Dennis W. Bennett

Understanding Single-Crystal X-Ray Crystallography

Related Titles

Guo, J. (ed.)

X-Rays in Nanoscience

Spectroscopy , Spectromicroscopy, and Scattering Techniques

2010

ISBN: 978-3-527-32288-6

Woollins, J. D. (ed.)

Inorganic Experiments

Third, Revised Edition

2010

ISBN: 978-3-527-32472-9

McPherson, A.

Introduction to Macromolecular Crystallography

2009

ISBN: 978-0-470-18590-2

Messerschmidt, A.

X-Ray Crystallography of Biomacromolecules

A Practical Guide

2007

ISBN: 978-3-527-31396-9

Scott, R. A. (ed.)

Applications of Physical Methods to Inorganic and Bioinorganic Chemistry

2007

ISBN: 978-0-470-03217-6

Dennis W. Bennett

Understanding Single-Crystal X-Ray Crystallography

WILEY-VCH Verlag GmbH & Co. KGaA

The Author

Professor Dennis W. Bennett
University of Wisconsin-Milwaukee
Department of Chemistry and Biochemistry
3210, N. Cramer Street
Milwaukee, WI 53211
USA

Library of Congress Card No.:
applied for

British Library Cataloguing-in-Publication Data
A catalogue record for this book is available from the British Library.

Bibliographic information published by the Deutsche Nationalbibliothek
The Deutsche Nationalbibliothek lists this publication in the Deutsche Nationalbibliografie; detailed bibliographic data are available on the Internet at <http://dnb.d-nb.de>.

Composition Uwe Krieg, Berlin
Printing and Binding betz-druck GmbH, Darmstadt
Cover Design Formgeber, Eppelheim

Printed in the Federal Republic of Germany
Printed on acid-free paper

ISBN: 978-3-527-32677-8 (HC)
978-3-527-32794-2 (SC)

To Vicki, the love of my life

Contents

Understanding Single-Crystal X-Ray Crystallography. Dennis W. Bennett

ISBN: 978-3-527-32677-8 (HC), 978-3-527-32794-2 (SC)

Foreword

In 1975, while in graduate school at the University of Utah, I was infected with the crystallography bug. As a nascent physical-inorganic chemist I had become frustrated with my inability to reconcile structural ambiguities that existed between my research and the literature that were the result of spectroscopic assignments (which the late F. Albert Cotton often referred to as "sporting methods"). Hoping to graduate at some date before the beginning of the next century, I sought the advice of one of the post-docs in my research group, who suggested that I attempt to grow some crystals and determine a crystal structure of one of the ambiguous compounds. Although I was aware that molecular structures could somehow be determined from crystals, having come to chemistry from biology, I hadn't the slightest notion of how this could be done. I signed up for Bill Cagle's two quarter X-ray crystallography sequence, and from my first day in class I knew that crystallography would forever be a central tool in my future research.

I began my academic career at Clemson University, hired there as an assistant professor in inorganic chemistry. During my interview I was shown a single-crystal X-ray diffractometer in the Physics Department that had been constructed entirely by the excellent machinists in Clemson's engineering department. I was informed that it had not been used in years, but would be mine if I accepted the position. After I began my work at Clemson, I soon became aware that this was a truly rudimentary instrument, operated with a DEC pdp-8 computer connected to a teletype and programmed by hand in binary. Instructions were introduced with paper tape, and there was virtually no software to drive the instrument to record intensity data from pre-selected reflections. Desperate to make it work, I dug into the literature to learn all that I could about diffractometers and X-ray diffraction, then created crude but sufficient code to generate an orientation matrix and collect data stored on paper tape. The data were transferred to the University's IBM 360 main-frame computer via a paper tape reader interfaced to a modem, and with the aid of *SHELX-76*,[1] I was able to solve the structure of a transition metal complex important to my research. Unfortunately, there were many outliers in the data, and the structure was of insufficient quality for publication. As an assistant professor, this was untenable, and I turned my attention to discovering the nature of the problem. After several months of troubleshooting I finally was able to demonstrate unequivocally that the chi circle was not a circle at all, but an ellipse! It had been machined out of aluminum – perhaps it self-annealed.

Clemson was facing a substantial financial shortfall at that moment, and I was informed that there was nothing that could be done for me. Faced with the possibility that I would be unable to earn tenure, I began seeking other employment. For-

Understanding Single-Crystal X-Ray Crystallography. Dennis W. Bennett

ISBN: 978-3-527-32677-8 (HC), 978-3-527-32794-2 (SC)

tunately, the Department of Chemistry at the University of Wisconsin-Milwaukee had just undergone an external graduate program review, and had been informed that their program was strong and growing, but a major deficiency was the lack of crystallography in the program. I was offered a position on the Chemistry faculty, based at least partly on my crystallography experience. Although the College of Letters and Science had insufficient funds to purchase a diffractometer for me, they had located a relatively new instrument at the Veterans Administration Hospital in Milwaukee that was used only sparingly, and that I would have virtually unlimited access to. I accepted the offer.

Shortly after I arrived at UWM the crystallographer at the VA Hospital unexpectedly moved to a new position in another state, and took his diffractometer with him. The College scrambled to help me out, but was still unable to come up with the funds necessary to buy an instrument. They were, however, able to allow me to purchase a used Picker circle and a hardware/ software package running on a DEC pdp-11 that appeared to work (although every time it ramped its stepper motors it sounded like a sawmill, much to the chagrin of my colleagues on the sixth floor!). While the hardware, with the exception of the tape drive, was adequate, the software was not (written in Dartmouth BASIC), and the DEC computer kept crashing. Data were stored on magnetic tape, and the tape drive failed constantly. Frustration again set in.

This occurred in the early 1980s, at the dawn of the IBM-PC age, and with the ability to store programs and data on floppy disks, the open architecture of the PC, and the availability of Microsoft QuickBASIC and MASM, I was able to create a complete diffractometer control package for the Picker circle that mimicked the functions on the Syntex diffractometers that I had become familiar with in my graduate work. In addition, the only readily available molecular graphics programs were to be found on commercial diffractometers, rendering it necessary for me create a molecular graphics program in QuickBASIC that would produce the ORTEP drawings needed for publication (the software has since evolved into the Microsoft Windows based program, *MOLXTL*[2]). I was finally in business! The software eventually found its way onto Picker diffractometers at a number of institutions, including Los Alamos National Laboratory, the University of Wisconsin at Madison, Georgetown University, and the University of Hawaii.

While my continuing (and often compulsive) pursuit of crystallography has often proved challenging, I would have been woefully unable to write this book without the thousands of hours that were spent struggling with the literature in order to gain the knowledge necessary to build and operate a diffractometer system. In the process of this pursuit I also became aware that others who had taken similar paths were diminishing rapidly in number, as crystallography hardware and software inevitably became more automated and less transparent. I feared that much of their knowledge was destined to be lost. This became very clear a few years ago when I set out to teach a crystallography course to our graduate students, many of whom had interests in biochemistry, but lacked a strong physical background. In seeking a crystallography text to provide this background it became clear that there were none. I taught the course without a textbook, and decided to use the resulting notes to create one. I naively estimated that it would take me no more than a year — but as I began writing I soon became aware of just how much I didn't know. Four years later the result is this book.

Preface

To know many things about something is laudable. To know sufficient things about something to understand it is exemplary. To know everything about something is impossible.

It is a testament to the ingenuity of the human species that mankind has been able to extend its collective senses, primarily designed for elementary survival, to encounter the universe on scales profoundly beyond the limits of those senses. X-ray crystallography has played a pivotal role in this enterprise, allowing us to observe and study entities which are unimaginably small. Over the past 50 years the use of X-ray diffraction to determine the structures of molecules and the components of extended solids has evolved from a very difficult and time-consuming undertaking to one that has become almost routine.

Prior to the ready accessibility of the digital computer, molecular structure determination using X-ray diffraction was a difficult and frequently exasperating venture. Several doctorates were often earned during the course of a single structural investigation. The combination of fast computers and new structural determination methods changed all that, and X-ray diffraction is now routinely utilized to determine the three-dimensional structures of molecules ranging from those containing a few atoms to macromolecules containing many thousand.

As X-ray crystallography became more accessible to the non-specialist, a number of books were written to describe the fundamentals of the experiment, and simultaneously, to serve as handbooks for the increasing number of investigators using this powerful technique. Many of those books necessarily compromised between an emphasis on understanding the underlying theory and a description of the basic experimental techniques, while others covered the theory rigorously, targeting those with a sufficient background in physics and mathematics. These books, when used in combination, generally met the needs of those involved in molecular structure determination, usually chemists and physicists with the requisite background knowledge.

Over the past 20 years, however, a revolution in crystallography has occurred, spurred on largely by the availability of "user-friendly" X-ray diffractometers, very efficient software packages for small-molecule structure determination and more recently the use of area detectors for both large and small molecule studies. The result has been a tremendous increase in the number of people undertaking crystallographic investigations, many with little or no understanding of the fundamental

Understanding Single-Crystal X-Ray Crystallography. Dennis W. Bennett

ISBN: 978-3-527-32677-8 (HC), 978-3-527-32794-2 (SC)

experiment, or the background needed to gain this understanding from most of the crystallography books currently available.

This book is designed to provide that background, not by attempting to explain the basis of X-ray structural determination in simple terms, but by presenting the material in a self-contained fashion, so that the necessary background is included and described as part of the explanation. Thus, the book is designed to serve primarily as a textbook, rather than as a handbook. Crystallography is a multi-faceted subject, and in writing this book I have made no attempt to be comprehensive. Rather, I have endeavored to cover the most fundamental concepts thoroughly, with an emphasis on *understanding*, foregoing many details which can be gleaned from nearly any of the books already available. A number of these are listed in the bibliography.*

The book is composed of eight chapters, and is designed to present a logical, step-by-step treatment of the elements of crystallography as they are encountered in a single-crystal structure determination.

Chapter 1 presents vectors and matrices in the context of the crystal lattice. While this discussion may be familiar to many readers who have a linear algebra background, many others do not, and many who do will will welcome the review. In addition, a substantial number of readers will not be familiar with the non-orthogonal coordinate systems common to crystallography that are covered in the first chapter.

Chapter 2 extends the lattice concept to include its contents, centering on the symmetry of the crystal lattice and the crystal structure itself. The concept of a mathematical group is introduced here and crystallographic point groups and space groups are discussed in detail.

Chapter 3 begins with a first-principles discussion of diffraction, followed by an extensive treatment of the diffraction of X-rays from single crystals. This portion of the chapter introduces the reciprocal lattice as the natural lattice of the diffraction pattern, and contains a determination of the relationships between the reciprocal lattice and the crystal lattice that are crucial to an understanding of diffraction theory. The chapter ends with an intuitive (but rigorous) derivation of the two fundamental equations of X-ray crystallography — the diffraction equation and the electron density equation.

Chapter 4 describes the theory and mathematics underlying the X-ray diffraction experiment and the details of X-ray data collection. The chapter uses the material in Chapter 1 to determine the means to transform locations (vectors) in the reciprocal lattice (the diffraction pattern) into vectors that can be utilized for the collection of the intensities in the diffraction pattern, thus providing the fundamental data for structure solution.

Chapter 5 describes the nature of both statistical (random) and systematic errors in the X-ray intensity data. The section on random errors includes a discussion of elementary statistics and probability distributions. This material is introduced in Chapter 5 because random errors in crystallography are important – giving rise to uncertainties in the intensity data. In addition, placing the material here provides the reader with an understanding of probability distributions before they are

*Several of the initial entries in the bibliography[3–18] are from Joseph Reibenspies, who listed them in the Spring, 2008 newsletter of the American Crystallographic Association as a "must read" list for X-ray crystallographers. I have added to the list.

encountered in the treatment of atom displacement errors later on in the chapter. Even more importantly, statistics forms the underlying basis for the use of probability methods in structure determination — the subject of Chapter 7. The second portion of the chapter discusses the systematic errors encountered in the diffraction experiment and the modification of the data to compensate for those errors.

Chapter 6 is based on the determination of the crystal structure from the structural information residing in the experimental X-ray intensity data. The focus of the chapter is on a mathematical function of the X-ray intensities known as the Patterson function. In theory, this function contains sufficient structural information to determine the locations of electron density maxima from the experimental data collected in the diffraction experiment. Various methods for extracting these locations from the Patterson function are described and illustrated throughout the chapter. Other methods based on the experimental intensities are also discussed.

Chapter 7 extends the statistical concepts introduced in Chapter 5 and applies them to the probability distributions of relationships between X-ray intensities determined in the diffraction experiment. These statistical relationships are intimately linked to the electron density distribution in the crystal and allow for the direct determination of the crystal structure from the intensity data. *Direct methods* are currently responsible for the solution of nearly all small molecule crystal structures, while the "experimental" methods of Chapter 6 are employed for the solution of macromolecular structures.

Chapter 8 completes the book with a treatment of the refinement of the crystal structure, which includes the final determination of the fundamental parameters that define the structure – the atomic positions and their displacements — and the statistical errors inherent in those parameters. It concludes with a discussion of the refinement of macromolecular crystal structures, a process that differs substantially from the refinement of crystal structures containing small molecules.

Each chapter concludes with several exercises designed to reinforce some of the concepts introduced in the chapter. These exercises are intended to serve as a starting point, rather than as a representation of "the important material" – much of which does not lend itself to a simple "question and answer" format. In all likelihood, instructors using the text will elect to add their own exercises, either as supplements or replacements.

Milwaukee, Wisconsin
August 2009

Dennis W. Bennett

Acknowledgements

I express my deepest gratitude to the late Robert W. Parry — the inspiration for my love of molecules — and to the late William F. Cagle, Jr., Robert J. Neustadt, Jack Simons, and Keith McDowell for introducing me to the thrill of elucidating their geometric and electronic structures. Don Miller, the physicist at Clemson University who constructed the diffractometer there, was also of tremendous assistance in my early years — teaching me to code in machine language and to interface the code with mechanical devices,.

I am especially thankful for three special colleagues — Dean Duncan, Eddy Tysoe, and Jim Otvos. Their knowledge, analysis, constructive criticism, and friendship were invaluable. In addition to Jim Otvos' individual contributions, he, and indeed all of my colleagues at LipoScience, not only kept me continually challenged intellectually, but also kept my family fed during the leaner years. Thanks also goes to Jung-Ja Kim, who taught me macromolecular crystallography during my sabbatical year with her at the Medical College of Wisconsin, and to Dan Haworth, who solved several structures while learning crystallography during his sabbatical year with me. One of those structures provided an excellent example used in several places in this book.

My former graduate students also deserve a special nod; most of the structures in my own research were solved by them. Among these, special thanks goes to Joe Guy, who solved the first publishable structure in my group, and to Jalal Siddiquee, who solved (by far) the most structures. *None* of these crystal structures would have been possible, however, without the uncanny expertise of Alan Thompson — our departmental instrumentation specialist. Were it not for Alan's continual insight, advice, and effort, I would never have been able to build and maintain the homemade instruments that were my mainstay for over 20 years. I am also appreciative of Chuck Campana of Bruker AXS, who helped me to obtain a used commercial diffractometer a few years ago and carted a low-temperature attachment across the border from Canada in his van — and of Bob Ponton, glassblower par excellence, who transported the attachment from Rochester to Milwaukee — on his own dime.

Finally I am profoundly indebted to my wife, Vicki, for her incredible patience, encouragement, and sacrifice — and to my son, Patrick, who turned out wonderfully, in spite of getting much less of my time than he deserved.

Understanding Single-Crystal X-Ray Crystallography. Dennis W. Bennett

ISBN: 978-3-527-32677-8 (HC), 978-3-527-32794-2 (SC)

Chapter 1

Crystal Lattices

1.1 The Solid State

X-ray crystallography deals with materials in the solid state, and it is there that we must begin. In principle, depending on the temperature, any material can exist as a solid, a liquid, or a gas — provided that it does not decompose before it attains a given state. Materials are comprised of either atoms, neutral molecules, or ions, which, in the gas phase, behave more or less independently, despite attractive forces between them. This is because these entities have sufficient kinetic energy that the attractive forces between them are comparatively very weak. While they may tug a bit on one another in passing, they readily escape from one another to continue their motion indefinitely.

As the material is cooled — as kinetic energy is removed from the atoms, molecules, or ions in the gas phase — a point is reached at which they no longer have sufficient kinetic energy to move freely, and the attractive forces begin to dominate, keeping the entities in close proximity to one another. While they have enough kinetic energy to move around, over, and under one another, they are now constrained to stay in the vicinity of one another, and the material is said to have condensed into the liquid state.

Finally, when the material has cooled sufficiently, the attractive forces become so dominant that the components hold onto one another tenaciously, tending to arrange themselves so that they can gain as much interaction as possible, resulting in a minimum energy configuration. In the process they lose their translational motion, and the material becomes rigid. We now consider the manner in which the components of these rigid materials arrange themselves in order to maximize the attractive forces between them.

Since the attractive forces increase as the distances between entities decrease it is probably not surprising that simple materials, consisting of spherical ions or atoms, will tend to pack the spheres together in an orderly arrangement in order to get as many spheres as possible touching one another. Fig. 1.1 shows the arrangement of the spherical atoms in metallic copper and the spherical ions in a typical salt, sodium chloride. Both provide an excellent illustration of the old idiom, "Nature abhors a vacuum." In both structures the organization is clearly evident – spheres

Understanding Single-Crystal X-Ray Crystallography. Dennis W. Bennett

ISBN: 978-3-527-32677-8 (HC), 978-3-527-32794-2 (SC)

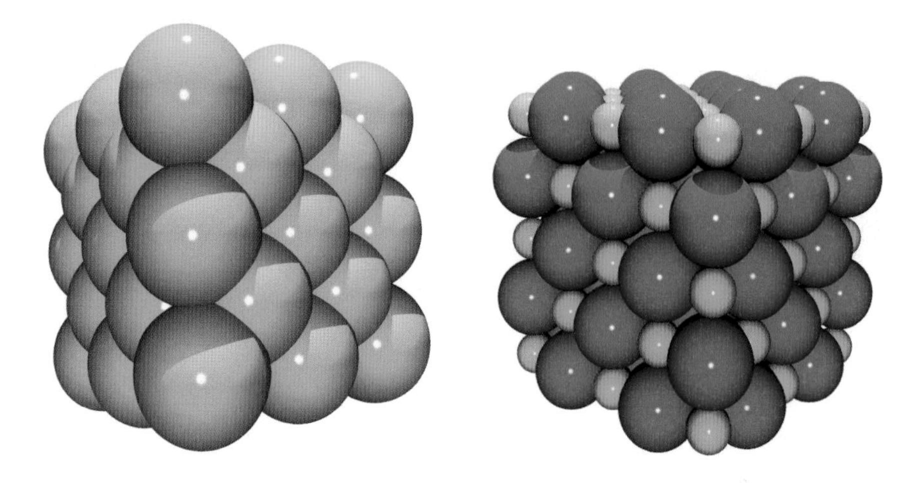

Figure 1.1 Arrangements of the atoms in copper and the ions in sodium chloride.

tend to arrange themselves into ordered three-dimensional arrays. The internal organization of such materials is reflected in their external appearance; both materials appear to the naked eye as cubes — or other polyhedra related to the cube — as shown in Fig. 1.2. In contrast to atoms and atomic ions, molecules and polyatomic ions are not spherical. Indeed, many molecules have virtually no symmetry of their own. Nevertheless, they generally form solid crystals with a symmetric appearance, suggesting that they also have ordered internal arrangements. In Fig. 1.3, a space-filling model of a sucrose molecule (table sugar) is compared to the macroscopic appearance of solid sucrose.

Figure 1.2 Naturally occurring crystals of metallic copper coated with tenorite and ionic sodium chloride (halite). Copper crystal specimen and photo courtesy of Rob Lavinsky; halite crystal photo courtesy of the Smithsonian Institution.

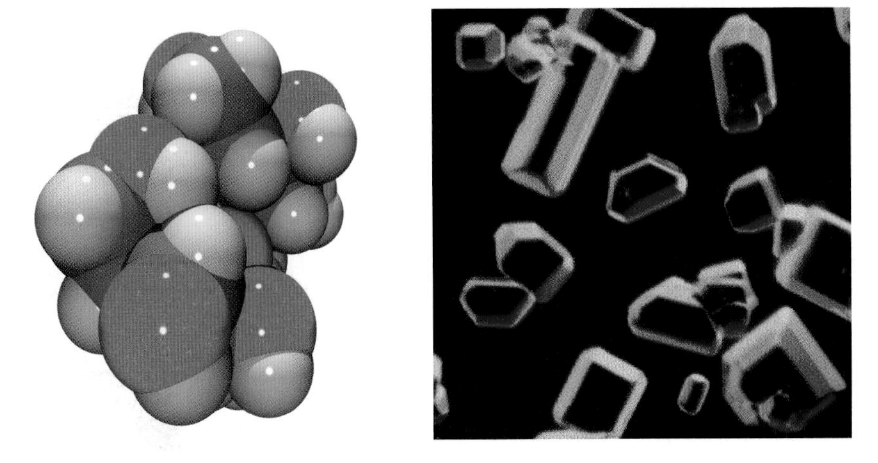

Figure 1.3 Space-filling model of a sucrose molecule and crystals of sucrose. Photo courtesy of Nicolas von Geijn and the Andrew van Hook Association.

The molecule has no observable symmetry, yet the solid material consists of symmetric crystals. A simpler molecule, 2-mercaptopyridine, illustrates how most molecules "fill the space," and pack next to one another in symmetric arrays. The space-filling representation of the molecule is shown on the left in Fig. 1.4. The arrangement of the molecules as they occur in the solid state indicates clearly that they arrange themselves in a periodically repeating sequence, which can be visualized as extending indefinitely throughout the three dimensions of the crystal. This periodicity is much easier to visualize if the space-filling representation of the molecule is replaced by a "ball and stick" model, indicating the locations of the atomic centers and their intramolecular connectivities (bonds), as illustrated in Fig. 1.5. Note that it is possible to construct a series of identical parallelepiped "cages" in

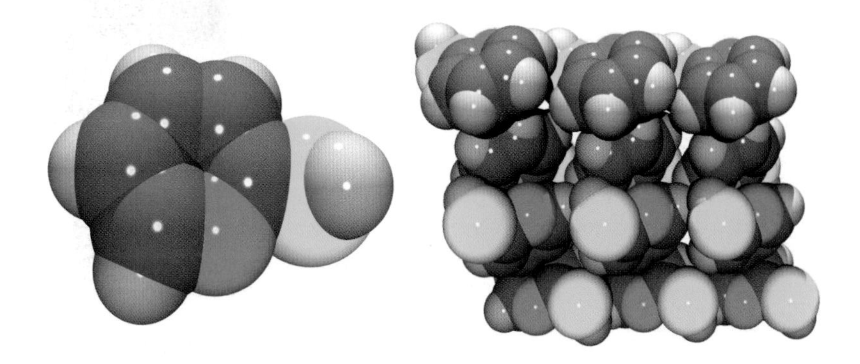

Figure 1.4 Space-filling model of 2-mercaptopyridine and the arrangement of the molecules in the solid state. The yellow atom is sulfur, the dark gray atoms are carbon, the light gray atoms are hydrogen, and the blue atom is nitrogen.

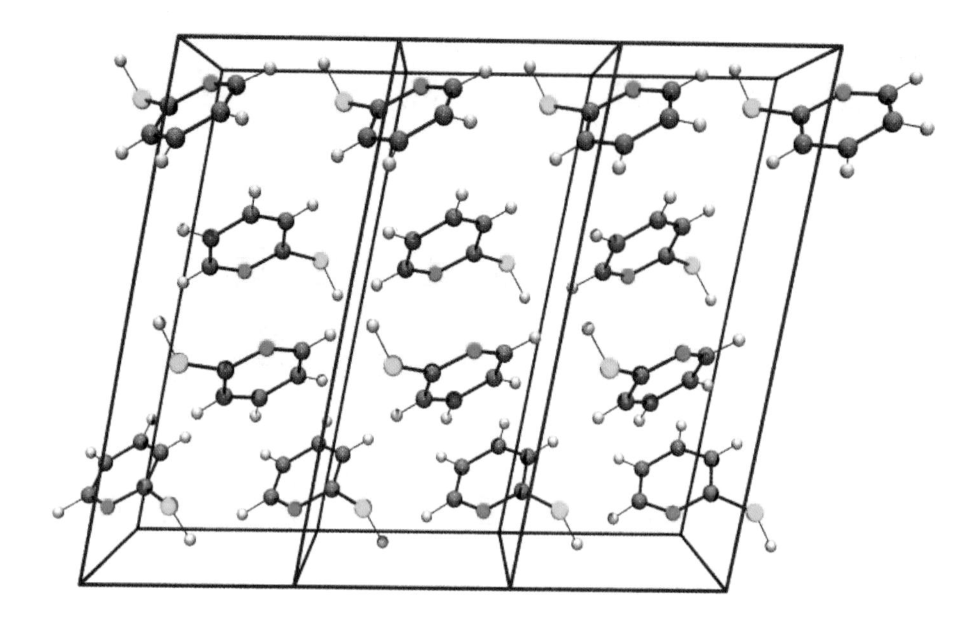

Figure 1.5 Crystallographic packing of 2-mercaptopyridine showing three unit cells translated along the *a* axis.

the structure, such that the contents and appearance of these cages are identical, again throughout the crystal.

We will come to know these parallelepipeds as the *unit cells* of the crystal and the framework that results from the entire array of these unit cells as the *crystal lattice*. Not all materials condense into these regular structures. Ideally, when a solid material is allowed to form slowly its component molecules or ions have sufficient opportunity to rearrange into a symmetric structure. However, if they lose kinetic energy too rapidly they will be trapped in less symmetric arrangements, and the rigid material will not exhibit the crystallinity of more ordered solids. These disordered materials are known as *glasses*. A common example is window glass, which is identical in chemical composition to quartz, except that window glass is formed from the rapid cooling of molten silicon dioxide, while quartz crystals are formed in nature from the same elements — slowly over millions of years. Fortunately for the crystallographer most substances, given the opportunity, tend to form crystalline solids.

1.2 The Crystal Lattice

1.2.1 Two-dimensional Lattices

It is often useful to begin a discussion in two dimensions, then extrapolate what we have learned to three dimensions. Consider a two-dimensional analog of the periodic arrangement of molecules described in Section 1.1 (Fig. 1.6), which we will call the *beta-structure*. The periodicity of this array, as in the previous three-

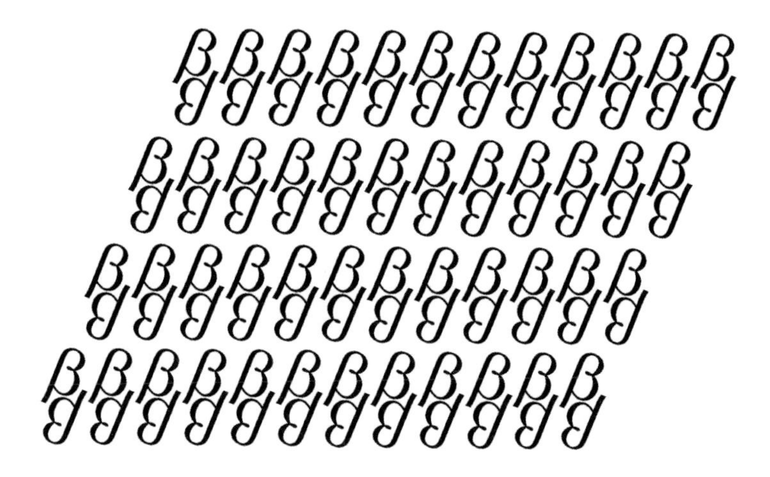

Figure 1.6 Two-dimensional periodic array.

dimensional case, can be represented by a two-dimensional lattice resulting from a series of parallelogram cells with identical contents (Fig. 1.7). The framework is created from two sets of equidistant parallel lines. The lattice is uniquely *defined* as the set of points at the intersections of these two sets of lines. The line segments

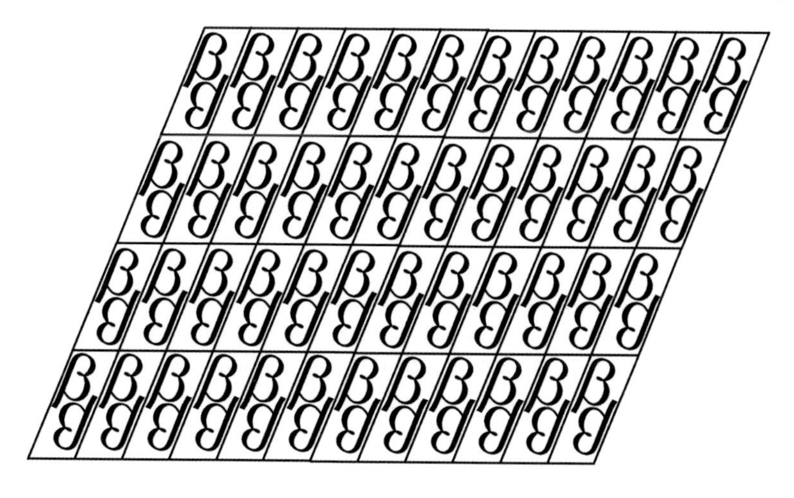

Figure 1.7 Lattice and contents of unit cells in a two-dimensional periodic array.

which make up the unit cell are called *axes*, since they will become the coordinate axes for locating points within the unit cell, much like the x and y axes in Cartesian coordinates. The unit cell is a parallelogram characterized by the lengths of its axes, a and b, and the angle between them, γ (Fig. 1.8). While the set of points defining

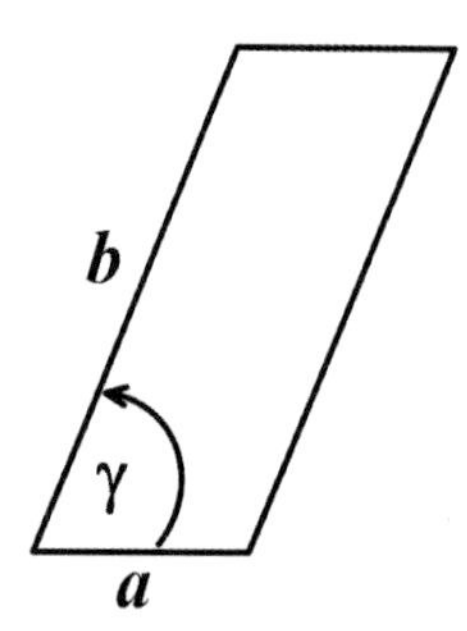

Figure 1.8 Unit cell for the beta-structure illustrating the crystallographic axes and the inter-axial angle.

the lattice is unique for a given structure, the sets of lines which define the unit cell, the repeating unit in the lattice, are the not the only sets of equidistant parallel lines which contain all of the points in the lattice, as illustrated in Fig. 1.9. Indeed, there are infinitely many sets of parallel lines in a given structure, each of which contains all of the points in the lattice. Since a given set of lines contains every point in the lattice and consists of equally spaced parallel lines, each set will divide a given unit cell axis into an integral number of equally spaced line segments.

In the example shown in Fig. 1.9, the red lines divide the a axis into 2 segments, and the b axis into 3 segments. The blue lines divide the a axis into 1 segment, and the b axis into 2 segments. Since each set of lines divides the axes differently, the two integers representing the number of divisions for both axes will be unique for a given set (subject to the caveat discussed below). It follows that each set of lines can be identified by these two integers, and that they can be used as *indices* to classify each set, provided that they are kept in "a, b" order. This scheme was devised by the Welsh crystallographer William Miller[19] (1801–1880), and the ordered integers are known as the *Miller indices* of the lines to which they correspond. They are generally placed between parentheses. In the example given in Fig. 1.9 the Miller indices of the red lines are (2 3), and(1 2) for the blue lines. By convention, the indices for a given set of lines are denoted (h k). This assignment is not unambiguous, since for every set of lines with a negative slope, there is another set of lines with indices of the same magnitude but with a positive slope, as indicated by the green lines in Fig 1.9. (which also divide a into one segment and b into two segments). We will find a convenient and formal way to deal with this ambiguity later on in the chapter by taking axial directions into account using vectors. For now we can deal with this informally by assigning positive directions along a and b to be to the right and up, and negative directions to be to the left and down. We select any lattice point as an origin and look for the next line away from the origin which crosses the axes. We note whether the crossing points are in positive or

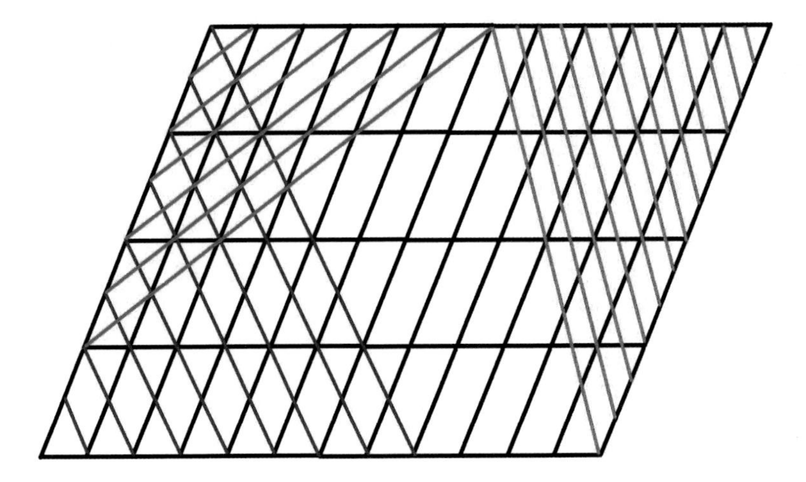

Figure 1.9 Three of the infinitely many sets of equidistant parallel lines that intersect every point in the beta-structure lattice.

negative directions along a and b, then assign the signs of those directions to the indices. Negative indices are indicated by placing a *bar* over the index. The lines in Fig. 1.9 serve as examples to clarify this. We find a blue line which intersects the axes one axial length to the *right* along a and, *for the same unit cell*, half of an axial length *up* along b. Indices $(h\ k) = (1\ 2)$ are therefore assigned to the blue lines. Note that there is also a blue line which crosses the a axis to the *left* and *down* along b, so that indices $(\bar{1}\ \bar{2})$ can also be assigned to the blue lines. While these represent the same lines, there are special cases in diffraction where *direction* is important, and we therefore keep them separate to account for this. Similarly, if we proceed in a positive direction along a to a green line ($h = +1$), we find that the line intersects the b axis halfway down — in a negative direction, and indices $(1\ \bar{2})$ are assigned to the green lines. Alternatively, if we proceed along a in a negative direction, we find a green line which intersects b in a positive direction, and we can equally assign indices $(\bar{1}\ 2)$ to the green lines — again retaining both sets of indices for the special cases that we will encounter later on. It is left to the reader to verify that the red lines have indices $(2\ 3)$ – and $(\bar{2}\ \bar{3})$. The original set of lines collinear with the a axis divide the b axis into 1 segment, and do not divide the a axis at all; thus they are assigned indices $(0\ 1)$ – and $(0\ \bar{1})$. Similarly, the lines collinear with the b axis are assigned indices$(1\ 0)$ – and $(\bar{1}\ 0)$. Note that the larger the magnitude of an index, the more times the lines cross the axis corresponding to that index, and the shorter the resulting line segments become. *It follows that as the magnitudes of the Miller indices increase, the spacings between the correspondingly indexed parallel lines must decrease.* We will encounter this *reciprocal* relationship again in the next section (in three dimensions) and throughout the text — it lies at the heart of the diffraction experiment!

1.2.2 Three-dimensional Lattices

Two-dimensional lattices are created by the intersection of two sets of equidistant parallel *lines;* three-dimensional lattices are the result of the intersection of three sets of intersecting parallel *planes* (Fig. 1.10). The unit cell thus becomes a parallelepiped, characterized by the length of its three axes, a, b, and c and the three corresponding inter-axial angles, α, β, and γ. By convention, α is the angle between the b and c axes, β is the angle between the a and c axes, and γ is the angle between the a and b axes, as illustrated for the 2-mercaptopyridine unit cell in Fig. 1.11(a).

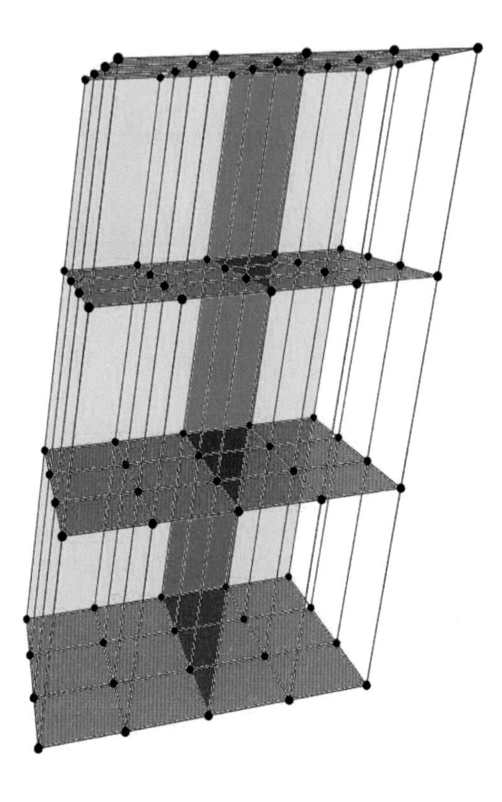

Figure 1.10 Three sets of intersecting equidistant parallel planes defining the lattice of the 2-mercaptopyridine structure. The lattice consists of the points indicated with small black spheres. Four a-b planes are shown in red, an a-c plane is shown in green, and a b-c plane is shown in blue.

The analogy with the two-dimensional lattice also holds for the Miller indices. Just as there are infinitely many sets of equidistant parallel lines intersecting every point in the two-dimensional lattice, there are an infinite number of sets of equidistant parallel planes which intersect every point in the three-dimensional lattice. Fig. 1.11(b) shows a set of planes that divides the a axis into two segments, the b axis into 2 segments and the c axis into 3 segments. As with each set of lines in two dimensions, this set of planes can be assigned Miller indices, $(h\ k\ l)$ — in this example, (2 2 3). There are seven other sets of planes with indices $(\bar{2}\ \bar{2}\ \bar{3})$, $(2\ 2\ \bar{3})$, $(\bar{2}\ \bar{2}\ 3)$, $(2\ \bar{2}\ 3)$, $(\bar{2}\ 2\ \bar{3})$, $(\bar{2}\ 2\ 3)$, and $(2\ 2\ \bar{3})$, determined by assigning directions along a, b, and c, just as we did in Sec. 1.2.1. Each set of planes is

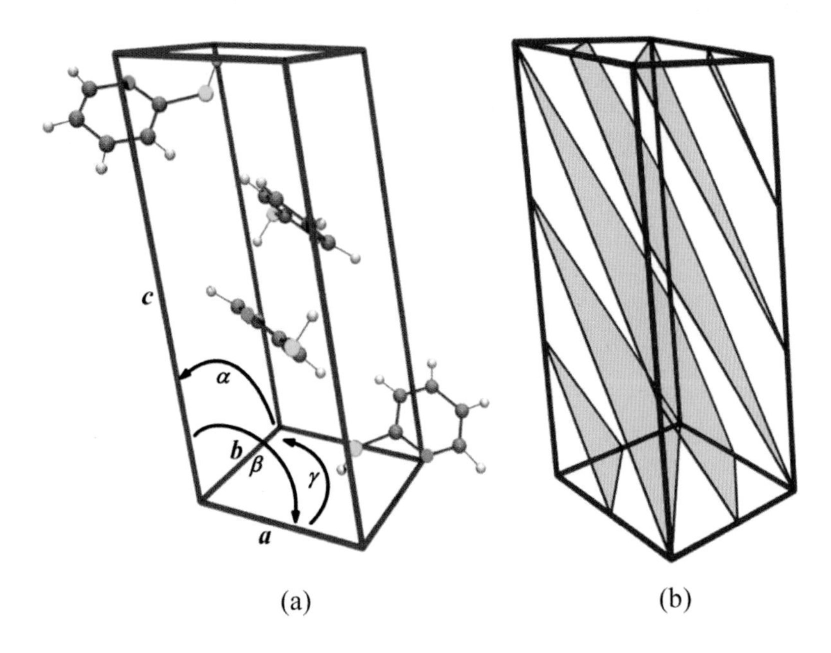

Figure 1.11 Unit cell for the 2-mercaptopyridine structure illustrating (a) conventional unit cell parameters and (b) intersection of the $(h\ k\ l) = (2\ 2\ 3)$ planes with the unit cell axes.

oriented differently in the lattice and is uniquely characterized by its indices.* As with the two-dimensional lattice, each set of planes with indices $(h\ k\ l)$ can also be indexed with $(\bar{h}\ \bar{k}\ \bar{l})$, so that there are actually four unique sets of planes; we retain the eight sets of indices to keep track of *direction.* The planes containing the b and c axes divide the a axis into one segment — and since they do not divide the b or c axes, they are assigned indices $(1\ 0\ 0)$ – and $(\bar{1}\ 0\ 0)$. Similarly, the a-c planes are assigned $(0\ 1\ 0)$— and $(0\ \bar{1}\ 0)$ indices, and the a-b planes are assigned $(0\ 0\ 1)$ — and $(0\ 0\ \bar{1})$ indices. As in the two-dimensional case, the larger a given index, the more the corresponding axis is divided, and the smaller the distance between the planes. *Again, there is a reciprocal relationship between the magnitude of the plane indices and the distances between the planes corresponding to those indices.*

*In the early days of crystallography, planes were indexed to describe crystal faces. For every set of planes with indices $(h\ k\ l)$, there is another set with indices $(2h\ 2k\ 2l)$, and another with indices $(3h\ 3k\ 3l)$, etc. All of these planes are parallel to one another, and macroscopically indistinguishable. For that reason Miller indices formally have no common factors, thus excluding the $(nh\ nk\ nl)$ planes for $n > 1$. Since we must include these planes, from this point on we will refer to the indices of all of the planes in the lattice as *general indices*, or more concisely, simply as *indices*.

1.3 Vectors in Crystallography

The discussion of indices in the previous section required us to consider both the *magnitudes* and *directions* of line segments in order to assign the indices unambiguously. Directed line segments are often referred to as *vectors*, although formally they are geometric *representations* of vectors. In order to develop the mathematics of crystallography and diffraction we will rely heavily on the use of vectors.

A vector is an ordered array of elements known as *components*. The number of components is referred to as the *dimension* of the vector. We will describe a vector by placing its components in square brackets: $[v_1\ v_2\ v_3 \ldots v_n]$, where v_i is the ith component of an n-dimensional vector. A boldface lower case letter will be used to indicate the vector (e.g., $\mathbf{v} = [v_1\ v_2\ v_3]$) *and* its geometric representation; the term *vector* will be used for both as well.

The sum of two vectors *of the same dimension* produces a third vector of that dimension,

$$\begin{aligned}\mathbf{v_i} + \mathbf{v_j} &= [v_{i,1}\ v_{i,2}\ v_{i,3} \ldots v_{i,n}] + [v_{j,1}\ v_{j,2}\ v_{j,3} \ldots v_{j,n}] \\ &= [(v_{i,1} + v_{j,1})\ (v_{i,2} + v_{j,2})\ (v_{i,3} + v_{j,3} \ldots (v_{i,n} + v_{j,n})]. \end{aligned} \tag{1.1}$$

A one-dimensional vector is known as a *scalar*. It is generally a single number which represents the *magnitude* of something, and is ordinarily not placed in brackets. The product of a vector of dimension n and a scalar results in a vector of the same dimension in which each component is multiplied by the scalar:

$$s\mathbf{v} = s[v_1\ v_2\ v_3 \ldots v_n] = [sv_1\ sv_2\ sv_3 \ldots sv_n]. \tag{1.2}$$

We have already encountered examples of two and three-dimensional vectors — the indices of sets of lines in two dimensions and sets of planes in three dimensions. The indices are descriptors, in the sense of identifying specific sets of planes. Indices are also vectors which occur commonly in the mathematics of diffraction. We will adopt the convention of placing indices in parentheses when referring specifically to sets of lattice planes, and in square brackets when they are to be employed as vectors.

Crystallography is chiefly concerned with directed line segments in two or three dimensions, and the discussion here will largely be confined to those vectors. A directed line segment can be described by specifying its length (*magnitude*) and direction, usually expressed as an angle or angles with respect to some standard direction, and indicated by an arrowhead at the leading end of the line segment. The leading end of the line segment is called its *head*, the trailing end is called its *tail*. A vector with its tail at point o and its head at point p will be indicated by placing an arrow over the symbols for the points, e.g., $\mathbf{v} = \overrightarrow{op}$. Unfortunately, there is no standard convention for indicating the magnitude of a vector. For example, the magnitude of the vector $\mathbf{v}$ can be found in various texts as $|\mathbf{v}|$, $\|\mathbf{v}\|$, or simply v. With the exception of cases where it might be confusing, a lower case script letter will be employed to indicate any scalar quantity, including the magnitude of a vector. When the vector is represented by the points which define it, its magnitude will be indicated by placing the symbol for the vector between vertical lines, e.g., $v = |\overrightarrow{op}|$.

1.3.1 Geometric Vector Addition and Multiplication

Geometric vectors are added and multiplied by an established set of rules, and we must consider those rules in order to link ordered pairs and triples of numbers with geometric vectors in two and three dimensions. Geometric vectors, $\mathbf{v_1}$ and $\mathbf{v_2}$, are added to one another by placing the tail of $\mathbf{v_2}$ at the head of $\mathbf{v_1}$, and creating a third *resultant* vector by connecting the tail of $\mathbf{v_1}$ to the head of $\mathbf{v_2}$ (Fig. 1.12). Note that this resultant vector is also obtained if we reverse the order in which we combine the vectors. For "regular" numbers (scalars) the order in which we add things does not alter the results, and we say that scalar addition is *commutative.* Vector addition behaves just like regular addition in this respect – vector addition is commutative.

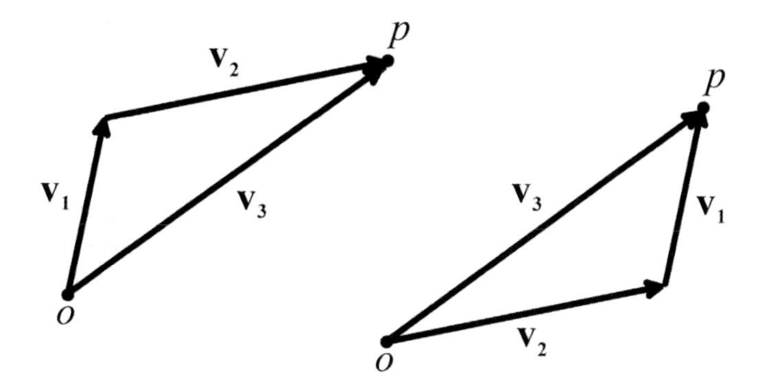

Figure 1.12 The addition of geometric vectors: $\mathbf{v_3} = \mathbf{v_1} + \mathbf{v_2} = \mathbf{v_2} + \mathbf{v_1}$. Note that merging the two representations creates a parallelogram with the resultant vector along the diagonal — providing a convenient method for vector addition.

Adding vectors in this manner makes intuitive sense if we consider our intent to make use of them in order to locate points inside the crystal lattice. Locating the point p relative to point o can be effected by traversing directly along $\mathbf{v_3}$, or by taking the path along $\mathbf{v_1}$, followed by $\mathbf{v_2}$. The combined effects of $\mathbf{v_1}$ and $\mathbf{v_2}$ are clearly additive in this respect, and equivalent to $\mathbf{v_3}$.

Vectors can be multiplied* in a number of ways. The simplest is multiplication by a scalar. Multiplication by a positive scalar produces a new vector in the same direction as the original with its magnitude multiplied by the scalar; multiplying by a negative scalar produces a new vector in the opposite direction. Fig. 1.13 illustrates a vector multiplied by the scalars 0.5 and -2.0, producing a new vector parallel to the original with half the length ($v_2 = 0.5v_1$) and an antiparallel vector with twice the length ($v_3 = -2.0v_1$).

Multiplying a vector by -1 produces the negative of that vector — a vector of equal magnitude but opposite direction. In order to subtract $\mathbf{v_2}$ from $\mathbf{v_1}$, we

---

* *Multiplication* is used here in a more abstract context to indicate a *combination* of entities in accordance with a specific set of rules. In this sense *addition* is also considered a form of *multiplication.*

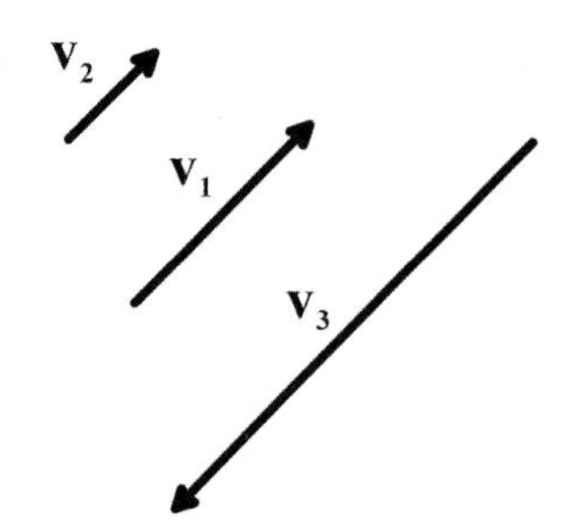

Figure 1.13 Geometric vector $\mathbf{v_1}$ multiplied by scalars: $\mathbf{v_2} = 0.5\mathbf{v_1}$; $\mathbf{v_3} = -2.0\mathbf{v_1}$

multiply $\mathbf{v_2}$ by -1 to negate it, then add it to $\mathbf{v_1}$. Note that vector subtraction is *not* commutative. As illustrated in Fig. 1.14, reversing the order of subtraction changes the sign of the resultant vector, but leaves the magnitude unchanged. As with addition, the subtraction of vectors parallels the non-commutative subtraction of ordinary numbers.

Vectors can also be multiplied together to produce a scalar. The *scalar product* of two vectors (also known as the *inner product* or *dot product*) is defined as the product of the magnitudes of the two vectors and the cosine of the angle between them:

$$\mathbf{v_1} \cdot \mathbf{v_2} = v_1 v_2 \cos\theta. \tag{1.3}$$

Since the magnitude of the original vectors and the angle between them does not change, $\mathbf{v_1} \cdot \mathbf{v_2} = \mathbf{v_2} \cdot \mathbf{v_1}$; the scalar product is commutative. The scalar product has three important properties which are very useful in vector analysis. First, the scalar product of a vector with itself gives us the square of the magnitude (length) of the vector:

$$\mathbf{v_1} \cdot \mathbf{v_1} = v_1 v_1 \cos(0) = v_1^2. \tag{1.4}$$

Second, the scalar product of two *orthogonal* (perpendicular) vectors is zero ($\theta = \pi/2 = 90°$) :

$$\mathbf{v_1} \cdot \mathbf{v_2} = v_1 v_2 \cos(\pi/2) = 0, \mathbf{v_1} \perp \mathbf{v_2}. \tag{1.5}$$

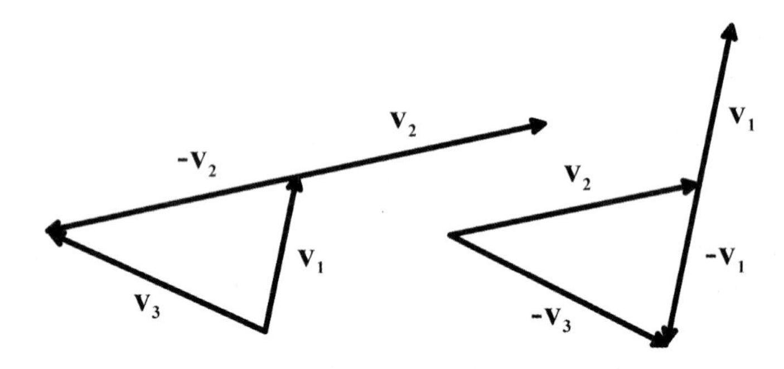

Figure 1.14 The subtraction of geometric vectors: $\mathbf{v_3} = \mathbf{v_1} - \mathbf{v_2}$; $\mathbf{v_3'} = -\mathbf{v_3} = \mathbf{v_2} - \mathbf{v_1}$.

The third property involves the *projection* of one vector onto another. This entails the construction of a perpendicular from one vector to the head of another, as illustrated in Fig. 1.15. $\mathbf{v_1}$ is said to have been projected onto $\mathbf{v_2}$. p_1 is the magnitude of the projection of $\mathbf{v_1}$ onto $\mathbf{v_2}$. p_1 can be determined from the scalar product $\mathbf{v_1} \cdot \mathbf{v_2}$ and the magnitude of $\mathbf{v_2}$:

$$\begin{aligned}
\cos\theta &= \frac{v_1}{p_1} \\
p_1 &= v_1 \cos\theta \\
\mathbf{v_1} \cdot \mathbf{v_2} &= v_1 v_2 \cos\theta \\
\mathbf{v_1} \cdot \mathbf{v_2} &= v_2 p_1 \\
p_1 &= \frac{\mathbf{v_1} \cdot \mathbf{v_2}}{v_2}. \qquad (1.6)
\end{aligned}$$

We will find all three of these properties very useful as we analyze the vectors in both the crystal lattice and the X-ray diffraction pattern.

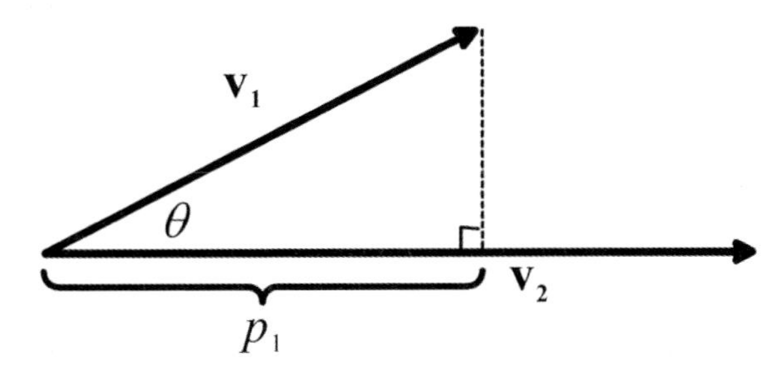

Figure 1.15 The projection of vector $\mathbf{v_1}$ onto the vector $\mathbf{v_2}$.

Finally, two vectors can be multiplied to produce a *vector product* (also known as an *outer product* or *cross product*) — a new vector created from the originals. The magnitude of the new vector is determined as the product of the magnitudes of the original two vectors and the sine of the angle between them:

$$\mathbf{v_3} = \mathbf{v_1} \times \mathbf{v_2}, \qquad (1.7)$$

$$v_3 = v_1 v_2 \sin\theta. \qquad (1.8)$$

This new vector is perpendicular to *both* of the original vectors, and is therefore perpendicular to the plane in which both vectors lie (Fig. 1.16). Since $\mathbf{v_3}$ can point in either of two directions, both perpendicular to the $\{\mathbf{v_1}, \mathbf{v_2}\}$ plane, we adopt the convention that it if we align the index finger of *the right hand* along $\mathbf{v_1}$, and the middle finger of the right hand along $\mathbf{v_2}$, then $\mathbf{v_3}$ will point in the direction of the right thumb. *This is referred to as the right hand rule.* Note that $\mathbf{v_2} \times \mathbf{v_1}$ produces a vector of the same magnitude pointing in the opposite direction to $\mathbf{v_1} \times \mathbf{v_2}$. Thus $\mathbf{v_2} \times \mathbf{v_1} = -(\mathbf{v_1} \times \mathbf{v_2})$; vector products are *not* commutative.

1.3.2 Basis Vectors and Coordinates

As the reader might suspect at this point, unit cell axes can be represented as vectors with tails sharing a common lattice point, which we will designate as the

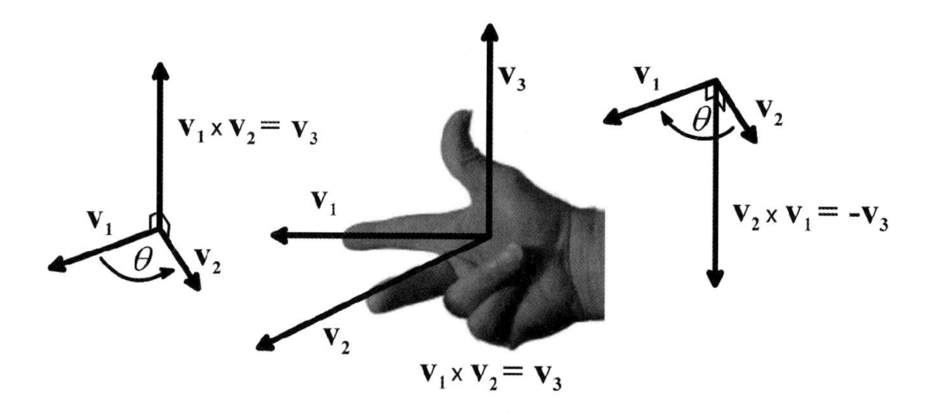

Figure 1.16 *Right hand rule* for the vector product: $\mathbf{v_3} = \mathbf{v_1} \times \mathbf{v_2}$, $-\mathbf{v_3} = \mathbf{v_2} \times \mathbf{v_1}$, $v_3 = v_1 v_2 \sin\theta$.

origin of the unit cell, illustrated in Fig. 1.17 for a two-dimensional unit cell. X-ray crystallography involves specification of locations inside the lattice, and the unit cell axes provide a convenient set of reference vectors for determining the location of a point anywhere in the lattice — as the end of a vector with its tail at the origin and its head coincident with the point. We can imagine taking a path to the point p along a vector $\mathbf{x}$ coincident with the $\mathbf{a}$ axis, then following a vector $\mathbf{y}$ parallel to the $\mathbf{b}$ axis to get to p — or alternatively taking the direct path along $\overrightarrow{\mathbf{op}}$ — which is clearly the vector sum of $\mathbf{x}$ and $\mathbf{y}$:

$$\overrightarrow{\mathbf{op}} = \mathbf{x} + \mathbf{y}$$

Because $\mathbf{x}$ is parallel to $\mathbf{a}$ it can be expressed as the vector $\mathbf{a}$ multiplied by a scalar, $\mathbf{x} = s_x\mathbf{a}$; similarly, $\mathbf{y} = s_y\mathbf{b}$:

$$\overrightarrow{\mathbf{op}} = s_x\mathbf{a} + s_y\mathbf{b}$$

Note that *any* vector in two-dimensions can be described as a vector sum of reference vectors multiplied by appropriate scalars. Since the sum contains only linear terms (no exponents), it is called a *linear combination*; the reference vectors are known as *basis* vectors, and the set of two reference vectors (in two-dimensions) is called a *basis set*, denoted $\{\mathbf{v_1}, \mathbf{v_2}\}$ — in our case $\{\mathbf{a}, \mathbf{b}\}$. In the general case, the selection of the initial basis set is arbitrary, provided that the two vectors are not parallel. We can express two parallel vectors as a linear combination of one another (by multiplying one of the vectors by a scalar and the other by zero). Such vectors are said to be *linearly dependent*. Basis vectors which are not parallel to one another cannot be expressed as linear combinations of one another, and are termed *linearly independent* vectors. We can now state these observations more formally — *any vector in two-dimensional space can be expressed as a linear combination of two linearly independent basis vectors.*

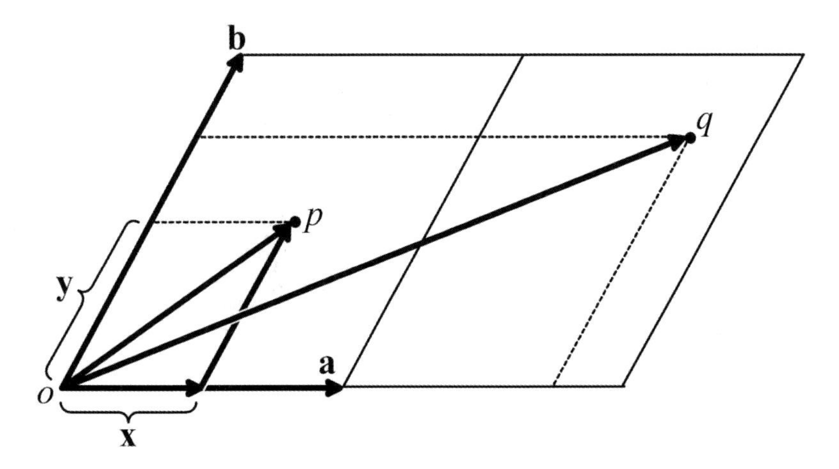

Figure 1.17 Two-dimensional unit cells illustrating the axes **a** and **b** as basis vectors for two-dimensional space and the vectors locating points p and q as linear combinations of the axial basis vectors.

A logical way to determine s_x and s_y is to express them as fractions of the lengths of the basis vectors:

$$\begin{aligned} s_x &= \frac{x}{a} = x_f \\ s_y &= \frac{y}{b} = y_f \\ \overrightarrow{\mathbf{op}} &= x_f\mathbf{a} + y_f\mathbf{b} \end{aligned}$$

Once we have selected our basis vectors, $\{\mathbf{a}, \mathbf{b}\}$, we can locate any point in the lattice by determining the ordered pair of numbers, $[x_f\ y_f]$, defining a vector from the origin of the unit cell to that point. The determination of a basis set and the expression of vectors as linear combinations of the basis vectors provides the link between vectors and their geometric representations in two dimensions:

$$\overrightarrow{\mathbf{op}} = [x_f\ y_f]. \tag{1.9}$$

The components of the vector, x_f and y_f, are called the *coordinates* of the vector, and when expressed as fractions of the magnitudes of the basis vectors are known as *fractional coordinates.* For example, in Fig. 1.17,

$$\begin{aligned} \overrightarrow{\mathbf{op}} = \tfrac{1}{2}\,\mathbf{a} + \tfrac{1}{2}\,\mathbf{b} &= [\tfrac{1}{2}\ \tfrac{1}{2}] \\ \overrightarrow{\mathbf{oq}} = 1\tfrac{3}{4}\,\mathbf{a} + \tfrac{3}{4}\,\mathbf{b} &= [1\tfrac{3}{4}\ \tfrac{3}{4}] \\ \mathbf{x} = \tfrac{1}{2}\,\mathbf{a} + 0\,\mathbf{b} &= [\tfrac{1}{2}\ 0] \\ \mathbf{y} = 0\,\mathbf{a} + \tfrac{1}{2}\,\mathbf{b} &= [0\ \tfrac{1}{2}] \\ \mathbf{a} = 1\,\mathbf{a} + 0\,\mathbf{b} &= [1\ 0] \qquad (1.10) \\ \mathbf{b} = 0\,\mathbf{a} + 1\,\mathbf{b} &= [0\ 1]. \qquad (1.11) \end{aligned}$$

The product of a vector and a scalar can now be expressed in terms of the components of the vector. Fig. 1.18 illustrates the formation of $\mathbf{v_2} = [v_{2x}\ v_{2y}]$ as the product of $\mathbf{v_1} = [v_{1x}\ v_{1y}]$ and a scalar, s. The triangles with sides (v_1, v_{1x}, v_{1y}) and (v_2, v_{2x}, v_{2y}) are similar triangles, which means that the ratios of the lengths of similar sides are a constant. Since $\mathbf{v_2} = s\mathbf{v_1}$,

$$\begin{aligned}
v_2 &= sv_1 \\
\frac{v_{1x}}{v_1} &= \frac{v_{2x}}{v_2} = \frac{v_{2x}}{sv_1} \qquad \frac{v_{1y}}{v_1} = \frac{v_{2y}}{v_2} = \frac{v_{2y}}{sv_1} \\
v_{2x} &= sv_{1x} \qquad v_{2y} = sv_{1y} \\
\mathbf{v_2} &= s\mathbf{v_1} = [v_{2x}\ v_{2y}] = s[v_{1x}\ v_{1y}] = [sv_{1x}\ sv_{1y}].
\end{aligned}$$

Multiplying a vector by a scalar produces a new vector formed by multiplying each component of the original vector by the scalar.

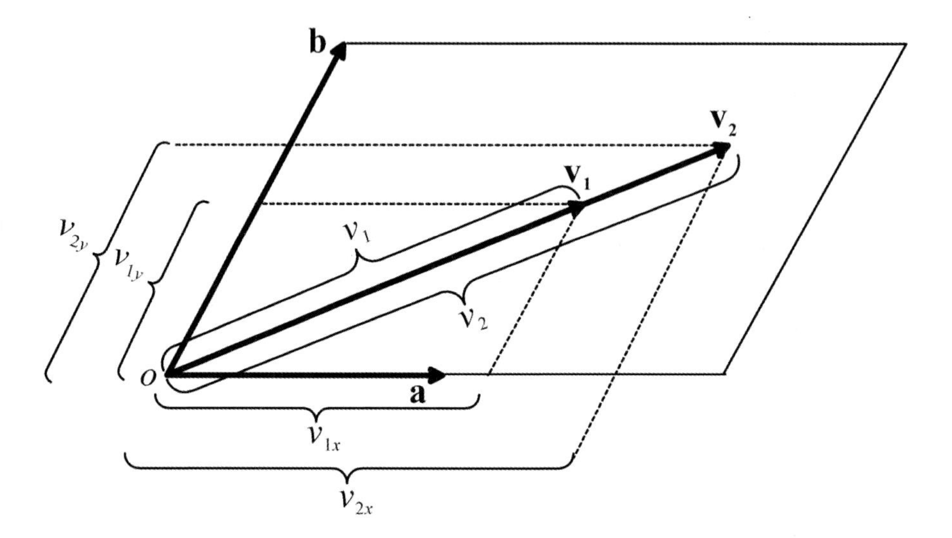

Figure 1.18 The components of the product of a vector and a scalar: $\mathbf{v_2} = s\mathbf{v_1}$.

Vector addition can also be expressed in terms of vector components. Fig. 1.19 illustrates the formation of $\mathbf{v_3} = [v_{3x}\ v_{3y}]$ as the sum of $\mathbf{v_1} = [v_{1x}\ v_{1y}]$ and $\mathbf{v_2} = [v_{2x}\ v_{2y}]$. In order to move the tail of $\mathbf{v_2}$ to the head of $\mathbf{v_1}$, $\mathbf{v_2}$ must be translated by adding v_{1x} to its $\mathbf{a}$ component and v_{1y} to its $\mathbf{b}$ component:

$$\begin{aligned}
\mathbf{v_1} &= v_{1x}\mathbf{a} + v_{1y}\mathbf{b} = [v_{1x}\ v_{1y}] \\
\mathbf{v_2} &= v_{2x}\mathbf{a} + v_{2y}\mathbf{b} = [v_{2x}\ v_{2y}] \\
\mathbf{v_3} &= v_{3x}\mathbf{a} + v_{3y}\mathbf{b} = [v_{3x}\ v_{3y}] \\
v_{3x} &= v_{1x} + v_{2x} \\
v_{3y} &= v_{1y} + v_{2y} \\
\mathbf{v_3} &= (v_{1x} + v_{2x})\mathbf{a} + (v_{1y} + v_{2y})\mathbf{b} = [(v_{1x} + v_{2x})\ (v_{1y} + v_{2y})] \qquad (1.12)
\end{aligned}$$

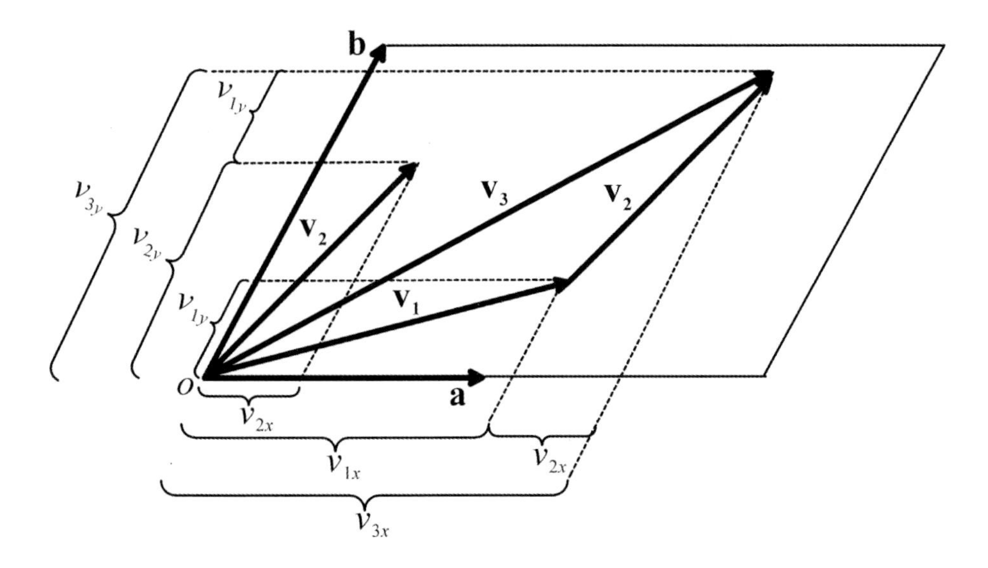

Figure 1.19 The components of the sum of two vectors: $\mathbf{v_3} = \mathbf{v_1} + \mathbf{v_2}$.

Finally, the subtraction of vectors can be accomplished using the vector components by combining scalar multiplication and vector addition:

$$\begin{aligned} -\mathbf{v_1} &= -1(v_{1x}\mathbf{a} + v_{1y}\mathbf{b}) = [-v_{1x} \ -v_{1y}] \\ \mathbf{v_2} - \mathbf{v_1} &= \mathbf{v_2} + (-\mathbf{v_1}) = (v_{2x} - v_{1x})\mathbf{a} + (v_{2y} - v_{1y})\mathbf{b} \\ &= [(v_{2x} - v_{1x}) \ (v_{2y} - v_{1y})] \end{aligned} \quad (1.13)$$

Adding or subtracting the components of two vectors produces a new vector which is the vector sum or difference of the original vectors.

Extrapolation of these concepts to a three-dimensional lattice is straightforward. *Any vector in three-dimensional space can be expressed as a linear combination of three linearly independent basis vectors.* The **c** axis becomes a third linearly independent basis vector, creating a three-dimensional basis set, $\{\mathbf{a}, \mathbf{b}, \mathbf{c}\}$, as illustrated in Fig. 1.20. A point p in the lattice is now found by traversing along vector x, coincident with the **a** axis, then along a vector y parallel to the **b** axis, and finally along z parallel to the **c** axis, such that

$$\overrightarrow{\mathbf{op}} = \mathbf{x} + \mathbf{y} + \mathbf{z}$$

$$\overrightarrow{\mathbf{op}} = s_x\mathbf{a} + s_y\mathbf{b} + s_z\mathbf{c}$$

$$s_x = \frac{x}{a} = x_f \quad (1.14)$$

$$s_y = \frac{y}{b} = y_f \quad (1.15)$$

$$s_z = \frac{z}{c} = z_f \quad (1.16)$$

$$\overrightarrow{\mathbf{op}} = x_f\mathbf{a} + y_f\mathbf{b} + z_f\mathbf{c} = [x_f \ y_f \ z_f]. \quad (1.17)$$

In Fig. 1.21(a) the vector $\overrightarrow{\mathbf{op}} = [\frac{1}{2} \ \frac{1}{2} \ \frac{1}{2}]$ defines the point at the center of the unit cell.

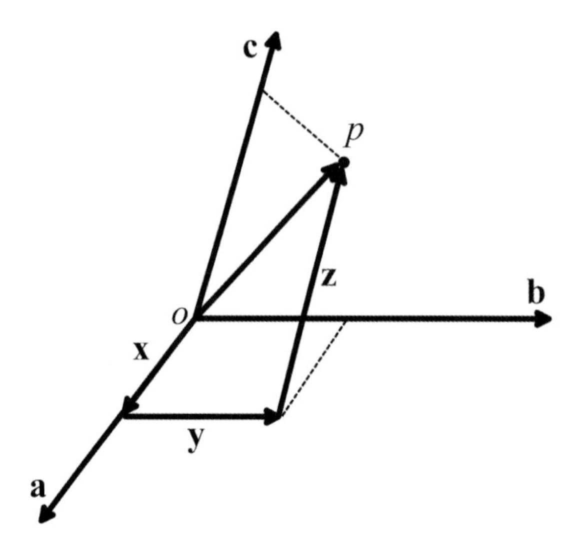

Figure 1.20 Three-dimensional unit cell illustrating the axes **a**, **b**, and **c** as basis vectors for three-dimensional space and the vector locating point p as a linear combination of the axial basis vectors.

The indices of a given set of lattice planes are conveniently determined from the fractional coordinates of the three vectors from the origin of the unit cell to the initial points of intersection of the planes along each coordinate axis. Recall that the indices were defined by the number of equal segments into which each axis was divided (Fig. 1.11). This led to an ambiguity in the assignment of the indices, since there were eight sets of planes which divided the axes into the same increments. These planes were differentiated by taking into account directions in the lattice, and the use of vectors provides a natural way to accomplish this. For a given set of planes, the **a**, **b**, and **c** axes are divided into h, k and l segments, respectively. The fractional coordinates to the first point of intersection along each axis are therefore $[(1/h\ 0\ 0]$, $[0\ 1/k\ 0]$, and $[0\ 0\ 1/l]$, as illustrated in Fig. 1.21(b). The signs of the indices are now taken as the signs of the vectors, determined from their directions in the right-handed unit cell coordinate system.

As we observed in two dimensions, multiplying a vector by a scalar in three dimensions produces a new vector formed by multiplying each component of the original vector by the scalar; adding or subtracting the components of two vectors produces the sum or difference of of the two vectors. For $\mathbf{v_1} = [v_{1x}\ v_{1y}\ v_{1z}]$ and $\mathbf{v_2} = [v_{2x}\ v_{2y}\ v_{2z}]$,

$$\mathbf{v_2} = s\mathbf{v_1} = [v_{2x}\ v_{2y}\ v_{2z}] = s[v_{1x}\ v_{1y}\ v_{1z}] = [sv_{1x}\ sv_{1y}\ v_{1z}] \tag{1.18}$$

$$\begin{aligned}\mathbf{v_3} &= \mathbf{v_1} + \mathbf{v_2} = (v_{1x} + v_{2x})\mathbf{a} + (v_{1y} + v_{2y})\mathbf{b} + (v_{1z} + v_{2z})\mathbf{c} \\ &= [(v_{1x} + v_{2x})\ (v_{1y} + v_{2y})\ (v_{1z} + v_{2z})]\end{aligned} \tag{1.19}$$

$$\begin{aligned}\mathbf{v_3'} &= \mathbf{v_2} - \mathbf{v_1} = (v_{2x} - v_{1x})\mathbf{a} + (v_{2y} - v_{1y})\mathbf{b} + (v_{2z} - v_{1z})\mathbf{c} \\ &= [(v_{2x} - v_{1x})\ (v_{2y} - v_{1y})\ (v_{2z} - v_{1z})].\end{aligned} \tag{1.20}$$

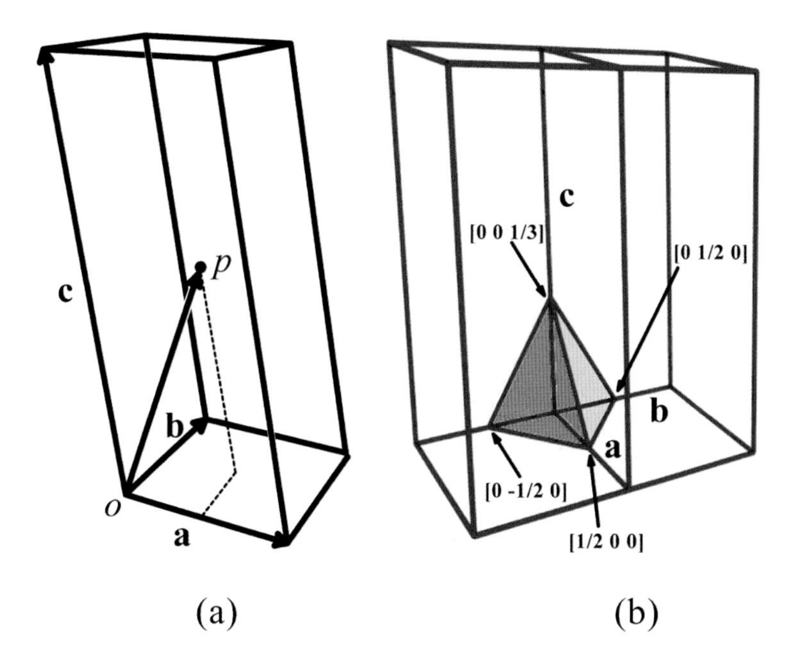

Figure 1.21 2-mercaptopyridine unit cell showing (a) p in the center of the cell at $[x_f\ y_f\ z_f] = [\frac{1}{2}\ \frac{1}{2}\ \frac{1}{2}]$ and (b) the fractional coordinates of the axial intersection vectors for the (2 2 3) planes (violet) and the (2 $\bar{2}$ 3) planes (red).

The multiplication of a vector by a scalar and the addition of vectors, both effected by operating on the vector components, is general for any basis set. This is *not* the case for the scalar and vector products. The formation of these two products from the components is transparent only in specific coordinate systems in which the basis vectors are vectors of unit length (*normalized*) which are perpendicular to one another(*orthogonal*). Such basis sets are known as *orthonormal* bases.

1.3.3 Orthonormal Bases

The unit cell axes are a natural choice for basis vectors when describing locations in the lattice — every coordinate consists of an integer (to define the specific unit cell) plus a fraction (to specify the position within that unit cell). Furthermore, since every unit cell is identical, the fractional coordinates alone uniquely define the internal structure of the crystal; the integers serve only to define the specific cell in which a targeted location exists. For example, the fractional coordinates [10.5 232.5 19.5] tell us that the point of interest lies in the center of a unit cell 10 a unit cell lengths along the **a** axial direction, 232 b units in the **b** direction, and 19 c units in the **c** direction. The periodic nature of the crystal lattice is reflected in this basis set, and we will see that its use greatly simplifies the mathematics of diffraction.

However, we will also need to perform numerical calculations, specifically concerned with molecular parameters such as bond lengths and angles. In addition, we will find the need to transform the natural coordinates of the crystal lattice into a laboratory reference frame in the form of a Cartesian coordinate system in order to collect diffraction data. This will require us to describe the vectors in the lattice with a basis set which will allow us to use the coordinates to compute scalar and vector products. As alluded to above, this necessitates the use of a unique basis set consisting of orthonormal vectors – vectors of unit length which are perpendicular to one another.

An orthonormal basis set is shown in Fig. 1.22. We will also refer to a basis set as a *coordinate system.* An orthonormal coordinate system is a special case of a Cartesian coordinate system in which the basis vectors are of unit length ($i = j = k = 1$), referred to as *unit vectors.* The **i**, **j**, and **k** vectors are mutually perpendicular and are chosen so that **k** points in the direction of $\mathbf{i} \times \mathbf{j}$. A point p is located in this coordinate system by traversing a distance x_c along **i**, then y_c along a vector parallel to **j**, and z_c along a vector parallel to **k**, exactly as was done in the general coordinate system; its location is determined by the vector $\overrightarrow{\mathbf{op}} = [x_c \; y_c \; z_c]$. We will use the subscript "c" to refer to a *Cartesian coordinate* throughout the text. A vector in this coordinate system is a linear combination of orthonormal basis vectors:

$$\overrightarrow{\mathbf{op}} = x_c\mathbf{i} + y_c\mathbf{j} + z_c\mathbf{k} = [x_c \; y_c \; z_c] \tag{1.21}$$

The multiplicative properties of vectors in an orthonormal coordinate system are unique, allowing for the determination of scalar and vector products from vector components, thus providing a powerful tool for determining distances and angles

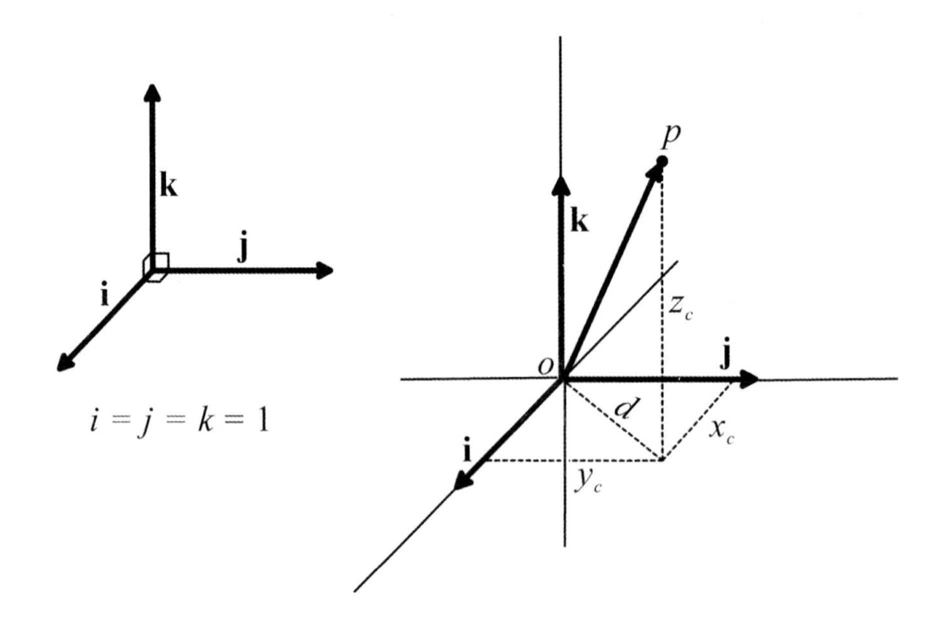

Figure 1.22 Orthonormal basis set — the point p is located at the end of the vector defined by $[x_c \; y_c \; z_c]$.

within the crystal lattice. This uniqueness arises from the multiplicative properties of the basis vectors themselves. The scalar products of **i**, **j**, and **k** are summarized below:

$$\begin{aligned} \mathbf{i}\cdot\mathbf{i} &= ii\cos(0) &= 1 \\ \mathbf{j}\cdot\mathbf{j} &= jj\cos(0) &= 1 \\ \mathbf{k}\cdot\mathbf{k} &= kk\cos(0) &= 1 \end{aligned} \tag{1.22}$$

$$\begin{aligned} \mathbf{i}\cdot\mathbf{j} &= ij\cos(\tfrac{\pi}{2}) &= 0 \\ \mathbf{i}\cdot\mathbf{k} &= ik\cos(\tfrac{\pi}{2}) &= 0 \\ \mathbf{j}\cdot\mathbf{k} &= jk\cos(\tfrac{\pi}{2}) &= 0. \end{aligned} \tag{1.23}$$

The vector product of **i**, **j**, or **k** with itself produces a vector of zero length (a point) known as the *null* vector:

$$\begin{aligned} |\mathbf{i}\times\mathbf{i}| &= ii\sin(0) &= 0 \implies \mathbf{i}\times\mathbf{i} &= \mathbf{0} \\ |\mathbf{j}\times\mathbf{j}| &= jj\sin(0) &= 0 \implies \mathbf{j}\times\mathbf{j} &= \mathbf{0} \\ |\mathbf{k}\times\mathbf{k}| &= kk\sin(0) &= 0 \implies \mathbf{k}\times\mathbf{k} &= \mathbf{0} \end{aligned} \tag{1.24}$$

The remaining vector products depend on the order in which the basis vectors are multiplied. We have defined the coordinate system such that the vector product $\mathbf{i}\times\mathbf{j}$ points in the direction of **k**. A coordinate system defined in this manner is termed a *right-handed coordinate system*. Since $|\mathbf{i}\times\mathbf{j}| = ij\sin(\frac{\pi}{2}) = 1$, the vector product of **i** and **j** is *exactly* **k**. Recalling that reversing the order of multiplication changes the sign of the vector product, the remaining vector products are determined in the same manner:

$$\begin{aligned} \mathbf{i}\times\mathbf{j} &= \mathbf{k} & \quad \mathbf{j}\times\mathbf{i} &= -\mathbf{k} \\ \mathbf{j}\times\mathbf{k} &= \mathbf{i} & \quad \mathbf{k}\times\mathbf{j} &= -\mathbf{i} \\ \mathbf{k}\times\mathbf{i} &= \mathbf{j} & \quad \mathbf{i}\times\mathbf{k} &= -\mathbf{j} \end{aligned} \tag{1.25}$$

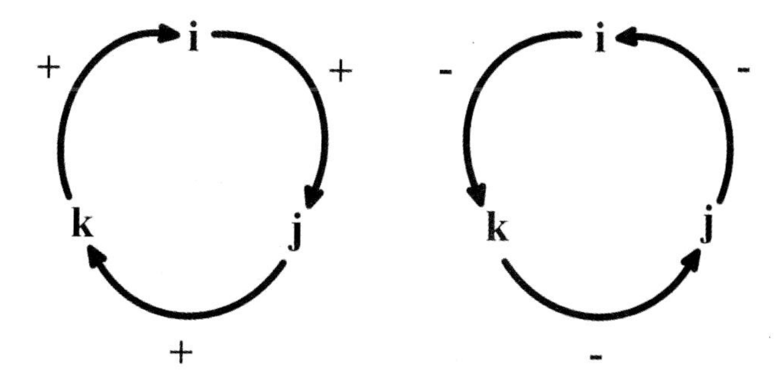

Figure 1.23 Mnemonic device for recalling the vector products of orthonormal basis vectors. Clockwise rotations are positive; counterclockwise rotations are negative. For example, $\mathbf{k}\times\mathbf{i} = \mathbf{j}$ and $\mathbf{j}\times\mathbf{i} = -\mathbf{k}$.

Although these vector products can always be generated by referring to the right-handed coordinate system illustrated in Fig. 1.22, a convenient mnemonic device for remembering them is shown in Fig. 1.23. Using this scheme, clockwise rotations produce positive vector products, counter-clockwise rotations produce negative vector products.

1.3.4 The Scalar Product in an Orthonormal Coordinate System

The right angles of the orthogonal coordinate system make it possible to use the Pythagorean theorem to determine the length of a vector with its tail at the origin from the components of the vector. Referring to Fig. 1.22, the diagonal d is the hypotenuse of a right triangle, as is the vector $\overrightarrow{\mathbf{op}}$:

$$\begin{aligned} d^2 &= x_c^2 + y_c^2 \\ |\overrightarrow{\mathbf{op}}|^2 &= d^2 + z_c^2 = x_c^2 + y_c^2 + z_c^2 \\ |\overrightarrow{\mathbf{op}}| &= (x_c^2 + y_c^2 + z_c^2)^{\frac{1}{2}}. \end{aligned} \tag{1.26}$$

Now, consider the arbitrary vector, $\mathbf{v_1}$ in Fig. 1.24. In order to determine its length we construct two vectors emanating from the origin, $\mathbf{v_2}$ with its head at the tail of $\mathbf{v_1}$, and $\mathbf{v_3}$ with its head at the head of $\mathbf{v_1}$. Clearly $\mathbf{v_3} = \mathbf{v_2} + \mathbf{v_1}$ and $\mathbf{v_1} = \mathbf{v_3} - \mathbf{v_2}$. The process of negating $\mathbf{v_2}$ and adding it to $\mathbf{v_3}$ translates $\mathbf{v_1}$ it to the origin. Indeed, this is just the reverse of the process of vector addition, where $\mathbf{v_1}$ is translated from the origin to the head of the vector $\mathbf{v_2}$ to which it is added. The components of the translated vector $\mathbf{v_1} = [v_{1x}\ v_{1y}\ v_{1z}]$ can therefore provide us with the length of the vector. Let $\mathbf{v_2} = [v_{2x}\ v_{2y}\ v_{2z}]$ and $\mathbf{v_3} = [v_{3x}\ v_{3y}\ v_{3z}]$. Then, $\mathbf{v_1} = [v_{1x}\ v_{1y}\ v_{1z}] = [(v_{3x} - v_{2x})\ (v_{3y} - v_{2y})\ (v_{3z} - v_{2z})]$. The subtraction of the coordinates of $\mathbf{v_2}$ from those of $\mathbf{v_3}$ translates $\mathbf{v_1}$ back to the origin, allowing for the determination of the squares of the lengths of all three vectors:

$$\begin{aligned} v_3^2 &= v_{3x}^2 + v_{3y}^2 + v_{3z}^2 \\ v_2^2 &= v_{2x}^2 + v_{2y}^2 + v_{2z}^2 \\ v_1^2 &= (v_{3x} - v_{2x})^2 + (v_{3y} - v_{2y})^2 + (v_{3z} - v_{2z})^2. \end{aligned} \tag{1.27}$$

The third equation allows us to compute distances between any two points in the lattice, provided that we have the coordinates for the two points in an orthonormal basis. We could, for example, imagine an atom with coordinates $[v_{3x}\ v_{3y}\ v_{3z}]$ and another at $[v_{2x}\ v_{2y}\ v_{2z}]$. The distance between the two atoms can be determined by subtracting the coordinates of the two points, squaring the differences, summing the squares, and taking the square root of the sum to give

$$v_1 = ((v_{3x} - v_{2x})^2 + (v_{3y} - v_{2y})^2 + (v_{3z} - v_{2z})^2)^{\frac{1}{2}}. \tag{1.28}$$

Finally, expanding Eqn. 1.27 provides an expression for the scalar product in terms of the coordinates *in an orthonormal basis*:

$$\begin{aligned} v_1^2 &= v_{3x}^2 - 2v_{3x}v_{2x} + v_{2x}^2 + v_{3y}^2 - 2v_{3y}v_{2y} + v_{2y}^2 + v_{3z}^2 - 2v_{3z}v_{2z} + v_{2z}^2 \\ v_1^2 &= (v_{2x}^2 + v_{2y}^2 + v_{2z}^2) + (v_{3x}^2 + v_{3y}^2 + v_{3z}^2) - 2(v_{2x}v_{3x} + v_{2y}v_{3y} + v_{2z}v_{3z}) \\ v_1^2 &= v_2^2 + v_3^2 - 2(v_{2x}v_{3x} + v_{2y}v_{3y} + v_{2z}v_{3z}). \end{aligned} \tag{1.29}$$

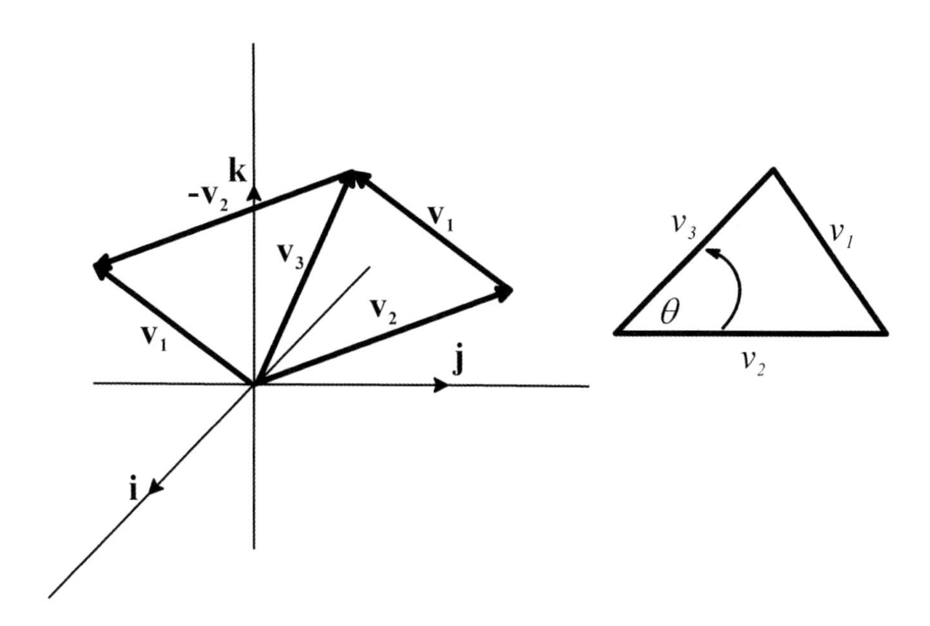

Figure 1.24 Vectors used to determine the length of a general vector in an orthonormal coordinate system. The Law of Cosines determines the length of $\mathbf{v_1}$, $v_1^2 = v_2^2 + v_3^2 - 2v_2v_3\cos\theta$.

Referring to the triangle to the right in Fig. 1.24, θ is the angle between $\mathbf{v_2}$ and $\mathbf{v_3}$. The Law of Cosines produces the square of the length of the third side of a triangle from the lengths of the other two sides and the angle between them. In this case we are interested in the square of the length of $\mathbf{v_1}$: $v_1^2 = v_2^2 + v_3^2 - 2v_2v_3\cos\theta$. Equating these two expressions for v_1^2 gives

$$\begin{aligned} v_2^2 + v_3^2 - 2v_2v_3\cos\theta &= v_2^2 + v_3^2 - 2(v_{2x}v_{3x} + v_{2y}v_{3y} + v_{2z}v_{3z}) \\ v_2v_3\cos\theta &= v_{2x}v_{3x} + v_{2y}v_{3y} + v_{2z}v_{3z}. \end{aligned} \tag{1.30}$$

We have arrived at a very useful result — $v_2v_3\cos\theta$ is the scalar product of $\mathbf{v_2}$ and $\mathbf{v_3}$:

$$\mathbf{v_2} \cdot \mathbf{v_3} = v_{2x}v_{3x} + v_{2y}v_{3y} + v_{2z}v_{3z}. \tag{1.31}$$

In an orthonormal coordinate system the scalar product of two vectors is given by the sum of the products of the components of each vector. The length of a vector with its tail at the origin is now readily determined by taking the scalar product of a vector with itself:

$$\begin{aligned} \mathbf{v_1} &= v_{1x}\mathbf{i} + v_{1y}\mathbf{j} + v_{1z}\mathbf{k} = [v_{1x}\ v_{1y}\ v_{1z}] \\ \mathbf{v_1}\cdot\mathbf{v_1} &= v_{1x}v_{1x} + v_{1y}v_{1y} + v_{1z}v_{1z} \\ v_1v_1\cos(0) &= v_1^2 = v_{1x}^2 + v_{1y}^2 + v_{1z}^2 \\ v_1 &= (v_{1x}^2 + v_{1y}^2 + v_{1z}^2)^{\frac{1}{2}} \\ v_1 &= (\mathbf{v_1}\cdot\mathbf{v_1})^{\frac{1}{2}}. \end{aligned} \tag{1.32}$$

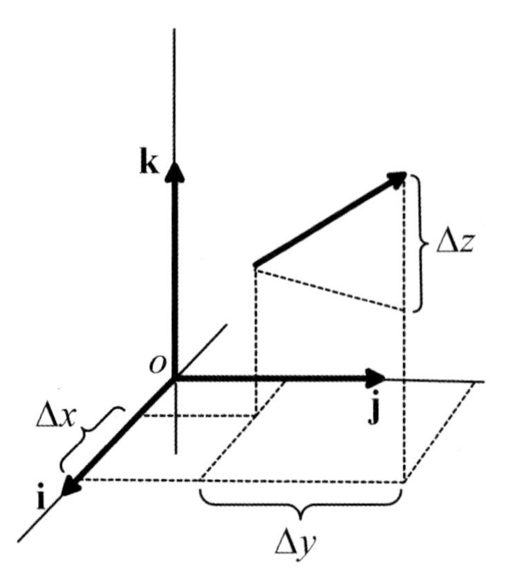

Figure 1.25 Vector in orthonormal coordinates illustrating the changes in the components obtained from subtracting the coordinates of the head of the vector from those of the tail. Translating the vector to the origin places the tail at [0 0 0] and the head at $[\Delta x\ \Delta y\ \Delta z]$.

The scalar product also yields the angle between two vectors emanating from the origin in an orthonormal system:

$$\begin{aligned}
\mathbf{v_1} &= v_{1x}\mathbf{i} + v_{1y}\mathbf{j} + v_{1z}\mathbf{k} = [v_{1x}\ v_{1y}\ v_{1z}] \\
\mathbf{v_2} &= v_{2x}\mathbf{i} + v_{2y}\mathbf{j} + v_{2z}\mathbf{k} = [v_{2x}\ v_{2y}\ v_{2z}] \\
v_1 &= (\mathbf{v_1} \cdot \mathbf{v_1})^{\frac{1}{2}} = (v_{1x}^2 + v_{1y}^2 + v_{1z}^2)^{\frac{1}{2}} \\
v_2 &= (\mathbf{v_2} \cdot \mathbf{v_2})^{\frac{1}{2}} = (v_{2x}^2 + v_{2y}^2 + v_{2z}^2)^{\frac{1}{2}} \\
\mathbf{v_1} \cdot \mathbf{v_2} &= v_1 v_2 \cos\theta = v_{1x}v_{2x} + v_{1y}v_{2y} + v_{1z}v_{2z} \\
\cos\theta &= \frac{\mathbf{v_1} \cdot \mathbf{v_2}}{v_1 v_2} = \frac{v_{1x}v_{2x} + v_{1y}v_{2y} + v_{1z}v_{2z}}{(v_{1x}^2 + v_{1y}^2 + v_{1z}^2)^{\frac{1}{2}}(v_{2x}^2 + v_{2y}^2 + v_{2z}^2)^{\frac{1}{2}}}.
\end{aligned} \tag{1.33}$$

The origin constraint is not as restrictive as it first appears, since we can translate a vector to the origin by subtracting the coordinates of its head and tail, as we did earlier to determine the length of a general vector. Fig. 1.25 provides an alternative way to envision this. Consider the head of the vector to have coordinates $[x_h y_h z_h]$, and the tail to have coordinates $[x_t\ y_t\ z_t]$ (These would be the components of $\mathbf{v_2}$ and $\mathbf{v_3}$ in the previous examples). Subtracting the components of the tail from those of the head gives the *change* in the coordinates along each of the axes: $[(x_h - x_t)\ (y_h - y_t)\ (z_h - z_t)] = [\Delta x\ \Delta y\ \Delta z]$. If we translate the vector by maintaining its magnitude and direction and moving its tail to the origin, the components of the vector will then be $[\Delta x\ \Delta y\ \Delta z]$, and the magnitude of the vector will be $((\Delta x)^2 + (\Delta y)^2 + (\Delta z)^2)^{\frac{1}{2}}$. For two vectors, $\mathbf{v_1}$ and $\mathbf{v_2}$, the scalar product will be obtained by translating both vectors to the origin: $\mathbf{v_1} \cdot \mathbf{v_2} = \Delta x_1 \Delta x_2 + \Delta y_1 \Delta y_2 + \Delta z_1 \Delta z_2$.

The parallel between the scalar product of vectors and the multiplication of ordinary numbers has already been established for the commutative property of multiplication, i.e., the product remains unchanged when the order of multiplication is changed. The associate property of multiplication also applies to ordinary numbers. In this case multiplications are always binary operations between pairs of numbers, and the property tells us that we can create any pairs that we wish to obtain their product: $abc = (ab)c = a(bc)$, etc. The analogous scalar product makes no sense, since the scalar product of any pair of vectors will result in a scalar, making a second scalar product impossible. However there is a close analogy in the multiplication of a scalar and a scalar product:

$$\begin{aligned} \mathbf{v_1} \cdot (s\mathbf{v_2}) &= v_1(sv_2)\cos\theta \\ &= s(v_1v_2)\cos\theta \ = \ s(\mathbf{v_1} \cdot \mathbf{v_2}) \end{aligned} \tag{1.34}$$

The scalar product is associative with respect to multiplication by a scalar. The multiplication of ordinary numbers is also distributive: $a(b + c) = ab + ac$. The scalar product analogy is $\mathbf{v_1} \cdot (\mathbf{v_2} + \mathbf{v_3})$. In orthonormal coordinates,

$$\begin{aligned} \mathbf{v_1} \cdot (\mathbf{v_2} + \mathbf{v_3}) &= [v_{1x}\ v_{1y}\ v_{1z}] \cdot [(v_{2x} + v_{3x})\ (v_{2y} + v_{3y})\ (v_{2z} + v_{3z})] \\ &= v_{1x}(v_{2x} + v_{3x}) + v_{1y}(v_{2y} + v_{3y}) + v_{1z}(v_{2z} + v_{3z}) \\ &= v_{1x}v_{2x} + v_{1x}v_{3x} + v_{1y}v_{2y} + v_{1y}v_{3y} + v_{1z}v_{2z} + v_{1z}v_{3z} \\ &= (v_{1x}v_{2x} + v_{1y}v_{2y} + v_{1z}v_{2z}) + (v_{1x}v_{3x} + v_{1y}v_{3y} + v_{1z}v_{3z}) \\ &= (\mathbf{v_1} \cdot \mathbf{v_2}) + (\mathbf{v_1} \cdot \mathbf{v_3}). \end{aligned} \tag{1.35}$$

The scalar product is distributive with respect to vector addition. We will find the use of scalar products invaluable in the calculation of interatomic distances and angles inside the crystal lattice.

A Note on Terminology: Scalar/Dot/Inner Product. The terms *scalar product*, *dot product*, and *inner product* are commonly used interchangeably. As demonstrated above, the scalar resulting from $\mathbf{v_1} \cdot \mathbf{v_2} = v_1v_2\cos\theta$ is equal to the scalar resulting from $\mathbf{v_1} \cdot \mathbf{v_2} = v_{1x}v_{2x} + v_{1y}v_{2y} + v_{1z}v_{2z}$, only if the components are defined in an orthonormal basis.

In most cases the use of various terms to describe both of these scalars is unambiguous, either because the coordinate system is orthonormal, or because there is no confusion in the context of their use. However, in crystallography, the transformation of vectors between orthonormal and non-orthonormal coordinate systems occurs regularly, and it is important to have a descriptive term for a scalar that does not vary when these transformations occur. We initially referred to this invariant term as *the scalar product*, and we will continue to do so (expanding the concept a bit when discussing reciprocal lattices).

On the other hand, the value of the sum of the products of the components of two vectors does vary with the coordinate system, but we will find this scalar useful in representing the sums of products of the components of vectors in the basis of the crystal lattice with the components of vectors defined in the basis of the lattice of the diffraction pattern. In addition, the scalar provides a convenient generalization of the matrix product of a vector and its transpose, encountered later on in this chapter. In order to differentiate this scalar from the other, we will adopt the convention that the sum of the products of the components is the *inner product* of the vectors. Whenever there is no need to distinguish between the two entities, the term *dot product* will apply.

1.3.5 The Vector Product in an Orthonormal Coordinate System

The vector product can also be expressed in terms of the coordinates in an orthonormal basis. Fig. 1.26 illustrates the vector product $\mathbf{v_1} = \mathbf{v_2} \times \mathbf{v_3}$ for two vectors with tails at the origin (translated as discussed previously, if necessary).

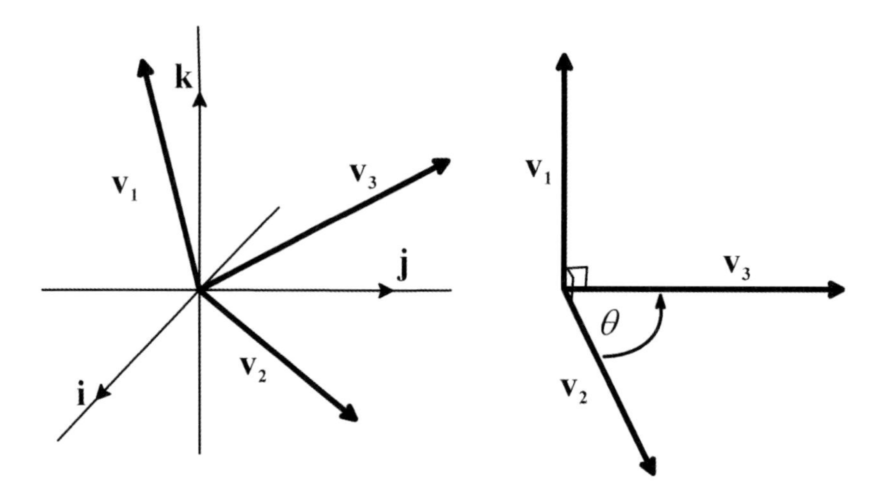

Figure 1.26 Vector product in orthonormal coordinates: $\mathbf{v_1} = \mathbf{v_2} \times \mathbf{v_3}$. $v_1 = v_2 v_3 \sin\theta$ with $\mathbf{v_1} \perp \mathbf{v_2}$ and $\mathbf{v_1} \perp \mathbf{v_3}$.

From the definition of the vector product,

$$\begin{aligned}
v_1 &= v_2 v_3 \sin\theta \\
v_1^2 &= v_2^2 v_3^2 \sin^2\theta \;=\; v_2^2 v_3^2 (1-\cos^2\theta) \\
&= v_2^2 v_3^2 - v_2^2 v_3^2 \cos^2\theta \;=\; v_2^2 v_3^2 - (v_2 v_3 \cos\theta)^2 \;=\; v_2^2 v_3^2 - (\mathbf{v_2}\cdot\mathbf{v_3})^2 \\
&= (v_{2x}^2 + v_{2y}^2 + v_{2z}^2)(v_{3x}^2 + v_{3y}^2 + v_{3z}^2) - (v_{2x}v_{3x} + v_{2y}v_{3y} + v_{2z}v_{3z})^2
\end{aligned}$$

Expansion and collection of the terms on the right gives

$$\begin{aligned}
v_1^2 &= (v_{2y}^2 v_{3z}^2 - 2v_{2y}v_{3y}v_{2z}v_{3z} + v_{2z}^2 v_{3y}^2) \\
&+ (v_{2z}^2 v_{3x}^2 - 2v_{2z}v_{3z}v_{2x}v_{3x} + v_{2x}^2 v_{3z}^2) \\
&+ (v_{2x}^2 v_{3y}^2 - 2v_{2x}v_{3x}v_{2y}v_{3y} + v_{2y}^2 v_{3x}^2) \\
v_1^2 &= (v_{2y}v_{3z} - v_{2z}v_{3y})^2 + (v_{2z}v_{3x} - v_{2x}v_{3z})^2 + (v_{2x}v_{3y} - v_{2y}v_{3x})^2. \qquad (1.36)
\end{aligned}$$

Since the square of the magnitude of $\mathbf{v_1}$ is the sum of the squares of its components, the square roots of the three terms in Eqn. 1.36 are clearly the components of the vector product of $\mathbf{v_2}$ and $\mathbf{v_3}$. Unfortunately, the equation does not tell us which coordinate axis each component corresponds to. To determine which is which we apply the added constraint that $\mathbf{v_1} \perp \mathbf{v_2}$ and $\mathbf{v_1} \perp \mathbf{v_3} \Longrightarrow \mathbf{v_1} \cdot \mathbf{v_2} = 0$ and $\mathbf{v_1} \cdot \mathbf{v_3} = 0$. In order for the terms in the scalar products to cancel to zero, every number triple in the scalar product sum must contain an x, y, and z contribution. Of the six possible ways of permuting the components, only when $(v_{2y}v_{3z} - v_{2z}v_{3y})$ is multiplied by v_{2x} or v_{3x}, $(v_{2z}v_{3x} - v_{2x}v_{3z})$ is multiplied by v_{2y} or v_{3y}, and $(v_{2x}v_{3y} - v_{2y}v_{3x})$ is multiplied by v_{2z} or v_{3z} will this be the case. We have thus arrived at an expression for the vector product of two vectors in terms of their orthonormal components:

$$\begin{aligned} \mathbf{v_2} \times \mathbf{v_3} &= (v_{2y}v_{3z} - v_{2z}v_{3y})\mathbf{i} + (v_{2z}v_{3x} - v_{2x}v_{3z})\mathbf{j} + (v_{2x}v_{3y} - v_{2y}v_{3x})\mathbf{k} \\ &= [(v_{2y}v_{3z} - v_{2z}v_{3y})\ (v_{2z}v_{3x} - v_{2x}v_{3z})\ (v_{2x}v_{3y} - v_{2y}v_{3x})]. \end{aligned} \tag{1.37}$$

In an orthonormal coordinate system the vector product of two vectors is a vector with components comprised of all the possible pairs of cross terms of the components of the original vectors; the vector product is often called the *cross* product.

It has already been established that the vector product is not commutative. It is instructive to revisit this in terms of the orthonormal components of the vector.

$$\begin{aligned} \mathbf{v_3} \times \mathbf{v_2} &= (v_{3y}v_{2z} - v_{3z}v_{2y})\mathbf{i} + (v_{3z}v_{2x} - v_{3x}v_{2z})\mathbf{j} + (v_{3x}v_{2y} - v_{3y}v_{2x})\mathbf{k} \\ &= -(v_{2y}v_{3z} - v_{2z}v_{3y})\mathbf{i} - (v_{2z}v_{3x} - v_{2x}v_{3z})\mathbf{j} - (v_{2x}v_{3y} - v_{2y}v_{3x})\mathbf{k} \\ &= -(\mathbf{v_2} \times \mathbf{v_3}). \end{aligned} \tag{1.38}$$

As demonstrated for the general coordinate system previously, the sign of the vector has changed, but the magnitude, $v_2v_3 \sin\theta$, remains the same. In addition to being commutative, the *scalar* product is also associative with respect to multiplication by a scalar, and distributive with respect to vector addition. Despite its non-commutative nature, does the vector product share any analogous properties with ordinary numbers? Consider the vector product $\mathbf{v_2} \times s\mathbf{v_3}$:

$$\begin{aligned} \mathbf{v_2} \times s\mathbf{v_3} &= (v_{2y}sv_{3z} - v_{2z}sv_{3y})\mathbf{i} + (v_{2z}sv_{3x} - v_{2x}sv_{3z})\mathbf{j} + (v_{2x}sv_{3y} - v_{2y}sv_{3x})\mathbf{k} \\ &= s(v_{2y}v_{3z} - v_{2z}v_{3y})\mathbf{i} + s(v_{2z}v_{3x} - v_{2x}v_{3z})\mathbf{j} + s(v_{2x}v_{3y} - v_{2y}v_{3x})\mathbf{k} \\ &= s((v_{2y}v_{3z} - v_{2z}v_{3y})\mathbf{i} + (v_{2z}v_{3x} - v_{2x}v_{3z})\mathbf{j} + (v_{2x}v_{3y} - v_{2y}v_{3x}))\mathbf{k} \\ &= s(\mathbf{v_2} \times \mathbf{v_3}). \end{aligned} \tag{1.39}$$

The vector product is associative with respect to scalar multiplication.

The *scalar product* provides for a convenient test for perpendicular vectors. If $\mathbf{v_1} \perp \mathbf{v_2}$ then $\mathbf{v_1} \cdot \mathbf{v_2} = v_1v_2\cos(\pi/2) = 0$. The associative property of the *vector product* allows for a convenient test for parallel vectors. Suppose $\mathbf{v_1} \| \mathbf{v_2} \Longrightarrow \mathbf{v_2} = s\mathbf{v_1}$. Thus $\mathbf{v_1} \times \mathbf{v_2} = \mathbf{v_1} \times s\mathbf{v_1} = s(\mathbf{v_1} \times \mathbf{v_1}) = \mathbf{0}$, the null vector, with components $[0\ 0\ 0]$. *If the scalar product is zero then the vectors are perpendicular. If the vector product is* **0** *then the vectors are parallel.* Finally, consider the vector product of a vector with the sum of two vectors:

$$\begin{aligned}
\mathbf{v_1} \times (\mathbf{v_2} + \mathbf{v_3}) &= (v_{1y}(v_{2z} + v_{3z})) - v_{1z}(v_{2y} + v_{3y}))\mathbf{i} \\
&+ (v_{1z}(v_{2x} + v_{3x})) - v_{1x}(v_{2z} + v_{3z}))\mathbf{j} \\
&+ (v_{1x}(v_{2y} + v_{3y})) - v_{1y}(v_{2y} + v_{3x}))\mathbf{k} \\
\mathbf{v_1} \times (\mathbf{v_2} + \mathbf{v_3}) &= (v_{1y}v_{2z} + v_{1y}v_{3z} - v_{1z}v_{2y} - v_{1z}v_{3y})\mathbf{i} \\
&+ (v_{1z}v_{2x} + v_{1z}v_{3x} - v_{1x}v_{2z} - v_{1x}v_{3z})\mathbf{j} \\
&+ (v_{1x}v_{2y} + v_{1x}v_{3y} - v_{1y}v_{2y} - v_{1y}v_{3x})\mathbf{k} \\
\mathbf{v_1} \times (\mathbf{v_2} + \mathbf{v_3}) &= \\
(v_{1y}v_{2z} - v_{1z}v_{2y})\mathbf{i} &+ (v_{1z}v_{2x} - v_{1x}v_{2z})\mathbf{j} + (v_{1x}v_{2y} - v_{1y}v_{2y})\mathbf{k} \\
+ (v_{1y}v_{3z} - v_{1z}v_{3y})\mathbf{i} &+ (v_{1z}v_{3x} - v_{1x}v_{3z})\mathbf{j} + (v_{1x}v_{3y} - v_{1y}v_{3x})\mathbf{k} \\
&= (\mathbf{v_1} \times \mathbf{v_2}) + (\mathbf{v_1} \times \mathbf{v_3}).
\end{aligned} \tag{1.40}$$

The vector product is distributive with respect to vector addition.

1.4 Matrices in Crystallography

As emphasized in the previous section, a basis set consisting of the unit cell axes is the natural coordinate system for the treatment of lattice symmetry, the topic of the next chapter, as well as for the mathematics of diffraction, treated in Chapter 3. On the other hand, computing distances and angles requires an orthonormal basis set. It follows that we will need a means to transform vectors expressed in one coordinate system to the other. It is also necessary to transform vectors in order to relocate coordinates when considering lattice symmetry. In general, these vector transformations are accomplished using matrices.

1.4.1 Matrix Definitions

A matrix is a rectangular array of numeric or algebraic *elements* arranged in m rows and n columns:

$$\mathbf{D} = \begin{bmatrix} d_{11} & d_{12} & d_{13} & \cdots & d_{1n} \\ d_{21} & d_{22} & d_{23} & \cdots & d_{2n} \\ d_{31} & d_{32} & d_{33} & \cdots & d_{3n} \\ d_{41} & d_{42} & d_{43} & \cdots & d_{4n} \\ \vdots & \vdots & \vdots & \vdots & \vdots \\ d_{m1} & d_{m2} & d_{m3} & \cdots & d_{mn} \end{bmatrix} \tag{1.41}$$

We describe the array as an $m \times n$ matrix, where the matrix element d_{ij} is the element at the intersection of the ith row and the jth column; m and n are called the *row dimension* and *column dimension*, respectively, of the matrix. In general, a boldface capital letter will be used to indicate a matrix, except in the special cases when $m = 1$ or $n = 1$.

When $m = 1$ or $n = 1$ the resulting matrices *are* vectors. When $m = 1$ the vector takes the form of a single row which we refer to as a *row vector*; when $n = 1$ the matrix takes the form of a single column referred to as a *column vector*:

$$\text{row vector } \mathbf{v} = \begin{bmatrix} v_1 & v_2 & v_3 & \cdots & v_n \end{bmatrix} \qquad \text{column vector } \mathbf{v} = \begin{bmatrix} v_1 \\ v_2 \\ v_3 \\ v_4 \\ \vdots \\ v_m \end{bmatrix}$$

It is often useful to express an $m \times n$ matrix as a matrix of n column vectors, each with m components:

$$\mathbf{D} = \begin{bmatrix} \mathbf{d_1} & \mathbf{d_2} & \mathbf{d_3} & \cdots & \mathbf{d_n} \end{bmatrix}, \qquad \mathbf{d_i} = \begin{bmatrix} d_{1i} \\ d_{2i} \\ d_{3i} \\ d_{4i} \\ \vdots \\ d_{mi} \end{bmatrix} \tag{1.42}$$

The *column rank* of the matrix is defined as the number of these vectors which are linearly independent (i.e., cannot be written as linear combinations of other column vectors in the matrix). The *row rank* of the matrix is defined as the number of row vectors in the matrix which are linearly independent. We will generally find that the rank of the matrices in crystallography will be $n = m = 3$. When $n = m$ the matrix is refereed to as a *square matrix*.

The *transpose* of an $m \times n$ matrix is generated by switching its rows and columns, creating an $n \times m$ matrix:

$$\mathbf{D}^T = \begin{bmatrix} d_{11} & d_{21} & d_{31} & \cdots & d_{m1} \\ d_{12} & d_{22} & d_{32} & \cdots & d_{m2} \\ d_{13} & d_{23} & d_{33} & \cdots & d_{m3} \\ d_{14} & d_{24} & d_{34} & \cdots & d_{m4} \\ \vdots & \vdots & \vdots & \vdots & \vdots \\ d_{1n} & d_{2n} & d_{3n} & \cdots & d_{mn} \end{bmatrix} = \begin{bmatrix} \mathbf{d_1}^T \\ \mathbf{d_2}^T \\ \mathbf{d_3}^T \\ \mathbf{d_4}^T \\ \cdots \\ \mathbf{d_n}^T \end{bmatrix}, \mathbf{d_i}^T = \begin{bmatrix} d_{1i} & d_{2i} & d_{3i} & d_{4i} & \cdots & d_{mi} \end{bmatrix} \tag{1.43}$$

Note that the column vectors in Eqn. 1.42 and the row vectors in Eqn. 1.43 are transposes of one another. Transposing a matrix switches d_{ij} and d_{ji}. A square matrix which remains unchanged when it is transposed (i.e. $d_{ij} = d_{ji}$ and $\mathbf{D} = \mathbf{D}^T$) appears identical above and below the diagonal elements of the matrix and is termed a *symmetric* matrix.

1.4.2 Matrix Operations

The product of a scalar and a matrix multiplies every element in the matrix by the scalar:

$$s\mathbf{D} = s\begin{bmatrix} d_{11} & d_{12} & d_{13} & \cdots & d_{1n} \\ d_{21} & d_{22} & d_{23} & \cdots & d_{2n} \\ d_{31} & d_{32} & d_{33} & \cdots & d_{3n} \\ d_{41} & d_{42} & d_{43} & \cdots & d_{4n} \\ \vdots & \vdots & \vdots & \vdots & \vdots \\ d_{m1} & d_{m2} & d_{m3} & \cdots & d_{mn} \end{bmatrix} = \begin{bmatrix} sd_{11} & sd_{12} & sd_{13} & \cdots & sd_{1n} \\ sd_{21} & sd_{22} & sd_{23} & \cdots & sd_{2n} \\ sd_{31} & sd_{32} & sd_{33} & \cdots & sd_{3n} \\ sd_{41} & sd_{42} & sd_{43} & \cdots & sd_{4n} \\ \vdots & \vdots & \vdots & \vdots & \vdots \\ sd_{m1} & sd_{m2} & sd_{m3} & \cdots & sd_{mn} \end{bmatrix} \quad (1.44)$$

The multiplication of a matrix and a scalar is a commutative process, since $sd_{ij} = d_{ij}s$.

The sum of two matrices is obtained by adding corresponding matrix elements:

$$\mathbf{D}+\mathbf{E} = \begin{bmatrix} d_{11} & d_{12} & \cdots & d_{1n} \\ d_{21} & d_{22} & \cdots & d_{2n} \\ d_{31} & d_{32} & \cdots & d_{3n} \\ d_{41} & d_{42} & \cdots & d_{4n} \\ \vdots & \vdots & \vdots & \vdots \\ d_{m1} & d_{m2} & \cdots & d_{mn} \end{bmatrix} + \begin{bmatrix} e_{11} & e_{12} & \cdots & e_{1n} \\ e_{21} & e_{22} & \cdots & e_{2n} \\ e_{31} & e_{32} & \cdots & e_{3n} \\ e_{41} & e_{42} & \cdots & e_{4n} \\ \vdots & \vdots & \vdots & \vdots \\ e_{m1} & e_{m2} & \cdots & e_{mn} \end{bmatrix}$$

$$= \begin{bmatrix} d_{11}+e_{11} & d_{12}+e_{12} & \cdots & d_{1n}+e_{1n} \\ d_{21}+e_{21} & d_{22}+e_{22} & \cdots & d_{2n}+e_{2n} \\ d_{31}+e_{31} & d_{32}+e_{32} & \cdots & d_{3n}+e_{3n} \\ d_{41}+e_{41} & d_{42}+e_{42} & \cdots & d_{4n}+e_{4n} \\ \vdots & \vdots & \vdots & \vdots \\ d_{m1}+e_{m1} & d_{m2}+e_{m2} & \cdots & d_{mn}+e_{mn} \end{bmatrix} \quad (1.45)$$

Because corresponding matrix elements are added the matrices must have identical row and column dimensions. Matrix addition is also commutative, since $d_{ij} + e_{ij} = e_{ij} + d_{ij}$.

Unlike matrix-scalar multiplication and matrix addition, multiplying matrices with one another has no simple numerical analogy, and is a bit more complicated and abstract. However, vectors are transformed by matrix multiplication and we will find the ability to create *matrix products* to be critical to the development of the mathematics of crystallography. As we shall soon observe, matrix products are not commutative, and we must keep track of the matrix "on the left" and the one "on the right". For the product **DE**, we say that **D** *premultiplies* **E** and that **E** *postmultiplies* **D**. In order to form the matrix product $\mathbf{F} = \mathbf{DE}$ the number of columns in **D** must equal the number of rows in **E**. For **D** $m \times q$ and **E** $q \times n$, the resulting product matrix **F** will be an $m \times n$ matrix with its elements defined by $f_{ij} = \sum_{k=1}^{q} d_{ik}e_{kj}$. The ij^{th} element in the product matrix is formed from the ith row of the matrix on the left and the jth column on the right by multiplying the first element in row i of **D** by the first element in column j of **E**, then multiplying the second elements, then the third ... and so forth until there are q such products. These products are then added together to form the ij^{th} element in **F**. While this may seem a bit confusing, this is precisely the process that was undertaken in the formation of the scalar product, which, for a general coordinate system becomes the *inner* product (see the previous section for a discussion of this terminology). The ij^{th} element of the product matrix is just the inner product of the ith row vector of the matrix on the left and the jth column vector of the matrix on the right, both of which contain q components. Thus,

$$\mathbf{DE} = \begin{bmatrix} d_{11} & d_{12} & \cdots & d_{1q} \\ d_{21} & d_{22} & \cdots & d_{2q} \\ d_{31} & d_{32} & \cdots & d_{3q} \\ d_{41} & d_{42} & \cdots & d_{4q} \\ \vdots & \vdots & \vdots & \vdots \\ d_{m1} & d_{m2} & \cdots & d_{mq} \end{bmatrix} \begin{bmatrix} e_{11} & e_{12} & \cdots & e_{1n} \\ e_{21} & e_{22} & \cdots & e_{2n} \\ e_{31} & e_{32} & \cdots & e_{3n} \\ e_{41} & e_{42} & \cdots & e_{4n} \\ \vdots & \vdots & \vdots & \vdots \\ e_{q1} & e_{q2} & \cdots & e_{qn} \end{bmatrix} \tag{1.46}$$

$$= \begin{bmatrix} (\sum_{k=1}^{q} d_{1k}e_{k1}) & (\sum_{k=1}^{q} d_{1k}e_{k2}) & \cdots & (\sum_{k=1}^{q} d_{1k}e_{kn}) \\ (\sum_{k=1}^{q} d_{2k}e_{k1}) & (\sum_{k=1}^{q} d_{2k}e_{k2}) & \cdots & (\sum_{k=1}^{q} d_{2k}e_{kn}) \\ (\sum_{k=1}^{q} d_{3k}e_{k1}) & (\sum_{k=1}^{q} d_{3k}e_{k2}) & \cdots & (\sum_{k=1}^{q} d_{3k}e_{kn}) \\ (\sum_{k=1}^{q} d_{4k}e_{k1}) & (\sum_{k=1}^{q} d_{4k}e_{k2}) & \cdots & (\sum_{k=1}^{q} d_{4k}e_{kn}) \\ \vdots & \vdots & \vdots & \vdots \\ (\sum_{k=1}^{q} d_{mk}e_{k1}) & (\sum_{k=1}^{q} d_{mk}e_{k2}) & \cdots & (\sum_{k=1}^{q} d_{mk}e_{kn}) \end{bmatrix}.$$

A simpler 3×3 example should serve to clarify this:

$$\mathbf{DE} = \begin{bmatrix} d_{11} & d_{12} & d_{13} \\ d_{21} & d_{22} & d_{23} \\ d_{31} & d_{32} & d_{33} \end{bmatrix} \begin{bmatrix} e_{11} & e_{12} & e_{13} \\ e_{21} & e_{22} & e_{23} \\ e_{31} & e_{32} & e_{33} \end{bmatrix} =$$

$$\begin{bmatrix} (d_{11}e_{11} + d_{12}e_{21} + d_{13}e_{31}) & (d_{11}e_{12} + d_{12}e_{22} + d_{13}e_{32}) & (d_{11}e_{13} + d_{12}e_{23} + d_{13}e_{33}) \\ (d_{21}e_{11} + d_{22}e_{21} + d_{23}e_{31}) & (d_{21}e_{12} + d_{22}e_{22} + d_{23}e_{32}) & (d_{21}e_{13} + d_{22}e_{23} + d_{23}e_{33}) \\ (d_{31}e_{11} + d_{32}e_{21} + d_{33}e_{31}) & (d_{31}e_{12} + d_{32}e_{22} + d_{33}e_{32}) & (d_{31}e_{13} + d_{32}e_{23} + d_{33}e_{33}) \end{bmatrix}.$$

For example,

$$\begin{aligned} \mathbf{DE} &= \begin{bmatrix} -1 & -2 & 3 \\ 2 & -1 & 2 \\ 0 & 1 & -2 \end{bmatrix} \begin{bmatrix} 1 & 1 & 2 \\ 3 & 2 & 0 \\ 2 & 1 & 2 \end{bmatrix} \\ &= \begin{bmatrix} (-1\cdot 1 - 2\cdot 3 + 3\cdot 2) & (-1\cdot 1 - 2\cdot 2 + 3\cdot 1) & (-1\cdot 2 - 2\cdot 0 + 3\cdot 2) \\ (2\cdot 1 - 1\cdot 3 + 2\cdot 2) & (2\cdot 1 - 1\cdot 2 + 2\cdot 1) & (2\cdot 2 - 1\cdot 0 + 2\cdot 2) \\ (0\cdot 1 + 1\cdot 3 - 2\cdot 2) & (0\cdot 1 + 1\cdot 2 - 2\cdot 1) & (0\cdot 2 + 1\cdot 0 - 2\cdot 2) \end{bmatrix} \\ &= \begin{bmatrix} -1 & -2 & 4 \\ 3 & 2 & 8 \\ -1 & 0 & -4 \end{bmatrix}. \end{aligned}$$

Finally, consider two column vectors with the same dimension, $\mathbf{v_a}$ and $\mathbf{v_b}$. Transposing $\mathbf{v_a}$ and premultiplying $\mathbf{v_b}$ $(n \times 1)$ by $\mathbf{v_a}^T$ $(1 \times n)$ results in a 1×1 matrix — a scalar. Thus the inner product, $\mathbf{v_a} \cdot \mathbf{v_b}$, can be represented as a matrix product, $\mathbf{v_a}^T\mathbf{v_b}$:

$$\mathbf{v_a}^T\mathbf{v_b} = \begin{bmatrix} v_{a1} & v_{a2} & \cdots & v_{an} \end{bmatrix} \begin{bmatrix} v_{b1} \\ v_{b2} \\ \vdots \\ v_{bn} \end{bmatrix} = (v_{a1}v_{b1} + v_{a2}v_{b2} + \cdots + v_{an}v_{bn}). \tag{1.47}$$

This provides a very convenient description of matrix multiplication. Expressing $\mathbf{D}$ as a column vector consisting of m row vectors of dimension q, and $\mathbf{E}$ as a row vector consisting of n column vectors of dimension q,

$$\mathbf{DE} = \begin{bmatrix} \mathbf{d_1}^T \\ \mathbf{d_2}^T \\ \mathbf{d_3}^T \\ \vdots \\ \mathbf{d_m}^T \end{bmatrix} \begin{bmatrix} \mathbf{e_1} & \mathbf{e_2} & \mathbf{e_3} & \cdots & \mathbf{e_n} \end{bmatrix} = \begin{bmatrix} \mathbf{d_1}^T\mathbf{e_1} & \mathbf{d_1}^T\mathbf{e_2} & \mathbf{d_1}^T\mathbf{e_3} & \cdots & \mathbf{d_1}^T\mathbf{e_n} \\ \mathbf{d_2}^T\mathbf{e_1} & \mathbf{d_2}^T\mathbf{e_2} & \mathbf{d_2}^T\mathbf{e_3} & \cdots & \mathbf{d_2}^T\mathbf{e_n} \\ \mathbf{d_3}^T\mathbf{e_1} & \mathbf{d_3}^T\mathbf{e_2} & \mathbf{d_3}^T\mathbf{e_3} & \cdots & \mathbf{d_3}^T\mathbf{e_n} \\ \vdots & \vdots & \vdots & \vdots & \vdots \\ \mathbf{d_m}^T\mathbf{e_1} & \mathbf{d_m}^T\mathbf{e_2} & \mathbf{d_m}^T\mathbf{e_3} & \cdots & \mathbf{d_m}^T\mathbf{e_n} \end{bmatrix}$$

$$= \begin{bmatrix} \mathbf{d_1}\cdot\mathbf{e_1} & \mathbf{d_1}\cdot\mathbf{e_2} & \mathbf{d_1}\cdot\mathbf{e_3} & \cdots & \mathbf{d_1}\cdot\mathbf{e_n} \\ \mathbf{d_2}\cdot\mathbf{e_1} & \mathbf{d_2}\cdot\mathbf{e_2} & \mathbf{d_2}\cdot\mathbf{e_3} & \cdots & \mathbf{d_2}\cdot\mathbf{e_n} \\ \mathbf{d_3}\cdot\mathbf{e_1} & \mathbf{d_3}\cdot\mathbf{e_2} & \mathbf{d_3}\cdot\mathbf{e_3} & \cdots & \mathbf{d_3}\cdot\mathbf{e_n} \\ \vdots & \vdots & \vdots & \vdots & \vdots \\ \mathbf{d_m}\cdot\mathbf{e_1} & \mathbf{d_m}\cdot\mathbf{e_2} & \mathbf{d_m}\cdot\mathbf{e_3} & \cdots & \mathbf{d_m}\cdot\mathbf{e_n} \end{bmatrix}. \quad (1.48)$$

1.4.3 Matrix Transformations

Premultipling an $n \times 1$ column vector by an $m \times n$ matrix produces a new $m \times 1$ column vector:

$$\mathbf{Dv} = \begin{bmatrix} d_{11} & d_{12} & \cdots & d_{1n} \\ d_{21} & d_{22} & \cdots & d_{2n} \\ \vdots & \vdots & \vdots & \vdots \\ d_{m1} & d_{m2} & \cdots & d_{mn} \end{bmatrix} \begin{bmatrix} v_1 \\ v_2 \\ \vdots \\ v_n \end{bmatrix} = \begin{bmatrix} (d_{11}v_1 + d_{12}v_2 + \cdots + d_{1n}v_n) \\ (d_{21}v_1 + d_{22}v_2 + \cdots + d_{2n}v_n) \\ \vdots \\ (d_{m1}v_1 + d_{m2}v_2 + \cdots + d_{mn}v_n) \end{bmatrix}$$

$$= \begin{bmatrix} \mathbf{d_1}^T \\ \mathbf{d_2}^T \\ \vdots \\ \mathbf{d_m}^T \end{bmatrix} \mathbf{v} = \begin{bmatrix} \mathbf{d_1}^T\mathbf{v} \\ \mathbf{d_2}^T\mathbf{v} \\ \vdots \\ \mathbf{d_m}^T\mathbf{v} \end{bmatrix} = \begin{bmatrix} \mathbf{d_1}\cdot\mathbf{v} \\ \mathbf{d_2}\cdot\mathbf{v} \\ \vdots \\ \mathbf{d_m}\cdot\mathbf{v} \end{bmatrix}. \quad (1.49)$$

Whenever the matrix is a square matrix, the dimension of the vector does not change and we say that we have *transformed* the original vector into a new one. In crystallography the square *transformation matrix* is nearly always a 3×3 matrix. By selecting appropriate elements for the transformation matrix we can perform specific *operations* on the vector to transform it in a specific manner. Transformation matrices are often called *matrix operators*. For example, consider the vector at point $(x_c\ y_c\ z_c)$ in an orthonormal basis in Fig. 1.27(a). Suppose that we wish to determine the coordinates of the point after it has been *reflected* across the xy (**ij**) plane. To accomplish this geometrically we project the head of the vector onto the plane, then extend it by the length of the perpendicular to the other side of the plane. The matrix equation that will accomplish the same task has the following form:

$$\begin{bmatrix} t_{11} & t_{12} & t_{13} \\ t_{21} & t_{22} & t_{23} \\ t_{31} & t_{32} & t_{33} \end{bmatrix} \begin{bmatrix} x_c \\ y_c \\ z_c \end{bmatrix} = \begin{bmatrix} x_c \\ y_c \\ -z_c \end{bmatrix}.$$

The x component in the new vector remains unchanged: $t_{11}x_c + t_{12}y_c + t_{13}z_c = x_c$. Thus $t_{11} = 1$, $t_{12} = 0$, and $t_{13} = 0$. The y coordinate also does not change. It

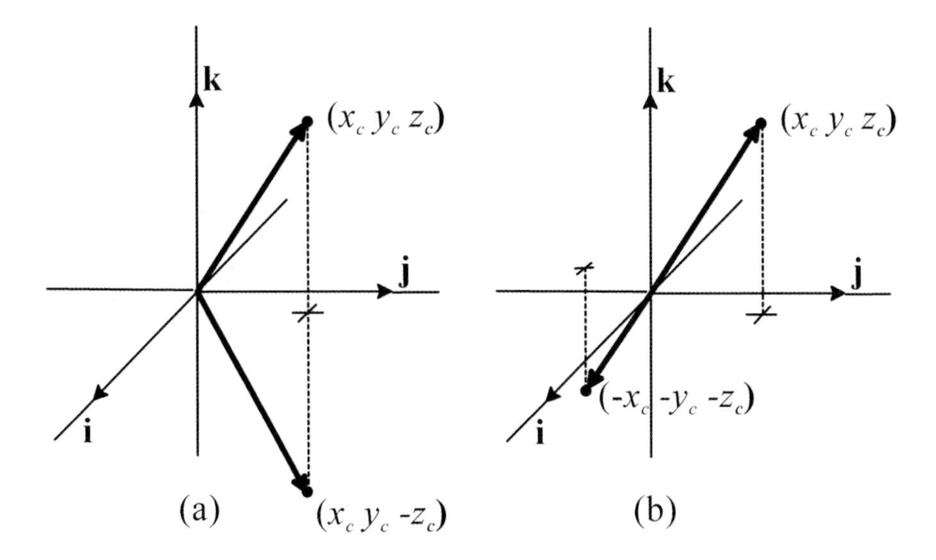

Figure 1.27 (a) Reflection of a vector at point $(x_c\ y_c\ z_c)$ across the xy plane. (b) Inversion of the vector $[x_c\ y_c\ z_c]$ through the origin.

follows that $t_{21} = 0$, $t_{22} = 1$, and $t_{23} = 0$. The z component changes in sign with no contributions from the original x and y components. Hence $t_{31} = 0$, $t_{32} = 0$, and $t_{33} = -1$:

$$\begin{bmatrix} 1 & 0 & 0 \\ 0 & 1 & 0 \\ 0 & 0 & -1 \end{bmatrix} \begin{bmatrix} x_c \\ y_c \\ z_c \end{bmatrix} = \begin{bmatrix} x_c \\ y_c \\ -z_c \end{bmatrix}. \tag{1.50}$$

The transformation matrices for reflections across the xz and yz planes are, respectively,

$$\begin{bmatrix} 1 & 0 & 0 \\ 0 & -1 & 0 \\ 0 & 0 & 1 \end{bmatrix} \text{ and } \begin{bmatrix} -1 & 0 & 0 \\ 0 & 1 & 0 \\ 0 & 0 & 1 \end{bmatrix}. \tag{1.51}$$

Operations such as *reflections* are necessary to describe lattice symmetry. Fig. 1.27(b) illustrates another important *symmetry operation* — the *inversion* of a vector through the origin. This transformation changes the sign of *all* three components, and therefore results in the following matrix equation:

$$\begin{bmatrix} -1 & 0 & 0 \\ 0 & -1 & 0 \\ 0 & 0 & -1 \end{bmatrix} \begin{bmatrix} x_c \\ y_c \\ z_c \end{bmatrix} = \begin{bmatrix} -x_c \\ -y_c \\ -z_c \end{bmatrix}. \tag{1.52}$$

In addition to reflections and inversions, a vector can be rotated by fixing its tail at the origin, maintaining its length, and moving its head to another position by rotating the vector around an axis. The *rotation* operation often takes place around one of the coordinate axes. Fig. 1.28(a) illustrates the rotation of a vector at point

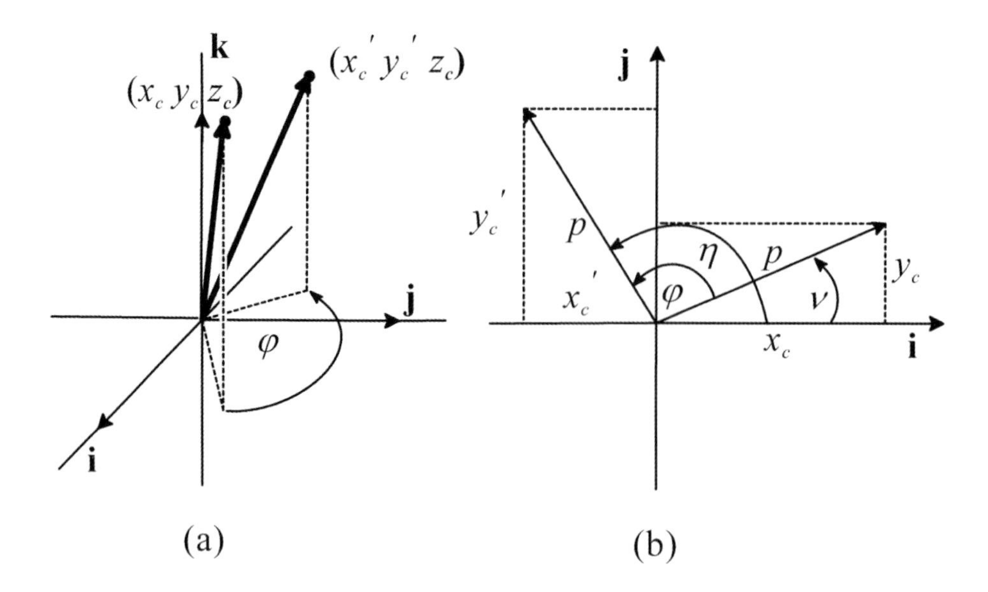

Figure 1.28 (a) Rotation of a vector at point $(x_c\ y_c\ z_c)$ through angle φ around the z axis. (b) Projection of the vector rotation onto the xy plane "looking down" the z axis. p is the length of the projected vector.

$(x_c\ y_c\ z_c)$ rotated around the z axis through an angle φ. The z coordinate remains constant in the rotation. Fig. 1.28(b) is the projection of the vector before and after rotation onto the xy plane, viewing down the z axis in the negative direction. Because we are operating in a right-handed coordinate system, pointing the thumb in the direction of the rotation axis, **k** in this case, assigns a positive rotation in the direction of the curl of the right hand fingers, counterclockwise when "looking down" the axis. The vector **p** is the rotated vector projected onto the plane, with magnitude p. The projected vector is rotated in the xy plane through the angle φ. If ν is the original angle of the projected vector with respect to the x axis, and η is the final angle, then

$$\begin{aligned}
x_c &= p\cos\nu \\
y_c &= p\sin\nu \\
x_c' &= p\cos\eta = p\cos(\nu+\varphi) \\
y_c' &= p\sin\eta = p\sin(\nu+\varphi). \\
z_c' &= z_c
\end{aligned}$$

From the trigonometric identities for angle sums,

$$\begin{aligned}
x_c' &= p\cos\nu\cos\varphi - p\sin\nu\sin\varphi = x_c\cos\varphi - y_c\sin\varphi \\
y_c' &= p\cos\nu\sin\varphi + p\sin\nu\cos\varphi = x_c\sin\varphi + y_c\cos\varphi.
\end{aligned}$$

The matrix equation to rotate a vector at $(x_c\ y_c\ z_c)$ through angle φ around the z axis is then

$$\begin{bmatrix} \cos\varphi & -\sin\varphi & 0 \\ \sin\varphi & \cos\varphi & 0 \\ 0 & 0 & 1 \end{bmatrix} \begin{bmatrix} x_c \\ y_c \\ z_c \end{bmatrix} = \begin{bmatrix} x'_c \\ y'_c \\ z_c \end{bmatrix}. \tag{1.53}$$

Rotation around the x axis produces essentially the same equations, this time transforming the y and z coordinates in exactly the same manner. Rotation about y, however, is positive when the x coordinates are reversed (see Fig. 1.29). In order to keep the signs of the trigonometric functions consistent in each of the quadrants the rotation must occur in the opposite direction; φ must be replaced by $-\varphi$ in the derivation. The resulting transformation matrices for x-rotation and y-rotation are therefore, respectively,

$$\begin{bmatrix} 1 & 0 & 0 \\ 0 & \cos\varphi & -\sin\varphi \\ 0 & \sin\varphi & \cos\varphi \end{bmatrix} \text{ and } \begin{bmatrix} \cos\varphi & 0 & \sin\varphi \\ 0 & 1 & 0 \\ -\sin\varphi & 0 & \cos\varphi \end{bmatrix}. \tag{1.54}$$

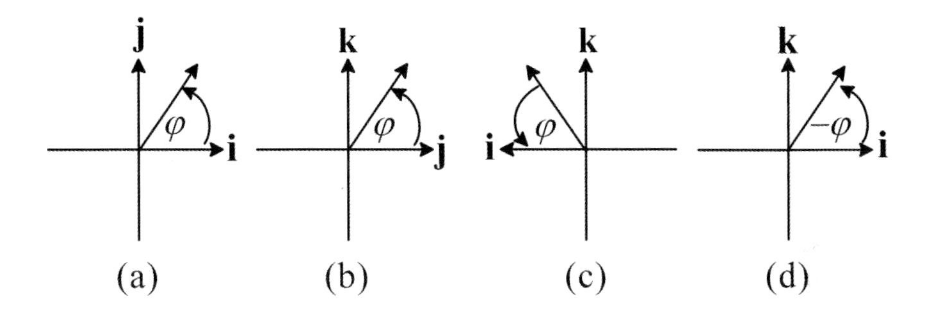

Figure 1.29 Views down the axes showing the signs of the rotational angle: (a)z axis, (b)x axis, and (c) y axis. (d) Inverted axes for the y-rotation showing the effect on the sign of the rotational angle.

It is important to note that each of the matrix operations described in this section maintained the magnitude of the vector transformed by them. If the columns of these matrices are treated as vectors, it can readily be shown that these vectors are orthogonal to one another. They also have unit magnitudes, and are therefore orthonormal vectors. A matrix with orthonormal columns is known as an *orthonormal matrix*. For example, consider the z-rotation matrix:

$$\mathbf{v_1} = \begin{bmatrix} \cos\varphi \\ \sin\varphi \\ 0 \end{bmatrix} \quad \mathbf{v_2} = \begin{bmatrix} -\sin\varphi \\ \cos\varphi \\ 0 \end{bmatrix} \quad \mathbf{v_3} = \begin{bmatrix} 0 \\ 0 \\ 1 \end{bmatrix}. \tag{1.55}$$

$$\begin{aligned}
v_1 &= (\cos^2\varphi + \sin^2\varphi + 0^2)^{\frac{1}{2}} \;=\; 1\\
v_2 &= (\sin^2\varphi + \cos^2\varphi + 0^2)^{\frac{1}{2}} \;=\; 1\\
v_3 &= (0^2 + 0^2 + 1^2)^{\frac{1}{2}} \;=\; 1\\
\mathbf{v_1}\cdot\mathbf{v_2} &= (-\cos\varphi\sin\varphi + \cos\varphi\sin\varphi + 0) \;=\; 0\\
\mathbf{v_1}\cdot\mathbf{v_3} &= (0+0+0) \;=\; 0\\
\mathbf{v_2}\cdot\mathbf{v_3} &= (0+0+0) \;=\; 0
\end{aligned}$$

Now, consider the transformation of a vector $\mathbf{v}$ by a general orthonormal transformation matrix, $\mathbf{T} = [\mathbf{t_1}\ \mathbf{t_2}\ \mathbf{t_3}]$:

$$\mathbf{Tv} = \mathbf{v}' = \begin{bmatrix} t_{11} & t_{12} & t_{13}\\ t_{21} & t_{22} & t_{23}\\ t_{31} & t_{32} & t_{33}\end{bmatrix}\begin{bmatrix} x_c\\ y_c\\ z_c\end{bmatrix} = \begin{bmatrix} x'_c\\ y'_c\\ z'_c\end{bmatrix} = \begin{bmatrix} t_{11}x_c + t_{12}y_c + t_{13}z_c\\ t_{21}x_c + t_{22}y_c + t_{23}z_c\\ t_{31}x_c + t_{32}y_c + t_{33}z_c\end{bmatrix}$$

$$\begin{aligned}
v^2 &= x_c^2 + y_c^2 + z_c^2\\
v'^2 &= (t_{11}x_c + t_{12}y_c + t_{13}z_c)^2 + (t_{21}x_c + t_{22}y_c + t_{23}z_c)^2 + (t_{31}x_c + t_{32}y_c + t_{33}z_c)^2\\
&= (t_{11}^2 + t_{21}^2 + t_{31}^2)x_c^2 + (t_{12}^2 + t_{22}^2 + t_{32}^2)y_c^2 + (t_{13}^2 + t_{23}^2 + t_{33}^2)z_c^2\\
&\quad + 2(t_{11}t_{12} + t_{21}t_{22} + t_{31}t_{32})x_c y_c\\
&\quad + 2(t_{11}t_{13} + t_{21}t_{23} + t_{31}t_{33})x_c z_c\\
&\quad + 2(t_{12}t_{13} + t_{22}t_{23} + t_{32}t_{33})y_c z_c\\
&= (\mathbf{t_1}\cdot\mathbf{t_1})x_c^2 + (\mathbf{t_2}\cdot\mathbf{t_2})y_c^2 + (\mathbf{t_3}\cdot\mathbf{t_3})z_c^2\\
&\quad + (\mathbf{t_1}\cdot\mathbf{t_2})x_c y_c + (\mathbf{t_1}\cdot\mathbf{t_3})x_c z_c + (\mathbf{t_2}\cdot\mathbf{t_3})y_c z_c\\
&= (1)x_c^2 + (1)y_c^2 + (1)z_c^2 + 0 + 0 + 0 \;=\; x_c^2 + y_c^2 + z_c^2 \;=\; v^2. \qquad (1.56)
\end{aligned}$$

When an orthonormal matrix operates on a vector it changes the direction of the vector but leaves its magnitude unchanged.

1.4.4 The Determinant of a Matrix

The *determinant* is a scalar that results from the signed products of the permutations of all the matrix elements in a matrix. Adding these products together produces a number with properties which tell us something about the matrix (hence the name). The determinant of a matrix is represented by placing the symbol of the matrix between vertical bars. For an $n \times n$ square matrix, $\mathbf{D}$,

$$|\mathbf{D}| = \begin{vmatrix} d_{11} & d_{12} & \cdots & d_{1n}\\ d_{21} & d_{22} & \cdots & d_{2n}\\ \vdots & \vdots & \vdots & \vdots\\ d_{n1} & d_{n2} & \cdots & d_{nn}\end{vmatrix} = \sum(-1)^k d_{1i_1}d_{2i_2}d_{3i_3}\cdots d_{ni_n}. \qquad (1.57)$$

The sum is over all products of n-fold permutations of the indices. $i_1, i_2, \cdots i_n$, is a permutation of the indices 1,2,$\cdots n$; k is the number of inversions ("switches") of indices necessary to achieve the permutation, e.g., $123 \xrightarrow{1} 132 \xrightarrow{2} 312 \xrightarrow{3} 321,\ k=3$. While this seems abstract, the example for $n=3$ illustrates that it is only tedious:

$$\begin{aligned}|\mathbf{D}| &= \begin{vmatrix} d_{11} & d_{12} & d_{13} \\ d_{21} & d_{22} & d_{23} \\ d_{31} & d_{32} & d_{33} \end{vmatrix} \\ &= (-1)^0 d_{11}d_{22}d_{33} + (-1)^1 d_{11}d_{23}d_{32} + (-1)^1 d_{12}d_{21}d_{33} \\ &\quad + (-1)^2 d_{12}d_{23}d_{31} + (-1)^2 d_{13}d_{21}d_{32} + (-1)^3 d_{13}d_{22}d_{31}.\end{aligned} \tag{1.58}$$

Four important properties of the determinant arise directly from the definition. Rather than cover the relatively abstract general proofs, the discussion here will be limited to the $n=3$ case, although in most instances the extrapolation to larger matrices is a logical one.

Determinant Property 1. *Switching two columns or rows in a matrix changes the sign of its determinant.*

$$\begin{aligned}|\mathbf{D}'| &= \begin{vmatrix} d_{11} & d_{13} & d_{12} \\ d_{21} & d_{23} & d_{22} \\ d_{31} & d_{33} & d_{32} \end{vmatrix} \\ &= (-1)^0 d_{11}d_{23}d_{32} + (-1)^1 d_{11}d_{22}d_{33} + (-1)^1 d_{13}d_{21}d_{32} \\ &\quad + (-1)^2 d_{13}d_{22}d_{31} + (-1)^2 d_{12}d_{21}d_{33} + (-1)^3 d_{12}d_{23}d_{31} \\ &= -|\mathbf{D}|.\end{aligned} \tag{1.59}$$

The permutations are the same, but the number of inversions for each has changed. In every case $(-1)^k$ has changed sign; the magnitude of the determinant has not changed, but the sign of the determinant has. *The sign of the determinant changes if a row or column in the matrix is switched with another.*

Determinant Property 2. *The determinant of a matrix that contains column or row vectors that are linearly dependent (not linearly independent) is exactly zero.* Consider the case where two columns (or rows) of the matrix are linearly dependent, that is, one column is a scalar multiple of the other (If the columns are treated as geometric vectors, the vectors would be parallel to one another):

$$\begin{aligned}|\mathbf{D}''| &= \begin{vmatrix} d_{11} & d_{12} & sd_{12} \\ d_{21} & d_{22} & sd_{22} \\ d_{31} & d_{32} & sd_{32} \end{vmatrix} \\ &= (-1)^0 d_{11}d_{22}sd_{32} + (-1)^1 d_{11}sd_{22}d_{32} + (-1)^1 d_{12}d_{21}sd_{32} \\ &\quad + (-1)^2 d_{12}sd_{22}d_{31} + (-1)^2 sd_{12}d_{21}d_{32} + (-1)^3 sd_{12}d_{22}d_{31} \\ &= s\{(-1)^0 d_{11}d_{22}d_{32} + (-1)^1 d_{11}d_{22}d_{32} + (-1)^1 d_{12}d_{21}d_{32} \\ &\quad + (-1)^2 d_{12}d_{22}d_{31} + (-1)^2 d_{12}d_{21}d_{32} + (-1)^3 d_{12}d_{22}d_{31}\} \\ &= s\begin{vmatrix} d_{11} & d_{12} & d_{12} \\ d_{21} & d_{22} & d_{22} \\ d_{31} & d_{32} & d_{32} \end{vmatrix}\end{aligned} \tag{1.60}$$

If we switch the second and thirds columns of the matrix the determinant clearly remains unchanged, yet, according to Eqn. 1.59 it must change sign. This is only possible if the determinant is zero:

$$|\mathbf{D}''| = -|\mathbf{D}''| = 0 \tag{1.61}$$

A matrix with a zero determinant is called a *singular* matrix. Eqn. 1.60 also illustrates another property of the determinant: *Multiplying a column or row of a determinant by a scalar multiplies the determinant by the scalar.*

Determinant Property 3. *The determinant of a matrix and its transpose are equal.* While this might appear intuitive, it is not obvious if one considers the definition of the determinant as a sum of permutations. The proof of this in the general case is based on arguments of the equivalency of various permutations, and would take us far afield. Fortunately, since we will only deal with 3×3 determinants, this can be demonstrated explicitly. Let $\mathbf{E} = \mathbf{D}^T$ and $e_{ij} = d_{ji}$. Then

$$\begin{aligned}
|\mathbf{D}| &= \begin{vmatrix} d_{11} & d_{12} & d_{13} \\ d_{21} & d_{22} & d_{23} \\ d_{31} & d_{n2} & d_{33} \end{vmatrix} = \sum (-1)^k d_{1i_1} d_{2i_2} d_{3i_3} \\
|\mathbf{E}| &= \begin{vmatrix} e_{11} & e_{12} & e_{13} \\ e_{21} & e_{22} & e_{23} \\ e_{31} & e_{32} & e_{33} \end{vmatrix} = \sum (-1)^k e_{1i_1} e_{2i_2} e_{3i_3} \\
&= \sum (-1)^k d_{i_1 1} d_{i_2 2} d_{i_3 3}
\end{aligned}$$

Since each term in the sum has one index for each row and column there will be a permutation in $|\mathbf{E}|$ matching every permutation in $|\mathbf{D}|$. Thus the only possible difference is in the signs of the matching permutations. The following table demonstrates that the signs do not change:

k	$\lvert\mathbf{D}\rvert$	$\lvert\mathbf{E}\rvert$	$\lvert\mathbf{D}^T\rvert$
0 ←	$d_{11}d_{22}d_{33}$ ←	$e_{11}e_{22}e_{33}$ ←	$d_{11}d_{22}d_{33}$
1 ←	$d_{11}d_{23}d_{32}$ ←	$e_{11}e_{23}e_{32}$ ←	$d_{11}d_{32}d_{23}$
1 ←	$d_{12}d_{21}d_{33}$ ←	$e_{12}e_{21}e_{33}$ ←	$d_{21}d_{12}d_{33}$
2 ←	$d_{12}d_{23}d_{31}$ ←	$e_{12}e_{23}e_{31}$	$d_{21}d_{32}d_{13}$
2 ←	$d_{13}d_{21}d_{32}$ ←	$e_{13}e_{21}e_{32}$	$d_{31}d_{12}d_{23}$
2 ←	$d_{13}d_{22}d_{31}$ ←	$e_{13}e_{22}e_{31}$ ←	$d_{31}d_{22}d_{13}$

Since every permutation in $|\mathbf{D}^T|$ has the same sign as the corresponding permutation in $|\mathbf{D}|$,

$$|\mathbf{D}^T| = |\mathbf{D}| \tag{1.62}$$

Determinant Property 4. *The determinant of the product of two matrices is the product of the determinants of the matrices.* This is another determinant property that appears deceptively intuitive on first glance. The proof demonstrates that it is not! Let $\mathbf{P} = \mathbf{DE}$.

$$
\begin{aligned}
|\mathbf{P}| &= |\mathbf{DE}| = \sum (-1)^k p_{1i_1} p_{2i_2} p_{3i_3}, \text{ where} \\
p_{1i_1} &= \sum_{j_1}^{3} d_{1j_1} e_{j_1 1}, \; p_{2i_2} = \sum_{j_2}^{3} d_{2j_2} e_{j_2 2}, \text{ and } p_{3i_3} = \sum_{j_3}^{3} d_{3j_3} e_{j_3 3}. \\
|\mathbf{P}| &= \sum (-1)^k \sum_{j_1}^{3} d_{1j_1} e_{j_1 1} \sum_{j_1}^{3} d_{2j_2} e_{j_2 2} \sum_{j_3}^{3} d_{3j_3} e_{j_3 3} \\
&= \sum_{j_1}^{3} \sum_{j_2}^{3} \sum_{j_3}^{3} d_{1j_1} d_{2j_2} d_{3j_3} \underbrace{\sum (-1)^k e_{j_1 1} e_{j_2 2} e_{j_3 3}}_{\boldsymbol{E}_{j_1 j_2 j_3}} .
\end{aligned}
$$

For every set of indices $j_1 j_2 j_3$, $\boldsymbol{E}_{j_1 j_2 j_3}$ is a signed sum of permutations — *a determinant*:

$$
\boldsymbol{E}_{j_1 j_2 j_3} = \sum (-1)^k e_{j_1 1} e_{j_2 2} e_{j_3 3} = \begin{vmatrix} e_{j_1 1} & e_{j_1 2} & e_{j_1 3} \\ e_{j_2 1} & e_{j_2 2} & e_{j_3 3} \\ e_{j_3 1} & e_{j_3 2} & e_{j_3 3} \end{vmatrix} .
$$

There will be a term in the complete sum for every set of indices $j_1 j_2 j_3$. However if $j_1 = j_2$ or $j_1 = j_3$ or $j_2 = j_3$ then two rows of the determinant for that term in the sum will be identical – and $\boldsymbol{E}_{j_1 j_2 j_3} = 0$. It follows that the only terms that survive in the complete sum are terms in which $j_1 \neq j_2 \neq j_3$, that is $(j_1 j_2 j_3)$ = (1,2,3), (2,3,1), etc. — all of the permutations of the indices $(j_1 j_2 j_3)$. Thus $\sum_{j_1}^3 \sum_{j_2}^3 \sum_{j_3}^3 d_{1j_1} d_{2j_2} d_{3j_3}$ is a sum which includes all the permutations of the indices — *it would be a determinant if the permutations were multiplied by the appropriate signs*:

$$
|\mathbf{P}| = \sum d_{1j_1} d_{2j_2} d_{3j_3} \boldsymbol{E}_{j_1 j_2 j_3} \tag{1.63}
$$

If $(j_1 j_2 j_3)$ = (1,2,3) then

$$
\boldsymbol{E}_{123} = \begin{vmatrix} e_{11} & e_{12} & e_{13} \\ e_{21} & e_{22} & e_{23} \\ e_{31} & e_{32} & e_{33} \end{vmatrix} = |\mathbf{E}|.
$$

Another permutation of $(j_1 j_2 j_3)$ generates the same matrix with one or more rows switched. For example, if $(j_1 j_2 j_3)$ = (1,3,2),

$$
\boldsymbol{E}_{132} = \begin{vmatrix} e_{11} & e_{12} & e_{13} \\ e_{31} & e_{32} & e_{33} \\ e_{21} & e_{22} & e_{23} \end{vmatrix} = -|\mathbf{E}|.
$$

For any permutation we must switch the rows to get back to (1,2,3) order. The sign changes for each switch. In general, where k' is the number of switches to restore the matrix to (1,2,3) order, $\boldsymbol{E}_{j_1 j_2 j_3} = (-1)^{k'} |\mathbf{E}|$. The number of row switches necessary to get $\boldsymbol{E}_{j_1 j_2 j_3}$ into (123) order is exactly the same as the number of switches necessary to get $d_{1j_1} d_{2j_2} d_{3j_3}$ into (123) order, and that is the number of switches necessary to create the signed permutation of $\mathbf{D}$ from (123) order. $(-1)^{k'}$ is the sign of the permutation:

$$\begin{aligned} |\mathbf{P}| &= \sum d_{1j_1} d_{2j_2} d_{3j_3} (-1)^{k'} |\mathbf{E}| \\ &= \sum (-1)^{k'} d_{1j_1} d_{2j_2} d_{3j_3} |\mathbf{E}| \\ &= |\mathbf{D}||\mathbf{E}| \\ |\mathbf{DE}| &= |\mathbf{D}||\mathbf{E}| \end{aligned} \tag{1.64}$$

The formal method for evaluating a determinant, which guarantees that all of the permutations with appropriate signs are obtained, is called *cofactor expansion.* It is undertaken in two steps. First, a row of the matrix is selected (or a column) — then a set of sub-matrices is formed by selecting each element in the row, "crossing out" the row and column that the element intersects, and evaluating the determinant of the remaining sub-matrix (called the *minor* of the element). In the second step each determinant is multiplied by $(-1)^{(i+j)}$, where i and j are the indices of the element which generates the minor. This new signed determinant,$\mathcal{D}_{ij}$, is called the *cofactor* of the matrix element, d_{ij}. Again, this apparently complex notion is clarified by an example. We select the first row of the determinant in Eqn. 1.58, determine the minors of each element in the row, and determine the cofactors. The cofactor of d_{11} is

$$\begin{vmatrix} \not{d}_{11} & \not{d}_{12} & \not{d}_{13} \\ \not{d}_{21} & d_{22} & d_{23} \\ \not{d}_{31} & d_{32} & d_{33} \end{vmatrix} \Longrightarrow \mathcal{D}_{11} = -1^{(1+1)} \begin{vmatrix} d_{22} & d_{23} \\ d_{32} & d_{33} \end{vmatrix} = \begin{vmatrix} d_{22} & d_{23} \\ d_{32} & d_{33} \end{vmatrix}. \tag{1.65}$$

Similarly,

$$\mathcal{D}_{12} = -1^{(1+2)} \begin{vmatrix} d_{21} & d_{23} \\ d_{31} & d_{33} \end{vmatrix} = - \begin{vmatrix} d_{21} & d_{23} \\ d_{31} & d_{33} \end{vmatrix} \tag{1.66}$$

and

$$\mathcal{D}_{13} = -1^{(1+3)} \begin{vmatrix} d_{21} & d_{22} \\ d_{31} & d_{32} \end{vmatrix} = \begin{vmatrix} d_{21} & d_{22} \\ d_{31} & d_{32} \end{vmatrix}. \tag{1.67}$$

The determinant is evaluated as the sum of the products of the matrix elements and their cofactors:

$$\begin{aligned} |\mathbf{D}| &= d_{11}\mathcal{D}_{11} + d_{12}\mathcal{D}_{12} + d_{13}\mathcal{D}_{13} \\ &= d_{11} \begin{vmatrix} d_{22} & d_{23} \\ d_{32} & d_{33} \end{vmatrix} - d_{12} \begin{vmatrix} d_{21} & d_{23} \\ d_{31} & d_{33} \end{vmatrix} + d_{13} \begin{vmatrix} d_{21} & d_{22} \\ d_{31} & d_{32} \end{vmatrix}. \end{aligned} \tag{1.68}$$

The cofactor determinants must now be evaluated in the same manner. For a large matrix this process is cumbersome, but for a 3×3 matrix we are left with only 2×2 determinants. The cofactors for these determinants contain only 1×1 determinants, which are obviously scalars. For a general 2×2 determinant:

$$\begin{aligned} |\mathbf{E}| &= \begin{vmatrix} e_{11} & e_{12} \\ e_{21} & e_{22} \end{vmatrix} = e_{11}(-1)^{(1+1)}|e_{22}| + e_{12}(-1)^{(1+2)}|e_{21}| \\ &= e_{11}e_{22} - e_{12}e_{21}, \end{aligned} \tag{1.69}$$

the difference of the products of the diagonal elements.

Evaluating each of the 2×2 determinants results in the evaluation of the determinant of the original matrix, which is identical to Eqn. 1.58:

$$\begin{aligned}|\mathbf{D}| &= d_{11}(d_{22}d_{33} - d_{23}d_{32}) - d_{12}(d_{21}d_{33} - d_{23}d_{31}) \\ &\quad + d_{13}(d_{21}d_{32} - d_{22}d_{31}) \\ &= d_{11}d_{22}d_{33} - d_{11}d_{23}d_{32} - d_{12}d_{21}d_{33} + d_{12}d_{23}d_{31} \\ &\quad + d_{13}d_{21}d_{32} - d_{13}d_{22}d_{31}. \end{aligned} \tag{1.70}$$

The determinant in the form of Eqn. 1.68 contains all of the possible permutations of the elements of the second and third rows of the matrix in the form of 2×2 determinants. Recall that the vector product contained all of these permutations as well. Writing the determinant as three row vectors gives us a convenient representation of the vector product in an orthonormal coordinate system:

$$\mathbf{v_2} \times \mathbf{v_3} = \begin{vmatrix} \mathbf{i} & \mathbf{j} & \mathbf{k} \\ v_{2x} & v_{2y} & v_{2z} \\ v_{3x} & v_{3y} & v_{3z} \end{vmatrix} = \mathbf{i}\begin{vmatrix} v_{2y} & v_{2z} \\ v_{3y} & v_{3z} \end{vmatrix} - \mathbf{j}\begin{vmatrix} v_{2x} & v_{2z} \\ v_{3x} & v_{3z} \end{vmatrix} + \mathbf{k}\begin{vmatrix} v_{2x} & v_{2y} \\ v_{3x} & v_{3y} \end{vmatrix}. \tag{1.71}$$

Expansion of Eqn. 1.71 results in the previously determined expression for the vector product (Eqn. 1.37). Note that $\mathbf{v_3} \times \mathbf{v_2}$ reverses the row vectors in the determinant, thus changing its sign! The representation of the vector product as a determinant provides a useful way to handle vector products – and will lead us to a simple method for computing the unit cell volume from the orthonormal coordinates of the unit cell axes.

1.4.5 The Inverse of a Matrix

In the previous section the cofactor expansion of a 3×3 matrix was shown to produce the determinant of the matrix. The first row of the matrix was selected to generate the expansion. It is left to the reader to verify that selection of the second or third row produces the same result. Formal proofs for general square matrices often turn out to be cumbersome. Proofs involving determinants and other properties of general matrices can be found in any number of linear algebra books (for example see Campbell, 1971[20]). For the remainder of the chapter we will take the simpler path of exhibiting properties for 3×3 matrices wherever the extrapolation to larger matrices appears logical.

In ordinary algebra, there is an inverse operation to multiplication which reverses the effect of the multiplication – the reciprocal. Multiplying a number by x, then by x^{-1} leaves the number unchanged: $x^{-1} \cdot (x \cdot s) = (x^{-1} \cdot x) \cdot s = (1)s = s$. The reciprocal *inverts* the original multiplication x, and is therefore also called the *inverse* of x. In number theory the number "1" is a multiplier which leaves the number that it multiplies unchanged, and is referred to as an *identity* element. Since we have discovered that matrices transform vectors, is there a matrix that will be analogous to the reciprocal in number theory which will undo a transformation? We begin to answer this question be determining the matrix that is analogous to the identity element in number theory, that is, a matrix that will multiply a vector and leave the vector unchanged. The only matrix that fulfills this function has diagonal elements of one, and off-diagonal elements of zero. The matrix is called *the identity matrix*, and is referred to by the symbol **I**:

$$\mathbf{Iv} = \begin{bmatrix} 1 & 0 & 0 \\ 0 & 1 & 0 \\ 0 & 0 & 1 \end{bmatrix} \begin{bmatrix} v_1 \\ v_2 \\ v_3 \end{bmatrix} = \begin{bmatrix} v_1 \\ v_2 \\ v_3 \end{bmatrix} \tag{1.72}$$

Consider the transformation $\mathbf{Dv} = \mathbf{v}'$. To reverse this transformation, we seek a matrix $\mathbf{D}'$ such that $\mathbf{D}'\mathbf{v}' = \mathbf{v}$, leaving $\mathbf{v}$ unchanged:

$$\begin{aligned} \mathbf{D}'\mathbf{v}' &= \mathbf{Iv} = \mathbf{v} \\ \mathbf{D}'(\mathbf{Dv}) &= (\mathbf{D}'\mathbf{D})\mathbf{v} = \mathbf{Iv} \\ \Longrightarrow \mathbf{D}'\mathbf{D} &= \mathbf{I}. \end{aligned} \tag{1.73}$$

Thus, presuming that matrix multiplication is associative (which we will demonstrate in the next section), the analog of the inverse in number theory is clearly the matrix which reverses the transform, $\mathbf{D}'$, which we will denote as the *inverse matrix* (or alternatively, just the *inverse*) of $\mathbf{D}$, $\mathbf{D}^{-1}$. Unfortunately, determining the inverse of a matrix is much less straightforward than determining the reciprocal from ordinary numbers. We begin the task by returning to the *cofactors* discussed in the last section. Every matrix element has a cofactor (Eqns. 1.65–1.67). It follows that we can generate the cofactor for each matrix element and create a new matrix which we call the *cofactor matrix* of the original matrix. For the matrix $\mathbf{D}$, we denote the cofactor matrix of $\mathbf{D}$ as $\mathbf{D_c}$:

$$\mathbf{D} = \begin{bmatrix} d_{11} & d_{12} & d_{13} \\ d_{21} & d_{22} & d_{23} \\ d_{31} & d_{32} & d_{33} \end{bmatrix} \qquad \mathbf{D_c} = \begin{bmatrix} \mathcal{D}_{11} & \mathcal{D}_{12} & \mathcal{D}_{13} \\ \mathcal{D}_{21} & \mathcal{D}_{22} & \mathcal{D}_{23} \\ \mathcal{D}_{31} & \mathcal{D}_{32} & \mathcal{D}_{33} \end{bmatrix} \tag{1.74}$$

We now evaluate the matrix product,

$$\mathbf{P} = \mathbf{DD_c^{\mathit{T}}} = \begin{bmatrix} d_{11} & d_{12} & d_{13} \\ d_{21} & d_{22} & d_{23} \\ d_{31} & d_{32} & d_{33} \end{bmatrix} \begin{bmatrix} \mathcal{D}_{11} & \mathcal{D}_{21} & \mathcal{D}_{31} \\ \mathcal{D}_{12} & \mathcal{D}_{22} & \mathcal{D}_{32} \\ \mathcal{D}_{13} & \mathcal{D}_{23} & \mathcal{D}_{33} \end{bmatrix} = \begin{bmatrix} p_{11} & p_{12} & p_{13} \\ p_{21} & p_{22} & p_{23} \\ p_{31} & p_{32} & p_{33} \end{bmatrix}.$$

$$\begin{aligned} p_{11} &= d_{11}\mathcal{D}_{11} + d_{12}\mathcal{D}_{12} + d_{13}\mathcal{D}_{13} = |\mathbf{D}| \\ p_{22} &= d_{21}\mathcal{D}_{21} + d_{22}\mathcal{D}_{22} + d_{23}\mathcal{D}_{23} = |\mathbf{D}| \\ p_{33} &= d_{31}\mathcal{D}_{31} + d_{32}\mathcal{D}_{32} + d_{33}\mathcal{D}_{33} = |\mathbf{D}|. \end{aligned} \tag{1.75}$$

p_{11} is the cofactor expansion of the determinant using the first row of the matrix, p_{22} is the cofactor expansion using the second row, and p_{33} is the cofactor expansion using the third row. The off-diagonal elements are sums of products of the elements of one row and the cofactors of another row. This creates a determinant with two identical rows, which must equal zero. Using p_{21} as an example,

$$\begin{aligned} p_{21} &= d_{21}\mathcal{D}_{11} + d_{22}\mathcal{D}_{12} + d_{23}\mathcal{D}_{13} \\ &= d_{21}\begin{vmatrix} d_{22} & d_{23} \\ d_{32} & d_{33} \end{vmatrix} - d_{22}\begin{vmatrix} d_{21} & d_{23} \\ d_{31} & d_{33} \end{vmatrix} + d_{23}\begin{vmatrix} d_{21} & d_{22} \\ d_{31} & d_{32} \end{vmatrix} \\ &= \begin{vmatrix} d_{21} & d_{22} & d_{23} \\ d_{21} & d_{22} & d_{23} \\ d_{31} & d_{32} & d_{33} \end{vmatrix} = 0. \end{aligned}$$

The remaining off-diagonal elements suffer the same fate. It follows that the product matrix consists of identical scalars, $|\mathbf{D}|$, along the diagonal, and zeros everywhere else:

$$\mathbf{P} = \mathbf{D}\mathbf{D}_\mathbf{c}^T = \begin{bmatrix} |\mathbf{D}| & 0 & 0 \\ 0 & |\mathbf{D}| & 0 \\ 0 & 0 & |\mathbf{D}| \end{bmatrix} = |\mathbf{D}| \begin{bmatrix} 1 & 0 & 0 \\ 0 & 1 & 0 \\ 0 & 0 & 1 \end{bmatrix} = |\mathbf{D}|\mathbf{I}. \tag{1.76}$$

Since $|\mathbf{D}|$ is a scalar, *provided that it is not zero,*

$$\mathbf{D}\frac{\mathbf{D}_\mathbf{c}^T}{|\mathbf{D}|} = \mathbf{I} \quad \Longrightarrow \quad \mathbf{D}^{-1} = \frac{\mathbf{D}_\mathbf{c}^T}{|\mathbf{D}|}. \tag{1.77}$$

In order to determine the inverse of a matrix we generate a cofactor matrix and transpose it (the transpose of the cofactor matrix is called the *adjoint* matrix), then divide each element of the resulting matrix by the determinant of the original matrix. It follows that *the inverse of a matrix can exist only if the determinant of the matrix is non-zero.* Thus, *if a matrix is singular, it does not have an inverse.* The inverse of a 3×3 matrix will be especially important throughout the remainder of the book. Given the determinant, $|\mathbf{D}|$, from Eqn. 1.70,

$$\mathbf{D}^{-1} = \frac{1}{|\mathbf{D}|} \begin{bmatrix} (d_{22}d_{33} - d_{32}d_{23}) & (d_{32}d_{13} - d_{12}d_{33}) & (d_{12}d_{23} - d_{22}d_{13}) \\ (d_{31}d_{23} - d_{21}d_{33}) & (d_{11}d_{33} - d_{31}d_{13}) & (d_{21}d_{13} - d_{11}d_{23}) \\ (d_{21}d_{32} - d_{31}d_{22}) & (d_{31}d_{12} - d_{11}d_{32}) & (d_{11}d_{22} - d_{21}d_{12}) \end{bmatrix}. \tag{1.78}$$

The inverse of the transformation matrices described earlier will be especially important in the discussion of symmetry at the end of this chapter. Recall that these matrices were orthonormal matrices. Consider the product of an orthonormal matrix $\mathbf{T} = [\mathbf{t_1}\ \mathbf{t_2}\ \mathbf{t_3}]$ and its transpose:

$$\begin{aligned} \mathbf{T}^T\mathbf{T} &= \begin{bmatrix} t_{11} & t_{21} & t_{31} \\ t_{12} & t_{22} & t_{32} \\ t_{13} & t_{23} & t_{33} \end{bmatrix} \begin{bmatrix} t_{11} & t_{12} & t_{13} \\ t_{21} & t_{22} & t_{23} \\ t_{31} & t_{32} & t_{33} \end{bmatrix} \\ &= \begin{bmatrix} (\mathbf{t_1}\cdot\mathbf{t_1}) & (\mathbf{t_1}\cdot\mathbf{t_2}) & (\mathbf{t_1}\cdot\mathbf{t_3}) \\ (\mathbf{t_2}\cdot\mathbf{t_1}) & (\mathbf{t_2}\cdot\mathbf{t_2}) & (\mathbf{t_2}\cdot\mathbf{t_3}) \\ (\mathbf{t_3}\cdot\mathbf{t_1}) & (\mathbf{t_3}\cdot\mathbf{t_2}) & (\mathbf{t_3}\cdot\mathbf{t_3}) \end{bmatrix} = \begin{bmatrix} 1 & 0 & 0 \\ 0 & 1 & 0 \\ 0 & 0 & 1 \end{bmatrix} = \mathbf{I}. \end{aligned} \tag{1.79}$$

The transpose of an orthonormal matrix is its inverse. Note also that the identity matrix is its own inverse:

$$\mathbf{I}\mathbf{I} = \mathbf{I} \Longrightarrow \mathbf{I}^{-1} = \mathbf{I}. \tag{1.80}$$

If a matrix is symmetric ($\mathbf{T} = \mathbf{T}^T$), its inverse will also be symmetric:

$$\begin{aligned} &\mathbf{T}\mathbf{T}^{-1} = \mathbf{I} \\ &(\mathbf{T}^{-1})^T\mathbf{T}^T = \mathbf{I} \\ &(\mathbf{T}^{-1})^T\mathbf{T} = \mathbf{I} \Longrightarrow \\ &(\mathbf{T}^{-1})^T = \mathbf{T}^{-1}. \end{aligned} \tag{1.81}$$

1.4.6 The Rules of Matrix Algebra

Matrix transformations are *linear* operations since they change the vectors upon which they operate linearly, i.e., they do not modify vectors with exponents or other functions. The matrix operations defined previously provide the basis for formulating the algebraic rules for combining and manipulating matrices. The rules of this *linear algebra* that are important to crystallography are summarized and rationalized in this section.

Rule 1. *Matrix additions are associative.* For $\mathbf{G} = (\mathbf{D}+\mathbf{E})+\mathbf{F}$, the $ijth$ matrix element of $\mathbf{G}$ is $g_{ij} = (d_{ij}+e_{ij})+f_{ij}$. For $\mathbf{H} = \mathbf{D}+(\mathbf{E}+\mathbf{F})$, the $ijth$ matrix element of $\mathbf{H}$ is $h_{ij} = d_{ij}+(e_{ij}+f_{ij})$. Clearly, $g_{ij} = h_{ij}$ and

$$(\mathbf{D}+\mathbf{E})+\mathbf{F} = \mathbf{D}+(\mathbf{E}+\mathbf{F}) \tag{1.82}$$

Rule 2. *Matrix additions are commutative.* For $\mathbf{F} = \mathbf{D}+\mathbf{E}$, the $ijth$ matrix element of $\mathbf{F}$ is $f_{ij} = d_{ij}+e_{ij}$. For $\mathbf{G} = \mathbf{E}+\mathbf{D}$, the $ijth$ matrix element of $\mathbf{G}$ is $g_{ij} = e_{ij}+d_{ij}$. Again, it is obvious that $f_{ij} = g_{ij}$ and

$$\mathbf{D}+\mathbf{E} = \mathbf{E}+\mathbf{D} \tag{1.83}$$

Rule 3. *Matrix additions are distributive with respect to scalar multiplication.* Let $\mathbf{F} = s(\mathbf{D}+\mathbf{E})$. Then $f_{ij} = s(d_{ij}+e_{ij}) = sd_{ij}+se_{ij}$. It follows that

$$s(\mathbf{D}+\mathbf{E}) = s\mathbf{D}+s\mathbf{E}. \tag{1.84}$$

Also, if $\mathbf{G} = (s_1+s_2)\mathbf{D}$, then $g_{ij} = (s_1+s_2)d_{ij} = s_1d_{ij}+s_2d_{ij}$ and

$$(s_1+s_2)\mathbf{D} = s_1\mathbf{D}+s_2\mathbf{D}. \tag{1.85}$$

Rule 4. *Matrix multiplications are associative.* For $\mathbf{G} = (\mathbf{DE})\mathbf{F}$ and $\mathbf{H} = \mathbf{D}(\mathbf{EF})$, Let $(\mathbf{DE})\mathbf{F} = \mathbf{PF}$ and $\mathbf{D}(\mathbf{EF}) = \mathbf{DQ}$. For simplicity, assume that $\mathbf{D}$ and $\mathbf{E}$ are 3×3 matrices and $\mathbf{F}$ is either a 3×3 matrix or a 3×1 column vector. Then

$$\begin{aligned}
g_{ij} &= \sum_{k=1}^{3} p_{ik}f_{kj} \\
p_{ik} &= \sum_{m=1}^{3} d_{im}e_{mk} \\
g_{ij} &= \sum_{k=1}^{3}\left(\sum_{m=1}^{3} d_{im}e_{mk}\right)f_{kj} \\
g_{ij} &= \sum_{m=1}^{3} d_{im}\left(\sum_{m=1}^{3} e_{mk}f_{kj}\right) \\
g_{ij} &= \sum_{m=1}^{3} d_{im}q_{mj} = h_{ij} \implies \mathbf{G} = \mathbf{H}.\ \text{Thus,} \\
(\mathbf{DE})\mathbf{F} &= \mathbf{D}(\mathbf{EF}).
\end{aligned} \tag{1.86}$$

Rule 5. *Matrix multiplications are, in general, **not** commutative.* Consider, for example, rotation of the vector $\mathbf{v}$ around the z axis by 90° with $\mathbf{R_Z}$, followed by a rotation around the x axis of 90° with $\mathbf{R_X}$, $\mathbf{R_X}(\mathbf{R_Z}\mathbf{v}) = (\mathbf{R_X}\mathbf{R_Z})\mathbf{v}$ (Rule 4). The matrix product creates a new transformation matrix, $\mathbf{R_1} = \mathbf{R_Z}\mathbf{R_X}$:

$$\begin{aligned}
\mathbf{R_1} &= \begin{bmatrix} \cos(\pi/2) & -\sin(\pi/2) & 0 \\ \sin(\pi/2) & \cos(\pi/2) & 0 \\ 0 & 0 & 1 \end{bmatrix} \begin{bmatrix} 1 & 0 & 0 \\ 0 & \cos(\pi/2) & -\sin(\pi/2) \\ 0 & \sin(\pi/2) & \cos(\pi/2) \end{bmatrix} \\
&= \begin{bmatrix} 0 & -1 & 0 \\ 1 & 0 & 0 \\ 0 & 0 & 1 \end{bmatrix} \begin{bmatrix} 1 & 0 & 0 \\ 0 & 0 & -1 \\ 0 & 1 & 0 \end{bmatrix} = \begin{bmatrix} 0 & 0 & 1 \\ 1 & 0 & 0 \\ 0 & 1 & 0 \end{bmatrix} \\
\mathbf{R_1}\mathbf{v} &= \mathbf{R_1} \begin{bmatrix} x_c \\ y_c \\ z_c \end{bmatrix} = \begin{bmatrix} z_c \\ x_c \\ y_c \end{bmatrix}
\end{aligned}$$

If we reverse the order of the operations, rotating around the x axis followed by rotation around the z axis we have $\mathbf{R_Z}(\mathbf{R_X}\mathbf{v}) = (\mathbf{R_Z}\mathbf{R_X})\mathbf{v}$. The new transformation matrix is now $\mathbf{R_2} = \mathbf{R_X}\mathbf{R_Z}$:

$$\begin{aligned}
\mathbf{R_2} &= \begin{bmatrix} 1 & 0 & 0 \\ 0 & \cos(\pi/2) & -\sin(\pi/2) \\ 0 & \sin(\pi/2) & \cos(\pi/2) \end{bmatrix} \begin{bmatrix} \cos(\pi/2) & -\sin(\pi/2) & 0 \\ \sin(\pi/2) & \cos(\pi/2) & 0 \\ 0 & 0 & 1 \end{bmatrix} \\
&= \begin{bmatrix} 1 & 0 & 0 \\ 0 & 0 & -1 \\ 0 & 1 & 0 \end{bmatrix} \begin{bmatrix} 0 & -1 & 0 \\ 1 & 0 & 0 \\ 0 & 0 & 1 \end{bmatrix} = \begin{bmatrix} 0 & -1 & 0 \\ 0 & 0 & 1 \\ 1 & 0 & 0 \end{bmatrix} \\
\mathbf{R_2}\mathbf{v} &= \mathbf{R_2} \begin{bmatrix} x_c \\ y_c \\ z_c \end{bmatrix} = \begin{bmatrix} -y_c \\ z_c \\ x_c \end{bmatrix}
\end{aligned}$$

Thus $\mathbf{R_Z}\mathbf{R_X} \neq \mathbf{R_X}\mathbf{R_Z}$. In general, for matrices $\mathbf{DE}$ and $\mathbf{ED}$,

$$\mathbf{DE} \neq \mathbf{ED}. \tag{1.87}$$

The order in which we multiply matrices matters! It follows that for the equality $\mathbf{D} = \mathbf{E}$, *pre-multiplying or post-multiplying both sides of the matrix equation by another matrix retains the equality*: $\mathbf{FD} = \mathbf{FE}$ and $\mathbf{DF} = \mathbf{EF}$, *but only in special cases does* $\mathbf{DF} = \mathbf{FE} = \mathbf{FD} = \mathbf{EF}$. When the order of multiplication of two matrices does not alter the product we say that the matrices *commute*.

Rule 6. *A matrix commutes with the identity matrix and its own inverse.* The identity matrix commutes with all matrices:

$$\mathbf{DI} = \begin{bmatrix} d_{11} & d_{12} & d_{13} \\ d_{21} & d_{22} & d_{23} \\ d_{31} & d_{32} & d_{33} \end{bmatrix} \begin{bmatrix} 1 & 0 & 0 \\ 0 & 1 & 0 \\ 0 & 0 & 1 \end{bmatrix} = \begin{bmatrix} d_{11} & d_{12} & d_{13} \\ d_{21} & d_{22} & d_{23} \\ d_{31} & d_{32} & d_{33} \end{bmatrix} = \mathbf{D}$$

$$\mathbf{ID} = \begin{bmatrix} 1 & 0 & 0 \\ 0 & 1 & 0 \\ 0 & 0 & 1 \end{bmatrix} \begin{bmatrix} d_{11} & d_{12} & d_{13} \\ d_{21} & d_{22} & d_{23} \\ d_{31} & d_{32} & d_{33} \end{bmatrix} = \begin{bmatrix} d_{11} & d_{12} & d_{13} \\ d_{21} & d_{22} & d_{23} \\ d_{31} & d_{32} & d_{33} \end{bmatrix} = \mathbf{D}$$

$$\mathbf{DI} = \mathbf{ID} = \mathbf{D}. \tag{1.88}$$

A matrix also commutes with its inverse:

$$\begin{aligned}
\mathbf{D}^{-1}\mathbf{D} &= \mathbf{I} \\
\mathbf{D}^{-1}\mathbf{D}^{-1}\mathbf{D} &= \mathbf{D}^{-1}\mathbf{I} = \mathbf{D}^{-1} \\
\mathbf{D}^{-1}\mathbf{D}\mathbf{D}^{-1} &= \mathbf{I}\mathbf{D}^{-1} = \mathbf{D}^{-1} \\
\mathbf{D}^{-1}\mathbf{D}^{-1}\mathbf{D} &= \mathbf{D}^{-1}\mathbf{D}\mathbf{D}^{-1} \\
\mathbf{D}^{-1}\mathbf{D} &= \mathbf{D}\mathbf{D}^{-1} = \mathbf{I}. \qquad (1.89)
\end{aligned}$$

Rule 7. *Matrix multiplications are distributive.* Let $\mathbf{S} = \mathbf{E} + \mathbf{F}$, $\mathbf{T} = \mathbf{D}(\mathbf{E} + \mathbf{F}) = \mathbf{DS}$, $\mathbf{V} = \mathbf{DE}$, and $\mathbf{W} = \mathbf{DF}$. Then

$$\begin{aligned}
t_{ij} &= \sum_{k=1}^{3} d_{ik}s_{kj} = \sum_{k=1}^{3} d_{ik}(e_{kj} + f_{kj}) \\
&= \sum_{k=1}^{3} d_{ik}e_{kj} + d_{ik}f_{kj} = \sum_{k=1}^{3} d_{ik}e_{kj} + \sum_{k=1}^{3} d_{ik}f_{kj} \\
&= v_{ij} + w_{ij} \Longrightarrow \mathbf{T} = \mathbf{V} + \mathbf{W}
\end{aligned}$$

$$\mathbf{D}(\mathbf{E} + \mathbf{F}) = \mathbf{DE} + \mathbf{DF}. \qquad (1.90)$$

In this case the matrix sum is pre-multiplied by a matrix, and this is formally called the *left distributive property of matrix multiplication.* By a similar argument, it is easy to demonstrate the *right distributive property of matrix multiplication,* in which the matrix sum is post-multiplied by a matrix: $(\mathbf{E} + \mathbf{F})\mathbf{D} = \mathbf{ED} + \mathbf{FD}$.

Rule 8. *The inverse of the product of an ordered array of matrices is the product of the inverses of the individual matrices* ***in reverse order****.*

$$\begin{aligned}
\mathbf{D}^{-1}\mathbf{D} &= \mathbf{D}^{-1}\mathbf{I}\mathbf{D} = \mathbf{D}^{-1}(\mathbf{E}^{-1}\mathbf{E})\mathbf{D} \\
&= (\mathbf{D}^{-1}\mathbf{E}^{-1})(\mathbf{ED}) = \mathbf{I} \\
&\Longrightarrow (\mathbf{ED})^{-1} = \mathbf{D}^{-1}\mathbf{E}^{-1} \\
(\mathbf{D}^{-1}\mathbf{E}^{-1})(\mathbf{ED}) &= (\mathbf{D}^{-1}\mathbf{E}^{-1})\mathbf{I}(\mathbf{ED}) \\
&= (\mathbf{D}^{-1}\mathbf{E}^{-1})(\mathbf{F}^{-1}\mathbf{F})(\mathbf{ED}) \\
&= (\mathbf{D}^{-1}\mathbf{E}^{-1}\mathbf{F}^{-1})(\mathbf{FED}) = \mathbf{I} \\
&\Longrightarrow (\mathbf{FED})^{-1} = \mathbf{D}^{-1}\mathbf{E}^{-1}\mathbf{F}^{-1}.
\end{aligned}$$

By induction,

$$(\mathbf{QR}\ldots\mathbf{FED})^{-1} = \mathbf{D}^{-1}\mathbf{E}^{-1}\mathbf{F}^{-1}\ldots\mathbf{Q}^{-1}\mathbf{R}^{-1}. \qquad (1.91)$$

Rule 9. *The transpose of the product of an ordered array of matrices is the product of the transposes of the individual matrices* ***in reverse order****.* Let $\mathbf{U} = \mathbf{DE}$,

$\mathbf{V} = (\mathbf{DE})^T$, $\mathbf{S} = \mathbf{E}^T, \mathbf{T} = \mathbf{D}^T$, and $\mathbf{W} = \mathbf{E}^T\mathbf{D}^T = \mathbf{ST}$. Again, for simplicity, assume that all matrices are 3×3.

$$\begin{aligned}
u_{ij} &= \sum_{k=1}^{3} d_{ik}e_{kj} \\
v_{ij} &= u_{ij}^T = u_{ji} = \sum_{k=1}^{3} d_{jk}e_{ki} \\
w_{ij} &= \sum_{i=1}^{3} s_{ik}t_{kj},\ s_{ik} = e_{ki} \text{ and } t_{kj} = d_{jk} \\
w_{ij} &= \sum_{i=1}^{3} e_{ki}d_{jk} = \sum_{i=1}^{3} d_{jk}e_{ki} = v_{ij} \Longrightarrow \mathbf{V} = \mathbf{W} \\
\mathbf{E}^T\mathbf{D}^T &= (\mathbf{DE})^T \\
\mathbf{F}^T\mathbf{E}^T\mathbf{D}^T &= \mathbf{F}^T(\mathbf{DE})^T = (\mathbf{DEF})^T
\end{aligned}$$

By induction,

$$(\mathbf{QR}\ldots\mathbf{FED})^T = \mathbf{D^T E^T F^T}\ldots\mathbf{Q^T R^T}. \tag{1.92}$$

Rule 10. *The transpose of the sum of two matrices is the sum of the transposes of the individual matrices.* Let $\mathbf{D} = \mathbf{A} + \mathbf{B}$.

$$\begin{aligned}
\begin{bmatrix} d_{11} & d_{12} & d_{13} \\ d_{21} & d_{22} & d_{23} \\ d_{31} & d_{32} & d_{33} \end{bmatrix} &= \begin{bmatrix} a_{11} & a_{12} & a_{13} \\ a_{21} & a_{22} & a_{23} \\ a_{31} & a_{32} & a_{33} \end{bmatrix} + \begin{bmatrix} b_{11} & b_{12} & b_{13} \\ b_{21} & b_{22} & b_{23} \\ b_{31} & b_{32} & b_{33} \end{bmatrix} \\
&= \begin{bmatrix} a_{11}+b_{11} & a_{12}+b_{12} & a_{13}+b_{13} \\ a_{21}+b_{21} & a_{22}+b_{22} & a_{23}+b_{23} \\ a_{31}+b_{31} & a_{32}+b_{32} & a_{33}+b_{33} \end{bmatrix}
\end{aligned} \tag{1.93}$$

$$\begin{aligned}
\mathbf{D}^T = \begin{bmatrix} d_{11} & d_{21} & d_{31} \\ d_{12} & d_{22} & d_{32} \\ d_{13} & d_{23} & d_{33} \end{bmatrix} &= \begin{bmatrix} a_{11}+b_{11} & a_{21}+b_{21} & a_{31}+b_{31} \\ a_{12}+b_{12} & a_{22}+b_{22} & a_{32}+b_{32} \\ a_{13}+b_{13} & a_{23}+b_{23} & a_{33}+b_{33} \end{bmatrix} \\
&= \begin{bmatrix} a_{11} & a_{21} & a_{31} \\ a_{12} & a_{22} & a_{32} \\ a_{13} & a_{23} & a_{33} \end{bmatrix} + \begin{bmatrix} b_{11} & b_{21} & b_{31} \\ b_{12} & b_{22} & b_{32} \\ b_{13} & b_{23} & b_{33} \end{bmatrix} \\
&= \mathbf{A}^T + \mathbf{B}^T.
\end{aligned} \tag{1.94}$$

Rule 11. *The transpose of the inverse of a matrix is the inverse of the transpose of the matrix.*

$$\begin{aligned} \mathbf{G}^{-1}\mathbf{G} &= \mathbf{I} \\ \mathbf{G}^T(\mathbf{G}^{-1})^T &= \mathbf{I}^T = \mathbf{I} \\ \mathbf{G}^T(\mathbf{G}^T)^{-1} &= \mathbf{I} \\ \mathbf{G}^T(\mathbf{G}^T)^{-1} &= \mathbf{G}^T(\mathbf{G}^{-1})^T \\ (\mathbf{G}^T)^{-1} &= (\mathbf{G}^{-1})^T. \end{aligned} \tag{1.95}$$

1.4.7 The Eigenvectors and Eigenvalues of a Matrix

In general, when a matrix operates on a vector, the vector changes in both direction and magnitude:

$$\mathbf{D}\mathbf{v} = \mathbf{v}'. \tag{1.96}$$

Among the infinitely many vectors that a matrix can transform, there is a subset of those vectors, *characteristic* of the matrix, that *may* change in magnitude, but either remain parallel or become antiparallel to the original vectors when operated on by the matrix. These characteristic vectors, $\{\mathbf{e}\}$, are known as *eigenvectors* (*eigen* in German roughly translates as *characteristic* or *innate*). When a matrix operates on its eigenvectors, the vector is stretched, shrunk, or unmodified by a scalar, but otherwise it either retains its direction or reverses it:

$$\mathbf{D}\mathbf{e}_i = \lambda_i \mathbf{e}_i, \quad ß = 1, 2, \ldots, n. \tag{1.97}$$

The scalar multiplier of each eigenvector, λ, is also characteristic of the matrix, and is known as the *eigenvalue* of the eigenvector.

Note that if an eigenvector is multiplied or divided by a constant the resulting vector is still an eigenvector of the matrix with the same eigenvalue:

$$\mathbf{D}(q\mathbf{e}_i) = \lambda_i(q\mathbf{e}_i). \tag{1.98}$$

Thus the "eigenvector" for a given eigenvalue is actually an infinite set of vectors with all possible magnitudes, all pointing in the same direction. The most useful eigenvector is the one of unit length, obtained by dividing any one of the set of eigenvectors by its length:

$$\begin{aligned} \mathbf{D}\frac{\mathbf{e}_i}{e_i} &= \lambda_i \frac{\mathbf{e}_i}{e_i} \\ \mathbf{D}\mathbf{e}_{\mathbf{u}i} &= \lambda_i \mathbf{e}_{\mathbf{u}i}, \quad e_{ui} = 1. \end{aligned} \tag{1.99}$$

For an $n \times n$ matrix there are n unit eigenvectors and n corresponding eigenvalues. For convenience we will focus on the three-dimensional case. The three resulting equations (Eqn. 1.99) can be represented by a single matrix equation:

$$\begin{aligned} \mathbf{D}\mathbf{E}_{\mathbf{u}} &= \mathbf{D}\begin{bmatrix} e_{11} & e_{12} & e_{13} \\ e_{21} & e_{22} & e_{23} \\ e_{31} & e_{32} & e_{33} \end{bmatrix} = \begin{bmatrix} e_{11} & e_{12} & e_{13} \\ e_{21} & e_{22} & e_{23} \\ e_{31} & e_{32} & e_{33} \end{bmatrix} \begin{bmatrix} \lambda_1 & 0 & 0 \\ 0 & \lambda_2 & 0 \\ 0 & 0 & \lambda_3 \end{bmatrix} \\ \mathbf{D}\mathbf{E}_{\mathbf{u}} &= \mathbf{E}_{\mathbf{u}}[\lambda]. \end{aligned} \tag{1.100}$$

The column vectors of $\mathbf{E_u}$ are the unit eigenvectors of $\mathbf{D}$, and the diagonal elements of $[\lambda]$ are the eigenvalues. The matrix consisting of the eigenvectors of $\mathbf{D}$, $\mathbf{E}$, is referred to as the *modal matrix* of $\mathbf{D}$. Eqn. 1.100 is equally valid for general eigenvectors, i.e., $\mathbf{DE} = \mathbf{E}[\lambda]$. Note that the equation must be written in this manner; $\mathbf{E}$ and $[\lambda]$ *do not commute.*

The treatment of atom displacements in Chapter 5 will involve the eigenvectors and eigenvalues of symmetric matrices. In preparation for this, we derive an important property for *symmetric* matrices: *If the matrix,* $\mathbf{D}$, *is symmetric, then its modal matrix is orthogonal.* To prove this we show that any pair of the eigenvectors that compose a modal matrix of $\mathbf{D}$, $\mathbf{e}_i$ and $\mathbf{e}_j$, with distinct eigenvalues $\lambda_i \neq \lambda_j$, have a scalar product of zero, given that $\mathbf{D} = \mathbf{D}^T$:

$$\mathbf{D}\,\mathbf{e}_i = \lambda_i\,\mathbf{e}_i$$

$$\mathbf{e}_j^T\mathbf{D}\,\mathbf{e}_i = \lambda_i\,\mathbf{e}_j^T\,\mathbf{e}_i \tag{1.101}$$

$$\mathbf{D}\,\mathbf{e}_j = \lambda_j\,\mathbf{e}_j$$

$$(\mathbf{D}\,\mathbf{e}_j)^T = \mathbf{e}_j^T\mathbf{D}^T = \mathbf{e}_j^T\mathbf{D} = \lambda_j\,\mathbf{e}_j^T$$

$$\mathbf{e}_j^T\mathbf{D}\,\mathbf{e}_i = \lambda_j\,\mathbf{e}_j^T\,\mathbf{e}_i \tag{1.102}$$

Subtracting Eqn. 1.101 from Eqn. 1.102 results in $(\lambda_j - \lambda_i)(\mathbf{e}_j^T\,\mathbf{e}_i) = 0$. Since $\lambda_j - \lambda_i \neq 0$, $\mathbf{e}_j^T\,\mathbf{e}_i = 0$; the eigenvectors of the modal matrix of $\mathbf{D}$ are orthogonal.

The eigenvectors and eigenvalues of matrix $\mathbf{D}$ (for $n = 3$) are determined by solving Eqn. 1.97:

$$\begin{bmatrix} d_{11} & d_{12} & d_{13} \\ d_{21} & d_{22} & d_{23} \\ d_{31} & d_{32} & d_{33} \end{bmatrix} \begin{bmatrix} e_1 \\ e_2 \\ e_3 \end{bmatrix} = \lambda \begin{bmatrix} e_1 \\ e_2 \\ e_3 \end{bmatrix}$$

$$d_{11}\,e_1 + d_{12}\,e_2 + d_{13}\,e_3 = \lambda\,e_1$$

$$d_{21}\,e_1 + d_{22}\,e_2 + d_{23}\,e_3 = \lambda\,e_2$$

$$d_{31}\,e_1 + d_{32}\,e_2 + d_{33}\,e_3 = \lambda\,e_3$$

The result is three homogenous (all equal to zero) linear equations:

$$(d_{11} - \lambda)\,e_1 + d_{12}\,e_2 + d_{13}\,e_3 = 0 \tag{1.103}$$

$$d_{21}\,e_1 + (d_{22} - \lambda)\,e_2 + d_{23}\,e_3 = 0 \tag{1.104}$$

$$d_{31}\,e_1 + d_{32}\,e_2 + (d_{33} - \lambda)\,e_3 = 0 \tag{1.105}$$

$$\begin{bmatrix} d_{11} - \lambda & d_{12} & d_{13} \\ d_{21} & d_{22} - \lambda & d_{23} \\ d_{31} & d_{32} & d_{33} - \lambda \end{bmatrix} \begin{bmatrix} e_1 \\ e_2 \\ e_3 \end{bmatrix} = \begin{bmatrix} 0 \\ 0 \\ 0 \end{bmatrix}$$

$$\mathbf{D}_\lambda \mathbf{e} = \mathbf{0}. \tag{1.106}$$

If $\mathbf{D}_\lambda$ has an inverse then $\mathbf{e} = \mathbf{D}_\lambda^{-1}\mathbf{0} = \mathbf{0}$, a trivial solution (correct, but useless). Thus for there to be a non-trivial solution, $\mathbf{D}_\lambda$ cannot have an inverse. Recall that in Sec. 1.4.5 it was shown that a matrix has an inverse if and only if its determinant is not singular (equal to 0). Thus the criterion for a nontrivial solution is that $|\mathbf{D}_\lambda| = 0$:

$$\begin{aligned}|\mathbf{D}_\lambda| &= (d_{11}-\lambda)(d_{22}-\lambda)(d_{33}-\lambda) - (d_{11}-\lambda)d_{23}d_{32} - d_{12}d_{21}(d_{33}-\lambda)\\ &\quad -d_{13}(d_{22}-\lambda)d_{31} + d_{12}d_{23}d_{31} + d_{13}d_{21}d_{32} = 0.\end{aligned}$$

Expanding and collecting the terms produces a third order polynomial (cubic) equation:

$$\begin{aligned}&\lambda^3 - (d_{11}+d_{22}+d_{33})\lambda^2\\ &\quad +(d_{22}d_{33} - d_{32}d_{23} + d_{11}d_{33} - d_{31}d_{13} + d_{11}d_{22} - d_{21}d_{12})\lambda\\ &\quad -(d_{11}d_{22}d_{33} - d_{11}d_{23}d_{32} - d_{12}d_{21}d_{33}\\ &\quad +d_{12}d_{23}d_{31} + d_{13}d_{21}d_{32} - d_{13}d_{22}d_{31}) = 0, \qquad (1.107)\end{aligned}$$

conveniently written in terms of the trace, the cofactors of the diagonal elements, and the determinant of $\mathbf{D}$:

$$\lambda^3 - (d_{11}+d_{22}+d_{33})\lambda^2 + (\mathcal{D}_{11}+\mathcal{D}_{22}+\mathcal{D}_{33})\lambda - |\mathbf{D}| = 0. \qquad (1.108)$$

The cubic equation, of the form $f(\lambda) = a\,\lambda^3 + b\,\lambda^2 + c\,\lambda + d = 0$, has three roots, λ_1, λ_2, and λ_3, which are the eigenvalues of $\mathbf{D}$.* Inserting each eigenvalue into Eqns. 1.103–1.105 allows us to solve for the corresponding eigenvector. The solutions are not unique — there are an infinite number of eigenvectors for each eigenvalue, differing by their magnitudes (Eqn. 1.99). Because of this each eigenvector solution will have one arbitrary component (there will be one vector in the infinite set that has this component). For example, setting $e_3 = 1$ and inserting λ_1 into Eqns. 1.103–1.105 finds e_2 in terms of e_1 from Eqn. 1.103. Substituting the expression for e_2 into Eqn. 1.104 then provides a value for e_1, and subsequently, e_2. The eigenvector has magnitude $e = \sqrt{(e_1^2 + e_2^2 + 1^2)}$, and $\mathbf{e}_u = [e_1/e\ e_2/e\ 1/e]^T$.* The remaining two unit eigenvectors are determined by substituting λ_2 and λ_3, respectively.

If a matrix is orthogonal, it does not change the magnitude of any vector that it transforms, including it eigenvectors. It follows that the real† eigenvalues of an orthogonal matrix must be ± 1:

$$\mathbf{D}\mathbf{e}_i = \pm\mathbf{e}_i, \quad i = 1, 2, \ldots, n. \qquad (1.109)$$

*There are analytical formulas for the three roots of a cubic equation,[21] just as there are for the two roots of a quadratic equation. The formulae are quite complex, involving a number of operations. The solutions for cubic and higher order equations are generally obtained numerically, by searching for values of the independent variable that set the function very close to zero (within some preselected tolerance limit).

*The representation of a vector in a text line requires it to be written as a row vector. Most of the vector operations throughout the book will be undertaken with column vectors, formally requiring the row vector in the text line to be written as its transpose. To avoid having to repeat this notation continually throughout the book, unless specifically indicated, *a row vector in a text line will be assumed to be a column vector.*

†An orthogonal rotation matrix can also have one real eigenvalue of +1 and two complex eigenvalues, $e^{i\theta} = \cos\theta + i\sin\theta$ and $e^{-i\theta} = \cos\theta - i\sin\theta$; the eigenvectors in these cases have imaginary components and do not change direction in "complex space."

1.5 Coordinate Systems in Crystallography

As discussed earlier, the ability to treat vectors in either the unit cell basis $\{\mathbf{a}\ \mathbf{b}\ \mathbf{c}\}$ or an orthonormal basis $\{\mathbf{i}\ \mathbf{j}\ \mathbf{k}\}$ requires a method for transforming the vector between the two bases. The matrix algebra described in the last section provides us with just that! We begin with the components of a vector defined in one coordinate system and set out to determine a transformation matrix that will generate the coordinates of the vector in a new coordinate system — a *change of basis.*

1.5.1 Change of Basis

Fig. 1.30 illustrates a vector $\mathbf{v}$ described in two different coordinate systems, one defined by the basis set $\{R\} = \{\mathbf{r_x}\ \mathbf{r_y}\ \mathbf{r_z}\}$ and the other defined by the basis set $\{S\} = \{\mathbf{s_x}\ \mathbf{s_y}\ \mathbf{s_z}\}$. In the $\{R\}$ basis $\mathbf{v} = [x\ y\ z]$ — in fractional coordinates $\mathbf{v} = [x_r\ y_r\ z_r]$ where $x_r = x/r_x$, $y_r = y/r_y$, and $z_r = z/r_z$. In the $\{S\}$ basis $\mathbf{v} = [x'\ y'\ z']$ — in fractional coordinates $\mathbf{v} = [x_s\ y_s\ z_s]$ where $x_s = x'/s_x$, $y_s = y'/s_y$, and $z_s = z'/s_z$.

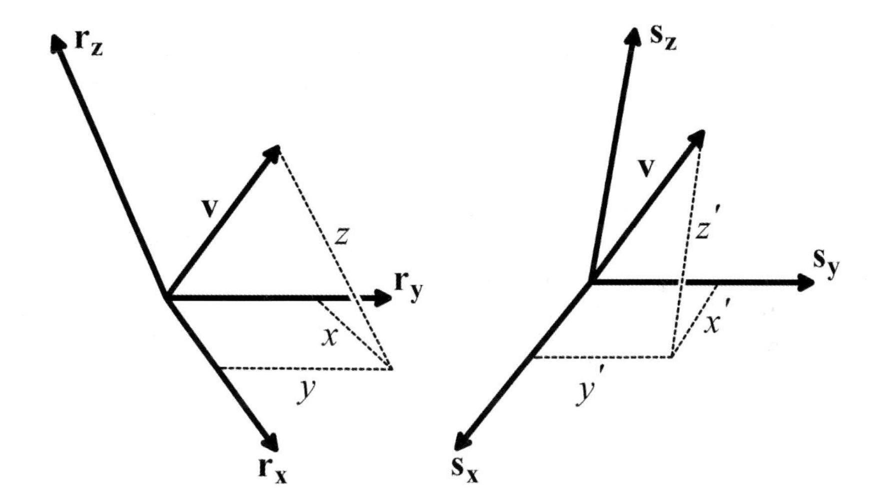

Figure 1.30 Vector $\mathbf{v}$ in two different coordinate systems, $\{R\} = \{\mathbf{r_x}\ \mathbf{r_y}\ \mathbf{r_z}\}$ and $\{S\} = \{\mathbf{s_x}\ \mathbf{s_y}\ \mathbf{s_z}\}$. $[x\ y\ z]$ are the displacement distances along each of the axes in $\{R\}$, $[x'\ y'\ z']$ are the displacements along each of the axes in $\{S\}$.

Suppose that we have described the vector in $\{R\}$, and wish to determine its coordinates in $\{S\}$. Thus we know $[x_r\ y_r\ z_r]$ and can express the vector in terms of the coordinates:

$$\mathbf{v} = x_r\mathbf{r_x} + y_r\mathbf{r_y} + z_r\mathbf{r_z}. \tag{1.110}$$

We wish to determine $[x_s\ y_s\ z_s]$ such that

$$\mathbf{v} = x_s\mathbf{s_x} + y_s\mathbf{s_y} + z_s\mathbf{s_z}. \tag{1.111}$$

In $\{R\}$ the components of the basis vectors are

$$\begin{aligned}\mathbf{r_x} &= 1\mathbf{r_x} + 0\mathbf{r_y} + 0\mathbf{r_z} &= [1\ 0\ 0]\\ \mathbf{r_y} &= 0\mathbf{r_x} + 1\mathbf{r_y} + 0\mathbf{r_z} &= [0\ 1\ 0]\\ \mathbf{r_z} &= 0\mathbf{r_x} + 0\mathbf{r_y} + 1\mathbf{r_z} &= [0\ 0\ 1].\end{aligned}$$

Since the basis vectors in $\{R\}$ are vectors in 3-space, each of them can also be described as a linear combination of the basis vectors in $\{S\}$:

$$\begin{aligned}\mathbf{r_x} &= r_{11}\mathbf{s_x} + r_{12}\mathbf{s_y} + r_{13}\mathbf{s_z}\\ \mathbf{r_y} &= r_{21}\mathbf{s_x} + r_{22}\mathbf{s_y} + r_{23}\mathbf{s_z}\\ \mathbf{r_z} &= r_{31}\mathbf{s_x} + r_{32}\mathbf{s_y} + r_{33}\mathbf{s_z}.\end{aligned}$$

Thus the vector in $\{R\}$ can be expressed as

$$\begin{aligned}\mathbf{v} = & \; x_r\,(r_{11}\mathbf{s_x} + r_{12}\mathbf{s_y} \;+\; r_{13}\mathbf{s_z}) \;+\; y_r\,(r_{21}\mathbf{s_x} + r_{22}\mathbf{s_y} + r_{23}\mathbf{s_z})\\ & + z_r\,(r_{31}\mathbf{s_x} + r_{32}\mathbf{s_y} + r_{33}\mathbf{s_z}).\end{aligned}$$

Expanding this expression and collecting terms gives

$$\begin{aligned}\mathbf{v} = & \; (x_r r_{11} + y_r r_{21} + z_r r_{31})\; s_x \;+\; (x_r r_{12} + y_r r_{22} + z_r r_{32})\; s_y\\ & + (x_r r_{13} + y_r r_{23} + z_r r_{33})\; s_z.\end{aligned}$$

This is $\mathbf{v}$ *expressed as a linear combination of the basis vectors in* $\{S\}$ We have determined the components of the vector in $\{S\}$ in terms of its components and the basis vectors in $\{R\}$, provided that we know the components of the basis vectors in $\{R\}$ in the $\{S\}$ basis:

$$\begin{aligned}x_s &= (x_r r_{11} + y_r r_{21} + z_r r_{31})\\ y_s &= (x_r r_{12} + y_r r_{22} + z_r r_{32})\\ z_s &= (x_r r_{13} + y_r r_{23} + z_r r_{33}).\end{aligned} \tag{1.112}$$

Eqns. 1.112 are easily recognized as a matrix equation:

$$\begin{bmatrix} r_{11} & r_{21} & r_{31}\\ r_{12} & r_{22} & r_{32}\\ r_{13} & r_{23} & r_{33}\end{bmatrix}\begin{bmatrix} x_r\\ y_r\\ z_r\end{bmatrix} = \begin{bmatrix} x_s\\ y_s\\ z_s\end{bmatrix} \tag{1.113}$$

Note that the matrix in this expression is the *transpose* of the matrix that consists of the coefficients of the components of the $\{R\}$ basis vectors in the $\{S\}$ basis.

1.5.2 Transformation from the Unit Cell Basis to an Orthonormal Basis

The strategy to formulate a method to transform vectors from a unit cell basis into an orthonormal basis should now be clear. We determine the components of the unit cell axes in the orthonormal basis and transpose the matrix of the components. This will provide a transformation matrix $\mathbf{B}$ such that $\mathbf{B}\mathbf{v_f} = \mathbf{v_c}$, where $\mathbf{v_f} = [x_f\ y_f\ z_f]$, in fractional coordinates in the unit cell basis, and $\mathbf{v_c} = [x_c\ y_c\ z_c]$, the Cartesian coordinates of the vector.

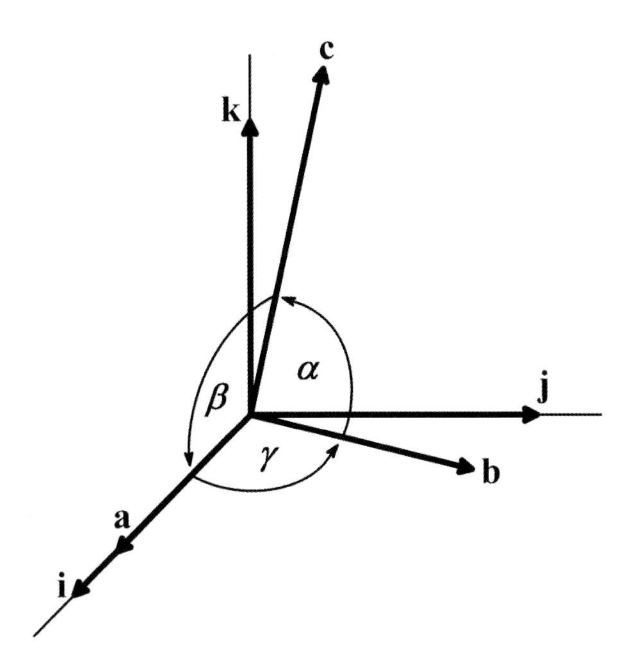

Figure 1.31 The standard orientation of a unit cell in a Cartesian coordinate system. The **a** axis is coincident with **i**, The **b** axis lies in the **ij** plane, and the z component of the **c** axis points in the same direction as **k**.

There is, however, one necessary step before the **B** matrix is determined. The orientation of the unit cell with respect to the Cartesian coordinate system is arbitrary. If we place the unit cell origin at the origin of the Cartesian system we can rotate the unit cell at any angle about the origin. Each orientation will, of course, create a different transformation matrix since the relation between the bases will change. We must therefore fix the orientation in some rational way. Fig. 1.31 illustrates the standard orientation of the unit cell with respect to the orthonormal basis. This orientation maintains right-handed coordinates for both systems. The **a** axis is coincident with the **i** axis, and points in the same direction (both are positive or negative simultaneously). The **b** axis is coplanar with **i** and **j** and points in a direction such that $\mathbf{a} \times \mathbf{b}$ is coincident with **i** and points in the same direction. The z component of the **c** axis points in the same direction as **k**.

We begin with a vector **v** with coordinates in the unit cell basis $\{\mathbf{a}\,\mathbf{b}\,\mathbf{c}\}$. Vectors will be considered column vectors in the discussion which follows. We will use the symbol $\mathbf{v_f} = [x_f\ y_f z_f]$ to denote a general vector with fractional coordinates in the unit cell basis:

$$\mathbf{v_f} = x_f\mathbf{a} + y_f\mathbf{b} + z_f\mathbf{c}. \tag{1.114}$$

We seek to determine the components *of the same vector* in Cartesian coordinates, which we will denote $\mathbf{v_c} = [x_c\ y_c\ z_c]$ such that

$$\mathbf{v_c} = x_c\mathbf{i} + y_c\mathbf{j} + z_c\mathbf{k}. \tag{1.115}$$

In the unit cell basis the components of the axial basis vectors are

$$\begin{aligned}
\mathbf{a} &= 1\mathbf{a} + 0\mathbf{b} + 0\mathbf{c} = [1\ 0\ 0] \\
\mathbf{b} &= 0\mathbf{a} + 1\mathbf{b} + 0\mathbf{c} = [0\ 1\ 0] \\
\mathbf{c} &= 0\mathbf{a} + 0\mathbf{b} + 1\mathbf{c} = [0\ 0\ 1].
\end{aligned} \tag{1.116}$$

The lengths of the unit cell axes, a, b, and c, and the angles between the axes, α, β, and γ are collectively known as *the unit cell parameters.* In the orthonormal basis the unit cell basis vectors have components $\mathbf{a_c} = [a_x\ a_y\ a_z]$, $\mathbf{b_c} = [b_x\ b_y\ b_z]$, and $\mathbf{c_c} = [c_x\ c_y\ c_z]$. The unit cell lengths and the components are expressed in the same units as the unit vectors of the Cartesian system (e.g., Å):

$$\begin{aligned}
\mathbf{a_c} &= a_x\mathbf{i} + a_y\mathbf{j} + a_z\mathbf{k} \\
\mathbf{b_c} &= b_x\mathbf{i} + b_y\mathbf{j} + b_z\mathbf{k} \\
\mathbf{c_c} &= c_x\mathbf{i} + c_y\mathbf{j} + c_z\mathbf{k}.
\end{aligned} \tag{1.117}$$

Determining these 9 components will provide the transformation matrix that we seek. Since $\mathbf{a}$ lies along the $\mathbf{i}$ axis,

$$a_x = a \tag{1.118}$$

$$b_x = 0 \tag{1.119}$$

$$c_x = 0. \tag{1.120}$$

Since $\mathbf{b_c}$ lies in the $\mathbf{ij}$ plane its z component is zero. The dot product of $\mathbf{a_c}$ and $\mathbf{b_c}$ gives us b_x:

$$\begin{aligned}
\mathbf{a_c} \cdot \mathbf{b_c} &= ab\cos\gamma \\
\mathbf{a_c} \cdot \mathbf{b_c} &= \mathbf{a_c}^T\mathbf{b_c} = \begin{bmatrix} a & 0 & 0 \end{bmatrix} \begin{bmatrix} b_x \\ b_y \\ 0 \end{bmatrix} = ab_x + 0 + 0 \Longrightarrow \\
ab_x &= ab\cos\gamma \\
b_x &= b\cos\gamma.
\end{aligned} \tag{1.121}$$

The dot product of $\mathbf{b_c}$ with itself provides b_y:

$$\begin{aligned}
\mathbf{b_c} \cdot \mathbf{b_c} &= bb\cos(0) = b^2 \\
\mathbf{b_c} \cdot \mathbf{b_c} &= \mathbf{b_c}^T\mathbf{b_c} = \begin{bmatrix} b\cos\gamma & b_y & 0 \end{bmatrix} \begin{bmatrix} b\cos\gamma \\ b_y \\ 0 \end{bmatrix} \\
&= b^2\cos^2\gamma + b_y^2 + 0 \Longrightarrow \\
b^2 &= b^2\cos^2\gamma + b_y^2 \\
b_y^2 &= b^2(1 - \cos^2\gamma) \\
&= b^2\sin^2\gamma
\end{aligned}$$

$$b_y = b\sin\gamma \tag{1.122}$$

$$b_z = 0. \tag{1.123}$$

The x component of $\mathbf{c_c}$ is determined from the dot product of $\mathbf{a_c}$ and $\mathbf{c_c}$:

$$\begin{aligned}
\mathbf{a_c} \cdot \mathbf{c_c} &= ac\cos\beta \\
\mathbf{a_c} \cdot \mathbf{c_c} &= \mathbf{a_c}^T\mathbf{c_c} \;=\; \begin{bmatrix} a & 0 & 0 \end{bmatrix} \begin{bmatrix} c_x \\ c_y \\ c_z \end{bmatrix} \\
&= ac_x + 0 + 0 \implies \\
ac_x &= ac\cos\beta \\
c_x &= c\cos\beta.
\end{aligned} \tag{1.124}$$

The y component of $\mathbf{c_c}$ is determined from the dot product of $\mathbf{b_c}$ and $\mathbf{c_c}$:

$$\begin{aligned}
\mathbf{b_c} \cdot \mathbf{c_c} &= bc\cos\alpha \\
\mathbf{b_c} \cdot \mathbf{c_c} &= \mathbf{b_c}^T\mathbf{c_c} \;=\; \begin{bmatrix} b\cos\gamma & b\sin\gamma & 0 \end{bmatrix} \begin{bmatrix} c\cos\beta \\ c_y \\ c_z \end{bmatrix} \\
&= bc\cos\beta\cos\gamma + c_y b\sin\gamma + 0 \implies \\
bc\cos\alpha &= bc\cos\beta\cos\gamma + c_y b\sin\gamma \\
c_y b\sin\gamma &= bc(\cos\alpha - \cos\beta\cos\gamma) \\
c_y &= \frac{c(\cos\alpha - \cos\beta\cos\gamma)}{\sin\gamma}
\end{aligned} \tag{1.125}$$

Finally, the z component of $\mathbf{c_c}$ is obtained from the length of the axis and the x and y components:

$$\begin{aligned}
c^2 &= c_x^2 + c_y^2 + c_z^2 \\
c_z^2 &= c^2 - c_x^2 - c_y^2 \\
&= c^2 - c^2\cos^2\beta - \frac{c^2(\cos\alpha - \cos\beta\cos\gamma)^2}{\sin^2\gamma} \\
&= c^2\left(\frac{\sin^2\gamma - \cos^2\beta\sin^2\gamma - (\cos\alpha - \cos\beta\cos\gamma)^2}{\sin^2\gamma}\right) \\
&= \frac{c^2}{\sin^2\gamma}(\sin^2\gamma - \cos^2\beta\sin^2\gamma - \cos^2\alpha - \cos^2\beta\cos^2\gamma + 2\cos\alpha\cos\beta\cos\gamma)
\end{aligned}$$

Substituting $-\cos^2\beta\sin^2\gamma - \cos^2\beta\cos^2\gamma = -\cos^2\beta(\sin^2\gamma + \cos^2\gamma) = -\cos^2\beta$ and $\sin^2\gamma = 1 - \cos^2\gamma$,

$$\begin{aligned}
c_z^2 &= \frac{c^2}{\sin^2\gamma}(1 - \cos^2\alpha - \cos^2\beta - \cos^2\gamma + 2\cos\alpha\cos\beta\cos\gamma). \\
c_z &= \frac{c(1 - \cos^2\alpha - \cos^2\beta - \cos^2\gamma + 2\cos\alpha\cos\beta\cos\gamma)^{\frac{1}{2}}}{\sin\gamma}.
\end{aligned} \tag{1.126}$$

Later in the chapter we will derive an expression for the unit cell volume (Eqn. 1.5.4): $V = abc(1 - \cos^2\alpha - \cos^2\beta - \cos^2\gamma + 2\cos\alpha\cos\beta\cos\gamma)^{\frac{1}{2}}$. Multiplying the expression for c_z by ab/ab results in

$$c_z = \frac{V}{ab\sin\gamma}. \tag{1.127}$$

We have now determined the components of the axial vectors in a Cartesian coordinate system, based on the unit cell parameters:

$$\begin{aligned}
\mathbf{a_c} &= a\,\mathbf{i} + 0\,\mathbf{j} + 0\,\mathbf{k} \\
\mathbf{b_c} &= b\cos\gamma\,\mathbf{i} + b\sin\gamma\,\mathbf{j} + 0\,\mathbf{k} \\
\mathbf{c_c} &= c\cos\beta\,\mathbf{i} + \left(\frac{c(\cos\alpha - \cos\beta\cos\gamma)}{\sin\gamma}\right)\mathbf{j} + \left(\frac{V}{ab\sin\gamma}\right)\mathbf{k}.
\end{aligned} \tag{1.128}$$

The matrix of coefficients is

$$\begin{aligned}
\mathbf{C} &= \begin{bmatrix} a_x & a_y & a_z \\ b_x & b_y & b_z \\ c_x & c_y & c_z \end{bmatrix} \\
&= \begin{bmatrix} a & 0 & 0 \\ b\cos\gamma & b\sin\gamma & 0 \\ c\cos\beta & \left(\frac{c(\cos\alpha - \cos\beta\cos\gamma)}{\sin\gamma}\right) & \frac{V}{ab\sin\gamma} \end{bmatrix}.
\end{aligned} \tag{1.129}$$

The matrix that will transform a vector in fractional coordinates based on a unit cell with cell parameters a, b, c, α, β γ, and V into an orthonormal basis is therefore:

$$\mathbf{B} = \mathbf{C}^T = \begin{bmatrix} a & b\cos\gamma & c\cos\beta \\ 0 & b\sin\gamma & \left(\frac{c(\cos\alpha - \cos\beta\cos\gamma)}{\sin\gamma}\right) \\ 0 & 0 & \frac{V}{ab\sin\gamma} \end{bmatrix}. \tag{1.130}$$

$$\mathbf{B}\begin{bmatrix} x_f \\ y_f \\ z_f \end{bmatrix} = \begin{bmatrix} x_c \\ y_c \\ z_c \end{bmatrix}. \tag{1.131}$$

In Chapter 3 a basis related to the diffraction pattern and known as a the *reciprocal basis* will be introduced. Each cell parameter in the unit cell basis is related to the cell parameters in the reciprocal basis, and we will later use these relationships to simplify **B**.

To transform a vector in Cartesian coordinates into unit cell coordinates we need only invert the **B** matrix:

$$\mathbf{B}\mathbf{v_f} = \mathbf{v_c} \tag{1.132}$$

$$\begin{aligned}
\mathbf{B}^{-1}\mathbf{B}\mathbf{v_f} &= \mathbf{B}^{-1}\mathbf{v_c} \\
\mathbf{I}\mathbf{v_f} &= \mathbf{B}^{-1}\mathbf{v_c}
\end{aligned}$$

$$\mathbf{B}^{-1}\mathbf{v_c} = \mathbf{v_f} \tag{1.133}$$

1.5.3 Determining Distances and Angles In the Unit Cell

A virtual plethora of crystal structures are now available to the scientific investigator, either in the published literature or archived in databases[22]. These structures contain the unit cell parameters and the $[x\ y\ z]$ coordinates of the atoms inside the unit cell. It is unlikely that there is a crystallographer alive who has not been confronted by frustrated colleagues who have taken the atomic coordinates from a published structure and attempted to calculate interatomic distances and angles from them. The reason, of course, is that they are treating the coordinates as Cartesian coordinates in an orthonormal basis, when the coordinates are actually listed as fractional coordinates in the unit cell basis. While there are software programs available which will accept fractional coordinates[2], allowing the user to determine molecular parameters, the use of the transform derived in the previous section is very straightforward, as an example using the 2-mercaptopyridine structure will illustrate.

The crystallographic data in the literature will ordinarily list the unit cell parameters, including the cell volume and the *space group* (the internal crystal symmetry – discussed in Sec. 2.4) along with other information related to data collection and refinement of the structure. Somewhere in the manuscript or archive* will be a list of the fractional coordinates of the atoms in the unit cell, labeled as x, y, and z (rather that x_f, etc.):

> The title compound, 2-mercaptopyridine, crystallized in the monoclinic space group $P2_1/n$, with $a = 6.112(5)$ Å, $b = 6.326(5)$ Å, $c = 14.314(5)$ Å, $\beta = 101.530(5)°$, $V = 542.3(6)$ Å^3, $Z = 4$, T = 293 K.

The numbers in parentheses after each of the unit cell parameters are the estimated standard deviations of the parameters. They are a measure of the uncertainty in the last digit (e.g., for a, the standard deviation of the axial length is 0.005 Å). Standard deviations will be discussed in detail in Chapters 5 and 8.

Positional Parameters of 2-Mercaptopyridine

Atom	x	y	z
S(1)	0.7403	0.0629	0.4073
H(1)	0.8714	0.0983	0.4799
N(1)	0.3705	0.2616	0.4294
C(1)	0.5501	0.2608	0.3860
C(2)	0.5630	0.4351	0.3256
H(2)	0.683	0.431	0.292
C(3)	0.4070	0.5920	0.3147
H(3)	0.423	0.693	0.271
C(4)	0.2289	0.5854	0.3617
H(4)	0.108	0.682	0.347
C(5)	0.2151	0.4162	0.4187
H(5)	0.083	0.368	0.453

Figure 1.32 displays a *displacement ellipsoid* plot[23] of the 2-mercaptopyridine molecule showing the atom labeling scheme. The displacement ellipsoids are informative in that they can tell us something about how the molecule vibrates (in good structures) or alternatively characterize the quality of the diffraction data (in not-so-good structures). For our purposes here we only note that this is typical of drawings in the literature and that the centers of the atoms lie at the centers of the ellipsoids. Note that not all of the unit cell parameters are listed. We will discover in Sec. 2.3.6 that some of the unit cell angles are fixed due to the lattice symmetry.

*Atomic positions were routinely published in the older literature. Current journals generally refer the reader to an archival database where the atomic parameters are stored.

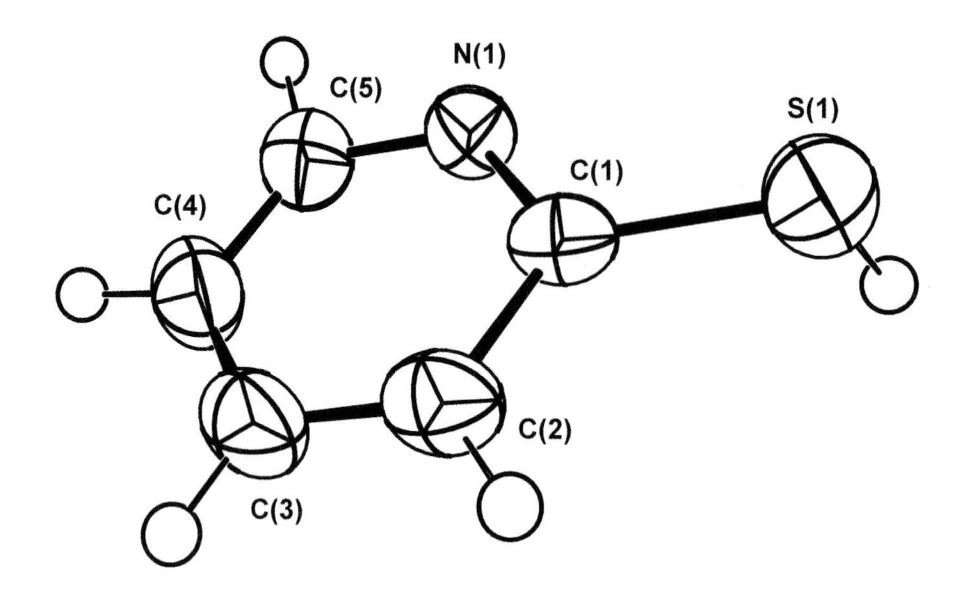

Figure 1.32 Displacement ellipsoid plot of the 2-mercaptopyridine molecule. Ellipsoids are plotted at the 50% level.

It is common practice to list only those parameters which are determined experimentally, rather than assigned by constraint. In this example these fixed angles are $\alpha = \gamma = 90°$. In a substantial majority of the structures found in the literature the unlisted angles will be 90°, *but not always.*

Suppose that we wish to know the interatomic distance (bond length) between S(1) and C(1), and the N(1)-C(1)-S(1) interatomic angle with C(1) at the vertex. To calculate these parameters we must first convert the fractional coordinates of the atom positions to Cartesian coordinates (in Å), requiring the **B** matrix:

$$\mathbf{B} = \begin{bmatrix} 6.112 & \{6.326\cos(90°)\} & \{14.314\cos(101.53°)\} \\ 0 & \{6.326\sin(90°)\} & \dfrac{14.314\{\cos(90°) - \cos(101.53°)\cos(90°)\}}{\sin(90°)} \\ 0 & 0 & \dfrac{542.3}{(6.112)(6.326)\sin(90°)} \end{bmatrix}.$$

$$= \begin{bmatrix} 6.112 & 0 & -2.861 \\ 0 & 6.326 & 0 \\ 0 & 0 & 14.025 \end{bmatrix}.$$

Using Eqn. 1.131 for the S(1) fractional coordinates transforms [(0.74025) (0.06290) (0.40731)] into Cartesian coordinates in Å:

$$\begin{bmatrix} 6.112 & 0 & -2.861 \\ 0 & 6.326 & 0 \\ 0 & 0 & 14.025 \end{bmatrix} \begin{bmatrix} 0.74025 \\ 0.06290 \\ 0.40731 \end{bmatrix} = \begin{bmatrix} 3.359 \\ 0.398 \\ 5.712 \end{bmatrix}.$$

The remainder of the coordinates, which we will call *crystal Cartesian coordinates*, allow for the calculation of any distance or angle in the molecule:

Crystal Cartesian Coordinates

Atom	x_c	y_c	z_c
S(1)	3.359	0.398	5.712
H(1)	3.953	0.622	6.731
N(1)	1.036	1.655	6.022
C(1)	2.258	1.650	5.414
C(5)	0.117	2.633	5.872
C(2)	2.510	2.752	4.567
C(4)	0.364	3.703	5.073
C(3)	1.587	3.745	4.414
H(5)	−0.789	2.328	6.353
H(2)	3.339	2.727	4.095
H(3)	1.810	4.384	3.801
H(4)	−0.333	4.31	4.867

The components of the $\overrightarrow{\mathrm{C(1)S(1)}}$ vector translated to the origin are

$$[(3.359)\ (0.398)(5.712)] - [(2.258)\ (1.650)\ (5.414)] = [(1.101)\ (-1.252)\ (0.298)].$$

The carbon-sulfur bond length is $|\overrightarrow{\mathrm{C(1)S(1)}}| = (1.101^2 + (-1.252)^2 + 0.298^2)^{\frac{1}{2}} =$ 1.694 Å. The components of the $\overrightarrow{\mathrm{C(1)N(1)}}$ vector translated to the origin are [(-1.222) (0.005) (0.608)], resulting in a carbon-nitrogen bond length of $|\overrightarrow{\mathrm{C(1)N(1)}}| = 1.365$ Å. The N(1)-C(1)-S(1) angle, ν, is determined from the dot product of these two vectors: $\overrightarrow{\mathrm{C(1)S(1)}} \cdot \overrightarrow{\mathrm{C(1)N(1)}} = [(1.101)(-1.222) + (-1.252)(0.005) + (0.298)(0.608)] =$ -1.170.

$$\overrightarrow{\mathrm{C(1)S(1)}} \cdot \overrightarrow{\mathrm{C(1)N(1)}} = \left(|\overrightarrow{\mathrm{C(1)S(1)}}|\right)\left(|\overrightarrow{\mathrm{C(1)N(1)}}|\right) \cos\nu = -1.170.$$

$$\cos\nu = \frac{-1.170}{\left(|\overrightarrow{\mathrm{C(1)S(1)}}|\right)\left(|\overrightarrow{\mathrm{C(1)N(1)}}|\right)} = \frac{-1.170}{(1.694)(1.365)} = -0.506.$$

$$\nu = 120.4^\circ.$$

1.5.4 Determining the Volume of the Unit Cell

The unit cell volume, which we have already used to simplify the **B** matrix, is also a useful parameter for the crystallographer during the early stages of structural investigation. With the cell volume in hand, the density of the crystal under investigation can be calculated and tested for consistency with the putative contents of the unit cell. A strange density usually means that the cell does not contain what it is thought to contain, or that the unit cell is incorrect. The general unit cell, shown in Fig. 1.33 is a parallelepiped, based on axes **a**, **b**, and **c**. The volume of a parallelepiped is determined as the area of its base, A_{bc}, times its height, h_a. The base, in turn, is a parallelogram with an area determined as the length of its

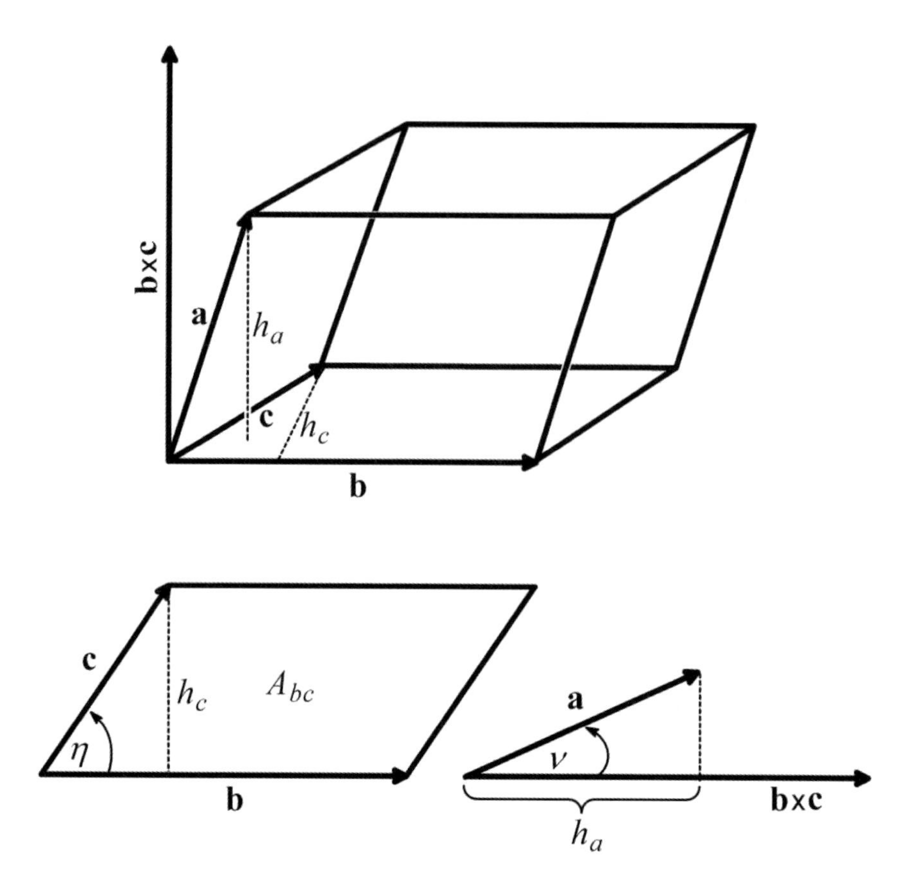

Figure 1.33 Unit cell parallelepiped. The **bc** vectors determine the base, with the perpendicular h_a the height. The b axis is the base of the parallelogram that makes up the base of the parallelepiped. The height of the parallelogram is the perpendicular, h_c.

base, b, times its height, h_c. The area of the base can be determined from the vector product of **b** and **c**:

$$\begin{aligned} A_{bc} &= bh_c \\ h_c &= c\sin\eta \\ |\mathbf{b}\times\mathbf{c}| &= bc\sin\eta = bh_c \\ A_{bc} &= |\mathbf{b}\times\mathbf{c}|. \end{aligned} \tag{1.134}$$

The height of the parallelepiped, h_a, is a perpendicular from the base, and is therefore parallel to $(\mathbf{b} \times \mathbf{c})$. The dot product of $\mathbf{a}$ and $(\mathbf{b} \times \mathbf{c})$ determines h_a, which is the projection of $\mathbf{a}$ onto $(\mathbf{b} \times \mathbf{c})$ (Eqn. 1.6):

$$\begin{aligned} V &= A_{bc} h_a \\ h_a &= \frac{\mathbf{a} \cdot (\mathbf{b} \times \mathbf{c})}{|\mathbf{b} \times \mathbf{c}|} \\ &= \frac{\mathbf{a} \cdot (\mathbf{b} \times \mathbf{c})}{A_{bc}} \\ V &= \mathbf{a} \cdot (\mathbf{b} \times \mathbf{c}). \end{aligned} \tag{1.135}$$

The scalar, $\mathbf{a} \cdot (\mathbf{b} \times \mathbf{c})$,* is known as a *scalar triple product*; the volume of the unit cell is the scalar triple product of unit cell vectors. The scalar triple product provides us with the means to determine the cell volume without having to transform the unit cell parameters into orthonormal coordinates. We begin with Eqn. 1.71:

$$\begin{aligned} \mathbf{b} \times \mathbf{c} &= \begin{vmatrix} \mathbf{i} & \mathbf{j} & \mathbf{k} \\ b_x & b_y & b_z \\ c_x & c_y & c_z \end{vmatrix} = \mathbf{i} \begin{vmatrix} b_y & b_z \\ c_y & c_z \end{vmatrix} - \mathbf{j} \begin{vmatrix} b_x & b_z \\ c_x & c_z \end{vmatrix} + \mathbf{k} \begin{vmatrix} b_x & b_y \\ c_x & c_y \end{vmatrix} \\ &= \left[\begin{vmatrix} b_y & b_z \\ c_y & c_z \end{vmatrix} \left(- \begin{vmatrix} b_x & b_z \\ c_x & c_z \end{vmatrix} \right) \begin{vmatrix} b_x & b_y \\ c_x & c_y \end{vmatrix} \right] \\ \mathbf{a} \cdot (\mathbf{b} \times \mathbf{c}) &= a_x \begin{vmatrix} b_y & b_z \\ c_y & c_z \end{vmatrix} - a_y \begin{vmatrix} b_x & b_z \\ c_x & c_z \end{vmatrix} + a_z \begin{vmatrix} b_x & b_y \\ c_x & c_y \end{vmatrix} \\ &= \begin{vmatrix} a_x & a_y & a_z \\ b_x & b_y & b_z \\ c_x & c_y & c_z \end{vmatrix}. \end{aligned} \tag{1.136}$$

The volume of the unit cell is the determinant of the matrix consisting of row vectors composed of the Cartesian coordinates of the unit cell axes. Thus,

$$\begin{aligned} V &= \begin{vmatrix} a_x & a_y & a_z \\ b_x & b_y & b_z \\ c_x & c_y & c_z \end{vmatrix} \\ V^2 &= \begin{vmatrix} a_x & a_y & a_z \\ b_x & b_y & b_z \\ c_x & c_y & c_z \end{vmatrix} \begin{vmatrix} a_x & a_y & a_z \\ b_x & b_y & b_z \\ c_x & c_y & c_z \end{vmatrix} \\ &= \begin{vmatrix} a_x & a_y & a_z \\ b_x & b_y & b_z \\ c_x & c_y & c_z \end{vmatrix} \begin{vmatrix} a_x & b_x & c_x \\ a_y & b_y & c_y \\ a_z & b_z & c_z \end{vmatrix} \quad \text{(Determinant Property 3)} \end{aligned} \tag{1.137}$$

*Note that $\mathbf{a} \cdot (\mathbf{b} \times \mathbf{c}) \equiv \mathbf{a} \cdot \mathbf{b} \times \mathbf{c}$ is unambiguous, since $(\mathbf{a} \cdot \mathbf{b}) \times \mathbf{c}$ is meaningless.

$$V^2 = \left| \begin{bmatrix} a_x & a_y & a_z \\ b_x & b_y & b_z \\ c_x & c_y & c_z \end{bmatrix} \begin{bmatrix} a_x & b_x & c_x \\ a_y & b_y & c_y \\ a_z & b_z & c_z \end{bmatrix} \right| \quad \text{(Determinant Property 4)}$$

$$= \begin{vmatrix} \mathbf{a}\cdot\mathbf{a} & \mathbf{a}\cdot\mathbf{b} & \mathbf{a}\cdot\mathbf{c} \\ \mathbf{b}\cdot\mathbf{a} & \mathbf{b}\cdot\mathbf{b} & \mathbf{b}\cdot\mathbf{b} \\ \mathbf{c}\cdot\mathbf{a} & \mathbf{c}\cdot\mathbf{b} & \mathbf{c}\cdot\mathbf{c} \end{vmatrix}$$

$$= \begin{vmatrix} a^2 & ab\cos\gamma & ac\cos\beta \\ ab\cos\gamma & b^2 & bc\cos\alpha \\ ac\cos\beta & bc\cos\alpha & c^2 \end{vmatrix}. \tag{1.138}$$

Expansion of the determinant gives

$$\begin{aligned} V^2 &= (a^2b^2c^2) + (a^2b^2c^2\cos\alpha\cos\beta\cos\gamma) + (a^2b^2c^2\cos\alpha\cos\beta\cos\gamma) \\ &\quad -(a^2b^2c^2\cos^2\beta) - (a^2b^2c^2\cos^2\alpha) - (a^2b^2c^2\cos^2\gamma). \\ &= a^2b^2c^2(1-\cos^2\alpha-\cos^2\beta-\cos^2\gamma+2\cos\alpha\cos\beta\cos\gamma). \\ V &= abc(1-\cos^2\alpha-\cos^2\beta-\cos^2\gamma+2\cos\alpha\cos\beta\cos\gamma)^{\frac{1}{2}}. \end{aligned} \tag{1.139}$$

This is the expression used to simplify the **B** matrix (Eqn. 1.126). The matrix of the determinant in Eqn. 1.138 contains all of the metrics of the unit cell and is known as *the metric tensor** of the lattice. We will encounter the metric tensor later on, as it is generally obtained experimentally without prior knowledge of the unit cell parameters, and provides a means for obtaining them!

1.5.5 Important Identities

In Chapter 3 a new lattice will be introduced which is reciprocal to the crystal lattice. The determination of the relationships between the basis vectors in this new lattice and those in the crystal lattice will require identities relating the basis vectors. These identities involve combinations of scalar and vector products, and will be developed here for later use.

The *scalar triple product*, $\mathbf{a}\cdot\mathbf{b}\times\mathbf{c}$ has already been introduced (Eqn. 1.136) and shown to be equal to the unit cell volume. It is useful to generate the remaining two scalar triple products from the first in order to determine the vector products that will yield positive cell volumes. We do this by switching rows in the determinant representation of $\mathbf{a}\cdot\mathbf{b}\times\mathbf{c}$, recalling that each switch changes the sign of the determinant (and therefore the volume):

$$\mathbf{a}\cdot(\mathbf{b}\times\mathbf{c}) = \begin{vmatrix} a_x & a_y & a_z \\ b_x & b_y & b_z \\ c_x & c_y & c_z \end{vmatrix} = -\begin{vmatrix} b_x & b_y & b_z \\ a_x & a_y & a_z \\ c_x & c_y & c_z \end{vmatrix} = \begin{vmatrix} b_x & b_y & b_z \\ c_x & c_y & c_z \\ a_x & a_y & a_z \end{vmatrix}$$

$$= \mathbf{b}\cdot(\mathbf{c}\times\mathbf{a}) \tag{1.140}$$

*The word *tensor* will be used here in its "physics" context, as an entity that characterizes the properties of a physical system. All of the tensors that we will encounter will be represented by 3×3 matrices and we will often use *tensor* and *matrix* interchangeably.

and

$$\mathbf{a}\cdot(\mathbf{b}\times\mathbf{c}) = \begin{vmatrix} a_x & a_y & a_z \\ b_x & b_y & b_z \\ c_x & c_y & c_z \end{vmatrix} = -\begin{vmatrix} c_x & c_y & c_z \\ b_x & b_y & b_z \\ a_x & a_y & a_z \end{vmatrix} = \begin{vmatrix} c_x & c_y & c_z \\ a_x & a_y & a_z \\ b_x & b_y & b_z \end{vmatrix}$$

$$= \mathbf{c}\cdot(\mathbf{a}\times\mathbf{b}). \qquad (1.141)$$

It follows that $V = \mathbf{b}\cdot(\mathbf{c}\times\mathbf{a})$ and $V = \mathbf{c}\cdot(\mathbf{a}\times\mathbf{b})$. Note that reversal of any of the vector products produces a negative volume for the unit cell. This is a useful diagnostic during the course of the structural investigation. If a negative unit cell volume is determined, one of the axes is pointing in a direction that renders the coordinate system left-handed. Reversing the direction of one of the axes in the vector product will correct this.

The *vector triple product* is the vector analog of the scalar triple product: $\mathbf{a}\times(\mathbf{b}\times\mathbf{c})$. The vector $\mathbf{b}\times\mathbf{c}$ is perpendicular to the bc plane. The vector triple product produces a vector that is perpendicular to that vector, and therefore lies in the bc plane. Its components are determined from the scalar products of $\mathbf{a}$ with $\mathbf{b}$ and $\mathbf{c}$:

$$\mathbf{a}\times(\mathbf{b}\times\mathbf{c}) = (\mathbf{a}\cdot\mathbf{c})\,\mathbf{b} - (\mathbf{a}\cdot\mathbf{b})\,\mathbf{c}. \qquad (1.142)$$

We prove this by showing that the expression on the right reduces to the vector triple product:

$$\begin{aligned}(\mathbf{a}\cdot\mathbf{c})\,\mathbf{b} - (\mathbf{a}\cdot\mathbf{b})\,\mathbf{c} &= (a_x c_x + a_y c_y + a_z c_z)(b_x\mathbf{i} + b_y\mathbf{j} + b_z\mathbf{k}) \\ &- (a_x b_x + a_y b_y + a_z b_z)(c_x\mathbf{i} + c_y\mathbf{j} + c_z\mathbf{k}).\end{aligned}$$

Expanding and collecting terms gives

$$\begin{aligned}(\mathbf{a}\cdot\mathbf{c})\,\mathbf{b} - (\mathbf{a}\cdot\mathbf{b})\,\mathbf{c} &= [a_y(b_x c_y - b_y c_x) - a_z(b_z c_x - b_x c_z)]\,\mathbf{i} \\ &+ [a_z(b_y c_z - b_z c_y) - a_x(b_x c_y - b_y c_x)]\,\mathbf{j} \\ &+ [a_x(b_z c_x - b_x c_z) - a_y(b_y c_z - b_z c_y)]\,\mathbf{k}.\end{aligned}$$

We define $\mathbf{v} = v_x\mathbf{i} + v_y\mathbf{j} + v_z\mathbf{k}$ such that $v_x = b_y c_z - b_z c_y$, $v_y = b_z c_x - b_x c_z$, and $v_z = b_x c_y - b_y c_x$, that is, $\mathbf{v} = \mathbf{b}\times\mathbf{c}$. Substituting these components:

$$\begin{aligned}(\mathbf{a}\cdot\mathbf{c})\,\mathbf{b} - (\mathbf{a}\cdot\mathbf{b})\,\mathbf{c} &= (a_y v_z - a_z v_y)\,\mathbf{i} \\ &+ (a_z v_x - a_x v_z)\,\mathbf{j} \\ &+ (a_x v_y - a_y v_x)\,\mathbf{k} \\ &= \mathbf{a}\times\mathbf{v} = \mathbf{a}\times(\mathbf{b}\times\mathbf{c}).\end{aligned}$$

The *scalar quadruple product* is the scalar product of two vector products, $(\mathbf{a}\times\mathbf{b})\cdot(\mathbf{c}\times\mathbf{d})$:

$$(\mathbf{a}\times\mathbf{b})\cdot(\mathbf{c}\times\mathbf{d}) = (\mathbf{a}\cdot\mathbf{c})(\mathbf{b}\cdot\mathbf{d}) - (\mathbf{a}\cdot\mathbf{d})(\mathbf{b}\cdot\mathbf{c}). \qquad (1.143)$$

The scalar triple product and the vector triple product are employed to prove this:

$$
\begin{aligned}
\text{Let } \mathbf{v} &= \mathbf{c} \times \mathbf{d}. \\
(\mathbf{a} \times \mathbf{b}) \cdot (\mathbf{c} \times \mathbf{d}) &= (\mathbf{a} \times \mathbf{b}) \cdot \mathbf{v} = \mathbf{v} \cdot (\mathbf{a} \times \mathbf{b}) \quad (\textit{scalar triple product}) \\
&= \begin{vmatrix} v_x & v_y & v_z \\ a_x & a_y & a_z \\ b_x & b_y & b_z \end{vmatrix} = - \begin{vmatrix} a_x & a_y & a_z \\ v_x & v_y & v_z \\ b_x & b_y & b_z \end{vmatrix} = \begin{vmatrix} a_x & a_y & a_z \\ b_x & b_y & b_z \\ v_x & v_y & v_z \end{vmatrix} \\
&= \mathbf{a} \cdot (\mathbf{b} \times \mathbf{v}) = \mathbf{a} \cdot [\mathbf{b} \times (\mathbf{c} \times \mathbf{d})] \quad (\textit{vector triple product}) \\
&= \mathbf{a} \cdot [(\mathbf{b} \cdot \mathbf{d})\, \mathbf{c} - (\mathbf{b} \cdot \mathbf{c})\, \mathbf{d}] \\
&= (\mathbf{a} \cdot \mathbf{c})(\mathbf{b} \cdot \mathbf{d}) - (\mathbf{a} \cdot \mathbf{d})(\mathbf{b} \cdot \mathbf{c}).
\end{aligned}
$$

The scalar quadruple product is conveniently represented as a 2×2 determinant:

$$
(\mathbf{a} \times \mathbf{b}) \cdot (\mathbf{c} \times \mathbf{d}) = \begin{vmatrix} \mathbf{a} \cdot \mathbf{c} & \mathbf{b} \cdot \mathbf{c} \\ \mathbf{a} \cdot \mathbf{d} & \mathbf{b} \cdot \mathbf{d} \end{vmatrix}. \tag{1.144}
$$

The *vector quadruple product* is the vector product of two vector products, $(\mathbf{a} \times \mathbf{b}) \times (\mathbf{c} \times \mathbf{d})$:

$$
\begin{aligned}
\text{Let } \mathbf{v} &= \mathbf{a} \times \mathbf{b}. \\
\mathbf{v} \times (\mathbf{c} \times \mathbf{d}) &= (\mathbf{v} \cdot \mathbf{d})\mathbf{c} - (\mathbf{v} \cdot \mathbf{c})\mathbf{d} \quad (\textit{vector triple product}) \\
&= (\mathbf{d} \cdot (\mathbf{a} \times \mathbf{b}))\mathbf{c} - (\mathbf{c} \cdot (\mathbf{a} \times \mathbf{b}))\mathbf{d} \\
&= (\mathbf{a} \cdot (\mathbf{b} \times \mathbf{d}))\mathbf{c} - (\mathbf{a} \cdot (\mathbf{b} \times \mathbf{c}))\mathbf{d} \\
\Longrightarrow (\mathbf{a} \times \mathbf{b}) \times (\mathbf{c} \times \mathbf{d}) &= (\mathbf{a} \cdot \mathbf{b} \times \mathbf{d})\mathbf{c} - (\mathbf{a} \cdot \mathbf{b} \times \mathbf{c})\mathbf{d}
\end{aligned} \tag{1.145}
$$

Exercises

1. The copper atoms depicted in Fig. 1.1 are arranged in a cubic unit cell. Each edge of the cell has the same length: $a = b = c = 3.6147$. Determine the distances between the planes with (a) (1 1 1), (b) (2 2 2), and (c) (3 3 3) indices.

2. Consider two 2-dimensional unit cells, each with the same axial lengths: $a = 2.40$ and $b = 3.20$. For unit cell A, $\gamma = 90°$; for unit cell B, $\gamma = 117°$. A point p is located in each unit cell at the end of the sum of a vector of magnitude 1.80 , parallel to the **a** axis and a vector of magnitude 2.40 , parallel to the **b** axis. (a) Determine the fractional coordinates of point p in each unit cell. (b) Determine the distance from the origin to point p in each unit cell.

3. (a) Derive a formula for the inverse of a 2×2 matrix and use matrix multiplication to demonstrate that your formula is correct ($\mathbf{DD}^{-1} = \mathbf{I}$). (b) Compute the inverse of the following matrix:

$$\mathbf{D} = \begin{bmatrix} 1.000 & 2.000 & 3.000 \\ 2.000 & 1.000 & 3.000 \\ 3.000 & 2.000 & 1.000 \end{bmatrix} .$$

   (c) Demonstrate that the matrix calculated in part (b) is $\mathbf{D}^{-1}$.

4. Show that (a) the inverse of a matrix for the rotation of angle φ about a coordinate axis (e.g., the x axis) is the rotation matrix for the $-\varphi$ rotation about the same axis, (b) the matrix for a reflection across a coordinate plane (e.g., the xz plane) is its own inverse, and (c) the inversion matrix is its own inverse.

5. Using matrices, show that sequential rotations about a coordinate axis of φ_1 followed by φ_2 is equivalent to a single rotation of $(\varphi_1 + \varphi_2)$ about the same axis.

6. Show that (a) the rotation matrices for rotation about the coordinate axes are orthonormal matrices, (b) the inverses of these matrices are their transposes, and (c) the rotation of a general vector effected by any of these matrices does not alter the length of the vector.

7. The monoclinic unit cell of CuO has the following parameters: $a = 4.6837(5)$ Å, $b = 3.4226(6)$ Å, $c = 5.1288(6)$ Å, $\alpha = 90.00°$, $\beta = 99.54°(1)$ and $\gamma = 90.00°$.* The fractional coordinates of the contents of the unit cell are

atom	x_f	y_f	z_f	atom	x_f	y_f	z_f
Cu1	0.2500	0.2500	0.0000	O1	0.0000	0.4184	0.2500
Cu2	0.7500	0.7500	0.5000	O2	0.5000	0.9184	0.2500
Cu3	0.2500	0.7500	0.5000	O3	0.0000	0.5816	0.7500
Cu4	0.7500	0.2500	0.5000	O4	0.5000	0.0816	0.7500

*Åsbrink, S. and Norrby, L.-J., *Acta. Cryst.*, B26, 8(1970).

(a) Determine the shortest (contact) distance between the copper(II) ions and the oxide ions in the unit cell. (b) Determine the volume of the unit cell. (c) Determine the mass of the unit cell in grams. (d) Determine the density of solid copper(II) oxide in g/cm^3.

8. The orthorhombic unit cell of $CuSO_4$ has the following parameters:* $a = 8.39$ Å, $b = 6.89$ Å $c = 4.83$ Å, $\alpha = 90.00°$, $\beta = 90.00°$ and $\gamma = 90.00°$

   The fractional coordinates of the basic unit in the unit cell (repeated in order to fill the cell) are

atom	x_f	y_f	z_f	atom	x_f	y_f	z_f
Cu1	0.000	0.000	0.893	O2	0.375	0.250	0.439
S1	0.185	0.250	0.445	O3	0.129	0.069	0.307
O1	0.141	0.250	0.755				

   (a) Determine the average sulfur-oxygen distance and the average O-S-O angle in the sulfate ion. (b) The experimentally measured density of anhydrous copper sulfate is 3.6 g/cm^3. How Many $CuSO_4$ units are in the unit cell? (c) The basic unit in the unit cell seems to be missing an oxygen atom. How can this be if the stoichiometry in the crystal is $CuSO_4$? Hint: The basic unit in the unit cell is called the *asymmetric* unit. You may have to read ahead in Chapter 2 to answer this question. The space group of the crystal is *Pnma*.

*Rao, B.R., *Acta. Cryst.*, 14, 321(1961).

Chapter 2

Crystal Symmetry

2.1 Symmetry

In the introduction to Chapter 1 the external appearance of a crystal was attributed to the internal *symmetry* of the crystal lattice. This internal symmetry is also intimately linked to the diffraction data that is used for structural determination. Incorporating the effects of crystal symmetry on these data is always useful, and very often essential for structural solution. It is virtually always necessary for structural refinement.

Nearly everyone has a qualitative notion of symmetry. If one were to ask whether the lower case "a" in *asymmetry* was more or less symmetric than the lower case "s" in *symmetry*, there would probably be a general consensus that the "s" seemed more symmetric than the "a". However, beyond voicing an intuitive sense that one appears more symmetric than the other, casual observers would probably be unable to explain precisely how they came to this conclusion. In order to deal with symmetry more formally, it will be necessary for us to define precisely what we mean by symmetry. *An object has symmetry if, after it has been moved in some specified manner, it is impossible to determine from its appearance that it has been moved.* In other words, the motion has either returned the object to its original state, or moved it to an equivalent one. Consider, for example, two "a"s superimposed on one another. We construct an axis perpendicular to the page through the center of the letters, then rotate one of the "a"s by an arbitrary angle. There is no angle (except 360°) that will generate an image identical to that of the original letters. Fig. 2.1 illustrates this for a 180° rotation. In contrast, beginning with two superimposed "s"s, the 180° rotation *appears* to have kept the letters in their original conformation (Fig 2.1(b)). This type of symmetry is called *rotational symmetry.* Each letter has the trivial (but essential) symmetry of a 360° rotation — but the "s" has 180° rotational symmetry as well — it is more symmetric. Note that the rotation occurred about an axis through a central point in each letter. If we relax this constraint, and allow one of the "a"s to translate after the 180° rotation, then the two letters *together* suddenly acquire rotational symmetry, with the rotation axis now located at the midpoint between the two letters, as shown in Figs. 2.1(c) and 2.1(d). Fig. 2.1(e) illustrates that this *translational symmetry* can

Understanding Single-Crystal X-Ray Crystallography. Dennis W. Bennett

ISBN: 978-3-527-32677-8 (HC), 978-3-527-32794-2 (SC)

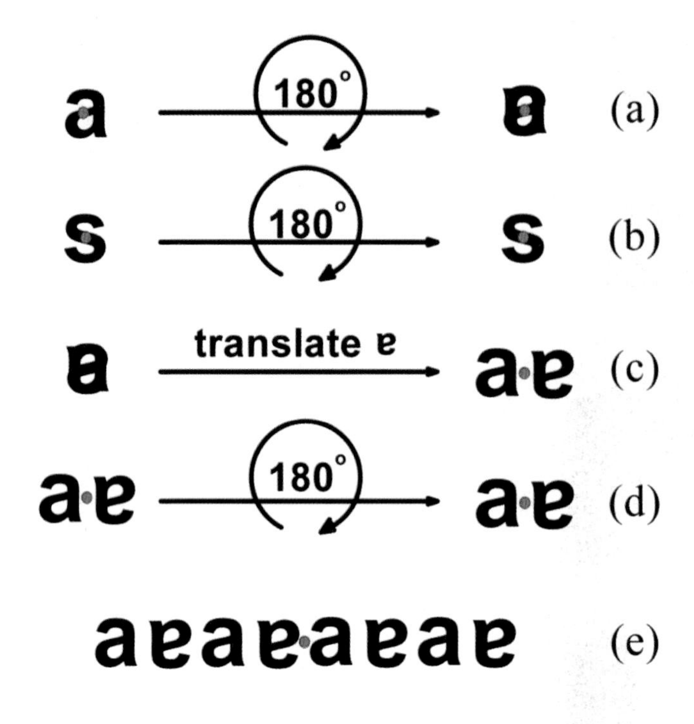

Figure 2.1 Rotational symmetry of the letters "a" and "s". The rotational axis passes through the center (red dot) and is perpendicular to the page.

be extended indefinitely, with the entire array of "a"s retaining the 180° rotational symmetry.

As illustrated in the previous example, an analysis of the symmetry of an object will require "moving" the object in a specific manner with respect to a geometric construct, in this case, rotation about a line (*axis*). The motion is referred to as an *operation*. When the operation results in a conformation of the object that appears identical to the original conformation, the motion is called a *symmetry operation*. The geometric construct is referred to as an *element* — a *symmetry element* when its associated operation is a symmetry operation. Another type of symmetry operation involves the *reflection* process discussed previously in Sec. 1.4.3 and illustrated in Fig. 1.27. The symmetry element in this case is a plane, termed a *reflection plane* or a *mirror plane*. The operation involves the construction of a vector to every point defining the object, followed by the projection of each vector to the plane and extension of each projection an equal distance to the other side of the plane. In Fig. 2.2, the chess knight viewed from the side exhibits no *mirror symmetry*, while the knight viewed from the front is clearly symmetric with respect to a mirror plane passing through its center. As observed previously for rotational symmetry, the mirror image of the side view can translated with respect to the original image, creating an array of such images that collectively possess mirror symmetry.

Figure 2.2 Mirror symmetry of the chess knight. Top: Side and front views showing mirror symmetry for the front view. Bottom: Array of side views exhibiting mirror symmetry.

The *inversion* process was also discussed previously in Sec. 1.4.3 and illustrated in Fig. 1.27. The inversion operation is accomplished by constructing a vector from every point defining the object to a single point, then extending the vector an equal distance beyond the point. When the point is a symmetry element it must lie at the center of the object, and is therefore referred to as an *inversion center* or a *center of symmetry*. Fig. 2.3(a) is a ball-and-stick model of the PF_6^- anion, which exhibits *inversion symmetry*; the inversion center lies in the center of the ion, at the intersection of the six $\overrightarrow{FP}$ vectors. Every (green) fluorine atom inverts through the center of the (orange) phosphorus atom to an equivalent atom on the opposing side. Fig. 2.3(b) is a model of the PF_5 molecule. While PF_5 has symmetry elements in the form of mirror planes and rotation axes, it does not have an inversion center. The hypothetical molecule, PFClBrHI, shown in Fig. 2.3(c), clearly does not have a center of symmetry, but if the molecule is inverted and then translated, the pair of molecules becomes *centrosymmetric* with respect to a point in the center of the pair (Fig. 2.3(d)). As before, this figure can be extended indefinitely into an array that will retain the inversion center.

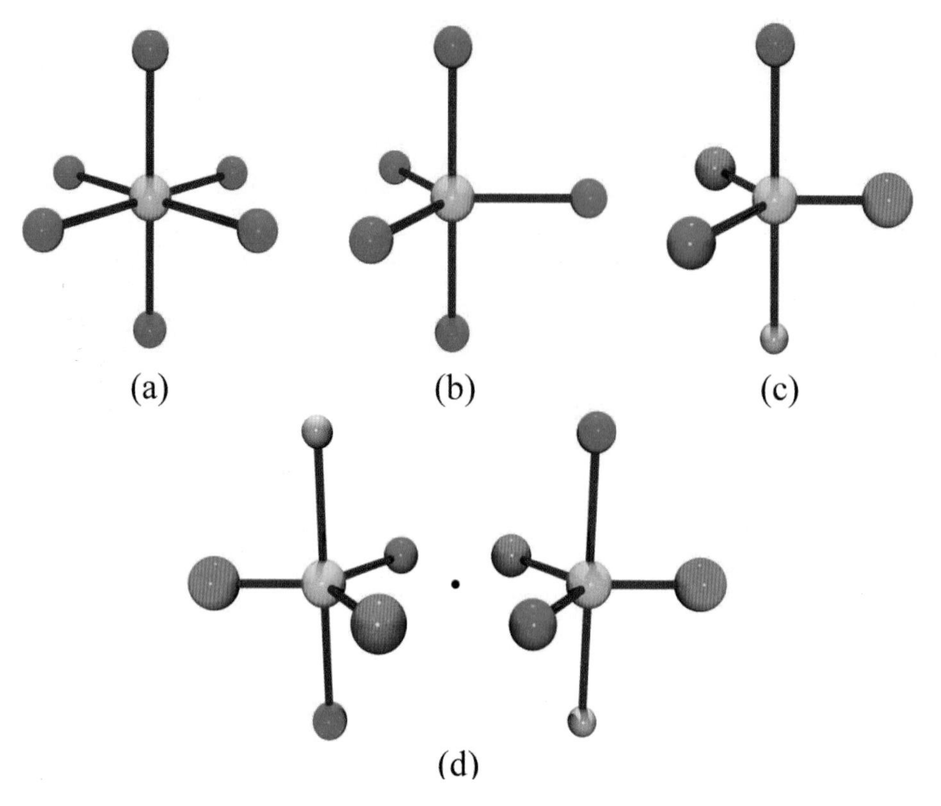

Figure 2.3 Inversion symmetry of the PF_6^- anion (a). The PF_5 molecule (b) and hypothetical PFClBrHI molecule (c) lack inversion centers. Two molecules of PFClBrHI (d), one inverted with respect to the other, can be arranged into a centrosymmetric pair.

2.2 Symmetry Group Theory

2.2.1 Sets of Symmetry Operations

It now appears reasonable to assess the symmetry of an object by listing all of its symmetry operations. The PF_5 molecule in Fig. 2.3(b) serves as an excellent illustrative example. We begin by looking for the symmetry elements in the molecule, then cataloging all of the independent symmetry operations corresponding to each element. It will be useful to note that there are two different location types for the fluorine atoms. There is a plane passing through three of the atoms, known as the *equatorial plane*, and the positions of these fluorine atoms will be referred to as *equatorial positions*. The other two fluorine atoms lie on an axis orthogonal to the equatorial plane; their positions will be referred to as *axial positions*. Fig. 2.4(a) is a view of the PF_5 molecule looking along the orthogonal axis at the plane containing the equatorial fluorine atoms. Fig. 2.4(b) is another view of the axis. The operations that produce a *symmetry-equivalent* conformation are rotations about this axis of $2\pi/3$ (120°) and $2 \cdot 2\pi/3$ (240°). Three rotations of $2\pi/3$ return the molecule to its original conformation; the axis is referred to as a *three-fold rotation*

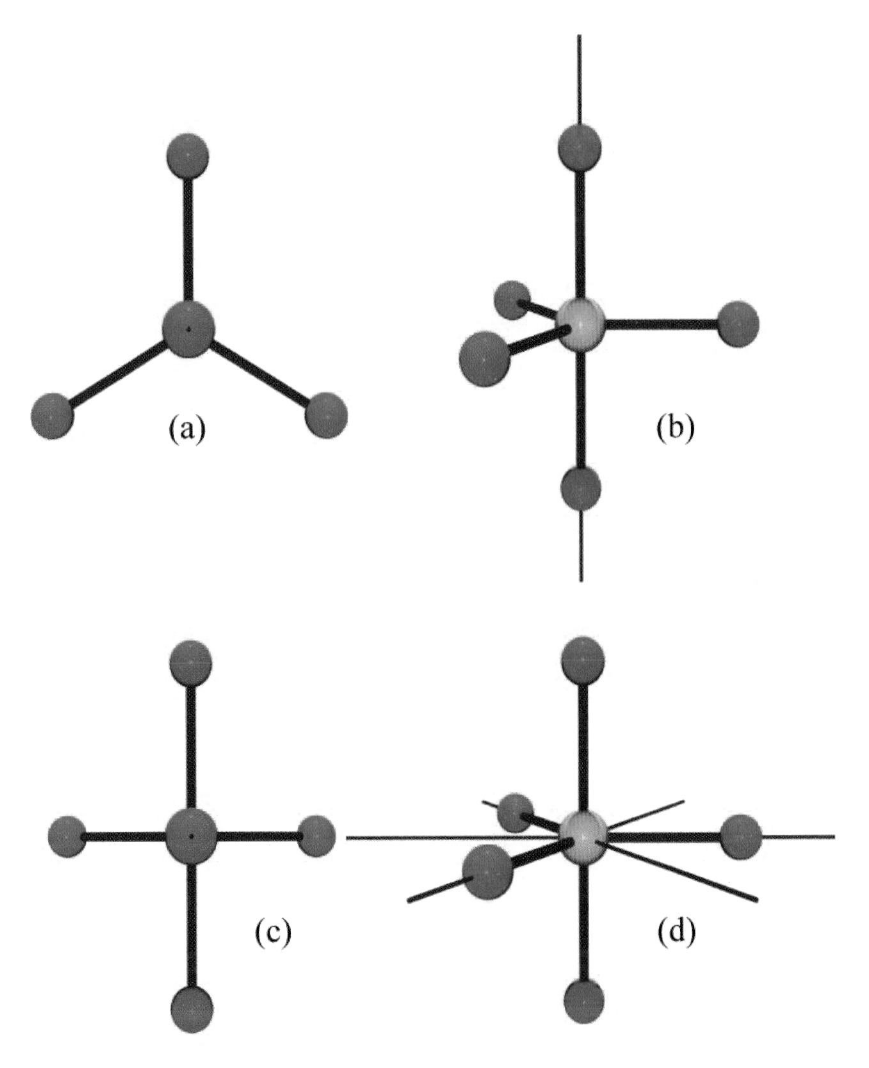

Figure 2.4 (a) Three-fold axis of the PF_5 molecule viewed along the axis. (b) Three-fold axis shown passing through the centers of the axial fluorine atoms. (c) Two-fold axis of PF_5 viewed along the axis. (d) The three two-fold axes in PF_5.

axis. Fig. 2.4(c) is a view along one of the $\overrightarrow{FP}$ vectors in the equatorial plane. A π rotation (180°) around this vector will switch the locations of the axial fluorine atoms with one another, and will similarly switch the equatorial atoms. Two such rotations will create the original configuration and the $\overrightarrow{FP}$ vector is therefore aligned along a *two-fold rotation axis.* It is common practice to align the *axis of highest symmetry* (the one generating the most rotation operations) with the z axis. Mirror planes that contain this axis are either *vertical mirror planes* or *dihedral mirror planes.* Vertical mirror planes contain two-fold axes that are perpendicular to the axis of highest symmetry; dihedral mirror planes (which are also "vertical") bisect the angle between these two-fold axes.

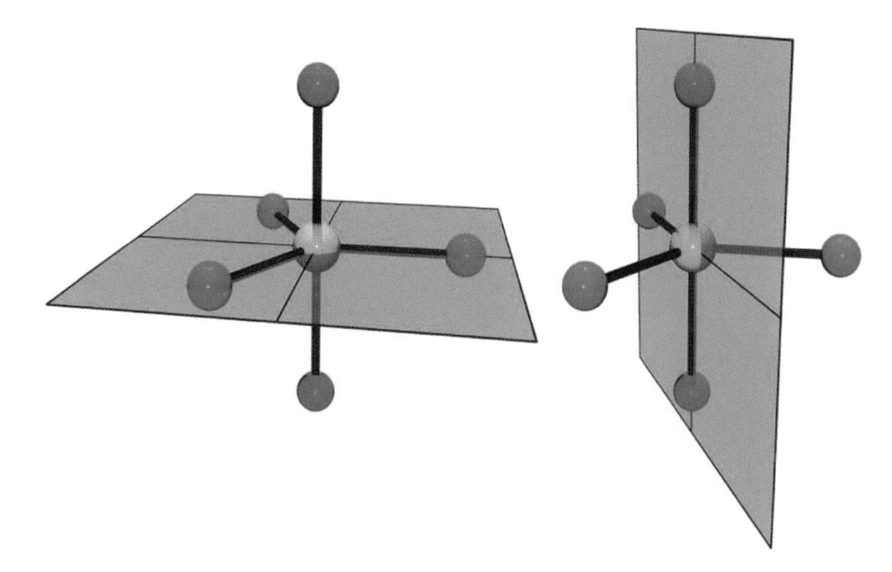

Figure 2.5 The horizontal mirror plane and one of three vertical mirror planes in the PF_5 molecule. The lines in the planes are xy and yz reference axes.

It is important to note here that symmetry axes and their corresponding operations are transformed along with the object whenever a symmetry operation is performed (or any other operation!). The three-fold rotations transform the two-fold axes into one another, and the three-fold axis into itself. Each two-fold rotation transforms the other two-fold axes into one another, and the three-fold axis into itself. However, *in no case is a three-fold axis transformed into a two-fold axis.* Because of this the two-fold operations are said to belong to one *symmetry class*, and the three-fold operations to another. In addition to rotational axes, the PF_5 molecule also contains horizontal and vertical mirror planes, illustrated in Fig. 2.5. There are three vertical mirror planes. A given vertical mirror plane reflects the two-fold axis that it contains into itself, and the other two-fold axes into one another. The three-fold axis is also reflected into itself. In addition the other vertical mirror planes are reflected into one another. The horizontal mirror plane is reflected into itself. The horizontal mirror plane reflects all of the other axes and planes into themselves. Again, *in no instance is a horizontal mirror plane transformed into a vertical mirror plane, or vice versa.* In addition the two-fold and three-fold rotations always transform the vertical mirror planes back into vertical planes and the horizontal mirror plane back into itself. It therefore follows that the vertical and horizontal mirror operations also belong to different symmetry classes since *there is no symmetry operation that will transform a member of one class into another.*

The *apparently* complete set of symmetry elements in the PF_5 molecule is shown in Fig. 2.6. They consist of a three-fold rotational axis, three two-fold rotational axes, a horizontal mirror plane, and three vertical mirror planes. The Schönflies symbols* for these elements, C_3, C_2, σ_h and σ_v, respectively, provide a convenient shorthand for representing these symmetry operations. The corresponding symme-

*named for their founder and discussed more thoroughly in the next section

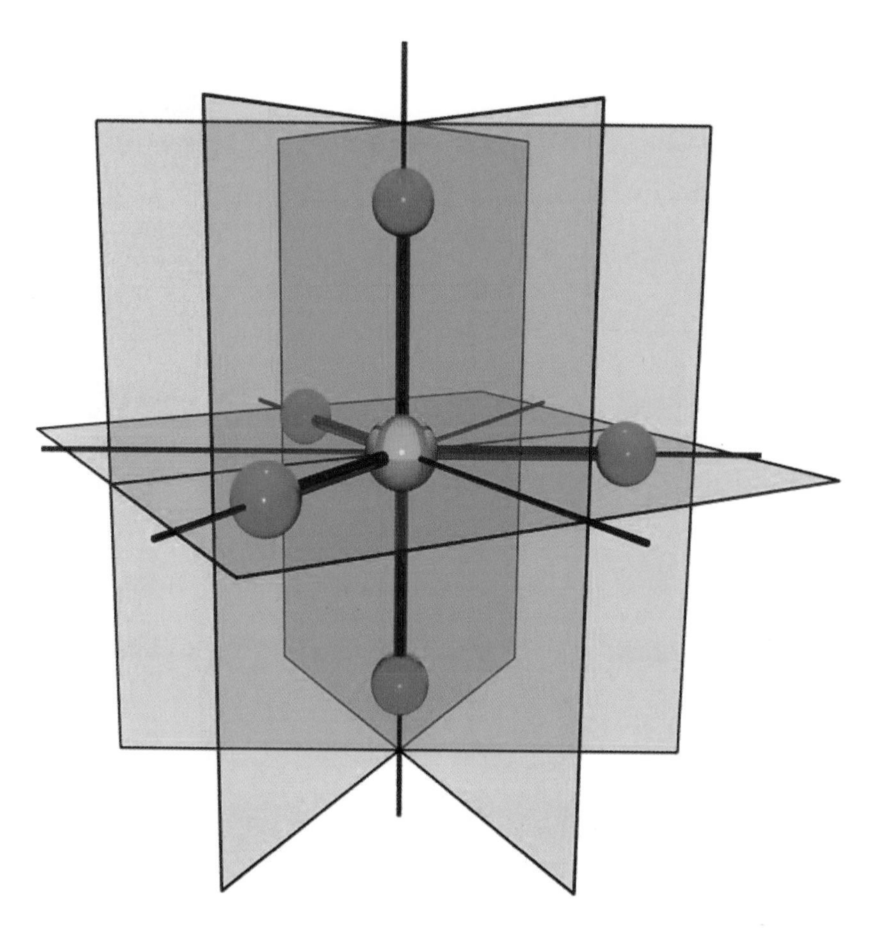

Figure 2.6 The symmetry elements in the PF_5 molecule.

try operations, grouped by classes, are: Two three-fold rotations about the same axis, symbolized as C_3^1 and C_3^2; three two-fold rotations, each about a separate axis, symbolized as C_2, C_2', and C_2''; reflection across a horizontal plane, σ_h; three reflection operations across each of three mirror planes, σ_v, σ_v', and σ_v''. In addition, we include the operation of *doing nothing*, since the object without transformation is one of the possible ways of arranging the atoms. For reasons that will become clear in the next section, we call this operation the *identity operation* and give it the symbol E.

It would appear that all of the possible symmetry operations for the PF_5 molecule have been catalogued. Fig. 2.7 shows all of the possible conformations of the fluorine atoms in the molecule, each with a different color in order to keep track of their locations in the original conformation — shown in Fig. 2.7(a). There are $3! = 6$ ways of arranging the equatorial fluorine atoms, and $2! = 2$ ways of arranging the axial atoms, for a total of 12 different arrangements. We have counted $E + 2C_3 + 3C_2 + \sigma_h + 3\sigma_v = 10$ operations. *It appears that there are two conformations of the molecule which are not accessible by any of the operations that we have listed* — shown in Figs. 2.7(e) and (f).

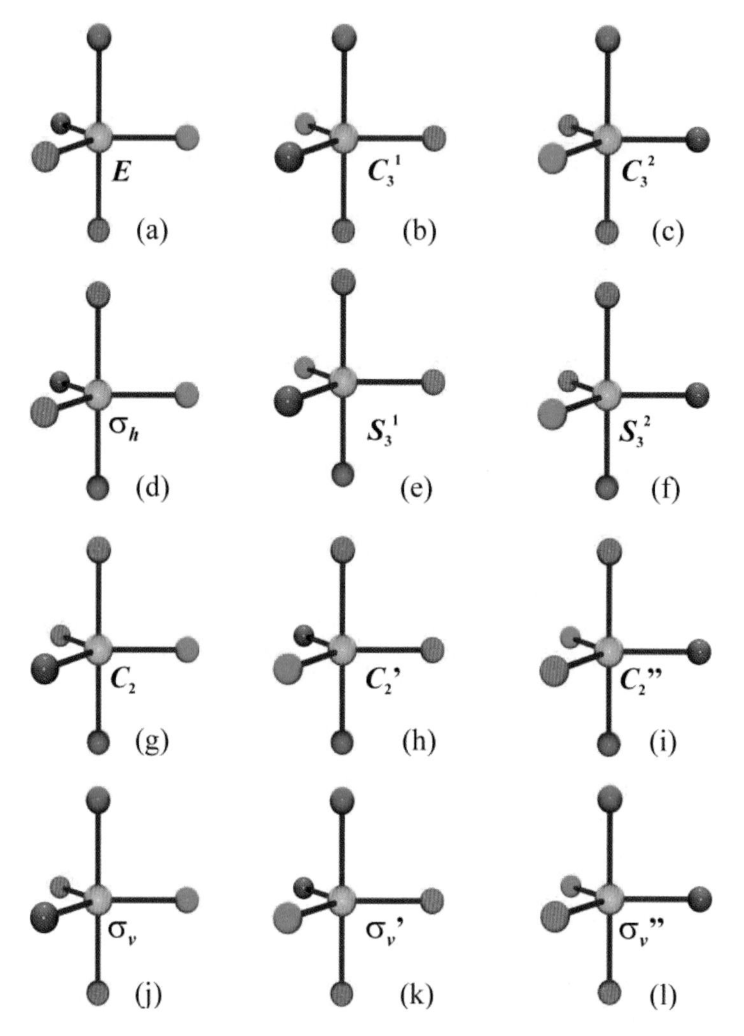

Figure 2.7 The 12 possible arrangements of the fluorine atoms in the PF_5 molecule and the symmetry operations that generate them.

In order to obtain each of these conformations we must create two new symmetry operations. The conformation in Fig. 2.7(e) can be obtained by a $2\pi/3$ rotation about the three-fold axis, *followed by* reflection across a perpendicular mirror plane. *This is a single operation – the rotation and reflection are inseparably linked.* The conformation in Fig. 2.7(f) is generated by a $4\pi/3$ rotation followed by a reflection. The symmetry element is composed of a rotation axis combined with a mirror plane perpendicular to the axis, and is therefore referred to as a *rotoreflection* axis.* The Schönflies notation for the rotoreflection axis in PF_5 is S_3. In this case the S_3 axis is collinear with the C_3 axis, and its mirror plane is coplanar with the σ_h mirror. The two symmetry operations are symbolized by the Schönflies symbols S_3^1 and S_3^2. While S_n axes are often collinear with C rotation axes and are also often perpendicular to σ_h mirror planes, the $C_2H_2Br_2F_2$ molecule shown in Fig. 2.8 has

*Unfortunately, the rotoreflection axis is also commonly known as an *improper* axis of rotation.

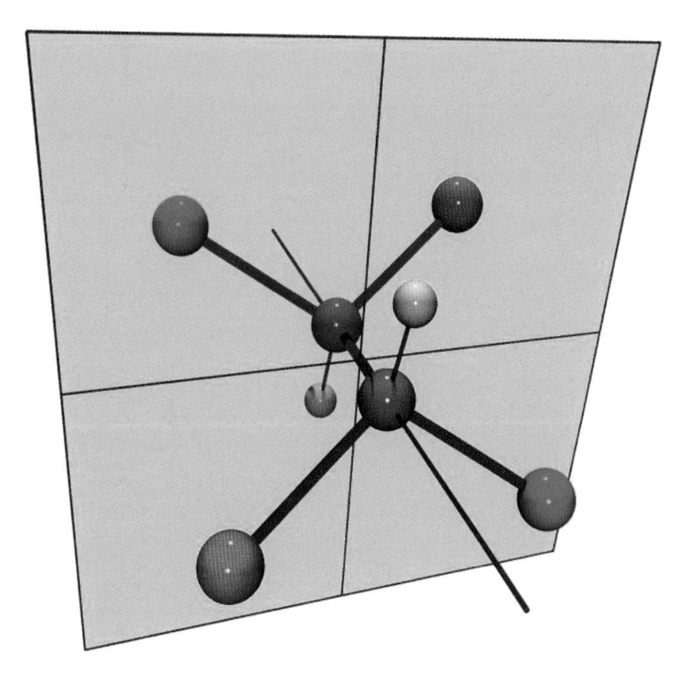

Figure 2.8 Two-fold rotoreflection (S_2) axis in $C_2H_2Br_2F_2$.

an S_2 axis (180° rotation) along the $\overrightarrow{CC}$ vector that does not coincide with either a rotational axis or horizontal mirror plane. The rotoreflection axis is an independent symmetry element and its associated rotation-reflection operations are independent symmetry operations.

The set of symmetry operations now appears complete. If this is the case, then any combination of symmetry operations should generate one of the 12 conformations in Fig. 2.7. It follows that any combination of two symmetry operations should be equivalent to a single symmetry operation. The combination of two operations is symbolized as (Operation2) × (Operation1), or simply (Operation2)(Operation1) and is read to indicate that the operation on the right is performed first, followed by the operation on the left. For example, $C_3^2\,\sigma_v$ indicates that the object is subjected to a reflection across the σ_v vertical mirror plane, followed by a $4\pi/3$ rotation around the three-fold axis. The two operations produce the conformation in Fig. 2.7(l) and we can equate the combined operations with the single operation that produces the same arrangement, i.e., $C_3^2\,\sigma_v = \sigma_v'$. Table 2.1, a multiplication* table, lists the results of all such combinations.

2.2.2 Symmetry Groups

An analysis of the multiplication table reveals that the symmetry operations constitute a set with very unique properties. Before considering the nature of this set we must first make note of an unfortunate ambiguity that is a consequence of the language used in discussing symmetry and the language used in set theory. The geometric constructs with which symmetry operations are performed are termed

*As discussed previously, *multiplication* here is used in the general sense of *combination*.

Table 2.1 Multiplication Table: Symmetry Operations for the PF_5 molecule.*

	E	C_3^1	C_3^2	C_2	C_2'	C_2''	σ_h	S_3^1	S_3^2	σ_v	σ_v'	σ_v''
E	E	C_3^1	C_3^2	C_2	C_2'	C_2''	σ_h	S_3^1	S_3^2	σ_v	σ_v'	σ_v''
C_3^1	C_3^1	C_3^2	E	C_2''	C_2	C_2'	S_3^1	S_3^2	σ_h	σ_v''	σ_v	σ_v'
C_3^2	C_3^2	E	C_3^1	C_2'	C_2''	C_2	S_3^2	σ_h	S_3^1	σ_v'	σ_v''	σ_v
C_2	C_2	C_2'	C_2''	E	C_3^1	C_3^2	σ_v	σ_v'	σ_v''	σ_h	S_3^1	S_3^2
C_2'	C_2'	C_2''	C_2	C_3^2	E	C_3^1	σ_v'	σ_v''	σ_v	S_3^2	σ_h	S_3^1
C_2''	C_2''	C_2	C_2'	C_3^1	C_3^2	E	σ_v''	σ_v	σ_v'	S_3^1	S_3^2	σ_h
σ_h	σ_h	S_3^1	S_3^2	σ_v	σ_v'	σ_v''	E	C_3^1	C_3^2	C_2	C_2'	C_2''
S_3^1	S_3^1	S_3^2	σ_h	σ_v''	σ_v	σ_v'	C_3^1	C_3^2	E	C_2''	C_2	C_2'
S_3^2	S_3^2	σ_h	S_3^1	σ_v'	σ_v''	σ_v	C_3^2	E	C_3^1	C_2'	C_2''	C_2
σ_v	σ_v	σ_v'	σ_v''	σ_h	S_3^1	S_3^2	C_2	C_2'	C_2''	E	C_3^1	C_3^2
σ_v'	σ_v'	σ_v''	σ_v	S_3^2	σ_h	S_3^1	C_2'	C_2''	C_2	C_3^2	E	C_3^1
σ_v''	σ_v''	σ_v	σ_v'	S_3^1	S_3^2	σ_h	C_2''	C_2	C_2'	C_3^1	C_3^2	E

*row × column (column operation performed first)

symmetry elements. A *set* is a collection of entities — these entities are called *elements* of the set. The symmetry operations for a given object constitute a set; the symmetry *operations* are elements of that set — *not* the symmetry elements.

The multiplication table illustrates that the elements of the set of symmetry operations belonging to the PF_5 molecule satisfy the following conditions:

1. *The product of any two elements in the set, including the product of an element with itself, produces another element in the set.* This is a consequence of having determined a symmetry operation for each of the possible configurations of the object (a molecule in this case). Performing multiple symmetry operations will always generate one of these conformations; which is accessible by an equivalent single symmetry operation.

2. *The product of any two elements of the set do not necessarily commute.* For example, $C_2\, S_3^2 = \sigma_v''$, while $S_3^2\, C_2 = \sigma_v'$.

3. *There is one element of the set, E, which commutes with every element of the set and leaves the element unchanged.* Thus we refer to the element E as an *identity* element. In general, for any element Y in the set, $EY = YE = Y$.

4. *Multiple combinations of symmetry operations are associative.* Provided that the order of the operations is retained, the associative law of multiplication holds: $X(YZ) = (XY)Z$. For example, $C_2(S_3^2\,\sigma_h) = C_2\,C_3^2 = C_2''$ and $(C_2\,S_3^2)\sigma_h = \sigma_v''\,\sigma_h = C_2''$.

5. *For every element in the set there is a commutative product which produces the identity element.* Every operation can be paired with one other operation which reverses the effect of the first, leaving the original conformation unchanged. That is, for every operation, Y, there is an operation, Z, such that $ZY = E$. Furthermore, the product commutes: $YZ = E$. It follows that Y and Z are inverse operations — *reciprocals* of one another. Thus *every element in the set has an inverse which is also an element of the set.* The inverse of Y will be indicated by Y^{-1}.

6. *The inverse of the product of a sequence of symmetry operations is the product of the inverses of those operations in reverse order.* Consider the product of symmetry operations YZ:

$$\begin{aligned} Z^{-1}Z &= Z^{-1}EZ = E \\ &= Z^{-1}(Y^{-1}Y)Z \\ &= (Z^{-1}Y^{-1})(YZ) = E \Longrightarrow \\ (Z^{-1}Y^{-1}) &= (YZ)^{-1}. \end{aligned} \tag{2.1}$$

   The argument can be extended to any number of symmetry operations, *as shown previously for matrices* (Eqn. 1.91).

Two important conclusions can be drawn from these observations. First, *the rules for the multiplication of symmetry operations are identical to those for matrix multiplication,* and second, the conditions listed above are precisely those conditions that *define* a special type of set known as a *group.* The set of symmetry operations constitute a mathematical group and the mathematics of *group theory* applies to them.

A mathematical group is determined by the elements of the set and a rule for combining them. The rule is called a *multiplication* rule, and the combinations determined by the application of that rule are termed *products.* For a *set* to be a *group*,

1. The product of any two elements of the set must be an element of the set.
2. The product of more than two elements in the set must be associative.
3. The set must contain an identity element.
4. Every element in the set must have an inverse that is also an element of the set.

For example, consider the set of integers with respect to the "+" multiplication operator; common addition is the multiplicative rule, and the sum of two integers constitutes a product. The sum (product) of any two integers generates another integer. The product of an integer and zero is the same integer — zero is the identity element. Every integer added to its negative produces the identity element — the inverse of an integer is its negative. Thus the set of integers constitute a group under the "+" operator — in this case an infinite one. Note that the set of non-negative integers is not a group with respect to addition, since its elements do not have inverses that are elements of the set. Also note, ironically, that the integers do not constitute a group if the multiplicative operator is chosen to be "×". In this case the identity element is one, but the inverses (e.g., $1/4$) are not elements of the set. The set of real numbers *is* a group under the "×" operator.

2.3 Point Groups

2.3.1 Molecular Point Symmetry

The previous discussion of symmetry made only brief mention of the symmetry of crystal lattices. It focused on the symmetry of objects in which the geometric or physical center was fixed. The sets of symmetry operations for a given symmetry type for which only rotations, reflections and inversions are allowed, referenced to a single fixed point are known as *symmetry point groups*, or more commonly simply as *point groups*. In particular, the point symmetry of molecules has been described, partly because it is easier to visualize the symmetry operations without the added complication of translation, and partly because many scientists experience their first introduction to symmetry in the study of the point symmetry of molecules. Furthermore, molecules are the constituents of unit cells, and molecular symmetry often plays a role in the determination of the symmetry of the crystal lattice.

Symmetry classification schemes are ordinarily based on the Schönflies notation. We have already encountered specific examples of the Schönflies notation for the

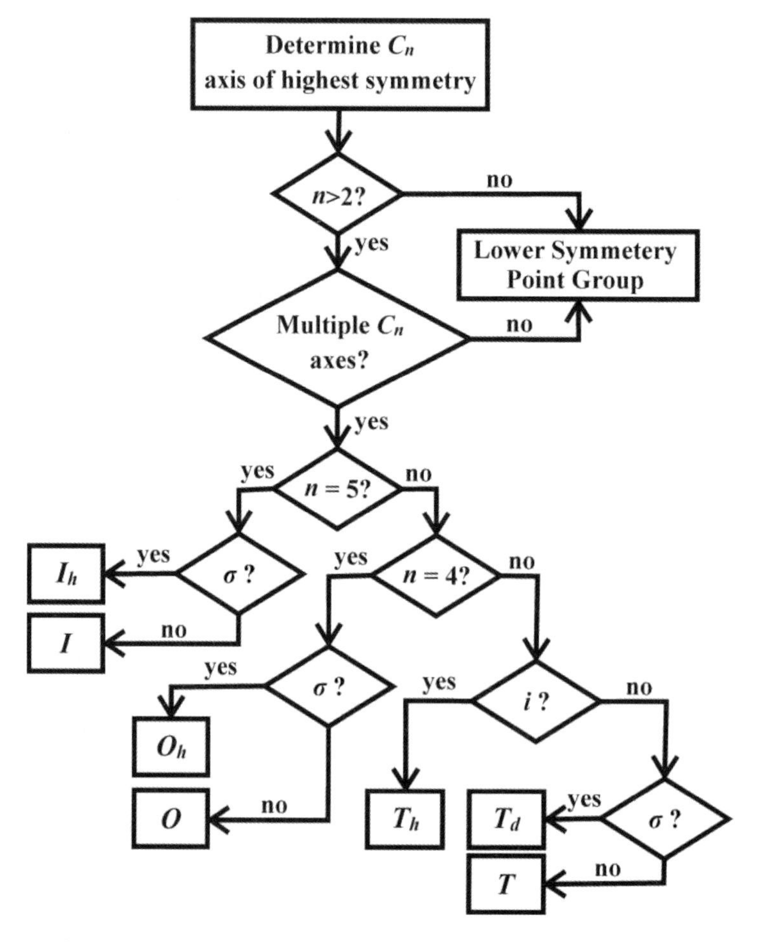

Figure 2.9 Scheme for the determination of higher symmetry point groups.

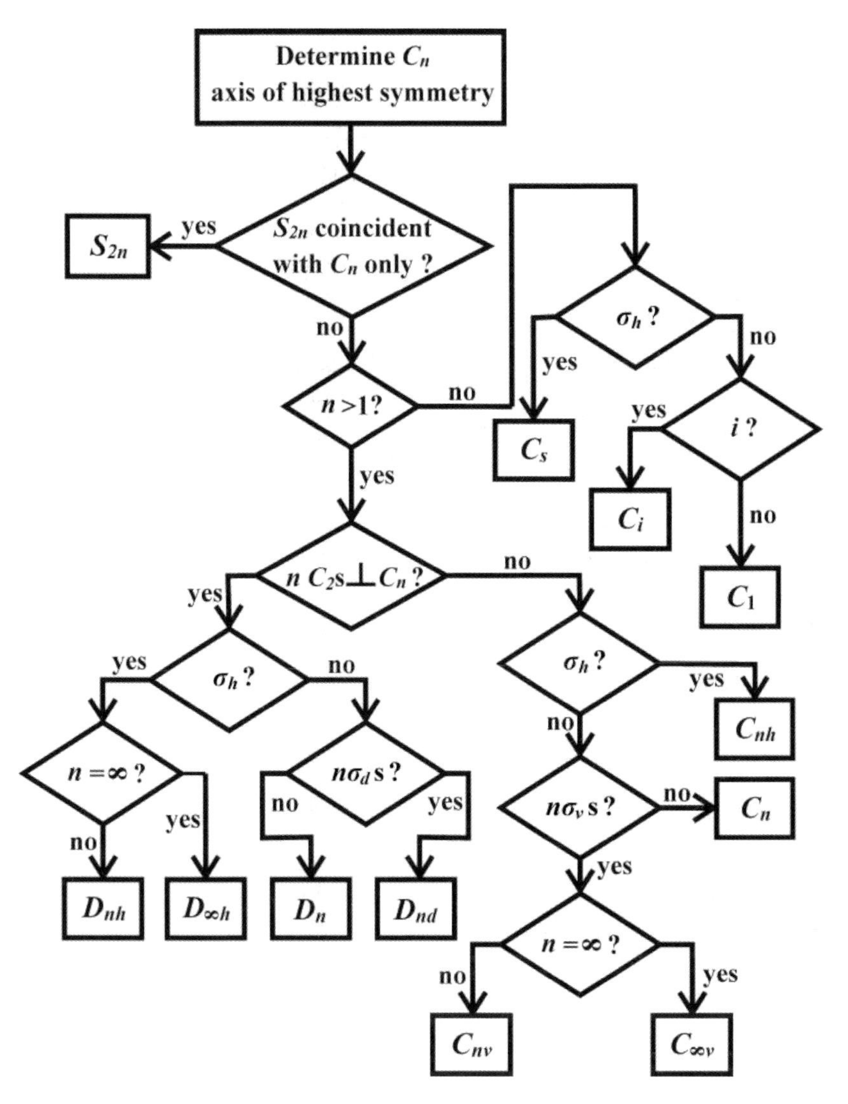

Figure 2.10 Scheme for the determination of lower symmetry point groups.

various symmetry elements and operations in point symmetry group theory. In general, mirror planes and their associated symmetry operations are assigned the same symbol, σ_x, where $x = h$, v, or d;* the inversion center *and* operator are assigned the symbol i. Rotation axes are represented with the symbol C_n for an n-fold axis; n-fold rotoreflection axes are given the symbol S_n. The symbols for the symmetry operations are $C_n^1, C_n^2, \ldots, C_n^{n-1}, C_n^n = E$ and $S_n^1, S_n^2, \ldots, S_n^n$ for rotation and rotoreflection operations about these axes. The symbols C_n^k and S_n^k refer to a rotation of $k \cdot (2\pi/n)$ radians $= k \cdot (360°/n)$ degrees.

Schönflies created this symbolic notation in the process of developing a scheme for classifying sets of symmetry operations into symmetry point groups. For a given point group the existence of certain symmetry operators requires that others must

*h, v, and d denote *horizontal*, *vertical*, and *dihedral* mirron planes, respectively.

also exist; Schönflies' classification scheme is based upon a determination of the subset of symmetry operations necessary to uniquely define the point group. The flowcharts shown in Figs. 2.9 and 2.10 provide a systematic approach to determining these subsets from the *symmetry elements* in a given molecule. The terminating boxes in the charts contain the Schönflies symbols for the various point groups. The process of assigning a point group begins with a determination of the highest symmetry C_n rotation axis. If there are *multiple* axes in the structure with $n > 2$ then the symmetry elements correspond to a higher symmetry point group and the scheme in Fig. 2.9 is utilized. In these cases the point group is often easily identified by inspection. If there are no multiple higher order axes, then the scheme in Fig. 2.10 is used. For example, the symmetry elements of the PF_5 molecule (Fig. 2.6) indicate that the rotation axis of highest symmetry is a C_3 axis, and that there is only one of them. Thus the point group is a "lower symmetry" point group. Referring to the flowchart in Fig. 2.10, $n = 3$ and is therefore greater than 1. For molecules, the question which follows is nearly always answered in the negative. A "yes" answer requires the existence of a C_n rotation axis, a S_{2n} rotoreflection axis ($2n$ is an even number) that is coincident with the C_n axis, and the absence of any other symmetry elements (with the possible exception of an inversion center). Very few molecules are assigned to S_{2n} point groups, but lattices – discussed in the next section – can exhibit S_4 or S_6 point group symmetry. In the PF_5 case, the answer is "no". Following this, the molecule contains three two-fold axes perpendicular to the three-fold axis; it contains a horizontal mirror plane; n is 3 rather than infinity — leading us to assign the D_{3h} point group to PF_5.

2.3.2 Matrix Representations of Groups

Each of the symmetry operations in the D_{3h} point group can be carried out with a transformation matrix. A general vector in the molecule will be transformed by each matrix into a symmetry-equivalent vector. The strong link between matrices and symmetry operations has already been discussed. In the previous section we discovered that the rules for the multiplication of the symmetry operations in a group and the rules for matrix multiplications were identical. In Sec. 1.24 transformation matrices were formulated that produced rotations, reflections, and inversions. Just as the product of two symmetry operations produces another symmetry operation, the product of two transformation matrices will result in another transformation matrix. Since symmetry operations are simply vector transformations for which symmetry-equivalent vectors are produced by rotating, reflecting or inverting, it should not be surprising that the product of two transformation matrices corresponding to two symmetry operations will produce a transformation matrix that corresponds to the product of the two symmetry operations. It follows that a multiplication table based on the transformation matrices rather than the symmetry operations themselves will look identical to Table 2.1. *The matrix multiplications parallel the multiplications of corresponding symmetry operations.* In group theory, any set that has multiplicative properties that parallel those of a group is known as a *representation* of the group. In the case of matrices, the set of symmetry transformation matrices that generate the symmetry operations of the group constitute a *matrix representation* of the group. The symmetry transformation matrices are not the only matrix representations – there are infinitely many more. Although a discussion of crystal symmetry does not require a detailed study of these matrix representations, they are intimately related to the physical properties of crystalline materials, and a discussion of crystal lattices would be incomplete without a brief discussion of their basic characteristics.

In chemistry and physics, we are often interested in the effects of the symmetry operations on specific mathematical functions which are related in some way to an object with a certain symmetry. For example, the behavior of the functions $f_1(x) = x$, $f_2(y) = y$, and $f_3(z) = z$ when the molecule undergoes all of its symmetry transformations is of interest in spectroscopy and quantum mechanics.[24] These functions behave the same as three vectors aligned alone each coordinate axis — or, for example, three dipoles aligned along the axes (the components of the dipole moment) — or the familiar p_x, p_y, and p_z atomic orbitals described in every general chemistry book. It follows that an understanding of the behavior of x, y, and z under the symmetry operations of a particular point group can provide information related to any object which exhibits the symmetry of that group An object that has been assigned a given point group is often said to *belong* to the group.

Returning to the PF_5 (D_{3h}) example, the molecule is oriented so that the C_3 axis is coincident with the z axis and one of the equatorial fluorine atoms lies on the x axis. The xz plane is selected as the σ_v mirror plane, and the x axis is the C_2 rotation axis. Recalling that $\cos(2\pi/3) = -1/2$, $\sin(2\pi/3) = \sqrt{3}/2$, $\cos(\pi) = -1$, and $\sin(\pi) = 0$, the compete set of transformation matrices for $[x \ y \ z]$ can be generated:

$$E \quad \begin{bmatrix} 1 & 0 & 0 \\ 0 & 1 & 0 \\ 0 & 0 & 1 \end{bmatrix} \qquad C_3^1 \quad \begin{bmatrix} -\frac{1}{2} & -\frac{\sqrt{3}}{2} & 0 \\ \frac{\sqrt{3}}{2} & -\frac{1}{2} & 0 \\ 0 & 0 & 1 \end{bmatrix} \qquad C_3^2 \quad \begin{bmatrix} -\frac{1}{2} & \frac{\sqrt{3}}{2} & 0 \\ -\frac{\sqrt{3}}{2} & -\frac{1}{2} & 0 \\ 0 & 0 & 1 \end{bmatrix}$$

$$\sigma_h \quad \begin{bmatrix} 1 & 0 & 0 \\ 0 & 1 & 0 \\ 0 & 0 & -1 \end{bmatrix} \qquad S_3^1 \quad \begin{bmatrix} -\frac{1}{2} & -\frac{\sqrt{3}}{2} & 0 \\ \frac{\sqrt{3}}{2} & -\frac{1}{2} & 0 \\ 0 & 0 & -1 \end{bmatrix} \qquad S_3^2 \quad \begin{bmatrix} -\frac{1}{2} & \frac{\sqrt{3}}{2} & 0 \\ -\frac{\sqrt{3}}{2} & -\frac{1}{2} & 0 \\ 0 & 0 & -1 \end{bmatrix}$$

$$C_2 \quad \begin{bmatrix} 1 & 0 & 0 \\ 0 & -1 & 0 \\ 0 & 0 & -1 \end{bmatrix} \qquad C_2' \quad \begin{bmatrix} -\frac{1}{2} & -\frac{\sqrt{3}}{2} & 0 \\ -\frac{\sqrt{3}}{2} & \frac{1}{2} & 0 \\ 0 & 0 & -1 \end{bmatrix} \qquad C_2'' \quad \begin{bmatrix} -\frac{1}{2} & \frac{\sqrt{3}}{2} & 0 \\ \frac{\sqrt{3}}{2} & \frac{1}{2} & 0 \\ 0 & 0 & -1 \end{bmatrix}$$

$$\sigma_v \quad \begin{bmatrix} 1 & 0 & 0 \\ 0 & -1 & 0 \\ 0 & 0 & 1 \end{bmatrix} \qquad \sigma_v' \quad \begin{bmatrix} -\frac{1}{2} & -\frac{\sqrt{3}}{2} & 0 \\ -\frac{\sqrt{3}}{2} & \frac{1}{2} & 0 \\ 0 & 0 & 1 \end{bmatrix} \qquad \sigma_v'' \quad \begin{bmatrix} -\frac{1}{2} & \frac{\sqrt{3}}{2} & 0 \\ \frac{\sqrt{3}}{2} & \frac{1}{2} & 0 \\ 0 & 0 & 1 \end{bmatrix}$$

The matrices for the mirror planes and diad axes that do not lie in the xz plane are generated by premultiplying the matrix that performs the same operation in the xz plane by the matrix performing the appropriate rotation, e.g., $\sigma_v'' = C_3^2 \sigma_v$.

If the matrices are grouped into symmetry classes, we discover a very important property of matrices corresponding to operations in the same class: *the sums of the diagonal elements of all the matrices corresponding to a class of symmetry operations are identical.* We have chosen these matrices to operate on $[x\ y\ z]$ as a general vector. Of the infinite number of general vectors there is a subset that change magnitude, but do not change direction when operated upon by a given matrix, i.e., the new vector is parallel to the original vector. For example,

$$\begin{bmatrix} -\frac{1}{2} & -\frac{\sqrt{3}}{2} & 0 \\ -\frac{\sqrt{3}}{2} & \frac{1}{2} & 0 \\ 0 & 0 & 1 \end{bmatrix} \begin{bmatrix} x_e \\ y_e \\ z_e \end{bmatrix} = \begin{bmatrix} \lambda x_e \\ \lambda y_e \\ \lambda z_e \end{bmatrix} = \lambda \begin{bmatrix} x_e \\ y_e \\ z_e \end{bmatrix}. \tag{2.2}$$

These special vectors are *characteristic* of the matrix, and are called its *eigenvectors* – *eigen* is German for "characteristic." (See Sec. 1.4.7) The scalar multipliers for the eigenvectors are also characteristic of the matrix and are termed its *eigenvalues.* The matrices for a given class of symmetry operations have the same eigenvalues, and are said to be *similar* to one another because of this. Importantly, the sum of the diagonal elements of a matrix, known as the *trace* or *character* of the matrix equals the sum of the eigenvalues of the matrix. It follows that *the matrices corresponding to a class of symmetry operations all have the same characters.*

As stated earlier, there are infinitely many possible matrix representations for a given point group. For example, in the vibrational spectroscopy of molecules, the symmetry analysis of vibrations begins with the placement of three coordinate vectors on each atom.[25] For n atoms this results in $3n$ vectors. The matrices that describe the transformation of these vectors for each operation in the group are $3n \times 3n$ matrices, which can become very large for molecules with many atoms. However, regardless of the size or complexity, the characters of the matrices corresponding to a given class of symmetry operations will be identical.

Another important feature of the $[x\ y\ z]$ transformation matrices is their form — they all consist of a 2×2 matrix in the upper left corner, and a 1×1 matrix in the lower right. The matrices are said to be *block-diagonalized.* This results in x and y being transformed as a pair (the resulting vector will have components that are linear combinations of x and y), and z being transformed independently, e.g.,

$$\begin{bmatrix} -\frac{1}{2} & -\frac{\sqrt{3}}{2} & 0 \\ -\frac{\sqrt{3}}{2} & \frac{1}{2} & 0 \\ 0 & 0 & -1 \end{bmatrix} \begin{bmatrix} x \\ y \\ z \end{bmatrix} = \begin{bmatrix} (-\frac{1}{2}x - \frac{\sqrt{3}}{2}y) \\ (-\frac{\sqrt{3}}{2}x + \frac{1}{2}y) \\ -z \end{bmatrix}.$$

The complete set of 2×2 matrices are the transformation matrices for the (x, y) pair, and are independent of the complete set of 1×1 matrices responsible for the transformation of z. The 3×3 matrix representation has been reduced to two independent smaller matrix representations. Both sets of smaller matrices are matrix representations of the groups, since each set parallels the multiplicative properties of the group. An important theorem in group theory (which we will not deal with here) states that the application of a suitable set of transformations (known as *similarity transformations*) will create a block-diagonalized matrix with each of the blocks constituting a representation of the group. Furthermore, the blocks will be as small as possible (they must be large enough to allow for any coordinates that are dependent on one another, such as x and y above)— they can be reduced no further. Each point group has a finite (and relatively small) number of these representations, known as the *irreducible* representations of the group.

Regardless of the size of the initial matrix representation, the block-diagonalized matrix will consist of one or more (often many!) of these irreducible representations.

Each irreducible representation for a group describes a different "behavior" when the symmetry operations of the group are performed. Thus a specific irreducible representation is said to constitute a *symmetry species* of the group. If a representation describes the behavior of an entity (e.g., a function) when the symmetry operators are applied, the entity is said to *belong* to the symmetry species defined by the representation.

2.3.3 Character Tables

The properties of symmetry point groups are conveniently summarized in *character tables.* Table 2.2, the D_{3h} character table is typical:

Table 2.2 D_{3h} Character Table

D_{3h}	E	$2C_3$	$3C_2$	σ_h	$2S_3$	$3\sigma_v$	linear	quadratic
A_1'	1	1	1	1	1	1		x^2+y^2, z^2
A_2'	1	1	-1	1	1	-1	R_z	
E'	2	-1	0	2	-1	0	(x,y)	(x^2-y^2, xy)
A_1''	1	1	1	-1	-1	-1		
A_2''	1	1	-1	-1	-1	1	z	
E''	2	-1	0	-2	1	0	(R_x, R_y)	(xz, yz)

The top row of the table consists of the symmetry operations of the group by symmetry class; the integer multiplier gives the number of symmetry operations in each class. The rows of numbers below the top line are the *characters* of the irreducible matrix representations of the group, each labeled in the column on the left with the symmetry species symbol for the representation. The entries in the columns to the right of the characters are functions that transform as a particular irreducible representation, placed in the rows of the symmetry species to which they belong. In the D_{3h} point group z belongs to the A_2'' representation. Since the matrix representation for the transformation of z is one-dimensional, the characters are the actual elements of the matrices, and the transformation of the function can be determined directly from the characters: z remains unchanged when the E, C_3, and σ_v operations are performed, and changes sign when the σ_h, C_2, and S_3 operations are performed. The pair, (x, y), belongs to the E' representation, consisting of two-dimensional matrices. In this case the characters do not give us the actual transformations of x and y. Once the transformation matrices for x, y, and z have been determined for a point group, the symmetry behavior of any other function of these three variables can be assessed. For instance, the function z can be either positive or negative, while z^2 is always positive. Thus $C_2z = -z$, but $C_2z^2 = (-z)^2 = [1]z^2$. Any symmetry operation on z^2 will leave it unchanged; z^2 belongs to the A_1' representation. The matrix representation consists of twelve one-dimensional matrices all containing a single element, 1. In this case the symmetry of the function is identical to the symmetry of the object (the PF_5 molecule in our example) since they both remain unchanged for all of the symmetry operations of the group. Because of this, the representation consisting of characters of +1 for all of the operations is termed *the totally symmetric representation.* As a second example, consider the function $x^2 + y^2$. The C_2 and C_3^1 operations performed on this function produce the following results:

$$\begin{aligned} C_2 x &= x \\ C_2 y &= -y \\ C_2 x^2 &= x^2 \\ C_2 y^2 &= (-y)^2 = y^2 \\ C_2(x^2+y^2) &= [1](x^2+y^2) \end{aligned} \tag{2.3}$$

$$\begin{aligned} C_3^1 x &= -\frac{1}{2}x - \frac{\sqrt{3}}{2}y \\ C_3^1 y &= \frac{\sqrt{3}}{2}x - \frac{1}{2}y \\ C_3^1 x^2 &= \left(-\frac{1}{2}x - \frac{\sqrt{3}}{2}y\right)^2 = \frac{1}{4}x^2 + \frac{\sqrt{3}}{4}xy + \frac{3}{4}y^2 \\ C_3^1 y^2 &= \left(\frac{\sqrt{3}}{2}x - \frac{1}{2}y\right)^2 = \frac{3}{4}x^2 - \frac{\sqrt{3}}{4}xy + \frac{1}{4}y^2 \\ C_3^1(x^2+y^2) &= \frac{1}{4}x^2 + \frac{3}{4}y^2 + \frac{3}{4}x^2 + \frac{1}{4}y^2 = [1](x^2+y^2) \end{aligned} \tag{2.4}$$

It is left to the reader to determine that the results are the same for the 10 remaining matrices; $x^2 + y^2$ belongs to the totally symmetric representation of the D_{3h} point group.

2.3.4 Lattice and Crystal Point Symmetry. The Hermann-Mauguin Notation.

The ordered array of points that make up a lattice can also be viewed as a single object by fixing one of the points in the lattice and observing the effects of point symmetry operations on the lattice as it is transformed about the point. Since the lattice point that we select is arbitrary, the same operations applied to any lattice point must produce the same results. We again turn to a two-dimensional lattice to begin an analysis of the consequences of these statements. Fig. 2.11 illustrates that lattices can exhibit 2-fold, 3-fold, 4-fold and 6-fold rotational symmetry. These operations rotate every lattice point into an equivalent lattice point. Since each lattice point contains a perpendicular rotation axis, these axes are also transformed into symmetry-equivalent rotation axes. *The four types of rotation axes shown, along with a 1-fold axis, are the only rotation axes possible for a two-dimensional lattice.*

To prove this assertion we select four collinear lattice points, separated from one another by a translation distance, t, as illustrated in Fig. 2.12. Assuming

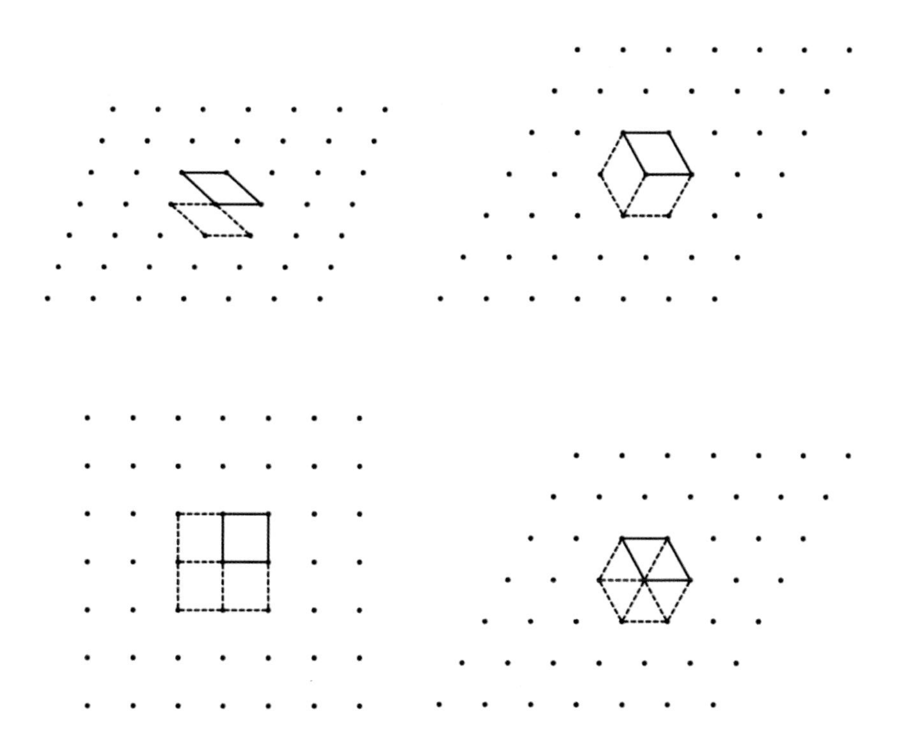

Figure 2.11 2-fold, 3-fold, 4-fold and 6-fold rotational axes in two-dimensional lattices. Each axis is perpendicular to the page. Solid parallelograms are unit cells.

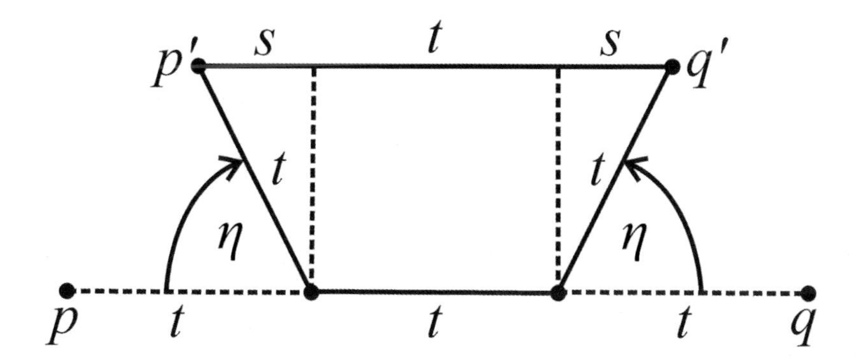

Figure 2.12 The results of two rotations around n-fold symmetry axes containing two adjacent lattice points.

Table 2.3 Possible Rotation Axes for a Two-Dimensional Lattice

m'	$\cos\eta$	η	n (C_n)
−3	−3/2	—	—
−2	−1	π (180°)	2
−1	−1/2	$2\pi/3$ (120°)	3
0	0	$\pi/2$ (90°)	4
1	1/2	$\pi/3$ (60°)	6
2	1	2π (360°)	1
3	3/2	—	—

that each lattice point contains an n-fold symmetry axis, we rotate $\eta = 2\pi/n$ in a counterclockwise direction about the point adjacent to the rightmost point, and $\eta = 2\pi/n$ in a clockwise direction about the point adjacent to the leftmost point. For each rotation, the points on each end of the four-point array are transformed onto other lattice points: point p is rotated onto point p' and point q is rotated onto point q'. Since there are an integral number of rotations allowed for each axis, both clockwise and counterclockwise rotations must be symmetry operations (e.g., for a 3-fold axis, $-C_3^1 = C_3^2$). The transformed lattice points are now separated by a distance of $2s + t$. However, since they remain lattice points — and lie on a line parallel to the original four points — they must be separated by an integral number of translation increments. Thus $2s + t = mt$, where m is an integer. Since $s = t\cos\eta$, $2t\cos\eta + t = mt \Longrightarrow \cos\eta = (m-1)/2 = m'/2$. Since $-1 \leqq \cos\eta \leqq 1$, and m' is an integer, there are only a few values of η possible, listed in Table 2.3. There is no 5-fold axis allowed in a two-dimensional lattice, nor is any axis with n greater than six possible. *All* two-dimensional lattices contain at least a two-fold axis (this is not true for three-dimensional lattices). Two-dimensional lattices can also exhibit mirror symmetry, and *every* lattice has inversion symmetry (Fig. 2.13). Every point in the lattice is an inversion center.

The Schönflies notation provides a concise shorthand for point symmetry. We will soon add translational symmetry operations to the point symmetry operations to create *space groups* from point groups. Schönflies added a superscript to the point group notation for each space group generated by the addition of specific translational symmetry operators. While symbols such as D_{2h}^{15} serve to differentiate between the space groups generated from a particular point group (there are 28 space groups with D_{2h} symmetry), the superscript contains no information about symmetry. Because of this Hermann and Mauguin developed an alternative system of symbolic notation that produces space group symbols which convey symmetry information. The Hermann-Mauguin (H-M) notation uses a single integer to describe rotation axes, i.e., 1, 2, 3, 4, and 6 for C_1, C_2, C_3, C_4 and C_6, respectively. The symbol for a mirror plane is m; mirror planes that are perpendicular to n-fold axes (σ_h) are symbolized by n/m, while those containing an n-fold axis (σ_v) are symbolized by nm. The most significant difference between the two schemes is the lack of rotoreflection axes in the H-M system. Instead, *rotoinversion* axes are used.

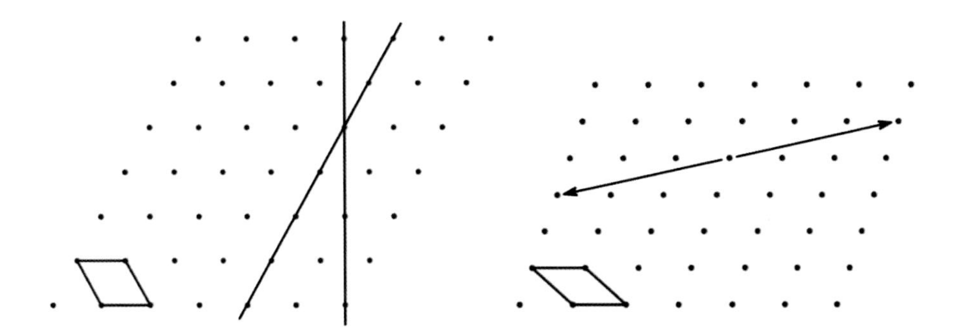

Figure 2.13 Mirror planes and an inversion center in two-dimensional lattices. Solid parallelograms are unit cells.

A rotoinversion axis consists of a rotation axis containing a center of inversion; like its rotoreflection analog, the line and point are inseparable. The rotoinversion operations about an n-fold rotoinversion axis involve a $k \cdot 2\pi/n$ rotation ($k = 1, 2, \ldots, n$) followed by an inversion. The symbol for a rotoinversion axis is the integer corresponding to the rotation axis with a bar over the number, i.e., $\overline{n} = \overline{1}$, $\overline{2}$, $\overline{3}$, $\overline{4}$, or $\overline{6}$. Since a 360° rotation followed by an inversion *is* an inversion, the Shoenflies symbol i becomes $\overline{1}$ in the H-M notation. Rotoinversion axes are described verbally as "n bar" axes. Since rotoreflections and rotoinversions are fundamentally different operations the reader may wonder how the notation schemes account for the same operations (which, of course, they must!). We will demonstrate shortly that each rotoinversion axis is equivalent to a symmetry element described by the Schönflies notation.

The addition of a third dimension to the lattices that we have just discussed is now relatively straightforward. If we consider a third translation the result will be to repeat the two dimensional lattice along the third dimension. We can envision looking down on the two dimensional lattice, then stacking identical and equidistant planes below the plane of the original two-dimensional lattice, as illustrated in Fig. 2.14. This places an added constraint on the rotational symmetry of the lattice. If the planes are stacked directly below (or above) the plane of origin, the rotational symmetry will remain intact. This is the case when the translation vector in the third dimension is perpendicular to the original plane. If, however, the added planes are translated in a non-orthogonal direction, the rotational symmetry is lost. It follows that three-dimensional lattices are also allowed only for 1, 2, 3, 4, and 6-fold rotational symmetry, and unlike two-dimensional lattices, are not required to have a 2-fold axis (everything has a 1-fold axis!). On the other hand, the non-orthogonal translation does preserve the center of symmetry; as in two-dimensions, *every lattice point in three dimensions is an inversion center.* The previous statement must be read with some caution here – we have not yet discussed the *contents* of the unit cells. Here we are considering only the symmetry of the *lattice*. To avoid confusion we will use the term *crystal symmetry* when discussing the symmetry of the lattice *and* its contents.

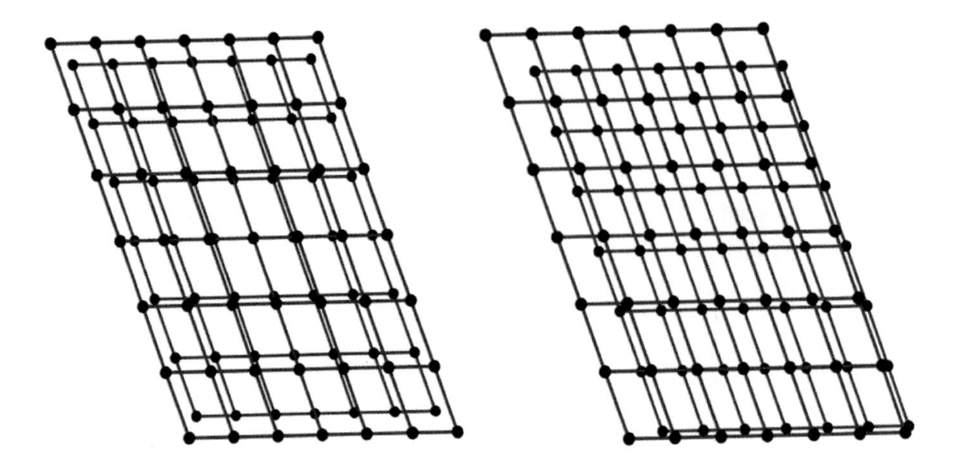

Figure 2.14 Translation of a two-dimensional lattice in an orthogonal direction(left) and in a non-orthogonal direction(right).

2.3.5 Stereographic Projections: Crystallographic Point Symmetry Elements.

Before crystallographers had access to modern X-ray diffraction techniques the internal symmetry of a crystal had to be inferred from its external appearance. Although the size and shape of the crystal faces depend on crystallization conditions, the angles between them depend on the internal symmetry of the crystal lattice. In 1809 William Hyde Wollaston, an English chemist, physicist, and physiologist developed a device, the *optical goniometer*, that measured the angle of reflection of light from the faces of a crystal. A "modern" version, designed near the end of the 19th century, is shown in Fig. 2.15. With the device shown, the crystal was rotated around horizontal and vertical axes until a collimated beam of light was reflected from a crystal face into the telescope. The horizontal and vertical rotation angles, in conjunction with the angle between the collimator and telescope, allowed the investigator to determine the direction of the vector normal to the plane of the crystal face.

As noted several times throughout the chapter, lattice symmetry is often easier to visualize in two dimensions. This was particularly true for the interpretation of the symmetry relationships between the plane normals that were determined by optical goniometry. In 1823 the German physicist and mineralogist, Franz E. Neumann adapted a method known to the ancient Greeks, the *stereographic projection*, to create a two-dimensional figure which facilitated the analysis of the underlying symmetry. The stereographic projection, illustrated in Fig. 2.16, is constructed by centering a sphere of radius r at the origin of an orthonormal coordinate system and projecting vectors of interest from the origin to the surface of the sphere.

If the vector lies on or above the xy plane, a line segment is created between the point of intersection of the vector with the sphere, (x, y, z), and the lowermost point on the sphere, $(0, 0, -r)$. If the vector lies below the xy plane the point of intersection of the projected vector with the sphere is connected with a line segment to the uppermost point on the sphere, $(0, 0, r)$. The point of intersection of the line

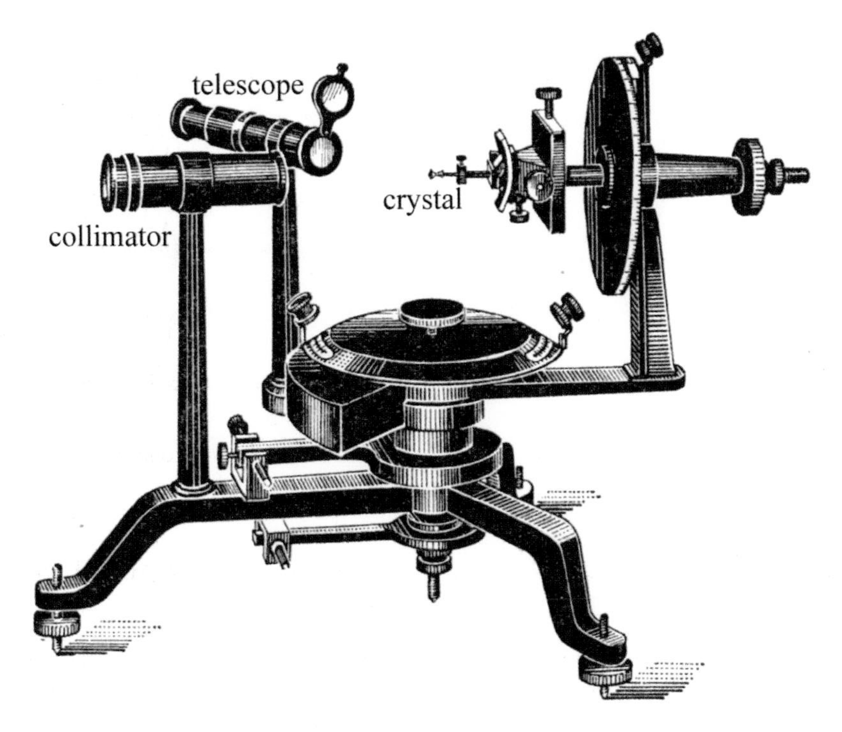

Figure 2.15 Two-circle optical goniometer designed near the end of the 19th century.

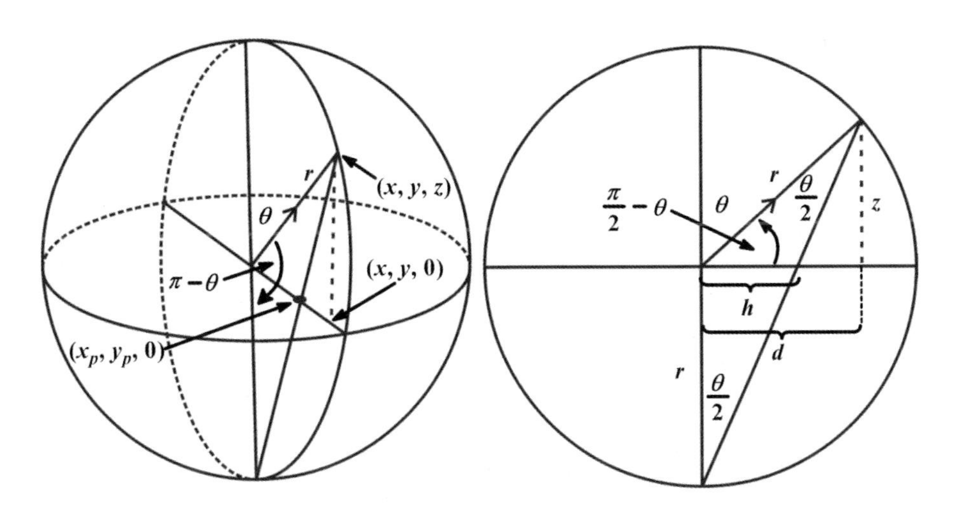

Figure 2.16 (left) Sphere with vector projected to its surface at (x, y, z) and connect to the "south pole" at $(0, 0, -r)$. The line segment connecting these points intersects the xy plane at $(x_p, y_p, 0)$. (right) The great circle containing (x, y, z) and $(0, 0, -r)$.

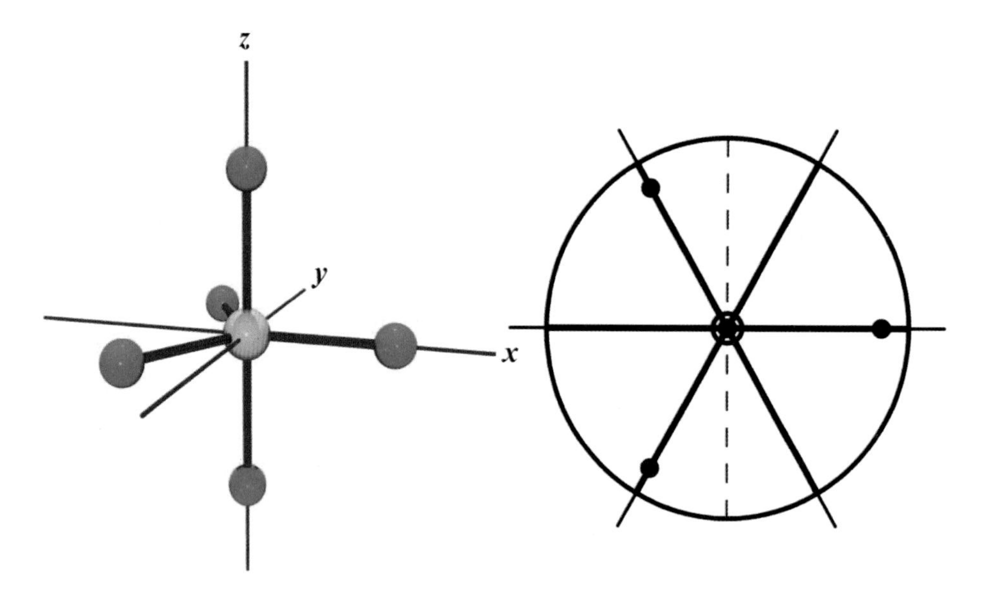

Figure 2.17 The PF_5 molecule in an orthonormal coordinate system and its stereographic projection. The three dots on the plane illustrate the three-fold axis. The axial point above the plane is shown as a projected dot, and the point below the plane as a projected ring. The heavier lines indicate vertical mirror planes, the lighter lines extending beyond the perimeter are the two-fold axes, and the solid circle indicates the horizontal mirror plane. The dashed line is a reference line for the y axis.

segment with the xy plane becomes the stereographic projection of the vector. It is located at $(x_p, y_p, 0)$, and can be determined by analyzing the great circle which contains the points $(0, 0, \pm r)$ and (x, y, z):

$$
\begin{aligned}
d &= (x^2 + y^2)^{\frac{1}{2}} \\
\frac{z}{d} &= \tan\left(\frac{\pi}{2} - \theta\right) \\
\theta &= \frac{\pi}{2} - \arctan\left(\frac{z}{d}\right) \\
h &= r \tan\left(\frac{\theta}{2}\right) \\
x_p &= \frac{x}{d} h \text{ and } y_p = \frac{y}{d} h.
\end{aligned}
\tag{2.5}
$$

The stereographic projection of the complete set of vectors is plotted on the equatorial great circle in the xy plane. The plot is referred to as a *stereogram*. Points of intersection for vectors above the plane are drawn with a filled dot; those from vectors below the plane are drawn with a circular ring. The equatorial circle is drawn with dashed lines unless it is coplanar with a mirror plane, in which case the circle is drawn with a solid line to indicate the presence of the mirror. The axis of highest symmetry is selected as the z axis. Conventionally, the order of the axis is indicated by a symbol (e.g., a three-fold axis is symbolized by a triangle). In cases where the order of the axis is obvious from the stereogram the axial symbols are often omitted. Axes in the xy plane are indicated with lines which extend

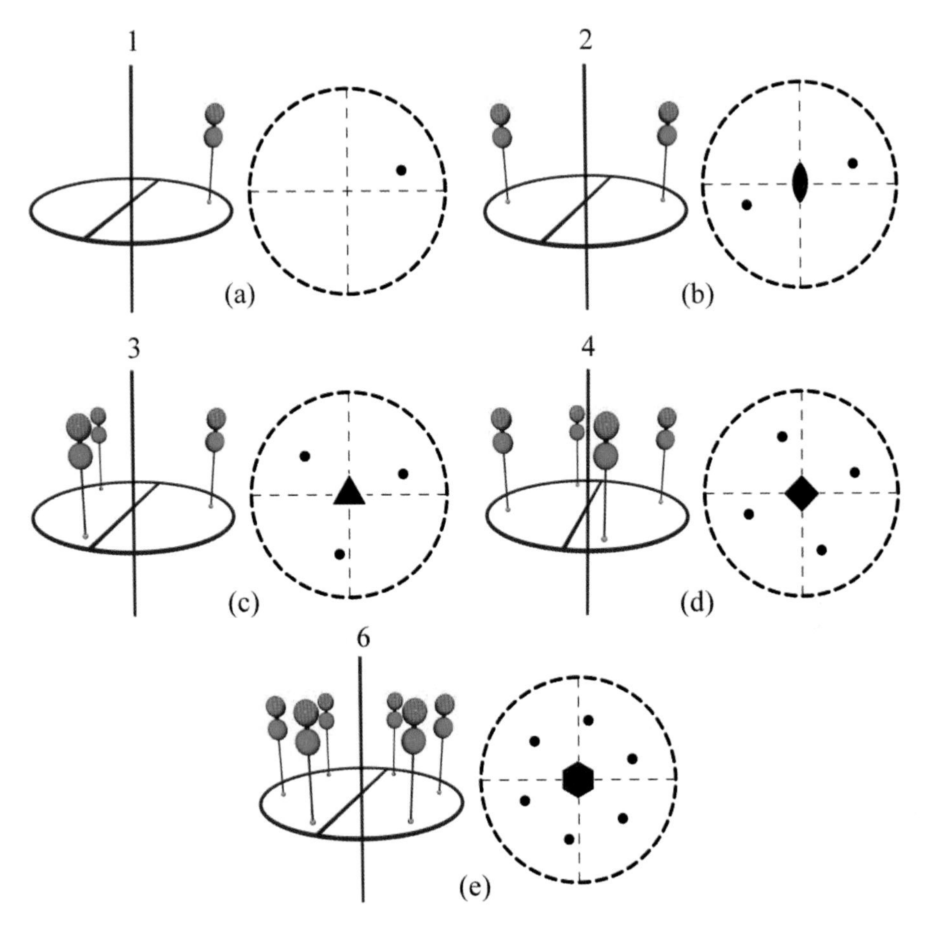

Figure 2.18 Rotation stereographic projections: (a) monad axis, $1 \equiv C_1$, (b) diad axis, $2 \equiv C_2$, (c) triad axis, $3 \equiv C_3$, (d) tetrad axis, $4 \equiv C_4$, (e) hexad axis, $6 \equiv C_6$.

beyond the perimeter of the circle.* Vertical mirror planes are shown as heavier lines corresponding to diameters of the circle. The x and y axes are indicated with light dashed horizontal lines and vertical lines respectively. The x axis is the reference axis for rotations. Thus, whenever there are mirror planes containing the z axis, or axes in the xy plane, at least one of these symmetry elements will be coincident with the x axis. Thus a light dashed line along x indicates the absence of such symmetry elements.

Stereographic projections can also be applied to molecules in order to evaluate their symmetry. The PF_5 molecule provides an example of a typical stereographic projection, illustrated in Fig. 2.17. The vectors that are projected to the sphere are those from the phosphorus atom in the center to the fluorine atoms. The simple two-dimensional figure clearly illustrates that rotation around any of the axes, or reflections across the planes will produce an identical projection.

*Traditionally, axial symbols are placed in a stereogram at the projected points where the axis intersects the sphere. Except in cases of high symmetry, this adds unnecessary complexity to the stereogram and will be avoided here — with the exception of cases where the axes do not intersect the sphere in one of the coordinate planes.

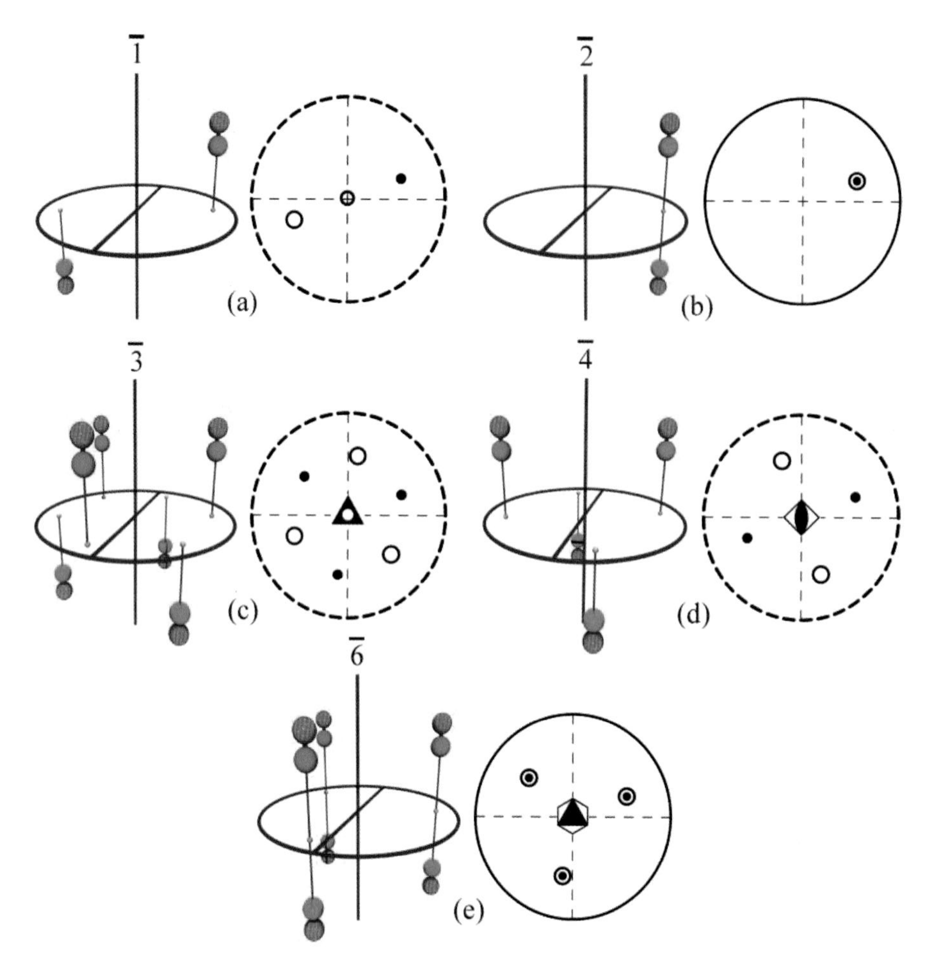

Figure 2.19 Rotoinversion stereographic projections: (a) $\bar{1} \equiv S_2 \equiv i$, (b) $\bar{2} \equiv S_1 \equiv m$, (c) $\bar{3} \equiv S_6$, (d) $\bar{4} \equiv S_4$, (e) $\bar{6} \equiv S_3 \equiv 3/m$.

In the next section we will make use of stereographic projections in our discussion of the crystallographic point groups. The stereographic projections of the allowed crystallographic symmetry operations will be combined to create the stereographic projections indicative of specific point groups. The projections of the five allowed rotational axes are shown in Fig. 2.18. The H-M notation and its Schönflies equivalent are given for each axis, along with the descriptive terms *monad*, *diad*, *triad*, *tetrad*, and *hexad* for 1-fold, 2-fold, 3-fold, 4-fold, and 6-fold axes, respectively. The geometric symbols for the axes are also shown.

The rotoinversion projections are shown in Fig. 2.19. The inversion, $\bar{1}$, is equivalent to S_2 in the Schönflies notation. The symbol for an inversion is the small open circle at the origin in the stereogram. The circle is often omitted since it is easy to confuse it with the symbol for a projected vector along the $-z$ axis; the existence of an inversion center is usually obvious from the appearance of the stereogram. The $\bar{2}$ axis is equivalent to S_1, which, in turn is just a complete rotation followed by a reflection — a reflection. Thus $\bar{2}$ is equivalent to m, symbolized by

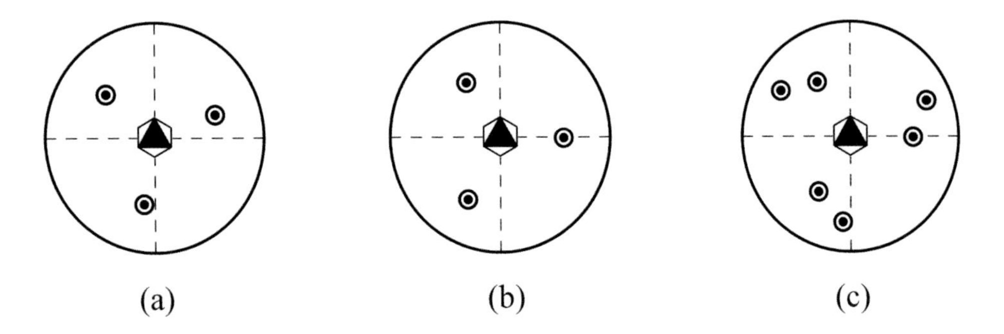

Figure 2.20 (a) Stereographic projection of the effects of $\bar{6}$ rotoinversion operations on a single point. (b) Apparent vertical mirror and diad symmetry generated by the rotoinversion operations. (c) Removal of the ambiguity by adding a second point.

the solid-line equatorial circle. The $\bar{3}$ axis is symbolized by a triangle containing an un-filled circle to represent the inversion center in the rotoinversion axis. The three-fold rotoinversion axis is equivalent to a six-fold rotoreflection axis, S_6. The $\bar{4}$ axis is equivalent to an S_4 axis.

A diad axis also exists as a consequence of the 4-fold rotoinversion; the $\bar{4}$ axis is therefore represented by a combination of the symbol for the tetrad and diad rotation axes. This is a general property of $\bar{n}$ axes when n is an even number — the existence of an even order rotoinversion axis requires the simultaneous existence of a rotation axis of order $n/2$. The $\bar{6}$ axis is equivalent to an S_3 axis, and because it is an even-ordered rotoinversion axis it is indicated with a combination of the triad and hexad axial symbols. The six-fold rotoinversion operations generate a horizontal mirror plane in addition to the triad axis so that $\bar{6} \equiv 3/m$ as well. *Each rotoinversion axis in the Hermann-Mauguin system is equivalent to a unique rotoreflection axis in the Schönflies system.* Note that odd rotoinversion axes impose a center of symmetry. It is also interesting to note that the set of rotation axes *and* rotoinversion axes constitute a complete collection of symmetry elements which includes the mirror plane and the inversion center. However, the importance of the mirror plane and center of symmetry in crystallography justifies their use as separate symmetry elements.

Stereographic projections ordinarily represent the effects of symmetry operations on a single point. This can give rise to apparent ambiguities, often implying the existence of more symmetry than there actually is. Consider, for example, the stereogram for the $\bar{6}$ rotoinversion axis, repeated in Fig. 2.20(a). If the initial point is located on the x axis, as indicated in Fig. 2.20(b), there appear to be vertical mirror planes in the xz plane and rotated 120° and 240°, respectively. The x axis also appears to contain a diad axis, with two more axes rotated by the same angles. This ambiguity is readily resolved by performing the operations on a second point, unrelated to the first, as illustrated in Fig. 2.20(c). The second point is related to its symmetry-equivalent points only by the $\bar{6}$ operations. Adding a second point to stereographic projections, however, renders them very difficult to interpret. Rather than nullifying the simplicity of the stereogram (the reason for using it), we simply agree that the initial point to be transformed will *not* be placed on the x axis.

After all symmetry-equivalent points are projected, if there are no vertical/dihedral mirror planes or rotation axes coincident with the x axis, the axis will continue to be indicated by a light dashed line.

2.3.6 The 32 Crystallographic Point Groups.

There are a total of ten symmetry elements available to a crystal lattice, and if a random combination of these symmetry elements was possible there would be a very large number of crystal symmetries to deal with. Fortunately, only certain combinations of these symmetry elements are possible — many combinations are redundant and the corresponding sets of symmetry operations must constitute a point group. Because of this the combination of certain symmetry elements will generate other symmetry elements. Several of these relationships warrant special attention:

1. ***The intersection of two mirror planes is a rotation axis.*** If the dihedral angle between the mirror planes is $2\pi/n$, then the axis is an n-fold axis. Fig. 2.21 illustrates the case when $n = 3$. The initial vector in Fig. 2.21(a) is reflected across an initial vertical mirror plane in Fig. 2.21(b). A second vertical mirror plane, rotated at an angle of $2\pi/3$ with respect to the first mirror plane, is introduced. The original vector and its original reflection are reflected across the second mirror (Fig. 2.21(c)). Finally all four vectors are reflected across the original mirror plane, generating a third pair of vectors (Fig. 2.21(d)). The stereogram in

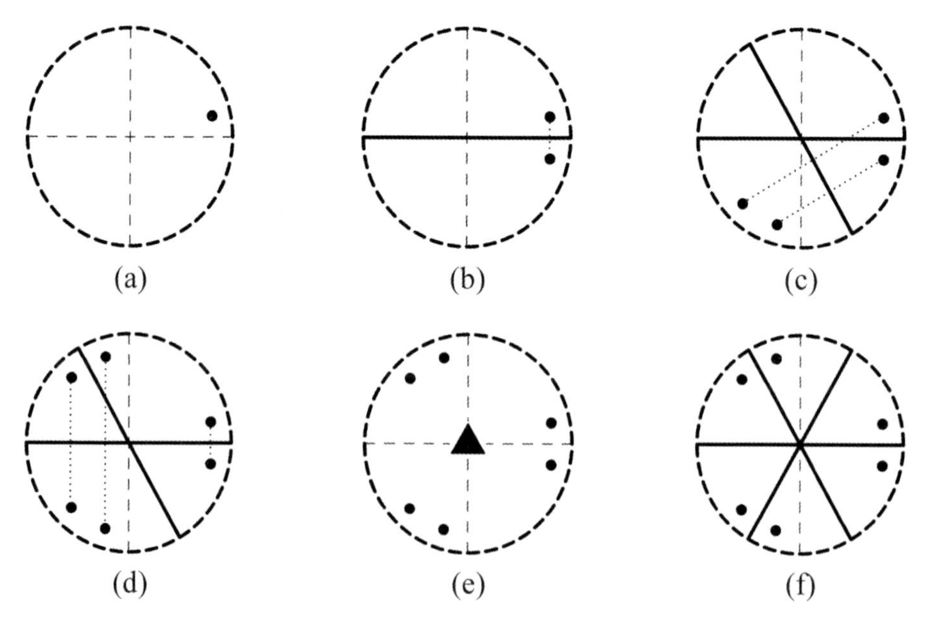

Figure 2.21 Stereograms: (a) Projected vector. (b) Reflection across an initial mirror plane. (c) Reflection across a second mirror plane at a $2\pi/3$ dihedral angle to the first. (d) Second reflection across the initial mirror plane. (e) Triad axis generated from the intersection of the two mirror planes. (f) Two additional mirror planes generated from the initial mirror plane containing the triad axis.

Fig. 2.21(e) clearly shows that the three pairs of vectors created by reflections across the two vertical mirror planes are related by a triad rotation axis.

2. ***If a mirror plane contains an n-fold axis, then there are n-1 additional mirror planes which also contain the axis.*** This is the converse of the previous relationship. The mirror planes will have dihedral angles of $k \cdot 2\pi/n,\ k = 1, 2, \ldots,\ n-1$, with respect to the original mirror plane. Referring to Fig. 2.21(b), an original mirror plane combined with a 3-fold rotation axis generates the stereogram in Fig. 2.21(f); the triad axis has created two additional mirror planes.

3. ***If there are two intersecting diad axes with an angle of 2π/n between them, then there must be an n-fold axis perpendicular to both of the diad axes.*** For example, Fig. 2.22 illustrates the creation of a triad axis perpendicular to two diad axes with an inter-axial angle between the diads of $2\pi/3$. The initial vector in Fig. 2.22(a) is rotated around an initial diad axis in Fig. 2.22(b). Rotation around a second diad axis at an angle of $2\pi/3$ with respect to the first creates a second pair of vectors as shown in Fig. 2.22(c). The four vectors resulting from these two rotations are now rotated around the first diad axis, generating a third pair of vectors (Fig. 2.22(d)). The three pairs of vectors are related by a triad axis perpendicular to the plane containing the two diad axes as indicated in Fig. 2.22(e). The two diad axes have generated a triad axis perpendicular to the plane containing the diad axes.

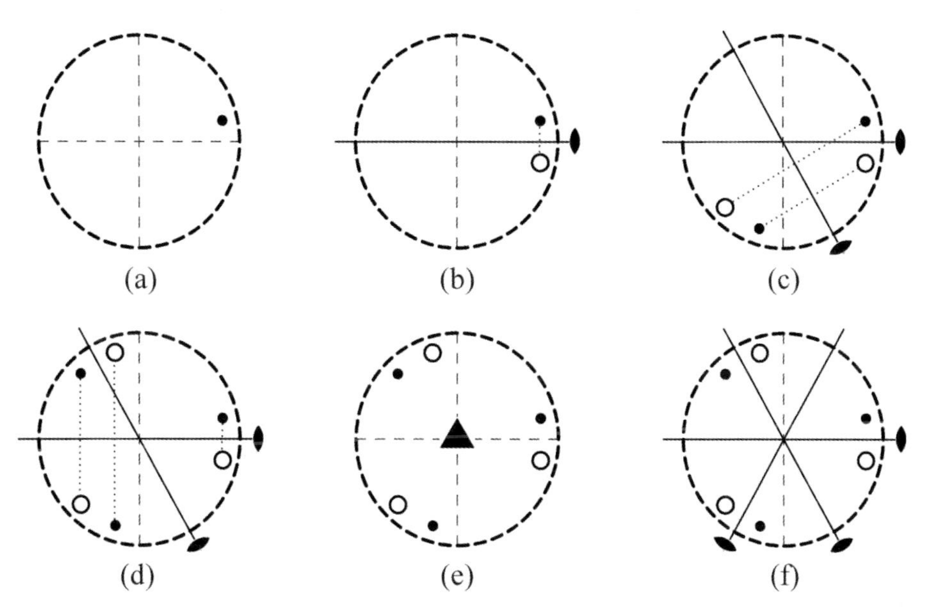

Figure 2.22 Stereograms: (a) Projected vector. (b) Rotation around an initial diad axis. (c) Rotation around a second diad axis at a $2\pi/3$ angle to the first. (d) Second rotation around the initial diad axis. (e) Triad axis generated from the intersection of the two diad axes. (f) Two additional diad axes generated from the initial diad axis and the perpendicular triad axis.

4. ***If there is an n-fold axis intersecting a perpendicular diad axis, then there are n-1 additional diad axes, separated from one another by angles of 2π/n.*** This is the converse of the third relationship. If the diad axis in

Fig. 2.22(b) intersects a perpendicular triad axis, then two more diad axes will be generated by the triad axis, as illustrated in Fig. 2.22(f).

5. *The presence of any two of the following: n-fold axis (n even), perpendicular mirror plane, inversion center — automatically implies the third.* Every rotational axis for which n is even generates a second symmetry-equivalent vector rotated 180° with respect to the original vector. Reflecting the pair of vectors across a perpendicular mirror plane creates a center of symmetry. Reversing these operations has the same effect; inverting a vector will create a vector rotated 180° with respect to the first but on the other side of the plane — a reflection of the second vector. Reflecting a vector and inverting it creates vectors rotated 180° with respect to one another, requiring n to be an even number.

These constraints, coupled with the limited number of allowed symmetry elements for the crystal lattice, result in a total of 32 possible crystallographic point groups. Five of these, the simplest, exhibit the symmetry of a single rotational axis exclusively. Table 2.4 lists the H-M and Schönflies notations (Sch), the classes of symmetry operations (Schönflies), and the figure in which the stereogram is illustrated for each of these point groups.

Table 2.4 Crystal Point Groups For Lattices With Single Rotational Axes.

H-M	Sch	Symmetry Operations						Stereogram
1	C_1	E						Fig. 2.18(a)
2	C_2	E	C_2					Fig. 2.18(b)
3	C_3	E	C_3	C_3^2				Fig. 2.18(c)
4	C_4	E	C_4	C_2	C_4^3			Fig. 2.18(d)
6	C_6	E	C_6	C_3	C_2	C_3^2	C_6^5	Fig. 2.18(e)

The point groups with the symmetry of a single rotoinversion axis constitute a second set of crystallographic point groups. The point group notations, symmetry operations, and stereogram locations for these point groups are listed in Table 2.5.

Table 2.5 Crystal Point Groups For Lattices With Single Rotoinversion Axes.

H-M	Sch	Symmetry Operations						Stereogram
$\bar{1}$	C_i	E	i					Fig. 2.19(a)
m	C_s	E	σ_h					Fig. 2.19(b)
$\bar{3}$	S_6	E	C_3	C_3^2	i	S_6^5	S_6	Fig. 2.19(c)
$\bar{4}$	S_4	E	S_4	C_2	S_4^3			Fig. 2.19(d)
$\bar{6}$	C_{3h}	E	C_3	C_3^2	σ_h	S_3^2	S_3^5	Fig. 2.19(e)

The Hermann-Mauguin symbols in Tables 2.4 and 2.5 are the "short forms" of the symbols. Each symbol lists those symmetry elements for which the set of corresponding symmetry operators is sufficient to generate the remainder of the symmetry elements for the point group. The "long form" of the point group symbol contains more information. Its structure depends on the type of crystal lattice — the *crystal system* — a topic to be covered in the next section. For lower-symmetry lattices, which are the more common, the long form of the symbol contains information related to the **a**, **b**, and **c** axes. For example, the long form of the $C_2 \equiv 2$ point group symbol is 1 2 1, when the **b** axis is chosen as the diad axis. The combination of a diad axis with an n-fold rotation axis (called the *principal* rotation axis) creates a new point group. Since a vector lying on the principal axis must transform into a symmetry-equivalent vector the diad axis must be perpendicular to the principal axis. Furthermore, relationship **4** requires that if there is one diad axis perpendicular to the n-fold axis then there must be n perpendicular diad axes.

The next four point groups are those formed from the intersection of an n-fold principal rotation axis with n diad axes perpendicular to the principal axis (Table 2.6). The diad axes in the 222 point group (Fig. 2.23(a)) cannot be transformed into one another by rotation around any of the three axes. The diad operations are therefore in different symmetry classes. The diad axes in the 32 point group (Fig. 2.23(b)) are all transformed into one another by the triad axis; the diad rotations in 32 are all in the same class. The long form of the point group symbol is 322, but since the triad and one diad are all that are necessary to generate the entire set of symmetry elements, the short form is 32. In the 422 point group the 4-fold axis transforms the diad axes at right angles into one another, but never transforms one axis into another rotated 45° from it. Referring to Fig. 2.23(c), the diad operations around axes marked with asterisks are members of one class; rotations around unmarked diads belong to another class. The 622 point group has similar properties. There are two sets of diad axes creating two classes of diad rotations, one set marked with asterisks in Fig. 2.23(d) and the other set unmarked.

Another four point groups are formed in a manner similar to that just described in which a vertical mirror plane is combined with an n-fold principal axis. According to relationship **2**, there must be n of these planes. Table 2.7 summarizes the results of these combinations. The two mirror planes in $2mm$ (Fig. 2.24(a)) cannot be rotated or reflected into one another; the reflections are in different classes. The three vertical mirrors in 3m (Fig. 2.24(b)) are related by the triad rotation, and are in the same class. There are two sets of mirror planes in $4mm$ and $6mm$ with the reflection operations in different classes. In Figs. 2.24(a) and (b), one set is marked with an asterisk and the other is unmarked. Because a vertical mirror plane corresponding to one class bisects the angle between two planes corresponding to the other class, the two classes of reflections are labeled σ_v and σ_d.

Three more point groups are formed by adding a horizontal mirror plane to diad, tetrad, and hexad axes (Table 2.8). The absence of the triad axis here occurs because the addition of a horizontal mirror plane to a triad axis creates a $\bar{6}$ rotoinversion axis (Fig. 2.19(e)), and we have already accounted for the $\bar{6}$ point group.

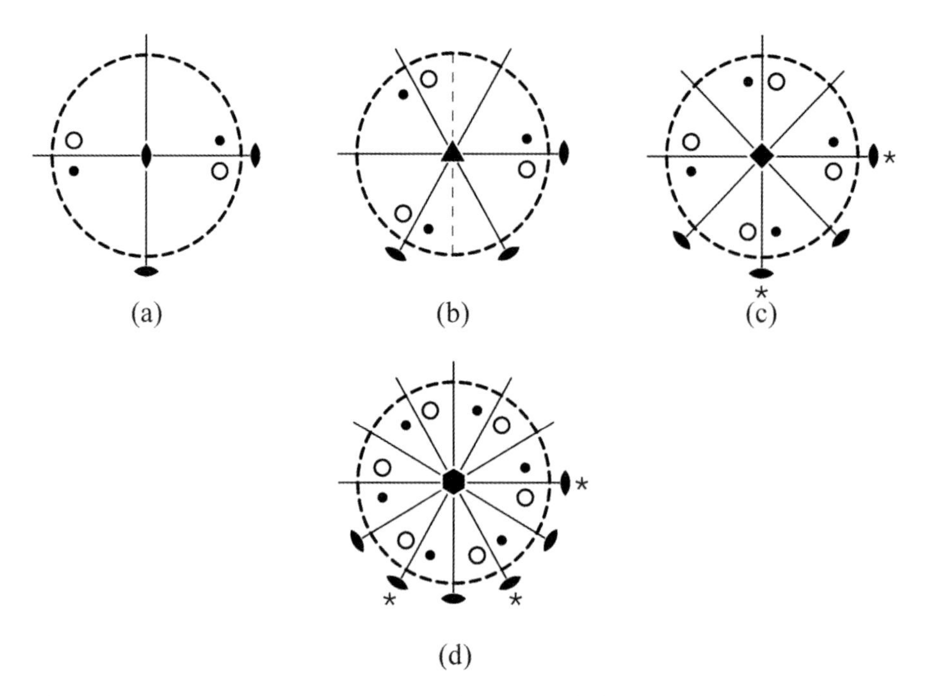

Figure 2.23 Point group stereograms: (a) 222 (D_2), (b) 32 (D_3), (c) 422 (D_4), and (d) 622 (D_6).

Table 2.6 Crystal Point Groups for Lattices with n Diad Axes Perpendicular to an n-Fold Principal Axis.

H-M	Sch	Symmetry Operations						Stereogram
222	D_2	E	$C_2(z)$	$C_2(y)$	$C_2(x)$			Fig. 2.23(a)
32	D_3	E	$2C_3$	$3C_2$				Fig. 2.23(b)
422	D_4	E	$2C_4$	C_4^2	$2C_2'$	$2C_2''$		Fig. 2.23(c)
622	D_6	E	$2C_6$	$2C_3$	C_2	$3C_2'$	$3C_2''$	Fig. 2.23(d)

The small circle in the center of each axial symbol in Fig. 2.25 indicates an inversion center resulting from the combination of an even order rotational axis and a mirror plane.

Up to this point we have generated a total of twenty-one point groups from a single axis, or a combination of axes and one other symmetry element. In order to generate additional point groups we must add another symmetry element to those already combined. This can be accomplished by combining a horizontal mirror plane with the symmetry elements in the D_n or C_{nv} groups *or* n vertical mirror planes or diad axes to the symmetry elements in the C_{nh}point group. An n-fold principal axis, a horizontal mirror plane, n diad axes and n vertical mirror planes are the resulting symmetry elements for each of the four point groups created by

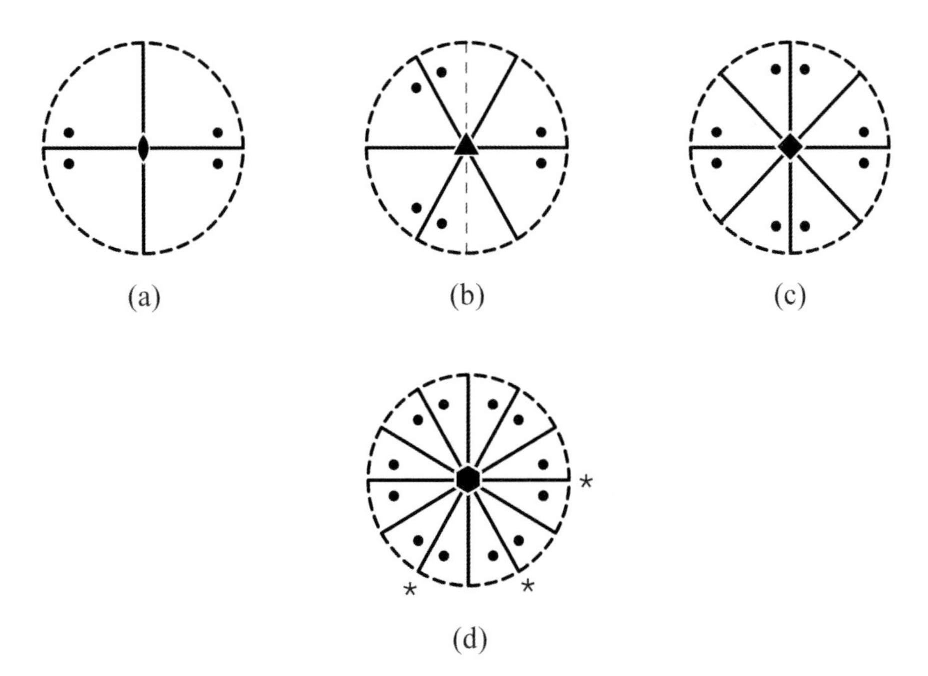

Figure 2.24 Point group stereograms: (a) $2mm$ (C_{2v}), (b) $3m$ (C_{3v}), (c) $4mm$ (C_{4v}), and (d) $6mm$ (C_{6v}).

Table 2.7 Crystal Point Groups for Lattices with n Vertical Mirror Planes Parallel to an n-Fold Principal Axis.

H-M	Sch	Symmetry Operations						Stereogram
$2mm$	C_{2v}	E	C_2	$\sigma_v(xz)$	$\sigma'_v(yz)$			Fig. 2.24(a)
$3m$	C_{3v}	E	$2C_3$	$3\sigma_v$				Fig. 2.24(b)
$4mm$	C_{4v}	E	$2C_4$	C_2	$2\sigma_v$	$2\sigma_d$		Fig. 2.24(c)
$6mm$	C_{6v}	E	$2C_6$	$2C_3$	C_2	$3\sigma_v$	$3\sigma_d$	Fig. 2.24(d)

any of these combinations (Table 2.9). The combination of a mirror plane and an n-fold principal axis creates an inversion center for $n = 2$, 4, and 6. The mirror plane perpendicular to the triad axis creates a 6-fold rotoinversion axis. For $n = 2$, there are three mirror planes at right angles to one another. All of the reflection operations are in different classes. There are also three perpendicular diads at right angles. The point group would formally be indicated as $2/mmm$, but since the selection of a principal axis in this case is arbitrary (any diad will do), the symbol is simply mmm. The point groups involving the tetrad and hexad principal axes each have two classes of vertical mirror planes, and are denoted $4/mmm$ and $6/mmm$, respectively. The combination of a triad axis and a mirror plane creates a $3/m \equiv \bar{6}$ rotoinversion axis. All of the vertical mirror reflections are in the same class in this

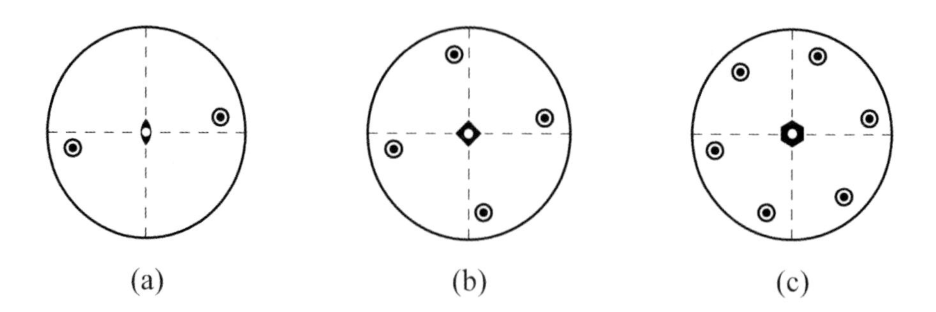

Figure 2.25 Point group stereograms: (a) $2/m$ (C_{2h}), (b) $4/m$ (C_{4h}), and (c) $6/m$ (C_{6h}).

Table 2.8 Crystal Point Groups for Lattices with a Horizontal Mirror Plane Perpendicular to an n-Fold Principal Axis.

H-M	Sch	Symmetry Operations	Stereogram
$2/m$	C_{2h}	E C_2 i σ_h	Fig. 2.25(a)
$4/m$	C_{4h}	E C_4 C_2 C_4^3 i S_4^3 σ_h S_4	Fig. 2.25(c)
$6/m$	C_{6h}	E C_6 C_3 C_2 C_3^2 C_6^5 i S_3^5 S_6^5 σ_h S_6 S_3	Fig. 2.25(d)

point group and its point group notation is $3/mm2 \equiv \bar{6}m2$. Although the diad axes are generated by the rotoinversion axis and the vertical mirror planes, they are important in the description of the crystal point symmetry, and are retained in the symbol.

At this stage it would appear that we have exhausted all of the possible combinations of mirror planes and diad axes. However, in the point groups already established the vertical mirror planes contain the diad axes. We can create another set of point groups by rotating the mirror planes by angles of $\pi/2n$ with respect to the diad axes such that the mirror planes bisect the angle between the diads. The diads also bisect the dihedral angles between the mirror planes, and the mirror planes are referred to as *dihedral mirror planes.* Four new point groups are generated, but only two of them are crystallographic point groups (Table 2.10). Reference to Fig. 2.27 provides an explanation. The addition of dihedral planes to a two-fold principal and two perpendicular diads produces the stereogram in Fig. 2.27(a). The relocation of the mirror planes has created another symmetry element — an S_4 rotoreflection axis. This is equivalent to a $\bar{4}$ rotoinversion axis, and the point group is labeled $\bar{4}2m$. The addition of dihedral planes to a three-fold axis with three perpendicular diad axes creates an S_6 rotoreflection axis, which is equivalent to a $\bar{3}$ rotoinversion axis (Fig. 2.27(b)). The vertical reflections are all in

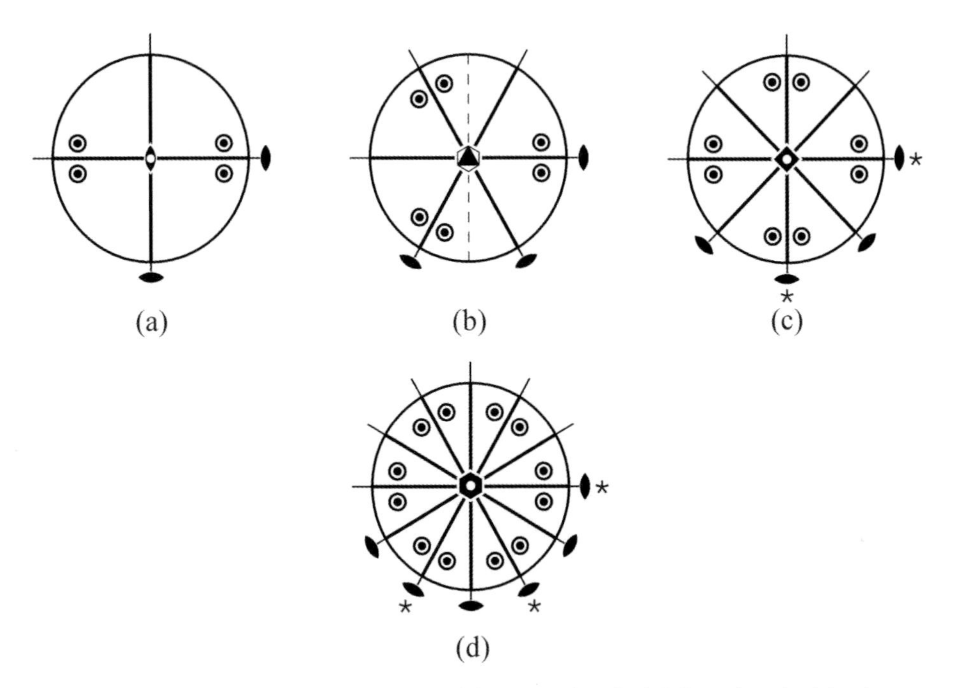

Figure 2.26 Point group stereograms: (a) mmm (D_{2h}), (b) $\bar{6}m2$ (D_{3h}), (c) $4/mmm$ (D_{4h}), and (d) $6/mmm$ (D_{6h}).

Table 2.9 Crystal Point Groups for Lattices with a Horizontal Mirror Plane and n Diad Axes Perpendicular to an n-Fold Principal Axis.

H-M	Sch	Symmetry Operations	Stereogram
mmm	D_{2h}	$E\,C_2(x)C_2(y)C_2(z)\quad i\quad \sigma(xy)\sigma(xz)\sigma(yz)$	Fig. 2.26(a)
$\bar{6}m2$	D_{3h}	$E\ 2C_3\ \ 3C_2\quad \sigma_h\ \ 2S_3\ \ 3\sigma_v$	Fig. 2.26(b)
$4/mmm$	D_{4h}	$E\ 2C_4\quad C_2\quad 2C_2'\ 2C_2''\quad i\quad 2S_4\quad \sigma_h\ \ 2\sigma_v 2\sigma_d$	Fig. 2.26(c)
$6/mmm$	D_{6h}	$E\ 2C_6\quad 2C_3\quad C_2\ \ 3C_2'\ 3C_2''\quad i\quad 2S_3\ 2S_6\ \sigma_h\ 3\sigma_d 3\sigma_v$	Fig. 2.26(d)

the same class, as are the diad rotations. The mirrors and the triad rotoinversion axis are sufficient to define the point group, $\bar{3}m$. Note that the $\bar{3}m$ point group is centrosymmetric (D_{nd} point groups are centrosymmetric for odd n). In general, the addition of n dihedral mirror planes to an n-fold principal axis and n perpendicular diads creates an S_{2n} rotoreflection axis. As illustrated in Figs. 2.27(c) and (d), the tetrad and hexad principal axes are collinear with 8-fold and 12-fold rototreflection axes. Since the rotational symmetry in a lattice is constrained to be six-fold or less, neither of these point groups can be a crystallographic point group. We have now exhausted possible combinations of diad, triad, tetrad and hexad axes with diad axes and mirror planes in creating 27 crystallographic point groups. An analysis of combinations of principal axes with triad axes is the next logical step. We have already analyzed the combination of triads and diads, result-

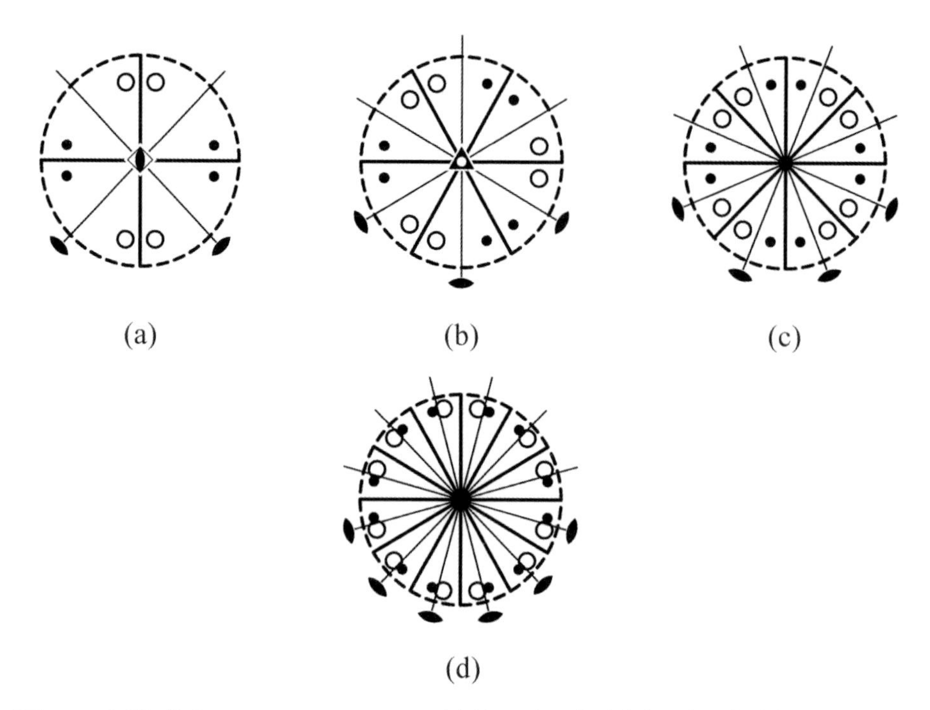

Figure 2.27 Point group stereograms: (a) $\bar{4}2m$ (D_{2d}), (b) $\bar{3}m$ (D_{3d}), (c) D_{4d}, and (d) D_{6d}.

Table 2.10 Crystal Point Groups for Lattices with n Diad Axes Perpendicular to an n-Fold Principal Axis and n Dihedral Mirror Planes.

H-M	Sch	Symmetry Operations						Stereogram
$\bar{4}2m$	D_{2d}	E	$2S_4$	C_2	$2C_2'$	$2\sigma_d$		Fig. 2.27(a)
$\bar{3}m$	D_{3d}	E	$2C_3$	$3C_2$	i	$2S_6$	$3\sigma_d$	Fig. 2.27(b)

ing in the 222, 32(2), 422, and 622 point groups. Fig. 2.28 shows stereograms of the combination of a triad principal axis with another triad axis in the xy plane rotated 0°, 30°, and 45°. These axes are obviously *not* symmetry axes. Attempts to find other angles in the xy plane prove equally fruitless, and we are compelled to conclude that the triad axes must be at some angle other than 90° with respect to one another. Furthermore, as the orders of the combined rotation axes increase, the possible location of symmetry-equivalent points becomes very constrained, since they must be in positions where rotations about multiple high order axes bring all of the vectors into one another. In other words, the orientation of the multiple three-fold axes will be critical to the generation of point groups containing them. The location of the triad axes must allow for the transformation of a general vector into a symmetry-equivalent one. To simplify the analysis, we consider the special case where the vector is collinear with one of the coordinate axes. It follows that symmetry-equivalent vectors also lie along the coordinate axes. If a rotation

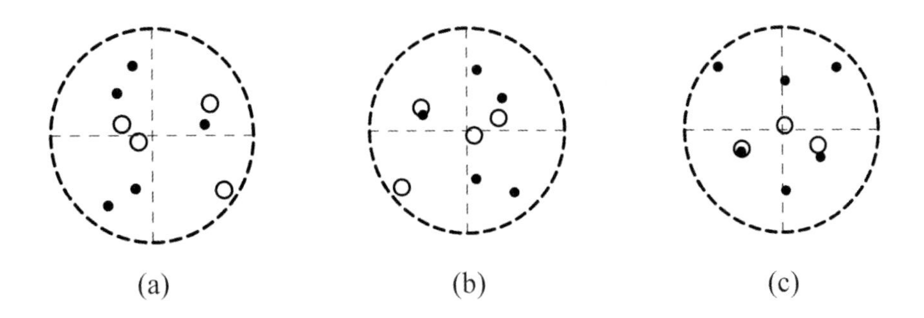

Figure 2.28 Stereograms of combinations of triad axes, one axis along z and the other rotated about z in the xy plane at (a) 0°, (b) 30°, and (c) 45°.

about a triad axis places the transformed vector at any other position, there exists no combination of three-fold rotations about any other axis or axes that will return the vector to its original location along the axis. Fig. 2.29(a) incorporates red spheres to represent the heads of the symmetry-equivalent vectors aligned with the coordinate axes. If the centers are connected with line segments, the resulting polygon is an *octahedron* as illustrated in the insert. Each of the eight faces is an equilateral triangle — connecting the center of any face with the origin creates a triad axis. Fig. 2.29(b) is a view along one of the four triad axes in the octahedron; a stereogram for this view illustrates the three-fold rotational symmetry. Note that the octahedron is reoriented so that the triad axis is coincident with the z axis in the stereographic projection. The construction of a cube with the vertices (red spheres) of the octahedron in the center of the faces provides the means for us to determine the orientation of the triad axis. The product of this construction is shown in Fig. 2.30(a). The stereogram is a view along any of the coordinate axes, indicating a tetrad axis along each coordinate axis. This view also illustrates that each of the coordinate planes is a mirror plane. Fig. 2.30(b) is a different view along the body diagonal of the cube, showing that the diagonal is coincident with a triad axis for the *combination* of the octahedron and cube. If the edges of the cube are of unit length, the diagonal of the cube is $\sqrt{2}$, and the length of the body diagonal is $(1^2 + (\sqrt{2})^2)^{\frac{1}{2}} = \sqrt{3}$. The cosine of the angle that the triad axis makes with the z axis is $1/\sqrt{3}$. This angle, $\approx 54.74°$, is very important in spectroscopy, and is often termed the *magic angle.*

The octahedron and cube apparently have the same symmetry: each of the coordinate axes is a four-fold axis, each coordinate plane is a mirror plane, and the four triad axes are the same in both polygons. Fig. 2.30(c) illustrates the construction of a third polygon with a triad axis at the magic angle. Removal of opposing vertices from the cube creates a *tetrahedron* oriented so that the triad axis passes through a vertex of the polygon and the center of its base — an equilateral triangle. The stereogram, viewed down any of the coordinate axes shows that there is a diad along each coordinate axis, but not a tetrad. The view and stereogram

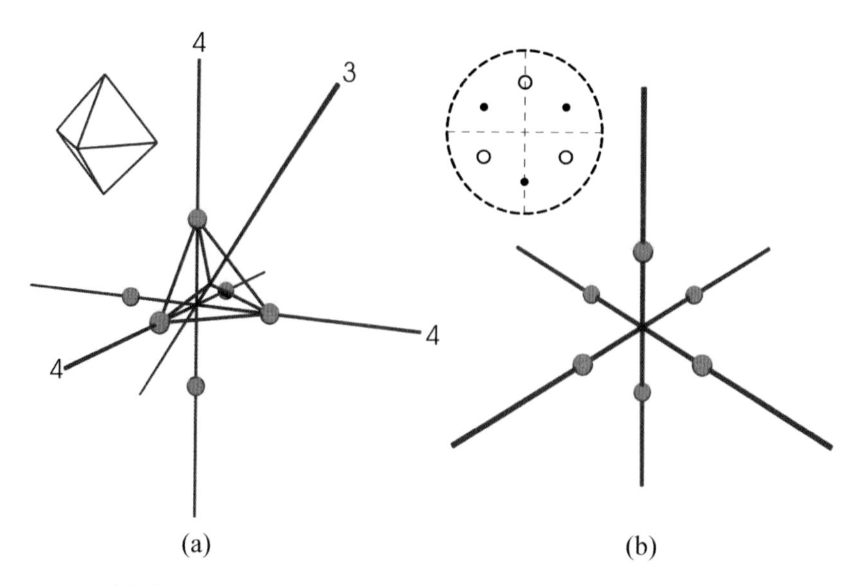

Figure 2.29 (a) Octahedron formed from symmetry-equivalent vectors along the Cartesian axes. (b) View and stereographic projection along the triad axis.

in Fig. 2.30(d) result from reorienting the tetrahedron so that the triad axis is coincident with the z axis.

The symmetry associated with the octahedron (and cube) is appropriately labeled *octahedral symmetry*; the symmetry associated with the tetrahedron is likewise referred to as *tetrahedral symmetry.* To analyze both of these in general, we now relax the constraint that the vector undergoing transformation must lie along a coordinate axis. Application of the diad operation for tetrahedral symmetry to a general vector, followed by rotation about the triad produces the stereogram in Fig. 2.31(a). The triad axes are indicated with symbols inside the great circle, since they intersect the sphere inside each of the octants of the coordinate system. Note that traditional stereograms place the symbols for all axes at their projected intersection points, usually on the great circle or in the center. This often makes the stereogram unnecessarily complicated, and has been avoided to this point, but the existence of multiple triads makes it necessary here. All of the triad axes are transformed into one another by the diad axes; the diads are all transformed into one another by the triads. Thus a single diad and triad are all that are necessary to determine the point group, 23. This should not be confused with the 32 point group. The principal axis, coincident with a coordinate axis is the initial number in the H-M short form. The stereogram for a general vector transformed with octahedral symmetry is shown in Fig. 2.31(b). In order to generate all of the symmetry elements a tetrad, triad, and diad axis are necessary, creating the 432 point group.

The final three point groups are created by replacing rotation axes in the 32 and 432 point groups with rotoinversion axes. Replacement of the triad axes in the 32 group with $\bar{3}$ axes generates mirror planes perpendicular to the two-fold axes in each of the coordinate planes, as illustrated in Fig. 2.11(c). The $\bar{3}$ axis creates a center of symmetry (Fig. 2.19(c)), and this, in turn produces the mirror planes. The same point group can be generated from the mirror planes and a triad axis since the mirror planes alone impose a center of symmetry. All of the

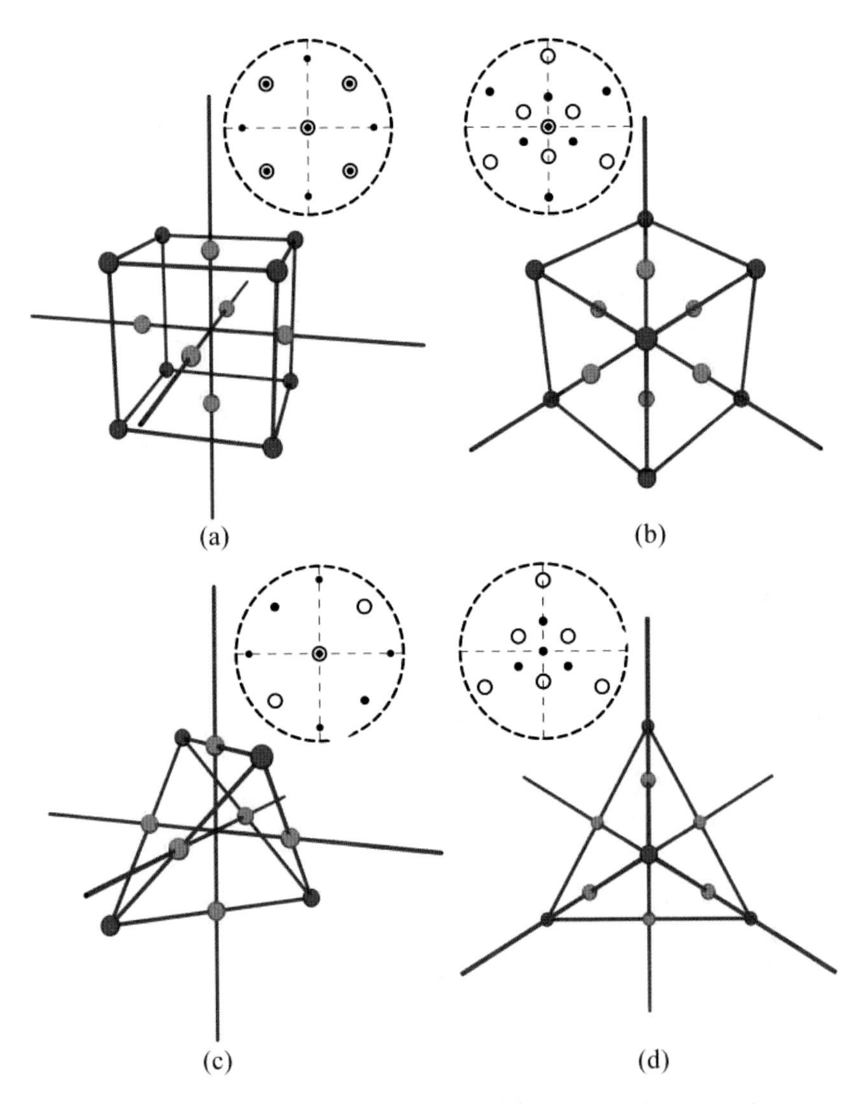

Figure 2.30 (a) Cube enclosing the octahedron. (b) View and stereographic projection along the body diagonal/triad axis of the cube. (c) Tetrahedron formed by removal of four opposing vertices of the cube. (d) View and stereographic projection along the triad axis in the tetrahedron.

planes and diads are transformed into one another by the threefold rotations. The principal axis is contained in the mirror so that a mirror plane and triad axis are sufficient to generate all of the symmetry elements — the 23 point group has been transformed into $m3$. If the diad axes in the 32 point group are replaced with $\bar{4}$ axes, two dihedral mirror planes are generated, bisecting the angle between the $\bar{4}$ axes (Fig. 2.31(d)). The principal axis is now a four-fold rotoinversion axis and the point group is labeled $\bar{4}3m$, which is the point group of the tetrahedron. Finally, the replacement of the triad axes in the 432 group with $\bar{3}$ axes produces horizontal mirror planes in the coordinate planes, and dihedral mirror planes bisecting the dihedral angles between them. The two sets of reflection operations are in different

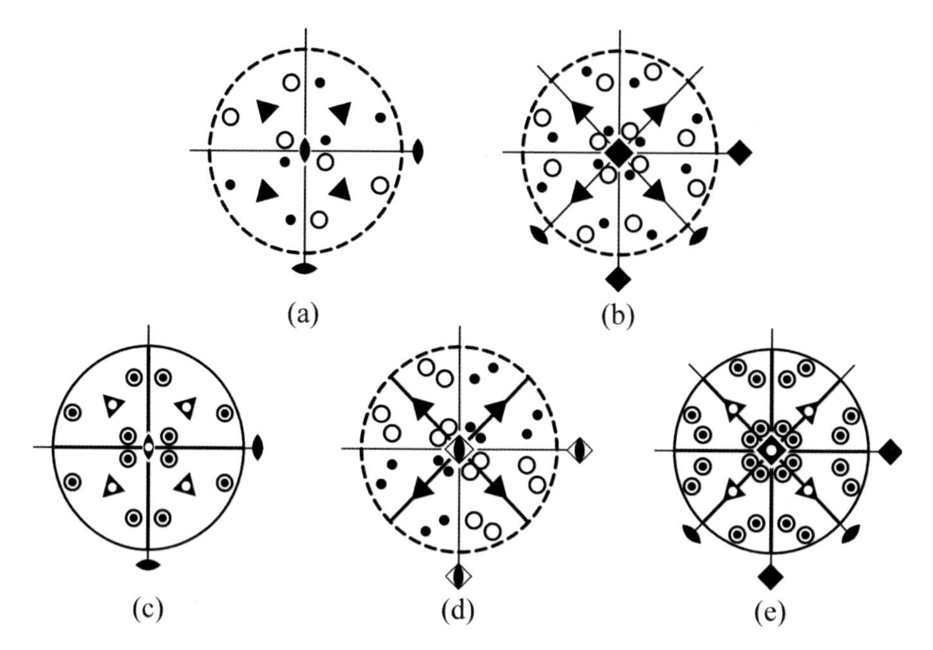

Figure 2.31 Point group stereograms: (a) 23 (T), (b) 432 (O), (c) $m3$ (T_h), (d) $\bar{4}3m$ (T_d), and (e) $m3m$ (O_h).

Table 2.11 Crystal Point Groups for Lattices with Multiple 3-fold Axes: The Cubic Point Groups.

H-M	Sch	Symmetry Operations	Stereogram
23	T	E $4C_3$ $4C_3^2$ $3C_2$	Fig. 2.31(a)
432	O	E $6C_4$ $3C_4^2$ $8C_3$ $6C_2$	Fig. 2.31(b)
$m3$	T_h	E $4C_3$ $4C_3^2$ $3C_2$ i $4S_6$ $4S_6^5$ $3\sigma_h$	Fig. 2.31(c)
$\bar{4}3m$	Td	E $8C_3$ $3C_2$ $6S_4$ $6\sigma_d$	Fig. 2.31(d)
$m3m$	O_h	E $8C_3$ $6C_2$ $6C_4$ $3C_4^2$ i $6S_4$ $8S_6$ $3\sigma_h$ $3\sigma_d$	Fig. 2.31(e)

classes. As with the $m3$ group, the $\bar{3}$ axis creates an inversion center. Again, the mirror planes alone also impose a center of symmetry. The symmetry elements can therefore be generated from a dihedral mirror, a horizontal mirror, and a triad axis (which becomes a $\bar{3}$ axis when the inversion center is created from the reflections). The point group symbol is therefore $m3m$. This is the point group of the cube *and* the octahedron. As we have just observed, these final five "high symmetry" point groups, giving us a total of 32, are all intimately related to the cube and are classified as the *cubic point groups*. The reader familiar with the packing of spheres will recognize the combination of the cube and octahedron in Fig. 2.30(a) as a *face-centered cubic unit cell* — a common unit cell for the elements. It is also the arrangement of the large anions in a significant number of ionic compounds.

2.3.7 The Symmetry Classification of Crystal Systems.

As discussed previously, a three-dimensional lattice can be constructed by stacking equidistant, parallel, two-dimensional lattice planes along a vector that is not parallel to the planes(Fig. 2.14). The two-dimensional lattice has two-fold and inversion symmetry, and while the inversion symmetry is retained when the stacking occurs, the two-fold symmetry is lost if the translation direction is not perpendicular to the two-dimensional planes. Since lattices are mathematical constructs represented by the points at the intersection of three sets of parallel planes, all lattices are centrosymmetric. However, the crystal may not reflect that symmetry if the contents of each unit cell are not related by a center of symmetry. Fig. 2.32 illustrates this for the two-dimensional case. If these two-dimensional lattices are stacked in a non-orthogonal direction the unit cells will have no special constraints placed on their geometries. In the general case all of the unit cell angles and axial lengths will be different; the resulting cell is said to belong to the *triclinic* crystal system and is called a *triclinic* unit cell (Fig. 2.33(a)). Neither structure will exhibit two-fold rotational symmetry. If the contents of the three-dimensional unit cell are related by a center of symmetry, then the entire structure will be centrosymmetric and will belong to the $\bar{1}$ point group. Otherwise the structure will have only the trivial symmetry of the 1 point group.

If the translation direction *is* perpendicular, a unit cell with an axis perpendicular to the other axes is created — *the geometry of the unit cell is determined by the symmetry of the lattice.* The resulting structure will have two-fold rotational symmetry, but will have additional symmetry only if the contents of the unit cells are related by that symmetry. In other words, *the lattice point symmetry is the limiting point symmetry of the structure, which may be decreased if the unit cell contents do not reflect that symmetry (the unit cell contents are related by lower symmetry than the lattice points).* The unit cell that results from a perpendicular translation has no constraints on its axial lengths, but now has a *unique axis* that is perpendicular to the plane containing the other two axes. The b axis is ordinarily assigned to the perpendicular axis, although in earlier times the c axis was selected

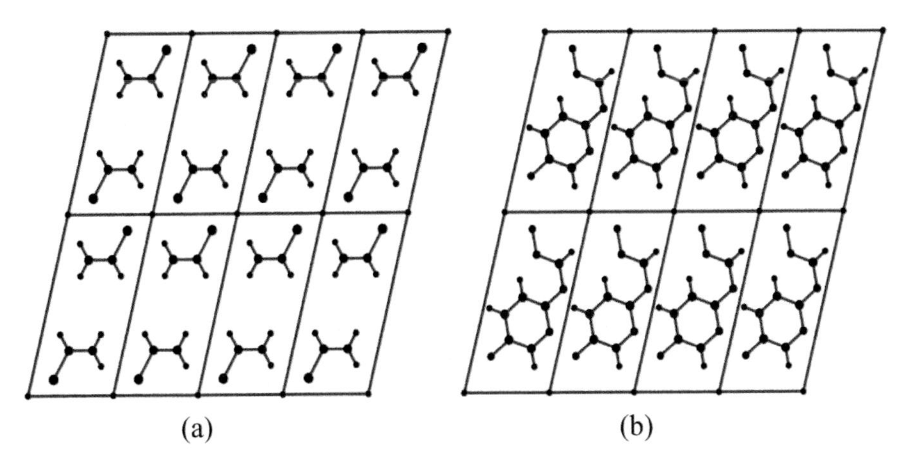

Figure 2.32 Two-dimensional lattices: (a) Unit cell contents related by a diad axis and inversion center. (b) Unit cell contents related only by a monad axis.

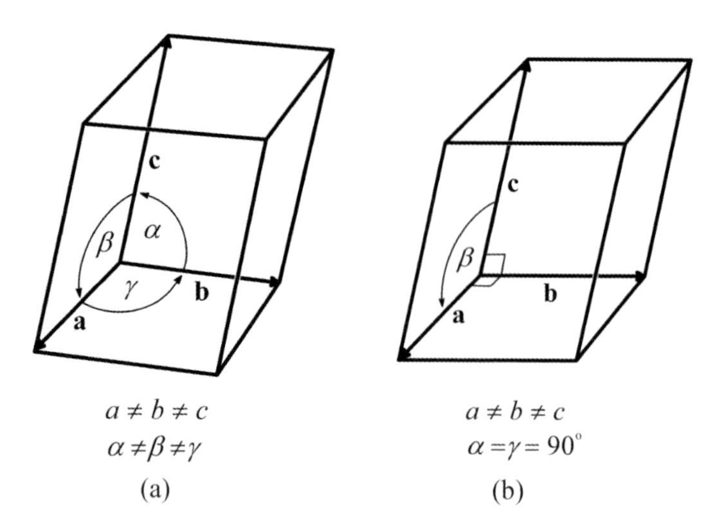

Figure 2.33 (a) Triclinic unit cell. (b) Monoclinic unit cell. The b axis is unique.

as the unique axis. The latter case is often referred to as the "first setting", while the former case is referred to as the "second setting." We will use the second setting exclusively; the interaxial angles α and γ in a monoclinic unit cell are both 90° (Fig. 2.33(b)). Structures with two-fold rotational symmetry are members of the *monoclinic* crystal system, provided that there is no additional two-fold or higher rotational symmetry. The *lattice* for the monoclinic system has two-fold rotational symmetry and inversion symmetry, and therefore must have mirror symmetry as well; the two-fold axes are perpendicular to the mirror planes. The monoclinic *crystal* system encompasses the 2, m, and $2/m$ point groups. The structure will belong to either the 2 or m point groups if it lacks a center of symmetry, depending on whether the axis is a 2 or $\bar{2}$ axis, and will belong to the $2/m$ point group if it is centrosymmetric.

If the axes in the original two-dimensional lattice are perpendicular to one another *and* the translation is also in an orthogonal direction, all three axial directions will have two-fold rotational symmetry. There are still no constraints on the axial lengths, but the interaxial angles are *all* now constrained to be 90° (Fig. 2.34(a)). Crystals with two-fold symmetry in all three axial directions but with no higher axial symmetry are members of the *orthorhombic* crystal system. Orthorhombic structures have 222 or $mm2$ symmetry if they are non-centrosymmetric, and have mmm symmetry if they are centrosymmetric.

If the original two-dimensional unit cell has axes that are *both* perpendicular *and* the same length, then the rotational axis is a tetrad axis, and perpendicular translation creates a unit cell with two equal axes, 90° interaxial angles(Fig. 2.34(b)), and four-fold symmetry along the translation axis. The third axial length is unconstrained, and is therefore the unique axis in this case. The crystal system is the *tetragonal* system, and accounts for the 4, $\bar{4}$, $4/m$, 422, $4mm$, $\bar{4}$2m, and $4/mmm$ point groups. The c axis is assigned as the unique axis in the tetragonal unit cell.

If, in addition to the above constraints, the translation distance is also equal to the length of the axes in the original two-dimensional lattice then the unit cell

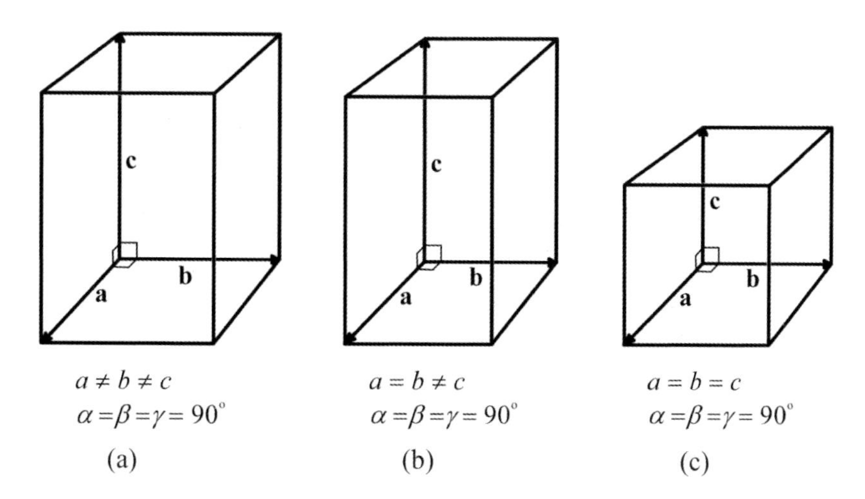

Figure 2.34 (a) Orthorhombic unit cell. (b) Tetragonal unit cell. The c axis is the unique tetrad axis. (c) Cubic unit cell.

becomes a cube (Fig. 2.34(c)) and is a member of the *cubic* crystal system, encompassing the 23, $m3$, 432, $\bar{4}3m$, and $m3m$ point groups.

The choice of the unit cells in a crystal lattice is based on the criterion that the contents of all of the unit cells in the lattice are identical. This choice is far from unique, as the two-dimensional examples in Fig. 2.35 illustrate. The lattice is the same as that depicted in Fig. 2.32(a). The unit cells in Fig. 2.35(a) have a longer axis than those in Fig. 2.32(a). Both unit cells have two-fold rotational symmetry, are centrosymmetric, and have origins at every lattice point. There is nothing to be gained by selecting the more complex cell with the longer axis, since both reflect the crystal symmetry, and the original unit cell would clearly be the preferable one. Unit cells with origins coincident with every lattice point are known as *primitive* unit cells. In addition to the primitive cells in the lattice, unit cells which have lattice points in the center of the cell are also possible choices, as shown in Fig. 2.35(b). As required, the cell contents of these larger unit cells are identical, and the two-fold and inversion symmetry of the original structure is also retained. Unit cells with lattice points inside the unit cell, usually in the center of a face or interior of the cell, are known as *centered* unit cells. In this example, there is nothing to be gained by choosing the larger unit cell, and the original primitive unit cell with the shortest independent axial vectors would remain the unit cell of choice. Since such a unit cell exists in every lattice (in three dimensions as well), one might surmise that this would always be the choice. However, in the mid-1800's, M.L. Frankenheim, a German physicist, made a systematic study of lattices, and determined that there were lattices for which the selection of a centered unit cell reflected more symmetry than the primitive unit cell. Fig. 2.36 illustrates this for a two-dimensional lattice. The primitive unit cells on the left exhibit no apparent symmetry, while the centered cell on the right has the full symmetry of the lattice — including mirror symmetry through the center of the unit cell. The selection of the higher symmetry unit cell is nearly always advisable. The determination of a crystal structure requires the location of electron density maxima (atoms) in the

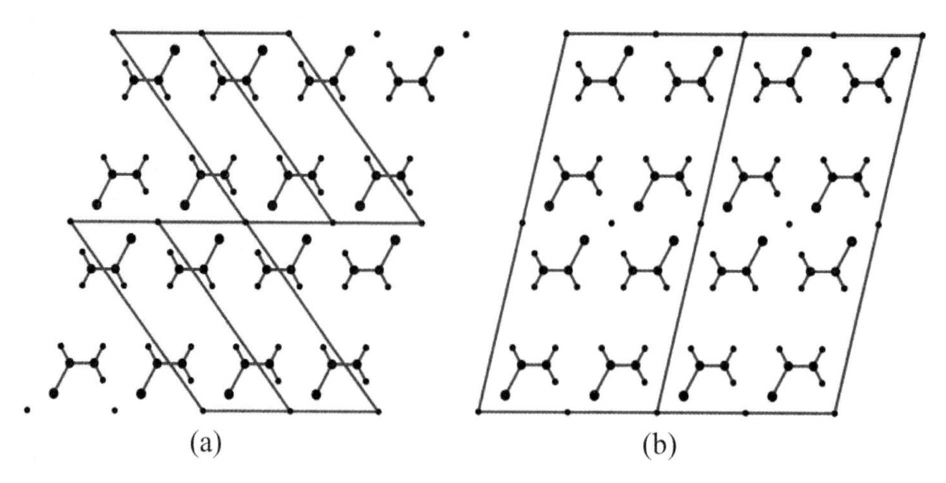

Figure 2.35 Alternative unit cell choices for a two-dimensional lattice: (a) Primitive unit cell. (b) Centered unit cell.

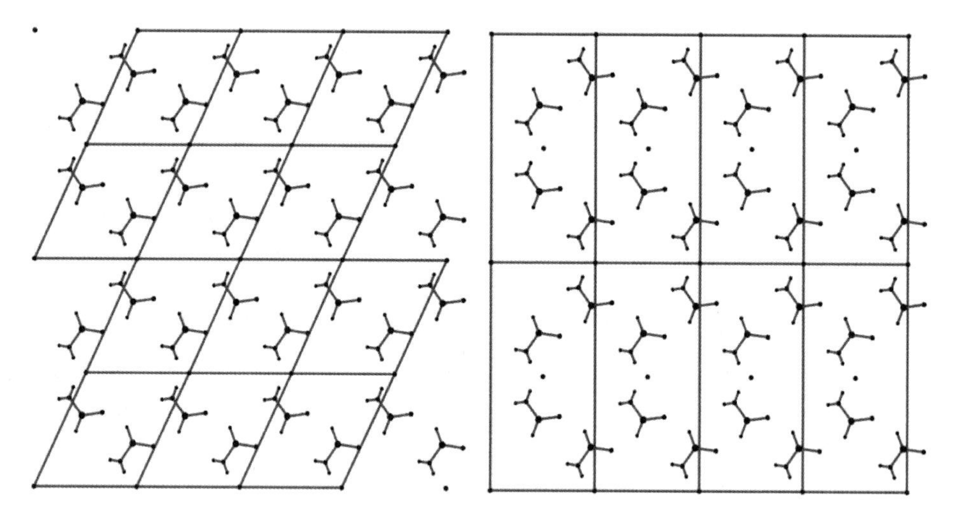

Figure 2.36 Two-dimensional lattice illustrating a lower-symmetry primitive unit cell on the left, and a higher-symmetry centered unit cell on the right.

unit cell, and the more of these atoms that are related to one another by symmetry, the fewer that we have to find. Perhaps of even more importance is the nature of the diffraction pattern which results from the interaction of X-radiation with the periodically arranged electron density in a crystal. The diffraction pattern contains the structural information that we must extract to solve the crystal structure. The crystal symmetry is reflected in the diffraction pattern, and it therefore makes sense to select a coordinate system which reflects the symmetry as well. Frankenheim determined that there were a total of 15 possible *space lattices* if centered unit cells were included. Unfortunately (for Frankenheim), in 1848 A. Bravais, a French physicist, determined that two of Frankenheim's lattices were redundant; the 14 possible space lattices are known as the *Bravais lattices*.

In the Bravais lattices, centered unit cells are symbolized by the location of the "internal" lattice points. If a lattice point is in the interior of the cell, specifically at $[x_f\ y_f\ z_f] = [\frac{1}{2}\ \frac{1}{2}\ \frac{1}{2}]$, then the cell is called an *I-centered* cell. This type of unit cell is also known as a *body-centered* unit cell. If the lattice point is in the center of only one of the three unique faces of a unit cell then one of the unit cell axes will be parallel to a vector connecting the lattice points in two opposing faces. The parallel axis defines the type of centering in the *base-centered unit cell.* For example, in a *C-centered* cell the line segment connecting the internal lattice points is parallel to the c axis; the points are located at $[x_f\ y_f\ z_f] = [\frac{1}{2}\ \frac{1}{2}\ 0]$. A- and B-centered unit cells have internal lattice points at $[x_f\ y_f\ z_f] = [\ 0\ \frac{1}{2}\ \frac{1}{2}]$ and $[\frac{1}{2}\ \ 0\ \frac{1}{2}]$, respectively. If there is a lattice point in the center of each face then the unit cell is termed an *F-centered* cell. The primitive unit cell is symbolized with a "P".

Triclinic and Monoclinic Crystal Systems. There are no unique lattices with centered triclinic unit cells. A larger unit cell with a centered lattice point will have no additional symmetry, and a smaller primitive cell will always be found in the triclinic lattice. In a monoclinic lattice, if an internal lattice point is in the ab plane, then the line segment connecting the points will parallel the c axis. As illustrated in Fig. 2.37(a), any primitive cell that is formed by placing these points at a unit cell origin will have no 90° interaxial angles. The choice of the a and c axes for the monoclinic unit cell is arbitrary, and a base-centered monoclinic unit cell can be described as either a C-centered unit cell or an A-centered unit cell, depending on whether the line segment connecting the points in the center of the bases parallels the c axis or the a. The centered unit cell is of higher symmetry than the primitive triclinic unit cell, and the lattice which defines it is one of the unique Bravais lattices. If the faces containing the internal lattice points lie in the ac plane then a B-centered unit cell would result. However, as shown in Fig. 2.37(b), the primitive unit cell formed by placing the lattice points at its origin is also a monoclinic unit cell; both cells share a common unique axis. It follows that there is nothing to be gained by selecting a B-centered unit cell. Similarly, as demonstrated in Figs.2.37(c) and (d), F-centered monoclinic unit cells and I-centered monoclinic unit cells can both be reduced to C-centered (or A-centered) monoclinic unit cells. There are therefore two unique unit cell types in the monoclinic system, P and C (or A).

Orthorhombic Crystal Systems. In the orthorhombic crystal system, base-centering can occur along any of the three unit cell axes. In Fig. 2.38(a) the primitive unit cell in such a lattice is shown to be a primitive monoclinic unit cell with its unique axis coincident with the "centering" axis. F- and I-centered orthorhombic unit cells contain base-centered monoclinic unit cells (Figs. 2.38(b) and (c)). The orthorhombic unit cell is of higher symmetry in all three cases; in addition to P unit cells, C (or A or B), F, and I unit cells are all unique in the orthorhombic system.

Tetragonal Crystal Systems. At first glance, it would appear that the tetragonal system would look very much like the orthorhombic system depicted in Fig. 2.38, with two of the axes in the orthorhombic unit cell becoming equal in length to form the tetragonal unit cell. In Fig. 2.39(a), this appears to be the case; the I-centered

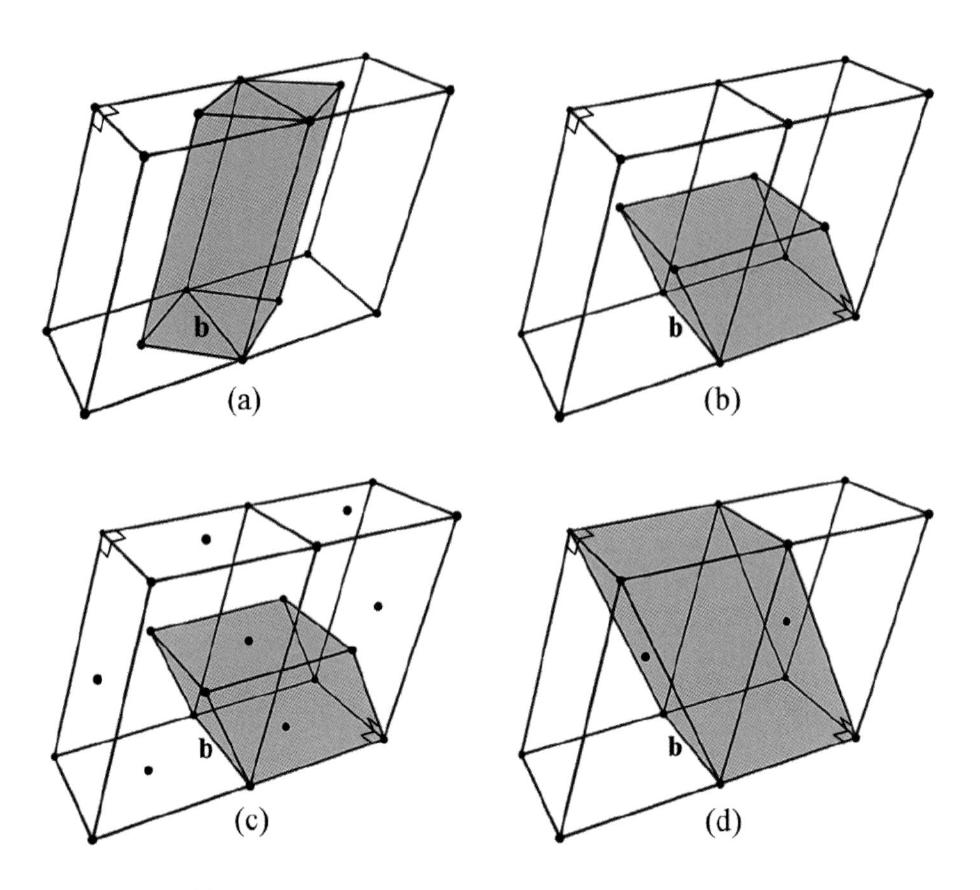

Figure 2.37 (a) C- or A-centered monoclinic unit cell with primitive triclinic unit cell. (b) B-centered monoclinic unit cell with primitive monoclinic unit cell. (c) F-centered monoclinic unit cell with C-centered (or A-centered) monoclinic unit cell. (d) I-centered monoclinic unit cell with C-centered (or A-centered) monoclinic unit cell.

tetragonal unit cell contains a base-centered monoclinic unit cell and is unique. However, the four-fold symmetry in the tetragonal system creates diagonals across the square base of the unit cell that are equal in length and perpendicular to one another. The result for the F-centered tetragonal unit cell is another tetragonal unit cell containing a single lattice point in its center, i.e., an I-centered unit cell, as illustrated in Fig. 2.39(b). The four-fold symmetry in the tetragonal unit cell requires that an A-centered unit cell must also be B-centered. If a unit cell is A-centered, addition of the vector $[\,0\ \frac{1}{2}\ \frac{1}{2}]$ to the coordinates of any lattice point will generate another lattice point. If the unit cell is B-centered, the addition of $[\frac{1}{2}\ 0\ \frac{1}{2}]$ to a lattice point will generate another lattice point. Beginning with the lattice point at the origin, $[0\ 0\ 0]$, the A-centering operation will generate a lattice point at $[\,0\ \frac{1}{2}\ \frac{1}{2}]$. Applying a B-centering operation to that lattice point will create another lattice point at $[\,0\ \frac{1}{2}\ \frac{1}{2}]+[\frac{1}{2}\ 0\ \frac{1}{2}] = [\frac{1}{2}\ \frac{1}{2}\ 1\,] \equiv [\frac{1}{2}\ \frac{1}{2}\ 0\,]$, in the center of the "$C$-centering" face. The result is an F-centered unit cell. It follows that the only unique unit cell types for the tetragonal system are P and I.

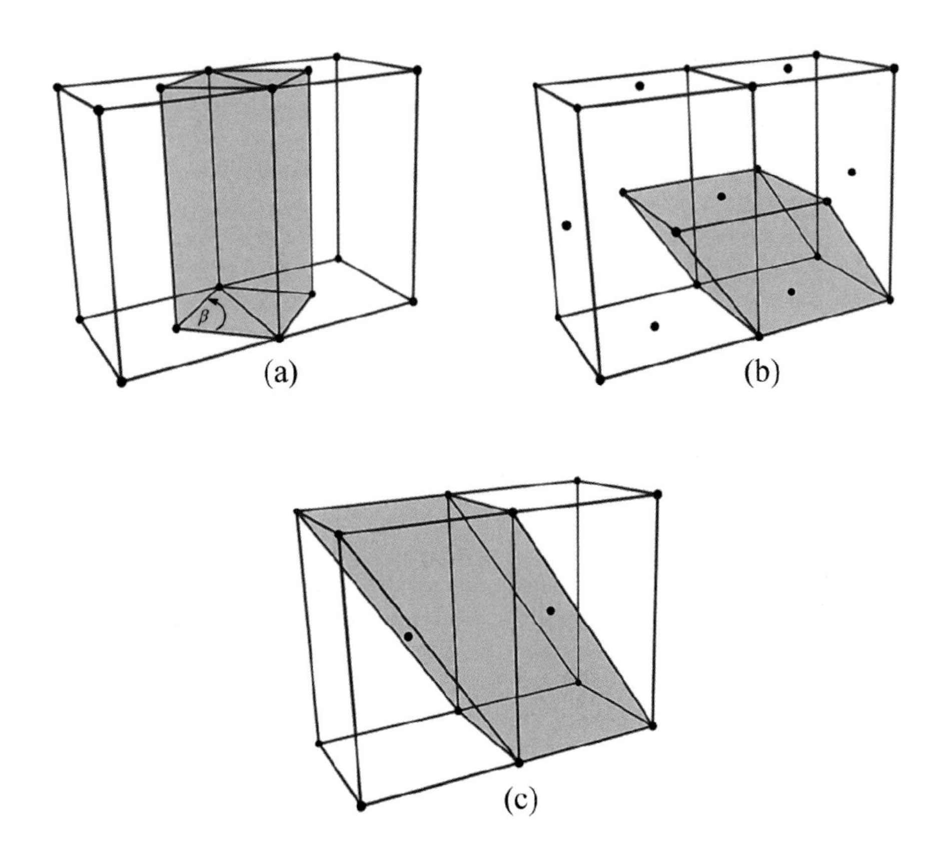

Figure 2.38 (a) C-, B-, or A-centered orthorhombic unit cell with primitive monoclinic unit cell ($\beta \neq 90^\circ$). (b) F-centered orthorhombic unit cell with base-centered monoclinic unit cell. (c) I-centered orthorhombic unit cell with base-centered monoclinic unit cell.

Cubic Crystal Systems. In the cubic system, the I-centered unit cell, like the tetragonal I-centered unit cell, contains a base-centered monoclinic unit cell (Fig. 2.39(c)). The F-centered unit cell is also similar to the tetragonal F-centered unit cell — with an I-centered *tetragonal* unit cell in the lattice(Fig. 2.39(d)). The cubic unit cell is higher in symmetry, and its lattice is a unique Bravais lattice. The symmetry of the cubic system requires that a lattice point in one face will be created in all of the other faces — F-centering is the only form of face-centering in cubic unit cells. There are P, I, and F unit cells in the cubic system. Both types of centered cubic unit cells occur commonly in higher symmetry crystals, and are often referred to as *bcc* (body-centered cubic) and *fcc* (face-centered cubic) unit cells.

Trigonal and Hexagonal Crystal Systems. Returning to the creation of three-dimensional lattices by the stacking of equidistant parallel two-dimensional lattices, we now consider the cases where the perpendicular translation axis is coincident with three-fold or six-fold rotation axes. Referring to Fig. 2.11, note that

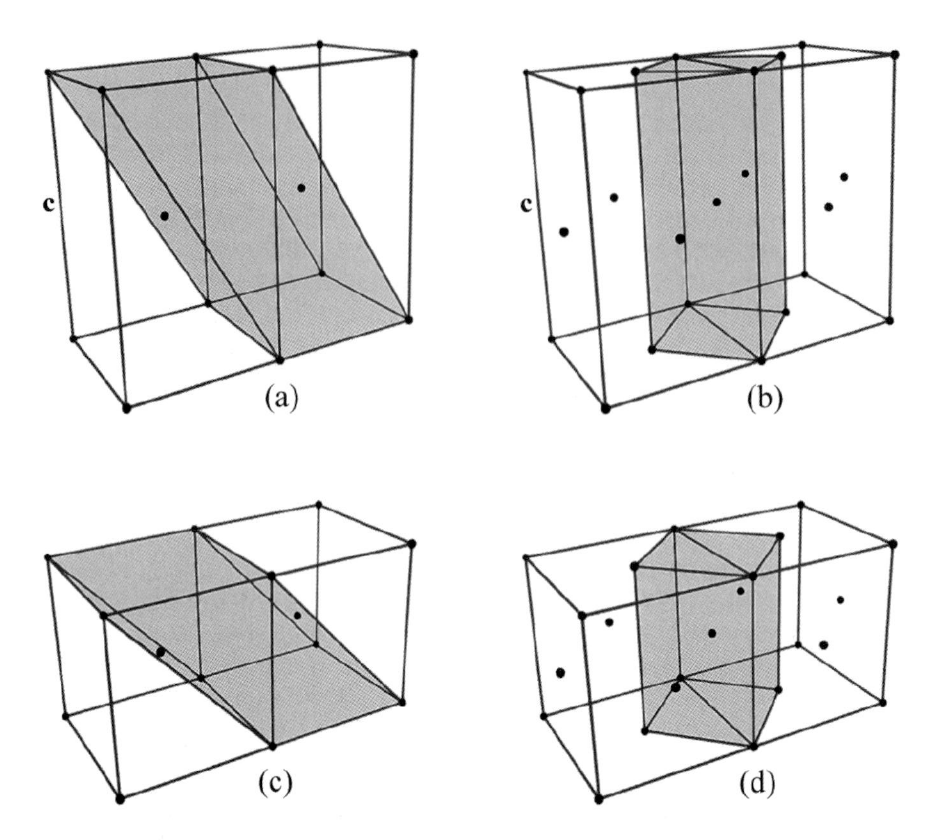

Figure 2.39 (a) I-centered tetragonal unit cell with base-centered monoclinic unit cell. (b) F-centered tetragonal unit cell with I-centered tetragonal unit cell. (c) I-centered cubic unit cell with base-centered monoclinic unit cell. (d) F-centered cubic unit cell with I-centered tetragonal unit cell.

the lattice with three-fold rotational symmetry is identical to that with six-fold symmetry! Translation in an orthogonal direction will not alter that — the lattice and unit cell for crystals with either axis are identical – the *lattice* always has six-fold (hexagonal) symmetry. This, of course, does not mean that trigonal and hexagonal crystal systems have identical symmetry – since this is defined by the *contents* of the unit cells in the lattice. As the reader might suspect, this ambiguity often leads to confusion — even among crystallographers! This confusion is compounded when we discuss the rhombohedral lattice next. Unfortunately, the terminology in crystallography is often as much historical as logical – the major source of the confusion arises from the use of the term *hexagonal axes* when describing a trigonal unit cell. *In order to avoid this problem we will use that term only when referring to unit cells with six-fold symmetry, and will use the term trigonal(H) axes when describing a unit cell with three-fold symmetry, even though the lattice and unit cell parameters are the same for both unit cells.* The reason for this will become apparent in the discussion of rhombohedral lattices. The trigonal(H)/hexagonal unit cell has two axes of equal length at an angle of 120° to one another in the plane perpendicular to the translation direction. These axes are assigned a and b

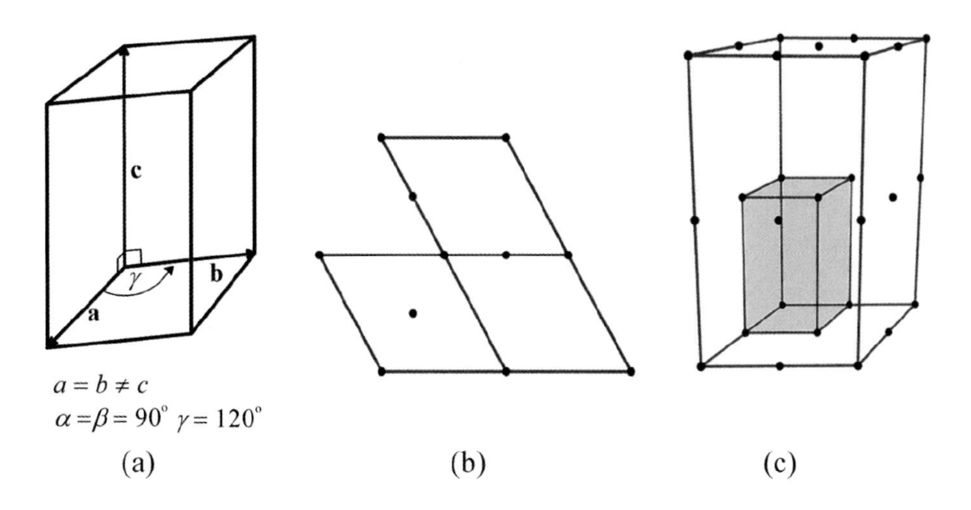

Figure 2.40 (a) Trigonal(H)/hexagonal unit cell. (b) Projection of a trigonal(H)/hexagonal unit cell showing three-fold symmetry of a centered lattice point. (c) Centered trigonal(H)/hexagonal unit cell with primitive trigonal(H)/hexagonal unit cell.

in the unit cell. The c axis is not constrained in its length, and is perpendicular to the ab plane (Fig. 2.40(a)).

To this point we have derived thirteen of the fourteen Bravais lattices. Based upon what we have already done, a centered lattice would seem an obvious choice for the last. The trigonal(H)/hexagonal unit cell might be viewed as a special case of a monoclinic unit cell, allowing it to be faced-centered along a non-unique axis. Fig 2.40(b) shows a projection of such a unit cell, looking down the c axis. One rotation about the axis places an internal lattice point in the center of the second face; a second rotation places a lattice point in the center of the unit cell. The three-fold symmetry produces lattice points at $[\,0\ \frac{1}{2}\ \frac{1}{2}]$, $[\frac{1}{2}\ 0\ \frac{1}{2}]$ and $[\frac{1}{2}\ \frac{1}{2}\ \frac{1}{2}]$ in the cell. As we have already observed in the tetragonal case, centering in two sets of parallel faces requires centering in the third set: $([\,0\ \frac{1}{2}\ \frac{1}{2}] + [\frac{1}{2}\ 0\ \frac{1}{2}] = [\frac{1}{2}\ \frac{1}{2}\ 1\,] \equiv [\frac{1}{2}\ \frac{1}{2}\ 0\,]$. The result is a combination of F- and I- centering. This creates an axis through the center of each unit cell intersecting a lattice point in increments of 1/2 along the axis. This axis must also exhibit three-fold symmetry and rotation about the axis generates lattice points at the mid-point of each unit cell axis (the projection is the same as that for the rotation of the initial face-centered lattice point — Fig 2.40(b)). These new lattice points define a new *primitive* unit cell, illustrated in Fig. 2.40(c). Thus the trigonal(H)/hexagonal unit cell must be primitive — we still have only 13 unique lattices.

Rhombohedral Crystal Systems. There is another way to stack two-dimensional lattices that introduces a three-fold axis — and results in the creation of a fourteenth Bravais lattice. Recall that a cubic unit cell has a triad axis that is not aligned with the plane faces defined by the unit cell axes. This axis is along the body-diagonal of the cube, making an equal angle of ≈ 54.74° with all three unit cell axes. If the cube is "distorted" by extending two opposing vertices in opposite directions along a body diagonal and constraining all of the edges to retain their equal lengths, the resulting geometric solid is a *rhombohedron*. The remaining six vertices move equally toward the body-diagonal when the two opposing vertices are extended, retaining the equality of the angles between the three axes and preserving the three-fold symmetry of the body diagonal. To create a lattice of rhombohedra we begin with a two-dimensional lattice with unit cell axes of equal length at angle α with respect to one another. The translation vector is selected so that the angle between the translation vector and each of the two inter-planar vectors is α and its length is equal to the lengths of the vectors in the plane. The stacking of lattice planes in this manner in illustrated in Fig. 2.41(a). Fig. 2.41(b) is a view down the three-fold axis of the *rhombohedral lattice*.

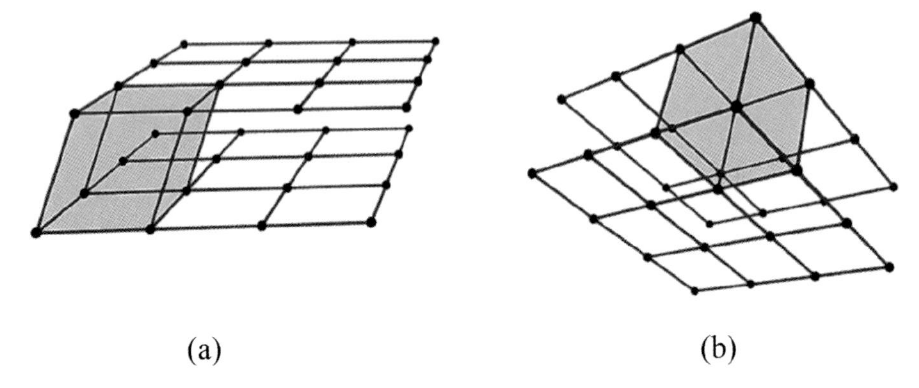

Figure 2.41 (a) Stacking of lattice planes in a rhombohedral lattice. (b) View down the triad axis in the rhombohedral lattice.

We therefore have two quite different unit cells with three-fold symmetry — the rhombohedral unit cell and the trigonal(H) unit cell. Together they make up the trigonal system, and account for point groups 3, $\bar{3}$, 32, $3m$, and $\bar{3}m$. The triad axis precludes centering in the rhombohedral unit cell as it does in the trigonal(H) unit cell; the rhombohedral unit cell is primitive. The existence of two types of trigonal unit cells, one corresponding to the hexagonal lattice, and the other to the rhombohedral lattice provides a dilemma of sorts, since all of the other crystal systems are described with single unit cell types with specific axial and angular constraints. A careful analysis of the rhombohedral lattice illustrates that there exists a centered* unit cell which has the same form as the trigonal(H)/hexagonal unit cell, i.e., it has two equal-length axes at 120° and a third axis perpendicular to the first two. This allows for the description of the rhombohedral lattice in terms of "hexagonal axes", even though there is nothing hexagonal about the unit cell – it has no six-fold symmetry. The unit cell is a trigonal unit cell, which we will designate as trigonal(R) to indicate that it is based on the rhombohedral lattice rather than the hexagonal lattice.

The body diagonal of the rhombohedron becomes the c axis in the trigonal(R) unit cell. This is the triad axis in the lattice, as is the c axis in the trigonal(H)

* *Centered* is used loosely here. There are internal lattice points in the unit cell, but they are not in the center of an axis, a face, or a diagonal.

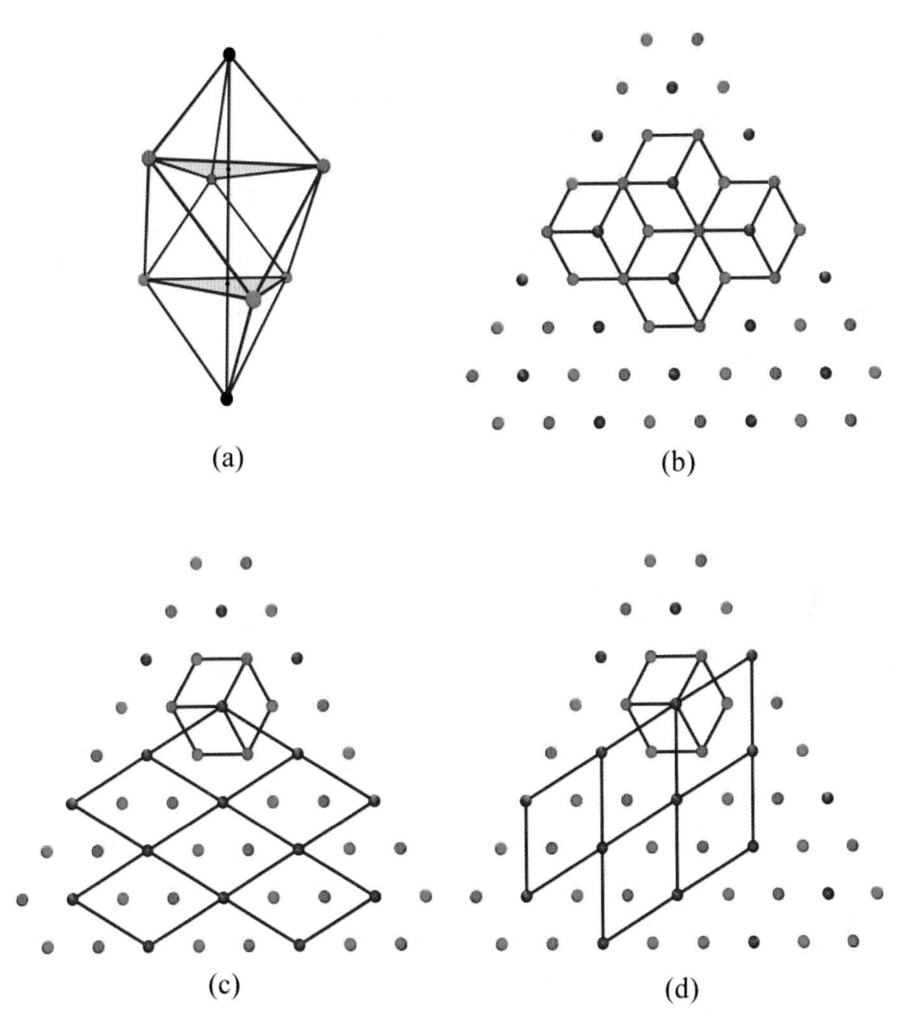

Figure 2.42 (a) Rhombohedral unit cell. Red spheres represent a layer of lattice points 1/3 of the distance along the body diagonal; blue spheres represent a layer of lattice points 2/3 of the distance along the body diagonal. (b) Projection of the rhombohedral lattice viewed down the body diagonal. (c) Projected trigonal(R) unit cells in the obverse setting. (d) Projected trigonal(R) unit cells in the reverse setting.

unit cell. Fig. 2.42(a) is a view of the rhombohedral unit cell with this axis shown vertically in the page. Along this axis there are two separate planes defined by the vertices that are not along the diagonal. These planes intersect the diagonal at 1/3 and 2/3 of its length; "upper" and "lower" planes are shown with red and blue spheres, respectively, and the apical vertices are shown as black spheres. Fig. 2.42(b) is a projection of the points in the lattice viewed down the diagonal three-fold axis. The projections of rhombohedral unit cells are also shown. We can envision the lattice as consisting of a set of three equidistant layers which repeats itself in increments of the length of the body diagonal of the rhombohedron. The first layer contains apical lattice points. The second layer contains the "upper" vertices and the third layer contains the "lower" vertices. The fourth layer is a repeat of the first, and so on ...

The layer containing the apical lattice points is perpendicular to the body diagonal, repeating in increments of the length of the body diagonal. Connecting the apical points in the manner shown in Fig 2.42(c) reveals a two-dimensional unit cell which is clearly the projection of the trigonal(R) unit cell described in the previous paragraph. Each trigonal(R) unit cell contains an "upper" lattice point at $[\frac{1}{3}\ \frac{2}{3}\ \frac{2}{3}]$ and a "lower" lattice point at $[\frac{2}{3}\ \frac{1}{3}\ \frac{1}{3}]$; both points are in the interior of the unit cell. Note that this unit cell is not unique. By connecting the apical lattice points as shown in Fig. 2.42(d), a second unit cell with the same dimensions as the first is created, but with the location of the interior lattice points reversed: the "upper" point is now at $[\frac{2}{3}\ \frac{1}{3}\ \frac{2}{3}]$ and the "lower" lattice point is at $[\frac{1}{3}\ \frac{2}{3}\ \frac{1}{3}]$. This second unit cell is essentially the reverse of the first and is said to be in the *reverse setting.* The first unit cell is (arbitrarily) chosen as the standard – in the *obverse setting.*

The primitive unit cell in the rhombohedral lattice is shown in Fig. 2.43(a). Fig. 2.43(b) illustrates the location (in the rhombohedral cell) of the internal lattice points and c axis of the trigonal(R) unit cell in the obverse setting. The trigonal(R) unit cell is shown in Fig. 2.43(c). The internal lattice points lie along the body diagonal of the trigonal(R) unit cell and share an edge with two other rhombohedra in the lattice. This is readily observed in Fig. 2.42(c) by connecting the points for the projected rhombohedra directly below and to the right of the one shown in the figure.

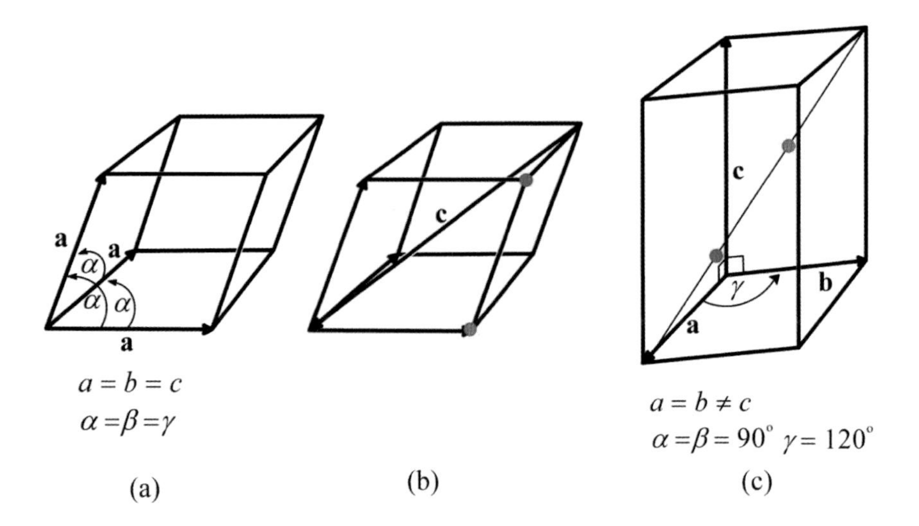

Figure 2.43 (a) Rhombohedral unit cell. (b) Location of trigonal(R) c axis and internal lattice points in the rhombohedral unit cell. (c). Trigonal(R) unit cell.

Unlike the previous examples of centering, the "centered" trigonal(R) unit cell has no more symmetry than the primitive rhombohedral unit cell in the lattice. However, the added convenience of a single unit cell type for the trigonal system makes trigonal(R) the unit cell of choice in most cases.

The hexagonal system completes the list of crystal systems, accounting for the 6, $\bar{6}$, $6/m$, 622, $6mm$, $\bar{6}m2$, and $6/mmm$ point groups. The hexagonal unit cell has already been discussed. Table 2.12 is a summary of the classification of the seven crystal systems according to their symmetry.

Table 2.12 The Point Symmetry Classification of Crystal Systems

Crystal System	Lattice Point Group	Crystal Point Groups	Bravais Lattices
Triclinic	$\bar{1}$	$1\ \bar{1}$	P
Monoclinic	$2/m$	$2\ m\ 2/m$	$P\ C$
Orthorhombic	mmm	$222\ mm2\ mmm$	$P\ C\ F\ I$
Trigonal	$6/mmm(P)$ or $\bar{3}m(R)$	$3\ \bar{3}\ 32\ 3m\ \bar{3}m$	$P^*\ R^{**}$
Tetragonal	$4/mmm$	$4\ \bar{4}\ 4/m\ 422$	$P\ I$
		$4mm\ \bar{4}2m\ 4/mmm$	
Hexagonal	$6/mmm$	$6\ \bar{6}\ 6/m\ 622$	P
		$6mm\ \bar{6}2m\ 6/mmm$	
Cubic	$m3m$	$23\ m3\ 432\ \bar{4}3m\ m3m$	$P\ F\ I$

*Same lattice as hexagonal lattice.

**Rhombohedral lattice.

2.4 Space Groups

2.4.1 Translational Symmetry

Centering. Crystal structure solution has its basis in the determination of the contents of the unit cell, and therefore the entire crystal. If the contents are related by symmetry, then it is only necessary to determine the minimal entity contained in the unit cell that will generate the remainder of the contents by the employment of symmetry operations. This entity is known as the *asymmetric unit* in the unit cell. For example, if the unit cell contains a mirror plane coplanar with the ac plane, an atom located in the asymmetric unit of the unit cell at $[x_f\ y_f\ z_f]$ will automatically place a symmetry-equivalent atom at $[x_f\ {-y_f}\ z_f]$. Up to this point we have considered only those symmetry operations determined by the point symmetry of the crystal. However, an analysis of centered unit cells indicates that point symmetry alone is not sufficient to define the contents of the unit cell based on the asymmetric unit. Consider the example in Fig. 2.44 – a lattice containing a C-centered orthorhombic unit cell and a primitive monoclinic unit cell, with the unique axis of the monoclinic unit cell coincident with the c axis of the centered unit cell. An atom is located in the asymmetric unit at $[x_f\ y_f\ z_f]_C$ (the subscript indicates that the coordinates are for the centered unit cell). The lattice point in the center of the ab face of the orthorhombic unit cell is an *origin* for the primitive unit cell, dictating that there must be an equivalent atom at the same location as in the primitive cell containing the original atom. The new atom must therefore

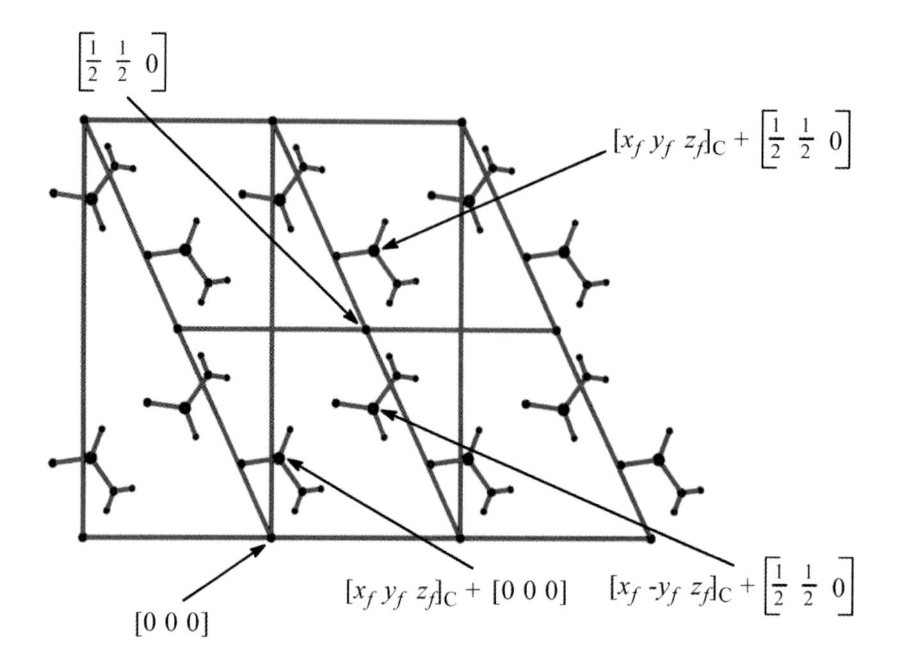

Figure 2.44 *C*-centered orthorhombic unit cells viewed along the *c* axis with primitive monoclinic unit cells. The primitive unit cell in the lattice is assumed to have its *b* axis coincident with the orthorhombic *c* axis. The vertical axis is the orthorhombic *b* axis. The orthorhombic *ac* plane is a mirror plane. A parallel mirror plane passes through the center of the orthorhombic unit cell.

be at $[x_f\ y_f\ z_f]_C$ with respect to the centered lattice point. Since the centered lattice point is at $[\frac{1}{2}\ \frac{1}{2}\ 0]$, the new atom must be at $[x_f\ y_f\ z_f]_C + [\frac{1}{2}\ \frac{1}{2}\ 0]$ from the origin of the centered unit cell. In order to "move" the asymmetric unit to a symmetry-equivalent position, we must *translate* it by adding the vector $[\frac{1}{2}\ \frac{1}{2}\ 0]$.

Furthermore, if the unit cell contents in the centered cell are also related by point symmetry operations, then applying a point symmetry operation to the vectors of the asymmetric unit, followed by translating them, will also constitute a symmetry operation. In this example, a reflection, followed by a translation will place a symmetry-equivalent atom at $[x_f\ -y_f\ z_f] + [\frac{1}{2}\ \frac{1}{2}\ 0]$. The symmetry of a centered crystal is not defined by translation operations alone, nor, as we have already observed, is it described by point group operations alone. *In order to determine every symmetry-equivalent location in the crystal we must combine both the point symmetry of the crystal and its translational symmetry.*

2.4.2 Crystal Space Symmetry

The Translation Group. In the structure just described, in addition to the point group operations and the "local" $[\frac{1}{2}\ \frac{1}{2}\ 0]$ translation operation, the periodic regularity of the crystal requires that a symmetry-equivalent location exists for every translation, of the form

$$[x_f\ y_f\ z_f]' = [x_f\ y_f\ z_f] + \mathbf{t}_j,$$
$$\mathbf{t}_j = m_1\,\mathbf{a} + m_2\,\mathbf{b} + m_3\,\mathbf{c} = [m_1\ m_2\ m_3], \tag{2.6}$$

where m_1, m_2, and m_3 ***are integers***. Although, this appears to be nothing more than a formal way of stating that the contents of every unit cell must be identical, we will see that the effects of combining this vector set with the operations of a point group are profound! The complete set of translations, $T_g = \{\mathbf{t}_j\}$, constitute a mathematical group if the multiplicative operator is vector addition. Two successive translations will always by equivalent to a translation by their vector sum, which is also an element of the set. Vector additions are always associative. The set of vectors contains an identity element, $[0\ 0\ 0] \equiv \mathbf{0}$, the null vector. Finally, the inverse operation is performed by adding the negative of the vector and producing the null vector — every element in the set has an inverse. The complete set of "global" translation operations given in Eqn. 2.6 is therefore a mathematical group; it is known as the *translation group*, which we will refer to as T_g. The combination of point group operations, "local" translation operations and the translation operations in the translation group form a new symmetry group which spans the space of the lattice and is known as a *space group*.

Before proceeding with a discussion of space groups we pause to describe some symbols that we will use throughout the remainder of the book. It should also be noted here that much of the material presented in this introductory section will become much clearer as the reader proceeds through the chapter. The symbol $\mathbf{R}(g)$ will be used to describe a point group operation and its corresponding matrix. The matrices will transform fractional coordinates, and we will discover that they are very simple in comparison to the matrices that transform Cartesian vectors. For example, a 120° rotation about a triad axis coincident with the c axis of a unit cell transforms the fractional coordinates of a general vector $[x_f\ y_f\ z_f]$ as follows:

$$\mathbf{R}(2)\begin{bmatrix} x_f \\ y_f \\ z_f \end{bmatrix} = \begin{bmatrix} 0 & \bar{1} & 0 \\ 1 & \bar{1} & 0 \\ 0 & 0 & 1 \end{bmatrix}\begin{bmatrix} x_f \\ y_f \\ z_f \end{bmatrix} = \begin{bmatrix} \bar{y}_f \\ x_f - y_f \\ z_f \end{bmatrix}. \tag{2.7}$$

In general, point group matrix operators for crystallographic bases will contain only 0, 1, or −1!

In addition to the centering translations described in the previous section we will find that point group operations can be combined with local translations to produce new space symmetry operators. We will refer to these combinations as *local operators*. They take the form $\mathbf{S}(g_i) = \mathbf{R}(g_i) + \mathbf{t}_L$, where $\mathbf{R}(g_i)$ is a point group operator and $\mathbf{t}_L$ is a local translation. If $\mathbf{t}_L = \mathbf{0}$, the null vector, then the local operator is simply a point symmetry operator. If an operator does have a translation component, then its successive application must produce a symmetry-equivalent position in the lattice — a translation into the next unit cell. This translation will be a member of the translation group, usually a unit cell translation along the translation direction of the local operator, e.g., $[0\ 1\ 0]$. Recall that one of the criteria for a group is that the product (combination) of any two elements in the group must be another element in the group — this is formally known as *closure.* There is no single operation in the set of local operations that will generate the combination of a local operation and a translation; the complete set of local symmetry operations for a crystal does not form a mathematical group. *However,* the application of two local symmetry operations *would* be equivalent to another single operation were it not for the addition of the translation group vector. *The product of any two local symmetry operations in the set of local operations will differ from a single symmetry operation in the set by (at most) the addition of a vector in the translation group.*

From the discussion above it would appear that *adding* the translation group to the set of local symmetry operators should form a space group. However, the set formed from the simple expansion of the set of local operations with the translation group is *not* itself a mathematical group. We consider a simple example – one that we will encounter again when we consider the $P2$ space group. The $P2$ space group consists of local diad rotational operators ($\mathbf{R}(2)$), combined with the translation group. The diad rotational axes are parallel to the b axis of the unit cell, and perpendicular to the ac plane. As we shall soon see, every rotation about a diad axis (2), followed by a translation from one of the vectors in the translation group that parallel the ac plane, will be equivalent to a diad rotation about a two-fold axis somewhere in the lattice (2'). The product of a rotation and translation is a rotation – satisfying the closure requirement. On the other hand, a diad rotation, followed by a translation in the b direction will also produce a symmetry-equivalent point, but there is no single symmetry operation in the set (rotational or translational) that will produce the same result.

In order to resolve this dilemma, new operators must be created that are combinations of local operations and translations, which we will refer to as "local+translation" operators (analogous to the rotoinversion and rotoreflection operators in point group theory). These new operators will be represented as $\mathbf{S}(g_i|\mathbf{t}_j)$, where g_i corresponds to an operator in the "local" set and $\mathbf{t}_j$ corresponds to a vector operator in the translation group. As an example, later on in the chapter we will find that a new operator can be formed by combining a diad rotation with a translation of half of the distance along one of the unit cell axes. The result is the transformation of a diad rotation operation, 2, into a diad rotation+translation operation, 2_1: $\mathbf{S}(2_1) = \mathbf{R}(2)+[0\ \frac{1}{2}\ 0]$. The operation in the space group that combines this operator with a translation group operator is then $\mathbf{S}(2_1)|\mathbf{t}_j) = \mathbf{S}(2_1)+\mathbf{t}_j = \mathbf{R}(2)+[0\ \frac{1}{2}\ 0]+\mathbf{t}_j$. The operation rotates a point about an axis parallel to the b axis, translates it by $\frac{1}{2}$ along the $\mathbf{b}$ direction, then translates it again along the vector $\mathbf{t}_j$. In general,

$$\mathbf{S}(g_i|\mathbf{t}_j) = \mathbf{S}(g_i) + \mathbf{t}_j = \mathbf{R}(g_i) + \mathbf{t}_L + \mathbf{t}_j. \tag{2.8}$$

The local operator set contains the identity element $\mathbf{S}(E) = \mathbf{R}(E) + [0\ 0\ 0]$; the translation group identity element is the null vector, $\mathbf{0}$. *The space group is the complete set of local+translation operators,*

$$\begin{array}{l}\{\mathbf{S}(E|\mathbf{0}), \mathbf{S}(g_2|\mathbf{0}), \ldots, \mathbf{S}(g_k|\mathbf{0}), \ldots\ldots, \mathbf{S}(E|\mathbf{t}_1), \mathbf{S}(g_2|\mathbf{t}_1), \ldots, \mathbf{S}(g_k|\mathbf{t}_1), \ldots\ldots,\\ \quad \mathbf{S}(E|\mathbf{t}_2), \mathbf{S}(g_2|\mathbf{t}_2), \ldots, \mathbf{S}(g_k|\mathbf{t}_2), \ldots\ldots, \mathbf{S}(E|\mathbf{t}_3), \mathbf{S}(g_2|\mathbf{t}_3), \ldots, \mathbf{S}(g_k|\mathbf{t}_3) \ldots,\\ \quad \vdots \qquad \vdots \qquad \vdots \qquad \vdots \qquad \vdots \qquad \vdots\\ \quad \mathbf{S}(E|\mathbf{t}_j), \mathbf{S}(g_2|\mathbf{t}_j), \ldots, \mathbf{S}(g_k|\mathbf{t}_j), \ldots\ldots\ldots\ldots\ldots\ldots\ldots\ldots\}.\end{array} \tag{2.9}$$

The complete set of local+translation symmetry operators is a mathematical group, satisfying the following criteria:

1. **Closure.** The product of any two symmetry operations in the set is another symmetry operation in the set. To prove this we consider the results of the symmetry operation product on a general vector, $\mathbf{r} = [x_f\ y_f\ z_f]$: $\mathbf{S}(g_i|\mathbf{t}_j)\mathbf{S}(g_k|\mathbf{t}_m)\mathbf{r}$. We begin by noting that the operation on a vector by the operator $(\mathbf{R}+\mathbf{t}_A+\mathbf{t}_B)$ is effected by first applying the point operation, $\mathbf{R}$, to the vector. This creates a new vector, which is then translated by adding $\mathbf{t}_A$ and $\mathbf{t}_B$. It follows that $(\mathbf{R}+\mathbf{t}_A+\mathbf{t}_B)\mathbf{r} = (\mathbf{R}+\mathbf{t}_A)\mathbf{r}+\mathbf{t}_B = \mathbf{R}\mathbf{r}+\mathbf{t}_A+\mathbf{t}_B$.* Thus,

$$
\begin{aligned}
&\mathbf{S}(g_i|\mathbf{t}_j)\mathbf{S}(g_k|\mathbf{t}_m)\mathbf{r} = \mathbf{S}(g_i|\mathbf{t}_j)(\mathbf{R}(g_k)\mathbf{r}+\mathbf{t}_K+\mathbf{t}_m)\\
&= (\mathbf{R}(g_i)+\mathbf{t}_I+\mathbf{t}_j)(\mathbf{R}(g_k)\mathbf{r}+\mathbf{t}_K+\mathbf{t}_m)\\
&= \mathbf{R}(g_i)(\mathbf{R}(g_k)\mathbf{r}+\mathbf{t}_K+\mathbf{t}_m)+\mathbf{t}_I+\mathbf{t}_j\\
&= \mathbf{R}(g_i)\mathbf{R}(g_k)\mathbf{r}+\mathbf{R}(g_i)\mathbf{t}_K+\mathbf{R}(g_i)\mathbf{t}_m+\mathbf{t}_I+\mathbf{t}_j\\
&= \mathbf{R}(g_n)\mathbf{r}+(\mathbf{R}(g_i)\mathbf{t}_K+\mathbf{t}_I)+(\mathbf{R}(g_i)\mathbf{t}_m+\mathbf{t}_j) \qquad (2.10)\\
&= \mathbf{R}(g_n)\mathbf{r}+\mathbf{t}_N+\mathbf{t}_o = (\mathbf{R}(g_n)+\mathbf{t}_N+\mathbf{t}_o)\mathbf{r}. \qquad (2.11)
\end{aligned}
$$

   The transformation of a vector in the translation group with a point group operator will generate another vector in the translation group since the operator matrices all have integer elements, and the components of the vector are all integers. Thus $\mathbf{t}_o$ is a vector in the translation group. Selecting the case when $\mathbf{t}_j = \mathbf{0}$ and $\mathbf{t}_m = \mathbf{0}$ gives $\mathbf{R}(g_n)+\mathbf{t}_N = \mathbf{S}(g_i)\mathbf{S}(g_k)$, which must differ from a local operation in the set of local operations by a vector in the translation group (which *can* be the null vector): $\mathbf{S}(g_i)\mathbf{S}(g_k) = \mathbf{S}(g_n)+\mathbf{t}_r$. Setting $\mathbf{t}'_o = \mathbf{t}_o+\mathbf{t}_r$,

$$
\begin{aligned}
\mathbf{S}(g_i|\mathbf{t}_j)\mathbf{S}(g_k|\mathbf{t}_m) &= \mathbf{R}(g_n)+\mathbf{t}_N+\mathbf{t}_o = \mathbf{S}(g_i)\mathbf{S}(g_k)+\mathbf{t}_o\\
&= \mathbf{S}(g_n)+\mathbf{t}_o+\mathbf{t}_r = \mathbf{S}(g_n|\mathbf{t}'_o), \qquad (2.12)
\end{aligned}
$$

   another operation in the set.

2. **Associativity.** Provided that the order of operations is not changed and that each operation occurs in only one pair, sequences of operations in the set can be performed pairwise with no constraints on the selections of the pairs. We consider the results of the product of three symmetry operations on a vector: $\mathbf{S}(g_i|\mathbf{t}_j)\mathbf{S}(g_k|\mathbf{t}_m)\mathbf{S}(g_p|\mathbf{t}_q)\mathbf{r}$,

$$
\begin{aligned}
&(\mathbf{S}(g_i)+\mathbf{t}_j)\Big((\mathbf{S}(g_k)+\mathbf{t}_m)(\mathbf{S}(g_p)\mathbf{r}+\mathbf{t}_q)\Big)\\
&= (\mathbf{S}(g_i)+\mathbf{t}_j)(\mathbf{S}(g_k)\mathbf{S}(g_p)\mathbf{r}+\mathbf{S}(g_k)\mathbf{t}_q+\mathbf{t}_m)\\
&= \mathbf{S}(g_i)\mathbf{S}(g_k)\mathbf{S}(g_p)\mathbf{r}+\mathbf{S}(g_i)\mathbf{S}(g_k)\mathbf{t}_q+\mathbf{S}(g_i)\mathbf{t}_m+\mathbf{t}_j \qquad (2.13)
\end{aligned}
$$

$$
\begin{aligned}
&\Big((\mathbf{S}(g_i)+\mathbf{t}_j)(\mathbf{S}(g_k)+\mathbf{t}_m)\Big)(\mathbf{S}(g_p)\mathbf{r}+\mathbf{t}_q)\\
&= \Big((\mathbf{S}(g_i)\mathbf{S}(g_k)+\mathbf{S}(g_i)\mathbf{t}_m+\mathbf{t}_j)\Big)(\mathbf{S}(g_p)\mathbf{r}+\mathbf{t}_q)\\
&= \mathbf{S}(g_i)\mathbf{S}(g_k)\mathbf{S}(g_p)\mathbf{r}+\mathbf{S}(g_i)\mathbf{S}(g_k)\mathbf{t}_q+\mathbf{S}(g_i)\mathbf{t}_m+\mathbf{t}_j. \qquad (2.14)
\end{aligned}
$$

*This is an important notion here. Otherwise the algebra is incomprehensible!

The results are identical and the symmetry operations in the set are associative.

3. **Identity.** There is an identity element in the set that commutes with every element of the set and leaves it unchanged:

$$\mathbf{S}(g_i|\mathbf{t}_j)\mathbf{S}(E|\mathbf{0}) = \mathbf{S}(E|\mathbf{0})\mathbf{S}(g_i|\mathbf{t}_j) = \mathbf{S}(g_i|\mathbf{t}_j), \tag{2.15}$$

   $\mathbf{S}(E|\mathbf{0})$ is the identity operator in the set.

4. **Inverse.** For every vector, $\mathbf{t}_j$, in the translation group, there is a vector in the opposite direction, $-\mathbf{t}_j$, and for every local operation, there is an operation that negates it, since every operation in the generating point group has an inverse in the group: $(\mathbf{S}(g_i))^{-1} = (\mathbf{R}(g_i))^{-1} - \mathbf{t}_L$. The set of local+translation operations therefore contains the inverse of each operation:

$$(\mathbf{S}(g_i|\mathbf{t}_j))^{-1} = (\mathbf{S}(g_i))^{-1} - \mathbf{t}_j = (\mathbf{R}(g_i))^{-1} - \mathbf{t}_L - \mathbf{t}_j, \tag{2.16}$$

   and $\mathbf{S}(g_i|\mathbf{t}_j)(\mathbf{S}(g_i|\mathbf{t}_j))^{-1} = (\mathbf{S}(g_i|\mathbf{t}_j))^{-1}\mathbf{S}(g_i|\mathbf{t}_j) = \mathbf{S}(E|\mathbf{0})$.

The set of local+translation operators is a mathematical group — a *crystallographic space group. Each operation in the space group produces a symmetry-equivalent position in the lattice — the structure remains invariant.*

Two subsets of the space group deserve special attention. The subset containing E as the local operator, $\{\mathbf{S}(E|\mathbf{t}_j)\}$, is simply the translation group, $\{\mathbf{0}, \mathbf{t}_1, \ldots, \mathbf{t}_j, \ldots\}$. Because the translation group is itself a group, it is referred to as a *subgroup* of the space group. The subset containing the null vector, $\{\mathbf{S}(g_i|\mathbf{0})\}$, consists of the set of local operations only. *The presence of these two subsets in the group allows us to treat local operations and translations independently whenever necessary.* Most importantly, every operation in the group is generated by the products of the operations in these two subsets. The symbol $\mathbf{S}(g_i|\mathbf{t}_j)$ indicates that a local operation is to be performed, $\mathbf{S}(g_i))$, followed by a translation, $\mathbf{t}_j$. The product $\mathbf{S}(E|\mathbf{t}_j)\mathbf{S}(g_i|\mathbf{0})$ performs the same operation: $\mathbf{S}(g_i|\mathbf{t}_j)$=$\mathbf{S}(E|\mathbf{t}_j)\mathbf{S}(g_i|\mathbf{0})$.

We will find it conceptually useful to treat the space group operations in terms of these products. To simplify the symbols, we express $\mathbf{S}(g_i|\mathbf{0})$ as $\mathbf{S}(g_i) = \mathbf{R}(g_i) + \mathbf{t}_L$, and $\mathbf{S}(E|\mathbf{t}_j)$ as $\mathbf{S}(\mathbf{t}_j)$. Thus the space group operations are conveniently expressed as

$$\mathbf{S}(g_i|\mathbf{t}_j) = \mathbf{S}(\mathbf{t}_j)\mathbf{S}(g_i), \tag{2.17}$$

which reads simply "perform local operation $\mathbf{S}(g_i)$, then translate the result of that operation along vector $\mathbf{t}_j$."

The local symmetry operator, $\mathbf{S}(g_i)$ combines a point symmetry operator with a local translation. If the translation vector is [0 0 0], then the local operator is a point group operator, e.g., for the $P2$ space group, $\mathbf{S}(2) = \mathbf{R}(2) + [0\ 0\ 0] \equiv \mathbf{R}(2)$. The identity is $\mathbf{S}(E) = \mathbf{R}(E) + [0\ 0\ 0] \equiv \mathbf{R}(E)$. The basic criterion for combining a point group operation with a local translation is that the application of the combined operation must result in identical unit cells in the lattice. It follows that a series of local operations that transform a general point in one unit cell into an adjacent unit cell must transform that point (vector) into a position that is symmetry-equivalent to one in the original unit cell. In other words, when performed sequentially, a finite number of these new point+translation operations must produce a result that is equivalent to a simple translation from one unit cell to another. Just as with the combination of local operations with the translation group operations, the combination of a point symmetry operation and a local translation is a new symmetry operation — the components are inseparable.

The creation of a space group begins with a point group. Local symmetry operations are created from the point group operators by the addition of local translation vector operators (which may be the null operator), that are then combined with the vector operators of the translation group to form the space group. The initial point group is often referred to as the "point group of the space group," even though the point symmetry has been modified by the inclusion of translations. This terminology has it roots in the study of crystalline materials. Many of their physical properties are independent of translational symmetry; solid materials often behave as if they have the point symmetry of the initial point group. The "point group" of the crystal can be inferred by removing the translations from all of the space group operators, leaving only those of the initial point group. The "point group of the space group" is more aptly referred to as the "generating point group" and we will use that terminology throughout the text.

In the late 1800's two mathematicians and an amateur scientist took very different approaches to establish that there were 230 possible crystallographic space groups. William Barlow, an English builder and amateur crystallographer made a comprehensive study of known crystal structures and deduced the possible space groups intuitively. At almost the same time, the Russian mathematician, E.S. Federov, accomplished this same task using a geometric approach, while in Germany A. Schönflies analyzed combinations of point symmetry operations and translations as part of his study of group theory to determine the crystallographic space groups that were feasible. As mentioned earlier, Schönflies created a notation for these new space groups based on the generating point groups, appended with superscripts which distinguished one group from another. Unfortunately the Schönflies symbols indicated nothing about the translational symmetry. The Hermann-Mauguin notation rectified this and became the standard notation for crystallography. The form of the H-M space group notation depends on the lattice type, the symmetry of the generating point group, and the specific symmetry operations. The H-M symbol for a specific space group begins with the Bravais lattice type (P, C (or A or B), F, I, or R) followed by symbols for specific symmetry operations which differ in meaning, depending on the crystal system.

In this brief introduction to space group theory, much has been left out, including the mechanics of the process of creating the space groups from crystallographic point groups and the translation group. Formally, this is accomplished by generating a group specific to the space group called a factor group that contains the point group operations, centering operations, etc. — then forming the space group as a product group of the factor group and the translation group. The reader interested in the details of space group generation is referred to the treatment in Zachariasen's 1945 book[26] with the caveat that the notation and treatment will take some time to decipher! We will take a much less rigorous approach here, but one that is significantly more intuitive. Beginning with the simplest of space groups, new space groups will by systematically "created" by adding and modifying local symmetry elements and operators. This process will continue until the complexity of the space groups requires us to consult the results from a more formal treatment — summarized in Volume A of *International Tables for Crystallography.*[27]

2.4.3 The Triclinic Space Groups

The $P1$ and $P\bar{1}$ Space Groups. The simplest space groups are derived from the simplest point groups — those of the triclinic system — 1 and $\bar{1}$. In the non-centrosymmetric case, the only translations possible are those that transform a vector in one unit cell into a symmetry-equivalent position in another unit cell — the operations in the translation group. The point group of the lattice and its contents is 1, with only a single local symmetry operation: $\mathbf{S}(E)$. Since the only local symmetry operator in the group is the identity operation the space group consists entirely of the translation group. The H-M symbol for the resulting space group, is $P1$. The lack of symmetry of a crystal in this space group places no constraints on the location of the origin of the unit cell, which can be chosen for convenience.

Although it is only a single symmetry element, the inversion center alters the symmetry of a crystal dramatically. The addition of an inversion operation, (which is its own inverse) to the identity operation creates a new point group, $\bar{1} \equiv \{E, \bar{1}\}$. Combining the translation group with this point group creates a new space group from E, $\bar{1}$, and T_g. The inversion operator is $\mathbf{S}(\bar{1}) = \mathbf{R}(\bar{1}) + [0\ 0\ 0]$. The inclusion of the translation group results in *many* new inversion operations in the group. The locations of the inversion centers can be deduced by treating the symmetry operations of the space group as the product of the inversion operations and translations via the vectors in the translation group: $\mathbf{S}(\mathbf{t}_j)\mathbf{S}(\bar{1})$.

Consider an atom at $[x_f\ y_f\ z_f]$ in the unit cell, with the origin selected so that it coincides with the $\bar{1}$ symmetry element. The inversion operation places a symmetry-equivalent atom at $[\bar{x}_f\ \bar{y}_f\ \bar{z}_f]$. Following the inversion, the translation group places equivalent atoms at $[\bar{x}_f\ \bar{y}_f\ \bar{z}_f] + [m_1\ m_2\ m_3] \equiv [(m_1 - x_f)(m_2 - y_f)(m_3 - z_f)]$. With the exception of the identity, inversion operations are the only local operations in the space group, and we now query whether there are other inversion centers in the lattice for which their associated inversion operations will produce the same result: $\mathbf{S}(\bar{1}') = \mathbf{S}(\mathbf{t}_j)\mathbf{S}(\bar{1})$, $\Rightarrow \mathbf{S}(\bar{1}')[x_f\ y_f\ z_f] = [(m_1 - x_f)(m_2 - y_f)(m_3 - z_f)]$. The general position, $[x_f\ y_f\ z_f]$, is arbitrary and we can select it at the location of the $\bar{1}'$ inversion center. Placing an atom *on the inversion center* at $[x_f^*\ y_f^*\ z_f^*]$ and performing the inversion operation generates a new atom at the same location as the original atom — back on to the inversion center. We have postulated here that the initial inversion and translation operations on an atom at $[x_f^*\ y_f^*\ z_f^*]$ places an atom at the same location — at $[(m_1 - x_f^*)(m_2 - y_f^*)(m_3 - z_f^*)]$. The $\bar{1}'$ inversion center is therefore located at

$$\begin{aligned} x_f^* &= m_1 - x_f^* \Longrightarrow x_f^* = \frac{m_1}{2} \\ y_f^* &= m_2 - y_f^* \Longrightarrow y_f^* = \frac{m_2}{2} \\ z_f^* &= m_3 - z_f^* \Longrightarrow z_f^* = \frac{m_3}{2}. \end{aligned} \tag{2.18}$$

Values of 0 or 1 for m_1, m_2, and m_3 place inversion centers at $[0\ 0\ 0]$, $[\frac{1}{2}\ 0\ 0]$, $[0\ \frac{1}{2}\ 0]$, $[0\ 0\ \frac{1}{2}]$, $[\frac{1}{2}\ \frac{1}{2}\ 0]$, $[\frac{1}{2}\ 0\ \frac{1}{2}]$, $[0\ \frac{1}{2}\ \frac{1}{2}]$, and $[\frac{1}{2}\ \frac{1}{2}\ \frac{1}{2}]$. *If the origin of the unit cell is*

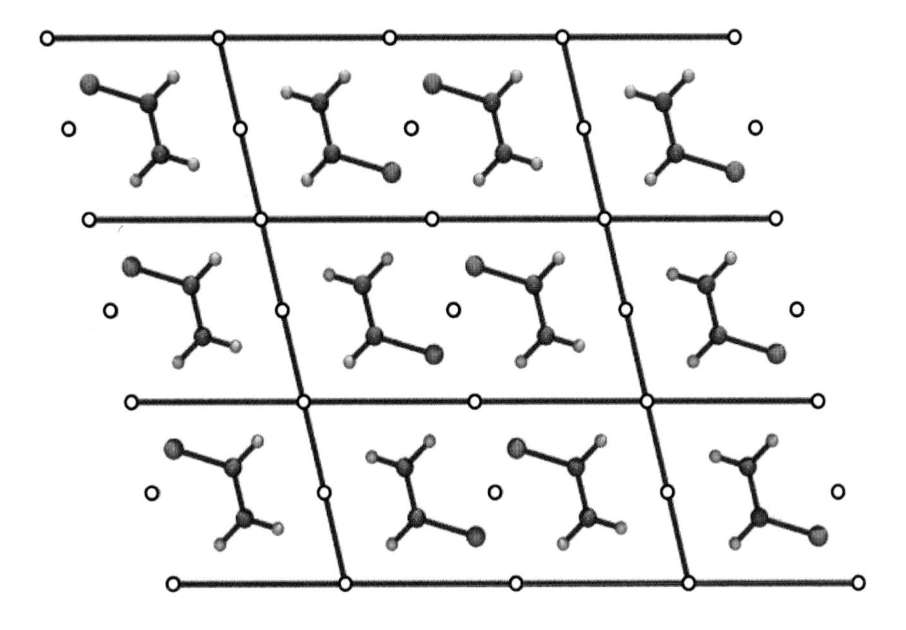

Figure 2.45 Projection of a hypothetical centrosymmetric triclinic structure onto the *ab* plane. Inversion centers are shown as open circles.

*coincident with an inversion center in the crystal, the inclusion of the translation group in the space group requires the existence of inversion centers in the center, the midpoint of each axis, and the center of each face of the unit cell.** Each of these inversion centers corresponds to an operation in the group. Values of m_1, m_2, and m_3 greater than 1 or less than 0 will place these inversion centers at equivalent locations in unit cells throughout the lattice. Fig. 2.45 shows the projection of a hypothetical centrosymmetric triclinic structure viewed in the direction of the *c* axis onto the *ab* plane. The centers of symmetry are indicated by open circles. The space group is created by combining the identity element, the translation group, and the complete set of inversion operations required by inclusion of the translation group; the inversion operations are their own inverses. The lattice of the space group is primitive, and its H-M symbol is $P\bar{1}$. As we will see in the discussions that follow, space group symbols have both long forms and short forms. For monoclinic and orthorhombic space groups, the long form of the space group symbol displays the symmetry element that is related to each of the axes, in *a*, *b*, *c* order. Applied to the two triclinic space groups, the long forms would be $P111$ and $P\bar{1}\bar{1}\bar{1}$. Because the

*If the origin is not made coincident with the inversion center, each unit cell will still have the same contents, and the crystal will still maintain its inversion symmetry. However, translating the origin away from the inversion center results in translation of the inversion operations to locations within the cell. This complicates matters considerably (see the $C2$ space group discussion for an example)! The origins of centrosymmetric unit cells are always chosen so that they are coincident with an inversion center.

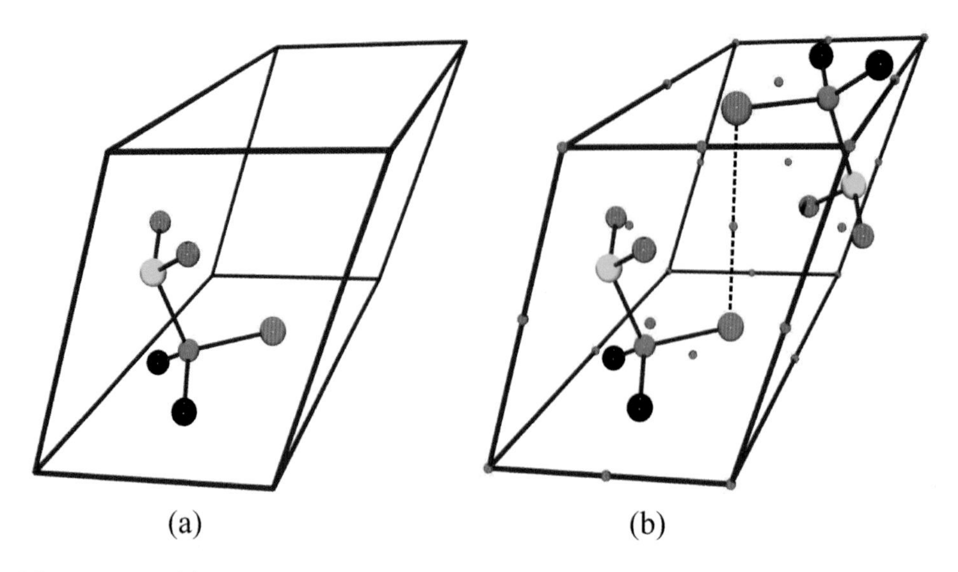

Figure 2.46 (a) $P1$ unit cell. The origin position is arbitrary; the "empty space" is for illustrative purposes. (b) $P\bar{1}$ unit cell. Inversion centers are illustrated with smaller red spheres.

long forms provide no additional information (the symmetry element is the same for each axis), only the short forms are used.

$P1$ and $P\bar{1}$ are the only two triclinic space groups, differentiated by the absence or presence of centers of symmetry; their unit cells are depicted in Fig. 2.46. It is interesting to note that while there are many molecules without symmetry, only a few choose to reside in a $P1$ unit cell, while many are found in the $P\bar{1}$ space group.

2.4.4 The Monoclinic Space Groups

Lattices for the monoclinic crystal system belong to the $2/m$ point group and exhibit two-fold rotational and mirror symmetry. The contents of the unit cells in these lattices can be related by a diad axis, a mirror plane, or both (in this case the unit cell also has an inversion center) – *and more.* The point symmetry operations available to the unit cell contents (asymmetric units) of monoclinic cells are E, 2, m, and $\bar{1}$:

$$\mathbf{R}(E) = \begin{bmatrix} 1 & 0 & 0 \\ 0 & 1 & 0 \\ 0 & 0 & 1 \end{bmatrix} \quad \mathbf{R}(2) = \begin{bmatrix} -1 & 0 & 0 \\ 0 & 1 & 0 \\ 0 & 0 & -1 \end{bmatrix}$$

$$\mathbf{R}(m) = \begin{bmatrix} 1 & 0 & 0 \\ 0 & -1 & 0 \\ 0 & 0 & 1 \end{bmatrix} \quad \mathbf{R}(\bar{1}) = \begin{bmatrix} -1 & 0 & 0 \\ 0 & -1 & 0 \\ 0 & 0 & -1 \end{bmatrix}$$

The $P2$ Space Group. We begin with the monoclinic space groups derived from the 2 point group. The unit cell contents in a structure with this point group are

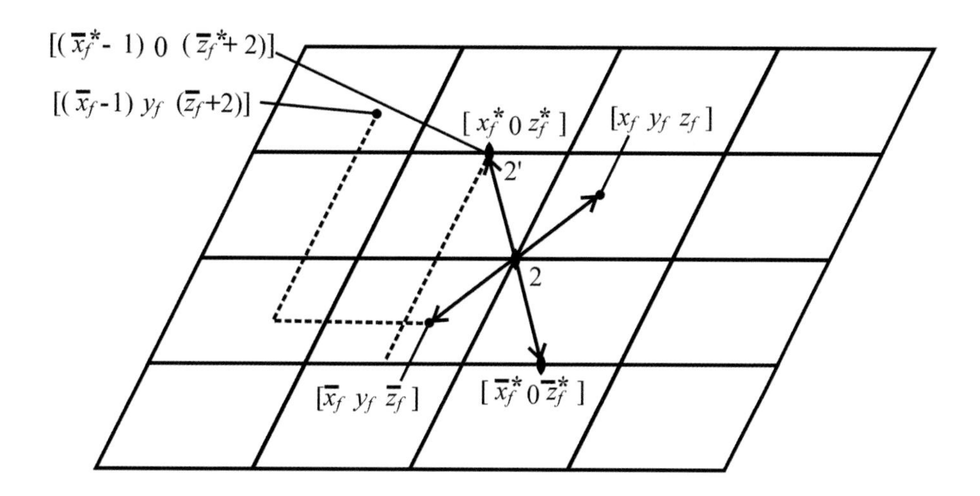

Figure 2.47 Results of a rotation about a diad axis (2), followed by a translation ([−1 0 2]), for a general point and a point at the intersection of another diad axis (2') with the *ac* plane.

related by a diad rotation about the unique axis.* A general vector at $[x_f\ y_f\ z_f]$ is transformed by the $\mathbf{R}(2)$ matrix:

$$\begin{bmatrix} -1 & 0 & 0 \\ 0 & 1 & 0 \\ 0 & 0 & -1 \end{bmatrix} \begin{bmatrix} x_f \\ y_f \\ z_f \end{bmatrix} = \begin{bmatrix} \bar{x}_f \\ y_f \\ \bar{z}_f \end{bmatrix}. \tag{2.19}$$

Addition of the translation group to create the space group results in symmetry-equivalent positions at $[\bar{x}_f\ y_f\ \bar{z}_f]+[m_1\ m_2\ m_3] \equiv [(m_1-x_f)(m_2+y_f)(m_3-z_f)]$. As we observed previously for the $P\bar{1}$ space group, the addition of the translation group requires there to be local symmetry elements throughout the lattice — diad axes in this case — and their corresponding symmetry operations: $\mathbf{S}(2') = \mathbf{R}(2') + [0\ 0\ 0]$. Again, we postulate that there is a diad operation that will be equivalent to the rotation and translation operations, $\mathbf{S}(2') = \mathbf{S}(\mathbf{t}_j)\mathbf{S}(2)$. To do so we look for diad axes with rotations that transform $[x_f\ y_f\ z_f]$ directly into $[(m_1-x_f)(m_2+y_f)(m_3-z_f)]$: $\mathbf{S}(2')[x_f\ y_f\ z_f] = [(m_1-x_f)(m_2-y_f)(m_3-z_f)]$.

As we noted previously, the $[x_f\ y_f\ z_f]$ position can be chosen on a symmetry element. To retain the two-fold symmetry of the lattice, the $2'$ axis must be parallel to the b axis, intersecting the ac plane at the point $[x_f^*\ 0\ z_f^*]$. Placing an atom at $[x_f^*\ 0\ z_f^*]$ *on the* $2'$ *rotation axis* and performing the rotation about the axis will place a symmetry-equivalent atom at the same point, $[x_f^*\ 0\ z_f^*]$. The initial rotation about the 2 axis followed by translation will place an atom at $[(m_1-x_f^*)(m_2+0)(m_3-z_f^*)]$,

*The origin of a monoclinic unit cell is selected so that it lies on a diad axis coincident with the unique axis, which we will always select as the b axis. As with the case of the origin in $P\bar{1}$, arbitrary placement of the origin at another position in the ac plane translates diad axes so that they intersect the ac plane inside the cell, serving only to complicate the rotation operators unnecessarily.

which must be the same position, as illustrated in Fig. 2.47 for $m_1 = -1$ and $m_3 = 2$. Thus, $[(m_1 - x_f^*)(m_2)(m_3 - z_f^*)] = [x_f^*\ 0\ z_f^*]$, and

$$\begin{aligned} x_f^* &= m_1 - x_f \Longrightarrow x_f^* = \frac{m_1}{2} \\ m_2 &= 0 \\ z_f^* &= m_3 - z_f \Longrightarrow z_f^* = \frac{m_3}{2}. \end{aligned} \tag{2.20}$$

The zero value for m_2 indicates that the y_f coordinate does not change. Values of 0 or 1 for m_1 and m_3 produce the points of intersection of the diad axes with the *ac* plane: $[0\ 0\ 0]$, $[\frac{1}{2}\ \ 0\ 0]$, $[0\ 0\ \frac{1}{2}]$ and $[\frac{1}{2}\ 0\ \frac{1}{2}]$, resulting in diad axes parallel to the unique *b* axis and passing through the origin, the centers of the *a* and *c* axes, and the center of the *ac* plane. The remaining integer values of m_1 and m_3 will place diad rotational axes (each with a corresponding two-fold rotational operation) at equivalent locations in unit cells throughout the lattice. A projection of four adjacent unit cells and their contents is shown in Fig. 2.48, illustrating the location of all the diad axes in the unit cell.

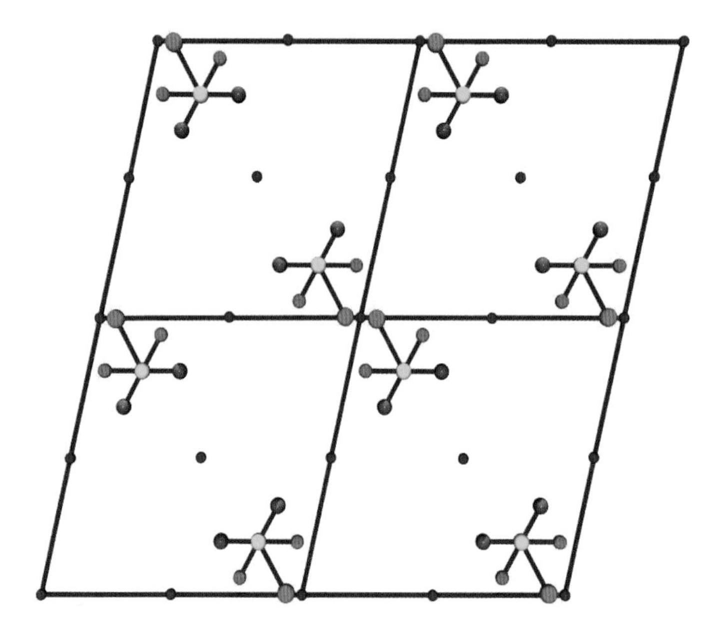

Figure 2.48 Projection of four unit cells with two-fold rotational symmetry. Diad rotation axes intersect the *ac* plane at the points marked by the blue spheres — at the origins, midpoints, and bisectors of *a* and *b*.

Fig. 2.49(a) illustrates a monoclinic unit cell with diad symmetry.* The unit cell contents are related by the diad axis passing through the center of the unit cell. While the origin is coincident with the *b* axis, there are no constraints on the

*In this section the monoclinic space groups will be studied by systematically analyzing possible symmetry combinations. In order to facilitate a comparison of unit cells corresponding to the various space groups, the same hypothetical asymmetric unit and unit cell with constant cell parameters are used in the illustrations. Because of this, unit cells with lower symmetry will appear to contain empty space – which would not be the case in an actual crystal structure.

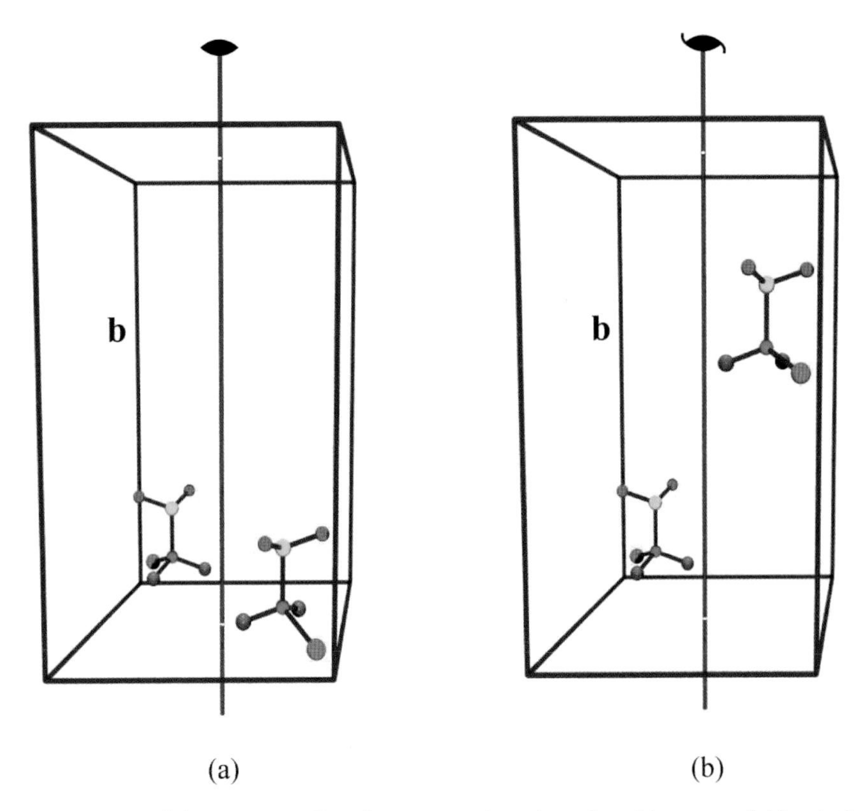

Figure 2.49 (a) $P2$ unit cell and asymmetric units related by a two-fold rotation axis. (b) $P2_1$ unit cell and asymmetric units related by a two-fold screw axis. The symbol for the diad screw axis is shown atop the axis.

location of this origin along the axis – and its position can be assigned arbitrarily. To clarify this, consider the structure in Fig. 2.49(a) and envision fixing the positions of the unit cell contents and translating the lattice (the unit cell and its origin) along the a axis or the c axis. The diad axis will no longer lie in the center of the cell, and the unit cell symmetry will not be coincident with the lattice symmetry. In contrast, consider the effects of translating the lattice along the b axis. In this case the two-fold axis relating the unit cell contents remain in the center of the unit cell, regardless of the position of the ac plane along the axis.

The space group formed from the identity operation, the translation operations in the translation group, and the complete collection of two-fold rotation operations (which are their own inverses) results in the first monoclinic space group (No. 3) listed in the *International Tables for Crystallography*[27]. The "full" H-M notation for monoclinic space groups is formed by adding three entities to the Bravais lattice symbol, each corresponding to the symmetry element related to each of the axes in the unit cell in a, b, c order. In the space group just discussed the b axis is a diad axis, and the long form of the space group symbol is $P121$. As with the H-M point group notation, the short form of the H-M symbol contains only the necessary symmetry to define the space group: $P2$. In the monoclinic system all of the symmetry elements are related to the unique axis — the short form will always contain only a single symmetry indicator.

The $P2_1$ Space Group. As previously discussed, specific local translations combined with point symmetry operations can generate new local symmetry operations. We now consider creating a new symmetry operation from a diad rotation coupled with a suitable translation. Fig. 2.49(b) shows a monoclinic unit cell in which asymmetric units are related by the *combined* operation of a two-fold rotation and a translation in the direction parallel to the b axis. A second diad rotation, followed by an identical translation, will place an asymmetric unit into an adjacent unit cell. In order for the contents of both unit cells to be the same, the new asymmetric unit and the original asymmetric unit must occupy identical positions in their respective unit cells. Thus they must be separated by the length of the b axis — related to one another by the translation group vector [0 1 0]. It follows that the translational portion of the rotation+translation operation must have a magnitude that is half of the length of the b axis, as illustrated in Fig. 2.50.

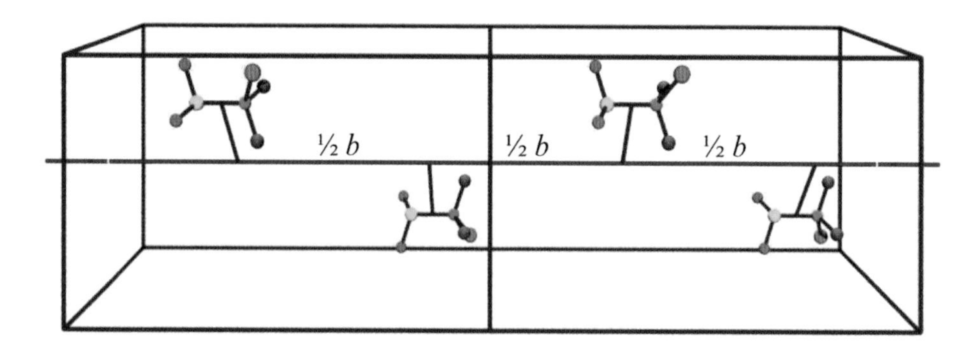

Figure 2.50 Two adjacent $P2_1$ unit cells sharing an ac face; the two-fold screw axis is parallel to the b axis.

When a screw is rotated, it also translates — the rotation-translation axis is called a *screw axis* — in this case a two-fold (or diad) screw axis. The screw axis and operation are designated by the symbol 2_1.* The graphical symbol for the two-fold screw axis is a modification of the symbol for the diad axis, as illustrated in Fig. 2.49(b). The two-fold screw operator is $\mathbf{S}(2_1) = \mathbf{R}(2_1) + \mathbf{t}_S$, where $\mathbf{t}_S = [0\ \frac{1}{2}\ 0]$, transforming the general vector $[x_f\ y_f\ z_f]$ into a symmetry-equivalent vector:

$$\begin{bmatrix} -1 & 0 & 0 \\ 0 & 1 & 0 \\ 0 & 0 & -1 \end{bmatrix} \begin{bmatrix} x_f \\ y_f \\ z_f \end{bmatrix} + \begin{bmatrix} 0 \\ \frac{1}{2} \\ 0 \end{bmatrix} = \begin{bmatrix} \bar{x}_f \\ \frac{1}{2} + y_f \\ \bar{z}_f \end{bmatrix}. \tag{2.21}$$

In order to form an element of the space group there must be an inverse operation for the diad screw operation: $(\mathbf{S}(2_1))^{-1} = \mathbf{R}(2_1) + \mathbf{t}'_S$, where $\mathbf{t}'_S = [0\ (-\frac{1}{2})\ 0]$. The

*In general, an N-fold screw axis in which the translation is k/N of the length of the axis parallel to the translation direction is designated by N_k.

application of both operations in either order brings an atom originally at $[x_f\ y_f\ z_f]$ back to its original position: $\mathbf{S}(2_1)(\mathbf{S}(2_1))^{-1} = (\mathbf{S}(2_1))^{-1}\mathbf{S}(2_1) = \mathbf{S}(E)$.

This new symmetry operation and its inverse, along with the identity operation and the translation group form elements in a space group generated from the 2 point group by *replacing* the diad rotation operations with diad screw operations. Similar to the creation of multiple diad axes in the $P2$ unit cell, inclusion of the translation group generates multiple diad screw axes in the unit cell of this new space group. Application of the 2_1 screw operation to an atom at $[x_f\ y_f\ z_f]$ generates a symmetry-equivalent atom at $[\bar{x}_f\ (y_f + \frac{1}{2})\ \bar{z}_f]$. The translation group vectors place additional equivalent atoms at $[\bar{x}_f\ (y_f + \frac{1}{2})\ \bar{z}_f] + [m_1\ m_2\ m_3] = [(m_1 - x_f)\ (m_2 + y_f + \frac{1}{2})\ (m_3 - z_f)]$. As we have noted previously, for every m_1, m_2, and m_3, there is a single local symmetry operation relating an atom at $[x_f\ y_f\ z_f]$ to each of these equivalent atoms. Here we are constrained to diad screw operations: $\mathbf{S}(2_1') = \mathbf{S}(\mathbf{t}_j)\mathbf{S}(2_1)$. The diad screw axes are parallel to the unique axis, intersecting the ac plane at $[x_f^*\ 0\ z_f^*]$. Placing an atom at $[x_f^*\ 0\ z_f^*]$ *on the* $2_1'$ *diad screw axis* and performing the diad screw operation generates a symmetry-equivalent atom that retains the original x_f^* and z_f^* coordinates, but is translated half the distance along the b axis to $[x_f^*\ \frac{1}{2}\ z_f^*]$. The diad screw operation on $[x_f^*\ 0\ z_f^*]$ about the original 2_1 axis, followed by translation, places an atom at the same position — at $[(m_1 - x_f^*)\ (m_2 + 0 + \frac{1}{2})\ (m_3 - z_f^*)]$, and

$$\begin{aligned} x_f^* &= m_1 - x_f \Longrightarrow x_f^* = \frac{m_1}{2} \\ \frac{1}{2} &= m_2 + \frac{1}{2} \Longrightarrow m_2 = 0 \\ z_f^* &= m_3 - z_f \Longrightarrow z_f^* = \frac{m_3}{2}. \end{aligned} \tag{2.22}$$

These equations are seen to be identical to Eqns. 2.20, indicating that the points of intersection of the diad screw axes with the ac plane in this space group are identical to the points of intersection of the diad rotation axes with the ac plane in the $P2$ space group: $[0\ 0\ 0]$, $[\frac{1}{2}\ 0\ 0]$, $[0\ 0\ \frac{1}{2}]$ and $[\frac{1}{2}\ 0\ \frac{1}{2}]$. The complete set of diad screw operations and their inverses, along with E and the translation group create the space group symbolized by the H-M notation $P\,1\,2_1\,1 \equiv P\,2_1$ (No. 4). The origin of the $P\,2_1$ unit cell lies at an arbitrary position along the b axis, just as it does in the $P2$ unit cell. Similarly, the b axis is selected so that it is coincident with a diad screw axis.

Local Symmetry Operations. If the set of local symmetry operations in a space group are all point group operations (without local translations) then the set is a mathematical group – the generating point group of the space group. If some of the members of the set are composed of point group operations coupled with local translations then the set of local operations is not a mathematical group – *but it behaves like one!*. The product of any two local operations on a location in a given unit cell will produce a symmetry-equivalent location that can be generated by a single operation, plus or minus a simple translation from the translation group, $[m_1\ m_2\ m_3]$, with m_1, m_2, and m_3 = -1, 0, or 1. If any one of the components in the vector is non-zero then the location will be in an adjacent unit cell. Its location can readily be transformed into a symmetry-equivalent position in the original unit cell by simply negating the translation (i.e., adding +1 to any fractional coordinates less than one, and -1 to any fractional coordinates greater than one). A simple example will serve to illustrate this. The local symmetry operators in the $P2$ space group are the identity operation $\mathbf{S}(E) \equiv \mathbf{R}(E)$ and the diad rotation operation $\mathbf{S}(2) \equiv \mathbf{R}(2)$. The product of any two of the operations in the set will produce another in the set. In particular, $\mathbf{S}(2)\mathbf{S}(2) = \mathbf{S}(E)$, satisfying the closure requirement for a group. The local operators in the $P2_1$ space group are the identity operation, $\mathbf{S}(E)$ and the diad screw operation, $\mathbf{S}(2_1)$. The products $\mathbf{S}(E)\mathbf{S}(2_1) = \mathbf{S}(E)\mathbf{S}(2_1) = \mathbf{S}(2_1)$ and $\mathbf{S}(E)\mathbf{S}(E) = \mathbf{S}(E)$ are consistent with the criteria for a group, but consider the results of $\mathbf{S}(2_1)\mathbf{S}(2_1)$:

$$\mathbf{S}(2_1)\begin{bmatrix} x_f \\ y_f \\ z_f \end{bmatrix} = \begin{bmatrix} \bar{1} & 0 & 0 \\ 0 & 1 & 0 \\ 0 & 0 & \bar{1} \end{bmatrix}\begin{bmatrix} x_f \\ y_f \\ z_f \end{bmatrix} + \begin{bmatrix} 0 \\ \frac{1}{2} \\ 0 \end{bmatrix} = \begin{bmatrix} \bar{x}_f \\ y_f + \frac{1}{2} \\ \bar{z}_f \end{bmatrix}$$

$$\mathbf{S}(2_1)\begin{bmatrix} \bar{x}_f \\ y_f + \frac{1}{2} \\ \bar{z}_f \end{bmatrix} = \begin{bmatrix} \bar{1} & 0 & 0 \\ 0 & 1 & 0 \\ 0 & 0 & \bar{1} \end{bmatrix}\begin{bmatrix} \bar{x}_f \\ y_f + \frac{1}{2} \\ \bar{z}_f \end{bmatrix} + \begin{bmatrix} 0 \\ \frac{1}{2} \\ 0 \end{bmatrix} = \begin{bmatrix} x_f \\ y_f + 1 \\ z_f \end{bmatrix}$$

$$= \begin{bmatrix} x_f \\ y_f \\ z_f \end{bmatrix} + \begin{bmatrix} 0 \\ 1 \\ 0 \end{bmatrix}. \tag{2.23}$$

Thus $\mathbf{S}(2_1)\mathbf{S}(2_1) = \mathbf{S}(E) + [0\ 1\ 0]$. The product of two $\mathbf{S}(2_1)$ operations is a new operation that is not in the set. However, negating the translation places the location back at its original location, as if it had been subjected to the $\mathbf{S}(E)$ operation. *If* we could ignore these translations (which we effectively do – by negating them), the set of local symmetry operations would be a mathematical group. *In general, every product of local operations will be equal to another operation in the set of local operations after translating the results of the product back into the original unit cell.* This will be implicit in the remaining discussions, when we state that there must be a single local symmetry operation that is the product of two local symmetry operations; from here on we will treat the set of local symmetry operations as if it was a group.

Space Group Tables (I). The four space groups already discussed, along with the remaining 226, are described in tabular detail in Volume A of the *International Tables for Crystallography*[27]. While it would be impossible to discuss all of the space groups and their properties in this volume, various components of those tables will be described whenever they pertain to particular aspects of our discussion in order for the reader to be able to make use of the vast amount of information contained in the *International Tables.* Each table describing a space group begins with a top line consisting of the H-M short symbol and the Schönflies symbol for the space group, the symbol for the generating point group* (without local translations), and the crystal system. A second line lists the sequential space group number, the full H-M symbol, and another symbol giving the symmetry of an important function used in structural solution, known as the *Patterson function,*

*The term "generating point group" is not in common usage. However it emphasizes that the inclusion of translation operators (always including the translation group) *replaces* the *point* symmetry with *space* symmetry.

which we will encounter in Chapter 6. For the $P2_1$ space group the first three lines appear as:

$\mathbf{P\,2_1}$	C_2^2	2	**Monoclinic**
No. 4	$P\,1\,2_1\,1$		Patterson symmetry $P\,1\,2/m\,1$

UNIQUE AXIS b

A number of projections of the unit cell follow this heading, showing the location of symmetry elements or symmetry-equivalent locations. For our purposes we will adopt a modified composite of such projections, using the notation employed in the *International Tables*. Fig. 2.51 illustrates this usage, showing projections of the $P2$ and $P2_1$ unit cells along the b axis onto the ac plane. The locations of the diad rotation and screw axes are shown where they intersect the ac plane. In projections where the diad axes lie in the plane (in an ab projection of a monoclinic unit cell, for example) the diad rotation axis is indicated with an arrow and the screw axis is represented with a "half-headed" arrow. The open circles represent a general position in the asymmetric unit in the unit cell, illustrating that there are two asymmetric units in the unit cell for each space group. The "+" signs next to the circles in the $P2$ projection indicate that the transformed general positions are all at the same (arbitrary) displacement, $+y_f$, along the b axis above the ac plane. The "1/2+" next to half of the circles in the $P2_1$ projection indicates that these locations are displaced half of the length of the b axis above the position indicated by the "+" sign as a result of the screw operation. In the *International Tables* a solid line represents a mirror plane in the projections containing the symmetry elements, but represents a reference line in those illustrating the positions in the unit cell. Since we have combined these projections we will use a heavy line to indicate a mirror plane.

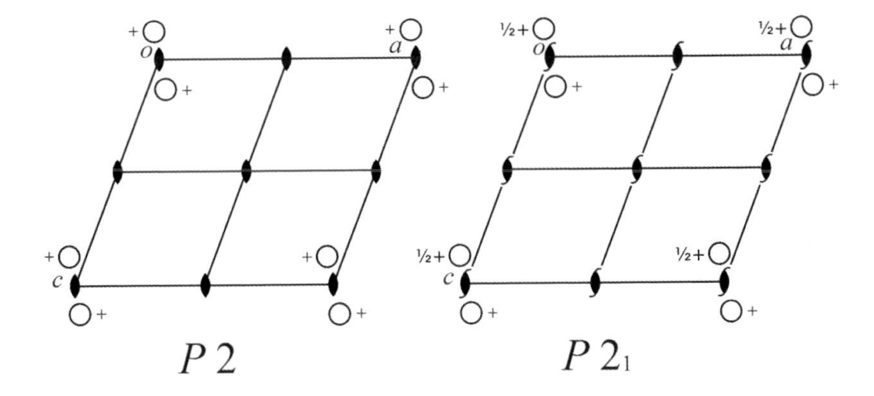

Figure 2.51 Orthogonal projections of $P2$ and $P2_1$ unit cells onto the ac plane. Diad rotational or screw axes intersect the ac plane at the origin and midpoint of the unit cell. In addition, diad axes intersect the midpoints of the a and c axes.

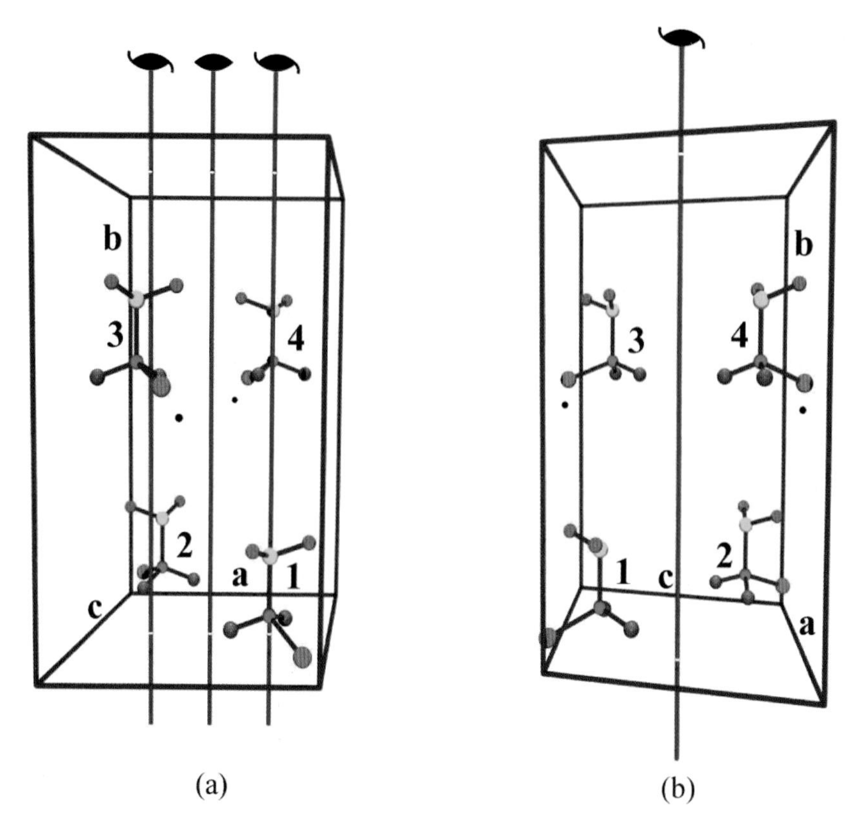

Figure 2.52 (a) $C2$ unit cell illustrating the location of diad rotation and screw axes. Centered lattice points are shown in the ab faces. (b) View parallel to the a axis showing a diad screw axis (all three axes are superimposed in this view).

The $C2$ Space Group. A third space group can be generated from the 2 point group by combining the identity and diad operations with the $C-$ or $A-$ centering operation.* Since the choice of the a and c axes are arbitrary, there is only one space group representing this combination of symmetry operations – we will use C-centering to illustrate it. As with the $P2$ space group, the origin is located on a diad axis coincident with the b axis. By an analysis identical to that given in the $P2$ discussion, diad rotation axes will intersect the ac plane in the same locations as they do in the $P2$ unit cell. Unlike the screw operator, the centering operation and the two-fold rotation are separate symmetry operations. In the centered unit cell the vectors related by the diad operation, $[x_f\ y_f\ z_f]$ and $[(1-x_f)\ y_f\ (1-z_f)] \equiv [\bar{x}_f\ y_f\ \bar{z}_f]$ are translated with the vector $\mathbf{t}_C = [\frac{1}{2}\ \ \frac{1}{2}\ 0]$, producing two more symmetry-equivalent positions at $[(x_f+\frac{1}{2})\ (y_f+\frac{1}{2})\ z_f]$ and $[(\bar{x}_f+\frac{1}{2})\ (y_f+\frac{1}{2})\ \bar{z}_f]$. Fig. 2.52(a) is a view of the centered unit cell looking at the ab plane. The four asymmetric units in the unit cell are labeled with the numbers **1** - **4** for reference. **1** and **2** are related by a two-fold rotation, as are **3** and **4**. The centering translation produces **3** from **1** and **4** from **2**. Thus **4** can be generated from **1** by a rotation followed by a translation. Treating the set of local symmetry operations as if it

*Recall that centering does not change the lattice point group.

was a group requires that there must be a single symmetry operation that is the product of the diad and centering operations. The illustration demonstrates clearly that this operation is not a diad rotation, and since the asymmetric units have different orientations, it can't be a centering translation either. Fig. 2.52(b) is a view parallel to the a axis, appearing to relate **1** to **4** through a rotation followed by a translation, a possible two-fold screw operation about an axis that is neither at the center, edge, or origin of the unit cell.

We hypothesize that the $\mathbf{S}(C)\mathbf{S}(2)$ product is $\mathbf{S}(2_1)$, and set out to find a way to test the hypothesis. An atom (the head of a vector) in **1** at $[x_f\ y_f\ z_f]$ is transformed into an atom in **4** at $[(\bar{x}_f + \frac{1}{2})\ (y_f + \frac{1}{2})\ \bar{z}_f]$ by a diad rotation and centering translation. The translation group operations place atoms at $[(m_1 - x_f + \frac{1}{2})\ (m_2 + y_f + \frac{1}{2})\ (m_3 - z_f)]$. The product of these symmetry operations must also be an element of the symmetry group, and we have proposed that the element is a diad screw operation: $\mathbf{S}(2_1) = \mathbf{S}(\mathbf{t}_g)\mathbf{S}(C)\mathbf{S}(2)$, where $\mathbf{S}(C) = \mathbf{R}(E) + [\frac{1}{2}\ \frac{1}{2}\ 0]$ is the C-centering operator. The hypothesis is tested in two steps. In the first step we *postulate* that the symmetry operation generating an equivalent atom at $[(m_1 - x_f + \frac{1}{2})\ (m_2 + y_f + \frac{1}{2})\ (m_3 - z_f)]$ from $[x_f\ y_f\ z_f]$ is a diad screw operation, allowing for the determination of the location of the putative axis in the unit cell. Following this we will test to see if this postulate is correct, by determining whether or not a general vector is transformed to the same position by the screw operation as it is by the product of the diad rotation and centering operation.

In order to accomplish the first step we position an atom onto our hypothesized screw axis at $[x_f^*\ 0\ z_f^*]$, the point where the screw axis would intersect the ac plane. As observed for the 2_1 axis in the $P2_1$ space group, the screw operation will place an equivalent atom at $[x_f^*\ \frac{1}{2}\ z_f^*]$. If screw operations are equivalent to the diad+centering+translation operations above then an atom at $[(m_1 - x_f^* + \frac{1}{2})\ (m_2 + 0 + \frac{1}{2})\ (m_3 - z_f^*)]$ will be at the same position:

$$\begin{aligned} x_f^* &= m_1 - x_f^* + \frac{1}{2} \Longrightarrow 2x_f^* = m_1 + \frac{1}{2} \Longrightarrow x_f^* = \frac{m_1}{2} + \frac{1}{4} \\ \frac{1}{2} &= m_2 + \frac{1}{2} \Longrightarrow m_2 = 0 \\ z_f^* &= m_3 - z_f^* \Longrightarrow z_f^* = \frac{m_3}{2}. \end{aligned} \tag{2.24}$$

If the hypothesized diad screw axes exist, then values of 0 or 1 for m_1 and m_3 will have them intersecting the ac plane at $[\frac{1}{4}\ 0\ \frac{1}{2}]$, $[\frac{3}{4}\ 0\ \frac{1}{2}]$, $[\frac{1}{4}\ 0\ 0]$, and $[\frac{3}{4}\ 0\ 0]$. The remaining integers will then generate the intersection points of equivalent screw axes throughout the lattice.

Selecting one of these screw axes as a test case, we now assess the validity of the hypothesis. Choosing the putative screw axis that intersects the ac plane at $[\frac{1}{4}\ 0\ \frac{1}{2}]$, we first transform an atom at the head of a general vector $[x_f\ y_f\ z_f]$ by rotating the vector about the axis, then translating it along this axis a distance of $\frac{1}{2}b$ to effect the screw operation. Since the axis is no longer coincident with the b axis, determining the transformation matrices to perform the rotation becomes complicated. An alternative way to accomplish the rotation is to translate the axis to the origin, moving the atom to be transformed along with it (Fig. 2.53). The two-fold rotation is then performed and the transformed point and axis are translated back, returning the axis to its original location. This rotation transforms $[x_f\ y_f\ z_f]$ into $[(\bar{x}_f + \frac{1}{2})\ y_f\ (\bar{z}_f + 1)]$. Since the transformed vector has a $(\bar{z}_f + 1)$ component, there must be an equivalent point in the original (adjacent) unit cell with a $\bar{z}$ component. That is, the rotation has transformed $[x_f\ y_f\ z_f]$ into $[(\bar{x}_f + \frac{1}{2})\ y_f\ \bar{z}_f]$. The rotation about the axis at $[\frac{1}{4}\ 0\ \frac{1}{2}]$ is effectively a rotation about the b axis, represented by $\mathbf{R}(2)$, followed by a translation of $\frac{1}{2}$ in the $\mathbf{a}$ direction. The operator therefore becomes $\mathbf{S}(2) = \mathbf{R}(2) + [\frac{1}{2}\ 0\ 0]$. Finally, adding a $\frac{1}{2}$ b translation results in the transformation of $[x_f\ y_f\ z_f]$ into $[(\bar{x}_f + \frac{1}{2})\ (y_f + \frac{1}{2})\ \bar{z}_f]$: $\mathbf{S}(2_1) = \mathbf{S}(2) + [0\ \frac{1}{2}\ 0]$:

$$\mathbf{S}(2_1)\begin{bmatrix} x_f \\ y_f \\ z_f \end{bmatrix} = \begin{bmatrix} \bar{1} & 0 & 0 \\ 0 & 1 & 0 \\ 0 & 0 & \bar{1} \end{bmatrix}\begin{bmatrix} x_f \\ y_f \\ z_f \end{bmatrix} + \begin{bmatrix} \frac{1}{2} \\ 0 \\ 0 \end{bmatrix} + \begin{bmatrix} 0 \\ \frac{1}{2} \\ 0 \end{bmatrix} = \begin{bmatrix} \bar{x}_f + \frac{1}{2} \\ y_f + \frac{1}{2} \\ \bar{z}_f \end{bmatrix}. \tag{2.25}$$

This is the same location as that generated when the centering translation and two-fold rotation are applied to $[x_f\ y_f\ z_f]$. The two operations are equivalent to a two-fold screw rotation about the axis at $[\frac{1}{4}\ 0\ \frac{1}{2}]$. The reader is encouraged to establish that the other axes are also diad screw axes.

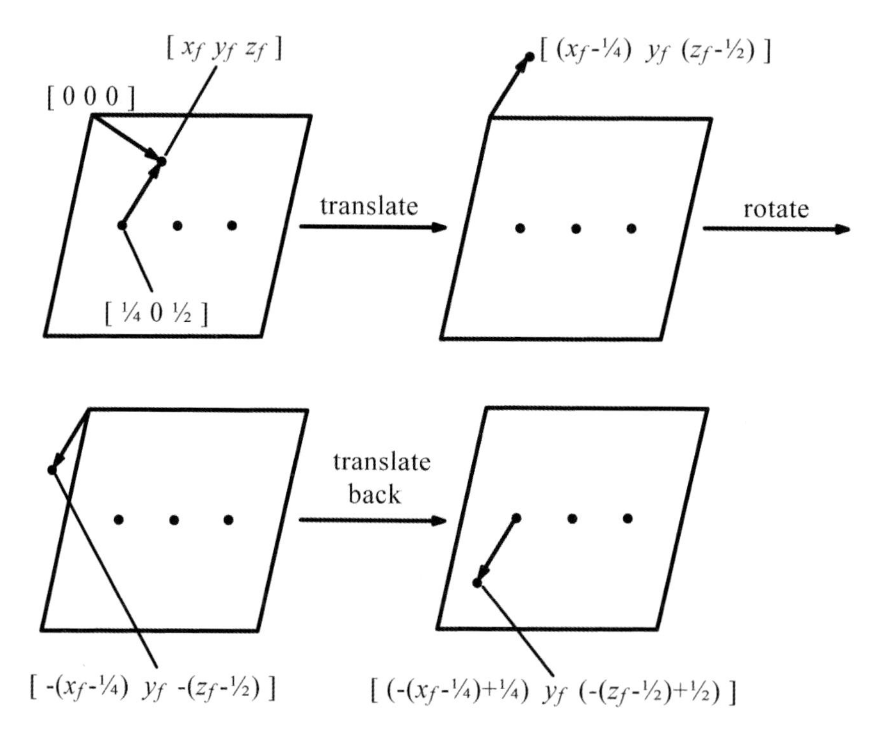

Figure 2.53 Two-fold rotation around an axis intersecting the ac plane at $[\frac{1}{4}\ 0\ \frac{1}{2}]$.

The centered space group created by combining the identity, the translation group, the complete set of diad operations, the complete set of centering translations and their inverses and all of the screw operations and their inverses, is the $C\,1\,2\,1 \equiv C\,2$ (No. 5) space group. The $P2$ space group contains the same diad rotation operations that are in the $C2$ space group, and is therefore a subgroup of the $C2$ group. The $C2$ space group is formed by adding centering operations to the elements of the $P2$ space group, with the presence of the centering and diad operations requiring the existence of diad screw operations in the new group.

It appears from the previous discussion that the $C2_1$ space group should be generated next, by combining the 2_1 operations with the centering translation (and its inverse). By arguments similar to those just presented, the screw axes will reside at the same locations as the diad axes in the $C2$ unit cell — and the combination of a centering translation and screw operation will produce a rotation about a diad rotation axis at $[\frac{1}{4}\ 0\ \frac{1}{2}]$, $[\frac{3}{4}\ 0\ \frac{1}{2}]$, $[\frac{1}{4}\ 0\ 0]$, or $[\frac{3}{4}\ 0\ 0]$. Fig. 2.54 is the projection of the $C2$ and $C2_1$ unit cells. For comparative purposes the general position has been selected at the same position relative to the two-fold rotation axis in each cell. Note that the location of the symmetry elements and the lattice contents occur at the same places in the lattice. The unit cells differ only by a selection of the origin. If the origin in the $C2_1$ unit cell is shifted by $[\frac{1}{4}\ 0\ 0]$, it will reside on the diad rotation axis rather than the screw axis and the unit cells will be indistinguishable. The $C2$ and $C2_1$ space groups are identical. Since the minimal symmetry operation necessary to generate the space group is the diad rotation, the space group is labeled with the $C2$ symbol. Note that the 2_1 axes are at the same locations as they are in the $P2_1$ unit cell before the origin is shifted.

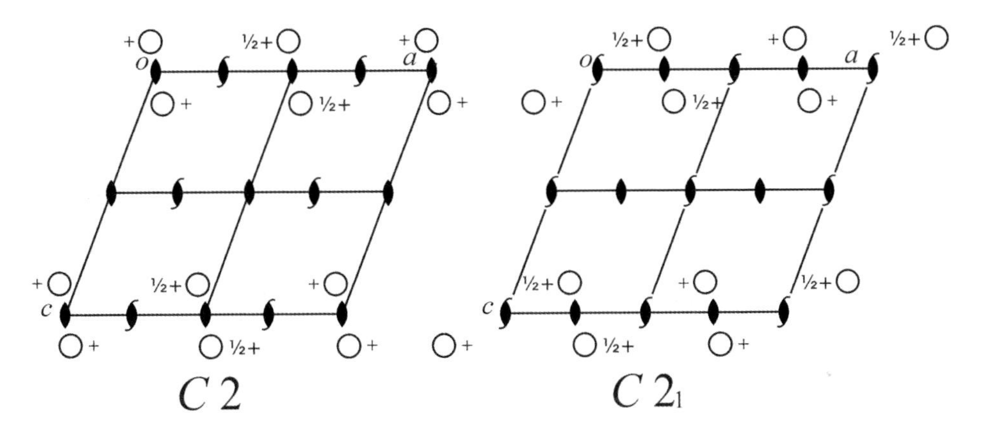

Figure 2.54 Orthogonal projections of $C2$ and $C2_1$ unit cells onto the *ac* plane.

Space Group Tables (II). The *International Tables* provides detailed information for each space group after the projections are plotted. Much of the information involves diffraction properties — which we will take up in the next chapter. We are concerned here with the most important information related to the lattice and its contents for the $P2$, $P2_1$, and $C2$ space groups, summarized in Table 2.13. The first entry under the **Positions** heading begins with a number, a letter and a symmetry symbol, followed by the position of a general vector at $x, y, z \equiv [x_f\ y_f\ z_f]$ and the positions generated by the local symmetry operations in the space group. The initial number is the number of symmetry-equivalent positions in the unit cell corresponding to a general vector. It is called the *multiplicity* of the general position – and corresponds to the number of asymmetric units in the unit cell (unless the *entire* asymmetric unit resides on a symmetry element). The letter is the *Wyckoff letter*, corresponding to a notation developed by R.W.G. Wyckoff. The symmetry symbol is the *site symmetry* of the position. The site symmetry is the point symmetry at that position; for the *general position* in the asymmetric unit it will always be 1. If there is an entry under the "Coordinates" heading, it indicates that translations are to be added to each of the positions listed. The two positions listed for a $C2$ unit cell must each have $(0, 0, 0)$ and $(\frac{1}{2},\frac{1}{2},0)$ added to them in order to generate the four symmetry-equivalent positions in the unit cell. Note that the multiplicity of the general position is 4 in the $C2$ unit cell. If there is no entry under "Coordinates" in the table, $(0, 0, 0)$ is assumed.

Table 2.13 Information in the *International Tables for Crystallography* for the $P2$, $P2_1$, and $C2$ Space Groups.

	$P2$	$P2_1$	$C2$
Origin on	2	2_1	2
Positions	2 e 1	2 a 1	4 c 1
(General)	(1) x, y, z	(1) x, y, z	(1) x, y, z
	(2) $\bar{x}, y, \bar{z}$	(2) $\bar{x}, y+\frac{1}{2}, \bar{z}$	(2) $\bar{x}, y, \bar{z}$
(Special)	1 d 2 $\frac{1}{2}, y, \frac{1}{2}$		
	1 c 2 $\frac{1}{2}, y, 0$		
	1 b 2 $0, y, \frac{1}{2}$		2 b 2 $0, y, \frac{1}{2}$
	1 a 2 $0, y, 0$		2 a 2 $0, y, 0$
Coordinates			$(0, 0, 0)+$ $(\frac{1}{2},\frac{1}{2},0)+$

The local matrix operators for the space group are readily deduced from the general positions. For example, from Table 2.13, the general positions for the $P2_1$ space group are listed as x, y, z and $\bar{x}, y+\frac{1}{2}, \bar{z}$. The matrices generate these positions from an arbitrary vector, $[x_f\ y_f\ z_f]$:

$$\mathbf{S}(E)\begin{bmatrix} x_f \\ y_f \\ z_f \end{bmatrix} = \begin{bmatrix} 1 & 0 & 0 \\ 0 & 1 & 0 \\ 0 & 0 & 1 \end{bmatrix}\begin{bmatrix} x_f \\ y_f \\ z_f \end{bmatrix} = \begin{bmatrix} x_f \\ y_f \\ z_f \end{bmatrix} \tag{2.26}$$

$$\mathbf{S}(2_1)\begin{bmatrix} x_f \\ y_f \\ z_f \end{bmatrix} = \begin{bmatrix} \bar{1} & 0 & 0 \\ 0 & 1 & 0 \\ 0 & 0 & \bar{1} \end{bmatrix}\begin{bmatrix} x_f \\ y_f \\ z_f \end{bmatrix} + \begin{bmatrix} 0 \\ \frac{1}{2} \\ 0 \end{bmatrix} = \begin{bmatrix} \bar{x}_f \\ y_f + \frac{1}{2} \\ \bar{z}_f \end{bmatrix}. \tag{2.27}$$

Following the description of the general position under the **Positions** heading, possible *special positions* are indicated. These are locations in the unit cell that are taken back onto themselves by a symmetry operation because *the point to be transformed resides on a symmetry element.* The special position is described with a series of four symbols indicating its multiplicity, its Wyckoff letter, its site symmetry, and the constraints on the vector that defines the special position. For example, a $P2$ unit cell has a general position denoted by 2 e 1; an atom located at $[x_f\ y_f\ z_f]$ generates another atom at $[\bar{x}_f\ y_f\ \bar{z}_f] \equiv [(1-x_f)\ y_f\ (1-z_f)]$, resulting in two atoms in the unit cell — the general position has a multiplicity of two.

The general position descriptors are followed by those corresponding to four different special positions in the unit cell. The multiplicity of each of these special positions is 1; the site symmetry is the point symmetry of the special position — two-fold rotational symmetry for $P2$ and $C2$. In the first special position in the list for $P2$, an atom located there will reside at an arbitrary location, y_f, on a diad rotational axis that intersects the *ac* plane in its center, at $[\frac{1}{2}\ 0\ \frac{1}{2}]$. The diad operation on the vector to an atom residing at some position along this axis will *not* generate an atom in another location, and *physically* it cannot generate an equivalent atom on top of itself!* The three remaining special positions are seen to be the diad axes that are coincident with the origin and the *a* and *c* axes. An atom located on a diad axis in the $P2$ unit cell will contribute only a single atom to the contents of the unit cell. The $P2_1$ unit cell has no special positions, since the screw operation always translates a vector to another position — an atom on a screw axis will never be transformed into a symmetry-equivalent atom back onto itself. The $C2$ unit cell has two special positions, each with a multiplicity of 2, at $[0\ y_f\ \frac{1}{2}]$ and $[0\ y_f\ 0]$, on diad axes at the origin and bisecting the *c* axis.

The reader may wonder why the special positions in the $C2$ unit cell have a multiplicity of 2, or why a position on one of the other diad rotational axes is not included in the $C2$ special position list. It is often instructive to consider a specific case, rather than dealing with the abstractions of vector algebra. Consider an atom located on the diad rotational axis at the origin — at $[0\ 0.2\ 0]$. This atom lies on the *b* axis, and is shared with the four unit cells that share the axis. Thus it contributes $1/4$ of an atom to the unit cell. Since the unit cell is bounded by four *b* axes, each contributing $1/4$ of an atom to the unit cell, the atom located on the *b* axis accounts for one atom in the unit cell. All of the two-fold axes will generate an atom at another origin, but, the screw axis intersecting the *ac* plane at $[\frac{1}{4}\ 0\ 0]$ generates a second atom at $[\frac{1}{2}\ 0.7\ 0]$, on the diad axis bisecting the *a* axis. This atom shares the *ab* plane with another unit cell, contributing half an atom to each cell. The screw axis intersecting the plane at $[\frac{1}{4}\ 0\ \frac{1}{2}]$ generates a second atom at $[\frac{1}{2}\ 0.7\ 1]$, the equivalent position on the opposite *ab* face, contributing another half of an atom to the unit cell. The remaining screw axes place equivalent atoms in other unit cells. It follows that an atom on the *b* axis results in a net of two atoms per unit cell, rather than the four that would correspond to an atom in the general position, giving the position a multiplicity of 2. The symmetry-equivalent atom resides on the diad rotational axis that bisects the *a* axis — it is in a special position there as well, but since one special position generates the other there is no need to include "1 *c* 2 $[\frac{1}{2}\ y_f\ 0]$" in the list (no additional atoms are accounted for by doing so). Now consider an atom on the diad rotational axis bisecting the *c* axis — at $[0\ 0.2\ \frac{1}{2}]$. The screw axis intersecting the *ac* plane at $[\frac{1}{4}\ 0\ \frac{1}{2}]$ generates a second atom at $[\frac{1}{2}\ 0.7\ \frac{1}{2}]$, on the diad rotational axis passing through the center of the unit cell. This contributes a second atom to the unit cell — corresponding to a multiplicity of 2. The other screw axes in the unit cell generate symmetry-equivalent atoms outside the cell from this atom, and because the symmetry-equivalent atom is also on a special position on the axis that intersects the *ac* plane in its center, there is no need for "1 *d* 2 $[\frac{1}{2}\ y_f\ \frac{1}{2}]$" in the special position list. It should be noted that the descriptors for the special positions in the list could be replaced by the two equivalent ones that are not included.

*When analyzing the symmetry, we *do* allow atoms to be placed on top of one another *conceptually*.

The *Pm* Space Group. We now turn our attention to the monoclinic space groups derived from the m point group. The space group with no local translation operators contains the identity, the reflection operator, $\mathbf{S}(m) = \mathbf{R}(m) + [0\ 0\ 0]$, and the translation group. The reflection operator transforms a general vector in the unit cell, $[x_f\ y_f\ z_f]$, into the symmetry-equivalent vector: $\mathbf{S}(m)[x_f\ y_f\ z_f] = [x_f\ \bar{y}_f\ z_f]$. As we have observed previously, the translation group generates equivalent positions at $[(m_1+x_f)(m_2-y_f)(m_3+z_f)]$, and there is a single local symmetry operation that transforms $[x_f\ y_f\ z_f]$ into each of these positions for each translation vector. Having only reflection operations available, we must therefore locate mirror planes and their reflection operations such that: $\mathbf{S}(m') = \mathbf{S}(\mathbf{t}_j)\mathbf{S}(m)$. The mirror planes will be perpendicular to the b axis — and we begin by placing an atom at the point of intersection of a mirror plane and the unique axis, at $[0\ y_f^*\ 0]$. The reflection operation, m', will generate a symmetry-equivalent atom in the same position, which must also be at the position generated by the original m operation after translation, $[(m_1+0)(m_2-y_f^*)(m_3+0)]$. Thus,

$$\begin{aligned} m_1 &= 0 \\ y_f^* &= m_2 - y_f^* \Longrightarrow y_f^* = \frac{m_2}{2} \\ m_3 &= 0. \end{aligned} \tag{2.28}$$

The atom and point of intersection of the mirror plane does not change its x_f and z_f positions; values of m_2 of 0 or 1 place the intersection point on the b axis at $[0\ 0\ 0]$ and $[0\ \frac{1}{2}\ 0]$. The remaining integers place mirror planes throughout the lattice at the same positions in each unit cell. A monoclinic unit cell with its contents related

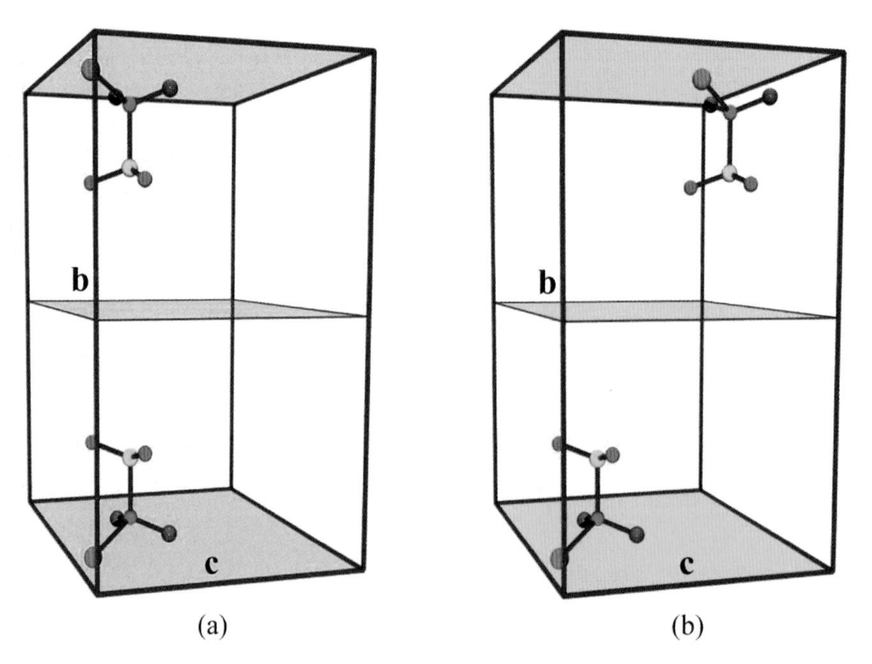

Figure 2.55 (a) Pm unit cell and asymmetric units related by a mirror plane (blue). (b) Pc unit cell and asymmetric units related by a c-glide plane (red).

by a mirror plane is illustrated in Fig. 2.55(a). The mirror planes, shown in blue, lie in the *ac* plane, or are parallel to it — passing through the center of the unit cell. Unit cells sharing an *ac* face will have the mirror image of the asymmetric unit on each side of the face, as illustrated in Fig. 2.56. The combination of the complete set of mirror operators (they are their own inverses) with the identity and translation group creates the $P1m1 \equiv Pm$ (No. 6) space group.

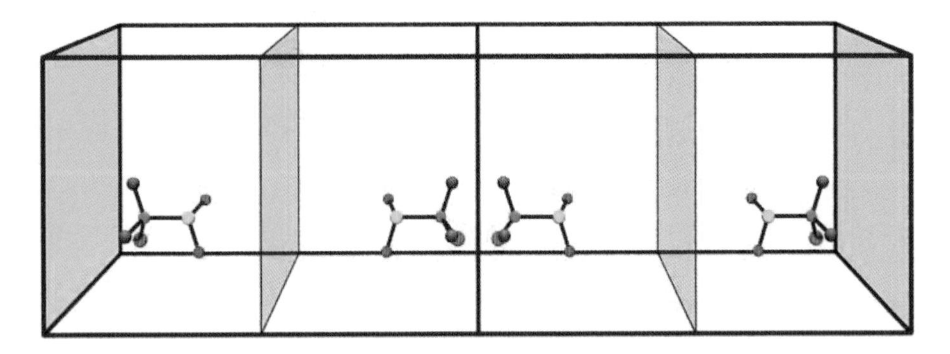

Figure 2.56 Two adjacent Pm unit cells sharing an *ac* face (center of the figure) and illustrating the mirror symmetry of the *ac* plane.

The *Pc* Space Group. Reflection operations behave in a similar manner to two-fold rotations, in that two sequential operations result in the identity operation. Previously, this allowed us to combine a diad rotation with a local translation halfway along the *b* axis to create a new symmetry operator and a new space group. In the same manner, we can combine a reflection operation with a translation. The translation in this case must be parallel to the mirror plane, and in a direction that moves the asymmetric unit into a symmetry-equivalent position in an adjacent unit cell. This effectively transforms the mirror plane in the Pm unit cell into a reflection-translation plane, a *glide plane*, as illustrated in red in Fig. 2.55(b). The reflection-translation operation, known as a *glide* operation, is demonstrated in Fig. 2.57. The asymmetric unit is reflected across the plane passing through the center of the cell exactly as in the reflection operation for the Pm unit cell. The reflected asymmetric unit is then translated a distance that is half of the length of the *c* axis, *along a vector parallel to the c axis.* A second reflection and translation places the asymmetric unit in the adjacent unit cell at a position the length of the *c* axis from its original location.

As with the screw operation, this reflection-translation operation retains the identical composition of the unit cell contents throughout the lattice. The operator for a reflection + *c*-translation is $\mathbf{S}(c) = \mathbf{R}(m) + \mathbf{t}_G$, where $\mathbf{t}_G = [0\ \ 0\ \ \frac{1}{2}]$, transforming the general vector $[x_f\ y_f\ z_f]$ into a symmetry-equivalent vector:

$$\begin{bmatrix} 1 & 0 & 0 \\ 0 & -1 & 0 \\ 0 & 0 & 1 \end{bmatrix} \begin{bmatrix} x_f \\ y_f \\ z_f \end{bmatrix} + \begin{bmatrix} 0 \\ 0 \\ \frac{1}{2} \end{bmatrix} = \begin{bmatrix} x_f \\ \bar{y}_f \\ \frac{1}{2} + z_f \end{bmatrix}. \tag{2.29}$$

The translation group creates equivalent positions at $[(m_1 + x_f)(m_2 - y_f)(m_3 + \frac{1}{2} + z_f)]$, and we now postulate single *c*-glide operations, $\mathbf{S}(c')$, that transform

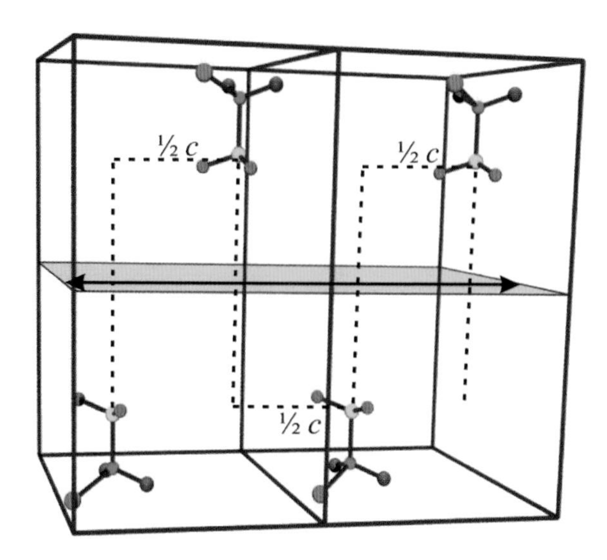

Figure 2.57 (a) Two adjacent Pc unit cells sharing an ab face; the translation arrow is parallel to the c axis.

$[x_f\ y_f\ z_f]$ into each of these positions. Placing an atom on the glide plane at its point of intersection with the b axis at $[0\ y_f^*\ 0]$ and performing the $\mathbf{S}(c')$ reflection-local translation operation generates an equivalent atom at $[0\ y_f^*\ \frac{1}{2}]$, remaining on the glide plane but translated in the c direction by $\frac{1}{2}c$, which must also be at a position generated by the original $\mathbf{S}(c)$ operation followed by translation: $[(m_1 + 0)(m_2 - y_f^*)(m_3 + \frac{1}{2} + 0)]$. Thus,

$$\begin{aligned} m_1 &= 0 \\ y_f^* &= m_2 - y_f \Longrightarrow y_f^* = \frac{m_2}{2} \\ m_3 &= 0. \end{aligned} \tag{2.30}$$

The glide planes intersect the b axis at the same locations as the mirror planes in the Pm space group, at $[0\ 0\ 0]$ and $[0\ \frac{1}{2}\ 0]$ — illustrated in Fig. 2.58. As before, the various values of m_2 place glide planes, each with a glide operator, throughout

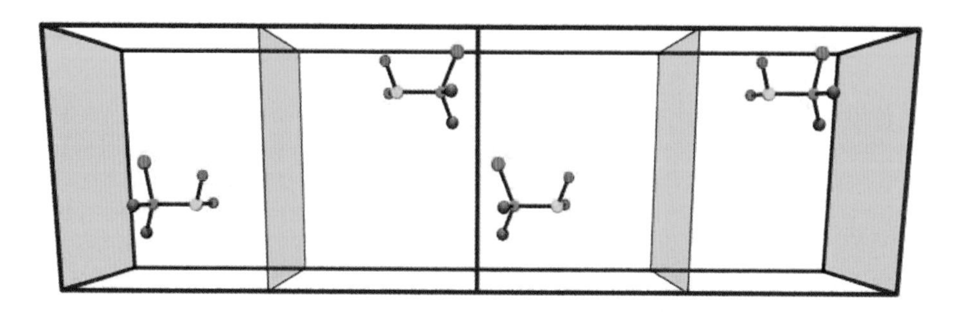

Figure 2.58 Two adjacent Pc unit cells sharing an ac face (center of the figure) and illustrating the glide symmetry of the ac plane.

the lattice at the same locations in each unit cell. In order to create an element in a mathematical group, each of the glide operators must have an inverse. For $m_2 = 0$, $(\mathbf{S}(c))^{-1} = \mathbf{R}(m) + \mathbf{t}'_G$, where $\mathbf{t}'_G = [0 \;\; 0 \;\; -\frac{1}{2}]$. The combination of the complete set of glide operators and their inverses, the translation group and the identity operation creates the $P\,1\,c\,1 \equiv P\,c$ (No. 7) space group, where the c refers specifically to the translation parallel to the c axis. In the long form of the symbol, the c is placed in the "b" position; as with m in $P\,1\,m\,1$, the position in the symbol indicates a reflection-translation operation across a plane perpendicular to the unique b axis.

Projections of the $P\,m$ and $P\,c$ unit cells are shown in Fig. 2.59. In the $P\,m$ unit cell the general position is amended with a dividing line and a comma in the left half of the circle. A point represented by an open circle (the general position) is considered to be part of an asymmetric unit that is right-handed. If the right-handed asymmetric unit is related to a left-handed image of itself through either a reflection operation, an inversion operation, or a rotoinversion operation, then a symmetry-equivalent point within the left-handed image is indicated by a circle with a comma in its center. The single symbol containing both an open half (right-handed) and a half containing a comma (left-handed) indicate that the projection of the asymmetric unit and its mirror image lie atop one another. The "+" sign next to the open half of the circle indicates that the general position in the asymmetric unit is at an arbitrary displacement, $+y_f$, along the b axis above the ac mirror plane, and the "$-$" sign next to the half containing the comma indicates that an equivalent position in its mirror image is located at an equal displacement, $-y_f$, along the b axis on the opposite side of the ac mirror plane.

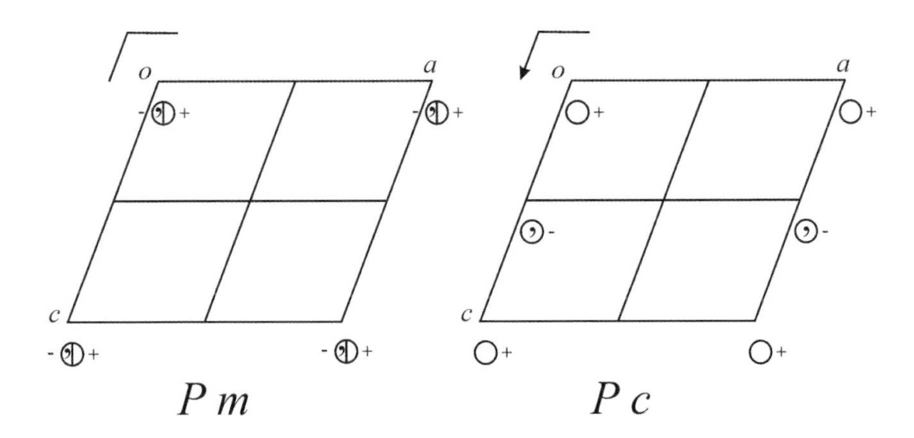

Figure 2.59 Orthogonal projections of $P\,m$ and $P\,c$ unit cells onto the ac plane.

Unlike the unit cells for space groups derived from the 2 point group — which allowed for an arbitrary placement of the origin along the b axis — a mirror plane in the $P\,m$ unit cell passes through the center of the unit cell , and the location of the origin is therefore constrained to lie in the ac plane — another mirror plane. The symbols in the $P\,c$ projection indicate that the mirror images are at equal distances above and below the ac plane, but that the mirror image of the asymmetric unit containing the general position has been translated by half of the distance along a vector parallel to the c axis. The intersecting lines in the upper left of the projection of the Pm unit cell indicate a mirror plane; the arrowhead at the end of the line parallel to the c axis of the projection of the Pc unit cell indicates a glide plane with a translation parallel to the axis. Since a glide plane passes through the center of the unit cell, the origin must lie on a glide plane.

Because the c axis is not unique in the Pc unit cell, it is also possible to translate along a vector parallel to the a axis. In order to differentiate these two glide planes from one another they are labeled with the axis that the translation vector parallels. Thus a c-glide operation couples a reflection across a plane perpendicular to the b axis with a $1/2\ c$ translation parallel to the c axis. An a-glide operation involves a reflection followed by a $1/2\ a$ translation parallel to the a axis. Since we can always initially assign the "translation axis" as the c axis, the a-glide does not create a new space group. The Pa and Pc space groups are equivalent. The c-glide is selected as the *standard* setting, but monoclinic space group symbols in which the unit cell has an a-glide plane occur commonly in the literature. Transforming from the Pa unit cell into a P_c unit cell is a relatively simple matter. The **a** and **c** basis vectors must be exchanged. This is accomplished by switching the a and c axial lengths and the x_f and z_f coordinates of the unit cell contents. It is also necessary to change the signs of the y_f coordinates, since the exchange of **a** and **c** turns $(\mathbf{c} \times \mathbf{a})$ into $(\mathbf{a} \times \mathbf{c}) = -(\mathbf{c} \times \mathbf{a})$. In a right-handed monoclinic unit cell, $(\mathbf{c} \times \mathbf{a})$ is a vector parallel to **b**. Switching the axes creates a $(\mathbf{c} \times \mathbf{a})$ vector in the opposite direction, indicating that **b** must be changed in direction in order to keep the unit-cell right-handed. The matrix transformation of the coordinates is

$$\begin{bmatrix} 0 & 0 & 1 \\ 0 & -1 & 0 \\ 1 & 0 & 0 \end{bmatrix} \begin{bmatrix} x_f \\ y_f \\ z_f \end{bmatrix} = \begin{bmatrix} x'_f \\ y'_f \\ z'_f \end{bmatrix}.$$

There is another possible glide operation for which it is not immediately obvious that the resulting space group will be equivalent to Pc. In this case the reflection is followed by a translation of $1/2\ a$ in the **a** direction and a second translation of $1/2\ c$ in the **c** direction. The translational vector, $(\mathbf{a} + \mathbf{c})/2$, is parallel to the *diagonal* connecting origins in the ac plane. A second reflection and translation will result in a translation of a parallel to the a axis, and c parallel to the c axis — clearly a unit translation to a symmetry-equivalent position in the unit cell located diagonally to the original cell. The resulting glide operation, $\mathbf{S}(n) = \mathbf{R}(m) + \mathbf{t}_G$, where $\mathbf{t}_G = [\frac{1}{2}\ 0\ \frac{1}{2}]$, is called a *diagonal* glide, identified by the letter n. It would seem reasonable to denote the n-glide as a d-glide, but the symbol d is reserved for another type of glide plane that occurs in the space groups with orthorhombic or higher symmetry, known as a *diamond glide*. The n derives from *net*, another name for a two-dimensional lattice. The unit cell diagonals connect the "net" points, just as the unit cell axes do. Indeed, this provides us with a clue for resolving whether or not the diagonal glide operation actually generates a new space group.

Fig. 2.60 shows a monoclinic axis projected along the b axis which intersects the ac plane in the center of the figure; we will assume that it contains an n-glide plane. The glide direction is parallel to the ac diagonal. The axes for the cell containing the diagonal glide plane are labeled **a** and **c**. As we have already discussed, the chosen unit cell is not unique — an alternative cell choice is shown in the figure. The **c** axis becomes the $\mathbf{a}'$ axis of the new cell, and the $\mathbf{c}'$ axis is chosen so that it

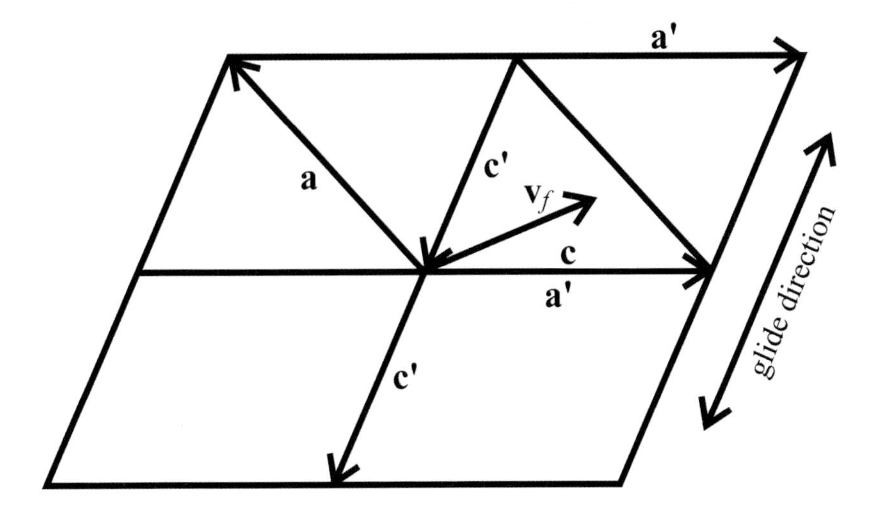

Figure 2.60 Projection of Pn unit cells with axes **a** and **c**. **b** is perpendicular to the projection plane. $\mathbf{v}_f$ is a general vector. $\mathbf{a}'$ and $\mathbf{c}'$ are axes for the Pc unit cell in the lattice.

is along the diagonal of the original unit cell; $\mathbf{c}'$ is the negative of the vector sum of **a** and **c**.* The alternative unit cell has a *c*-glide plane, and the space group is Pc. The Pn and Pc space groups are therefore equivalent.

Two decades ago standard practice required that the coordinates of atoms located in a crystal structure determination were reported in terms of the standard unit cell, which, in monoclinic systems meant the unit cell with a *c*-glide plane. For monoclinic unit cells, this practice has largely fallen by the wayside and space groups with *n* and *a* glide planes are routinely reported in the literature. Despite this trend, it is often useful to be able to transform between unit cells, and the Pn to Pc transformation provides an instructive example.

The unit cell vectors for the Pn, $[\mathbf{a}\ \mathbf{b}\ \mathbf{c}]$, and Pc unit cells, $[\mathbf{a}'\ \mathbf{b}'\ \mathbf{c}']$ are related as follows: $\mathbf{a}' = \mathbf{c}$, $\mathbf{b}' = \mathbf{b}$, and $\mathbf{c}' = -(\mathbf{a} + \mathbf{c})$. In order to determine the Pc unit cell parameters from the Pn unit cell, we determine the Cartesian coordinates of the unit cell vectors in the Pn basis from the Cartesian coordinates of the unit cell vectors in the Pc unit cell by creating the **B** matrix (Eqns. 1.129 and 1.130) from

*The diagonal vector of the Pn unit cell is $\mathbf{a} + \mathbf{c}$. Choosing this vector as $\mathbf{c}'$ changes the direction of $\mathbf{a} \times \mathbf{c}$, creating a left-handed unit cell. To keep the unit cell right handed, $\mathbf{c}'$ must be chosen in the reverse direction: $\mathbf{c}' = -(\mathbf{a} + \mathbf{c})$.

the Pc unit cell parameters, a, b, c, β, and $\alpha = \gamma = 90°$: $\mathbf{B} = [\mathbf{a}_c\ \mathbf{b}_c\ \mathbf{c}_c]$. The unit cell parameters for the new unit cell are obtained from the relationships above:

$$\begin{aligned}
\mathbf{a}'_c = \mathbf{c}_c &= [c_x\ c_y\ c_z] \\
\mathbf{b}'_c = \mathbf{b}_c &= [b_x\ b_y\ b_z] \\
\mathbf{c}'_c = -(\mathbf{a}_c + \mathbf{c}_c) &= [(-(a_x + c_x))\ \ (-(a_y + c_y))\ \ (-(a_z + c_z))] \\
a' &= c \\
b' &= b \\
c' = (\mathbf{c}'_c \cdot \mathbf{c}'_c)^{\frac{1}{2}} &= ((a_x + c_x)^2 + (a_y + c_y)^2 + (a_z + c_z)^2)^{1/2} \\
\cos\beta' &= \frac{\mathbf{a}'_c \cdot \mathbf{c}'_c}{a'c'}.
\end{aligned}$$

The vector $\mathbf{v}_f$, shown in projection in Fig. 2.60, is a general vector in the lattice. It has fractional coordinates of $[x_f\ y_f\ z_f]$ in the Pn unit cell, and $[x'_f\ y'_f\ z'_f]$ in the Pc unit cell. The transformation of the vector is effected by expressing it in both coordinate bases:

$$\begin{aligned}
\mathbf{v}_f &= x_f\mathbf{a}_c + y_f\mathbf{b}_c + z_f\mathbf{c}_c \\
&= x'_f\mathbf{a}' + y'_f\mathbf{b}' + z'_f\mathbf{c}' \\
&= x'_f\mathbf{c}_c + y'_f\mathbf{b}_c - z'_f(\mathbf{a}_c + \mathbf{c}_c) \\
&= -z'_f\mathbf{a}_c + y'_f\mathbf{b}_c + (x'_f - z'_f)\mathbf{c}_c \Longrightarrow \\
z'_f &= -x_f \\
y'_f &= y_f \\
x'_f - z'_f &= x'_f + x_f \ = \ z_f \\
x'_f &= z_f - x_f.
\end{aligned}$$

This provides a matrix for transforming the contents of the Pn unit cell into the Pc unit cell:

$$\begin{bmatrix} -1 & 0 & 1 \\ 0 & 1 & 0 \\ -1 & 0 & 0 \end{bmatrix} \begin{bmatrix} x_f \\ y_f \\ z_f \end{bmatrix} = \begin{bmatrix} x'_f \\ y'_f \\ z'_f \end{bmatrix}. \tag{2.31}$$

Because there are only a limited number of *alternate cell choices* in the monoclinic space groups, such transformations are needed only occasionally. On the other hand, there is no obvious unique axis in an orthorhombic unit cell. There are six possible ways to label the axes in the unit cells of each of the 58 orthorhombic space groups. The symmetry labels for each of the axes change for each of the six permutations, and it is important to be able to transform the many possibilities into a limited number of *standard settings*. The "P_n to P_c" transformation serves as an instructive example of how this is done.

The *Cm* Space Group. In a manner similar to the formation of the $C2$ space group from the 2 point group, a new space group can be generated from the m point group by combining the identity and reflection operations with the $C-$ (or $A-$) centering operation. The centering translation adds two symmetry-equivalent positions to $[x_f\ y_f\ z_f]$ and $[x_f\ \bar{y}_f\ z_f]$: $[(x_f + \frac{1}{2})\ (y_f + \frac{1}{2})\ z_f]$ and $[(x_f + \frac{1}{2})\ (\bar{y}_f + \frac{1}{2})\ z_f]$.

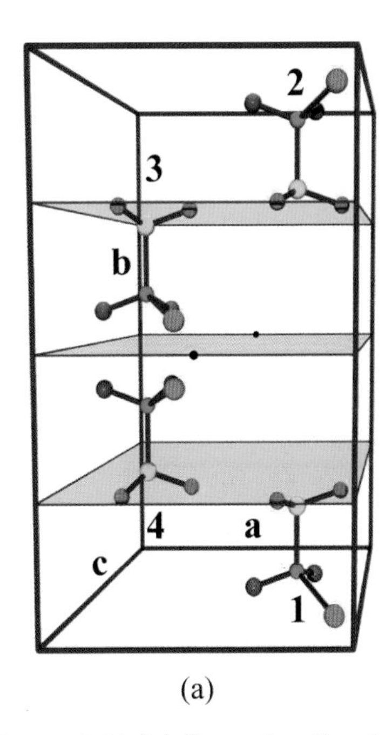

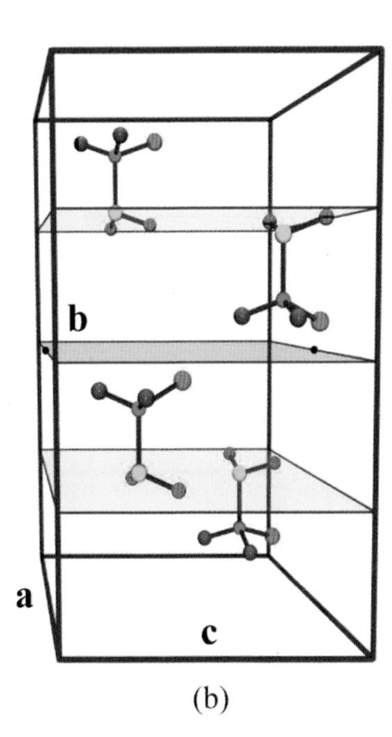

(a) (b)

Figure 2.61 (a) Cm unit cell and asymmetric units related by a mirror plane (blue) and a-glide planes (red). (b) Cc unit cell and asymmetric units related by a c-glide plane (red) and n-glide planes (yellow).

Fig. 2.61(a) illustrates the results of these operations; the asymmetric units are labeled **1** - **4** for reference. **1** and **2** are related by a reflection operation, as are **3** and **4**. The centering operation produces **3** from **1** and **4** from **2**. Thus **4** can be generated from **1** by a reflection followed by a centering translation. In order for these local operations to behave as if they were elements in a mathematical group there must be a single symmetry operation that is the product of the reflection and centering operations. The illustration indicates that this operation is neither another reflection nor a translation, and indeed appears to be a reflection-translation operation with the translation vector parallel to the a axis. As we have done previously, we determine whether or not the operation that generates **4** from **1** involves an a-glide plane by first postulating its existence in order to determine its location. We then test the correctness of the postulate by attempting to demonstrate that the product of the reflection and centering operations is equivalent to an a-glide operation in which the reflection is across the proposed glide plane.

The reflection operator will transform an atom at $[x_f\ y_f\ z_f]$ in asymmetric unit **1** into an equivalent atom in **2** at $[x_f\ \bar{y}_f\ z_f]$. A subsequent centering operation will generate an equivalent atom in **4** at $[(x_f + \frac{1}{2})\ (\frac{1}{2} - y_f)\ z_f]$. The translation group places atoms at $[(m_1 + \frac{1}{2} + x_f)\ (m_2 + \frac{1}{2} - y_f)\ (m_3 + z_f)]$. We begin by postulating that the single operation in the group that is the product of the reflection+centering+translation operations for a given m_1, m_2, and m_3 is an a-glide operation: $\mathbf{S}(a) = \mathbf{S}(\mathbf{t}_g)\mathbf{S}(C)\mathbf{S}(m)$. Placing an atom at $[0\ y_f^*\ 0]$ on an a-glide plane at its point of intersection on the b axis and performing the glide operation generates an equivalent atom that remains on the glide plane at $[\frac{1}{2}\ y_f^*\ 0]$.

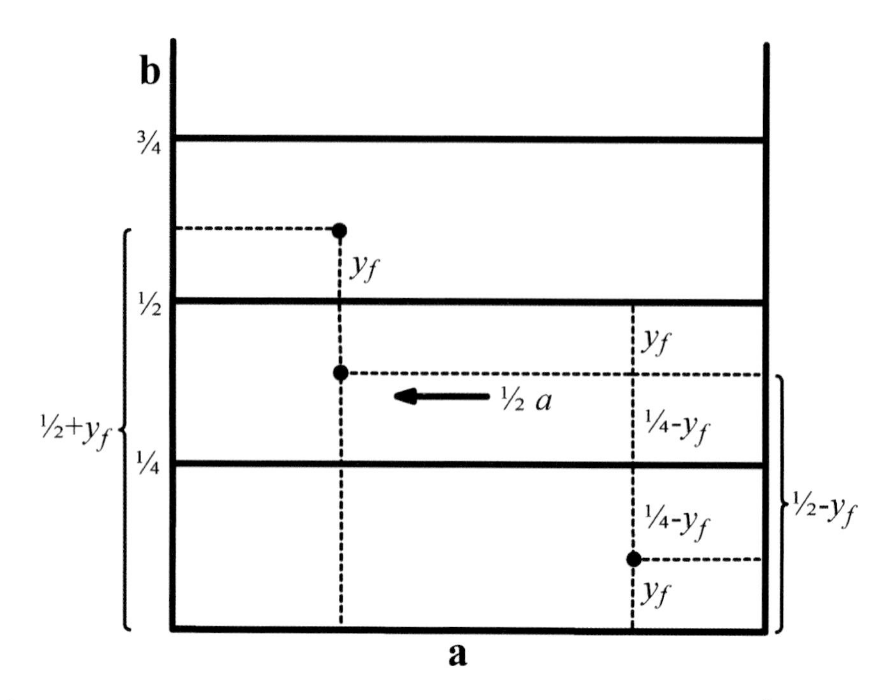

Figure 2.62 The results of an a-glide operation with a glide plane at $[0\ \frac{1}{4}\ 0]$ and a mirror plane at $[0\ \frac{1}{2}\ 0]$.

We have postulated that this atom must be in the same location as an atom at $[(m_1 + \frac{1}{2} + 0)\ (m_2 + \frac{1}{2} - y_f^*)\ (m_3 + 0)]$, resulting in

$$\begin{aligned} m_1 &= 0 \\ y_f^* &= m_2 + \frac{1}{2} - y_f \Longrightarrow y_f^* = \frac{m_2}{2} + \frac{1}{4} \\ m_3 &= 0. \end{aligned} \tag{2.32}$$

Values of m_2 of 0 and 1 place the points of intersection of our hypothesized glide planes at $[0\ \frac{1}{4}\ 0]$ and $[0\ \frac{3}{4}\ 0]$, midway between the mirror planes passing through the origin and center of the unit cell. The remaining integers place these glide planes throughout the lattice.

We now establish that an a-glide operation across one of these planes is equivalent to a reflection across the mirror plane in the center of the unit cell, followed by a centering operation. Referring to Fig. 2.62, an atom in the general position at $[x_f\ y_f\ z_f]$ is at displacement y_f above the ac plane. To perform the glide operation we first reflect the atom by projecting it to the glide plane, a distance of $1/4$-y_f, and extending it the same distance above the plane. The new position in the b direction is $y_f + 2(1/4 - y_f) = 1/2 - y_f$. The operation is effectively a reflection across the ac plane, represented by $\mathbf{R}(m)$, followed by a translation of $\frac{1}{2}$ in the $\mathbf{b}$

direction: $\mathbf{S}(m) = \mathbf{R}(m) + [0\ \frac{1}{2}\ 0]$. The glide operation adds 1/2 to x_f placing the atom at $[(x_f + \frac{1}{2})\ (\frac{1}{2} - y_f)\ z_f]$: $\mathbf{S}(a) = \mathbf{S}(m) + [\frac{1}{2}\ 0\ 0]$:

$$\mathbf{S}(a)\begin{bmatrix} x_f \\ y_f \\ z_f \end{bmatrix} = \begin{bmatrix} 1 & 0 & 0 \\ 0 & \bar{1} & 0 \\ 0 & 0 & 1 \end{bmatrix}\begin{bmatrix} x_f \\ y_f \\ z_f \end{bmatrix} + \begin{bmatrix} 0 \\ \frac{1}{2} \\ 0 \end{bmatrix}\begin{bmatrix} \frac{1}{2} \\ 0 \\ 0 \end{bmatrix} + = \begin{bmatrix} \bar{x}_f + \frac{1}{2} \\ \frac{1}{2} - y_f \\ z_f \end{bmatrix}. \tag{2.33}$$

This is the same location that the combined mirror and centering operations generate; the plane that crosses the b axis at $[0\ \frac{1}{4}\ 0]$ is an a-glide plane, and the a-glide operation belongs to the space group. Further reflection of the atom at this position across the mirror plane in the center of the unit cell places its mirror image at an equal distance above the mirror plane, at $1/2 + y_f$. It is left to the reader to establish that an a-glide operation across the plane at $[0\ \frac{3}{4}\ 0]$ is also equivalent to a mirror operation followed by a centering operation. Combining the reflection operations, the a-glide operations and their inverses, the C centering operations and their inverses, with identity and translations in the translation group creates the $C\,1\,m\,1 \equiv C\,m$ space group (No. 8). As we have observed previously, the centered space group can be generated from the primitive space group by adding the centering operation and its inverse.

The *Cc* Space Group. The addition of the centering operation and its inverse to the $P\,c$ space group creates new symmetry-equivalent locations from $[x_f\ y_f\ z_f]$ and $[x_f\ \bar{y}_f\ (z_f + \frac{1}{2})]$: $[(x_f + \frac{1}{2})\ (y_f + \frac{1}{2})\ z_f]$ and $[(x_f + \frac{1}{2})\ (\bar{y}_f + \frac{1}{2})\ (z_f + \frac{1}{2})]$. In comparing these positions with those for $C\,m$ we see that the only difference is an additional translation of $1/2\ c$ parallel to the c axis. As can be seen in Fig. 2.61(b) this transforms the *mirror* plane in the $C\,m$ unit cell into a c-glide plane (in red) and adds a *translation* along $\mathbf{c}$ to that along $\mathbf{a}$, converting the a-glide planes into diagonal n-glide planes (in yellow). It is left to the reader as an exercise to establish the existence and location of these planes. The addition of the C-centering operation to the symmetry operations in the $P\,c$ space group generates the $C\,1\,c\,1 \equiv C\,c$ space group (No. 9).

Projections for the $C\,m$ and $C\,c$ space groups are illustrated in Fig. 2.63. There are two symmetry planes indicated by the symbols in the upper left of each figure.

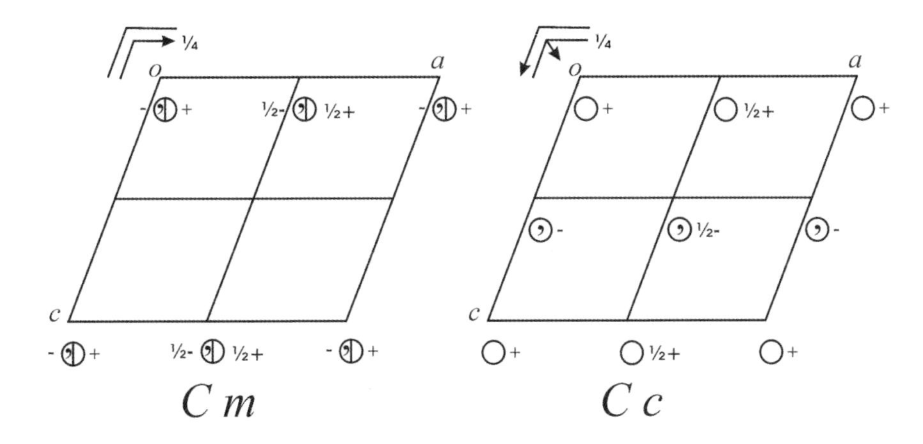

Figure 2.63 Orthogonal projections of $C\,m$ and $C\,c$ unit cells onto the ac plane.

For the Cm projection the symbols refer to a mirror plane in the ac plane (and therefore in the center of the unit cell), and an a-glide plane at $1/4\ b$ above the ac plane (and therefore at $-1/4\ b \equiv 3/4\ b$). The corresponding symbols for the Cc unit cell show that there is a c-glide plane in the ac plane (and center), and a diagonal n-glide plane at $1/4\,b$ (and $3/4\,b$). The plus and minus signs indicate equal displacements above and below the ac plane. The $\pm 1/2$ next to the symbols tell us that the glide-equivalent positions are at $1/2\ b$ plus or minus the displacement of the general position "+" at $+y_f$ above the ac plane (see Fig. 2.62). A summary of the results of the analysis of space groups generated from the m point group, as listed in the *International Tables* for the Pm, Pc, and Cm and Cc space groups, is given in Table 2.14. In addition to the general positions, the Pm unit cell has two special positions of multiplicity 1, on the mirror planes at $[x_f\ \frac{1}{2}\ z_f]$ and $[x_f\ 0\ z_f]$. As it was with the $P2_1$ unit cell, the translation in the Pc unit cell precludes any special positions, since nothing will be reflected onto itself. The Cm unit cell has one special position with a multiplicity of 2, on the mirror plane at $[x_f\ 0\ z_f]$. The centering operation creates another position at $[x_f\ \frac{1}{2}\ z_f]$, and just as in the $C2$ unit cell, it is not a separate special position. For the same reasons as for the Pc unit cell, the Cc unit cell also has no special positions since it contains no mirror planes.

Table 2.14 Information in the *International Tables for Crystallography* for the Pm, Pc, and Cm and Cc Space Groups.

	Pm	Pc	Cm	Cc
Origin on	m	c-glide	m	c-glide
Positions	$2\ c\ 1$	$2\ a\ 1$	$4\ b\ 1$	$4\ a\ 1$
(General)	$(1) x, y, z$	$(1) x, y, z$	$(1) x, y, z$	$(1) x, y, z$
	$(2) x, \bar{y}, z$	$(2) x, \bar{y}, \frac{1}{2}+z$	$(2) x, \bar{y}, z$	$(2) x, \bar{y}, \frac{1}{2}+z$
Coordinates			$(0,0,0)+\ (\frac{1}{2},\frac{1}{2},0)+$	$(0,0,0)+\ (\frac{1}{2},\frac{1}{2},0)+$

A generalization of the symmetry tools that have been described throughout the discussion of space groups to this point can now be used to create all 230 space groups from (a) the operators in the crystallographic point groups, (b) C, F, and I centering operators and their inverses, (c) screw and glide operators and their inverses, (d) the identity operator, and the (e) translation operators in the translation group. This approach becomes more difficult as the symmetry of the point groups increase, and it is here that we take note of two observations gleaned from the previous discussion that allow us to generate the remaining monoclinic space groups in a much more straightforward manner: (1) New space groups can be created by replacing point operations with combined point+translation operations. (2) It is possible to expand the symmetry operations in a "lower symmetry" space group by adding specific symmetry operations to create a new space group.

The seven monoclinic space groups that we have discussed in some detail have been created from point groups without a center of symmetry, and therefore are all non-centrosymmetric. The remaining monoclinic space groups are derived from the point group $2/m$, which is centrosymmetric. It therefore seems reasonable to create the centrosymmetric monoclinic space groups by adding the inversion operation to the operations in the non-centrosymmetric groups. As we established in the centrosymmetic triclinic space group, $P\bar{1}$, which was created by adding an inversion center to the operations in the noncentrosymmetric space group, $P1$, an inversion center is always coincident with the origin, and the translation group establishes inversion centers at the midpoints of all of the unit cell axes and in the center of each unit cell.

The $P2/m$ Space Group. Recall that the combination of a diad axis and an inversion center creates a mirror plane perpendicular to the diad axis, and combining an inversion center with a mirror plane creates a perpendicular diad axis. The general positions for a $P2$ unit cell are $[x_f\ y_f\ z_f]$ and $[\bar{x}_f\ y_f\ \bar{z}_f]$. Applying the inversion operator to each of these positions generates two new positions in the unit cell: $\mathbf{S}(\bar{1})[x_f\ y_f\ z_f] = [\bar{x}_f\ \bar{y}_f\ \bar{z}_f]$ and $\mathbf{S}(\bar{1})[\bar{x}_f\ y_f\ \bar{z}_f] = [x_f\ \bar{y}_f\ z_f]$. The $[x_f\ \bar{y}_f\ z_f]$ position is clearly the result of a reflection across the *ac* plane (see the discussion of the $P\,m$ space group) indicating the presence of mirror planes intersecting the b axis at $[0\ 0\ 0]$ and $[0\ \frac{1}{2}\ 0]$. The space group generated by combining an inversion operation with a two fold rotation operation, or the equivalent, combining a mirror operation with a rotation operation, again with the inclusion of the translation group, creates the $P\,1\,2/m\,1 \equiv P\,2/m$ space group (No. 10). The $P\,2/m$ unit cell is depicted in Fig 2.64(a). The origin of the $P\,2/m$ unit cell lies on an inversion center which is, simultaneously, the point of intersection of a diad axis and a mirror plane.

The $P2_1/m$ Space Group. The addition of an inversion operator to the two general positions in the $P\,2_1$ space group creates a new space group with two new symmetry-equivalent positions in the unit cell at $\mathbf{S}(\bar{1})[x_f\ y_f\ z_f] = [\bar{x}_f\ \bar{y}_f\ \bar{z}_f]$ and $\mathbf{S}(\bar{1})[\bar{x}_f\ (y_f + \frac{1}{2})\ \bar{z}_f] = [x_f\ (\bar{y}_f - \frac{1}{2})\ z_f]$. As we have just observed, the inclusion of inversion operations with the diad rotation operations in the $P2$ space group generates reflection operations with corresponding mirror planes passing through the origin and midpoints of the $P2/m$ unit cell. In the analogous space group created from $P2_1$, there is no symmetry-equivalent $[x_f\ \bar{y}_f\ z_f]$ position in the unit cell. There is therefore no mirror plane passing through the origin or center of the unit cell, and it appears that the combination of the diad screw operations in the $P2_1$ space group with inversion operations might *not* require that the space group also contains reflection operations. *If* there are reflection operations in the set of local operations, then there must be an equivalent product of operations with $\mathbf{S}(m) = \mathbf{S}(\bar{1})\mathbf{S}(2_1)$ and/or $\mathbf{S}(m) = \mathbf{S}(2_1)\mathbf{S}(\bar{1})$ being the most likely candidates. Beginning with an atom at $[x_f\ y_f\ z_f]$, the $\mathbf{S}(\bar{1})\mathbf{S}(2_1)$ product generates an atom at $[x_f\ (\bar{y}_f - \frac{1}{2})\ z_f]$. To test this conjecture, we adopt the strategy undertaken previously in determining whether or not additional symmetry operations existed in the centered non-centrosymmetric space groups. That is, we first determine the location of potential mirror planes in the unit cell, and then attempt to demonstrate that reflection operations with respect to these planes are equivalent to the product of other operations in the space group. Placing an atom at $[0\ y_f^*\ 0]$, at the point of location of a potential mirror plane with the b axis, and performing a mirror operation, generates a symmetry-equivalent atom at the same position. If the mirror plane is the product of a diad screw rotation and inversion, then the presence of the

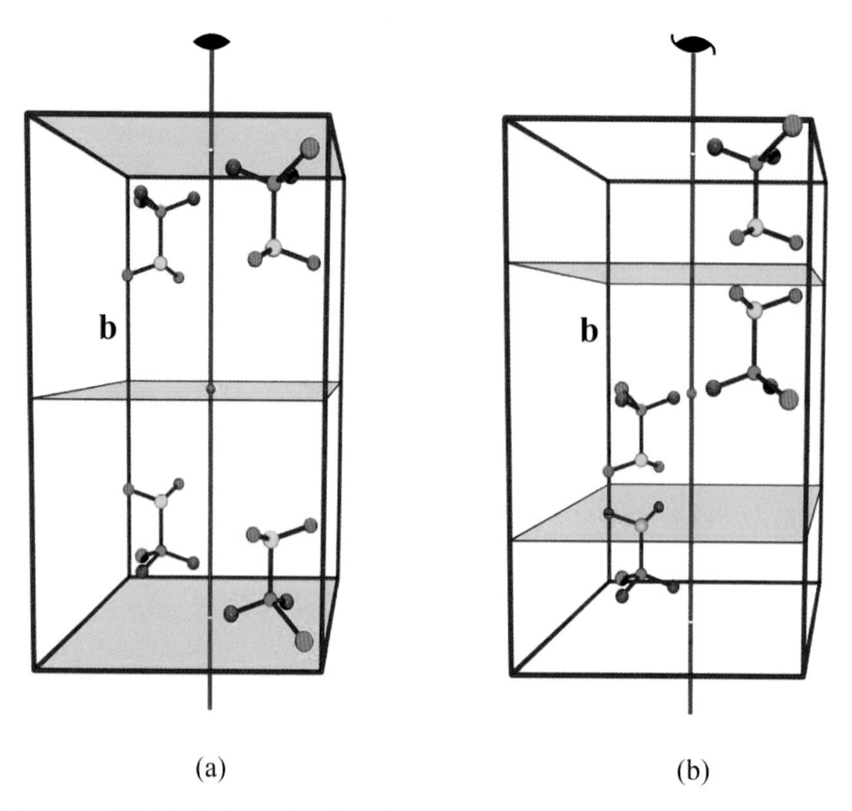

Figure 2.64 (a) $P2/m$ unit cell and asymmetric units related by a two-fold rotation axis and mirror plane at $[0\ \frac{1}{2}\ 0]$. (b) $P2_1/m$ unit cell and asymmetric units related by a two-fold screw axis and mirror planes at $[0\ \frac{1}{4}\ 0]$ and $[0\ \frac{3}{4}\ 0]$. The inversion centers in the center of the unit cells are depicted as red spheres.

translation group requires the same atom to be at $[(m_1+0)\ (m_2-\bar{y}_f^*-\frac{1}{2})\ (m_3+0)]$, giving

$$\begin{aligned} m_1 &= 0 \\ y_f^* &= m_2 - \frac{1}{2} - y_f^* \Longrightarrow y_f^* = \frac{m_2}{2} - \frac{1}{4} \\ m_3 &= 0. \end{aligned} \tag{2.34}$$

If there are mirror planes in the unit cell, then they intersect the b axis at $[0\ \frac{1}{4}\ 0]$ and $[0\ \frac{3}{4}\ 0]$, midway between the center of the b axis and the origin. Referring to Fig. 2.62, reflecting an atom at $[x_f\ y_f\ z_f]$ across the mirror plane that intersects the b axis at $[0\ \frac{1}{4}\ 0]$ generates an atom at $[x_f\ (\frac{1}{2}-y_f)\ z_f]$, illustrating that the reflection operation across such a plane is equivalent to the product of a screw operation and an inversion. The combination of the diad screw operation and inversion retains mirror symmetry, but relocates the symmetry elements inside the unit cell. The space group created by the addition of the inversion operation to the local operations in $P2_1$ is the $P1\,2_1/m\,1 \equiv P2_1/m$ (No. 11) space group. The origin in the $P2_1/m$ unit cell, shown in Fig. 2.64(b), is on an inversion center and intersects a two-fold screw axis, *but it does not lie on a mirror plane.* The projections of the $P2/m$ and $P2_1/m$ unit cells in Fig. 2.65 are seen to be similar

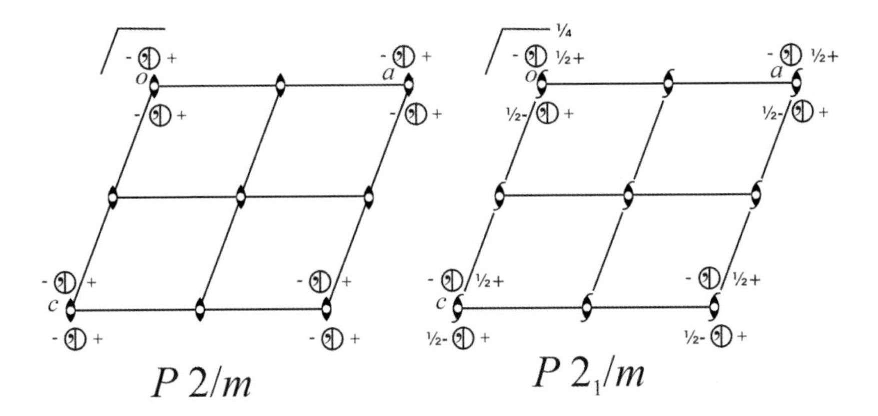

Figure 2.65 Orthogonal projections of $P\,2/m$ and $P\,2_1/m$ unit cells onto the *ac* plane. Inversion centers and diad rotational or screw axes intersect the midpoint of each unit cell, and the *ac* plane at the origin and center of the *ac* face and *a* and *b* axes. The mirror planes in the $P\,2_1/m$ unit cell intersect the *b* axis at $[0\ \frac{1}{4}\ 0]$ and $[0\ \frac{3}{4}\ 0]$.

to those for the $P\,2$ and $P\,2_1$ projections in Fig. 2.51, with the addition of mirror planes and inversion centers (indicated with open circles).

The $C2/m$ Space Group. The addition of the C-centering operation to the $P\,2/m$ space group operations generates a new space group with four more positions in the unit cell — resulting in a multiplicity of 8 for the general position. The space group can also be created by adding an inversion operation to the Cm space group, since the inversion center and the mirror plane generate the diad axis. As with the Cm unit cell, there are mirror planes in the *ac* plane and through the center of the unit cell, and *a*-glide planes intersecting the *b* axis at $[0\ \frac{1}{4}\ 0]$ and $[0\ \frac{3}{4}\ 0]$. As a third alternative, the space group can be generated by adding an inversion center to the $C2$ space group, since the inversion center and diad axis create a mirror plane. There are diad rotation axes at the origin, at the center of the unit cell, and at the midpoint of the *a* and *b* axes — and diad screw axes at $[\frac{1}{4}\ 0\ 0]$, etc., in the same locations as in the $C2$ unit cell. The origin of the unit cell is at the intersection of a diad axis and a mirror plane. As shown in Fig. 2.66, the unit cell for this space group, $C\,1\,2/m\,1 \equiv C\,2/m$ (No. 12), appears to be a combination of the unit cells for $C2$ (Fig. 2.52 and Cm in Fig. 2.61(a). However, as indicated in the projections in Fig. 2.67, the new space group is centrosymmetic, and every diad rotation axis and screw axis contains an inversion center, indicated by the open circles. The inversion centers are located on the diad screw axes are at the point where the axes intersect the glide planes; the "1/4" beside the circles representing these inversion centers indicate that they are at $1/4b$ above the *ac* plane. We have repeatedly discussed strategies for determining the existence and location of the symmetry elements in a unit cell throughout the chapter and the reader is encouraged to apply those strategies to the $C\,2/m$ space group as a useful exercise.

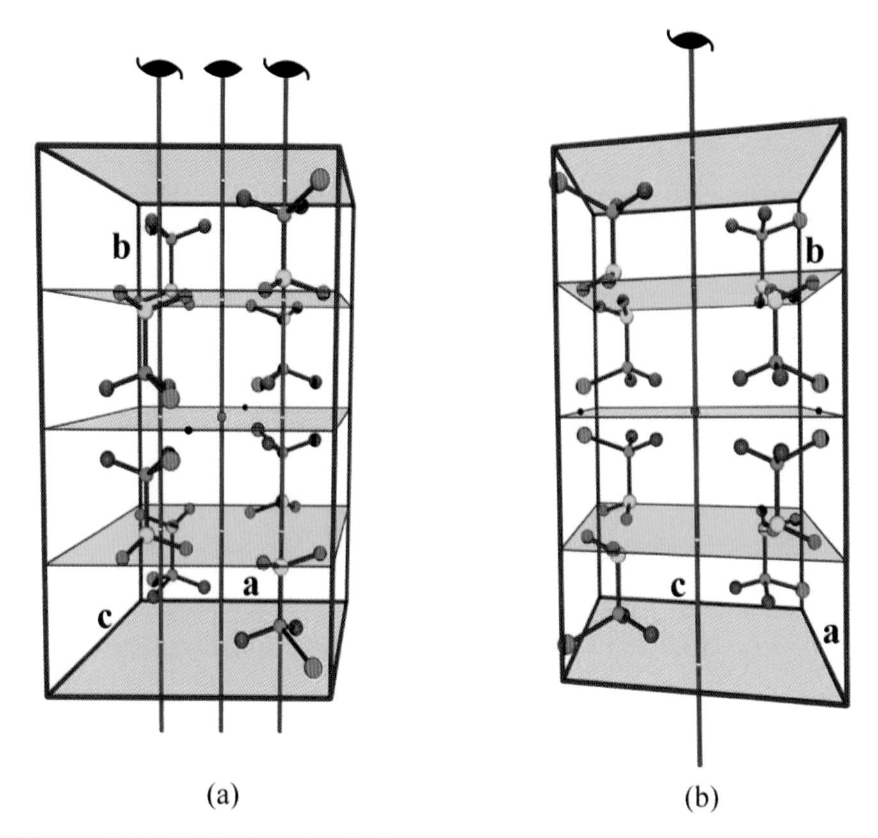

Figure 2.66 (a) $C\,2/m$ unit cell illustrating the location of diad rotation axes, screw axes, mirror planes (blue), and a-glide planes (red). The inversion center in the center of the unit cell is shown as a red sphere. Centered lattice points are shown in the ab faces. (b) View parallel to the a axis showing a diad screw axis (all three axes are superimposed in this view).

The $C2_1/m$ Space Group. The $C\,2_1/m$ space group can be created by adding the C-centering operation to the $P\,2_1/m$ space group, or by adding an inversion operation to the $C2 \equiv C2_1$ space group. As illustrated in Fig. 2.67, the results are the same as they were when considering $C\,2$ and $C\,2_1$; the $C\,2_1/m$ unit cell is identical to the $C\,2/m$ unit cell, differing only by the selection of the origin. As with $C2$ the space groups are identical, and the selected symbol is $C\,2/m$. Table 2.15 summarizes the results of the analysis of the centrosymmetric $P\,2/m$, $P\,2_1/m$, and $C\,2/m$ space groups.

The $P\,2/m$ unit cell has a plethora of special positions! There are two special positions of multiplicity 2 on the mirror planes, one at $[x\ \frac{1}{2}\ z]$, with the inversion center generating a second position at $[\bar{x}_f\ \frac{1}{2}\ \bar{z}_f] \equiv [\bar{x}_f\ \ (1-\frac{1}{2})\ \bar{y}_f] \equiv [\bar{x}_f\ \frac{1}{2}\ \bar{z}_f]$, the other at $[x_f\ 0\ z_f]$, inverted to $[\bar{x}_f\ 0\ \bar{z}_f]$. There are four special positions on the two-fold axes which intersect the origin, unit cell axes and center of the unit cell. They also have multiplicities of 2, since the inversion changes the sign of y_f, e.g., $[\frac{1}{2}\ y_f\ \frac{1}{2}] \longrightarrow [\frac{1}{2}\ \bar{y}_f\ \frac{1}{2}]$. Finally, each inversion center is a special position of multiplicity 1. The unit cell has eight unique inversion centers, one on each face, one in the center, one at the origin, and one on each unit cell axis (including the

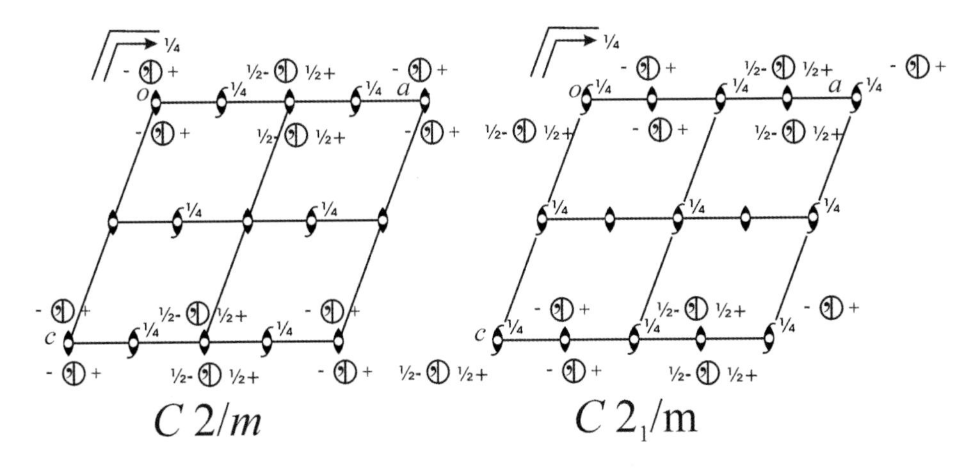

Figure 2.67 Orthogonal projections of $C\,2/m$ and $C\,2_1/m$ unit cells onto the ac plane.

Table 2.15 Information in the *International Tables for Crystallography* for the $P\,2/m$, $P\,2_1/m$, and $C\,2/m$ Space Groups.

	$P\,2/m$	$P\,2_1/m$	$C\,2/m$
Origin on	$\bar{1} \equiv 2/m$	$\bar{1}$ on 2_1	$\bar{1} \equiv 2/m$
Positions	4 o 1	4 f 1	8 j 1
(General)	$(1) x, y, z$	$(1) x, y, z$	$(1) x, y, z$
	$(2) \bar{x}, y, \bar{z}$	$(2) \bar{x}, y + \frac{1}{2}, \bar{z}$	$(2) \bar{x}, y, \bar{z}$
	$(3) \bar{x}, \bar{y}, \bar{z}$	$(3) \bar{x}, \bar{y}, \bar{z}$	$(3) \bar{x}, \bar{y}, \bar{z}$
	$(4) x, \bar{y}, z$	$(4) x, \bar{y} + \frac{1}{2}, z$	$(4) x, \bar{y}, z$
Coordinates			$(0, 0, 0)+$ $(\frac{1}{2}, \frac{1}{2}, 0)+$

b axis). An atom on an origin will contribute 1/8 of an atom per unit cell, and with 8 origins, this results in 1 atom in the cell. An atom on an a axis shares four unit cells and contributes 1/4 of an atom to the cell for a net of 1 atom per cell, since there are four a axes surrounding the unit cell — likewise for b and c. An atom on an ab face shares two unit cells, and there are two ab faces per unit cell — the same for ac and bc. Finally, an atom in the center of the unit cell belongs solely to the unit cell. An atom on any inversion center results in a contribution of one atom per unit cell. The $P\,2_1/m$ unit cell has fewer special positions, all of multiplicity 2. One is on a mirror plane. Since the mirror plane is offset from the ac plane by 1/4 b, the inversion generates the second position on the other mirror plane: $[x_f\ \frac{1}{4}\ z_f] \longrightarrow [\bar{x}_f\ -\frac{1}{4}\ \bar{z}_f] \equiv [\bar{x}_f\ \frac{3}{4}\ \bar{z}_f]$. There are four special positions on each of the inversion centers, each inverted to a second inversion center, (each inversion center contributing one atom per unit cell). $C\,2/m$ has special positions of multiplicity 4, one on a mirror plane, two on diad axes and two on the inversion centers lying on the glide planes. It also has four special positions of multiplicity 2 on the inversion centers lying on the mirror planes. The reader is encouraged to

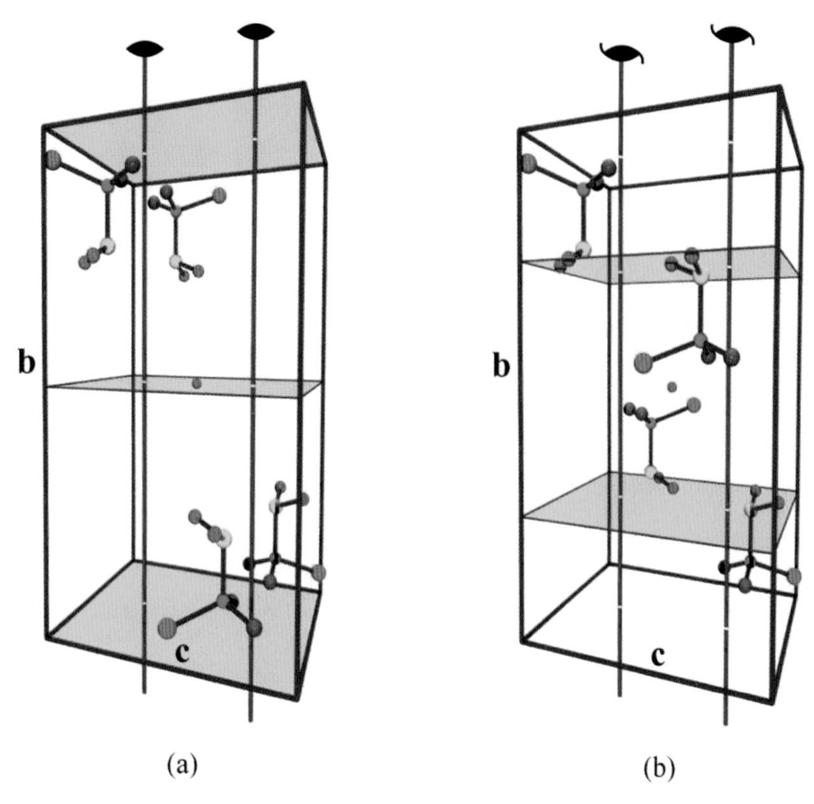

Figure 2.68 (a) $P\,2/c$ unit cell and asymmetric units related by a two-fold rotation axis and c-glide plane at $[0\ \frac{1}{2}\ 0]$. (b) $P\,2_1/c$ unit cell and asymmetric units related by a two-fold screw axis and c-glide planes at $[0\ \frac{1}{4}\ 0]$ and $[0\ \frac{3}{4}\ 0]$. The inversion centers in the center of the unit cells are depicted as red spheres.

establish the location of these special positions and rationalize their multiplicities as another useful exercise.

The $P2/c$ Space Group. Adding an inversion operation to the $P\,m$ space group introduces a diad operation and creates the $P\,2/m$ space group, which we have already encountered. However, adding an inversion operation to the $P\,c$ space group generates a new space group, $P\,1\,2/c\,1 \equiv P\,2/c$ (No. 13). The general positions for the Pc unit cell are $[x_f\ y_f\ z_f]$ and $[x_f\ \bar{y}_f\ (z_f + \frac{1}{2})]$. The inversion operation generates two new general positions at $[\bar{x}_f\ \bar{y}_f\ \bar{z}_f]$ and $[\bar{x}_f\ y_f\ (\bar{z}_f - \frac{1}{2})]$. The $(\bar{z}_f - \frac{1}{2})$ component is equivalent to $(\bar{z}_f - \frac{1}{2} + 1)$, *since we can always add a positive or negative integer to any coordinate*, giving $[\bar{x}_f\ y_f\ (\bar{z}_f + \frac{1}{2})]$.

Using arguments similar to those made for the $C2$ space group (the reader is encouraged to do so – see Exercise 2.6), the inversion operation also adds diad rotations to the space group. The diad axes are not coincident with the b axis, but intersect the ac plane at $[0\ 0\ \frac{1}{4}]$, $[0\ 0\ \frac{3}{4}]$, $[\frac{1}{2}\ 0\ \frac{1}{4}]$ and $[\frac{1}{2}\ 0\ \frac{3}{4}]$. To affirm this here, we recall that rotation about diad axes at $[\frac{1}{2}\ 0\ \frac{1}{4}]$ in the $C2$ unit cell in the presence of an a-glide plane is equivalent to a diad rotation about the b axis, followed by a translation of $\frac{1}{2}a$ in the **a** direction. By analogy, we postulate here that diad

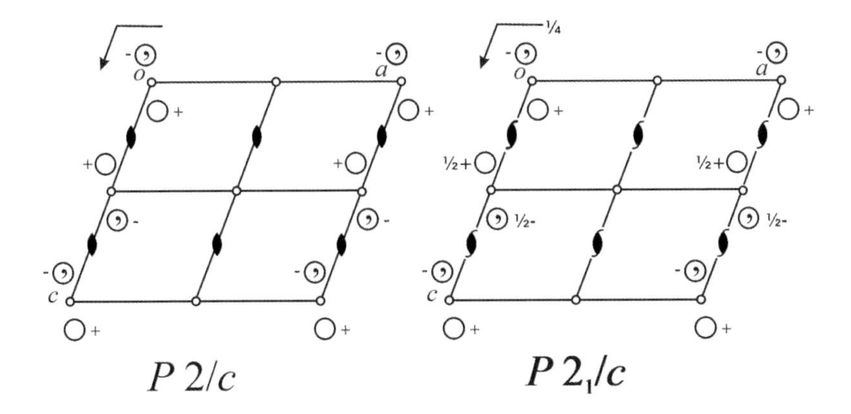

Figure 2.69 Orthogonal projections of $P\,2/c$ and $P\,2_1/c$ unit cells onto the ac plane. Inversion centers intersect the midpoint of each unit cell, and the ac plane at the origin and center of the ac face and a and b axes. The diad axes in both unit cells *do not* intersect the inversion centers. The glide planes in the $P\,2_1/c$ unit cell intersect the b axis at $[0\ \frac{1}{4}\ 0]$ and $[0\ \frac{3}{4}\ 0]$.

rotation will effectively be a diad rotation about b, followed by a translation of $\frac{1}{2}c$ in the **c** direction: $\mathbf{S}(2) = \mathbf{R}(2) + [0\ 0\ \frac{1}{2}]$:

$$\begin{bmatrix} \bar{1} & 0 & 0 \\ 0 & 1 & 0 \\ 0 & 0 & \bar{1} \end{bmatrix} \begin{bmatrix} x_f \\ y_f \\ z_f \end{bmatrix} + \begin{bmatrix} 0 \\ 0 \\ \frac{1}{2} \end{bmatrix} = \begin{bmatrix} \bar{x}_f \\ y_f \\ \bar{z}_f + \frac{1}{2} \end{bmatrix}. \tag{2.35}$$

Application of the inversion operator generates $[\bar{x}_f\ \bar{y}_f\ \bar{z}_f]$ and $[x_f\ \bar{y}_f\ (z_f + \frac{1}{2})]$, verifying the existence and location of the diad axes. The $P2/c$ unit cell is depicted in Fig. 2.68(a). The inversion centers are at the origins, centers of the unit cell axes and midpoints of the faces.

The $P2_1/c$ Space Group. At this stage we have exhausted all of the primitive non-centrosymmetric monoclinic space groups to which we can add inversion symmetry to generate new space groups. In cataloguing these space groups, we note that for every space group with a diad rotational axis, there is another that can be generated from it by converting the rotational axis to a screw axis. So far we have the pairs $P\,2$, $P\,2_1$ and $P\,2/m$, $P\,2_1/m$. Converting the diad rotation axis in $P\,2/c$ will complete the set, forming the space group $P\,1\,2_1/c\,1 \equiv P\,2_1/c$ (No. 14). This adds a $1/2$ translation to the y_f coordinate of the $[\bar{x}_f\ y_f\ (\bar{z}_f + \frac{1}{2})]$ and $[x_f\ \bar{y}_f\ (z_f + \frac{1}{2})]$ positions in $P\,2/c$, transforming them to $[\bar{x}_f\ (y_f + \frac{1}{2})\ (\bar{z}_f + \frac{1}{2})]$ and $[x_f\ (\bar{y}_f + \frac{1}{2})\ (z_f + \frac{1}{2})]$. The $P\,2_1/c$ unit cell is illustrated in Fig. 2.68(b), and its projection, along with the projection of the $P\,2/c$ unit cell is shown in Fig. 2.69. As we have observed several times in the past, the transformation of diad axes into screw axes effectively moves glide and mirror planes from the ac plane and midpoint of the unit cell to points at $1/4\ b$ and $3/4\ b$ along the b axis. The existence of these glide planes can be verified by assuming their existence and determining whether or not the resulting symmetry operations are identical to those above. Since we

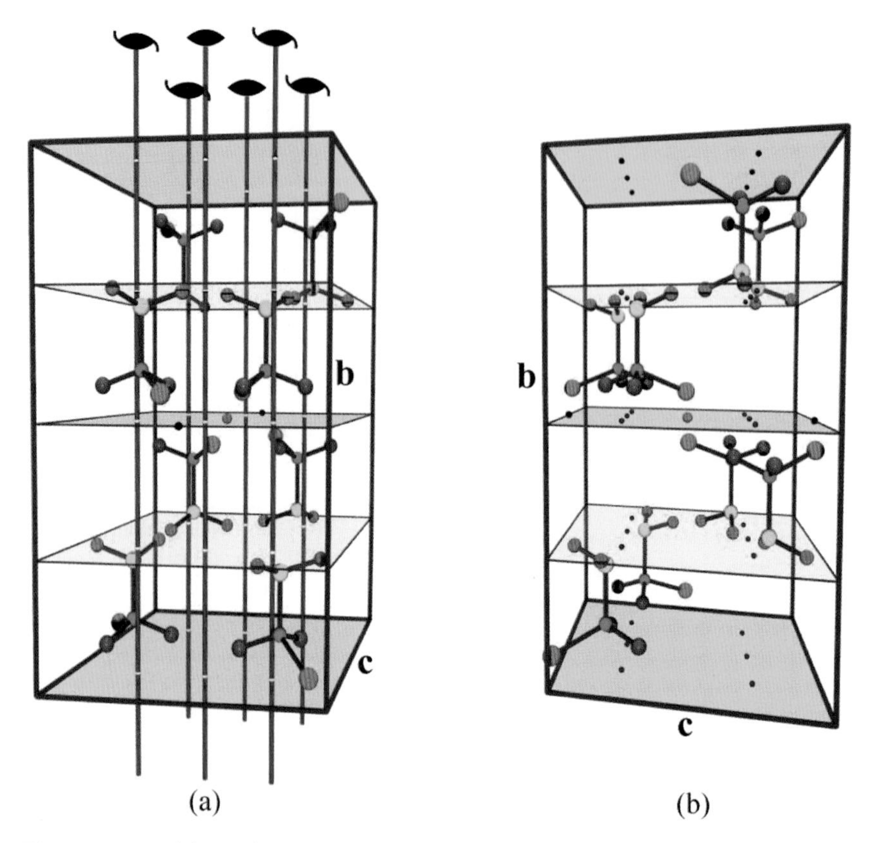

Figure 2.70 (a) $C\,2/c$ unit cell illustrating the location of diad rotation axes, screw axes, c-glide planes (red), and n-glide planes (yellow). The inversion center in the center of the unit cell is shown as a red sphere. Centered lattice points are shown in the ab faces. (b) View parallel to the a axis (the intersection points of the rotational and screw axes with the planes are indicated with blue spheres).

have selected the c glide direction, by analogy with Eqn. 2.33, the symmetry operator for reflection across the glide plane intersecting b at $[0\ \frac{1}{4}\ 0]$, followed by a $\frac{1}{2}c$ translation, is postulated to be $\mathbf{S}(c) = \mathbf{S}(m) + [0\ 0\ \frac{1}{2}]$:

$$\mathbf{S}(c)\begin{bmatrix} x_f \\ y_f \\ z_f \end{bmatrix} = \begin{bmatrix} 1 & 0 & 0 \\ 0 & \bar{1} & 0 \\ 0 & 0 & 1 \end{bmatrix}\begin{bmatrix} x_f \\ y_f \\ z_f \end{bmatrix} + \begin{bmatrix} 0 \\ \frac{1}{2} \\ 0 \end{bmatrix} + \begin{bmatrix} 0 \\ 0 \\ \frac{1}{2} \end{bmatrix} = \begin{bmatrix} x_f \\ \bar{y}_f + \frac{1}{2} \\ z_f + \frac{1}{2} \end{bmatrix}. \tag{2.36}$$

Applying the inversion operator produces the two additional positions, verifying the existence of the glide planes in the unit cell. The origin of the $P\,2_1/c$ unit cell is on an inversion center that does not intersect *either* the glide planes or the 2_1 axis. $P2_1/c$ is the most commonly reported space group in X-ray crystallography.

The $C2/c$ Space Group. The thirteenth — and final — monoclinic space group is created by adding the C-centering operator to $P\,2/c$ or $P\,2_1/c$, creating the $C\,1\,2/c\,1 \equiv C\,2/c$ (No. 15) space group. Four new general positions are created: $[(x_f + \frac{1}{2})\ (y_f + \frac{1}{2})\ z_f]$, $[(\bar{x}_f + \frac{1}{2})\ (y_f + \frac{1}{2})\ (\bar{z}_f + \frac{1}{2})]$, $[(\bar{x}_f + \frac{1}{2})\ (\bar{y}_f + \frac{1}{2})\ \bar{z}_f]$, and

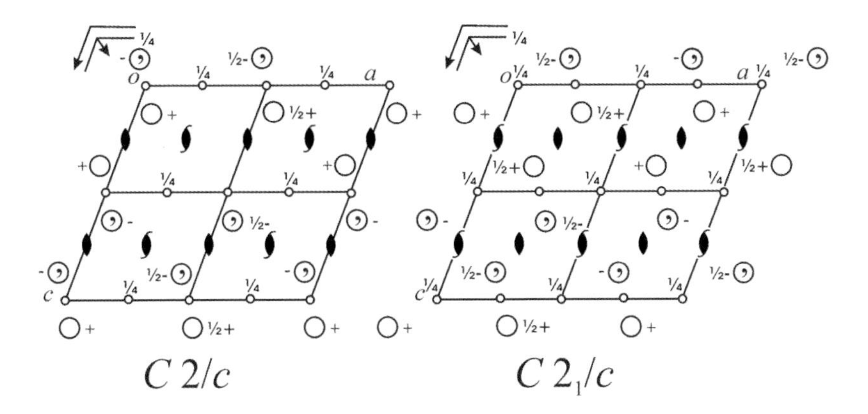

Figure 2.71 Orthogonal projections of $C\,2/c$ and $C\,2_1/c$ unit cells onto the ac plane.

$[(x_f + \frac{1}{2})\ (\bar{y}_f + \frac{1}{2})\ z_f + \frac{1}{2}]$. The combination of the screw axes, glide planes, and centering operations places all of the axes at positions of $1/4\ c$ and $3/4\ c$ along the c axis, and $1/4\ a$ (2_1), $1/2\ a$ (2), and $3/4\ a$ (2_1) along the a axis. c-glide planes are present in the ac plane and midpoint of the unit cell, with n-glide planes at $1/4\ b$ and $3/4\ b$ along the b axis. The origin is on an inversion center intersecting a c-glide plane. The diad axes and inversion centers do *not* intersect. As with $C\,2$ and $C\,2/m$, $C\,2/c$ differs from $C\,2_1$ solely by an origin shift as illustrated in the projections in Fig. 2.71 — and the space group is referred to as $C\,2/c$.

Table 2.16 summarizes the results of the analysis of the $P\,2/c$, $P\,2_1/c$, and $C\,2/c$ space groups. The $P\,2$ unit cell has two special positions of multiplicity 2 on the diad axes, and four positions of multiplicity 2 on the inversion centers. The $P\,2_1/c$ unit cell has four positions of multiplicity 2 on the inversion centers. The $C\,2/c$ unit cell has one special position of multiplicity 4 on a rotational axis, and four positions of multiplicity 4 on the inversion centers. Again, the reader will find it useful to verify these assignments.

Table 2.16 Information in the *International Tables for Crystallography* for the $P\,2/c$, $P\,2_1/c$, and $C\,2/c$ Space Groups.

	$P\,2/c$	$P\,2_1/c$	$C\,2/c$
Origin on	$\bar{1}$ on c-glide	$\bar{1}$	$\bar{1}$ on c-glide
Positions	$4\ g\ 1$	$4\ e\ 1$	$8\ f\ 1$
(General)	$(1) x, y, z$	$(1) x, y, z$	$(1) x, y, z$
	$(2) \bar{x}, y, \bar{z}+\frac{1}{2}$	$(2) \bar{x}, y+\frac{1}{2}, \bar{z}+\frac{1}{2}$	$(2) \bar{x}, y, \bar{z}+\frac{1}{2}$
	$(3) \bar{x}, \bar{y}, \bar{z}$	$(3) \bar{x}, \bar{y}, \bar{z}$	$(3) \bar{x}, \bar{y}, \bar{z}$
	$(4) x, \bar{y}, z+\frac{1}{2}$	$(4) x, \bar{y}+\frac{1}{2}, z+\frac{1}{2}$	$(4) x, \bar{y}, z+\frac{1}{2}$
Coordinates			$(0,0,0)+$ $(\frac{1}{2},\frac{1}{2},0)+$

This completes a rather detailed analysis of the monoclinic space groups. In doing so most of the important relationships have been developed for mirror planes, glide planes and diad rotational and screw axes and the basic tools for generating all of the remaining space groups have been established. The only new translational operators are higher-order screw axes, and we will summarize the remainder of the space groups in the context of the specific contributions of higher symmetry and the effects of these additional operators.

2.4.5 The Orthorhombic Space Groups

Lattices in the orthorhombic crystal system have mmm symmetry. The only additional symmetry element in the orthorhombic system is a relatively rare and unique type of diagonal glide plane known as a *diamond glide plane.* This type of glide plane can occur if the unit cell is F-centered. The symmetry-equivalent position produced by the centering translation is also a simple diagonal lattice translation for the primitive unit cell in the lattice. Thus if there is a diagonal glide glide plane in the primitive lattice, involving a reflection followed by a translation of half the primitive diagonal, in the face centered unit cell this will be a $1/4$ translation along its diagonal (see Fig. 2.44). Thus the diamond glide involves a reflection, followed by a translation of $1/4$ of the diagonal length (e.g., $(\mathbf{a}+\mathbf{b})/4$). Diamond glide planes occur in pairs separated by $1/4$ of the length of the unit cell axis to which they are perpendicular, with the glide vector pointing along opposite diagonals in the unit cell (see Fig. 2.92 in the section discussing cubic space groups for an example of diamond glide planes in the diamond unit cell). In the orthorhombic system there are only two F-centered unit cells with diamond glide planes. In spite of this, the additional lattice symmetry in the orthorhombic system results in 59 orthorhombic space groups — Nos. 16-74. The point symmetry operations available to the asymmetric units of the orthorhombic system are E, 2, m, and $\bar{1}$, the same as those in the monoclinic system, but they can occur along any of the unit cell axes. In the general case, no axis is unique in the orthorhombic unit cell (it is perhaps more accurate to state that orthorhombic unit cells exhibit *three* unique axes). In cases where the point group does have a unique axis, such as $mm2$, the standard setting is chosen such that this axis is the c axis, but there is no simple general rule for assigning the axes in an orthorhombic system. Just as in the monoclinic system the H-M notation for orthorhombic space groups is formed by adding three entities to the Bravais lattice symbol, each corresponding to the symmetry element related to each of the axes in the unit cell in a, b, c order. However, unlike the monoclinic symbol, all three entities are needed, since each of the three axes of the unit cell is orthogonal to the plane defined by the other two axes; orthorhombic unit cells have *at least* diad symmetry in all three axial directions. As an example, a unit cell with a diagonal glide plane perpendicular to the a axis, a c glide plane perpendicular to the b axis, and a diad screw axis parallel to the c axis represents the $Pna2_1$ space group.

The $P\,222$ Space Group. The simplest space group in the orthorhombic system is the $P222$ space group, created from the 222 point group, with a diad axis coincident with each of the unit cell axes. For a general position at $[x_f\ y_f\ z_f]$, rotation about each axis generates symmetry-equivalent positions at $[x_f\ \bar{y}_f\ \bar{z}_f]$, $[\bar{x}_f\ y_f\ \bar{z}_f]$, and $[\bar{x}_f\ \bar{y}_f\ z_f]$. The three symmetry operators, $\mathbf{S}(2)_x^a \equiv \mathbf{R}(2)_x$, $\mathbf{S}(2)_y^b \equiv \mathbf{R}(2)_y$, and $\mathbf{S}(2)_z^c \equiv \mathbf{R}(2)_z$,* along with the translation group and the identity, generate a mathematical group – application of the local symmetry operators to any of the general positions will produce one of those positions, indicating that any combination of operations will be equivalent to a single operation. The origin of the $P222$ unit cell is at the intersection of the diad axes coincident with each unit cell axis.

The $P\,2_122 \equiv P222_1$ Space Group. Converting the a axis in the $P222$ space group into a diad screw axis generates the $P2_122$ space group (the standard setting is $P222_1$, consistent with the c axis as the unique axis – we adopt the $P2_122$ setting here for a systematic development of the $P2_12_12_1$ space group). The screw operation, $\mathbf{S}(2_1)_x^a = \mathbf{R}(2)_x + [\frac{1}{2}\ 0\ 0]$, transforms $[x_f\ y_f\ z_f]$ into $[(x_f + \frac{1}{2})\ \bar{y}_f\ \bar{z}_f]$. There is no constraint on the location of the first diad axis, which we make coincident with the c axis. The product of the $\mathbf{S}(2_1)_x^a$ and $\mathbf{S}(2)_z^c$ operations produces

$$\mathbf{S}(2)_z^c\mathbf{S}(2_1)_x^a \begin{bmatrix} x_f \\ y_f \\ z_f \end{bmatrix} = \mathbf{S}(2)_z^c \begin{bmatrix} x_f + \frac{1}{2} \\ \bar{y}_f \\ \bar{z}_f \end{bmatrix} = \begin{bmatrix} \bar{x}_f + \frac{1}{2} \\ y_f \\ \bar{z}_f \end{bmatrix}. \tag{2.37}$$

For this product to behave as a product in a symmetry group there must be a single operation that is its equivalent. A third diad axis, coincident with the b axis would transform $[x_f\ y_f\ z_f]$ into $[\bar{x}_f\ y_f\ \bar{z}_f]$, which is a new position that is *not* generated by the product of the screw operation and the diad rotation: $\mathbf{S}(2)_y^b \neq \mathbf{S}(2)_z^c\mathbf{S}(2_1)_x$. To determine the single operator that is equivalent to the $\mathbf{S}(2)_z^c\mathbf{S}(2_1)_x$ product, we refer to the discussion of the $C2$ space group. The resulting coordinates in Eqn. 2.37 are generated by a diad rotation about an axis parallel to the b axis, but intersecting the ac plane at $[\frac{1}{4}\ 0\ \frac{1}{2}]$, effectively a diad rotation about b followed by a translation of $\frac{1}{2}a$ in the **a** direction: $\mathbf{S}(2)_y = \mathbf{R}(2)_y + [\frac{1}{2}\ 0\ 0]$. The second set of diad axes in the $P2_122$ unit cell intersect the ac plane at $[\frac{1}{4}\ 0\ 0]$, $[\frac{3}{4}\ 0\ 0]$, $[\frac{1}{4}\ 0\ \frac{1}{2}]$ and $[\frac{3}{4}\ 0\ \frac{1}{2}]$. The $\mathbf{S}(2_1)_x^a$, $\mathbf{S}(2)_y$, and $\mathbf{S}(2)_z^c$ operators, along with the identity and translation group generate the $P2_122$ space group. The origin of the $P2_122$ unit cell is chosen at the point of intersection of the diad axis coincident with the c axis and the 2_1 axis coincident with the a axis. The location of the origin on the b axis is arbitrary.

The $P\,2_12_12$ Space Group. Converting the diad axis in the $P2_122$ space group that parallels the b axis and intersects the ac plane at $[\frac{1}{4}\ 0\ \frac{1}{2}]$ creates the $P2_12_12$ space group: $\mathbf{S}(2_1)_y = \mathbf{S}(2)_y + [0\ \frac{1}{2}\ 0] = \mathbf{R}(2)_y + [\frac{1}{2}\ \frac{1}{2}\ 0]$, and $\mathbf{S}(2_1)_y\,[x_f\ y_f\ z_f] = [(\bar{x}_f + \frac{1}{2})\ (y_f + \frac{1}{2})\ \bar{z}_f]$. The product of this operation with the diad rotation about c is

$$\mathbf{S}(2)_z^c\mathbf{S}(2_1)_y \begin{bmatrix} x_f \\ y_f \\ z_f \end{bmatrix} = \mathbf{S}(2)_z^c \begin{bmatrix} \bar{x}_f + \frac{1}{2} \\ y_f + \frac{1}{2} \\ \bar{z}_f \end{bmatrix} = \begin{bmatrix} x_f + \frac{1}{2} \\ \bar{y}_f + \frac{1}{2} \\ \bar{z}_f \end{bmatrix}. \tag{2.38}$$

*These operators will be assumed to involve diad rotations about axes coincident with the a, b, and c axes, respectively. Diad rotations about axes that are not coincident with the unit cell axes will be represented without the axial superscript — as $\mathbf{S}(2)_x$ (for example) — to indicate that they can be effected by a diad rotation about a unit cell axis, followed by a translation. Diad screw operations about a unit cell axis will also be labeled with a superscript defining the axis (e.g., $\mathbf{S}(2_1)_x^a$); those lacking the superscript will be considered to occur about axes not coincident with the unit cell axes.

The diad screw axis coincident with the a axis in The $P2_122$ unit cell transforms $[x_f\ y_f\ z_f]$ into $[(x_f + \frac{1}{2})\ \bar{y}_f\ \bar{z}_f]$ — $\mathbf{S}(2_1)^a_x$ is clearly not equivalent to the product of the operations related to the other two axes. To determine the single operation that is the product of these operations we again return to the discussion of the $C2$ space group. By analogy, a diad rotation about an axis parallel to the a axis but intersecting the bc plane at $[0\ \frac{1}{4}\ \frac{1}{2}]$ will transform $[x_f\ y_f\ z_f]$ to $[x_f\ (\bar{y}_f + \frac{1}{2})\ \bar{z}_f]$. A diad screw operation about this axis will produce the results in Eqn. 2.38. The diad screw operator for the axis parallel to the a axis is $\mathbf{S}(2_1)_x = \mathbf{R}(2)_x + [\frac{1}{2}\ \frac{1}{2}\ 0]$ and the 2_1 axes intersect the bc plane at $[0\ \frac{1}{4}\ 0]$, $[0\ \frac{3}{4}\ 0]$, $[0\ \frac{1}{4}\ \frac{1}{2}]$, and $[0\ \frac{3}{4}\ \frac{1}{2}]$. The $\mathbf{S}(2_1)_x$, $\mathbf{S}(2_1)_y$, and $\mathbf{S}(2)^c_z$ operators, along with the identity and the translation group, generate the $P2_12_12$ space group. The 2_1 axes both lie in the same plane, and the origin of the $P2_12_12$ unit cell is selected at the point of intersection of the plane containing the 2_1 axes and the diad axis coincident with the c axis.

The $P2_12_12_1$ Space Group. The $P2_12_12_1$ space group is the most commonly reported non-centrosymmetric space group in X-ray crystallography. It is created by converting the diad axis coincident with the c axis in the $P2_12_12$ space group into a diad screw axis parallel to the c axis. Since we expect the $\mathbf{S}(2_1)_z$ operation to be equivalent to the product of the $\mathbf{S}(2_1)_x$ and $\mathbf{S}(2_1)_y$ operations, this product must generate a rotation about an axis parallel to the c axis and a translation of $\frac{1}{2}c$ in the $\mathbf{c}$ direction. The product of these operators, as they exist in the $P2_12_12$ space group, cannot do this, since they result in a rotation about the c axis ($\mathbf{S}^c_z \equiv \mathbf{R}(2)_z$). Using similar arguments to those made in the $C2$ space group discussion, it is readily shown that the rotation of $[x_f\ y_f\ z_f]$ about an axis parallel to the b axis at $[\frac{1}{2}\ 0\ \frac{1}{4}]$ generates $[\bar{x}_f\ y_f\ (\bar{z}_f + \frac{1}{2})]$, producing the necessary translation for a diad screw operation in the direction of $\mathbf{c}$. A diad screw operation about the same axis produces $[\bar{x}_f\ (y_f + \frac{1}{2})\ (\bar{z}_f + \frac{1}{2})]$: $\mathbf{S}(2_1)_y = \mathbf{R}(2)_y + [0\ \frac{1}{2}\ \frac{1}{2}]$. The $\mathbf{S}(2_1)_x$ operator is the same one as in the $P2_12_12$ space group, $\mathbf{S}(2_1)_x = \mathbf{R}(2)_x + [\frac{1}{2}\ \frac{1}{2}\ 0]$, and the product of $\mathbf{S}(2_1)_x$ and $\mathbf{S}(2_1)_y$ now results in

$$\mathbf{S}(2_1)_x\mathbf{S}(2_1)_y \begin{bmatrix} x_f \\ y_f \\ z_f \end{bmatrix} = \begin{bmatrix} \bar{x}_f + \frac{1}{2} \\ \bar{y}_f + 1 \\ z_f + \frac{1}{2} \end{bmatrix} = \begin{bmatrix} \bar{x}_f + \frac{1}{2} \\ \bar{y}_f \\ z_f + \frac{1}{2} \end{bmatrix}. \tag{2.39}$$

Setting $\mathbf{S}(2_1)_z = \mathbf{S}(2_1)_y\mathbf{S}(2_1)_y = \mathbf{R}(2)_z + [\frac{1}{2}\ 0\ \frac{1}{2}]$ results in the operators necessary, along with the identity and translation group, to define the $P2_12_12_1$ space group. The rotation portion of the operator, $\mathbf{S}(2)_z = \mathbf{R}(2)_z + [\frac{1}{2}\ 0\ 0]$ indicates a diad rotation about an axis intersecting the ab plane at $[\frac{1}{4}\ \frac{1}{2}\ 0]$.

The $P2_12_12_1$ unit cell is illustrated in Fig. 2.72. The diad screw axes parallel to the a axis intersect the bc plane at $[0\ \frac{1}{4}\ 0]$, $[0\ \frac{3}{4}\ 0]$, $[0\ \frac{1}{4}\ \frac{1}{2}]$, and $[0\ \frac{3}{4}\ \frac{1}{2}]$; the diad screw axes parallel to the b axis intersect the ac plane at $[0\ 0\ \frac{1}{4}]$, $[0\ 0\ \frac{3}{4}]$, $[\frac{1}{2}\ 0\ \frac{1}{4}]$ and $[\frac{1}{2}\ 0\ \frac{3}{4}]$; the diad screw axes parallel to the c axis intersect the ab plane at $[\frac{1}{4}\ 0\ 0]$, $[\frac{1}{4}\ \frac{1}{2}\ 0]$, $[\frac{1}{4}\ \frac{1}{2}\ 0]$, and $[\frac{3}{4}\ \frac{1}{2}\ 0]$. All of the 2_1 axes parallel to a given unit cell axis lie in a plane; the three resulting orthogonal planes intersect in a single point, which is chosen as the origin.

For non-centrosymmetric structures the origin is typically chosen at a point of highest symmetry. The systematic addition of mirror and glide planes to the $P222$, $P222_1$, $P2_12_12$, and $P2_12_12_1$ unit cells results in the formation of a number

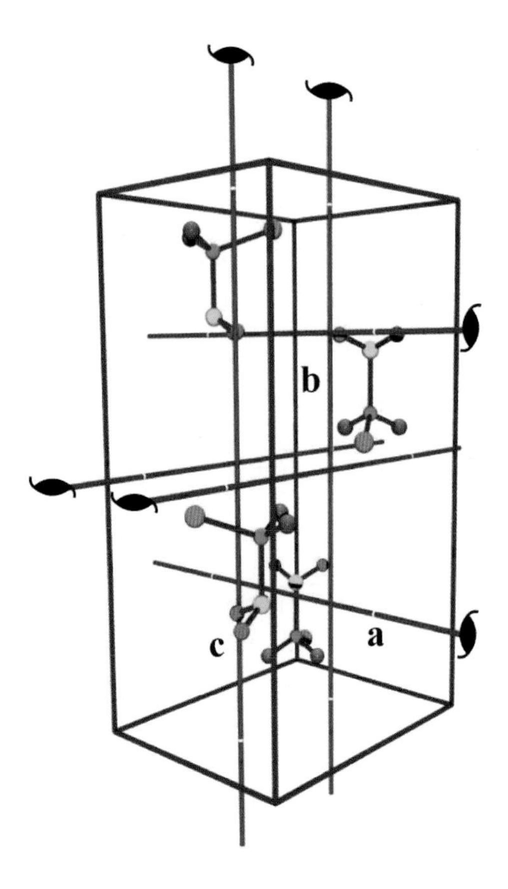

Figure 2.72 $P2_12_12_1$ unit cell showing diad screw axes.

of non-centrosymmetric orthorhombic space groups. The resulting space group descriptions are found in the *International Tables for Crystallography.*

The *P*mmm Space Group. The simplest centrosymmetric space group in the orthorhombic system is the *Pmmm* space group, created from the *mmm* point group. For a general position at $[x_f\ y_f\ z_f]$, reflection across each mirror plane generates symmetry-equivalent positions at $[\bar{x}_f\ y_f\ z_f]$, $[x_f\ \bar{y}_f\ z_f]$, and $[x_f\ y_f\ \bar{z}_f]$. Further reflection of these positions generates $[x_f\ \bar{y}_f\ \bar{z}_f]$, $[\bar{x}_f\ y_f\ \bar{z}_f]$, and $[\bar{x}_f\ \bar{y}_f\ z_f]$, which can also be produced by diad rotations about the a, b, and c axes, respectively. The presence of a mirror plane and a perpendicular diad axis (there are three here) produces an inversion center, clearly evident by comparing the six positions generated by the reflections. There is therefore an eighth symmetry-equivalent location at $[\bar{x}_f\ \bar{y}_f\ \bar{z}_f]$. The *Pmmm* space group is formed from the identity, the three mirror operators, the three rotation operators, the inversion operator, and the translation group. The origin for the unit cell is chosen at the center of symmetry, which must lie at the intersection of the three mirror planes in order to remain invariant when reflected across each of them.

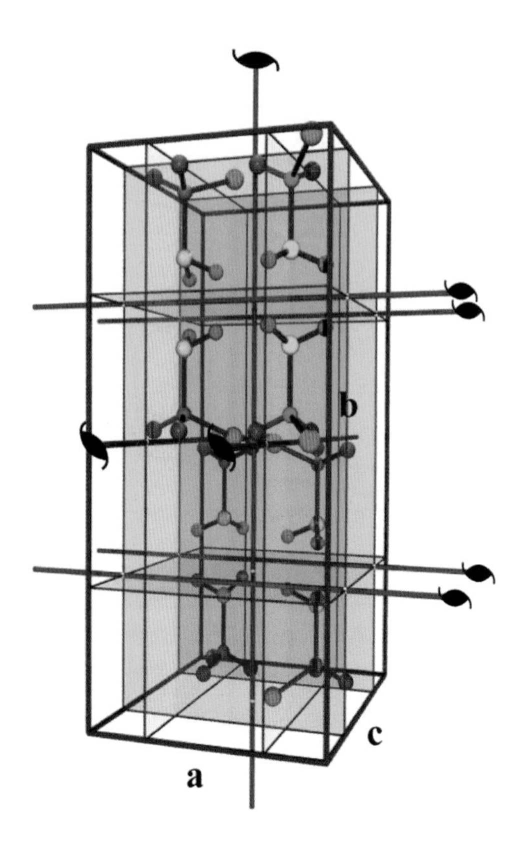

Figure 2.73 $Pnma$ unit cell showing screw axes, mirror planes (blue), a-glide planes (red) and n-glide planes (yellow).

The remaining orthorhombic space groups are generated by systematically converting the mirror planes in $Pmmm$ into glide planes. The presence of a large number of symmetry elements in the unit cells of these space groups makes this a non-trivial process, as illustrated in the unit cell for the $Pnma$ space group in Fig. 2.73. The origin of the $Pnma$ unit cell is on an inversion center that intersects the 2_1 axis coincident with the b axis. It contains three sets of symmetry planes, all at $1/4$ and $3/4$ of the distance along an axis: n-glide planes perpendicular to a, mirror planes perpendicular to b and a-glide planes perpendicular to c. It also contains screw axes parallel to all three unit cell axes. The "full" H-M symbol, $P\,2_1/n\,2_1/m\,2_1/a$, indicates the presence of all of these symmetry elements. The unit cell for the $Pnma$ space group illustrates the general complexity of orthorhombic unit cells. Monoclinic unit cells are less complicated because they are lower in symmetry, and the remaining crystal systems have unit cells which are less complicated because they are higher in symmetry! Fig. 2.74 illustrates projections of the symmetry elements of the $P2_12_12_1$ and $Pnma$ unit cells onto the ab plane. Unlike monoclinic lattices, and those with higher axial symmetry, the possibility for different symmetry elements in all three dimensions makes it very difficult to

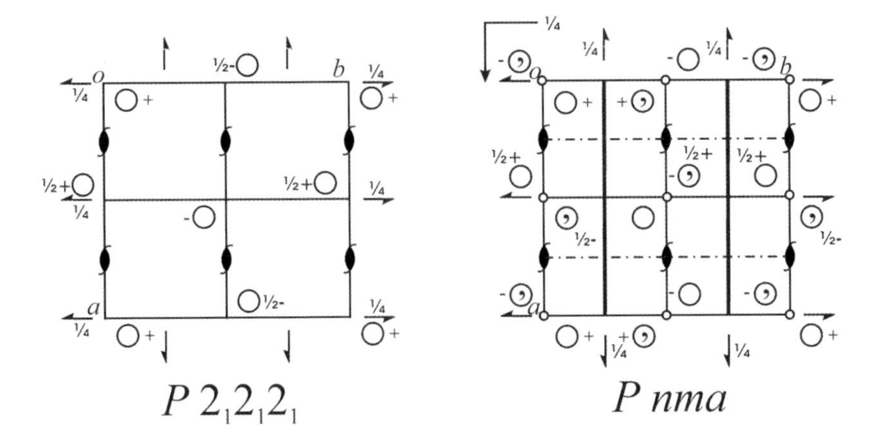

Figure 2.74 Orthogonal projections of $P\,2_12_12_1$ and $P\,nma$ unit cells onto the ab plane.

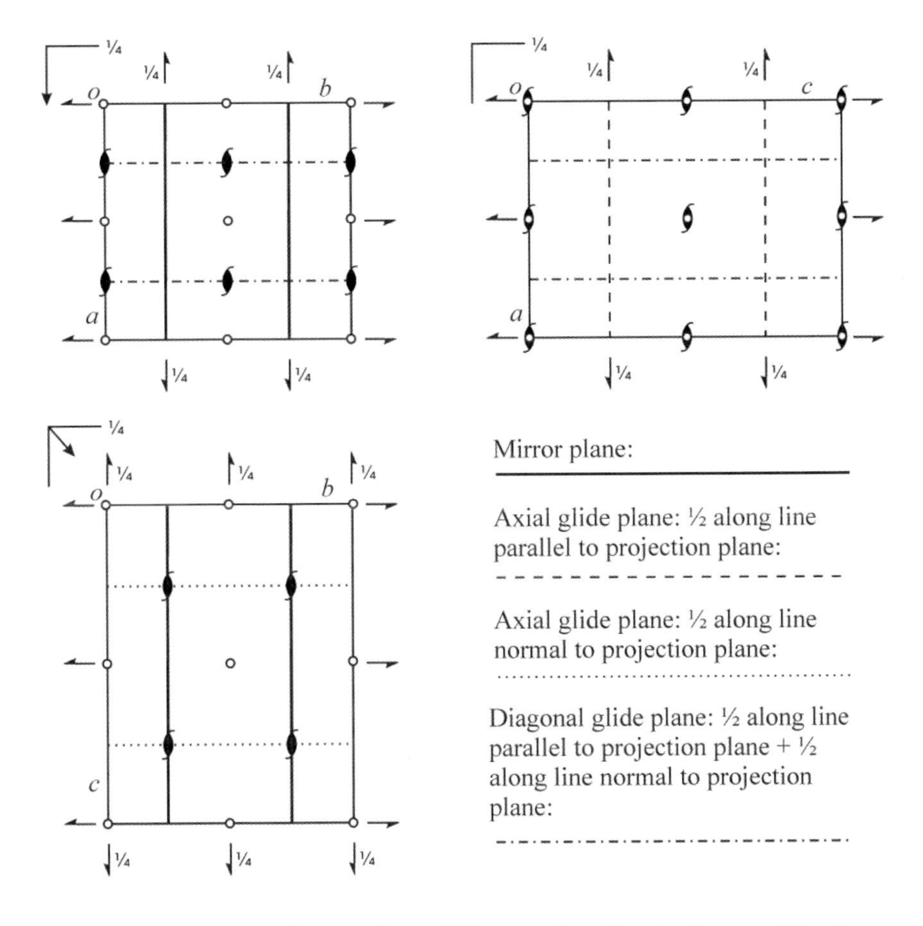

Figure 2.75 Orthogonal projections of $P\,nma$ unit cell onto ab, ac and bc planes.

envision the symmetry components of the orthorhombic unit cell, either in three dimensions or in a single projection. Views of the orthorhombic unit cell along each of the crystallographic axes are especially helpful in this regard; the *International Tables for Crystallography* includes projections onto all three unit cell facial planes for orthorhombic (and monoclinic) unit cells. Fig. 2.75 illustrates these projections for the $Pnma$ space group as they appear in the *International Tables* and provides a key for the common graphical symbols for glide and mirror planes. A complete list of symbols is included in the *International Tables.*

Transforming Orthorhombic Unit Cells. In the course of ordinary structure determination unit cell axes are assigned without knowledge of the internal symmetry of the crystal, which can generally be determined only after data collection, and in some cases after structural solution. Because of their often complex symmetry, orthorhombic crystals are especially problematic. For example, it is possible that a structural solution will yield a structure with a 2_1 axis parallel to the a axis, a c-glide plane perpendicular to the b axis, and a diagonal glide plane perpendicular to the c axis. We would assign the space group symbol $P2_1cn$. A cursory look at this symbol would not indicate that it refers to the same space group as $Pna2_1$, even though it does — with the axes renamed. There are four more permutations of the unit cell axis assignments, $P2_1nb$, $Pc2_1n$, $Pbn2_1$ and $Pn2_1a$, creating six different

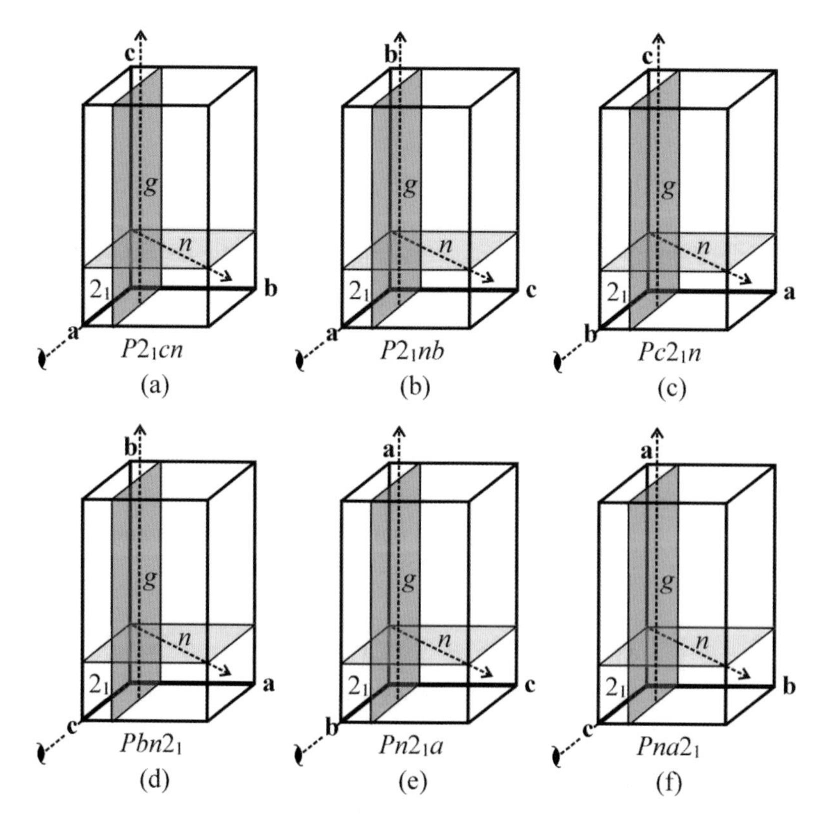

Figure 2.76 Permutations of the unit cell axes of a typical orthorhombic unit cell and the space group symbols for each setting.

symbols for the same space group, illustrated in Fig. 2.76. Since crystal structure determination and analysis requires a knowledge of the symmetry-equivalent positions in the unit cell, we generally must consult a source such as the *International Tables for Crystallography* for the locations of the general and special positions in the unit cell. Unfortunately, the coordinates contained in most sources are those for the standard setting, and we are faced either with transforming those coordinates so that they are compatible with the *non-standard* setting of the unit cell, *or* transforming the unit cell into the standard setting (the preferable choice).

As indicated in Figs. 2.76(a) and 2.76(f), the transformation of the $P2_1cn$ unit cell into a $Pna2_1$ unit cell — the standard setting — requires exchanging the a and c axes. To effect this change we reassign the axial lengths of **c** to **a** and **a** to **c** (the angles are all 90°), and swap the x_f coordinates of the unit cell contents with the z_f coordinates. The axial switch changes the direction of $\mathbf{a} \times \mathbf{c}$ (the **b** direction), creating a left-handed unit cell. This is corrected by changing the direction of **c** prior to the swap – by changing the signs of the z_f coordinates of the components of the unit cell. If two pairs of axes are switched, no sign reversal is required. The standard settings of the orthorhombic space groups and their possible permutations are listed in Volume *A* of the *International Tables for Crystallography.* The permutations are in columns with headings such as $\bar{\mathbf{c}}\mathbf{ba}$, which indicates that the direction of the c axis should be reversed, followed by an exchange of the c and a axes. The $\bar{\mathbf{c}}\mathbf{ba}$ transformation is the transformation necessary to convert the $P2_1cn$ unit cell and its contents into a $Pna2_1$ unit cell.

2.4.6 The Trigonal Space Groups

The $P3_221$ and $P3_121$ Space Groups. With the trigonal system we begin a discussion of the so-called "higher symmetry" space groups. To the E, 2, m, and $\bar{1}$ point symmetry operations of the monoclinic and orthorhombic systems, also available to the trigonal system, we add 3^1, 3^2, $\bar{3}^1$ and $\bar{3}^2$. The transformation matrices for E, 2, m, and $\bar{1}$ are identical in orthogonal and non-orthogonal coordinate systems since they do not "mix" the basis vectors. The only effect of each of these operations for a given vector component is to leave it unchanged, or to change its sign. However, there is no general matrix for rotations that become linear combinations of the basis vectors, since each fractional coordinate is specific to a given basis vector. In a general (fractional) coordinate system, a vector must first be transformed to orthonormal coordinates, rotated, then transformed back to fractional coordinates. Fortunately, for rotations in space lattices, the basis vectors involved in 3-, 4-, and 6-fold rotations are of the same length, and the rotational angles for the symmetry operators are the same as the angles between the basis vectors. This allows for a determination of the lengths of the components along each of the unit cell axes, and hence the fractional coordinates.

The $2\pi/3$ rotation of a general vector in a trigonal unit cell is illustrated in Fig. 2.77. The figure is a projection of a trigonal unit cell on the ab plane with x and y the lengths of the vector components in the unit cell basis along the a and b axes, respectively. These lengths correspond to the fractional coordinates $x_f = x/a$ and $y_f = y/a$, since both unit cell axes have a magnitude of a. The 3-fold operation rotates the vector and the parallelogram that defines it through $\gamma = 120°$ such that the x component of the original vector along the a axis is now aligned with the b axis and therefore with the y component of the rotated vector. Extending a parallel

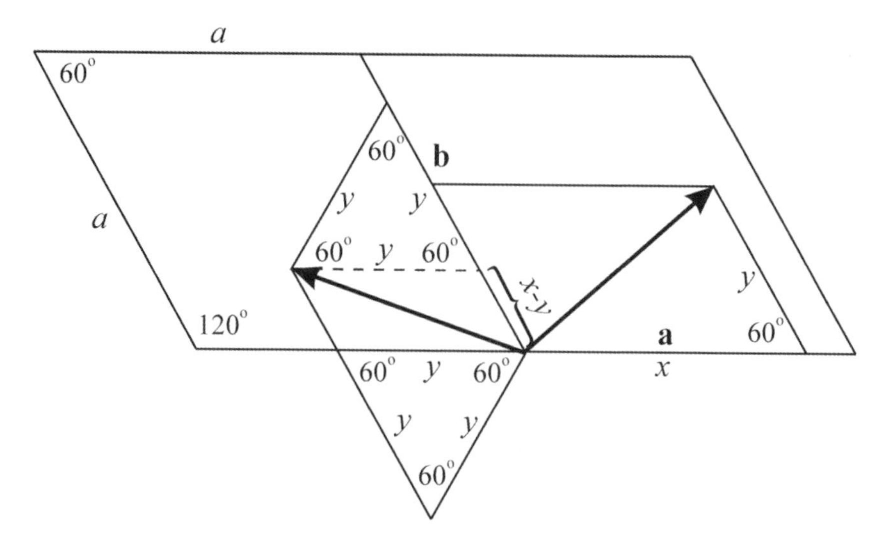

Figure 2.77 $2\pi/3$ rotation of a general vector about the c axis in a trigonal unit cell.

line from the head of the rotated vector to the b axis creates an equilateral triangle with edge length y. The intersection of the line with the b axis defines the distance along the b axis for the fractional coordinate of the rotated vector, $x - y$. The length of the a component of the rotated vector is y, in the negative direction. Thus $x'_f = -y/a = -y_f$ and $y'_f = (x-y)/a = x/a - y/a = x_f - y_f$; the coordinates of the rotated vector in the trigonal unit cell are $[x'_f\ y'_f\ z_f'] = [\bar{y}_f\ (x_f - y_f)\ z]$. The second rotation operation is effected by a $\pi/3$ rotation of $[x'_f\ y'_f\ z_f']$, giving $[x''_f\ y''_f\ z_f''] = [\bar{y}'_f\ (x'_f - y'_f)\ z'] = [(y_f - x_f)\ \bar{x}_f\ z]$. Creating matrix operators for the 3^1 and 3^2 operations, then multiplying them by $\mathbf{R}(\bar{1})$ to create the matrices for $\bar{3}^1$ and $\bar{3}^2$ provides the necessary point symmetry operators for the trigonal system (this assumes that rhombohedral unit cells are transformed to trigonal(R) unit cells.):

$$\mathbf{R}(3^1) = \begin{bmatrix} 0 & -1 & 0 \\ 1 & -1 & 0 \\ 0 & 0 & 1 \end{bmatrix} \quad \mathbf{R}(3^2) = \begin{bmatrix} -1 & 1 & 0 \\ -1 & 0 & 0 \\ 0 & 0 & 1 \end{bmatrix}$$

$$\mathbf{R}(\bar{3}^1) = \begin{bmatrix} 0 & 1 & 0 \\ -1 & 1 & 0 \\ 0 & 0 & -1 \end{bmatrix} \quad \mathbf{R}(\bar{3}^2) = \begin{bmatrix} 1 & -1 & 0 \\ 1 & 0 & 0 \\ 0 & 0 & -1 \end{bmatrix}$$

There are also local translations available to the trigonal system. If a triad rotation about the c axis is coupled with a translation of $1/3c$, then followed by two more identical rotations and translations the result will be a complete 2π rotation and translation by the length of c into the next unit cell. The symmetry element is a triad screw axis, and the operation just described is illustrated in Fig. 2.78(a). Fig. 2.78(b) is the mirror image of Fig. 2.78(a). *The figures cannot be superimposed.* They are said to be *enantiomorphs* of one another (we will have more to say about

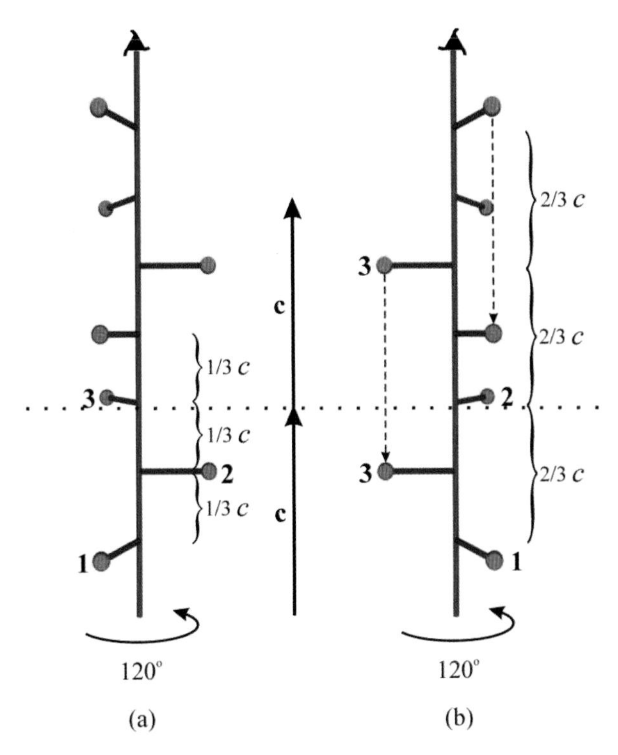

Figure 2.78 (a) 3_1 screw axis. (b) 3_2 screw axis. Symbols for both axes indicating opposite rotations are shown atop each axis.

this later). Thus the addition of translational symmetry to the trigonal system adds two symmetry elements instead of one. Note that the screw axis in Fig. 2.78(a) appears to behave like a right-handed screw or helix – much like a wood screw — if it is rotated clockwise it will advance. The mirror image of this axis behaves like a left-handed screw. If rotated counterclockwise it will advance. How do we distinguish between them? We begin by adopting the convention that all*screw operations must be right-handed.* This forces us to find a way to perform a series of right-handed rotations for the left-handed screw axis that will duplicate the effects of the left-handed rotations. Fig. 2.78(b) illustrates how this is accomplished. Following a clockwise (right-handed) $\pi/3$ rotation, the translation is $2/3c$ rather than $1/3c$. This skips the second position and generates the third. A second operation results in a translation of $4/3c$ – or $1/3c$ in the next unit cell. This is symmetry-equivalent to the "skipped" position – every point along the axis will be generated, just as if it had been rotated counterclockwise and advanced by $1/3c$. These two screw elements and associated operations are now distinguished by the different translational increments. An N-fold screw axis with translation of k/N is called an N_k screw axis. Thus Fig. 2.78(a) illustrates a 3_1 screw axis, and Fig. 2.78(b) illustrates a 3_2 screw axis.

Combinations of point group operators, the local translation operators $[0\ 0\ \frac{1}{3}]$ and $[0\ 0\ \frac{2}{3}]$, inverses of all new operations, and the translation group results in 25 trigonal space groups, Nos. 143-167. The triad symmetry elements all parallel the unique c axis. The diad symmetry elements are all constrained to the ab plane, either along axes parallel to the unit cell axes in the plane (and all of their symmetry-equivalents generated by the triad operations), or along the [1 $\bar{1}$ 0] direction (the long diagonal in the unit cell). Glide or mirror planes are perpendicular to one of these two directions. There are no mirror planes perpendicular to the c axis in the trigonal system. Recall that a triad axis and a horizontal mirror plane generates a six-fold rotoinversion axis — and the trigonal system has no six-fold symmetry. The combination belongs to the hexagonal system. The H-M notation for the trigonal system reflects the unique placement of its symmetry elements. The general symbol is $Pqrs$, but may be abbreviated if certain components are unnecessary. The q element is always present, and indicates the nature of the triad axis. The second element describes the symmetry with respect to the axial directions, and the third the symmetry with respect to the [1 $\bar{1}$ 0] direction. Space groups with 3-fold screw axes can exist as enantiomorphs of one another – and are described by *different* space groups. The crystal structure of quartz (SiO_2) provides an interesting example.[28] α-Quartz crystallizes in the trigonal system, as

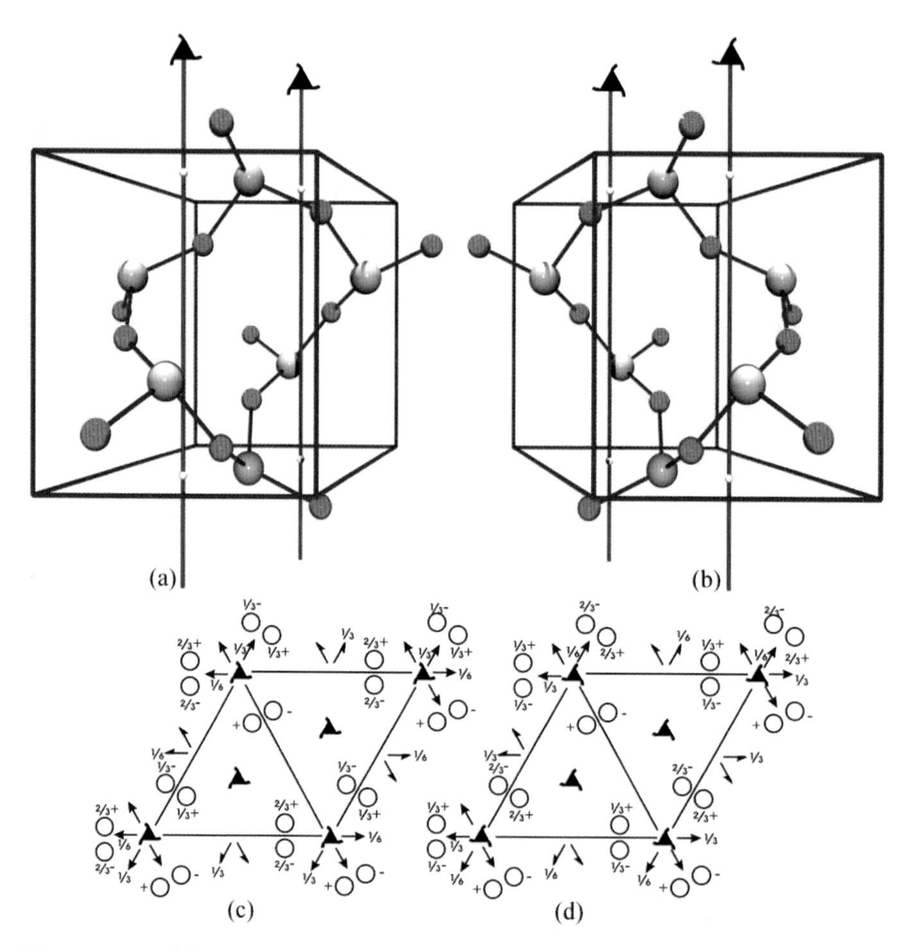

Figure 2.79 (a) *D*-quartz unit cell in the $P3_121$ space group. (b) *L*-quartz unit cell in the $P3_221$ space group. (c) Orthogonal projection of $P3_121$ unit cell onto the ab plane. d) Orthogonal projection of $P3_221$ unit cell onto the ab plane.

either *D*–quartz or *L*–quartz. Each structure is the mirror image of the other as shown in Fig. 2.79. Although it is difficult (impossible) to ascertain from the figure, there are diad rotational and screw axes parallel to the *a* and *b* axes. The space groups are assigned as $P3_121$ and $P3_221$, respectively, indicating the left- and right-hand screw axes in each unit cell. The "1" for the third subscript might seem unnecessary, but there are two trigonal space groups with both types of screw axes and diad symmetry in the [1 1 0] direction, denoted $P3_112$ and $P3_212$ — the third subscript is added to avoid confusion. On the other hand, the $P3$ space group is unequivocal, and its full form is $P3$.

The $R\bar{3}$ Space Group. To the uninitiated, the first encounter with a rhombohedral space group can prove daunting! The *International Tables* shows overlapping projections for the unit cells in these space groups, based on "rhombohedral" and "hexagonal" axes. The figure appears as a hexagon atop a rhombus — the rhombus is a projection of the "hexagonal" unit cell and the hexagon is a projection of the rhombohedral unit cell. This is the major reason for the use of "trigonal(R)" axes, rather than "hexagonal" axes here. Fig. 2.80 illustrates the relationship between the rhombohedral unit cell and the trigonal unit cell in the rhombohedral lattice. Early crystal structures involving rhombohedral lattices were reported with rhombohedral unit cells, but today they are rarely used, current practice favoring the trigonal(R) unit cell even though it is not primitive. It is therefore useful to be able to transform a rhombohedral unit cell into a trigonal(R) unit cell. In addition this provides an excellent exercise in unit cell transformation — we will use a practical example to demonstrate it.

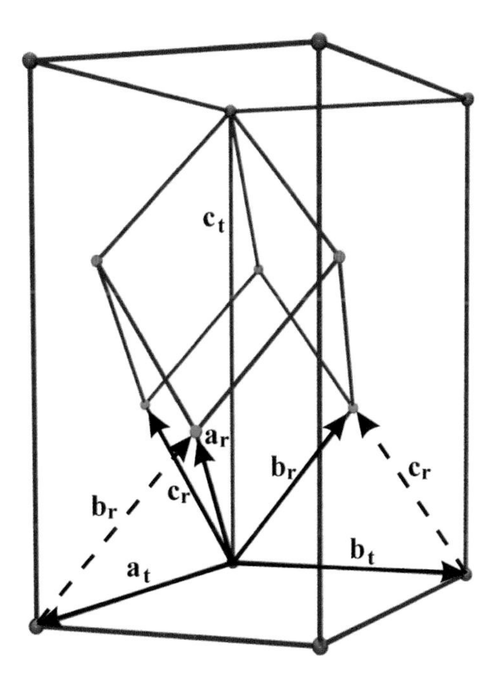

Figure 2.80 Trigonal(R) unit cell and rhombohedral unit cell (obverse) in the rhombohedral lattice.

In 1964 E.B. Fleischer[29] published the crystal structure of the novel cubane molecule, consisting of eight carbon atoms and eight hydrogen atoms arranged into a cube. Although the molecule had cubic symmetry, it crystallized in the rhombohedral space group $R\bar{3}$. The cell parameters were $a = 5.340(2)$ Å and $\alpha = 72.26(5)°$. The asymmetric unit consisted of two carbon atoms and two hydrogen atoms. One carbon atom, C_s, at $[x_f\ y_f\ z_f] = (0.1155, 0.1155, 0.1155)$ and one hydrogen atom, H_s, at (0.2100, 0.2100, 0.2100) were located on the threefold (diagonal) axis of the rhombohedral unit cell in special positions. The other carbon atom, C_g, and hydrogen atom, H_g were found in general positions at (-0.1871, 0.1952,0.1071) and (-0.3246, 0.3468, 0.1848), respectively. We will use these data to transform the rhombohedral unit cell and its contents into a trigonal(R) unit cell. Referring to Fig. 2.80, the relationships between the axes are

$$\begin{aligned}
\mathbf{a}_r &= \mathbf{a}_t + \mathbf{b}_r \Longrightarrow \mathbf{a}_t = \mathbf{a}_r - \mathbf{b}_r \\
\mathbf{b}_r &= \mathbf{b}_t + \mathbf{c}_r \Longrightarrow \mathbf{b}_t = \mathbf{b}_r - \mathbf{c}_r \\
\mathbf{c}_t &= \mathbf{a}_r + \mathbf{b}_r + \mathbf{c}_r.
\end{aligned} \tag{2.40}$$

In order to transform these axes we determine the Cartesian components of the rhombohedral unit cell vectors by determining the cell volume (Eqn. 1.139), $V = a^3(1 - 3\cos^2\alpha + 2\cos^3\alpha)^{\frac{1}{2}}$ — then creating the $\mathbf{B}$ matrix:

$$\mathbf{B} = \begin{bmatrix} a & a\cos\alpha & a\cos\alpha \\ 0 & a\sin\alpha & \dfrac{a(\cos\alpha - \cos\alpha\cos\alpha)}{\sin\alpha} \\ 0 & 0 & \dfrac{V}{a^2\sin\alpha} \end{bmatrix} = \begin{bmatrix} a_{rx} & b_{rx} & c_{rx} \\ a_{ry} & b_{ry} & c_{ry} \\ a_{rz} & b_{rz} & c_{rz} \end{bmatrix} \tag{2.41}$$

$$= \begin{bmatrix} 5.340 & 1.627 & 1.627 \\ 0 & 5.086 & 1.188 \\ 0 & 0 & 4.945 \end{bmatrix}.$$

The Cartesian components of the axial vectors in the trigonal(R) basis are then computed and the unit cell parameters determined:

$$\begin{aligned}
\mathbf{a}_{tc} &= ((5.340 - 1.627), (0 - 5.086), (0 - 0)) = (3.713, -5.086, 0) \\
\mathbf{b}_{tc} &= ((1.627 - 1.627), (5.086 - 1.188), (0 - 4.945)) = (0, 3.898, -4.945) \\
\mathbf{c}_{tc} &= ((5.340 + 1.627 + 1.627), (0 + 5.086 + 1.188), (0 + 0 + 4.945)) \\
&= (8.594, 6.274, 4.945) \\
a_t &= (\mathbf{a}_{tc} \cdot \mathbf{a}_{tc})^{\frac{1}{2}} = 6.297 \\
b_t &= (\mathbf{b}_{tc} \cdot \mathbf{b}_{tc})^{\frac{1}{2}} = 6.297 \\
c_t &= (\mathbf{c}_{tc} \cdot \mathbf{c}_{tc})^{\frac{1}{2}} = 11.733
\end{aligned}$$

$$\begin{aligned}
\cos\alpha_t &= (\mathbf{b}_{tc} \cdot \mathbf{c}_{tc})/(b_t c_t) = 0 \Longrightarrow \alpha_t = 90° \\
\cos\beta_t &= (\mathbf{c}_{tc} \cdot \mathbf{a}_{tc})/(c_t a_t) = 0 \Longrightarrow \beta_t = 90° \\
\cos\gamma_t &= (\mathbf{a}_{tc} \cdot \mathbf{b}_{tc})/(a_t b_t) = -0.5 \Longrightarrow \gamma_t = 120°
\end{aligned}$$

In order to transform the fractional coordinates of a general vector in the rhombohedral unit cell, $[x_r\ y_r\ z_r]$, into fractional coordinates in the trigonal(R) unit cell, $[x_t\ y_t\ z_t]$, we express the vector in both bases to develop the relationship between the fractional coordinates in each basis:

$$\begin{aligned}
\mathbf{v}_f &= x_r\mathbf{a}_r + y_r\mathbf{b}_r + z_r\mathbf{c}_r \\
&= x_t\mathbf{a}_t + y_t\mathbf{b}_t + z_t\mathbf{c}_t \\
&= x_t(\mathbf{a}_r - \mathbf{b}_r) + y_t(\mathbf{b}_r - \mathbf{c}_r) + z_t(\mathbf{a}_r + \mathbf{b}_r + \mathbf{c}_r) \\
&= (x_t + z_t)\mathbf{a}_r + (-x_t + y_t + z_t)\mathbf{b}_r + (-y_t + z_t)\mathbf{c}_r \Longrightarrow
\end{aligned}$$

$$x_r = x_t + z_t \tag{2.42}$$

$$y_r = -x_t + y_t + z_t \tag{2.43}$$

$$z_r = -y_t + z_t \tag{2.44}$$

Adding Eqns. 2.42, 2.43 and 2.44 gives $z_t = 1/3x_r + 1/3y_r + 1/3z_r$. Inserting this result into Eqns. 2.44 and 2.42 results in $y_t = 1/3x_r + 1/3y_r - 2/3z_r$ and $x_t = 2/3x_r - 1/3y_r - 1/3z_r$. The transformation to trigonal coordinates can now be accomplished using the matrix equation:

$$\begin{bmatrix} \frac{2}{3} & -\frac{1}{3} & -\frac{1}{3} \\ \frac{1}{3} & \frac{1}{3} & -\frac{2}{3} \\ \frac{1}{3} & \frac{1}{3} & \frac{1}{3} \end{bmatrix} \begin{bmatrix} x_r \\ y_r \\ z_r \end{bmatrix} = \begin{bmatrix} x_t \\ y_t \\ z_t \end{bmatrix}.$$

Applying this transform to the rhombohedral coordinates of the four atoms in the cubane asymmetric unit results in trigonal(R) fractional coordinates for C_s at (0, 0, 0.1155), H_s, at (0, 0, .2100), C_g at (−0.2255, −0.0687, 0.0384), and H_g at (−0.3936, −0.1158, 0.0690). In order to generate the coordinates for the complete molecule or the contents of the unit cell, the general positions in the *International Tables* are consulted. The rhombohedral space group for cubane is $R\bar{3}$. The general positions are given for both rhombohedral and hexagonal (trigonal(R)) axes. By applying the symmetry transformations for each unit cell, including the "centering" translations for the trigonal(R) unit cell and adding ±1 to each coordinate (including the symmetry-equivalent positions) the contents of each unit cell and the unit cells surrounding it can be determined. Fig. 2.81 shows the results of this exercise. The graphical representation of the unit cells and their contents are referred to as *packing diagrams.* The hydrogen atoms in the cubane molecules have been removed in the packing diagrams for clarity. A projection of the unit cell as it appears in the *International Tables for Crystallography* is also shown. Note that the space group is centrosymmetric, and that there are alternating right- and left-handed screw axes in the unit cells.

2.4.7 The Tetragonal Space Groups

The $P4_2/nnm$ Space Group. The tetragonal space groups do not have three-fold symmetry, but they do incorporate the two-fold symmetry of the "lower symmetry" space groups. The additional symmetry for the tetragonal space groups arises from the four-fold rotation axis of the tetragonal point groups. The point symmetry operators are 4^1, $4^2 \equiv 2$, 4^3, $\bar{4}^1$, $\bar{4}^2 \equiv 2$, and $\bar{4}^3$. Since the axes are at right angles to one another and the **a** and **b** basis vectors are the same length, the

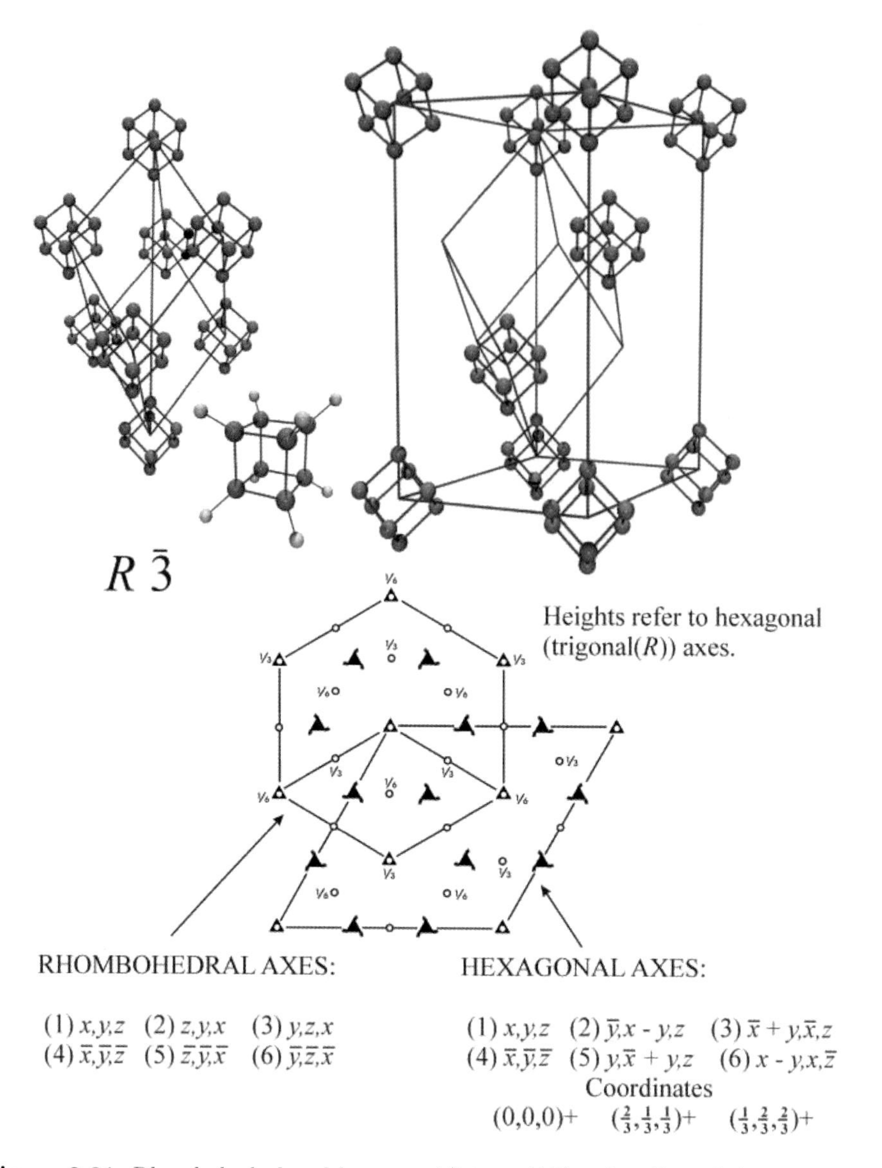

Figure 2.81 Rhombohedral and hexagonal/trigonal(R) unit cells and their contents for cubane.

transformation matrices will be the same in an orthonormal basis and the unit cell basis. The $\pi/2$ rotation of a general vector in a tetragonal unit cell is illustrated in Fig. 2.82. The figure is a projection of the unit cell on the ab plane with x and y the lengths of the vector components in the unit cell basis along the a and b axes, respectively. As we have previously noted in the trigonal unit cell, these lengths correspond to the fractional coordinates $x_f = x/a$ and $y_f = y/a$. The results are much more straightforward for the tetragonal case, however. For each rotation, the magnitudes of the vector components simply exchange with one another. It follows that $x'_f = -y/a = -y_f$ and $y'_f = x/a = x_f$ for the first rotation. For the second

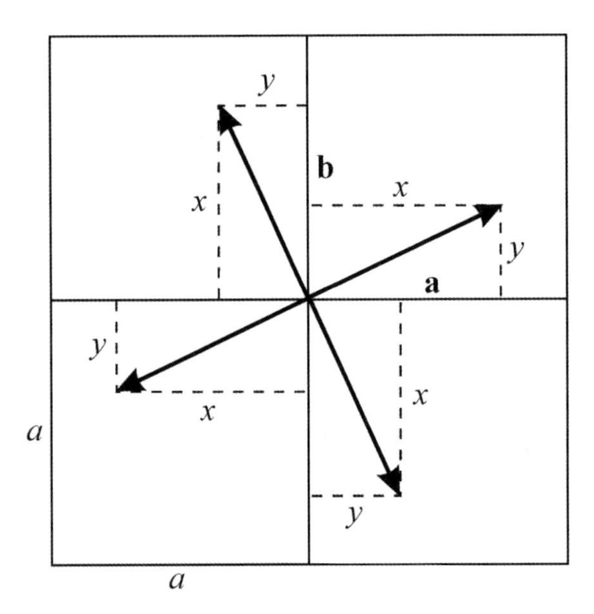

Figure 2.82 Three $\pi/2$ rotations of a general vector about the c axis in a tetragonal unit cell.

rotation, $x''_f = -x_f$ and $y''_f = -y_f$ — a two fold rotation — and for the third, $x'''_f = x_f$ and $y'''_f = -x_f$. We already have operators for the diad operations; there are therefore four new point group operations in this space group. As we did with the trigonal space group, we create matrix operators for the 4^1 and 4^3 operations, then multiply them by $\mathbf{R}(\bar{1})$ to create the matrices for $\bar{4}^1$ and $\bar{4}^3$:

$$\mathbf{R}(4^1) = \begin{bmatrix} 0 & -1 & 0 \\ 1 & 0 & 0 \\ 0 & 0 & 1 \end{bmatrix} \quad \mathbf{R}(4^3) = \begin{bmatrix} 0 & 1 & 0 \\ -1 & 0 & 0 \\ 0 & 0 & 1 \end{bmatrix}$$

$$\mathbf{R}(\bar{4}^1) = \begin{bmatrix} 0 & 1 & 0 \\ -1 & 0 & 0 \\ 0 & 0 & -1 \end{bmatrix} \quad \mathbf{R}(\bar{4}^3) = \begin{bmatrix} 0 & -1 & 0 \\ 1 & 0 & 0 \\ 0 & 0 & -1 \end{bmatrix}$$

In addition to tetrad axes, local translations of $1/4c$, $1/2c$, and $3/4$c also make four-fold screw axes possible in this space group. The 4_1 axis undergoes a complete rotation and translation into the next unit cell after four rotation-translations (Fig. 2.83(a)). As shown in Fig. 2.83(b) the 4_2 screw axis behaves much like the 3_2 screw axis in the trigonal system, in that the second symmetry-equivalent position is skipped after the first rotation, but "recovered" on the next rotation. The result is an axis with two-fold symmetry. However, it is a "special" two-fold axis, with alternating positions along the axis rotated 90° with respect to one another. The 4_3 operations skip the first *two* symmetry-equivalent positions on the initial rotation, then recover the positions in subsequent operations. Fig. 2.83(c) demonstrates that this is the same result that would be attained if the 4_3 axis was rotated counter-clockwise followed by a translation of $1/4c$ — the 4_3 screw axis is left-handed and

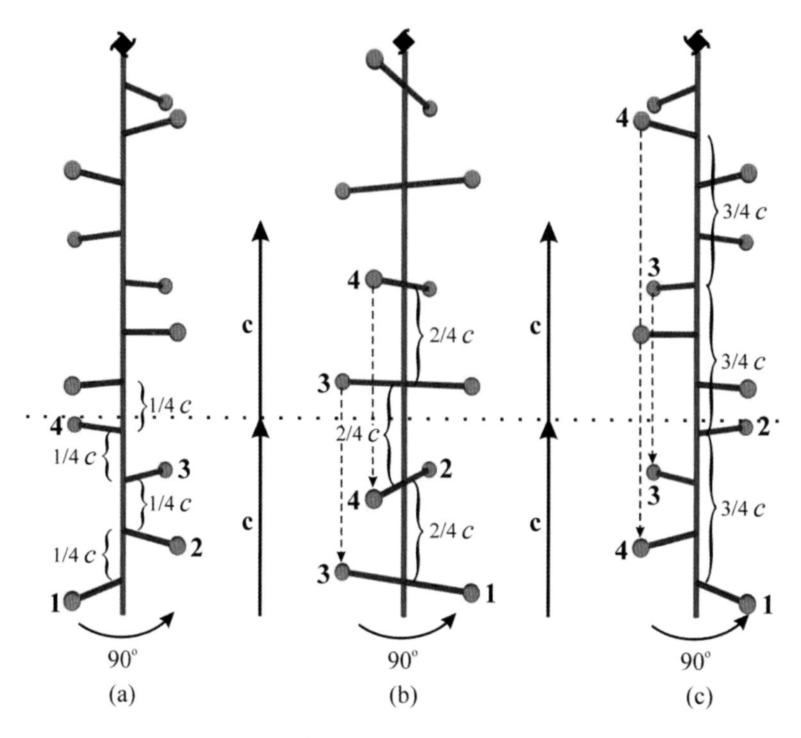

Figure 2.83 (a) 4_1 screw axis. (b) 4_2 screw axis. (c) 4_3 screw axis. Symbols for axes indicating opposite rotations for the 4_1 and 4_3 axes are shown atop each axis.

is the enantiomorph of the 4_1 screw axis. In addition to the typical local translation operations, I-centered unit cells in the tetragonal system can also contain diamond glide planes, as we observed in two of the F-centered orthorhombic space groups (recall that the I-centered lattice can be viewed alternately as an F-centered lattice).

Combinations of the diad operations (including mirror planes and glide planes), the inversion operation, the local tetrad symmetry operations (and their inverses), and the translation group results in 68 tetragonal space groups (Nos. 75-142). The H-M notation is very similar to that for the trigonal space groups. It is also of the general form $P\,qrs$. The first descriptor, q, indicates the nature of the tetrad axis, the second the symmetry with respect to the axial directions, and the third the symmetry with respect to the $[1\ \bar{1}\ 0]$ direction.

The crystal structure of a unique form of elemental boron provides an informative example of a tetragonal space group.[30] The chemistry of boron is extraordinarily unique – so unique that W. N. Lipscomb was awarded the Nobel Prize for elucidating the nature of the structure and bonding in boron compounds in 1976. Much of the chemistry of boron centers around its elemental structure – where it occurs as *icosahedra* in several allotropic forms, one of which is tetragonal. The H-M notation for the space group of tetragonal boron is $P\,4_2/nnm$; the long form is $P\,4_2/n\ 2/n\ 2/m$. The first element tells us that there is a 4_2 axis parallel to the c axis, and perpendicular to a diagonal glide plane, the second that there are diagonal glide planes perpendicular to the a and b axes, and the third indicates

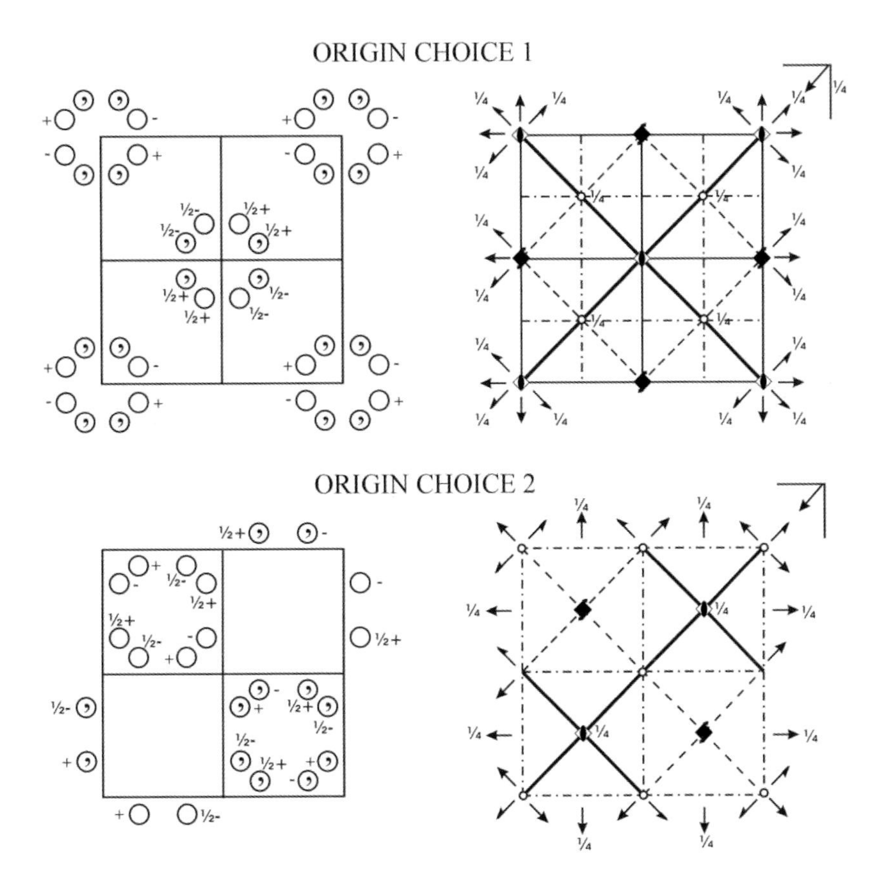

Figure 2.84 Orthogonal projections of $P4_2/nnm$ with two different choices of origin. The first choice places the origin at the intersection of the $\bar{4}$ axis parallel to c, diad axes aligned with the a and b axes, and a mirror plane perpendicular to the $[1\ \bar{1}\ 0]$ (diagonal) direction. The second choice places the origin at the center of symmetry, (1/4, -1/4, 1/4) from the origin in the first choice.

the presence of mirror planes perpendicular to the diagonals of the unit cell. The symmetry operations with respect to these elements give rise to inversion centers, $\bar{4}$ axes and axial glide planes. *The International Tables* lists two origin choices for the unit cell, as do several other centrosymmetric tetragonal space groups. This occurs when the inversion center is not coincident with a point of high site symmetry. The projections of the $P4_2/nnm$ unit cells showing both origin choices as they appear in the *International Tables* are shown in Fig. 2.84. Origin choice 1, at (-1/4, 1/4, -1/4) from the inversion center, is at the intersection of a $\bar{4}$ axis, a diad axis, and a mirror plane. The projection appears to provide a better picture of the unit cell contents and symmetry elements than origin choice 2, which apparently illustrates a unit cell with two empty quadrants. Alternative choices of origin for centrosymmetric structures may be of value when analyzing a solved structure, but they is never a good reason to use them for structure solution. The mathematics of diffraction is greatly simplified for centrosymetric structures with the inversion center at the origin, as are the methods for structure solution. Furthermore, many

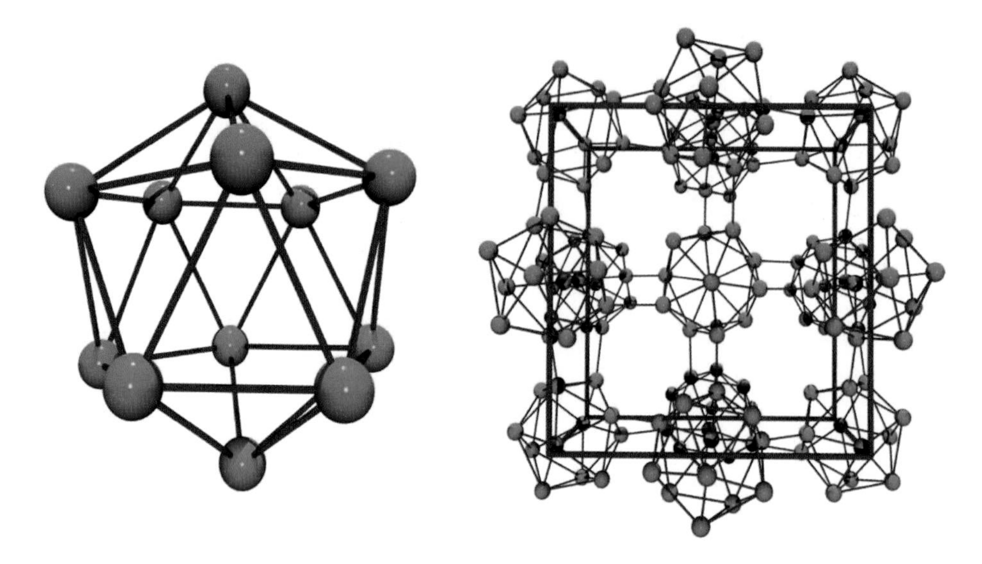

Figure 2.85 Boron icosahedron (slightly distorted) and packing diagram for origin choice 2 viewed down the c axis.

crystallography software programs *assume* the presence of the $[\bar{x}_f\ \bar{y}_f\ \bar{z}_f]$ coordinate if the input indicates that the structure is centrosymmetric. Displacing the origin from the inversion center also displaces this vector — and modifies its coordinates.

Fig. 2.84 also illustrates another important point when viewing unit cell projections. As unit cells increase in symmetry, their projections can often become deceptive. It is important to remember that the open circles do not represent the contents of the unit cell, but rather the location of a *single point* and its symmetry-equivalents — a *single atom* would produce another atom at every point in the diagram. Fig. 2.85 is a packing diagram of the boron unit cell with the second origin choice — clearly the symmetry operators of the unit cell and its neighbors have left no quadrants empty!

2.4.8 The Hexagonal Space Groups

The $P\,6_3/mmc$ Space Group. The hexagonal system combines the diad symmetry of the lower symmetry space groups, the triad symmetry of the trigonal system and the complete point symmetry of the hexagonal lattice. The new symmetry elements are the hexad axes; the resulting symmetry operations are 6^1, $6^2 \equiv 3$, $6^3 \equiv 2$, $6^4 \equiv 3^2$, 6^5, $\bar{6}^1$, $\bar{6}^2 \equiv \bar{3}$, $\bar{6}^3 \equiv \bar{2}$, $\bar{6}^4 \equiv \bar{3}^2$, and $\bar{6}^5$. As with the triad operators, the orthogonal transformation matrices will differ from those necessary to transform the fractional coordinates. The $2\pi/6$ rotation of a general vector in a hexagonal unit cell is illustrated in Fig. 2.86. The figure is a projection of a hexagonal unit cell on the ab plane with x and y the lengths of the vector components in the unit cell basis along the a and b axes, corresponding to the fractional coordinates $x_f = x/a$ and $y_f = y/a$. The figure on the left illustrates that the 6-fold operation rotates the vector and the parallelogram that defines it through 60° such that the y component of the original vector along the b axis is now aligned with the a axis, pointing in the

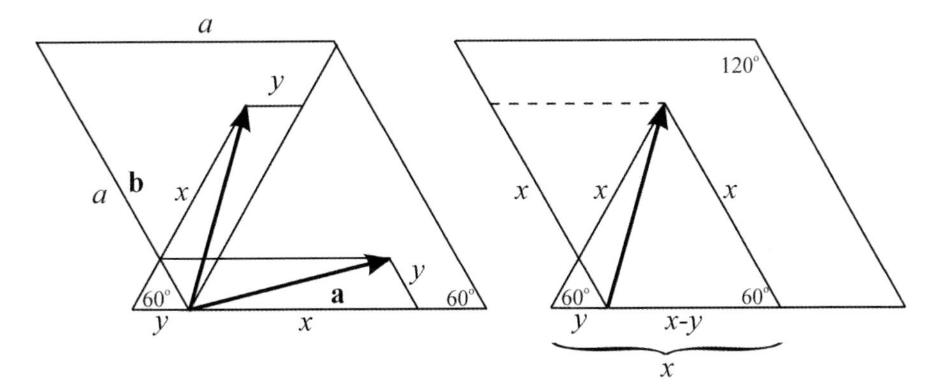

Figure 2.86 $2\pi/6$ rotation of a general vector about the c axis in a hexagonal unit cell.

-**a** direction. The figure on the right shows the y^I component of the rotated vector, parallel to the b axis. It is at an angle of 60° to the a axis, as is the "x" edge of the rotated parallelogram. Thus the triangle created from the "x" edge of the rotated parallelogram, the y^I component, and the line consisting of the rotated "y" edge of the parallelogram and the x^I component of the rotated vector forms an equilateral triangle. It follows that $x^I = (x - y)/a = x/a - y/a = x_f - y_f$ and $y^I = x/a = x_f$. The coordinates of the rotated vector are $[x_f^I \; y_f^I \; z_f^I] = [(x_f - y_f) \; x_f \; z_f]$. Applying this operation sequentially provides the remaining four symmetry-equivalent positions:

$$\begin{array}{lll}
x_f^{II} = x_f^I - y_f^I = -y_f & y_f^{II} = x_f^I = x_f - y_f & z_f^{II} = z_f^I = z_f \\
x_f^{III} = x_f^{II} - y_f^{II} = -x_f & y_f^{III} = x_f^{II} = -y_f & z_f^{III} = z_f^{II} = z_f \\
x_f^{IV} = x_f^{III} - y_f^{III} = y_f - x_f & y_f^{IV} = x_f^{III} = -x_f & z_f^{IV} = z_f^{III} = z_f \\
x_f^{V} = x_f^{IV} - y_f^{IV} = y_f & y_f^{V} = x_f^{IV} = y_f - x_f & z_f^{V} = z_f^{IV} = z_f
\end{array}$$

The matrix operators for the rotations are

$$\mathbf{R}(6^1) = \begin{bmatrix} 1 & -1 & 0 \\ 1 & 0 & 0 \\ 0 & 0 & 1 \end{bmatrix} \qquad \mathbf{R}(6^2) = \begin{bmatrix} 0 & -1 & 0 \\ 1 & -1 & 0 \\ 0 & 0 & 1 \end{bmatrix} \equiv \mathbf{R}(3^1)$$

$$\mathbf{R}(6^3) = \begin{bmatrix} -1 & 0 & 0 \\ 0 & -1 & 0 \\ 0 & 0 & 1 \end{bmatrix} \equiv \mathbf{R}(2) \qquad \mathbf{R}(6^4) = \begin{bmatrix} -1 & 1 & 0 \\ -1 & 0 & 0 \\ 0 & 0 & 1 \end{bmatrix} \equiv \mathbf{R}(3^2)$$

$$\mathbf{R}(6^5) = \begin{bmatrix} 0 & 1 & 0 \\ -1 & 1 & 0 \\ 0 & 0 & 1 \end{bmatrix}.$$

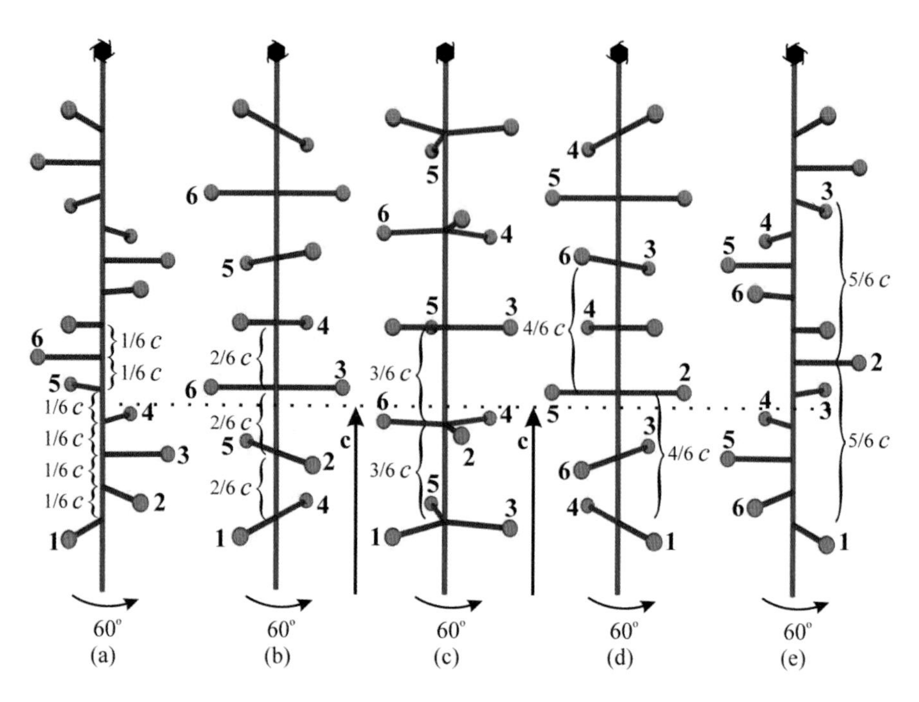

Figure 2.87 (a) 6_1 screw axis. (b) 6_2 screw axis. (c) 6_3 screw axis. (d) 6_4 screw axis. (e) 6_5 screw axis. Symbols for axes are shown atop each axis.

The rotoinversion operators, $\mathbf{R}(\bar{6}^n)$, are obtained by negating the elements in each of the matrices. The only *new* operators in this set are 6^1, 6^5, $\bar{6}^1$, and $\bar{6}^5$. However each operator can by combined with translations of $n/6\,c$ to create the symmetry operations resulting from 6_1, 6_2, 6_3, 6_4, and 6_5 screw axes, illustrated in Fig. 2.87. The 6_1 and 6_5 axes are enantiomorphs of one another – right and left handed six-fold screw axes. The 6_2 and 6_4 axes are also 3_1 and 3_2 axes — and therefore are also enantiomorphs. The 6_3 axis is also a diad rotation axis.

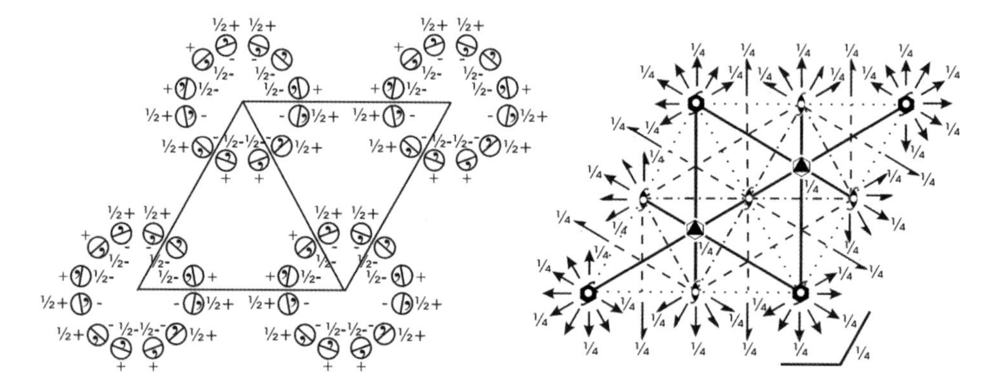

Figure 2.88 Orthogonal projections of the $P6_3/mmc$ unit cell onto the ab plane.

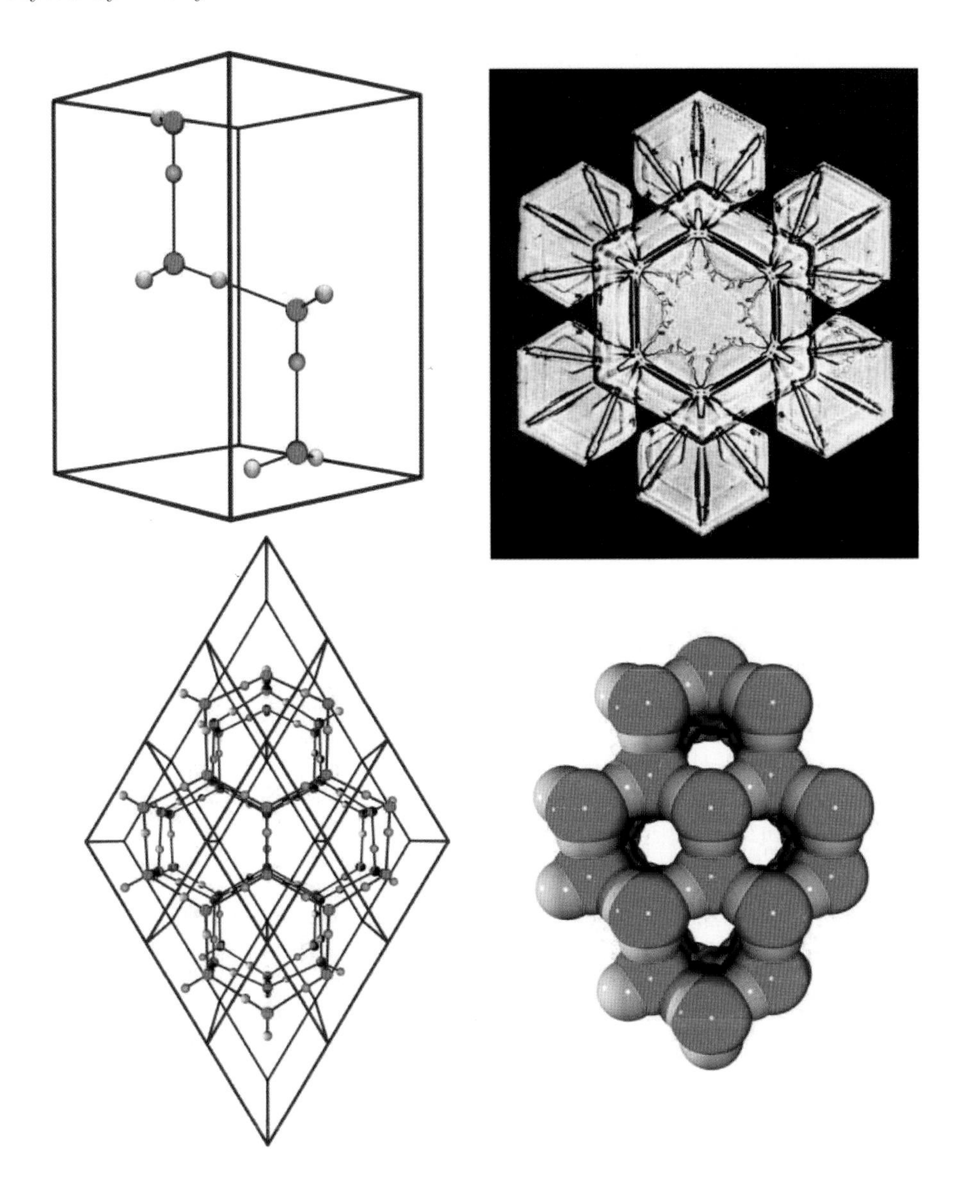

Figure 2.89 Upper left: Ice(I) unit cell. Lower left: Ice packing diagram viewed down the c axis. Lower right: Space-filling model of the packing diagram on the left. Upper right: Snowflake.

Combinations of these new operations with those of the trigonal and lower-symmetry space groups results in 27 hexagonal space groups (Nos. 168-194). The H-M notation for the hexagonal space groups follow essentially the same rules as for the trigonal system: for $P\,qrs$, q gives the hexad symmetry, r provides the axial symmetry, and s refers to the $[1\ \bar{1}\ 0]$ diagonal.

The crystal structure of ice provides an excellent example of a hexagonal crystal structure.[31] The structure is unique in that it is one of the few in which the crystal contains substantial empty space. The result is a solid less dense than the liquid, which allows it to float, rather than sink in its own liquid. The space group for ice is $P6_3/mmc$, indicating a 6_3 axis along c perpendicular to a mirror in the ab plane, mirror planes perpendicular to the axes, and a c glide plane perpendicular to the "long" unit cell diagonal in the ab plane. The long form of the space group symbol is $P6_3/m\,2/m\,2/c$.

This combination of symmetry elements generates many more, as the $P6_3/m\,m\,c$ projections in Fig. 2.88 demonstrate. Near the beginning of Chapter 1 it was noted that crystals adopt macroscopic appearances that reflect the underlying symmetry of the crystal. No material does this better than water! Fig. 2.89 shows the hexagonal unit cell for water. There are four water molecules in the unit cell — forming an extended structure through hydrogen bonding (the "longer" bonds in the figure). The structure is also viewed along the c axis, revealing the hexagonal symmetry. A space-filling model of the same view clearly reveals the empty space in the lattice. Finally, the photograph in the upper right is a sample of the many exquisite snowflake photographs taken by Wilson A. Bentley (known as "Snowflake Bentley") over the course of his life.

One minor perturbation regarding hexagonal and trigonal unit cells must be addressed before we finish with the space groups. In much of the older literature – and some of the current literature (in materials science, geology, and mineralogy in particular), an apparently odd fourth index is assigned to the lattice planes. Because of the symmetry of the hexagonal lattice, the short diagonal in the ab plane of the trigonal(H)/hexagonal unit cell is an alternative unit cell axis. Taking this axis in the direction of $-(\mathbf{a}+\mathbf{b})$ provides a third axis, rotated 120° to the initial axes. The axis is redundant, since it is a linear combination of the first two, but determining the intersection of a given set of lattice planes with this axis as well as the original three axes often aids in the discussion of the complex symmetry of the hexagonal and trigonal systems. The "new" indices are then $(h, k, -(h+k), l)$. However, these indices are not used for structural determination; the fourth index is not only redundant, it is formally unnecessary.

2.4.9 The Cubic Space Groups

The $P2_13$, $Fd\bar{3}m$, and $F\,\mathrm{m}\bar{3}m$ Space Groups. The cubic system adds no new symmetry elements, but it does have additional symmetry operators (otherwise there would be no new space groups!). The new point symmetry operators are for the triad rotations along the body diagonal of the cube. Since the triad axes do not align with any coordinate axis, we must do some work to generate the transformation matrices for the three-fold operations — around axes in the [1 1 1] direction of the unit cell. Fortunately, the fractional coordinates of the axes in the cubic unit cell are identical to the unit vectors in an orthonormal coordinate system; the transformation matrices can be determined in a Cartesian system. There are a number of ways to determine the transformation matrix for rotation about an arbitrary axis. The most intuitive approach is to use the transformations that we already have, namely rotations around the coordinate axes. Fig. 2.90 illustrates how this is done.

The point (x, y, z) is to be rotated through an angle θ about the rotation axis $\mathbf{r}$. The point can be considered to be attached to $\mathbf{r}$ with a line segment lying in the rotation plane and perpendicular to $\mathbf{r}$. The $\mathbf{r}$ vector, along with the point, is first

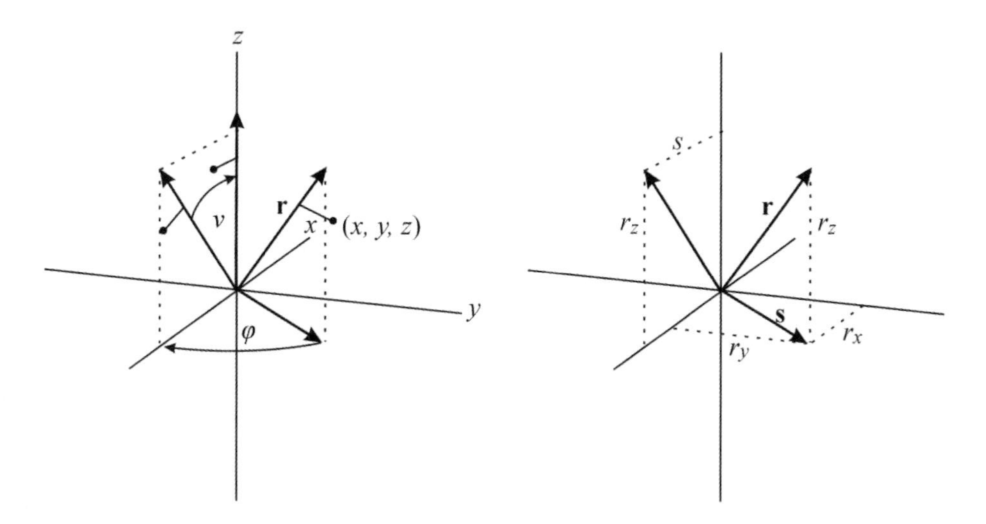

Figure 2.90 Transformation of an arbitrary rotation axis to the z axis.

rotated about the z axis through angle φ into the xz plane using the rotation matrix $\mathbf{R}_z$. This rotates the projection of $\mathbf{r}$, $\mathbf{s}$, onto the x axis, so that s, the length of the projected vector, becomes the x component of the rotated $\mathbf{r}$ vector. The y component of the rotated vector is zero, and the z component remains unchanged. The rotated $\mathbf{r}$ vector in the xy plane is rotated a second time, around the y axis through angle ν using the rotation matrix $\mathbf{R}_y$, making it coincident with the z axis. The "attached" point and its perpendicular line segment now lie in a plane perpendicular to the z axis. The rotation of the point is accomplished by rotating it around the z axis with the rotation matrix $\mathbf{R}_\theta$. The axis transformation process is then reversed by applying $\mathbf{R}_y^{-1}$, then $\mathbf{R}_z^{-1}$, taking $\mathbf{r}$ back to its original position, along with the rotated point. The entire process is accomplished with a series of matrix multiplications: $\mathbf{R_{rot}v} = \mathbf{v}' \Rightarrow \mathbf{R}_z^{-1}\mathbf{R}_y^{-1}\mathbf{R}_\theta\mathbf{R}_y\mathbf{R_z v} = \mathbf{v}'$. Since the matrices are all orthogonal we can compute $\mathbf{R}_{yz} = \mathbf{R}_y\mathbf{R}_z$ so that $\mathbf{R}_{rot} = \mathbf{R}_{yz}^{-1}\mathbf{R}_\theta\mathbf{R}_{yz} = \mathbf{R}_{yz}^T\mathbf{R}_\theta\mathbf{R}_{yz}$. The components of the projection of the rotation axis give us the cosine and sine of the z rotation angle: $\cos\varphi = r_x/s$ and $\sin\varphi = r_y/s$, where s is the length of the projected vector, $s = (r_x^2 + r_y^2)^{\frac{1}{2}}$. The components of the rotated vector in the xz plane provide the sine and cosine of the y rotation angle: $\cos\nu = r_z/r$ and $\sin\nu = s/r$, where $r = (r_x^2 + r_y^2 + r_z^2)^{\frac{1}{2}}$. $\mathbf{R}_{yz}$ can now be determined:

$$\mathbf{R}_{yz} = \mathbf{R}_y\mathbf{R}_z = \begin{bmatrix} r_x/s & r_y/s & 0 \\ -r_y/s & /r_x/s & 0 \\ 0 & 0 & 1 \end{bmatrix} \begin{bmatrix} r_z/r & 0 & -s/r \\ 0 & 1 & 0 \\ s/r & 0 & r_z/r \end{bmatrix}$$

$$= \begin{bmatrix} (r_x r_z)/(rs) & (r_y r_z)/(rs) & -s/r \\ -r_y/s & r_x/s & 0 \\ r_x/r & r_y/r & r_z/r \end{bmatrix}. \tag{2.45}$$

The matrix for a rotation about the axis $\mathbf{r}$ is determined by forming the matrix product, $\mathbf{R}_{yz}^T \mathbf{R}_\theta \mathbf{R}_{yz}$:

$$\mathbf{R}_{rot} = \mathbf{R}_{yz}^T \begin{bmatrix} \cos\theta & \sin\theta & 0 \\ -\sin\theta & \cos\theta & 0 \\ 0 & 0 & 1 \end{bmatrix} \mathbf{R}_{yz} = \tag{2.46}$$

$$\begin{bmatrix} \frac{r_x^2+(r_y^2+r_z^2)\cos\theta}{r^2} & \frac{r_x r_y(1-\cos\theta)-r_z r\sin\theta}{r^2} & \frac{r_x r_z(1-\cos\theta)+r_y r\sin\theta}{r^2} \\ \frac{r_x r_y(1-\cos\theta)+r_z r\sin\theta}{r^2} & \frac{r_y^2+(r_x^2+r_z^2)\cos\theta}{r^2} & \frac{r_y r_z(1-\cos\theta)-r_x r\sin\theta}{r^2} \\ \frac{r_x r_z(1-\cos\theta)-r_y r\sin\theta}{r^2} & \frac{r_y r_z(1-\cos\theta)+r_x r\sin\theta}{r^2} & \frac{r_z^2+(r_x^2+r_y^2)\cos\theta}{r^2} \end{bmatrix}.$$

This matrix may appear forbidding, but the triad diagonal rotation axis in the cubic unit cell (and the rhombohedral unit cell as well!) simplifies it tremendously. The rotation axis has components $[r_x\ r_y\ r_z] = [1\ 1\ 1]$. Thus $r^2 = 1^2 + 1^2 + 1^2 = 3$. For a $3^1 \equiv 2\pi/6$ rotation, $\cos\theta = -1/2$ and $\sin\theta = \sqrt{3}/2$, and for a $3^2 \equiv 4\pi/6$ rotation, $\cos\theta = -1/2$ and $\sin\theta = -\sqrt{3}/2$. Because of the equivalency of r_x, r_y, and r_z, there are only three different terms in the matrix, the diagonal terms, which are equivalent, and those which either add or subtract the $\sin\theta$ term. The diagonal terms become $(1 + 2\cos\theta)/3 = (1 + 2(-1/2))/3 = 0$. For the 3^1 operation the terms with a negative $\sin\theta$ term become $(1(1 + (1/2)) - (\sqrt{3}(\sqrt{3}/2)))/3 = 0$,while those with a positive $\sin\theta$ term become $(1(1 + (1/2)) + (\sqrt{3}(\sqrt{3}/2)))/3 = 1$. The sign of the sine terms change for the 3^2 operation, so that when the $\sin\theta$ term is subtracted, the matrix element is equal to one, and when added it is equal to zero. The new rotation operators for the cubic space groups are therefore

$$\mathbf{R}_{111}(3^1) = \begin{bmatrix} 0 & 0 & 1 \\ 1 & 0 & 0 \\ 0 & 1 & 0 \end{bmatrix} \quad \mathbf{R}_{111}(3^2) = \begin{bmatrix} 0 & 1 & 0 \\ 0 & 0 & 1 \\ 1 & 0 & 0 \end{bmatrix}.$$

The rotoinversion operations are obtained by negating the matrix elements. Thus the three fold axis transforms $[x_f\ y_f\ z_f]$ into $[z_f\ x_f\ y_f]$ and $[y_f\ z_f\ x_f]$. These "out of plane" rotations combine with other symmetry operations to generate other operations that are inclined to the planes containing the unit cell axes, in particular 2, 2_1, 3_1, and 3_2. Combinations of these symmetry operations result in 36 cubic space groups, Nos. 195-230. The H-M space group symbols for the cubic groups are again of the form $P\,qrs$, where q is the diad or tetrad symmetry of the unit cell axes, r is the symmetry with respect to the [1 1 1] (body diagonal) triad, and s corresponds to diad symmetry along [1 1 0] (face diagonal).

The cubic space groups fall into two general sub-classes, those with tetrahedral symmetry, and those with octahedral symmetry. The crystal structure of carbon monoxide serves to illustrate the tetrahedral space groups. The space group is (approximately) $P\,2_1 3$.* For a high symmetry space group it is relatively simple.

*$P\,2_1 3$ is the original space group assigned to solid carbon monoxide, based on Raman spectroscopy studies.[32] X-ray diffraction studies at high pressure later revealed that the carbon monoxide molecules were probably statically disordered, packing in a hexagonal space group, with $P6_3m3$ or $P6_3/mmc$ the most likely candidates.[33]

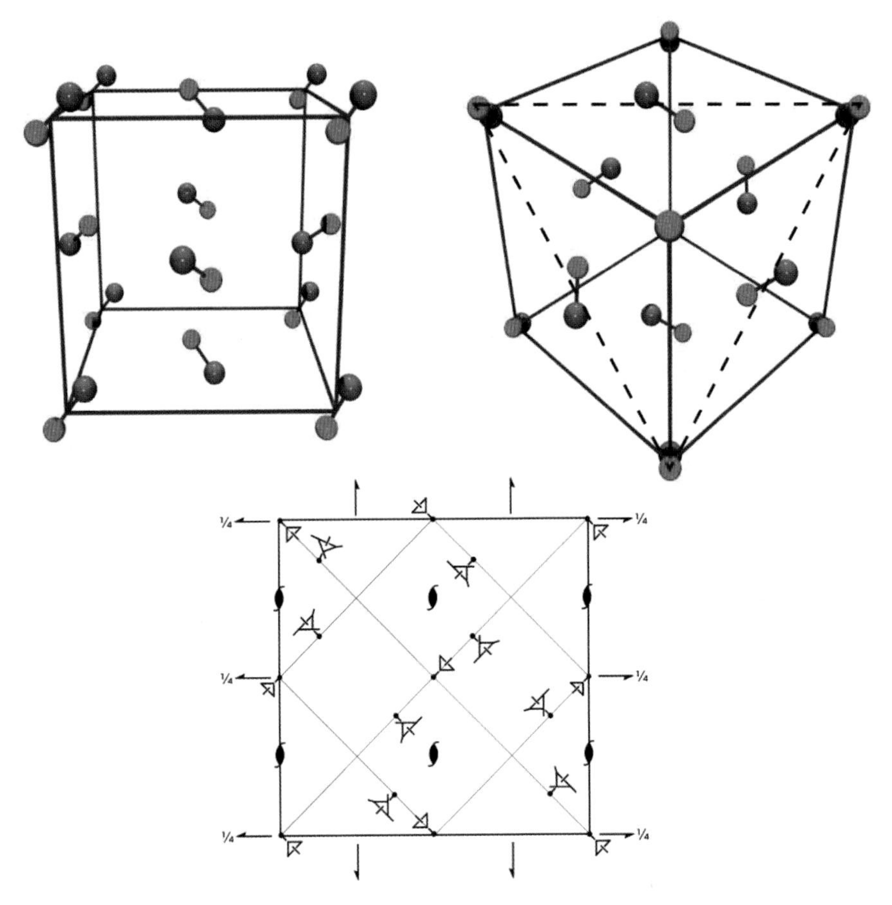

Figure 2.91 Upper left: Carbon monoxide unit cell. Upper right: View down the three fold axis. Lower: Orthogonal projection of the $P2_13$ unit cell.

The projection in Fig. 2.91 illustrates that the symmetry elements are diad screw axes parallel to the unit cell axes; like the $P2_12_12_1$ unit cell, the origin is at the center of the three perpendicular sets of screw axes. The triad axis along the diagonal is depicted as an open triangle with a line and "point of entry" to indicate that the axis does not lie in a plane. The form of this symbol is typical for symmetry elements not lying in the projection plane (which occur only with cubic space groups). The combination of the triad axis and the screw axes generates triad screw axes inside the cell. The view on the left illustrates the screw operations parallel to each axis; the one on the right shows the tetrahedral symmetry of the unit cell.

Cubic space groups with tetrahedral symmetry are not frequently observed; those with octahedral symmetry are far more common. Several of these have unit cells that contain diamond glide planes. The F-centered groups with diamond glide planes have them in alternating pairs separated by $1/4$ of the length of the unit cell axis. Each glide plane in a pair generates a reflection followed by a $1/4$ translation along the cell diagonal, with the glide directions along opposite diagonals for each glide plane in the pair. This is typical of diamond glide planes, and is illustrated

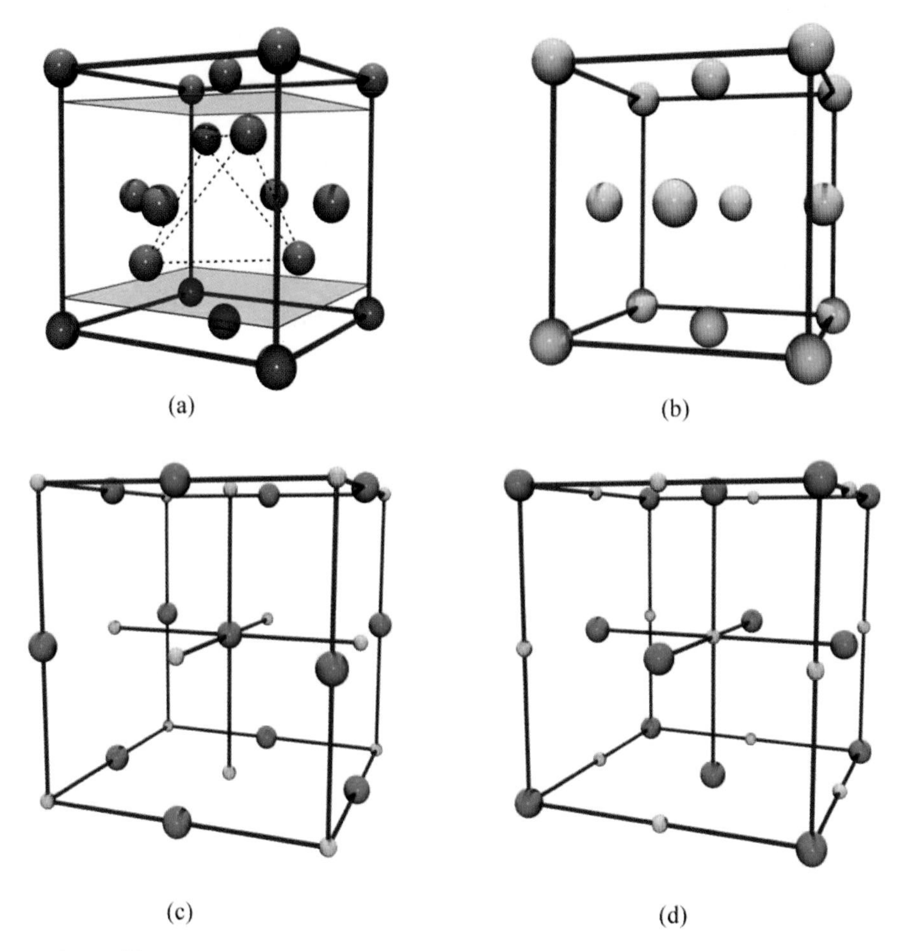

Figure 2.92 Unit cells in the F-centered cubic space groups.: (a) The diamond unit cell, space group $F\,d\bar{3}m$. The diamond glide planes are shown in blue. (b) Face-centered cubic unit cell of copper in the $F\,m\bar{3}m$ space group. (c) Sodium chloride unit cell ($F\,m\bar{3}m$) with a sodium ion at the origin. (d) Sodium chloride unit cell with a chloride ion at the origin.

in the diamond unit cell in Fig. 2.92(a).[34] The diamond glide planes, shown in blue, are at 1/8 and -1/8 along the unit cell axis. The diamond unit cell — in space group $F\,d\bar{3}m$ — exhibits virtually all of the cubic point symmetry. There are four carbon atoms at the corners, creating a cube. The atoms in the faces create an octahedron, and the internal atoms lie at 1/4 of the distance along each body diagonal — forming a tetrahedron (indicated by the dotted lines in the figure). The glide directions are aligned with the edges of the tetrahedron. In addition to these "conventional" diamond glide planes, I-centered unit cells have the centering translation vector along the body diagonal of the cube, and the glide direction must occur in this direction. Thus the glide plane must contain the body diagonal, and is therefore out of the projection plane. Instead it is the plane containing two opposing edges of the cube passing through the center of the unit cell. The glide translation in the body diagonal direction is therefore 1/4 along all three axes, i.e., $(\mathbf{a}+\mathbf{b}+\mathbf{c})/4$.

Finally, we end this discussion of space groups where we began the discussion of crystals in the early part of Chapter 1. Figs. 1.1 and 1.2 were presented to point out the internal and external symmetry of two substances related to the cube — copper,[35] and sodium chloride.[36] Not only are these two very different materials cubic in the solid state, but both materials pack in the same space group, $F\,m\bar{3}m$ (no. 225). It is the most common space group for the close-packing of spheres. The full H-M notation is $F\,4/m\bar{3}2/m$, indicating tetrad axes perpendicular to mirror planes aligned with each unit cell axis, a triad rotoinversion axis along the body diagonal, and diad axes along the face diagonals perpendicular to mirror planes, which must therefore also be aligned with the face diagonals. Figs. 2.92 (b), (c), and (d) show the $F\,m\bar{3}m$ unit cells for copper and sodium chloride. The copper unit cell is correctly described as a face-centered cubic (fcc) unit cell. Unfortunately, the sodium chloride unit cell is also often described as "face-centered cubic", and it is clearly not. The bottom of the figure displays the unit cell with two different origin choices – since the placement of the origin could equally be on a chloride ion or a sodium ion. In each case the face is occupied by the ion that is not at the origin. If only the chloride ions are considered — or the sodium ions — the *sublattice* can be considered to be fcc — and the entire lattice can be described as the combination of two inter-penetrating fcc sublattices. Each ion is surrounded by an octahedral arrangement of counter-ions. Since the chloride ions are much larger than the sodium ions the structure is often described as a close-packed (fcc) array of chloride ions containing sodium ions in the resulting "octahedral holes." This is a simple example of the way many minerals are characterized — as arrays of anions containing cations in octahedral or tetrahedral "holes."

The combination of mirror symmetry, inversion symmetry, and diad, triad, and tetrad symmetry creates a very complex projection for this unit cell — it is typical of those in the cubic system. The reader is encouraged to consult the *International Tables for Crystallography* for views of the projections of the cubic space groups.

2.4.10 General Considerations

Despite the existence of 230 crystallographic space groups, only a small subset of these are routinely observed. Although neither domain is exclusive, molecular species, which are generally relatively asymmetric, are predominantly found in the triclinic, monoclinic and orthorhombic systems, while symmetric species such as atoms and ions (quite often spheres) are discovered most often in the higher symmetry space groups. The space groups selected by molecular species are generally even more limited, with molecules packing in space groups with screw axes and/or glide planes which allow them to translate with respect to one another in order to attain maximum intramolecular contact. Indeed, five space groups account for three-fourths of all the known experimentally determined molecular space groups,[37] $P\bar{1}$, $P\,2_1$, $P\,2_1/c$, $C\,2/c$ and $P\,2_12_12_1$. Of these, $P\,2_1/c$ accounts for over a third of all the know space groups for molecular compounds. On the other hand, crystals of elements and ionic compounds are generally found with orthorhombic or higher symmetry, with $F\,m\bar{3}m$, $F\,d\bar{3}m$, $P\,6_3/mmc$, $P\,m\bar{3}m$, and $P\,2_1/c$ constituting the five most common space groups found for these materials. $P\,2_1/c$ shows up in this group as well!

In molecular crystal structures, including large molecular ions, the asymmetric unit is usually a molecule or part of a molecule (related by symmetry to another part) in the unit cell. In order for an asymmetric unit to occupy a unit cell with a mirror plane or inversion center it must be accompanied by its mirror or inversion image. It follows that if two molecules are enantiomers* of one another *both* enantiomers must be present in space groups that are centrosymmetric or contain mirror planes. The converse of this is also true: If there is only one enantiomer present in the unit cell, the unit cell will not contain either mirror planes or inversion centers. Biological macromolecules are intriguing in this respect, since they are all enantiomorphic. Naturally occurring nucleic acids are composed exclusively of *D*- enantiomers of the nucleotides, while proteins consist entirely of *L*- amino acids. Since the other enantiomers do not exist; all biological molecules are found in non-centrosymmetric unit cells without mirror planes.

Finally, while the point symmetry of the structure is lost with the addition of local translation operators, the generating point group is still important in describing the physical properties of crystalline materials, since most physical properties are not influenced by the translational symmetry. For example, the dipole moment of a material depends on the addition of the dipole moments of the components of the material. The relative translation of one dipole to another does not affect this sum. *The generating point group can always be determined by converting all screw axes to rotation axes and all glide planes to mirror planes.* One important property which depends on the generating point group is known as *piezoelectricity*. Crystals in point groups without centers of symmetry can contain axes which are not symmetric in the reverse direction. Such axes are called *polar axes*. When crystals with polar axes (quartz is an example) are undisturbed their internal electric dipoles are arranged so that they cancel one another. Compression of the crystal along a polar axis alters the direction of these dipoles, and since there is no center of symmetry (in which case the distortions would cancel one another), the compression induces a dipole moment in the crystal. With enough compression an electric discharge (spark!) can result. This phenomenon, and another related phenomenon in which the induced dipole moment varies with temperature (known as *pyroelectricity*), can be useful in determining whether or not a crystal is centrosymmetric – an often essential prerequisite for space group determination. Few other physical properties are sensitive to differences between centrosymmetric and non-centrosymmetric crystals. The measurement of piezoelectrically was more common in the early days of crystallography than it is today; the method has largely been replaced with statistical methods related to the diffraction of X-rays from crystals — crystal diffraction — our next topic.

*The term *enantiomer* is ordinarily used to describe a molecule that cannot be superimposed on its inverted image; the term *enantiomorph* is employed to describe a crystal lattice that cannot be superimposed on its inverted image, i.e., a non-centrosymmetric lattice.

Exercises

1. A ball and stick model of the POF_3 molecule is shown below:

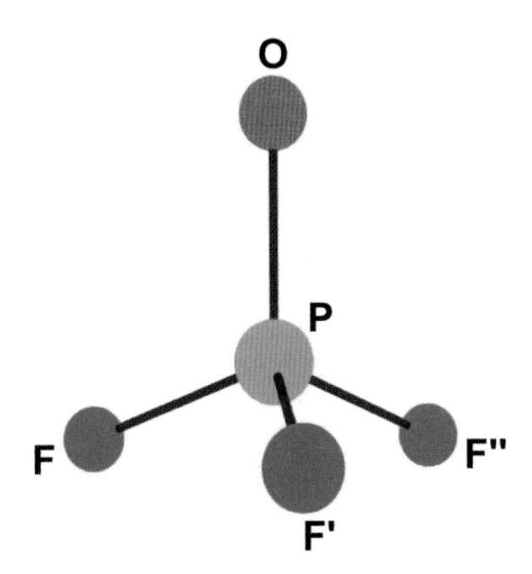

The fluorine atoms are all equivalent by symmetry. (a) Determine the symmetry elements and operations for the molecule. (b) Determine the point symmetry group of the molecule. (c) Create a multiplication table for the symmetry operations and show that the operations constitute a mathematical group (the point group in (b)) (d) Compare your multiplication table to Table 2.1. What can you conclude about the relationship between the point group of POF_3 and the point group of PF_5 from this comparison?.

2. Generate the matrix products for the (a)$C_2'S_3^2$ and (b) $S_3^2C_2'$ operation products for the D_{3h} point group. (c) Assess whether or not the matrix products produce the same results as the products of the operations given in Table 2.1. (d) Is the complete set of matrices for the D_{3h} point group a mathematical group? Explain your answer (why or why not?).

3. Using the scheme outlined in Figs. 2.9 and 2.10, verify the assignment of the Schönflies symbols for the point groups of lattices with a single rotoinversion axis.

4. Using the scheme outlined in Figs. 2.9 and 2.10, determine Schönflies and Hermann-Mauguin notations for the point groups of the Bravais lattices for each crystal system. Figs. 2.11 and 2.42 may prove useful.

5. The unit cell for the 2-mercaptopyridine crystal structure (Sec. 1.5.3) is described in a non-standard space group – $P2_1/n$. The standard setting is $P2_1/c$. (a) Determine the unit cell parameters for the $P2_1/c$ unit cell. (b) Determine the fractional coordinates of the S(1), C(1), and N(1) atoms in the $P2_1/c$ unit cell. (c) Verify the transformed coordinates in (b) by calculating the C(1)–S(1) and C(1)–N(2) interatomic distances and the N(1)–C(1)–S(1) interatomic angle in the $P2_1/c$ unit cell.

6. Recall that the product of local operations of a space group must generate another local operation if the translations generated by the product are negated (or ignored!). (a) For the $P2/c$ space group, diad rotation axes are postulated to result from a combination of c-glide operations, inversion operations, and translation group operations: $\mathbf{S}(2) = \mathbf{S}(t_j)\mathbf{S}(\bar{1})\mathbf{S}(c)$. Determine the locations of the diad axes in the $P2/c$ unit cell. (b) Verify that the $\mathbf{S}(2)$ operator in the $P2/c$ space group is given by Eqn. 2.35. (c) For the $P2_1/c$ space group, c-glide planes are postulated to result from a combination of diad screw operations (about axes at the locations determined in part (a)), inversion operations, and translation group operations: $\mathbf{S}(c) = \mathbf{S}(\mathbf{t}_j)\mathbf{S}(\bar{1})\mathbf{S}(2_1)$. Determine the locations of the glide planes in the $P2_1/c$ unit cell. (d) Verify that the $\mathbf{S}(c)$ operator in the $P2_1/c$ space group is given by Eqn. 2.36.

7. The transformation of unit cells from one setting to another does not affect the locations of the atoms, and therefore does not affect the locations of the symmetry elements in the lattice. (a)Use the locations of the points of intersection of the screw axes with the ac plane and the glide planes with the b axis in the $P2_1/c$ unit cell to determine the location of these symmetry elements in the $P2_1/n$ unit cell.* (b) Determine the symmetry operators, $\mathbf{S}(n)$ and $\mathbf{S}(2_1)$, for the n-glide and diad screw operations in the $P2_1/n$ unit cell. (c) Given the general position, $[x_f\ y_f\ z_f]$, in the $P2_1/n$ unit cell, determine the symmetry-equivalent general positions within the unit cell. (d) Determine the location of all the S(1)-C(1) bonded pairs in the 2-mercaptopyridine crystal structure (Sec. 1.5.3). Verify that all of the transformed pairs have the same internuclear distance.

8. Referring to Fig. 2.47, it was postulated that the symmetry element located at the 2' location was a diad rotational axis, but this assertion was not proved (any symmetry element that would transform an atom onto itself would have worked). (a) Show that an axis located at the position in the figure is a two-fold rotational axis. (b) Prove that any axis with a location resulting from a general product of a diad rotational operation about an axis at the origin followed by a translation from a vector in the translation group, $\mathbf{S}(\mathbf{t}_j)\mathbf{S}(2)$, $\mathbf{t}_j = [m_1\ m_2\ m_3]$, is a diad rotational axis. m_1, m_2, and m_3 are arbitrary integers.

9. The general positions for the $P2_12_12_1$ space group are listed in the *International Tables for Crystallography* as: (1) $x,\ y,\ z$ (2) $\bar{x}+\frac{1}{2},\ \bar{y},\ z+\frac{1}{2}$ (3) $\bar{x},\ y+\frac{1}{2},\ \bar{z}+\frac{1}{2}$ (4) $x+\frac{1}{2},\ \bar{y}+\frac{1}{2},\ \bar{z}$. (a) Determine the matrix operators for the local operators in $P2_12_12_1$ from these general positions. (b) Demonstrate that the set of local operations resulting from the application of these operators to a general vector *behaves like* a mathematical group if symmetry-equivalent positions due to translations from vectors in the translation group are transformed to equivalent positions without the translations.

10. Referring to Exercise 8 at the end of Chapter 1, the $CuSO_4$ crystal structure was reported to pack in the $Pnma$ space group. This is space group No. 62 in the *International Tables for Crystallography*. Although it is more common to transform a unit cell in a non-standard setting into one in the standard setting (in this case $Pnma$), consider transforming the unit cell and its contents into

*You can also locate these points using the "group theory" approach. You might wish to do it both ways!

the non-standard *Pmcn* unit cell. (a) Determine the cell parameters of the transformed unit cell and develop a matrix that will transform the fractional coordinates of the Cu, S, and O atoms from the *Pnma* unit cell to the *Pmcn* unit cell. (b) Look up the general and special positions in the *Pnma* unit cell and determine their locations in the *Pmcn* unit cell. (c) Determine the coordinates of the atoms in the asymmetric unit of the *Pmcn* unit cell. Are the contents of the cell consistent with $CuSO_4$ stoichiometry? (e) When the unit cell is transformed from *Pnma* to *Pmcn*, what happens to the indices of reflections from each set of lattice planes in the crystal? (You may have to read ahead in Chapter 3 to answer this question).

Chapter 3

Crystal Diffraction: Theory

3.1 Electromagnetic Radiation

The interaction between electromagnetic radiation and material objects allows us to see them. If the objects are too small, our radiation detectors (eyes) are incapable of sensing the wavelengths of the radiation necessary to visualize them directly, and we must resort to indirect methods to create the microscopic images that we are interested in. The interaction of electromagnetic radiation with the regularly repeated arrangement in crystalline solids amplifies the images of small entities (atoms and molecules) sufficiently for us to observe them, provided that we can somehow mimic the image formation process that our detectors and complex computers (brains) do automatically for larger objects.

3.1.1 The Electric Field.

Objects with certain properties are able to influence one another when separated in space. Indeed, we define these properties based on our observations of this influence (it is how we observe them!). One such property is known as *charge*, and we begin by considering how objects with charge influence one another. Qualitatively, objects with the same charge exert a repulsive force on one another, while objects with opposite charges exert an attractive force.

In order to "explain" the ability of one charged object to influence another we invoke the notion of an *electric field*. Consider the sphere with charge q_1 depicted in Fig. 3.1, which we will call the *source charge*. We envision an electric field

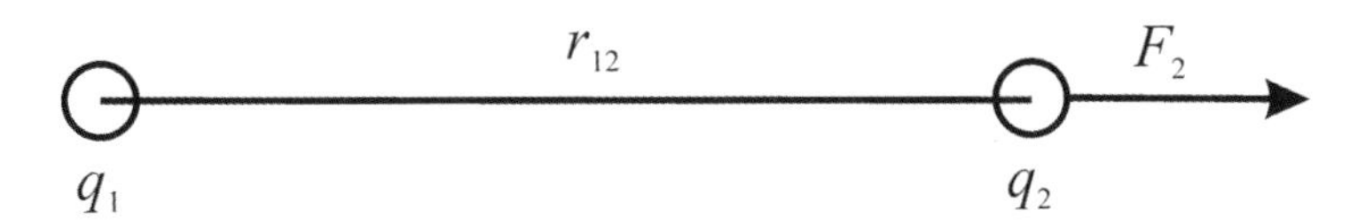

Figure 3.1 The electric field surrounding an object with charge q_1 exerts a force, F_2 on an object with charge q_2 at a distance r_{12}.

Understanding Single-Crystal X-Ray Crystallography. Dennis W. Bennett

ISBN: 978-3-527-32677-8 (HC), 978-3-527-32794-2 (SC)

surrounding the charged sphere which exerts a force on a second sphere with*test charge* q_2. We now wish to determine the magnitude of the strength of this field – that is – we seek a number that will indicate how the field resulting from charge q_1 influences a general charge placed somewhere else in space. In order to do this, we use the force exerted on the sphere with charge q_2 as a measure of this influence. However, as might be expected, the greater the magnitude of the test charge, q_2, the greater the force: $F_2 \alpha q_2$. Thus to make the magnitude of the electric field at the location of the test charge independent of the specific charge, we define it as $\mathcal{E} = F_2/q_2$.

By performing a few experiments, we determine two important properties of the electric field and its ability to influence another charge. First, the force, and therefore the magnitude of the electric field decreases as the distance, r_{12}, increases — in inverse proportion: $F \alpha 1/r_{12}$. Second, the influence is not instantaneous. That is, if the sphere containing the source charge is suddenly moved toward the sphere with the test charge (we assume here that both charges are negative), the second sphere will respond to an additional repulsive force from the first sphere, since they are now closer to one another. However, it will do so only after a time delay of r_{12}/c has occurred, where c is the speed of light. By responding, we mean that its speed has changed, which means it has gained energy from the first sphere, after we have increased the energy of the first sphere by moving it. Apparently this energy is transmitted from the source charge to the test charge at the speed of light.

3.1.2 Waves.

We now consider a *gedanken* experiment — a thought experiment — one that would be impractical to perform, but will allow us to discuss the essential elements of electromagnetic radiation without getting bogged down in experimental details. We make use of a hypothetical generator, a device which alternatively removes electrons from one electrode and moves them to a second electrode making it negative and leaving the first electrode positive, then reverses so that the first electrode becomes

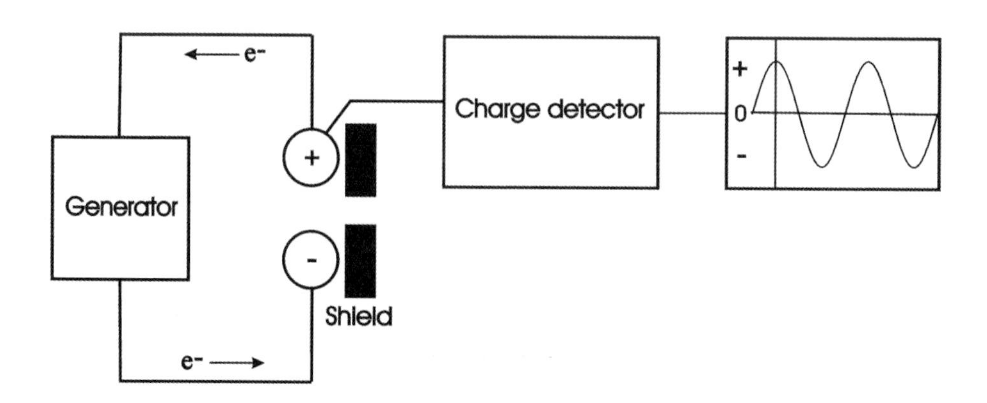

Figure 3.2 Hypothetical generator.

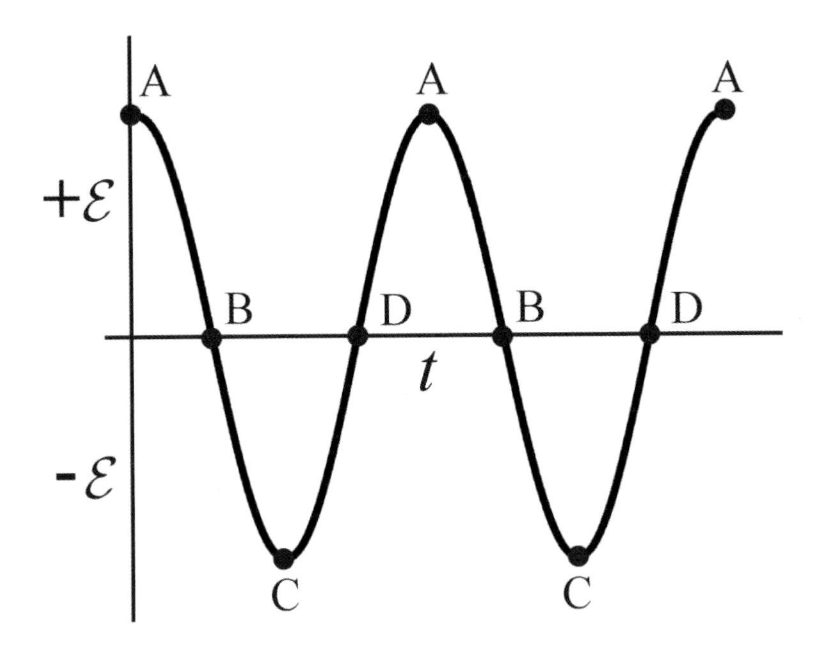

Figure 3.3 Electron velocity at different values of the electric field in the hypothetical generator: (A) $v = 0$. (B) $v = +v_{max}$. (C) $v = 0$. (D) $v = -v_{max}$.

positive and the second becomes negative. When switched on, the generator will charge each electrode in a sinusoidal manner, as depicted in Fig. 3.2. Our generator is an ideal one, subject to no constraints — it can shuttle the charges back and forth between the two electrodes at any rate (frequency) that we desire. The electrodes are shielded so that the electric field of the charges on the electrodes can only be sensed in the region directly between then.

An electron is now placed between the electrodes with the generator turned off. We restrict the electron to "up and down" motion between the electrodes – as in a wire or an electron-sized tube (remember, this is a gedanken experiment!). The electron is initially at rest. The generator has been designed so that the electron will initially be exposed to the maximum value of the electric field. When the generator is switched on, the electron is subjected to a force from the electric field, $\mathcal{E}$, between the electrodes — and it begins to move (points (A) in Fig. 3.3). *If* this force remained constant the electron would accelerate, gaining kinetic energy as a function of time. We could then determine the kinetic energy of the electron as a function of time using Newton's law, $F = m_e a$, with $F = \mathcal{E}q_e$, where q_e is the charge on the electron, m_e is its mass, and a is its acceleration. After time t the speed of the electron would be $v = at = Ft/m_e = \mathcal{E}q_e t/m_e$. The kinetic energy of the electron would then be $\mathrm{E}_k = 1/2 m_e v^2 = \mathcal{E}^2(q_e^2 t^2/2m_e)$. However, the force changes as the electric field changes, and since the electric field is the sum of the individual fields of identical charges on each electrode, the electric field will vary sinusoidally, in step with the charges on the electrodes. The electron will continue to be accelerated (by lesser amounts) until the electric field reaches zero, at points (B) in Fig. 3.3, at which time the velocity of the electron will be at its maximum value in the upward direction, $+v_{max}$. At these points the electron will see a reversal in the sign of the electric field, and will begin to decelerate until the electric field has reached its minimum at points (C) in the figure. The electron will now have a zero velocity and will change its direction. It will be accelerated in the downward direction until the electric field is again at zero. At these points (D) the electron will have its maximum downward velocity, $-v_{max}$. The electron will then be decelerated until the electric field reaches the next point, (A), where the velocity of the electron will again be zero and it will again change direction. Although the electron is constantly changing in speed, during each cycle of this oscillation the electron does have an *average* speed, and therefore has some average kinetic energy, which we can determine by calculating the average of the electric field over the time it takes for one oscillation, known as the period, T: $\langle \mathrm{E}_k \rangle = 1/2m_e\langle v^2 \rangle = \langle \mathcal{E} \rangle^2 (q_e^2 T^2/2m_e)$. To determine $\langle \mathrm{E}_k \rangle$, we assume that $\mathcal{E}$ can be represented by a sine or cosine wave. For our experiment the sine wave is easier to average and we will describe $\mathcal{E}(t)$ as $\mathcal{E}(t) = \mathcal{E}_o \sin 2\pi\nu t$, where ν is the frequency of the oscillation — the number of oscillations per second, and $\mathcal{E}_o$ is the maximum magnitude of the electric field. Thus the period T is the number of seconds per oscillation, $1/\nu$. Since the electric field changes sign, and the speed of the electron depends only on the magnitude of the field, we will average the magnitude of the electric field over half of an oscillation (since the average value of the electric field *vector* is zero over the complete oscillation). We do this by adding all of the values of $\mathcal{E}(t)$ over the time interval $T/2$, then dividing by the time interval. Since the values are continuous, the sum becomes an integral:

$$\begin{aligned} \langle \mathcal{E}(t) \rangle &= \frac{\int_0^{T/2} \mathcal{E}_o \sin(2\pi\nu t)\, dt}{T/2} = \frac{2}{T} \int_0^{T/2} \mathcal{E}_o \sin(2\pi\nu t)\, dt \\ &= 2\nu \int_0^{1/(2\nu)} \mathcal{E}_o \sin(2\pi\nu t)\, dt = \frac{2\mathcal{E}_o}{\pi}. \end{aligned} \tag{3.1}$$

The average magnitude of the electric field over a single oscillation is therefore $4\mathcal{E}_o/\pi$, and the average kinetic energy of the electron over one oscillation is

$$\langle \mathrm{E}_k \rangle = \mathcal{E}_o^2 (8e^2 T^2)/(\pi m_e). \tag{3.2}$$

Since the electron will continue to oscillate in the same manner so long as the generator remains on, we can assume that the average speed of the electron will remain constant, and that its average kinetic energy, initially delivered by the electric field, is proportional to the *square* of the magnitude of the electric field.

As noted above, the electron stops and reverses whenever the electric field reaches a maximum or minimum value. At each of these times, the electron's speed and kinetic energy are zero. It has somehow lost the energy that it gained in its interaction with the electric field, and the electric field must provide a new supply of energy to the electron to get it moving through another half-oscillation. Thus the electron is constantly absorbing energy from the generator and giving it up somewhere else, apparently *radiating* it away into space. Since our electron has its own electric field, which is moving up and down as the electron moves, we design a detector to see if we can determine the nature of this radiating energy. The detector is nothing more than a test charge — another electron — between two electrodes which are connected to one another through a wire, at a distance r from the source electron, depicted in Fig. 3.4. We also constrain this test electron to "up and down" motion.

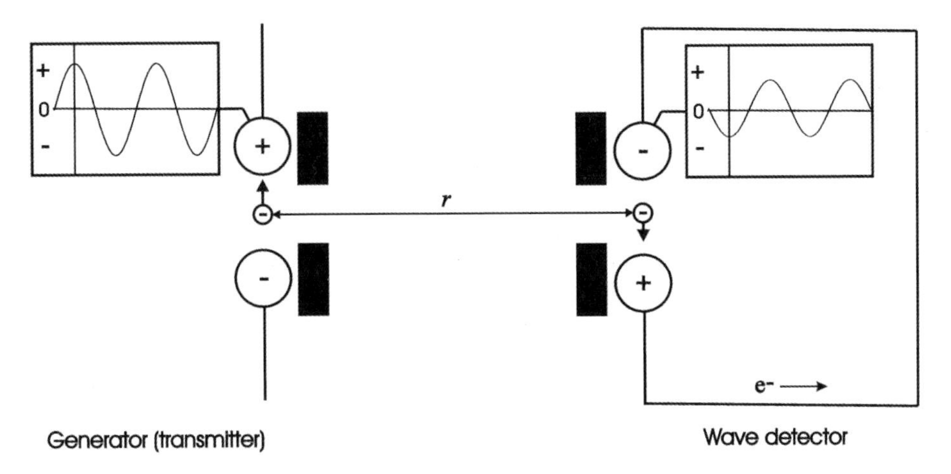

Figure 3.4 Electron oscillator and electron wave-detector.

Before the generator is switched on, both charges are at rest. When the switch is thrown the source electron begins to move, and after a delay of r/c seconds, the test charge also begins to move, but in the opposite direction to the source electron, since they repel one another. As the test electron approaches an electrode it repels the electrons on the electrode, which migrate through the wire to the other electrode. If we measure the charge on an electrode we find that it also varies sinusoidally, but in an opposite manner to the oscillating charge on the source electrode. We describe the two oscillations as being *180° out of phase* with respect to one another (we will soon discuss *phase* more rigorously).

The test electron has apparently been provided with energy emitted by the oscillating source electron, giving us an answer as to where the energy that is continuously radiating away from the source is going. We can therefore envision our oscillating electron as a *primary* source of radiating energy, creating an oscillating electric field — *a wave* — which moves through space at the speed of light, able to transfer energy to a secondary test electron. Furthermore, we can undertake the same exercise with respect to the second electron as we did with the source electron to determine its average kinetic energy, which *must also depend on square of the average magnitude of the electric field due to the source electron at a distance of r from it.* In addition, just as the primary electron does, the test electron continues to oscillate "in step" (actually 180° out of step) with the source electron after a single cycle, gaining no additional energy, so that it must also be reradiating any additional energy that it picks up over the course of an oscillation. In other words, the test electron becomes a secondary source of radiating energy.

3.1.3 Particles.

Continuing our gedanken experiment — we remove the test electron, our wave detector — and replace it (in the same location) with a device that can detect particles and count them. Photomultiplier tubes and Geiger counters are examples of practical particle detectors; our hypothetical device can detect and determine the energy of any and all particles. As in the previous experiment, we switch the generator on and watch our detector to see if there are any effects. After r/c

seconds the detector informs us that it has captured its first particle, and as long the generator remains on, these particles are detected in a continuous stream. Thus our oscillating electron appears to be emitting particles, just as it is creating waves, and we seek to discover if these waves and particles are related to one another.

The particles are moving at the speed of light (just like the waves!), and must therefore have kinetic energy. Indeed, we discover that for a given oscillating frequency (the frequency of the oscillating electric field), all of these particles have the *same* energy. Furthermore, the energy of these particles is found to be proportional to the frequency of the oscillating field, with the proportionality constant the same for all frequencies: $E_{particle} = h\nu$, where h is known as *Planck's constant*, after Max Planck, its discoverer.

Are these particles another way for the oscillating electron to emit energy? To test this we determine the average kinetic energy of the source electron over a complete cycle as we did previously, noting that it is the amount of energy released during one oscillation. If we create a sphere of radius r around the source electron, and measure the average kinetic energy of our "wave detecting" test electron at every point on the sphere, then sum all of the energies (since it is continuous this would actually be an integral), we find that we have accounted for *all* of the energy emitted over a complete oscillation. We then replace the wave detector with our particle counter and count N particles emitted after sampling every point on the sphere, each time over a single oscillation. Summing the individual particle energies gives $E = Nh\nu$, which, perhaps surprisingly, *is the same amount of energy that we determined to be carried by the wave – the energy emitted by the source electron!*

We have come face-to-face with one of the most profound dilemmas in modern physics. We have observed that energy can be transmitted through space in the form of a traveling wave, and that the same energy appears to be transmitted through space in a stream of particles, small packets of energy known as light quanta or *photons*. We cannot resolve this dilemma here, as a quote from Einstein underscores: "All these fifty years of constant brooding have brought me no nearer to the answer to the question 'What are light quanta?' ". On the other hand, it is possible to give the issue some context, by noting that wherever the square of the electric field is large, there are many photons, and where the field is weak there are few. Quantitatively, we find that the number of photons found in a given region of space is proportional to the square of the electric field at that point (see below), and we can envision the photons as transporters of energy, with the electric field dictating where these photons *are most likely* to appear. This way of thinking is in the spirit of that espoused by Eyring, *et.al.*[38]: "Since we always observe photons, and not light waves, we must logically conclude that light is 'really' a stream of photons. The waves are the mathematical expressions of the way in which the photons move. The photons of a beam of light do not obey Newton's laws of motion but the laws of wave motion." An alternative way of thinking is to visualize photons as entities with varying degrees of localization, delocalized as waves when left undisturbed (in which case we observe the effects of the wave) and localized as particles when we interact with them in order to detect them. The former description is the easiest to discuss, and we will incorporate it throughout the remainder of the text. In reality, we are unable to distinguish among these and other explanations, and we must be content with knowing that light has both wave and particle properties, and that we must use both to describe it appropriately.

We speak of the energy radiating from a source simply as *radiation*. If we replace the test electron with a small magnet we discover the existence of an oscillating magnetic field as well as an electric field, and therefore this radiating energy is referred to as *electromagnetic radiation*. Experimentally, we count the number of photons in various regions of space as a measure of the *intensity* of the electromagnetic radiation in those regions. Formally, the intensity at a given point in space is a measure of the number of photons that pass through a unit area at that point per unit time. Since each photon carries a specific amount of energy, the intensity is a measure of the energy that passes through the unit area per unit time. Recall that the test electron at some point in space is subject to an accelerating force due to the electric field of radiation emitted by the oscillator, thus acquiring kinetic energy that it subsequently releases in the form of radiation. This is the energy measured when we measure the intensity. The average kinetic energy of the oscillating electron is proportional to the square of the magnitude of the electric field of the radiation, $\langle \mathrm{E}_k \rangle \ \alpha \ \mathcal{E}^2$ (Eqn. 3.2). From this we conclude that *the intensity of the electromagnetic radiation at a point in space is proportional to the square of the magnitude of the electric field at that point:* $I \ \alpha \ \mathcal{E}^2$.

3.1.4 Interference.

We continue our experiment by placing a second "source" electron, oscillating at the same frequency, in front of the first oscillating electron — closer to the test electron. Since the second oscillator is closer, its electric field at the point of the test electron will be slightly greater, but we can assume that if the distance to the test electron is much greater than the distance between the source electrons the difference will be negligible. We might anticipate that the electric field at the test electron, measured by the amplitude of its oscillation, will be proportional to the sum of the amplitudes of the two oscillators — but we observe something quite different. As expected, the test electron continues to oscillate at the same frequency, but the magnitude of the oscillation, which measures the magnitude of the electric field at the point of the test electron, varies tremendously, depending of the placement of the second oscillator. At certain placements of the second oscillator the test electron stops oscillating completely, indicating that the electric field is zero at that point. This would occur if the electric fields of the oscillators were equal and opposite to one another, since the test electron senses the field of each source electron. If the field from one oscillator is going "up", while the other is going "down" the test electron will see both fields changing in opposite directions, and the effects of those fields on the test electron will tend to cancel one another. On the other hand, if the fields from both oscillators are changing in the same direction the effects of the two oscillating electric fields will be additive. It follows that we must determine exactly how the propagating electromagnetic waves combine in order to assess their net effect on the test electron.

In order to make the analysis general, we allow the waves emanating from the two oscillators to have different amplitudes, depicted in Fig. 3.5. The oscillators are separated by a distance L, and we assume that they are oscillating at the same frequency in step with one another, that is, they both go through their maxima and minima at the same time (we say that they are *in phase* with one another). At some time t, the test electron "sees" the electric field from the first oscillator, which varies as the cosine (or sine) as a function of time: $\mathcal{E}_1 = \mathcal{E}_{o,1} \cos(2\pi\nu t)$. At the same time,

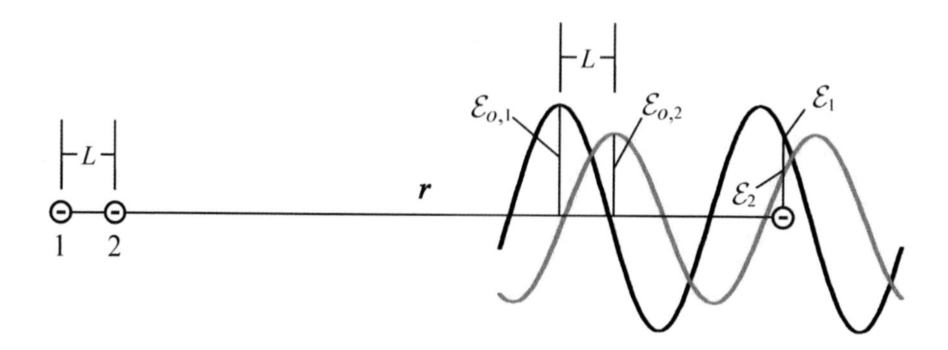

Figure 3.5 Propagating electric field waves from two oscillating electrons.

the test electron is subjected to the electric field from the second oscillator. Since this oscillator is closer, its wave arrives at the test electron L/c seconds earlier than the wave from the first oscillator. It is L/c seconds ahead of the first wave, and the field from the second oscillator at time t is therefore $\mathcal{E}_2 = \mathcal{E}_{o,2}\cos(2\pi\nu(t + (L/c))$. The resultant electric field at the test electron is the sum of the electric fields from both oscillators; $\mathcal{E}_r = \mathcal{E}_1 + \mathcal{E}_2 = \mathcal{E}_{o,1}\cos(2\pi\nu t) + \mathcal{E}_{o,2}\cos(2\pi\nu(t + (L/c))$. In order to determine how this resultant electric field varies in time we must develop a method to combine (*superposition*) the electric field waves. To do this, we represent each wave as a vector in the **ij** plane, with its magnitude equal to its maximum amplitude, and the angle with respect to the **i** axis equal to the argument in the cosine function, illustrated in Fig. 3.6.

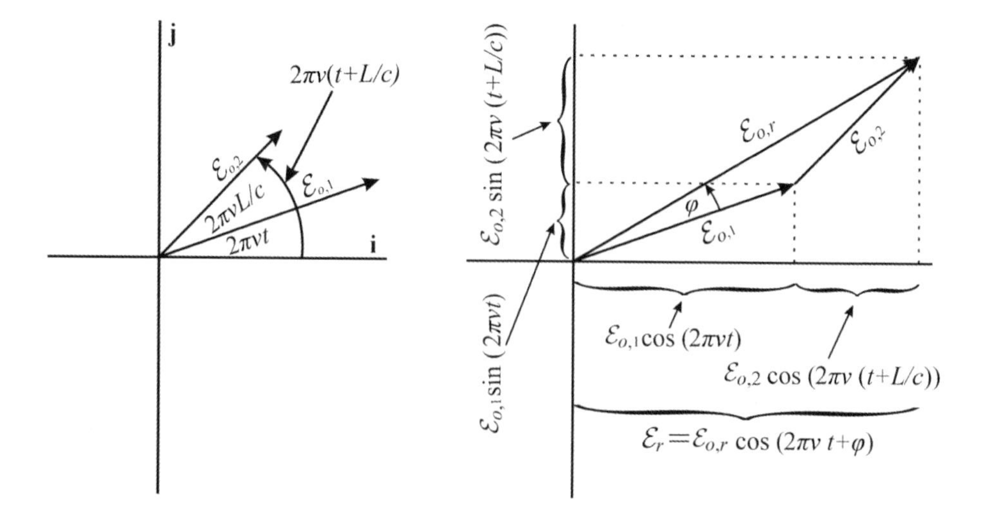

Figure 3.6 Representation of propagating waves as rotating vectors and superposition of the waves with a vector sum.

Referring to the figure, consider the vector representing the wave propagating from the first oscillator, $\overrightarrow{\mathcal{E}}_1$. The vector rotates with time, and makes a complete revolution when $t_0 = T = 1/\nu$ (in which case the cosine is equal to 1 and the vector is aligned along the **i** axis). The projection of the vector on the **i** axis at time t is $\mathcal{E}_1 = \mathcal{E}_{o,1}\cos(2\pi\nu t)$, the magnitude of the electric field due to the first oscillator at the test electron. When $t_0 = T$ the vector representing the wave propagating from the second oscillator, $\overrightarrow{\mathcal{E}}_2$, is at an angle of $2\pi L/c$ with respect to the **i** axis, and the projection of the vector at time t is $\mathcal{E}_2 = \mathcal{E}_{o,2}\cos(2\pi\nu(t+(L/c))$, the magnitude of the electric field at the test electron due to the second oscillator. Thus the net electric field at the test electron at time t is the sum of the projections of the two vectors along the **i** axis. These vectors are separated by an angle $2\pi L/c$, and will rotate in tandem with one another retaining that angle, which we call the *phase shift** between the two vectors (waves). Furthermore, the vector sum produces a resultant vector with a phase shift of φ with respect to the first vector, rotating in tandem with $\overrightarrow{\mathcal{E}}_1$ and $\overrightarrow{\mathcal{E}}_2$. Since the resultant vector makes a complete revolution over the period, T, the wave that it represents is oscillating at the same frequency as the constituent waves. The projection of the vector on the **i** axis is the sum of the projections of $\overrightarrow{\mathcal{E}}_1$ and $\overrightarrow{\mathcal{E}}_2$ — the net electric field at the test electron, which is another cosine function: $\mathcal{E}_r = \mathcal{E}_{o,r}\cos(2\pi\nu(t+\varphi))$. *The test electron does not "see" two oscillating electric fields, but only one, resulting from the superposition of the waves from the two oscillators and oscillating at the same frequency.* The placement of a third oscillator, etc. will produce the same results — the net electric field wave will always be sinusoidal. The amplitude and the phase angle of the resultant wave will depend on the relative amplitudes and phase angles of the constituent waves. The **i** component of the resultant vector is $\mathcal{E}_r = \mathcal{E}_i = \mathcal{E}_{o,1}\cos(2\pi\nu t) + \mathcal{E}_{o,2}\cos(2\pi\nu(t+(L/c))$, the **j** component of the vector is $\mathcal{E}_j = \mathcal{E}_{o,1}\sin(2\pi\nu t) + \mathcal{E}_{o,2}\sin(2\pi\nu(t+(L/c))$. The magnitude of the resultant, $\mathcal{E}_{o,r}$, is $\sqrt{\mathcal{E}_j^2 + \mathcal{E}_i^2}$ and the rotation angle with respect to the **i** axis, $2\pi\nu(t+\varphi)$, is the angle with tangent $\mathcal{E}_j/\mathcal{E}_i$. The phase shift with respect to the first wave, φ, is readily obtained by evaluating $\mathcal{E}_j/\mathcal{E}_i$ at $t_0 = 0$.

It is now possible to rationalize the variation in the electric field at the detector as the relative distance between the oscillators is varied. The phase shift between the two waves will change as L/c changes, and the resultant wave will vary in amplitude, depending on how the waves superposition. Fig. 3.7 illustrates four scenarios resulting from different placements of the second oscillator, which we now make identical to the first oscillator, with the same maximum amplitudes for the electric fields propagating from each oscillator. The length of a wave over one oscillation is known as its *wavelength*, and is ordinarily denoted by the symbol λ. To determine this length, consider an observer at the test electron. The observer notes that the electric field reaches a maximum every $T = 1/\nu$ seconds, where T is the period of the oscillation. Since the wave is moving past the test electron at c, the speed of light, it will travel a distance $cT = c/\nu$ during a single oscillation. This is the length of the wave between maxima — its wavelength: $\lambda = c/\nu$.

*The use of the term *phase* is often ambiguous, as it is used variably to indicate the difference in the rotation angles between two rotating vectors *or* as the angle of rotation with respect to a reference vector, in our case the **i** axis. The term *phase shift* will be used here to indicate the former — and *phase angle* to indicate the latter.

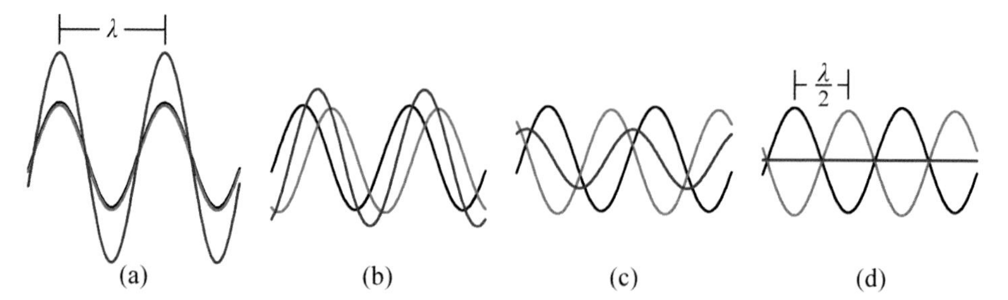

Figure 3.7 Interference of waves at the location of the test electron from two equal-amplitude oscillators separated by different distances (resultant wave shown in blue, $\mathcal{E}_{o,2} = \mathcal{E}_{o,1}$): (a) In-phase constructive interference. (b) Constructive interference. (c) Destructive interference. (d) 180° out-of-phase destructive interference.

In Fig. 3.7(a), $L = \lambda = c/\nu$ (or some integer multiple of the wavelength). The electric field of the second oscillator at the test electron is $\mathcal{E}_2 = \mathcal{E}_{o,2}\cos(2\pi\nu(t + ((c/\nu)/c)) = \mathcal{E}_{o,2}\cos(2\pi\nu t + 2\pi)$. The phase shift with respect to the first oscillator is $2\pi \equiv 0$; the waves are *in phase* with one another. Thus their resultant is the maximum value possible – the sum of the electric fields of each wave. The waves behave as if they have interfered with one another, and the combination of waves is known as *interference*. Note, however, that the electric fields of the waves are independent of one another (or we would not be able to simply add them to get a resultant wave), and the "interference" is really just an observation of the net effect of both electric fields acting on the test electron at the same place and time. When the interference results in a net electric field that is larger than the electric field of the two constituent waves, we call the interference *constructive interference*, illustrated in Fig. 3.7(b). When the interference results in an electric field that is smaller than the constituent fields as in Fig. 3.7(c) we call the interference *destructive interference*. If the second oscillator is placed so that $L = \lambda/2$ (or an odd integer multiple of half of the wavelength), $\mathcal{E}_2 = \mathcal{E}_{o,2}\cos(2\pi\nu(t + ((c/2\nu)/c)) = \mathcal{E}_{o,2}\cos(2\pi\nu t + \pi)$. The oscillators are now separated by a phase shift of π, and we say that they are *180° out of phase* with one another. In this case, as shown in Fig. 3.7(d), equal-amplitude waves cancel one another completely, and the net electric field at the test electron is zero.

We just noted that the two oscillators are constantly radiating energy, producing photons independently of one another, yet there appears to be no indication of this at the test electron when the waves are 180° out of phase. Replacing the wave detector (electron) with a particle counter verifies the observation – *there are no photons captured by the detector when the magnitude of the electric field is zero at the detector location.* Since the photons are still being produced in the same quantities, they must have gone elsewhere!

3.2 Diffraction

In the initial setup of our experiment we placed the oscillators and detector on the same line. However the oscillators are radiating in all directions. Since we have restricted their motions to "up and down", these directions are limited to a horizontal plane (perpendicular to the oscillation direction) in our gedanken experiment. We

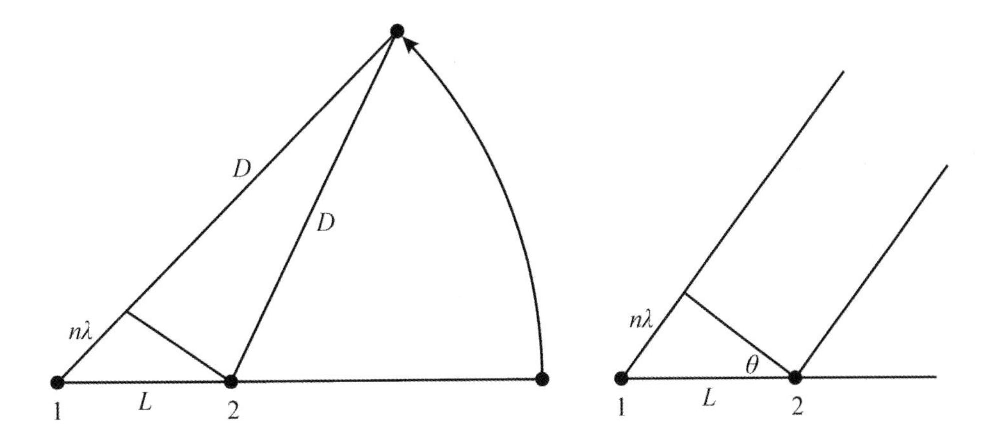

Figure 3.8 Test electron moved to a position where the difference in the distances each wave must travel is an integer multiple of the wavelength (left). The paths are essentially parallel in "real world" situations (right).

can therefore search for the missing photons by rotating a vector in the horizontal plane, attached to the midpoint of the oscillators on one end, and the wave detecting electron on the other. For a given displacement of the oscillators, L, the electric field will vary as we rotate the detector, and will reach a maximum each time that the difference between the distances that the two waves must travel to reach the detecting electron is an integer multiple of the wavelength, $n\lambda$, as shown in the drawing on the left in Fig. 3.8. In this case the waves will be in phase. If the rotation is continued the difference between the wave paths will at some point be half a wavelength less than at the maximum point, $(n - 1/2)\lambda$, and the electric field will be zero at that point. Continued rotation will find another maximum magnitude for the electric field at a path difference of $(n - 1)\lambda$, followed by another minimum at $(n - 3/2)\lambda$, and so on. As we expect, the placement of a particle detector at various points along the path registers photon counts that follow the square of the magnitude of the electric field exactly. This alternating constructive and destructive interference is known as *diffraction*, and the repeated pattern of maxima and minima is called a *diffraction pattern.* In practical situations, the distance between the detector and the oscillators is orders of magnitude greater than indicated in the figure on the left. In addition, the width of the detector will actually be orders of magnitude greater than the distance between the oscillators. Because of this, the waves can be considered to be radiating from the oscillators along parallel paths toward the detector, as depicted in the drawing on the right in Fig. 3.8.

When the waves are in phase the angle between the wave paths and the line connecting the oscillators is $\pi/2 - \theta$, where $\sin\theta = n\lambda/L$; L is the distance between the oscillators. A maximum in the diffraction pattern will occur for every $n = L\sin\theta/\lambda$. Since $-1 \leq \sin\theta \leq 1$, n can take on integer values of $-L/\lambda \leq n \leq L/\lambda$. The number of diffraction maxima is therefore proportional to L; as the distance between the oscillators increases, the number of diffraction maxima increases and the spacing between the diffraction maxima decreases — *there is an inverse relationship between the separation of the oscillators and the distance between the diffraction maxima.*

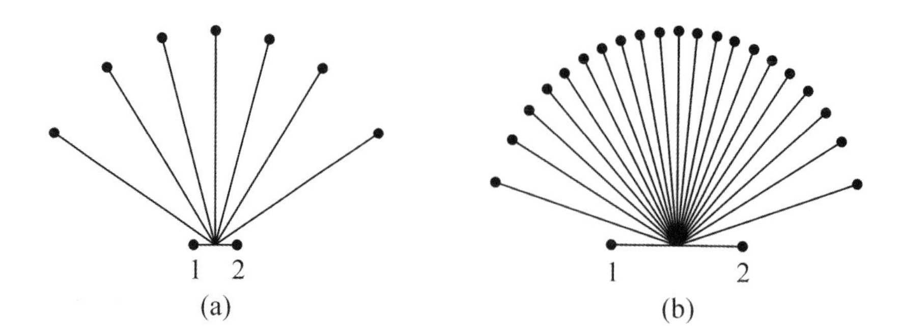

Figure 3.9 (a) Location of diffraction maxima for oscillators 1 and 2 separated by 3.5 times the wavelength. (b) Location of diffraction maxima for oscillators 1 and 2 separated by 10.5 times the wavelength.

This reciprocal relationship is illustrated in Fig. 3.9. Fig. 3.9(a) shows the location of the detector electron at positions where electric field maxima are observed when the distance between the oscillators is 3.5 times the wavelength of the oscillations. Fig. 3.9b represents the case in which the wavelength is unchanged, but the distance between the oscillators has been increased to 10.5 times the wavelength.

While our thought experiment has provided the physical basis for observing a diffraction pattern, the number of photons emitted from two oscillating electrons would be impossible to detect. However, if there are a large number of equally spaced oscillators, as illustrated in Fig. 3.10, at the angle where constructive interference occurs, the path to the detector of the wave from each oscillator differs from the paths of all the other oscillators by an integer number of wavelengths — *they are all in phase with one another.* The superposition of the waves from all of the oscillators, and the resulting square of the electric field and number of photons — *the intensity* — becomes large enough to be detected. *This is the essence of the X-ray diffraction experiment — the regular arrangement of the atoms in the crystal, with the nucleus of each atom surrounded by electrons, provides a very large number of symmetrically arranged oscillators, and an observable number of photons in the direction of the diffraction maxima. Furthermore, since $L > n\lambda$, diffraction maxima will occur only if the oscillators are separated by at least one wavelength.* The last statement is of extreme importance in crystallography. The wavelength of the radiation used to "observe" the crystal structure must be shorter than the distance between repeating points in the crystal lattice. It follows that the wavelengths of visible light are much too long for diffraction from arrays of atoms or molecules and we must employ shorter wavelength *X-radiation.*

The type of diffraction that we have been discussing is called *Fraunhofer diffraction*, after its discoverer, Joseph von Fraunhofer. The oscillators in this type of diffraction are in phase with one another. One way to induce a row of equally spaced electrons to oscillate in phase is to shine a laser light source on them in a perpendicular direction to the row. The electromagnetic waves emanating from the laser all have the same wavelengths and are in phase with one another, thus setting the electron oscillators in motion in phase with one another.

In X-ray crystallography we will find it necessary to expose crystals to radiation at angles which are *not* perpendicular to the oscillators in the crystal. The waves

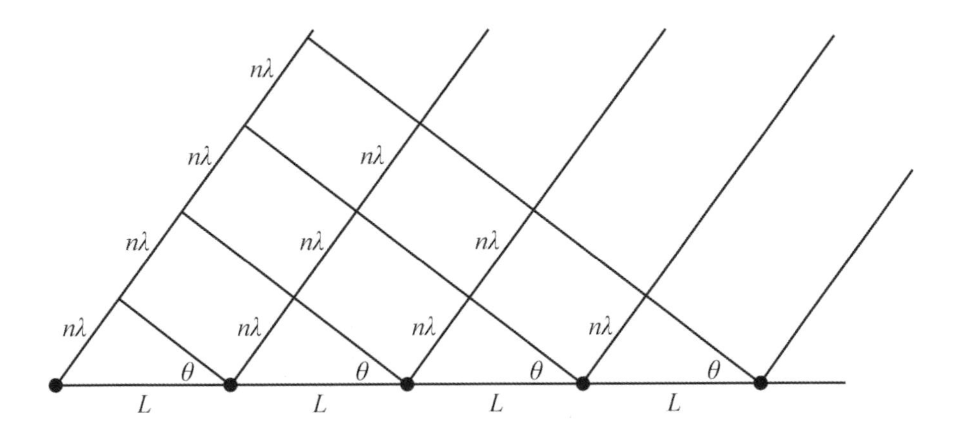

Figure 3.10 Fraunhofer diffraction from a row of equally spaced oscillators.

from the radiation source will arrive at the secondary oscillators with each oscillator experiencing a phase shift which depends on the distance the incident wave must travel to arrive at a specific oscillator. We must now establish the conditions for diffraction when the incident radiation which drives a row of equally spaced secondary oscillators is no longer perpendicular to the row. This type of diffraction is known as *Laue diffraction*, named for its discoverer, the German physicist, Max von Laue.[39] Laue first predicted that crystals could serve as diffraction gratings for X-rays, despite the objections of eminent physicists such as Arnold Sommerfeld. Fortunately Laue persevered, and was awarded the Nobel Prize in Physics in 1914 as *the discoverer of X-ray diffraction from crystals.* Laue formulated the conditions for diffraction based on *the reciprocal lattice*, which we will encounter later in the chapter. This type of diffraction is also known as *Bragg diffraction*, named after the father and son team, William Henry Bragg and William Lawrence Bragg, who developed the conditions for diffraction based on the planes in the *direct lattice* rather then the reciprocal lattice.[40] In 1915 they were jointly awarded the Nobel Prize for their contributions to crystallography.

In order to determine the conditions necessary for Laue/Bragg diffraction we modify our hypothetical device by replacing the row of primary oscillators with electrons (without the generators) to serve as secondary oscillators, illustrated in Fig. 3.11. The electrons will be set in motion by a beam of incoming source radiation that has a single frequency (*monochromatic*) and is in phase (*coherent*). For now the oscillators will be constrained to "up and down" motion so that the analysis can be restricted to the plane. The new gedanken experiment involves changing the direction of the source radiation and rotating a detector to locate diffraction maxima. The incident waves are 180° out of phase with respect to the waves emitted from the secondary oscillators, but this is universal, and will not affect the arguments that follow, so we will ignore this phase shift for simplicity. The radiation emitted from the secondary oscillators is sent out in all directions (in our special case all directions in the plane), and is referred to as *scattered radiation.* The secondary oscillators are often referred to as *scatterers.* Since the waves must travel different distances to get to each scatterer, the scatterers are no longer oscillating

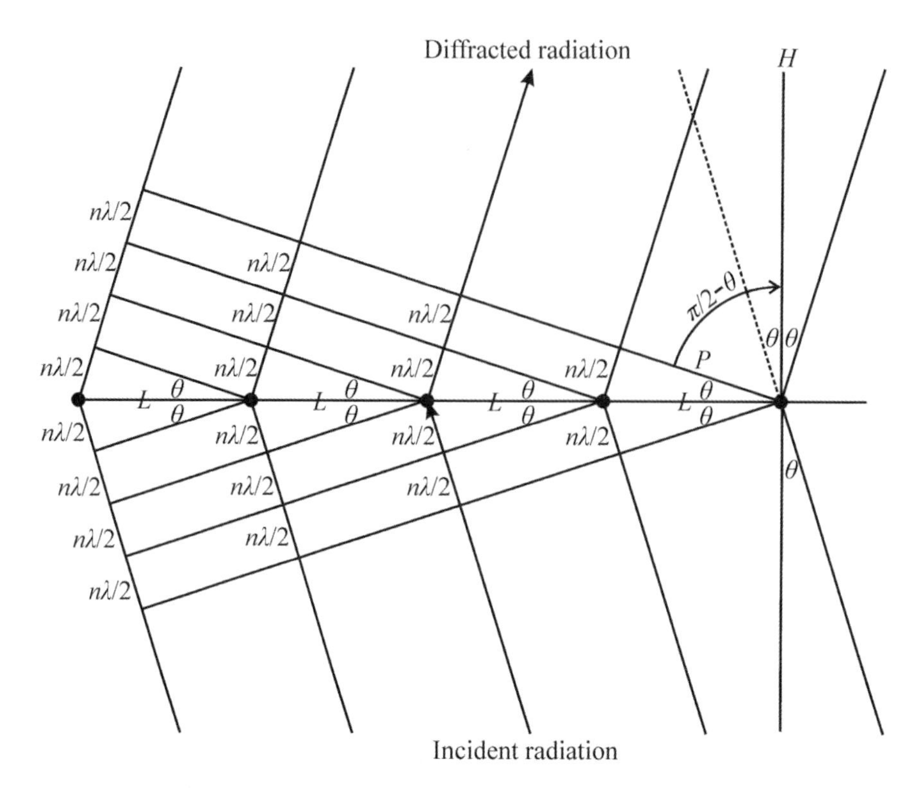

Figure 3.11 Laue/Bragg diffraction from a row of equally spaced secondary oscillators.

in phase with respect to one another. At an arbitrary angle this will result in destructive interference, and diffraction maxima will *not* be observed at *any* angle of the detector. However, if the angle of the incident radiation is adjusted so that the waves arrive at each oscillator along paths that differ by an integral multiple of *half* of a wavelength, then moving the detector to a position where the paths to the detector differ by the same distance will result in a net path difference that is an integral multiple of a *whole* wavelength; the diffracted waves will be in phase and a diffraction maximum will be observed. As shown in Fig. 3.11, this occurs when the path to the detector is at the same angle with respect to the scatterers as the path from the radiation source.The conditions for diffraction can now be readily deduced, giving us the *Bragg Law for diffraction*:

$$\sin\theta = \frac{n\lambda/2}{L} = \frac{n\lambda}{2L} \tag{3.3}$$

The figure also illustrates that the angle that the incident and diffracted beams of radiation make with the *perpendicular* to the row of scatterers is the same angle, θ. To prove this we consider the line segment P, connecting the diffraction vectors from two adjacent scatterers and perpendicular to the vectors. The angle between P and the line perpendicular to the row of scatterers, H, is $\pi/2 - \theta$. Since P and the diffracted bean are perpendicular, the angle between the diffracted beam and H is $\pi/2 - (\pi/2 - \theta) = \theta$. The same argument can be made for the angle between

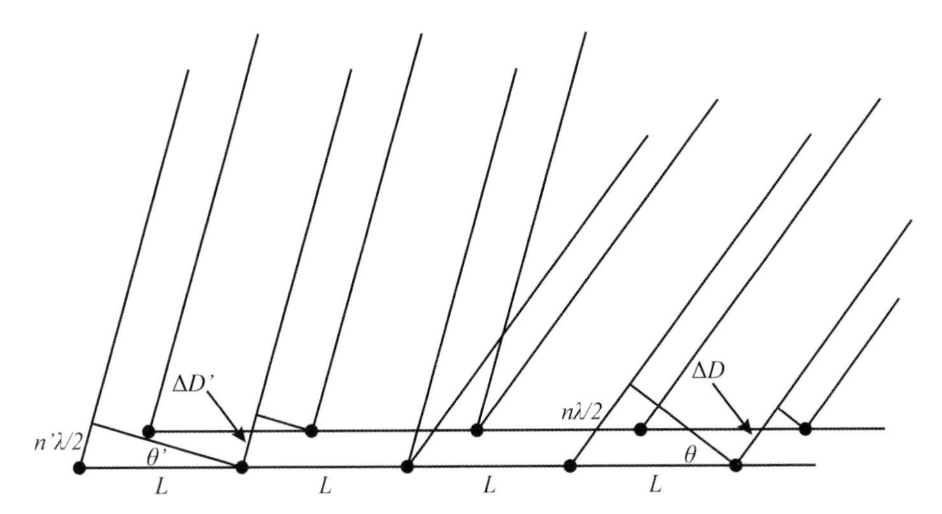

Figure 3.12 Diffraction from two rows of equally spaced oscillators for two different diffraction angles.

the incident beam direction and H, which is also θ, and it follows *that the angle between the incident and diffracted beams is the angle* 2θ.

The row of scatterers that we have been discussing is a simple analog of equally spaced entities in the "unit cells" of a one-dimensional lattice. If the lattice is two- or three-dimensional, there will be other sets of scatterers (i.e., other components of an asymmetric unit) which are separated by the same spacing (in adjacent unit cells), but offset from the first set of scatterers. In order to attain a sense of the effects of the other oscillators on the diffraction pattern, we add a second row of scatterers with the same spacing, offset from the first and closer to the detector as illustrated in Fig. 3.12. This row of scatterers will also produce diffraction maxima at the same angle as the first. However the waves from each oscillator and its new neighbor will now undergo interference, and the net intensity at the detector at the position of each diffraction maximum will be altered. Furthermore, the path difference, ΔD, and consequently the phase shift between the two sets of waves will differ for each diffraction angle, altering the intensity for each diffraction maximum differently. The relative locations of the two sets of scatterers will determine the phase differences and resultant intensities of the waves at each location in the diffraction pattern. *If we are able to determine these phase differences and intensities then we should be able to determine the relative locations of different sets of scatterers. This is what X-ray crystal structure determination is all about — determining the intensities in the diffraction pattern and phase shifts of the waves from scatterers in the crystal lattice in order to determine the relative locations of those scatterers in the lattice.*

3.2.1 The Reciprocal Lattice.

The relative location of scatterers, defined by the lattice coordinate system, determines the relative location of diffraction maxima and their intensities. Conversely, the location and intensities of diffraction maxima in the diffraction pattern determine the location of scatterers in the lattice. Chapters 1 and 2 described the representation of the locations of the scatters in terms of a coordinate system that reflected the periodicity of their locations — the crystal lattice. For reasons that will soon become apparent, we refer to this lattice as the *real lattice* or *the direct lattice*. In the simple one-dimensional case the direct lattice defines the location of the scatterers, as shown in the upper half of Fig. 3.13. The picture has been made a little more realistic by treating the scatterers at each point as "one dimensional hydrogen atoms", allowing the electron on each "atom" to move about its nucleus in the $\pm d$ directions. The bound electrons spend most of their time at or near the nucleus, with the chance of finding an electron away from the nucleus diminishing exponentially. Intuitively, we might expect each atom to behave, on the average, as if its scattering was emanating from the center of the atom. We will come to identify (in three dimensions) the concentration of the electrons at a location as the *electron density* at that location. Thus a plot of scatterer concentration (electron density) versus distance reveals maxima that occur in equal increments of the distance between scatterers, L, along the d axis of the plot.

The variable that determines the occurrence of intensity maxima in the diffraction pattern is the diffraction angle. The maxima occur every time that $2\sin\theta/\lambda$ is an integral multiple of the *reciprocal* of the spacing between scatterers, $1/L$. If the

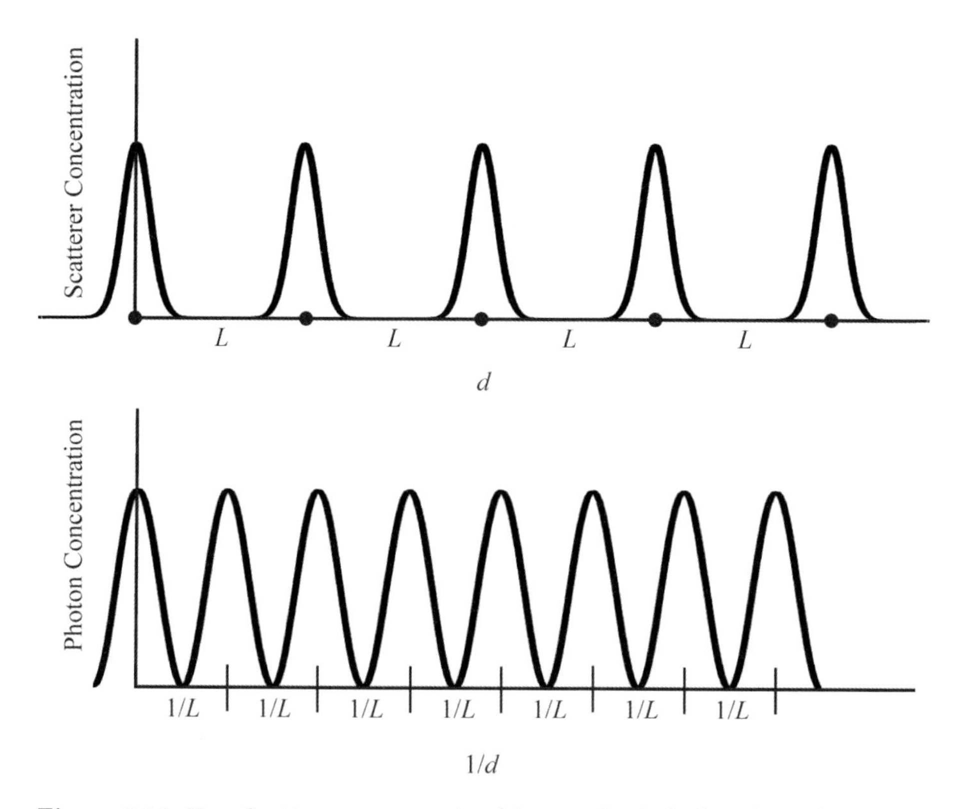

Figure 3.13 Top: Scatterer concentration (electron density) plotted as a function of distance in direct space. Bottom: Photon concentration (intensity) plotted as a function of the reciprocal of distance in reciprocal space.

intensity (number of photons) at each angle for our source and detector is plotted versus $2\sin\theta/\lambda = 1/d$ as illustrated in the bottom half of Fig. 3.13, we will observe a maximum on the plot every time $1/d = n/L$. Thus a natural coordinate system for observing the intensity of photons from our scatterers has spacings of $1/L$ in $1/d$ space — which we call *reciprocal space.* The lattice (in this case one-dimensional) with $1/L$ spacings is known as *the reciprocal lattice — it is the natural lattice of the diffraction pattern in reciprocal space.*

Functions in direct and reciprocal space are linked by related independent variables (each variable is independent in its own space) and the functions constitute what is known as a *Fourier pair.* They can be converted into one another by a powerful mathematical tool known as a *Fourier transform.* We will take up this important topic later in the chapter; for now we note that the location of the diffraction maxima in reciprocal space can be determined from the location of the scatterers in direct space, and the locations of the scatterers in direct space can be determined from the diffraction maxima in reciprocal space, both by way of the Fourier transform.

While we are unable to "see" in reciprocal space, the mathematics of crystallography is simplified tremendously if we treat interference phenomena in reciprocal space and scatterer location in real space *simultaneously.* Because of this we must discover how the reciprocal lattice and the direct lattice are related to one another. In order to accomplish this task we introduce a construct initially developed by the physicist Paul Peter Ewald.[41] In three dimensions this construct is known as the *Ewald sphere*; we will begin by limiting the Ewald sphere to the plane — creating an "Ewald circle."

Before constructing the circle, we select a scatterer in the real lattice as an origin for both the real lattice and the reciprocal lattice. Since the lattice is one-dimensional the real and reciprocal axes will be collinear. This is illustrated in the upper portion of Fig. 3.14. With the incident beam of radiation perpendicular to the lattices, a circle is constructed *in reciprocal space* with radius $1/\lambda$ (the reciprocal of the wavelength of the incident radiation) that is tangent to the lattice lines at the origin, bisected by the incident beam. Note that the relative scaling of the real and reciprocal lattices in the drawing is arbitrary. The only constraint is that the circle in reciprocal space has the same scaling as the reciprocal lattice.

The incident beam is now pivoted at the origin point with the Ewald circle remaining "attached" to the beam as depicted in the bottom portion of Fig. 3.14. The circle intersects the reciprocal axis, creating a triangle inscribed in a circle, which must therefore be a right triangle. The beam and circle are now pivoted until the Laue/Bragg diffraction conditions are met and $\sin\theta = n\lambda/2L$. The inscribed right triangle now has a segment of the reciprocal axis as one of its edges, another edge of length $2/\lambda$, and the third edge perpendicular to the direct and reciprocal axes. Thus the angle subtended by the diameter of the Ewald circle and the perpendicular edge is the diffraction angle, θ. Representing the length of the opposite edge as u, $\sin\theta = u/(2/\lambda) = u\lambda/2$. Since we are at a diffraction angle, $u\lambda/2 = n\lambda/2L$ and $u = n/L = n(1/L)$. *When the conditions for diffraction are met the Ewald circle intersects the reciprocal lattice at one of its lattice points.* The converse is equally true. The condition for Laue/Bragg diffraction can be met by rotating the beam until the Ewald circle intersects a reciprocal lattice point.

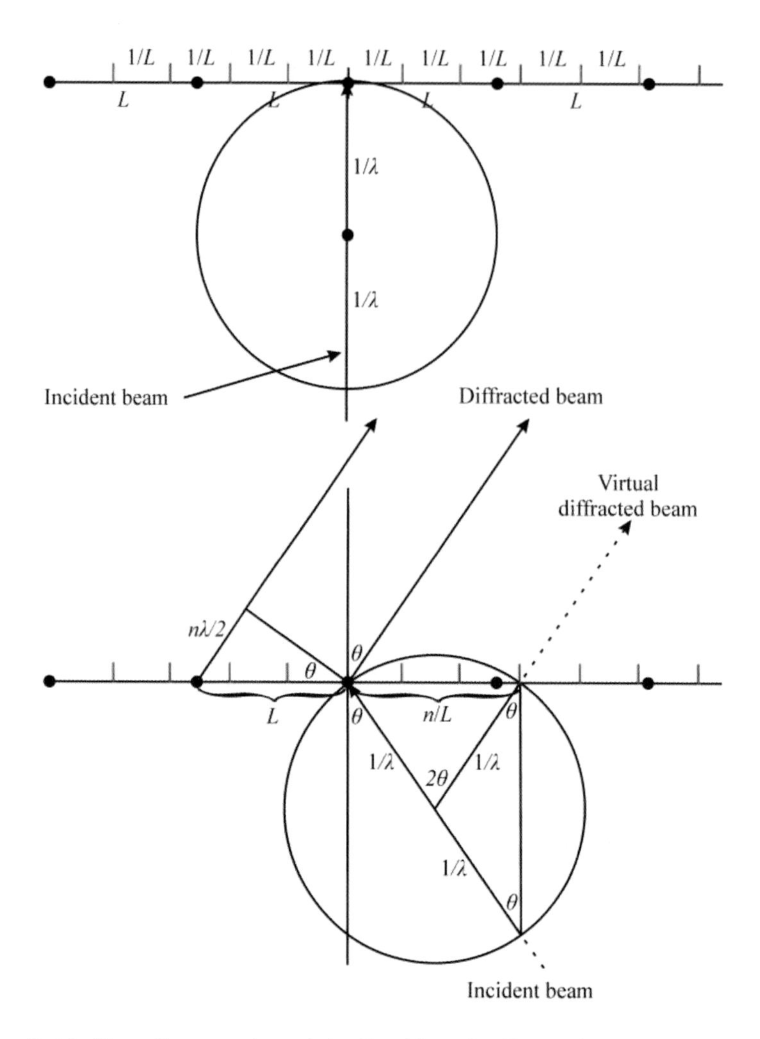

Figure 3.14 Top: Construction of the Ewald circle. Direct lattice points are shown as dark circles, reciprocal lattice points are indicated by red lines. Bottom: Rotation of the Ewald circle to an angle where Laue diffraction conditions are met.

An equivalent way to create the conditions for diffraction is to rotate the lattice about the origin, leaving the beam (and Ewald circle) in a fixed position. Referring to Fig. 3.14, note that when a reciprocal lattice point intersects the circle, the radius vector through the reciprocal lattice point creates an isosceles triangle such that the third angle in the isosceles triangle must be $\pi - 2\theta$ and the radius vector makes an angle of 2θ with the incident beam. *Thus when a reciprocal lattice point crosses the circle, diffraction conditions are met, and a vector from the center of the circle through the point of contact will parallel the direction of the diffracted radiation.* This allows for a convenient way to create an image of the diffraction pattern from the reciprocal lattice. If we replace our "point" detector with an "area detector" such as a fluorescent screen, photographic plate, or charge-couple device (CCD) the photons that arrive will create an image of the diffraction pattern. The experiment is illustrated in Fig. 3.15. The actual diffracted beams are emanating

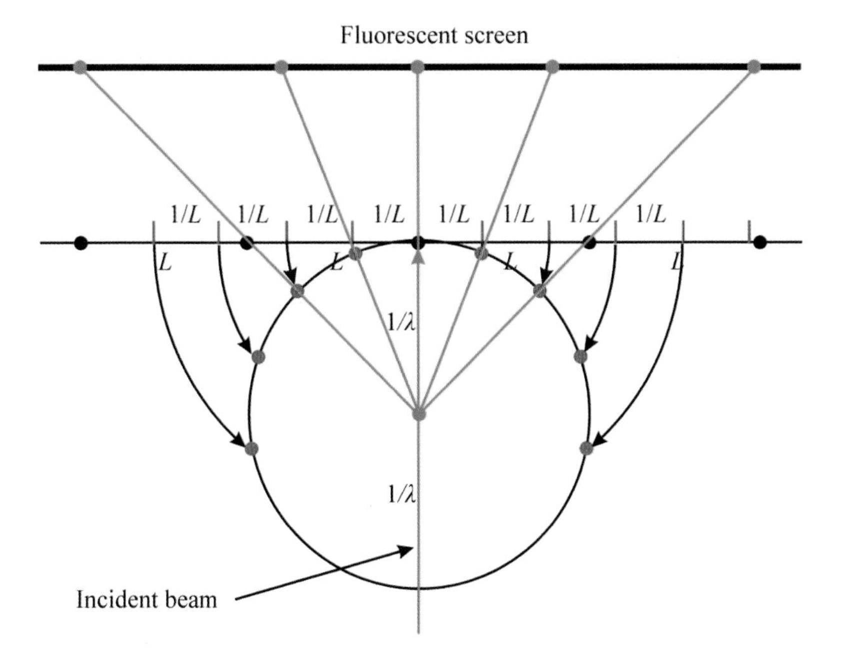

Figure 3.15 Creation of a distorted image of the reciprocal lattice by rotating the reciprocal lattice through the Ewald circle.

from the scatterers in the direct lattice, but the scale of a real experiment is such that we can consider them to be emanating from the center of the Ewald circle along radii that intersect the reciprocal lattice points whenever they intersect the circle — represented by the "virtual" diffracted beam in Fig. 3.14. The image created on the screen is seen to be a magnified, but distorted representation of the reciprocal lattice, with the diffraction maxima spreading out from the center, rather than remaining equally spaced. However, the figure has been grossly exaggerated for illustrative purposes. Note that if there were several reciprocal lattice points near the center of the figure they would intersect the circle with only a small amount of rotation, and would appear effectively evenly spaced on the screen — creating a scaled image of the reciprocal lattice near the center. A more realistic situation in which the scatterers are separated by 10.5 times the wavelength is illustrated in Fig. 3.16. The spacings on the screen near its center are virtually equal, separated by a constant, K, times the spacing in the reciprocal lattice, $1/L$; K depends on the incident radiation wavelength and the distance between the scatterers and the screen. It can be chosen arbitrarily for graphical purposes. Actual spacings on the screen are generally orders of magnitude greater than the distances between scatterers in the lattice. The reciprocal relationship linking the distance between scatterers and the distance between diffraction maxima is clearly seen here, since the distance between diffraction maxima, K/L, varies inversely with L.

Consider placing rows of scatters equally spaced above and below the initial row, and offset from one another as illustrated on the left in Fig. 3.17. The scatterers are again restricted to "up and down" motion. Unlike the situation depicted in

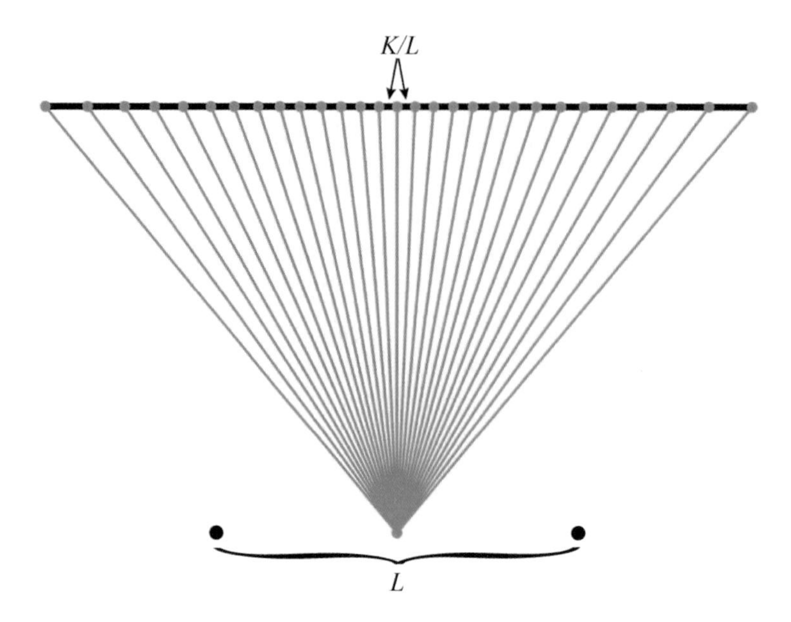

Figure 3.16 Diffraction pattern for Laue/Bragg diffraction from scatterers separated by 10.5 times the wavelength.

Fig. 3.12, the scatterers are all at the same distance from the detector/screen. The resulting diffraction pattern is illustrated on the right in Fig. 3.17. The diffraction "spots" from each row are aligned in columns on the screen that are perpendicular to the rows of scatterers. *The spots are not offset from one another since their location on the screen depends only on the diffraction angle, which in turn depends only on the spacing between the scatterers.*

Now that we have determined the general conditions for diffraction from single electrons constrained to "up and down" oscillations we are in a position to consider more realistic scatterers – the electrons in atoms. We replace the electron scatterers with hydrogen atoms, consisting of a single proton in the nucleus and one electron, which moves at about one-tenth of the speed of light about the nucleus. While its actual trajectory cannot be determined, on the average, it can be found at some location within a sphere with the nucleus at the center. The electron has an equal probability of being at any direction from the center of the sphere, and at the same time that it is set into oscillation by the incident electric field, it is also moving to a new location. This combination of motions causes the electron to radiate in a direction other than the plane — indeed, *its motion is so rapid that on the average an electron "bound" to a nucleus radiates in all directions.*

The hydrogen atom scatterers radiate in all directions, but diffraction conditions still must be met in order to observe intensity maxima. Thus, the waves scattered from each atom must still constructively interfere, but these waves are no longer constrained to the plane. If the circular path that the wave detector originally took (Fig. 3.8) can now be tilted to sample regions above and below the plane, diffraction maxima will be observed at the same locations around the circle as it is tilted. If the detector is replace with a fluorescent screen, the result will be a series of parallel lines, perpendicular to the row of scatterers, depicted in Fig. 3.18(a). If a is the

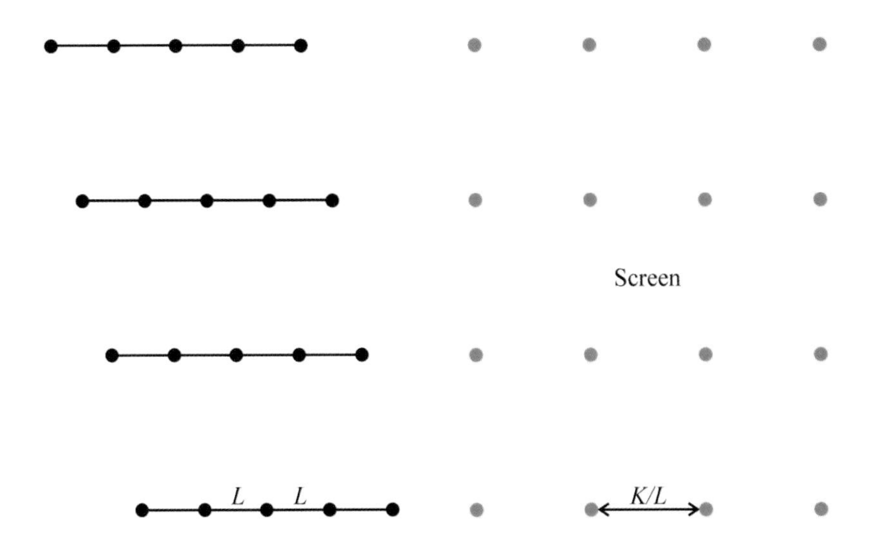

Figure 3.17 Diffraction pattern (right) from rows of equally spaced scatterers (left). The incident beam is normal to the page.

distance between the lines, they will be spaced by a distance of K/a in reciprocal space. Furthermore, referring to Fig. 3.17, addition of rows above and below the initial row will create no new lines, since the diffraction angles will be the same.

Now, suppose a second set of the oscillators with a different spacing, b are placed along a line perpendicular to the incident beam, but at an angle γ with respect to the original row. An Ewald circle is constructed so that it is parallel to the direction of the new row of oscillators. As the row of scatterers is rotated in the plane of the Ewald circle, lines of diffraction maxima will occur which are perpendicular to the row, spaced by a distance of K/b in reciprocal space (Fig. 3.18(b)). Again, addition of parallel rows of offset scatterers separated by the same distance will result in no new lines.

If the two rows are combined into a single array, we have a two-dimensional lattice with scatterers at each origin, depicted in Fig. 3.19(a). The diffraction pattern will now exhibit maxima when both rows of oscillators produce waves that constructively interfere simultaneously — at the points of intersection of the diffraction "lines" produced by each row of oscillators. These points of mutual constructive interference are often called diffraction *nodes*. The result observed on the screen is a scaled version of the two dimensional lattice of diffraction maxima — the reciprocal lattice in two dimensions, illustrated in Fig. 3.19(b). We can now identify a *reciprocal unit cell* in the diffraction pattern with reciprocal axial lengths denoted by a^* and b^*, scaled with the constant K, at an angle γ^* with respect to one another. The labels are chosen so that each reciprocal axis closely parallels the corresponding direct axis (i.e., if γ^* is 90° then $\mathbf{a}$ and $\mathbf{a}^*$ are parallel).

There are infinitely many sets of rows of scatterers in the two dimensional lattice, each one independently capable of producing a set of perpendicular diffraction lines. The diffraction from the direct lattice lines with indices (1 1) are shown in Fig. 3.19(c). Since diffraction from each oscillator has already been accounted

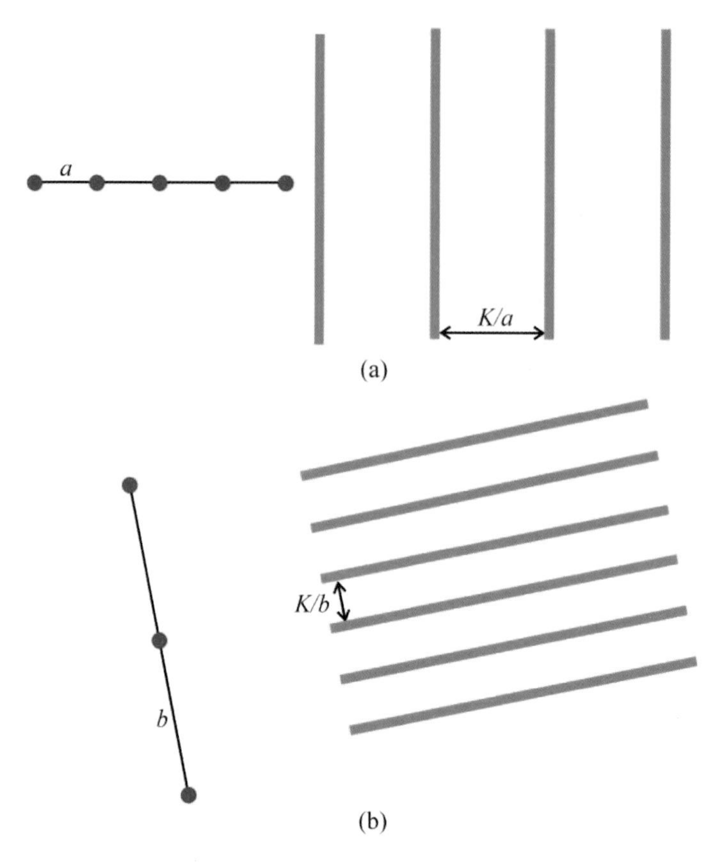

Figure 3.18 Diffraction pattern from rows of scatterers emitting radiation in all directions.

for with the a and b rows of scatterers, the other rows of scatterers will produce parallel "diffraction" lines that will constructively interfere simultaneously at the same points as the a and b rows of scatterers. In other words, *selection of a different direct lattice unit cell will result in the same diffraction pattern.* It follows that the diffraction lines from each row must intersect the reciprocal lattice points, just as all of the lines in the direct lattice intersect the points in the direct lattice. The parallel green lines in Fig. 3.19(b) illustrate this for diffraction due to the oscillators on the (1 1) lines of the direct lattice. The (1 1) lines in the direct lattice have generated a perpendicular set of diffraction lines, which have become the $(\bar{1}^*\ 1^*)^*$ lines in the reciprocal lattice. Note that the scatterers along the (0 1) lines in the direct lattice containing the a axis have produced the perpendicular set of diffraction lines, $(\bar{1}^*\ 0^*)$, containing the b^* axis. Similarly, the (1 0) lines in the direct lattice containing the b axis have produced the perpendicular set of diffraction lines, $(0^*\ 1^*)$, containing the a^* axis. In the two-dimensional reciprocal lattice the $a*$ axis is perpendicular to the b axis in the direct lattice, and the b^* axis is perpendicular to the a axis in the direct lattice. When we extend this concept

*The asterisk on the integers indicates that they are the integer indices for a set of lines in the reciprocal lattice.

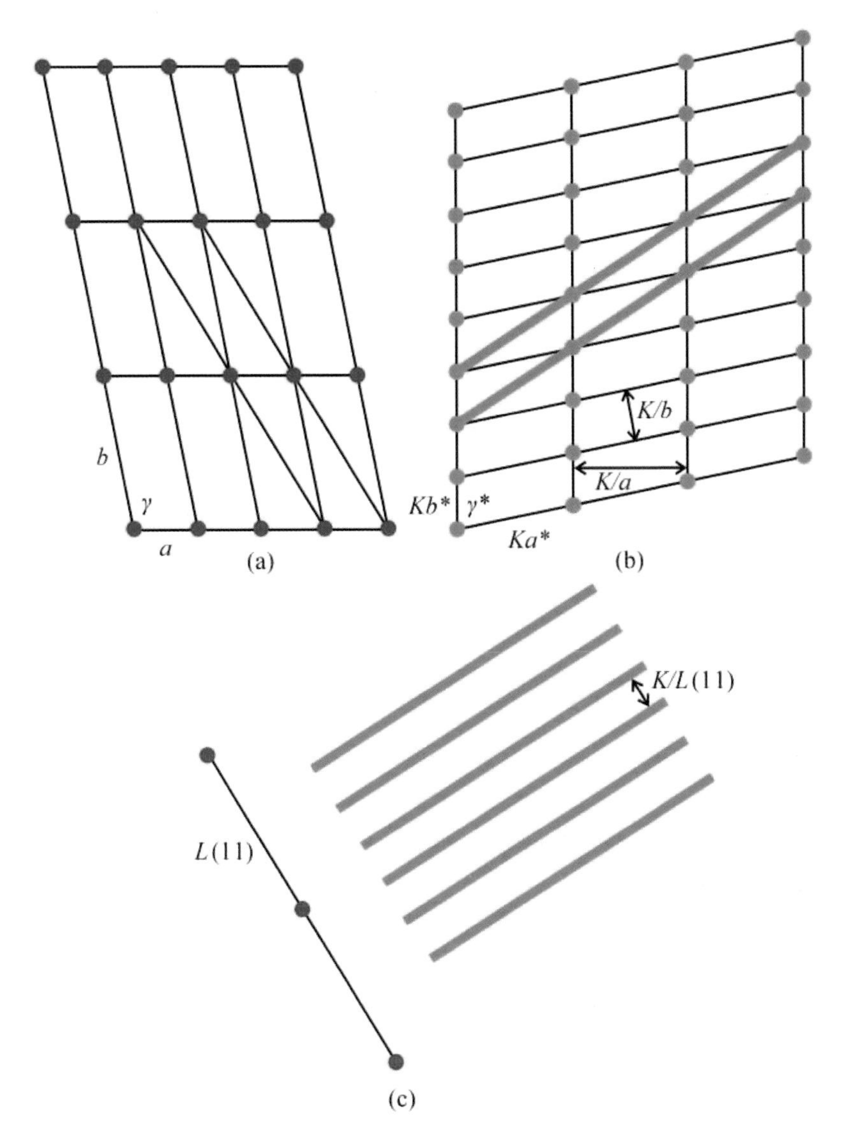

Figure 3.19 (a) Two-dimensional direct lattice containing atomic scatterers at each origin. (b) Diffraction pattern of the two-dimensional lattice showing the scaled reciprocal unit cell. (c) Contribution to the diffraction pattern of the (1 1) lines in the direct lattice.

two three dimensions we will discover that the a^* axis is perpendicular to the b axis *and* the c axis, and therefore is perpendicular to the bc plane. In general, a set of lattice lines in the two-dimensional direct lattice with indices $(h\ k)$ will produce a set of perpendicular lattice lines in the reciprocal lattice with indices $(\bar{k}^*\ h^*)$. The converse of this is also true — a set of lattice lines in the reciprocal lattice with indices $(h^*\ k^*)$ will produce a set of perpendicular lattice lines in the direct lattice with indices $(\bar{k}\ h)$.

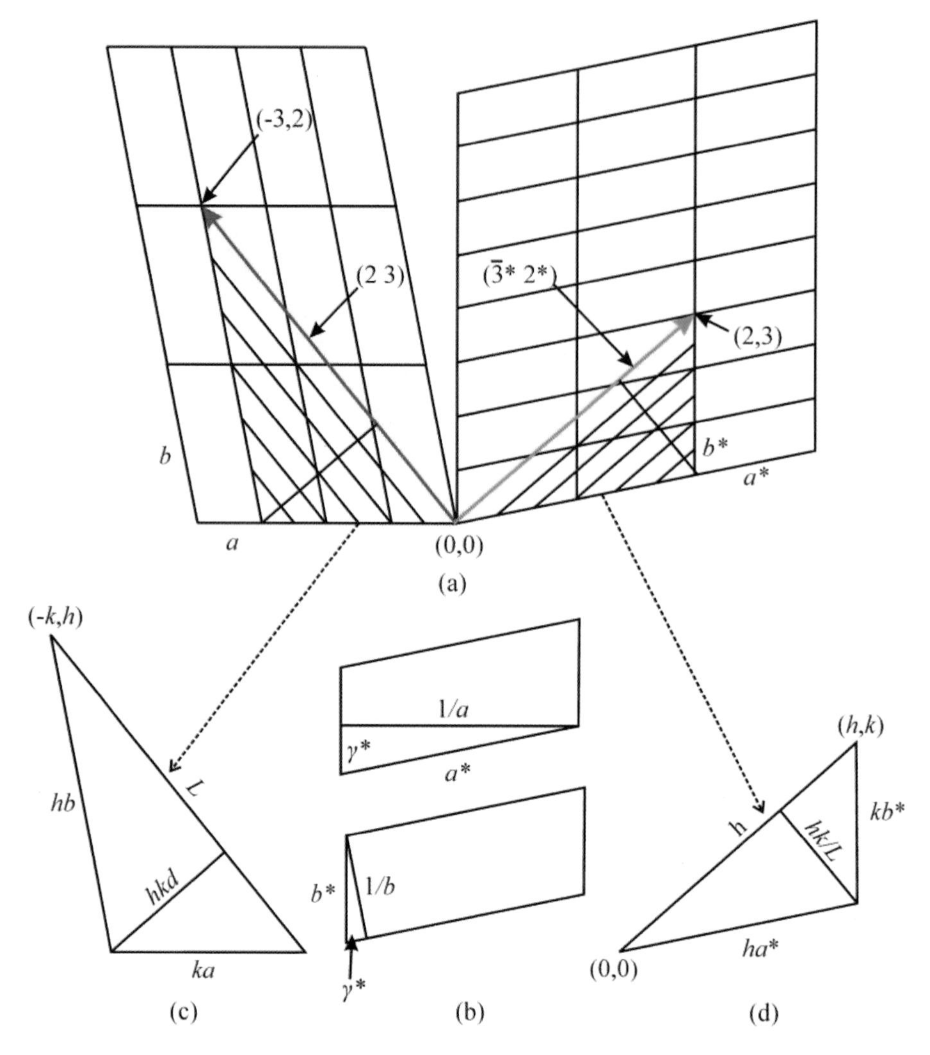

Figure 3.20 (a) Direct (left) and reciprocal (right) lattices sharing a common origin. The (2 3) lines and their resulting perpendicular ($\bar{3}^*$ 2^*) diffraction lines are also shown. (b) Reciprocal unit cell showing the relationship between the spacing of diffraction lines and reciprocal cell axes. (c) Triangle created from the unit cell axes and the vector from the direct lattice origin to the $(-k, h)$ lattice point in the direct lattice. (d) Triangle created from the reciprocal cell axes and the vector from the reciprocal cell origin to the (h, k) lattice point in the reciprocal lattice.

To determine the relationship between the direct and reciprocal lattices we "shrink" the reciprocal lattice to its "original size", by setting K to 1, then join the direct and reciprocal lattices at a common origin as pictured in Fig. 3.20(a). The (2 3) lattice lines in the direct lattice, and the resulting ($\bar{3}^*$ 2^*) lines in the reciprocal lattice are also shown. Note that the reciprocal vector to the (2,3) lattice point in the reciprocal lattice is perpendicular to the (2 3) planes in the real lattice. Such vectors are of considerable importance in the development of the formal mathematics of diffraction. In general, a reciprocal lattice point, (h, k) can be generated by creating a vector from the common lattice origin that is perpendicular to the $(h\ k)$ lines in the direct lattice. The length of the vector is related reciprocally to the spacing between the direct lattice lines, just as the distances between lines (diffraction lines) in the reciprocal lattice are related inversely to the distance between direct lattice points — the lengths of vectors from the common origin to those points.

In order to prove this we first derive a relationship between the direct and reciprocal unit cell axes. Referring to Fig. 3.20(b) we can readily determine that $\sin\gamma^* = (1/a)/a^* = (1/b)/b^* \Rightarrow a^*/b^* = b/a$. Beginning at the $(h\ k)$ plane that contains two lattice points, we now construct two triangles, one consisting of edges along the unit cell axes and the vector to the $(-k, h)$ point in the direct lattice (Fig. 3.20(c)), the other consisting of edges along the reciprocal cell axes and the vector to the (h, k) reciprocal lattice point, $\mathbf{h}$ (Fig. 3.20(d). The ratio of the "axial edge" lengths of the direct lattice triangle is ka/hb; the ratio for the reciprocal lattice triangle is $kb^*/ha^* = ka/hb$ — *the triangles are similar triangles.* It follows that the ratio of their base lengths and heights must also be equal. Starting at the origin of each triangle, we note that there will always be hk parallel lines dividing the triangles. In the direct lattice triangle these lines are separated by a distance of d. Treating the $[-h\ k]$ vector as the base, the height of the triangle is hkd and the length of its base is L. The reciprocal lattice triangle has hk parallel lattice lines each separated by a reciprocal distance of $1/L$, giving it a height of hk/L. Thus, $h/(hk/L) = L/hkd$ and $h = 1/d$. *The vector from the origin to a point* (h, k) *in the reciprocal lattice is perpendicular to the* $(h\ k)$ *lines in the direct lattice with a reciprocal length of* $1/d$, *where* d *is the distance between the direct lattice lines.*

The reciprocal lattice vectors provide a compact representation of the sets of direct lattice lines that generate them — giving their location, direction, and spacing in the direct lattice. The converse is also true — a direct lattice vector contains all of the information necessary to construct the corresponding sets of lines in the reciprocal lattice. Each point in the reciprocal lattice can be generated from a vector perpendicular to the lines in the direct lattice with indices corresponding to the coordinates of the points, with a length equal to the reciprocal of the distance between the lines. Again, the converse is true for the direct lattice.

We now extend the discussion to a three-dimensional lattice — by adding parallel equidistant ab planes along a translation in the direction of a c axis. If this lattice is rotated around the a axis until the ac plane is perpendicular to the incident beam, an identical treatment will generate a c^* axis which is perpendicular to the a axis. The a^* axis will also be perpendicular to the c axis. If the lattice is oriented so that the bc plane is perpendicular to the incident beam, we will find that the c^* axis is perpendicular to the b axis, and the b^* axis is perpendicular to the c axis. It follows that the $\mathbf{a}^*$ vector is perpendicular to the bc plane, the $\mathbf{b}^*$ vector is perpendicular to the ac plane, and the $\mathbf{c}^*$ vector is perpendicular to the ab plane, illustrated in Fig. 3.21. Similarly, $\mathbf{a}$ is perpendicular to the b^*c^* plane and so forth. The volume of the direct unit cell is the area of its base, the bc plane, times its height, which is d_{100}, the distance between the $bc \equiv (1\ 0\ 0)$ planes. The area of the base is the magnitude of the vector product of $\mathbf{b}$ and $\mathbf{c}$, $A_{bc} = |\mathbf{b} \times \mathbf{c}|$ (Eqn. 1.134). The volume of the unit cell is also given by the scalar triple product of the basis vectors: $V = \mathbf{a} \cdot (\mathbf{b} \times \mathbf{c}) \equiv \mathbf{a} \cdot \mathbf{b} \times \mathbf{c}$ (Eqn. 1.135). The $\mathbf{a}^*$ axis is the $\mathbf{h}_{100}$ vector in

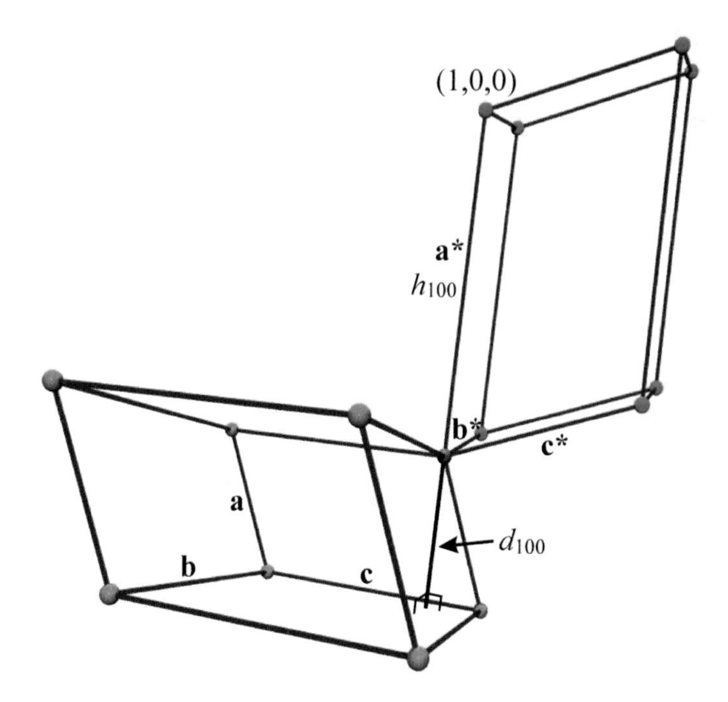

Figure 3.21 Direct unit cell (blue spheres) and reciprocal unit cell (green spheres) sharing a common origin.

the reciprocal lattice with length $1/d_{100}$. The relationship between the $\mathbf{a}^*$ axis and the direct unit cell axial vectors follows:

$$\begin{aligned} V &= A_{bc}d_{100} = |\mathbf{b} \times \mathbf{c}|\, d_{100} \\ d_{100} &= \frac{V}{|\mathbf{b} \times \mathbf{c}|} = \frac{\mathbf{a} \cdot \mathbf{b} \times \mathbf{c}}{|\mathbf{b} \times \mathbf{c}|} \\ h_{100} &= a^* = \frac{|\mathbf{b} \times \mathbf{c}|}{\mathbf{a} \cdot \mathbf{b} \times \mathbf{c}}. \end{aligned}$$

$\mathbf{a}^*$ is a vector perpendicular to the bc plane and parallel to $\mathbf{b} \times \mathbf{c}$, with magnitude h_{100}. Thus,

$$\mathbf{a}^* = \frac{\mathbf{b} \times \mathbf{c}}{\mathbf{a} \cdot \mathbf{b} \times \mathbf{c}} = \frac{\mathbf{b} \times \mathbf{c}}{V}. \tag{3.4}$$

Treating the $\mathbf{b}^*$ and $\mathbf{c}^*$ axes in the same manner completes the expressions for the reciprocal cell basis in terms of the direct cell vectors:

$$\mathbf{b}^* = \frac{\mathbf{c} \times \mathbf{a}}{\mathbf{a} \cdot \mathbf{b} \times \mathbf{c}} = \frac{\mathbf{c} \times \mathbf{a}}{V}. \tag{3.5}$$

$$\mathbf{c}^* = \frac{\mathbf{a} \times \mathbf{b}}{\mathbf{a} \cdot \mathbf{b} \times \mathbf{c}} = \frac{\mathbf{a} \times \mathbf{b}}{V}. \tag{3.6}$$

Expressions for the direct unit cell basis in terms of the reciprocal axial vectors are determined in a similar fashion. The reciprocal unit cell volume is determined in the same manner as that for the direct cell volume (Section 1.5.4):

$$V^* = \mathbf{a}^* \cdot \mathbf{b}^* \times \mathbf{c}^* \quad (3.7)$$

$$= a^*b^*c^*(1 - \cos^2\alpha^* - \cos^2\beta^* - \cos^2\gamma^* + 2\cos\alpha^*\cos\beta^*\cos\gamma^*)^{\frac{1}{2}} \quad (3.8)$$

The spacing between b^*c^* planes is $1/a$ and

$$V^* = A^*_{b^*c^*}\frac{1}{a} = |\mathbf{b}^* \times \mathbf{c}^*|\,\frac{1}{a}$$

$$a = \frac{|\mathbf{b}^* \times \mathbf{c}^*|}{\mathbf{a}^* \cdot \mathbf{b}^* \times \mathbf{c}^*}.$$

a is a vector perpendicular to the b^*c^* plane and parallel to $\mathbf{b}^* \times \mathbf{c}^*$; similar treatments for **b** and **c** result in the direct lattice expressions:

$$\mathbf{a} = \frac{\mathbf{b}^* \times \mathbf{c}^*}{\mathbf{a}^* \cdot \mathbf{b}^* \times \mathbf{c}^*} = \frac{\mathbf{b}^* \times \mathbf{c}^*}{V^*} \quad (3.9)$$

$$\mathbf{b} = \frac{\mathbf{c}^* \times \mathbf{a}^*}{\mathbf{a}^* \cdot \mathbf{b}^* \times \mathbf{c}^*} = \frac{\mathbf{c}^* \times \mathbf{a}^*}{V^*} \quad (3.10)$$

$$\mathbf{c} = \frac{\mathbf{a}^* \times \mathbf{b}^*}{\mathbf{a}^* \cdot \mathbf{b}^* \times \mathbf{c}^*} = \frac{\mathbf{a}^* \times \mathbf{b}^*}{V^*}. \quad (3.11)$$

In Sec. 1.5.5 identities involving scalar and vector products were developed. We will now make use of these identities to determine a number of important relationships. The first of these are the scalar products of the direct and reciprocal cell basis vectors:

$$\mathbf{a} \cdot \mathbf{a}^* = \mathbf{a} \cdot \frac{\mathbf{b} \times \mathbf{c}}{\mathbf{a} \cdot \mathbf{b} \times \mathbf{c}} = \frac{\mathbf{a} \cdot \mathbf{b} \times \mathbf{c}}{\mathbf{a} \cdot \mathbf{b} \times \mathbf{c}} = 1 \quad (3.12)$$

$$\mathbf{b} \cdot \mathbf{b}^* = \mathbf{b} \cdot \frac{\mathbf{c} \times \mathbf{a}}{\mathbf{a} \cdot \mathbf{b} \times \mathbf{c}} = \frac{\mathbf{b} \cdot \mathbf{c} \times \mathbf{a}}{\mathbf{b} \cdot \mathbf{c} \times \mathbf{a}} = 1 \quad (3.13)$$

$$\mathbf{c} \cdot \mathbf{c}^* = \mathbf{c} \cdot \frac{\mathbf{a} \times \mathbf{b}}{\mathbf{a} \cdot \mathbf{b} \times \mathbf{c}} = \frac{\mathbf{c} \cdot \mathbf{a} \times \mathbf{b}}{\mathbf{c} \cdot \mathbf{a} \times \mathbf{b}} = 1 \quad (3.14)$$

$$\mathbf{a} \cdot \mathbf{b}^* = \mathbf{a} \cdot \frac{\mathbf{c} \times \mathbf{a}}{\mathbf{a} \cdot \mathbf{b} \times \mathbf{c}} = \frac{\mathbf{c} \cdot \mathbf{a} \times \mathbf{a}}{\mathbf{a} \cdot \mathbf{b} \times \mathbf{c}} = \frac{\mathbf{c} \cdot \mathbf{0}}{\mathbf{a} \cdot \mathbf{b} \times \mathbf{c}} = 0 \quad (3.15)$$

$$\mathbf{a} \cdot \mathbf{c}^* = \mathbf{a} \cdot \frac{\mathbf{a} \times \mathbf{b}}{\mathbf{a} \cdot \mathbf{b} \times \mathbf{c}} = \frac{\mathbf{b} \cdot \mathbf{a} \times \mathbf{a}}{\mathbf{a} \cdot \mathbf{b} \times \mathbf{c}} = \frac{\mathbf{b} \cdot \mathbf{0}}{\mathbf{a} \cdot \mathbf{b} \times \mathbf{c}} = 0 \quad (3.16)$$

$$\mathbf{b} \cdot \mathbf{c}^* = \mathbf{b} \cdot \frac{\mathbf{a} \times \mathbf{b}}{\mathbf{a} \cdot \mathbf{b} \times \mathbf{c}} = \frac{\mathbf{a} \cdot \mathbf{b} \times \mathbf{b}}{\mathbf{a} \cdot \mathbf{b} \times \mathbf{c}} = \frac{\mathbf{a} \cdot \mathbf{0}}{\mathbf{a} \cdot \mathbf{b} \times \mathbf{c}} = 0 \quad (3.17)$$

Similarly, $\mathbf{a}^* \cdot \mathbf{b} = \mathbf{a}^* \cdot \mathbf{c} = \mathbf{b}^* \cdot \mathbf{c} = 0$. These relationships make geometric sense as well. For example the cosine of the angle, η, between **a** and $\mathbf{a}^*$ is $(1/a)/a^* = 1/(aa^*)$ (see Fig. 3.20(b)), so that $\mathbf{a} \cdot \mathbf{a}^* = aa^* \cos\eta = aa^*/aa^* = 1$. $\mathbf{b}^*$ is perpendicular to the ac plane requiring $\mathbf{a} \cdot \mathbf{b}^* = 0$, and so forth.

The reciprocal relationship exhibited by the scalar products of the unit cell axes in the direct and reciprocal unit cells is also reflected in their unit cell volumes, which are the inverses of one another:

$$\begin{aligned} VV^* &= (\mathbf{a}\cdot\mathbf{b}\times\mathbf{c})(\mathbf{a}^*\cdot\mathbf{b}^*\times\mathbf{c}^*) \\ &= \begin{vmatrix} a_x & a_y & a_z \\ b_x & b_y & b_z \\ c_x & c_y & c_z \end{vmatrix} \begin{vmatrix} a_x^* & b_x^* & c_x^* \\ a_y^* & b_y^* & c_y^* \\ a_z^* & b_z^* & c_z^* \end{vmatrix} = \begin{vmatrix} \mathbf{a}\cdot\mathbf{a}^* & \mathbf{a}\cdot\mathbf{b}^* & \mathbf{a}\cdot\mathbf{c}^* \\ \mathbf{b}\cdot\mathbf{a}^* & \mathbf{b}\cdot\mathbf{b}^* & \mathbf{b}\cdot\mathbf{c}^* \\ \mathbf{c}\cdot\mathbf{a}^* & \mathbf{c}\cdot\mathbf{b}^* & \mathbf{c}\cdot\mathbf{c}^* \end{vmatrix} \\ &= \begin{vmatrix} 1 & 0 & 0 \\ 0 & 1 & 0 \\ 0 & 0 & 1 \end{vmatrix} = 1. \end{aligned} \tag{3.18}$$

The relationship between these two coordinate systems is further underscored if we create a reciprocal basis *from the reciprocal basis* by constructing $(\mathbf{a}^*)^*$, etc. We make use of the scalar triple product (Eqn. 1.136) and the vector quadruple product (Eqn. 1.145):

$$\begin{aligned} \mathbf{a}^* &= \frac{\mathbf{b}\times\mathbf{c}}{V} \\ (\mathbf{a}^*)^* &= \frac{\mathbf{b}^*\times\mathbf{c}^*}{V^*} = \frac{(\mathbf{c}\times\mathbf{a})\times(\mathbf{a}\times\mathbf{b})}{VV^*V} \\ &= \frac{(\mathbf{c}\cdot\mathbf{a}\times\mathbf{b})\,\mathbf{a}-(\mathbf{a}\cdot\mathbf{a}\times\mathbf{b})\,\mathbf{c}}{V} \\ &= \frac{(\mathbf{c}\cdot\mathbf{a}\times\mathbf{b})\,\mathbf{a}}{V}-\frac{(\mathbf{b}\cdot\mathbf{a}\times\mathbf{a})\,\mathbf{c}}{V} = \frac{(\mathbf{c}\cdot\mathbf{a}\times\mathbf{b})\,\mathbf{a}}{V} \\ &= \frac{\mathbf{c}\cdot\mathbf{a}\times\mathbf{b}}{\mathbf{c}\cdot\mathbf{a}\times\mathbf{b}}\,\mathbf{a} = \mathbf{a}. \end{aligned}$$

Similarly, $(\mathbf{b}^*)^* = \mathbf{b}$ and $(\mathbf{c}^*)^* = \mathbf{c}$; the bases are truly reciprocals of one another!.

We now have the tools to obtain the reciprocal unit cell parameters from the direct cell parameters and vice versa. The axial lengths are the lengths of the vectors that define them:

$$a^* = \frac{|\mathbf{b}\times\mathbf{c}|}{V} = \frac{bc\sin\alpha}{V} \qquad a = \frac{|\mathbf{b}^*\times\mathbf{c}^*|}{V^*} = \frac{b^*c^*\sin\alpha^*}{V^*} \tag{3.19}$$

$$b^* = \frac{|\mathbf{c}\times\mathbf{a}|}{V} = \frac{ac\sin\beta}{V} \qquad b = \frac{|\mathbf{c}^*\times\mathbf{a}^*|}{V^*} = \frac{a^*c^*\sin\beta^*}{V^*} \tag{3.20}$$

$$c^* = \frac{|\mathbf{a}\times\mathbf{b}|}{V} = \frac{ab\sin\gamma}{V} \qquad c = \frac{|\mathbf{a}^*\times\mathbf{b}^*|}{V^*} = \frac{a^*b^*\sin\gamma^*}{V^*} \tag{3.21}$$

To determine the interaxial angle relationships we create unit vectors $\mathbf{a_1}$, $\mathbf{b_1}$, and $\mathbf{c_1}$ along $\mathbf{a}$, $\mathbf{b}$, and $\mathbf{c}$, respectively (e.g., $\mathbf{a_1} = \mathbf{a}/a$, etc.), then define the vector products $\mathbf{v_c} = \mathbf{a_1}\times\mathbf{b_1}$ and $\mathbf{v_a} = \mathbf{b_1}\times\mathbf{c_1}$. $\mathbf{c}^*$ is perpendicular to $\mathbf{a}$ and $\mathbf{b}$, as is $\mathbf{v_c}$. $\mathbf{v_c}$ is therefore parallel to $\mathbf{c}^*$ with magnitude $|\mathbf{a_1}\times\mathbf{b_1}| = a_1b_1\sin\gamma = \sin\gamma$. $\mathbf{v_a}$ is parallel to $\mathbf{a}^*$ with magnitude $|\mathbf{b_1}\times\mathbf{c_1}| = b_1c_1\sin\alpha = \sin\alpha$. Thus the angle between $\mathbf{v_c}$ and $\mathbf{v_a}$ is the angle between $\mathbf{c}^*$ and $\mathbf{a}^*$, β^*, and

$$\begin{aligned} \mathbf{v_c}\cdot\mathbf{v_a} &= v_cv_a\cos\beta^* \\ (\mathbf{a_1}\times\mathbf{b_1})\cdot(\mathbf{b_1}\times\mathbf{c_1}) &= \sin\gamma\sin\alpha\cos\beta^*. \end{aligned}$$

The expression $(\mathbf{a_1} \times \mathbf{b_1}) \cdot (\mathbf{b_1} \times \mathbf{c_1})$ is a scalar quadruple product (Eqn. 1.143), allowing for the determination of $\mathbf{v_c} \cdot \mathbf{v_a}$ solely in terms of the direct cell angles:

$$\begin{aligned}(\mathbf{a_1} \times \mathbf{b_1}) \cdot (\mathbf{b_1} \times \mathbf{c_1}) &= \begin{vmatrix} \mathbf{a_1} \cdot \mathbf{b_1} & \mathbf{a_1} \cdot \mathbf{c_1} \\ \mathbf{b_1} \cdot \mathbf{b_1} & \mathbf{b_1} \cdot \mathbf{c_1} \end{vmatrix} \\ &= \begin{vmatrix} (1)(1)\cos\gamma & (1)(1)\cos\beta \\ (1)(1)\cos 0 & (1)(1)\cos\alpha \end{vmatrix} \\ &= \cos\gamma\cos\alpha - \cos\beta.\end{aligned}$$

The two expressions for $\mathbf{v_c} \cdot \mathbf{v_a}$ are set equal to one another,

$$\sin\gamma\sin\alpha\cos\beta^* = \cos\gamma\cos\alpha - \cos\beta,$$

and $\cos\beta^*$ can be expressed in terms of the direct cell angles. The same approach yields expressions for the remaining reciprocal and direct cell angles:

$$\cos\alpha^* = \frac{\cos\beta\cos\gamma - \cos\alpha}{\sin\beta\sin\gamma} \qquad \cos\alpha = \frac{\cos\beta^*\cos\gamma^* - \cos\alpha^*}{\sin\beta^*\sin\gamma^*} \tag{3.22}$$

$$\cos\beta^* = \frac{\cos\alpha\cos\gamma - \cos\beta}{\sin\alpha\sin\gamma} \qquad \cos\beta = \frac{\cos\alpha^*\cos\gamma^* - \cos\beta^*}{\sin\alpha^*\sin\gamma^*} \tag{3.23}$$

$$\cos\gamma^* = \frac{\cos\alpha\cos\beta - \cos\gamma}{\sin\alpha\sin\beta} \qquad \cos\gamma = \frac{\cos\alpha^*\cos\beta^* - \cos\gamma^*}{\sin\alpha^*\sin\beta^*} \tag{3.24}$$

Perhaps the most important relationship between the direct and reciprocal lattices is the relationship between interplanar spacings in the direct lattice and the lengths of vectors to lattice points in the reciprocal lattice. In the two dimensional lattice the reciprocal lattice vector to specific coordinates (h, k) in the reciprocal lattice was found to be perpendicular to the lattice lines in the direct lattice with indices $(h\ k)$ with a length equal to the reciprocal of the distance between the $(h\ k)$ lines. A general vector in the three-dimensional reciprocal lattice will have fractional coordinates x_f^*, y_f^*, z_f^* such that

$$\mathbf{v}^* = x_f^*\,\mathbf{a}^* + y_f^*\,\mathbf{b}^* + z_f^*\,\mathbf{c}^*. \tag{3.25}$$

A vector to a *lattice point* (often referred to as a *node*) in the three-dimensional reciprocal lattice will have integer coordinates:

$$\mathbf{h}_{hkl} = h\,\mathbf{a}^* + k\,\mathbf{b}^* + l\,\mathbf{c}^*. \tag{3.26}$$

While the extrapolation from lines to planes in going from two to three dimensions seems intuitive, we must rigorously establish analogous relationships for interplanar spacings and reciprocal lattice directions and lengths for the three dimensional lattices. Indeed, we might simply have *defined* $\mathbf{a}^*$, $\mathbf{b}^*$ and $\mathbf{c}^*$ with Eqns. 3.4, 3.5, and 3.6, and sought to determine the relationships between the lattices directly. Fig. 3.22 represents a unit cell in direct space intersected by planes with indices $(h\ k\ l)$ at the origin and at distances a/h along $\mathbf{a}$, b/k along $\mathbf{b}$, and c/l along $\mathbf{c}$. The perpendicular distance between planes is d_{hkl}. The reciprocal lattice is assumed to be joined with the direct lattice at the origin, and we consider a vector from the origin to the reciprocal lattice point (h, k, l), $\mathbf{h}_{hkl}$ (Eqn. 3.26). Two vectors, $\mathbf{v_1}$ and $\mathbf{v_2}$, are created from the intersection points of the $(h\ k\ l)$ plane

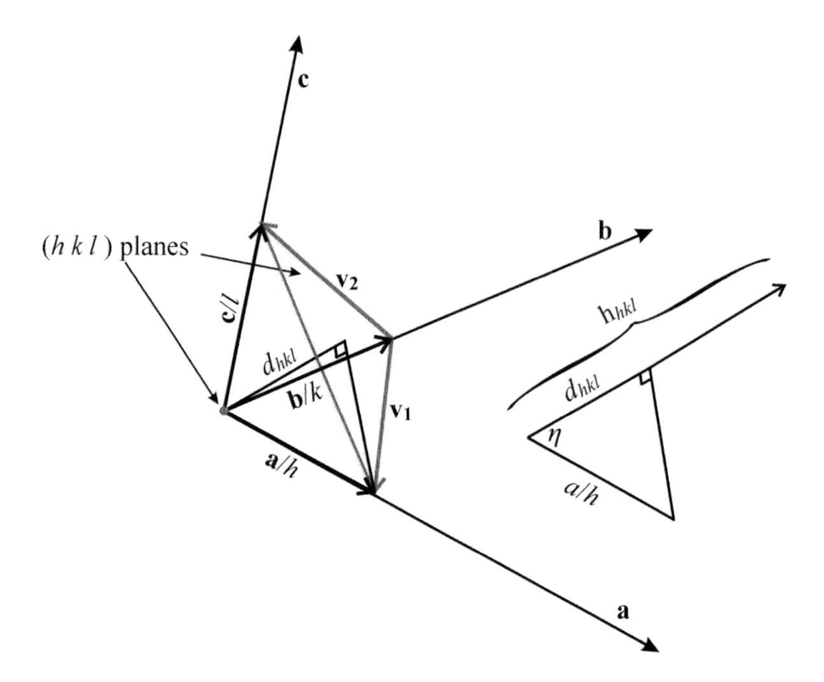

Figure 3.22 Unit cell in direct space intersected by parallel planes with indices $(h\ k\ l)$. The red dot indicates that one plane intersects at the origin. Insert: Reciprocal lattice vector $[h\ k\ l]$ from a common origin to the reciprocal lattice point (h, k, l).

along each of the direct unit cell axes. Both of these vectors lie in the $(h\ k\ l)$ plane, and are perpendicular to the reciprocal lattice vector $\mathbf{h}_{hkl}$:

$$\begin{aligned}
\mathbf{v_1} &= \left(\frac{\mathbf{a}}{h} - \frac{\mathbf{b}}{k}\right) \\
\mathbf{v_1} \cdot \mathbf{h}_{hkl} &= \frac{\mathbf{a}}{h} \cdot (h\,\mathbf{a}^* + k\,\mathbf{b}^* + l\,\mathbf{c}^*) - \frac{\mathbf{b}}{k} \cdot (h\,\mathbf{a}^* + k\,\mathbf{b}^* + l\,\mathbf{c}^*) \\
&= \frac{h}{h} + 0 + 0 - 0 - \frac{k}{k} - 0 = 0. \\
\mathbf{v_2} &= \left(\frac{\mathbf{c}}{l} - \frac{\mathbf{b}}{k}\right) \\
\mathbf{v_2} \cdot \mathbf{h}_{hkl} &= \frac{\mathbf{c}}{l} \cdot (h\,\mathbf{a}^* + k\,\mathbf{b}^* + l\,\mathbf{c}^*) - \frac{\mathbf{b}}{k} \cdot (h\,\mathbf{a}^* + k\,\mathbf{b}^* + l\,\mathbf{c}^*) \\
&= 0 + 0 + \frac{l}{l} - 0 - \frac{k}{k} - 0 = 0.
\end{aligned}$$

Since the reciprocal lattice vector $\mathbf{h}_{hkl}$ *is perpendicular to two non-collinear vectors in the* $(h\ k\ l)$ *plane, it is perpendicular to the plane.* The insert in Fig. 3.22

illustrates the relationship between this vector and the direct unit cell. If η is the angle between the a axis and $\mathbf{h}_{hkl}$, then

$$\begin{aligned}
\frac{\mathbf{a}}{h}\cdot\mathbf{h}_{hkl} &= \frac{\mathbf{a}}{h}\cdot(h\,\mathbf{a}^* + k\,\mathbf{b}^* + l\,\mathbf{c}^*) = \frac{h}{h} + 0 + 0 = 1 \\
\frac{\mathbf{a}}{h}\cdot\mathbf{h}_{hkl} &= \frac{a}{h}\mathrm{h}_{hkl}\cos\eta \\
\cos\eta &= \frac{d_{hkl}}{(a/h)} = \frac{h\,d_{hkl}}{a} \Longrightarrow \frac{a\,\mathrm{h}_{hkl}\,h\,d_{hkl}}{ha} = 1 \\
\mathrm{h}_{hkl}d_{hkl} &= 1 \\
\mathrm{h}_{hkl} &= \frac{1}{d_{hkl}} \ . \qquad (3.27)
\end{aligned}$$

As we observed in the two-dimensional case, *the reciprocal lattice vector from a common origin to the reciprocal lattice point* (h, k, l) *is normal to the* $(h\ k\ l)$ *planes in the direct lattice with a magnitude equal to the reciprocal of the distance between the planes.*

3.2.2 X-ray Diffraction: The Diffraction Equation.

A crystal is composed of a periodic arrangement of atomic nuclei surrounded by electrons. These electrons scatter X-radiation, which has wavelengths short enough to allow for the determination of the location of the electron scatterers in the crystal lattice, provided that we are able to observe the collective effects from the scattering of a large number of these scatterers in effectively identical environments. This is accomplished by establishing the conditions for the diffraction of X-rays from a crystal, and determining the relative intensities of the diffracted radiation for a number of diffraction maxima. In order to establish the conditions necessary for the diffraction from crystals we must take a closer look at the interaction of X-radiation with electrons, specifically electrons "bound" to atomic nuclei.

Electrons bound to nuclei are constrained to specific energy states. These electronic energy states are know as *stationary states*, and quantum mechanics requires that only these states are allowed. The lowest energy state is known as the *electronic ground state*, in which the electrons, on a time-average basis, are as close to the nuclei as possible. The other electronic states are known as *excited states*, and require the absorption of specific amounts of energy, *quanta*, resulting in the electrons (again on a time-average basis) residing farther away from the nuclei. From a quantum mechanical point of view, when the electric field of a photon interacts with an electron in an atom the energy of the photon is transferred to the atom and the photon is considered to have been "absorbed." If this energy is sufficient to increase the kinetic energy of the electron enough to allow it to escape from the nucleus the electron becomes a *photoelectron*, and either the entire photon is absorbed in the process, or a photon of lower energy is emitted. The former process is known as the *photoelectric effect*, and the latter is known as the *Compton effect*.

If the electron remains bound to the atom (as most do) it must end up either in the electronic ground state, or an electronic excited state. The electron has absorbed the energy of the photon in this process, and since it is bound to the nucleus through coulombic attractive forces it will tug on the nucleus as its motion changes, thus transferring a portion of the added momentum to the much more massive nucleus. The partitioning of this added momentum depends on the degree of attraction between the nucleus and electron, which in turn depends on the distance of the electron from the nucleus when the photon is absorbed and the effective nuclear charge experienced by the electron. If this attraction is very strong then most of the momentum will go toward moving the electron and nucleus together, with only a small fraction of the added momentum affecting their relative motions. The electron will have absorbed insufficient energy to either exit the atom or create a stationary excited state, and the energy must be released from the atom – *in the form of an emitted photon of the same frequency as the incident photon.* This type of scattering is known as *elastic scattering.* If the coupling is weak, then the electron can gain a larger portion of the momentum and either be ejected from the atom, or excited to an electronic excited state. Both processes result in scattered photons of lower frequency. This type of scattering is known as *inelastic scattering.*

Since diffraction conditions are based upon constructive interference, those photons that are scattered inelastically will be unlikely to add to diffraction maxima, although they will be scattered at all angles, and will arrive at the detector along with the elastically scattered photons. We must therefore consider whether or not they will influence our ability to detect the photons resulting from diffraction. To do so we consider a typical crystal 0.3 mm $\times$ 0.3 mm $\times$ 0.3 mm containing approximately 10^{18} electrons bound to nuclei. Assuming that scattered photons have an oscillating electric field with maximum amplitude ϵ, the individual electric fields in a specific direction for the inelastically scattered photons will contribute ϵ^2 to the intensity, and since they are at different frequencies and phases, interference effects will tend to average to zero (some waves will interfere constructively, others destructively). Because of this the trajectories of the incoherent photons will be spread out in all directions and their contributions to the electric field at given point in space will simply be additive; the intensity at a point in space will be proportional to the number of scatterers. We denote the contribution of inelastic scattering to the net magnitude of the electric field from n inelastic scatterers observed at a specific point in space as $I_{incoh} \alpha n \epsilon^2$. Under diffraction conditions, n elastically scattered photons have electric fields that constructively interfere *at the locations of diffraction maxima.* Interference effects in this case do not average out, and the trajectories of the photons are directed toward these locations, and away from those locations where their waves destructively interfere. The net electric field at the locations of the diffraction maxima is determined by superpositioning each of the waves. The result is an electric field at these locations that is proportional to the number of scatterers, $\mathcal{E} \alpha n \epsilon$, resulting in a net intensity proportional to the *square* of the number of scatterers, $I_{coh} \alpha \mathcal{E}^2 \alpha n^2 \epsilon^2$. For the example above, under diffraction conditions, $I_{incoh} \alpha 10^{18}$, while $I_{coh} \alpha (10^{18})^2 \alpha 10^{36}$! *Under diffraction conditions the contribution to observed X-ray intensities from inelastic scattering is orders of magnitude less than the contribution from elastic scattering, and while inelastic scattering contributes to the background radiation it can be ignored as a contributor to the intensities observed at diffraction maxima. We will therefore focus solely on the elastically scattered radiation, the radiation at the same wavelength as the incident radiation, in the treatments that follow.*

X-radiation is scattered from electrons which are in constant motion on a time scale in which we are able to observe only their average behavior. Before considering the conditions for diffraction from moving electrons, we begin with the simplification depicted in Fig. 3.23, in which a tightly bound electron is fixed at the origin, o, and another at point p. A coherent monochromatic source of X-radiation is directed toward these electrons at an arbitrary angle, and a photon-counting detector is placed at another arbitrary angle to capture any photons scattered in its direction. At the points q and q' the oscillating electric fields are in phase. The radiation is scattered from the electrons at o and p, undergoing a 90° phase shift.* We are interested only in the relative phases of the waves scattered from each electron, and since the radiation scattered from both electrons undergoes the same 90° phase shift we can ignore it. If $\mathcal{E}_s$ is the maximum amplitude of the electric field due to the scattering by a single electron, then each electron will scatter a certain fraction of the incident radiation, $f_e = (\mathcal{E}_s/\mathcal{E}_I)$. This fraction, termed the *electron scattering factor*, is a measure of the ability of an electron to scatter radiation — its *scattering power*.

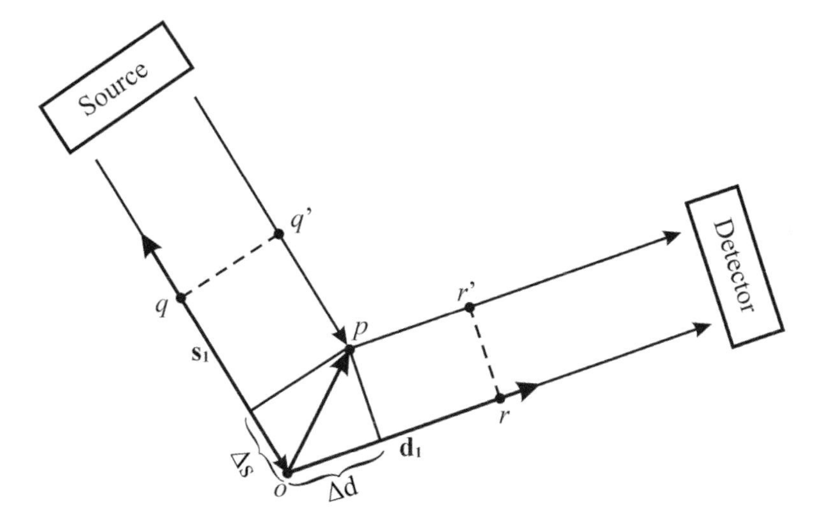

Figure 3.23 Elastic scattering of X-radiation from electrons at the origin (o) and at the point p.

For reference we await a maximum in the electric field to arrive at the points q and q' and somehow "mark" the maxima on the coherent incident waves. We then monitor the scattered wave at point r, setting the time to $t_o = 0$ when a maximum in the wave passes the point. At some later time, t, the electric field of the scattered radiation from the electron at o is $\mathcal{E}_r(t) = \mathcal{E}_s \cos(2\pi\nu t)$. At the same time t, the wave arriving at r' from point p has had to travel a smaller distance, less by $L = \Delta s + \Delta d$. We also observe that the mark made at q' arrives at point r' before the mark made at q arrives at r. After scattering, the wave from p appears to be traveling ahead of the wave from o — at point r' and time t the wave appears as if it had been traveling for $t + (L/c)$ seconds relative to the wave from o, where c is the speed of light: $\mathcal{E}_{r'}(t) = \mathcal{E}_s \cos(2\pi\nu(t + (L/c))$. The phase difference between waves scattered from electrons at points o and p is therefore $\varphi_{op} = 2\pi\nu t - (2\pi\nu(t + (L/c)) = -2\pi\nu(L/c)$. If we create *unit* vectors, $\mathbf{s_1}$ and $\mathbf{d_1}$,

*Because electrons in atoms are subjected to restoring forces, the phase shift is different from that of a free electron. We will deal with this more formally in Section 5.1.2.

directed toward the source and detector, respectively, then Δs is the projection of the vector $\overrightarrow{op}$ onto $\mathbf{s_1}$ and Δd is the projection of $\overrightarrow{op}$ onto $\mathbf{d_1}$. According to Eqn. 1.6,

$$\begin{aligned}\Delta s &= \frac{\overrightarrow{op}\cdot\mathbf{s_1}}{s_1} = \overrightarrow{op}\cdot\mathbf{s_1}\ ,\\ \Delta d &= \frac{\overrightarrow{op}\cdot\mathbf{d_1}}{d_1} = \overrightarrow{op}\cdot\mathbf{d_1}\ \text{ and}\\ L &= \overrightarrow{op}\cdot\mathbf{s_1}+\overrightarrow{op}\cdot\mathbf{d_1}\ .\end{aligned}$$

Since $\lambda = c/\nu$, the phase difference between a wave scattered from the origin and one scattered from the point p is related to the wavelength of the incident radiation by

$$\varphi_{op} = -2\pi\nu L/\lambda = -2\pi\left[\ \overrightarrow{op}\cdot(\mathbf{s_1}+\mathbf{d_1})\ \right]/\lambda\ . \tag{3.28}$$

Since the two scattered waves are out of phase with one another by φ_{op} they must be superpositioned in order to determine the resultant electric field and the photon intensity at the detector. This is accomplished by treating the two waves as rotating vectors as we have done previously. Referring to Fig. 3.6, the resultant vector is the sum of the vectors representing each wave:

$$\begin{aligned}\overrightarrow{\mathcal{E}_o(t)} &= [(\mathcal{E}_s\cos 2\pi\nu t)\ \ (\mathcal{E}_s\sin 2\pi\nu t)]\\ \overrightarrow{\mathcal{E}_p(t)} &= [(\mathcal{E}_s\cos(2\pi\nu t+\varphi_{op}))\ \ (\mathcal{E}_s\sin(2\pi\nu t+\varphi_{op}))]\\ \overrightarrow{\mathcal{E}_{op}(t)} &= [(\mathcal{E}_s\cos 2\pi\nu t+\mathcal{E}_s\cos(2\pi\nu t+\varphi_{op}))\ (\mathcal{E}_s\sin 2\pi\nu t+\mathcal{E}_s\sin(2\pi\nu t+\varphi_{op}))]\end{aligned}$$

Since photons are counted over many cycles the measured intensity will be proportional to the square of the magnitude of the resultant vector, $|\overrightarrow{\mathcal{E}_{op}(t)}|^2$, which is independent of time. The vectors above can therefore be simplified by setting $t = 0$. This aligns the vector representing the wave scattered from the origin along the $\mathbf{i}$ axis with a zero phase angle.

$$\begin{aligned}\overrightarrow{\mathcal{E}_o} &= [\mathcal{E}_s\ \ 0]\\ \overrightarrow{\mathcal{E}_p} &= [(\mathcal{E}_s\cos\varphi_{op})\ \ (\mathcal{E}_s\sin\varphi_{op})]\\ \overrightarrow{\mathcal{E}_{op}} &= [(\mathcal{E}_s+\mathcal{E}_s\varphi_{op})\ (0+\mathcal{E}_s\sin\varphi_{op})]\end{aligned}$$

The experiment is now modified by placing n “fixed” electrons at various points with respect to the origin in order to determine their effect on the intensity measured at the detector. The electron at the origin is removed, but a hypothetical wave scattered from the origin is used as a *scattering reference*. Since the intensity measured from the resultant wave depends only on relative phases of the waves scattered from each point this phase reference is arbitrary. However, we will soon discover that using a lattice point origin as the scattering reference will greatly simplify the mathematics that we are about to develop. Fig. 3.24 illustrates the vector superposition of waves scattered from “fixed electrons” at three points in space. The extrapolation to n electrons is obvious and the resulting wave vector is

$$\overrightarrow{\mathcal{E}_d} = \left[\left(\sum_{j=1}^{n}\mathcal{E}_s\cos\varphi_j\right)\ \left(\sum_{j=1}^{n}\mathcal{E}_s\sin\varphi_j\right)\right], \tag{3.29}$$

where $\varphi_j = -2\pi\left[\ \overrightarrow{op_j}\cdot(\mathbf{s_1}+\mathbf{d_1})\ \right]/\lambda$.

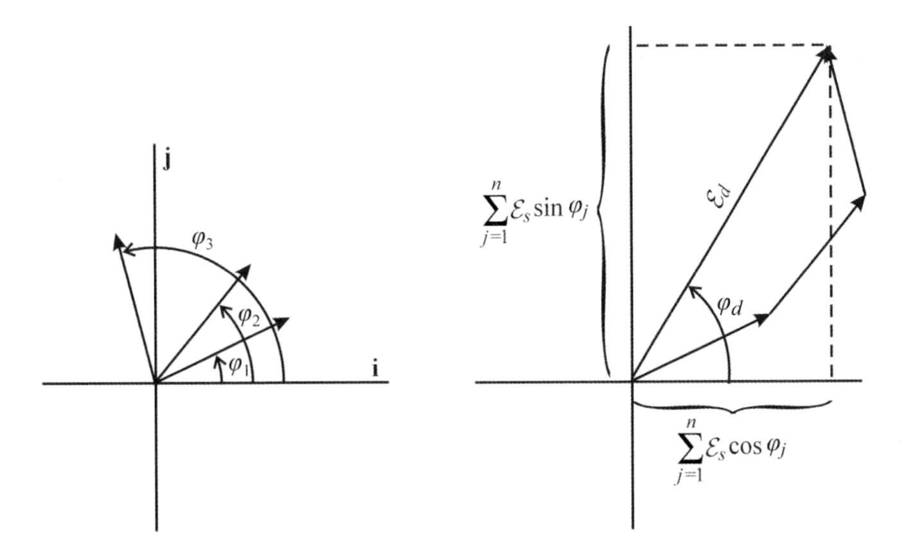

Figure 3.24 Vector superposition of scattered waves from $n = 3$ fixed electrons.

The resultant vector that represents the net electric field of the wave propagating toward the detector is an ordered pair of numbers generated from the sum of the components of the vectors representing each scattered wave, each an ordered pair of numbers. The superposition of waves requires that we keep the **i** and **j** components of these vectors separate, but handling ordered pairs of numbers is cumbersome, especially for creating equations that contain the vectors. A convenient way to keep the components of a vector separate is to write them as a complex number, e.g., $[x\ y] \equiv x + iy$. Note that the rules for complex number addition require that the "real" parts of the number are added separately from the "imaginary" parts, just as the rules for adding vectors require that the x and y components are added separately. Thus if we write our component wave vectors as $[(\mathcal{E}_s \cos\varphi_j)\ (\mathcal{E}_s \sin\varphi_j)] \equiv (\mathcal{E}_s \cos\varphi_j) + i(\mathcal{E}_s \sin\varphi_j)$ the resultant vector can be represented as a complex number:

$$\overrightarrow{\mathcal{E}_d} = \left(\sum_{j=1}^{n} \mathcal{E}_s \cos\varphi_j\right) + i\left(\sum_{j=1}^{n} \mathcal{E}_s \sin\varphi_j\right). \tag{3.30}$$

The major advantage of representing vectors as complex numbers is that we can make use of the tools of complex arithmetic. One of the most powerful of these tools was developed by the mathematician Leonard Euler in the 18th century. By expanding $\sin\theta$, $\cos\theta$, and $e^{i\theta}$ as infinite series, Euler showed that

$$\cos\theta + i\sin\theta = e^{i\theta}. \tag{3.31}$$

The vector representing a scattered wave can now be represented as

$$\overrightarrow{\mathcal{E}_j} = \mathcal{E}_s(\cos\varphi_j + i\sin\varphi_j) = \mathcal{E}_s e^{i\varphi_j}. \tag{3.32}$$

The resultant wave vector is then

$$\overrightarrow{\mathcal{E}_d} = \mathcal{E}_s \sum_{j=1}^{n}(\cos\varphi_j + i\sin\varphi_j) = \mathcal{E}_s \sum_{j=1}^{n} e^{i\varphi_j}. \tag{3.33}$$

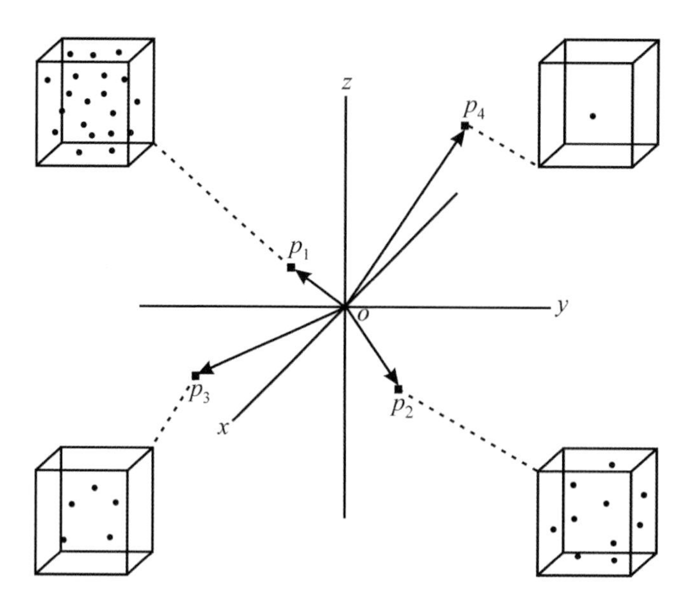

Figure 3.25 Electron locations in volume elements at four different points in space determined over a specified time period.

The intensity observed at the detector is proportional to the square of amplitude of the resultant wave, $I \,\alpha\, \mathcal{E}_d^2$, where

$$\mathcal{E}_d = \left[\left(\sum_{j=1}^{n} \mathcal{E}_s \cos\varphi_j\right)^2 + \left(\sum_{j=1}^{n} \mathcal{E}_s \sin\varphi_j\right)^2\right]^{\frac{1}{2}}. \tag{3.34}$$

The phase of the resultant wave, relative to the phase of a wave scattered from the origin, is determined from its tangent:

$$\varphi_d = \arctan\left(\frac{\sum_{j=1}^{n} \mathcal{E}_s \sin\varphi_j}{\sum_{j=1}^{n} \mathcal{E}_s \cos\varphi_j}\right). \tag{3.35}$$

We have now developed the tools to handle scattering from electrons residing at any number of locations, but *the electrons in a real material are in constant motion. Scattering will only occur from a specific location when an electron is at that location.* To extend our analysis to moving electrons, we divide the crystal — the space where electrons are moving — into very small volume elements. If we were able to locate the electrons as they moved in and out of these volumes we could sample each element in equal time increments, marking the location of any electrons found over a certain period of time. For simplicity, consider a system with only a single electron. Fig. 3.25 is an example of what might be expected from such an exercise at four locations in space. Qualitatively, we would expect the most scattering from the volume element at p_1, and the least from the element at p_4.

To make this quantitative we count the number of dots in each volume, assuming that the probability that a photon will be scattered from a given volume element will be equal to the probability that the electron is in that volume element, and that will be proportional to the relative number of dots. The number itself does not give us this probability directly, since it depends on the volume of the element. This dependence is removed if we divide by the volume, which will then give us the number of dots in a unit volume. This is a measure of the concentration of dots at a specific point – the "dot density." Since these dots represent the location of an electron, the dot density at a point (x, y, z) is a measure of the relative probability of finding the electron in a volume element at that point: $f_p(x, y, z) = (\# \; of \; dots \; in \; volume \; element)/(volume \; of \; element)$. The relative probability is actually a density, and is therefore referred to as a *probability density.* $f_p(x, y, z)$ is a relative number since the number of dots in a volume element depends on the time interval over which the electron locations are determined.

To put the relative probability density, $f_p(x, y, z)$, on an absolute scale, we note that the probability of finding the electron somewhere in space is unity. We call the absolute probability density for an electron its *electron density,* $\rho(x, y, z)$. $\rho(x, y, z)$ is a measure of the probability per unit volume of finding an electron at a certain point in space. If we make the volume element infinitesimally small, with volume dV, then the probability of finding an electron in a specific volume element at (x, y, z) is the probability per unit volume times the volume, $\rho(x, y, z)\, dV$, $dV \equiv dxdydz$. For a given "experiment" $\rho(x, y, z) \,\alpha\, f_P(x, y, z) \Rightarrow \rho(x, y, z) = K f_P(x, y, z)$. The sum of the probabilities of finding the electron in each volume element must equal the overall probability of finding the electron somewhere in space, which is one. With infinitesimal volume elements the sum becomes an integral over all space:

$$\int_{-\infty}^{\infty}\int_{-\infty}^{\infty}\int_{-\infty}^{\infty} \rho(x, y, z)\, dxdydz = 1\,, \tag{3.36}$$

$$\int_{-\infty}^{\infty}\int_{-\infty}^{\infty}\int_{-\infty}^{\infty} K f_P(x, y, z)\, dxdydz = 1 \text{ and}$$

$$K = \frac{1}{\int_{-\infty}^{\infty}\int_{-\infty}^{\infty}\int_{-\infty}^{\infty} f_P(x, y, z)\, dxdydz}\,,$$

providing the constant that puts the probability density on an absolute scale. For a system with m electrons,

$$\int_{-\infty}^{\infty}\int_{-\infty}^{\infty}\int_{-\infty}^{\infty} \rho(x, y, z)\, dxdydz = m. \tag{3.37}$$

Assuming that the volume elements into which we have divided the crystal are very small, we can ignore the tiny amounts of destructive interference that would occur from scattering at different locations within the volume (we will soon correct this by making the volume elements infinitesimally small). Each volume element will have volume $\Delta V_{xyz} = \Delta x \Delta y \Delta z$. The probability, $\rho(x, y, z)\Delta V_{xyz}$, represents the fraction of time that an electron is present in the volume element ΔV_{xyz} at the location $p_j \equiv (x, y, z)$, and therefore represents the fraction of the photons scattered from that volume element. If, on the average, there is always an electron in ΔV_{xyz}, then $\rho(x, y, z)\Delta V_{xyz} = 1$ and the wave scattered from ΔV_{xyz} will be the same as that from a "fixed" electron, $\overrightarrow{\mathcal{E}_j} = \mathcal{E}_s e^{i\varphi_j}$. If, on the average, an electron is found only a third of the time in ΔV_{xyz}, then $\rho(x,y,z)\Delta V_{xyz} = 1/3$ and $\overrightarrow{\mathcal{E}_j} = (1/3)\mathcal{E}_s e^{i\varphi_j}$. *In general, the amplitude of the wave scattered from a volume element at point* $p_j \equiv (x,y,z)$ *is proportional to the probability of finding the electron in that volume element,* $\overrightarrow{\mathcal{E}_j} = \rho(x,y,z)\Delta V_{xyz}\mathcal{E}_s e^{i\varphi_j}$.

The scattered wave from n volume elements in a crystal can now be determined in the same manner as the scattered wave from n fixed electrons, since the relative phases depend only on the locations of the volume elements, and the amplitudes of the waves scattered from each element depends on the electron density in those elements. It follows from Eqn. 3.33 that the resultant wave scattered toward the detector from n selected volume elements is

$$\overrightarrow{\mathcal{E}_d} = \mathcal{E}_s \sum_{j=1}^{n} \rho_j \Delta V_j e^{i\varphi_j} = \mathcal{E}_s \sum_{j=1}^{n} \rho_j \Delta V_j e^{-2\pi i[\ \overrightarrow{op_j}\cdot(\mathbf{s_1}+\mathbf{d_1})\ ]/\lambda},$$

where $p_j = (x_j, y_j, z_j)$.

To determine the resultant wave scattered toward the detector *from the entire crystal* we make the volume elements very very small — and superposition the waves from every volume element in the crystal. The volume elements now have volume dV, and the sum becomes an integral over the volume of the crystal, V_x:

$$\overrightarrow{\mathcal{E}_x} = \mathcal{E}_s \int_0^{V_x} \rho_j e^{i\varphi_j}\, dV = \mathcal{E}_s \int_0^{V_x} \rho_j e^{-2\pi i[\ \overrightarrow{op_j}\cdot(\mathbf{s_1}+\mathbf{d_1})\ ]/\lambda}\, dV, \tag{3.38}$$

where $p_j = (x,y,z)$, $\rho_j = \rho(x,y,z)$, $\varphi_j = \varphi(x,y,z)$ and $dV = dxdydz$.

If we select a lattice point as the scattering reference, o, in the crystal, then a vector to the general point p_j, $\overrightarrow{op_j}$, can be described in terms of its location in the crystal using the unit cell basis:

$$\overrightarrow{op_j} = L_j\mathbf{a} + M_j\mathbf{b} + N_j\mathbf{c} + x_j\mathbf{a} + y_j\mathbf{b} + z_j\mathbf{c}, \tag{3.39}$$

where L_j, M_j, and N_j are integer increments from the scattering origin to the origin of the unit cell containing the point p_j, and x_j, y_j and z_j are the *fractional coordinates* of p_j in that unit cell.

The wave scattered toward the detector can now be described in terms of locations in the crystal:

$$\overrightarrow{\mathcal{E}_x} = \mathcal{E}_s \int_0^{V_x} \rho_j e^{-2\pi i[\ (L_j\mathbf{a}+M_j\mathbf{b}+N_j\mathbf{c}+x_j\mathbf{a}+y_j\mathbf{b}+z_j\mathbf{c})\cdot(\mathbf{s_1}+\mathbf{d_1})\ ]/\lambda}\, dV. \tag{3.40}$$

While the equation above uses a crystal coordinate system, with the selection of a more general coordinate system it effectively describes the scattering from any material. Diffraction from a crystalline material, however, demands constructive interference resulting from the periodic repetition of the scatterers in the lattice. This periodicity is reflected in identical unit cell contents — the electron density distribution in any given unit cell is identical to the electron density distribution of any other unit cell. The relative locations of the scattering electrons is identical in each cell, and it follows that the scattered wave vector from each unit cell will be identical to that from any other unit cell. These wave vectors will interfere with one another, depending on the relative location of each unit cell. If $\overrightarrow{\mathcal{E}_m}$ is the wave scattered from the mth unit cell, then for a crystal with K unit cells, these waves superposition to form the resultant wave propagating toward the detector:

$$\overrightarrow{\mathcal{E}_x} = \sum_{m=1}^{K} \overrightarrow{\mathcal{E}_m}. \tag{3.41}$$

Note that this is a vector sum, *not* a scalar sum.

The contribution to the overall scattering from the mth unit cell is determined by integrating the waves scattered from each point in the unit cell over the volume of the unit cell:

$$\begin{aligned}\overrightarrow{\mathcal{E}_m} &= \mathcal{E}_s \int_0^{V_c} \rho_j e^{-2\pi i[\ (L_j\mathbf{a}+M_j\mathbf{b}+N_j\mathbf{c}+x_j\mathbf{a}+y_j\mathbf{b}+z_j\mathbf{c})\cdot(\mathbf{s_1}+\mathbf{d_1})\ ]/\lambda}\, dV \\ &= \mathcal{E}_s \int_0^{V_c} \rho_j e^{-2\pi i[\ (L_j\mathbf{a}+M_j\mathbf{b}+N_j\mathbf{c})\cdot(\mathbf{s_1}+\mathbf{d_1})\ ]/\lambda} \\ &\qquad \times\ e^{-2\pi i[(x_j\mathbf{a}+y_j\mathbf{b}+z_j\mathbf{c})\cdot(\mathbf{s_1}+\mathbf{d_1})\ ]/\lambda}\, dV.\end{aligned}$$

L_j, M_j, and N_j are constants for a specific unit cell; there are no terms in the first exponential expression that depend on the fractional coordinates in the unit cell, and the expression can be factored out of the integral. The terms remaining inside the integral depend only on the fractional coordinates of the unit cell, *and are the same for every unit cell.* The subscripts can therefore be dropped inside the integral — we replace them with x_f, y_f, and z_f to emphasize that the integration variables are fractional coordinates referenced to the origin of the unit cell. *The result is the basic equation of X-ray diffraction:*

$$\begin{aligned}\overrightarrow{\mathcal{E}_m} &= \mathcal{E}_s e^{-2\pi i[\ (L_j\mathbf{a}+M_j\mathbf{b}+N_j\mathbf{c})\cdot(\mathbf{s_1}+\mathbf{d_1})\ ]/\lambda} \\ &\quad \times \int_0^{V_c} \rho(x_f, y_f, z_f) e^{-2\pi i[(x_f\mathbf{a}+y_f\mathbf{b}+z_f\mathbf{c})\cdot(\mathbf{s_1}+\mathbf{d_1})\ ]/\lambda}\, dV.\end{aligned} \tag{3.42}$$

The equation describing the scattering contribution of a general unit cell has been factored into a product of two expressions. In the analysis that follows we will discover that the first expression, when summed over the entire crystal, will give us the necessary conditions for diffraction. Indeed, the analysis will result in an independent derivation of the Laue/Bragg diffraction law! The first expression creates the *necessary* conditions for the observation of intensities at the detector, but *not sufficient* conditions. *The second expression, which depends on the contents of the unit cell, determines those conditions.* The first expression defines the criteria for the observation of diffraction maxima; the second expression determines the intensities that will be observed at those diffraction maxima. It is this second expression that we will use to ascertain the electron density distribution in the unit cell.

Each unit cell in the crystal is identified by the unique set of integers, L, M, and N, where L varies from L_{min} to L_{max}, M varies from M_{min} to M_{max}, and M varies from N_{min} to N_{max}. Eqn. 3.41 then becomes

$$\overrightarrow{\mathcal{E}_x} = \sum_L \sum_M \sum_N \overrightarrow{\mathcal{E}}_{LMN}. \tag{3.43}$$

The integral in Eqn. 3.42 is contained in every term in the sum and can be factored out. The wave scattered from the crystal is the superposition of the waves summed over all of the integers. The range of each of these integers is very large — *the sum will be over more that 10^{15} integers along each axis.* Thus the net resultant wave vector becomes

$$\begin{aligned}\overrightarrow{\mathcal{E}_x} &= \mathcal{E}_s \sum_L \sum_M \sum_N e^{-2\pi i[\ (L\mathbf{a}+M\mathbf{b}+N\mathbf{c})\cdot(\mathbf{s_1}+\mathbf{d_1})\ ]/\lambda} \\ &\times \int_0^{V_c} \rho(x_f, y_f, z_f) e^{-2\pi i[(x_f\mathbf{a}+y_f\mathbf{b}+z_f\mathbf{c})\cdot(\mathbf{s_1}+\mathbf{d_1})\ ]/\lambda}\, dV. \qquad (3.44)\end{aligned}$$

The sum in Eqn. 3.44 can be written as the sum of the products of three exponentials, which in turn can be written as the product of three sums (since the sum of products and the product of sums each generate a sum containing all possible combinations of the three exponentials):

$$\begin{aligned}&\sum_L \sum_M \sum_N e^{-2\pi i[(L\mathbf{a}+M\mathbf{b}+N\mathbf{c})\cdot(\mathbf{s_1}+\mathbf{d_1})]/\lambda} \\ &= \sum_L \sum_M \sum_N e^{-2\pi i[(L\mathbf{a})\cdot(\mathbf{s_1}+\mathbf{d_1})\ ]/\lambda} e^{-2\pi i[(M\mathbf{b})\cdot(\mathbf{s_1}+\mathbf{d_1})]/\lambda} e^{-2\pi i[(N\mathbf{c})\cdot(\mathbf{s_1}+\mathbf{d_1})]/\lambda} \\ &= \sum_L e^{-2\pi i[(L\mathbf{a})\cdot(\mathbf{s_1}+\mathbf{d_1})\ ]/\lambda} \sum_M e^{-2\pi i[(M\mathbf{b})\cdot(\mathbf{s_1}+\mathbf{d_1})\ ]/\lambda} \sum_N e^{-2\pi i[(N\mathbf{c})\cdot(\mathbf{s_1}+\mathbf{d_1})\ ]/\lambda}\end{aligned}$$

Recall that a major reason for constructing the reciprocal lattice was to allow for the treatment of interference phenomena in reciprocal space and scattering locations in direct space simultaneously. We are now in a position to do just that! The "interference vector" is composed of unit vectors toward the source and detector, *divided* by the wavelength — *a vector in reciprocal space.* We can therefore express it as a vector in terms of the reciprocal basis of our crystal:

$$\frac{\mathbf{s_1}+\mathbf{d_1}}{\lambda} = p\mathbf{a}^* + q\mathbf{b}^* + r\mathbf{c}^*, \qquad (3.45)$$

where p, q, and r are fixed for a given orientation of the X-ray source and detector since they are determined by $\mathbf{s_1}$ and $\mathbf{d_1}$.

Recalling that $\mathbf{a}\cdot\mathbf{a}^* = 1$, $\mathbf{a}\cdot\mathbf{b}^* = 0$, etc., Eqn. 3.44 becomes

$$\begin{aligned}\overrightarrow{\mathcal{E}_x} &= \mathcal{E}_s \sum_L e^{-2\pi i pL} \sum_M e^{-2\pi i qM} \sum_N e^{-2\pi i rN} \\ &\times \int_0^{V_c} \rho(x_f, y_f, z_f) e^{-2\pi i(px_f+qy_f+rz_f)}\, dV. \qquad (3.46)\end{aligned}$$

The exponential term in any one of the three sums in the equation can be viewed as a vector. For example $e^{-2\pi i pL} \equiv [(\cos(-2\pi pL))\ (\sin(-2\pi pL))]$ is a vector with magnitude 1 and phase angle $-2\pi pL$, where p depends on the orientation of the source and detector. Fig. 3.26 demonstrates the results of the addition of these vectors for p chosen arbitrarily as 1/9. The results should not be surprising. The scattering for an arbitrary orientation of the source and detector produces waves from one unit cell in the crystal that are out of phase with waves scattered from other unit cells. For such a configuration the three sums in the equation will add to zero — only background radiation will be observed at the detector. The observation is perfectly general. If the scattering origin is selected in the center of the lattice so that an equal number of positive and negative integers enter the sum, it will add

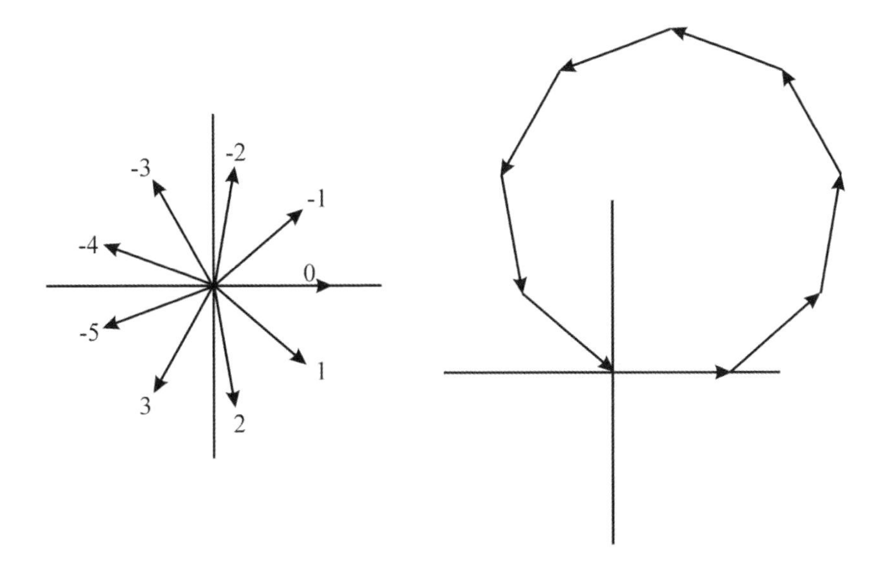

Figure 3.26 $e^{-2\pi ipL}$ vector plotted for $p = 1/9$, $L = 3$ to -5. The vector sum is shown on the right.

to zero for an arbitrarily selected value of p. As another example, the resultant vector with $p = 0.9$ and L varying in integer increments from -999 to 1000 has a magnitude of 0.000002.

Under what conditions will these sums not add to zero? Clearly, these will be the conditions that will allow for constructive interference of the scattered radiation. Suppose that p is an integer, say j. Then $jL = n$, another integer, generating the vectors $e^{-2\pi in} \equiv \cos(-2\pi n) + i\sin(-2\pi n) = 1$. All of the vectors will be in phase, pointing in the same direction. Since each value of L adds 1 to the sum, the magnitude of the resultant will be a very large number. An instructive example illustrates the sensitivity of this superposition (in double precision arithmetic). If $p = 0.999$ rather than 0.9 in the example given above, the resultant has a magnitude of 0.0003. For $p = 0.99999$ the resultant is 1998.6. It is not until $p = 0.999999$ that the value is 2000.

The same argument holds for the other two sums and the necessary conditions for diffraction from a crystal have now been determined: *the detector and source must be oriented so that the components of the vector* $(\mathbf{s_1} + \mathbf{d_1})/\lambda$ in the reciprocal cell basis are integers:

$$\frac{\mathbf{s_1} + \mathbf{d_1}}{\lambda} = j\mathbf{a}^* + j'\mathbf{b}^* + j''\mathbf{c}^*$$

This is a vector that we have already encountered! It is a reciprocal lattice vector from the origin to a point in the lattice, perpendicular to a set of planes in the direct lattice with indices identical to the integer components of the vector. We can immediately identify j, j', and j'' with the indices of these planes:

$$\frac{\mathbf{s_1} + \mathbf{d_1}}{\lambda} = h\,\mathbf{a}^* + k\,\mathbf{b}^* + l\,\mathbf{c}^* = \mathbf{h}_{hkl} \quad \text{and} \tag{3.47}$$

$$\mathbf{s_1} + \mathbf{d_1} = \lambda\,\mathbf{h}_{hkl}. \tag{3.48}$$

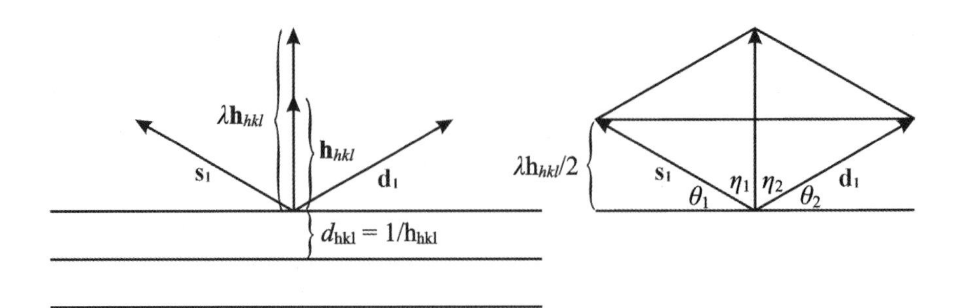

Figure 3.27 Orientation of unit vectors toward the source and detector to align their vector sum with the reciprocal lattice vector, $\mathbf{h}_{hkl}$.

Eqn. 3.48 defines the conditions for diffraction based on the reciprocal lattice, first formulated by Max von Laue.[39] They are often referred to as *the Laue conditions. To observe diffraction we must move the radiation source and detector to positions where the vector sum of* $\mathbf{s_1}$ *(pointing toward the source) and* $\mathbf{d_1}$ *(pointing toward the detector) is parallel to* $\mathbf{h}_{hkl}$*, the reciprocal lattice vector perpendicular to the* $(h\ k\ l)$ *planes in the direct lattice!* Referring to Fig. 3.27, since the unit vectors have the same length, the parallelogram that generates their sum, $\lambda\mathbf{h}_{hkl}$, must be a rhombus in which the resultant vector bisects the angle between them. Thus $\eta_1 = \eta_2 = \eta$. Since $\lambda\mathbf{h}_{hkl}$ is perpendicular to the $(h\ k\ l)$ planes, $\theta_1 = \theta_2 = (\pi/2) - \eta$. In addition, the line segment that connects the heads of the two unit vectors is a diagonal for the rhombus, and therefore bisects the other diagonal, $\lambda\mathbf{h}_{hkl}$. It follows that

$$\sin\theta_{hkl} = \frac{\lambda \mathrm{h}_{hkl}/2}{s_1} = \frac{\lambda \mathrm{h}_{hkl}}{2} \quad \text{and}$$

$$\lambda \mathrm{h}_{hkl} = 2\sin\theta_{hkl}. \tag{3.49}$$

$$\text{Since } \mathrm{h}_{hkl} = \frac{1}{d_{hkl}},\ \frac{\lambda}{d_{hkl}} = 2\sin\theta_{hkl} \quad \text{and}$$

$$\lambda = 2d_{hkl}\sin\theta_{hkl}. \tag{3.50}$$

Eqns. 3.49 and 3.50 are expressions of the Bragg diffraction law[40] in terms of the length of the reciprocal lattice vector* and the interplanar spacings, respectively, corresponding to the planes in the lattice with indices $(h\ k\ l)$. *The Bragg law was derived here assuming only that the electron density distribution in each unit cell is identical to the electron density in any other cell.* It follows that the electron density distribution along each set of planes will be periodic. Diffraction from these planes is more complex, but directly analogous to the two dimensional rows of point scatterers discussed previously, in which each set of lines in the direct lattice gave rise to diffraction, observed in an image of the reciprocal lattice (Figs. 3.19 and 3.20).

When diffraction conditions are met, a beam of photons is directed toward a set of lattice planes at the angle θ_{hkl}, and a beam of photons is subsequently scattered at the same angle away from the planes. It is as if the lattice planes were

*Eqn. 3.49 is a summary of the Laue diffraction conditions in reciprocal space, which is equivalent to Eqn. 3.50 in direct space. It would probably be more appropriate to characterize the equations as the "Laue/Bragg" law, but common usage refers to either equation as "the Bragg Law."

acting as mirrors, with X-rays being reflected from them. In many texts discussing crystallography the Bragg law is derived by invoking this notion of reflection (See Appendix A). While the derivations are simple and straightforward, they do little justice to the underlying basis of the law. However, most early treatments describe diffraction phenomena as reflections from planes; the beam of photons resulting from diffraction from a given set of planes was described as a reflection of the incident beam from those planes. This terminology remains in general use today; we will use the term *reflection* to refer to the photons observed at diffraction maxima.

Under diffraction conditions each exponential term in the three sums in Eqn. 3.46 will contribute 1 to the sum. The result of summing over the entire crystal creates the product: $(L_{max} - L_{min})(M_{max} - M_{min})(N_{max} - N_{min})$. Since the three terms depend on the number of unit cells along each crystallographic axis, the product scales with the crystal volume. Thus the intensity observed for a given reflection depends directly on the volume of the crystal. Recall that the electron scattering factor, the fraction of the incident radiation scattered by an electron toward the detector, was given by $f_e = (\mathcal{E}_s/\mathcal{E}_I)$. f_e represents the scattering power of an electron and is independent of the intensity of the incident beam. Thus if $\mathcal{E}_I$ increases, $\mathcal{E}_s$ increases proportionally. The intensity observed at the detector therefore depends directly on the incident beam intensity as well. Substituting $f_e\mathcal{E}_{\mathcal{I}}$ for $\mathcal{E}_s$ reduces Eqn. 3.46 to

$$\begin{aligned}\overrightarrow{\mathcal{E}_x} &= f_e\mathcal{E}_I(L_{max} - L_{min})(M_{max} - M_{min})(N_{max} - N_{min}) \\ &\quad \times \int_0^{V_c} \rho(x_f, y_f, z_f)e^{-2\pi i(hx_f + ky_f + lz_f)}\, dV. \end{aligned} \tag{3.51}$$

For a given experiment, the size of the crystal and the intensity of the incident radiation are constant and the terms outside of the integral in the equation are either zero if diffraction conditions are not met, or a constant, D, when they are met:

$$\overrightarrow{\mathcal{E}_x} = D\int_0^{V_c} \rho(x_f, y_f, z_f)e^{-2\pi i(hx_f + ky_f + lz_f)}\, dV. \tag{3.52}$$

In order to observe a reflection, the source and detector are adjusted until D is nonzero. Since D is a constant for all sets of planes, the amplitude of this wave depends only on the value of the integral in Eqn. 3.52, which can be seen to be a superposition of waves scattered from every volume element in the unit cell, weighted by the electron density of each volume element. Thus the amplitude depends on the electron density distribution in the unit cell and the indices of the planes from which the diffraction is occurring. The indices dictate the direction of the beam entering the unit cell. The inner product[†] of the reciprocal lattice vector $\mathbf{h}_{hkl} = [h\ k\ l]$ with the direct lattice vector, $\mathbf{r}_f = [x_f\ y_f\ z_f]$, generates the relative phase of a wave scattered from the volume element at (x_f, y_f, z_f). The electron density at that location determines the amount of scattering, the intensity, of the wave scattered from there. The more electron density at a specific location, the greater the contribution of the wave to the overall superposition. Thus the electron distribution determines the relative interference resulting from unequal scattering

[†]See Sec. 1.3.4 for a discussion of the distinction between scalar product and inner product adopted here.

from different regions in the cell. *The intensity of the reflection from each set of planes contains information about the electron density distribution in the unit cell — and it is our task to extract that information.*

Finally, *the diffraction equation* is conveniently represented by replacing the phase term in the exponential with the inner product of the position vector in direct space and the *diffraction vector** in reciprocal space:

$$\vec{\mathcal{E}}_{\mathbf{h_{hkl}}} = D \int_0^{V_c} \rho(\mathbf{r}_f) e^{-2\pi i(\mathbf{h}_{hkl} \cdot \mathbf{r}_f)} \, dV. \tag{3.53}$$

The $\mathbf{h}_{hkl}$ subscript for the net electric field *vector* representing the diffracted wave indicates that the amplitude and phase (relative to a hypothetical wave scattered from the lattice origin) of the diffracted beam from a given set of planes will be unique to that set of planes and determined by the value of the integral corresponding to those planes.

3.2.3 X-ray Diffraction: The Electron Density Equation.

The diffraction equation allows for the calculation of the amplitudes and phases of the diffraction maxima for all of the possible reflections in the diffraction pattern from the electron density function, $\rho(x_f, y_f, z_f)$, for a *single unit cell.* While this will be an important exercise in the process of solving a crystal structure, *our goal is to reverse the process — to calculate the electron density function from the diffraction pattern.*

Unfortunately, unlike the waves in the diffraction pattern, which are described by trigonometric functions, there is no general analytical form for the electron density function. This is a common problem in mathematics and is often solved by representing the function as an infinite series of known functions. The electron density function repeats itself periodically (just like the diffraction waves), with the period identified with the unit cell dimensions. An infinite series representation of $\rho(x_f, y_f, z_f)$ would therefore necessarily involve periodic functions such as the sine and cosine. From the mid-1700s to the early 1800s a number of famous mathematicians, including Leonhard Euler, Jean Le Rond d'Alembert, and Joseph Fourier investigated the expansion of specific functions as trigonometric series. In 1807 Fourier proposed the trigonometric series expansion of general functions in order to describe the mathematics of heat transfer.[42] Although the theory was harshly criticized by the mathematicians of the day, including Fourier's mentor, Joseph Lagrange, the *Fourier series* is one of today's most powerful and extensively utilized mathematical constructs.

In order to develop the Fourier series expansion of the electron density function we begin in one dimension and consider a series of cosine functions, $\cos(n\pi x)$, where x varies from -1 to 1 and n is an integer varying from $-\infty$ to ∞. Suppose that we wish to use a series of cosine functions to represent a periodic function $f(x')$ with $-d <= x' < d$ defining the period of the function (the repeating "unit" lies between $-d$ and d). In order to match the period of $f(x')$ with the range of x for the cosine functions, we define $x = x'/d$ so that $-1 < x < 1$ is the period for $f(x)$. By adding cosine functions of varying wavelengths *and amplitudes* we

*The reciprocal lattice vector for a reflection is traditionally described as the *diffraction vector*. The vector in the direction of the diffracted beam is referred to as the *reflection vector*.

hope to model the function. When n is small, the cosine wavelengths are long, and the overall gross structure of the function (its "envelope") is approximated. As n increases, the cosine functions introduce more "wiggles" and increasingly finer features of the function are added in as the wavelengths of the cosine functions decrease. In principle, an infinite number of cosines should reproduce the function exactly, providing that its periodic behavior is the same as the periodic behavior of the cosine functions.

The cosine function is a periodic *even function.* An even function is symmetric with respect to the origin, that is, $f(x) = f(-x)$. It follows that any superposition of cosine functions will be an even function, and we assert that an even function, f_{ev}, can be represented by an infinite series of cosine functions:

$$f_{ev}(x) = \sum_{n=-\infty}^{\infty} A_n \cos(n\pi x) \tag{3.54}$$

In order to create the series, we must determine the weighting factors, A_n, from $f_{ev}(x)$. To do this we multiply both sides of Eqn. 3.54 by $\cos(m\pi x)$, where m is an integer, and integrate both sides of the equation over the range $x = -1$ to $x = 1$:

$$\begin{aligned} \int_{-1}^{1} f_{ev}(x)\cos(m\pi x)\,dx &= \int_{-1}^{1} \left(\cos(m\pi x) \sum_{n=-\infty}^{\infty} A_n \cos(n\pi x) \right) dx \\ &= \sum_{n=-\infty}^{\infty} A_n \int_{-1}^{1} \cos(m\pi x)\cos(n\pi x)\,dx. \end{aligned} \tag{3.55}$$

To evaluate the integral on the right we make use of the following identities (the proofs are readily found in elementary textbooks):

$$\int \cos(u)\,du = \sin(u) + C \tag{3.56}$$

$$\int \cos^2(u)\,du = \frac{u}{2} + \frac{\sin 2u}{4} + C \tag{3.57}$$

$$\cos\alpha\cos\beta = \frac{1}{2}\cos(\alpha+\beta) + \frac{1}{2}\cos(\alpha-\beta). \tag{3.58}$$

Substituting the formula for the cosine product and setting $k = m+n$ and $l = m-n$ $(m \neq n)$ gives

$$\begin{aligned} &\int_{-1}^{1} \cos(m\pi x)\cos(n\pi x)\,dx \\ &= \frac{1}{2}\int_{-1}^{1} \cos((m+n)\pi x)\,dx + \frac{1}{2}\int_{-1}^{1} \cos((m-n)\pi x)\,dx \\ &= \frac{1}{2}\left(\frac{\sin k\pi x}{k\pi} + \frac{\sin l\pi x}{l\pi} \right)\Bigg|_{-1}^{1} = 0 + 0 - 0 - 0 = 0, \quad m \neq n. \end{aligned}$$

If m and n are not the same then the corresponding term in the sum is zero. If $m = n$ then:

$$\begin{aligned} &\int_{-1}^{1} \cos(n\pi x)\cos(n\pi x)\,dx = \int_{-1}^{1} \cos^2(n\pi x)\,dx \\ &= \frac{1}{n\pi}\left(\frac{n\pi x}{2} + \frac{\sin(2n\pi x)}{4} \right)\Bigg|_{-1}^{1} = \frac{1}{2} + 0 - \left(-\frac{1}{2}\right) - 0 = 1, \quad m = n. \end{aligned}$$

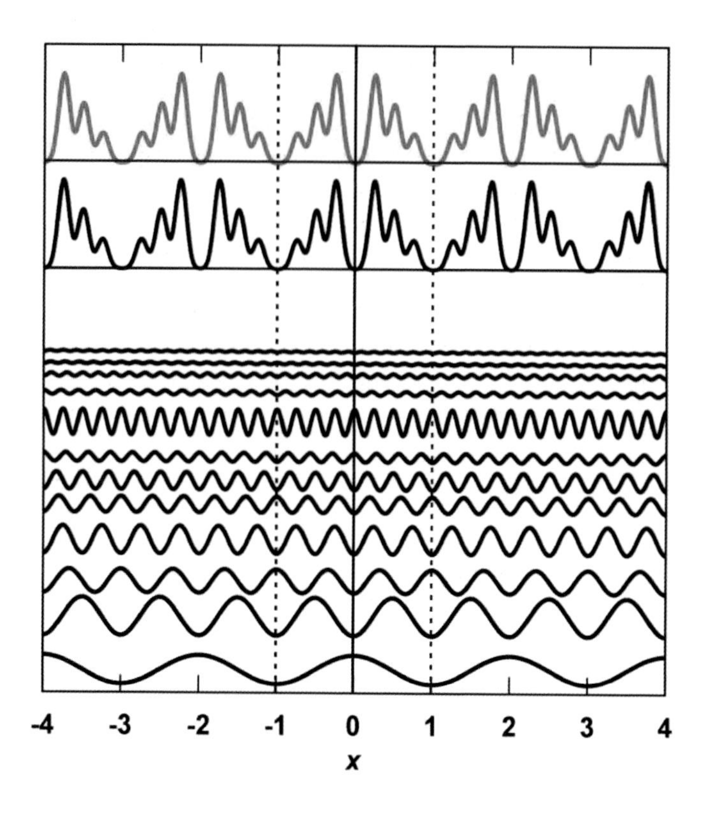

Figure 3.28 Cosine series representation of an even periodic function. The function is plotted in red. The individual cosine functions and their sum are plotted in black.

The only term that survives in the sum is where $m=n$ and

$$\int_{-1}^{1} f_{ev}(x)\cos(n\pi x)\,dx = A_n \int_{-1}^{1} \cos^2(n\pi x)\,dx;$$

$$A_n = \int_{-1}^{1} f_{ev}(x)\cos(n\pi x)\,dx. \tag{3.59}$$

This makes perfect sense! The integral is a measure of the "overlap" of the two functions. If, for the same values of x, a given cosine function has values similar to those of f_{ev}, then the product of those values will tend to add "constructively" to the integral, while a cosine function with little in common with f_{ev} will generate products which tend to cancel one another. In a loose sense, the cosine function is weighted in proportion to how much it has in common with the function that we are attempting to model.

Fig. 3.28 is an example of the cosine series representation of an even function. The function, shown in red, was created using a sum of six Gaussian functions taking the form $f_{ev}(x) = \sum_{i=1}^{6} C_i e^{(x-X_i)^2}$ over the interval $-1 < x < 1$, where X_i locates the Gaussian's at $X_i = -0.75, -0.5, -0.25, 0.25, 0.5,$ and 0.75 with magnitudes given by $C_i = 5, 10, 15, 15, 10,$ and 5, respectively. The values of A_n were determined by numerical integration over the interval between -1 and 1, indicated by the dashed lines in the figure. The contributing cosine functions, $A_n \cos(n\pi x)$ are plotted in black, along with their sum, illustrating clearly that a *finite* number of terms in the sum are sufficient to model this relatively simple one-dimensional function.

The electron density distribution is an even function if the unit cell has a center of symmetry, since $\rho(x_f, y_f, z_f) = \rho(-x_f, -y_f, -z_f)$. *If the unit cell is centrosymmetric we should be able to extrapolate the treatment above to three dimensions and model the electron density function using a cosine series expansion.* However, the electron density distribution for a non-centrosymmetric unit cell is *not* an even function, and cannot be represented by a cosine series. In order to resolve this dilemma we consider a trigonometric series of sine functions. The sine function is an odd function, antisymmetric with respect to the origin, such that $f(x) = -f(-x)$. The same arguments invoked for the cosine series can be invoked for the sine series; an odd function can be represented as an infinite series of sine functions:

$$f_{od}(x) = \sum_{n=-\infty}^{\infty} B_n \sin(n\pi x) \tag{3.60}$$

To determine the weighting factors, B_n, from $f_{od}(x)$ we multiply both sides of Eqn. 3.60 by $\sin(m\pi x)$ and integrate both sides of the equation over the range $x = -1$ to $x = 1$:

$$\begin{aligned} \int_{-1}^{1} f_{od}(x) \sin(m\pi x)\, dx &= \int_{-1}^{1} \left(\sin(m\pi x) \sum_{n=-\infty}^{\infty} B_n \sin(n\pi x) \right) dx \\ &= \sum_{n=-\infty}^{\infty} B_n \int_{-1}^{1} \sin(m\pi x) \sin(n\pi x)\, dx. \end{aligned} \tag{3.61}$$

Employing the same approach as before, we use the following identities to evaluate the integral on the right:

$$\int \sin(u)\, du = -\cos(u) + C \tag{3.62}$$

$$\int \sin^2(u)\, du = \frac{u}{2} - \frac{\sin 2u}{4} + C \tag{3.63}$$

$$\sin\alpha \sin\beta = \frac{1}{2}\cos(\alpha - \beta) - \frac{1}{2}\cos(\alpha + \beta). \tag{3.64}$$

The reader can easily verify that the integral is zero when $m \neq n$, and unity when $m = n$. By identical arguments,

$$\begin{aligned} \int_{-1}^{1} f_{od}(x) \sin(n\pi x)\, dx &= B_n \int_{-1}^{1} \sin^2(n\pi x)\, dx; \\ B_n &= \int_{-1}^{1} f_{od}(x) \sin(n\pi x)\, dx. \end{aligned} \tag{3.65}$$

Fig. 3.29 is an example of the sine series representation of an odd function. The function, shown in red, was created using the previous sum-of-Gaussians function for the cosine series and changing the sign of the Gaussian weighting coefficients for the negative values of X_i: $f_{od}(x) = \sum_{i=1}^{6} C_i e^{(x - X_i)}$, $X_i = -0.75, -0.5, -0.25, 0.25, 0.5, 0.75$ and $C_i = -5, -10, -15, 15, 10, 5$. As before the values of B_n were

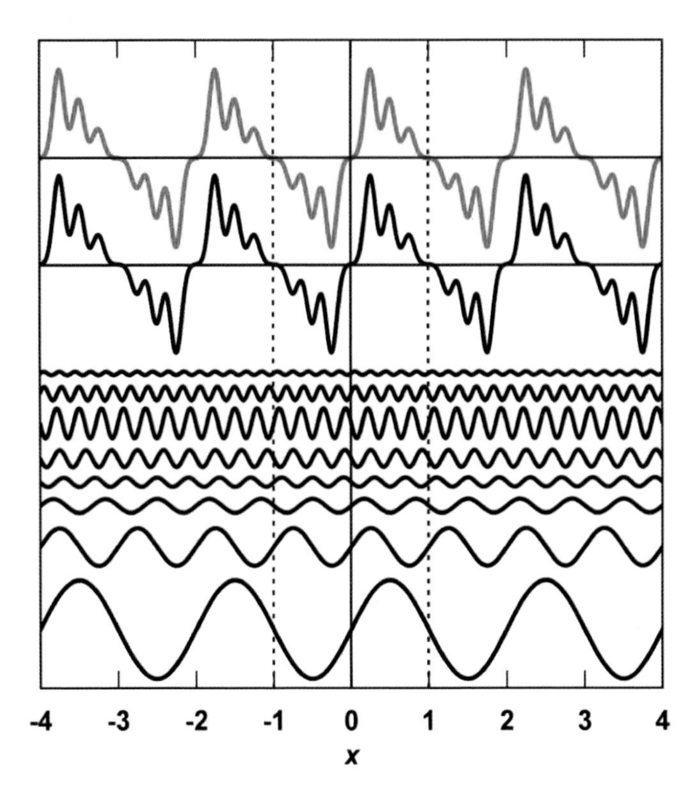

Figure 3.29 Sine series representation of an odd periodic function. The function is plotted in red. The individual sine functions and their sum are plotted in black.

determined by numerical integration over the interval between -1 and 1, indicated by the dashed lines in the figure. The contributing sine functions, $B_n \sin(n\pi x)$ are plotted in black, along with their sum.

Both even and odd functions can be represented with trigonometric series, the first using cosine functions, the second using sine functions. However, the electron density distribution in a non-centrosymmetric unit cell is neither symmetric *nor* antisymmetric with respect to the origin – *not even* – *not odd*! Fourier argued that *any* periodic function could be represented by an infinite series of sine and cosine functions. The general function, $f(x)$, can always be expressed as a sum of an even and odd function:

$$\begin{aligned} f(x) &= \frac{1}{2}f(x) + \frac{1}{2}f(x) = \frac{1}{2}f(x) + \frac{1}{2}f(x) + \frac{1}{2}f(-x) - \frac{1}{2}f(-x) \\ &= \frac{1}{2}[f(x) + f(-x)] + \frac{1}{2}[f(x) - f(-x)]. \end{aligned}$$

Let $g(x) = \frac{1}{2}[f(x) + f(-x)]$ and $h(x) = \frac{1}{2}[f(x) - f(-x)]$. Then $g(-x) = \frac{1}{2}[f(-x) + f(x)] = g(x)$ and $h(-x) = \frac{1}{2}[f(-x) - f(x)] = -h(x)$. $g(x)$ is an even function and

$h(x)$ is an odd function; $f(x) = g(x) + h(x)$, and can be represented by the sum of a cosine series and a sine series, known as a *Fourier series*:

$$f(x) = \sum_{n=-\infty}^{\infty} A_n \cos(n\pi x) + \sum_{n=-\infty}^{\infty} B_n \sin(n\pi x). \tag{3.66}$$

The electron density function in one dimension is periodic, with the length of the period equal to the axial length, a. In this case the period of the function would be $-a/2 < x' < a/2$ with $x = x'/(a/2) = 2x'/a$, the period covering half a unit cell on each side of the origin. Since x'/a is the fractional coordinate, x_f, the independent variable is $x = 2x_f$. Thus, in terms of the fractional coordinates of the one-dimensional unit cell,

$$\rho(x_f) = f(2x_f) = \sum_{n=-\infty}^{\infty} A_n \cos(2\pi n x_f) + \sum_{n=-\infty}^{\infty} B_n \sin(2\pi n x_f). \tag{3.67}$$

$\rho(x_f)$ is the Fourier series representation of a periodic function in which the independent variable is the fractional coordinate in the unit cell. It is the function that we now employ to model the electron density in the unit cell. Fig. 3.30 illustrates the Fourier series representation of the electron density in a hypothetical non-centrosymmetric one-dimensional unit cell containing the BeClF molecule as the asymmetric unit. Sines and cosines of increasingly higher frequency are added from bottom to top in the figure.

In order to get Eqn. 3.67 into exponential form, we again make use of Euler's relation (Eqn. 3.31):

$$e^{i\phi} = \cos\phi + i\sin\phi$$

$$e^{-i\phi} = \cos(-\phi) + i\sin(-\phi) = \cos\phi - i\sin\phi \tag{3.68}$$

$$e^{i\phi} + e^{-i\phi} = 2\cos\phi \Longrightarrow \cos\phi = \frac{e^{i\phi} + e^{-i\phi}}{2} \tag{3.69}$$

$$\begin{aligned} e^{i\phi} - e^{-i\phi} = 2i\sin\phi \Longrightarrow \sin\phi &= \frac{e^{i\phi} - e^{-i\phi}}{2i} = \frac{(e^{i\phi} - e^{-i\phi})i}{2i^2} \\ &= i\frac{(e^{-i\phi} - e^{i\phi})}{2} \end{aligned} \tag{3.70}$$

$$\begin{aligned} \rho(x_f) &= \sum_{n=-\infty}^{\infty} [A_n \cos(2\pi n x_f) + B_n \sin(2\pi n x_f)] \\ &= \sum_{n=-\infty}^{\infty} \left[\frac{1}{2}A_n(e^{-2\pi i n x_f} + e^{2\pi i n x_f}) + \frac{1}{2}iB_n(e^{-2\pi i n x_f} - e^{2\pi i n x_f})\right] \end{aligned}$$

$$\rho(x_f) = \sum_{n=-\infty}^{\infty} \left[\frac{1}{2}(A_n + iB_n)e^{-2\pi i n x_f} + \frac{1}{2}(A_n - iB_n)e^{2\pi i n x_f}\right]. \tag{3.71}$$

In the superposition of wave vectors, the vectors were conveniently represented by complex numbers (Eqns. 3.30–3.34). Conversely, complex numbers are conveniently represented by vectors, as illustrated in Fig. 3.31. $A_n + iB_n$ and $A_n - iB_n$ are complex conjugates of one another: $A_n + iB_n \equiv [(A_n)\ (B_n)] = \vec{C}_n$ and $A_n - iB_n \equiv [(A_n)\ (-B_n)] = \vec{C}_n^*$. $C_n = (A_n^2 + B_n^2)^{1/2}$ is the magnitude of both

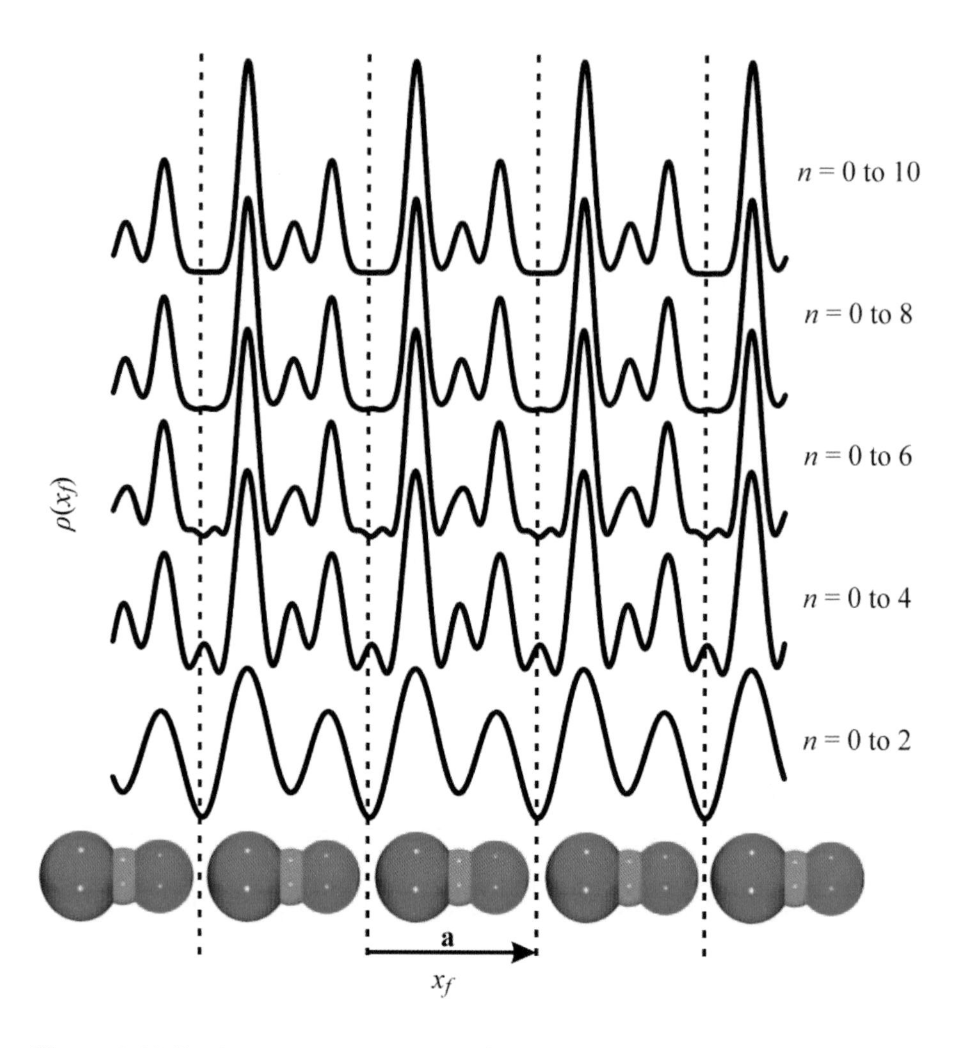

Figure 3.30 Fourier series representation of the electron density in a hypothetical one-dimensional crystal containing the linear BeClF molecule as the asymmetric unit of the unit cell. n is the index of the cosine *and* sine terms in the Fourier sums. As n increases the frequencies of the sinusoidal components increase.

vectors. The phase angle of $\overrightarrow{C_n}$ is $\varphi_n = \arctan(B_n/A_n)$; the phase angle of $\overrightarrow{C}^*_n$ is $\arctan(-B_n/A_n) = -\varphi_n$. The electron density function can now be expressed in terms of a superposition of these vectors:

$$\rho(x_f) = \frac{1}{2}\sum_{n=-\infty}^{\infty} \vec{C}_n e^{-2\pi i n x_f} + \frac{1}{2}\sum_{n=-\infty}^{\infty} \vec{C}^*_n e^{2\pi i n x_f}. \tag{3.72}$$

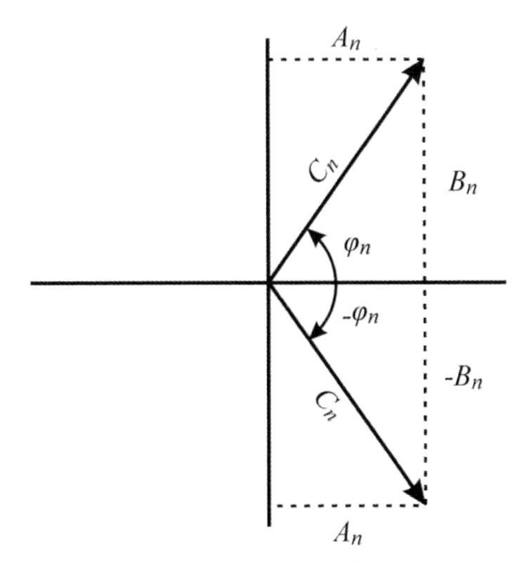

Figure 3.31 Vector representation of the complex numbers $A_n + iB_n$ and $A_n - iB_n$.

Since the sums are over the range $n = -\infty$ to $n = \infty$, for every n there is a $-n$:

$$\begin{aligned} A_n &= \int \rho(x_f)\cos(2\pi n x_f)\,dx_f \\ A_{-n} &= \int \rho(x_f)\cos(-2\pi n x_f)\,dx_f = \int \rho(x_f)\cos(2\pi n x_f)\,dx_f = A_n \\ B_n &= \int \rho(x_f)\sin(2\pi n x_f)\,dx_f \\ B_{-n} &= \int \rho(x_f)\sin(-2\pi n x_f)\,dx_f = -\int \rho(x_f)\sin(2\pi n x_f)\,dx_f = -B_n \end{aligned}$$

For every $|n|$, the first sum in Eqn. 3.72 will contain the terms $(A_n + iB_n)e^{-2\pi n x_f}$ for n, and $(A_n - iB_n)e^{2\pi n x_f}$ for $-n$. The second sum will contain the terms $(A_n - iB_n)e^{2\pi n x_f}$ for n, and$(A_n + iB_n)e^{-2\pi n x_f}$ for $-n$. Both sums are identical, and

$$\rho(x_f) = \sum_{n=-\infty}^{\infty} \vec{C}_n e^{-2\pi i n x_f} = \sum_{n=-\infty}^{\infty} \vec{C}^*_n e^{2\pi i n x_f}. \tag{3.73}$$

The expansion of the Fourier series to two dimensions will require sinusoidal functions in two dimensions. These functions take the form $\cos(m\pi x + n\pi y)$ and $\sin(m\pi x + n\pi y)$, where $x = 2x_f$ and $y = 2y_f$ and are waves which propagate in specific directions in the xy plane. To consider how these waves propagate we consider the cosine function in the special cases where $x = 0$ and $y = 0$. When $x = 0$ the function along the y axis is $\cos(n\pi y)$, with the period defined by n; n cycles of the function will occur between $y = -1$ and $y = 1$. This is identical to the cosine functions in the one-dimensional cosine series. Similarly, when $y = 0$, the function is $\cos(m\pi x)$ along the x-axis, with m cycles occurring between $x = -1$ and $x = 1$. If the points on the axes where a minimum or maximum of the function occurs are connected with a line, the direction of propagation of the wave will be perpendicular to the line. Fig. 3.32 illustrates two-dimensional cosine waves for

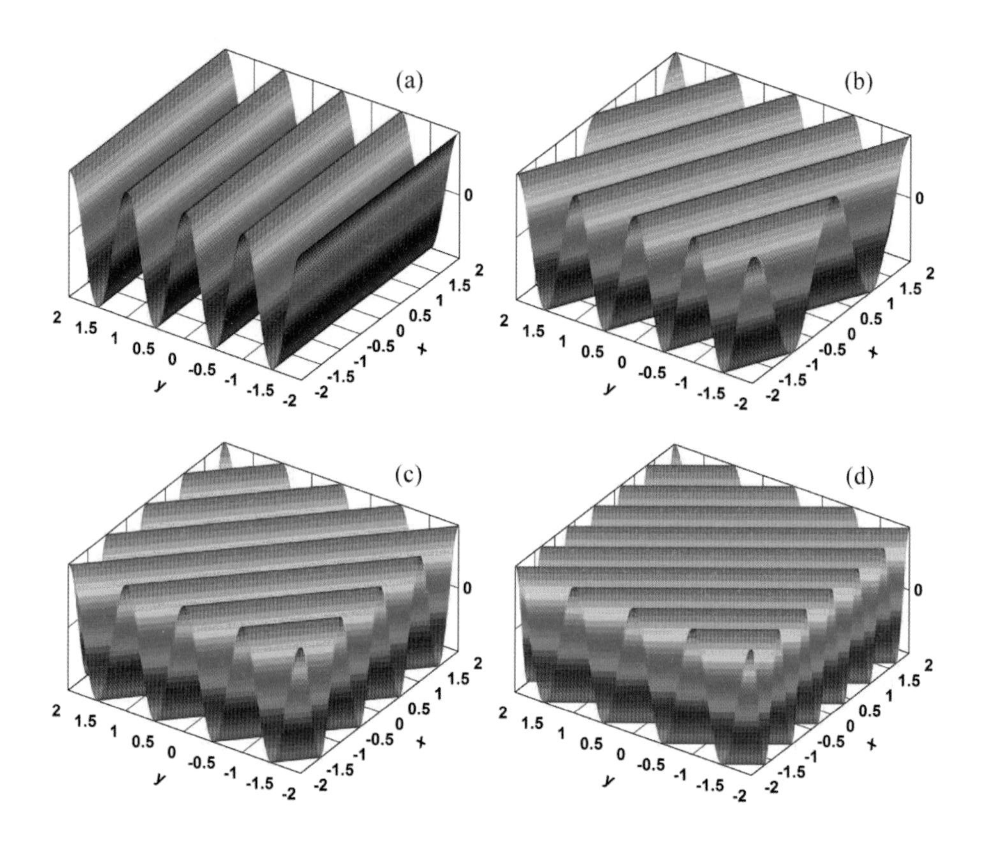

Figure 3.32 Two-dimensional cosine waves: (a) $f(x,y) = \cos(\pi(0x + 2y))$, (b) $f(x,y) = \cos(\pi(1x + 2y))$, (c) $f(x,y) = \cos(\pi(2x + 2y))$, and (d) $f(x,y) = \cos(\pi(3x + 2y))$; $x = 2x_f$ and $y = 2y_f$.

$n = 2$ and m varying from 0 to 3. As m and n change the periods along each axis change, altering the frequency of the cosine wave and its direction of propagation — which will be determined by the ratio of m and n. Since m and n take on values from minus infinity to infinity, every possible direction will be covered. For all cosine waves with the same $m : n$ ratio, there will be a complete set of waves of increasing frequency corresponding to m, n, $2m, 2n$, $3m, 3n$ and so forth, all propagating in the same direction — there is a cosine series for every direction in the unit cell. The same arguments hold for the sine series, and the Fourier series representation for the electron density in a two-dimensional crystal is

$$\rho(x_f, y_f) = \sum_{m=-\infty}^{\infty} \sum_{n=-\infty}^{\infty} A_{mn} \cos(2\pi(mx_f + ny_f)) + \sum_{m=-\infty}^{\infty} \sum_{n=-\infty}^{\infty} B_{mn} \sin(2\pi(mx_f + ny_f)). \tag{3.74}$$

In exponential form,

$$\rho(x_f, y_f) = \sum_{m=-\infty}^{\infty} \sum_{n=-\infty}^{\infty} \vec{C}_{mn} e^{-2\pi i(mx_f + ny_f)}. \tag{3.75}$$

The extrapolation to the three-dimensional electron density function is straightforward:

$$\begin{aligned}\rho(x_f, y_f, z_f) &= \sum_{m=-\infty}^{\infty} \sum_{n=-\infty}^{\infty} \sum_{o=-\infty}^{\infty} A_{mno} \cos(2\pi(mx_f + ny_f + oz_f)) \\ &+ \sum_{m=-\infty}^{\infty} \sum_{n=-\infty}^{\infty} \sum_{o=-\infty}^{\infty} B_{mno} \sin(2\pi(mx_f + ny_f + oz_f)) \qquad (3.76) \\ &= \sum_{m=-\infty}^{\infty} \sum_{n=-\infty}^{\infty} \sum_{o=-\infty}^{\infty} \vec{C}_{mno} e^{-2\pi i(mx_f + ny_f + oz_f)}. \qquad (3.77)\end{aligned}$$

Now that we have a functional form for the electron density, we return to the diffraction equation, Eqn. 3.53. The magnitude of the electric field vector depends on the intensity of the incident beam and the volume of the crystal, manifest in the constant multiplier, D. The net wave is obtained by multiplying D by an integral evaluated over the unit cell electron density. It is from this integral that we wish to extract the electron density function and determine the crystal structure. The integral, which is a wave vector — a superposition of the waves scattered from a unit cell — is denoted $\vec{F}_{hkl}$, and is known as the *structure factor* for the $(h\ k\ l)$ reflection:

$$\begin{aligned}\vec{F}_{hkl} &= \frac{\vec{\mathcal{E}}_{hkl}}{D} \\ &= \int_0^{V_c} \rho(x_f, y_f, z_f) e^{-2\pi i(hx_f + ky_f + lz_f)}\, dV. \qquad (3.78)\end{aligned}$$

Note that dividing the experimental electric field vector by D renders the structure factor vector independent of the incident beam intensity and size of the crystal.

Substituting the Fourier series representation for the electron density into the structure factor expression gives

$$\begin{aligned}\vec{F}_{hkl} &= \sum_{m=-\infty}^{\infty} \sum_{n=-\infty}^{\infty} \sum_{o=-\infty}^{\infty} \int_0^{V_c} \vec{C}_{mno} e^{-2\pi i(mx_f + ny_f + oz_f)} e^{-2\pi i(hx_f + ky_f + lz_f)}\, dV \\ &= \sum_{m=-\infty}^{\infty} \sum_{n=-\infty}^{\infty} \sum_{o=-\infty}^{\infty} \vec{C}_{mno} \int_0^{V_c} e^{-2\pi i[(m+h)x_f + (n+k)y_f + (o+l)z_f]}\, dV. \qquad (3.79)\end{aligned}$$

If the unit cell axes are orthogonal we can evaluate the integral in each term in the sum by replacing the fractional coordinates with their Cartesian equivalents and setting $L = m + h$, $M = n + k$ and $N = o + l$:

$$\vec{F}_{hkl} = \sum_{m=-\infty}^{\infty} \sum_{n=-\infty}^{\infty} \sum_{o=-\infty}^{\infty} C_{mno} \int_0^a \int_0^b \int_0^c e^{-2\pi i[L(x_c/a) + M(y_c/b) + N(z_c/c)]} dx_c dy_c dz_c,$$

since $dV = dx_c dy_c dz_c$. If the unit cell axes are not orthogonal, then dV will not be equal to $dx_c dy_c dz_c$, but we can scale dV by noting that the ratio of the actual volume differential, dV, to the product of the differentials of the axial lengths, $dx_c dy_c dz_c$ (the orthogonal differential volume), is the same as the ratio of the actual volume, V_c, to the product of axial lengths, abc (the orthogonal volume): $dV/(dx_c dy_c dz_c) = V_c/(abc)$ and $dV = V_c(dx_c dy_c dz_c)/(abc)$. Note that this gives $dV = dx_c dy_c dz_c$ for an orthogonal unit cell. Evaluation of the integral for a general term in the sum becomes:

$$\int_0^a \int_0^b \int_0^c e^{-2\pi i L x_c/a} e^{-2\pi i M y_c/b} e^{-2\pi i N z_c/c} \frac{V_c}{abc}\, dx_c dy_c dz_c$$
$$= \int_0^a \int_0^b e^{-2\pi i L x_c/a} e^{-2\pi i M y_c/b} \left[-\frac{c}{2\pi i N} e^{-2\pi i N z_c/c} \right]_0^c \frac{V_c}{abc}\, dx_c dy_c.$$

Since $e^{-2\pi i N c/c} = e^{-2\pi i N} = \cos(-2\pi N) + i\,\sin(-2\pi N) = 1 + 0 = 1$, the expression in the brackets becomes $-c/2\pi i N(1-1) = 0$. This will clearly be the case for any integral in which at least one of the integers, L, M, or N is non-zero. It follows that the only possibility for a non-zero term in the sum is when $L = M = N = 0$, that is, when $h = -m,\ k = -n\ and\ l = -o$. In this case the integral becomes

$$\int_0^a \int_0^b \int_0^c e^0 e^0 e^0 \frac{V_c}{abc}\, dx_c dy_c dz_c$$
$$= \frac{V_c}{abc} \int_0^a \int_0^b \int_0^c dx_c dy_c dz_c$$
$$= \frac{V_c}{abc} abc \;=\; V_c. \tag{3.80}$$

This is the only non-zero term in the sum in Eqn. 3.79, and the net structure factor becomes

$$\overrightarrow{F}_{hkl} = V_c \overrightarrow{C}_{\bar{h}\bar{k}\bar{l}}. \tag{3.81}$$

This has led us to an elegant relationship between the unit cell wave vectors scattered from each (h k l) plane in the crystal and the coefficients in the Fourier series representation of the electron density. It is the link that we seek between the diffraction pattern and the electron density! The result is the *electron density equation*:

$$\overrightarrow{C}_{\bar{h}\bar{k}\bar{l}} = \frac{1}{V_c} \overrightarrow{F}_{hkl} \text{ and} \tag{3.82}$$

$$\rho(x_f, y_f, z_f) = \frac{1}{V_c} \sum_{h=-\infty}^{\infty} \sum_{k=-\infty}^{\infty} \sum_{l=-\infty}^{\infty} \overrightarrow{F}_{hkl} e^{-2\pi i (h x_f + k y_f + l z_f)} \tag{3.83}$$

$$\rho(\mathbf{r}_f) = \frac{1}{V_c} \sum_{\mathbf{h}} \overrightarrow{F}_{\mathbf{h}} e^{-2\pi i (\mathbf{h} \cdot \mathbf{r}_f)}. \tag{3.84}$$

The diffraction equation and the electron density equation constitute a *Fourier pair** that will allow us to iterate toward a structural solution — alternatively computing structure factors from electron densities, and electron densities from structure factors.

*These equations can be derived concisely from formal Fourier transform theory, but the derivations are much less intuitive than the approach taken here. For the interested reader these derivations can be found in Appendix B.

3.2.4 X-ray Diffraction: The Spherical Atom Approximation.

Although the structure factors are discrete entities that can be incorporated directly into the electron density equation, the electron density function required to compute structure factors is continuous — the structure factors are determined by integration. While in principle it should be possible to "digitize" the electron density by selecting values at specific points in the unit cell and numerically integrating over the electron density at those points, such an approach would be tedious and cumbersome. More importantly, we will be required to approximate the electron density function with a model during the iterative structural solution, and this would require us to assign an electron density value to each point, based on our model.

Fortunately, the unit cell is filled with atoms – localized regions of electron density that are approximately spherical. In this approximation, the electron density in the unit cell is modeled as a collection of overlapping spheres of electron density centered on the atomic nuclei in the cell; each atom is assumed to retain its electrons. Since electrons in atoms move at approximately $1/10$ of the speed of light, scattering from the electrons in a given atom will occur independently from those of another. Electron transfer, electron repulsions, and covalent bonding result in minor (but important!) deviations from this model. However, the approximation is adequate for structural solution and we will defer a discussion of these perturbations until we consider structural refinement.

Fig. 3.33(a) is a representation of a hypothetical two-dimensional unit cell containing two molecules of carbon monoxide. The molecules are approximated as overlapping spheres of electron density. If each atom is replaced by a single electron at its center, we have created a set of four fixed scattering centers, and the net scattered wave, $\overrightarrow{\mathcal{E}}_{hkl}$, for a reflection from a unit cell containing these four fixed electrons, would be (Eqn. 3.33):

$$\overrightarrow{\mathcal{E}}_{hkl} = \mathcal{E}_s \sum_{j=1}^{4} e^{i\varphi_j}.$$

$\mathcal{E}_s$ is the magnitude of the electric field due to scattering from a single electron, and φ_j is the phase of the wave scattered from an electron at location j relative to that from a hypothetical reference electron chosen at an arbitrary location (e.g., the origin of the unit cell). Recall that $\overrightarrow{F}_{hkl}$ was rendered independent of the magnitude of the incident beam electric field, $\mathcal{E}_I$, by dividing by the constant factor D (Eqn. 3.79). Since $\mathcal{E}_s$ is proportional to $\mathcal{E}_I$, we must divide by $\mathcal{E}_I$ to remove the dependency on the magnitude of incident beam:

$$\overrightarrow{\mathcal{E}'}_{hkl} = \frac{\overrightarrow{\mathcal{E}}_{hkl}}{\mathcal{E}_I} = \frac{\mathcal{E}_s}{\mathcal{E}_I} \sum_{j=1}^{4} e^{i\varphi_j} = f_e \sum_{j=1}^{4} e^{i\varphi_j},$$

where $f_e = \mathcal{E}_s/\mathcal{E}_I$ is the scattering factor for a single electron — a measure of its "scattering power," and $\mathcal{E}'$ is the relative magnitude of the electric field vector of the wave scattered from the unit cell.

If the single electron at each point is now replaced with the electrons in the atom centered at that point — as if the electron density for each atom was collapsed to

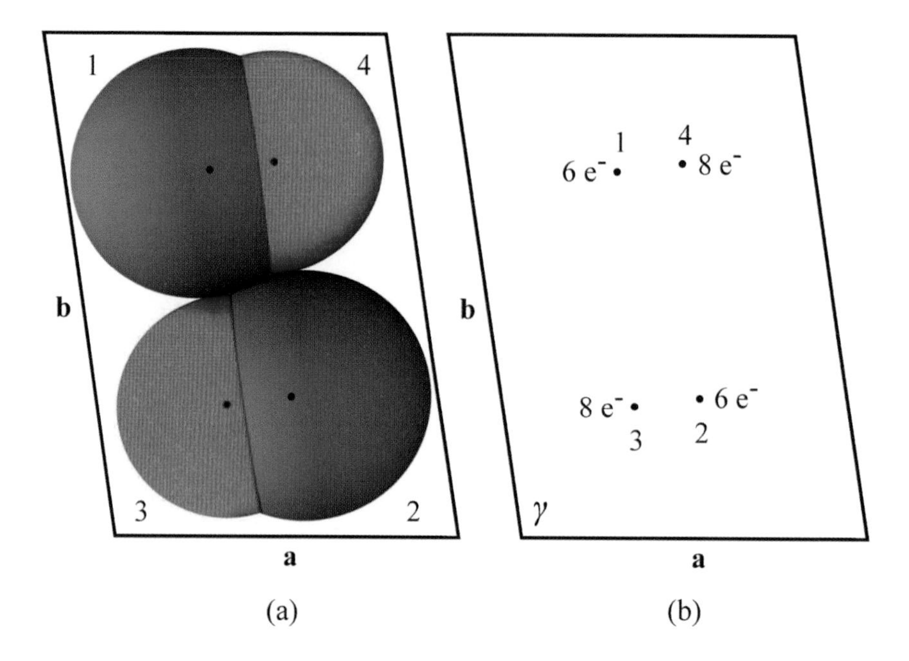

Figure 3.33 (a) Hypothetical two-dimensional unit cell containing two molecules of carbon monoxide. (b) Unit cell in which overlapping atomic spheres of electron density have been replaced with fixed electrons at the centers of the atoms.

its center (Fig. 3.33(b)) — the waves scattered from each electron at a specific point will scatter in phase with all the other electrons at that point since the phase depends on the location of the scatterer with respect to the origin. These waves will superposition, and the net amplitude of the wave scattered from each center will be the sum of the amplitudes of the waves scattered from each electron. The relative electric field vector for the net wave scattered from the unit cell will be the superposition of the waves scattered from each center:

$$\overrightarrow{\mathcal{E}'}_{hkl} = 6f_e e^{i\varphi_1} + 6f_e e^{i\varphi_2} + 8f_e e^{i\varphi_3} + 8f_e e^{i\varphi_4}.$$

In general, for a unit cell consisting of n atoms, if the electrons were concentrated at the center of the atom the relative electric field vector factor would be

$$\overrightarrow{\mathcal{E}'}_{hkl} = \sum_{j=1}^{n} f_e Z_j e^{i\varphi_j}, \tag{3.85}$$

where Z_j is the atomic number of atom j (the number of electrons in the neutral atom).

In order to transform the relative electric field vector into the structure factor for our hypothetical "point atom" unit cell, we first consider the actual integrated structure factor for a real unit cell:

$$\overrightarrow{F}_{hkl} = \int_0^{V_c} \rho(x_f, y_f, z_f) e^{-2\pi i(hx_f + ky_f + lz_f)}\, dV.$$

If all of the electrons were concentrated to a point at the scattering origin, the structure factor would reflect the lack of destructive interference,

$$\vec{F}_{hkl} = \int_0^{V_c} \rho(x_f, y_f, z_f) e^{-2\pi i(h0+k0+l0)}\, dV = \int_0^{V_c} \rho(x_f, y_f, z_f)\, dV.$$

Integrating over the electron density in the unit cell for this hypothetical case gives the total number of electrons in the unit cell:

$$\vec{F}_{hkl} = \sum_{j=1}^{n} Z_j. \tag{3.86}$$

If the electrons in our "fixed electron" unit cell are concentrated at the scattering origin ($\varphi_j = 0$) then

$$\begin{aligned} \vec{\mathcal{E}'}_{hkl} &= \sum_{j=1}^{n} f_e Z_j e^{i0} = f_e \sum_{j=1}^{n} Z_j \quad \text{and} \\ \frac{\vec{\mathcal{E}'}_{hkl}}{f_e} &= \sum_{j=1}^{n} Z_j \;=\; \vec{F}_{hkl}. \end{aligned} \tag{3.87}$$

It follows that the structure factor for the fixed electron unit cell is

$$\vec{F}_{hkl} = \frac{\vec{\mathcal{E}'}_{hkl}}{f_e} = \sum_{j=1}^{n} Z_j e^{i\varphi_j}. \tag{3.88}$$

If the electrons for each atom could be treated as if they were concentrated at the center of the atom the calculation of the structure factor would be reduced to a straightforward summation.

Our model, of course, requires that the fixed electrons at the atomic centers are actually"spread out" into the spherical atoms to which they belong. The waves scattered from each electron no longer scatter in phase; their relative phases now depend on the instantaneous location of each electron when the scattering event occurs. Fig. 3.34 illustrates the consequences of scattering from different regions in each atom — the result is a diminution of the magnitude of the structure factor resulting from destructive interference. Furthermore, as shown in the figure, the phase difference — and the destructive interference — increases as the diffraction angle increases. If we are to recover the simple form of Eqn. 3.88 we must find a way to determine the resultant wave scattered from each spherical atom so that the structure factor calculation can become a simple superposition of the scattered waves from the atoms in the unit cell.

In order to determine the results of scattering from a single atom, we consider a crystal in the diffraction condition. For simplicity we place the scattering reference at the center of an atom with atomic number Z. The coordinate system is constructed so that the z axis is aligned with the diffraction vector, as illustrated in Fig. 3.35. Following the same arguments that were used to determine the net scattered wave from the crystal (Eqn. 3.38), the net scattered wave from the atom will result from the superposition of the waves scattered from each location in the atom:

$$\vec{\mathcal{E}_a} = \mathcal{E}_s \int_0^{\infty} \rho(\mathbf{r}) e^{i\varphi(\mathbf{r})}\, dV. \tag{3.89}$$

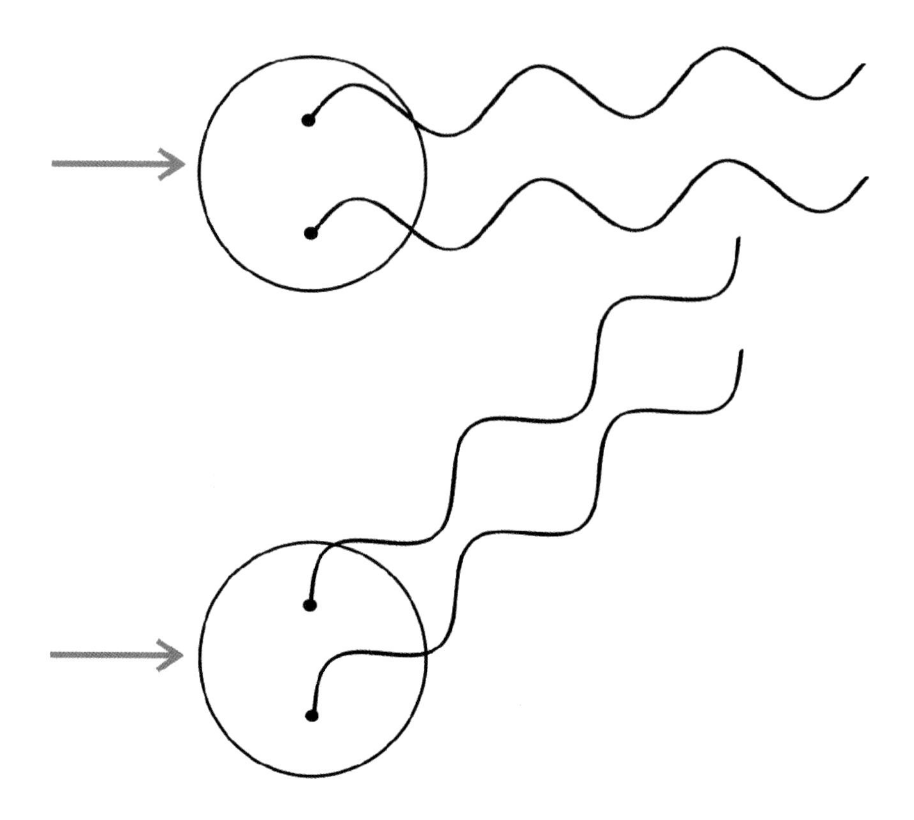

Figure 3.34 Top: X-rays scattering from an electron in two different locations in an atom at $2\theta = 6°$. Bottom: X-rays scattering from an electron at two different locations in an atom at $2\theta = 40°$.

To remove incident beam dependence, we again divide by $\mathcal{E}_I$:

$$\overrightarrow{\mathcal{E}'_a} = \frac{\mathcal{E}_s}{\mathcal{E}_I} \int_0^\infty \rho(\mathbf{r}) e^{i\varphi(\mathbf{r})}\, dV = f_e \int_0^\infty \rho(\mathbf{r}) e^{i\varphi(\mathbf{r})}\, dV. \tag{3.90}$$

Since the scattering from the atom is occurring while the crystal is diffracting, $\varphi(\mathbf{r}) = -2\pi i\mathbf{r} \cdot \mathbf{h}$. Since the atom is spherical, the integral can be evaluated in spherical polar coordinates, where $dV = dx\, dy\, dz = d\eta\, \sin\nu\, d\nu\, r^2\, dr$:

$$\overrightarrow{\mathcal{E}'_a} = f_e \int_0^\infty \int_\pi^0 \int_0^{2\pi} \rho(r,\nu,\eta) e^{-2\pi i(\mathbf{r}\cdot\mathbf{h})}\, d\eta\, \sin\nu\, d\nu\, r^2\, dr. \tag{3.91}$$

The electron density in the atom is assumed to distributed as it is in the free atom, with spherical symmetry independent of η and ν, so that $\rho(r,\nu,\eta) = \rho(r)$. With $\mathbf{h}$ aligned with the z axis, $\mathbf{r} \cdot \mathbf{h} = \mathrm{h}r \cos\nu$ and

$$\overrightarrow{\mathcal{E}'_a} = f_e \int_0^\infty \int_\pi^0 \int_0^{2\pi} \rho(r) e^{-2\pi i(\mathrm{h}r\cos\nu)}\, d\eta\, \sin\nu\, d\nu\, r^2\, dr. \tag{3.92}$$

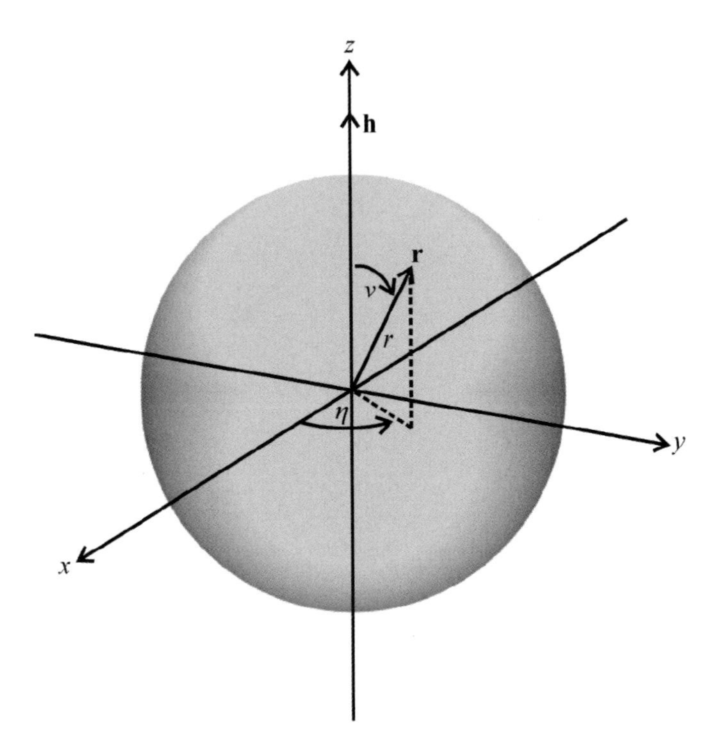

Figure 3.35 Spherical atom coordinate system with diffraction vector aligned along the z axis.

This integral is evaluated by integrating first over η:

$$\int_0^{2\pi} d\eta = 2\pi.$$

The integral over ν is evaluated by expressing the integrand with Euler's relations:

$$\int_\pi^0 e^{-2\pi i(\mathrm{h}r\cos\nu)} \sin\nu\, d\nu =$$
$$\int_\pi^0 \cos(-2\pi \mathrm{h}r\cos\nu)\sin\nu\, d\nu + i\int_\pi^0 \sin(-2\pi \mathrm{h}r\cos\nu)\sin\nu\, d\nu.$$

The imaginary expression evaluates to zero, *setting the phase to zero*, and the real expression becomes $2\sin(2\pi \mathrm{h}r)/(2\pi \mathrm{h}r)$. Substituting the magnitude of the diffraction vector, h$= 2\sin\theta/\lambda$,

$$\begin{aligned}\overrightarrow{\mathcal{E}'_a}(\theta) &= f_e\int_0^\infty r\rho(r)\frac{2\sin(4\pi r\sin\theta/\lambda)}{\sin\theta/\lambda}\, dr \\ &= \left(f_e\int_0^\infty r\rho(r)\frac{2\sin(4\pi r\sin\theta/\lambda)}{\sin\theta/\lambda}\, dr\right)e^{i0}.\end{aligned} \tag{3.93}$$

The expression $e^{i0} = 1$ has been added to the expression for the relative amplitude of the wave scattered from the atom to illustrate that it has become a scalar –

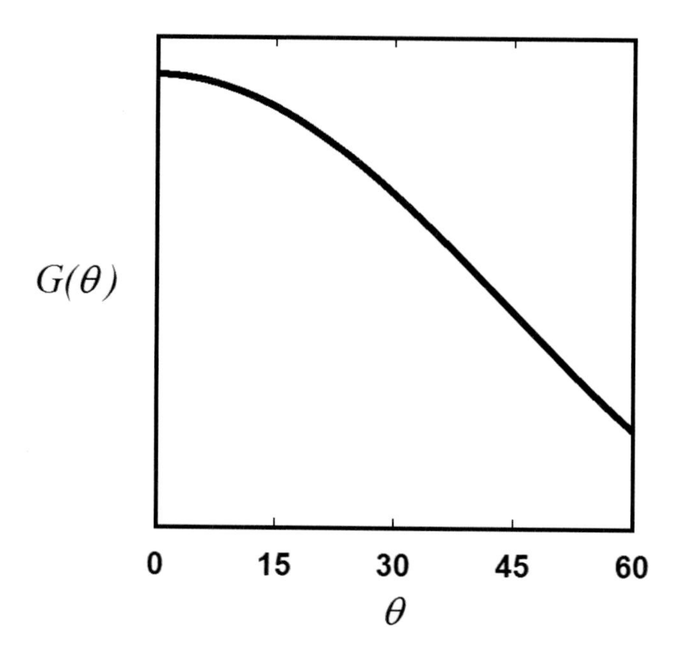

Figure 3.36 Plot of $G(\theta) = (2\sin(4\pi r \sin\theta/\lambda))/(\sin\theta/\lambda)$ versus θ.

that the net wave has a phase angle of zero with respect to a wave emanating from the center of the atom (the scattering origin). *Thus the net scattered wave from a spherical atom can be considered to be a wave scattered from the atomic center with amplitude* $\mathcal{E}'_a(\theta)$.

Although the integral is relatively complicated, for a given value of 2θ it can be computed if the electron density function for the spherical atom, $\rho(r)$, is known. It is instructive to observe the behavior of the portion of the integrand that depends on θ, illustrated in Fig. 3.36 for fixed values of r and λ. For any value of r, the function decreases as θ increases, which results in a smaller value for the integral and a subsequent attenuation of $\mathcal{E}'_a(\theta)$. It is this expression that accounts for the effects of destructive interference as a result of scattering from different regions within the atom.

In the limit, at $\theta=0$, $\mathcal{E}'_a(\theta)$ is a maximum, and represents the case where $\varphi = 0$ and the scattered waves from every portion of the atom are constructively interfering (Fig. 3.34):

$$\overrightarrow{\mathcal{E}}'_a(\theta = 0) = f_e \int_0^{\infty} \rho(\mathbf{r}) e^{i0}\, dV = f_e \int_0^{\infty} \rho(\mathbf{r}) = f_e Z_a, \tag{3.94}$$

where Z_a is the number of electrons in the atom. The structure factor at $2\theta = 0$ can now be determined by treating the scattering from the cell as the superposition of waves from the atomic electrons located at the center of the atoms (Eqn. 3.88):

$$\overrightarrow{F}_{hkl}(\theta = 0) = \frac{\mathcal{E}'_{hkl}}{f_e} = \sum_{j=1}^{n} \frac{\mathcal{E}'_j(\theta = 0)}{f_e} e^{i\varphi_j} = \sum_{j=1}^{n} Z_j e^{i\varphi_j}. \tag{3.95}$$

When $2\theta > 0$ the scattering can be considered to be occurring from the electrons concentrated at the atomic centers with the scattering efficiency of those electrons lowered, due to destructive interference:

$$\vec{F}_{hkl}(\theta) = \frac{\mathcal{E}'_{hkl}}{f_e} = \sum_{j=1}^{n} \frac{\mathcal{E}'_j(\theta)}{f_e} e^{i\varphi_j} < \sum_{j=1}^{n} Z_j e^{i\varphi_j}. \tag{3.96}$$

The phase of the wave scattered from each atom depends only on the fractional coordinates of the center of the atom: $\varphi_j = -2\pi\, i\,(hx_j + ky_j + lz_j)$. For each atom the factor

$$\frac{\mathcal{E}'_j(\theta)}{f_e} = \int_o^\infty r\rho(r)G(\theta, r)\, dr = f_j(\theta), \tag{3.97}$$

where $G(\theta, r) = (2\sin(4\pi r \sin\theta/\lambda))/(\sin\theta/\lambda)$, represents the scattering efficiency of atom j, $\mathcal{E}'_j(\theta)$, relative to the scattering efficiency of a hypothetical electron at its center (f_e). This factor is the known as the *atomic scattering factor* for atom j, and is given the symbol $f_j(\theta)$.

Computing the scattering factor requires $\rho(r)$ in Eqn. 3.97, which can be determined by solving the Schrödinger equation in spherical polar coordinates for the electronic state function (often referred to as a "wave equation") of the multi-electronic, spherically symmetric atom:

$$-\frac{\hbar^2}{2m_e}\left[\frac{1}{r^2}\frac{\partial}{\partial r}\left(r^2\frac{\partial\psi}{\partial r}\right) + \frac{1}{r^2\sin^2\nu}\frac{\partial^2\psi}{\partial\eta^2} + \frac{1}{r^2\sin\nu}\frac{\partial}{\partial\nu}\left(\sin\nu\frac{\partial\psi}{\partial\theta}\right)\right] + V\psi = E\psi,$$

where m_e is the (reduced) mass of the electron and $V \equiv V(\mathbf{r}_1, \mathbf{r}_2, \ldots, \mathbf{r}_Z)$, the electrostatic potential energy of the Z electrons in the atom. The equation is a differential equation; the solution is the state function $\psi(r)$. The Schrödinger equation can be solved rigorously only for one-electron atoms or ions, but it can be solved to a very high degree of accuracy for multi-electron atoms by employing approximation methods based on the self-consistent field method developed initially by Douglas R. Hartree in 1928[43] and modified (independently) in 1930 by Vladimir A. Fock[44] and John C. Slater.[45] The details of these solutions would take us *far* beyond the scope of the text.* The state function for the electrons in the atom is the analog of the electric field function of the photons in a beam of light; both are based on the wave properties of the particles involved. Squaring the magnitude of the electric field of the light wave at a specific point in space gives the intensity of the light at that point — the *photon density*. Squaring the magnitude of the electronic state function at a point in space gives the *electron density* at that point: $\rho(r) = \psi^2(r)$.

Accurate electronic state functions for all of the atoms in the periodic table are available from sophisticated quantum chemistry calculations, and the atomic scattering factors can be determined for all of the atoms in the periodic table of the elements:

$$f_j(\theta) = \int_o^\infty r\,\psi_j^2(r)G(\theta, r)\, dr. \tag{3.98}$$

*Interested readers can find a plethora of discussions of the Hartree-Fock-Slater method. An excellent treatment is found in *Modern Quantum Chemistry* by Szabo and Ostlund[46].

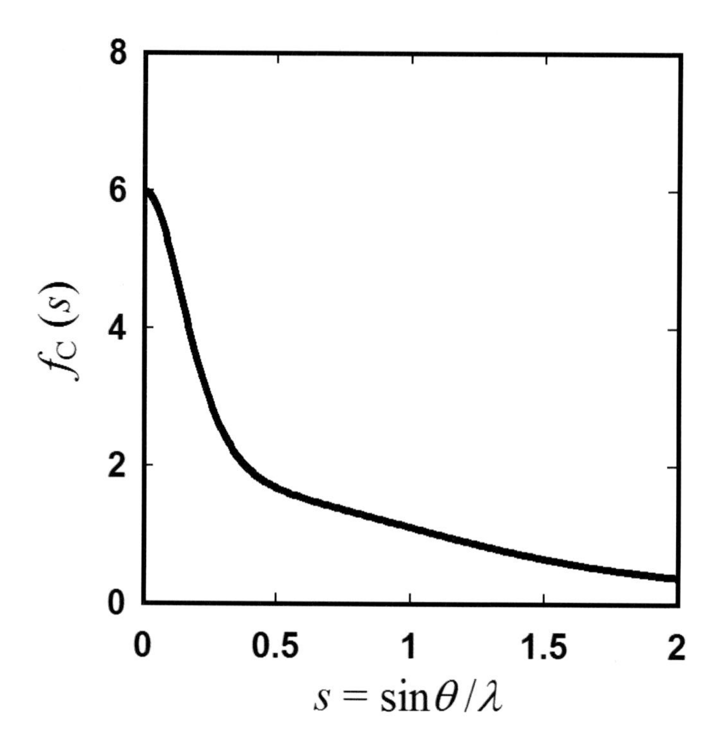

Figure 3.37 Plot of the carbon atom scattering factor.

θ does not vary linearly in reciprocal space since the natural dimension is $1/d = \mathrm{h} = 2\sin\theta/\lambda$ (Fig. 3.13). Because of this f_j is generally evaluated for values of $s = \mathrm{h}/2 = \sin\theta/\lambda$ rather than θ. For atom j,

$$f_j(s) = \int_0^\infty r\psi_j^2(r)\frac{2\sin(4\pi r\, s)}{s}\, dr. \tag{3.99}$$

The atomic scattering factor calculations provide a table of values of $f(s)$, usually computed in s increments of 0.01 from 0 to 2. For use in computer programs the resulting curves can be conveniently and accurately modeled with a sum of four Gaussian functions and a baseline constant,

$$f_j(s) = \sum_{j=1}^{4} a_i e^{-b_i s^2} + c. \tag{3.100}$$

The nine parameters for each atom are determined by a least-squares fit of the values from the model function in Eqn. 3.100 and the computed values determined from Eqn. 3.99. The resulting scattering factor function for the carbon atom is plotted in Fig. 3.37. Tables of atomic scattering factors and functional parameters to compute them can be found in the *International Tables for Crystallography.*

The structure factor for a reflection can now be expressed as a sum of waves scattering from the atomic centers in the unit cell with scattering efficiencies given by the scattering factors for each atom at the diffraction angle:

$$\overrightarrow{F}_{hkl} = \sum_{j=1}^{n} f_j(hkl) e^{-2\pi\, i(hx_j+ky_j+lz_j)}. \tag{3.101}$$

The electron density equation, also a sum, constitutes the other half of the Fourier pair:

$$\rho(x,y,z) = \frac{1}{V_c} \sum_{hkl} \overrightarrow{F}_{hkl} e^{-2\pi i(hx+ky+lz)}. \tag{3.102}$$

In Eqn. 3.101, (x_j, y_j, z_j) are the fractional coordinates of the center of atom j in the unit cell, and (x, y, z) are the fractional coordinates of a point in the unit cell in Eqn. 3.102.*

3.2.5 Calculating Structure Factors and Electron Density.

Calculating Structure Factors. If we know the atom locations, then the structure factor *vectors*, with amplitudes and relative phases for each reflection, can be readily calculated from Eqn. 3.101. Setting $\xi_j(hkl) = -2\pi(hx_j + ky_j + lz_j)$,

$$\begin{aligned} \overrightarrow{F}_{hkl} &= \sum_{j=1}^{n} f_j(hkl) e^{i\,\xi_j(hkl)} = F_{hkl} e^{i\,\varphi_{hkl}} \\ &= \sum_{j=1}^{n} f_j(hkl)\cos(\xi_j(hkl)) + i \sum_{j=1}^{n} f_j(hkl)\sin(\xi_j(hkl)) \\ &= A_{hkl} + i\, B_{hkl}. \end{aligned} \tag{3.103}$$

Fig. 3.38(a) shows the result of the superposition of the scattered waves from the atoms in the unit cell for the reflection with indices hkl, creating the structure factor, $\overrightarrow{F}_{hkl}$. With $A_{hkl} = \sum_{j=1}^{n} f_j(hkl)\cos\xi_j(hkl)$ and $B_{hkl} = \sum_{j=1}^{n} f_j(hkl)\sin\xi_j(hkl)$,

$$F_{hkl} = \left(A_{hkl}^2 + B_{hkl}^2\right)^{\frac{1}{2}} \tag{3.104}$$

$$\varphi_{hkl} = \arccos\left(\frac{A_{hkl}}{F_{hkl}}\right) = \arcsin\left(\frac{B_{hkl}}{F_{hkl}}\right) = \arctan\left(\frac{B_{hkl}}{A_{hkl}}\right). \tag{3.105}$$

Calculating Electron Density. Similarly, given the set of amplitudes and relative phases of the scattered waves from a crystal, Eqn. 6.2 provides locations in the unit cell where there are high concentrations of electron density — the locations of

*Up to this point fractional coordinates have been labeled with the f subscript to differentiate them from Cartesian coordinates. We will adopt the convention here that unlabeled coordinates are considered to be fractional coordinates.

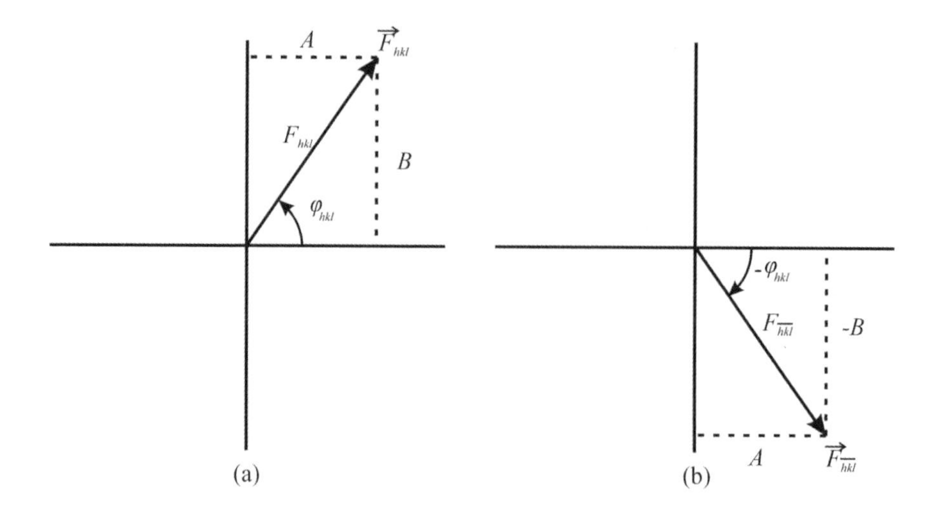

Figure 3.38 (a) Structure factor wave vector for hkl reflection. (b) Structure factor wave vector for $\bar{h}\bar{k}\bar{l}$ reflection.

the atoms! In order to calculate the electron density we first consider the structure factor magnitude and phase for the negative reflection, $-\mathbf{h} = [\bar{h}\ \bar{k}\ \bar{l}\ ]$.

$$\begin{aligned}\vec{F}_{\bar{h}\bar{k}\bar{l}} &= \sum_{j=1}^{n} f_j(\bar{h}\bar{k}\bar{l}) e^{-2\pi\, i(-hx_j - ky_j - lz_j)} \\ &= \sum_{j=1}^{n} f_j(\bar{h}\bar{k}\bar{l}) e^{-\, i\,\xi_j(hkl)} \end{aligned} \tag{3.106}$$

In the reciprocal lattice, the vector $-\mathbf{h}$ is equal in magnitude, h, and opposite in direction to the vector $\mathbf{h}$. Since $f_j(hkl)$ depends on h (h$= 2s$ in Eqn. 3.99), $f_j(\bar{h}\bar{k}\bar{l}) = f_j(hkl)$ *,

$$\begin{aligned}\vec{F}_{\bar{h}\bar{k}\bar{l}} &= \sum_{j=1}^{n} f_j(\bar{h}\bar{k}\bar{l}) \cos\xi_j(\bar{h}\bar{k}\bar{l}) + i\sum_{j=1}^{n} f_j(\bar{h}\bar{k}\bar{l}) \sin\xi_j(\bar{h}\bar{k}\bar{l}) \\ &= \sum_{j=1}^{n} f_j(hkl)\cos(-\xi_j(hkl)) + i\sum_{j=1}^{n} f_j(hkl)\sin(-\xi_j(hkl)) \\ &= \sum_{j=1}^{n} f_j(hkl)\cos(\xi_j(hkl)) - i\sum_{j=1}^{n} f_j(hkl)\sin(\xi_j(hkl)) \\ &= A_{hkl} - i\,B_{hkl}. \end{aligned} \tag{3.107}$$

*The spherical scattering model derived in this chapter must be modified slightly to account for a phenomenon known as *anomalous dispersion*, which causes $f_j(\bar{h}\bar{k}\bar{l})$ to deviate slightly from $f_j(hkl)$ (see Sec. 5.1.2).

Fig. 3.38(b) illustrates the structure factor vector, $\overrightarrow{F}_{\bar{h}\bar{k}\bar{l}}$, with magnitude and phase

$$\varphi_{\bar{h}\bar{k}\bar{l}} = \arctan\left(\frac{-B_{hkl}}{A_{hkl}}\right) = -\varphi_{hkl} \text{ and} \tag{3.108}$$

$$F_{\bar{h}\bar{k}\bar{l}} = \left(A_{hkl}^2 + (-B_{hkl})^2\right)^{\frac{1}{2}} = F_{hkl}. \tag{3.109}$$

$\overrightarrow{F}_{\bar{h}\bar{k}\bar{l}}$ has the same magnitude as $\overrightarrow{F}_{hkl}$ (subject to the footnote below); their phases are negatives of one another. Since the intensity of the reflection is proportional to F_{hkl}^2, both reflections have the same intensity. This observation is known as the *Friedel Law*, which will be treated more formally in Sec. 4.6.3.

There is an infinite number of sets of parallel planes in the direct lattice, corresponding to an infinite number of lattice points in the reciprocal lattice. The electron density function is a sum over all the indices:

$$\rho(\mathbf{r}) = \frac{1}{V_c}\sum_{\mathbf{h}} \overrightarrow{F}_{hkl} e^{-2\pi i(\mathbf{r}\cdot\mathbf{h})} =$$

$$\rho(x,y,z) = \frac{1}{V_c}\sum_{h=-\infty}^{\infty}\sum_{k=-\infty}^{\infty}\sum_{l=-\infty}^{\infty} F_{hkl} e^{i\varphi_{hkl}} e^{-2\pi i(hx+ky+lz)}. \tag{3.110}$$

For every set of hkl indices in Eqn. 3.110, there is a set with indices $\bar{h}\bar{k}\bar{l}$ and the equation can be factored into two sums. Setting $\phi_{hkl} = -2\pi(hx + ky + lz)$ and noting that $\phi_{\bar{h}\bar{k}\bar{l}} = -2\pi(-hx - ky - lz) = -\phi_{hkl}$,

$$\begin{aligned}
\rho(x,y,z) &= \frac{1}{V_c}\sum_{hkl} F_{hkl} e^{i\varphi_{hkl}} e^{i\phi_{hkl}} + \frac{1}{V_c}\sum_{\bar{h}\bar{k}\bar{l}} F_{\bar{h}\bar{k}\bar{l}} e^{i\varphi_{\bar{h}\bar{k}\bar{l}}} e^{i\phi_{\bar{h}\bar{k}\bar{l}}} \\
&= \frac{1}{V_c}\sum_{hkl} F_{hkl} e^{i\varphi_{hkl}} e^{i\phi_{hkl}} + \frac{1}{V_c}\sum_{\bar{h}\bar{k}\bar{l}} F_{hkl} e^{-i\varphi_{hkl}} e^{-i\phi_{hkl}} \\
&= \frac{1}{V_c}\sum_{hkl} F_{hkl} e^{i\Phi_{hkl}} + \frac{1}{V_c}\sum_{\bar{h}\bar{k}\bar{l}} F_{hkl} e^{-i\Phi_{hkl}},
\end{aligned}$$

where $\Phi_{hkl} = \varphi_{hkl} + \phi_{hkl}$. Using Euler's relations,

$$\begin{aligned}
\rho(x,y,z) &= \frac{1}{V_c}\sum_{hkl} F_{hkl}\cos\Phi_{hkl} + i\frac{1}{V_c}\sum_{hkl} F_{hkl}\sin\Phi_{hkl} \\
&\quad + \frac{1}{V_c}\sum_{\bar{h}\bar{k}\bar{l}} F_{hkl}\cos(-\Phi_{hkl}) + i\frac{1}{V_c}\sum_{\bar{h}\bar{k}\bar{l}} F_{hkl}\sin(-\Phi_{hkl}) \\
&= \frac{1}{V_c}\sum_{hkl} F_{hkl}\cos\Phi_{hkl} + i\frac{1}{V_c}\sum_{hkl} F_{hkl}\sin\Phi_{hkl} \\
&\quad + \frac{1}{V_c}\sum_{\bar{h}\bar{k}\bar{l}} F_{hkl}\cos\Phi_{hkl} - i\frac{1}{V_c}\sum_{\bar{h}\bar{k}\bar{l}} F_{hkl}\sin\Phi_{hkl} \\
&= \frac{1}{V_c}\sum_{hkl} F_{hkl}\cos\Phi_{hkl} + \frac{1}{V_c}\sum_{\bar{h}\bar{k}\bar{l}} F_{hkl}\cos\Phi_{hkl}.
\end{aligned}$$

Since the sum over hkl is the same as the sum over $\bar{h}\bar{k}\bar{l}$ for the imaginary terms they add to zero — and since $\cos\Phi_{hkl} = \cos(-\Phi_{hkl})$,

$$\rho(x,y,z) = \frac{1}{V_c}\sum_{h=-\infty}^{\infty}\sum_{k=-\infty}^{\infty}\sum_{l=-\infty}^{\infty} F_{hkl}\cos(2\pi(hx+ky+lz) - \varphi_{hkl}). \tag{3.111}$$

Centrosymmetric Structures. The presence of a center of symmetry imposes constraints on the structure factor and electron density expressions that warrant special consideration. Referring to Eqn. 3.103, if the structure is centrosymmetric, for every $\xi_j(hkl) = -2\pi(hx_j + ky_j + lz_j)$, there is a $-\xi_j(hkl) = -2\pi(h\bar{x}_j + k\bar{y}_j + l\bar{z}_j)$, and

$$\begin{aligned}
\vec{F}_{hkl} &= \sum_{j=1}^{n/2} f_j(hkl)\cos(\xi_j(hkl)) + i\sum_{j=1}^{n/2} f_j(hkl)\sin(\xi_j(hkl)) \\
&+ \sum_{j=1}^{n/2} f_j(hkl)\cos(-\xi_j(hkl)) + i\sum_{j=1}^{n/2} f_j(hkl)\sin(-\xi_j(hkl)) \\
&= 2\sum_{j=1}^{n/2} f_j(hkl)\cos(\xi_j(hkl)).
\end{aligned} \tag{3.112}$$

The imaginary term vanishes, and the centrosymmetric structure factor vector is restricted to lie along the real axis, with a phase angle of $\varphi_{hkl} = 0$ if the structure factor is positive, or $\varphi_{hkl} = \pi$ if it is negative. Thus,

$$\vec{F}_{hkl} = 2\sum_{j=1}^{n/2} f_j(hkl)\cos(-2\pi(hx_j + ky_j + lz_j)) = F_{hkl}S_{hkl}, \tag{3.113}$$

where S_{hkl} is the sign of the structure factor.

The constraint on phases also affects the electron density equation for centrosymmetric structures. Applying the trigonometric identity, $\cos(\alpha - \beta) = \cos\alpha\cos\beta - \sin\alpha\sin\beta$, the cosine term in Eqn. 3.111 can be written as

$$\begin{aligned}
&\cos(2\pi(hx + ky + lz) - \varphi_{hkl}) = \\
&\cos(2\pi(hx + ky + lz))\cos\varphi_{hkl} - \sin(2\pi(hx + ky + lz)\sin\varphi_{hkl}
\end{aligned}$$

Since $\sin\varphi_{hkl} = 0$ and $\cos\varphi_{hkl} = \pm 1 = S_{hkl}$, the electron density equation for the centrosymmetric structure becomes

$$\begin{aligned}
\rho(x,y,z) &= \frac{1}{V_c}\sum_{h=-\infty}^{\infty}\sum_{k=-\infty}^{\infty}\sum_{l=-\infty}^{\infty} F_{hkl}S_{hkl}\cos(2\pi(hx + ky + lz)) \\
&= \frac{1}{V_c}\sum_{h=-\infty}^{\infty}\sum_{k=-\infty}^{\infty}\sum_{l=-\infty}^{\infty} \vec{F}_{hkl}\cos(2\pi(hx + ky + lz)).
\end{aligned} \tag{3.114}$$

The Phase Problem. Given the magnitudes of the diffracted waves from a crystal in the form of structure factor amplitudes, F_{hkl}, and their relative phases, φ_{hkl}, Eqn. 3.111 allows us to calculate the electron density distribution in the unit cell. It is here that we encounter a dilemma fundamental to the determination of X-ray crystal structures. The intensities of the various reflections, proportional to $\mathcal{E}^2$, are obtained by counting the photons at the location of the diffraction maxima for each set of planes. The amplitudes of the scattered waves can then be obtained as the square roots of these intensities, which, after appropriate scaling, *provide the magnitudes of the structure factors, F_{hkl}, but not the relative phases. Unfortunately, the phase information is not available experimentally and we are confronted with* *the daunting task of obtaining information that appears to be irrevocably lost — the* ***phase problem.***

As the reader might suspect, determining the relative phases of the structure factors will require significant effort. Chapters 6 and 7 are devoted entirely to various approaches to solving the phase problem. Solutions to the problem require an experimental determination of the lattice parameters and symmetry in addition to a complete set of intensities from the lattice planes — we must first consider the details of the diffraction experiment itself.

Exercises

1. Consider four equally spaced hydrogen atoms, arranged in a line along the x axis of a Cartesian coordinate system:

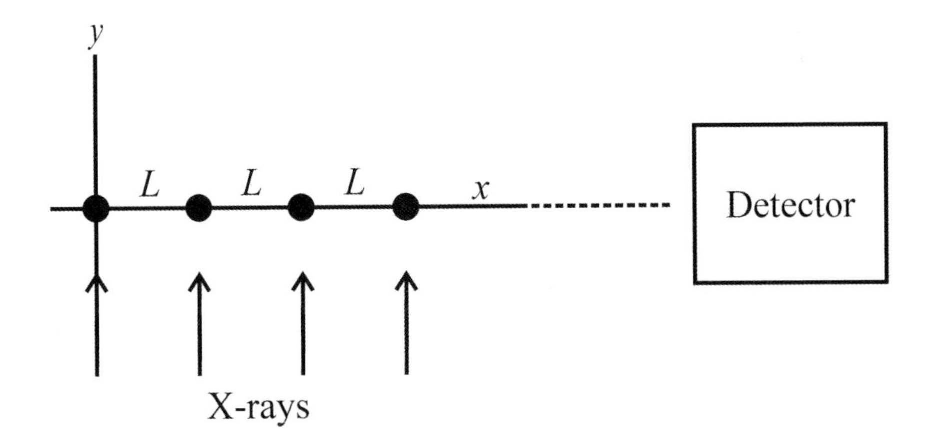

   A monochromatic, in-phase X-ray beam with $\lambda = 1.500$ is directed parallel to the y axis; the atom at the origin is selected as the scattering (phase) reference. A wave detector is placed at a distance of one meter from the origin. The maximum electric field due to each atom observed at the detector (measured in the absence of the other atoms) is $\mathcal{E}_a$. (a) Determine general formulae for the maximum amplitude of the resultant wave observed at the detector, $\mathcal{E}_{o,r}$, and its relative phase, φ_r, with respect to a wave scattered from the origin. (b) Determine $\mathcal{E}_{o,r}$ when $L = 1.125$, $L = 2.250$, and $L = 4.500$. Note that since the frequencies of all of the waves and the resultant wave are the same, all of the wave vectors rotate together in time. Thus $\mathcal{E}_{o,r}$ and φ_r remain constant, and are conveniently evaluated at $t = 0$.

2. The detector in Exercise 1 can be rotated in the xy plane so that its distance from the origin remains constant. Because the distances between the atoms are orders of magnitude different from the distance to the detector, the waves emanating from each atom can be considered to be traveling along parallel vectors toward a point on the detector in any direction (See Fig. 3.8). (a) Determine general formulae for the maximum amplitude of the resultant wave observed at the

detector, $\mathcal{E}_{o,r}$, and its relative phase, φ_r. These formulae will be similar to those derived in Exercise 1, but will contain the detector angle, θ, between a vector to the center of the detector and the y axis. (b) Determine the smallest non-zero angle through which the detector must be rotated to observe a maximum intensity at the detector (a diffraction maximum) and $\mathcal{E}_{o,r}$ and φ_r for this angle for $L = 1.125$, $L = 2.250$, and $L = 4.500$. (c) Determine the smallest angle through which the detector must be rotated to observe a minimum intensity at the detector (a diffraction minimum) and $\mathcal{E}_{o,r}$ and φ_r for this angle for $L = 1.125$, $L = 2.250$, and $L = 4.500$.

3. The detector and the X-ray source in Exercise 2 are modified so that vectors in the direction of the X-ray beam and vectors toward the detector both make equal angles with the y axis (Fig. 3.11). (a) Determine general formulae for the maximum amplitude of the resultant wave observed at the detector, $\mathcal{E}_{o,r}$, and its relative phase, φ_r. (b) Determine the smallest angle through which the detector must be rotated to observe a maximum intensity at the detector (a diffraction maximum) and $\mathcal{E}_{o,r}$ and φ_r for this angle for $L = 1.125$, $L = 2.250$, and $L = 4.500$. (c) Determine the smallest angle through which the detector must be rotated to observe a minimum intensity at the detector (a diffraction minimum) and $\mathcal{E}_{o,r}$ and φ_r for this angle for $L = 1.125$, $L = 2.250$, and $L = 4.500$.

4. X-rays for crystal structure determination are commonly generated by bombarding specific metals with high energy electrons. The metals most often used are copper and molybdenum, which emit average "monochromatized" wavelengths of 1.5418 Å and 0.7107 Å, respectively. Referring to Fig. 3.15, with the area detector (screen) at a distance of 0.1 meters from the crystal, consider a 1-dimensional crystal lattice with $a = 10$ Å — typical of a small-molecule crystal structure. The spots on the screen are finite in size – with widths on the order of a millimeter. To measure the intensity of a single spot, it must not overlap with adjacent spots. (a) Determine the number of diffraction maxima (spots on the screen) that it is possible to observe for each wavelength. (b) Determine the distance between the spots on the screen for $h = 1$ and $h = 2$. (c) Repeat (a) for a unit cell with $a = 100$ Å, typical of a macromolecular crystal structure. (d) Repeat (b) for the 100 Å unit cell.

5. In Chapter 1 an expression for the unit cell volume in terms of the unit cell parameters was derived for a general (triclinic) unit cell. In Chapter 3 relationships between the direct and reciprocal unit cell parameters were derived for the general case. (a) Derive explicit expressions for the unit cell volumes of monoclinic, orthorhombic, tetragonal, cubic, trigonal/hexagonal, and rhombohedral unit cells. (b) Derive expressions for the direct↔reciprocal unit cell relationships for each of the unit cell types in part (a).

6. In Chapter 1 the **B** matrix was derived to transform a direct lattice vector, $[x_f\ y_f\ z_f]$, into Cartesian coordinates, $[x_c\ y_c\ z_c]$. Using identical arguments (see Fig. 4.14), a vector in reciprocal fractional coordinates, $[x_f^*\ y_f^*\ z_f^*]$, can be transformed into a vector in reciprocal Cartesian coordinates, $[x_c^*\ y_c^*\ z_c^*]$ by the $\mathbf{B}^*$ matrix for the structure. (a) Using the cell parameters for the mercaptopyridine unit cell in Sec. 1.5.3, determine the reciprocal cell parameters and the $\mathbf{B}^*$ matrix for the structure. (b) Determine the distance between the planes with

$(h\ k\ l) = (1\ 2\ 3)$ in the mercaptopyridine lattice. (c) Determine the distance between the planes with $(h\ k\ l) = (4\ 5\ 6)$. (d) Calculate the Bragg diffraction angles for the reflections from both sets of planes for copper and molybdenum radiation (the wavelengths are given in Exercise 4).

7. The hypothetical two-dimensional carbon monoxide structure modeled in Fig. 3.33 has $a = 4.855$ Å, $b = 6.953$ Å, and $\gamma = 98.8°$. (a) Referring to Fig. 3.20, determine expressions for the two-dimensional reciprocal cell parameters — a^*, b^*, and γ^*. (b) Adopting the convention that the a and a^* axes are coincident with the unit vector $\mathbf{i}$ in their respective orthonormal coordinate systems, derive the $\mathbf{B}$ and $\mathbf{B}^*$ matrices that convert vectors in direct and reciprocal fractional coordinates to Cartesian coordinates. (c) Determine a^*, b^*, and γ^* for the CO unit cell. (d) Determine the distances between the (1 2) and (2 3) lines in the two-dimensional lattice.

8. The hypothetical carbon monoxide unit cell in Exercise 7 has the following contents:

atom	x	y
C1	0.567	0.278
O1	0.377	0.262
C2	0.433	0.722
O2	0.623	0.738

The parameters for Eqn. 3.100 for the scattering factors of carbon and oxygen are:

	a_1	b_1	a_2	b_2	a_3	b_3	a_4	b_4	c
C	2.3100	20.8439	1.0200	10.2075	1.5886	0.5687	0.8650	51.6512	0.2156
O	3.0485	13.2771	2.2868	5.7011	1.5463	0.3239	0.8670	32.9089	0.2508

Determine the amplitudes and phase angles of the structure factors for the (1 2) and (2 3) reflections (Hint: Do Exercise 7 first).

9. Given the following "experimental" structure factor amplitudes and phases for the hypothetical carbon monoxide crystal in Exercises 7 and 8, calculate the electron density at (a) the origin of the unit cell, (b) the center of O1, and (c) the midpoint between C1 and O1.

h	k	F_{hk}	φ_{hk}	h	k	F_{hk}	φ_{hk}	h	k	F_{hk}	φ_{hk}
0	0	28.000	0	1	0	19.496	π	2	0	4.865	0
0	1	3.065	π	1	1	3.510	π	2	1	4.782	0
0	2	21.171	π	1	2	14.640	0	2	2	3.580	π
0	3	5.727	0	1	3	0.210	0	2	3	2.779	π
0	4	11.950	0	1	4	8.757	π	2	4	2.128	0
0	5	5.699	π	1	5	1.439	0	2	5	1.317	0
3	0	3.884	0	4	0	6.155	π	5	0	5.034	0
3	1	1.270	π	4	1	2.836	π	5	1	5.075	0
3	2	3.937	π	4	2	6.581	0	5	2	5.940	π
3	3	1.353	0	4	3	1.427	0	5	3	3.372	π
3	4	3.587	0	4	4	6.370	π	5	4	6.210	0
3	5	1.432	π	4	5	0.060	π	5	5	1.640	0

Chapter 4

Crystal Diffraction: Experiment

Methods for the collection and analysis of X-ray diffraction data are many and varied. Rather than attempting a comprehensive treatment of the various methods, we will limit the discussion here to the collection of X-ray diffraction data from single crystals. The emphasis here will be on underlying principles rather than experimental details — the reader is referred to any of the standard texts in the bibliography for discussions of the more practical aspects of X-ray data collection.

4.1 The Sphere of Reflection.

In the typical single crystal X-ray diffraction experiment a crystal with linear dimensions of less than a millimeter (e.g., 0.3 mm $\times$ 0.3 mm $\times$ 0.3 mm) is mounted onto a device known as a goniometer head which has adjustable axes so that the crystal can be positioned in the center of an incident X-ray beam. The cross sectional area of the beam is such that the crystal is exposed to nearly monochromatic X-radiation of uniform intensity. The goniometer head is attached to a goniometer, a device that rotates the crystal through various angles, keeping the crystal in the center of the incident beam at all times. A photon detection device is attached to the goniometer, allowing for the measurement of the intensities from the various reflections as the crystal is rotated into the orientations necessary to satisfy the diffraction conditions for each reflection.

In Section 3.2 we determined that the diffraction conditions for a one-dimensional lattice were met when a reciprocal lattice point intersected a circle with a radius of $1/\lambda$ — the Ewald circle. The *Ewald sphere*, more descriptively termed *the sphere of reflection*, is the three-dimensional extension of the Ewald circle. As shown in Fig. 4.1, a sphere of radius $1/\lambda$ is constructed with the incident beam passing through the center of the sphere and the origin of the reciprocal lattice. The equatorial great circle is assumed to be the xy plane of the "laboratory reference frame" (e.g., parallel to the floor). The crystal is rotated about the origin until a reciprocal lattice point intersects the sphere at some point p. The great circle that contains the origin and point p is shown in the upper right of the figure.

Understanding Single-Crystal X-Ray Crystallography. Dennis W. Bennett

ISBN: 978-3-527-32677-8 (HC), 978-3-527-32794-2 (SC)

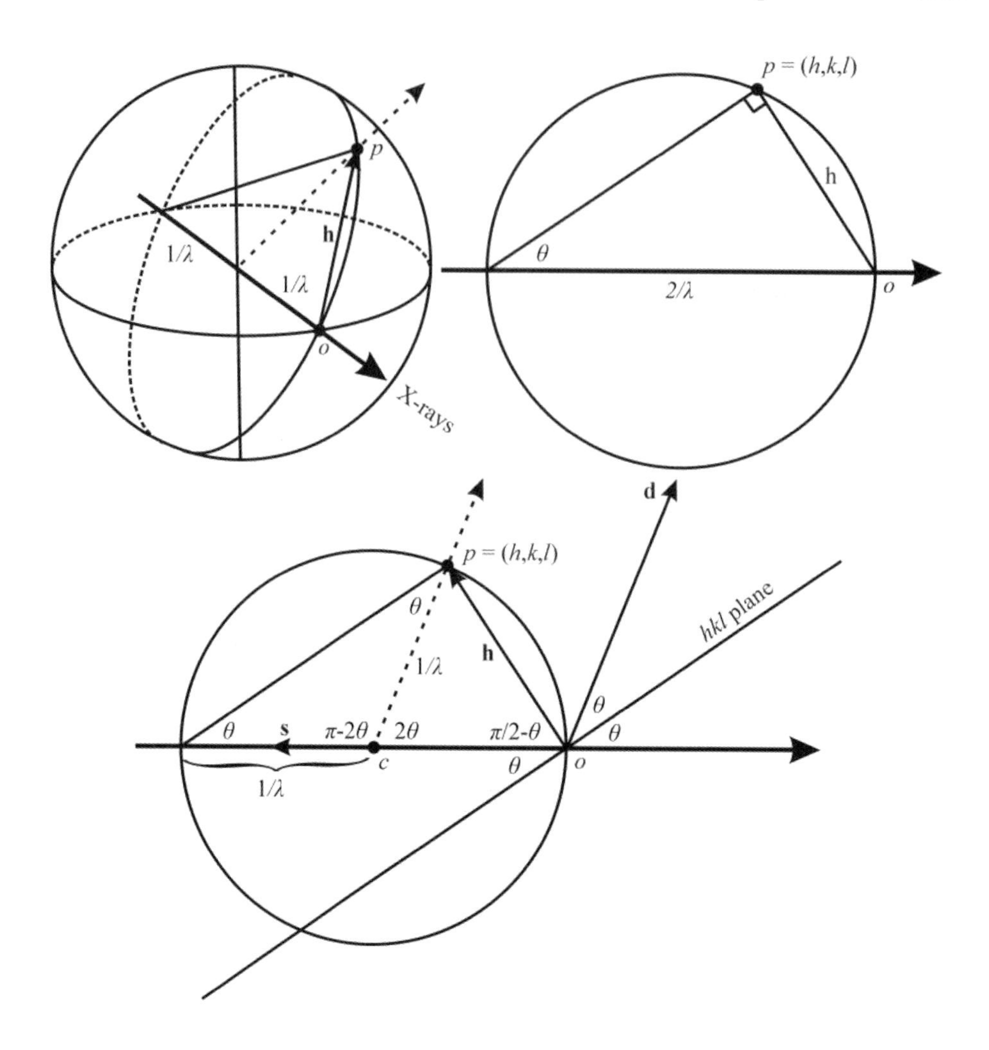

Figure 4.1 Intersection of a reciprocal lattice point with the sphere of reflection.

The triangle in the figure is inscribed in a circle, and is therefore a right triangle. It follows that the sine of the angle θ is the length of the reciprocal lattice vector, h, divided by the diameter of the circle, $2/\lambda$: $\sin\theta = \mathrm{h}/(2/\lambda) = \lambda\mathrm{h}/2$, and $\lambda\mathrm{h} = 2\sin\theta$ — Eqn. 3.49 — *the Bragg diffraction law.* θ is the angle that the incident beam must make with the planes perpendicular to the $\overrightarrow{op} \equiv \mathbf{h}_{hkl}$ reciprocal lattice vector.

The drawing at the bottom of Fig. 4.1 shows that the angle, θ, *is* the angle between the incident beam and the *hkl* planes, and is therefore the angle necessary for diffraction to occur. Just as in the one-dimensional case, diffraction conditions are met when a three-dimensional reciprocal lattice point intersects the sphere of reflection. The angle between the diffracted beam and the incident beam is 2θ. Although the actual diffraction occurs from the scatterers in the direct lattice, the vector from the center of the sphere of reflection through the intersecting reciprocal lattice point (dashed arrow in the figure) is in the same plane as the diffracted vector, also at an angle of 2θ with respect to the incident beam. The two vectors are parallel and the scale of the experiment is such that we can assume that the diffracted beam from a given set of *hkl* planes emanates from the center of the sphere of reflection and passes through the (h, k, l) reciprocal lattice point — the head of the $\mathbf{h}_{hkl}$ reciprocal lattice vector perpendicular to the planes.

4.2 Recording the Diffraction Pattern: Film Methods.

In section 3.2.1 a distorted image of the one-dimensional reciprocal lattice was created by rotating the lattice line about the origin — generating a reflection that was observed as a "spot" on a screen or other area detector each time a reciprocal lattice point intersected the Ewald circle (Figs. 3.15 and 3.16). Similarly, rotation of a three dimensional crystal about an axis perpendicular to the incident beam will create a distorted projected image of the reciprocal lattice whenever a reciprocal lattice point intersects the sphere of reflection, schematically illustrated in Fig. 4.2.

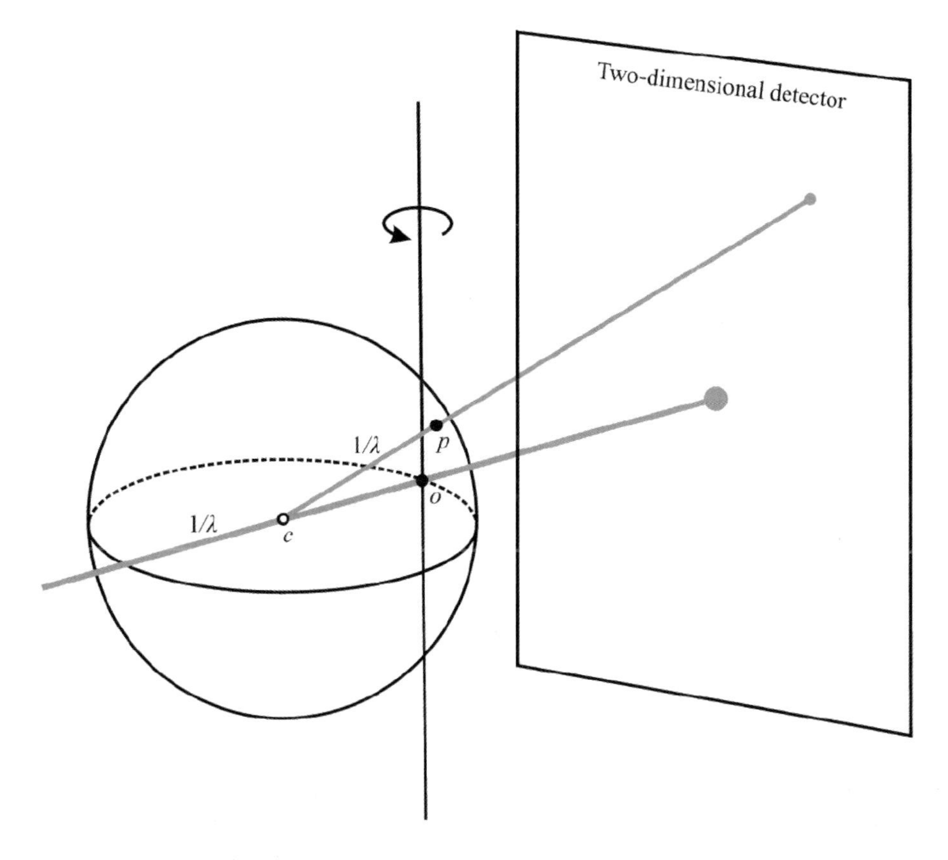

Figure 4.2 Rotation of a crystal about an axis perpendicular to the incident beam producing a reflection each time a reciprocal lattice point, p, intersects the sphere of reflection. Point o represents a common origin for the direct and reciprocal lattices; c is the center of the Ewald sphere. Relative scales are distorted for illustrative purposes.

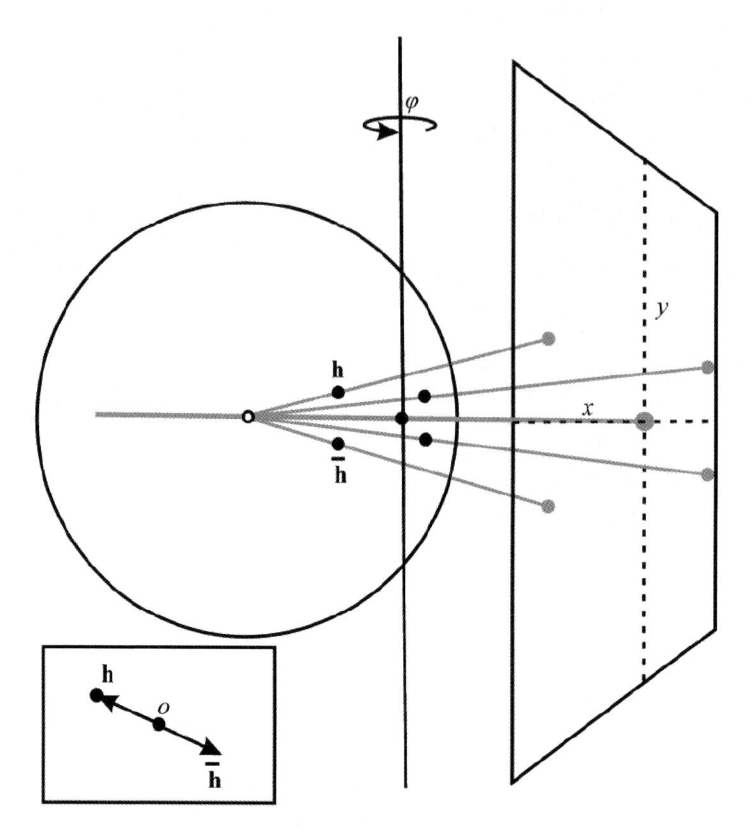

Figure 4.3 Rotation of the crystal around an axis (*the* φ *axis*) perpendicular to the incident beam results in the intersection of reciprocal lattice points and their negatives with the sphere of reflection, creating an image with x and y mirror symmetry.

The direction of the diffracted beam is parallel to the vector from the center of the sphere of reflection to the reciprocal lattice point that intersects the sphere. Since the relative scaling of the direct and reciprocal lattices is arbitrary, the sphere can be constructed so that its size is orders of magnitude smaller than the distance to the screen, allowing us to use this vector as the actual reflection vector. The scaling in the figure is exaggerated for graphical purposes.

As the crystal is rotated though 360° a given reciprocal lattice point intersects the sphere of reflection twice. If the beam is considered to be aligned along the z axis with the rotation axis aligned with the y axis,* then the vector $[h\ k\ l]$ will intersect the sphere at a point (x, y, z) on the sphere, and again at $-x, y, z$. In addition, The negative of the reciprocal lattice vector, $[\bar{h}\ \bar{k}\ \bar{l}]$, will intersect the sphere at $(x, -y, z)$ and $(-x, -y, z)$. The xz and yz planes are therefore mirror planes for the points of intersection. This mirror symmetry is reflected in the image of the reciprocal lattice on the screen as depicted in Fig. 4.3.

If the two-dimensional detector is a sheet of photographic film, then the image can be recorded permanently. The *rotation photograph* shown on the left in Fig. 4.4

*We will refer to this Cartesian coordinate system as *the laboratory reference frame.*

Figure 4.4 Rotation photographs of a ruby crystal about an arbitrary axis (left) and a sucrose crystal about the b unit cell axis (right).

is from a rhombohedral ruby (aluminum oxide) crystal which is centered at the origin of the laboratory reference frame, but is at an arbitrary orientation with respect to the x, y, and z axes.

The photograph on the right is from a sucrose crystal (Fig. 1.3). The monoclinic crystal has been aligned so that the b axis is coincident with the y rotation axis of the laboratory reference frame, creating an *axial rotation photograph.* Recall that the a^*c^* plane is perpendicular to the b axis — the reciprocal lattice points are now in planes which are parallel to the xz plane. The reciprocal lattice therefore intersects the sphere in parallel layers, creating parallel lines of reflections on the film. The line of reflections in the center of the photograph results from the $(h\ 0\ l)$ layer in the reciprocal lattice. The layers above the line are for the $(h\ 1\ l)$, $(h\ 2\ l)$, etc. reflections and those below the line are for the $(h\ \bar{1}\ l)$, $(h\ \bar{2}\ l)$, etc. reflections, as shown on the left in Fig. 4.5*

The measurement of distances between the *layer lines* in axial rotation photographs allows us to determine the lengths of the unit cell axes. In the drawing on the right in Fig. 4.5, D is the distance from the crystal to the film, L is the distance between layer lines, and d^* is the distance between a^*c^* planes in the reciprocal lattice. The reflection vector intersects the sphere of reflection at a distance of $1/\lambda$ from the center of the sphere at a height of nd^* from the *zero layer line*, where n is the value of the k index of the reciprocal lattice point intersecting the sphere. As shown in the figure, the reflection vector creates similar triangles with $\sin\eta = d^*/(1/\lambda) = \lambda d^*$ and $d^* = (\sin\eta)/\lambda$, where $\eta = \arctan(L/D)$. d^* is the projection of $\mathbf{b}^*$ on $\mathbf{b}$. From Eqn. 1.6, $d^* = (\mathbf{b}^* \cdot \mathbf{b})/b$. Since $\mathbf{b}^* \cdot \mathbf{b} = 1$, $d^* = 1/b$ and $b = \lambda/(\sin\eta)$. The measurement of the interlayer spacings from ro-

*The sucrose unit cell is monoclinic and the b^* axis is parallel to the unique b axis. The figure depicts the general (triclinic) case.

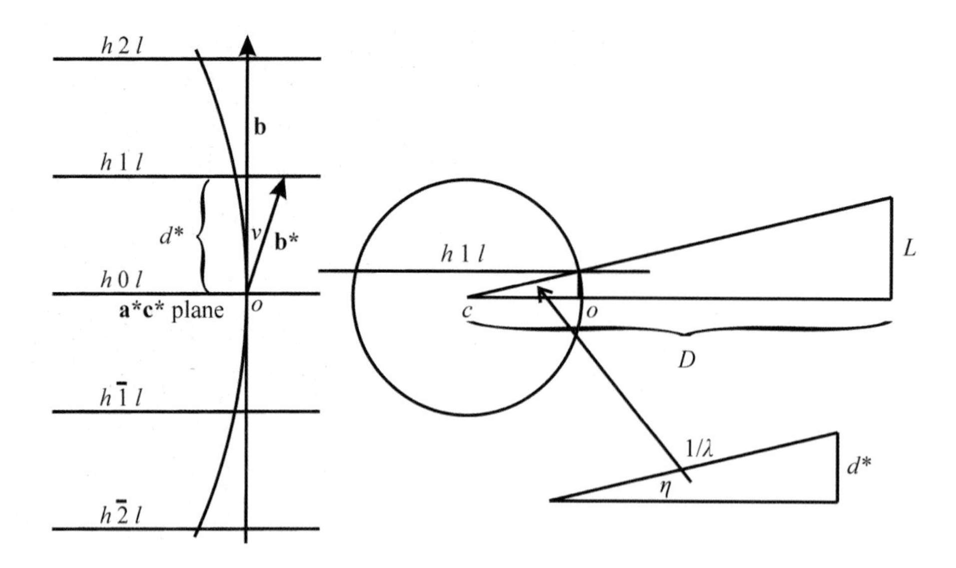

Figure 4.5 Left: Intersection of reciprocal lattice planes with the sphere of reflection during axial rotation. Right: Similar triangles created by reflection vectors for the $(h\ 0\ l)$ and $(h\ 1\ l)$ planes.

tation photographs around the a and c axes will provide the lengths of a and c as well.

The brightness of the spots in the photographs in Fig. 4.4 illustrates the varying intensities of the reflections from each reciprocal lattice layer. If we were somehow able to measure the degree of exposure of each spot on the film we could determine the relative intensities of the reflections creating each spot. This, in turn, would give us the magnitudes of the structure factors for each reflection. Unfortunately, while we are able to assign a reciprocal lattice layer to a specific line in the axial photograph on the right in the figure, we are unable to assign indices to the reflections (a process known as *indexing*), since we do not know which reciprocal lattice vector created a given spot. Furthermore, if the lengths of the reciprocal lattice vectors in the layer plane are similar, the reflections will overlap, and the exposures due to coincident reflections will combine. If we were able determine the rotation angle, φ, corresponding to each reflection, we would know the orientation of the crystal when the reflection occurred, and would be able to index the reflections in the photograph, provided that we could find a way to avoid overlapping reflections. Measurement of the relative intensity would then give us F_{hkl} for each reflection.

In the early days of single crystal structure determination, *film methods* provided the only practical means to the measurement of structure factor amplitudes, and a number of strategies evolved to measure the relative exposure and index the spots on film in various configurations. The first approach involved a simple modification of the *rotation method* known as *the oscillation method.* Although straightforward, the method is time consuming and tedious. It begins with the mounting and alignment of a crystal on a goniometer head equipped with two mutually perpendicular arcs so that the crystal can be rotated about its center, and translational adjustments so that the crystal remains oriented in the center of the incident beam as it

Figure 4.6 Goniometer head with lower translational adjustments and upper angular adjustments. Insert: Crystal glued to a glass capillary and mounted on the goniometer head.

rotates (Fig. 4.6). The crystal is commonly glued to a glass fiber and attached to a small post which is inserted in the goniometer head. The alignment exercise alone often takes a number of trials, since the crystal must be oriented so that a crystallographic axis is coincident with the rotation axis. An alternating series of rotation photographs and goniometer head adjustments is performed until the crystal axis is aligned along the φ axis and an axial rotation photograph is obtained. Once the crystal is aligned along an axis, it is oscillated back and forth through a small angle repeatedly (since a single oscillation will fail to expose the film significantly). Only those reflections occurring within a limited range of φ will be observed, allowing for the assignment of indices to the limited set of reflections.* A number of photographs are taken at various increments of φ angles until the full 360° rotation has taken place.

A major disadvantage of the rotation method was the large number of photographs necessary to unambiguously assign the indices of individual reflections, since it was necessary to include redundant reflections for the different increments of φ in order to compare the exposed spots on one photograph to another. In

*The details of the index assignment for various film methods will not be discussed here. Nevertheless, a careful study of these methods is pedagogically useful and historically important and the interested reader is encouraged to refer to[47], [48], and [49] for comprehensive treatments of the rotation/oscillation, Weissenberg, and precession methods, respectively.

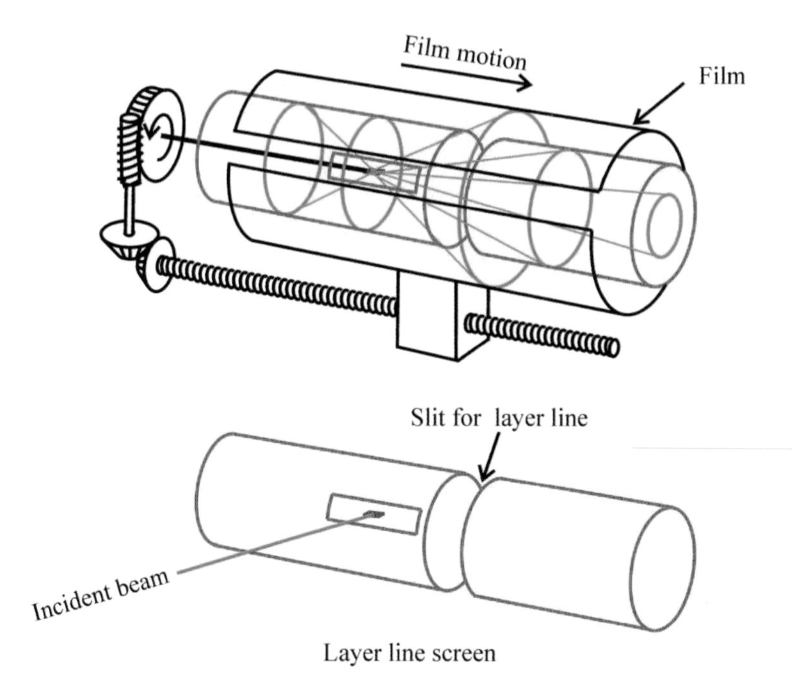

Figure 4.7 Schematic representation of the Weissenberg goniometer. Reflections from various reciprocal lattice layers are indicated by the cones shown in green. The layer line screen allows reflections from a single reciprocal lattice layer to reach the film, which is translated as the crystal rotates.

1924 the physicist/chemist/mathematician/engineer/crystallographer, Karl Weissenberg, published a paper[50] describing an instrument that removed the ambiguities discussed above — by *translating the film* while rotating the crystal. The *Weissenberg goniometer* is illustrated schematically in Fig. 4.7. The film encircles the crystal in a cylindrical holder, with the crystal rotating around an axis passing through the center of the cylinder. The film holder is attached to a gear mechanism that couples the crystal rotation with the film translation. A movable screen with a slit is placed so that a specific layer line of reflections is allowed to reach the film, with all other reflections blocked from doing so. The location of a given spot on the film resulting from a specific reflection is a function of the crystal rotation angle *and* film translation distance — each reflection can be identified and indexed from its position on the film.

The motion of the film in the *Weissenberg method* is different from the motion of the recorded layers of reciprocal lattice, resulting in a distorted image (Fig. 4.9). As we noted earlier, the *symmetry* of the reciprocal lattice is important in space group determination, and although the Weissenberg method allowed for the determination of reflection intensities, the distortion of the reciprocal lattice image made it difficult to assess its symmetry. In the late 1930s, W.F. de Jong and J. Bouman[51] developed another *moving film method* in which the film rotated synchronously with the crystal, creating an undistorted image of the reciprocal lattice. The *de*

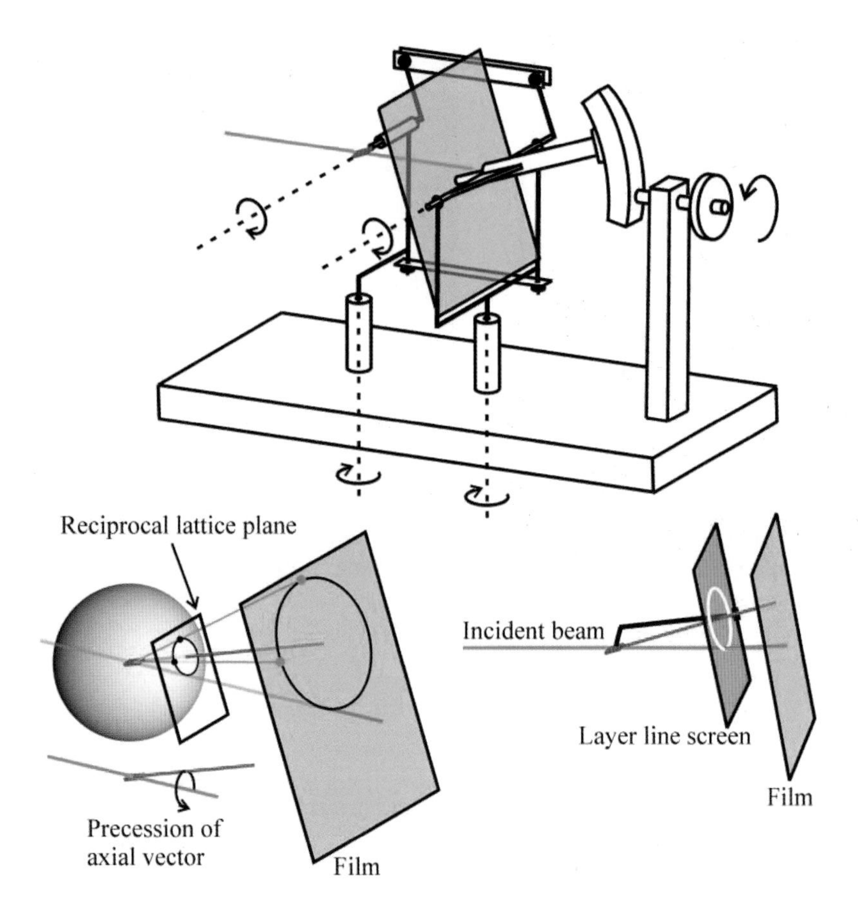

Figure 4.8 Top: Schematic representation of the Buerger precession goniometer. Bottom left: Synchronous precession of a tilted reciprocal lattice plane and film produces an undistorted image of the reciprocal lattice plane. Bottom right: Layer line screen allows for the recording of a single reciprocal lattice layer.

Jong–Bouman method was a modification of the rotation method, and was difficult to implement since it required simultaneous rotation of the crystal and film. In the early 1940s the chemist/mining engineer/mineralogoist/crystallographer, Martin J. Buerger, developed an ingenious moving film method[48,49] in which the same effect could be accomplished with a synchronous precession of the film and unit cell axis — obviating the need to rotate the film around the axis. The principle is illustrated at the bottom left of Fig. 4.8. Unlike the rotation method, in which the unit cell axis is tangent to the sphere of reflection and perpendicular to the incident beam, the cell axis is aligned with the incident beam, with the perpendicular reciprocal lattice planes also perpendicular to the beam. The unit cell axis is then tilted at an angle with respect to the incident beam; each reciprocal lattice layer now intersects the sphere of reflection in a circle. The film is simultaneously tilted *at the same angle* so that the reciprocal lattice layers and the film remain parallel (normal to the unit cell axis). The rotation of the axial vector around the beam axis is known as *precession*. *If* the crystal and film precess in sync through 360° around the in-

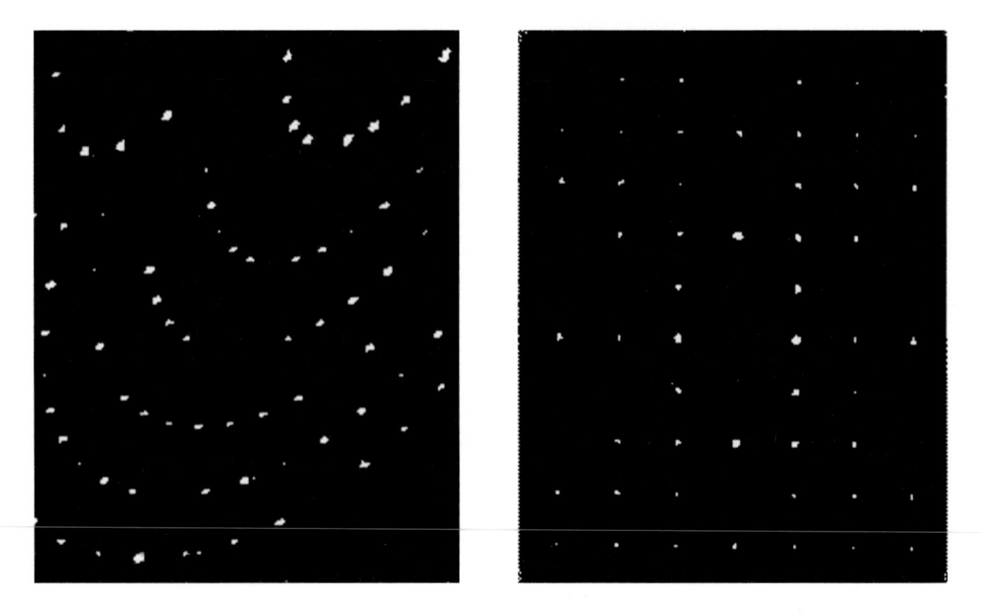

Figure 4.9 Left: Distorted image of a reciprocal lattice plane in a Weissenberg photograph. Right: Undistorted image of a reciprocal lattice plane in a precession photograph.

cident beam axis the tilt angle is maintained; the circle of intersection moves with the axis and is projected onto the film, which simultaneously moves with the axis. Each time a reciprocal lattice point crosses the edge of the circle, a reflection is "projected" to an equivalent point on the film. Because the film motion parallels the motion of the reciprocal lattice, each spot on the film is at the same relative position on the film and as it is in the reciprocal lattice; *creating an undistorted image of the lattice* (Fig. 4.9). While the direct rotation of the film and crystal provide no advantages over the de Jong–Bouman method, Buerger noted that gyroscopes are able to precess without undergoing complete rotation by employing a complex series of motions effected by the use of rotating forks known as *gimbals*. Based on this principle he developed the *Buerger precession goniometer*, illustrated schematically at the top of Fig. 4.8. The film and crystal are coupled in such a manner that the film always remains normal to the unit cell axis and parallel to the reciprocal lattice layers as they cross the sphere of reflection during the precession motion. The motion is effected by rotating an "off axis" arc (which establishes the tilt angle), inducing the horizontal and vertical rotations indicated in the figure, which combine to create the simultaneous precession of the crystal and film. The zero-layer of the reciprocal lattice intersects the sphere with a relatively small circle; higher-level layers intersect with larger circles. Thus each circle creates a *reflection cone* of a different size, and a specific reciprocal lattice layer can be recorded by blocking the reflections from other layers with a screen containing an annular opening of the appropriate radius, as depicted at the bottom right of the figure. The layer line screen is attached to the horizontal rotation axis of the crystal – it too remains normal to the unit cell axis. All three entities, the circle of intersection, the annular opening of the layer line screen, and the projected circle onto the film lie on the perimeter of the reflection cone, parallel to its base.

Although they have largely been replaced by more modern methods, the importance of photographic methods to the development of single crystal structure determination cannot be overstated. At their best, however, the determination of reflection intensities from film data required a significant effort, much of which was involved in the necessary alignment of the crystal along a unit cell axis prior to the collection of the data. The measurement of the intensities required an additional effort. The degree of film exposure was either estimated visually or recorded with an optical device known as a densitometer. Neither method produced especially accurate intensities, and both were dependent on the quality of the photograph, especially for the measurement of weaker reflections. In the early days of crystallography X-ray crystal structure determination was not an undertaking for the faint-hearted, and the *relatively* routine determination of X-ray crystal structures had to await the development of methods in which indices and accurate intensities from an arbitrarily aligned crystal could be determined. The advent of the digital computer, photon counting devices, and the automated diffractometer made such a task possible, and it is there that we now focus our discussion.

4.3 Recording the Diffraction Pattern: Counter Methods.

4.3.1 Serial Detectors

In the late 1940s the measurement of the diffraction patterns from powdered solids had developed to a point where commercial instruments were available that employed Geiger-Müller counters rather than film to record diffraction intensities. Geometric constraints rendered the recording of intensities from the diffraction of single crystals more difficult. In 1952 M. J. Buerger devised the first apparatus designed specifically for the measurement of intensities from single crystals. The instrument incorporated a modification of the rotation method to accommodate a Geiger-Müller counter that could be moved in both horizontal and vertical planes. The intensity of each reflection was measured by manually rotating the crystal and moving the counter to calculated positions where the diffracted beam could enter the counter aperture, then oscillating the crystal through a small angle while counting the diffracted photons. The device constrained the crystal to its rotational axis in the equatorial (xy) plane of the laboratory reference frame, with the counter moving in three dimensions so that it could point toward the various reflections emanating from the crystal. Yet another ingenious device created by Buerger, the three dimensional motion of the detector and its attached wiring proved somewhat cumbersome, and an alternative apparatus, one that moved the crystal and constrained the detector to the equatorial plane, was devised by Thomas C. Furnas and David Harker in 1953.[52] This new apparatus was the prototype for today's modern single-crystal diffractometer.

In the modern diffractometer system a crystal is mounted onto a goniometer head with translational adjustments as illustrated in Fig. 4.10. These adjustments allow for a crystal with an arbitrary orientation to be centered in the X-ray beam so that its center of mass remains at a single point for any rotation of the crystal about its center. The goniometer is constructed so that a given reflection vector, which depends on the orientation of the unit cell in the crystal, can point toward

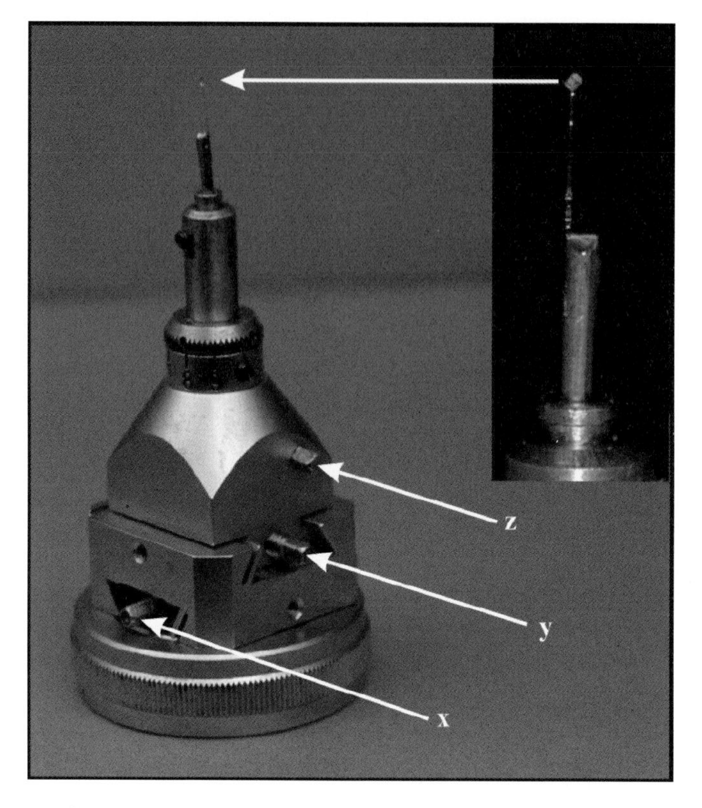

Figure 4.10 Goniometer head with x, y, and z translational adjustments. Insert: Crystal glued to a glass fiber and mounted on the goniometer head

the detector, with coordinates in the laboratory (goniometer) reference frame. Until relatively recently the detector was almost universally a scintillation counter, designed to count photons for individual reflections. Since the detector measured the intensities of each reflection in serial order, it has been called a *serial detector* to differentiate it from the increasingly common *area detector*. The mathematics for diffractometers was derived for serial detectors, then modified for area detectors, and we will take that approach here.

The operation that orients each reflection so that its intensity can be recorded is depicted graphically in Fig. 4.11. Fig. 4.11(a) shows a crystal centered in the laboratory reference frame, defined by the Cartesian vectors $\mathbf{x}_l$, $\mathbf{y}_l$, and $\mathbf{z}_l$, with the xy plane parallel to the "laboratory floor". $\mathbf{h}$ is the reciprocal lattice vector responsible for the reflection; a perpendicular $(h\ k\ l)$ plane in the direct lattice is shown in blue. The instrument is designed so that the reciprocal lattice vector can be rotated into the xy plane with the reflection vector at an angle 2θ with respect to the incident X-ray beam (Fig. 4.11(b)). Fig. 4.11(c) is a view down the $\mathbf{z}_l$ axis (pointing toward the viewer). The $(h\ k\ l)$ planes are now perpendicular to the xy plane of the goniometer, oriented so that a detector at angle 2θ can record the intensity under diffraction conditions, $\lambda\mathbf{h} = \mathbf{s_1} + \mathbf{d_1}$ (Fig. 4.11(d)).

Since the crystal is not translated, the orientation of the reciprocal lattice vector is effected by rotation of the crystal. The most convenient rotation would be about

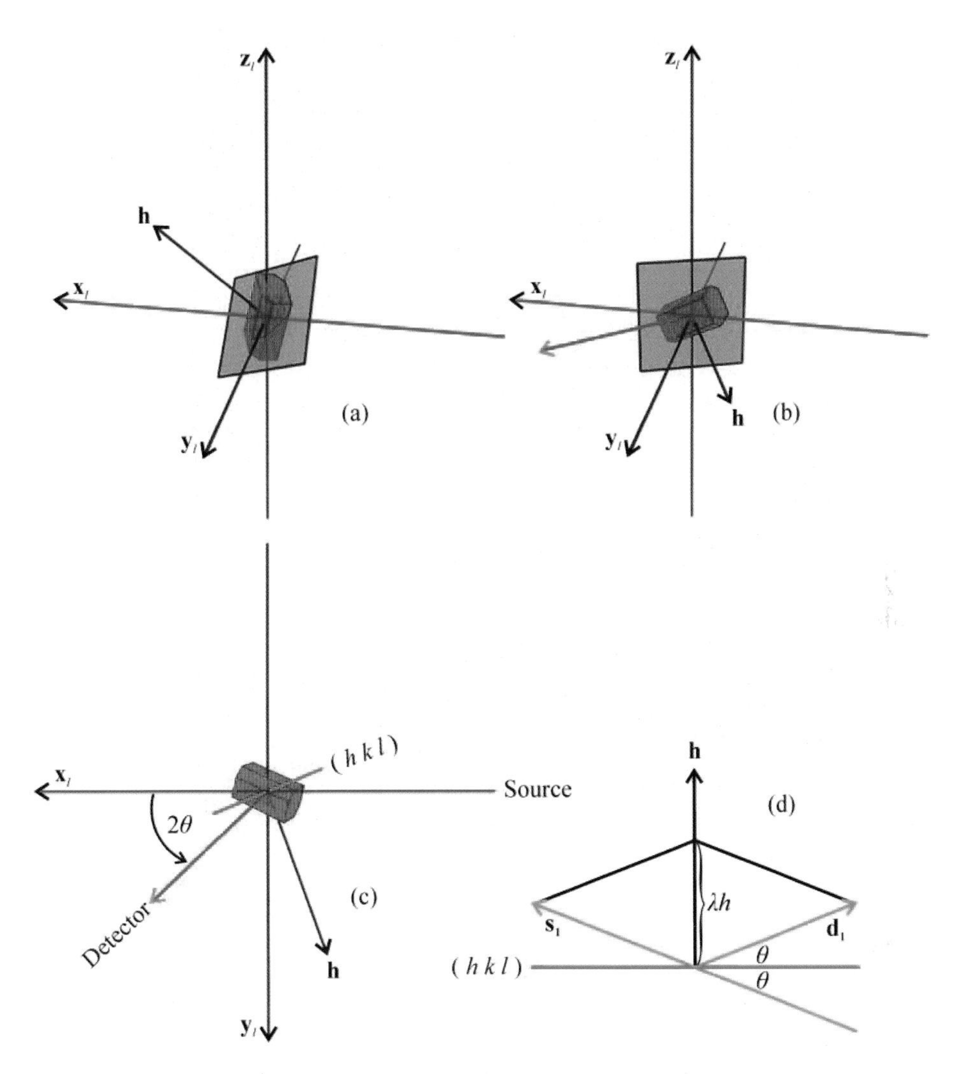

Figure 4.11 (a) Crystal and reciprocal lattice vector in an arbitrary orientation with respect to the laboratory reference frame. (b) Crystal rotated into the diffraction condition in the equatorial plane of the reference frame. (c) View of the diffracting crystal along the $\mathbf{z}_l$ axis of the reference frame. (d) Conditions for diffraction.

an axis perpendicular to the plane containing the vector before and after rotation, but since each reciprocal lattice vector points in a different direction it would be very difficult to construct an instrument that would perform single rotations about rotation axes pointing in virtually every direction in three dimensions. Rotational systems are common in classical mechanics, and a transformation initially derived by Euler is especially useful for the rotation of rigid bodies (with their own internal coordinate systems), in terms of a fixed external coordinate system. In our case the rigid body is the crystal, and the external system is the laboratory reference frame. The Euler transformation utilizes three rotations; the angles of these rotations are known as the *Eulerian angles*. There are a number of conventions for the Euler rotations — we will adopt the following: The laboratory reference frame is a right-handed coordinate system with the x_l axis in the direction of the incident beam, pointing *away* from the source. The first rotation, φ, is a clockwise (left-handed) rotation around the z_l axis; the second, χ, is a clockwise rotation about the x_l axis; the third, ω, is a counter-clockwise (right-handed) rotation about the z_l axis.*

Fig. 4.12(a) represents a reciprocal lattice vector **h** and its perpendicular $(h\ k\ l)$ plane, shown in blue. A logical way to rotate **h** into the diffraction condition in the x_l, y_l plane is depicted in Figs. 4.12 (b)–(d). In this case φ_0 is chosen so that **h** rotates into the y_l, z_l plane. χ_0 then rotates the vector onto the y_l axis, followed by an ω_0 rotation in the x_l, y_l plane to an angle of $\pi/2 - \theta$ with respect to the x_l axis. The $(h\ k\ l)$ plane is then at the diffraction angle θ with respect to the incident beam along the x_l axis. For reasons soon to become apparent we will call this specific orientation of the crystal the *bisecting position.* This is not the only way to rotate the vector into the diffraction condition. As shown in Figs. 4.12 (e)–(g), φ_1 can be an arbitrary rotation, provided that χ_1 is selected so that **h** is rotated into the x_l, y_l plane. This allows ω_1 to rotate the vector into the diffraction condition. Note that the only difference between Fig. 4.12(d) and Fig. 4.12(g) is the orientation of the $(h\ k\ l)$ plane. In each case the plane is perpendicular to the x_l, y_l plane, and at an angle θ with respect to the x_l axis — but the plane — and therefore the crystal — has rotated around the reciprocal lattice vector, **h**. It follows that different combinations of the rotation angles can rotate the crystal around the reciprocal lattice vector with the crystal remaining in the diffraction condition. The rotational angle is given the symbol ψ, with $\psi = 0$ corresponding to orientation of the crystal in the bisecting position.

While it might appear that it would be unnecessary to record intensities in any orientation other than the simplest – the bisecting position – the ability to record intensities by rotating the crystal about ψ provides two distinct advantages. First, the actual goniostat must be constructed so that its various components move in three dimensions, and the mechanical constraints imposed can make it difficult to record certain reflections in the bisecting position. Second, the actual intensities are attenuated as the incident and diffracted X-rays travel through the crystal due to a phenomenon know as *absorption*, which we will discuss in the next chapter. The longer the path of travel, the more the reflection is attenuated, and determining the variation in intensities from *the same* $(h\ k\ l)$ planes in different orientations (and path lengths) can assist in correcting for these changes in absorption.

Furnas and Harker constructed a goniometer that allowed for independent rotations of the Eulerian angles. The goniometer head was mounted onto a semi-circular track; the χ rotation occurred when the goniometer head moved along the track. The goniometer head was also able to rotate independently about its own longitudinal axis, the φ rotation, and the track and goniometer head assembly could be rotated simultaneously about the z_l axis — the ω rotation. The cradle-like shape of the χ semicircle, along with the Eulerian rotations, prompted the inventors to describe their goniometer as an *Eulerian Cradle.* This term became general for similar goniometers, even though the "cradle" was soon replaced with a complete circle for the χ rotation. The mechanisms for rotating the crystal through the Eulerian

*This is only one of several alternative conventions for the definitions of the Eulerian angles and laboratory (goniometer) coordinate system, including another in the *International Tables for Crystallography.* In addition to differences in angle sign and axis direction, ω is often defined with respected to a rotating reference frame rather than the fixed (laboratory) frame.

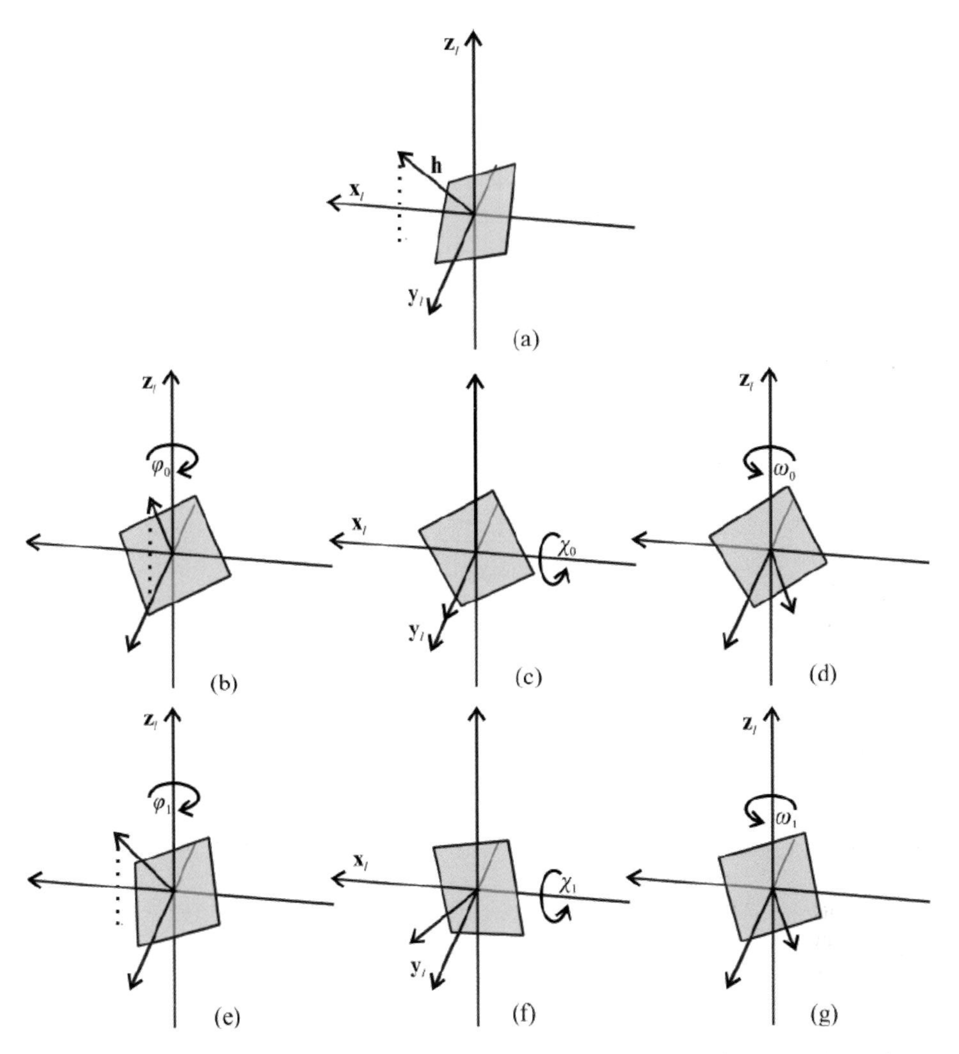

Figure 4.12 (a) Crystal and reciprocal lattice vector, **h**, in an arbitrary orientation. The perpendicular $(h\ k\ l)$ plane is shown in blue. (b) φ_0 rotation of **h** onto the y_l, z_l plane. (c) χ_0 rotation of **h** onto the y_l axis. (d) ω_0 rotation of **h** into the diffraction condition. (e) φ_1 rotation of **h**. (f) χ_1 rotation of **h** onto the x_l, y_l plane. (g) ω_1 rotation of **h** into the diffraction condition.

angles are generally referred to as the *phi*, *chi*, and *omega* "circles." In addition to these three, a fourth circle that rotates the detector in the x_l, y_l plane about the z_l axis to the diffraction angle is referred to as the *two-theta* circle. The 2θ rotation angle is right-handed, in the same direction as the ω rotation.

The *four-circle goniometer*, illustrated schematically in Fig. 4.13, has been the mainstay of single crystal diffraction studies for over four decades.* While goniome-

*An often-used alternative to the Eulerian cradle replaces the *chi* circle with an arm that rotates about an axis at fixed angle with respect to the z_l axis. The rotation angle, κ, is a linear combination of χ and ω; the goniostat incorporating *kappa* geometry is less geometrically constrained than the Eulerian cradle (See Sec. 4.7.5).

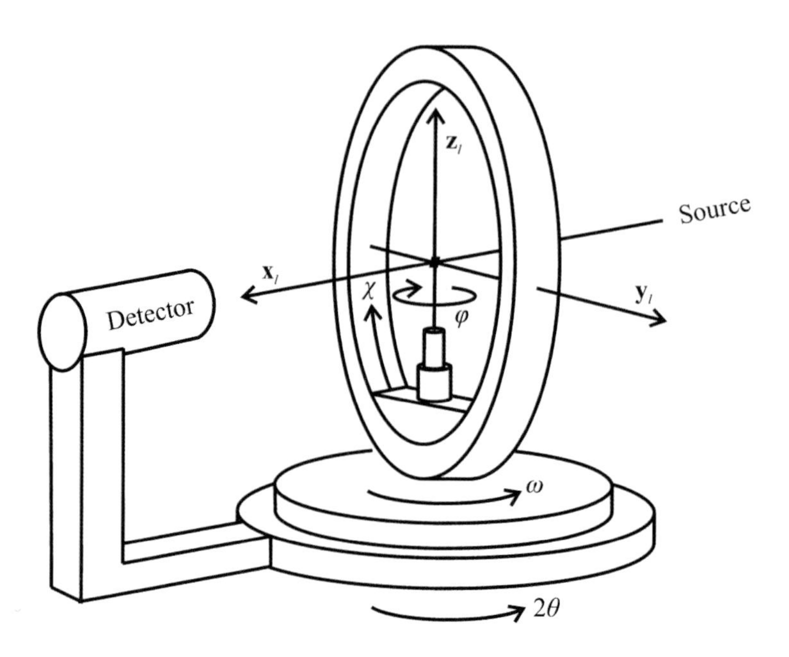

Figure 4.13 Schematic representation of a four-circle Eulerian goniometer. φ and χ are clockwise rotations, ω and 2θ are counter-clockwise rotations. When the angles are set to zero, the χ circle is coincident with the y_l, z_l plane; the rotational axis of the χ circle is coincident with the x_l axis, pointing toward the detector at $2\theta = 0$, and the rotational axis of the φ circle is aligned with the z_l axis. The rotational axes of the ω and 2θ circles are always aligned with the z_l axis.

ter configurations are now being modified to accommodate the new area detectors the underlying principles providing the basis for their design remain those originally established for the four-circle goniometer. The determination of intensities from a *four-circle diffractometer* requires a reorientation of the reciprocal lattice vector for each reflection into the diffraction condition in the x_l, y_l plane. In order to determine the Eulerian angles necessary to accomplish this task the vector must first be transformed into Cartesian coordinates based on the crystal. These coordinates can then be transformed into coordinates in the laboratory (diffractometer) reference frame and rotated through φ, χ, and ω into the appropriate position to observe diffraction.

The transformation of a general reciprocal lattice vector, $\mathbf{v}_\mathbf{f}^* = x_f^*\mathbf{a}^* + y_f^*\mathbf{b}^* + z_f^*\mathbf{c}^*$, into reciprocal Cartesian coordinates is accomplished in exactly the same manner as the transformation of a direct lattice vector into direct lattice Cartesian coordinates (Sec. 1.5.2). The reciprocal Cartesian coordinate system, illustrated in Fig. 4.14, is chosen so that the $\mathbf{a}^*$ axis is coincident with the $\mathbf{i}$ axis, the $\mathbf{b}^*$ axis is coplanar with $\mathbf{i}$ and $\mathbf{j}$, and the z component of the $\mathbf{c}^*$ axis points in the same direction as $\mathbf{k}$. The components of the reciprocal lattice vector in Cartesian coordinates are given by $\mathbf{v}_\mathbf{c}^* = [x_c^* \; y_c^* \; z_c^*]$ such that $\mathbf{v}_\mathbf{c}^* = x_c^*\mathbf{i} + y_c^*\mathbf{j} + z_c^*\mathbf{k}$. Replacing a, b, c, α, β, and γ in Eqns. 1.116–1.129 with a^*, b^*, c^*, α^*, β^*, and γ^* results in

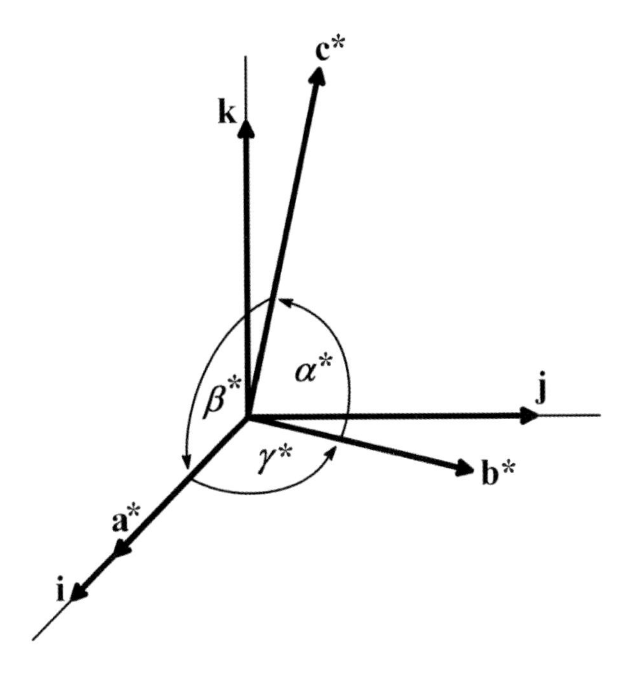

Figure 4.14 The standard orientation of a reciprocal unit cell in a Cartesian coordinate system. The unit vectors have dimensions of 1/length.

the matrix that transforms $\mathbf{v}_\mathbf{f}^*$ into $\mathbf{v}_\mathbf{c}^*$:

$$\mathbf{B}^* = \begin{bmatrix} a^* & b^* \cos\gamma^* & c^* \cos\beta^* \\ 0 & b^* \sin\gamma^* & \frac{c^*(\cos\alpha^* - \cos\beta^* \cos\gamma^*)}{\sin\gamma^*} \\ 0 & 0 & \frac{V^*}{a^* b^* \sin\gamma^*} \end{bmatrix}. \tag{4.1}$$

$$\mathbf{B}^* \begin{bmatrix} x_f^* \\ y_f^* \\ z_f^* \end{bmatrix} = \begin{bmatrix} x_c^* \\ y_c^* \\ z_c^* \end{bmatrix}. \tag{4.2}$$

The relationships between direct and reciprocal lattices allow us to simplify the Cartesian transformation matrices for real and reciprocal lattice vectors:

$$\begin{aligned} \frac{c^*(\cos\alpha^* - \cos\beta^* \cos\gamma^*)}{\sin\gamma^*} &= -\frac{c^* \sin\beta^*(\cos\beta^* \cos\gamma^* - \cos\alpha^*)}{\sin\beta^* \sin\gamma^*} \\ &= -c^* \sin\beta^* \cos\alpha \end{aligned} \tag{4.3}$$

$$\frac{V^*}{a^* b^* \sin\gamma^*} = \frac{1}{c} \tag{4.4}$$

$$\mathbf{B} = \begin{bmatrix} a & b\cos\gamma & c\cos\beta \\ 0 & b\sin\gamma & -c\sin\beta\cos\alpha^* \\ 0 & 0 & 1/c^* \end{bmatrix} \tag{4.5}$$

$$\mathbf{B}^* = \begin{bmatrix} a^* & b^*\cos\gamma^* & c^*\cos\beta^* \\ 0 & b^*\sin\gamma^* & -c^*\sin\beta^*\cos\alpha \\ 0 & 0 & 1/c \end{bmatrix}. \tag{4.6}$$

$\mathbf{B}^*\mathbf{v}_\mathbf{f}^* = \mathbf{v}_\mathbf{c}^*$ expresses the reciprocal lattice vector in terms of a Cartesian coordinate system based on the crystal. In order to rotate the vector in the laboratory reference frame we must determine the change-of-basis transformation matrix that will convert the reciprocal lattice Cartesian coordinates into "laboratory" coordinates. This is the matrix that relates the orientation of the crystal unit cell (specifically, the reciprocal unit cell) to the diffractometer coordinate system — it is ordinarily given the symbol **U** and is known as the *orientation matrix*. At this stage we have no knowledge of the orientation of the reciprocal lattice; for now we will assume that we are able to determine the orientation matrix and delay the details of its determination until later on in the chapter. **U** transforms the coordinates of the vector $\mathbf{v}_\mathbf{c}^*$ into coordinates in the laboratory basis, $\{\mathbf{x}_l\ \mathbf{y}_l\ \mathbf{z}_l\}$:

$$\mathbf{U}\mathbf{v}_\mathbf{c}^* = \mathbf{U}\,\mathbf{B}^*\mathbf{v}_\mathbf{f}^* = \mathbf{v}_\mathbf{l}^*. \tag{4.7}$$

The Eulerian rotations can now be employed to reorient the vector in the laboratory reference frame. Each set of rotations begin with the diffractometer angles set to zero, i.e., the crystal begins in the same orientation each time. The clockwise φ rotation about the z_l axis is followed by a clockwise χ rotation around the x_l axis, then a counterclockwise ω rotation, again about z_l. The corresponding rotation matrices are denoted $\mathbf{\Phi}$, $\mathbf{X}$, and $\mathbf{\Omega}$, transforming the vector $\mathbf{v_l}^*$ into the rotated vector, $\mathbf{v}_\mathbf{r}^*$: $\mathbf{\Omega}\,\mathbf{X}\,\mathbf{\Phi}\,\mathbf{v}_\mathbf{l}^* = \mathbf{v}_\mathbf{r}^*$, where

$$\mathbf{\Phi} = \begin{bmatrix} \cos\varphi & \sin\varphi & 0 \\ -\sin\varphi & \cos\varphi & 0 \\ 0 & 0 & 1 \end{bmatrix} \tag{4.8}$$

$$\mathbf{X} = \begin{bmatrix} 1 & 0 & 0 \\ 0 & \cos\chi & \sin\chi \\ 0 & -\sin\chi & \cos\chi \end{bmatrix}, \quad \text{and} \tag{4.9}$$

$$\mathbf{\Omega} = \begin{bmatrix} \cos\omega & -\sin\omega & 0 \\ \sin\omega & \cos\omega & 0 \\ 0 & 0 & 1 \end{bmatrix}. \tag{4.10}$$

A general reciprocal lattice vector in the crystal can now be transformed in the laboratory reference frame to a vector rotated through the Eulerian angles:

$$\mathbf{\Omega}\,\mathbf{X}\,\mathbf{\Phi}\,\mathbf{U}\,\mathbf{B}^*\mathbf{v}_\mathbf{f}^* = \mathbf{v}_\mathbf{r}^*. \tag{4.11}$$

The reciprocal lattice vectors responsible for diffraction are those to the lattice points (nodes) in the reciprocal lattice, with integer components, $\mathbf{h} = [h\ k\ l]$, and length h. For a reflection to be observed by a serial detector this vector must

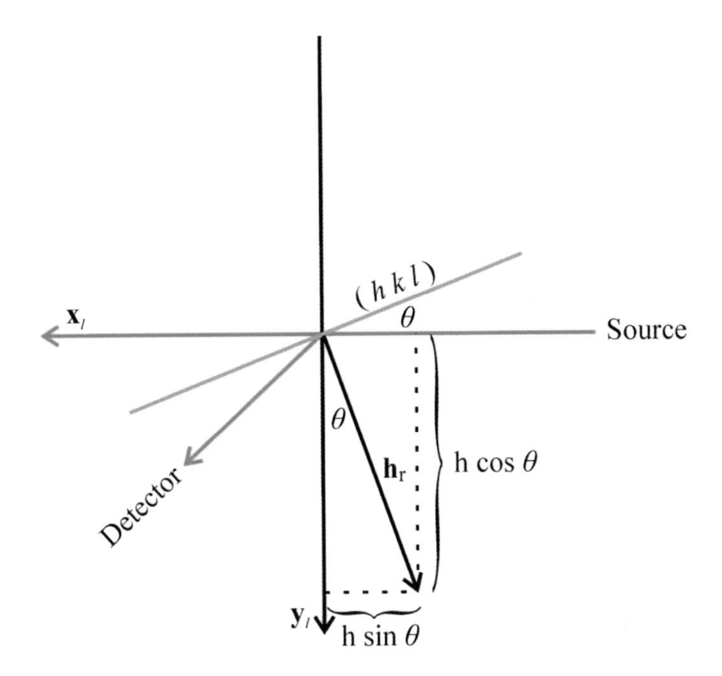

Figure 4.15 Reciprocal lattice vector in the diffraction condition in the x_l, y_l plane. Its components in the diffractometer coordinate system are $[(-\mathrm{h}\sin\theta)\ (\mathrm{h}\cos\theta)\ \ 0]$.

be rotated into the x_l, y_l plane so that the perpendicular lattice planes are at the diffraction angle θ with respect to the incident beam along the x_l axis, as shown in Fig. 4.15. It follows that the angle between the perpendicular reciprocal lattice vector and the y_l axis is also θ. Eqn. 3.50 (the Bragg diffraction law), $\lambda\mathrm{h} = 2\sin\theta$, relates the length of the reciprocal lattice vector to the diffraction angle and the X-ray wavelength: $\mathrm{h} = 2\sin\theta/\lambda$ (the orthogonal transformations of the reciprocal lattice vector have not altered its length). Thus, under diffraction conditions, our vector equation becomes

$$\mathbf{\Omega}\,\mathbf{X}\,\mathbf{\Phi}\,\mathbf{U}\,\mathbf{B}^* \begin{bmatrix} h \\ k \\ l \end{bmatrix} = \begin{bmatrix} -\mathrm{h}\sin\theta \\ \mathrm{h}\cos\theta \\ 0 \end{bmatrix} = \frac{2\sin\theta}{\lambda} \begin{bmatrix} -\sin\theta \\ \cos\theta \\ 0 \end{bmatrix} = \mathbf{h_r}. \tag{4.12}$$

Our aim here is to develop an equation that relates the indices of a specific reflection to the Eulerian angles necessary to rotate the vector into the diffraction condition. We therefore rearrange the equation by premultiplying both sides sequentially by $\mathbf{\Omega}^{-1}$, $\mathbf{X}^{-1}$, and $\mathbf{\Phi}^{-1}$.

$$\mathbf{\Phi}^{-1}\mathbf{X}^{-1}\mathbf{\Omega}^{-1}\mathbf{\Omega}\,\mathbf{X}\,\mathbf{\Phi}\,\mathbf{U}\,\mathbf{B}^* \begin{bmatrix} h \\ k \\ l \end{bmatrix} = \frac{2\sin\theta}{\lambda}\,\mathbf{\Phi}^{-1}\mathbf{X}^{-1}\mathbf{\Omega}^{-1} \begin{bmatrix} -\sin\theta \\ \cos\theta \\ 0 \end{bmatrix}.$$

Since all three matrices are orthogonal, their transposes are their inverses:

$$\mathbf{UB}^* \begin{bmatrix} h \\ k \\ l \end{bmatrix} = \frac{2\sin\theta}{\lambda}\, \mathbf{\Phi}^T\, \mathbf{X}^T\, \mathbf{\Omega}^T \begin{bmatrix} -\sin\theta \\ \cos\theta \\ 0 \end{bmatrix}. \tag{4.13}$$

The matrix $\mathbf{A}^* = \mathbf{UB}^*$ encompasses the reciprocal lattice parameters and the orientation of the lattice with respect to the diffractometer coordinate system. The expression on the right contains the angular information necessary to observe diffraction:

$$\begin{aligned} \mathbf{\Omega}^T \begin{bmatrix} -\sin\theta \\ \cos\theta \\ 0 \end{bmatrix} &= \begin{bmatrix} \cos\omega & \sin\omega & 0 \\ -\sin\omega & \cos\omega & 0 \\ 0 & 0 & 1 \end{bmatrix} \begin{bmatrix} -\sin\theta \\ \cos\theta \\ 0 \end{bmatrix} \\ &= \begin{bmatrix} \sin\omega\cos\theta - \cos\omega\sin\theta \\ \sin\omega\sin\theta + \cos\omega\cos\theta \\ 0 \end{bmatrix} \\ &= \begin{bmatrix} \sin(\omega-\theta) \\ \cos(\omega-\theta) \\ 0 \end{bmatrix}. \end{aligned} \tag{4.14}$$

The final vector expression in Eqn. 4.14 results from the application of the trigonometric identities for the sine and cosine of the difference between two angles. The expression on the right in Eqn. 4.13 becomes the "un-rotated" diffraction vector, $\mathbf{h}_\mathbf{l}$, determined by the diffractometer angles in the diffraction condition*:

$$\begin{aligned} \mathbf{A}^* \begin{bmatrix} h \\ k \\ l \end{bmatrix} &= \frac{2\sin\theta}{\lambda} \begin{bmatrix} \cos\varphi & -\sin\varphi & 0 \\ \sin\varphi & \cos\varphi & 0 \\ 0 & 0 & 1 \end{bmatrix} \begin{bmatrix} 1 & 0 & 0 \\ 0 & \cos\chi & -\sin\chi \\ 0 & \sin\chi & \cos\chi \end{bmatrix} \begin{bmatrix} \sin(\omega-\theta) \\ \cos(\omega-\theta) \\ 0 \end{bmatrix} \\ &= \frac{2\sin\theta}{\lambda} \begin{bmatrix} \cos\varphi\sin(\omega-\theta) - \sin\varphi\cos\chi\cos(\omega-\theta) \\ \sin\varphi\sin(\omega-\theta) + \cos\varphi\cos\chi\cos(\omega-\theta) \\ \sin\chi\cos(\omega-\theta) \end{bmatrix} \\ &= \mathbf{h}_\mathbf{l}. \end{aligned} \tag{4.15}$$

For a given reflection, θ is known, and $\mathbf{h}_\mathbf{l}$ is determined by the angles ω, χ, and φ. ω is the rotation of the reciprocal lattice vector in the x_l, y_l plane into the diffraction condition; φ and χ are the rotations that bring the vector onto the x_l, y_l plane. As noted previously, if a value of ω is specified, then the φ rotation and χ rotations are also determined. Furthermore, different values of ω will result in different ψ rotations of the crystal around the reciprocal lattice vector. The simplest case occurs when the φ and χ rotations align the reciprocal lattice vector along the y_l axis, which we now can see is also in the plane of the χ circle (Fig. 4.13). Thus, in this special case, the ω rotation angle is θ and the χ circle is at an angle of $(90° - \theta)$

---

*$\mathbf{h}_\mathbf{l} = \mathbf{A}^*\mathbf{h}$ is the reciprocal lattice vector $\mathbf{h} = [h\ k\ l]$ defined in Cartesian coordinates in the laboratory reference frame. $\mathbf{\Phi}^T\, \mathbf{X}^T\, \mathbf{\Omega}^T \mathbf{h}_\mathbf{r} = \mathbf{h}_\mathbf{l}$ because the vector in the diffraction condition, $\mathbf{h}_\mathbf{r}$, is the vector $\mathbf{h}_\mathbf{l}$ rotated through the angles ω, χ, and φ. That is, $\mathbf{\Omega}\,\mathbf{X}\,\mathbf{\Phi}\mathbf{h}_\mathbf{l} = \mathbf{h}_\mathbf{r}$.

to both the incident and diffracted beams (Fig. 4.15); it bisects the angle between them. This is the *bisecting position* that we previously alluded to. Recall that it is also the position where $\psi = 0$ (by definition). In this position $\omega = \theta$ and $\omega - \theta = 0$. Since $\cos 0 = 1$ and $\sin 0 = 0$ Eqn. 4.15 is simplified considerably:

$$\mathbf{A}^*\mathbf{h} = \begin{bmatrix} a_{11} & a_{12} & a_{13} \\ a_{21} & a_{22} & a_{23} \\ a_{31} & a_{32} & a_{33} \end{bmatrix} \begin{bmatrix} h \\ k \\ l \end{bmatrix} = \frac{2\sin\theta}{\lambda} \begin{bmatrix} -\sin\varphi\cos\chi \\ \cos\varphi\cos\chi \\ \sin\chi \end{bmatrix} = \mathbf{h}_{\mathrm{l}}. \tag{4.16}$$

If we could determine the elements of $\mathbf{A}^*$ we would be in a position to calculate the diffraction angles in the bisecting position for every reflection from its indices alone. Counting photons at this position would give us intensities, and therefore *indexed* structure factor magnitudes. The treatment that follows can be extended to the general case, where $\omega \neq \theta$, but we will restrict ourselves to determining the diffractometer angles for a given reflection in the bisecting position. Eqn. 4.16 provides three simultaneous equations:

$$a_{11}h + a_{12}k + a_{13}l = (2\sin\theta/\lambda)(-\sin\varphi\cos\chi) \tag{4.17}$$
$$a_{21}h + a_{22}k + a_{23}l = (2\sin\theta/\lambda)(\cos\varphi\cos\chi) \tag{4.18}$$
$$a_{31}h + a_{32}k + a_{33}l = (2\sin\theta/\lambda)\sin\chi \tag{4.19}$$

These relationships provide the diffractometer angles as a function of the indices and X-ray wavelength for $\omega = \theta$:

$$\frac{a_{11}h + a_{12}k + a_{13}l}{a_{21}h + a_{22}k + a_{23}l} = \frac{(2\sin\theta/\lambda)(-\sin\varphi\cos\chi)}{(2\sin\theta/\lambda)(\cos\varphi\cos\chi)} = -\frac{\sin\varphi}{\cos\varphi} = -\tan\varphi$$

$$\varphi = \arctan\left[-\left(\frac{a_{11}h + a_{12}k + a_{13}l}{a_{21}h + a_{22}k + a_{23}l}\right)\right] \tag{4.20}$$

$$\frac{a_{31}h + a_{32}k + a_{33}l}{a_{21}h + a_{22}k + a_{23}l} = \frac{(2\sin\theta/\lambda)(\sin\chi)}{(2\sin\theta/\lambda)(\cos\varphi\cos\chi)} = \frac{\sin\chi}{\cos\varphi\cos\chi} = \frac{\tan\chi}{\cos\varphi}$$

$$\chi = \arctan\left[\left(\frac{a_{31}h + a_{32}k + a_{33}l}{a_{21}h + a_{22}k + a_{23}l}\right)\cos\varphi\right] \tag{4.21}$$

$$\frac{2\sin\theta}{\lambda} = \frac{a_{31}h + a_{32}k + a_{33}l}{\sin\chi}$$

$$\theta = \omega = \arcsin\left[\left(\frac{a_{31}h + a_{32}k + a_{33}l}{2\sin\chi}\right)\lambda\right]. \tag{4.22}$$

4.3.2 Area Detectors

While the use of serial counters revolutionized X-ray crystallographic data collection, film methods still offered the advantage of obtaining a number of intensities more-or-less simultaneously. Although the data-collection process using a serial detector is automated with computer-driven motors and counting devices, the process can be time-consuming since the crystal and detector must be oriented and the intensity measured one reflection at a time. Crystal structure determinations usually take several days for small-molecule structures, which require on the order of 2000–5000 reflections. For large-molecule structures, the number of reflections

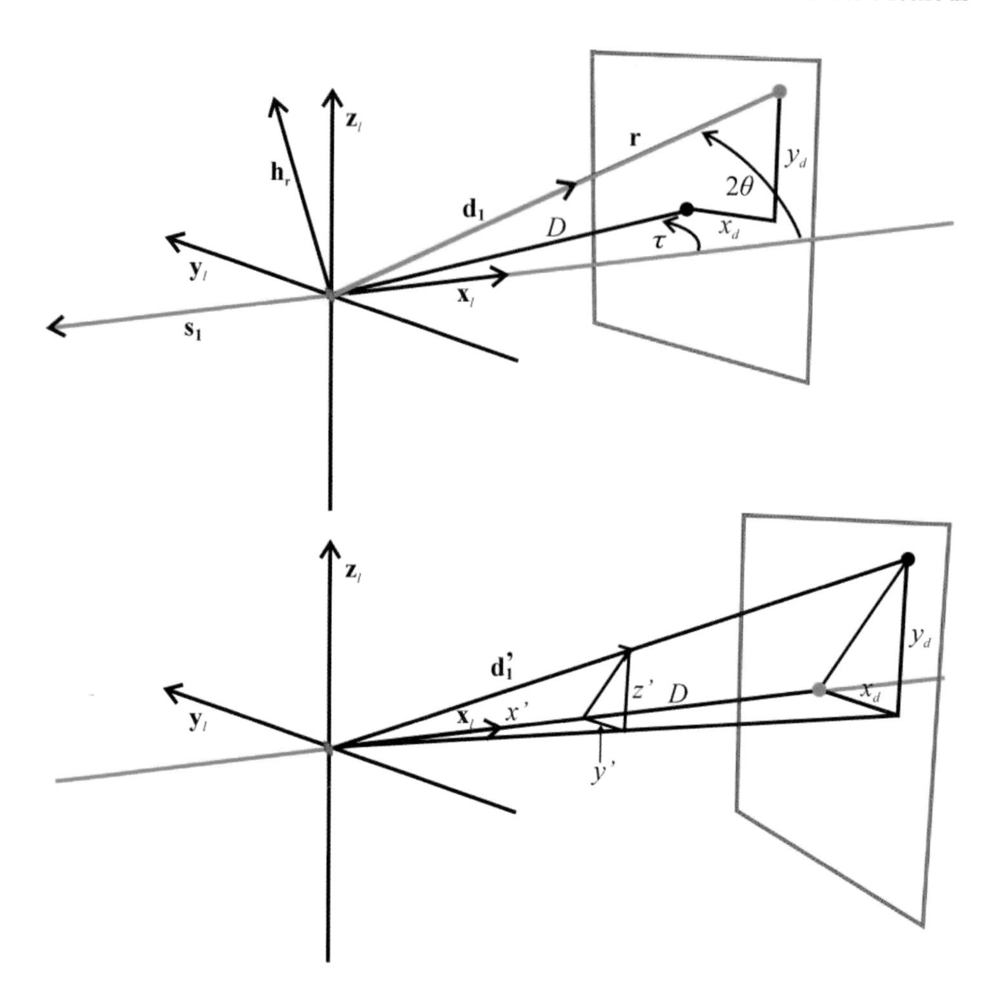

Figure 4.16 Top: A reflection, **r**, observed at x_d, y_d on the planar surface of an area detector rotated an angle τ in the equatorial plane. Bottom: The "virtual" position of the reflection if it had occurred at $\tau = 0$, with the detector perpendicular to the incident beam along the x_l axis of the laboratory reference frame.

needed for structural determination is at least an order of magnitude greater, demanding a tremendous amount of time for data collection. This is compounded by the ephemeral nature of many macromolecular crystals, which are easily damaged in the X-ray beam and are unable to remain exposed to it for the long periods necessary to collect this large quantity of data with a serial detector.

The second revolution in X-ray data collection came about in order to solve the problems associated with macromolecular structures — by employing devices that served essentially as "electronic film." These devices, know as *area detectors*, are position-sensitive photon counters. They combine the advantages of film methods with those of counter methods, including the ability to determine a structure from an arbitrarily aligned crystal. The most commonly used area detector employs a charge-coupled device (CCD) as a photon counter. The initial photons are detected by a planar X-ray sensitive phosphorescent screen coupled to a solid state CCD detector with a fiber-optic bundle; the actual photons detected are those emitted by the phosphor. *The CCD detector (similar to those in digital cameras) not only counts these photons, but detects where they are emitted from the screen.* It is mounted on the two-theta circle, in place of the serial detector (Fig. 4.13), and can be rotated to various positions around the circle. The angle of rotation is the detector angle rather than the diffraction angle, and is expressed as τ rather than 2θ here to avoid confusion. This is illustrated schematically in Fig. 4.16. In the drawing at the top of the figure, a planar detector, originally perpendicular to the incident beam has been rotated through the angle τ about the z_l axis in the equatorial plane. At a specific orientation of the crystal (φ, χ, ω) a reflection is observed at point (x_d, y_d) on the detector surface where we can record its location and measure its intensity.

With a serial detector, intensities are measured by specifying the indices for a reflection, then applying the orientation matrix and determining the diffractometer angles required to observe the reflection in the equatorial plane. In contrast, with the area detector φ, χ, ω and τ are set; the reflections are measured by fixing τ, χ and φ, and scanning through a small range in omega (alternatively, ω is fixed and φ is scanned). In order to measure the intensity of a specific reflection we must know its location on the surface of the detector at these angles. While Eqn. 4.11 was introduced in order to reorient the reciprocal lattice vector into the equatorial plane, it is valid for any orientation of the reciprocal lattice vector:

$$\mathbf{\Omega}\,\mathbf{X}\,\mathbf{\Phi}\,\mathbf{U}\,\mathbf{B}^*\mathbf{h} \;=\; \mathbf{\Omega}\,\mathbf{X}\,\mathbf{\Phi}\,\mathbf{A}^*\mathbf{h} \;=\; \mathbf{h_r}, \tag{4.23}$$

$$\mathbf{A}^*\begin{bmatrix} h \\ k \\ l \end{bmatrix} \;=\; \mathbf{\Phi}^T\,\mathbf{X}^T\,\mathbf{\Omega}^T\mathbf{h_r} \;=\; \mathbf{\Upsilon}\mathbf{h_r} \;=\; \mathbf{h_l} \;\; \text{where} \tag{4.24}$$

$$\mathbf{\Upsilon} = \mathbf{\Phi}^T\,\mathbf{X}^T\,\mathbf{\Omega}^T \;= \tag{4.25}$$

$$\begin{bmatrix} (\cos\varphi\cos\omega + \sin\varphi\cos\chi\sin\omega) & (\cos\varphi\sin\omega - \sin\varphi\cos\chi\cos\omega) & (\sin\varphi\sin\chi) \\ (\sin\varphi\cos\omega - \cos\varphi\cos\chi\sin\omega) & (\sin\varphi\sin\omega + \cos\varphi\cos\chi\cos\omega) & (-\cos\varphi\sin\chi) \\ (-\sin\chi\sin\omega) & (\sin\chi\cos\omega) & \cos\chi \end{bmatrix}$$

and $\mathbf{h_r}$ is the reciprocal lattice vector in the Cartesian coordinates of the laboratory reference frame after rotation through (φ, χ, and ω). The orientation matrix, $\mathbf{A}^*$, determines the Cartesian coordinates of the reciprocal lattice vector in the laboratory reference frame prior to rotation, and is independent of the final position of the vector; it is the same for serial and area detection.*

If a reflection $\mathbf{h}$ with component indices (h, k, l) has been rotated through (φ, χ, and ω) into the diffraction condition, then a reflection vector defined in the laboratory coordinate system, $\mathbf{r}$, intersects the area detector at (x_d, y_d). Since the crystal is in the diffraction condition, the vector sum of a unit vector along $\mathbf{r}$, $\mathbf{d_1}$ — and a unit vector directed toward the source, $\mathbf{s_1} = [-1\ 0\ 0]$, will equal $\lambda\mathbf{h_r}$ (Eqn. 3.48 and Fig. 4.11(d)), again in the laboratory coordinate system. We

*The $\mathbf{U}$ matrix relates the orientation of vectors in the reciprocal Cartesian basis to the laboratory reference frame, while $\mathbf{A}^*$ relates the orientation of vectors in the reciprocal unit cell basis to the laboratory reference frame. Both matrices are "orientation matrices".

make use of this relationship to determine the components of $\mathbf{d_1}$ in the laboratory reference frame:

$$\mathbf{\Omega X \Phi A^* h} = \mathbf{h_r} = \begin{bmatrix} h_x \\ h_y \\ h_z \end{bmatrix} \tag{4.26}$$

$$\mathbf{s_1} + \mathbf{d_1} = \lambda \mathbf{h_r}$$

$$\mathbf{d_1} = \lambda \mathbf{h_r} - \mathbf{s_1} = \begin{bmatrix} \lambda h_x + 1 \\ \lambda h_y \\ \lambda h_z \end{bmatrix} = \begin{bmatrix} x \\ y \\ z \end{bmatrix} \tag{4.27}$$

The reflection vector begins at the origin and ends at location (x_d, y_d) on the detector plane. Determining the point of intersection on the detector plane is facilitated by treating the vector in a common coordinate system. Since we have the coordinates of a unit vector in the direction of $\mathbf{r}$, a straightforward way to accomplish this task is to rotate $\mathbf{d_1}$ and the detector through the angle $-\tau$ about the z_l axis, as if the reflection had occurred at $\tau = 0$. This is illustrated at the bottom of Fig. 4.16. D is the distance from the crystal to the center of the area detector. (x', y', z') are the coordinates of $\mathbf{d_1}$ after rotation:

$$\mathbf{d_1'} = \begin{bmatrix} \cos\tau & \sin\tau & 0 \\ -\sin\tau & \cos\tau & 0 \\ 0 & 0 & 1 \end{bmatrix} \begin{bmatrix} x \\ y \\ z \end{bmatrix} = \begin{bmatrix} x' \\ y' \\ z' \end{bmatrix} \tag{4.28}$$

The head of the vector $\mathbf{d_1'}$ and its extension to the end of the rotated reflection vector create a set of similar triangles as shown in the figure. The direction of x_d is in the $-y_l$ direction such that

$$\frac{x_d}{D} = \frac{-y'}{x'} \implies x_d = D\frac{-y'}{x'} \tag{4.29}$$

$$\frac{y_d}{x_d} = \frac{z'}{-y'} \implies y_d = x_d\frac{z'}{-y'} = D\frac{z'}{x'} \tag{4.30}$$

Given the orientation matrix, crystal rotation angles, position of the detector and indices, we can now determine the location on the detector surface for any reflection that is in the diffraction condition; *only those indices that satisfy the conditions for diffraction will send photons to the detector.* This requires that the $(h\ k\ l)$ planes are at an angle of θ with respect to the incident beam and the reciprocal lattice vector $\mathbf{h_r}$ is at an angle of $(90° - \theta)$ to the beam in the direction of $\mathbf{s_1}$ (Fig. 4.1); theta is given by the length of $\mathbf{h_r}$: $\mathrm{h}_r = 2\sin\theta/\lambda$. The dot product of $\mathbf{s_1}$ and $\mathbf{h_r}$ is $-h_x$ since $\mathbf{s_1} = [-1\,0\,0]$. Thus $\mathrm{h}_r s_1 \cos(90° - \theta) = (2\sin\theta/\lambda)(1)(\sin\theta) = -h_x$ and $h_x = -2\sin^2\theta/\lambda$. For a given detector position and crystal orientation, only those reciprocal lattice vectors oriented so that their x_l component is equal to $-2\sin^2\theta/\lambda$ can be observed on the detector. The area detector is set at a specific angle, τ — the crystal is oriented to φ, χ and ω — and omega is scanned over a small angular range. The detector measures and stores a response for each location on the detector surface, proportional to the number of X-ray photons that contact it. The result is the equivalent of a quantitative photographic image; the image for a given set of angles, φ, χ, ω and τ, is called a *frame*. A subset of indices that meet the criteria for diffraction are selected, the locations of their reflections are computed, and the recorded intensity at each of these locations is determined. The result is a set of indexed structure factor amplitudes for each orientation of the crystal and detector.

4.4 Determining the Orientation Matrix and Unit Cell.

The determination of the matrix elements for $\mathbf{A}^*$ begins with the measurement of the diffraction angles for a number of reflections — usually those of higher intensity — and the determination of the Cartesian components of the reciprocal lattice vector for each reflection in the laboratory reference frame prior to rotation of the crystal, $\mathbf{h_l}$. When serial detectors are employed the reciprocal lattice vector, $\mathbf{h_r}$, is constrained to the equatorial plane, and is uniquely determined by the diffractometer angles (Fig. 4.15 and Eqn. 4.12). Since $\mathbf{\Phi}^T\,\mathbf{X}^T\,\mathbf{\Omega}^T\mathbf{h_r} = \mathbf{h_l}$, so is $\mathbf{h_l}$. For serial detectors there are several ways to obtain diffractometer angles for a number of reflections, including both random and more-or-less systematic searches through the diffractometer angles, looking for high photon counts observed in the detector. A more systematic approach is the use of a rotation photograph. The diffraction image is ordinarily collected with a rotation of φ through 360° with the other diffractometer angles set at zero. The images in Fig. 4.4 were collected in this manner. Fig. 4.17 illustrates the relationship between the four symmetric spots on the film from a reflection and the diffractometer angles needed to observe the reflection with a serial detector. ω is set to *zero*; the χ circle is perpendicular to the incident beam axis (x_l); 2θ is the angle between the beam axis and the vector from the center of the crystal to a spot on the film; χ is the angle necessary to rotate the spot into the x_l, y_l plane at the diffraction angle. These angles can be determined directly from distances measured on the film, x and y, and the distance between the crystal and the film, D:

$$q = (x^2 + y^2)^{\frac{1}{2}}$$

$$2\theta = \arctan\left(\frac{q}{D}\right) = \arctan\left[\frac{(x^2 + y^2)^{\frac{1}{2}}}{D}\right] \tag{4.31}$$

$$\chi = \arctan\left(\frac{y}{x}\right). \tag{4.32}$$

For a given reflection, the angle φ is still unknown. Let us suppose for a moment that we knew φ for a reflection. To observe it with a serial detector we would move the detector to 2θ, then rotate the crystal about the φ axis until the reciprocal lattice point contacted the sphere of reflection — causing a diffracted beam to point to the position where it was observed on the film. Finally, we would rotate the crystal through χ. Since the χ rotation is around the incident beam axis at $\omega = 0$, the reciprocal lattice point would follow a circular path along the surface of the sphere of reflection onto the x_l, y_l plane. This would place the reflection vector at an angle of 2θ in the plane, pointing directly at the detector.

Since we do not know φ, we establish the same conditions by rotating the crystal to χ and the detector to 2θ, then rotating the crystal about the φ axis (no longer coincident with z_l) until a large number of photons are counted in the detector.

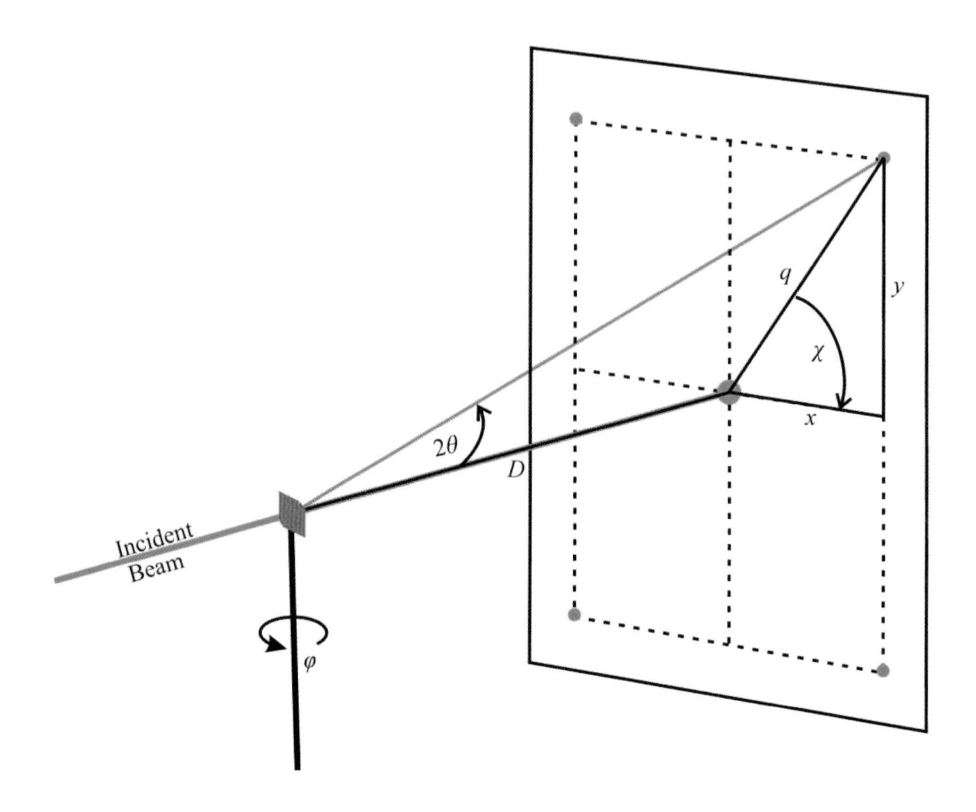

Figure 4.17 Relationship between the reflections observed on a photograph and the diffractometer angles required to observe the reflection in the x_l, y_l plane at $\omega = 0$ (*not* θ). The film is normal to the incident beam at a distance D from the crystal.

The diffracted beam has a finite cross-section; once located it is necessary to determine the central point of the reflection. In the case of the serial detector, this is accomplished by sequentially refining ω, 2θ, and χ — a process known as *centering.* φ remains fixed, since only two of the three Eulerian angles are necessary to move the crystal into the diffraction condition (as we previously observed by fixing ω, then rotating through χ and φ). The process begins by rotating ω (the χ circle) a small amount, moving the crystal out of the diffraction condition. ω is then scanned slowly through the diffraction condition while observing the intensity of the diffracted beam in the detector. A relatively small change in omega results in destructive interference and a rapid diminution of the observed intensity; the omega-scan peaks are usually sharp, with a maximum intensity when diffraction conditions are optimal. The value of ω for the reflection is determined at this maximum. The 2θ scan is quite different. The detector has a collimator in front of it to limit the portion of the detector measuring the intensities to a small cross-sectional area (otherwise the reflection would be observed continuously as the detector moved through a very large range of angles). Since the crystal remains in the diffraction condition, the intensity is monitored as the collimator opening moves past the diffracted beam; the reflection strikes different points on the detector surface within the window of the collimator opening. When the collimator opening reaches the edge of the diffracted beam the intensity decreases rapidly. The center of the reflection in the 2θ scan is therefore measured by determining the two angles where the intensity fall-off occurs, usually when the intensity has diminished by 25% — 2θ is taken as the mid-point of these two angles. The χ scan is similar to the 2θ scan. In this case the crystal essentially remains in the diffraction condition (the reciprocal lattice point follows a nearly circular path along the surface of the sphere of reflection, affected only by small deviations of ω); the reflection passes across the collimator window as the reciprocal lattice vector moves in and out of the x_l, y_l plane. As with 2θ, the "centered" position of χ for the reflection is the midpoint of the two angles where the intensity tails off. The centering process is repeated until the angles in two successive measurement sequences are "identical" within specified tolerance limits, usually a few hundredths of a degree.

When employing an area detector the determination of diffractometer angles for a number of reflections is more straightforward, but less accurate. In order to collect a set of the reflections the crystal must be rotated so that the reciprocal lattice points pass through the sphere of reflection. The pixel locations on the CCD detector provide the location of the reflection on the phosphorescent screen, but do not indicate the angle of rotation (the situation is the same for other area detectors). This was the difficulty encountered with the rotation method — requiring oscillation through a small angle – the oscillation method. For film methods this was a cumbersome process, but with automation and "electronic film," many oscillation images (frames) can be collected in a short period of time. For each frame, φ and χ are fixed, and ω is rotated over a small angular range. For small-molecule crystallography this is on the order of 0.5 degrees, less than the width of a peak in the diffraction pattern. Omega is considered to be the angle in the center of this rotational range. As described below, the unit cell will be based on the angles, and it might appear that the area detector will not provide accurate unit cells. The advantage of the area detector lies in its ability to collect a large number of reflections in a short period of time and the errors in omega cancel one another since they are expected to be randomly distributed. With a serial detector, the unit cell is usually determined with 10–15 reflections, and refined with a few more. With an area detector the initial unit cell can be determined with many more reflections and refined with a very large number. Furthermore the individual reflections from an area detector are generally observed in more than one frame (usually successive frames). Since the positions in one frame would be accurately predicted from the position in another frame if omega was known accurately for each, the "best" omegas for redundant reflections will tend to be the those that minimize the deviations between predicted and measured positions. This results in highly accurate unit cells *once the errors in omega are taken care of.*

For the area detector we collect a number of frames, commonly between 50 and 100, for which we know φ, χ, and τ accurately, and ω within the limits of its scan range. While the crystal rotation angles are known, in contrast to the serial detector, $\mathbf{h_r}$ is *not* confined to the equatorial plane. We must therefore determine it from the intersection of the reflection vector $\mathbf{r}$ and the detector plane — at (x_d, y_d) — subsequently allowing us to compute $\mathbf{h_l}$. To do so we reverse the process described in the previous section. We begin by selecting a number of more intense reflections from each frame and determining (x_d, y_d) for each "spot". If the reflection vector has been observed at $\tau = 0$ then its coordinates would be $(D, -x_d, y_d)$

(Fig. 4.16); as before, the x_d direction is in the $-y_l$ direction. Rotating this vector about z_l through the angle τ gives the components of the actual reflection vector in the laboratory reference frame:

$$\mathbf{r} = \begin{bmatrix} r_1 \\ r_2 \\ r_3 \end{bmatrix} = \begin{bmatrix} \cos\tau & -\sin\tau & 0 \\ \sin\tau & \cos\tau & 0 \\ 0 & 0 & 1 \end{bmatrix} \begin{bmatrix} D \\ -x_d \\ y_d \end{bmatrix} = \begin{bmatrix} D\cos\tau + x_d\sin\tau \\ D\sin\tau - x_d\cos\tau \\ y_d \end{bmatrix}. \quad (4.33)$$

Calculating $r = (D^2 + x_d^2 + y_d^2)^{1/2}$, the unit vector along the reflection in the laboratory reference frame is $\mathbf{d_1} = \mathbf{r}/r = [r_1/r \; r_2/r \; r_3/r]$. The unit vector toward the source points in the negative x_l direction: $\mathbf{s_1} = [-1 \; 0 \; 0]$. Since $\lambda\mathbf{h_r} = \mathbf{s_1} + \mathbf{d_1}$, $\mathbf{h_r}$ can be determined,

$$\mathbf{h_r} = \frac{1}{\lambda}\begin{bmatrix} (r_1/r) - 1 \\ r_2/r \\ r_3/r \end{bmatrix}, \quad (4.34)$$

and $\mathbf{h_l} = \mathbf{\Phi}^T \mathbf{X}^T \mathbf{\Omega}^T \mathbf{h_r}$ calculated. In addition, 2θ, the angle between the incident and diffracted beam, is the angle between $\mathbf{d_1}$ and $-\mathbf{s_1} = [1 \; 0 \; 0]$, and can be determined from their scalar product:

$$-\mathbf{s_1} \cdot \mathbf{d_1} = (1)(1)\cos 2\theta \quad (4.35)$$

$$\cos 2\theta = r_1/r. \quad (4.36)$$

Armed with an array of diffractometer angles for n reflections, we are now in a position to attempt the determination of the orientation matrix and unit cell.* If the reflections were collected using a serial detector we begin by creating an $\mathbf{h_l}$ vector for each reflection using Eqn. 4.15; if an area detector was used, $\mathbf{h_l}$ has already been determined for each reflection from its area detector coordinates, as described above. Recall that $\mathbf{A}^*\mathbf{h}_i = \mathbf{UB}^*\mathbf{h}_i = \mathbf{h_l}_i = \mathbf{z}_i{}^\dagger$; these vectors are reciprocal lattice vectors for each reflection, converted to Cartesian coordinates in the diffractometer reference frame before rotation of the crystal. Each of these vectors terminates at a unit cell origin in the reciprocal lattice, and therefore constitutes an axis for one of the infinitively many unit cells in the lattice. In principle, we could select any three of these vectors that do not lie in the same plane as a reciprocal unit cell. Each of the remaining vectors would then be expressed in terms of the unit cell as $\mathbf{v} = h\mathbf{a}^* + k\mathbf{b}^* + l\mathbf{c}^*$ in the selected basis, with integer components consisting of the indices of the reflections in the basis. In the selected basis the axial vectors have indices (1 0 0), (0 1 0), and (0 0 1) for $\mathbf{a}^*$, $\mathbf{b}^*$, and $\mathbf{c}^*$, respectively. However, arbitrarily choosing three vectors will result in an arbitrary reciprocal unit cell, and a correspondingly arbitrary direct unit cell. The appropriate unit cell for the lattice is a specific unit cell – the reduced cell – which we encounter in Sec. 4.6.1. In order to obtain the reduced unit cell, it is best to begin with a unit cell that is close to

*The uncertainty in omega and the potential absence of low-index reflections located with an area detector can require a more sophisticated approach than the one outlined here[53]. The underlying principles are the same.

†In the treatment that follows the use of the symbol $\mathbf{h_l}$ becomes somewhat cumbersome as we will need to apply a number of subscripts and superscripts to its components. We will temporarily redefine the vector as $\mathbf{z}$.

it, and our goal here is to determine *all* possible unit cell axes that are consistent with the reciprocal lattice vectors generated from the reflection array.

The vectors are sorted by length $(2\sin\theta/\lambda)$, and, beginning with the shortest vector, the vector array is tested sequentially, searching for the three shortest *linearly independent* vectors. This is accomplished by creating the matrix $\mathbf{Z} = [\mathbf{z}_q\ \mathbf{z}_r\ \mathbf{z}_s]$ and evaluating its determinant. If the determinant is non-zero then $\mathbf{Z}^{-1}$ exists and the three column vectors are linearly independent. *If* we know the indices of these three reflections, as might be the case if the crystal had been previously oriented using film data, then the matrix $\mathbf{H} = [\mathbf{h}_q\ \mathbf{h}_r\ \mathbf{h}_s]$ can be formed and $\mathbf{A}^*$ and its inverse can be determined directly from $\mathbf{A}^*\mathbf{H} = \mathbf{Z}$:

$$\mathbf{A}^* = \mathbf{Z}\mathbf{H}^{-1} \tag{4.37}$$

$$\mathbf{A}^{*-1} = (\mathbf{Z}\mathbf{H}^{-1})^{-1} = \mathbf{H}\mathbf{Z}^{-1} \tag{4.38}$$

In the general case the indices of the centered reflections are unknown, and we must find a way to get them if we are to determine $\mathbf{A}^*$ from an arbitrarily oriented crystal about which we may know absolutely nothing. Eqn. 4.38 provides the means for doing so. Denoting the matrix elements of $\mathbf{A}^{*-1}$ and $\mathbf{Z}^{-1}$ as a^i_{ij} and z^i_{ij}, respectively,

$$\begin{bmatrix} a^i_{11} & a^i_{12} & a^i_{13} \\ a^i_{21} & a^i_{22} & a^i_{23} \\ a^i_{31} & a^i_{32} & a^i_{33} \end{bmatrix} = \begin{bmatrix} h_q & h_r & h_s \\ k_q & k_r & k_s \\ l_q & l_r & l_s \end{bmatrix} \begin{bmatrix} z^i_{11} & z^i_{12} & z^i_{13} \\ z^i_{21} & z^i_{22} & z^i_{23} \\ z^i_{31} & z^i_{32} & z^i_{33} \end{bmatrix}. \tag{4.39}$$

The first row vector of $\mathbf{A}^{*-1}$, $\mathbf{a}^i_1$, is then given by

$$\begin{aligned} a^i_{11} &= h_q z^i_{11} + h_r z^i_{21} + h_s z^i_{31} \\ a^i_{12} &= h_q z^i_{12} + h_r z^i_{22} + h_s z^i_{32} \\ a^i_{13} &= h_q z^i_{13} + h_r z^i_{23} + h_s z^i_{33}. \end{aligned} \tag{4.40}$$

The second and third row vectors, $\mathbf{a}^i_2$ and $\mathbf{a}^i_3$, are identical, except that the *integer* triple (h_q, h_r, h_s) is replaced with (k_q, k_r, k_s) for $\mathbf{a}^i_2$ and (l_q, l_r, l_s) for $\mathbf{a}^i_3$. Each integer triple is a set of three indices; the magnitude of each integer will be relatively small since we have determined $\mathbf{Z}$ for the three shortest non-coplanar vectors. The number of possible integer triples is therefore limited. Since we do not know any of the indices, we *guess* a set of three small integers and compute a trial row vector, $\mathbf{a}^i$ (it does not matter which one), and test it to determine whether or not it is a candidate for one of the rows of the matrix, based on the following argument: We know that *for all n reflections*,

$$\mathbf{A}^* \begin{bmatrix} h_j \\ k_j \\ l_j \end{bmatrix} = \begin{bmatrix} z_{1j} \\ z_{2j} \\ z_{3j} \end{bmatrix}, \ j = 1 \text{ to } n.$$

$$\begin{bmatrix} h_j \\ k_j \\ l_j \end{bmatrix} = \mathbf{A}^{*-1} \begin{bmatrix} z_{1j} \\ z_{2j} \\ z_{3j} \end{bmatrix} = \begin{bmatrix} a^i_{11} & a^i_{12} & a^i_{13} \\ a^i_{21} & a^i_{22} & a^i_{23} \\ a^i_{31} & a^i_{32} & a^i_{33} \end{bmatrix} \begin{bmatrix} z_{1j} \\ z_{2j} \\ z_{3j} \end{bmatrix}, \ j = 1 \text{ to } n.$$

In order for the trial vector to be a legitimate row vector for $\mathbf{A^{*-1}}$, *the inner product of* $\mathbf{z}_j$ *and the trial vector,* $\mathbf{a}^i = [a_1^i \; a_2^i \; a_3^i]$, *must be an index for the jth reflection* — an *integer* — *for all n reflections*:

$$a_1^i z_{1j} + a_2^i z_{2j} + a^i z_{3j} = h_j \text{ or } k_j \text{ or } l_j \quad = \quad \text{an integer.} \tag{4.41}$$

Since the range of possible integer triples is limited, the strategy for determining the $\mathbf{A}^*$ matrix is to test all the integer triples within a small range, e.g., (-5,-5,-5) to (5,5,5), computing trial row vectors and determining whether or not they produce integers from Eqn. 4.41. There will be a number of solutions, all of which are legitimate row vectors in the matrix. We must choose three of them to create $\mathbf{A}^*$, which consists of the product of the orientation matrix and the Cartesian transformation matrix for the reciprocal basis. In order to make this choice we consider the composition of $\mathbf{A}^*$, the matrix that transforms a reciprocal lattice vector into Cartesian coordinates in the laboratory reference frame: $\mathbf{A^*h} = \mathbf{v_l}$. In particular, the axial reciprocal lattice vectors, $\mathbf{a}^* = [1 \; 0 \; 0]$, $\mathbf{b}^* = [0 \; 1 \; 0]$, and $\mathbf{a}^* = [0 \; 0 \; 1]$, are transformed into Cartesian vectors in the laboratory frame:

$$\begin{aligned} \mathbf{A^*a^*} &= \mathbf{a_l^*} \\ \mathbf{A^*b^*} &= \mathbf{b_l^*} \\ \mathbf{A^*c^*} &= \mathbf{c_l^*} \Rightarrow \end{aligned}$$

$$\mathbf{A}^* \begin{bmatrix} 1 & 0 & 0 \\ 0 & 1 & 0 \\ 0 & 0 & 1 \end{bmatrix} = [\mathbf{a_l^*} \; \mathbf{b_l^*} \; \mathbf{c_l^*}] = \begin{bmatrix} a_{xl}^* & b_{xl}^* & c_{xl}^* \\ a_{yl}^* & b_{yl}^* & c_{yl}^* \\ a_{zl}^* & b_{zl}^* & c_{zl}^* \end{bmatrix} = \mathbf{A}^*. \tag{4.42}$$

The columns of $\mathbf{A}^*$ *are the Cartesian components of the reciprocal cell axes in the laboratory reference frame.* It follows that

$$\mathbf{A^*}^T\mathbf{A}^* = \begin{bmatrix} \mathbf{a_l^*}\cdot\mathbf{a_l^*} & \mathbf{a_l^*}\cdot\mathbf{b_l^*} & \mathbf{a_l^*}\cdot\mathbf{c_l^*} \\ \mathbf{b_l^*}\cdot\mathbf{a_l^*} & \mathbf{b_l^*}\cdot\mathbf{b_l^*} & \mathbf{b_l^*}\cdot\mathbf{c_l^*} \\ \mathbf{c_l^*}\cdot\mathbf{a_l^*} & \mathbf{c_l^*}\cdot\mathbf{b_l^*} & \mathbf{c_l^*}\cdot\mathbf{c_l^*} \end{bmatrix}. \tag{4.43}$$

The symmetric matrix in Eqn. 4.43, $\mathbf{G}^* = \mathbf{A^*}^T\mathbf{A}^*$, *gives us the lengths and angles of the reciprocal unit cell.* It is referred to as the *reciprocal metric tensor*:

$$\mathbf{G}^* = \begin{bmatrix} a^{*2} & a^*b^*\cos\gamma^* & a^*c^*\cos\beta^* \\ a^*b^*\cos\gamma^* & b^{*2} & b^*c^*\cos\alpha^* \\ a^*c^*\cos\beta^* & b^*c^*\cos\alpha^* & c^{*2} \end{bmatrix} \tag{4.44}$$

Any three linearly independent row vectors selected from the set of solutions — those that yield integers from Eqn. 4.41 — can be used to create the $\mathbf{A}^*$ matrix (from $\mathbf{A^{*-1}}$); each set will generate a different reciprocal unit cell (and therefore a different direct unit cell). While we could use the relationships between real and reciprocal unit cells to calculate the direct cell for each choice, there is a simpler, and more elegant method for doing this. We begin by constructing an analog to $\mathbf{A}^*$ with the direct lattice parameters — identical with the exception of the replacement of the reciprocal lattice parameters with direct lattice parameters *and the transposition of the matrix*:

$$\mathbf{A} = \begin{bmatrix} a_{xl} & a_{yl} & a_{zl} \\ b_{xl} & b_{yl} & b_{zl} \\ c_{xl} & c_{yl} & c_{zl} \end{bmatrix}. \tag{4.45}$$

The product of this matrix and its transpose would yield the *direct lattice metric tensor*, from which we could readily determine the direct cell parameters:

$$\mathbf{AA}^T = \begin{bmatrix} \mathbf{a}_l \cdot \mathbf{a}_l & \mathbf{a}_l \cdot \mathbf{b}_l & \mathbf{a}_l \cdot \mathbf{c}_l \\ \mathbf{b}_l \cdot \mathbf{a}_l & \mathbf{b}_l \cdot \mathbf{b}_l & \mathbf{b}_l \cdot \mathbf{c}_l \\ \mathbf{c}_l \cdot \mathbf{a}_l & \mathbf{c}_l \cdot \mathbf{b}_l & \mathbf{c}_l \cdot \mathbf{c}_l \end{bmatrix} = \mathbf{G} = \begin{bmatrix} a^2 & ab\cos\gamma & ac\cos\beta \\ ab\cos\gamma & b^2 & bc\cos\alpha \\ ac\cos\beta & bc\cos\alpha & c^2 \end{bmatrix}. \tag{4.46}$$

Now consider $\mathbf{AA}^*$ and $\mathbf{GG}^*$:

$$\begin{aligned} \mathbf{AA}^* &= \begin{bmatrix} a_{xl} & a_{yl} & a_{zl} \\ b_{xl} & b_{yl} & b_{zl} \\ c_{xl} & c_{yl} & c_{zl} \end{bmatrix} \begin{bmatrix} a^*_{xl} & b^*_{xl} & c^*_{xl} \\ a^*_{yl} & b^*_{yl} & c^*_{yl} \\ a^*_{zl} & b^*_{zl} & c^*_{zl} \end{bmatrix} \\ &= \begin{bmatrix} \mathbf{a}_l \cdot \mathbf{a}^*_l & \mathbf{a}_l \cdot \mathbf{b}^*_l & \mathbf{a}_l \cdot \mathbf{c}^*_l \\ \mathbf{b}_l \cdot \mathbf{a}^*_l & \mathbf{b}_l \cdot \mathbf{b}^*_l & \mathbf{b}_l \cdot \mathbf{c}^*_l \\ \mathbf{c}_l \cdot \mathbf{a}^*_l & \mathbf{c}_l \cdot \mathbf{b}^*_l & \mathbf{c}_l \cdot \mathbf{c}^*_l \end{bmatrix} = \begin{bmatrix} 1 & 0 & 0 \\ 0 & 1 & 0 \\ 0 & 0 & 1 \end{bmatrix} \Rightarrow \\ \mathbf{A}^{*-1} &= \mathbf{A}. \end{aligned} \tag{4.47}$$

$$\begin{aligned} \mathbf{GG}^* &= \mathbf{AA}^T\mathbf{A}^{*T}\mathbf{A}^* \\ \mathbf{A}^T\mathbf{A}^{*T} &= (\mathbf{A}^*\mathbf{A})^T = \mathbf{I} \text{ and} \\ \mathbf{GG}^* &= \mathbf{AA}^* = \mathbf{I} \Rightarrow \\ \mathbf{G}^{*-1} &= \mathbf{G}. \end{aligned} \tag{4.48}$$

$\mathbf{A}$ is the inverse of $\mathbf{A}^*$ *and* $\mathbf{G}$ is the inverse of $\mathbf{G}^*$, dramatically underscoring the reciprocal nature of the relationship between the direct and reciprocal lattices. *The selection of three row vectors for* $\mathbf{A}^{*-1} \equiv \mathbf{A}$ *immediately provides the direct and reciprocal cell parameters from* $\mathbf{G} = \mathbf{AA}^T$ *and* $\mathbf{G}^* = (\mathbf{AA}^T)^{-1}$.

Finally, consider the determinant of $\mathbf{G}$ (or $\mathbf{G}^*$):

$$\begin{aligned} |\mathbf{G}| &= \begin{vmatrix} \mathbf{a}_l \cdot \mathbf{a}^*_l & \mathbf{a}_l \cdot \mathbf{b}^*_l & \mathbf{a}_l \cdot \mathbf{c}^*_l \\ \mathbf{b}_l \cdot \mathbf{a}^*_l & \mathbf{b}_l \cdot \mathbf{b}^*_l & \mathbf{b}_l \cdot \mathbf{c}^*_l \\ \mathbf{c}_l \cdot \mathbf{a}^*_l & \mathbf{c}_l \cdot \mathbf{b}^*_l & \mathbf{c}_l \cdot \mathbf{c}^*_l \end{vmatrix} \\ &= (a^2b^2c^2) + (ab\cos\gamma\, bc\cos\alpha\, ac\cos\beta) + (ab\cos\gamma\, bc\cos\alpha\, ac\cos\beta) \\ &\quad -(ac\cos\beta\, b^2 ac\cos\beta) - (bc\cos\alpha\, bc\cos\alpha\, a^2) - (ab\cos\gamma\, ab\cos\gamma\, c^2) \\ &= (a^2b^2c^2) + (a^2b^2c^2)(\cos\alpha\cos\beta\cos\gamma) + (a^2b^2c^2)(\cos\alpha\cos\beta\cos\gamma) \\ &\quad -(a^2b^2c^2\cos^2\alpha) - (a^2b^2c^2\cos^2\beta) - (a^2b^2c^2\cos^2\gamma) \\ &= a^2b^2c^2(1 - \cos^2\alpha - \cos^2\beta - \cos^2\gamma + 2\cos\alpha\cos\beta\cos\gamma) \\ &= V^2. \end{aligned} \tag{4.49}$$

A parallel derivation gives $|\mathbf{G}^*| = V^{*2}$. The determinant of the direct metric tensor provides the unit cell volume; the determinant of the reciprocal metric tensor provides the reciprocal cell volume.

Using Eqn. 4.15, all n centered reflections can now be indexed based on a unit cell generated by the selection of three row vectors:

$$\mathrm{h} = \mathbf{A}^{*-1}\mathbf{z} = \mathbf{A}\mathbf{z}. \tag{4.50}$$

The problem now is to select the "best" unit cell on which to base these indices. *The row vectors of* **A** *are the unit cell vectors in Cartesian coordinates* (Eqn. 4.45). *Each trial row vector that produces an integer solution from Eqn. 4.41 is a legitimate axial vector for the direct lattice.* We generate a set of such *axial solution* vectors, $\{\mathbf{a}_j^i\}$, then determine their magnitudes from $\mathbf{a}_j^i \cdot \mathbf{a}_j^i$ and the cosines of the angles between them from $\mathbf{a}_j^i \cdot \mathbf{a}_k^i$. The initial choice of the three unit cell axes would ordinarily consist of the three shortest non-coplanar vectors, determined by sorting the axial solutions and selecting the shortest vector as the first solution, the next shortest vector *not collinear* with the first (cosine $\neq \pm 1$) as the second solution, and the third vector *not in the same plane as the first two* (not perpendicular to the vector product of the first two solutions) as the third solution. The resulting primitive unit cell will be the unit cell of choice only if there is not higher symmetry in the lattice. This is usually indicated by interaxial angles of 90° — those with cosines of zero. The initial choice then usually consists of the three shortest non-coplanar vectors, unless there are at least two longer vectors perpendicular to a third. It is often the case that the three shortest vectors are also perpendicular to one another. It is possible that the three shortest non-coplanar solutions determined from the reflections are not the three shortest non-coplanar vectors in the lattice, since they are based on an arbitrary set of reflections. In addition, there may still be higher symmetry than the initial unit cell choice indicates. A procedure known as *cell reduction* will assist us in finding the shortest axial solutions, from which we can assess whether or not there is additional lattice symmetry in making a final unit cell choice. We will first consider the refinement of our initial unit cell choice before addressing the topic of cell reduction.

4.5 Refining the Orientation Matrix and Unit Cell.

Once the unit cell and orientation matrix have been selected the integer indices of all of the reflections can be computed from Eqn. 4.50, and once indexed, any three linearly independent reflections can be used to determine the orientation matrix and unit cell:

$$\mathbf{A}^* \begin{bmatrix} h_q & h_r & h_s \\ k_q & k_r & k_s \\ l_q & l_r & l_s \end{bmatrix} = \mathbf{A}^*\mathbf{H} = \begin{bmatrix} z_{1q} & z_{1r} & z_{1s} \\ z_{2q} & z_{2r} & z_{2s} \\ z_{3q} & z_{3r} & z_{3s} \end{bmatrix} = \mathbf{Z}$$

$$\mathbf{A}^* = \mathbf{Z}\mathbf{H}^{-1}. \tag{4.51}$$

The **Z** matrix is composed of experimentally measured angles. Although each orientation matrix calculated from the above equation should be identical to any other, measurement errors will cause one solution to deviate from another, the degree of deviation depending on the magnitude of the error in the measurements. It is imperative that the orientation matrix be as accurate as possible, since it determines the diffractometer angles for the measurement of reflection intensities with serial detectors, and the indices of reflections for reflections measured with area detectors. Even more importantly, it determines the unit cell, which is the coordinate system on which the location of atoms in the crystal structure is based. An inaccurate unit cell translates to inaccurate bond lengths and angles and a poor crystal structure. Since we have a number of experimental observations, we must find a way to utilize *all* of them to determine the orientation matrix and unit cell. If the errors in the angular measurements are random then averaging all of the possible orientation matrices would tend to cancel these errors — since they are equally distributed on both sides of the average. *This is especially important for the angular measurements from area detectors, since each individual ω can lie anywhere within its scan range.* However, determining the averages directly is impractical since computing all possible matrices and averaging them is a formidable task — even for a relatively small number of reflections. Fortunately, the *method of least squares* provides an optimal orientation matrix and cell parameters in a straightforward and efficient manner.*

The indices for every one of our n reflections provide three equations, each equation *calculating* a component of the $\mathbf{z}$ vector for the reflection (hence the superscript c):

$$\mathbf{A}^*\mathbf{h}_j = \mathbf{z}_j^c, \quad j = 1, 2, \ldots, n.$$

$$\begin{bmatrix} a_{11} & a_{12} & a_{13} \\ a_{21} & a_{221} & a_{23} \\ a_{31} & a_{32} & a_{33} \end{bmatrix} \begin{bmatrix} h_j \\ k_j \\ l_j \end{bmatrix} = \begin{bmatrix} z_{1j}^c \\ z_{2j}^c \\ z_{3j}^c \end{bmatrix}$$

$$a_{11}h_j + a_{12}k_j + a_{13}l_j = z_{1j}^c \tag{4.52}$$

$$a_{21}h_j + a_{22}k_j + a_{23}l_j = z_{2j}^c \tag{4.53}$$

$$a_{31}h_j + a_{32}k_j + a_{33}l_j = z_{2j}^c. \tag{4.54}$$

Each equation is a separate one, generating n calculated values, z_{ij}^c, $i = 1, 2, 3;\ j = 1, 2, \ldots, n$. For each of these calculated values there is a corresponding *observed* value, z_{ij}^o, determined from the measured angles for each reflection. If the orientation matrix was exact, the calculated and observed values would be identical and $z_{ij}^c - z_{ij}^o$ would be zero. Since it is not exact, there is an error in the expected value versus the actual value: $z_{ij}^c - z_{ij}^o = \epsilon_{ij}$. We seek the most accurate orientation matrix — the one that will produce calculated values as close as possible to the ones that we observe. In other words, we wish to determine those values of a_{ij}, the elements of the orientation matrix, that will minimize the errors, or more precisely, the average error. Attempting to minimize the average value of the error proves problematic; we must square the values to remove problems with signs – we instead minimize the square root of the average *squared* value of the error (known as the *root-mean-square* error). This is accomplished by summing the squares of the n errors for each equation, dividing by n to compute the average, then taking the

*There are many variations of the method described here, including "weighting" schemes that can be applied in the event that some parameters are known with more certainty than others. See Chapter 8 for a more comprehensive discussion of the method of least squares.

square root of the result. Since the root mean square error will be minimized if the sum of the squares of the errors is minimized, the *least-squares orientation matrix* is determined by finding the elements of $\mathbf{A}^*$ that minimize $\sum \epsilon_{ij}^2 = \sum (z_{ij}^c - z_{ij}^o)^2$.

Since Eqns. 4.52–4.54 are independent of one another we can treat them separately. We illustrate the determination of the least-squares matrix elements for the first row of the matrix, Eqn. 4.52; the second and third rows are treated in exactly the same manner. The expression that we wish to minimize is

$$\begin{aligned}\sum_{j=1}^{n} \epsilon_{1,j}^2 &= \sum_{j=1}^{n} (z_{1j}^c - z_{1j}^o)^2 \\ &= \sum_{j=1}^{n} \left[(a_{11}h_j + a_{12}k_j + a_{13}l_j) - z_{1j}^o \right]^2\end{aligned}$$

$$\begin{aligned}= \sum_{j=1}^{n} \Big[& h_j^2 a_{11}^2 + k_j^2 a_{12}^2 + l_j^2 a_{13}^2 \\ & + 2h_j\, k_j\, a_{11}a_{12} + 2h_j\, l_j\, a_{11}a_{13} + 2k_j\, l_j\, a_{12}a_{13} \\ & - 2h_j\, z_{1j}^o\, a_{11} - 2k_j\, z_{1j}^o\, a_{12} - 2l_j\, z_{1j}^o\, a_{13} - (z_{1j}^o)^2 \Big].\end{aligned} \tag{4.55}$$

The variables in the expression are the matrix elements, a_{11}, a_{12} and a_{13}. It is at a minimum when its partial derivative with respect to each of the variables is zero:

$$\frac{\partial \sum \epsilon_{1,j}^2}{\partial a_{11}} = 2\sum_{j=1}^{n} \left(h_j^2\, a_{11} + h_j\, k_j\, a_{12} + h_j\, l_j\, a_{13} - h_j\, z_{1j}^o \right) = 0 \tag{4.56}$$

$$\frac{\partial \sum \epsilon_{1,j}^2}{\partial a_{12}} = 2\sum_{j=1}^{n} \left(h_j\, k_j\, a_{11} + k_j^2\, a_{12} + k_j\, l_j\, a_{13} - k_j\, z_{1j}^o \right) = 0 \tag{4.57}$$

$$\frac{\partial \sum \epsilon_{1,j}^2}{\partial a_{13}} = 2\sum_{j=1}^{n} \left(h_j\, l_j\, a_{11} + k_j\, l_j\, a_{12} + l_j^2\, a_{13} - l_j\, z_{1j}^o \right) = 0. \tag{4.58}$$

This results in three simultaneous linear equations in the three variables to be determined:

$$\left(\sum_{j=1}^{n} h_j^2\right) a_{11} + \left(\sum_{j=1}^{n} h_j\, k_j\right) a_{12} + \left(\sum_{j=1}^{n} h_j\, l_j\right) a_{13} = \sum_{j=1}^{n} h_j\, z_{1j}^o \tag{4.59}$$

$$\left(\sum_{j=1}^{n} h_j\, k_j\right) a_{11} + \left(\sum_{j=1}^{n} k_j^2\right) a_{12} + \left(\sum_{j=1}^{n} k_j\, l_j\right) a_{13} = \sum_{j=1}^{n} k_j\, z_{1j}^o \tag{4.60}$$

$$\left(\sum_{j=1}^{n} h_j\, l_j\right) a_{11} + \left(\sum_{j=1}^{n} k_j\, l_j\right) a_{12} + \left(\sum_{j=1}^{n} l_j^2\right) a_{13} = \sum_{j=1}^{n} l_j\, z_{1j}^o. \tag{4.61}$$

Evaluating the coefficient sums on the left, which we denote t_{mn} (e.g., $t_{23} = \sum k_j l_j$), and the sums on the right, denoted w_{1m}, results in a straightforward matrix equation:

$$\begin{bmatrix} t_{11} & t_{12} & t_{13} \\ t_{21} & t_{22} & t_{23} \\ t_{31} & t_{32} & t_{33} \end{bmatrix} \begin{bmatrix} a_{11} \\ a_{12} \\ a_{13} \end{bmatrix} = \begin{bmatrix} w_{11} \\ w_{12} \\ w_{13} \end{bmatrix}. \tag{4.62}$$

The second and third rows of $\mathbf{A}^*$ are determined in the same manner. Note that the t_{mn} coefficients are the same for all three rows of the matrix (Eqns. 4.52–4.54); the entire least-squares orientation matrix can be evaluated with one equation:

$$\begin{bmatrix} t_{11} & t_{12} & t_{13} \\ t_{21} & t_{22} & t_{23} \\ t_{31} & t_{32} & t_{33} \end{bmatrix} \begin{bmatrix} a_{11} & a_{21} & a_{31} \\ a_{12} & a_{22} & a_{32} \\ a_{13} & a_{21} & a_{33} \end{bmatrix} = \begin{bmatrix} w_{11} & w_{21} & w_{31} \\ w_{12} & w_{22} & w_{32} \\ w_{13} & w_{21} & w_{33} \end{bmatrix}. \tag{4.63}$$

$$\begin{aligned} \mathbf{TA}^{*T} &= \mathbf{W} \\ \mathbf{A}^* &= (\mathbf{T}^{-1}\mathbf{W})^T. \end{aligned} \tag{4.64}$$

The least-squares orientation matrix can now provide the least squares direct and reciprocal unit cells from $\mathbf{G}$ and $\mathbf{G}^*$. The least squares method also provides standard errors (estimated standard deviations) for the lattice parameters — the details are given in Sec. 8.1.2.

4.6 Determining the Bravais Lattice.

4.6.1 Reduction of the Unit Cell

The unit cell that is selected and refined from measured reflections may not be the unit cell corresponding to the Bravais lattice — or the one with the three shortest non-coplanar axial vectors. We must therefore find a way to systematically determine the unit cell with the Bravais lattice symmetry from our selected unit cell — so that we are able to exploit the symmetry for structural solution and refinement.

In 1929, the Swiss mineralogist and crystallographer, Paul Niggli, published an extensive study of lattice geometry[54] in which he demonstrated that the unit cell for every Bravais lattice could be *reduced* to a *unique* primitive unit cell consisting of the three shortest non-coplanar vectors in the lattice (which we will call the *Niggli cell*), and that this unit cell exhibited mathematical relationships that would allow for its transformation into the unit cell of the Bravais lattice. In principle, any set of unit cell axes for a given lattice can be reduced into the Niggli cell for that lattice. Niggli created a comprehensive list of these reduced cells, including the relationships that linked each Niggli cell to a specific Bravais lattice, manifest in a matrix of the scalar products of the reduced cell axial vectors. This matrix is defined as follows (the subscript $\mathbf{r}$ signifies "reduced"):

$$\mathbf{S} = \begin{bmatrix} \mathbf{a_r}\cdot\mathbf{a_r} & \mathbf{b_r}\cdot\mathbf{b_r} & \mathbf{c_r}\cdot\mathbf{c_r} \\ \mathbf{b_r}\cdot\mathbf{c_r} & \mathbf{a_r}\cdot\mathbf{c_r} & \mathbf{a_r}\cdot\mathbf{b_r} \end{bmatrix} = \begin{bmatrix} s_{11} & s_{22} & s_{33} \\ s_{23} & s_{13} & s_{12} \end{bmatrix} \tag{4.65}$$

The top row of the scalar product matrix consists of the squares of the axial lengths and are termed the *symmetric* scalars. The bottom row consists of the inter-axial scalar products, referred to as the *asymmetric scalars.* (Note that these are simply the unique elements in the direct metric tensor for the reduced cell).

The scalar product matrix for a given Bravais lattice is determined from the geometric relationship between the Bravais lattice and the Niggli cell — as illustrated in the next section. There are 44 different forms for these matrices, and therefore 44 different Niggli cells. Depending on the relative lengths of the axes, a given Bravais lattice type can give rise to more than one Niggli cell. However, every Niggli cell will correspond to a single Bravais lattice type. In some cases a Bravais lattice will exhibit more than one reduced unit cell consisting of the three shortest non-coplanar vectors (which can be redundant), and it is therefore necessary to establish criteria that would lead to a single standard reduced cell in order to allow for the unequivocal determination of the Bravais lattice type; these are the criteria that were established by Niggli. They begin with the classification of the unit cell basis into one of two types, based on the interaxial angles. If a reduced cell has an interaxial angle less than 90°, then the axes are altered (by changing their directions) until all of the angles are $< 90°$. If the cell has one angle $\geq 90°$ then the axes are adjusted until all of the angles are $\geq 90°$. The standard reduced cell is constrained to be right-handed and its axes are labeled so that $a \leq b \leq c \Rightarrow s_{11} \leq s_{22} \leq s_{33}$. Reduced cells with acute angles are referred to as Type I reduced cells in the *positive reduced form.* Those with right angles or obtuse angles are referred to as Type II reduced cells in the *negative reduced form.*

The constraints on the asymmetric scalars are a consequence of the nature of the reduced cell. Since it consists of the three shortest translations in the lattice, there can be no diagonal in the reduced cell shorter than the shortest of the reduced cell axes. For acute interaxial angles this requires that the asymmetric scalar product for two axes will be less than or equal to half of the square of the magnitude of the shorter axis. This is illustrated at the top of Fig. 4.18. In Fig. 4.18(a) the length of the **b** axis is the same as the length of the diagonal. In this case, since the **b** axis and the diagonal are candidates for one of the shortest three axes, a special condition exists, and one of the two must be selected for the standard cell. The magnitude of the projection of **b** onto **a**, p_a, is $\frac{1}{2}a$ since the triangle consisting of **a**, **b**, and the diagonal is isosceles. Thus

$$p_a = b\cos\gamma = \frac{\mathbf{a}\cdot\mathbf{b}}{a} = \frac{1}{2}a \text{ and } \mathbf{a}\cdot\mathbf{b} = \frac{1}{2}a^2 = \frac{1}{2}(\mathbf{a}\cdot\mathbf{a}) \Rightarrow s_{12} = \frac{1}{2}s_{11}. \quad (4.66)$$

In Fig. 4.18(b) the acute angle is larger, and p_a is less than $\frac{1}{2}a$. In this situation the diagonal is longer than the length of the **b** axis. Since **a**, **b**, and the diagonal are coplanar, the diagonal cannot be a reduced cell axis. In this case $\mathbf{a}\cdot\mathbf{b} < \frac{1}{2}(\mathbf{a}\cdot\mathbf{a})$. On the other hand, in Fig. 4.18(c) the acute angle is smaller and the diagonal is shorter than the **b** axis, — **b** *cannot be a reduced cell axis.* When this occurs, $\mathbf{a}\cdot\mathbf{b} > \frac{1}{2}(\mathbf{a}\cdot\mathbf{a})$. Thus, given that **a** is the shortest axis in the lattice, **b** can be a reduced cell axis only if $\mathbf{a}\cdot\mathbf{b} \leq \frac{1}{2}(\mathbf{a}\cdot\mathbf{a}) \Rightarrow s_{12} \leq \frac{1}{2}s_{11}$. Since **a** is also shorter than **c**, **c** can be a reduced cell axis only if $s_{13} \leq \frac{1}{2}s_{11}$. Furthermore, since **b** is shorter than **c**, $\mathbf{b}\cdot\mathbf{c} \leq \frac{1}{2}(\mathbf{b}\cdot\mathbf{b}) \Rightarrow s_{23} \leq \frac{1}{2}s_{22}$. When special cases occur (when there is more than one potential reduced cell axis with the same length), further constraints, originally established by Niggli, are applied to chose between redundant

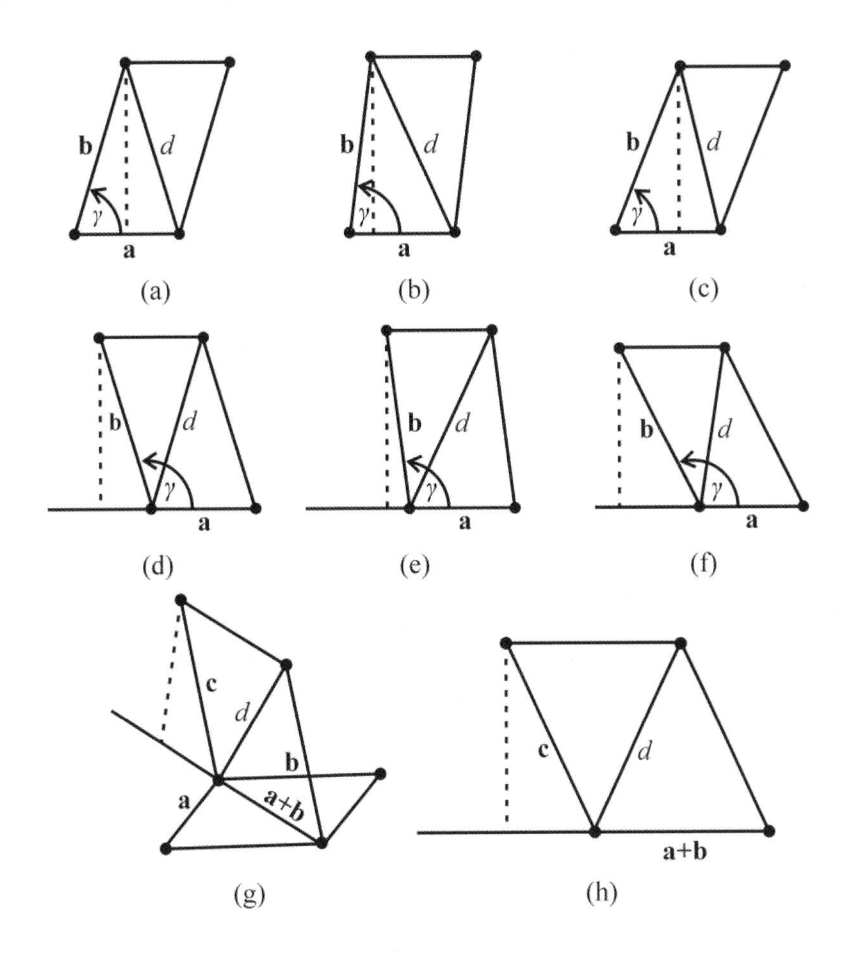

Figure 4.18 Acute interaxial angles: (a) $\mathbf{a} \cdot \mathbf{b} = \frac{1}{2}(\mathbf{a} \cdot \mathbf{a})$; $b = d$. (b) $\mathbf{a} \cdot \mathbf{b} < \frac{1}{2}(\mathbf{a} \cdot \mathbf{a})$; $b < d$. (c) $\mathbf{a} \cdot \mathbf{b} > \frac{1}{2}(\mathbf{a} \cdot \mathbf{a})$; $d < b$. Obtuse interaxial angles: (d) $-(\mathbf{a} \cdot \mathbf{b}) = |\mathbf{a} \cdot \mathbf{b}| = \frac{1}{2}(\mathbf{a} \cdot \mathbf{a})$; $b = d$. (e) $-(\mathbf{a} \cdot \mathbf{b}) = |\mathbf{a} \cdot \mathbf{b}| < \frac{1}{2}(\mathbf{a} \cdot \mathbf{a})$; $b < d$. (f) $-(\mathbf{a} \cdot \mathbf{b}) = |\mathbf{a} \cdot \mathbf{b}| > \frac{1}{2}(\mathbf{a} \cdot \mathbf{a})$; $d < b$. (g) Relationship between the body diagonal and the longest axis, **c**. (f) $-\mathbf{c} \cdot (\mathbf{a} + \mathbf{b}) = \frac{1}{2}(\mathbf{a} + \mathbf{b}) \cdot (\mathbf{a} + \mathbf{b})$; $c = d$.

axes in order to generate a standard reduced cell. The general criteria and special conditions for Type I reduced cells are outlined in Table 4.1.

The situation for potential reduced cell axes with obtuse angles is depicted in Fig. 4.18(d)–(h). In Fig. 4.18(d) the length of the **b** axis is the same as the length of the diagonal. The magnitude of the projection is the same as that in Fig. 4.18(a), $\frac{1}{2}a$, but since the angle is obtuse the cosine is negative and p_a is given by $-b \cos \gamma = -(\mathbf{a} \cdot \mathbf{b})/a = |(\mathbf{a} \cdot \mathbf{b})|/a$. Using arguments identical to those for Type I reduced cells, the general criteria for the asymmetric scalar products for Type II reduced cells are $|\mathbf{a} \cdot \mathbf{b}| \leq \frac{1}{2}(\mathbf{a} \cdot \mathbf{a})$, $|\mathbf{a} \cdot \mathbf{c}| \leq \frac{1}{2}(\mathbf{a} \cdot \mathbf{a})$, and $|\mathbf{b} \cdot \mathbf{c}| \leq \frac{1}{2}(\mathbf{b} \cdot \mathbf{b})$ — that is $|s_{12}| \leq \frac{1}{2}s_{11}$. $|s_{13}| \leq \frac{1}{2}s_{11}$, and $|s_{23}| \leq \frac{1}{2}s_{22}$. If the angles of the reduced cell are all acute then the body diagonal will always be longer than any of the unit cell axes. If, however, the angles are obtuse, the angle between the **c** axis — the longest axis – and the **ab** plane can increase to a point where the body diagonal of the

Table 4.1 Criteria for Reduced Cell Type I in the *International Tables for Crystallography.*

General conditions:

$$s_{11} \leq s_{22} \leq s_{33}; s_{23} \leq \tfrac{1}{2}s_{22}; s_{13} \leq \tfrac{1}{2}s_{11}; s_{12} \leq \tfrac{1}{2}s_{11}$$

Special conditions:

if $s_{11} = s_{22}$	then	$s_{23} < s_{13}$
if $s_{22} = s_{33}$	then	$s_{13} < s_{12}$
if $s_{23} = \frac{1}{2}s_{22}$	then	$s_{12} < 2s_{13}$
if $s_{13} = \frac{1}{2}s_{11}$	then	$s_{12} < 2s_{23}$
if $s_{12} = \frac{1}{2}s_{11}$	then	$s_{13} < 2s_{23}$

unit cell is shorter than the **c** axis. The relationship between the body diagonal and the **c** axis is illustrated in Fig. 4.18g. For Type II reduced cells, the length of the **c** axis must remain less than or equal to the body diagonal. The projection to determine this is along the **ab** face diagonal, **a**+**b**, illustrated in Fig. 4.18(g). The angle is obtuse and employing an identical argument leads to the conclusion that the magnitude of the projection must be less than half of the magnitude of **a**+**b**:

$$\begin{aligned}
-\mathbf{c}\cdot(\mathbf{a}+\mathbf{b}) &\leq \frac{1}{2}[(\mathbf{a}+\mathbf{b})\cdot(\mathbf{a}+\mathbf{b})]\\
-\mathbf{c}\cdot(\mathbf{a}+\mathbf{b}) &\leq \frac{1}{2}[(\mathbf{a}+\mathbf{b})\cdot(\mathbf{a}+\mathbf{b})]\\
-(\mathbf{c}\cdot\mathbf{a})-(\mathbf{c}\cdot\mathbf{b}) &\leq \frac{1}{2}[(\mathbf{a}\cdot\mathbf{a})+2(\mathbf{a}\cdot\mathbf{b})+(\mathbf{b}\cdot\mathbf{b})]\\
-(\mathbf{c}\cdot\mathbf{a})-(\mathbf{c}\cdot\mathbf{b})-(\mathbf{a}\cdot\mathbf{b}) &\leq \frac{1}{2}[(\mathbf{a}\cdot\mathbf{a})+(\mathbf{b}\cdot\mathbf{b})]\\
(|\mathbf{c}\cdot\mathbf{a}|+|\mathbf{c}\cdot\mathbf{b}|+|\mathbf{a}\cdot\mathbf{b}|) &\leq \frac{1}{2}[(\mathbf{a}\cdot\mathbf{a})+(\mathbf{b}\cdot\mathbf{b})] \Rightarrow\\
(|s_{23}|+|s_{13}|+|s_{12}|) &\leq \frac{1}{2}(s_{11}+s_{22}).
\end{aligned} \tag{4.67}$$

The general criteria and special conditions for Type II reduced cells are outlined in Table 4.2.

The strategy for determining the Bravais lattice for a specific crystal is therefore a matter of determining the Niggli cell, generating the scalar product matrix, and selecting the corresponding Bravais lattice, subsequently transforming the unit cell and reflection indices and generating a new orientation matrix by performing least squares with the newly indexed reflections.

There is, unfortunately, no simple algorithm for determining the Niggli cell directly from a primitive un-reduced cell in the lattice. Instead, a reduced cell consisting of the three shortest non-coplanar vectors is established, followed by their transformation (if necessary) into a reduced cell that satisfies the criteria in Table 4.1 or Table 4.2. In the late 1950s Martin Buerger developed several approaches to generate a unit cell consisting of the three shortest non-coplanar

Table 4.2 Criteria for Reduced Cell Type II in the *International Tables for Crystallography.*

General conditions:		
$s_{11} \leq s_{22} \leq s_{33}; \lvert s_{23}\rvert \leq \frac{1}{2}s_{22}; \lvert s_{13}\rvert \leq \frac{1}{2}s_{11}$		
$(\lvert s_{23}\rvert + \lvert s_{13} + \lvert s_{12}\rvert) \leq \frac{1}{2}(s_{11} + s_{22})$		
Special conditions:		
if $s_{11} = s_{22}$	then	$\lvert s_{23}\rvert < \lvert s_{13}\rvert$
if $s_{22} = s_{33}$	then	$\lvert s_{13}\rvert < \lvert s_{12}\rvert$
if $s_{23} = \frac{1}{2}s_{22}$	then	$s_{12} = 0$
if $s_{13} = \frac{1}{2}s_{11}$	then	$s_{12} = 0$
if $s_{12} = \frac{1}{2}s_{11}$	then	$s_{13} = 0$
if $(\lvert s_{23}\rvert + \lvert s_{13} + \lvert s_{12}\rvert)$ $= \frac{1}{2}(s_{11} + s_{22})$	then	$s_{11} < 2\lvert s_{13}\rvert + \lvert s_{12}\rvert$

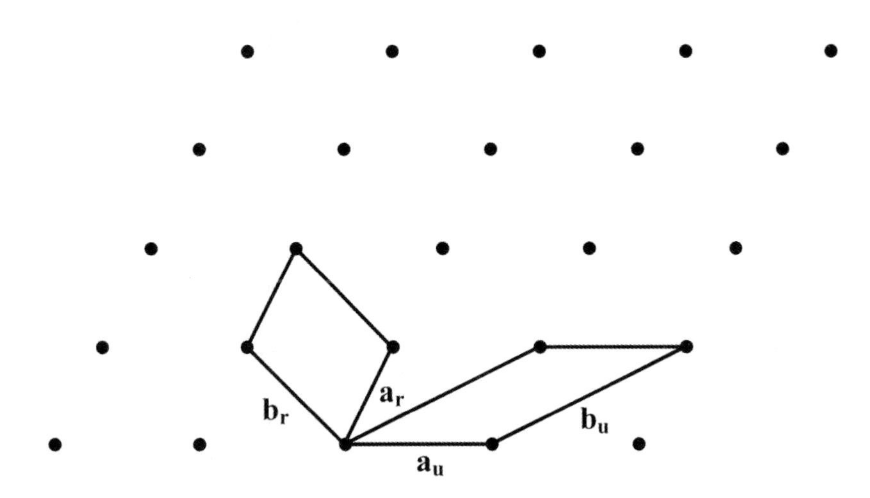

Figure 4.19 Two-dimensional lattice illustrating a primitive un-reduced cell with axes $\mathbf{a_u}$ and $\mathbf{b_u}$ and reduced cell with axes $\mathbf{a_r}$ and $\mathbf{b_r}$. The reduced unit cell axes can be described as integer combinations of the un-reduced cell axes: $\mathbf{a_r} = -1\,\mathbf{a_u} + 1\,\mathbf{b_u}$ and $\mathbf{b_r} = -2\,\mathbf{a_u} + 1\,\mathbf{b_u}$.

vectors from a general set of *primitive* unit cell axes for a given lattice. The simplest of these approaches was also the one that leant itself to computation. As illustrated in Fig. 4.19, any primitive unit cell in the lattice can be described as a linear combination of the unit cell axes for any other primitive unit cell, *with integer coefficients.* In particular, a unit cell with the three shortest non-coplanar axes, $\{\mathbf{a_r}, \mathbf{b_r}, \mathbf{c_r}\}$, can be described in terms of any un-reduced unit cell, $\{\mathbf{a_u}, \mathbf{b_u}, \mathbf{c_u}\}$, *provided that it is primitive*:

$$\begin{aligned}\mathbf{a_r} &= p_1\mathbf{a_u} + q_1\mathbf{b_u} + r_1\mathbf{c_u}\\ \mathbf{b_r} &= p_2\mathbf{a_u} + q_2\mathbf{b_u} + r_2\mathbf{c_u}\\ \mathbf{c_r} &= p_3\mathbf{a_u} + q_3\mathbf{b_u} + r_3\mathbf{c_u},\end{aligned} \tag{4.68}$$

where p_i, q_i, and r_i are integers. Since we already have the Cartesian components of our selected unit cell at hand it is a simple matter to generate trial linear combinations, $\mathbf{t_r} = p\mathbf{a_u} + q\mathbf{b_u} + r\mathbf{c_u}$, beginning at zero and testing with integers p, q, and r with increasing absolute value. $\mathbf{t} \cdot \mathbf{t}$ will provide the trial vector lengths; the process stops when the selected integers result exclusively in longer vectors. The trial vectors can be sorted by length and the three shortest of these that are non-collinear (the determinant of their $\mathbf{A}$ matrix is non-zero) will constitute a reduced cell. We call this reduced cell a *Buerger cell*, which may or may not be identical to the Niggli reduced unit cell. For a given Bravais lattice every Buerger cell is a reduced unit cell, but only one Buerger cell will satisfy the Niggli cell criteria — that Niggli cell will uniquely identify the Bravais lattice.

Once a Buerger unit cell is established for the lattice, its scalar product matrix is computed. If the Niggli criteria are met then the cell reduction is complete and the Buerger cell *is* the standard reduced cell. If the scalar product criteria are not met, then the Buerger cell is transformed by generating integral linear combinations of unit cell axes. Since the Buerger cell is always within a diagonal of the Niggli cell, the integers are restricted to -1, 0, and $+1$. Suppose, for example, that a Buerger reduced cell generates the following scalar product matrix after labeling and transforming the axes until the general conditions for a Type I reduced cell are met (the subscript signifies "Buerger"):

$$\mathbf{S} = \begin{bmatrix} \mathbf{a_b}\cdot\mathbf{a_b} & \mathbf{b_b}\cdot\mathbf{b_b} & \mathbf{c_b}\cdot\mathbf{c_b} \\ \mathbf{b_b}\cdot\mathbf{c_b} & \frac{\mathbf{a_b}\cdot\mathbf{a_b}}{2} & \mathbf{a_b}\cdot\mathbf{b_b} \end{bmatrix} = \begin{bmatrix} s_{11} & s_{22} & s_{33} \\ s_{23} & \frac{s_{11}}{2} & s_{12} \end{bmatrix}.$$

This is a special case, $s_{13} = s_{11}/2$, one in which the $\mathbf{c_b}$ axis and the $\mathbf{a_b c_b}$ diagonal are the same length. If $\mathbf{a} \cdot \mathbf{b} \leq 2\mathbf{b} \cdot \mathbf{c}$ then the Buerger cell is the Niggli cell. If it is not — if $2\mathbf{b} \cdot \mathbf{c} < \mathbf{a} \cdot \mathbf{b}$ — then the diagonal must become the Niggli cell $\mathbf{c}$ axis, which we will label $\mathbf{c_n}$, and the $\mathbf{c_b}$ axis becomes the diagonal: $\mathbf{c_b} = \mathbf{a_b} + \mathbf{c_n}$ and $\mathbf{c_n} = \mathbf{c_b} - \mathbf{a_b}$. In order to keep the unit cell right-handed and the angles acute we must also change the signs of the $\mathbf{a_b}$ and $\mathbf{b_b}$ axes. Thus the transformation from the Buerger cell to the Niggli cell is:

$$\begin{aligned}\mathbf{a_n} &= -\mathbf{a_b}\\ \mathbf{b_n} &= -\mathbf{b_b}\\ \mathbf{c_n} &= -\mathbf{a_b} + \mathbf{c_b}\end{aligned}$$

As a matrix equation this becomes

$$\begin{bmatrix} \bar{1} & 0 & 0 \\ 0 & \bar{1} & 0 \\ \bar{1} & 0 & 1 \end{bmatrix} \begin{bmatrix} \mathbf{a_b} \\ \mathbf{b_b} \\ \mathbf{c_b} \end{bmatrix} = \begin{bmatrix} \mathbf{a_n} \\ \mathbf{b_n} \\ \mathbf{c_n} \end{bmatrix}.$$

In order to determine if the transformed cell is the Niggli cell, we evaluate the scalar products, noting that $s_{13} = s_{11}/2 = \mathbf{a_b} \cdot \mathbf{c_b} = (\mathbf{a_b} \cdot \mathbf{a_b})/2$:

$$\begin{aligned}
\mathbf{a_n} \cdot \mathbf{a_n} &= -\mathbf{a_b} \cdot -\mathbf{a_b} = \mathbf{a_b} \cdot \mathbf{a_b} \Rightarrow s'_{11} = s_{11}. \\
\mathbf{b_n} \cdot \mathbf{b_n} &= -\mathbf{b_b} \cdot -\mathbf{b_b} = \mathbf{b_b} \cdot \mathbf{b_b} \Rightarrow s'_{22} = s_{22}. \\
\mathbf{c_n} \cdot \mathbf{c_n} &= (-\mathbf{a_b} + \mathbf{c_b}) \cdot (-\mathbf{a_b} + \mathbf{c_b}) = \mathbf{a_b} \cdot \mathbf{a_b} - 2\,\mathbf{a_b} \cdot \mathbf{c_b} + \mathbf{c_b} \cdot \mathbf{c_b} \\
&= \mathbf{a_b} \cdot \mathbf{a_b} - 2\left(\frac{\mathbf{a_b} \cdot \mathbf{a_b}}{2}\right) + \mathbf{c_b} \cdot \mathbf{c_b} = \mathbf{c_b} \cdot \mathbf{c_b} \Rightarrow s'_{33} = s_{33}. \\
\mathbf{b_n} \cdot \mathbf{c_n} &= -\mathbf{b_b} \cdot (-\mathbf{a_b} + \mathbf{c_b}) = \mathbf{a_b} \cdot \mathbf{b_b} - \mathbf{b_b} \cdot \mathbf{c_b} \Rightarrow s'_{23} = s_{12} - s_{23}. \\
\mathbf{a_n} \cdot \mathbf{c_n} &= -\mathbf{a_b} \cdot (-\mathbf{a_b} + \mathbf{c_b}) = \mathbf{a_b} \cdot \mathbf{a_b} - \mathbf{a_b} \cdot \mathbf{c_b} \\
&= \mathbf{a_b} \cdot \mathbf{a_b} - \left(\frac{\mathbf{a_b} \cdot \mathbf{a_b}}{2}\right) = \frac{\mathbf{a_b} \cdot \mathbf{a_b}}{2} \Rightarrow s'_{13} = s_{13}. \\
\mathbf{a_n} \cdot \mathbf{b_n} &= -\mathbf{a_b} \cdot -\mathbf{b_b} = \mathbf{a_b} \cdot \mathbf{b_b} \Rightarrow s'_{12} = s_{12}.
\end{aligned}$$

$s'_{23} = s_{12} - s_{23}$ is the only element that has changed in the scalar product matrix. The largest value for s_{12} is $s_{11}/2$ and the smallest value of s_{23} approaches zero (since the original angles are acute and s_{23} must remain positive). Thus the largest value possible for s'_{23} is $s_{11}/2 = s'_{11}/2$ and the general criteria are satisfied. The special condition remains to be satisfied — we must determine if $s'_{12} < 2\,s'_{23}$. Since $s'_{12} = s_{12}$ and $s_{12} > 2\,s_{23}$, $s'_{12} > 2\,s'_{12} - 2\,s'_{23}$ and $2\,s'_{23} > s'_{12}$ — the transformed cell is the Niggli cell.

There are a total of eleven such special cases to consider, and therefore eleven similar treatments; the diagnostic relationship between the scalar products and the transformation matrix for conversion to the standard Niggli cell are tabulated for each case in *Volume I** of the *International Tables for Crystallography.* A detailed discussion of the criteria for the standard reduced cell has been added to *Volume A* of the *International Tables.* In 1976 Krivey and Gruber published a very efficient algorithm that determines the Buerger and Niggli cells from an arbitrary primitive unit cell[56].

4.6.2 Searching for Higher Lattice Symmetry.

The determination of the standard reduced unit cell leads directly to the determination of the Bravais lattice since each of the 44 Niggli cell types corresponds to a specific Bravais lattice. The example in Fig. 4.20 illustrates that a specific Bravais lattice type can lead to more than one type of reduced cell, depending on the relative lengths of the axes. In this example, the *ab* planes of two different *C*-centered monoclinic unit cells are shown, along with the corresponding reduced unit cells. The *c* axis is shared by the centered and reduced cells in each case. In Fig. 4.20(a) the centered and reduced cells also share the *a* axis. The primitive unit cell shown

*The table, also published in a paper by Santoro and Mighell[55], was not included in *Volume A* of the *International Tables.*

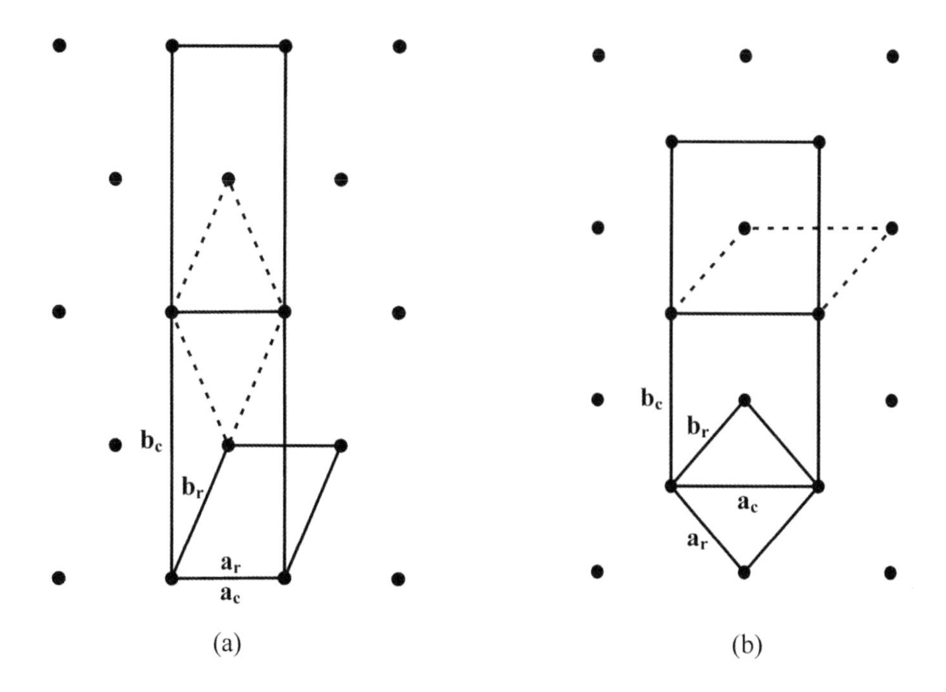

Figure 4.20 (a) The *ab* plane of a *C*-centered monoclinic lattice. The centered cell and the reduced cell share *a* and *c* axes. (b) The *ab* plane of a *C*-centered monoclinic unit cell for which the reduced cell and the centered cell share only the *c* axis.

with dashed lines is not the reduced cell, since its *a* axis is longer than its diagonal, which is the *a* axis for the reduced cell. Using the **c** subscript for the centered cell of the Bravais lattice, and **r** for the reduced cell, the cells are related as follows:

$$\begin{aligned}\mathbf{a_r} &= \mathbf{a_c}\\ \mathbf{b_r} &= \frac{1}{2}\mathbf{a_c} + \frac{1}{2}\mathbf{b_c}\\ \mathbf{c_r} &= \mathbf{c_c}.\end{aligned}$$

The form of the scalar products for the standard reduced cell are uniquely determined by this type of *C*-centered monoclinic lattice:

$$\begin{aligned}s_{11} &= \mathbf{a_r}\cdot\mathbf{a_r} = \mathbf{a_c}\cdot\mathbf{a_c} = a_c^2\\ s_{22} &= \mathbf{b_r}\cdot\mathbf{b_r} = \frac{1}{2}(\mathbf{a_c}+\mathbf{b_c})\cdot\frac{1}{2}(\mathbf{a_c}+\mathbf{b_c})\\ &= \frac{1}{4}(\mathbf{a_c}\cdot\mathbf{a_c}+2\mathbf{a_c}\cdot\mathbf{b_c}+\mathbf{b_c}\cdot\mathbf{b_c}) \quad (\mathbf{a_c}\cdot\mathbf{b_c}=0 \text{ since } \gamma=90^\circ.)\\ &= \frac{1}{4}(a_c^2+b_c^2)\\ s_{33} &= \mathbf{c_r}\cdot\mathbf{c_r} = \mathbf{c_c}\cdot\mathbf{c_c} = c_c^2\end{aligned}$$

$$
\begin{aligned}
s_{13} &= \mathbf{a_r} \cdot \mathbf{c_r} = \mathbf{a_c} \cdot \mathbf{c_c} \\
s_{23} &= \mathbf{b_r} \cdot \mathbf{c_r} = \frac{1}{2}(\mathbf{a_c} + \mathbf{b_c}) \cdot \mathbf{c_c} \\
&= \frac{1}{2}(\mathbf{a_c} \cdot \mathbf{c_c}) + \frac{1}{2}(\mathbf{b_c} \cdot \mathbf{c_c}) = \frac{1}{2}(\mathbf{a_c} \cdot \mathbf{c_c}) = \frac{s_{13}}{2} \quad (\ \mathbf{b_c} \cdot \mathbf{c_c} = 0 \ \text{since} \ \alpha = 90^\circ.) \\
s_{12} &= \mathbf{a_r} \cdot \mathbf{b_r} = \mathbf{a_c} \cdot \frac{1}{2}(\mathbf{a_c} + \mathbf{b_c}) \\
&= \frac{1}{2}(\mathbf{a_c} \cdot \mathbf{a_c}) + \frac{1}{2}(\mathbf{b_c} \cdot \mathbf{a_c}) = \frac{1}{2}a_c^2 = \frac{s_{11}}{2}.
\end{aligned}
$$

The unique scalar product matrix for the reduced cell corresponding to the C-centered Bravais lattice illustrated in Fig. 4.20(a) is

$$
\begin{bmatrix} \mathbf{a_r} \cdot \mathbf{a_r} & \mathbf{b_r} \cdot \mathbf{b_r} & \mathbf{c_r} \cdot \mathbf{c_r} \\ \mathbf{b_r} \cdot \mathbf{c_r} & \mathbf{a_r} \cdot \mathbf{c_r} & \mathbf{a_r} \cdot \mathbf{b_r} \end{bmatrix} = \begin{bmatrix} s_{11} & s_{22} & s_{33} \\ \frac{s_{13}}{2} & s_{13} & \frac{s_{11}}{2} \end{bmatrix}. \tag{4.69}
$$

The transformation matrix to convert from the reduced cell to the C-centered monoclinic cell is determined from the relationship between the unit cells:

$$
\begin{aligned}
\mathbf{a_c} &= \mathbf{a_r} \\
\mathbf{b_c} &= 2\mathbf{b_r} - \mathbf{a_c} = 2\mathbf{b_r} - \mathbf{a_r} \\
\mathbf{c_c} &= \mathbf{c_r}.
\end{aligned}
$$

The reduced cell has an acute angle, and therefore three acute angles – it is a Type I reduced cell. The conventional monoclinic unit cell is chosen so that β is non-acute, requiring a change in the direction of the c axis. In order to keep the unit cell right handed, the b axis must also be reversed. This results in the matrix equation for the transformation of the standard reduced unit cell to the conventional C-centered monoclinic unit cell:

$$
\begin{bmatrix} 1 & 0 & 0 \\ 1 & \bar{2} & 0 \\ 0 & 0 & \bar{1} \end{bmatrix} \begin{bmatrix} \mathbf{a_r} \\ \mathbf{b_r} \\ \mathbf{c_r} \end{bmatrix} = \begin{bmatrix} \mathbf{a_c} \\ \mathbf{b_c} \\ \mathbf{c_c} \end{bmatrix}. \tag{4.70}
$$

The monoclinic unit cell illustrated in Fig. 4.20(b) shares only the c axis with the reduced cell. In this case the primitive cell sharing the a axis with the centered cell, shown with dashed lines, has a diagonal shorter than the a axis and the diagonal is the reduced cell axis. The relationship between the centered cell and the reduced cell is now

$$
\begin{aligned}
\mathbf{a_r} &= \frac{1}{2}\mathbf{a_c} - \frac{1}{2}\mathbf{b_c} \\
\mathbf{b_r} &= \frac{1}{2}\mathbf{a_c} + \frac{1}{2}\mathbf{b_c} \\
\mathbf{c_r} &= \mathbf{c_c}.
\end{aligned} \tag{4.71}
$$

The reduced cell scalar product matrix, unique for this type of C-centered lattice, is given by

$$\begin{aligned}
s_{11} &= \mathbf{a_r}\cdot\mathbf{a_r} = \frac{1}{2}(\mathbf{a_c}-\mathbf{b_c})\cdot\frac{1}{2}(\mathbf{a_c}-\mathbf{b_c})\\
&= \frac{1}{4}(\mathbf{a_c}\cdot\mathbf{a_c}-2\mathbf{a_c}\cdot\mathbf{b_c}+\mathbf{b_c}\cdot\mathbf{b_c}) = \frac{1}{4}(a_c^2+b_c^2)\\
s_{22} &= \mathbf{b_r}\cdot\mathbf{b_r} = \frac{1}{2}(\mathbf{a_c}+\mathbf{b_c})\cdot\frac{1}{2}(\mathbf{a_c}+\mathbf{b_c})\\
&= \frac{1}{4}(\mathbf{a_c}\cdot\mathbf{a_c}+2\mathbf{a_c}\cdot\mathbf{b_c}+\mathbf{b_c}\cdot\mathbf{b_c}) = \frac{1}{4}(a_c^2+b_c^2) = s_{11}\\
s_{23} &= \mathbf{b_r}\cdot\mathbf{c_r} = \frac{1}{2}(\mathbf{a_c}+\mathbf{b_c})\cdot\mathbf{c_c} = \frac{1}{2}\mathbf{a_c}\cdot\mathbf{c_b}\\
s_{13} &= \mathbf{a_r}\cdot\mathbf{c_r} = \frac{1}{2}(\mathbf{a_c}-\mathbf{b_c})\cdot\mathbf{c_c} = \frac{1}{2}\mathbf{a_c}\cdot\mathbf{c_c} = s_{23}\\
s_{12} &= \mathbf{a_r}\cdot\mathbf{b_r} = \frac{1}{2}(\mathbf{a_c}-\mathbf{b_c})\cdot\frac{1}{2}(\mathbf{a_c}+\mathbf{b_c}) = \frac{1}{4}(a_c^2-b_c^2).
\end{aligned}$$

The unique scalar product matrix is therefore

$$\begin{bmatrix}\mathbf{a_r}\cdot\mathbf{a_r} & \mathbf{b_r}\cdot\mathbf{b_r} & \mathbf{c_r}\cdot\mathbf{c_r}\\ \mathbf{b_r}\cdot\mathbf{c_r} & \mathbf{a_r}\cdot\mathbf{c_r} & \mathbf{a_r}\cdot\mathbf{b_r}\end{bmatrix} = \begin{bmatrix}s_{11} & s_{11} & s_{33}\\ s_{23} & s_{23} & s_{12}\end{bmatrix}. \tag{4.72}$$

Since the reduced cell contains an obtuse angle (Type II), the transformation matrix does not require reorientation of the axes:

$$\begin{aligned}
\mathbf{a_c} &= \mathbf{a_r}+\mathbf{b_r}\\
\mathbf{b_c} &= -\mathbf{a_r}+\mathbf{b_r}\\
\mathbf{c_c} &= \mathbf{c_r}
\end{aligned}$$

$$\begin{bmatrix}1 & 1 & 0\\ \bar{1} & 1 & 0\\ 0 & 0 & 1\end{bmatrix}\begin{bmatrix}\mathbf{a_r}\\ \mathbf{b_r}\\ \mathbf{c_r}\end{bmatrix} = \begin{bmatrix}\mathbf{a_c}\\ \mathbf{b_c}\\ \mathbf{c_c}\end{bmatrix}. \tag{4.73}$$

The scalar product and transformation matrices for the 44 Niggli cells can be found in *Volume I/A of the International Tables for Crystallography.* They provide the means for determining the conventional unit cell in the Bravais lattice from a primitive unit cell in the lattice, based solely on the metrics of the lattice.* In most cases this corresponds to the actual symmetry of the crystal. However, it is possible that the symmetry of the contents of the lattice may not reflect the symmetry of the Bravais lattice determined from the reduced cell. This occurs when the actual symmetry is lower than the apparent symmetry. For example, a monoclinic unit cell may have β very close or equal to 90° "by accident," and the standard reduced cell will predict an orthorhombic unit cell. In order to ascertain whether or not the

*An alternative to the approach outlined here is to generate a Buerger unit cell and search the lattice for two-fold symmetry axes[57]. Although not completely rigorous, the method is computationally more efficient since it only requires the generation of the three shortest non-coplanar vectors in the lattice, and does not require a "table search." The method generally produces the correct Bravais lattice.

Bravais lattice determined from cell reduction reflects the actual symmetry of the crystal we must examine the crystal symmetry directly — by probing the symmetry of the reciprocal lattice.

4.6.3 Symmetry of the Reciprocal Lattice.

The Friedel Law

In direct space the crystal structure is referenced to its natural coordinate system — the direct lattice. In reciprocal space it is the diffraction pattern that is referenced to its natural coordinate system — the reciprocal lattice. Since the "nodes" in the lattice — the reciprocal lattice points — are those responsible for diffraction, we observe intensities only in the vicinity of those points; the fractional coordinates of interest in the reciprocal lattice are the integer indices. Assigning the corresponding diffraction intensity to each lattice point in reciprocal space is analogous to assigning electron density to each point in direct space — the crystal structure. We call the reciprocal space analog of the crystal structure the *intensity-weighted reciprocal lattice.* The diffraction pattern is an image of the intensity-weighted reciprocal lattice.

According to Eqn. 3.79, the reflection vector for a given reciprocal lattice vector, $\mathbf{h} = [h\ k\ l]$ is a scalar multiple of the wave scattered from a unit cell in the lattice (the structure factor):

$$\overrightarrow{\mathcal{E}}_{hkl} = D\overrightarrow{F}_{hkl} = D\int_0^{V_c} \rho(x_f, y_f, z_f)e^{-2\pi i(hx_f+ky_f+lz_f)}\,dV. \tag{4.74}$$

Recall that the integral is a superposition of waves scattered from each point in the unit cell, resulting in a net wave with amplitude $\mathcal{E}$ and phase, φ; the phase is determined by the values in the exponent, $-2\pi i(hx_f + ky_f + lz_f)$. The reflection vector for the reciprocal lattice vector in the opposite direction, $\bar{\mathbf{h}} = [\bar{h}\ \bar{k}\ \bar{l}]$, is therefore

$$\overrightarrow{\mathcal{E}}_{\overline{hkl}} = D\overrightarrow{F}_{\overline{hkl}} = D\int_0^{V_c} \rho(x_f, y_f, z_f)e^{-2\pi i(-hx_f-ky_f-lz_f)}\,dV. \tag{4.75}$$

The net scattered wave has an amplitude of $\mathcal{E}$ and a phase of $-\varphi$, as illustrated in Fig. 4.21.

The intensities of the scattered waves from the hkl planes are given by $\overrightarrow{\mathcal{E}}^{*}_{hkl}\overrightarrow{\mathcal{E}}_{hkl} = \mathcal{E}e^{-i\varphi}\mathcal{E}e^{i\varphi} = \mathcal{E}^2$, and $\overrightarrow{\mathcal{E}}^{*}_{\overline{hkl}}\overrightarrow{\mathcal{E}}_{\overline{hkl}} = \mathcal{E}e^{i\varphi}\mathcal{E}e^{-i\varphi} = \mathcal{E}^2$ for the $\overline{hkl}$ planes. *The intensities for hkl and $\overline{hkl}$ are the same.* This introduces a center of symmetry into the intensity-weighted reciprocal lattice. The proposition that all diffraction patterns are centrosymmetric was originally put forth in 1913 by the French crystallographer Georges Friedel[58] in order to explain the connection between the symmetry of the diffraction pattern and the symmetry of the crystal. The equivalency of the intensities from $\mathbf{h}$ and $\bar{\mathbf{h}}$ is known as the *Friedel Law,* although we will find that it is only a close approximation (we will soon encounter a small deviation from the treatment given here due to a phenomenon known as *anomalous dispersion*). Since the structure factors differ from the scattered waves by a constant factor, the structure factor amplitudes also exhibit this centrosymmetry: $F_{hkl} = F_{\overline{hkl}}$. The Friedel Law is often expressed in terms of the amplitudes of the structure factors.

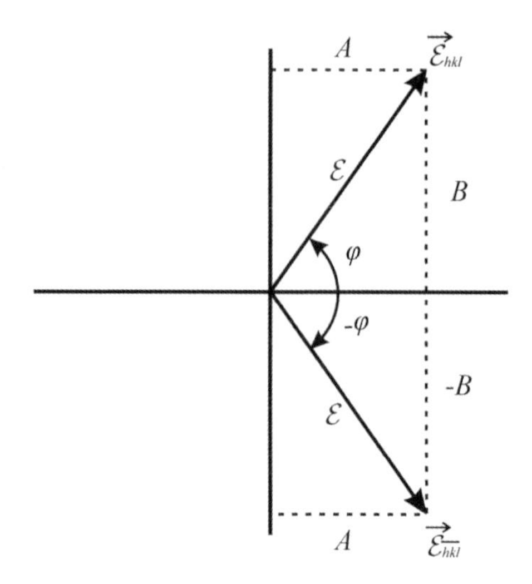

Figure 4.21 Diffracted waves for the $\mathbf{h} = [h\ k\ l]$ and $\bar{\mathbf{h}} = [\bar{h}\ \bar{k}\ \bar{l}]$ reflections. $\vec{\mathcal{E}}_{hkl} = \mathcal{E}e^{i\varphi} = \mathcal{E}(\cos\varphi + i\sin\varphi) = A + iB$. $\vec{\mathcal{E}}_{\overline{hkl}} = \mathcal{E}e^{-i\varphi} = \mathcal{E}(\cos\varphi - i\sin\varphi) = A - iB$.

Laue Symmetry

The intensities in the diffraction pattern depend on the electron density distribution in the crystal, and it should not be surprising that the symmetry of the crystal is reflected in those intensities. Consider, for example, a monoclinic crystal with two-fold rotational symmetry about the b axis. The electron density at (x_f, y_f, z_f) is identical to that at $(-x_f, y_f, -z_f)$. The scattered wave for the reflection with reciprocal lattice vector $\mathbf{h}' = [h'\ k'\ l']$, integrating over the symmetry-equivalent unit cell, is

$$\begin{aligned}\vec{\mathcal{E}}_{h'k'l'} &= D\vec{F}_{h'k'l} = D\int_0^{V_c} \rho(-x_f, y_f, -z_f)e^{-2\pi i(-h'x_f + k'y_f - l'z_f)}\,dV \\ &= D\int_0^{V_c} \rho(x_f, y_f, z_f)e^{-2\pi i(-h'x_f + k'y_f - l'z_f)}\,dV. \end{aligned} \tag{4.76}$$

Replacing $h' = \bar{h}$, $k' = k$ and $l' = \bar{l}$,

$$\vec{\mathcal{E}}_{\bar{h}k\bar{l}} = D\vec{F}_{\bar{h}k\bar{l}} = D\int_0^{V_c} \rho(x_f, y_f, z_f)e^{-2\pi i(hx_f + ky_f + lz_f)}\,dV = \vec{\mathcal{E}}_{hkl}. \tag{4.77}$$

It follows that $\mathcal{E}^2_{\bar{h}k\bar{l}} = \mathcal{E}^2_{hkl}$; The two-fold axis creates symmetry-equivalent reflections in the diffraction pattern: $I_{hkl} = I_{\bar{h}k\bar{l}}$ and $F_{hkl} = F_{\bar{h}k\bar{l}}$, where $I_{hkl} = \mathcal{E}^2_{hkl}$.

If the two-fold rotational axis is replaced with a diad screw axis, the electron density at (x_f, y_f, z_f) is identical to that at $(-x_f, y_f + 1/2, -z_f)$, and

$$\begin{aligned}\vec{\mathcal{E}}_{h'k'l'} &= D\vec{F}_{h'k'l} = D\int_0^{V_c} \rho(-x_f, y_f + 1/2, -z_f)e^{-2\pi i(-h'x_f + k'(y_f + 1/2) - l'z_f)}\,dV \\ &= De^{-\pi ik}\int_0^{V_c} \rho(x_f, y_f, z_f)e^{-2\pi i(-h'x_f + k'y_f - l'z_f)}\,dV. \end{aligned} \tag{4.78}$$

Again replacing $h' = \bar{h}$, $k' = k$ and $l' = \bar{l}$,

$$\vec{\mathcal{E}}_{\bar{h}k\bar{l}} = D\vec{F}_{\bar{h}k\bar{l}} = e^{-\pi ik}\vec{\mathcal{E}}_{hkl}. \tag{4.79}$$

The scattered wave has a modified phase, due to the presence of the translational symmetry in the crystal. However, the intensity is proportional to $\mathcal{E}^2_{\bar{h}k\bar{l}}$:

$$\mathcal{E}^2_{\bar{h}k\bar{l}} = \vec{\mathcal{E}}_{\bar{h}k\bar{l}}\vec{\mathcal{E}}^*_{\bar{h}k\bar{l}} = e^{-\pi ik}\vec{\mathcal{E}}_{hkl}\,e^{\pi ik}\vec{\mathcal{E}}^*_{hkl} = e^0\mathcal{E}^2_{hkl} = \mathcal{E}^2_{hkl}. \tag{4.80}$$

Thus, $I_{hkl} = I_{\bar{h}k\bar{l}}$ and $F_{hkl} = F_{\bar{h}k\bar{l}}$. The diad screw axis produces the same reflection symmetry as the diad rotational axis.

A monoclinic crystal can also have a mirror plane perpendicular to the unique b axis. For this symmetry element the electron density at $(x_f, -y_f, z_f)$ is the same as that at (x_f, y_f, z_f). By a similar argument, substituting $(x_f, -y_f, z_f)$ for the electron density and $h' = h$, $k' = \bar{k}$ and $l' = l$ for the indices gives $\vec{\mathcal{E}}_{h\bar{k}l} = \vec{\mathcal{E}}_{hkl}$. For a mirror plane in the direct lattice, $\mathcal{E}^2_{h\bar{k}l} = \mathcal{E}^2_{hkl} \Rightarrow I_{hkl} = I_{h\bar{k}l}$ and $F_{hkl} = F_{h\bar{k}l}$ in the intensity-weighted reciprocal lattice. As with the screw axis above, the substitution of a glide plane for mirror plane will produce the same intensities for the same reflections. *In general, the symmetry of the diffraction pattern reflects the point symmetry of the crystal, but not its translational symmetry.*

Finally, if the monoclinic crystal is centrosymmetric the inversion center requires symmetry-equivalent electron density at $(-x_f, -y_f, -z_f)$. With a similar treatment, substituting $h' = \bar{h}$, $k' = \bar{k}$ and $l' = \bar{l}$ results in $I_{hkl} = I_{\bar{h}\bar{k}\bar{l}}$ and $F_{hkl} = F_{\bar{h}\bar{k}\bar{l}}$. Thus the center of symmetry is reflected in the diffraction pattern as a center of symmetry.* Unfortunately, the Friedel Law precludes us from distinguishing from diffraction patterns resulting from centrosymmetric structures and those with inversion symmetry imposed by the law; *all* diffraction patterns are centrosymmetric. The diffraction pattern can therefore reflect the point symmetry of only 11 of the 32 point groups, those that have centers of symmetry. The symmetry of the diffraction pattern was first observed from Laue photographs, and the 11 characteristic centrosymmetric point groups are known as the *Laue groups*.†

A diffraction pattern from a triclinic crystal is centrosymmetric, with no other symmetry. As a result triclinic crystals exhibit the intensity pattern of four pairs of reflections, indicative of the Laue group $\bar{1}$: $I_{hkl} = I_{\bar{h}\bar{k}\bar{l}}$, $I_{\bar{h}kl} = I_{h\bar{k}\bar{l}}$, $I_{h\bar{k}l} = I_{\bar{h}k\bar{l}}$, and $I_{hk\bar{l}} = I_{\bar{h}\bar{k}l}$. These centrosymmetric pairs of reflections are often termed *Friedel pairs*. For a given set of indices, $|h|$, $|k|$, and $|l|$, a triclinic crystal will, in general, exhibit four different reflection intensities, one for each Friedel pair. The only centrosymmetric monoclinic point group is $2/m$. The combination of the two-fold axis (unique axis b) and the mirror plane results in two sets of four equivalent intensities: $I_{hkl} = I_{\bar{h}k\bar{l}} = I_{\bar{h}\bar{k}\bar{l}} = I_{h\bar{k}l}$ and $I_{\bar{h}\bar{k}l} = I_{\bar{h}kl} = I_{hk\bar{l}} = I_{h\bar{k}\bar{l}}$.

The only centrosymmetric orthorhombic point group is mmm. The mirror plane perpendicular to the a axis results in $I_{hkl} = I_{\bar{h}kl}$, the mirror plane perpendicular to the b axis results in $I_{hkl} = I_{h\bar{k}l}$, and the mirror plane perpendicular to c results

*It might appear from these examples that the point symmetry operations in the direct lattice always correspond to identical operations in the intensity-weighted reciprocal lattice. However, this is not always the case. The diffraction symmetry should always be derived from the symmetry in the direct lattice.

†The Laue method for observing the diffraction pattern differs from other methods — the crystal remains stationary and white X-radiation provides a spectrum of wavelengths that bring the various crystal planes into the diffraction condition.

in $I_{hkl} = I_{hk\bar{l}}$. Addition of the center of symmetry generates the four remaining symmetry-equivalent intensities: $I_{\bar{h}\bar{k}\bar{l}}$, $I_{\bar{h}\bar{k}l}$,$I_{\bar{h}k\bar{l}}$ and $I_{h\bar{k}\bar{l}}$. Again, these equivalencies are often expressed in terms of the structure factor amplitudes: $F_{hkl} = F_{hk\bar{l}} = F_{h\bar{k}l} = F_{\bar{h}kl} = F_{\bar{h}\bar{k}\bar{l}} = F_{\bar{h}\bar{k}l} = F_{\bar{h}k\bar{l}} = F_{h\bar{k}\bar{l}}$.

There are two centrosymmetric trigonal point groups, $\bar{3}$ and $\bar{3}m$. The threefold axis requires identical electron density at (x_f, y_f, z_f), $(-y_f, x_f - y_f, z_f)$, and $(y_f - x_f, -x_f, z_f)$. Integrating over the symmetry-equivalent unit cell generated from the $\mathbf{R}(3^1)$ operation for the general reciprocal lattice vector $\mathbf{h}' = [h'\ k'\ l']$ results in

$$\vec{\mathcal{E}}_{h'k'l'} = D\vec{F}_{h'k'l} = D\int_0^{V_c} \rho(-y_f, x_f - y_f, z_f) e^{-2\pi i(h'(-y_f)+k'(x_f-y_f)+l'z_f)}\, dV.$$

$$= D\int_0^{V_c} \rho(x_f, y_f, z_f) e^{-2\pi i(k'x_f+(-h'-k')y_f+l'z_f)}\, dV. \tag{4.81}$$

Replacing $h' = -k - h$, $k' = h$ and $l' = l$,

$$\vec{\mathcal{E}}_{(\bar{h}+\bar{k})hl} = D\vec{F}_{(\bar{h}+\bar{k})hl} = D\int_0^{V_c} \rho(x_f, y_f, z_f) e^{-2\pi i(hx_f+ky_f+lz_f)}\, dV = \vec{\mathcal{E}}_{hkl}. \tag{4.82}$$

A similar treatment for the $\mathbf{R}(3^2)$ operation results in the symmetry-equivalent intensities arising from the three-fold axis: $I_{hkl} = I_{(\bar{h}+\bar{k})hl} = I_{k(\bar{h}+\bar{k})l}$. The inversion symmetry creates three more: $I_{\bar{h}\bar{k}\bar{l}}$, $I_{(h+k)\bar{h}\bar{l}}$ and $I_{\bar{k}(h+k)\bar{l}}$. The diffraction symmetry for the Laue group $\bar{3}$ in terms of the structure factor amplitudes is $F_{hkl} = F_{(\bar{h}+\bar{k})hl} = F_{k(\bar{h}+\bar{k})l} = F_{\bar{h}\bar{k}\bar{l}} = F_{(h+k)\bar{h}\bar{l}} = F_{\bar{k}(h+k)\bar{l}}$. The Laue groups and the crystallographic point groups that produce their characteristic diffraction patterns are listed in Table 4.3. The reader is encouraged to apply the symmetry operations for each Laue group to determine the diffraction symmetry of the remaining point groups.

Table 4.3 Laue Groups and Corresponding Crystallographic Point Groups.

Laue Group	Point Groups	Crystal System	Degeneracy*
$\bar{1}$	$1, \bar{1}$	triclinic	2
$2/m$	$2, m, 2/m$	monoclinic	4
mmm	$222, mm2, mmm$	orthorhombic	8
$\bar{3}$	$3, \bar{3}$	trigonal	6
$\bar{3}m$	$32, 3m, \bar{3}m$	trigonal	12
$4/m$	$4, \bar{4}, 4/m$	tetragonal	8
$4/mmm$	$422, 4mm, \bar{4}2m, 4/mmm$	tetragonal	16
$6/m$	$6, \bar{6}, 6/m$	hexagonal	12
$6/mmm$	$622, 6mm, \bar{6}m2, 6/mmm$	hexagonal	24
$m3$	$23, m3$	cubic	24
$m3m$	$432, \bar{4}3m, m3m$	cubic	48

*The number of symmetry-equivalent intensities for the general reflection with indices *hkl*.

Each Laue group exhibits a characteristic symmetry in the diffraction pattern. Combining the results of an analysis of the lattice metrics via cell reduction and the analysis of the diffraction symmetry – by measuring the intensities of reflections corresponding to the crystal system determined from cell reduction – nearly always results in the assignment of the correct Bravais lattice. We now turn to the details of measuring the intensities.

4.7 The Measurement of Integrated Intensities.

4.7.1 Reflections

Up to this point we have been treating diffraction maxima as one-dimensional vectors, arriving at the detector at a single point (serial detectors are often referred to as *point detectors*). Experimentally observed reflections, however, are finite in size; several phenomena combine to create diffracted beams that have a finite, variable, and often non-uniform cross-section. The "spots" on an area detector surface are further distorted because the detector plane is not generally orthogonal to the reflection vector.

Diffraction from the crystal itself gives the reflection a finite size, since the scattering occurs from different regions inside the crystal. Furthermore, as with any beam of light, the diffracted beam will *diverge* (spread out) as it emanates from the crystal. In addition, the diffraction conditions are not satisfied instantaneously. As p, q, and r in Eqn. 3.46 approach integers, that is, as the crystal orientation approaches diffraction conditions, the waves from individual unit cells will begin to constructively interfere toward a maximum when the planes are at the diffraction angle. While these effects are usually small, the diffraction peaks are also broadened because X-radiation is not actually monochromatic. While most of the extraneous wavelengths emitted from X-ray sources can be removed using suitable filters or a monochromator, the incident X-ray beam actually consists of two inseparable components from the K atomic shell of the target material producing the X-rays. These components, known as the $K\alpha_1$ and $K\alpha_2$, are slightly different in wavelength, and contribute unequally to the incident radiation — the $K\alpha_2$ contribution is about half of the $K\alpha_1$ contribution. This results in two different diffraction maxima (in position and intensity) for each reflection — at two different diffraction angles: $\theta_1 = \arcsin(\lambda_1/2d_{hkl})$ and $\theta_2 = \arcsin(\lambda_2/2d_{hkl})$ such that $\Delta \sin\theta = (\lambda_2 - \lambda_1)/2d_{hkl}$. As the interplanar distance decreases (and θ increases) the difference between the maxima increases. Fig. 4.22 illustrates the spread in the diffraction maxima as θ increases when the target material is copper. At low Bragg angles (2θ) the peaks due to each wavelength are essentially superimposed, while at high angles they become almost completely resolved. In this case $K\alpha_1 = 1.54051$ Å and $K\alpha_2 = 1.54433$ Å. The other common target material is molybdenum, for which $K\alpha_1 = 0.70926$ Å and $K\alpha_2 = 0.71354$ Å. The actual value of the wavelength used for most crystallographic calculations is a weighted average of the two contributing wavelengths, with each component weighted in proportion to its contribution to the total intensity, 1.5418 Å for Cu $K\alpha$ and 0.7107 Å for Mo $K\alpha$.

The most variable contributor to the width of the observed peak in the diffraction pattern results from the nature of the crystal itself. To this point the crystal has been assumed to consist of a "perfect" lattice containing precisely aligned unit cells. As crystals grow from supersaturated solutions, however, local gradients in the solution, fluctuations in temperature, small impurities, etc. cause the growth

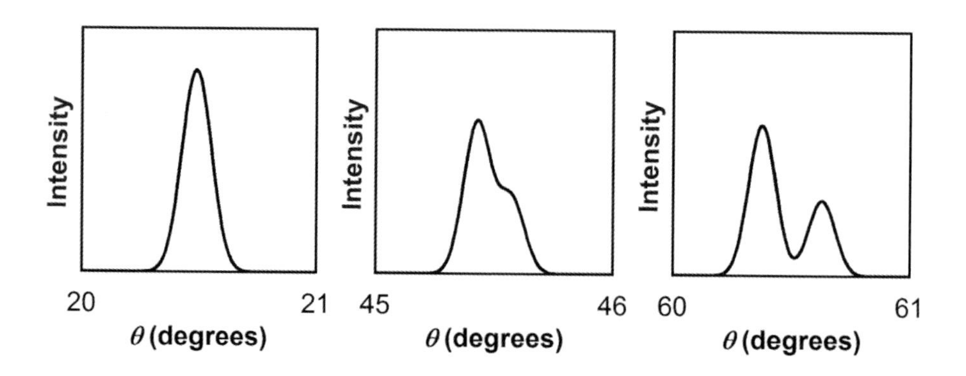

Figure 4.22 Diffraction maxima locations for reflections from planes with interplanar spacings of 2.2, 1.1 and 0.9 Å, respectively (left to right) with Cu $K\alpha$ radiation. The higher-angle maxima are due to the $K\alpha_2$ component of the incident radiation.

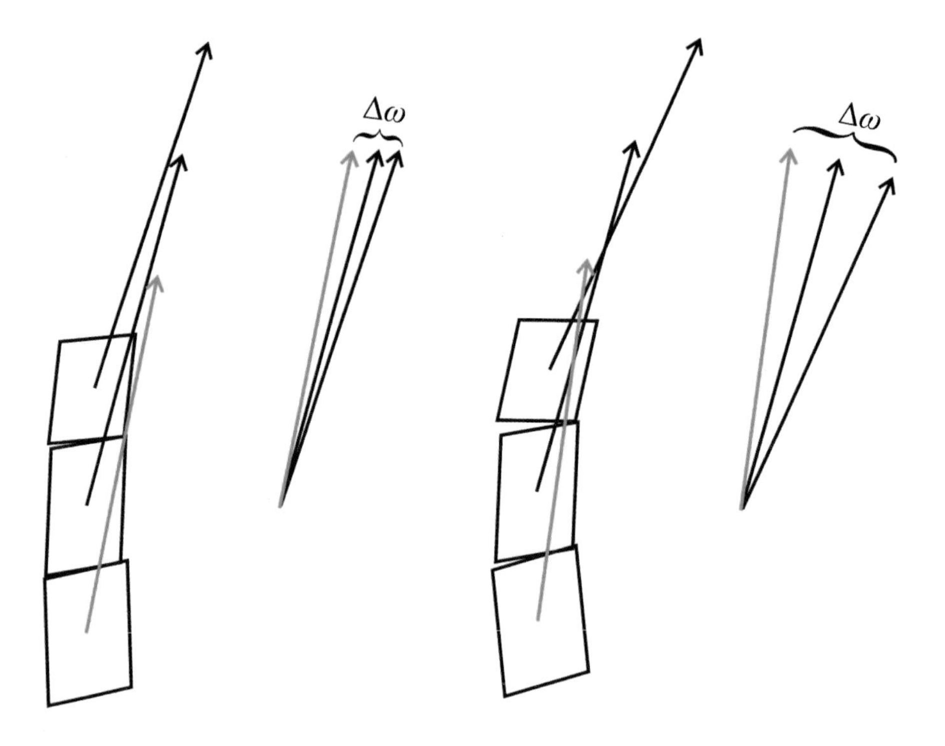

Figure 4.23 Left: Diffraction from slightly misaligned mosaic blocks. Right: Diffraction from significantly misaligned mosaic blocks. (This hypothetical example is limited to two dimensions and the misalignments are exaggerated.)

pattern to alter slightly. Mechanical stress can also cause micro-cracks along lattice planes. The result is that crystals ordinarily consist of small blocks on the order of 10^{-7} mm in length that are slightly misaligned from one another. The blocks form a *mosaic* pattern, and are known as mosaic blocks. The degree of block misalignment is termed the *mosaicity* of the crystal. This misalignment is usually very small, on the order of 0.1°–0.2°, but in extreme cases can be as great as 2°–3°. The results of the misalignment are illustrated in Fig. 4.23 (The displacements have been exaggerated for graphical purposes). On the left are blocks rotated only slightly. If the block at the bottom is in the diffraction condition, the two blocks above must each be rotated through an additional increment of ω in order to intersect their diffraction vectors with the sphere of reflection. The blocks on the right suffer a greater mosaicity, and the difference in the values of omega for each block increases. Crystals that are highly mosaic require scans through large increments of ω in order to measure the net intensity of a reflection — resulting in broad peaks when measured with a serial detector. Those with only small block displacements exhibit sharp peaks as omega is scanned. It would appear that the ideal crystal would be the "perfect crystal" — one with no misalignments. However, the lack of mosaicity results in another phenomenon known as *extinction* (discussed in the next chapter), which can result in a significant diminution of intensities. Thus the best crystals for diffraction studies are those with a large number of slightly misaligned mosaic blocks.

4.7.2 The Integrated Intensity.

Fig. 4.24 depicts a small crystal subjected to an incident beam of intensity I_0 directed away from the source along the unit vector $\mathbf{s_1}$. The beam encounters atom n in the crystal; some of the radiation is scattered toward a detector. A unit vector, $\mathbf{r_1}$ points toward a point p on the detector, with some of the coherently scattered radiation arriving at point p. The *integrated intensity* is measured by counting the total number photons arriving in an area on the detector surface over the time interval necessary to pass the entire reciprocal lattice node through the sphere of reflection. It is measure of total energy — rather than intensity — since intensity is a measure of the energy per cross-sectional area per unit time. The term "integrated intensity" is used because the energy is determined from the integration of the intensity over area and time, but it is somewhat of a misnomer and it might better be referred to as "integrated energy". However the usage is universal in the language of crystallography, and we will continue to use it. Whenever it becomes important to distinguish the integrated intensity from the actual intensity we will refer to the integrated intensity by the symbol $\mathcal{I}$. To derive an explicit expression for the integrated intensity, we first determine an expression for the energy per unit time per unit area arriving at p — the intensity, I_p.

Atom n is located at $\mathbf{v_j} = x_j\mathbf{a} + y_j\mathbf{b} + z_j\mathbf{c}$ inside a unit cell in the crystal (atom n is the *jth* atom in the unit cell). The vector from the crystal origin, o, to the origin of this unit cell has components that are integer numbers of unit cells in the direction of each unit cell axis; the net vector from the crystal origin to the atom is $\mathbf{v_n} = L\mathbf{a} + M\mathbf{b} + N\mathbf{c} + \mathbf{v_j}$. As the incident beam passes the origin its wave vector is described by

$$\overrightarrow{\mathcal{E}_o(t)} = \mathcal{E}_I e^{-2\pi i\nu t} = \mathcal{E}_I e^{2\pi i(c/\lambda)t}. \tag{4.83}$$

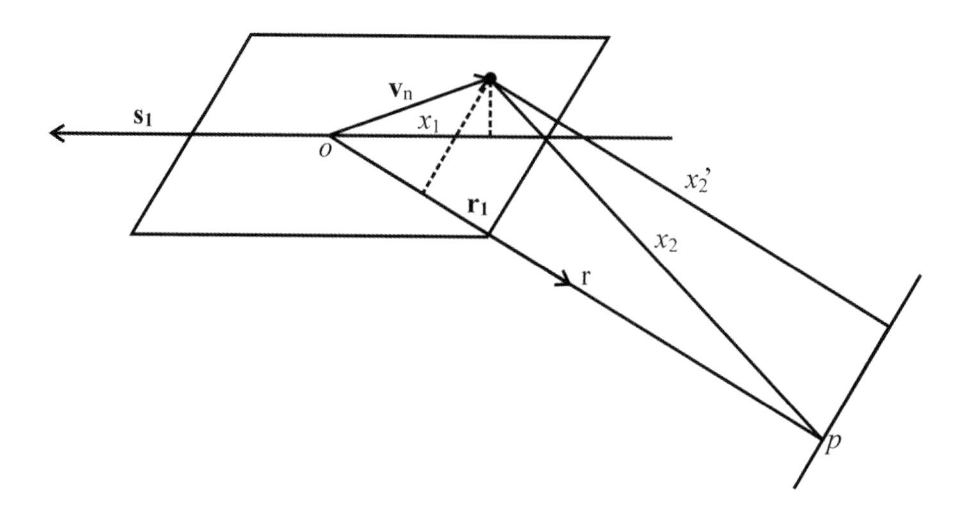

Figure 4.24 Elastic scattering from an atom inside a small crystal.

The beam passes an additional distance, x_1 to reach the atom, and the scattered radiation travels another distance, x_2, to arrive at point p. The total distance, $x = x_1 + x_2$, is traveled in time $t' = x/c$, where x_1 is the projection of $\mathbf{v_n}$ onto $-\mathbf{s_1}$, and x_2' is the distance from the origin to point p, less the projection of $\mathbf{v_n}$ in the direction of $\mathbf{r_1}$. Since the detector is very far from crystal, x_2 is assumed to be equal to x_2':

$$\begin{aligned} x_1 &= -\mathbf{v_n} \cdot \mathbf{s_1}, && (4.84)\\ x_2 &= r - \mathbf{v_n} \cdot \mathbf{r_1} \quad \text{and} && (4.85)\\ x &= r - \mathbf{v_n} \cdot (\mathbf{s_1} + \mathbf{r_1}). && (4.86) \end{aligned}$$

Suppose that the atom was replaced by a fixed electron at the same location. The force on the electron (the force that will set it into oscillation – perpendicular to the direction of the incident beam) is the product of the magnitude of the electric field of the incident radiation and the charge on the electron (q_e), allowing us to determine the acceleration of the electron as a function of the magnitude of the electric field:

$$\begin{aligned} F &= m_e a = \mathcal{E}_I q_e \\ a &= \frac{\mathcal{E}_I q_e}{m_e}. && (4.87) \end{aligned}$$

The magnitude of the electric field at point p, $\mathcal{E}_p$, is directly proportional to the acceleration, directly proportional to the charge on the particle being accelerated, and inversely proportional to the distance at which the radiation is detected, r:

$$\mathcal{E}_p \; \alpha \; \frac{q_e}{r} a.$$

The proportionality constant for this relationship is $1/4\pi\epsilon_0 c^2$ [59]:

$$\begin{aligned}\mathcal{E}_p &= \frac{1}{4\pi\epsilon_0 c^2}\frac{q_e}{r}a \\ &= \frac{1}{4\pi\epsilon_0 c^2}\frac{\mathcal{E}_I q_e^2}{r\, m_e}.\end{aligned} \tag{4.88}$$

Thus, $\mathcal{E}_p = \kappa_e \mathcal{E}_I$, where $\kappa_e = q_e^2/4\pi\epsilon_o r m_e c^2$. According to the treatment in Sec. 3.2.4 the atom scatters as if there were f_n electrons scattering from the center of the atom, where f_n is the scattering factor for the atom. Thus the amplitude of the scattered wave is $\mathcal{E}_p = \kappa_e f_n \mathcal{E}_I$. Ignoring the phase shift due to scattering (since it is the same for all atoms), the scattered wave vector arriving at point p is

$$\begin{aligned}\overrightarrow{\mathcal{E}_p(t)} &= \kappa_e f_n \mathcal{E}_I e^{-2\pi i[(c/\lambda)(t-t')]} = \kappa_e f_n \mathcal{E}_I e^{-2\pi i[(c/\lambda)(t-(x/c))]} \\ &= \kappa_e f_n \mathcal{E}_I e^{-2\pi i[(c/\lambda)t-(1/\lambda)(r-\mathbf{v_n}\cdot(\mathbf{s_1}+\mathbf{r_1}))]}.\end{aligned} \tag{4.89}$$

For mathematical convenience, the crystal (scattering) origin is now placed at an extreme point, so that L, M, and N are all non-negative integers, with maximum values of L_x, M_x, and N_x. The net wave vector arriving at point p from the entire crystal is the superposition of the waves from each unit cell, summed over all of the unit cells:

$$\begin{aligned}\overrightarrow{\mathcal{E}_p(t)} &= \kappa_e \mathcal{E}_I \sum_{L=0}^{(L_x-1)}\sum_{M=0}^{(M_x-1)}\sum_{N=0}^{(N_x-1)}\sum_j f_j\, e^{-2\pi i[(c/\lambda)t-(1/\lambda)(r-\mathbf{v_n}\cdot(\mathbf{s_1}+\mathbf{r_1}))]} \\ &= \kappa_e \mathcal{E}_I e^{-2\pi i[\nu t-(r/\lambda)]}\sum_j f_j e^{-2\pi i(\mathbf{q}\cdot\mathbf{v_j})} \\ &\quad\times \sum_{L=0}^{(L_x-1)} e^{-2\pi i(\mathbf{q}\cdot\mathbf{a})L}\sum_{M=0}^{(M_x-1)} e^{-2\pi i(\mathbf{q}\cdot\mathbf{b})M}\sum_{N=0}^{(N_x-1)} e^{-2\pi i(\mathbf{q}\cdot\mathbf{c})N},\end{aligned} \tag{4.90}$$

where $\mathbf{q} = (\mathbf{s_1}+\mathbf{r_1})/\lambda$. The last three sums in Eqn. 4.90 are geometric series of the form

$$\sum_{m=0}^{m_x-1} az^m = az^0 + az^1 + \cdots az^{m_x-1} = \frac{zaz^{m_x-1}-1}{z-1} = \frac{az^{m_x}-1}{z-1}. \tag{4.91}$$

In these sums $a = 1$ and

$$\begin{aligned}\overrightarrow{\mathcal{E}_p(t)} &= \kappa_e \mathcal{E}_I e^{-2\pi i[\nu t-(r/\lambda)]}\sum_j f_j e^{-2\pi i(\mathbf{q}\cdot\mathbf{v_j})} \\ &\quad\times\left(\frac{e^{-2\pi i(\mathbf{q}\cdot\mathbf{a})L_x}-1}{e^{-2\pi i(\mathbf{q}\cdot\mathbf{a})}-1}\right)\left(\frac{e^{-2\pi i(\mathbf{q}\cdot\mathbf{b})M_x}-1}{e^{-2\pi i(\mathbf{q}\cdot\mathbf{b})}-1}\right)\left(\frac{e^{-2\pi i(\mathbf{q}\cdot\mathbf{c})N_x}-1}{e^{-2\pi i(\mathbf{q}\cdot\mathbf{c})}-1}\right).\end{aligned} \tag{4.92}$$

The intensity at point p, I_p, is proportional to the square of the magnitude of the electric field, $I_p = \epsilon_o c\mathcal{E}_p^2 = \epsilon_o c\,(\overrightarrow{\mathcal{E}_p^*(t)}\cdot\overrightarrow{\mathcal{E}_p(t)})$. Using Euler's relation, $e^{iv} = \cos v + i\sin v$, and the trigonometric identity, $\sin^2 v = (1-\cos 2v)/2$, each of the last three terms in Eqn. 4.92, times its complex conjugate, has the form

$$\left(\frac{e^{imy}-1}{e^{iy}-1}\right)\left(\frac{e^{-imy}-1}{e^{-iy}-1}\right) = \frac{2-2\cos(my)}{2-2\cos y} = \frac{\sin^2(my/2)}{\sin^2(y/2)}, \tag{4.93}$$

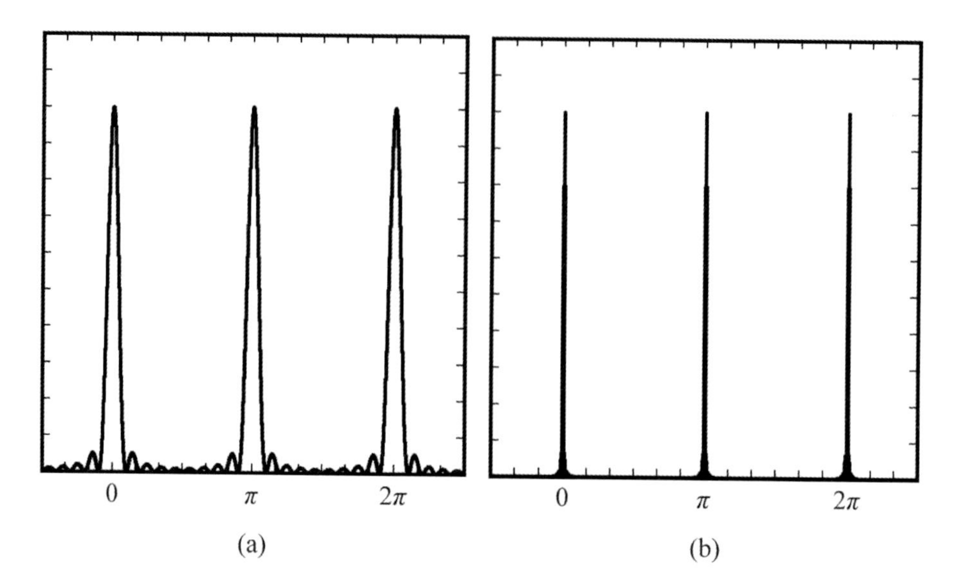

Figure 4.25 Plot of $\sin^2(mz)/\sin^2 z$ for (a) $m = 10$ and (b) $m = 100$; $z = -\pi/2$ to $5\pi/2$.

where $y = -2\pi i(\mathbf{q}\cdot \mathbf{a})$, and $z = y/2 = \pi i(\mathbf{q}\cdot \mathbf{a})$, etc. The function $\sin^2(mz)/\sin^2(z)$ for $m = 10$ and $m = 100$ is plotted in Figs. 4.25 (a) and (b), respectively. The function maximizes around integer increments of π, and as m increases even moderately, the function has values other than zero only when z is extremely close to $n\pi$, where n is an integer. For maximum intensity to be observed at point p, $\pi i(\mathbf{q}\cdot \mathbf{a}) = n\pi \Rightarrow \mathbf{q}\cdot \mathbf{a}$ is an integer. Since $\mathbf{q}$ is the vector sum of two vectors in reciprocal space, $\mathbf{q} = q_1\mathbf{a}^* + q_2\mathbf{b}^* + q_3\mathbf{c}^*$. The following conditions must therefore hold for significant intensity to be observed:

$$\begin{aligned} \mathbf{q}\cdot \mathbf{a} &= q_1 = n_1 \\ \mathbf{q}\cdot \mathbf{b} &= q_2 = n_2 \\ \mathbf{q}\cdot \mathbf{c} &= q_3 = n_3. \end{aligned} \tag{4.94}$$

$\mathbf{q}$ is a reciprocal vector with integer indices *and is therefore a diffraction vector*, $\mathbf{h}$, for some set of hkl planes. Thus $\mathbf{r_1}$ is a unit vector pointing in the direction of a diffraction node ($\mathbf{r_1} = \mathbf{d_1}$) and $\mathbf{q} = \mathbf{h} = (\mathbf{s_1} + \mathbf{d_1})/\lambda$. We have arrived at the Laue conditions (Eqn. 3.48) by yet another route!

The remaining sum to be multiplied by its complex conjugate in Eqn. 4.90 becomes

$$\sum_j f_j e^{-2\pi i(\mathbf{h}\cdot \mathbf{v_j})}, \tag{4.95}$$

which we recognize from Eqn. 3.101 as the structure factor, $\overrightarrow{F}_{hkl}$. The product is therefore $\overrightarrow{F^*}_{hkl}\cdot \overrightarrow{F}_{hkl} = F^2_{hkl}$, the square of the amplitude of the structure factor for the reflection satisfying the Laue conditions. The exponential term, $e^{-2\pi i[\nu t-(r/\lambda)]}$, is of the form e^{iz}, and the product is $e^{iz}e^{-iz} = e^0 = 1$.

Combining these results,

$$\mathcal{E}_{hkl,p}^2 = \kappa_e^2 \mathcal{E}_I^2 F_{hkl}^2 \left(\frac{\sin^2[\pi(\mathbf{h} \cdot \mathbf{a}) L_x]}{\sin^2[\pi(\mathbf{h} \cdot \mathbf{a})]} \right) \left(\frac{\sin^2[\pi(\mathbf{h} \cdot \mathbf{b}) M_x]}{\sin^2[\pi(\mathbf{h} \cdot \mathbf{b})]} \right) \times \left(\frac{\sin^2[\pi(\mathbf{h} \cdot \mathbf{c}) N_x]}{\sin^2[\pi(\mathbf{h} \cdot \mathbf{c})]} \right). \tag{4.96}$$

Multiplying both sides of the equation by the proportionality constant $\epsilon_o c$ gives the intensity at point p:

$$I_{hkl,p} = \left(\frac{q_e^2}{4\pi\epsilon_o r m_e c^2} \right)^2 I_0 \, F_{hkl}^2 \left(\frac{\sin^2[\pi(\mathbf{h} \cdot \mathbf{a}) L_x]}{\sin^2[\pi(\mathbf{h} \cdot \mathbf{a})]} \right) \times \left(\frac{\sin^2[\pi(\mathbf{h} \cdot \mathbf{b}) M_x]}{\sin^2[\pi(\mathbf{h} \cdot \mathbf{b})]} \right) \left(\frac{\sin^2[\pi(\mathbf{h} \cdot \mathbf{c}) N_x]}{\sin^2[\pi(\mathbf{h} \cdot \mathbf{c})]} \right). \tag{4.97}$$

The reader may have noted that the substitution of $\mathbf{h}$ for $\mathbf{q}$ in Eqn. 4.92 would set each exponential in the last three sums to $e^0 = 1$, creating the product $L_x M_x N_x$ and greatly simplifying the equation:

$$\overrightarrow{\mathcal{E}_{hkl,p}} = \left(\frac{q_e^2}{4\pi\epsilon_o r m_e c^2} \right) \mathcal{E}_I e^{-2\pi i[\nu t - (r/\lambda)]} L_x M_x N_x \sum_j f_j e^{-2\pi i(\mathbf{h} \cdot \mathbf{v_j})}. \tag{4.98}$$

Comparing this with Eqn. 3.53,

$$\overrightarrow{\mathcal{E}_{hkl}} = D \int_0^{V_c} \rho(\mathbf{r}_f) e^{-2\pi i(\mathbf{h} \cdot \mathbf{r}_f)} \, dV = D \sum_j f_j e^{-2\pi i(\mathbf{h} \cdot \mathbf{v_j})}, \tag{4.99}$$

provides an explicit formulation of the constant, D.*

We did not, however, make this obvious simplification since we are interested in what happens to the intensity as the crystal is rotated through the diffraction condition and a reciprocal lattice node of finite dimensions passes through the sphere of reflection. For that we will need the intensity relationship described by Eqn. 4.97.

When the crystal is in the diffraction condition for a given reflection, the reciprocal lattice node corresponding to the reflection intersects the sphere at different points within the node, resulting in diffracted beams emanating from the crystal in slightly different directions. This is depicted in Fig. 4.26. The view is along the z_l axis in the laboratory reference frame; the maximum diffraction is consider to be taking place in the equatorial x_l, y_l plane. The detector plane is perpendicular to the diffracted beam. It is convenient to consider the point p on the detector to be the point of maximum intensity — for which d_1 exactly satisfies the Laue conditions, treating the displacement vector to points p' in the vicinity of p as the integration variable, $\mathbf{\Delta D}$. The integrated intensity is the sum of the photon energy collected over all the intervals between $\mathbf{\Delta D}$ and $\mathbf{\Delta D} + d\mathbf{\Delta D}$.

Orthogonal components of the displacement vector can be formed by rotating the reflection vector at point p through the angle η in the x_l, y_l plane, followed by a

*This assumes that all of the unit cells corresponding to each value of L, M, and N are accounted for, i.e., that the crystal is a parallelepiped. In other cases, $L_x N_x M_x$ would be replaced by the total number of unit cells in the crystal.

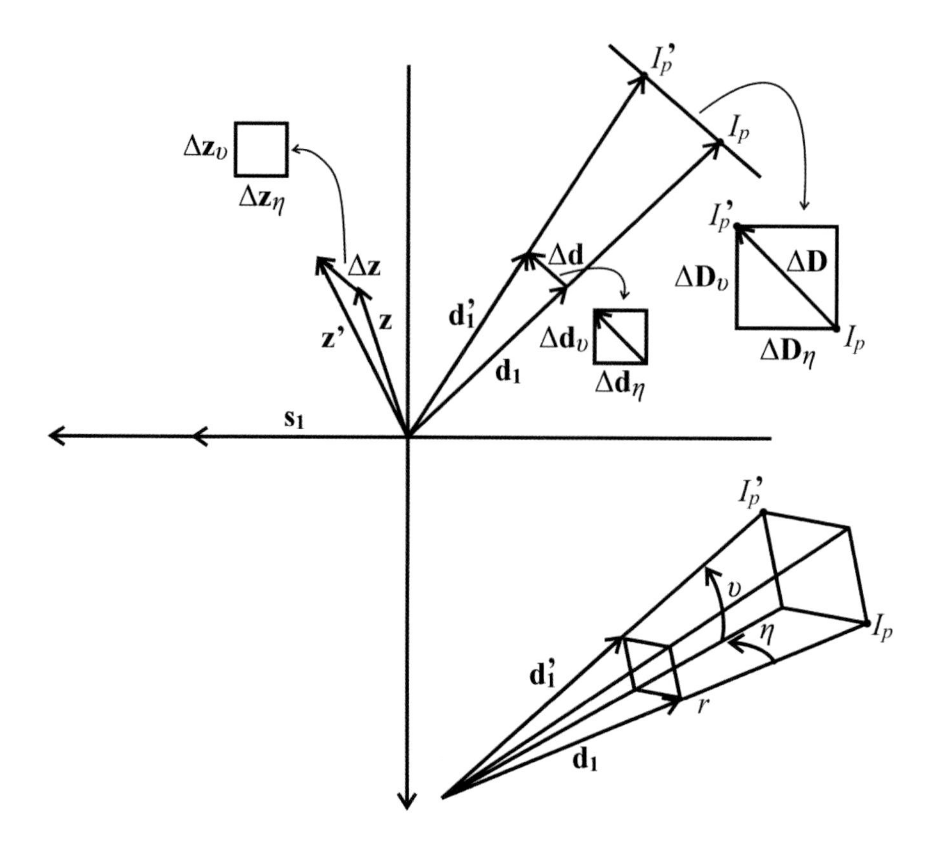

Figure 4.26 Diffraction resulting from different regions in a reciprocal lattice node intersecting the sphere of reflection.

perpendicular υ rotation to point p'. A differential change in $\Delta\mathbf{D}$ then corresponds to a differential area with perpendicular edges $d\Delta\mathbf{D}_\eta = r\,d\eta$ in the x_l, y_l plane and $d\Delta\mathbf{D}_\upsilon = r\,d\upsilon$, parallel to the z_l axis, where r is the distance from the crystal to the detector. If we count photons for a small time interval, dt, then the energy detected will be the energy per unit time per unit area, I_p, multiplied by the area and the time: $dE = I_p\,dt\,dA = I_p\,dt\,r^2\,d\eta\,d\upsilon$.

The crystal is now rotated through the angle ω, at constant angular velocity, υ_ω, about the z_l axis. ω is sufficiently large to pass the entire reciprocal lattice node through the surface of the sphere of reflection. The total number of photons incident on the detector are counted as the crystal is rotated, resulting in the integrated intensity:

$$\mathcal{I} = \int dE = \int_t \int_A I_p\,dt\,dA = \int_{t(\omega)} \int_\eta \int_\upsilon I_p\,dt\,r^2\,d\eta\,d\upsilon. \tag{4.100}$$

Since υ_ω is constant, $\upsilon_\omega = d\omega/dt$ and $dt = d\omega/\upsilon_\omega$:

$$\mathcal{I} = \frac{r^2}{\upsilon_\omega} \int_\omega \int_\eta \int_\upsilon I_p\,d\omega\,d\eta\,d\upsilon. \tag{4.101}$$

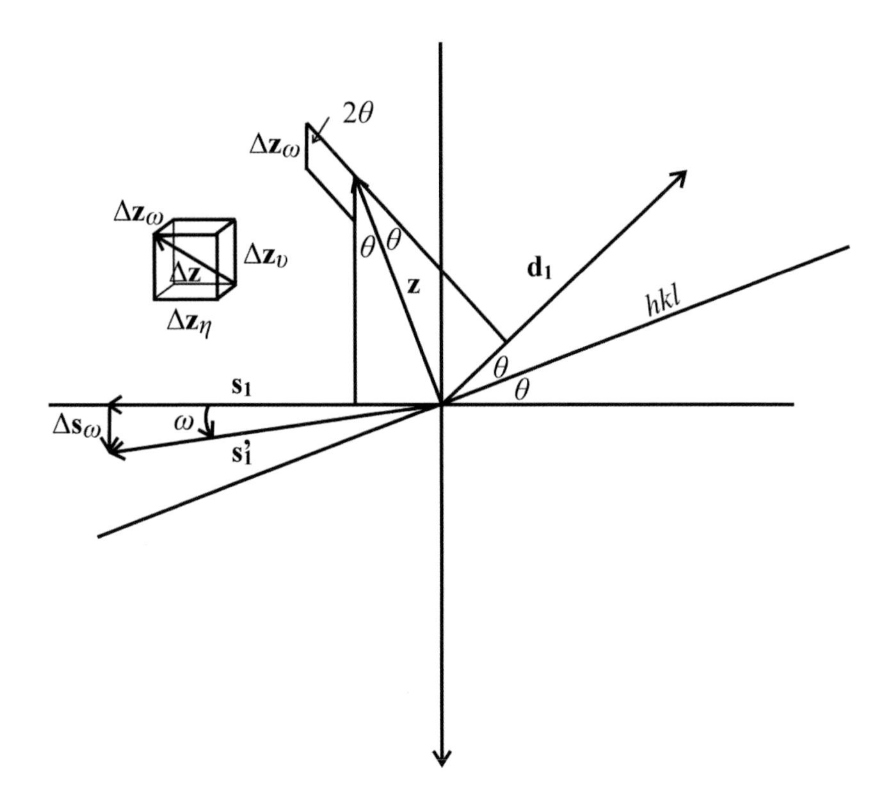

Figure 4.27 Displacement volume created by rotation of the source.

Substitution of Eqn. 4.97 into the equation and evaluating the integral will result in an equation for the integrated intensity. However, the expression for the intensity at point p contains the reciprocal lattice vector components as variables — in the reciprocal lattice basis — rather than the variables in Eqn. 4.101, which are in the reciprocal Cartesian basis. Before we can evaluate the integral we must determine the differential in the appropriate basis.

The integrated intensity is based on changes in the displacement vector at the detector surface, while changes in the intensity from Eqn. 4.97 at a specific point p' depends on $\mathbf{h}'$ at that point. Since the differential at the detector is actually the change in the displacement vector between p and p', corresponding to unit reflection vectors $\mathbf{d_1}$ and $\mathbf{d}_1'$, we must relate it to changes in the displacement vector between $\mathbf{h}$ and $\mathbf{h}'$. The displacement vector between $\mathbf{d_1}$ and $\mathbf{d}_1'$, $\Delta\mathbf{d}$, is parallel to $\Delta\mathbf{D}$, with components $|\Delta\mathbf{d}_\eta| = 1\,\eta$ and $|\Delta\mathbf{d}_\upsilon| = 1\,\upsilon$. Since $(\mathbf{s_1} + \mathbf{d_1}) = \lambda\mathbf{h} = \mathbf{z}$ and $(\mathbf{s_1} + \mathbf{d}_1') = \lambda\mathbf{h}' = \mathbf{z}'$, $(\mathbf{z}' - \mathbf{z}) = (\mathbf{d_1}' - \mathbf{d_1}) = \Delta\mathbf{d} = \Delta\mathbf{z}$. There is a displacement vector at the end of the vector $\lambda\mathbf{h} = \mathbf{z}$ that is identical to the displacement vector, $\Delta\mathbf{d}$, with orthogonal components $|\Delta\mathbf{z}_\eta| = \eta$ and $|\Delta\mathbf{z}_\upsilon| = \upsilon$.

To determine the relationship between the ω rotation of the crystal and the $\mathbf{z}$ vector it is convenient to leave the crystal in a fixed position and rotate the source in the x_l, y_l plane through the angle, ω, depicted in Fig. 4.27. This results in a displacement vector, $\Delta\mathbf{s_1}$, with magnitude $1\,\omega$. Again, $\Delta\mathbf{z}_\omega = (\mathbf{s_1}' + \mathbf{d_1}) - (\mathbf{s_1} +$

$\mathbf{d_1}) = \Delta\mathbf{s_1}$. Substituting $\mathbf{d_1}'$ for $\mathbf{d_1}$ results in the same displacement for $\mathbf{z}'$. The effect of the rotation is to translate the area element with edge vectors $\Delta\mathbf{z}_\eta$ and $\Delta\mathbf{z}_\upsilon$ in a direction perpendicular to the source vector, creating a third edge, $\Delta\mathbf{z}_\omega$. This creates a parallelepiped with a parallelogram face in the x_l, y_l plane defined by the $\Delta\mathbf{z}_\eta$ and $\Delta\mathbf{z}_\omega$ vectors, at an angle of 2θ with respect to one another — completed by a third vector, $\Delta\mathbf{z}_\upsilon$, perpendicular to this face. The parallelepiped has edge lengths of ω, η, and υ. Differential changes $d\omega$, $d\eta$, and $d\upsilon$ will result in a differential volume in the shape of a parallelepiped with the same angles, defined by the differential vectors $d\Delta\mathbf{z}_\omega$, $d\Delta\mathbf{z}_\eta$, and $d\Delta\mathbf{z}_\upsilon$ with magnitudes $d\omega$, $d\eta$, and $d\upsilon$. From Eqn. 1.135, the volume of the differential parallelepiped is the triple scalar product of these vectors:

$$\Delta V = d\Delta\mathbf{z}_\upsilon \cdot (d\Delta\mathbf{z}_\omega \times d\Delta\mathbf{z}_\eta). \tag{4.102}$$

$d\Delta\mathbf{z}_\omega \times d\Delta\mathbf{z}_\eta$ creates a vector that is perpendicular to the x_l, y_l plane with magnitude $|d\Delta\mathbf{z}_\omega||d\Delta\mathbf{z}_\eta| \sin 2\theta = d\omega\, d\eta\, \sin 2\theta$. This vector is therefore parallel to $d\Delta\mathbf{z}_\upsilon$, and the scalar product of these two vectors gives the volume:

$$\Delta V = d\omega\, d\eta\, d\upsilon\, \sin 2\theta. \tag{4.103}$$

$\Delta\mathbf{z}$ can also be expressed in terms of displacement vectors in the reciprocal basis: $\Delta\mathbf{z} = \lambda\,\Delta\mathbf{h} = \lambda(p_1\mathbf{a}^* + p_2\mathbf{b}^* + p_3\mathbf{c}^*)$. λp_1, λp_2, and λp_3 are the components along the three reciprocal axes and the magnitudes of the differential vectors are then $\lambda\, dp_1$, $\lambda\, dp_2$ and $\lambda\, dp_3$ along the same three reciprocal cell axes, selected so that the volume of the differential element is the same as that in Eqn. 4.103. The volume is again determined from the triple scalar product of the component vectors:

$$\begin{aligned}\Delta V &= d\Delta\mathbf{z}_{a*} \cdot (d\Delta\mathbf{z}_{b*} \times d\Delta\mathbf{z}_{c*}) \\ &= \lambda\, dp_1\mathbf{a}^* \cdot (\lambda\, dp_2\mathbf{b}^* \times \lambda\, dp_3\mathbf{c}^*) \;=\; \lambda^3\, dp_1 dp_2 dp_3\, (\mathbf{a}^* \cdot (\mathbf{b}^* \times \mathbf{c}^*)) \\ &= \lambda^3 V_c^* dp_1 dp_2 dp_3 \;=\; \frac{\lambda^3}{V_c}\, dp_1 dp_2 dp_3.\end{aligned} \tag{4.104}$$

where V_c^* and V_c are the reciprocal and direct unit cell volumes, respectively. Setting the expressions for the volume equal to one another gives

$$d\omega\, d\eta\, d\upsilon = \frac{\lambda^3}{V_c \sin 2\theta}\, dp_1 dp_2 dp_3. \tag{4.105}$$

The integral to be evaluated is now

$$\mathcal{I} = \frac{\lambda^3}{V_c \sin 2\theta} \frac{r^2}{\upsilon_\omega} \int_{p_1}\int_{p_2}\int_{p_3} I_p\, dp_1 dp_2 dp_3, \tag{4.106}$$

where I_p is the intensity at a general point on the detector surface corresponding to $\mathbf{h}' = \mathbf{h} + \Delta\mathbf{h}$. Noting that $\mathbf{h} \cdot \mathbf{a} = h$, making $L_x\pi\mathbf{h} \cdot \mathbf{a} = n\pi$ (where n is an integer) — that $\sin^2(n\pi + \alpha) = \sin^2 \alpha$ — and that $(p_1\mathbf{a}^* + p_2\mathbf{b}^* + p_3\mathbf{c}^*) \cdot \mathbf{a} = p_1$, the $\sin^2$ terms in I_p become, for example,

$$\frac{\sin^2(L_x\pi\mathbf{h} \cdot \mathbf{a} + L_x\pi\Delta\mathbf{h} \cdot \mathbf{a})}{\sin^2(\pi\mathbf{h} \cdot \mathbf{a} + \pi\Delta\mathbf{h} \cdot \mathbf{a})} = \frac{\sin^2(p_1\pi\, L_x)}{\sin^2(p_1\pi)}. \tag{4.107}$$

The intensity is maximum when $\mathbf{h}' = \mathbf{h}$ and $p_1 = p_2 = p_3 = 0$. The integral has values other than zero only when p_1, p_2 and p_3 are very small numbers — when $\mathbf{h}' \approx \mathbf{h}$. The integrations can therefore be performed from $p_1 = -\infty$ to ∞, etc., with significant contributions to the integrals occurring for only a small range of small values of p_1, p_2 and p_3. Furthermore, when x is small, $\sin x \approx x$ $\Rightarrow \sin^2(p_1\pi) \approx (p_1\pi)^2$. Thus,

$$\begin{aligned} \mathcal{I}_{hkl} &= \frac{\lambda^3 r^2}{v_\omega V_c \sin 2\theta}\left(\frac{q_e^2}{4\pi\epsilon_o r m_e c^2}\right)^2 I_0\, F_{hkl}^2 \\ &\times \int_{-\infty}^{\infty} \frac{\sin^2(p_1\pi\, L_x)}{(p_1\pi)^2}\, dp_1 \int_{-\infty}^{\infty} \frac{\sin^2(p_2\pi\, M_x)}{(p_2\pi)^2}\, dp_2 \\ &\times \int_{-\infty}^{\infty} \frac{\sin^2(p_3\pi\, N_x)}{(p_3\pi)^2}\, dp_3. \end{aligned} \tag{4.108}$$

Each of the integrals is a definite integral of the form

$$\int_{-\infty}^{\infty} \frac{\sin^2 u}{u^2}\, du = \pi. \tag{4.109}$$

Setting $u = p_1\pi L_x$, etc., the integrals evaluate to L_x, M_x, and N_x, and

$$\mathcal{I}_{hkl} = \frac{\lambda^3}{v_\omega V_c \sin 2\theta}\left(\frac{q_e^2 F_{hkl}}{4\pi\epsilon_o m_e c^2}\right)^2 I_0 L_x M_x N_x. \tag{4.110}$$

If the crystal is in the shape of a parallelepiped, so that all values of L, M, and N are represented, then $L_x M_x N_x$ = the total number of unit cells in the crystal, and the volume of the crystal is $V_x = L_x M_x N_x V_c$. A crystal of arbitrary shape can be considered to be composed of mosaic blocks approximated as parallelepipeds, in which case Eqn. 4.110 holds for V_x = the volume of the mosaic block. In either case,

$$\mathcal{I}_{hkl} = \frac{\lambda^3}{v_\omega \sin 2\theta}\left(\frac{q_e^2 F_{hkl}}{4\pi\epsilon_o m_e c^2 V_c}\right)^2 I_0 V_x. \tag{4.111}$$

Dividing both sides of Eqn. 4.111 by the intensity of the incident beam and multiplying by the angular velocity of the ω rotation provides an integrated intensity that is independent of the intensity of the source and the time of the intensity measurement, i.e., it depends only on the crystal (or mosaic block). It represents the ability of the crystal to diffract radiation for each specific reflection and is referred to as the "reflecting power" of the crystal/block. It is ordinarily given the symbol R_{hkl}. Further division by the volume of the crystal/block gives its reflecting power per unit volume, Q_{hkl}:

$$Q_{hkl} = \frac{\mathcal{I}_{hkl} v_\omega}{I_0 V_x} = \frac{\lambda^3}{\sin 2\theta}\left(\frac{q_e^2 F_{hkl}}{4\pi\epsilon_o m_e c^2 V_c}\right)^2 \tag{4.112}$$

$$R_{hkl} = Q_{hkl} V_x. \tag{4.113}$$

For a given crystal Eqn. 4.111 can be written as a simple function of the square of the structure factor:

$$\mathcal{I}_{hkl} = \frac{1}{v_\omega \sin 2\theta_{hkl}} D' F_{hkl}^2. \tag{4.114}$$

The $1/\sin 2\theta$ term is known as the *Lorentz factor*, L, for a lattice node passing through the sphere of reflection in the equatorial plane. It is discussed in detail in Sec. 5.1.2. The intensity is also attenuated by polarization of the X-ray beam, and the equation will include a polarization factor, P. The polarization factor is derived in Sec. 5.1.2 for unpolarized incident radiation ($P = (1 + \cos^2 2\theta)/2$); and in *The Rotation Method in Crystallography*[47] for the partially polarized radiation from a monochromator. In practice the integrated intensity is divided by the scan time (multiplied by the scan rate), giving a relative quasi-normalized integrated intensity:

$$I_{hkl} = \mathcal{I}_{hkl} v_{\omega} = (D'LP)F_{hkl}^2. \tag{4.115}$$

Although I_{hkl} is still not rigorously an intensity (the cross sectional area of the diffracted beam has not been taken into account), its square root provides a relative structure factor (the structure factor multiplied by a constant) and the magnitude of the structure factor can be determined by scaling $\sqrt{I_{hkl}}$. Since the aim of data collection is the determination of the structure factors needed to solve the crystal structure, we will refer to I_{hkl} as the experimentally observed intensity.

4.7.3 Intensities From Serial Detectors

In addition to the nearly monochromatic photons that enter the detector from the reflection vector, the surrounding space contains a large contribution from the incoherently scattered radiation. These photons make up the major portion of the *background radiation*; they arise from every point in reciprocal space. Because they also arrive at the detector when the crystal is in the diffraction condition, their contribution to the measured intensity must be subtracted from the total intensity in order to account for the photons arising solely from diffraction.

If reflections were one-dimensional the collection of intensity data with a serial detector would be a simple matter of rotation of the crystal into the diffraction condition in the equatorial plane, moving the detector to the appropriate 2θ, and counting the photons entering the detector for a set period of time. The background radiation could then be measured by moving the crystal by a small angle out of the diffraction condition and counting the background photons for an equal period of time in the vicinity of the reflection. In doing so we would make the assumption that the background intensity has the same value at the nearby reflection location in reciprocal space. The net number of counts due to diffraction, the integrated intensity, would then be the total number of counts minus the number of counts for the background. In order to "normalize" the intensities for different counting periods (see above), the total number of counts would be divided by the counting time, resulting in an intensity in *counts per second (cps)*.

In the actual measurement of the intensity with a serial detector, the crystal must be rotated through the entire width of the peak and the photons counted for each value of ω, either by stopping in increments and counting (*step scan*) or counting "on the fly" (*continuous scan*). The orientation matrix determines the goniometer angles for the reflection. The crystal is rotated into the diffraction condition, the detector is rotated to 2θ, omega is moved to one side of the peak outside of the diffraction condition and the background intensity is determined by counting the photons, C_{B1}, for time t_{B1}: $I_{B1} = C_{B1}/t_{B1}$. Omega is then scanned

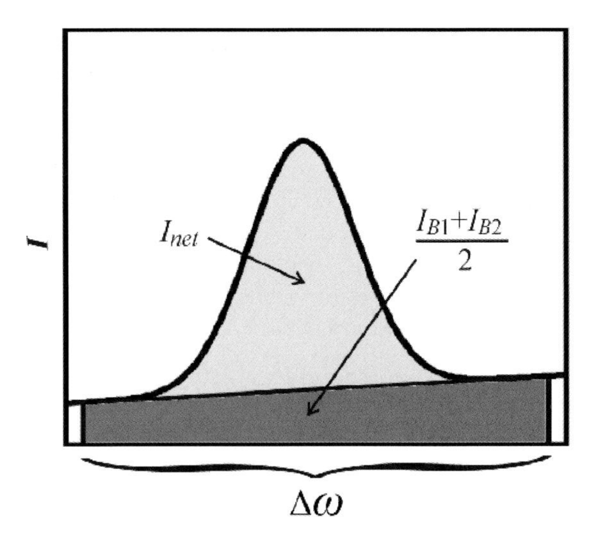

Figure 4.28 Intensity determined by the background-peak-background (BPB) method.

slowly through the peak. The photons can either be counted and summed, or counted in steps and stored as a profile of the reflection. The peak intensity is determined by dividing the total number of counts, C_P by the total counting time, t_P: $I_P = C_P/t_P$. Since the background intensities may differ through the scan, the background on the other side of the peak, $I_{B2} = C_{B2}/t_{B2}$, is also measured. Because it is impossible to determine the exact manner in which the background varies as the scan proceeds, the variation is assumed to be linear — the two backgrounds are averaged. The net intensity in counts per second is determined by subtracting the background contribution from the total intensity:

$$I_{net} = I_P - \frac{I_{B1} + I_{B2}}{2} \tag{4.116}$$

Fig. 4.28 demonstrates the results of this process, known as the *background-peak-background* (BPB) method.

The determination of exactly where the background ends and the peak begins depends on the width of the peak, which in turn varies for each crystal and is dependent on the diffraction angle. The peak widths must therefore be determined empirically, preferably by measuring peak widths for a number of reflections, and determining an empirical function of θ that will provide a reasonable estimate of the widths of the peaks at various Bragg angles. The BPB method requires an accurate orientation matrix since the entire peak must lie within the scan window.

If the peak profiles are stored, an alternative to this approach is to determine the intensities by analyzing the peaks either during or after data collection, and fitting the peaks to model profiles of strong and well-resolved reflections[60]. These model profiles are created by selecting a number of stronger reflections, preferably with approximately the same Bragg angles as those to be fit, and averaging their profiles after background subtraction. In this method the edges of the model peaks are determined at the point where the intensity divided by its standard deviation (see Sec. 5.1.1) reaches a minimum on each side of the peak. All of the reflections in that region are then considered to have the same peak shape, differing from the model reflection profile only by a scale factor, and the determination of their integrated intensities is accomplished by determining this scale factor for each reflection. This can be accomplished using the method of least squares. If p_i is the intensity for the ith step in the known profile, then the predicted intensity for the ith value in the reflection is $I_c = fp_i + b_i$, where b_i is the predicted value for the background, based on a linear interpolation of backgrounds measured on each side of the peak. If the observed value for the ith step is I_0, then the constant, f, can be determined by minimizing the sum of the squares of the differences of the observed and calculated values for all of the n steps in the profile:

$$\begin{aligned} \epsilon_i^2 &= (I_c - I_o)^2 = f^2p_i^2 + b_i^2 + I_{oi}^2 + 2fp_ib_i - 2fp_iI_{oi} - 2b_iI_{oi}. \\ \frac{\partial \sum_{i=1}^{n} \epsilon_i^2}{\partial f} &= \sum_{i=1}^{n}(2fp_i^2 + 2p_ib_i - 2p_iI_{oi}) = 0. \\ f &= \frac{\sum_{i=1}^{n} p_i(I_{oi} - b_i)}{\sum_{i=1}^{n} p_i^2} \end{aligned} \tag{4.117}$$

The profile can be allowed to vary in its relative position in the reflection window to further improve the fit. The major advantage of profile analysis lies in its ability to separate the background from the peak, decreasing the statistical uncertainty in the intensity. As we will see in the next section, it has become an important part of the peak integration routines for area detector data collected from crystals containing macromolecules.

There are two additional characteristic frequencies arising from the X-ray source that are very close to the $K\alpha$ frequencies, $K\beta_1$ and $K\beta_2$. Prior to the advent of graphite monochromators, filters consisting of specific metals (nickel for radiation from a copper target; zirconium for radiation from a molybdenum target) were routinely utilized to remove most of the extraneous radiation from the incident beam, including the $K\beta$ component. The metal filters block *most* of the $K\beta$ radiation, and are known as *beta filters.* Since a portion of this radiation remains in the incident beam it will diffract at slightly different Bragg angles from those due to the $K\alpha$ radiation. In order to count the photons from the $K\beta$ component of filtered radiation, the detector must also be moved. This is accomplished by rotating the detector to $2\theta - \Delta\omega$ and the crystal to $\omega - \Delta\omega/2$, then scanning through the peak by moving the detector through $2\Delta\omega$ while the crystal rotates through $\Delta\omega$. This type of scan is known as an "$\omega - 2\theta$ scan." When the detector remains in a fixed position the scan is termed an "ω scan." Graphite monochromators "pre-diffract" the radiation using a strong reflection from a large single crystal of graphite. The monochromator crystal is set at the Bragg angle for $K\alpha$ diffraction, effectively eliminating the $K\beta$ component. This renders the simultaneous motion of the detector unnecessary unless filtered radiation is used.*

4.7.4 Intensities From Area Detectors

The two-dimensionality of the area detector surface demands a more sophisticated approach to the determination of integrated intensities. The reflection is observed

*The $\omega - 2\theta$ scan method is also useful for very high Bragg angles using Cu radiation because of the separation of the $K\alpha_1$ and $K\alpha_2$ components of the diffracted beam (Fig. 4.22). The separation is much smaller with Mo radiation, for which ω scans are nearly always appropriate.

as a "spot" on the surface of the detector. As with serial detection it is necessary to distinguish the background radiation from that arising from diffraction. For a number of reasons this becomes more difficult for the area detector.

The natural broadening of the diffracted beam resulting from divergence and the approach to the diffraction condition tends to create a cylindrical beam, which becomes elliptical due to polarization (discussed in the next chapter). If this was the only factor involved in determining the shape of the spot, it could be modeled by assuming that the reciprocal lattice "point" was actually an ellipsoid which produced a cone of reflection vectors with an approximately elliptical cross-section; the spot on the detector would be elliptical in shape. However, there are other factors which broaden the diffracted beam and distort the shape of the spot. The $K\alpha_1/K\alpha_2$ separation effectively "thickens" the surface of the sphere of reflection (with radii at the maxima of $1/\lambda_1$ and $1/\lambda_2$). Since the spheres contain a common origin at a surface point (the origin of the reciprocal lattice), this thickening is not uniform, and the relative directions of the two diffracted vectors depends on the locations of their intersection with each of their Ewald spheres. This tends to broaden the spot profile, but not in an easily predictable manner. The effects of crystal mosaicity on the reflection can be envisioned by extending the illustration in Fig. 4.23 to three dimensions. The result will be a cone of reflection vectors rotating into the diffraction condition at different values of ω. This will correspond to a cone of diffraction vectors (reciprocal lattice vectors) centered about an average reciprocal lattice vector. For a given frame, the effect on the spot will again be difficult to predict, since the spread in the ω rotation direction cannot be observed in a single frame. Since the reciprocal lattice vector from each mosaic block will produce a reflection at the same Bragg angle, the spot will be distorted in an arc, where each reflection vector is at the Bragg angle, but rotated slightly around the x_l axis in the laboratory coordinate system.

In principle, it should be possible to model all of these distortions by imagining a reciprocal lattice node shaped in such a manner that its points of contact with the sphere of reflection will produce the reflection vectors necessary to create the spot observed on the detector. However, the nature of the distortions is such that there is, to date, no rigorous model for predicting the shapes of these virtual reciprocal lattice "points" — or their corresponding spots on area detector surfaces. For our purposes we will take the approach that the shape of the observed reflection on the detector can be modeled with a three dimensional solid centered at each reciprocal lattice node that is more or less ellipsoidal. This would produce a set of reflection vectors with an "elliptoid" (more or less elliptical) cross-section, resulting in an "elliptoidal" spot shape.

There are two general strategies for area detector data collection, based to some degree on whether or not the crystals are composed of "small molecules" or macromolecules. For macromolecules, with large unit cells, the diffraction maxima are numerous; for a given value of omega there can be many reflections in the diffraction condition. Furthermore, these maxima are densely packed and in close proximity to one another — unit cells for macromolecules are very large, resulting in many long interplanar spacings and a dense reciprocal lattice as a consequence. Indeed many of the observed reflections can overlap, and it is necessary to sample the reciprocal lattice in small increments. Omega scans are usually over narrow ranges (e.g., 0.1°), much less than the width of a single reflection. Because of this, the

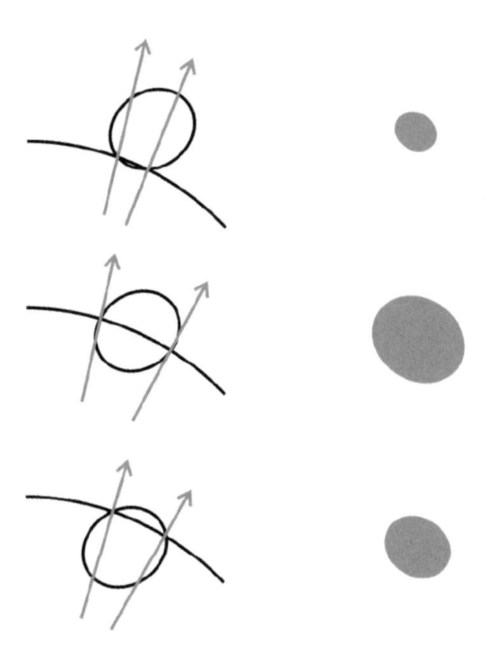

Figure 4.29 Graphical representation of the creation of diffraction spots for three separate frames on an area detector surface as the reflection passes through the sphere of reflection. The position of the nodal solid is assumed to be at the end of an incremental scan of $\Delta\omega$ in each drawing.

intensity due to a single reflection is observed in a number of sequential frames as illustrated schematically in Fig. 4.29.

Each pixel in a frame records an intensity, creating a two-dimensional array for each increment of $\Delta\omega$.* Each reflection spans several frames, and the contribution to the intensity from each frame must be combined to create the integrated intensity for the reflection. The data are therefore effectively three dimensional, with the rotation angle constituting the third dimension. The data is ordinarily analyzed by creating a "box" in the predicted vicinity of the reflection in $\{x_d, y_d, \omega\}$ space, where x_d and y_d are commonly pixel coordinates on the detector surface.† The size of the box is such that the reflection is contained within the box. As with integration from a serial detector, the background is determined by selecting portions of the box outside of the predicted envelope of the reflection and averaging those values. The background is then considered to vary linearly with respect to the box coordinates, allowing for the assignment of a background contribution to each pixel.

In the simplest case the integrated intensity can be obtained by subtracting the background contribution and summing the net intensity for each pixel inside the box, then normalizing the intensity to counts per second by dividing by the time taken to record the frame. Unfortunately, for macromolecular data, this approach

*While the omega scan is very common, depending on the experimental parameters the actual rotation axis — often referred to as the *spindle axis* — may be an axis other than the ω axis.

†In data analysis these coordinates are often mapped into other coordinate systems.

assumes that there is only one reflection in the box, and this is often not the case. As with data collected with a serial detector, the integration process can be improved by profile analysis, providing better counting statistics *and* a way to deal with overlapping reflections. In the case of the area detector, reflections within a region surrounding the reflection to be integrated are background-corrected and averaged to create a representative 3-dimensional profile in $\{x_d, y_d, \omega\}$ space. The remaining reflections in the subset are fit with this profile by assuming that the profile of the reflections to be fit differ from the model profile by a scale factor. The ith step in Eqn. 4.117 becomes the ith pixel in the three-dimensional data set, and the scale factor is determined by the method of least squares in the same manner as that for serial detector data.† In addition to improving the quality of the data set for individual reflections, overlapping profiles can be incorporated in order to determine the relative contributions to the intensities when two peaks overlap. Thus profile fitting has become an important part of the peak integration routines for macromolecular data.

The reflections from crystals containing "small molecules" are generally well-resolved from one another, precluding the need for small increments in omega in the data collection scheme. Omega scans in this case cover a larger interval, usually in the range of $0.3°-0.5°$. There are fewer frames per reflection, decreasing the data-collection time substantially. Although reflection overlap is rarely a problem, profile analysis is performed routinely to improve the quality of the data set. Since all of the diffraction data can be collected without an orientation matrix, it is possible to collect a complete data set using the area detector and handle the process of unit cell determination, peak integration and structural solution completely "off-line."

4.7.5 Limits to the Collection of Intensity Data

The Limiting Sphere

The reciprocal lattice extends infinitely, but the number of reflections available for data collection does not. In order for a reflection to be observed, a reciprocal lattice point must cross the sphere of reflection. As shown in Fig. 4.30, as the crystal is rotated, only those reciprocal lattice vectors with lengths less than $2/\lambda$ are able to cross the Ewald sphere. The reciprocal lattice points that can produce reflections are therefore limited to those that lie within a sphere centered at the origin of the reciprocal lattice with a radius of $2/\lambda$, which we refer to as *the limiting sphere.* The number of reciprocal lattice points that lie within this sphere can be calculated by noting that each corner of the reciprocal unit cell shares a reciprocal lattice point with seven other unit cells. There is $1/8$ of a lattice point per corner and 8 corners — one reciprocal lattice point per reciprocal unit cell. The number of reciprocal unit cells in the limiting sphere, M, is equal to the volume of the sphere divided by the volume of the reciprocal cell:

$$M = \frac{V_s}{V^*} = \frac{4}{3}\pi\left(\frac{2}{\lambda}\right)^3\left(\frac{1}{V^*}\right) = \frac{32\pi}{3\lambda^3 V^*} = \frac{32\pi}{3\lambda^3}V, \tag{4.118}$$

since the volume of the direct unit cell is the reciprocal of the volume of the reciprocal unit cell. The cube of the wavelength has a profound effect on the number of

†This is a representative example of the approach to profile analysis. There are a number of profile fitting schemes currently in use; some differ significantly from this approach[61].

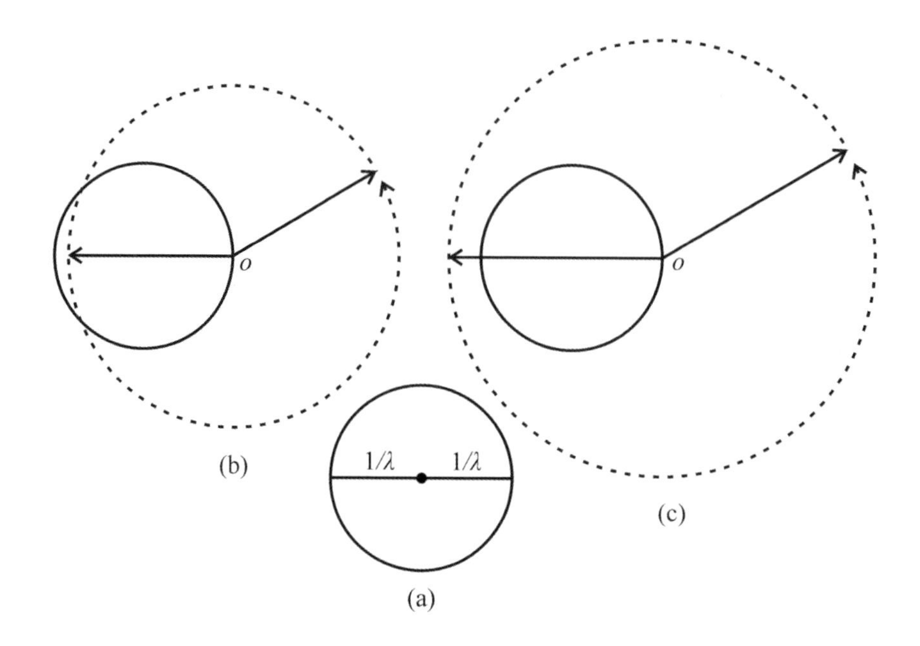

Figure 4.30 (a) Sphere of reflection with a diameter of $2/\lambda$. (b) A reciprocal lattice vector with length h$\leq 2/\lambda$ can intersect the sphere of reflection. (c) A reciprocal lattice vector with h $> 2/\lambda$ cannot intersect the sphere of reflection.

reflections available for data collection. For any two wavelengths the ratio of the number of reflections possible for a given crystal is

$$\frac{M_1}{M_2} = \frac{32\pi V/3\lambda_1^3}{32\pi V/3\lambda_2^3} = \left(\frac{\lambda_2}{\lambda_1}\right)^3. \tag{4.119}$$

For molybdenum versus copper radiation, the ratio is $(1.5418/0.7107)^3 = 10.2$. The use of molybdenum radiation provides *many* more reflections.

The additional data comes with a price, however. The shorter wavelength of Mo radiation decreases the distance between reflections in the diffraction pattern, increasing the potential for overlap. This is ordinarily not a problem with "small molecule" crystals, but macromolecular crystals pack in large unit cells, resulting in densely packed diffraction patterns that require longer wavelength radiation to limit reflection overlap.

Resolution

The Fourier series representation of the electron density,

$$\rho(\mathbf{r}) = \frac{1}{V_c}\sum_{\mathbf{h}} \vec{F}_{\mathbf{h}} e^{-2\pi i(\mathbf{h}\cdot\mathbf{r})}, \tag{4.120}$$

constructs an image of the unit cell contents by superpositioning sinusoidal functions (see Fig. 3.32) of increasingly higher frequencies as **h** becomes larger, i.e., as the distance between the lattice planes becomes smaller. As indicated in the simple one-dimensional example illustrated in Fig. 3.30, the higher frequency terms in the series provide the detail in the electron density function. The ability to discern one maximum in the electron density function from another depends on the smallest distance between maxima in the terms in the Fourier series. Qualitatively, the limit of our ability to discern between peaks (atoms) in an electron density map, the *resolution* of the map, is expected to be proportional to $1/\mathrm{h}_{max} = d_{min}$, since the separation between the maxima become smaller as h becomes larger. Ironically, the reflections that are the most important for resolving the locations of the atoms in a unit cell are those reflections that occur at the highest diffraction angles — with intensities that are the most difficult to obtain accurately.

Applying Raleigh scattering theory for optical diffraction to the three dimensional amplitude function given in Eqn. 4.120, the resolution can be shown to be approximately $0.917/\mathrm{h}_{max} = 0.917\, d_{min}$.[62] If the crystal is of high quality, so that all possible reflections can be collected, then $\mathrm{h} = 2/\lambda$ and the resolution limit is

$$L_{res}(min) = 0.917/\mathrm{h}_{max} = 0.459\,\lambda. \tag{4.121}$$

For CuK_α radiation, $L_{res}(min) = 0.71$; for MoK_α radiation, $L_{res}(min) = 0.33$. The distance between a carbon atom and a bonded hydrogen atom is about 1 Å; the distance between two bonded carbon atoms is about 1.5 Å. If all the reflections are available, both X-ray wavelengths can provide *resolution at the atomic level.* However, if the data set is limited to lower angle reflections, this might not be the case. As the data become more and more limited, the atomic peaks in the Fourier map overlap more and more, eventually to a point where they can no longer be distinguished from one another. Because of this, the quality of the diffraction data can profoundly influence the determination of the locations of the atoms in a crystal and their refined positions.

For most crystals, the number of observable reflections is less than indicated by Eqn. 4.118. If $2\sin\theta_{lim}/\lambda = \mathrm{h}_{lim} = 2/\lambda$ then a crystal that diffracts to a maximum value of $2\theta_{max} < 2\theta_{lim}$ will result in a limiting sphere (Fig. 4.30) with radius $\mathrm{h}_{max} < 2/\lambda$. The maximum number of observable reflections is then

$$M = \frac{V_s}{V*} = \frac{4}{3}\pi(\mathrm{h}_{max})^3\left(\frac{1}{V*}\right) = \frac{4}{3}\pi V\left(\frac{1}{d_{min}}\right)^3. \tag{4.122}$$

The "resolving power" of the X-ray data is reflected in a comparison of the number of unique observations made in the experiment to the number of parameters that must be determined in order to define the structure. For *any* experiment, there must be at least one *unique* observation made for every parameter to be determined — the ratio of the number of data collected to the number of parameters must be *at least* unity. Small molecule structures generally have *data-to-parameter ratios* greater than unity, and are said to be *over-determined.* For example, the mercaptopyridine structure described in Sec. 1.5.3 was determined from data collected with MoK_α radiation to $2\theta_{max} = 50°$. Substituting $\mathrm{h}_{max} = 2\sin 25°/0.71073$ Å and the 542.2 Å cell volume into Eqn. 4.122 indicates that there are 3820 observable reflections. The Laue group is $2/m$, resulting in $3820/4 = 955$ unique reflections. There are seven non-hydrogen atoms in the asymmetric unit — the hydrogen atoms are ordinarily placed in their calculated positions. Each atom requires three positional parameters to describe its location, and another six displacement parameters to describe (approximately) its motion. The data-to-parameter ratio is therefore 955/63=15.1.

On the other hand, macromolecular asymmetric units generally contain atoms numbering in the thousands. A typical structure might have 2500 non-hydrogen atoms in the asymmetric unit, in a unit cell with a volume of 100,000 Å^3. Even though the much larger direct unit cell results in a much smaller reciprocal cell — and many more reflections, the crystals often diffract poorly, with resolution limits of 2.5–3.0 Å commonly observed. For the example given above, at 2.5 Å resolution, $h_{max} = 0.917/2.5 = 0.367\ ^{-1}$. Substitution in Eqn. 4.122 results in 20,671 reflections. If the space group is $P2_1$, a common macromolecular space group, the Laue group is $2/m$ and the number of unique reflections is $20,667/4 = 5167$. To define the structure at the atomic level three positional parameters and a single isotropic displacement parameter are required — 10,000 parameters. At 2.5 Å resolution the data-to-parameter ratio is 0.5 — the structure is significantly *under-determined*. As we shall see in Chapter 8, the refinement of under-determined crystal structures requires the restraint/constraint of a number of the parameters in order to accommodate for the lack of data necessary to define them uniquely.

Instrument Constraints

The actual number of reflections collected is always less than the number theoretically available. In many cases instrument geometry limits the ability to measure certain reflections. The major obstacle in the Eulerian goniostat is the χ circle (Fig. 4.13), which must be constrained from collision with the source and detector collimators, and may block some reflections from reaching the detector. In order to increase the freedom of motion of the crystal, an alternative to the Eulerian cradle is often employed. The arrangement replaces the χ circle with a rotational axis fixed at an angle of about 50° called the κ axis. The entire goniometer head assembly is rotated about the angle κ as illustrated in Fig. 4.31. The arrangement effectively works like a robotic arm, with movement similar to the motion of the detector devised by Buerger prior to the development of the Eulerian cradle (Sec. 4.3.1).

Although the motion of the kappa goniometer is more complex, the angles required to observe a reflection are easily mapped into the traditional Eulerian angles and the equations derived in this chapter are equally applicable to both geometric arrangements.

X-ray Scattering From Light Atoms

The derivation of the scattering factor expression in the previous chapter illustrated that intensities become much weaker at greater diffraction angles. This tends to render more reflections unobservable as 2θ increases. Recall that the electrons in atoms or ions are in constant motion around the nuclei (moving about 1/10 of the speed of light), tending to occupy an approximately spherical region of space over time. Bonding interactions distort the spherical symmetry of the electron "cloud" a bit, but to a good approximation a crystal can be considered to be composed of overlapping spheres of electrons. X-rays scatter from different regions inside each sphere, depending on the instantaneous location of the electrons, and therefore interfere with one another differently. As the angle between the incident and diffracted beams becomes larger the phase differences increase, resulting in more de-

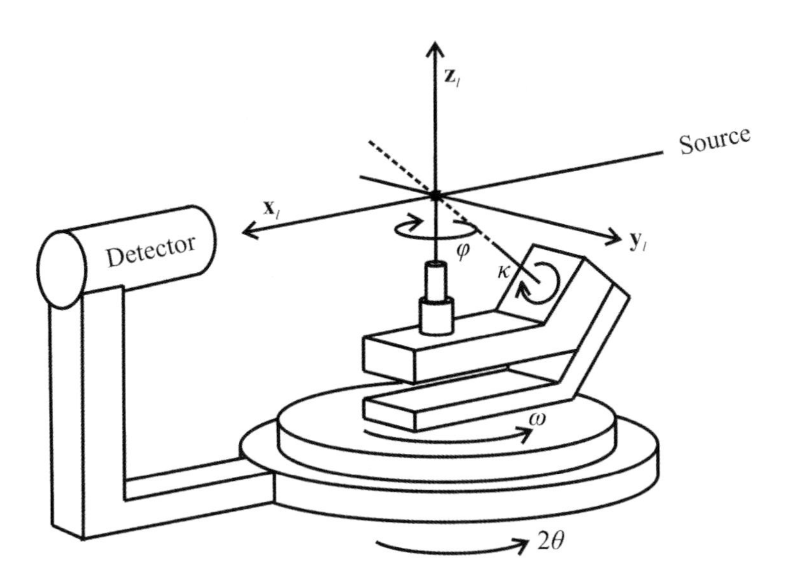

Figure 4.31 Schematic representation of a four-circle goniometer with κ geometry.

structive interference and a weaker intensity. For practical purposes the diffraction angle limits the number of reflections to be collected, depending on the scattering ability of the crystal. The ability of a particular atom to scatter X-rays depends on the number of electrons in the atom. Crystals containing elements with fewer electrons (*light atoms*), exhibit less scattering than those containing elements with large numbers of electrons (*heavy atoms*), and the limits of 2θ for data collection depend to some degree on the composition of the crystal. This is illustrated dramatically in Fig. 4.32. The atomic scattering factors, indicating the scattering efficiency of the atoms, are plotted as a function of $\sin\theta/\lambda$ for three atoms in Group IV of the periodic table: carbon (6 electrons), germanium (32 electrons) and lead (82 electrons). At large values of 2θ reflections from a crystal containing only carbon and other light atoms will become too weak to observe, while crystals containing heavier atoms will tend to retain their "scattering power."

Crystal Quality

Crystals often vary considerably in their ability to provide measurable intensities. In some cases the crystals are simply too thin or small to produce a sufficient number of diffracted photons. In others, the composition or physical state of the crystal may cause problems. The diffraction experiment demands periodic electron density, and the less periodic this becomes, the less constructive the interference. If the atomic arrangements are disordered in a somewhat random fashion, due either to thermal motion or varied arrangements in the crystal, intensities will tend to decrease. For example, if the crystal is near its melting point, or contains randomly distributed solvent molecules, it will tend to diffract poorly. Thermal motion, in

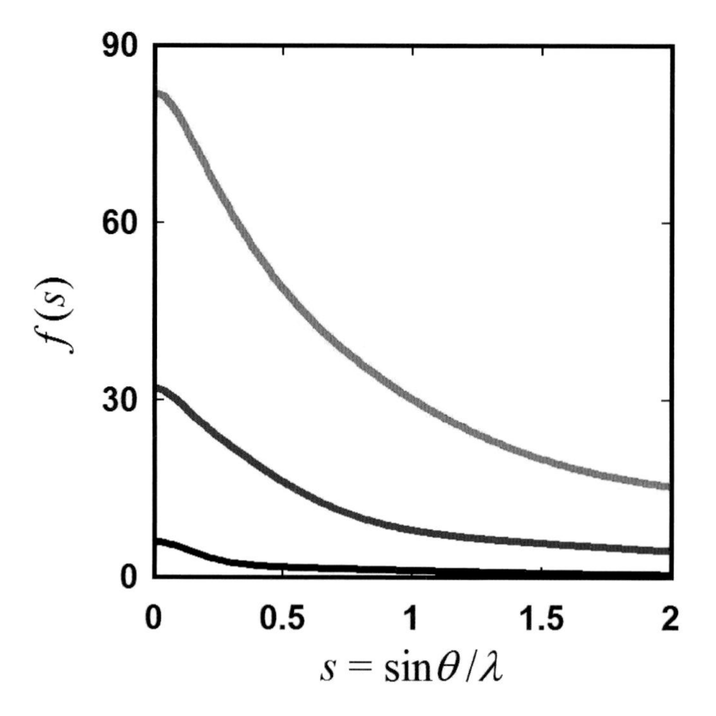

Figure 4.32 Plot of atomic scattering factors for carbon (black), germanium (blue), and lead (red).

general, tends to increase the effective size of the atoms in the crystal, and therefore the destructive interference described above. It is possible to reduce this effect by collecting the data at lower temperatures.

Exercises

1. A crystal of mercaptopyridine (Sec. 1.5.3) is aligned so that its a axis is perpendicular to the direction of the incident X-ray beam ($\lambda = 0.7107$ Å) along the z axis of the laboratory reference frame. An area detector is positioned so that its plane is perpendicular to the incident beam at a distance of 9 cm from the crystal. (a) Predict the distance between the horizontal rows of diffraction maxima (spots) on the detector surface. (b) Predict the location of the spots on the detector surface from the (0 1 0) and (0 0 1) reflections.

2. The orientation matrix for the mercaptopyridine crystal used to determine the structure discussed in Sec 1.5.3, employing Mo $K\alpha$ radiation (0.71073 Å), was

$$\mathbf{A}^* = \begin{bmatrix} 0.02650 & -0.04430 & 0.06840 \\ -0.04094 & -0.14859 & -0.02004 \\ 0.15986 & -0.03066 & -0.00154 \end{bmatrix}.$$

(a) Determine the Eulerian angles and detector angle for the $(2\,\bar{4}\,11)$ reflection for the rotation of the crystal and the observation of a diffraction maximum in the bisecting position. (b) A reflection is located with the following Eulerian and detector angles: $2\theta = 21.673°$, $\omega = 0.114°$, $\varphi = 119.921°$,$\chi = 60.878°$, $2\theta = 21.673°$. What are its indices?

3. In part (a) of the previous exercise the reflection vector, $\mathbf{h_r}$, was oriented with the angles ω, χ, and φ into the diffraction condition in the bisecting position with Eulerian angles specific for each reflection. When an area detector is employed, χ, φ, and the detector angle, τ, are fixed; $\mathbf{h_r}$ is oriented with ω alone. From Eqn. 4.26, for a given reflection, $\mathbf{X\Phi A^*h}$ is a specified vector, $\mathbf{v}$, oriented by ω: $\mathbf{\Omega v} = \mathbf{h_r}$. (a) Derive an expression for ω in terms of $\mathbf{h_r} = [h_x\ h_y\ h_z]$ and $\mathbf{v} = [v_x\ v_y\ v_z]$. (b) For $\chi = 54.73°$ and $\phi = 270°$, determine the ω angle that will orient the $(2\,\bar{4}\,11)$ reflection vector of the mercaptopyridine crystal in the previous exercise into the diffraction condition. (c) Determine the location of the $(2\,\bar{4}\,11)$ "spot" on an area detector surface for the settings of χ and φ in (b), a crystal-to-detector distance of 6 cm, and a detector angle of 26°.

4. The mercaptopyridine crystal in Exercise 2 provided 14 reflections with the following refined diffractometer angles:

Reflection #	2θ	ω	ϕ	χ
1	7.951	−0.633	229.286	54.806
2	9.389	0.113	318.467	56.348
3	10.222	0.174	294.654	49.341
4	13.746	−0.272	89.943	75.436
5	14.861	0.069	310.896	77.474
6	15.929	0.315	269.490	3.521
7	16.010	0.327	215.404	18.527
8	17.496	0.286	245.219	358.839
9	18.831	−0.213	309.305	57.254

Reflection #	2θ	ω	ϕ	χ
10	19.954	0.257	70.621	43.299
11	21.673	0.114	119.921	60.878
12	22.477	0.079	19.438	73.882
13	23.399	−0.208	239.967	65.376
14	24.324	0.132	285.315	69.025

In the search for axial solutions using integer triples ranging from $(n_1,\ n_2,\ n_3) = (-5, -5, -5)$ to $(5,\ 5,\ 5)$, 57 potential unit cell axes, $\mathbf{v} = [v_x\ v_y\ v_z]$, with lengths of 25 Å or less were determined using Eqn. 4.40 — based on $\mathbf{Z}^{-1}$ created from the first three reflections in the list:

$$\begin{aligned} v_x &= n_1 z^i_{11} + n_2 z^i_{21} + n_3 z^i_{31} \\ v_y &= n_1 z^i_{12} + n_2 z^i_{22} + n_3 z^i_{32} \\ v_z &= n_1 z^i_{13} + n_2 z^i_{23} + n_3 z^i_{33}. \end{aligned}$$

These trial solutions were tested using Eqn. 4.41, which must provide an integer (in practice, a number close to an integer) for $\mathbf{h}_\mathrm{l} \equiv \mathbf{z}$ for each reflection:

$$v_x z_x + v_y z_y + v_z z_z = m.$$

The candidates that meet this test are the Cartesian components of each potential unit cell axis; the selection of the "best" unit cell is facilitated by computing their lengths and angles with respect to one another, looking for short axes with angles approaching 90°. Although the solutions for this crystal provided the monoclinic unit cell described in the text, we consider another possible solution from the 57 candidates (the "**u**" subscript indicates that the unit cell is not the reduced unit cell for the lattice — see Exercise 5):

axis	n_1	n_2	n_3	v_x	v_y	v_z
$\mathbf{a_u}$	1	1	1	−0.1980	−1.1664	5.9618
$\mathbf{b_u}$	−1	−2	−2	−1.5685	−4.7848	−7.1966
$\mathbf{c_u}$	2	2	3	15.1534	1.4349	4.0624

(a) Determine the parameters for the unit cell. (b) Determine the indices of the reflections in the list based on the unit cell.

5. When refined by the method of least squares, the unreduced cell in the previous problem results in the following orientation matrix:

$$\mathbf{A}^* = \begin{bmatrix} -0.01781 & 0.02410 & 0.06840 \\ -0.18953 & -0.16863 & -0.02004 \\ 0.12920 & -0.03220 & -0.00154 \end{bmatrix}.$$

Eqns. 4.68 can be utilized to produce linear combinations of the Cartesian components of the refined unit cell axes in order to produce a "Buerger" reduced cell. For the unreduced unit cell defined by the orientation matrix above, the integers (p, q, r) ranging from $(-2, -2, -2)$ to (2,2,2) produce 124 lattice vectors, 62 of which are unique (for every vector created from (p, q, r), its negative will be created with $(-p, -q, -r)$). The shortest 20 of these are tabulated below:

axis	p	q	r	length	axis	p	q	r	length
$\mathbf{v}_1$	1	0	0	6.1071	$\mathbf{v}_{11}$	1	0	-1	15.6495
$\mathbf{v}_2$	1	1	0	6.3268	$\mathbf{v}_{12}$	1	2	1	15.6555
$\mathbf{v}_3$	2	1	0	8.7917	$\mathbf{v}_{13}$	0	0	1	15.7231
$\mathbf{v}_4$	0	1	0	8.7952	$\mathbf{v}_{14}$	2	2	1	15.7272
$\mathbf{v}_5$	2	0	0	12.2141	$\mathbf{v}_{15}$	1	-1	-1	16.6519
$\mathbf{v}_6$	2	2	0	12.6537	$\mathbf{v}_{16}$	2	1	1	16.8559
$\mathbf{v}_7$	1	-1	0	13.7577	$\mathbf{v}_{17}$	0	2	0	17.5904
$\mathbf{v}_8$	1	2	0	14.0525	$\mathbf{v}_{18}$	2	0	-1	17.8098
$\mathbf{v}_9$	0	1	1	14.3168	$\mathbf{v}_{19}$	0	2	1	17.8168
$\mathbf{v}_{10}$	1	1	1	14.3962	$\mathbf{v}_{20}$	1	0	1	18.0033

(a) Determine the Cartesian components of the refined unit cell axes. (b) Determine "Buerger" cell parameters for the lattice. (c) Use the Buerger unit cell to determine the "Niggli" reduced cell for the lattice. (c) Determine the indices for the reflections listed in Exercise 4 based on the reduced cell.

6. According to Eqn. 4.37, any three indexed reflections that are linearly independent (so that $\mathbf{H}^{-1}$ has an inverse) are sufficient to define the orientation matrix and unit cell parameters. Reflections 11, 13, and 7 in the array in Exercise 2 have indices of $(2\ 0\ \bar{6})$, $(3\ \bar{1}\ 2)$, and $(1\ 1\ 4)$, respectively. (a) Demonstrate that the reciprocal lattice vectors for the reflections are linearly independent. (b) Determine the orientation matrix based solely on the three reflections. (c) Calculate the metric tensor and determine the unit cell parameters based solely on the three reflections.

7. Show that a c-glide plane perpendicular to the b axis has the same effect on the symmetry of the intensity weighted reciprocal lattice as a mirror plane perpendicular to the b axis (without the translational symmetry).

8. The simplest tetragonal space group is $P4$, which represents the $4/m$ Laue group. (a) Referring to Sec. 2.4.7, determine the symmetry-equivalent positions in the $P4$ space group for the general direct lattice vector, $[x_f\ y_f\ z_f]$. (b) Determine the symmetry-equivalent intensities in the intensity-weighted reciprocal lattice for a general reciprocal lattice vector, $[h\ k\ l]$, in the $P4$ space group.

9. Consider crystals of a small molecule (S) and a protein (P), each rectangular solids with dimensions 0.3 mm × 0.4 mm × 0.5 mm. Both molecules pack in orthorhombic space groups. The unit cell dimensions for the small molecule lattice are a = 8 Å, b = 10 Å, and c = 12 Å, while the unit cell dimensions

for the protein lattice are $a = 48$ Å, $b = 50$ Å, and $c = 52$ Å. Suppose that a structure factor, F_{0k0}, fortuitously has the same value for both structures. Using the integrated intensity of this reflection as an estimate of the comparative "diffracting power" of the crystals, calculate the ratio $\mathcal{I}^{\mathrm{S}}_{0k0}/\mathcal{I}^{\mathrm{P}}_{0k0}$. For simplicity assume that $\sin 2\theta_{\mathrm{P}}/\sin 2\theta_{\mathrm{S}} \simeq \sin\theta_{\mathrm{P}}/\sin\theta_{\mathrm{S}}$.

10. A data-to-parameter ratio in the vicinity of 10:1 is often cited as a criterion for a "good" (atomic resolution) crystal structure. Referring to Exercise 9, assume that the small molecule asymmetric unit contains seven non-hydrogen atoms, the protein asymmetric unit contains 1800 non-hydrogen atoms, and that data are collected with Cu $K\alpha$ radiation ($\lambda = 1.5418$ Å). (a) Determine the maximum value of the diffraction angle, 2θ, necessary to attain a 10:1 data-to-parameter ratio for the small molecule and protein structures. (b) Determine the data-to-parameter ratios for the protein crystal at 1 Å (atomic) resolution, 2.5 Å resolution, and 5 Å resolution.

Chapter 5

Crystal Diffraction: Data

5.1 Experimental Error

The experimental intensities collected from X-ray diffraction eventually become the coefficients in the Fourier series representation of the electron density (Eqn. 3.83). Since they are measured quantities they are subject to errors — the quality of the final structure clearly depends on these errors.

In all experiments the experimentalist encounters two types of errors. The first type involves random fluctuations in the experiment that are beyond the control of the experimentalist to model or remove. These *random errors* are statistical* in nature, and while they cannot be removed from the experiment, their effects must be evaluated and considered in the analysis of the data. Random errors determine the reproducibility of a measurement. If the magnitude of the fluctuations is small the experimentalist can be confidant that a series of measurements will produce results that are close to one another, while the results of repeated experiments may vary considerably if the fluctuations are large. The degree of fluctuation is a measure of the *precision* of the measurement. In general, repeated measurements will tend to cluster around an average value, and with enough repetitions an accurate (see below) average can be obtained. However, in most experiments it is impractical to repeat an experiment enough times to completely "average out" the random error, and we must instead assess the magnitude of the random error in order to determine the degree of confidence that we have in data obtained with a limited number of observations.

The second type of error involves phenomena that are inherent in the experimental system – and is known as *systematic error*. For example, crystals absorb radiation which tends to lower the observed intensities. The amount of absorption depends on the amount of material that the incident and diffracted beams are exposed to on their way through the crystal. Since the path length varies, depending on the Bragg angle and the orientation of the crystal, the effect on the intensities will vary. The relative intensity of a given reflection depends on this path length, and if we fail to take it into account the relative intensities will have values that

*Probability and statistics play major roles in many aspects of crystallography. They will be encountered often throughout the remainder of the text, and will therefore be treated in some detail here.

Understanding Single-Crystal X-Ray Crystallography. Dennis W. Bennett

ISBN: 978-3-527-32677-8 (HC), 978-3-527-32794-2 (SC)

differ from those that would be observed if this error in their measurements did not exist. The degree to which a measurement differs from its known or expected value is a measure of the *accuracy* of the measurement. In this particular case the accuracy of our measurements can be improved if we can somehow predict the variations in absorption and correct the observed values accordingly. Systematic errors must be identified by the experimentalist, and either modeled into the data or removed from it.

5.1.1 Random Error

Even in the most carefully controlled experiment the removal of systematic error does not guarantee that a measurement will be exact. All measurements are subject to random fluctuations that cannot be removed, even by the most careful experimentalist or sophisticated model. Since we can never control these *random errors* completely, we must be able to assess their effects on the observed data — in our case intensity data. For example, if a measured intensity is 10,000 counts per second, with random fluctuations in the neighborhood of 100 counts per second, then we might expect repeated measurements of the same reflection to provide values around 1% of one another — any one of the measured intensities will have a high degree of "certainty." If, on the other hand, the intensity that we measure is 10 counts per second, and the fluctuations are around 3 counts per second, then a second measurement would be expected to lie within about 30% of the first— much less "certain." While we cannot completely remove these fluctuations in our experiments, quantifying the degree of uncertainty in a given intensity will assist us in determining how much *weight* to give to it in the determination and refinement of the crystal structure. In order to assess how we might go about determining this quantification we turn to a simple example.

Consider two trucks full of apples, one type grown in a hothouse under controlled conditions (truck A), the other type grown in a field where environmental conditions cause large variations in growth rates (truck B). We remove an apple from each truck and weigh it, wishing to know the degree to which the weight of each apple represents an average apple in the truck. In order to make this assessment we must weigh more than one apple. We assume here that our scale is calibrated properly; that it is void of systematic errors. Table. 5.1 gives the results of the experiment for samples of apples taken from trucks A and B. The average weight of the apples in each truck is 5.29 *oz*. If the owners of trucks A and B wished to sell you 1,000 apples by weighing a single apple and subsequently "counting" the apples by weighing them in bulk, you would be much more *confidant* that you are actually getting that number from truck A than from truck B. 1000 apples from truck A *or* truck B would weigh something very close to 5290 *oz*. (the average times a thousand). If apple #5 from truck A above was selected as the "experimental apple", the experiment would predict that there would be 1000 apples in 5260 *oz*.— you would get almost the number that you paid for. The selection of any apple from truck A would produce essentially the same result. Selecting apple #5 from truck B would predict that there would be 1000 apples in 4310 *oz*. — and you would get far fewer apples than you bargained for. On the other hand you might get lucky and select apple #4 from truck B, but in either case you would not be nearly as confidant of your choice.

Table 5.1 Weights of Apples from Truck A and Truck B.

Truck A		Truck B	
$w(oz.)$	$w - \bar{w}$	$w(oz.)$	$w - \bar{w}$
5.23	−0.06	4.39	−0.90
5.27	−0.02	5.09	−0.20
5.29	0.00	4.79	−0.50
5.30	0.01	6.10	0.81
5.26	−0.03	4.31	−0.98
5.36	0.07	5.55	0.26
5.33	0.04	5.32	0.03
5.34	0.05	5.98	0.69
5.27	−0.02	5.90	0.60
5.24	−0.05	5.49	0.20

The column to the right of each column of apple weights is the difference between the weight of each apple and the average weight of an apple. The difference between a measurement and the average value of a number of measurements is known as the *deviation* or *residual* of the measurement. The deviations are small when the measurements lie close to the average value (known as the *mean* value), and large when they do not. Since the deviations vary, we might consider an average deviation as a good measure of the degree of uncertainty in an experiment. However, note that the deviations are distributed evenly on both sides of the mean – they will tend to average to zero since they can be either negative or positive: $(1/n)\sum_{i=1}^{n}(w_i - \bar{w}) \approx 0$, where n is the number of measurements. To avoid the negative numbers we might consider squaring the deviations and taking the average value of the squares of the deviations. The resulting *averaged squared deviation* is obtained by summing the squares of the deviations and dividing by the number of observations (measurements). For statistical reasons we define a parameter that is nearly identical to the averaged squared deviation, known as the *sample variance*:

$$s^2 = \frac{1}{n-1}\sum_{i=1}^{n}(w_i - \bar{w})^2. \tag{5.1}$$

The symbol for the variance includes the superscript "2"; the "sample" qualifier indicates that this is the variance of a sample of the entire apple *population*. The division by $n-1$ rather than by n occurs because the sum of the deviations will equal zero as n becomes large. Thus the sum of $n-1$ of the deviations allows us to calculate the last one, i.e., one of the deviations is determined by the remaining $n-1$ of them — it is not actually an independent deviation. The variance is therefore somewhat larger than the averaged squared deviation — often negligibly so.

Since the variance is essentially the average *squared* deviation, we might consider a more representative number for the uncertainty in the data (often referred to as the *scatter* in the data) to be the square root of the variance. This number is known as the *root-mean-square* deviation, or more commonly, the *standard deviation*, in this case the *sample standard deviation*:

$$s = \left(\frac{1}{n-1}\sum_{i=1}^{n}(w_i - \bar{w})^2\right)^{\frac{1}{2}}. \quad (5.2)$$

In order to obtain the actual variances and standard deviations for our apple experiment we would need to know the real averages of the apple weight, which would require weighing all of the apples in each truck. The actual variance and standard deviation of the entire population are given the symbols σ^2 and σ, respectively. Determining these values for our apple population would be difficult and tedious; most populations in experimental science are so large that such a task is virtually impossible. If μ is the actual mean of the entire population, then, in general

$$\sigma^2 = \lim_{n\to\infty}\frac{1}{n}\sum_{i=1}^{n}(w_i - \mu)^2 = \langle (w_i - \mu)^2\rangle, \quad (5.3)$$

where $\langle (w_i - \mu)^2\rangle$ is the average value of the mean square deviation for the entire population. In order to determine how best to estimate the population variance, we imagine selecting a number of samples of the same size, determining the variance of each sample, and averaging the variances:

$$\begin{aligned}
\langle s^2\rangle &= \frac{1}{n-1}\left\langle \sum_{i=1}^{n}(w_i - \bar{w})^2\right\rangle = \frac{1}{n-1}\left\langle \sum_{i=1}^{n} w_i^2 - 2\bar{w}\sum_{i=1}^{n} w_i + \sum_{i=1}^{n}\bar{w}^2\right\rangle \\
&= \frac{1}{n-1}\left\langle \sum_{i=1}^{n} w_i^2 - 2n\bar{w}^2 + n\bar{w}^2\right\rangle = \frac{1}{n-1}\left\langle \sum_{i=1}^{n}\langle w_i^2\rangle - n\langle \bar{w}^2\rangle\right\rangle \\
&= \frac{n}{n-1}(\langle w_i^2\rangle - \langle \bar{w}^2\rangle). \quad (5.4)
\end{aligned}$$

The average value of w_i^2 can be evaluated from the population variance:

$$\begin{aligned}
\sigma^2 &= \langle (w_i - \mu)^2\rangle = \langle w_i^2\rangle - 2\mu\langle w_i\rangle + \langle \mu^2\rangle = \langle w_i^2\rangle - \mu^2 \Longrightarrow \\
\langle w_i^2\rangle &= \sigma^2 + \mu^2. \quad (5.5)
\end{aligned}$$

The average value of $\bar{w}^2$ is given by

$$\begin{aligned}
\langle \bar{w}^2\rangle &= \left\langle \left(\frac{w_1 + w_2 + \cdots + w_n}{n}\right)^2\right\rangle \\
&= \frac{1}{n^2}\left(\left\langle \sum_{i=1}^{n} w_i^2\right\rangle + \left\langle \sum_{i=1}^{n}\sum_{j\neq i}^{n} w_i w_j\right\rangle\right). \quad (5.6)
\end{aligned}$$

There are n of the "w_i^2" terms in Eqn. 5.6, and $n(n-1)$ of the "$w_i w_j$" cross terms. Since w_i and w_j are independent,

$$\langle w_i w_j\rangle = \langle w_i\rangle\langle w_j\rangle = \mu^2. \quad (5.7)$$

Thus

$$\begin{aligned}
\langle \bar{w}^2 \rangle &= \frac{1}{n^2}(n(\sigma^2 + \mu^2) + n(n-1)\mu^2) = \frac{\sigma^2}{n} + \mu^2 \quad \text{and} \\
\langle s^2 \rangle &= \frac{n}{n-1}\left((\sigma^2 + \mu^2) - \left(\frac{\sigma^2}{n} + \mu^2\right)\right) \\
&= \left(\frac{n}{n-1}\right)\sigma^2\left(\frac{n-1}{n}\right) = \sigma^2.
\end{aligned} \tag{5.8}$$

It follows that the sample variances will average to the population variance. The average value is often called the *expected value.* Thus the expected value of the sample variance is the population variance,* and we can assume that *the sample variance and standard deviation are the best single estimators of the population variance and standard deviation.*

Experimentally determined sample standard deviations are referred to as *estimated standard deviations* (abbreviated as *e.s.d.*s). The estimated standard deviations of the weights of apples in trucks A and B are 0.04 and 0.61, respectively. Although it is rigorously more correct to use the symbol s for the estimated standard deviation, it is common practice in experimental science to use the symbols σ^2 and σ for the estimated variance and standard deviation, and we will do so throughout the text.

For our intensity experiment we are rarely even afforded the luxury of estimating the standard deviation of our intensity measurements from a sample population of such measurements. We generally record each intensity only once, for a relatively brief period of time. If we were able to record intensities repeatedly and average them, or alternatively to record the intensity for a long period of time, we would expect the value obtained to be a reasonable estimate of the average intensity. Because we must measure a large number of intensities, we wish to know approximately how closely a single intensity measured in a practical period of time represents the value from repeated measurements or long counting times. The standard deviation is an indication of the average deviation of a measurement from the mean value of many measurements, and we must find a way to estimate it without actually knowing the mean value!

The Binomial Distribution

A reflection wave arrives at the detector with an electric field of magnitude $\mathcal{E}$, with the *probability* that a photon will arrive at the detector proportional to $\mathcal{E}^2$. As we discussed in Chapter 3, whether or not a photon arrives at the detector (a phenomenon that we will term an *event*) depends on statistical probability rather than certainty. Even though a given intensity may be measured as 9865 counts per second in a given time interval, another measurement may yield 9645. Although ***extremely*** unlikely, it is even *possible* that we will measure 3 counts per second in a particular experiment. If we knew the standard deviation of the measurement we would have some idea of just how probable 9645, or 8250, or 3 counts per second would be for a given measurement. The fluctuation in counts is the major source of random (statistical) error in the X-ray diffraction experiment, and assessment of

*Note that this justifies the $n-1$ divisor in Eqn. 5.1.

this error requires that we have a way of determining the probability that a certain number of counts will occur in a given experiment.

We have already encountered probabilities in the discussion of electron density in the previous chapter. Suppose that a very weak reflection, one with less than a count per second, is being recorded at the detector (we will ignore the background here). We wish to know the probability that, in a given time period, n photons will arrive at the detector. We begin by selecting a time increment that is small enough that there is a probability that a photon will *not* arrive at the detector — in this case we choose an increment of 1 second. We now monitor the detector in increments of 1 second, the monitoring of a single second constituting a *trial.* Our first goal is to determine the probability that we will observe a photon during a single trial. We do this by "keeping score" – placing a red ball in a jar every time a photon is observed, and a blue one ever time one is not (we assume here that the occurrence of two photons is very unlikely – if not we would simply change the time increment until that was the case). After 5000 of these observations we stop and count the balls, discovering 1000 red balls and 4000 blue balls. An event has been observed in $\frac{1}{5}$ of the trials, and we conclude that the probability that an event will occur in a single trial is $p = \frac{1}{5}$. The probability that an event will *not* occur (blue ball) is $q = \frac{4}{5} = 1 - p$. The probability that an event will either occur *or* not occur is $p + q = 1$, certainty.

Given the probability of an event occurring in a single trial, we now wish to assess the probability that n events will occur in N trials for a range of values of n. For example, suppose that in our experiment we want to know the probability that a 10 increment (10 second) count will produce 10 counts, or 9, or 8, . . . down to 0 counts. In order to perform this experiment we set up a large number of arrays (W) of 10 boxes, one box for each increment (trial):

1	2	3	4	5	6	7	8	9	10

For a given 10 second count, a red ball or a blue ball will be placed in the box corresponding to the trial, depending on whether or not a photon is observed (whether or not an event has occurred). We conduct W independent experiments, then sort all of the arrays into groups, depending on the number of red balls (counts) that we find in the array. In order to determine the probability that n counts will occur in 10 seconds, we need only count the number of arrays with n counts and divide by the total number of arrays. The resulting fraction is the probability, which we denote $f_N(n) = f_{10}(n)$, where $N = 10$ represents the number of trials, and n is the number of events observed in N trials.

Rather than perform this tedious experiment, perhaps there is a way to calculate these probabilities. Let us attempt to calculate the probability of observing 10 counts in 10 seconds. A representative array would look like this:

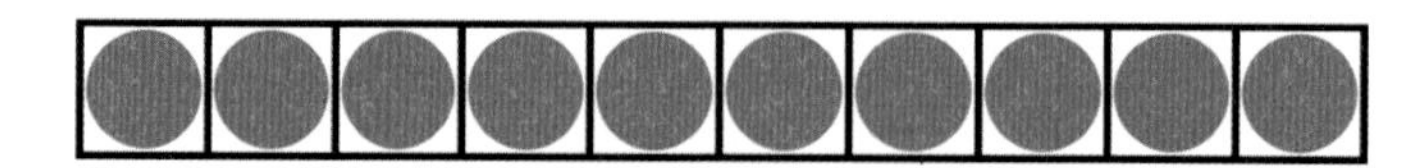

Since there is a $p = \frac{1}{5}$ chance that the first box will contain a red ball, the number of arrays with a red ball in the first box will be $\frac{1}{5}W$. $\frac{1}{5}$ of these, $\frac{1}{5}\left(\frac{1}{5}W\right)$, will have a red ball in the second box, $\frac{1}{5}\left(\frac{1}{5}\left(\frac{1}{5}W\right)\right)$ will have a red ball in the third box, and so on. Thus the number of arrays with red balls in all 10 boxes is $\left(\frac{1}{5}\right)^{10} W$, and $f_{10}(10) = \left[\left(\frac{1}{5}\right)^{10} W\right]/W = \left(\frac{1}{5}\right)^{10} = p^{10}$. By an identical argument, the probability of observing *zero* counts in 10 seconds (each box containing a blue ball) is clearly $f_{10}(0) = \left(\frac{4}{5}\right)^{10} = (1-p)^{10}$.

The two extreme cases are easy to evaluate, since there is only one array configuration that will exclusively contain red balls or blue balls. For the intermediate cases we must deal with a number of arrangements with the same number of counts. Consider the arrays with three counts, for example. We may observe a count in the second trial, the third trial, and the 6th trial for some of the arrays:

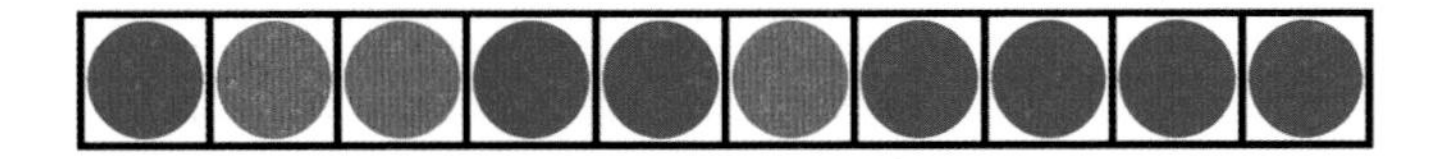

The number of arrays with a blue ball in box one will be $\frac{4}{5}W$. $\frac{1}{5}$ of these, $\frac{1}{5}\left(\frac{4}{5}W\right)$ will have a red ball in the second box, etc. The number of arrays with this particular configuration (which we will call configuration a) is $\left(\frac{1}{5}\right)^3\left(\frac{4}{5}\right)^7 W$, and $f_{10}^{a}(3) = \left(\frac{1}{5}\right)^3\left(\frac{4}{5}\right)^{(10-3)}$.

This is, of course, not the only configuration representing three counts. Another, which we call configuration b, looks like this:

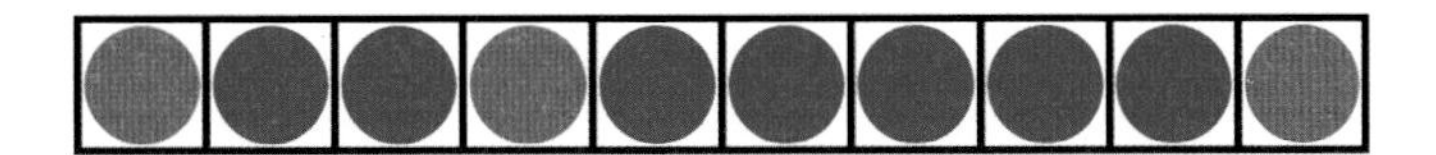

The number of arrays with this configuration is also $\left(\frac{1}{5}\right)^3\left(\frac{4}{5}\right)^7 W$, and $f_{10}^{b}(3) = \left(\frac{1}{5}\right)^3\left(\frac{4}{5}\right)^{(10-3)}$. This doubles the number of arrays representing three counts, and therefore doubles the probability of observing 3 counts in 10 intervals. A third configuration will triple the probability, and so forth. There are C such configurations representing three counts, such that $f_{10}(3) = C\left(\frac{1}{5}\right)^3\left(\frac{4}{5}\right)^{(10-3)}$. C is the number of different configurations containing three red balls, and we must determine it in order to obtain the probability.

We do this in a somewhat roundabout manner, beginning by temporarily labeling the three red balls with the numbers 1, 2, and 3. There are 10 boxes in which to place ball 1, after which there are 9 boxes available for ball 2, then 8 boxes for ball 3. Thus there are a total of $L = 10 \times 9 \times 8$ different ways to arrange three labeled red balls in 10 boxes. However, since we cannot actually distinguish between the balls, a number of these arrangements are redundant, as illustrated in Fig. 5.1.

We are not looking for the number of different configurations with 3 labeled balls, but the number with 3 indistinguishable balls. If we label the boxes containing the red balls with the letters A, B, and C in configuration a (or any other with three red balls), we can choose any one of the three labeled balls to place in box A, leaving two choices for box B, and only one choice for box C. Thus there are $3 \times 2 \times 1 = 3! = 6$ ways of arranging the labeled balls *for every unlabeled configuration.* By labeling the red balls we have counted each unlabeled configuration 3! times: $L = C \times 3!$ and $C = L/3! = (10 \times 9 \times 8)/3!$. For the reader familiar with combinations and

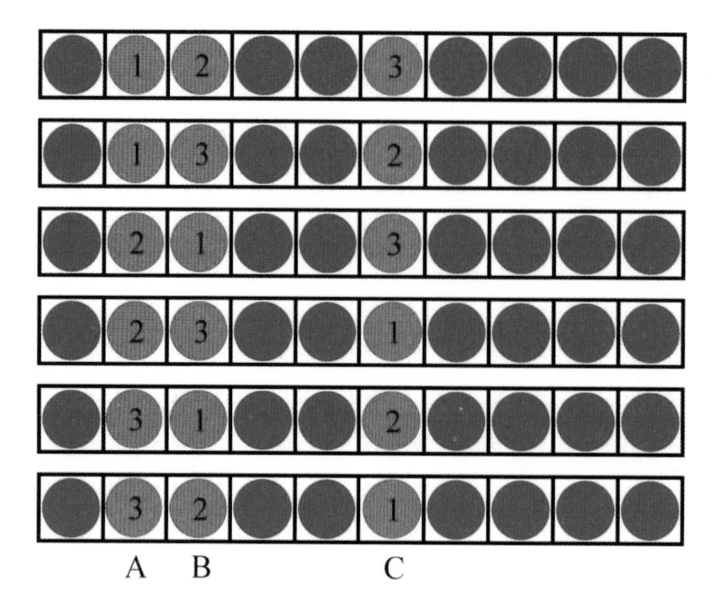

Figure 5.1 Redundant arrangements of three labeled balls.

permutations, 3! is the number of permutations of three distinguishable objects and C is the number of combinations of 10 objects taken 3 at a time, which we denote C_3^{10}. The probability of observing 3 events in 10 trials can now be computed:

$$f_{10}(3) = \frac{10 \times 9 \times 8}{3!}\left(\frac{1}{5}\right)^3\left(\frac{4}{5}\right)^{(10-3)} = C_3^{10}\left(\frac{1}{5}\right)^3\left(\frac{4}{5}\right)^{(10-3)}.$$

Using the same arguments, the probability of observing 4 events in 10 trials is

$$f_{10}(4) = \frac{10 \times 9 \times 8 \times 7}{4!}\left(\frac{1}{5}\right)^4\left(\frac{4}{5}\right)^{(10-4)} = C_4^{10}\left(\frac{1}{5}\right)^4\left(\frac{4}{5}\right)^{(10-4)}.$$

In general, if p is the probability of observing an event (count) in a single trial (time increment) and N is the number of trials, then the probability that n events will occur in N trials is

$$\begin{aligned} f_{N,p}(n) &= \frac{N \times (N-1) \times \ldots \times (N-n+2) \times (N-n+1)}{n!} p^n (1-p)^{(N-n)} \\ &= C_n^N p^n (1-p)^{(N-n)}. \end{aligned} \tag{5.9}$$

Counting probabilities are a member of a large family of probabilities that are "binary" in nature — either the event happens, or it does not — and are uniquely related to the binomial expansion. While we will not deal rigorously with the link between counting probabilities and the binomial expansion; the expansion of the binomial $[p + (1-p)]^{10} = (p+q)^{10}$ will serve to illustrate it:

$$\begin{aligned} (p+q)^{10} &= (p+q)(p+q)(p+q)(p+q)(p+q)(p+q)(p+q)(p+q)(p+q)(p+q) \\ &= 1p^{10} + 10p^9q^1 + 45p^8q^2 + 120p^7q^3 + 210p^6q^4 + 252p^5q^5 \\ &\quad +210p^4q^6 + 120p^3q^7 + 45p^2q^8 + 10p^1q^9 + 1q^{10}. \end{aligned}$$

Consider the term $120p^3q^7$. The coefficient is the number of times a product containing 3p terms and 7q terms occurs in the expansion, entirely analogous to the number of times that an array contains 3 red balls and 7 blue balls, when any arrangement is equally probable. Thus 120 must be C_3^{10} : $(10 \times 9 \times 8)/3! = 720/6 = 120$. The term corresponding to $p^n q^{N-n}$ in the binomial expansion of $(p+q)^N$ is $f_{N,p}(n)$, the probability of observing n events in N trials if the probability for a single trial is p.

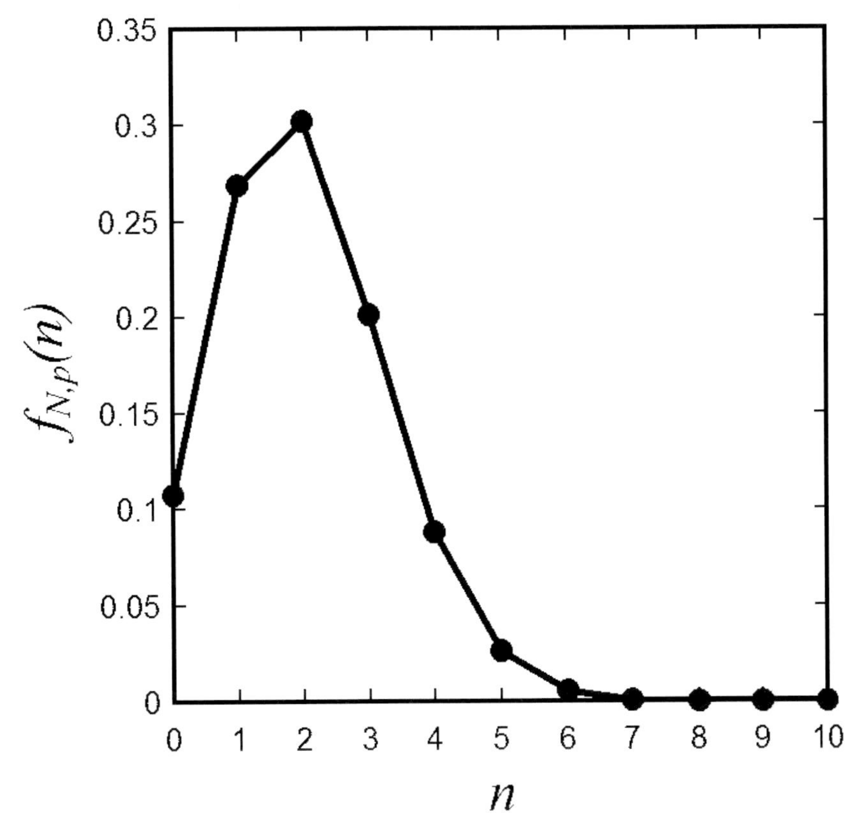

Figure 5.2 Binomial probability distribution for the probability of n events in $N = 10$ trials if the probability for an event in a single trial is $p = 1/5$.

Fig. 5.2 is a plot of $f_{N,p}(n)$ versus n for our example. It illustrates how the probabilities are distributed among the possible events for a set number of trials, and is known as a *probability distribution.* Because of its relationship to the binomial expansion it is referred to as a *binomial distribution.* $f_{N,p}(n) = C_n^N p^n (1-p)^{(N-n)}$ is known as the *binomial distribution function.* Note that the plot makes intuitive sense. The 5000 trial experiment resulted in 1000 counts — on the average two counts per 10 trials (10 seconds) — and this would be expected to be the most probable number. Even so, it is almost as likely that one or three counts will be observed. The low count rate also indicates a large probability that no counts at all will be observed in ten seconds, while it is very unlikely that more than 6 will be observed in that period of time.

From the distribution above, we may wish to know the probability that *either* one, or two, or three counts will be observed in the 10 second interval, $f_{10}(1or2or3)$. To answer this question we return to our arrays. The number of arrays with one count is $f_{10}(1)W$, those with two counts is $f_{10}(2)W$, and those with three counts is $f_{10}(3)W$. The total number of arrays that contain one, two or three counts is the sum of these numbers: $f_{10}(1or2or3)W = f_{10}(1)W + f_{10}(2)W + f_{10}(3)W = [f_{10}(1) + f_{10}(2) + f_{10}(3)]W$, and $f_{10}(1or2or3) = f_{10}(1) + f_{10}(2) + f_{10}(3)$, the sum of the probabilities for each count. The sum of *all* of the probabilities in the probability distribution is the probability that somewhere between zero and ten counts will be observed. Since that is a certainty, the sum of the probabilities must clearly equal unity: $\sum_{n=0}^{10} f_{10}(n) = 1$. This is general for the binomial distribution:

$$\sum_{n=0}^{N} f_{N,p}(n) = \sum_{n=0}^{N} C_n^N p^n (1-p)^{(N-n)} = 1 \tag{5.10}$$

Recall that all of this work has, as its final aim, the determination of the standard deviation for our observed intensities. In order to accomplish this we will need to derive another probability distribution, known as the *Poisson distribution*, as a limiting form of the binomial distribution. In preparation for this we first determine an expression for the average number of counts observed from the W experiments, based on the binomial distribution. In our collection of arrays, for every n there are $f_{N,p}(n)W$ arrays, and a contribution of $nf_{N,p}(n)W$ counts to the total number of counts (red balls) in all the arrays. Thus the total number of counts in all the arrays is $\sum_{n=0}^{N} nf_{N,p}(n)W$. The average number of counts in an array, $\bar{n}$,* is the total number of counts in all the arrays divided by the number of arrays:

$$\bar{n} = \frac{\sum_{n=0}^{N} nf_{N,p}(n)W}{W} = \sum_{n=0}^{N} nf_{N,p}(n) = \sum_{n=0}^{N} nC_n^N p^n (1-p)^{(N-n)} \tag{5.11}$$

As expected, this is just an average of all n counts, weighted by the probability that each n will be observed.

The Poisson Distribution

Although a very low count rate was selected for the derivation of the binomial probability distribution, the distribution is perfectly general. To deal with realistic counting rates, however, the time increment for a given trial will have to be very much smaller (to restrict the interval to a single event) and N will subsequently be larger. As a practical example, we consider a reflection with an average of 10,000 counts per second, counted over a 100 second time period. The time increments will need to be on the order of 10^{-6} seconds, making N on the order of 10^8. Numbers of this magnitude make the binomial distribution function unwieldy, and we seek a more convenient representation in order to determine an expression for the variance and standard deviation of our intensity measurements. In order to accomplish this we must determine an expression for the variance of the binomial distribution, and this in turn requires us to simplify the expression for the mean that we derived

*The average value is alternately referred to as the *mean value* or the *expected value*, and is denoted either with a bar over the variable, e.g., $\bar{x}$ — or its enclosure in bra-kets, e.g, $\langle x \rangle$. Both notations will be used throughout the text.

above. Both of these tasks require a bit of trickery.† To develop the expression for $\bar{n}$ we begin by differentiating both sides of Eqn. 5.10 with respect to p and rearranging the results:

$$\sum_{n=0}^{N} C_n^N (np^{(n-1)}(1-p)^{(N-n)} - (N-n)p^n(1-p)^{(N-n-1)}) = 0$$

$$\sum_{n=0}^{N} C_n^N np^{(n-1)}(1-p)^{(N-n)} = \sum_{n=0}^{N} C_n^N (N-n)p^n(1-p)^{(N-n-1)}$$

$$\sum_{n=0}^{N} nC_n^N (p^{(n-1)}(1-p)^{(N-n)} + p^n(1-p)^{(N-n-1)}) = N\sum_{n=0}^{N} C_n^N p^n(1-p)^{(N-n-1)}.$$

Multiplying both sides by $p(1-p)$:

$$\sum_{n=0}^{N} nC_n^N ((1-p)p^n(1-p)^{(N-n)} + pp^n(1-p)^{(N-n)}) = Np\sum_{n=0}^{N} C_n^N p^n(1-p)^{(N-n)}.$$

The expression on the right is Np times the sum of all the probabilities (Eqn. 5.10) $= Np \times 1$. The terms in the expression on the left combine to give $\bar{n}$:

$$\sum_{n=0}^{N} nC_n^N p^n(1-p)^{(N-n)} = Np \tag{5.12}$$

$$\bar{n} = Np. \tag{5.13}$$

This simple and elegant result makes intuitive sense. The average number of events (counts) is the probability of observing an event in a single trial times the number of trials. To obtain the variance, which is the average squared deviation, we return to our collection of W arrays, and determine the number of times each $(n-\bar{n})^2$ occurs. This is, of course, the same number as the number of occurrences of n, $f_{N,p}(n)W$. The variance is then the sum of the squared deviations for each array, divided by the number of arrays:

$$\sigma^2 = \frac{\sum_{n=0}^{N}(n-\bar{n})^2 f_{N,p}(n)W}{W} = \sum_{n=0}^{N}(n-\bar{n})^2 f_{N,p}(n) \tag{5.14}$$

$$= \sum_{n=0}^{N}(n^2 - 2n\bar{n} + \bar{n}^2) f_{N,p}(n).$$

$$= \sum_{n=0}^{N} n^2 f_{N,p}(n) - 2\bar{n}\sum_{n=0}^{N} n f_{N,p}(n) + \bar{n}^2\sum_{n=0}^{N} f_{N,p}(n).$$

As with $\bar{n}$, which is a weighted average of the number of counts (n), the variance is a weighted average of the squared deviations for each n. Substituting $\bar{n} = Np$

†The derivations of $\bar{n}$ and σ that follow are modified versions of those found in Appendix B of the excellent book on experimental statistics by H.D. Young[63].

and noting that the second sum is $\bar{n}$ while the third sum is unity (the sum of all the probabilities) gives

$$\sigma^2 = \sum_{n=0}^{N} n^2 f_{N,p}(n) - 2N^2p^2 + N^2p^2 = \sum_{n=0}^{N} n^2 f_{N,p}(n) - N^2p^2. \tag{5.15}$$

We now differentiate Eqn. 5.12 with respect to p, multiply each side by $p(1-p)$, and rearrange the equation so that the expression for σ^2 in Eqn. 5.15 is on the left:

$$\sum_{n=0}^{N} nC_n^N (np^{(n-1)}(1-p)^{(N-n)} - (N-n)p^n(1-p)^{(N-n-1)} = N$$

$$\sum_{n=0}^{N} n^2C_n^N p^n(1-p)^{(N-n)} - Np\sum_{n=0}^{N} nC_n^N p^n(1-p)^{(N-n)} = Np(1-p)$$

The terms in the first sum are $n^2 f_{N,p}(n)$ and the second sum is the mean, $\bar{n} = Np$. Thus

$$\sum_{n=0}^{N} n^2 f_{N,p}(n) = N^2p^2 + Np(1-p),$$

$$\sum_{n=0}^{N} n^2 f_{N,p}(n) - N^2p^2 = Np(1-p) \text{ and}$$

$$\sigma^2 = Np(1-p) \Longrightarrow \tag{5.16}$$

$$\sigma = \sqrt{Np(1-p)}. \tag{5.17}$$

Now that we have obtained simple expressions for the mean and standard deviation of the binomial distribution, we return to the practical example that we began this section with. As with the "low count" experiment, we count for W intervals of 100 seconds per interval and determine (experimentally) that the average number of counts per second for an interval is 10,000. Thus $\bar{n}$ for the experiment is $100 \times 10000 = 10^6$ counts. In order to divide the 100 second interval into increments (trials) small enough that the probability of a photon being counted in a given trial is much less than 1 we select a microsecond ($10^{-6}s$) as the time increment for a single trial. Thus the number of trials will be $N = 100 \times 10^6 = 10^8$. Since $Np = \bar{n}$, $p = 0.01$; the probability that a count will be observed in a single trial is $1/100$.

The binomial distribution now becomes

$$\begin{aligned} f_{N,p}(n) &= C_n^N (0.01)^n (0.99)^{(N-n)} \\ &= \frac{N(N-1)(N-2)\dots(N-n+1)}{n!}(0.01)^n(0.99)^{(N-n)} \end{aligned}$$

The mean value for n is 10^6 and we expect that the values of n that will produce significant probabilities will be of the same order of magnitude, approximately centered around the mean value. Furthermore, since $p \ll 1$, p^n will approach zero rapidly as n gets very much larger than its mean value, just as we observed in the "low count" experiment (Fig. 5.2). In the numerator in the expression for C_n^N the largest term is $N = 10^8$, and the smallest term is $N-n+1 = 10^8-(\sim 10^6)+1 \approx 10^8$.

Thus each of the n terms in the product can be replaced by N and the product becomes N^n. Similarly, the exponent $(N-n)$ can be replaced by N. This result is general when p is small and N is large, and the probability distribution function becomes

$$f_{N,p}(n) = \frac{N^n}{n!} p^n (1-p)^N = \frac{(Np)^n}{n!} (1-p)^N. \tag{5.18}$$

Substituting $\bar{n} = Np$ and $N = \bar{n}/p$,

$$f_{N,p}(n) = \frac{(\bar{n})^n}{n!} (1-p)^{\bar{n}/p} = \frac{(\bar{n})^n}{n!} [(1-p)^{(1/p)}]^{\bar{n}}. \tag{5.19}$$

Note that the equation is now independent of N and we can select p as small as we like (by selecting shorter time increments for each trial). The only expression containing p is $(1-p)^{(1/p)}$. As p gets smaller we note that this expression is one commonly evaluated in the theory of limits in elementary calculus:

$$\lim_{p \to 0} (1-p)^{(1/p)} = e^{-1} \implies \tag{5.20}$$

$$f_{\bar{n}}(n) = \frac{(\bar{n})^n e^{-\bar{n}}}{n!}. \tag{5.21}$$

This limiting form of the binomial distribution, which depends only on $\bar{n} = Np$ instead of N and p independently, is known as the *Poisson distribution.* The variance and standard deviation for this distribution follow from the variance of the binomial distribution and the small magnitude of p (so that $(1-p) \approx 1$):

$$\sigma^2 = Np(1-p) = \bar{n}(1-p) = \bar{n} \text{ and} \tag{5.22}$$

$$\sigma = \sqrt{\bar{n}}. \tag{5.23}$$

The variance of the Poisson distribution is the mean and the standard deviation is the square root of the mean!

Just as we did with the binomial distribution, to get the probability that some value of n will be observed in a specified region, we add up all of the probabilities in that region. However, for large numbers of events the values of n are very close together (since $\bar{n}$ is so large) and often we can (must!) treat n as a continuous variable, $n = x$. This presents us with a bit of a dilemma. While it is reasonable to ascertain the probability of observing a number, n, as the fraction of the total number of observations — with an infinite number of continuous variables, x, the probability of observing one of those variables *exactly* is meaningless.

To resolve this dilemma we consider a hypothetical experiment: A beam of photons is passed through a small hole, impinging on a wire of infinite length at a distance away from the hole, perpendicular to the direction of the incident beam. The photons spread out as they pass from the hole to the wire, with the largest number of photons arriving at a point on the wire along a line in the direction of the incident photons. The number of photons striking the wire diminishes in each direction away from this point and gradually falls to zero. If we allow the wire to be represented by the x axis, then, while it is inappropriate to ask for the probability of observing a photon at some point along the wire, x_i, it is perfectly reasonable to ask for the probability of observing a photon *somewhere* within an interval along the wire between x_i and some other point, x_j. Indeed, in a specified time increment, the probability is simply the number of photons that fall within the interval, divided by the total number of photons that strike the wire.

To make the probability determination general, we investigate how this fraction varies as we move from one point to another along the wire. To accomplish this we employ a detector with a small width, w_d, to count photons. Locating the detector at point x_i and counting for a set time interval, we obtain m_i, the number of photons that fall within the width of the detector along the x axis at that point. Dividing m_i by w_d gives the number of photons per unit length (the *number density*) that contact the wire at (or near) x_i:

$$\rho(x_i) = \frac{m_i}{w_d}. \tag{5.24}$$

The *fraction* of the number of photons that strike the detector is m_i/n_T, where n_T is the total number of photons that contact the wire in the same time interval. The fraction of the number of photons per unit length at x_1 (the *fractional density*) is therefore

$$f(x_i) = \frac{m_i}{n_T\, w_d}, \tag{5.25}$$

where $f_(x_i)$ represents the probability that a photon will be observed within an interval of unit length at(or near) x_i, and is therefore called a *probability density.*

We now divide the wire into equal increments of $\Delta\, x$, such that x_i and x_j each lie on the left side of one of the increments (for convenience). If $\Delta\, x$ is relatively large, then the number of counts observed with the detector will vary at different locations within the interval. However, if $\Delta\, x$ is small enough, we can assume that the number of counts measured by the detector at any location within an interval will be about the same, i.e., the probability density within an interval will remain approximately constant.

The probability that a photon will be observed in the interval between x and $x + \Delta\, x$ is the fraction of the number of photons that strike the wire per unit length in the interval times the length of the interval:

$$Pr(x_i) \approx f(x_i)\Delta\, x. \tag{5.26}$$

Similarly, the probability that a photon will be observed in the interval between x_{i+1} and $x_{i+1} + \Delta\, x$ is

$$Pr(x_{i+1}) \approx f(x_{i+1})\Delta\, x. \tag{5.27}$$

The fraction of the total number of photons that strike the wire in the region between x_i and x_j is the sum of the fractions for each interval:

$$Pr(x_i \cdots x_j) \approx \sum_{k=i}^{j} f(x_k)\, \Delta x. \tag{5.28}$$

The error in this approximation lies in the assumption that $f_x(k)$ remains constant throughout the interval. The approximation is improved by making $\Delta\, x$ smaller and smaller, and in the limit, as $\Delta\, x \longrightarrow dx$, the probability expression becomes exact:

$$Pr(x_i \cdots x_j) = \int_{x_i}^{x_j} f(x)\, dx. \tag{5.29}$$

Integrating the probability of finding a photon over the entire wire results in

$$Pr(-\infty \cdots \infty) = \int_{-\infty}^{\infty} f(x)\, dx = 1, \tag{5.30}$$

a necessary condition for a *probability density function.* The probability of finding a photon between a specific x_1 and $x_1 + dx$ is the probability per unit distance times the distance:

$$Pr_{x_1} = f(x_1)\, dx. \tag{5.31}$$

We now return to the Poisson probability function. In order to deal with $n!$, which is a discrete variable that becomes gigantic as n takes on even moderate levels (10000! is a number well beyond the limits of any computer to handle), we approximate it in Eqn. 5.21 with Stirling's approximation[64] for $n!$:

$$n! \approx n^n e^{-n} \sqrt{2\pi n} \quad \text{and, setting } n = x, \tag{5.32}$$

$$f_P(x) = \frac{1}{\sqrt{2\pi x}} \left(\frac{\bar{x}}{x}\right)^x e^{-(\bar{x}-x)}. \tag{5.33}$$

The subscript "P" has been added to indicate that the probability belongs to an effectively continuous Poisson probability distribution. Fig. 5.3 shows a plot of $f_P(x)$ (in black) for $\bar{x} = 5$. The discrete Poisson distribution for $\bar{n} = 5$ is shown in red. The probability for the discrete distribution in the region between 3 and 8 would be

$$Pr(n = 3 \cdots n = 8) = \sum_{k=3}^{5} f_P(n_k), \; n_k = k. \tag{5.34}$$

For the continuous distribution, since $\Delta x = 1$, the probability is approximated by

$$Pr(x = 3 \cdots x = 8) \approx \sum_{k=3}^{5} f_P(x_k)\Delta x = \sum_{k=3}^{5} f_P(x_k), \; x_k = k. \tag{5.35}$$

In the limit,

$$Pr(x = 3 \cdots x = 8) = \int_{3}^{5} f_P(x)\, dx. \tag{5.36}$$

In the general case the integral becomes:

$$Pr(x_1 \cdots x_2) = \int_{x_1}^{x_2} f_P(x)\, dx = \int_{x_1}^{x_2} \frac{1}{\sqrt{2\pi x}} \left(\frac{\bar{x}}{x}\right)^x e^{-(\bar{x}-x)}\, dx \tag{5.37}$$

As expected, the sum of all the probabilities is unity:*

$$Pr(-\infty \cdots \infty) = \int_{-\infty}^{\infty} f_P(x)\, dx = 1, \tag{5.38}$$

illustrating clearly that $f_P(x)$ is a probability density function. See Sec. 3.2.2 for a discussion of the electron density function as a concrete example of another probability density function.

*The integrals in Eqns. 5.38–5.41 are readily evaluated numerically between $\bar{n} - 5\sigma$ and $\bar{n} + 5\sigma$ since the probabilities are negligible outside of those limits.

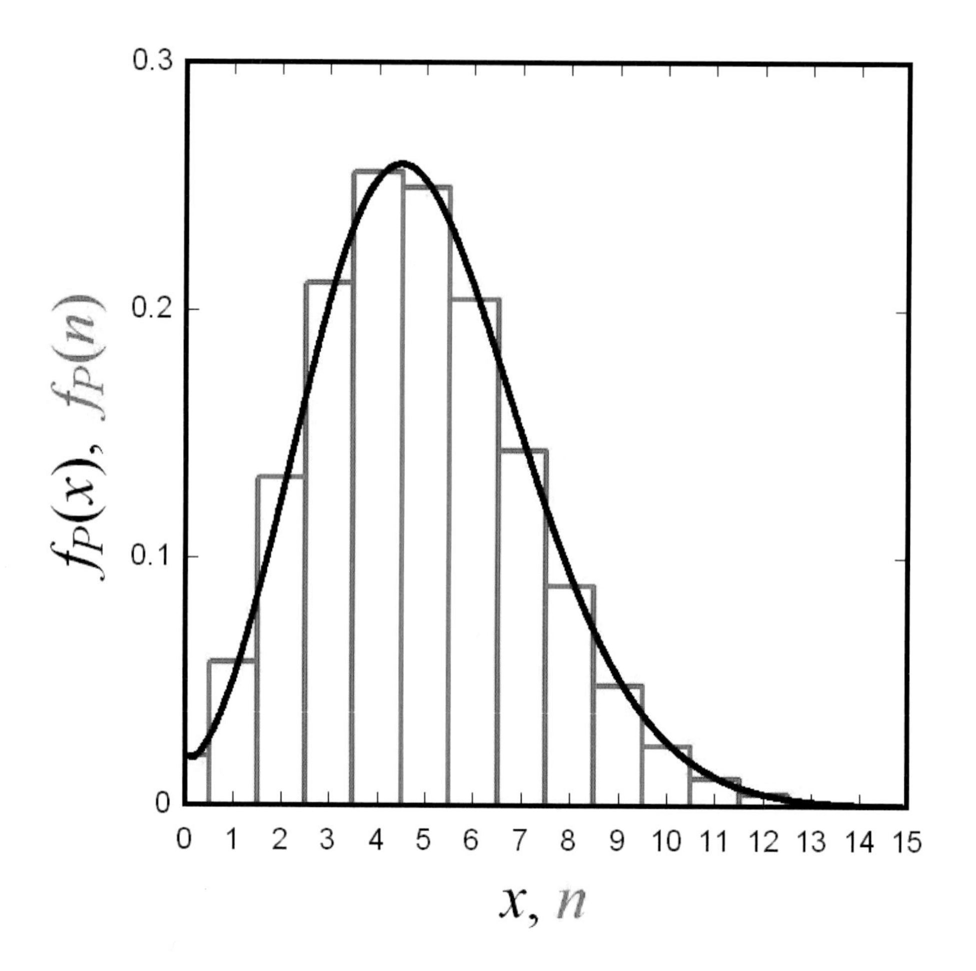

Figure 5.3 Black: Continuous Poisson probability distribution for $\bar{x} = 5$ Red: Discrete Poisson probability distribution for $\bar{n} = 5$.

The average value of a function of x is the sum of each value of the function for each x, weighted by the probability of observing that value of x, and therefore the probability of observing the corresponding value of the function:

$$\langle g(x) \rangle = \int_{-\infty}^{\infty} g(x) f_P(x)\, dx. \tag{5.39}$$

The mean value of x and variance of the Poisson distribution are the weighted averages of x and the squared deviations, respectively:

$$\bar{x} = \int_{-\infty}^{\infty} x f_P(x)\, dx \tag{5.40}$$

$$\sigma^2 = \int_{-\infty}^{\infty} (x - \bar{x})^2 f_P(x)\, dx \tag{5.41}$$

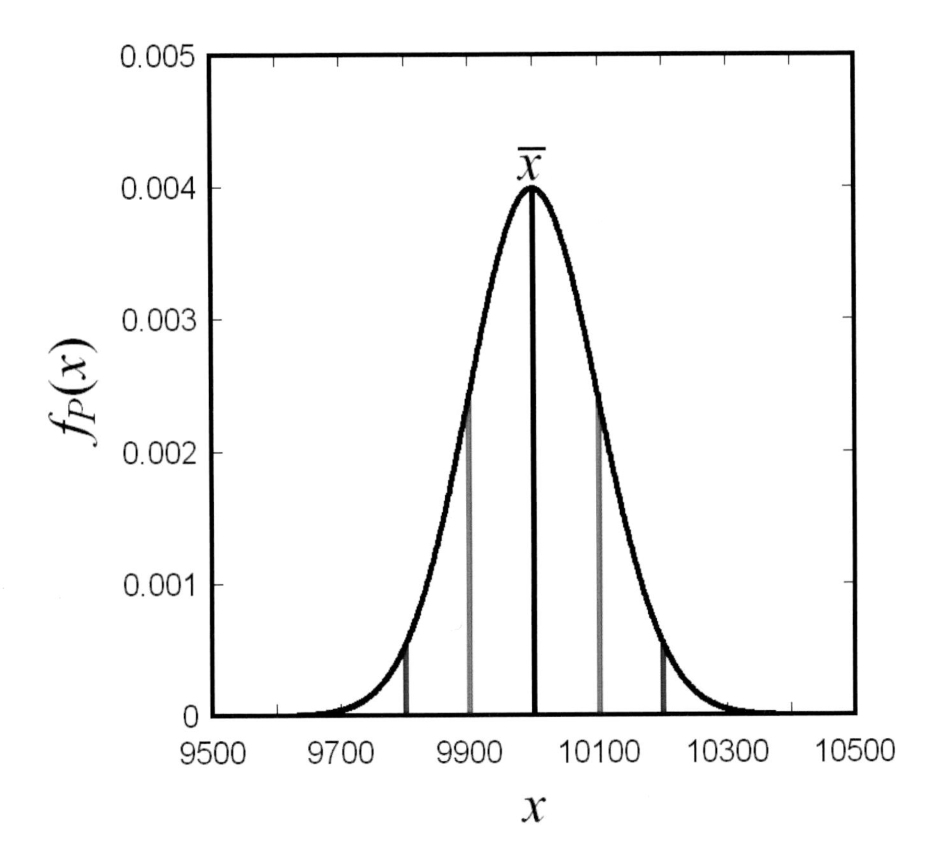

Figure 5.4 Poisson probability distribution for $\bar{x} = 10,000$. The red lines indicate the region of probabilities for values of x that lie within one standard deviation of the mean. The blue lines mark the region of probabilities for values of n that lie within two standard deviations of the mean.

The Poisson distribution for $\bar{n} = \bar{x} = 10,000$ is illustrated in Fig. 5.4. The standard deviation is 100 and the red bars indicate the portion of the probabilities that lie within one standard deviation of $\bar{x}$. The blue lines indicate all of the probabilities that lie within two standard deviations, and nearly all of the probabilities lie within three standard deviations of the mean. The integral over the interval $\bar{x} - \sigma$ to $\bar{x} + \sigma$ has a value of 0.683. A large number of identical counting experiments would yield intensities between 9900 and 10100 counts 68.3% of the time. The integral between $\bar{x} - 2\sigma$ to $\bar{x} + 2\sigma$ is 0.954; 95.4% of the measurements would yield intensities between 9800 and 10200 counts. The integral encompassing $\pm 3\sigma$ has a value of 0.997 and it is almost certain that any intensity measured will fall between 9700 and 10300 counts. The limits $\pm c\sigma$ are often referred to as *confidence limits*. The fractions above depend on the nature of the distribution, and will remain the same for all values of $\bar{n}$ (except when $\bar{n}$ gets very small — well out of the range of the counts recorded in the diffraction experiment).

What does this mean for the actual counting of photons? To obtain the actual standard deviation we still need the mean value, which we can obtain only from repeated measurements. However, to get an *estimate* of the standard deviation, we can assume that the observed intensity is close to its mean value, at least close enough for its square root to provide a reasonable estimate of its standard deviation. For a given intensity, I_{hkl}, we assume that $I_{hkl} \approx \bar{I}_{hkl}$, that $\sigma^2(I_{hkl}) \approx I_{hkl}$ and that $\sigma(I_{hkl}) \approx \sqrt{I_{hkl}}$. *The estimated standard deviation of the intensity is the square root of the intensity.* We will find this very useful in the determination of weighting schemes for the solution and refinement of crystal structures.

For the actual measurement of intensities, the standard deviation has often been used to decide which reflections are too weak to be considered observed (referred to as *unobserved reflections* or *less-thans*). These are reflections that are flagged as unreliable because their measured intensities are close to the standard deviations of those intensities. In essence, their measured intensities are smaller than the uncertainty in those intensities. We can be fairly confident that the measured intensity of a reflection will lie within two standard deviations of its "actual value" (the average value obtained from a large number of repeated measurements), and almost certain that it will lie within three standard deviations. Those reflections that have intensities that are less than the confidence limits set by the experimentalist, usually somewhere between $\sigma(I)$ and $3\sigma(I)$, are often considered to be unobserved (See Sec. 8.2 for a discussion).

The Gaussian Distribution

The reader familiar with the *Gaussian* (*normal*) distribution may have noted that the Poisson distribution plotted in Fig. 5.4 appears very similar to a plot of the Gaussian distribution, and that the probabilities integrate to the same fractions. For smaller values of $\bar{n}$ the Poisson distribution is decidedly skewed (e.g., Fig. 5.3), but as $\bar{n}$ increases, the probability of observing small numbers of events becomes less and less likely, and the Poisson distribution becomes effectively indistinguishable from the Gaussian distribution. As with the Poisson distribution, the Gaussian distribution is a limiting approximation to the binomial distribution for large values of N — but it is symmetric about the mean, while the Poisson distribution need not be. The Poisson distribution describes the probabilities for events that are limited on one side of the mean value, and not on the other. Counting statistics follow this distribution since there is an absolute limit of zero counts on the low end of the distribution, but no absolute limit on the high end. Many other physical events, however, are distributed equally about the mean value, and are more appropriately described by the Gaussian distribution. We will make use of the Gaussian probability distribution on several occasions throughout the remainder of this chapter and it will provide the basis for the statistical solution of the phase problem – the subject of Chapter 7.

To develop the Gaussian distribution we return to the binomial distribution, giving the probability that n events will occur in N trials:

$$
\begin{aligned}
f_{N,p}(n) &= C_n^N p^n (1-p)^{(N-n)} \\
&= \frac{N \times (N-1) \times \ldots \times (N-n+2) \times (N-n+1)}{n!} p^n q^{(N-n)} \\
&= \frac{N!}{n!(N-n)!} p^n q^{(N-n)}.
\end{aligned} \tag{5.42}
$$

As before, p is the probability of observing an event in a single trial and $q = (1-p)$ is the probability of *not* observing the event in a single trial; $q + p = 1$. The variance of the distribution is $\sigma^2 = Np(1-p) = Npq$ and the mean value of the number of events observed is $\bar{n} = Np$. We wish to evaluate the distribution as N gets very large, and must therefore approximate the large factorials using Stirling's approximation (Eqn. 5.32):

$$
\begin{aligned}
N! &\approx N^N e^{-N} \sqrt{2\pi N} = N^{(N+\frac{1}{2})} e^{-N} \sqrt{2\pi}, \quad \text{etc.}, \\
C_n^N &= \frac{N!}{n!(N-n)!} \\
&= \frac{N^{(N+\frac{1}{2})} e^{-N} \sqrt{2\pi}}{(n^{(n+\frac{1}{2})} e^{-n} \sqrt{2\pi})((N-n)^{((N-n)+\frac{1}{2})} e^{-(N-n)}) \sqrt{2\pi})} \\
&= \frac{1}{\sqrt{2\pi}} \frac{N^{(N+\frac{1}{2})} e^{-N}}{n^{(n+\frac{1}{2})} (N-n)^{(N-n+\frac{1}{2})} e^{-n-N+n}} \\
&= \frac{1}{\sqrt{2\pi}} \frac{N^{(N+\frac{1}{2})}}{n^{(n+\frac{1}{2})} (N-n)^{(N-n+\frac{1}{2})}}.
\end{aligned} \tag{5.43}
$$

Multiplying the expression by $N^{\frac{1}{2}}/N^{\frac{1}{2}}$ results in

$$
C_n^N = \frac{1}{\sqrt{2\pi\, N}} \left(\frac{N}{n}\right)^{(n+\frac{1}{2})} \left(\frac{N}{N-n}\right)^{(N-n+\frac{1}{2})}. \tag{5.44}
$$

To put $p^n q^{(N-n)}$ in the same form we multiply it by $(pq)^{\frac{1}{2}}/(pq)^{\frac{1}{2}}$, giving

$$
p^n q^{(N-n)} = \frac{p^{(n+\frac{1}{2})}}{\sqrt{p}} \frac{q^{(N-n+\frac{1}{2})}}{\sqrt{q}}. \tag{5.45}
$$

Substituting Eqns. 5.44 and 5.45 into Eqn. 5.42 provides the expression that we will use to determine the probability distribution:

$$
\begin{aligned}
f_{N,p}(n) &= \frac{1}{\sqrt{2\pi\, Npq}} \frac{p^{(n+\frac{1}{2})}}{(n/N)^{(n+\frac{1}{2})}} \frac{q^{(N-n+\frac{1}{2})}}{((N-n)/N)^{(n+\frac{1}{2})}} \\
&= \frac{1}{\sqrt{2\pi\, Npq}} \left(\left(\frac{n}{Np}\right)^{(n+\frac{1}{2})}\right)^{-1} \left(\left(\frac{N-n}{Nq}\right)^{(N-n+\frac{1}{2})}\right)^{-1}.
\end{aligned} \tag{5.46}
$$

Fortunately, this expression can be simplified by taking the logarithm of both sides and defining a new variable, $s = (n - Np)/\sqrt{Npq}$ that will allow us to approximate the resulting logarithms with power series expansions:

$$\ln\left(f_{N,p}(n)\right) = -\ln\left(\sqrt{2\pi\, Npq}\right) - \left(n + \frac{1}{2}\right)\ln\left(\frac{n}{Np}\right) - \left(N - n + \frac{1}{2}\right)\ln\left(\frac{N-n}{Nq}\right). \tag{5.47}$$

We now describe n, $N - n$, and the arguments of the logarithms in terms of s:

$$n = Np + s\sqrt{Npq} \tag{5.48}$$

$$\frac{n}{Np} = 1 + s\frac{\sqrt{Npq}}{Np} = 1 + s\sqrt{\frac{q}{Np}} \tag{5.49}$$

$$N - n = N - Np - s\sqrt{Npq} = N(1-p) - s\sqrt{Npq} = Nq - s\sqrt{Npq} \quad \text{and} \tag{5.50}$$

$$\frac{N-n}{Nq} = 1 - \frac{s\sqrt{Npq}}{Nq} = 1 - s\sqrt{\frac{p}{Nq}}. \tag{5.51}$$

Since we are examining the case where N is very large, $s\sqrt{q/Np}$ and $s\sqrt{p/Nq}$ are assumed to be less than 1. For situations where $|z| < 1$, $\ln(1+z)$ can be expanded as a Taylor series (Appendix I):

$$\ln(1+z) = z - \frac{1}{2}z^2 + \frac{1}{3}z^3 - \frac{1}{4}z^4 + \cdots. \tag{5.52}$$

Thus,

$$\ln\left(\frac{n}{Np}\right) = s\sqrt{\frac{q}{Np}} - s^2\frac{1}{2}\frac{q}{Np} \quad \text{and} \tag{5.53}$$

$$\ln\left(\frac{N-n}{Nq}\right) = -s\sqrt{\frac{p}{Nq}} - s^2\frac{1}{2}\frac{p}{Nq}. \tag{5.54}$$

The higher-order terms have been ignored here since they contain increasing powers of N in the denominator, and are negligible in comparison with the first two terms, since N is large. Substituting Eqns. 5.48, 5.50, 5.53, and 5.54 into Eqn. 5.47,

$$\ln\left(f_{N,p}(n)\right) = -\ln\left(\sqrt{2\pi\, Npq}\right) - \left(Np + s\sqrt{Npq} + \frac{1}{2}\right)\left(s\sqrt{\frac{q}{Np}} - s^2\frac{1}{2}\frac{q}{Np}\right) - \left(Nq - s\sqrt{Npq} + \frac{1}{2}\right)\left(-s\sqrt{\frac{p}{Nq}} - s^2\frac{1}{2}\frac{p}{Nq}\right). \tag{5.55}$$

Expanding the expression and collecting terms inside the square roots gives

$$\ln\left(f_{N,p}(n)\right) = -\ln\left(\sqrt{2\pi\, Npq}\right) - s\sqrt{Npq} + \frac{1}{2}s^2q - s^2q + \frac{1}{2}s^3\sqrt{\frac{q^3}{Np}} + \frac{1}{2}s\sqrt{\frac{q}{Np}} - \frac{1}{4}s^2\frac{q}{Np} + s\sqrt{Npq} + \frac{1}{2}s^2p - s^2p - \frac{1}{2}s^3\sqrt{\frac{p^3}{Nq}} + \frac{1}{2}s\sqrt{\frac{p}{Nq}} + \frac{1}{4}s^2\frac{p}{Nq}. \tag{5.56}$$

Since N is large, all of the terms containing it in the denominator are negligible in comparison with those that don't; the remaining coefficients of s cancel one another. The equation reduces to

$$\begin{aligned} \ln\left(f_{N,p}(n)\right) &= -\ln\left(\sqrt{2\pi\, Npq}\right) - \frac{1}{2}s^2(p+q) \\ &= -\ln\left(\sqrt{2\pi\, Npq}\right) - \frac{1}{2}s^2, \end{aligned} \tag{5.57}$$

since $(p+q) = 1$. Substituting the expression for s back into the equation yields

$$\ln\left(f_{N,p}(n)\right) = -\ln\left(\sqrt{2\pi\, Npq}\right) - \frac{1}{2}\frac{(n-Np)^2}{Npq}. \tag{5.58}$$

The variance of the distribution is $\sigma^2 = Npq$, the standard deviation is $\sigma = \sqrt{Npq}$, and the mean is $\bar{n} = Np$. Substituting and exponentiating both sides gives

$$\ln\left(f_{N,p}(n)\right) = -\ln\left(\sigma\sqrt{2\pi}\right) - \frac{1}{2}\frac{(n-\bar{n})^2}{\sigma^2} \quad \text{and} \tag{5.59}$$

$$f_{N,p}(n) = \frac{1}{\sigma\sqrt{2\pi}}\, e^{-(n-\bar{n})^2/(2\sigma^2)}. \tag{5.60}$$

The magnitude of N renders n essentially continuous, and as with the Poisson distribution, we can replace n with the continuous variable x:

$$f_G(x) = \frac{1}{\sigma\sqrt{2\pi}}\, e^{-(x-\bar{x})^2/(2\sigma^2)}. \tag{5.61}$$

The subscript "G" has been employed to indicate a Gaussian probability distribution.

Comparing Eqn. 5.21 with Eqn. 5.61 reveals the major difference in the Poisson and Gaussian distributions. For a given x less than $\bar{x}$, the value of x' symmetrically disposed on the other side of $\bar{x}$ is $x' = \bar{x} + (\bar{x} - x) = 2\bar{x} - x$. The exponent in the Poisson function is $(\bar{x} - x)$, $(\bar{x} - x') = -(\bar{x} - x)$ — and it follows that $f_P(x) \neq f_P(x')$. The other terms in the Poisson distribution are also asymmetric with respect to x, but with some significant effort it can be shown that the Poisson distribution becomes identical to the Gaussian distribution in the limit (compare the curves for $\sigma = 100$ in Fig. 5.4 and Fig. 5.5). The exponent in the Gaussian function contains $(x - \bar{x})$, the only term containing the independent variable, and again, $(x' - \bar{x}) = -(x - \bar{x})$. However, the term is squared in the Gaussian function making $(x' - \bar{x})^2 = (x - \bar{x})^2$, and $f_G(x) = f_G(x')$; the Gaussian distribution is symmetric and centered on its mean value. Because of its relative simplicity, the Gaussian distribution is often employed as an excellent approximation in cases where a rigorous treatment requires Poisson statistics.

Fig. 5.5 illustrates two Gaussian distributions with different standard deviations. One distribution has the same mean (10000) and standard deviation (σ=100) as the Poisson distribution in Fig. 5.4, while the other distribution has the same mean, but a larger standard deviation (σ=175), i.e., a higher probability that values of x will occur farther from the mean. Qualitatively, both distributions appear to have the same area under their respective curves, but in order to establish that $f_G(x)$ is a general probability distribution, we must demonstrate that it integrates to unity for any value of the standard deviation.

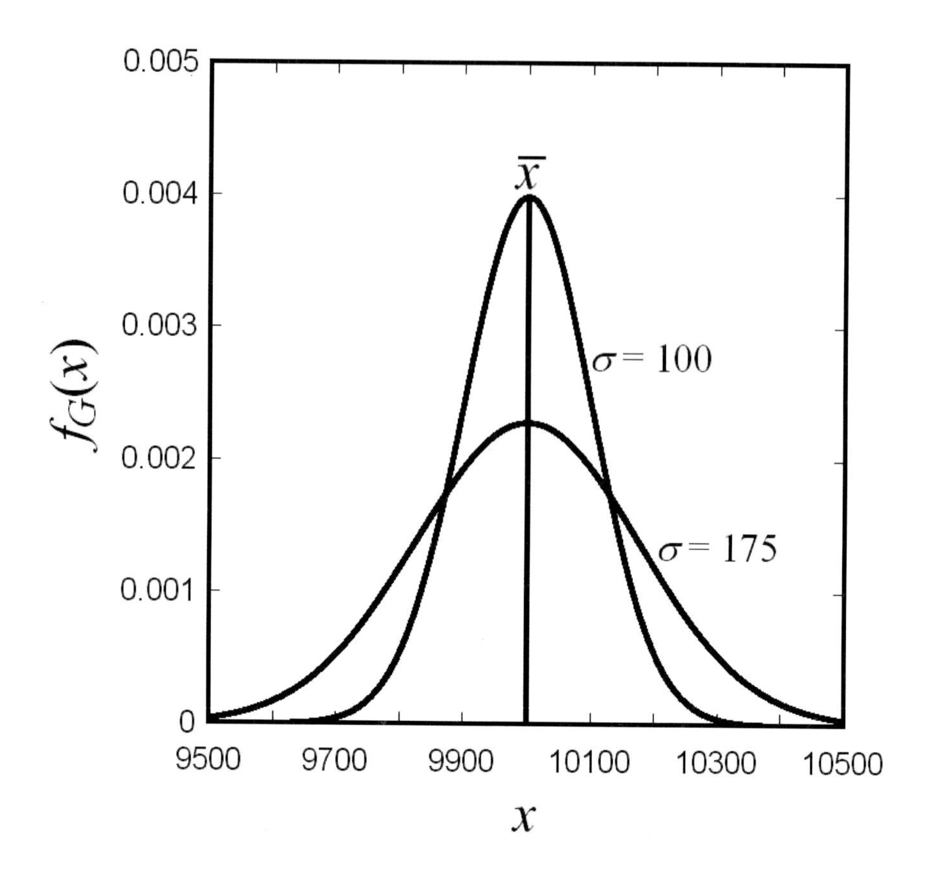

Figure 5.5 Gaussian probability distribution for $\bar{x} = 10,000$ for two different standard deviations.

The exponential in Eqn. 5.61 is of the form $e^{-a^2y^2}$, where $a = 1/(\sigma\sqrt{2})$ and $y = (x - \bar{x})$. To evaluate the integral we first consider a simpler one:

$$I(u) = \int_{-\infty}^{\infty} e^{-u^2}\, du \quad \text{and} \quad I(v) = \int_{-\infty}^{\infty} e^{-v^2}\, dv. \tag{5.62}$$

Since the variables cover the same range, the integrals are identical, and

$$I^2 = I(u)I(v) = \int_{v=-\infty}^{\infty}\int_{u=-\infty}^{\infty} e^{-(u^2+v^2)}\, dudv. \tag{5.63}$$

If the variables are expressed in polar coordinates, $u = r\cos\theta$,$v = r\sin\theta$ and $u^2 + v^2 = r^2$. The differential in the plane in polar coordinates is $dudv = r\,dr\,d\theta$, and the integral over the xy $(r\theta)$ plane is

$$\begin{aligned} I^2 &= \int_{r=0}^{\infty}\int_{\theta=0}^{2\pi} e^{-r2} r\,dr\,d\theta = \int_{r=0}^{\infty}\left(\int_{\theta=0}^{2\pi} d\theta\right) e^{-r2} r\,dr \\ &= 2\pi \int_{r=0}^{\infty} e^{-r^2} r\,dr. \end{aligned}$$

Setting $z = r^2 \Rightarrow dz = 2dr$,

$$I^2 = 2\pi \int_{r=0}^{\infty} \frac{1}{2} e^{-z}\, dz = 2\pi \frac{1}{2}(1) = \pi \quad \text{and}$$
$$I = \int_{-\infty}^{\infty} e^{-u^2}\, du = \sqrt{\pi}. \tag{5.64}$$

In the Gaussian integral, $u = ay$ and $du = a\,dy$:

$$\int_{-\infty}^{\infty} e^{-a^2y^2} a\,dy = \sqrt{\pi} \quad \text{and} \quad \int_{-\infty}^{\infty} e^{-a^2y^2}\, dy = \frac{1}{a}\sqrt{\pi}. \tag{5.65}$$

Since $dx = dy$,

$$\int_{-\infty}^{\infty} e^{-(x-\bar{x})^2/(2\sigma^2)}\, dx = \sigma\sqrt{2}\sqrt{\pi} = \sigma\sqrt{2\pi} \quad \text{and}$$
$$\int_{-\infty}^{\infty} f_G(x)\, dx = \frac{1}{\sigma\sqrt{2\pi}} \int_{-\infty}^{\infty} e^{-(x-\bar{x})^2/(2\sigma^2)}\, dx = 1. \tag{5.66}$$

Thus the sum of the probabilities is unity for any value of σ, and the Gaussian function is indeed a general probability density function. The Gaussian probability distribution is commonly referred to as a *normal distribution*, and variables that are characterized by a Gaussian distribution are said to be *normally distributed*.

The Error Function. As with the Poisson integral, the probability of observing a value of x between two limits is obtained by integrating the Gaussian integral between the limits:

$$Pr(x_j \le x \le x_k) = \int_{x_j}^{x_k} f_G(x)\, dx = \frac{1}{\sigma\sqrt{2\pi}} \int_{x_j}^{x_k} e^{-(x-\bar{x})^2/(2\sigma^2)}\, dx, \tag{5.67}$$

illustrated in Fig. 5.6(a). While the Gaussian integral can be evaluated from $x = -\infty$ to $x = \infty$, it cannot be evaluated analytically between finite limits. Explicit integrals depend on σ and $\bar{x}$, and while it is possible to determine specific probabilities by numerical integration, it is a cumbersome process that can be avoided by combining σ and $\bar{x}$ into a single variable and recasting the Gaussian integral into a function of that variable. Specific values of the function can then be computed numerically and tabulated for use in the determination of probabilities between specified limits. The function is known as the *error function*, $\mathrm{erf}(t_k)$, where t_k is determined by x_k, σ and $\bar{x}$.

The error function is developed by considering the special case in which the integration limits are symmetrically disposed about the mean, as illustrated in Fig. 5.6(b). Substituting $x' = x - \bar{x}$ effectively shifts the integral so that it is symmetrically disposed about the origin: $x'_k = x_k - \bar{x}$, $\bar{x}' = \bar{x} - \bar{x} = 0$, and $x'_j = x_j - \bar{x} = -x'_k$, illustrated in Fig. 5.6(c). Since $dx'/dx = 1$, $dx = dx'$, and

$$Pr(x_j \le x \le x_k) = \frac{1}{\sqrt{\pi}(\sigma\sqrt{2})} \int_{-x'_k}^{x'_k} e^{-(x'/(\sigma\sqrt{2}))^2}\, dx'. \tag{5.68}$$

A second change of variables produces the error function. Defining $t = x'/(\sigma\sqrt{2}) = (x - \bar{x})/(\sigma\sqrt{2})$ yields $dt/dx' = 1/(\sigma\sqrt{2}) \Rightarrow dx' = \sigma\sqrt{2}\, dt$, and

$$Pr(x_j \le x \le x_k) = \frac{1}{\sqrt{\pi}} \int_{-t_k}^{t_2} e^{-t^2}\, dt. \tag{5.69}$$

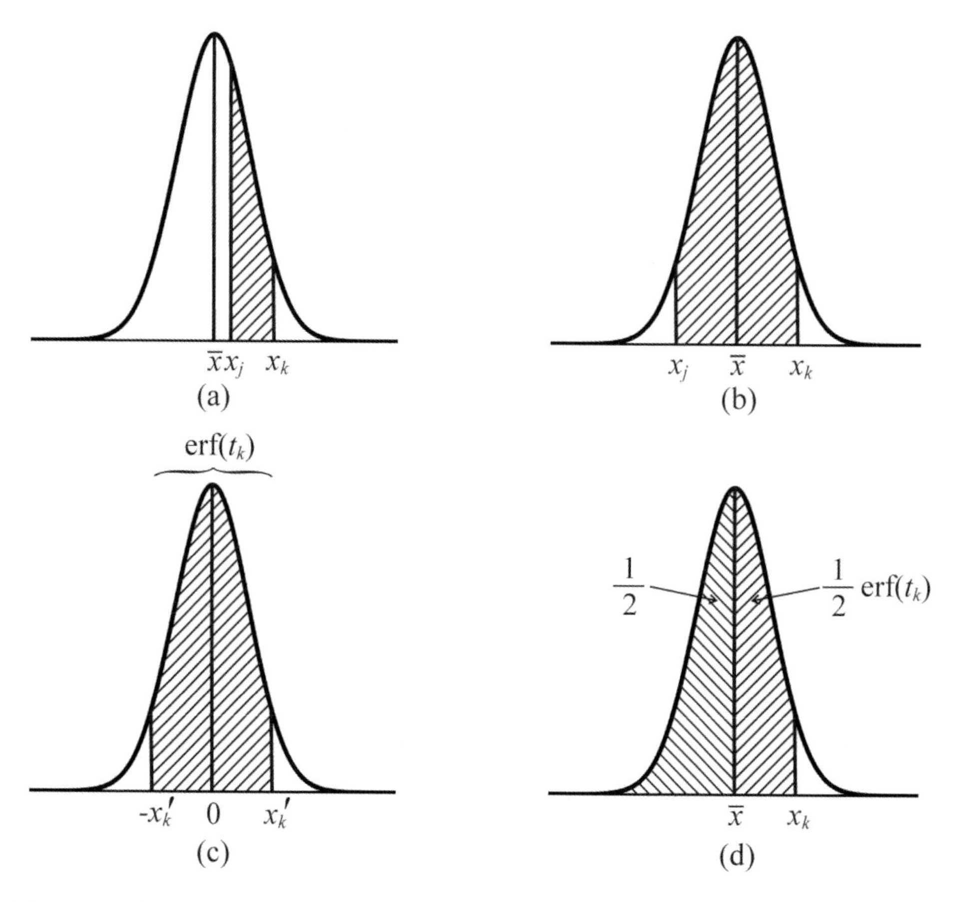

Figure 5.6 (a) Probability integral for $Pr(x_j \leq x \leq x_k)$. (b) Probability integral for x_j and x_k symmetric about the mean. (c) Probability integral determined from $\mathrm{erf}(t_k)$. (d) Probability integral for $Pr(x \leq x_k)$.

The symmetric Gaussian integral in this form, as a function of $t_k = (x_k - \bar{x})/(\sigma\sqrt{2})$, is known as the error function because of its common use in the treatment of random errors equally distributed about a mean value. Because of its symmetry it needs only to be evaluated from zero to t_k:

$$\mathrm{erf}(t_k) = \frac{2}{\sqrt{\pi}} \int_0^{t_k} e^{-t^2}\, dt. \tag{5.70}$$

For a specified mean and standard deviation, the probability that x will be found between $\bar{x} \pm \Delta x$, where $\Delta x = |x_k - \bar{x}|$ is given directly by $\mathrm{erf}(t_k)$. The error function has been numerically integrated with high precision for small increments of t_k and tabulated in a number of collections of mathematical tables and software packages. An excellent approximation to the error function is given by[65]

$$\mathrm{erf}(t_k) \simeq (1 - e^{q(t_k)})^{1/2}, \tag{5.71}$$

where

$$q(t_k) = \left(-t_k^2 \frac{(4/\pi) + at_k^2}{1 + at_k^2}\right), \quad a = \left(\frac{8}{3\pi}\right)\left(\frac{\pi - 3}{4 - \pi}\right) = 0.1400. \tag{5.72}$$

The error function is defined as an odd function, so that $\text{erf}(-t_k)=-\text{erf}(t_k)$. Using this property, Fig. 5.6(d) shows the way in which the error function is used to determine the probability that x lies between *any* limits, which is where our discussion began. The probability that x will lie between the mean and x_k is $\text{erf}(t_k)/2$. The probability that x will be less than or equal to the mean is 1/2. Thus the probability that x is less than or equal to x_k is $1/2 + \text{erf}(t_k)/2$. The probability that x is less than or equal to x_j is $1/2 + \text{erf}(t_j)/2$. Subtracting the two integrals from one another will result in the integral illustrated in Fig. 5.6(a):

$$Pr(x_j \leq x \leq x_k) = \frac{1}{2} + \frac{1}{2}\text{erf}(t_k) - \frac{1}{2} - \frac{1}{2}\text{erf}(t_j) = \frac{\text{erf}(t_k) - \text{erf}(t_j)}{2}. \tag{5.73}$$

We will encounter the error function in three dimensions when we discuss displacement ellipsoids later in the chapter.

The Multivariate Gaussian Distribution

In the discussion of vibration later in this chapter we will discover that when an atom vibrates, the displacement *vectors* from its mean position are approximately normally distributed. To describe the *displacement volume* of the atom we will need to ascertain the probability of observing a specific vector, $\mathbf{v} = [x_1^o, x_2^o, x_3^o]$. This is the probability of *simultaneously* observing a specific x_1^o, x_2^o and x_3^o between x_1^o and $x_1^o + dx_1$, x_2^o and $x_2^o + dx_2$, and x_3^o and $x_3^o + dx_2$, respectively. (x_1^o, x_2^o, x_3^o) is a specific ordered triplet selected from three sets of normally distributed random variables, $\{x_1\}$, $\{x_2\}$, and $\{x_3\}$.* In the treatment that follows we consider the simpler two-variable (*bivariate*) case first, and will extrapolate to three dimensions at the appropriate time.

The probability of observing x_1^o between x_1^o and $x_1^o + dx_1$ for *any* value of x_2 is simply $Pr(x_1^o) = f_G(x_1^o)\, dx_1$ with mean $\bar{x}_1$, and variance $\sigma_1^2 = \langle(x_1 - \bar{x}_1)^2\rangle$. The probability of observing x_2^o between x_2^o and $x_2^o + dx_2$ for *any* value of x_1 is $Pr(x_2^o) = f_G(x_2^o)\, dx_2$ with mean $\langle x_2\rangle$, and variance $\sigma_1^2 = \langle(x_2 - \bar{x}_2)^2\rangle$.† The probability of these two observations occurring simultaneously is $Pr(x_1^o, x_2^o) = f_G(x_1^o, x_2^o)\, dx_1 dx_2$; $f_G(x_1^o, x_2^o)$ is the probability per unit area (in x_1, x_2 space) and is known as a *joint probability density function.* Because of the simultaneity constraint, we might anticipate that, in addition to the squared residuals in the variances, a term involving both residuals could find its way into the joint probability distribution:

$$\sigma_{12} = \langle(x_1 - \bar{x}_1)(x_2 - \bar{x}_2)\rangle\, . \tag{5.74}$$

σ_{12} is the average product of the residuals, and is known as the *covariance* of the distributions of $\{x_1\}$ and $\{x_2\}$.

Variance and Covariance. The variance of a distribution represents the average squared deviation from the mean, and is an indicator of how closely the variables remain in the vicinity of their average values. The covariance, however, is an

*In the previous section x_1, etc., referred to specific values from the set of random variables $\{x\}$. Here the subscripts describe *sets* of variables; superscripts signify specific members of those sets.

†$Pr(x_1^o)$ and $Pr(x_2^o)$ are known as *marginal probabilities* since they occur in the margins of the x_1, x_2 plane.

indicator of the relative simultaneous behavior of the variables. Suppose we observe that, in most cases, whenever a specific x_1^o deviates from its mean with a certain magnitude and sign, a corresponding x_2^o deviates with about the same magnitude and sign — the variables seem to be linked (*correlated*). The covariance will then contain many pairs of positive numbers in the sum created to generate the average, and will therefore be a relatively large positive number. If the members of each pair of residuals tend to have about the same magnitude but are opposite in sign, then the covariance will tend to be a relatively large negative number. If the signs and magnitudes vary randomly, then the product of the residuals will average to zero. In the first instance we say that the variables are *positively correlated*, in the second instance we describe the variables as being *negatively correlated*, and in the third instance, when $\sigma_{12} = 0$, we describe them as *uncorrelated*. Variables that are completely independent of one another are uncorrelated.

Since the magnitude of the covariance depends on the relative sizes of the residuals, it is useful to "normalize" the covariance by defining a measure of correlation that is independent of the scaling of the variables. This quasi-normalized covariance is known as the *correlation coefficient*:

$$\rho_{12} = \frac{\sigma_{12}}{\sigma_1\,\sigma_2} = \frac{\langle (x_1 - \bar{x}_1)(x_2 - \bar{x}_2)\rangle}{\left(\langle (x_1 - \bar{x}_1)^2\rangle\right)^{1/2}\left(\langle (x_2 - \bar{x}_2)^2\rangle\right)^{1/2}}. \tag{5.75}$$

When the variables are uncorrelated, $\sigma_{12} = 0$ and $\rho_{12} = 0$. When x_1 and x_2 behave identically they have the same mean and variance, and

$$\rho_{12} = \frac{\left(\langle (x_1 - \bar{x}_1)^2\rangle\right)^2}{\left(\langle (x_1 - \bar{x}_1)^2\rangle\right)^2} = 1. \tag{5.76}$$

When the variables have the same magnitudes but opposite signs then

$$\rho_{12} = \frac{\left(\langle -(x_1 - \bar{x}_1)^2\rangle\right)^2}{\left(\langle (x_1 - \bar{x}_1)^2\rangle\right)^2} = -1. \tag{5.77}$$

Thus the correlation coefficient varies from $+1$ (perfect positive correlation) through zero (no correlation) to -1 (perfect negative correlation).

We will find it useful to derive several variance/covariance identities. Noting that the average value of a variable times a constant (such as $\bar{x}$) is the product of the constant times the average value of the variable, and that the average value of a sum is the sum of the average values of the components of the sum, we can derive the following relationships:

Identity 1. Given the variance, $\sigma^2(x) = \langle (x - \bar{x})^2\rangle$,

$$\begin{aligned} \sigma^2(x) &= \langle x^2 - 2x\bar{x} + \bar{x}^2\rangle = \langle x^2\rangle - 2\bar{x}\bar{x} + \bar{x}^2 \\ &= \langle x^2\rangle - \bar{x}^2. \end{aligned} \tag{5.78}$$

Identity 2. For $y = a + bx$, a and b constants,

$$\begin{aligned} \sigma^2(y) &= \left\langle ((a + bx) - \langle a + bx\rangle)^2 \right\rangle = \langle (a + bx)^2\rangle - (\langle a + bx\rangle)^2 \\ &= a^2 + 2ab\bar{x} + b^2\langle x^2\rangle - a^2 - 2ab\bar{x} - b^2\bar{x}^2 = \\ &= b^2\,\sigma^2(x). \end{aligned} \tag{5.79}$$

Identity 3. For $y = ax_1 + bx_2$, a and b constants,

$$\begin{aligned}\sigma^2(y) &= \left\langle \left((ax_1 + bx_2) - \langle ax_1 + bx_2\rangle\right)^2 \right\rangle \\ &= a^2\langle x_1^2\rangle + 2ab\langle x_1x_2\rangle + b^2\langle x_2^2\rangle - a^2\bar{x}_1^2 - 2ab\langle x_1x_2\rangle - b^2\bar{x}_2^2 \\ &= a^2(\langle x_1^2\rangle - \bar{x}_1^2) + b^2(\langle x_2^2\rangle - \bar{x}_2^2) \\ &= a^2\,\sigma^2(x_1) + b^2\,\sigma^2(x_2). \end{aligned} \tag{5.80}$$

Identity 4. Given the covariance, $\sigma_{12}(x_1, x_2) = \langle (x_1 - \bar{x}_1)(x_2 - \bar{x}_2)\rangle$,

$$\begin{aligned}\sigma_{12}(x_1, x_2) &= \langle x_1x_2\rangle - \langle x_1\bar{x}_2\rangle - \langle x_2\bar{x}_1\rangle + \bar{x}_1\bar{x}_2 \\ &= \langle x_1x_2\rangle - \bar{x}_1\bar{x}_2 - \bar{x}_1\bar{x}_2 + \bar{x}_1\bar{x}_2 \\ &= \langle x_1x_2\rangle - \bar{x}_1\bar{x}_2. \end{aligned} \tag{5.81}$$

Identity 5. for $y_1 = a + bx_1$ and $y_2 = c + dx_2$,

$$\begin{aligned}\sigma_{12}(y_1, y_2) &= \langle y_1\, y_2\rangle - \bar{y}_1\bar{y}_2 \\ &= \langle (a + bx_1)(c + dx_2)\rangle - \langle (a + bx_1)\rangle\langle (c + dx_2)\rangle \qquad (5.82) \\ &= ac + ad\bar{x}_2 + bc\bar{x}_1 + bd\langle x_1x_2\rangle \\ &\quad -ac - ad\bar{x}_2 - bc\bar{x}_1 - bd\bar{x}_1\bar{x}_2 \\ &= bd(\langle x_1x_2\rangle - \bar{x}_1\bar{x}_2) \;=\; bd\,\sigma_{12}(x_1, x_2). \end{aligned} \tag{5.83}$$

The *bivariate normal distribution* function that we are working toward involves continuous variables, and a simpler example incorporating discrete values will serve to illustrate some of ideas discussed above. Consider the components of the 8 vectors in Figs. 5.7(a) and 5.7(b). The vectors are described with two different sets of variables. In Fig. 5.7(a) the variable sets $\{u_1\}$ and $\{u_2\}$ are the Cartesian coordinates of the vectors, based on an orthogonal coordinate system, while the variable sets $\{x_1\}$ and $\{x_2\}$ in Fig. 5.7(b) are the fractional coordinates of the vectors based on a two dimensional unit cell. The mean of $\{u_2\}$ is zero, and for every residual product in the covariance sum there is a negative term of equal magnitude — the sum averages to zero, e.g., $(u_2^3)(u_1^4 - \bar{u}_1) + (-u_2^3)(u_1^4 - \bar{u}_1) = 0$. The variables $\{u_1\}$ and $\{u_2\}$ are uncorrelated. On the other hand, there are no such cancelations in the covariance sum for $\{x_1\}$ and $\{x_2\}$, e.g., $(x_2^3)(x_1^A - \bar{x}_1)) + (-x_2^3)(x_1^B - \bar{x}_1) \neq 0$, and the variables are correlated.

The differences in the uncorrelated variables, $\{u_1\}$ and $\{u_2\}$, and the correlated variables, $\{x_1\}$ and $\{x_2\}$, become apparent if we consider some function of the vectors. For example, suppose we define a "vector length function," $f_L = (c_i^2 + c_j^2)^{1/2}$, where c_i and c_j are the projections of the vector along the **i** and **j** axes, respectively. If the vector is described with the uncorrelated variables, then

$$f_L(u_1, u_2) = \sqrt{(u_1)^2 + (u_2)^2}. \tag{5.84}$$

If, on the other hand, the vector is described with the correlated variables, then clearly

$$f_L(x_1, x_2) \neq \sqrt{(x_1)^2 + (x_2)^2}. \tag{5.85}$$

In order to determine the length of the vector using these variables we must determine the projections along **i** and **j** from the fractional coordinates. This can be

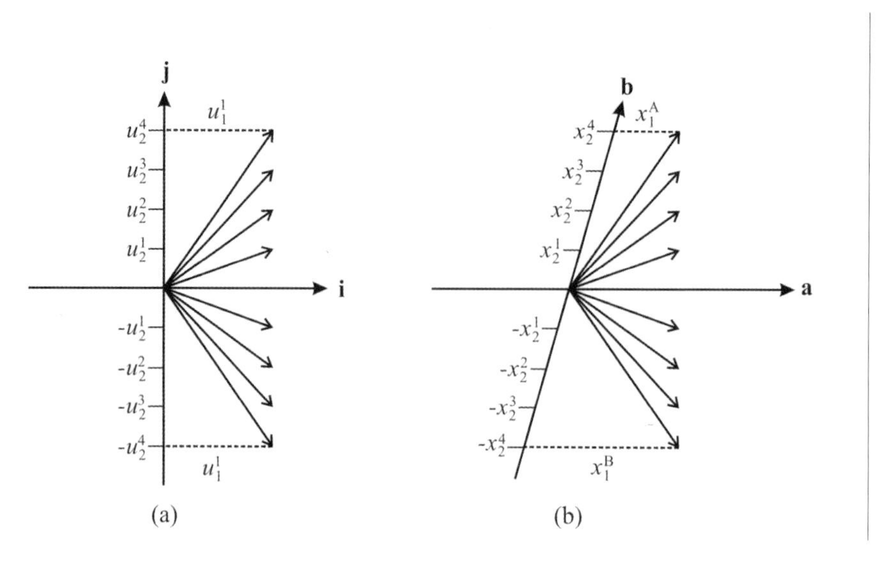

Figure 5.7 (a) Vector set described with orthogonal coordinates $\{u_1\}$ and $\{u_2\}$. (b) Vector set described with fractional coordinates $\{x_1\}$ and $\{x_2\}$.

accomplished by transforming (x_1, x_2) using the two-dimensional analog of the **B** matrix, $\mathbf{B_2}$:

$$\mathbf{B_2}\begin{bmatrix} x_1 \\ x_2 \end{bmatrix} = \begin{bmatrix} b_{11} & b_{12} \\ b_{21} & b_{22} \end{bmatrix}\begin{bmatrix} x_1 \\ x_2 \end{bmatrix} = \begin{bmatrix} u_1 \\ u_2 \end{bmatrix}$$

$$u_1 = b_{11}x_1 + b_{12}x_2$$

$$u_2 = b_{21}x_1 + b_{22}x_2$$

$$f_L(x_1, x_2) = \sqrt{(b_{11}x_1 + b_{12}x_2)^2 + (b_{21}x_1 + b_{22}x_2)^2}.$$

For clarity, the vector set in Fig. 5.7 was plotted with only one value for u_1. Consider the case where the number of vectors is increased by adding additional sets of 8 vectors for u_1^2, u_1^3 and u_1^4 and so forth. Now, suppose that we wish to search for all of the vectors with a specific projection along the **i** axis that have a specific length (or lie within a certain range). To screen the vector lengths we begin by selecting a value of $\{u_1\}$ that corresponds to the projection along the **i** axis, and enter it into in Eqn. 5.84. This selection is independent of the values of the variables in $\{u_2\}$. We now randomly select any value of $\{u_2\}$ in order to determine the length of a "test vector." The selection does not depend in any way of the values of the variables in $\{u_1\}$ — the variables in $\{u_1\}$ and $\{u_2\}$ are *independent* of one another. We now repeat the vector search with the same vectors, but in the "crystal" coordinate system. In order to determine the distance along the **i** axis, we can no longer select a value of $\{x_1\}$ independently of any specific value of $\{x_2\}$. Only those vectors with specific combinations of the variables will be candidates for testing, since both variables are required to determine the projection along the **i** axis — and only after we have determined those vectors with the appropriate projections can we proceed to determine the lengths along the **j** axis needed to calculate the vector lengths. Again, this requires knowing both variables simultaneously. The variables cannot be selected independently.

There is an important subtlety here that arises in this simple example. If the coordinate axes are rotated in Fig. 5.7(a), then the variables will appear to be correlated. In such cases, however, we can always find a suitable rotation of the coordinate system that will "uncorrelate" the variables. In contrast, there is no possible rotation of the coordinate axes in Fig. 5.7(b) that will render the variables independent of one another. We will encounter this later in the chapter in the treatment of displacement ellipsoids.

Variable Transformation. Our goal is to determine the Gaussian joint probability distribution for any set of variables, given the distribution in terms of uncorrelated variables. Specifically, we wish to determine $f_G(x_1, x_2)\,dx_1 dx_2$, given $f_G(u_1, u_2)\,du_1 du_2$, where (x_1, x_2) are fractional coordinates, and (u_1, u_2) are independent Cartesian coordinates.* To simplify matters, we adjust $\{u_1\}$ and $\{u_2\}$ so that each set of variables has a mean of zero, and a variance of 1. This is accomplished by placing a set of variables into *standard form.* Suppose for example, that the variable set $\{u'\}$ has a mean of $\bar{u}'$ and a variance of $\sigma^2(u') = \langle (u' - \bar{u}')^2 \rangle$. We define a new set of variables, $\{u\}$:

$$u = \frac{u' - \bar{u}'}{\sigma(u')} \tag{5.86}$$

This does not alter the variables fundamentally; they are simply shifted equally by a constant, $\bar{u}'$, to create a mean of zero, and scaled by another constant,$1/\sigma^2(u')$ to create a unit variance:

$$\begin{aligned} \bar{u} &= \frac{\langle u' - \bar{u}' \rangle}{\sigma(u')} = \frac{\langle \bar{u}' - \bar{u}' \rangle}{\sigma(u')} = 0; \\ \sigma^2(u) &= \langle (u' - \bar{u}')^2 \rangle = \langle u^2 \rangle = \frac{\langle (u' - \bar{u}')^2 \rangle}{\sigma^2(u')} = 1. \end{aligned} \tag{5.87}$$

Both sets of variables cover the xy plane, *and there is a one-to-one correspondence between them,* since each point in the plane is defined with a unique pair from each set.

The transformation of the variables themselves from $\{u_1\}$ and $\{u_2\}$ to $\{x_1\}$ and $\{x_2\}$ is straightforward. We consider the general case, when the coordinate systems do not have a common origin, in which case the origin of the u_1, u_2 system lies at point (q_1, q_2) from the x_1, x_2 origin. If $\mathbf{L} = \mathbf{B_2}^{-1}$, then

$$\begin{bmatrix} x_1 \\ x_2 \end{bmatrix} = \begin{bmatrix} l_{11} & l_{12} \\ l_{21} & l_{22} \end{bmatrix} \begin{bmatrix} u_1 \\ u_2 \end{bmatrix} + \begin{bmatrix} q_1 \\ q_2 \end{bmatrix}, \quad \text{and}$$

$$x_1 = f_1(u_1, u_2) = q_1 + l_{11}u_1 + l_{12}u_2 \tag{5.88}$$

$$x_2 = f_2(u_1, u_2) = q_2 + l_{21}u_1 + l_{22}u_2. \tag{5.89}$$

The inverse transformations are then

$$u_1 = g_1(x_1, x_2) = \frac{l_{22}x_1' - l_{12}x_2'}{l_{11}l_{22} - l_{12}l_{21}} \tag{5.90}$$

$$u_2 = g_2(x_1, x_2) = \frac{-l_{21}x_1' + l_{11}x_2'}{l_{11}l_{22} - l_{12}l_{21}}, \tag{5.91}$$

where $x_1' = x_1 - q_1$ and $x_2' = x_2 - q_2$.

---

*Although we will develop the bivariate normal distribution based on a specific application, the derivations here will be completely general.

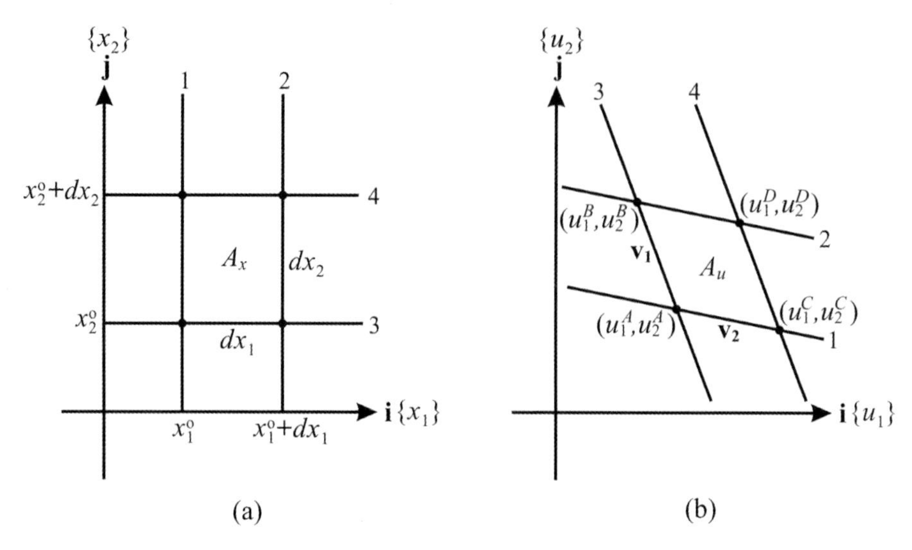

Figure 5.8 (a) Plot of $\{x_1\}$ and $\{x_2\}$ along orthogonal axes. (b) Plot of $\{u_1\}$ and $\{u_2\}$ along orthogonal axes.

The transformation of the differentials presents a more challenging problem. We begin by plotting the variable sets, $\{x_1\}$ and $\{x_2\}$, along orthogonal axes, as illustrated in Fig. 5.8(a) (since we are concerned only with the variables, they can be represented in any coordinate system). For a given pair, (x_1^o, x_2^o), we construct lines at $x_1 = x_1^o$, $x_1 = x_1^o + dx_1$, $x_2 = x_2^o$, and $x_2 = x_2^o + dx_2$. The intersection of these lines creates a rectangle with area, $A_x = dx_1 dx_2$.

Substitution of $x_1 = x_1^o$ and $x_1 = x_1^o + dx_1$ into Eqn. 5.88 and solving for u_1 creates two parallel lines in the u_1, u_2 plane:

$$u_2 = -\frac{l_{11}}{l_{12}} u_1 + (x_1^o - q_1) \quad \text{(Line 1);} \tag{5.92}$$

$$u_2 = -\frac{l_{11}}{l_{12}} u_1 + (x_1^o - q_1 + dx_1) \quad \text{(Line 2).} \tag{5.93}$$

Substitution of $x_2 = x_2^o$ and $x_2 = x_2^o + dx_2$ into Eqn. 5.89 and solving for u_2 creates another set of parallel lines:

$$u_2 = -\frac{l_{21}}{l_{22}} u_1 + (x_2^o - q_2) \quad \text{(Line 3);} \tag{5.94}$$

$$u_2 = -\frac{l_{21}}{l_{22}} u_1 + (x_2^o - q_2 + dx_2) \quad \text{(Line 4).} \tag{5.95}$$

The lines intersect to form a parallelogram with area A_u. Because of the one-to-one mapping of (x_1, x_2) and (u_1, u_2), whenever a point falls within the area A_x, a corresponding point will fall within the area A_u. The probability that a point will be found in A_x is the probability per unit area times the area:

$$Pr(x_1, x_2) = f_G(x_1, x_2) A_x = f_G(x_1, x_2)\, dx_1 dx_2. \tag{5.96}$$

The probability that a corresponding point will be found in A_u is also the probability per unit area times the area:

$$Pr(u_1, u_2) = f_G(u_1, u_2)A_u. \tag{5.97}$$

The probability of a point falling within either area is the same:

$$f_G(x_1, x_2)\, dx_1 dx_2 = f_G(u_1, u_2)A_u, \tag{5.98}$$

and we are left with the determination of the area, A_u — the differential in the u_1, u_2 system. From Eqn. 1.134, the area of a parallelogram is the magnitude of the vector product of the vectors corresponding to two connected edges, $\mathbf{v_1}$ and $\mathbf{v_2}$ in Fig. 5.8. Denoting the coordinates of the vertices of the parallelogram that define these vectors as (u_1^A, u_2^A), (u_1^B, u_2^B), and (u_1^C, u_2^C),

$$\mathbf{v_1} = \begin{bmatrix} (u_1^B - u_1^A) \\ (u_2^B - u_2^A) \end{bmatrix} \qquad \mathbf{v_2} = \begin{bmatrix} (u_1^C - u_1^A) \\ (u_2^C - u_2^A) \end{bmatrix}.$$

The vectors are in the **ij** plane, and the sole component of the perpendicular vector product is along the **k** axis. The component along the **k** axis is therefore the magnitude of the vector product and thus the area of A_u. From Eqn .1.71,

$$\begin{aligned} \mathbf{v_1} \times \mathbf{v_2} &= \begin{vmatrix} \mathbf{i} & \mathbf{j} & \mathbf{k} \\ (u_1^B - u_1^A) & (u_2^B - u_2^A) & 0 \\ (u_1^C - u_1^A) & (u_2^C - u_2^A) & 0 \end{vmatrix} \\ &= [(u_1^B - u_1^A)(u_2^C - u_2^A) - (u_1^C - u_1^A)(u_2^B - u_2^A)]\,\mathbf{k} \end{aligned} \tag{5.99}$$

$$\begin{aligned} A_u &= |\mathbf{v_1} \times \mathbf{v_2}| \\ &= u_1^B u_2^C + u_1^C u_2^A + u_1^A u_2^B - u_1^B u_2^A - u_1^C u_2^B - u_1^A u_2^C. \end{aligned} \tag{5.100}$$

The terms in Eqn. 5.100 result from the expansion of a determinant with 1's in the first column, and the coordinates of the three vertices in the second and third columns:

$$A_u = \begin{vmatrix} 1 & u_1^A & u_2^A \\ 1 & u_1^B & u_2^B \\ 1 & u_1^C & u_2^C \end{vmatrix}. \tag{5.101}$$

The four lines in each plot in Fig. 5.8 are the same lines in two different coordinate systems, and their points of intersection must therefore correspond to one another. The vertex coordinates can therefore be expressed in terms of $\{x_1\}$ and $\{x_2\}$:

$$\left.\begin{aligned} u_1^A &= g_1(x_1^o, x_2^o) \\ u_2^A &= g_2(x_1^o, x_2^o) \end{aligned}\right\} \quad \text{vertex } A \tag{5.102}$$

$$\left.\begin{aligned} u_1^B &= g_1(x_1^o + dx_1, x_2^o) \\ u_2^B &= g_2(x_1^o + dx_1, x_2^o) \end{aligned}\right\} \quad \text{vertex } B \tag{5.103}$$

$$\left.\begin{aligned} u_1^C &= g_1(x_1^o, x_2^o + dx_2) \\ u_2^C &= g_2(x_1^o, x_2^o + dx_2) \end{aligned}\right\} \quad \text{vertex } C. \tag{5.104}$$

The coordinates for vertex A can be obtained in terms of x_1^o and x_2^o from Eqns. 5.90 and 5.91, but since we do not know how a small differential change in either variable will change the function at that point, we have no direct way of determining the coordinates for the vertices containing the differentials in terms of x_1^o, x_2^o, dx_1, and dx_2 (which is the aim of this exercise!). In other words, if we know the value of some function $f(x)$, evaluated at x_o, we need to know how that value of the function will change when x_o becomes $x_o + dx$. If the change is small we can differentiate the function with respect to x, and evaluate the derivative at x_o, $(\partial(f(x))/\partial x)_{x_o}$. This tells us how the function is changing as the variable changes at (or near) x_o – it is the change in the function per unit change of the variable at the point x_o. Assuming that this rate of change is constant between x_o and $x_o + dx$, the change in the function is then the change in the function per unit change in the variable times the change in the variable:

$$\begin{aligned} \Delta f(x) &= \left(\frac{\partial f(x)}{\partial x}\right)_{x_o} dx \\ f(x_o + dx) &= f(x_o) + \Delta f(x) = f(x_o) + \left(\frac{\partial f(x)}{\partial x}\right)_{x_o} dx. \end{aligned} \tag{5.105}$$

All of the vertices can now be described in terms of x_1^o, x_2^o, dx_1, and dx_2:

$$u_1^B = g_1(x_1^o, x_2^o) + \left(\frac{\partial g_1(x_1, x_2)}{\partial x_1}\right)_{x_1^o, x_2^o} dx_1 \tag{5.106}$$

$$u_2^B = g_2(x_1^o, x_2^o) + \left(\frac{\partial g_2(x_1, x_2)}{\partial x_1}\right)_{x_1^o, x_2^o} dx_1 \tag{5.107}$$

$$u_1^C = g_1(x_1^o, x_2^o) + \left(\frac{\partial g_1(x_1, x_2)}{\partial x_2}\right)_{x_1^o, x_2^o} dx_2 \tag{5.108}$$

$$u_2^C = g_2(x_1^o, x_2^o) + \left(\frac{\partial g_2(x_1, x_2)}{\partial x_2}\right)_{x_1^o, x_2^o} dx_2 \tag{5.109}$$

The differential, A_u, is created from Eqn. 5.101. Setting $g_1 \equiv g_1(x_1^o, x_2^o)$ and $g_2 \equiv g_2(x_1^o, x_2^o)$,

$$A_u = \begin{vmatrix} 1 & g_1 & g_2 \\ 1 & \left(g_1 + \frac{\partial g_1}{\partial x_1} dx_1\right) & \left(g_2 + \frac{\partial g_2}{\partial x_1} dx_1\right) \\ 1 & \left(g_1 + \frac{\partial g_1}{\partial x_2} dx_2\right) & \left(g_2 + \frac{\partial g_2}{\partial x_2} dx_2\right) \end{vmatrix}. \tag{5.110}$$

Expanding the determinant gives an expression with only the differentials and derivatives remaining, which can be expressed in the form of another determinant:*

$$\begin{aligned} A_u &= \left(\frac{\partial g_1}{\partial x_1}\frac{\partial g_2}{\partial x_2} - \frac{\partial g_1}{\partial x_2}\frac{\partial g_2}{\partial x_1} \right) dx_1 dx_2 \\ &= \begin{vmatrix} \dfrac{\partial g_1}{\partial x_1} & \dfrac{\partial g_2}{\partial x_1} \\ \\ \dfrac{\partial g_1}{\partial x_2} & \dfrac{\partial g_2}{\partial x_2} \end{vmatrix} dx_1 dx_2 \end{aligned} \tag{5.111}$$

The probability equation (Eqn. 5.98) is now

$$f_G(x_1, x_2)\, dx_1 dx_2 = f_G(u_1, u_2) \begin{vmatrix} \dfrac{\partial g_1}{\partial x_1} & \dfrac{\partial g_2}{\partial x_1} \\ \\ \dfrac{\partial g_1}{\partial x_2} & \dfrac{\partial g_2}{\partial x_2} \end{vmatrix} dx_1 dx_2, \tag{5.112}$$

and the relationship between the bivariate probability density functions expressed in two different variable spaces is

$$f_G(x_1, x_2) = f_G(u_1, u_2) \begin{vmatrix} \dfrac{\partial g_1}{\partial x_1} & \dfrac{\partial g_2}{\partial x_1} \\ \\ \dfrac{\partial g_1}{\partial x_2} & \dfrac{\partial g_2}{\partial x_2} \end{vmatrix}. \tag{5.113}$$

The Bivariate Gaussian Distribution. If Pr_a is the probability of observing some variable, u_a, and Pr_b is the probability of observing another *independent* variable, u_b, then the probability of observing u_a and u_b jointly is $Pr_a Pr_b$ (See Sec. 5.1.1 for a discussion). Recall that the variables $\{u_1\}$ and $\{u_2\}$ are independent, with zero means and unit variances. It follows that the probability of observing specific variables u_1 and u_2 *simultaneously* is

$$Pr(u_1, u_2) = f_G(u_1, u_2)\, du_1\, du_2 = f_G(u_1)\, du_1\; f_G(u_2)\, du_2, \quad \text{and} \tag{5.114}$$

$$f_G(u_1, u_2) = \frac{1}{\sqrt{2\pi}} e^{-u_1^2/2} \frac{1}{\sqrt{2\pi}} e^{-u_2^2/2} = \frac{1}{2\pi} e^{-(u_1^2+u_2^2)/2}. \tag{5.115}$$

Using the methods developed in the previous section, we are now in a position to derive the general bivariate Gaussian probability density function, $f_G(x_1, x_2)$. We will find it convenient to place both variable sets at a common origin and shift the origin after the derivation is complete. Thus we define

$$x'_1 = x_1 - q_1 = l_{11}u_1 + l_{12}u_2 \tag{5.116}$$

$$x'_2 = x_2 - q_2 = l_{21}u_1 + l_{22}u_2. \tag{5.117}$$

If u_1 and u_2 are normally distributed, then x'_1 and x'_2 will also be normally distributed. The mean values of x'_1 and x'_2 are zero:

$$\langle x'_1 \rangle = l_{11}\langle u_1 \rangle + l_{12}\langle u_2 \rangle = l_{11}(0) + l_{12}(0) = 0 \text{ , etc..} \tag{5.118}$$

*The matrix of the first derivatives of a multivariate function is known as the *Jacobian Matrix*. Its determinant — the *Jacobian Determinant* (more commonly referred to as simply "the Jacobian") — is used often to generate a change in variables.

Because they have unit variances and zero means, $\langle u_1^2 \rangle = \langle u_2^2 \rangle = 1$. Because they are independent, $\sigma_{12}(u_1, u_2) = \langle u_1 u_2 \rangle - \bar{u}_1 \bar{u}_2 = \langle u_1 u_2 \rangle = 0$. From Eqns. 5.80 and 5.81, the variances and covariance of x_1' and x_2' are

$$\begin{aligned}
\sigma_1^2(x_1') &= l_{11}^2 \sigma^2(u_1) + l_{12}^2 \sigma^2(u_2) \;=\; l_{11}^2(1) + l_{12}^2(1) = l_{11}^2 + l_{12}^2 && (5.119)\\
\sigma_2^2(x_2') &= l_{21}^2 \sigma^2(u_1) + l_{22}^2 \sigma^2(u_2) \;=\; l_{21}^2(1) + l_{22}^2(1) = l_{21}^2 + l_{22}^2 && (5.120)\\
\sigma_{12}(x_1', x_2') &= \langle x_1' x_2' \rangle - \bar{x}_1' \bar{x}_2' \;=\; \langle x_1' x_2' \rangle \\
&= \langle\, l_{11} l_{21}\, u_1^2 + l_{12} l_{22}\, u_2^2 + l_{11} l_{22}\, u_1 u_2 + l_{12} l_{21}\, u_1 u_2 \rangle \\
&= l_{11} l_{21} \,\langle u_1^2 \rangle + l_{12} l_{22} \,\langle u_2^2 \rangle + l_{11} l_{22} \,\langle u_1 u_2 \rangle + l_{12} l_{21} \,\langle u_1 u_2 \rangle \\
&= l_{11}\, l_{21} + l_{12}\, l_{22}. && (5.121)
\end{aligned}$$

The inverse functions for x_1' and x_2', $u_1 = g_1(x_1', x_2')$ and $u_2 = g_2(x_1', x_2')$ are given in Eqns. 5.90 and 5.91. To simplify, let $\eta = l_{12} l_{22} - l_{12} l_{21}$ and ρ_{12} be the correlation coefficient for x_1' and x_2'. Then

$$\begin{aligned}
u_1 &= g_1(x_1', x_2') = \frac{l_{22}\, x_1' - l_{12}\, x_2'}{\eta} && (5.122)\\
u_2 &= g_2(x_1', x_2') = \frac{l_{21}\, x_1' - l_{11}\, x_2'}{\eta} && (5.123)\\
\rho_{12} &= \frac{\sigma_{12}(x_1', x_2')}{\sigma_1(x_1')\, \sigma_2(x_2')} = \frac{l_{11}\, l_{21} + l_{12}\, l_{22}}{\sqrt{l_{11}^2 + l_{12}^2}\, \sqrt{l_{21}^2 + l_{22}^2}}. && (5.124)
\end{aligned}$$

The "variable transformation determinant" (the Jacobian Determinant, Eqn. 5.113) is generated by differentiating $g_1(x_1', x_2')$ and $g_2(x_1', x_2')$:

$$\begin{vmatrix} \dfrac{\partial g_1}{\partial x_1} & \dfrac{\partial g_2}{\partial x_1} \\[2ex] \dfrac{\partial g_1}{\partial x_2} & \dfrac{\partial g_2}{\partial x_2} \end{vmatrix} = \begin{vmatrix} \dfrac{l_{22}}{\eta} & \dfrac{-l_{21}}{\eta} \\[2ex] \dfrac{-l_{12}}{\eta} & \dfrac{l_{11}}{\eta} \end{vmatrix} = \frac{l_{22} l_{11} - l_{12} l_{21}}{\eta^2} = \frac{1}{\eta}. \qquad (5.125)$$

Eqn. 5.113 now gives us the bivariate Gaussian distribution function:

$$f_G(x_1', x_2') = \frac{1}{2\pi}\, e^{-(u_1^2 + u_2^2)/2} \left(\frac{1}{\eta} \right) = \frac{1}{2\pi\, \eta}\, e^{-(u_1^2 + u_2^2)/2}. \qquad (5.126)$$

The exponent in the function is

$$\begin{aligned}
-\frac{1}{2}(u_1^2 + u_2^2) &= -\frac{1}{2} \left[\left(\frac{l_{22}\, x_1' - l_{12}\, x_2'}{\eta} \right)^2 + \left(\frac{l_{21}\, x_1' - l_{11}\, x_2'}{\eta} \right)^2 \right] \\
&= -\frac{1}{2\eta^2} [(l_{22}^2 + l_{21}^2) {x_1'}^2 - 2(l_{11} l_{21} + l_{12} l_{22}) x_1' x_2' + (l_{11}^2 + l_{12}^2) {x_1'}^2], && (5.127)
\end{aligned}$$

which can be expressed in terms of the variances and covariances by noting that $\rho\, \sigma_1\, \sigma_2 = \sigma_{12} = l_{11} l_{21} + l_{12} l_{22}$ and

$$\begin{aligned}
\sigma_1^2\, \sigma_2^2 (1 - \rho_{12}^2) &= (l_{11}^2 + l_{12}^2)(l_{21}^2 + l_{22}^2) \left(1 - \frac{(l_{11} l_{21} + l_{12} l_{22})^2}{(l_{11}^2 + l_{12}^2)(l_{21}^2 + l_{22}^2)} \right) \\
&= (l_{11} l_{22} - l_{12} l_{21})^2 = \eta^2. && (5.128)
\end{aligned}$$

Thus,

$$-\frac{1}{2}(u_1^2+u_2^2) = -\frac{1}{2\sigma_1^2\,\sigma_2^2(1-\rho_{12}^2)}(\sigma_2^2 x_1'^2 - 2\rho_{12}\,\sigma_1\,\sigma_2 x_1' x_2' + \sigma_1^2 x_2'^2)$$
$$= -\frac{1}{2(1-\rho_{12}^2)}\left(\frac{x_1'^2}{\sigma_1^2} - 2\rho_{12}\frac{x_1'\,x_2'}{\sigma_1\,\sigma_2} + \frac{x_1'^2}{\sigma_2^2}\right). \qquad (5.129)$$

The expression is simplified considerably if the expression in parentheses in Eqn. 5.129 is written in matrix form:

$$-\frac{1}{2}(u_1^2+u_2^2) = -\frac{1}{2}\frac{1}{(1-\rho_{12}^2)}\begin{bmatrix} x_1' & x_2' \end{bmatrix}\begin{bmatrix} \frac{1}{\sigma_1^2} & -\frac{\rho_{12}}{\sigma_1\,\sigma_2} \\ -\frac{\rho_{12}}{\sigma_1\,\sigma_2} & \frac{1}{\sigma_2^2} \end{bmatrix}\begin{bmatrix} x_1' \\ x_2' \end{bmatrix}. \qquad (5.130)$$

It is readily shown that

$$\frac{1}{(1-\rho_{12}^2)}\begin{bmatrix} \frac{1}{\sigma_1^2} & -\frac{\rho_{12}}{\sigma_1\,\sigma_2} \\ -\frac{\rho_{12}}{\sigma_1\,\sigma_2} & \frac{1}{\sigma_2^2} \end{bmatrix} = \begin{bmatrix} \sigma_1^2 & \rho_{12}\,\sigma_1\,\sigma_2 \\ \rho_{12}\,\sigma_1\,\sigma_2 & \sigma_2^2 \end{bmatrix}^{-1}$$
$$= \begin{bmatrix} \sigma_1^2 & \sigma_{12} \\ \sigma_{12} & \sigma_1^2 \end{bmatrix}^{-1} = \mathbf{S}^{-1}. \qquad (5.131)$$

S is known as the *variance-covariance matrix*, or more commonly as the *covariance* matrix (since the variance is the covariance of a variable with itself).

Since x_1 and x_2 differ from x_1' and x_2' only by the constants, q_1 and q_2, Eqns. 5.79 and 5.83 (with $b = d = 1$) tell us that the variances and covariances are the same for both variables. The vector $[x_1'\ x_2'] = \mathbf{x}'^T$ can therefore be replaced by $[(x_1-q_1)\ (x_2-q_2)] = [x_1\ x_2] - [q_1\ q_2] = (\mathbf{x}-\mathbf{q})^T$. Furthermore, the determinant of the covariance matrix,

$$|\mathbf{S}| = \sigma_1^2\,\sigma_2^2 - \sigma_1^2\,\sigma_2^2\,\rho_{12}^2 = \sigma_1^2\,\sigma_2^2(1-\rho_{12}^2) = \eta^2, \qquad (5.132)$$

gives us the pre-exponential factor,

$$\frac{1}{2\pi\,\eta} = \frac{1}{2\pi\,|\mathbf{S}|^{1/2}}, \qquad (5.133)$$

and the resulting distribution function:

$$f_G(x_1,x_2) = f_G(\mathbf{x}) = \frac{1}{2\pi\,|\mathbf{S}|^{1/2}}\,e^{-\frac{1}{2}(\mathbf{x}-\mathbf{q})^T\,\mathbf{S}^{-1}\,(\mathbf{x}-\mathbf{q})}. \qquad (5.134)$$

If x_1 and x_2 are uncorrelated, then setting $\rho_{12} = 0$ in Eqn. 5.134 results in the expected product of the marginal (independent) distributions,

$$f_G(x_1,x_2) = \frac{1}{\sqrt{2\pi}\,\sigma_1}\,e^{-(x_1-q_1)^2/(2\sigma_1^2)}\,\frac{1}{\sqrt{2\pi}\,\sigma_2}\,e^{-(x_2-q_2)^2/(2\sigma_2^2)}, \qquad (5.135)$$

and we can immediately identify q_1 and q_2 with $\bar{x}_1$ and $\bar{x}_2$, resulting in the general form for the bivariate Gaussian distribution:

$$f_G(\mathbf{x}) = \frac{1}{2\pi\,|\mathbf{S}|^{1/2}}\,e^{-\frac{1}{2}(\mathbf{x}-\bar{\mathbf{x}})^T\,\mathbf{S}^{-1}\,(\mathbf{x}-\bar{\mathbf{x}})}. \qquad (5.136)$$

The Trivariate Gaussian Distribution. Eqn. 5.136 is readily expanded to any number of dimensions. There is a $1/\sqrt{2\pi}$ term for each of n dimensions, resulting in

$$f_G(x_1, x_2, \ldots x_n) = f_G(\mathbf{x}) = \frac{1}{(2\pi)^{n/2}\,|\mathbf{S}|^{1/2}}\, e^{-\frac{1}{2}\,(\mathbf{x}-\bar{\mathbf{x}})^T\,\mathbf{S}^{-1}\,(\mathbf{x}-\bar{\mathbf{x}})}. \tag{5.137}$$

Note that Eqn. 5.137 reduces to Eqn. 5.61 in the univariate case ($n = 1$). We are especially interested in the *trivariate Gaussian distribution*, which will be essential for a characterization of the three dimensional displacement vectors of the vibrating atoms in the crystal:

$$f_G(x_1, x_2, x_3) = f_G(\mathbf{x}) = \frac{1}{(2\pi)^{3/2}\,|\mathbf{S}|^{1/2}}\, e^{-\frac{1}{2}\,(\mathbf{x}-\bar{\mathbf{x}})^T\,\mathbf{S}^{-1}\,(\mathbf{x}-\bar{\mathbf{x}})}, \tag{5.138}$$

where the three-dimensional covariance matrix is a logical extension of its two-dimensional analog in the bivariate distribution:

$$\mathbf{S} = \begin{bmatrix} \sigma_1^2 & \rho_{12}\,\sigma_1\,\sigma_2 & \rho_{13}\,\sigma_1\,\sigma_3 \\ \rho_{12}\,\sigma_1\,\sigma_2 & \sigma_2^2 & \rho_{23}\,\sigma_2\,\sigma_3 \\ \rho_{13}\,\sigma_1\,\sigma_3 & \rho_{23}\,\sigma_2\,\sigma_3 & \sigma_2^2 \end{bmatrix} = \begin{bmatrix} \sigma_1^2 & \sigma_{12} & \sigma_{13} \\ \sigma_{12} & \sigma_2^2 & \sigma_{23} \\ \sigma_{13} & \sigma_{23} & \sigma_3^2 \end{bmatrix}. \tag{5.139}$$

The Central Limit Theorem

It is difficult to overstate the significance of the Gaussian probability distribution. Indeed, we have already noted that as the number of independent random variables* increases, the Poisson distribution becomes indistinguishable from a Gaussian distribution. The binomial distribution also approaches the Gaussian distribution in the limit. While we will forego the formal proofs of these assertions, they are based upon an important theorem — the *central limit theorem* — that we will need when we formally discuss the statistical distributions of experimentally measured intensities. *The central limit theorem tells us that, given a population of values, if we repeatedly collect* n *random samples from this population, the sums (and averages) of each set of the* n *values will create a new population that will become normally distributed as* n *increases — regardless of the nature of the probability distribution of the original population.*

To prove the central limit theorem, we take a closer look at the mean and variance of a general probability distribution. The mean and variance are characteristic of the distribution, and we discover that there are similar higher order terms that are collectively unique to the distribution. For a random variable, x, these terms, known as *moments* of the distribution, take on the general form

$$\mu_k = \int_{-\infty}^{\infty} (x - \langle x \rangle)^k f(x)\, dx = \left\langle (x - \langle x \rangle)^k \right\rangle, \tag{5.140}$$

where μ_k is termed "the kth moment of the distribution about the mean." For $k = 0$, $u_0 = 1$, the integral of the normal probability distribution; for $k = 1$, u_1 is

*A random variable is a variable that is not uniquely determined each time that it is observed, but can take on different values; it is characterized by its probability distribution. Experimentally measured intensities are random variables.

the mean value of the residual (zero for a normal distribution). $u_2 = \left\langle (x - \langle x \rangle)^2 \right\rangle$, which is the variance, $\sigma^2(x)$; the variance is often referred to as the *second moment of the distribution.* When $k = 3$, u_3 approaches zero for symmetric distributions, and will be non-zero when the distribution is skewed. The third moment therefore characterizes the "skewness" of the distribution. The values of μ_4 determine whether the distribution will exhibit a sharp peak or a broad peak, providing a measure of the distribution's "peakedness." Most importantly, the moments of a distribution characterize it; they are conveniently determined by a function that is also uniquely characteristic of the distribution, known as its *moment generating function.* This function takes the form

$$g(t) = \left\langle e^{t(x - \langle x \rangle)} \right\rangle . \tag{5.141}$$

Since e^y can be expanded as a power series,

$$\begin{aligned}
e^y &= \sum_{n=0}^{\infty} \frac{y^n}{n!} = \frac{y^0}{1!} + \frac{y^1}{1!} + \frac{y^2}{2!} + \frac{y^3}{3!} + \cdots \Longrightarrow \\
e^{t(x - \langle x \rangle)} &= 1 + t(x - \langle x \rangle) + \frac{t^2 (x - \langle x \rangle)^2}{2!} + \frac{t^3 (x - \langle x \rangle)^3}{3!} + \cdots , \quad \text{and} \\
g(t) &= 1 + \langle t(x - \langle x \rangle) \rangle + \frac{t^2 \left\langle (x - \langle x \rangle)^2 \right\rangle}{2!} + \frac{t^3 \left\langle (x - \langle x \rangle)^3 \right\rangle}{3!} + \cdots , \\
&= \mu_0 + t\,\mu_1 + \frac{t^2}{2!}\,\mu_2 + \frac{t^3}{3!}\,\mu_3 + \cdots = \sum_{k=0}^{\infty} \frac{\mu_k}{k!}\, t^k .
\end{aligned} \tag{5.142}$$

The rth moment of the distribution is obtained from this function by differentiating it with respect to t r times $(d^r(g(t))/dt^r)$, then setting t to zero.

For the normal distribution, with x in standard form (Eqn. 5.86), $\langle x \rangle = 0$, $\sigma^2(x) = 1$, and the kth moment of the distribution is

$$\mu_k = \frac{1}{\sqrt{2\pi}} \int_{-\infty}^{\infty} x^k \, e^{-x^2/2} \, dx . \tag{5.143}$$

When k is odd, the integral is an odd function that evaluates to zero. When k is even, for $k = 2m$, μ_{2m} can be determined from the definite integral

$$\int_{-\infty}^{\infty} x^{2m} e^{-a^2 x^2} \, dx = 2 \left(\frac{1 \cdot 3 \cdot 5 \cdots (2m - 1)}{2^{m+1} a^{2m}} \frac{\sqrt{\pi}}{a} \right) . \tag{5.144}$$

Setting $a = 1/\sqrt{2}$ and multiplying the expression on the right by $(2^m\, m!)/(2^m\, m!)$ gives

$$\mu_{2m} = \frac{(2m)!}{2^m\, m!} \quad \text{and} \tag{5.145}$$

$$\begin{aligned}
g(t) &= \sum_{k=0}^{\infty} \frac{\mu_k}{k!}\, t^k = \sum_{m=0}^{\infty} \frac{\mu_{2m}}{(2m)!}\, t^{2m} \\
&= \sum_{m=0}^{\infty} \frac{1}{2^m\, m!}\, t^{2m} .
\end{aligned} \tag{5.146}$$

The power series expansion of $e^{t^2/2}$ is

$$\begin{aligned} e^{t^2/2} &= 1 + \frac{t^2}{2} + \frac{(t^2/2)^2}{2!} + \frac{(t^2/2)^3}{3!} + \cdots \\ &= 1 + \frac{t^2}{2} + \frac{(t^2)^2}{2^2 2!} + \frac{(t^2)^3}{2^3 3!} + \cdots \\ &= \sum_{m=0}^{\infty} \frac{1}{2^m\, m!} t^{2m}. \end{aligned} \tag{5.147}$$

Thus the moment generating function for the normal distribution is

$$g(t) = e^{t^2/2}. \tag{5.148}$$

The proof of the central limit theorem is based on a determination of the moment generating function for a sum of independent random variables. Consider the variables x_1 and x_2 with distributions described by moment generating functions $g(1)$ and $g(2)$, respectively. Let $S = x_1 + x_2$, where x_1 and x_2 are sampled n times. The distribution of the resulting values of S has a moment generating function, $g(t)$, such that

$$\begin{aligned} g(t) &= \left\langle e^{\,t(S - \langle S \rangle)} \right\rangle \\ &= \left\langle e^{\,t(x_1 + x_2 - \langle x_1 \rangle - \langle x_2 \rangle)} \right\rangle \\ &= \left\langle e^{\,t(x_1 - \langle x_1 \rangle)}\, e^{\,t(x_2 - \langle x_2 \rangle)} \right\rangle. \end{aligned} \tag{5.149}$$

The average value of the product of two independent variables, $y_1 y_2$ is the sum of each product times the fraction of times that the product occurs, i.e., it is the weighted sum of the probability of obtaining a certain product multiplied by that product:

$$\begin{aligned} \langle (y_1 y_2) \rangle &= \sum_{y_1} \sum_{y_2} (y_1 y_2) Pr(y_1 y_2) \\ &= \sum_{y_1} \sum_{y_2} y_1 y_2 Pr(y_1) Pr(y_2) \\ &= \sum_{y_1} y_1 Pr(y_1) \sum_{y_2} y_2 Pr(y_2) \\ &= \langle y_1 \rangle \langle y_2 \rangle. \end{aligned} \tag{5.150}$$

The mean of the product of two independent variables is the product of their means. Because x_1 and x_2 are independent,

$$g(t) = \left\langle e^{\,t(x_1 - \langle x_1 \rangle)} \right\rangle \left\langle e^{\,t(x_2 - \langle x_2 \rangle)} \right\rangle = g(1)\, g(2). \tag{5.151}$$

The moment generating function of a sum of independent random variables is the product of their individual moment generating functions.

We now consider a population $\{X\}$ of random independent variables characterized by an arbitrary probability distribution with mean $\langle X \rangle$ and variance $\sigma^2(X)$.

We repeatedly collect n samples of the population, each time generating their sum: $S = x_1 + x_2 + \cdots + x_n$. S has variance, σ_S^2 and mean $\langle S \rangle$, where

$$\begin{aligned} \langle S \rangle &= \langle x_1 + x_2 + \cdots + x_n \rangle = \langle x_1 \rangle + \langle x_2 \rangle + \cdots + \langle x_n \rangle \quad \text{and} & (5.152) \\ \sigma_S^2 &= \sigma_1^2(x_1) + \sigma_2^2(x_2) + \cdots + \sigma_n^2(x_n). & (5.153) \end{aligned}$$

Eqn. 5.153 follows from Eqn. 5.80. In order to determine the nature of the distribution of S, we place it in standard form — as we did with the variable x in the treatment of the normal distribution above (Eqn. 5.86):

$$\begin{aligned} S' &= \frac{S - \langle S \rangle}{\sigma_S} = \frac{(x_1 - \langle x_1 \rangle) + (x_2 - \langle x_2 \rangle) + \cdots + (x_n - \langle x_n \rangle)}{\sigma_S} \\ &= \frac{(x_1 - \langle x_1 \rangle)}{\sigma_S} + \frac{((x_2 - \langle x_2 \rangle)}{\sigma_S} + \cdots + \frac{((x_n - \langle x_n \rangle)}{\sigma_S}. \end{aligned} \quad (5.154)$$

The moment generating function of each term in the sum is given by:

$$\begin{aligned} g_i(t) &= \left\langle e^{t(x_i - \langle x_i \rangle)/\sigma_S} \right\rangle \\ &= 1 + \mu_{1i}\left(\frac{t}{\sigma_S}\right) + \frac{\mu_{2i}}{2!}\left(\frac{t}{\sigma_S}\right)^2 + \frac{\mu_{3i}}{3!}\left(\frac{t}{\sigma_S}\right)^3 + \cdots \end{aligned} \quad (5.155)$$

Since σ_S is the sum of the variances of the n samples, it will be a large number, and will get larger as n increases. Thus we expect the third term to be small with respect to 1, and the higher order terms, containing $1/\sigma^3$, $1/\sigma^4$, etc., to be negligible. The average deviation from the mean will be approximately zero (zero for symmetric distributions), and dividing by σ_S renders it even smaller, allowing us to ignore the second term. Since $\mu_{2i} = \sigma_i^2(x_i)$, the moment generating function for a term in the sum reduces to

$$g_i(t) = 1 + \frac{\sigma_i^2}{2}\left(\frac{t}{\sigma_S}\right)^2 = 1 + \frac{\sigma_i^2}{2}\left(\frac{t^2}{\sigma_S^2}\right). \quad (5.156)$$

The moment generating function for S' is the product of the moment generating functions of the sum:

$$\begin{aligned} g_S(t) &= g_1(t)\, g_2(t) \cdots g_n(t) \\ &= \left(1 + \frac{\sigma_1^2}{2}\frac{t^2}{\sigma_S^2}\right)\left(1 + \frac{\sigma_2^2}{2}\frac{t^2}{\sigma_S^2}\right) \cdots \left(1 + \frac{\sigma_n^2}{2}\frac{t^2}{\sigma_S^2}\right). \end{aligned} \quad (5.157)$$

Taking the logarithm of both sides,

$$\begin{aligned} \ln(g_S(t)) &= \ln\left(1 + \frac{\sigma_1^2}{2}\frac{t^2}{\sigma_S^2}\right) + \ln\left(1 + \frac{\sigma_2^2}{2}\frac{t^2}{\sigma_S^2}\right) \cdots + \ln\left(1 + \frac{\sigma_n^2}{2}\frac{t^2}{\sigma_S^2}\right) \\ &= \sum_{i=1}^{n} \ln\left(1 + \frac{\sigma_i^2}{2}\frac{t^2}{\sigma_S^2}\right). \end{aligned} \quad (5.158)$$

Again, because σ_S^2 is a large number for large values of n, the second term in the argument for each logarithm is very small compared to 1, and we can apply the well known approximation from elementary calculus, $\ln(1 + y) \approx y$ for small y. Thus

$$\ln(g_S(t)) = \sum_{i=1}^{n}\left(\frac{\sigma_i^2}{2}\frac{t^2}{\sigma_S^2}\right) = \left(\frac{1}{2}\frac{t^2}{\sigma_S^2}\right)\sum_{i=1}^{n}\sigma_i^2 = \frac{t^2}{2}. \quad (5.159)$$

The moment generating function for the sum of random variables from an arbitrary distribution is therefore

$$g_S(t) = e^{t^2/2}, \tag{5.160}$$

which is identical to Eqn. 5.148, *the moment generating function for the Gaussian distribution. The sum of a large number of independent random variables is characterized by a normal probability distribution with mean and variance equal to the sum of the means and variances of the independent variables,* Q.E.D.

5.1.2 Systematic Error

The accuracy of the results of a crystal structure determination depends upon the accuracy of the intensities measured in the diffraction experiment. There are several phenomena in the experiment that result in systematic errors in the observed intensities that must be dealt with before the reflection data are used for structural solution and refinement. None of these factors are easily removed from the experiment, and therefore must be handled by correcting the intensity data with appropriate models.

The magnitudes of the integrated intensities in the X-ray diffraction experiment depend on the size of the crystal and the intensity of the incident radiation. On the other hand, the magnitudes on the structure factors must not depend on specific experimental details since they will be used to model the electron density in the crystal. The measured intensity magnitudes are therefore relative numbers that must be scaled to "electron density" units before structural determination and analysis can be undertaken. In the treatment that follows we develop several correction factors that will modify the measured diffraction intensity, I_d: $I_{rel} = (correctionfactor) * \langle I_d \rangle$, where I_{rel} is the corrected relative intensity, and $\langle I_d \rangle$ is the time-averaged value for the intensity of the reflection. *Because I_{rel} will be scaled, we can incorporate any constants in a correction factor into the scale factor* (see Sec. 5.2).

The Polarization Correction

When a beam of unpolarized light is reflected from a surface, the reflected beam of light is less intense than the incident beam, due in part to a *polarization* of the reflected radiation. The electric field vector of the incident radiation is usually *unpolarized*, meaning that it has an equal probability at any single instant of being at any arbitrary angle in a plane perpendicular to the direction of propagation of the incident beam. On the average, the beam has an equal intensity in any (perpendicular) direction. The reflected beam does *not* have an equal intensity in each direction perpendicular to its propagation direction, and is therefore said to be *polarized*.

X-radiation from a diffracted beam also undergoes polarization. When photons are emitted from an X-ray source each photon carries with it a unit of angular momentum. The classical analog of this phenomenon is a spinning bullet leaving the barrel of a rifle. The "spin" is either clockwise or counter-clockwise about the direction of propagation of the photon. If all of the photons are spinning in the same direction (we cannot, of course, know that they are actually "spinning," but we do know that each has a *spin angular momentum*), the resulting electric field vector for the radiation will be oscillating in a direction perpendicular to the direction of propagation, and simultaneously rotating as a result of this spin. This oscillating, rotating electric field can act on a test electron by accelerating it in the direction of the oscillating electric field (perpendicular to the propagation direction), as discussed in Chapter 3, and simultaneously subjecting it to a "twisting force" (*torque*).

The beam of photons emitted from a conventional X-ray source (e.g., the anode of an X-ray tube), consists of an equal number of "clockwise" and "counterclockwise" photons. The rotational effects cancel one another and the net result is an electric field vector that is oscillating in a direction perpendicular to the propagation direction, but is not rotating. Even though the electric field vector is not rotating, the direction of the net electric field vector is arbitrary; there is an equal probability at any one instant that this vector will be at any given angle. At any instant, the electric field of a beam of photons from a conventional X-ray source can be treated as a vector perpendicular to the direction of propagation – at an arbitrary angle in a plane perpendicular to the propagation direction. When this electric field acts on the test electron, it accelerates it in the direction of the electric field vector, but does not subject it to torque.

Fig. 5.9 depicts the interaction of the electric field of the incident radiation with an electron inside a crystal that has been oriented into the diffraction condition. Fig. 5.9(a) represents the case in which the electric field vector is parallel to the plane containing the incident and diffracted beams. The electron experiences a force due to the electric field, and a subsequent acceleration in the direction of the electric field vector due to this force: $F = m_e a$, where m_e is the mass of the electron. The acceleration vector can be resolved into two components, one perpendicular to the direction of the diffracted beam, and one parallel to it. The perpendicular component of the acceleration, $a' = a \sin \eta$, creates an oscillating electric field at some distance r perpendicular to the diffracted beam with magnitude $\mathcal{E}_{d\|}$ (the "$\|$" symbol is used here to indicate that the electric field vectors of the incident and diffracted beams both lie in the same plane as the beams themselves). The intensity of the diffracted beam is proportional to the square of the magnitude of the electric field perpendicular to the direction of propagation: $I_d \; \alpha \; \mathcal{E}_{d\|}^2$.

The force on the electron is the product of the magnitude of the electric field of the incident radiation and the charge on the electron (q_e), allowing us to determine the acceleration of the electron as a function of the magnitude of the electric field:

$$\begin{aligned} F &= m_e a = \mathcal{E}_{I\|} q_e \\ a &= \frac{\mathcal{E}_{I\|} q_e}{m_e}. \end{aligned} \tag{5.161}$$

The magnitude of $\mathcal{E}_{d\|}$ is directly proportional to the perpendicular component of the acceleration, directly proportional to the charge on the particle being accelerated, and inversely proportional to the distance at which the radiation is detected, r:

$$\mathcal{E}_{d\|} \; \alpha \; \frac{q_e}{r} a \sin \eta.$$

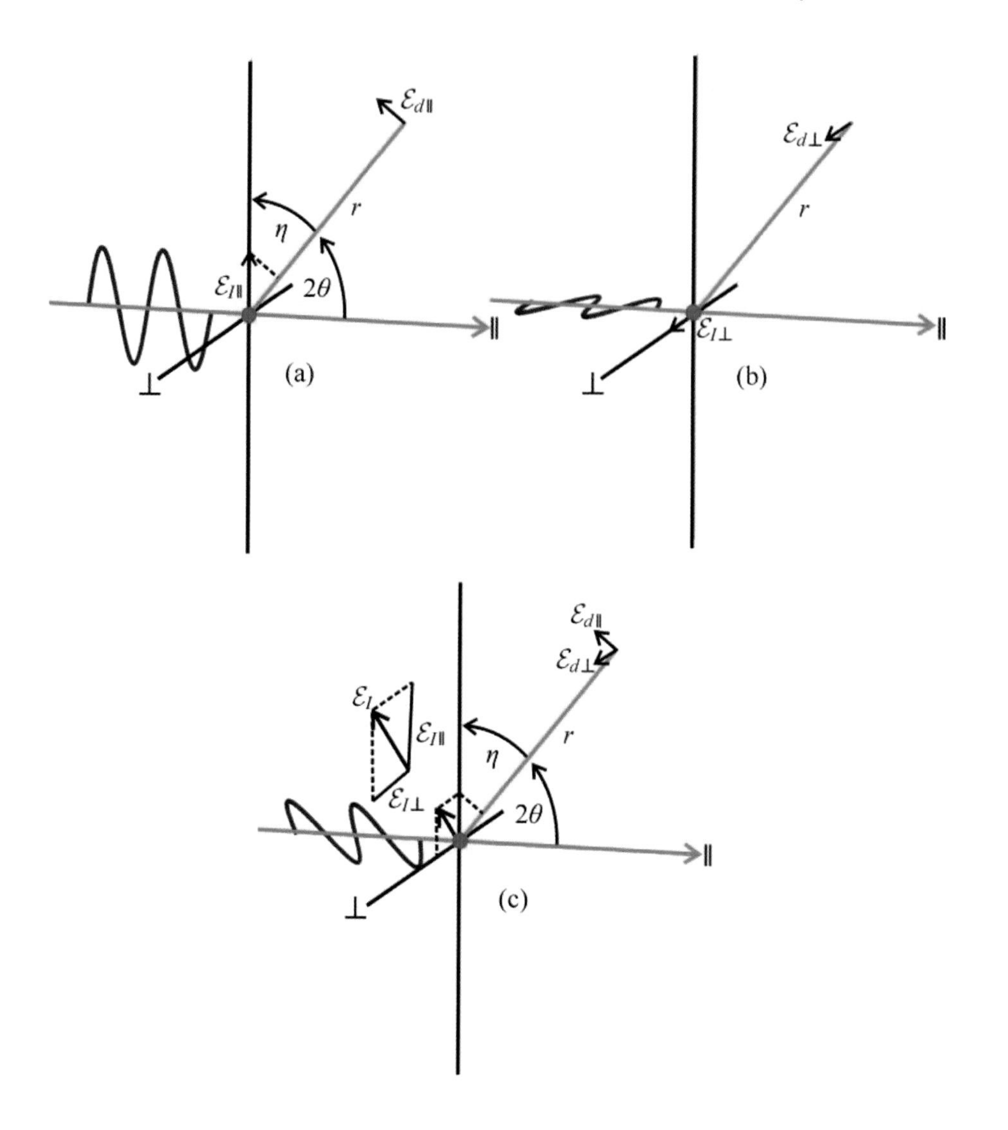

Figure 5.9 Effect on the electric field of the radiation scattered from an electron when the electric field vector of the incident radiation is (a) parallel to the diffraction plane, (b) perpendicular to the diffraction plane or (c) at an arbitrary angle to the diffraction plane.

The proportionality constant for this relationship is $\kappa = 1/4\pi\epsilon_0 c^2$ [59]:

$$\begin{aligned}\mathcal{E}_{d\|} &= \kappa \frac{q_e}{r} a \sin\eta \\ &= \kappa \frac{\mathcal{E}_{I\|} q_e^2}{r\, m_e} \sin\eta. \end{aligned} \tag{5.162}$$

Fig. 5.9(b) represents the case where the electric field vector of the incident radiation is perpendicular to the plane containing the incident and diffracted beams.

In this case the angle between the diffracted beam and the electric field vector of the incident radiation is $\eta' = 90°$. By an identical treatment,

$$\begin{aligned}
\mathcal{E}_{d\perp} &= \kappa \frac{q_e}{r} a \sin \eta' \\
&= \kappa \frac{\mathcal{E}_{I\perp} q_e^2}{r\, m_e} \sin(90°) = \kappa \frac{\mathcal{E}_{I\perp} q_e^2}{r\, m_e}.
\end{aligned} \tag{5.163}$$

In these two special cases, $\mathcal{E}_{I\|} = \mathcal{E}_{I\perp}$, and since $\sin \eta < 1$, the radiation from the beam polarized in the diffraction plane is less intense than the beam polarized in a direction perpendicular to the plane.

In Fig. 5.9(c), the electric field vector is at an arbitrary angle with respect to the plane containing the incident and diffracted beams. The intensity of the incident radiation is proportional to the square of the magnitude of the electric field vector: $I_0 \; \alpha \; \mathcal{E}_I^2$. The constant of proportionality is $\epsilon_0 c$[66]: $I_0 = \epsilon_0 c\, \mathcal{E}_I^2$. The electric field of the incident radiation can be resolved into components parallel and perpendicular to the diffraction plane, $\mathcal{E}_{I\|}$ and $\mathcal{E}_{I\perp}$. Each component contributes to the net intensity of the incident beam: $I_{I\|} = \epsilon_0 c\, \mathcal{E}_{I\|}^2$ and $I_{I\perp} = \epsilon_0 c\, \mathcal{E}_{I\perp}^2$. Since these components are perpendicular, $\mathcal{E}_{I\|}^2 + \mathcal{E}_{I\perp}^2 = \mathcal{E}_I^2$, and the net intensity is the sum of the intensities from each component: $I_0 = I_{I\|} + I_{I\perp}$.

In a similar fashion, the electric field of the diffracted beam can be resolved into parallel and perpendicular components, $\mathcal{E}_{d\|}$ and $\mathcal{E}_{d\perp}$. The parallel component of the diffracted beam results from the acceleration in the diffraction plane due to the $\mathcal{E}_{I\|}$, and the perpendicular component results from $\mathcal{E}_{I\perp}$. Again, the net diffracted intensity is the sum of the intensities from each component:

$$\begin{aligned}
I_d &= \epsilon_0 c\, \mathcal{E}_d^2 = \epsilon_0 c\, \mathcal{E}_{d\|}^2 + \epsilon_0 c\, \mathcal{E}_{d\perp}^2 \\
&= \epsilon_0 c \left(\kappa \frac{\mathcal{E}_{I\|} q_e^2}{r\, m_e} \sin \eta \right)^2 + \epsilon_0 c \left(\kappa \frac{\mathcal{E}_{I\perp} q_e^2}{r\, m_e} \right)^2 \\
&= \left(\frac{\kappa\, q_e^2}{r\, m_e} \right)^2 \left(\epsilon_0\, c\, \mathcal{E}_{I\|}^2 \sin^2 \eta + \epsilon_0\, c\, \mathcal{E}_{I\perp}^2 \right) \\
&= \left(\frac{\kappa\, q_e^2}{r\, m_e} \right)^2 \left(I_{I\|} \sin^2 \eta + I_{I\perp} \right).
\end{aligned} \tag{5.164}$$

Since the electric field vector of the incident radiation has an equal probability of being at any angle, we can average over all the angles to determine the average incident beam intensity: $\langle I_0 \rangle = \langle I_{I\|} \rangle + \langle I_{I\perp} \rangle$. Furthermore, $I_{I\|}$ and $I_{I\perp}$ can take on all possible values so that $\langle I_{I\|} \rangle = \langle I_{I\perp} \rangle$. Thus,

$$\begin{aligned}
\langle I_0 \rangle &= 2\langle I_{I\|} \rangle = 2\langle I_{I\perp} \rangle \text{ and} \\
\langle I_{I\|} \rangle &= \langle I_{I\perp} \rangle = \frac{1}{2} \langle I_0 \rangle.
\end{aligned} \tag{5.165}$$

The average value of the diffracted intensity, *the intensity that we measure*, is therefore

$$\begin{aligned}\langle I_d\rangle &= \left(\frac{\kappa\, q_e^2}{r\, m_e}\right)^2 \left(\langle I_{I\parallel}\rangle \sin^2\eta + \langle I_{I\perp}\rangle\right)\\ &= \left(\frac{\kappa\, q_e^2}{r\, m_e}\right)^2 \left(\frac{1}{2}\langle I_0\rangle \sin^2\eta + \frac{1}{2}\langle I_0\rangle\right)\\ &= \langle I_0\rangle \left(\frac{\kappa\, q_e^2}{r\, m_e}\right)^2 \left(\frac{1}{2}\sin^2\eta + \frac{1}{2}\right). \qquad (5.166)\end{aligned}$$

Referring to Fig. 5.9, the diffraction angle, $2\theta = 90° - \eta$, and $\sin\eta = \sin(90° - 2\theta) = \cos 2\theta$. Setting the constant term $\kappa\, q_e^2/r\, m_e$ in Eqn. 5.166 equal to κ_e results in an equation that models the effects of polarization as a function of the diffraction angle:

$$\langle I_d\rangle = \kappa_e^2\langle I_0\rangle \left(\frac{1}{2}\cos^2 2\theta + \frac{1}{2}\right). \qquad (5.167)$$

where the expression in parentheses is known as the polarization factor, P *:

$$P = \left(\frac{1+\cos^2 2\theta}{2}\right). \qquad (5.169)$$

As 2θ approaches zero, the polarization effect also approaches zero, since the perpendicular and parallel components of the diffracted beam become equal. As 2θ goes to zero, p becomes unity, and in the limit, $\langle I_d\rangle = \kappa_e^2\langle I_0\rangle$. In general, $\kappa_e^2\langle I_0\rangle$ would be the intensity of the diffracted beam if it was not affected by polarization, and multiplying $\kappa_e^2\langle I_0\rangle$ by the polarization factor diminishes the intensity by that factor. It therefore follows that *the diffracted beam intensity must be multiplied by the reciprocal of the polarization factor in order to correct it for the polarization of the diffracted beam:*

$$\frac{\langle I_d\rangle}{P} = \kappa_e^2\langle I_0\rangle$$

The relative intensity, corrected for polarization of the X-radiation is therefore

$$I_{rel}(P) = \frac{\langle I_d\rangle}{P}. \qquad (5.170)$$

The Lorentz Correction

As discussed in the previous chapter, reciprocal lattice points must actually be considered to be finite in size, with diffraction occurring as long as a portion of the

*If the incident radiation is polarized, as it is when it emanates from a synchrotron source and/or diffracts from a monochromator crystal, then P must include these additional polarization effects as well[47]. For a monochromator crystal the polarization factor is

$$P = \frac{(|\cos 2\theta_M|\cos^2\epsilon + \sin^2\epsilon)\cos^2 2\theta + |\cos 2\theta_M|\sin^2\epsilon + \cos^2\epsilon}{1+|\cos 2\theta_M|}, \qquad (5.168)$$

where θ_M is the diffraction angle for the monochromator reflection, and ϵ is the angle between the diffraction planes of the monochromator crystal and the specimen crystal.

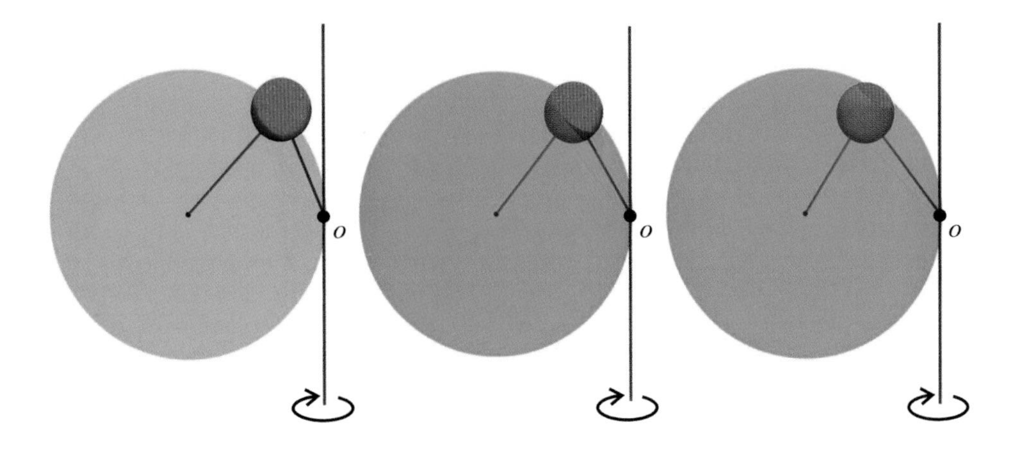

Figure 5.10 The rate of penetration of a reciprocal lattice node (greatly exaggerated) into the sphere of reflection is the rate at which it approaches the center of the sphere.

volume of a reciprocal lattice node is intersecting the sphere of reflection (Fig. 4.29). As the crystal is rotated at a constant rate in the diffraction experiment the time that the reciprocal lattice node remains in contact with the sphere depends on the location of the node (at the end of the reciprocal lattice vector) on the surface of the sphere. This, in turn depends on the rate at which the node is penetrating the sphere of reflection. Fig. 5.10 illustrates a reciprocal lattice node, approximated as a sphere, as it passes through the Ewald sphere. The size of the node has been magnified (tremendously!) for graphical purposes. The degree of penetration can be measured by the distance between the center of the node and the center of the sphere of reflection. The rate of penetration of the node into the sphere is therefore the rate at which the center of the node approaches the center of the Ewald sphere.

The number of photons observed for a given reflection will depend on the amount of time that its reciprocal lattice node remains in the diffraction condition. Even though the crystal is rotating at a constant rate, the rate at which the penetration varies will alter the measurement time, and we must find a way to correct for the variation of the integrated intensity obtained from different points of intersection on the sphere of reflection. The time that a node remains in contact with the sphere of reflection is inversely proportional to the rate at which it approaches the center of the sphere, and it is that rate that must be determined for each reflection.

Fig. 5.11(a) depicts a reciprocal lattice point, p, corresponding to a general reciprocal lattice vector $\mathbf{h}$, that has been rotated into contact with the sphere of reflection: $\boldsymbol{\Omega}\,\mathbf{X}\,\boldsymbol{\Phi}\,\mathbf{A}^*\mathbf{h} \;=\; \mathbf{h_r} = [h_x\ h_y\ h_z]$. Point p has coordinates (h_x, h_y, h_z) in the reciprocal Cartesian basis of the laboratory reference frame, defined with the X-ray beam along $+x_l$ and the origin of the reciprocal lattice at o, contacting the sphere of reflection at the point where the X-ray beam intersects the sphere as it exits. The reflection is observed by rotating ω at a constant angular velocity (*scan rate*), $v_\omega = \Delta\omega/\Delta t$. The rotation occurs about the $\mathbf{z}_l$ axis in a plane parallel to the $\mathbf{x}_l, \mathbf{y}_l$ plane intersecting the $\mathbf{z}_l$ axis at $o' \equiv (0, 0, h_z)$. The vector that is rotated in this plane is the projection of $\mathbf{h_r}$ onto the plane, consisting of the line segment of length h_p — connecting o' and p. The rate at which the point p passes through the sphere in this plane is given by its linear velocity in the plane, $v_p = h_p v_\omega$. If we construct a line segment from p to the center of the sphere, c, the penetration velocity of the reciprocal lattice point, v_c, will be the component of $\mathbf{v_p}$ (its projection) along that direction (toward the center of the sphere).

Fig. 5.11(b) illustrates the relationship between the linear velocity vector and the "penetrating vector." If η is the angle between the two vectors then $v_c = v_p \cos\eta$. In order to determine $\cos\eta$, we construct a perpendicular from the center of the small circle (formed from the intersection of the sphere and the rotation plane), c_c, to point q along the linear velocity vector, $\mathbf{v_p}$. Because $\mathbf{v_p}$ is perpendicular $\overrightarrow{o'p}$, $\overrightarrow{c_cq}$ is parallel to $\overrightarrow{o'p}$. The line segment between the center of the sphere and q creates a triangle (c, c_c, q) that is perpendicular to the rotation plane, so that $\overrightarrow{cq}$ is also perpendicular to $\overrightarrow{pq}$. Since $\overrightarrow{pc}$ is a radius of the sphere, its length is $1/\lambda$. Letting s be the length of the op segment, $\cos\eta = s/(1/\lambda) = s\lambda$.

Fig. 5.11(c) is a view looking down the $\mathbf{z}_l$ axis onto the small circle in the rotation plane. The dashed circle is a projection of the great circle in the $\mathbf{x}_l\mathbf{y}_l$ plane. The coordinates of p in the rotation plane are h_x, h_y. If ξ is the angle between $\overrightarrow{c_co'}$ and $\overrightarrow{o'p}$, then

$$\begin{aligned} v_c &= v_p \cos\eta = h_p v_\omega \cos\eta \\ &= h_p v_\omega s\lambda \\ \sin\xi &= \frac{h_y}{h_p} = \frac{s}{(1/\lambda)} = s\lambda \\ v_c &= v_\omega h_p \frac{h_y}{h_p} \\ &= v_\omega h_y. \end{aligned} \tag{5.171}$$

The penetration velocity of the lattice point depends only on the scan rate, v_ω, and the y coordinate of the reciprocal lattice point in contact with the sphere of reflection. The time of penetration is inversely proportional to the penetration velocity:

$$t_p = C\frac{1}{v_\omega h_y}. \tag{5.172}$$

Since t_p is the time that a reciprocal lattice point remains in contact with the sphere, the measured intensity, $\langle I_d \rangle$, is proportional to t_p. It therefore follows that *in order to correct the measured intensity for variations due to different penetration times, we must divide each measured intensity by its penetration time*:

$$I_{corrected} = \frac{\langle I_d \rangle}{t_p} = (\langle I_d \rangle)\frac{1}{C} v_\omega h_y. \tag{5.173}$$

Since we are concerned with relative intensities we can ignore the constant, C, which can be included in the scale factor. If the scan rate is constant it can also be included in the scale factor. However, if the scan rate varies for different reflections we must correct the penetration time to account for the variation. The measured intensities, $\langle I_d \rangle$, are determined by dividing integrated intensities by their measurement times, effectively normalizing them to a constant scan rate. Since the penetration time, t_p, is inversely proportional to the scan rate, it must also be normalized to a constant scan rate. This is accomplished by multiplying the penetration time for

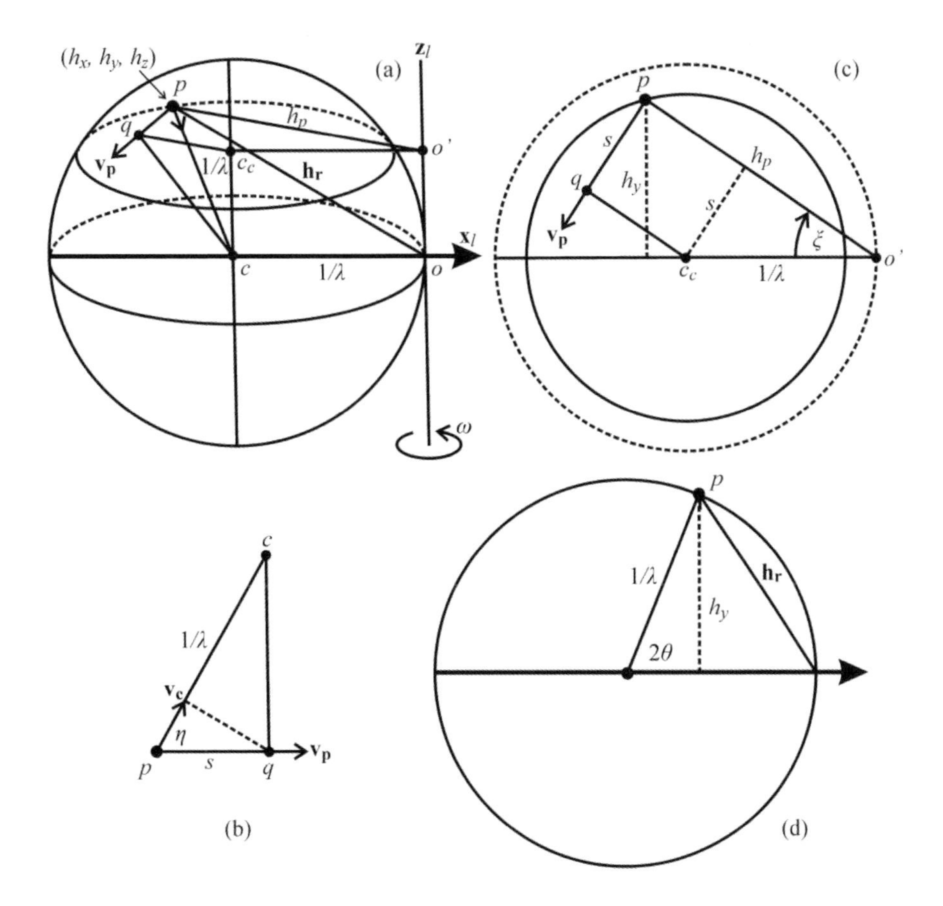

Figure 5.11 (a) Reciprocal lattice point p crossing the sphere of reflection at velocity v_p. (b) Relationship between linear velocity, v_p and the velocity of penetration into the sphere of reflection, v_c. (c) View down the $\mathbf{z}_l$ axis of the small circle resulting from the intersection of the sphere of reflection with the rotation plane. (d) Reciprocal lattice point crossing the sphere of reflection on the great circle in the equatorial plane.

each reflection by its scan rate. The normalized penetration time, which we denote L (see below), is given by

$$L = t_p v_\omega = \frac{1}{h_y} \quad \text{and} \tag{5.174}$$

$$I_{rel}(L) = \frac{\langle I_d \rangle}{L} = \langle I_d \rangle h_y. \tag{5.175}$$

This effect on the intensities of diffracted X-radiation was first observed by Nobel laureate Hendrik A. Lorentz,* the Dutch physicist responsible for the discovery of the Zeeman Effect (along with his student, Pieter Zeeman), the Lorentz contraction in relativity theory, and numerous contributions to the theory of electromagnetic radiation. The correction factor L is known as *the Lorentz factor*.

*The observation was made by Lorentz during one of his classroom lectures.

For diffractometers employing an area detector, the Lorentz factor is a relatively complex function of the diffractometer angles and the orientation matrix elements. From Eqn. 4.24, $\boldsymbol{\Upsilon}^T\mathbf{A}^*\mathbf{h} = \mathbf{h_r}$:

$$\begin{bmatrix} \upsilon_{11} & \upsilon_{21} & \upsilon_{31} \\ \upsilon_{12} & \upsilon_{22} & \upsilon_{32} \\ \upsilon_{13} & \upsilon_{23} & \upsilon_{33} \end{bmatrix} \begin{bmatrix} a_{11} & a_{12} & a_{13} \\ a_{21} & a_{22} & a_{23} \\ a_{31} & a_{32} & a_{33} \end{bmatrix} \begin{bmatrix} h \\ k \\ l \end{bmatrix} = \begin{bmatrix} h_x \\ h_y \\ h_z \end{bmatrix}$$

$$\begin{aligned} h_y = \; & \upsilon_{12}(a_{11}h + a_{12}k + a_{13}l) + \\ & \upsilon_{22}(a_{21}h + a_{22}k + a_{23}l) + \\ & \upsilon_{32}(a_{31}h + a_{32}k + a_{33}l), \end{aligned} \tag{5.176}$$

where $\upsilon_{12} = (\cos\varphi\sin\omega - \sin\varphi\cos\chi\cos\omega)$, $\upsilon_{22} = (\sin\varphi\sin\omega + \cos\varphi\cos\chi\cos\omega)$, and $\upsilon_{32} = (\sin\chi\cos\omega)$.

The Lorentz factor takes on a much simpler form when the reflections are measured in the equatorial plane with a serial detector. Fig. 5.11(d) illustrates this case. Since h_y lies in the diffraction plane,

$$\begin{aligned} \sin 2\theta &= \frac{h_y}{(1/\lambda)} = h_y\lambda \\ h_y &= \frac{\sin 2\theta}{\lambda} = \frac{2\sin\theta\cos\theta}{\lambda}. \end{aligned} \tag{5.177}$$

The wavelength can be incorporated into the scale factor and the Lorentz factor for reflections collected with a serial detector becomes

$$L = \frac{1}{2\sin\theta\cos\theta}. \tag{5.178}$$

The Absorption Correction

Although we have been treating the intensity of the incident X-ray beam as a constant, it diminishes as the beam passes through the crystal. The intensity of the diffracted X-ray beam is proportional to the intensity of the incident beam, and the changes in incident beam intensity will affect the magnitudes of the observed intensities in the diffraction experiment. Since we are interested in relative intensities, *if* the loss in intensity in the incident beam was a constant, this loss could simply be incorporated into the intensity scale factor. Unfortunately, intensity loss, referred to as *absorption*, depends on the distance that the beam travels as it passes through the crystal, which in turn depends on the shape of the crystal and the diffraction angle. Consequently, it is necessary to correct for the effects of varying intensity loss in the measured intensity data.

In order to understand the absorption process we must focus on the quantum-mechanical nature of the atom. The electronic state function of the atom (see Sec. 3.2.4) is rigorously a multielectronic function that depends on all of the electrons simultaneously, but it can be described with a high degree of accuracy as a complex product (actually a sum of products – a determinant) of one-electron functions known as orbitals. These functions are the familiar *s*, *p*, *d* and *f* orbitals described qualitatively in numerous elementary chemistry and physics textbooks.

The atom is often loosely envisioned as being composed of overlapping spherical "shells", with the energy range of the electron in a given shell determined by an integer known as the *principal quantum number*, n. Each shell contains a set of orbitals corresponding to two more integer quantum numbers, l and m_l, such that the energies of the orbitals vary slightly within a given principle quantum level. The lowest lying energy level corresponds to $n = 1$ and is known as the K level for the atom. The next level, for $n = 2$ corresponds to the L level, and so forth. The letters corresponding to the principal quantum levels for the atom begin with K because the energy levels for atoms were first observed empirically, and it was not known whether or not lower levels might later be discovered.

When an X-ray photon encounters an electron in an atom its electric field interacts with the electron. Quantum mechanically, the photon is instantaneously absorbed by the electron, raising it to a higher energy level, placing the atom into an excited state. If the excited state is not a stationary (time independent) state a new photon is created and ejected — the atom returns instantly to its ground state (this time-dependent state is often referred to as a *virtual state*, since it does not correspond to an observable atomic state). The photon travels from the atom in a new direction, but retains all of the energy of the incident photon, i.e., the photon is *elastically scattered*. This is the radiation that is observed in the diffraction experiment.

If the atom instead finds itself in an excited stationary state during the absorption process, the atom usually remains in that state for an observable period of time (the *lifetime* of the state). If the energy of the photon corresponds exactly to the difference in energies between the ground and excited states of the atom then the entire photon is absorbed. If it does not, another photon is emitted — one with a lower energy than the incident photon. This photon is also emitted in a new direction – it is *inelastically scattered*. The atom ordinarily returns to the ground state through non-radiative processes.

The most important loss of photons arises from the creation of *photoelectrons* from the constituent atoms in the crystal. In this case, the incident photon has sufficient energy to excite an electron to a point where it is removed from the atom, producing a photoelectron. In the process, the photon is completely absorbed and the ejected electron has a kinetic energy that is the difference between the energy of the absorbed photon and the energy required to remove the electron. This is the well-known *photoelectric effect*. Albert Einstein was awarded the first of his two Nobel prizes for its discovery. Fig. 5.12 is a plot of the relative number of X-ray photons absorbed as a function of wavelength. The discontinuous nature of the plot is a direct consequence of the quantum nature of the atom. The probability that a photon will be absorbed is optimal when its energy matches the energy required to remove an electron from the atom, its *ionization energy*. The maxima in the figure, referred to as *absorption edges*, correspond to the ionization energies of the electrons from various quantum levels (K, L, M, etc.) of the atom. Once an ionization energy is reached, the ionization probability decreases until another photon-ionization energy match is achieved and the absorption jumps to a new maximum. The complex nature of the absorption profile is an important consideration in determining the effects of absorption – *the probability that an X-ray photon will be absorbed by an atom depends intimately on both the energy of the photon and the element-specific quantum-mechanical nature of the atom.*

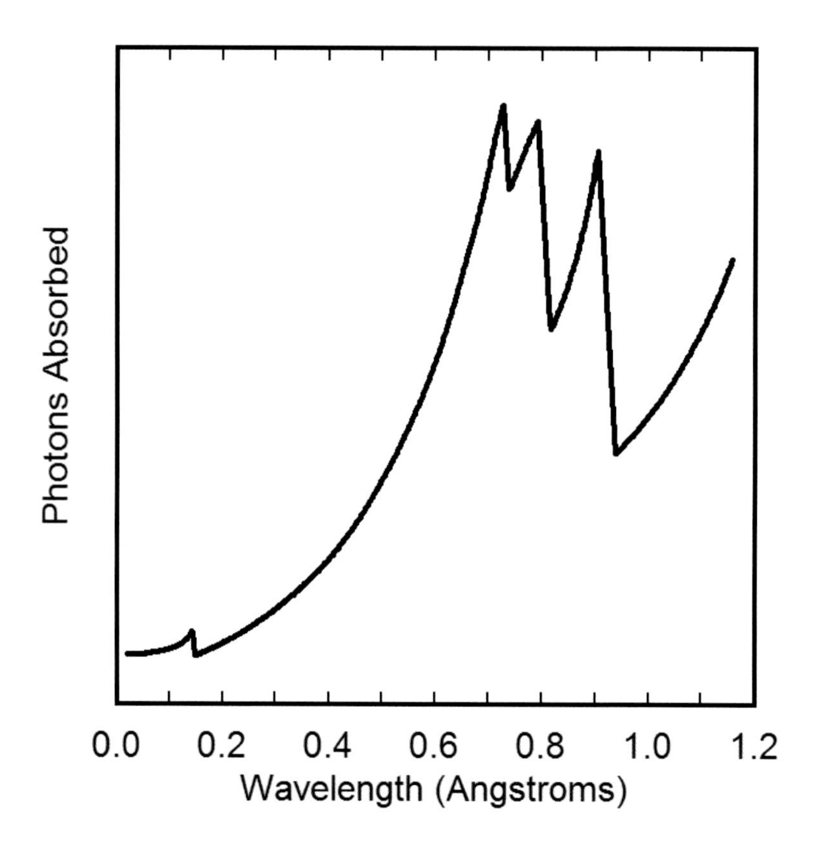

Figure 5.12 Relative absorption of X-ray photons by a sample of lead as a function of X-ray wavelength.

We begin an analysis of the absorption of X-rays by considering a single atom, atom type A, subjected to a beam of monochromatic X-radiation of a specific wavelength. Of the photons that impinge upon the atom, a certain fraction, μ_{A} will be absorbed due to the processes discussed above and the remaining photons will be transmitted. The *atomic absorption coefficient*, μ_A, is a measure of the probability that a photon will be absorbed by a single "type A" atom, and is characteristic of the atom type and the wavelength.

Fig. 5.13 depicts an X-ray beam with a unit cross-sectional area passing a distance x through a crystal composed of compound X consisting of two types of atoms, atom type A and atom type B. $A = 1$ because we wish to determine the effects of absorption on the *intensity* of the beam. Intensity is a measure of the energy per unit area and is therefore *proportional to the number of photons in a unit area of the beam.*

For the moment, we ignore the "B" atoms, and focus on the effects of the "A" atoms on the intensity of the beam as it passes through the crystal. Each time an atom is encountered by the beam it loses a fraction of its intensity, $\Delta I/I = \mu_A$.

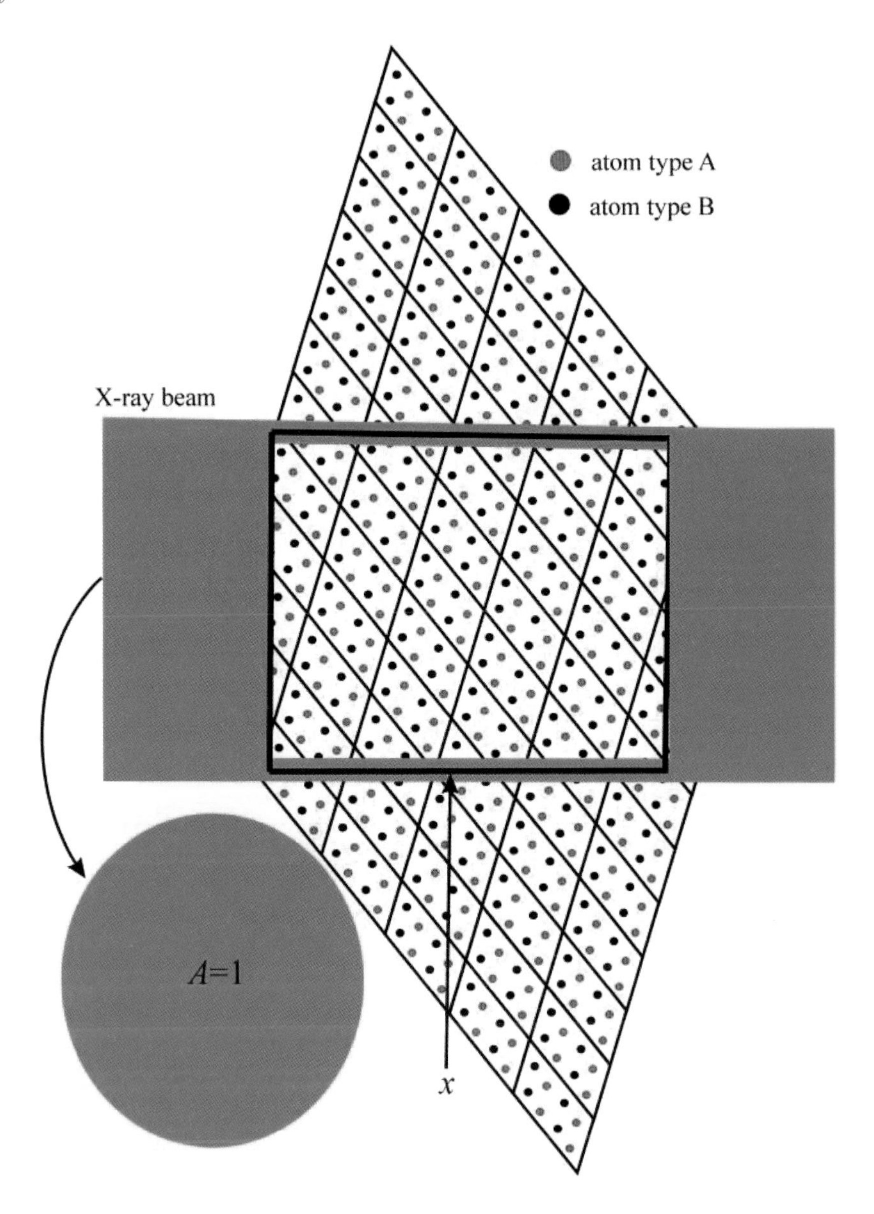

Figure 5.13 X-ray beam with unit cross-sectional area passing a distance x through a hypothetical crystal consisting of two atom types.

If n atoms are encountered *simultaneously*, then $\Delta I/I = n\mu_A$. If the interactions are not simultaneous, then this relationship will not be exact, since the intensity will change as the beam passes from one atom to the next. For a small change in distance, Δx, the interactions are *nearly* simultaneous and the relationship can be considered to be a reasonable approximation. If C_A is the concentration of atom type A in the crystal — the number of "A" atoms in a unit volume — and V is the volume of the crystal that the beam passes through, then $n = C_A V$ and $\Delta I/I \approx C_A V \mu_A$. If the beam travels a small distance Δx, then $V = A\Delta x = 1\,\Delta x$

and $\Delta I/I \approx C_A \Delta x \mu_A$. Since C_A and σ_A are constants in the experiment, we set $C_A \mu_A = \mu_l^A$ and rearrange the expression to give the *rate* of intensity loss with distance (the negative sign indicates a decrease in intensity for a positive change in distance):

$$\frac{\Delta I}{\Delta x} \approx -\mu_l^A I. \tag{5.179}$$

μ_l^A is known as the *linear absorption coefficient* since it is related to the linear distance traversed by the beam through the crystal. μ_l^A has units of *area/volume* = 1/*distance*, usually expressed in mm^{-1} or cm^{-1}. Eqn. 5.179 indicates that the rate at which the intensity decreases is directly proportional to the intensity. Qualitatively, this tells us that the rate of intensity loss diminishes as the beam passes through the crystal, since the intensity diminishes. In Fig. 5.14(a) this approximation is illustrated for several intervals of Δx. Focusing on the interval that begins with intensity I_1 and ends with I_2, note that the rate of intensity loss in the previous interval is larger, and the rate in the subsequent interval is smaller.

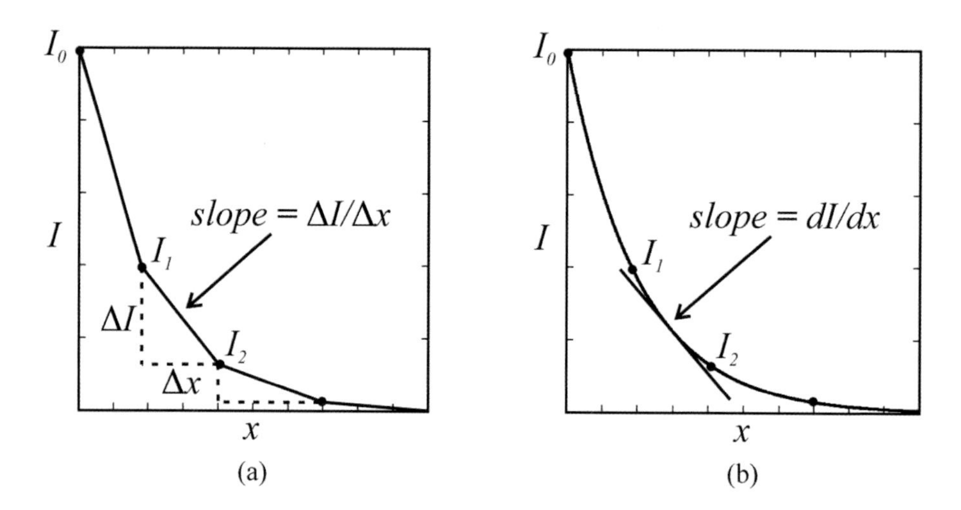

Figure 5.14 (a) The rate of change of intensity with distance approximated with linear changes, $\Delta x/\Delta I$. (b) The rate of change of intensity with distance in the limit as $\Delta x \to dx$ and $\Delta I \to dI$.

Because the intensity is continually changing over the interval, a reasonable approximation to this rate can be determined by using the average of I_1 and I_2 over the interval:

$$\frac{I_2 - I_1}{x_2 - x_1} = \frac{\Delta I}{\Delta x} \approx -\mu_l^A \left(\frac{I_1 + I_2}{2} \right). \tag{5.180}$$

The error in this approximation occurs because the actual rate of change in intensity that results when the intensity is $(I_1 + I_2)/2$ differs from our linear approximation of the rate of change. The error is increased if Δx is increased, and it is decreased if Δx is made smaller — as x_2 approaches x_1 and I_2 approaches I_1. In the limit Δx

becomes the differential dx, ΔI becomes dI, $I_1 = I_2 = I$, and Eqn. 5.180 becomes exact (Fig. 5.14(b)):

$$\frac{dI}{dx} = -\mu_l^A \left(\frac{I+I}{2} \right) = -\mu_l^A I. \tag{5.181}$$

By rearranging the variables and integrating we obtain an equation that gives the intensity of an X-ray beam with an initial intensity, I_0, as it passes a distance x through the crystal (recall that we are considering only the effects of atom type A here):

$$\begin{aligned} \int_{I_0}^{I} \frac{dI}{I} &= -\mu_l^A \int_0^x dx \\ \ln I - \ln I_0 &= -\mu_l^A x \\ \ln \frac{I}{I_0} &= -\mu_l^A x \\ I &= I_0 e^{-\mu_l^A x} \end{aligned} \tag{5.182}$$

Eqn. 5.182 was determined empirically by Pierre Bouguer before 1729.[67]. The relationship is ordinarily referred to as the *Lambert-Beer Law*, mis-attributed to Johann Heinrich Lambert, who cited and quoted from Bouguer's *Essai d'Optique sur la Gradation de la Lumière.*[68] In 1852, August Beer included the concentration of solutions in the absorption coefficient;[69] since the relationship depends only on absorption, it is equally valid for photons of any wavelength and is commonly used in absorption spectroscopy to measure concentrations of molecules in solution (recall that μ_l^A contains the concentration of atom type A).

If we now include the absorption contributions for atom type B, then $\mu_l^B = C_B \mu_B$, and the contribution to the intensity loss from both types of atoms will be additive:

$$\frac{dI}{I} = -\mu_l^A dx - \mu_l^B dx = -(\mu_l^A + \mu_l^B)\, dx = -\mu_l^X dx \quad \text{and} \tag{5.183}$$

$$\frac{I}{I_0} = e^{-\mu_l^X x}, \tag{5.184}$$

where $\mu_l^X = \mu_l^A + \mu_l^B$ is the linear absorption coefficient of the material X composed of atoms of type A and B.

Eqn. 5.184 provides a straightforward relationship between the relative intensity, I/I_0, and the path length of the beam, x. *In order to make use of the equation we must know the path length for each reflection through the crystal and the linear absorption coefficient as well.* μ_l^X depends on the various types of atoms in the material and the number of each that are encountered per unit distance. To determine μ_l^X experimentally it would be necessary to measure the intensity loss for an X-ray beam passing a distance x through a crystal of material X. Unfortunately, most crystals used for structural studies are far too small for accurate measurements of μ_l^X, and we must find a way to determine it from the composition of the crystal.

Given the density of the crystal, each atom type constitutes a specific fraction of the mass and density of the crystal. If we determine the effects of absorption of a specific mass of each atom type we can determine the linear absorption coefficient, since a specific mass of each atom type will interact with the X-ray beam over a given distance. In order to determine the effects of absorption for this mass, we begin with a crystal of the elemental material composed entirely of atom type A. Instead of focusing on the effects of absorption as the beam traverses a certain distance dx through the crystal, we note that it simultaneously encounters a certain mass of the material, dm_A, containing n_A atoms — so that $dm_A = n_A\times$ the mass of a single atom. In this case the fractional change in the intensity is given by:

$$\frac{dI}{I} \alpha \, dm_A = -\mu_m^A \, dm_A, \tag{5.185}$$

where μ_m^A is the *mass absorption coefficient.* The volume containing the mass dm_A is $V = Adx = 1\,dx$. If ρ_A is the density of the material, then $\rho_A = dm_A/V = dm_A/dx$, and $dm_A = \rho_A dx$. Thus,

$$\begin{aligned} \frac{dI}{I} &= -\rho_A \mu_m^A dx \Rightarrow \\ \mu_l^A &= \rho_A \, \mu_m^A \;\; \text{and} \qquad (5.186) \\ \frac{I}{I_0} &= e^{-\mu_l^A x} = e^{-\mu_m^A \rho_A x}. \qquad (5.187) \end{aligned}$$

In order to determine the mass absorption coefficient for an element, a beam with intensity I_0 is passed through a thickness x of the material and the attenuated intensity, I is measured. From these data μ_l^A is determined and $\mu_m^A = \mu_l^A/\rho_A$ is calculated from the density of the element.

A crystal containing more than one type of atom can be treated in exactly the same manner. If ρ_X is the density of the material then

$$\begin{aligned} \frac{dI}{I} &= -\mu_m^X \, dm_X = -\mu_m^X \, \rho_X \, dx \;\; \text{and} \qquad (5.188) \\ \mu_l^X &= \rho_X \, \mu_m^X. \qquad (5.189) \end{aligned}$$

For the material containing atom types A and B, the contribution of each atom to dm_X is dependent on the fraction of the mass that each atom type represents. If f_A is the fraction of the mass of the material represented by atom A, then the beam encounters a mass of $dm_A = f_A \, dm_X$ due to the presence of atom type A. Similarly, the beam encounters a mass of $dm_B = f_B \, dm_X$ due to the presence of atom type B. Each atom type contributes proportionately to the intensity loss:

$$\begin{aligned} \frac{dI_A}{I} &= -\mu_m^A \, dm_A = -\mu_m^A \, f_A dm_X \\ \frac{dI_B}{I} &= -\mu_m^B \, dm_B = -\mu_m^B \, f_B dm_X \\ \frac{dI}{I} &= -(f_A \mu_m^A + f_B \mu_m^B) dm_X \\ &= -(f_A \mu_m^A + f_B \mu_m^B) \rho_X dx \\ &= -\mu_l^X dx. \end{aligned}$$

It follows that

$$\mu_l^X = (f_A \mu_m^A + f_B \mu_m^B) \rho_X.$$

This argument can be extended to all of the elements in a crystal:

$$\mu_l = (f_A \mu_m^A + f_B \mu_m^B + f_C \mu_m^C + \cdots)\rho. \tag{5.190}$$

If the volume of the unit cell and its contents are known, then the density of the crystal and the fractional mass of each element can be easily computed. Mass absorption coefficient tables for the elements covering X-ray wavelengths from the commonly used X-ray target materials can be found in the *International Tables for Crystallography*. *The linear absorption coefficient can be calculated from the contents of the unit cell and the mass absorption coefficients of its component elements.*

An example of the determination of the linear absorption coefficient for a crystal containing "light, medium, and heavy" elements at three common X-ray wavelengths serves to illustrate several important considerations in the structural solution process. The molecular compound, $C_{21}H_{16}N_6O_9S_3Mn_2W$, packs in a monoclinic unit cell with cell parameters $a = 16.876$ Å, $b = 9.675$ Å, $c = 17.926$ Å, $\beta = 106.96°$, and $V = 2799.65$ Å^3. The molar mass of the compound is 886.32 g/mol. Table 5.2 lists the fractional masses of the elements in this compound and the mass absorption coefficients of each of the elements for Cr$K\alpha$ ($\lambda = 2.2909$ Å), Cu$K\alpha$ ($\lambda = 1.5418$ Å), and Mo$K\alpha$ ($\lambda = 0.7107$ Å) radiation.

Table 5.2 Mass Absorption Coefficients and Fractional Masses of Elements in $C_{21}H_{16}N_6O_9S_3Mn_2W$.

	$\mu_m = \mu_l/\rho$ (cm^2/g)*			Elemental	Fractional
Element	Cr $K\alpha$	Cu $K\alpha$	Mo $K\alpha$	Mass (g/mol)	Composition
H	0.545	0.435	0.380	16.13	0.018
C	14.5	4.60	0.625	252.231	0.285
N	23.9	7.52	0.916	84.04	0.095
O	36.6	11.5	1.31	143.99	0.162
S	272	89.1	9.55	96.20	0.109
Mn	93.0	285	34.7	109.88	0.124
W	458	172	99.1	183.85	0.207

**International Tables for Crystallography.*

The mass of a formula unit (molecule) of the compound is 886.32 g/N_o = 1.472×10^{-21} g, where $N_o = 6.022 \times 10^{23}$ (Avogadro's number). The volume of the unit cell is 2.800×10^{-21} cm^3. Monoclinic unit cells ordinarily contain either 2, 4, or 8 asymmetric units; a molecule is often an asymmetric unit. Two formula units per unit cell would result in a density of $(2 \times 1.472)/2.800 = 1.051$ g/cm^3. This is consistent with a compound containing only light atoms (e.g., H,C,N,O, etc.), which generally exhibit densities around 1 g/cm^3. However, the heavier atoms in the molecule (S, Mn, W) constitute over 40% of its mass, and the density is expected to be substantially greater. Four formula units doubles the density to 2.102 g/cm^3 — a reasonable value. Eight formula units would require the material to have a density of 4.204 g/cm^3, which is too large. We therefore predict that the structure has four molecules per unit cell, with $\rho = 2.102$ g/cm^3. This exercise is always useful during the early stages of structure determination. *If a consistent density can't be found for the unit cell, then the unit cell determination should be carefully scrutinized before continuing.*

The linear absorption coefficient for the three wavelengths can now be determined:

$$\mu_l = (f_{\mathrm{H}}\mu_m^{\mathrm{H}} + f_{\mathrm{C}}\mu_m^{\mathrm{C}} + f_{\mathrm{N}}\mu_m^{\mathrm{N}} + f_{\mathrm{O}}\mu_m^{\mathrm{O}} + f_{\mathrm{S}}\mu_m^{\mathrm{S}} + f_{\mathrm{Mn}}\mu_m^{\mathrm{Mn}} + f_{\mathrm{W}}\mu_m^{\mathrm{W}})\rho$$

$$\mu_l(\mathrm{Cr}\,K\alpha) = 311.9\ \mathrm{cm}^{-1}$$

$$\mu_l(\mathrm{Cu}\,K\alpha) = 177.8\ \mathrm{cm}^{-1}$$

$$\mu_l(\mathrm{Mo}\,K\alpha) = \ \ 55.4\ \mathrm{cm}^{-1}$$

This example illustrates a general trend — X-ray absorption increases as the wavelength of the X-radiation increases. It follows that errors due to absorption from different path lengths through the crystal will be greater for less energetic X-rays. This is a major advantage of Mo $K\alpha$ radiation over Cu $K\alpha$ or Cr $K\alpha$ radiation. The discontinuous nature of the X-ray absorption profile for an atom (see Fig. 5.12) complicates this picture to some degree — if the X-ray wavelength happens to be near an absorption edge for a particular atom in the crystal, then the absorption will be significantly increased. Note the value of μ_m for manganese in Table 5.2. The absorption of Cu $K\alpha$ radiation for manganese is over three times that of the longer wavelength Cr $K\alpha$ radiation.

The elements in Table 5.2 are listed in order of increasing atomic number, illustrating another general trend: elements that contain more electrons absorb more X-ray photons, just as they scatter more X-ray photons. Not only do heavy elements contribute proportionately more of the scattered radiation in the diffraction pattern — they also absorb more of the X-rays as they pass through the crystal. Absorption edges also affect this trend — note that sulfur absorbs over three times as much radiation as manganese when exposed to Cr $K\alpha$ radiation.

Absorption Corrections From Crystal Measurements. The magnitude of the absorption of X-rays is different for each reflection, and is determined by the amount of material that the incident and diffracted beams encounter as they pass through the crystal on their way to the detector. This, in turn, depends on the diffraction angle and the shape and orientation of the crystal in the diffraction condition for each reflection.

If the crystal has well-defined faces its image can be magnified and measured using an optical device attached to the diffractometer. The optical device is oriented on the instrument so that its local coordinate system, x_o, y_o, z_o, can be easily transformed into the diffractometer (laboratory) reference frame, x_l, y_l, z_l. In a typical (but not exclusive) setting the longitudinal axis (viewing axis) of the device is oriented so that a rotation of the diffractometer Eulerian angles, φ_o and χ_o (or alternatively, φ_o and ω_o), will realign the y_l axis along the viewing axis, y_o; the x_l axis will be parallel to a reference line on the optical device, x_o; the z_l axis will be perpendicular to the reference line, z_o. A vector oriented in the optical coordinate system can then readily be transformed back into the laboratory reference frame by $-\varphi_o$ and $-\chi_o$ rotations. The optical device is ordinarily equipped with a graduated reticle, calibrated so that its graduations can provide measurements of the crystal along the z_o axis of the telescope, as illustrated in Fig. 5.15(a). The crystal is measured by rotating φ and χ until each plane face of the crystal is "edge on"

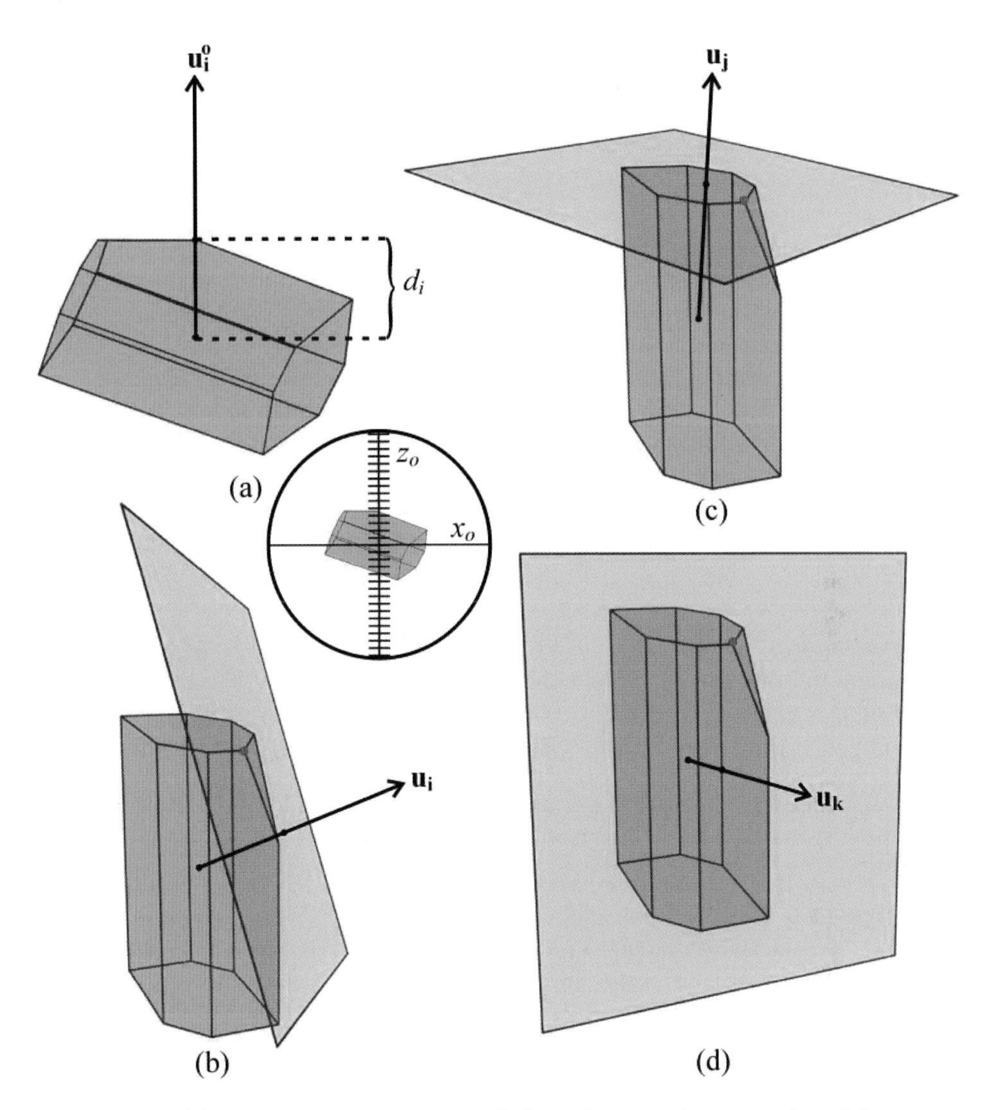

Figure 5.15 (a) Crystal face at a distance d_i from the crystal center oriented in an optical device with the unit vector $\mathbf{u}_i^o$ normal to the facial plane and aligned with the z_o axis. (b)–(d) Three plane faces defining a vertex of the crystal polyhedron(marked with a red dot). Unit vectors, $\mathbf{u}_i$, $\mathbf{u}_j$, and $\mathbf{u}_k$ are plane normals at distances d_i, d_j, and d_k, described in the laboratory reference frame at $\omega = \chi = \varphi = 0$.

(i.e., it appears as a line) and is parallel to the x_o axis. The measured distance from the center of mass of the crystal (the origin), d_m, and the rotation angles, φ_m and χ_m, are recorded for each face. This information is sufficient to define the shape of the crystal and its orientation in the laboratory reference frame. The angles determine the direction of a vector normal to a plane, and the distance specifies its displacement from the center of the crystal. The crystal is considered to be a convex polyhedron here, * and the set of n planes defined by their normal vectors

*A convex polyhedron is a polyhedron for which a line connecting any two points on its surface will contain points exclusively within the polyhedron. Crystal polyhedra that are not convex

intersect so that the bound plane surfaces of the crystal are uniquely determined. Each vertex is created by the intersection of three of these planes, as shown in Fig. 5.15 (b)–(d). The three planes, denoted with i, j, and k, are defined by unit perpendicular vectors $\mathbf{u_i}$, $\mathbf{u_j}$, and $\mathbf{u_k}$ with their surfaces at distances d_i, d_j, and d_k from the center of the crystal. The red dot in the figure identifies the vertex created by the intersection of these three planes.

In the reference frame of the optical device the vector normal to the ith facial plane is aligned with the z_o axis, with components $\mathbf{u_i^o} = [0\ 0\ 1]$, observed in this position at diffractometer angles φ_m and χ_m. Rotating this vector through angles $-\varphi_o$ and $-\chi_o$ will place it back in the laboratory reference frame, now aligned with the z_l axis, with components $[0\ 0\ 1]$. The goniometer will now be at angles $\omega_i = 0$, $\chi_i = \chi_m - \chi_o$ and $\varphi_i = \varphi_m - \varphi_o$. If the diffractometer angles are set to zero (the reference position of the crystal), then φ_i and χ_i are the angles that will rotate the unit vector, $\mathbf{u_i}$, to align it along the z_l axis:

$$\begin{bmatrix} 1 & 0 & 0 \\ 0 & \cos\chi_i & \sin\chi_i \\ 0 & -\sin\chi_i & \cos\chi_i \end{bmatrix} \begin{bmatrix} \cos\varphi_i & \sin\varphi_i & 0 \\ -\sin\varphi_i & \cos\varphi_i & 0 \\ 0 & 0 & 1 \end{bmatrix} \begin{bmatrix} u_{i1} \\ u_{i2} \\ u_{i3} \end{bmatrix} = \begin{bmatrix} 1 \\ 0 \\ 0 \end{bmatrix}$$

The measurement of φ_m and χ_m for all of the crystal faces allows us to determine the normal unit vectors in the reference orientation of the crystal for all n faces of the crystal:

$$\begin{bmatrix} \cos\varphi_i & -\sin\varphi_i & 0 \\ \sin\varphi_i & \cos\varphi_i & 0 \\ 0 & 0 & 1 \end{bmatrix} \begin{bmatrix} 1 & 0 & 0 \\ 0 & \cos\chi_i & -\sin\chi_i \\ 0 & \sin\chi_i & \cos\chi_i \end{bmatrix} \begin{bmatrix} 1 \\ 0 \\ 0 \end{bmatrix} = \begin{bmatrix} u_{i1} \\ u_{i2} \\ u_{i3} \end{bmatrix} = \mathbf{u_i} \tag{5.191}$$

When crystals form, they generally terminate along low-index lattice planes, forming planar faces. The normal unit vectors determined here are therefore parallel to the reciprocal lattice vectors describing the faces. The normal vectors are in reciprocal Cartesian coordinates, but they can be transformed into coordinates in the reciprocal basis using the orientation matrix. Since $\mathbf{A}^*\mathbf{v_f^*} = \mathbf{UB}^*\mathbf{v_f^*} = \mathbf{v_l^*}$ transforms a vector in the reciprocal basis into a vector in the laboratory reference frame,

$$\mathbf{u_i^*} = (\mathbf{A}^*)^{-1}\mathbf{u_i} = u_{i1}^*\mathbf{a}^* + u_{i2}^*\mathbf{b}^* + u_{i3}^*\mathbf{c}^*, \tag{5.192}$$

describes the unit vector perpendicular to the ith crystal face in the reciprocal basis. The reciprocal lattice vector for the facial plane, $\mathbf{h_i}$, is parallel to $\mathbf{u_i^*}$ with integer components that can be generated by dividing each component of $\mathbf{u_i^*}$ by the smallest component, then multiplying each resultant component by the smallest integer necessary to create integer indices. These indices will be those of one set of lattice planes parallel to the crystal face. This exercise serves two purposes. First, we find it useful to express the normal unit vectors corresponding to each crystal face in the reciprocal basis, and second, constraining the components of the vector parallel to each $\mathbf{u_i^*}$ to have integer components corrects measurement errors. The normal unit vector, $\mathbf{u_i^*}$, can be refined by transforming $\mathbf{A}^*\mathbf{h_i} = \mathbf{h_{il}}$, calculating $\mathbf{u_i} = \mathbf{h_{il}}/\mathrm{h}_{il}$ in the Cartesian laboratory reference frame, then transforming $\mathbf{u_i}$ back into the reciprocal basis with Eqn. 5.192.

contain re-entrant angles, which allow a connecting line to exit the crystal and re-enter it at another point. Such crystals are often indicative of a phenomenon known as *twinning*.

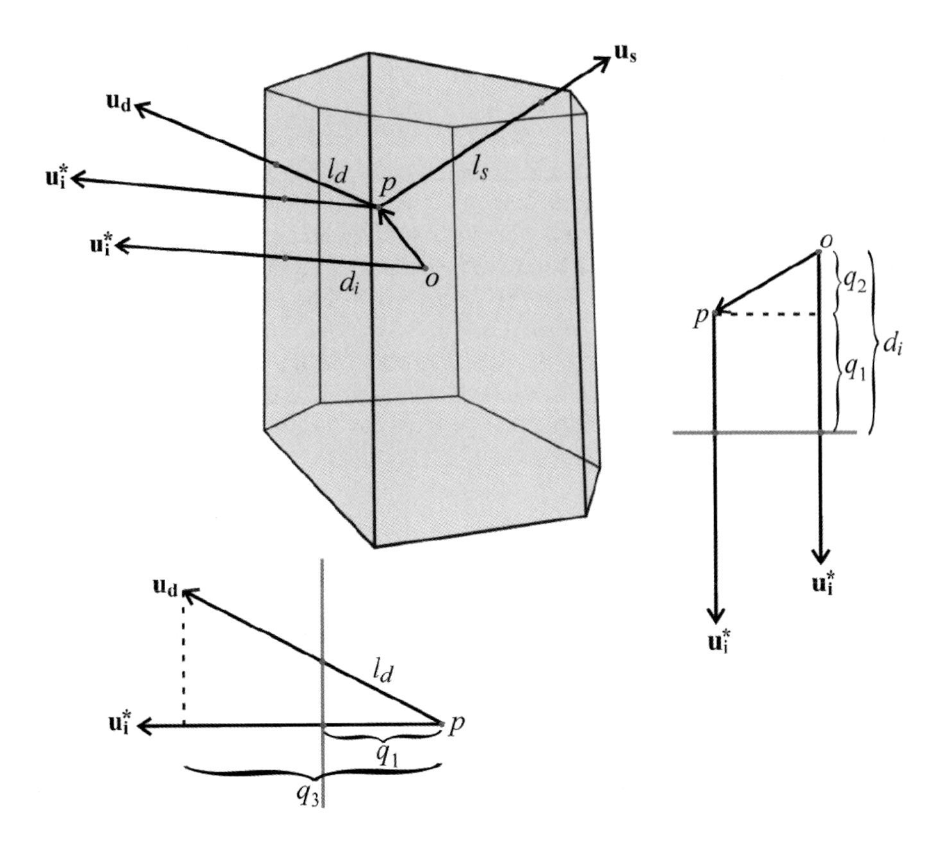

Figure 5.16 Path of the diffracted X-ray beam from a differential volume located at point p inside the crystal. $\mathbf{u_d}$ and $-\mathbf{u_s}$ are unit vectors in the directions of the paths of the diffracted and incident beams. $l = l_d + l_s$ is the length of the path through the crystal.

In order to calculate the absorption correction for a reflection, we determine the absorption due to diffraction from a small volume element in the crystal centered about a point p — then sum the contributions from all the volume elements in the crystal to obtain the overall absorption. In the limit this sum becomes an integral. Fig. 5.16 shows the path of the incident and diffracted beams from a small volume element, ΔV_p, located at point p inside the crystal. The overall length of the path through the crystal is $l = l_d + l_s$, and

$$I_p = I_0 e^{-\mu_l (l_d + l_s)_p}. \tag{5.193}$$

$\mathbf{u_i^*}$ is the unit vector perpendicular to the ith face of the crystal from which the beam exits, $\mathbf{u_d}$ is a unit vector pointing in the direction of the path taken by the diffracted beam, and $\mathbf{u_s}$ is a unit vector pointing in the negative direction of the path of the incident beam. The origin of the crystal, o, is the origin of the laboratory coordinate system (at the center of the crystal), and the distance d_i is the measured perpendicular distance from the origin to the ith face.

A "copy" of the normal unit vector to the ith face, $[u_{i1}^* \; u_{i2}^* \; u_{i3}^*]$, is translated to p. It remains parallel to the line segment d_i and the distance from p to the face is therefore $q_1 = d_i - q_2$, where q_2 is the projection of $\overrightarrow{op}$ onto the plane normal: $q_2 = \overrightarrow{op} \cdot \mathbf{u_i^*}$. l_d is the line segment from the point p to the ith face along the diffracted beam path. The projection of l_d onto $\mathbf{u_i^*}$ is the perpendicular distance from the point p to the face, q_1. These two line segments create a right triangle with the line segment connecting the points of intersection at the face forming the opposite side. The projection of $\mathbf{u_d}$ onto $\mathbf{u_i^*}$ forms a similar right triangle with an adjacent side of length $q_3 = \mathbf{u_i^*} \cdot \mathbf{u_d}$. The ratio of hypotenuse to adjacent side is the same for both triangles, allowing for the determination of the length of the path of the diffracted beam from point p to the exit point on the ith face:

$$\begin{aligned} \frac{l_d}{q_1} &= \frac{u_1^*}{q_3} \\ l_d &= (d_i - q_2)\frac{1}{q_3} \\ &= \frac{d_i - (\overrightarrow{op} \cdot \mathbf{u_i^*})}{\mathbf{u_i^*} \cdot \mathbf{u_d}}. \end{aligned} \tag{5.194}$$

If the incident beam enters the crystal through the jth face, then

$$l_s = \frac{d_j - (\overrightarrow{op} \cdot \mathbf{u_j^*})}{\mathbf{u_j^*} \cdot \mathbf{u_s}}. \tag{5.195}$$

For every reflection, unit vectors in the directions of the incident and diffracted beams, $\mathbf{s_1}$ (toward the source) and $\mathbf{d_1}$, define the diffraction condition: $\mathbf{s_1} + \mathbf{d_1} = \lambda\mathbf{h_r}$. The vectors are described in the reciprocal Cartesian coordinates of the laboratory reference frame. For nearly all configurations, $\mathbf{s_1}$ is directed along the x_l axis toward the source with components $[-1\ 0\ 0]$. For diffractometers equipped with a serial detector, $\mathbf{d_1}$ lies in the equatorial plane with components $[(\cos 2\theta)\ (\sin 2\theta)\ 0]$:

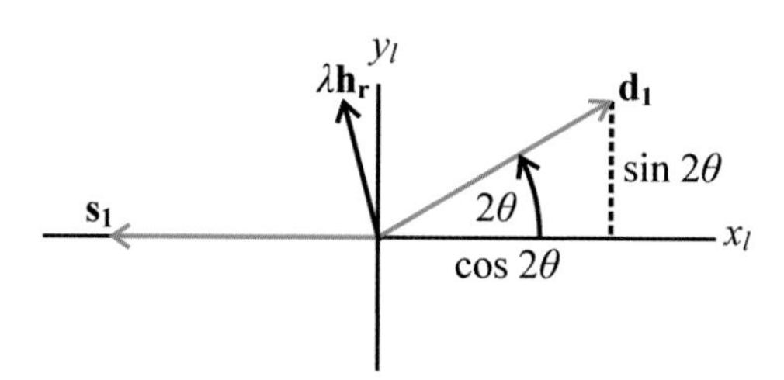

For area detector systems, the reflection vector $\mathbf{r}$ is determined from the coordinates of the reflection spot on the detector plane with Eqn. 4.33; $\mathbf{d_1} = \mathbf{r}/r$.

$\mathbf{s_1}$ and $\mathbf{d_1}$ are defined in the laboratory reference frame *after* rotation of the crystal from its reference position at $\omega = \chi = \varphi = 0$:

$$\mathbf{\Omega}\,\mathbf{X}\,\mathbf{\Phi}\,\mathbf{v_l} = \mathbf{v_r},$$

where $\mathbf{v_l}$ is a vector in reciprocal Cartesian coordinates in the laboratory reference frame. Since the unit vectors normal to the faces,$\mathbf{u_i^*}$, are defined with the crystal in the reference position, $\mathbf{s_1}$ and $\mathbf{d_1}$ must be "rotated back" to the reference position as well:

$$\mathbf{u_d^l} = \mathbf{\Phi}^T\,\mathbf{X}^T\,\mathbf{\Omega}^T\,\mathbf{d_1} \tag{5.196}$$
$$\mathbf{u_s^l} = \mathbf{\Phi}^T\,\mathbf{X}^T\,\mathbf{\Omega}^T\,\mathbf{s_1}. \tag{5.197}$$

Recalling the relationships between direct and reciprocal lattice scalar products ($\mathbf{a}^* \cdot \mathbf{a} = 1$, $\mathbf{a}^* \cdot \mathbf{b} = 0$, etc.), the scalar products in Eqns. 5.194 and 5.195 are conveniently evaluated if $\overrightarrow{op}$, $\mathbf{u_d}$ and $\mathbf{u_s}$ are expressed as fractional coordinates in the direct lattice basis. Since $\overrightarrow{op}$ is the vector between two points in the crystal, it is readily expressed in the direct basis. However, $\mathbf{u_d^l}$ and $\mathbf{u_s^l}$ are expressed in the reciprocal Cartesian basis and we must find a way to express them as direct lattice vectors.

A vector in the reciprocal lattice basis has fractional components, x_f^*, y_f^*, and z_f^* such that $\mathbf{v}^* = x_f^* \mathbf{a}^* + y_f^* \mathbf{b}^* + z_f^* \mathbf{c}^*$. Expressing the vector in the direct lattice basis,

$$x_f^* \mathbf{a}^* + y_f^* \mathbf{b}^* + z_f^* \mathbf{c}^* = x_f \mathbf{a} + y_f \mathbf{b} + z_f \mathbf{c}.$$

The scalar product $\mathbf{v}^* \cdot \mathbf{a}^*$ gives x_f in terms of the reciprocal vector components.

$$\begin{aligned} \mathbf{v}^* \cdot \mathbf{a}^* &= x_f^* \mathbf{a}^* \cdot \mathbf{a}^* + y_f^* \mathbf{b}^* \cdot \mathbf{a}^* + z_f^* \mathbf{c}^* \cdot \mathbf{a}^* \\ &= x_f \mathbf{a} \cdot \mathbf{a}^* + y_f \mathbf{b} \cdot \mathbf{a}^* + z_f \mathbf{c} \cdot \mathbf{a}^* \\ &= x_f \end{aligned} \tag{5.198}$$

Repeating the operation with $\mathbf{b}^*$ and $\mathbf{c}^*$ provides y_f and z_f:

$$\begin{aligned} x_f^* \mathbf{a}^* \cdot \mathbf{b}^* + y_f^* \mathbf{b}^* \cdot \mathbf{b}^* + z_f^* \mathbf{c}^* \cdot \mathbf{b}^* &= y_f \\ x_f^* \mathbf{a}^* \cdot \mathbf{c}^* + y_f^* \mathbf{b}^* \cdot \mathbf{c}^* + z_f^* \mathbf{c}^* \cdot \mathbf{c}^* &= z_f \end{aligned}$$

As a matrix equation,

$$\begin{bmatrix} x_f \\ y_f \\ z_f \end{bmatrix} = \begin{bmatrix} \mathbf{a}^* \cdot \mathbf{a}^* & \mathbf{a}^* \cdot \mathbf{b}^* & \mathbf{a}^* \cdot \mathbf{c}^* \\ \mathbf{b}^* \cdot \mathbf{a}^* & \mathbf{b}^* \cdot \mathbf{b}^* & \mathbf{b}^* \cdot \mathbf{c}^* \\ \mathbf{c}^* \cdot \mathbf{a}^* & \mathbf{c}^* \cdot \mathbf{b}^* & \mathbf{c}^* \cdot \mathbf{c}^* \end{bmatrix} \begin{bmatrix} x_f^* \\ y_f^* \\ z_f^* \end{bmatrix} \tag{5.199}$$

The matrix is readily identified as the reciprocal metric tensor (Eqn. 4.43), derived from the orientation matrix: $\mathbf{A}^{*T}\mathbf{A}^* = \mathbf{G}^*$. To express $\mathbf{u_d^l}$ and $\mathbf{u_s^l}$ in the direct lattice basis we first transform the Cartesian vectors into the reciprocal basis with $(\mathbf{A}^*)^{-1}$ (see Eqn. 5.192), then into the direct lattice basis with $\mathbf{G}^*$. Noting that $\mathbf{G}^*(\mathbf{A}^*)^{-1} = \mathbf{A}^{*T}\mathbf{A}^*(\mathbf{A}^*)^{-1} = \mathbf{A}^{*T}$,

$$\mathbf{u_d} = \mathbf{G}^*(\mathbf{A}^*)^{-1}\mathbf{u_d^l} = \mathbf{A}^{*T}\mathbf{\Phi}^T \mathbf{X}^T \mathbf{\Omega}^T \mathbf{d_1} \tag{5.200}$$

$$\mathbf{u_s} = \mathbf{G}^*(\mathbf{A}^*)^{-1}\mathbf{u_s^l} = \mathbf{A}^{*T}\mathbf{\Phi}^T \mathbf{X}^T \mathbf{\Omega}^T \mathbf{s_1}. \tag{5.201}$$

The path length for a reflection can now be determined if we know the entrance and exit faces of the incident and diffracted beams for the reflection. This problem is easier to analyze with the hypothetical two-dimensional crystal represented in Fig. 5.17; the extrapolation to a third dimension is straightforward. The unit "beam" vector with its tail at point p is either directed toward the source, in which case the path length is l_s, or it points in the direction of the diffracted beam with path length l_d. The points of intersection of all the face "planes" are shown Fig. 5.17(a). Eqn. 5.194 or Eqn. 5.195 determines the distance from the point p to each of the points $i \ldots l$, and $q \ldots t$. In order for the beam to intersect a face in the positive direction along the vector, the scalar product of the plane normal vector $\mathbf{u}^*$ and the beam vector must be positive since they must be at an acute

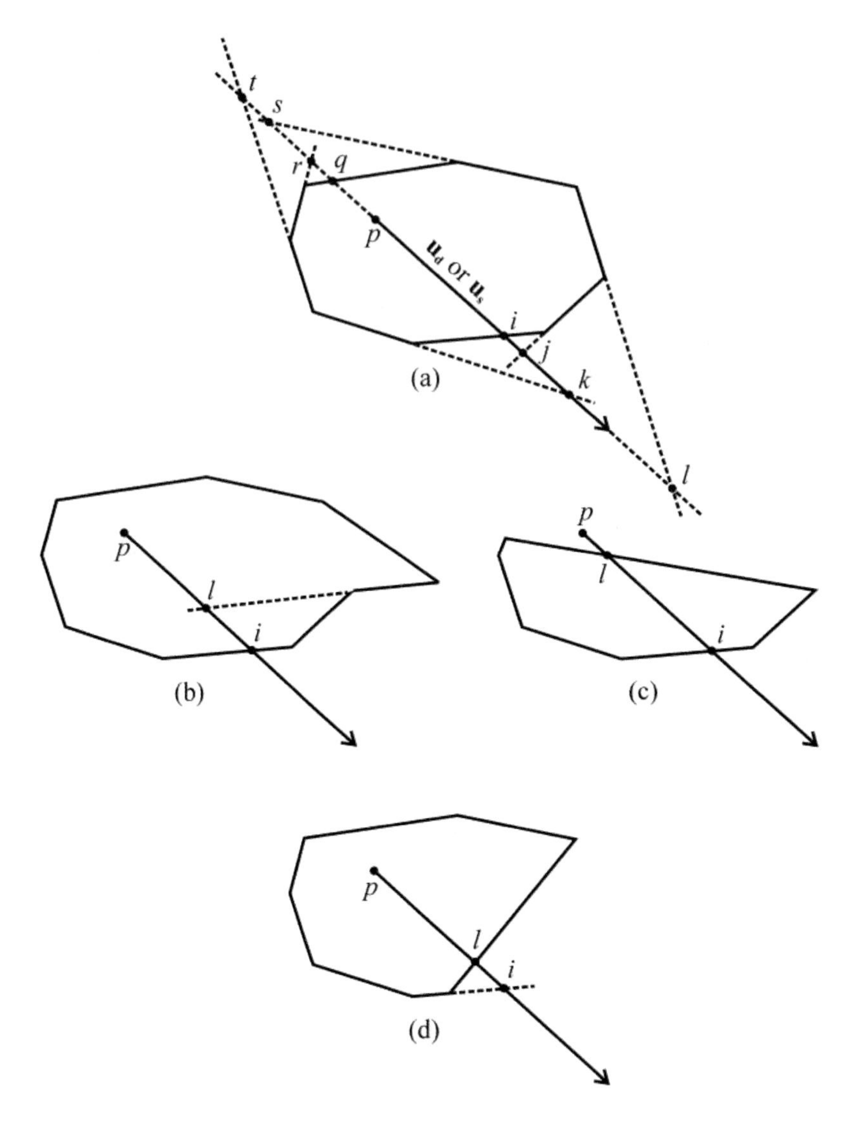

Figure 5.17 (a) Points of intersection of a unit vector in the direction of the beam vector (incident beam or diffracted beam) on all face "planes" of a hypothetical two-dimensional crystal. (b)–(d) Possible rotations of the crystal face plane containing intersection point l.

angle with respect to one another. The positive scalar product results in a positive number for the path length. Points $i \ldots l$ all have positive path lengths. In contrast, all intersections in the antiparallel direction of the beam (points $q \ldots t$) result from obtuse angles between the plane normal and the beam vector. Their scalar products are negative and the path lengths are subsequently calculated as negative numbers. Since these points lie in a direction opposite to the direction of the beam vector they are not entry/exit points and we can disregard them.

We now consider the points of intersection along the beam in the positive direction. Focussing on the plane that the beam intersects at point i, the entry/exit point, consider rotating one of the other planes at a vertex in order to determine the limits on the intersection points of the remaining planes. Selecting the plane with intersection point l as an example, rotation around one vertex is limited to a rotation angle that in one extreme brings it parallel to the plane with which it shares a vertex, and the other in which the plane is parallel to the beam vector. In the former case further rotation would create a "re-entrant angle" and the polygon would no longer be convex (Fig. 5.17(b)). In the latter case continuing the rotation would intersect the vector on the negative side of p until p itself was intersected; further rotation would then exclude point p from the crystal (Fig. 5.17(c)). If the plane rotates around the other vertex then it can be rotated until it intersects at a shorter distance in the positive direction, but then its point of intersection becomes the entry/exit point rather than point i (Fig. 5.17(c)). The results of these constraints are such that the remaining planes have their points of intersection, j, k, and l, at greater distances along the beam vector than the entry/exit point, i. Based on these considerations we conclude that *in order to determine the path length for a reflection, l_s and l_d are computed for all of the planes. The incident beam path length is the shortest of all of the l_s lengths that are greater than zero; the diffracted beam path length is the shortest of all the l_d lengths that are greater than zero.*

In order to correct the measured intensity of a reflection, $\langle I_d \rangle$, for absorption, we consider the ratio of the measured intensity to the corrected intensity, $I_{rel}(T)$, the intensity that would be observed in the absence of absorption:

$$T = \frac{I_{ab}}{I_{noab}} = \frac{\langle I_d \rangle}{I_{rel}(T)}. \tag{5.202}$$

A certain fraction of the incident beam intensity, f_s, will be elastically scattered, then altered by some factor, C, due to interference. The contribution to the intensity is also proportional to the size of the volume element at point p, ΔV_p, with some proportionality constant D. In the absence of absorption the intensity due to diffraction from volume element ΔV_p at point p is

$$I^p_{noab} = f_s I_0 C D \Delta V_p. \tag{5.203}$$

The intensity due to scattering from the entire crystal volume, V_x, is then summed over the contribution from each volume element:

$$I_{noab} = \sum_p C D f_s I_0 \Delta V_p = C D f_s I_0 \sum_p \Delta V_p = C D f_s I_0 V_x, \tag{5.204}$$

where V_x is the volume of the crystal. The intensity due to diffraction from the volume element at p is actually attenuated due to absorption, and I_0 becomes $I_0 e^{-\mu_l (l_s + l_d)_p}$:

$$I_{ab} = \sum_p C D f_s I_0 e^{-\mu_l (l_s + l_d)_p} \Delta V_p = C D f_s I_0 \sum_p e^{-\mu_l (l_s + l_d)_p} \Delta V_p. \tag{5.205}$$

The summation above is approximate, since it assumes (incorrectly) that the path length does not vary within the volume element, ΔV_p. The absorption correction ratio for the reflection, T, known as its *transmission factor*, can now be approximated from the path lengths for each volume element in the crystal and the linear attenuation coefficient:

$$\begin{aligned} T &= \frac{I_{ab}}{I_{noab}} = \frac{CDf_sI_0 \sum_p e^{-\mu_l(l_s+l_d)_p} \Delta V_p}{CDf_sI_0V_x} \\ &= \frac{1}{V_x} \sum_p e^{-\mu_l(l_s+l_d)_p} \Delta V_p. \end{aligned} \tag{5.206}$$

The approximation can be improved by making ΔV_p smaller and smaller; in the limit, as $\Delta V_p \to dV$,

$$I_{noab} = \int_0^{V_x} CDf_sI_0 dV = CDf_sI_0 \int_0^{V_x} dV = CDf_sI_0V_x, \tag{5.207}$$

$$I_{ab} = \int_0^{V_x} CDf_sI_0e^{-\mu_l(l_s+l_d)}\, dV, \text{ and} \tag{5.208}$$

$$\begin{aligned} T &= \frac{I_{ab}}{I_{noab}} = \frac{CDf_sI_0 \int_0^{V_x} e^{-\mu_l(l_s+l_d)}\, dV}{CDf_sI_0V_x} \\ &= \frac{1}{V_x} \int_0^{V_x} e^{-\mu_l(l_s+l_d)}\, dV. \end{aligned} \tag{5.209}$$

Referring to Eqn. 5.202, the intensity of the reflection can now be corrected for absorption by dividing the measured intensity by the transmission factor:

$$I_{rel}(T) = \frac{\langle I_d \rangle}{T}. \tag{5.210}$$

The integral and sum give the same results for the case where there is no absorption, since I_0 does not change within a volume element. Although the path lengths in Eqn. 5.209 can be determined for any point p in the crystal, each crystal is different and each reflection for a given crystal has its own unique set of path lengths. *There is therefore no analytical solution for the integral and it must be evaluated numerically for each reflection.*

Because both expressions for the transmission factor require summations in practice, the most straightforward approach to determining T is to make use of the simpler summation in Eqn. 5.206. This requires the division of the crystal into volume elements centered around the points, p. Since we wish to express $\overrightarrow{op}$ in terms of the direct lattice basis, a logical way to accomplish this is to create a parallelepiped with lengths in integral multiples of the unit cell basis vectors, centered about p, as illustrated in Fig. 5.18(a). If $K\mathbf{a}$, $L\mathbf{b}$, and $M\mathbf{c}$ are the lengths of the edges of this volume element, then its volume is $V_p = (KLM)V_{cell}$. A grid is then established by extending this element from the origin of the crystal in the direction of each of the unit cell axes:

$$\overrightarrow{og} = k(K\mathbf{a}) + l(L\mathbf{b}) + m(M\mathbf{c}), \tag{5.211}$$

where k, l, and m are positive and negative integers of sufficient magnitude to encompass the volume of the crystal. The point p is located at $K\mathbf{a}/2$, $L\mathbf{b}/2$, and $M\mathbf{c}/2$ within the volume element, and

$$\begin{aligned} \overrightarrow{op} &= (k(K\mathbf{a}) + K\mathbf{a}/2) + (l(L\mathbf{b}) + L\mathbf{b}/2) + (m(M\mathbf{c}) + M\mathbf{c}/2) \\ &= \left(\frac{(2k+1)K}{2}\right)\mathbf{a} + \left(\frac{(2l+1)L}{2}\right)\mathbf{b} + \left(\frac{(2m+1)M}{2}\right)\mathbf{c} \qquad (5.212) \\ &= p_1\mathbf{a} + p_2\mathbf{b} + p_3\mathbf{c}. \qquad (5.213) \end{aligned}$$

Since u_i^*, u_j^*, u_s, and u_d are unit vectors, $\overrightarrow{op} \cdot \mathbf{u}^*$ will ordinarily be in Å units. In order to incorporate $\overrightarrow{op} \cdot \mathbf{u}^*$ into Eqns. 5.194 and 5.195, it must be converted to the same units as d_i (e.g., cm).

To establish the limits of k, l, and m, we consider a point on the surface of one of the crystal planes as in Fig. 5.18(b). The projection of any vector to a point on the plane to the plane normal, $\mathbf{u_i^*}$, will be the distance from the origin to the plane, d_i:

$$\overrightarrow{op} \cdot \mathbf{u_i^*} = d_i. \qquad (5.214)$$

Because the crystal polyhedron is convex, the projection of the $\overrightarrow{op}$ vector *for any point inside the crystal* on any of the plane normal vectors will be less than the distance from the origin to the plane. This is depicted in the hypothetical two dimensional unit cell in Fig. 5.18(c). The reader can easily verify that the projection on any one of the plane normal vectors for the point inside the crystal will either be negative (e.g., $\overrightarrow{op} \cdot \mathbf{u_n^*}$) or positive and inside the crystal (e.g., $\overrightarrow{op} \cdot \mathbf{u_l^*}$). If the point lies on a face, the projection *is* the perpendicular distance to the face (d_l in the figure). *If the point lies outside of the crystal, there will be least one projection that will be greater than the perpendicular distance to a face.* Thus, the criterion that establishes whether or not a point is inside a crystal or on a face is

$$\overrightarrow{op} \cdot \mathbf{u_i^*} \leq d_i, \quad i = 1, 2 \ldots, n \qquad (5.215)$$

for the n faces of the crystal.

The points inside the crystal are determined by testing $p(klm)$ for each value of the integers and rejecting those that do not simultaneously satisfy the n inequalities above. The limits of the search are established by determining the minimum (most negative) and maximum $\overrightarrow{op}$ vectors, which will be vectors to points at the vertices of the crystal. The vertices are unique points, in that they share three faces (Fig. 5.15). For a vertex sharing the i, j, and k faces in Fig. 5.18(b), Eqn. 5.214, where $\overrightarrow{op} = \mathbf{v_{ijk}} = [v_1 \; v_2 \; v_3]$, must hold for all three faces:

$$\begin{aligned} \mathbf{v_{ijk}} \cdot \mathbf{u_i^*} &= v_1\mathbf{a} \cdot u_{i1}\mathbf{a}^* + v_2\mathbf{b} \cdot u_{i2}\mathbf{b}^* + v_3\mathbf{c} \cdot u_{i3}\mathbf{c}^* \\ &= v_1 u_{i1} + v_2 u_{i2} + v_3 u_{i3} \;=\; d_i \qquad (5.216) \\ \mathbf{v_{ijk}} \cdot \mathbf{u_j^*} &= v_1 u_{j1} + v_2 u_{i2} + v_3 u_{j3} \;=\; d_j \qquad (5.217) \\ \mathbf{v_{ijk}} \cdot \mathbf{u_k^*} &= v_1 u_{k1} + v_2 u_{k2} + v_3 u_{k3} \;=\; d_k. \qquad (5.218) \end{aligned}$$

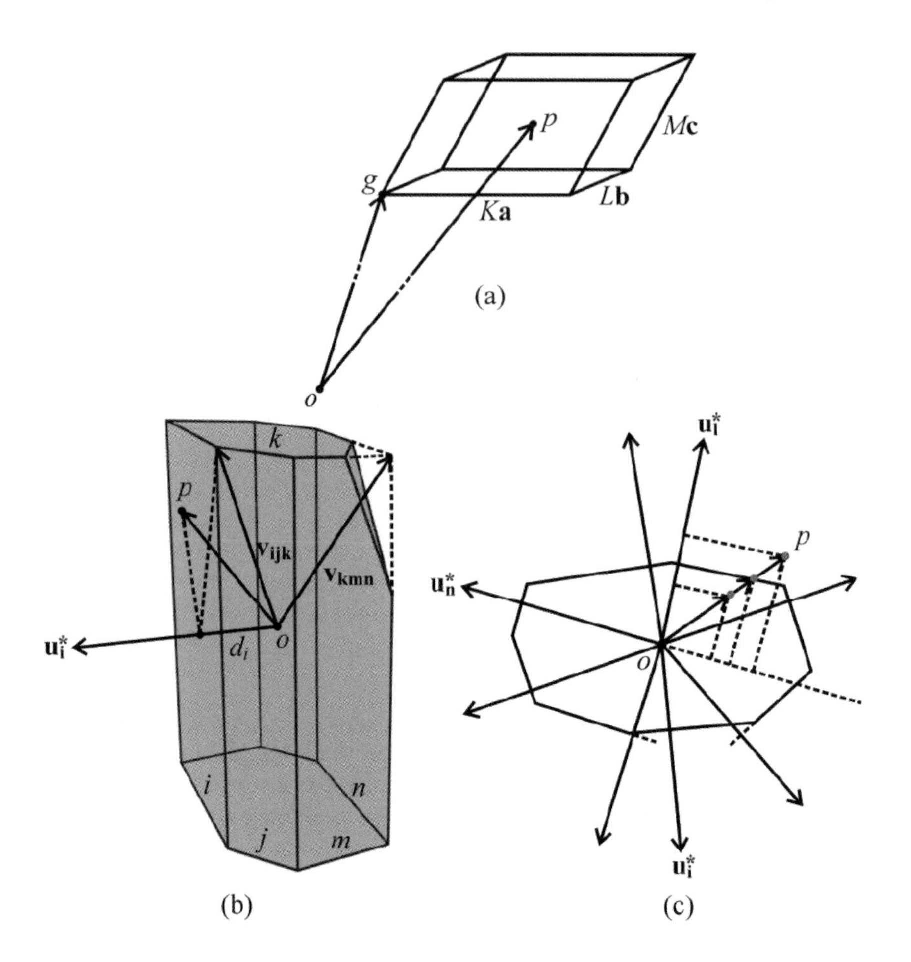

Figure 5.18 (a) Grid volume element in terms of the direct lattice basis. (b) Vectors to a point in a face and points at real and virtual vertices. (c) Projections of vectors to points inside the crystal, in a face, and outside the crystal onto the plane normal vectors.

These three simultaneous equations allow for the solution of the coordinates of the vertex. As a matrix equation,

$$\begin{bmatrix} u_{i1} & u_{i2} & u_{i3} \\ u_{j1} & u_{j2} & u_{j3} \\ u_{k1} & u_{k2} & u_{k3} \end{bmatrix} \begin{bmatrix} v_1 \\ v_2 \\ v_3 \end{bmatrix} = \begin{bmatrix} d_i \\ d_j \\ d_k \end{bmatrix}$$

$$\begin{bmatrix} v_1 \\ v_2 \\ v_3 \end{bmatrix} = \begin{bmatrix} u_{i1} & u_{i2} & u_{i3} \\ u_{j1} & u_{j2} & u_{j3} \\ u_{k1} & u_{k2} & u_{k3} \end{bmatrix}^{-1} \begin{bmatrix} d_i \\ d_j \\ d_k \end{bmatrix}. \tag{5.219}$$

All of the vertices of the crystal can be established in this manner. Each solution will generate a "real" vertex unless the crystal is truncated by another face, as depicted by $\mathbf{v_{kmn}}$ in Fig. 5.18(b). The vertex is a point outside of the crystal – a "virtual" vertex, and will therefore have at least one projection onto a plane normal vector longer than the distance to that plane. As with any point, a vertex must satisfy the condition,

$$\mathbf{v_{ijk}} \cdot \mathbf{u_i^*} \leq d_i, \quad i = 1, 2 \ldots, n. \tag{5.220}$$

The solution of the coordinates for the vertices provides an added bonus — the coordinates can be used to create a graphical image of the crystal, allowing for a test of the initial measurements.

This "simple" method for determining T now requires only the unit vectors, $\mathbf{s_1}$ and $\mathbf{d_1}$, for each reflection. Unfortunately, the method fails to accurately treat the contributions from points lying close to a face plane. If the point is close enough to a surface that part of the volume element extends outside the crystal, then the contribution to the absorption from that point will be overestimated. If, on the other hand, the point lies just outside of the crystal, then the contribution from the portion of the volume element surrounding the point that remains inside the crystal will be ignored, underestimating its contribution. If the size of the volume element is very small, then these effects will also be small, and should effectively cancel one another. When the software for numerical absorption corrections was originally developed, computers had very limited memories and processor speeds and practical grids were constrained to a few thousand points. Dividing the crystal into a relatively course $16 \times 16 \times 16$ grid generates nearly 5000 points, around the limit for computers of the time. To resolve the grid problem in the 1950's William R. Busing and Henri A. Levy[70]* employed a more sophisticated summation method to evaluate Eqn. 5.209, based on Gauss's formula of numerical integration[72]. This method establishes a point sampling method that avoids the errors inherent in the use of an arbitrarily fixed volume element. *Gaussian quadrature* remains an important component of most of the current computer software available for numerical absorption correction calculations.†

Absorption Corrections From Intensity Measurements. Many crystals grow with poorly defined faces; many others require protection from the environment and must be encased in glass or quartz capillary tubes, making it difficult so see them clearly in an optical device. Such crystals do not lend themselves to the absorption correction method just discussed, and a logical alternative is to *measure* the absorption of the crystal directly in order to determine the transmission factors empirically. During the early development of the four-circle diffractometer, Thomas C. Furnas, one of the pioneers in modern single crystal diffractometry, suggested[74] a surprisingly simple method for obtaining a relatively accurate estimate of the transmission factor for a reflection, based solely on its φ rotation angle. Since the orientation of the crystal depends on ω, χ, and φ, this notion appears counter-intuitive.

However, consider a hypothetical (and impractical) experiment with the setup shown in Fig. 5.19(a). With ω and χ set to zero, the crystal is rotated incrementally through φ and the intensity for each value of φ, which varies with the path length

*Busing and Levy were also responsible for the initial matrix description of the setting angles for four-circle diffractometers.[71]

†Other interesting approaches for calculating T include dividing the crystal polyhedron into tetrahedra; the absorption contribution for each tetrahedron can be solved analytically.[73]

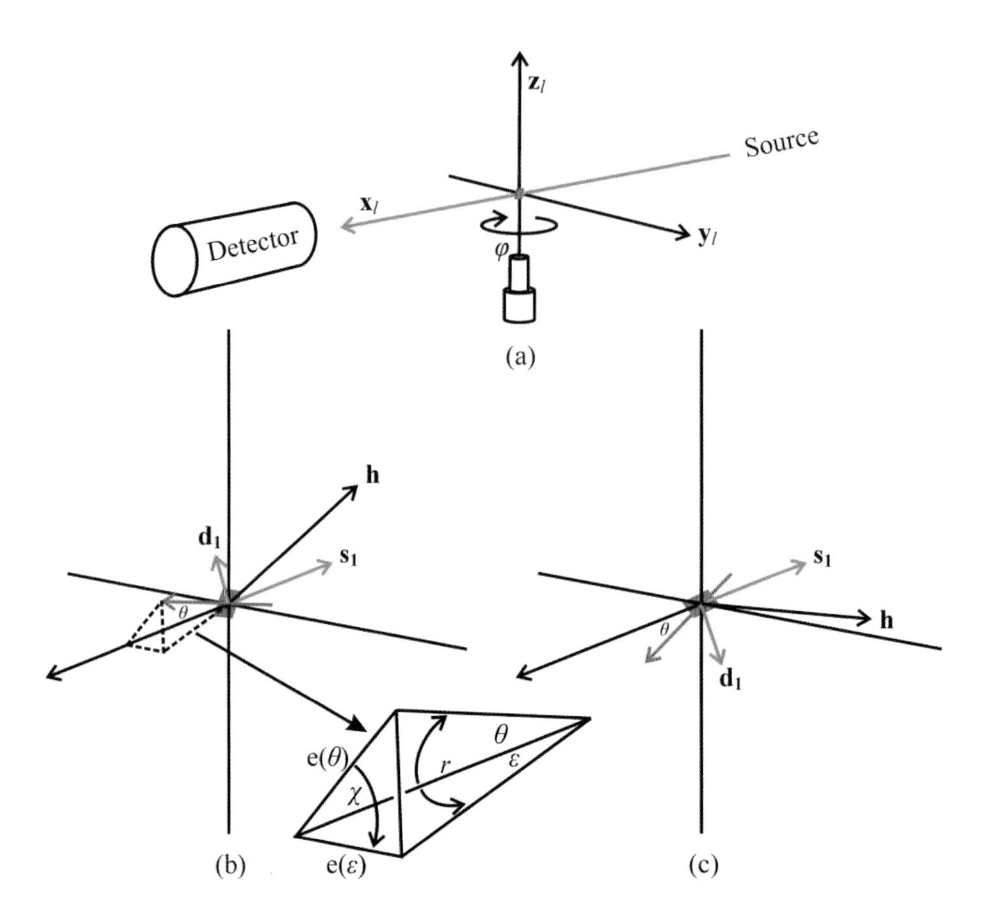

Figure 5.19 (a) Hypothetical arrangement to measure the transmission factor as a function of the φ rotational angle. (b) Diffraction occurring from an arbitrary set of *hkl* planes as φ rotates through 360°. (c) Rotation of the reflection around the incident beam (x_l) axis, placing the diffraction and reflection vectors into the equatorial plane.

through the crystal, is measured with the detector. The transmission factor for each φ is then calculated as $T(\varphi) = I(\varphi)/I_{noabs}$ — where I_{noabs} is the intensity in the absence of absorption. Although we do not know I_{noabs}, it is a constant which can be included in the intensity scale factor; we are only required to determine relative values of the transmission factor. The intensities, $I(\varphi)$, could themselves serve this purpose, but in order to keep the transmission factors on approximately the same scale, we divide each intensity by the maximum intensity:

$$T(\varphi) = \frac{I(\varphi)}{I(\varphi)_{max}}, \quad \varphi = 0^\circ \rightarrow 360^\circ. \tag{5.221}$$

In a complete 360° rotation of the crystal, every set of planes will be at the correct angle to diffract X-rays at some value of φ. The plane containing the reflection vector, $\mathbf{d_1}$ and diffraction vector, $\mathbf{h}$, will point in a direction determined by the rotation of the planes with respect to the equatorial plane (Fig. 5.19(b)). If the crystal is rotated through φ until a reflection occurs, as shown in the figure, the transmission factor for the reflection could be determined if the path length through the crystal could be determined. If the crystal at this value of φ is rotated around the beam axis, the angle between the planes and the incident beam will remain constant, the crystal will remain in the diffraction condition, and the path of the incident and diffracted beams through the crystal will not change. *Thus the transmission factor measured for a reflection with a specific φ rotation at $\chi = 0$ will be the same as that for the reflection when it rotates through χ into the equatorial plane, where its intensity can be measured with a detector* (Fig. 5.19(c)).

The transmission function, $T(\varphi)$, measured in our hypothetical experiment, depends on the path of the incident beam, but if we assume that the path of the diffracted beam does not vary significantly from the incident beam, then the function could be used to assign a transmission factor to every reflection from the value of φ of the reflection in the diffraction condition. This value of φ will be the angle for which the path of the incident beam passes through the crystal, with the diffracted beam traveling in another direction. This is depicted in Fig. 5.20(a). The incident beam vector, $\mathbf{s_1}$, the diffraction vector $\mathbf{h}$, and the reflection vector $\mathbf{d_1}$ lie in the same plane, *the diffraction plane* (Fig. 3.27). The *hkl* plane vector (in red) lies along the intersecting line between the *hkl* plane and the diffraction plane. The *hkl* plane is perpendicular to the diffraction plane and the *hkl* plane vector is at an angle of θ with respect to the incident beam, bisecting the 2θ angle between the reflection vector and the beam. The projection of the *hkl* plane vector onto the equatorial plane is at an angle, ε, with respect to the incident beam. A view of the projection onto the equatorial plane (along the z_l axis) is shown in Fig. 5.20a. Rather than using the value of φ for the reflection in the diffraction condition, Furnas' method used the value that would align the projection of the *hkl* plane vector with the incident beam, depicted in Fig. 5.20(b). $T(\varphi)$ would then correspond (approximately) to the average path taken by the incident and reflected beams (splitting the difference). The approximation arises because the diffracted beam does not lie in the equatorial plane.

In 1967, 10 years after Furnas suggested this empirical approach in an instrument instruction manual, A.C. North, D.C. Phillips, and F. Scott Mathews[75] extended the treatment to account for differences in the directions of the incident beam and the projection of the diffracted beam. Using the value of φ that aligns the plane vector along the incident beam as a reference, φ_{hkl} (Fig. 5.20(b)), the value of φ that corresponds to the incident beam path is $\varphi_{hkl} - \varepsilon$, and the transmission factor for the incident beam absorption for a specific reflection is then $T(\varphi_{hkl} - \varepsilon)$. The value of φ that aligns the projection of the path of the diffracted beam along the incident beam is $\varphi_{hkl} + \varepsilon$, as shown in (Fig. 5.20(c)). The value of the transmission factor for the diffracted beam projection is then $T(\varphi_{hkl} + \varepsilon)$. The transmission factor for the reflection is taken as the average of the incident and diffracted beam transmission factors:

$$T = \frac{T(\varphi_{hkl} - \varepsilon) + T(\varphi_{hkl} + \varepsilon)}{2}. \tag{5.222}$$

Referring to Fig. 5.19(a), ε can be determined from the diffractometer angles for the reflection by noting that the triangles with angles θ and ε share a common edge, r. If r is rotated through each angle, two arcs of length $r\theta$ and $r\varepsilon$ will be created. The ratios of the arc lengths will be equal to the ratios of their secants, and these

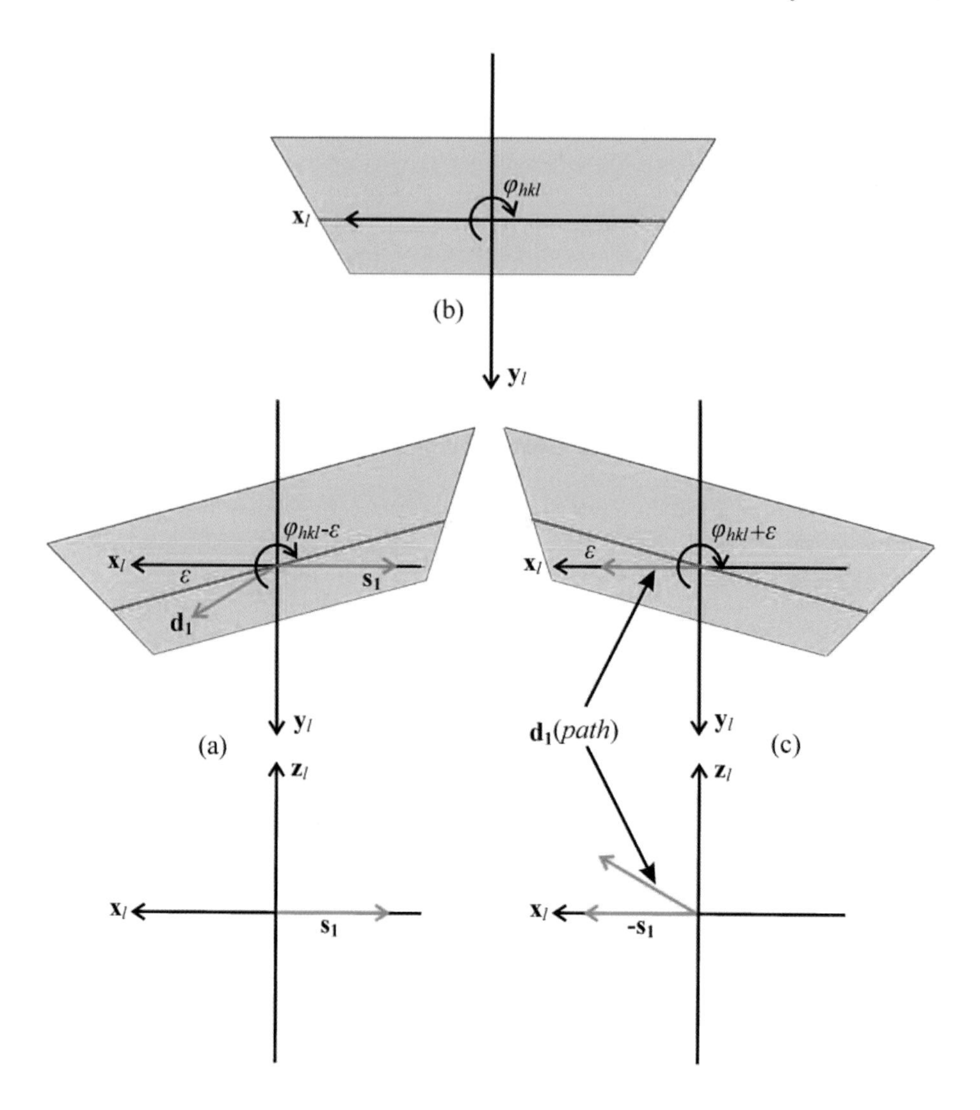

Figure 5.20 (a) Projection along the z_l and y_l axes of an hkl plane in the diffraction condition. (b) Rotation of the hkl plane about the z_l axis, aligning the plane vector projection with the incident beam. (c) Rotation of the hkl plane about the z_l axis, aligning the path of the reflection vector with the incident beam.

secants will be very close to the lengths of the opposite edges of both triangles, $e(\theta)$ and $e(\varepsilon)$. Thus,

$$\begin{aligned} \cos\chi &= \frac{e(\varepsilon)}{e(\theta)} = \frac{r\varepsilon}{r\theta} = \frac{\varepsilon}{\theta} \\ \varepsilon &= \theta\cos\chi. \end{aligned} \tag{5.223}$$

We are still left with the task of determining $T(\varphi)$. The experiment to generate $T(\varphi)$ directly is impractical because the actual fraction of the incident beam that is absorbed is extremely difficult to measure due to the small size of the crystals used in the diffraction experiment. The solution to this dilemma was alluded to in Chapter 4. Consider the crystal in Figs. 4.11(b) and (c), in the diffraction condition with the diffraction vector in the equatorial plane, and the *hkl* planes perpendicular to the equatorial plane. If the diffractometer had a mechanism to rotate the crystal around the diffraction vector (which, in general it does not), rotation of the crystal would keep the planes at the same angle with respect to the incident beam and the reflection vector in the equatorial plane at the same angle – pointing toward the detector. Now, suppose that we are able to align the crystal on the goniometer head so that, in the reference position (with all the angles set to zero), a diffraction vector for some reflection is aligned along the φ rotation axis (coincident with z_l). Referring to Fig. 4.12, *any* φ rotation will leave the diffraction vector aligned with the z_l axis, and the angle to rotate the vector into the equatorial plane, χ, will be 90°. After ω rotation into the diffraction condition, rotation about φ will also be rotation about **h**, and the crystal will remain in the diffraction condition. Rotation about φ in increments and determination of the integrated intensities for each φ will provide the relative transmission factors, $T(\varphi)$ which can be *quasi-normalized** by dividing by the maximum observed intensity (Eqn. 5.221).

In the early days of single crystal diffraction, crystals were routinely aligned so that a crystallographic axis was coincident with the vertical axis of the goniometer head — generally using film methods. This involved a tedious and time-consuming iterative trial and error process requiring the collection and development of a number of diffraction photographs. For crystals with monoclinic or higher symmetry, alignment along a perpendicular direct unit cell axis was also alignment along a parallel reciprocal unit cell axis and with some significant effort an axial reflection (e.g., 0 h 0) could provide the desired transmission function.

Prior to the development of relatively routine methods for structural solution, the effort expended in crystal alignment was justifiable, since it was negligible in comparison to the effort required to solve and refine the structure. However, as the time for structural solution decreased, crystals were no longer oriented prior to intensity data collection. This made it necessary to *rotate the crystal around a diffraction vector that was not aligned with the φ rotation axis* in order to determine $T(\varphi)$.

Referring again to Fig. 4.12, each rotation of the diffraction vector about φ, followed by a χ rotation onto the equatorial plane and a subsequent rotation through the appropriate ω places the crystal in the diffraction condition. For each set of angles, $\{\omega, \chi, \varphi\}$, the crystal will be rotated around the diffraction vector at a different angle. This angle, denoted by the symbol, ψ, is the analog of the azimuth of a celestial body, and the rotation is often referred to as an azimuthal rotation. Rotating around the vector in increments of ψ and recording the values of the intensity and φ for each increment will allow us to determine relative transmission factors as a function of φ from a single reflection — which we can then apply to all other reflections! The intensity scan for a reflection at a given value of ψ is known as an *azimuthal scan*, or more commonly, as a *psi scan*. It should be noted here that for a reflection with $\chi = 90°$, ψ is determined solely by φ rotation; as reflections vary increasingly from 90°, ω becomes increasingly important in the

*The transmission factors are scaled by dividing each intensity by the maximum intensity so that the maximum transmission coefficient is equal to 1. True normalization would require the sum of the transmission coefficients to equal 1.

determination of ψ. The omega rotation requires the movement of the entire chi circle, and access to some values of ψ (and therefore φ) may not be possible due to instrument geometry constraints.[†] Thus, the reflections for psi scans are optimally chosen with χ as close to 90° as possible.

The values of ψ for a reflection are determined by the values of ω, χ, and φ. We arbitrarily choose the orientation of the crystal in the bisecting position to assign $\psi = 0$. A mathematically convenient way to analyze rotation around the diffraction vector is to rotate the crystal through φ_o and χ_o in preparation for a final rotation through ω_o in order to place the reflection in the bisecting position in the equatorial plane:

$$\mathbf{X_o \Phi_o h_l} = \begin{bmatrix} 0 \\ h \\ 0 \end{bmatrix}. \tag{5.224}$$

These are the diffractometer angles (for a serial detector) that are determined from the orientation matrix for the bisecting position; prior to ω_o rotation the diffraction vector points along the y_l axis with components $(0\ h\ 0)$. Since the reflecting planes are perpendicular to the equatorial plane at this point, rotation through ω_o will have no further effect on the plane rotation about the diffraction vector. In other words, after φ_o and χ_o rotation the reflecting planes planes lie parallel to the x_l, z_l plane and the value of ψ will be its value in the bisecting position, which we have defined as $\psi = 0$. Before moving the crystal into the diffraction condition, we first rotate it about the diffraction vector aligned with the y_l axis through an angle ψ using the appropriate rotations of ω, χ, and φ. The crystal is finally rotated an *additional* increment of ω_o ($\omega' = \omega + \omega_o$) into the diffraction condition.

The ψ rotation is about the y_l axis, and it leaves the diffraction vector unchanged:

$$\begin{bmatrix} \cos\psi & 0 & \sin\psi \\ 0 & 1 & 0 \\ -\sin\psi & 0 & \cos\psi \end{bmatrix} \mathbf{X_o \Phi_o h_l} = \begin{bmatrix} 0 \\ h \\ 0 \end{bmatrix}$$

$$\mathbf{\Psi X_o \Phi_o h_l} = \mathbf{R}_\psi \mathbf{h_l} = \begin{bmatrix} 0 \\ h \\ 0 \end{bmatrix}. \tag{5.225}$$

From Eqns. 4.8 and 4.9, $\mathbf{R}_\psi =$

$$\begin{bmatrix} (\cos\psi\cos\varphi_o + \sin\psi\sin\chi_o\sin\varphi_o) & (\cos\psi\sin\varphi_o - \sin\psi\sin\chi_o\cos\varphi_o) & (\sin\psi\cos\chi_o) \\ (-\cos\chi_o\sin\varphi_o) & (\cos\chi_o\cos\varphi_o) & (\sin\chi_o) \\ (-\sin\psi\cos\varphi_o + \cos\psi\sin\chi_o\sin\varphi_o) & (-\sin\psi\sin\varphi_o - \cos\psi\sin\chi_o\cos\varphi_o) & (\cos\psi\cos\chi_o) \end{bmatrix}$$

$$= \begin{bmatrix} r_{11} & r_{12} & r_{13} \\ r_{21} & r_{22} & r_{23} \\ r_{31} & r_{32} & r_{33} \end{bmatrix}. \tag{5.226}$$

[†]Although diffractometers with kappa geometry do not have a chi circle, they suffer from other constraints.

The actual rotation is accomplished by moving the diffractometer to an ω, χ, and φ, where

$$\mathbf{\Omega X \Phi h_l} = \mathbf{R h_l} = \begin{bmatrix} 0 \\ h \\ 0 \end{bmatrix}. \tag{5.227}$$

From Eqns. 4.8, 4.9, and 4.10, $\mathbf{R} =$

$$\begin{bmatrix} (\cos\omega\cos\varphi + \sin\omega\cos\chi\cos\varphi) & (\cos\omega\sin\varphi - \sin\omega\cos\chi\cos\varphi) & (-\sin\omega\sin\chi) \\ (\sin\omega\cos\varphi - \cos\omega\cos\chi\sin\varphi) & (\sin\omega\sin\varphi + \cos\omega\cos\chi\cos\varphi) & (\cos\omega\sin\chi) \\ (\sin\chi\sin\varphi) & (-\sin\chi\cos\varphi) & (\cos\chi) \end{bmatrix}$$

Since both matrices accomplish the same rotation they must be identical; $\mathbf{R}_\psi \equiv \mathbf{R}$. The values of φ_o and χ_o are constant for a given reflection, and the matrix $\mathbf{R}_\psi$ is evaluated for each value of ψ for which an intensity is to be measured. Since the matrices are equivalent, the diffractometer settings necessary to effect each ψ rotation can be determined from the matrix elements of $\mathbf{R}_\psi$:

$$\begin{aligned} r_{31}^2 + r_{32}^2 &= \sin^2\chi\sin^2\varphi + \sin^2\chi\cos^2\varphi \\ &= \sin^2\chi(\sin^2\varphi + \cos^2\varphi) = \sin^2\chi \end{aligned}$$

$$\tan\chi = \frac{\sin\chi}{\cos\chi} = \frac{(r_{31}^2 + r_{32}^2)^{\frac{1}{2}}}{r_{33}}; \tag{5.228}$$

$$\frac{r_{31}}{-r_{32}} = \frac{\sin\chi\sin\varphi}{\sin\chi\cos\varphi} = \frac{\sin\varphi}{\cos\varphi}$$

$$\tan\varphi = \frac{r_{31}}{-r_{32}}; \tag{5.229}$$

$$\frac{-r_{13}}{r_{23}} = \frac{\sin\omega\sin\chi}{\cos\omega\sin\chi} = \frac{\sin\omega}{\cos\omega}$$

$$\tan\omega = \frac{-r_{13}}{r_{23}}. \tag{5.230}$$

For a given value of ψ the diffractometer angles are determined from the arctangents of the expressions above, ω_o is added to ω, and the crystal is rotated by ψ about $\mathbf{h}$ and moved into the diffraction condition. The quasi-normalized transmission coefficients, $T(\psi)$, are reorganized as a function of the φ angle corresponding to each value of ψ to create $T(\varphi)$ for the reflection.

The use of intensities from a single reflection introduces another error into the model: the path of the diffracted beam for the reference reflection is no longer in the direction of the incident beam. Thus $T(\varphi + \varepsilon)$ no longer represents the transmission factor for the projection of $\mathbf{d_1}$ for each reflection. However, if $T(\varphi)$ is determined for several reflections with different orientations of their reflection vectors *and averaged*, the average transmission coefficient for the diffracted beam should approximate that of a diffracted beam in the average direction – in the direction of the incident beam.

$T(\varphi)$ corrects for the *anisotropic absorption* due to the shape of the crystal, but since it is based on the incident beam direction, it does not account for the change in path length as the diffraction angle increases. Ideally, a second transmission

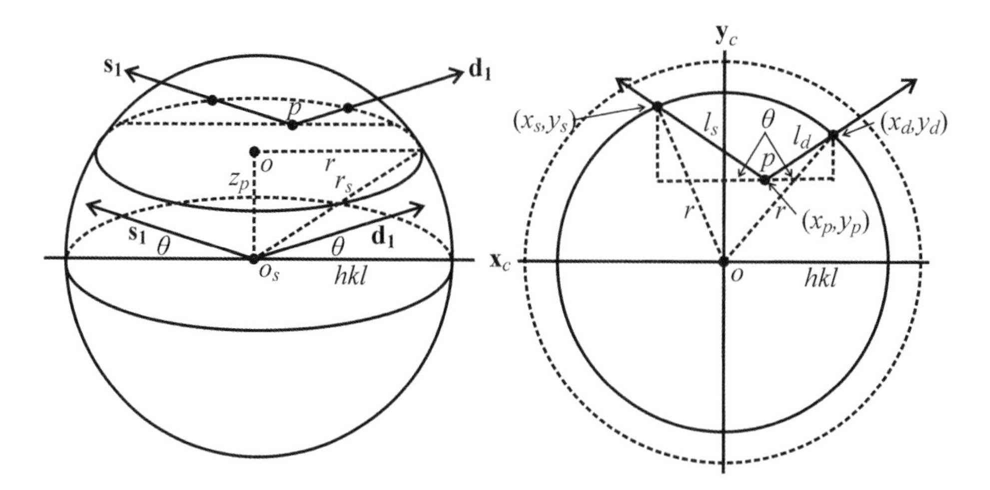

Figure 5.21 Path of an X-ray beam entering a spherical crystal from the left, diffracting at angle 2θ from a volume element at point p in the sphere and exiting the sphere to the right. The hkl planes are parallel to the x_c, z_c plane of a Cartesian coordinate system in direct space.

coefficient, $T(\theta)$, that depends on this variation of path length with the diffraction angle but *not* variations in crystal shape, would serve to correct for the path length variations resulting solely from differences in 2θ. This *isotropic absorption* correction can be approximated by treating the crystal as a sphere with the same volume as the crystal. Fig. 5.21 depicts diffraction from a set of hkl planes at diffraction angle 2θ from a spherical crystal with radius r_s. A Cartesian coordinate system is established with the hkl planes parallel to the x_c, z_c plane. The scattering plane is therefore parallel to the x_c, y_c plane. Scattering from the point p at (x_p, y_p, z_p) occurs from an incident beam path of length l_s and a diffracted beam l_d, entering and exiting the sphere at points on the small circle at height z_p with radius $r = (r_s^2 - z_p^2)^{1/2}$. As we did with the crystal in the previous section, we will determine the scattering path length, $l_s + l_d$, for a general volume element at point p in the sphere, then numerically integrate over the volume of the sphere to determine $T(\theta)$.

The drawing on the right is a projection onto the diffraction plane, looking down the z_c axis. The coordinates of the exit point on the small circle, (x_d, y_d), are given by

$$\begin{aligned} x_d &= x_p + l_d \cos\theta \\ y_d &= y_p + l_d \sin\theta. \end{aligned}$$

Since the exit point lies on the circle of radius r,

$$x_d^2 + y_d^2 = r^2$$
$$(x_p + l_d \cos\theta)^2 + (y_p + l_d \sin\theta)^2 = r^2.$$

Expanding, collecting terms, and setting $(\cos^2\theta + \sin^2\theta) = 1$,

$$l_d^2 + 2(x_p \cos\theta + y_p \sin\theta) l_d + (x_p^2 + y_p^2 - r^2) = 0. \quad (5.231)$$

The length of the path of the diffracted beam from the point p is the positive square root of this quadratic equation:

$$l_d = \frac{-2(x_p \cos\theta + y_p \sin\theta) \pm \sqrt{4(x_p \cos\theta + y_p \sin\theta)^2 - 4(x_p^2 + y_p^2 - r^2)}}{2}.$$

Expanding, collecting terms, and setting $(\cos^2\theta - 1) = -\sin^2\theta$ and $(\sin^2\theta - 1) = -\cos^2\theta$,

$$l_d = -x_p \cos\theta - y_p \sin\theta + (r^2 - (x_p \sin\theta - y_p \cos\theta)^2)^{\frac{1}{2}}. \tag{5.232}$$

The determination of l_s is identical, except that

$$\begin{aligned} x_s &= x_p - l_s \cos\theta \\ y_s &= y_p + l_s \sin\theta, \end{aligned}$$

resulting in

$$l_s = x_p \cos\theta - y_p \sin\theta + (r^2 - (x_p \sin\theta + y_p \cos\theta)^2)^{\frac{1}{2}}. \tag{5.233}$$

The overall path length is therefore

$$l_s + l_d = -2y_p \sin\theta + (r^2 - (x_p \sin\theta + y_p \cos\theta)^2)^{\frac{1}{2}} + (r^2 - (x_p \sin\theta - y_p \cos\theta)^2)^{\frac{1}{2}}.$$

For points in the third and fourth quadrants,

$$\begin{aligned} x_d &= x_p + l_d \cos\theta \\ y_d &= y_p - l_d \sin\theta \\ x_s &= x_p - l_s \cos\theta \\ y_s &= y_p - l_s \sin\theta, \quad \text{and} \\ l_s + l_d &= 2y_p \sin\theta + (r^2 - (x_p \sin\theta + y_p \cos\theta)^2)^{\frac{1}{2}} + (r^2 - (x_p \sin\theta - y_p \cos\theta)^2)^{\frac{1}{2}}. \end{aligned}$$

Since y_p is positive in the first two quadrants, and negative in the second two, for a general point,

$$\begin{aligned} l_s + l_d = {} & -2|y_p| \sin\theta + \\ & (r^2 - (x_p \sin\theta + y_p \cos\theta)^2)^{\frac{1}{2}} + (r^2 - (x_p \sin\theta - y_p \cos\theta)^2)^{\frac{1}{2}}. \end{aligned} \tag{5.234}$$

The numerical integration is identical to that for the general crystal described in the previous section, except that the coordinate system for the sphere is a Cartesian system.

$$T(\theta) \approx \frac{1}{V_{sph}} \sum_p e^{-\mu_l (l_s(\theta) + l_d(\theta))_p} \Delta V_p\,; \tag{5.235}$$

$$T(\theta) = \frac{1}{V_{sph}} \int_0^{V_{sph}} e^{-\mu_l (l_s(\theta) + l_d(\theta))}\, dV. \tag{5.236}$$

Because r is defined by z_p, the integration is easily performed in incremental slices along the z_c axis.* The points in the summation are constrained to lie inside the sphere, and therefore must satisfy $x_p^2 + y_p^2 + z_p^2 \leq r_s^2$.

*If r is held constant rather than being allowed to vary with z_p, the transmission factor for a cylinder can be determined.

The *International Tables for Crystallography* contains tabulated values of the transmission factor for spheres of unit radius for a range of linear absorption coefficients.[†] The path length increases linearly with the sphere radius. Since the transmission factor is the fraction of the radiation transmitted, an increase in the radius and path lengths will result in a proportional decrease in the transmission coefficient. To obtain the transmission coefficient for a sphere of a specific radius, the values in the table must be *divided* by the radius. Since $T(\theta)$ is incorporated into the overall transmission factor – a relative number – the tabulated values are suitable for a *theta correction* to the transmission factor determined from psi scans:

$$T_{hkl} = T(\varphi)T_{sph}(\theta). \tag{5.237}$$

An expanded table of *absorption factors* (the inverses of the transmission factors), giving the fraction of the radiation absorbed for a sphere of unit radius, is also found in the *International Tables*.

Despite the apparent limitations of this simple and straightforward empirical method, it approximates numerical absorption corrections computed from accurately measured crystals remarkably well — provided that the linear absorption coefficient of the crystal is not too large. As μ_l increases, the deviation in the direction of the diffracted beam path has a more pronounced effect on the absorption, and the model tends to break down. Ideally, a transmission function of all of the diffractometer angles, $T(\theta, \omega, \chi\varphi)$, would provide transmission factors that are not based on these approximations. In principle, if the intensities of a sufficient number of sets of equivalent reflections can be measured ("equivalent" in the sense that they would, in the absence of absorption, have identical intensities), and these reflections cover the complete range of angles involved in data collection, then it should be possible to generate $T(\theta, \omega, \chi\varphi)$, provided that we can represent it in functional form. Such a multivariate function is represented by a multidimensional surface and $T(\theta, \omega, \chi\varphi)$ is often called a *transmission surface* in 4-dimensional reciprocal space. For a serial detector, equivalent intensities are collected with psi scans — the number of equivalent intensities can be expanded to include reflections at various χ angles. In addition, in order to cover the angular ranges, symmetry-equivalent reflections can be included. This, of course, necessitates significantly more data, requiring significantly more time for data collection when a serial detector is employed. An area detector, on the other hand, routinely provides a large number of intensities from equivalent reflections, including those related by symmetry. Area detector data are therefore especially suited for the generation of a multidimensional transmission function.

The problem here, of course, is that we know nothing of the functional form of $T(\theta, \omega, \chi, \varphi)$. Indeed, the function will be different for every crystal! We will find it convenient to develop the transmission factor, $T(\theta, \omega, \chi, \varphi)$, in terms of its reciprocal, the absorption factor, $A(\theta, \omega, \chi, \varphi)$; the corrected intensity then becomes $I_{rel}(T) = A\langle I_d \rangle$, where $\langle I_d \rangle$ is the measured intensity. Since the diffraction angles

[†] If the crystal is chemically and mechanically stable it can be shaped into a sphere using a device that employs a stream of air to tumble the crystal against a very fine grit abrasive.

cycle periodically (e.g., $\omega = \omega + 360°$), $A(\theta, \omega, \chi, \varphi)$ is periodic as well, and *it should be possible to expand the function as a Fourier Series*[76]:

$$\begin{aligned} A(\theta, \omega, \chi, \varphi) &= \sum_m \sum_n \sum_o \sum_p G_{mnop} \cos(m\theta + n\omega + o\chi + p\varphi) \\ &+ \sum_m \sum_n \sum_o \sum_p H_{mnop} \sin(m\theta + n\omega + o\chi + p\varphi), \end{aligned} \quad (5.238)$$

where m, n, o, and p are integers varying from $-m_{max}$ to m_{max}, etc. This expansion is entirely analogous to the Fourier series expansion of the electron density function (Eqn. 3.76) developed in Chapter 3. The problem is now "reduced" to determining the Fourier coefficients, G_{mnop} and H_{mnop}.

Consider a set of n_h equivalent reflections, where h represents $(|h|, |k|, |l|)$ for the set. The absorption factor for the ith reflection of the set, $A_{hi}(\theta, \omega, \chi, \varphi)$, generates the corrected intensity, $A_{hi}I_{hi}$, where $I_{hi} \equiv \langle I_d \rangle_{hi}$. If the absorption is *isotropic*, as it would be for a spherical crystal, we anticipate that all of the n reflections in the set will have the same intensity, and that each intensity, I_{hi}, will be equal to the average intensity of the set, $\langle I_h \rangle$. If the absorption is *anisotropic*, then $I_{hi} \neq I_{hj}$, and we seek absorption correction factors such that $A_{hi}I_{hi} = A_{hj}I_{hj}$ and $A_{hi}I_{hi} = \langle A_h I_h \rangle$ for all of the reflections in the set. Ideally, the Fourier coefficients would provide absorption factors that would make this exact, resulting in $A_{hi}I_{hi} - \langle A_h I_h \rangle = 0$ for all n reflections. Practically, the "best" Fourier coefficients will be those that minimize the differences between the corrected values and their averages. These are generated by the method of least squares, which was introduced in Sec. 4.5 as a method to refine the unit cell and orientation matrix. In an analogous manner, we wish to determine the Fourier coefficients that will minimize the sum of the squares of the differences between the corrected intensities and their average value, commonly referred to as χ^2 (not to be confused with the diffractometer angle χ!). For a single set of n reflections this becomes

$$\chi^2 = \sum_{i=1}^{n} (A_{hi}I_{hi} - \langle A_h I_h \rangle)^2. \quad (5.239)$$

For multiple sets of reflections the function to minimize is

$$\chi^2 = \sum_h \sum_{i=1}^{n_h} w_{hi}(A_{hi}I_{hi} - \langle A_h I_h \rangle)^2. \quad (5.240)$$

The weighting factor in the sum, w_{hi}, is designed to weight the squared intensities in the sum according to our "confidence" in the intensity values, and is therefore based on its standard deviation, $\sigma(I_{hi})$. Since the intensity is squared in the sum, the weight is determined by the variance of the intensity, $\sigma^2(I_{hi})$. $w_{hi} = 1/\sigma^2(I_{hi})$ — the smaller the variance the greater the confidence.

The Fourier coefficients that minimize χ^2 are determined by differentiating with respect to each Fourier coefficient, setting the derivatives to zero (at the minimum), and solving the resulting system of equations for the coefficients:*

$$\frac{\partial \chi^2}{\partial G_{mnop}} = 0 \quad \text{and} \quad \frac{\partial \chi^2}{\partial H_{mnop}} = 0, \quad m = -m_{max} \text{ to } m_{max}, \text{ etc.}. \quad (5.241)$$

*The actual least-squares solutions are facilitated with some trickery in the form of a change of variables, etc. Various restraints and constraints are also applied to improve the behavior of the fitting algorithm.

Once the Fourier coefficients are determined, the transmission factor, $T(\theta,\omega,\chi,\varphi) = 1/A(\theta,\omega,\chi,\varphi)$, is computed for every reflection in the data set. Just as it was with $T(\varphi)$, $T(\theta,\omega,\chi,\varphi)$ corrects only for the absorption anisotropy; the complete transmission factor for the reflection is determined by multiplying the anisotropic transmission factor by the transmission factor for a sphere:

$$T_{hkl} = T(\theta,\omega,\chi\varphi)T_{sph}(\theta). \tag{5.242}$$

In practice, determination of the number of Fourier terms (the range of values of m, n, o and p) proves to be rather difficult. The use of the sine and cosine functions in the Fourier series expansion are not the only wave forms that can be used to expand a periodic function — an alternative approach is to expand the anisotropic absorption factor using *real spherical harmonic functions.*[77]. The spherical harmonics are functions that are "the angular part" of the solutions to a differential equation known as Laplace's equation:

$$\frac{\partial^2 f(x,y,z)}{\partial x^2} + \frac{\partial^2 f(x,y,z)}{\partial y^2} + \frac{\partial^2 f(x,y,z)}{\partial z^2} = 0. \tag{5.243}$$

The solution to a differential equation is a function that satisfies it. For this particular equation, if a function is differentiated twice with respect to each of the variables, and the sum of the results of those differentiations add to zero, then the function is a solution to Laplace's equation. The solutions to Laplace's equation are found by transforming the coordinates to spherical polar coordinates. The relationship between spherical coordinates and Cartesian coordinates is shown on the lower right in Fig. 5.22. It is easily shown that a point determined by a vector $\mathbf{r}$, originally aligned along the z axis, rotated from z by an angle of η, then about z at an angle of υ, with spherical coordinates (r,η,υ), has (x,y,z) coordinates of $(r\sin\eta\cos\upsilon,\ r\sin\eta\sin\upsilon,\ r\cos\upsilon)$. Laplace's equation becomes

$$\frac{1}{r^2}\frac{\partial}{r}\left(r^2\frac{\partial f(r,\eta,\upsilon)}{\partial r}\right) + \frac{1}{r^2\sin\eta}\frac{\partial}{\partial\eta}\left(\sin\eta\frac{\partial f(r,\eta,\upsilon)}{\partial\eta}\right) + \frac{1}{r^2\sin^2\eta}\frac{\partial^2 f(r,\eta,\upsilon)}{\partial\upsilon^2}.$$

Because the variables in the equation are independent of one another the solutions consist of a radial part, $R(r)$ and an angular part, $Y(\eta,\upsilon)$:

$$Y_l^m(\eta,\upsilon) = Ne^{im\upsilon}P_l^m(\cos\eta), \tag{5.244}$$

where l takes on positive integer values, $0,1,2,\ldots$; for every value of l, $m = -l\ldots 0\ldots + l$. The spherical harmonics are orthogonal functions; N is a normalization constant such that

$$\int_0^{\pi}\int_0^{2\pi} Y_{l'}^{m'}Y_l^{m*}\sin\eta\,d\upsilon d\eta = \delta_{ll'}\delta_{mm'}, \tag{5.245}$$

unity if $l = l'$ and $m = m'$ and zero if either $l \neq l'$ or $m \neq m'$. $P_l^m(\cos\eta)$ is a function of $\cos\eta$ known as an associated Legendre function:

$$P_l^m(\cos\eta) = \frac{(-1)^m}{2^l l!}(1-\cos^2\eta)^{m/2}\frac{d^{(l+m)}}{d\cos\eta^{(l+m)}}(\cos^2\eta - 1)^l. \tag{5.246}$$

These functions are indeed messy, but they are equally elegant and useful.* Spherical harmonics are three dimensional analogs of the sine and cosine functions, often conceptualized as "waves on a flooded planet,"and can be used in the same manner as the sinusoidal solutions are used in the Fourier series. Spherical harmonics with even l are even functions — i.e., they are centrosymmetric — those with l odd are non-centrosymmetric.

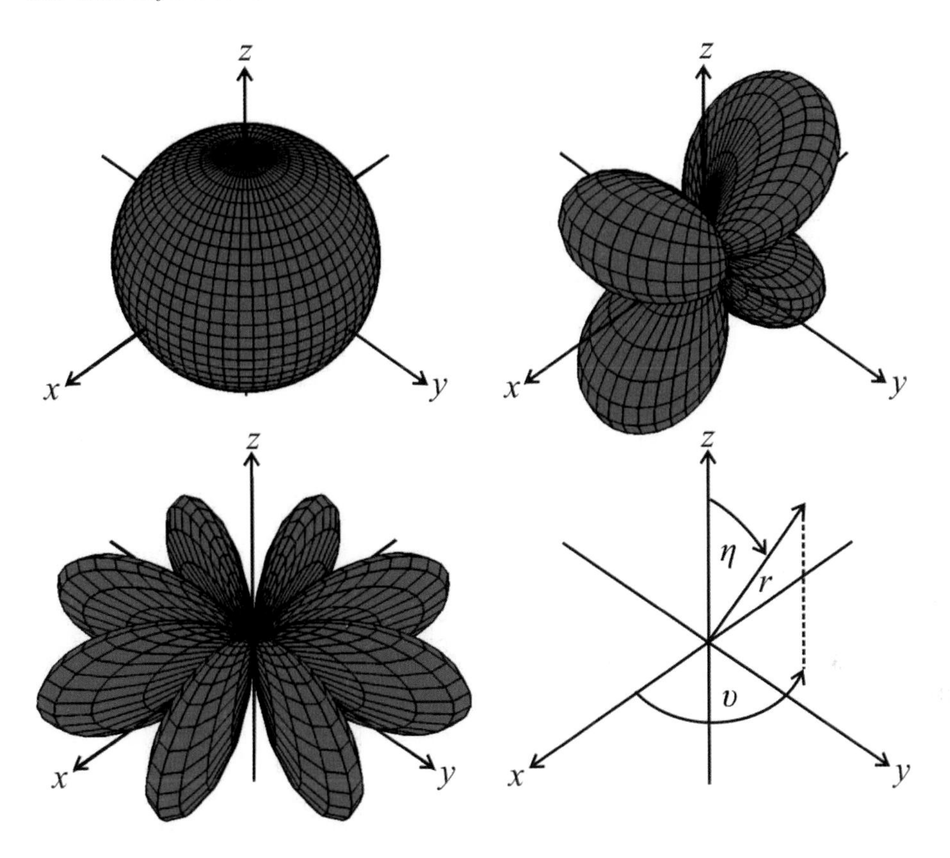

Figure 5.22 Surface plots of even real spherical harmonic functions: y_0^0, y_2^1, and y_4^4. Lower right: Relationship between spherical polar coordinates and Cartesian coordinates.

The infinite set of spherical harmonics, each generated from a unique combination of l and m, is known as a *complete set.* A complete set is the functional analog of a vector basis set; any function of the same variables can be expanded as a linear combination of the *basis functions* of a complete set. For the spherical harmonics,

$$f(\eta, \upsilon) = \sum_{l=0}^{\infty} \sum_{m=-l}^{+l} C_l^m Y_l^m(\eta, \upsilon). \tag{5.247}$$

There is one small complication here to consider — the spherical harmonic solutions have both amplitude and phase — they are complex numbers. However,

*They are, for example, the angular solutions to the Schrödinger equation (Sec. 3.2.5) for atomic state functions.

by using Euler's relations, $\cos \upsilon = (e^{i\upsilon} + e^{-i\upsilon})/2$ and $\sin \upsilon = (e^{i\upsilon} - e^{-i\upsilon})/2i$, the complex spherical harmonics can be combined into a complete set of real spherical harmonic functions:

$$y_l^m(\eta, \upsilon) = \begin{cases} \frac{1}{\sqrt{2}}(Y_l^m + Y_l^{-m}) = \sqrt{2}N\cos(m\upsilon)P_l^m(\cos\eta) & m > 0 \\ Y_l^0 & m = 0 \\ \frac{1}{\sqrt{2}i}(Y_l^{|m|} - Y_l^{-|m|}) = \sqrt{2}N\sin(|m|\upsilon)P_l^{|m|}(\cos\eta) & m < 0. \end{cases} \quad (5.248)$$

The real spherical harmonic functions for $l = 0$, 1, 2 and, 3 determine the shapes of the familiar s, p, d, and f atomic orbitals illustrated in elementary chemistry and physics texts.

Although the crystal itself is not, in general, centrosymmetric, the absorption surface is; for a reflection diffracting at 2θ, rotating the crystal by 180° will result in diffraction at -2θ *along the same path.* The anisotropic absorption surface can therefore be modeled as a linear combination of the even real spherical harmonic functions:

$$A_{aniso}(\eta, \upsilon) = 1 + \sum_{l=0}^{\infty} \sum_{m=-l}^{+l} c_l^m y_l^m(\eta, \upsilon), \quad l = 0, 2, 4, \ldots . \quad (5.249)$$

The basis functions in this expansion are orthonormal, and it is much easier to decide on the number of functions to incorporate in a finite approximation to the absorption surface, thus offering distinct advantages over the sinusoidal functions in the Fourier series expansion. The leading "1" is added to the expansion since the θ dependence is not included; if the crystal is spherical then $A_{aniso} = 1$ and $A_{hkl} = A_{aniso}A_{sph}(\theta) = A_{sph}(\theta)$. Surface plots of examples of real even spherical harmonics are shown in Fig. 5.22 for $l = 0$, $l = 2$ and $l = 4$.

The arguments in the spherical harmonics must be converted to diffractometer coordinates. For serial detector data the Eulerian angles are readily converted to the equivalent polar coordinate angles for the orientation of the reflection vector, but for an area detector, the reflection vector is determined directly from the spot locations on the detector plane, and it is more convenient to employ the reciprocal Cartesian coordinates of the reflection vector, $\mathbf{d_1}$ (this is also true for the Fourier expansion). This requires normalized Cartesian representations of the real spherical harmonics, which have been tabulated by Coppens and Paturle[78] for the functions through $l = 8$. For $\mathbf{d_1} = [x_d \; y_d \; z_d]$,*

$$A_{aniso} = 1 + \sum_{l=0}^{l_{max}} \sum_{m=-l}^{+l} c_l^m y_l^m(x_d, y_d, z_d) \quad l = 0, 2, 4, \ldots . \quad (5.251)$$

The intensities of sets of equivalent reflections are collected, A_{aniso} is calculated for each reflection, and the coefficients in the expansion are determined by the

*The components of the incident beam vector are also often included, equally weighted with the diffracted beam vector:

$$A_{aniso} = 1 + \sum_{l=0}^{l_{max}} \sum_{m=-l}^{+l} c_l^m [y_l^m(x_s, y_s, z_s) + y_l^m(x_d, y_d, z_d)], \quad (5.250)$$

where $\mathbf{s_1} = [x_s \; y_s \; z_s]$ in the negative direction of the incident beam.

least-squares method described above for the Fourier coefficients (Eqn. 5.240). The transmission factor for a general reflection is then given by

$$T_{hkl} = \frac{1}{A_{aniso}(x_d, y_d, z_d)A_{sph}(\theta)}. \tag{5.252}$$

Absorption Corrections From The Structure Model. The reader may be surprised at the space devoted here to absorption corrections, but they are arguably the most important source of error in the intensity data. Modern area detector systems provide excellent views of the crystal, allowing for the determination of numerical absorption corrections for many crystals. The amount of redundant data collected by area detectors allows for empirical absorption corrections when this is not possible. Thus, with modern instrumentation, absorption corrections should be a routine part of the handling of the intensity data set.

However, with data collected from serial detectors, absorption corrections often present a more difficult challenge. Because of the importance of the absorption correction, methods have been developed to create absorption surfaces from the intensity data set itself (without redundant data). In order to describe these methods we must first briefly discuss the subjects of the next three chapters.

The structural solution process results in a model based on the locations of the atoms in the unit cell. From this it is possible to calculate the magnitudes of the structure factors, F_c, for the reflections in the data set. These are then compared to the observed structure factor magnitudes, F_o, determined from the intensity data. The model (atom locations, etc.) can then be refined using the method of least squares, minimizing the weighted sum of the squares of the residuals:

$$R = \sum_{hkl} w_{hkl}(F_{c,hkl} - F_{o,hkl})^2. \tag{5.253}$$

If each of the observed structure factors is modified with an absorption correction factor such that

$$F_{m,hkl} = A(\omega, \chi, \varphi)A_{sph}(\theta)F_{o,hkl}, \tag{5.254}$$

then the absorption correction factors can be determined by minimizing the differences between the modified-observed and calculated structure factors with respect to the correction factors rather than the atom locations, etc. Various methods[79,80] differ in the way that the absorption correction function is defined. One approach is to expand $A(\omega, \chi, \varphi)$ as a Fourier series as in Eqn. 5.238, and to minimize

$$R = \sum_{hkl} w_{hkl}(F_{c,hkl} - A(\omega, \chi, \varphi)A_{sph}(\theta)F_{o,hkl})^2, \tag{5.255}$$

such that $\partial R/\partial G_{hkl}$=0 and $\partial R/\partial H_{hkl}$ =0.

Since the refinement of the absorption surface depends on the model, it is difficult to distinguish between the effects of model refinement and absorption refinement in the final refinement of the structure. The basic assumption in the use of these methods is that the errors in atom displacement vary more-or-less smoothly throughout reciprocal space, while absorption effects are more local. Unfortunately, it is virtually impossible to determine when this assumption holds, and when it does not. There has been substantial debate in the crystallographic community regarding the validity of using a model to modify the data upon which it is based. The following quote is taken from the documentation of the popular structural software package authored by George Sheldrick[1]:

> The program SHELXA has been kindly donated to the system by an anonymous user. This applies "absorption corrections" by fitting the observed to the calculated intensities as in the program DIFABS. SHELXA is intended for EMERGENCY USE ONLY, eg., when the world's only crystal falls off the diffractometer before there is time to make proper absorption corrections by indexing crystal faces or by determining an absorption surface experimentally by measuring equivalent reflections at different azimuthal angles, etc. The following restrictions apply to the use of SHELXA:
>
> (a) The structure should not be twinned (racemic twinning is allowed), the data should have been collected from one crystal (inter-batch scale factors should not have been refined), and there may not be a re-orientation matrix on the HKLF instruction. Otherwise there are no restrictions on the type of structure (SHELXA is equally (un)suitable for proteins) or the instructions used in the SHELXL refinement.
>
> (b) It is understood that any structure determined by means of this scientifically dubious procedure WILL NEVER BE PUBLISHED! The anonymous author of SHELXA has no intention of ever writing a paper about it that could be cited and thereby ruin his reputation.

The Extinction Correction

In the early 1900s pioneers in single crystal X-ray diffraction were intent on taking advantage of the diffraction experiment in order to locate electrons within the crystal. In 1914 Charles C. Darwin[81] published a theory that related the intensities of X-ray reflections to the "actual positions which the electrons occupy in the atom." In attempting to match experiment with theory, Darwin noted that the diminution of reflection intensities appeared to be greater than could be accounted for by absorption effects alone.

Darwin coined the term *extinction** to describe the loss in intensity due to the phenomena responsible for this observation. Since absorption coefficients are measured in the absence of diffraction, he concluded that any additional attenuation of the beam must arise from diffraction. In a second paper[82] Darwin set out to describe extinction in terms of energy losses that occur to the incident and diffracted beams as they pass through the crystal while the crystal is in the diffraction condition. He treated each beam separately, identifying extinction in the reflected beam as *primary extinction*, and extinction in the incident beam as *secondary extinction*.

Primary extinction was seen to arise from multiple diffraction within the crystal. Multiple diffraction occurs because the reflected beam is at the appropriate

*Unfortunately, this term implies that the incident beam is "extinguished". In practice, extinction results in a diminution of the beam that is negligible in many cases. However, in some crystals the effect is pronounced, resulting in systematic errors in the measured intensities from those crystals.

angle to be re-diffracted, resulting in a subsequent energy loss from the diffracted beam. The primary extinction effect was expected to be largest in a "perfect" single crystal, that is, when all of the lattice planes are simultaneously in the diffraction condition, maximizing the effects of multiple diffraction. The effect was expected to be minimized for crystals consisting of small, slightly misaligned *mosaic* blocks, in which case only a fraction of the blocks are undergoing diffraction at a given angle. These predictions were (and are) qualitatively consistent with experimental observations. Secondary extinction was considered to be due to a loss in intensity of the transmitted beam as it transfers energy into the diffracted beam. The loss in intensity at a given angle of the crystal (as it is rotated through the diffraction condition) would then depend on the intensity of the reflection and the fraction of the mosaic blocks underdoing diffraction at that angle. The phenomenon would diminish the incident beam intensity as if the radiation was being absorbed, and would thus appear as a reflection-specific addition to the absorption coefficient.

These two types of extinction are treated separately in the current literature, although it is experimentally difficult to determine when one begins and the other ends. The confusion has arisen partly because of the fundamental nature of extinction. As we will show in the following treatment, extinction arises from the interference of the X-rays waves as they traverse through the crystal. Up to this point we have ignored these effects, assuming that the waves scattered from each plane in the crystal are independent of one another and the incident beam wave as well; the description is based on geometric considerations alone. Kinematics is the branch of physics which involves the description of motion without examining the forces which produce the motion; the geometric theory is a *kinematical theory.* Dynamics, on the other hand, involves a description of motion *and* the forces which produce it. Darwin postulated that extinction resulted from the interference of the waves as they passed through the crystal and that it was necessary to treat these interactions explicitly. Darwin's theory was the first *dynamical theory* of X-ray diffraction.

Dynamical Diffraction. In order to understand the effects of dynamical interference on reflection intensities we return to the hypothetical oscillator and detector introduced in Chapter 3 to describe electromagnetic radiation and diffraction. Referring to Fig. 3.4, the generator electron produces a wave with instantaneous amplitude $\overrightarrow{\mathcal{E}_p(t)} = \mathcal{E}_p \cos(2\pi\nu t) = \mathcal{E}_p e^{2\pi i\nu t}$. If the detector electron is located at a distance, r, from the generator, it "sees" the field that was at the generator r/c seconds in the past:

$$\overrightarrow{\mathcal{E}_p(t)} = \mathcal{E}_p e^{2\pi i\nu(t-r/c)}. \tag{5.256}$$

Now, suppose that we introduce a thin plate of material between the generator and the detector. When some of the photons encounter the material they are absorbed into a virtual state, then scattered coherently. Of these some will be scattered toward the detector. Since this process is not instantaneous these photons take longer to get to the detector. The wave due to these scattered photons will also be delayed, and will no longer be in phase with the wave from the generator. (Classically, the bound electrons become secondary oscillators, with the delay in their oscillations resulting from their resistance to being accelerated – their effective inertia). Because of this additional retardation, the superposition of the scattered wave with the transmitted wave produces a phase shift that will be observed at the detector; it will *appear* as if the wave slows down as it passes through the material (when, in reality, it is a new wave, still moving at the speed of light). The ratio of the actual speed of the wave, c, to its apparent speed through a material, c_{app}, is known as the refractive index of the material, $n_r = c/c_{app}$, and $c_{app} = c/n_r$ (the change in phase is responsible for refraction, the "bending" of light rays as they pass through the material). If the plate is of thickness, T, then the wave will appear to take $(T/c)_{app}$ seconds rather than T/c seconds to pass through the material, resulting in a delay of $\Delta t = (T/c)_{app} - (T/c) = (n_r - 1)(T/c)$ seconds for the wave to reach the detector. The electric field vector of this retarded wave observed at the detector is therefore

$$\begin{aligned} \overrightarrow{\mathcal{E}'_p(t)} &= \mathcal{E}_p e^{2\pi i\nu(t-r/c-(n_r-1)(T/c))} \\ &= \mathcal{E}_p e^{2\pi i\nu(t-r/c)} e^{-2\pi i\nu(n_r-1)(T/c)}. \end{aligned} \tag{5.257}$$

e^q can be expanded as a power series,

$$e^q = \sum_{n=0}^{\infty} \frac{q^n}{n!} = \frac{1}{1} + \frac{q}{1} + \frac{q^2}{2} + \frac{q^3}{6} + \cdots. \tag{5.258}$$

If q is small, then $q^2 \ll z$ and $e^q \approx 1 + q$. Since T is small,

$$e^{-2\pi i\nu(n_r-1)(T/c)} \approx 1 - 2\pi i\nu(n_r - 1)(T/c) \quad \text{and}$$

$$\overrightarrow{\mathcal{E}'_p(t)} \approx \mathcal{E}_p e^{2\pi i\nu(t-r/c)} - i\left(\frac{2\pi\nu(n_r - 1)T}{c}\right)\mathcal{E}_p e^{2\pi i\nu(t-r/c)} \tag{5.259}$$

$$= \overrightarrow{\mathcal{E}_p(t)} - i\overrightarrow{\mathcal{E}_s(t)}. \tag{5.260}$$

Fig. 5.23(a) is a plot of the wave vector when $t - r/c = 0$. The first term is the electric field at the detector observed in the absence of scattering from the material (Eqn. 5.256), and the second term must therefore be the contribution due to scattering. The resultant electric field vector at the detector has components $(\mathcal{E}_p, -\mathcal{E}_s)$, as plotted in Fig. 5.23(a). The wave resulting from scattering is seen to be $-90°$ out phase with respect to the source radiation. Eqn. 5.260 describes the resultant wave in the direction of the incident beam, which is a superposition of the incident and scattered waves. Since the diffracted beam travels in a different direction, its phase is determined solely by the imaginary term. However, as illustrated in Fig. 5.23(b), a small fraction of the diffracted beam can be re-diffracted from a second *hkl* plane. *The scattering from each plane results in a $-90°$ phase shift, sending the beam resulting from secondary diffraction in the same direction as the incident beam — but 180° out of phase with it. This secondary diffraction therefore destructively interferes with the incident beam, diminishing its intensity.* The transmitted beam also undergoes a slight phase shift, γ, (the relative magnitude of $\overrightarrow{\mathcal{E}_s}$ is greatly exaggerated in the figure), resulting in a small difference in phase between the scattered waves from each plane (the kinematical model assumes that the phases between reflections from each plane are identical). The net amplitude of the diffracted beam therefore depends on multiple scattering and sequential phase shifts as the X-ray beam traverses through the crystal.

In order to focus on scattering from a specific set of planes, we consider a hypothetical crystal consisting of layers of atoms with their centers residing at each

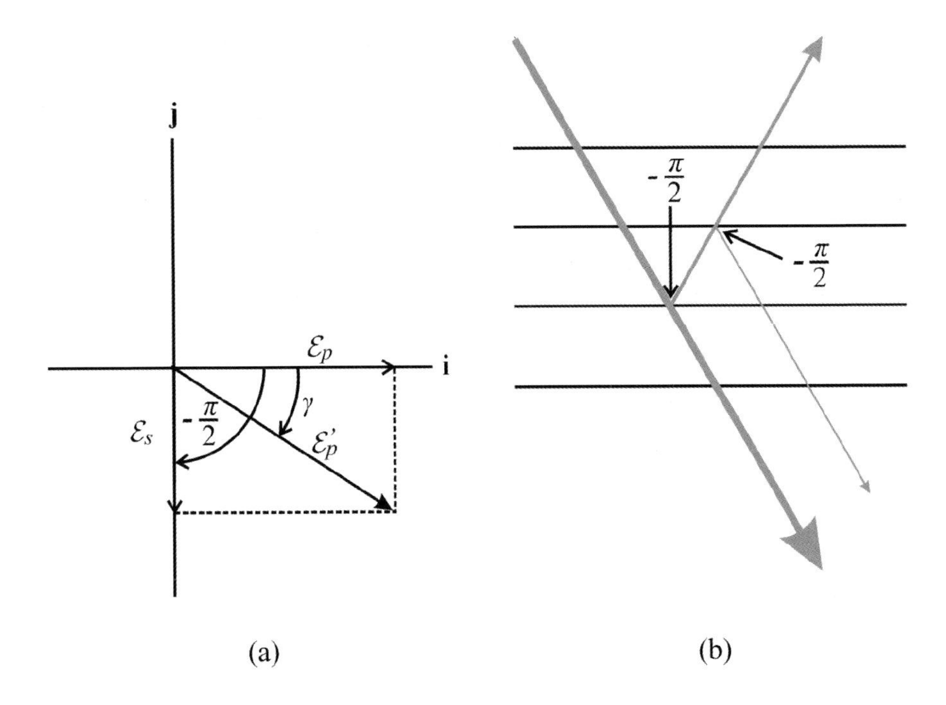

Figure 5.23 (a) Phase relationship between incident beam radiation and scattered radiation. (b) Attenuation of the incident beam by secondary diffraction.

origin. When it is necessary, we will ignore deviations in electron distribution in the crystal, treating each plane as a thin homogeneous layer of electron density, ρ_e, with an effective value that approximates the effects of the atomic electrons on the x-ray beam as it passes through each plane of atoms. This is justifiable on the basis that the distance between the source and detector in the diffraction experiment is orders of magnitude greater than the distance between the scattering centers; the crystal can be viewed as a uniform distribution of scatterers. Fig. 5.24 depicts the phase relationships between the transmitted incident beam and the reflected beam from two adjacent planes, where $\overrightarrow{\mathcal{E}_p} \equiv \mathbf{T}$ and $\overrightarrow{\mathcal{E}_s} \equiv \mathbf{S}$. $\mathbf{T}_j$ and $\mathbf{T}_{j+1}$ are the wave vectors for the transmitted beam just above the j and $j+1$ planes; $\mathbf{S}_j$ and $\mathbf{S}_{j+1}$ are the wave vectors for the reflected beam just above the j and $j+1$ planes.

At point q, just below the j plane, the transmitted beam experiences a small phase shift due to a superposition of the wave from the small number of photons scattered coherently in its direction (90° out of phase): $\mathbf{T}_j' = \mathbf{T}_j e^{-i\gamma}$. The perpendicular dashed line to the left in the figure corresponds to points at which the transmitted wave vectors are in phase. The wave at point p has the same amplitude as it does at point t, just above the $j+1$ plane, but lags behind by a phase angle, φ. If the crystal is exactly in the diffraction condition then $\varphi = \pi$ (the transmitted and reflected waves have phase differences from successive planes that total a complete wavelength, 2π, in order for the reflected waves to constructively interfere). To obtain the integrated intensity we must consider phase angles as the crystal rotates through the diffraction condition so that $\varphi = \phi + \delta$, where delta is

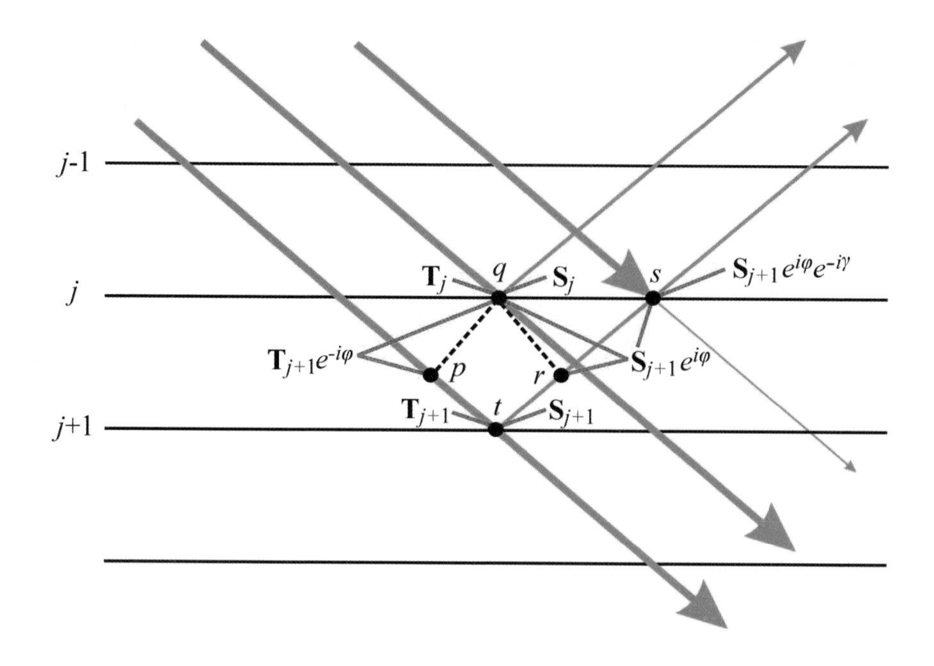

Figure 5.24 Phase relationships between transmitted and reflected waves from two adjacent planes in the crystal.

a small fraction of π. The transmitted wave just below the j plane at point q has the same amplitude and phase as the transmitted wave at point p, $\mathbf{T}_{j+1}e^{-i\varphi}$.

Each time a transmitted wave crosses a plane, the reflected wave is shifted by $-90°$ and a fraction of the radiation, σ, is reflected, $\mathbf{S} = -i\sigma\mathbf{T}$. At point r the wave reflected from the $j+1$ plane has the same amplitude as the reflected wave at point t, but leads by a phase angle of φ, $\mathbf{S}_{j+1}e^{i\varphi}$. The reflected waves just below the plane at points q and s have the same amplitude and phase as the wave at point r. As the reflected beam interacts with the electrons as it passes back through layer j it experiences a γ phase shift to $\mathbf{S}_{j+1}e^{i\varphi}e^{-i\gamma}$. The net reflected wave from the j plane is the superposition of the waves,

$$\mathbf{S}_j = -i\sigma\mathbf{T}_j + \mathbf{S}_{j+1}e^{i\varphi}e^{-i\gamma}. \tag{5.261}$$

In addition, a fraction (the same fraction) of the wave is re-diffracted in the direction of the transmitted wave, experiencing a second $-90°$ phase shift, depicted in the figure as the small vector emanating from point s. The superposition of the transmitted wave with the wave from secondary diffraction results in destructive interference and a diminution of the net wave transmitted through the j plane:

$$\mathbf{T}_{j+1}e^{-i\varphi} = \mathbf{T}'_j - i\sigma\mathbf{S}_{j+1}e^{i\varphi} = \mathbf{T}_j e^{-i\gamma} - i\sigma\mathbf{S}_{j+1}e^{i\varphi} \tag{5.262}$$

$$\mathbf{T}_j = \mathbf{T}_{j+1}e^{-i\varphi}e^{i\gamma} + i\sigma\mathbf{S}_{j+1}e^{i\varphi}e^{i\gamma}. \tag{5.263}$$

Ignoring small changes in phase, $\mathbf{T}'_j \simeq \mathbf{T}_j$ and $\mathbf{S}_{j+1}e^{i\varphi} \simeq \mathbf{S}_j \simeq i\sigma\mathbf{T}_j$, and

$$\mathbf{T}_{j+1}e^{-i\varphi} \simeq \mathbf{T}_j - (i\sigma)(-i\sigma\mathbf{T}_j) \simeq \mathbf{T}_j(1-\sigma^2), \tag{5.264}$$

reducing the amplitude of the wave from approximately T_j to approximately $(1 - \sigma^2)T_j$. Since each crystal has a very large number of parallel planes, the decrease in the amplitude of the incident beam can be substantial if the crystal is effectively "perfect" — with all its planes aligned.

The Amplitude Reflectivity. To determine the effect of these interactions on the integrated intensity we must first determine an expression for the ratio of the wave vector of the reflected wave as it leaves the crystal and the transmitted wave as it enters the crystal, the *amplitude reflectivity*, $\mathbf{R}_A$. We can envision determining the transmitted beam wave vector as it exits the crystal with k planes and the reflected beam wave vector from the kth plane, then applying Eqns. 5.261 and 5.263 sequentially until we have $\mathbf{T}_0$ and $\mathbf{S}_0$ with

$$\mathbf{R}_A = \frac{\mathbf{S}_0}{\mathbf{T}_0}. \tag{5.265}$$

If the crystal was exactly in the diffraction condition and there was no phase shift as the wave vectors crossed each plane, φ would equal π, γ would be zero, and $\mathbf{S}_j$ would be in phase with $\mathbf{S}_{j+1}e^{i\pi}$:

$$\begin{aligned} \mathbf{S}_{j+1}e^{i\pi} &= \mathbf{S}_j \\ \mathbf{S}_{j+1} &= \mathbf{S}_j e^{-i\pi} = \mathbf{S}_j e^{i\pi}. \end{aligned} \tag{5.266}$$

When the crystal is out of the diffraction condition $\varphi = \pi + \delta$ and this phase shift, coupled with the phase shift that results from the transmitted and reflected beams as they cross the planes puts $\mathbf{S}_j$ and $\mathbf{S}_{j+1}e^{i\pi}$ slightly out of phase and differing slightly in magnitude,

$$\mathbf{S}_{j+1} = \mathbf{S}_j e^{i\pi} e^{-\eta}, \tag{5.267}$$

since η will be a complex number with both real and imaginary parts (Eqn. 5.261). Substituting Eqn. 5.267 into Eqn. 5.261,

$$\mathbf{S}_j = -i\sigma\mathbf{T}_j + \mathbf{S}_j e^{i\pi} e^{-\eta} e^{i\varphi} e^{-i\gamma} = -i\sigma\mathbf{T}_j + \mathbf{S}_j e^{i\pi} e^{-\eta} e^{i\pi} e^{i\delta} e^{-i\gamma}. \tag{5.268}$$

Noting that $e^x \approx (1 - x)$ for small x, that the product of any of the small variables η, δ, and γ are negligible, and that $e^{i2\pi} = 1$, Eqn. 5.268 becomes

$$\begin{aligned} \mathbf{S}_j &= -i\sigma\mathbf{T}_j + \mathbf{S}_j(1-\eta)(1+i\delta)(1-i\gamma) \quad \text{and} \\ \frac{\mathbf{S}_j}{\mathbf{T}_j} &= \frac{-i\sigma}{1-(1-\eta)(1+i\delta)(1-i\gamma)} = \frac{-i\sigma}{\eta - i\delta + i\gamma} \\ &= \frac{\sigma}{i\eta + \delta - \gamma}. \end{aligned} \tag{5.269}$$

The phase variable, η depends on σ, δ, and γ. It is determined by decoupling Eqns. 5.261 and 5.262, obtaining an equation containing only reflected wave vectors, and incorporating Eqn. 5.267 in order to solve for $i\eta$. The derivation is a bit lengthy, and is given in Appendix C. The result is an expression for η^2:

$$\eta^2 = \sigma^2 - (\delta - \gamma)^2. \tag{5.270}$$

To determine $i\eta$, we note that $(i\eta)^2 = -\eta^2$,

$$\begin{aligned} (i\eta)^2 &= (\delta-\gamma)^2 - \sigma^2 \quad \text{and} \\ i\eta &= \pm\sqrt{(\delta-\gamma)^2 - \sigma^2}. \end{aligned} \tag{5.271}$$

The amplitude reflectivity becomes

$$\mathbf{R}_A = \frac{\mathbf{S}_j}{\mathbf{T}_j} = \frac{\mathbf{S}_0}{\mathbf{T}_0} = \frac{\sigma}{\pm\sqrt{(\delta-\gamma)^2 - \sigma^2} + \delta - \gamma}. \tag{5.272}$$

In order to simplify the expression we define a new phase shift variable, $\epsilon = \delta - \gamma$:

$$\mathbf{R}_A = \frac{\sigma}{\epsilon \pm \sqrt{\epsilon^2 - \sigma^2}}. \tag{5.273}$$

The Integrated Intensity. To determine the integrated intensity we will integrate the ratio of the reflected intensity, $I_S = \mathbf{S}_0\mathbf{S}_0^*$, and the incident beam intensity, $I_0 = \mathbf{T}_0\mathbf{T}_0^*$, as a function of δ, the deviation of the wave vector phase from π as the crystal is rotated through the diffraction condition. The intensity ratio is known as the *intensity reflectivity*,

$$R_I = \frac{I_S}{I_0} = \left(\frac{\mathbf{S}_0}{\mathbf{T}_0}\right)\left(\frac{\mathbf{S}_0}{\mathbf{T}_0}\right)^* = \mathbf{R}_A\mathbf{R}_A^*. \tag{5.274}$$

To describe the intensity reflectivity curve that we intend to integrate (known as the Darwin reflectivity curve) we further simplify Eqn. 5.273 by defining $x = \epsilon/\sigma$. Dividing the numerator and denominator of the equation by σ,

$$\begin{aligned} \mathbf{R}_A &= \frac{1}{x \pm \sqrt{x^2-1}} \\ &= \left(\frac{1}{x \pm \sqrt{x^2-1}}\right)\left(\frac{x \mp \sqrt{x^2-1}}{x \mp \sqrt{x^2-1}}\right) \\ &= x \mp \sqrt{x^2-1}. \end{aligned} \tag{5.275}$$

This creates three unique regimes for the amplitude reflectivity, illustrated in the graph on the left in Fig. 5.25. For $x \geq 1$ the red curve represents the value of the function $x - \sqrt{x^2-1}$, which approaches zero as x increases. The black curve represents $x + \sqrt{x^2-1}$, which increases as x increases. Since x increases as ϵ increases — as the crystal moves out of the diffraction condition — we expect the reflectivity to decrease to zero, and we conclude that for $x \geq 1$, the reflectivity function is $x - \sqrt{x^2-1}$. Following similar arguments, the amplitude reflectivity function for $x \leq -1$ is $x + \sqrt{x^2-1}$. In the region $-1 < x < 1$ the square root is imaginary, and the function is $x \mp i\sqrt{1-x^2}$.

The intensity reflectivity curve, illustrated in graph on the right in Fig. 5.25, is created from the product of the amplitude reflectivity function and its complex conjugate for each region:

$$R_I(x) = \begin{cases} (x - \sqrt{x^2-1})^2 & x \geq 1 \\ (x + i\sqrt{1-x^2})(x - i\sqrt{1-x^2}) = 1 & -1 < x < 1 \\ (x + \sqrt{x^2-1})^2 & x \leq -1 \end{cases} \tag{5.276}$$

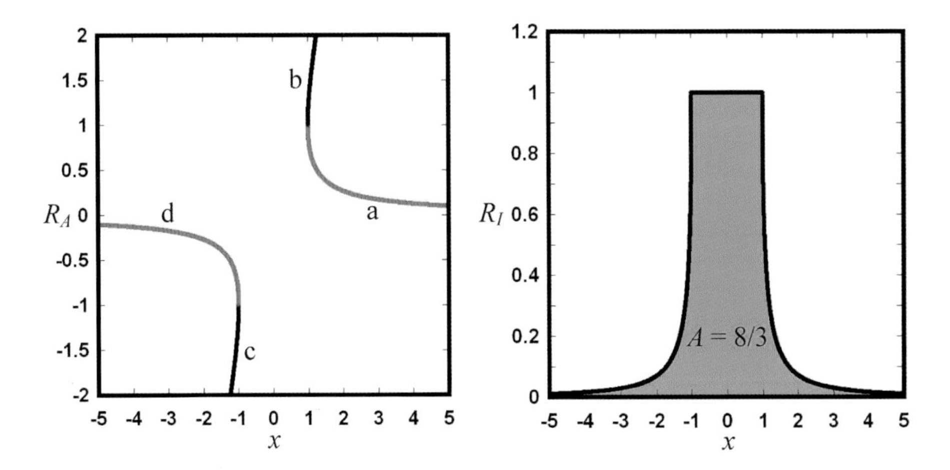

Figure 5.25 Left: Amplitude reflectivity curve. (a) $x - \sqrt{x^2 - 1}$. (b) $x + \sqrt{x^2 - 1}$. (c) $x - \sqrt{x^2 - 1}$. (d) $x + \sqrt{x^2 - 1}$. Right: Intensity reflectively curve.

Integrating over the three regions gives the integrated intensity reflectivity:*

$$\begin{aligned} A &= \int_{-1}^{-\infty} (x + \sqrt{x^2 - 1})^2 dx + (2)(1) + \int_{1}^{\infty} (x - \sqrt{x^2 - 1})^2 dx \\ &= 2 + 2\int_{1}^{\infty} (x - \sqrt{x^2 - 1})^2 dx = 2 + 2\left(\frac{1}{3}\right) = \frac{8}{3}. \end{aligned} \tag{5.277}$$

This result provides the relative integrated intensity as the crystal is rotated and x changes. A change in x corresponds to a change in the diffraction angle, θ; to determine the integrated intensity in the form that we are accustomed to we must determine the same integral with respect to changes in θ. That is, we have

$$\int_x R_I(x)\, dx,$$

and we seek

$$\int_\theta R_I(\theta)\, d\theta.$$

The integrals for each variable will be identical, but expressed in different units. To convert the integral from "x" to "θ" units we convert the differentials upon which they are based. The incident radiation is only approximately monochromatic, consisting of a narrow band of wavelengths. Because of this, as the crystal is rotated so that the average phase at points p and r in Fig. 5.24 varies from π to $\pi + \delta$,

*The integrals for $|x| \geq 1$ are evaluated as

$$\begin{aligned} \int (x \quad - \quad \sqrt{x^2 - 1})^2 dx &= \int 2x^2 dx - \int 1 dx - \int 2x(x^2 - 1)^{1/2}\, dx \\ &= \left(\frac{2}{3}x^3 - x - \frac{2}{3}(x^2 - 1)^{3/2}\right)\Big|_1^\infty = 0 - -\frac{1}{3} = \frac{1}{3}. \end{aligned}$$

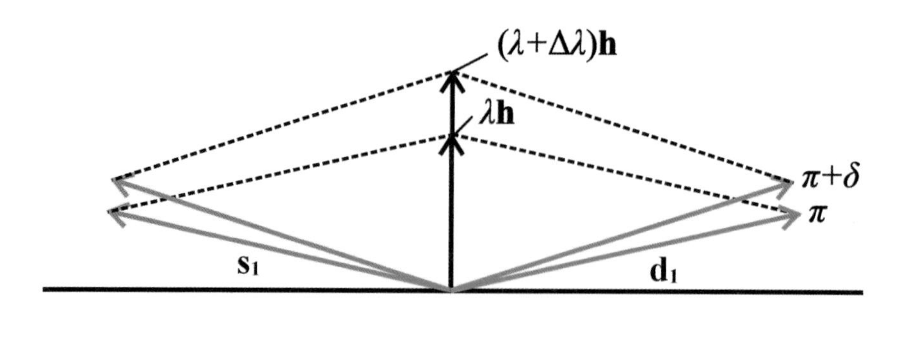

Figure 5.26 Change in the reflection vector phase and corresponding change in the diffraction vector magnitude as the crystal rotates through the diffraction condition.

the diffraction vector varies from $\lambda\mathbf{h}$ to $(\lambda + \Delta\lambda)\mathbf{h}$, as depicted in Fig. 5.26. For small changes, the relative change in the phase is equal to the relative change in the magnitude of the diffraction vector, a vector in reciprocal space, which in turn can be related to changes in θ:

$$\frac{\Delta\lambda\mathbf{h}}{\lambda\mathbf{h}} = \frac{\Delta\lambda}{\lambda} = \frac{\delta}{\pi} = \xi \tag{5.278}$$

$$\delta = \pi\xi \tag{5.279}$$

$$x = \frac{\epsilon}{\sigma} = \frac{\delta - \gamma}{\sigma} = \frac{\pi\xi}{\sigma} - \frac{\gamma}{\sigma}. \tag{5.280}$$

γ and σ are constant, and

$$\frac{dx}{d\xi} = \frac{\pi}{\sigma} \quad \text{and} \quad dx = \frac{\pi}{\sigma}\, d\xi. \tag{5.281}$$

λ, ξ, and θ are related by the Bragg equation:

$$\begin{aligned}
\lambda &= 2d\sin\theta \\
\frac{d\lambda}{d\theta} &= 2d\cos\theta \\
\frac{d\lambda}{\lambda} &= \frac{2d\cos\theta\, d\theta}{2d\sin\theta} = \frac{d\theta}{\tan\theta_B} \\
\frac{\Delta\lambda}{\lambda} &= \frac{\Delta\theta}{\tan\theta_B} = \xi \quad \text{and} \quad \Delta\theta = \xi\tan\theta_B \\
\frac{d\Delta\theta}{d\xi} &= \frac{d\theta}{d\xi} = \tan\theta_B \\
d\xi &= \frac{d\theta}{\tan\theta_B}.
\end{aligned} \tag{5.282}$$

It follows that

$$dx = \frac{\pi}{\sigma} \frac{d\theta}{\tan\theta_B} \quad \text{and} \quad d\theta = \frac{\sigma}{\pi} \tan\theta_B \, dx. \tag{5.283}$$

The relative integrated intensity, now integrated over the diffraction angle, becomes

$$\begin{aligned} \int_0^{2\pi} R_I(\theta)\, d\theta &= \int_0^{2\pi} R_I(\theta) \frac{\sigma}{\pi} \tan\theta_B \, dx \\ &= \frac{\sigma}{\pi} \tan\theta_B \int_{-\infty}^{\infty} R_I(x)\, dx = \frac{8}{3}\frac{\sigma}{\pi} \tan\theta_B. \end{aligned} \tag{5.284}$$

The subscript on the tangent argument indicates that it is a constant at the Bragg angle, θ_B, for the reflection. The reduction of the variables has given us a ratio that is independent of the volume of the crystal. Clearly the ratio must depend on the crystal volume, since the intensity of the incident beam as it enters the crystal, I_0, remains constant, while the intensity of the diffracted beam as it leaves the crystal, $I_S \equiv I_{\mathbf{h}}$, is proportional to the number and area of the reflecting planes (i.e., the number of unit cells in the crystal). Thus, the integrated intensity ratio above must be multiplied by the volume of the crystal or mosaic block undergoing diffraction:

$$\frac{I_{\mathbf{h}}}{I_0} = \frac{8}{3}\frac{\sigma}{\pi} \tan\theta \, V_x. \tag{5.285}$$

From Appendix D (P is the polarization factor, $(1 + \cos^2 2\theta)/2$),

$$\sigma = \frac{\lambda\, d}{\sin\theta} \left(\frac{q_e^2}{4\pi\epsilon_o m_e c^2 V_c} \right) P^{1/2} F_{\mathbf{h}}, \tag{5.286}$$

and

$$\frac{I_{\mathbf{h}}}{I_0} = \frac{8}{3} \frac{\tan\theta}{\pi} \frac{\lambda\, d}{\sin\theta} \left(\frac{q_e^2}{4\pi\epsilon_o m_e c^2 V_c} \right) P^{1/2} F_{\mathbf{h}} V_x. \tag{5.287}$$

Substituting $d = \lambda/2\sin\theta$ and recalling that $2\cos\theta\sin\theta = \sin 2\theta$ gives the dynamical integrated intensity

$$I_{\mathbf{h}}^D = \left(\frac{8}{3\pi} \right) \frac{\lambda^2}{\sin 2\theta} \left(\frac{q_e^2}{4\pi\epsilon_o m_e c^2 V_c} \right) P^{1/2} F_{\mathbf{h}} I_0 V_x. \tag{5.288}$$

Comparing the dynamical integrated intensity resulting from extinction to the kinematical integrated intensity determined earlier (Eqn. 4.111) produces a surprising result (v_ω is the scan speed):

$$I_{\mathbf{h}}^K = \mathcal{I}_{hkl} v_\omega = \frac{\lambda^3}{\sin 2\theta} \left(\frac{q_e^2}{4\pi\epsilon_o m_e c^2 V_c} \right)^2 P F_{\mathbf{h}}^2 I_0 V_x. \tag{5.289}$$

In the absence of extinction the intensity of a reflection depends on the square of the magnitude of the structure factor. This is the premise upon which crystal structure solutions are based. *When extinction is taken into account the intensity of the reflection is proportional to the structure factor instead of its square.* It would appear that extinction would have a dramatic effect on the intensities measured in the diffraction experiment, requiring a significant correction before they could be used to solve the crystal structure. The ratio of the kinematical and dynamical intensities is

$$\frac{I_{\mathbf{h}}^K}{I_{\mathbf{h}}^D} = \frac{3\pi\lambda}{8V_c}\left(\frac{q_e^2}{4\pi\epsilon_o m_e c^2}\right) P^{1/2} F_{\mathbf{h}}. \tag{5.290}$$

The equation above for the dynamical integrated intensity was originally derived by Darwin;[83] its derivation assumed no limit on the number of planes in the crystal. Ewald,[84–86] employing a more general approach based on propagating wave fields and Maxwell's equations, obtained essentially the same result. Using an approach that began with the one outlined here, Darwin extended his treatment to derive a correction factor for a crystal with a limited number of planes:

$$\frac{I_{\mathbf{h}}^K}{I_{\mathbf{h}}^D} = \frac{m\sigma}{\tanh(m\sigma)}, \tag{5.291}$$

where m is the number of planes in the crystal. Its derivation still depends on assuming a crystal of infinite width, and it is difficult to apply in most situations. *More importantly, although some crystals exhibit pronounced extinction effects, most crystals do not.* In order to explain this we must delve into the nature of the crystal itself.

Extinction in Mosaic Crystals. The measured intensities from most crystals are much closer to those predicted by the kinematical model than they are by those predicted from the dynamical model. The dynamical model for diffraction requires the *simultaneous* interaction of transmitted and diffracted waves as the crystal rotates through the diffraction condition. The treatment in the previous section implicitly assumed that diffraction was occurring from a single "perfect" crystal, with the transmitted and reflected waves from every portion of the crystal involved at the same time. In an attempt to explain the observation that only intensities from "perfect" crystals seemed to suffer from extinction, Darwin concluded that the deviations must arise because most crystals are not perfect; the imperfections in some way retarding the extinction effect.

In examining crystal surfaces Darwin noted that the surfaces of crystals examined microscopically appeared to be "wrinkled". Based on this observation he constructed the mosaic model of the crystal that we referred to in Chapter 4. In Darwin's construction, designed more for mathematical convenience than a representation of physical reality, crystals consist of slightly misaligned mosaic blocks (Fig. 4.23). At any instant during the measurement of intensities, only a fraction of these blocks are in the diffraction condition since only a small number of the blocks are aligned simultaneously at any specific angle as the crystal is rotated. The incident beam intensity is therefore diminished minimally by extinction at any specific angle, and the overall effect on the intensity is minimized. For example, suppose that a perfect crystal and a mosaic crystal are rotated by angular increments through the diffraction condition and that transmitted beam intensity for the perfect crystal undergoes a 50% intensity loss due to all of the planes being in registry at the same time. If, at each increment, only 10% of the mosaic blocks are oriented at an angle where they are able to diffract, then the intensity loss for the mosaic crystal will be 5% rather than 50%.*

Darwin described a crystal consisting of mosaic blocks small enough to render extinction effects negligible as an *ideally imperfect crystal.* While the actual physical structure of a crystal is likely to be considerably more complex, crystals consist of many imperfections resulting in multiple misalignments, and the mosaic model can serve to mimic these imperfections. The size of the mosaic blocks will, to a large extent, determine the degree to which extinction affects the observed intensity. Darwin extended his treatment even further by limiting the width and breadth of crystal, thus modeling the extinction correction for a mosaic block. The extinction factor derived by Darwin is expressed in terms of the reflecting power per unit volume (Eqn. 4.112):

$$\frac{I_{\mathbf{h}}^K}{I_{\mathbf{h}}^D} = \frac{\sqrt{2Q_{\mathbf{h}}x^2/\lambda \tan\theta}}{\tanh(\sqrt{2Q_{\mathbf{h}}x^2/\lambda \tan\theta})}, \tag{5.292}$$

where x is the mean depth of a mosaic block. Unfortunately, the expression is difficult to use as an extinction correction factor for real crystals since the average size of a mosaic block is generally unknown. To date there is no method for correcting rigorously for extinction, and we will take an empirical approach.

Referring to Eqn. 5.267,

$$\mathbf{S}_{j+1} = \mathbf{S}_j e^{i\pi} e^{-\eta} = -\mathbf{S}_j e^{-\eta}. \tag{5.293}$$

The negative sign occurs because the transmitted and reflected waves at point t in Fig. 5.24 are 180° out of phase with the waves at point q. Recalling its original definition, $\eta = \eta_r + i\,\eta_i$ is a complex number and

$$\mathbf{S}_{j+1} = -\mathbf{S}_j e^{-\eta_r} e^{-i\eta_i}. \tag{5.294}$$

In a derivation that parallels the one for the determination of η in Eqn. 5.267, we find an identical expression for the transmitted beam,

$$\mathbf{T}_{j+1} = \mathbf{T}_j e^{i\pi} e^{-\eta} = -\mathbf{T}_j e^{-\eta}, \tag{5.295}$$

with the same expression for η. Thus

$$\mathbf{T}_{j+1} = -\mathbf{T}_j e^{-\eta_r} e^{-i\eta_i}. \tag{5.296}$$

Each time the transmitted beam passes to the next plane the transmitted and reflected beams shift slightly in phase by $e^{-i\eta_i}$, and diminish in amplitudes from T_j and S_j to $T_j e^{-\eta_r}$ and $S_j e^{-\eta_r}$. Taking the product of the amplitude vectors and their complex conjugates, it follows that the intensities of both beams are diminished by a factor of $e^{-2\eta_r}$ for each layer. Applying this factor for m layers gives

$$I_{j+m} = I_j(e^{-2\eta_r} e^{-2\eta_r} e^{-2\eta_r} e^{-2\eta_r} \ldots e^{-2\eta_r}) = I_j e^{-2m\eta_r}. \tag{5.297}$$

*It has also been argued that the loss in intensity is diminished because the probability for multiple diffraction becomes smaller as the block size decreases. The argument is based on the probability that a diffracted photon will encounter an electron (only a small fraction do) and be coherently scattered a second time before it leaves the mosaic block. Because of this, as the blocks become very small, the probability of multiple diffraction within a block also becomes very small. While this is true, secondary diffraction can result from an encounter with a second block in alignment with the first; the phenomenon still depends on the number of aligned mosaic blocks.

If d is the distance between planes then the length that the beam travels as it traverses m planes is $l = md/\sin\theta$, and

$$I_{j+m} = I_j e^{-2(\eta_r \sin\theta/d)l}. \tag{5.298}$$

In particular, if l is the path length from the surface of the crystal, then

$$I = I_0 e^{-2(\eta_r/d)l}. \tag{5.299}$$

Comparing this equation to Eqn. 5.184, we see that the quantity, $2(\eta_r \sin\theta/d)$ is behaving as if it was a linear absorption coefficient:

$$I = I_0 e^{-\mu_e(\mathbf{h})\, l}. \tag{5.300}$$

Indeed, we might refer to $\mu_e(\mathbf{h})$ as a *linear extinction coefficient.* Unlike the absorption coefficient, each reflection has its own extinction coefficient; the amount of extinction depends on the magnitude of the structure factor for each reflection.

Although we cannot determine these reflection-specific extinction coefficients rigorously, we can establish a relationship that will allow us to deal with them practically, (and somewhat crudely). The theory is developed by considering the "reflecting power" of the mosaic blocks discussed earlier: $R_{hkl} = Q_{hkl} P V_x$, where V_x is the volume of the block, Q_{hkl} is the reflecting power per unit volume of the block given by Eqn. 4.112, and P is the correction factor for polarization of the diffracted beam. For simplicity we will denote the polarization-corrected reflecting power per unit volume as Q_{hkl}. The intensity of the secondary diffraction is proportional to Q_{hkl}; it follows that the degree of destructive interference in the primary beam will also depend on Q_{hkl}. Since the integrated intensity is actually an energy,

$$\begin{aligned} Q_{hkl} &= \frac{\mathcal{I}_{hkl} v_\omega}{I_0 V} = \frac{E v_\omega}{I_0 V}, \\ E &= Q_{hkl} I_0 \frac{1}{v_\omega} V = Q_{hkl} I_0\, t\, V \quad \text{and} \\ dE &= Q_{hkl} I_0\, t\, dV. \end{aligned} \tag{5.301}$$

The time, t, in which the photons are collected at the detector is the reciprocal of the scan speed, v_ω. Since $\mathcal{I}_{hkl}$ is the energy observed at the detector, we will refer to it as E_d, and use it as the indicator of energy loss in the X-ray beam (the detector is where the effects of any losses are observed). Let us first consider absorption in the crystal. Since the intensity is proportional to Q_{hkl}, it diminishes as the X-ray beam travels through the block, due to absorption. If $Q_{0,hkl}$ is its value when the beam enters the block, then it will be attenuated to $Q_{hkl} = Q_{0,hkl} e^{-\mu_l l}$ after traveling a distance l. Consider an incident beam with unit cross sectional area as depicted in Fig. 5.27(a). The differential energy observed at the detector after absorption as the beam travels the distance dl is

$$dE_d = Q_{0,hkl} e^{-\mu_l l} I_0\, t\, dV = Q_{0,hkl} e^{-\mu_l l} I_0\, t\, (1)\, dl. \tag{5.302}$$

If A_b is the average cross sectional area of the entire beam as it passes through the block (and is diffracted toward the detector), then, for the entire mosaic block

$$dE_d = Q_{0,hkl} e^{-\mu_l l} I_0\, t\, A_b\, dl. \tag{5.303}$$

The total energy of the observed beam, E_a, attenuated due to absorption loss, is then the integral over the total path length of the beam, l_t:

$$E_a = \int dE_d = \int_0^{l_t} Q_{0,hkl} e^{-\mu_l l} I_0 \, t \, A_b \, dl = Q_{0,hkl} I_0 \, t \, A_b \int_0^{l_t} e^{-\mu_l l} \, dl. \quad (5.304)$$

Since E_a is proportional to I_0, varying the incident beam intensity will also vary the energy. However, the relationship that we are aiming for is determined by the ratio of total energies with the same incident beam intensity, and the ratio will not depend on I_0. We therefore consider the case where I_0 has been decreased to a point where the last photon is absorbed just as the diffracted beam is exiting the block. This will allow for integration over l from zero to infinity, since there will be no contributions to the integral beyond $l = l_t$ (i.e., $I_0 e^{-\mu_l} \approx 0$):

$$E_a = Q_{0,hkl} I_0 \, t \, A_b \int_0^{\infty} e^{-\mu_l l} \, dl = \frac{Q_{0,hkl} I_0 \, t \, A_b}{\mu_l}. \quad (5.305)$$

To determine the energy loss due to extinction we make several approximations. First, we assume that the scattering power per unit volume remains close enough to its original value, $Q_{0,hkl}$, that we can consider it constant. A more rigorous treatment would replace Q_{hkl} with its average value. Second, we *assume that there are no absorption losses* so that we are able to assess the energy loss resulting from extinction alone. Finally, we assume that each reflection loses the same fraction of its energy each time diffraction occurs from a unit volume of a mosaic block. In other words, we are (crudely) assuming that the reflection-specific loss for a given reflection depends linearly on its reflecting power. Based on these considerations, if u_e is the fractional energy loss per unit volume, then

$$\frac{\Delta E}{\Delta V} = \frac{\Delta E}{A_b \Delta l} = -u_e \, Q_{hkl} I_0 \, t \approx -u_e \, Q_{0,hkl} I_0 \, t, \quad (5.306)$$

and we approximate the differential energy loss due to extinction as

$$dE = -u_e \, Q_{hkl} I_0 \, t \, A_b \, dl \approx -u_e \, Q_{0,hkl} I_0 \, t \, A_b \, dl. \quad (5.307)$$

Now consider the energy loss due to absorption. For a beam of unit cross sectional area,

$$\frac{dI}{I_0} = -\mu_l \, dl \quad \text{and} \quad dI = -I_0 \mu_l \, dl. \quad (5.308)$$

The intensity, I, is the energy of the beam per unit area per unit time. For the mosaic block with average beam area A_b, the energy of the beam is $E = I \, A_b \, t$, and the differential energy loss due to absorption is

$$dE = dI \, A_b \, t = -\mu_l I_0 t \, A_b \, dl. \quad (5.309)$$

Comparing Eqn. 5.309 with Eqn. 5.307 implies that $u_e \, Q_{0,hkl}$ behaves as if it was a linear absorption coefficient, with the net energy loss from the volume element being

$$dE = -(\mu_l + u_e \, Q_{0,hkl}) I_0 \, t \, A_b \, dl. \quad (5.310)$$

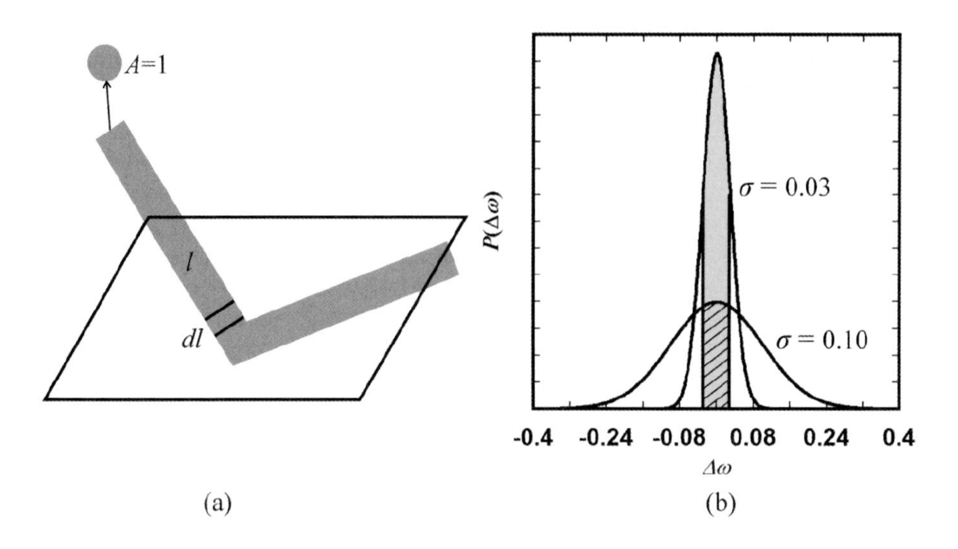

Figure 5.27 (a) Incident and diffracted beams of unit area passing through a mosaic block. (b) Normal probability distributions for mosaic block alignments.

This suggests that the total energy of the observed beam, attenuated by both absorption and extinction, is given by (cf. Eqn. 5.305)

$$E_{ae} = Q_{0,hkl} I_0\, t\, A_b \int_0^\infty e^{-(\mu_l + u_e Q_{0,hkl})l}\, dl = \frac{Q_{0,hkl} I_0\, t\, A_b}{\mu_l + u_e Q_{0,hkl}}. \tag{5.311}$$

If the mosaic blocks are all perfectly aligned, then they behave effectively as one block, and A_b becomes the average cross sectional area of the beam for the entire crystal; E_{ae} is observed at the detector as the diffracted energy for the crystal. If only a fraction of the blocks, g, are aligned at any given value of the scan angle, ω, then the effect will be as if $u_e Q_{0,hkl}$ in the exponential in Eqn. 5.311 was replaced with $g u_e Q_{0,hkl}$:

$$E_{ae} = \frac{Q_{0,hkl} I_0\, t\, A_b}{\mu_l + g u_e Q_{0,hkl}}. \tag{5.312}$$

If $g = 1$ then the crystal is "perfect" and the energy loss due to extinction is at its maximum. If $g = 0$ (or very small) then the only energy loss is due to absorption, — the crystal is "ideally imperfect". For situations between these extremes, at a given value of the rotational angle, ω, some of the blocks will be aligned "exactly" for diffraction to occur. Others will be misaligned slightly so that some diffraction will occur, still others will be misaligned to an extent where they are out of the diffraction condition. The deviation from ω we will refer to as $\Delta\omega$. If we assume that the mosaic block alignment is statistically distributed about a mean value, we can approximate the distribution with a Gaussian probability distribution (Sec. 5.1.1) in order to determine the fraction of the blocks, g, that are capable of diffraction for a given rotational angle:

$$Pr(\Delta\omega) = \frac{1}{\sigma\sqrt{2\pi}} e^{-(\Delta\omega)^2/(2\sigma^2)}, \tag{5.313}$$

where σ is the standard deviation of the distribution of $\Delta\omega$ (See Sec. 5.1.1), and $\langle\Delta\omega\rangle = 0$. Fig. 5.27(b) is a plot of $Pr(\Delta\omega)$ versus $\Delta\omega$. The probability distribution is normalized so that

$$\int_{-\infty}^{\infty} Pr(\Delta\omega)\, d\Delta\omega = \frac{1}{\sigma\sqrt{2\pi}} \int_{-\infty}^{\infty} e^{-(\Delta\omega)^2/(2\sigma^2)}\, d\Delta\omega = 1. \tag{5.314}$$

The integral of the normalized Gaussian distribution over all values is the sum of all of the probabilities of a specific misalignment occurring between some $\Delta\omega$ and $\Delta\omega + d\Delta\omega$ — which is obviously unity. Suppose that diffraction can be observed from blocks with misalignments between $\Delta\omega = -q$ and $\Delta\omega = +q$, indicated by the vertical lines in Fig. 5.27(b). The area under the curve between these limits will then be the probability that the blocks will be aligned somewhere within these limits, i.e., the integral is the fraction of the number of blocks aligned so that diffraction can occur:

$$g = \frac{1}{\sigma\sqrt{2\pi}} \int_{-q}^{q} e^{-(\Delta\omega)^2/(2\sigma^2)}\, d\Delta\omega. \tag{5.315}$$

The two curves in Fig. 5.27(b) are for two different degrees of misalignment, indicated by the width of each peak, which in turn depends on its standard deviation, σ. As the standard deviation decreases (sharper peak), the integral (in blue) increases, indicating that the fraction of "diffracting blocks" is increasing. If σ gets small enough, the curve will lie completely within the limits and the integral will equal 1 — all of the blocks will diffract. As σ increases (broader peak) the integral (indicated with cross-hatches) decreases, as does the fraction of the mosaic blocks that are aligned to diffract.

In order to correct for extinction for a specific reflection, we seek a correction factor, y_e, that will modify the observed intensity attenuated by both absorption and diffraction, I_{ea}, so that it is an intensity attenuated only by absorption I_a (which will be further modified with an absorption correction): $I_a = y_e I_{ea}$ and $y_e = I_a/I_{ea}$. To determine the intensity ratio, we convert E_a and E_b to relative intensities by dividing the energies by the area of the beam and the collection time to obtain energy per unit area per unit time. We also divide by the incident beam intensity to place the intensities on a relative scale:

$$I_a = \frac{E_a}{I_0 A_b t} = \frac{Q_{0,hkl}}{\mu_l} \quad \text{and} \quad I_{ae} = \frac{E_{ae}}{I_0 A_b t} = \frac{Q_{0,hkl}}{\mu_l + g u_e Q_{0,hkl}}. \tag{5.316}$$

The extinction correction factor can be determined from the ratio of these intensities:

$$y_e = \frac{I_a}{I_{ae}} = \frac{\mu_l + g u_e Q_{0,hkl}}{\mu_l} = 1 + \frac{g u_e Q_{0,hkl}}{\mu_l} = 1 + gueI_a \tag{5.317}$$

This ratio contains the corrected intensity in the correction factor, but rearrangement gives

$$\begin{aligned} I_a &= I_{ae} + I_a g u_e I_{ae} \quad \text{and} \quad I_a(1 - g u_e I_{ae}) = I_{ae} \Rightarrow \\ y_e &= \frac{I_a}{I_{ae}} = \frac{1}{1 - g u_e I_{ae}}. \end{aligned} \tag{5.318}$$

The approach taken here has been to give the reader a conceptual basis for the underlying principles responsible for extinction phenomena, and to provide at least some of the ingredients that must go into the formulation of rigorously derived extinction correction expressions. Unfortunately, the assumptions made in the derivation above, and the difficulty in assessing the mosaic parameter, g, or the energy loss parameter, u_e, make it virtually impossible to determine the extinction correction factors directly. An alternative to doing so is to await the solution and refinement of the crystal structure. The calculated intensities, I_c, are based solely on the structural model, and therefore if exact, would correspond to observed intensities corrected for all experimental errors. The expression $I_a = y_e I_{ae}$ indicates that I_a has been corrected for extinction, and I_{ae} has not. If the actual observed intensities, I_o, have been scaled and corrected for everything *except* extinction, then, by analogy, $I_c/I_o = I_a/I_{ae}$. Substituting in Eqn. 5.317,

$$y_e = \frac{I_c}{I_o} = 1 + gu_e I_c = gu_e I_c + 1. \tag{5.319}$$

The ratio of the calculated to observed intensities increases as the intensities increase; the magnitude of the increase depends on the magnitude of gu_e. This is a useful indicator — if I_c/I_o values are consistently larger than 1 for the stronger reflections in the data set then there is a high probability that the data set has suffered extinction effects. T*he extinction effect is most pronounced for the strongest reflections*, and the simplest approach to the extinction problem is to eliminate the reflections that are the most intense from the data set. This is often difficult in practice since the (approximate) linear nature of the increase in I_c/I_o makes it difficult to determine an intensity threshold for data elimination.

It follows that it is better to attempt to correct for extinction effects whenever they appear significant. Note that Eqn. 5.319 is of the form $y = mx + b$ — the equation of a line. After accurate absorption corrections have been made and the structure has been refined with as much accuracy as possible, a plot of $y_e = I_c/I_o$ versus I_c should yield a least-squares line with a slope of gu_e, allowing us to determine gu_e and correct the observed intensities: $I_o' = y_e I_o$. An evaluation of gu_e in this manner serves as a useful diagnostic. If gu_e is near zero then extinction effects can be ignored. If it is large,* then an attempt to correct for extinction is warranted. The crude nature of the treatment above, and the lack of other rigorous approaches, suggests that a better alternative is to attempt to modify the crystal physically — by mechanical or thermal methods. In many cases, plunging a crystal into liquid nitrogen will thermally "shock" the crystal, breaking it up internally into very small blocks and rendering it effectively ideally imperfect (kinematical), reducing g to a point where it is unnecessary to correct for extinction. *Fortunately, most crystals are sufficiently mosaic to allow the investigator to ignore extinction effects.*

The reader may have noted that a distinction between "primary extinction" and "secondary extinction" was not made in the previous discussion. Secondary extinction was suggested by Darwin in order to rationalize energy loss in the beam that would occur even if interference (primary extinction) was not responsible. It was based on the notion that the transfer of energy from the incident beam to the

*"Large" is relative here. Since I_o/I_c is a ratio, and I_c is an intensity, they differ by about three orders of magnitude. Thus, to assess gu_e, multiplying it by the intensity of the strongest reflection will give "large" values around 1.

diffracted beam from diffraction alone depleted the incident beam intensity. As with primary extinction, secondary extinction depended on the mosaic distribution and reflecting power of each individual reflection. Darwin proposed that the power gained by the diffracted beam must be lost by the diffracted beam, thus conserving energy. Based upon this he formulated a set of kinematical energy transfer equations to determine corrections for secondary extinction[83]. Other investigators[87–90] refined Darwin's treatment, using the transfer equations to determine factors putatively correcting for both primary and secondary extinction. The two approaches have often been discussed as being compatible, with the kinematical model a limiting case for the dynamical model; the difficulty in employing the dynamical model is ordinarily cited as the reason for using the kinematical corrections. *However, the models are mutually contradictory.* As discussed in Appendix E, "secondary extinction" effects are fully accounted for by the experimentally determined absorption coefficients, and the apparent failure of the dynamical model to conserve energy is based on a flawed assumption. Extinction must be modeled dynamically in order to quantitatively determine its effects on reflection intensities. There is currently no practical theory that accomplishes this task for small single crystals of arbitrary shape.

The Dispersion Correction

In the treatment of extinction in the previous section we determined that the coherently scattered radiation from a thin crystal was approximately 90° out of phase with respect to the incident radiation. Only a small portion of the radiation is scattered, and a relatively small fraction of that radiation is scattered back in the direction of the incident beam. The small amplitude electromagnetic wave due to these scattered photons superpositions with the much larger amplitude wave of the transmitted beam wave, producing a small shift in the phase of the transmitted beam. At the detector the resultant phase delay results in an *apparent* change in the speed of light. The ratio of the speed of light to its apparent speed as it passes through a material is known as the refractive index, n_r, of the material. The refractive index depends on the electronic nature of the material and the wavelength of the radiation. Formally, the variation of the refractive index with wavelength is known as *dispersion.* The term arises from the observation in optics that the various wavelengths in a polychromatic beam of light will be subjected to a unique refractive index for each wavelength, causing each wavelength to take a different path through the material – the beam is *dispersed* (spread out) into its component wavelengths. For visible light this produces the familiar color spectrum. For monochromatic X-rays the wavelength dependence of the refractive index results in very small changes in the phase shift of the diffracted beam in most circumstances, usually with only small effects on the observed intensity. However, there are circumstances when the dispersion effect is pronounced for a specific atom type in the crystal, causing a phase shift differing substantially from the usual 90° from the atoms of that type. Since the relative locations of the scatterers in the crystal are determined by the relative phases of the diffracted radiation, this anomalous scattering, known as *anomalous dispersion*, must be corrected for, since it produces shifts in the phase that are *not* a function of atom locations.

To determine the effects of dispersion on the diffracted radiation, we again consider a thin crystal subjected to monochromatic incident radiation, radiating from

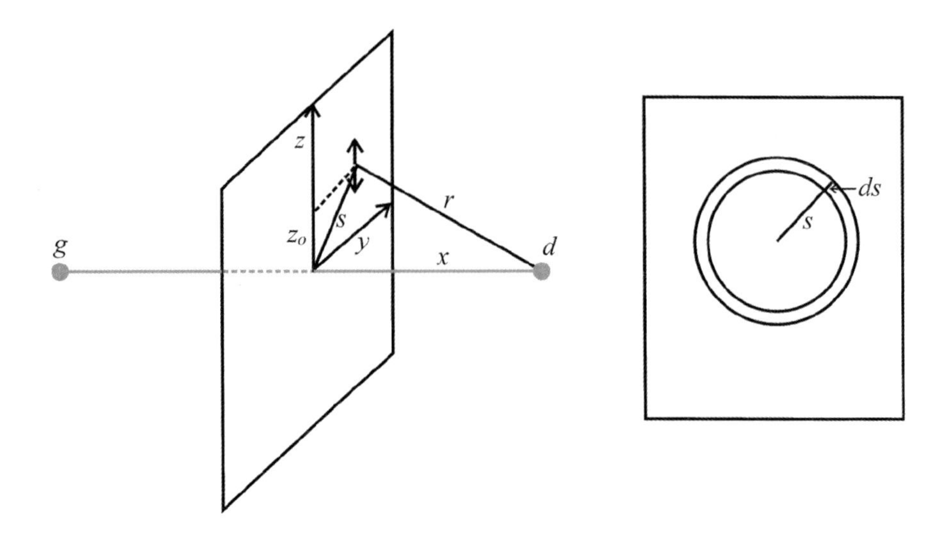

Figure 5.28 Oscillation of bound electrons resulting from the oscillating electric field of the incident radiation.

a "generator" and ultimately arriving at a detector after interacting with the material. This is depicted graphically in Fig. 5.28. For simplicity we will consider the incident beam to be plane polarized in the z direction and to be traveling in the x direction, away from the generator; we constrain the secondary radiation emitted from the electron oscillators in the material to remain plane polarized (since polarization effects will be accounted for in the polarization factor). Each electron in the material is subjected to an oscillating electric field with frequency, ν and amplitude $\mathcal{E}_I$: $\overrightarrow{\mathcal{E}_I(t)} = \mathcal{E}_I e^{i2\pi\nu t} = \mathcal{E}_I e^{i\omega t}$, where $\omega = 2\pi\nu$. As a result each bound electron in the material experiences a force that sets it into motion – turning it into a secondary oscillator. A given oscillator at position z_o experiences a sinusoidal displacement,

$$z(t) = z_o e^{i\omega t}, \tag{5.320}$$

a velocity in the z direction,

$$v(t) = \frac{dz(t)}{dt} = i\omega z_o e^{i\omega t}, \tag{5.321}$$

and an acceleration,

$$a(t) = \frac{dv(t)}{dt} = i^2\omega^2 z_o e^{i\omega t} = -\omega^2 z_o e^{i\omega t}. \tag{5.322}$$

The magnitude of the electric field observed at the detector, $\mathcal{E}_d$, is directly proportional to the acceleration, directly proportional to the charge on the particle being accelerated, and inversely proportional to the distance at which the radiation is detected, r. It has the same phase as the electric field emanating from the oscillating electron at time $t - r/c$.

For electron j this is

$$\overrightarrow{\mathcal{E}_{d,\,j}(t)} = \kappa\,\frac{q_e}{r}\,a(t) = -\kappa\,\frac{q_e}{r}\,\omega^2 z_o e^{i\omega(t-r/c)}, \tag{5.323}$$

where $\kappa = 1/(4\pi\epsilon_o c^2)$.

The net wave observed at the detector is the superposition of the waves emanating from each oscillator. If we let ρ_A be the number of electron oscillators per unit area in Fig. 5.28, we can consider a ring consisting of two concentric circles, one at radius s, and the other at radius $s + ds$. The area of the ring is $A_{ring} = 2\pi s\, ds$, and the number of electron oscillators in the ring is $A_{ring}\rho_A = 2\pi s\, ds\, \rho_A$. The contribution to the electric field at the detector from the oscillators in the ring is then $2\pi s\, ds\, \rho_A \overrightarrow{\mathcal{E}_{d,j}(t)}$; the net wave at the detector resulting from each ring is therefore

$$\overrightarrow{\mathcal{E}_{ring,\,d}} = -\kappa\,\frac{q_e}{r}\omega^2 z_o e^{i\omega(t-r/c)}\pi\rho_A\, 2s\, ds. \tag{5.324}$$

In order to determine the electric field observed at the detector we superposition the resultant waves from each ring:

$$\overrightarrow{\mathcal{E}_d} = \int_o^{s(max)} -\kappa\,\frac{q_e}{r}\omega^2 z_o e^{i\omega(t-r/c)}\pi\rho_A\, 2s\, ds \tag{5.325}$$

$$= -\kappa q_e \omega^2 z_o e^{i\omega t}\pi \int_o^{\infty} \rho_A \frac{e^{-i\omega(r/c)}}{r} 2s\, ds, \tag{5.326}$$

with $s(max)$ the maximum value of s.

The differential can be expressed in terms of r by noting that $r^2 = s^2 + x^2$:

$$\frac{d(r^2)}{dr} = \frac{d(s^2)}{dr} + \frac{d(x^2)}{dr} = \frac{d(s^2)}{dr},$$

$$2r = 2s\frac{ds}{dr} \quad \text{and} \quad 2r\, dr = 2s\, ds. \tag{5.327}$$

When $s = 0$, $r = x$, and the integral is evaluated from $r = x$ to $r = r(max)$:

$$\overrightarrow{\mathcal{E}_d} = -\kappa q_e \omega^2 z_o e^{i\omega t} 2\pi \int_x^{r(max)} \rho_A e^{-i\omega(r/c)}\, dr \tag{5.328}$$

$$= -\kappa q_e \omega^2 z_o e^{i\omega t} 2\pi\, \overrightarrow{\eta}. \tag{5.329}$$

The integral in the expression is represented by the symbol $\overrightarrow{\eta}$ to emphasize that it is a vector resulting from a superposition of the vector contributions from each oscillator. While the formal evaluation of this integral is straightforward, it appears to produce an arbitrary result, with the final phase of the resultant wave dependent on $r(max)$. The reason for this can be seen if we attempt to evaluate the integral numerically, beginning with $r = z$ and summing small rectangles with length Δr:

$$\overrightarrow{\eta} = \int_x^{r(max)} \rho_A e^{-i\,\omega(r/c)}\, dr \approx \sum_{n=0}^{n(max)} \rho_A e^{-i\,\omega[(z+n\Delta r)/c]}\Delta r. \tag{5.330}$$

Letting $\phi = -\omega x/c$ and $\Delta\phi = -\omega\Delta r/c$, the first three terms in the sum are

$$\overrightarrow{\eta_0} = \rho_A \Delta r(e^{-i\phi})$$

$$\overrightarrow{\eta_1} = \rho_A \Delta r(e^{-i(\phi+\Delta\phi)})$$

$$\overrightarrow{\eta_2} = \rho_A \Delta r(e^{-i(\phi+\Delta\phi+\Delta\phi)}), \quad \text{etc.}$$

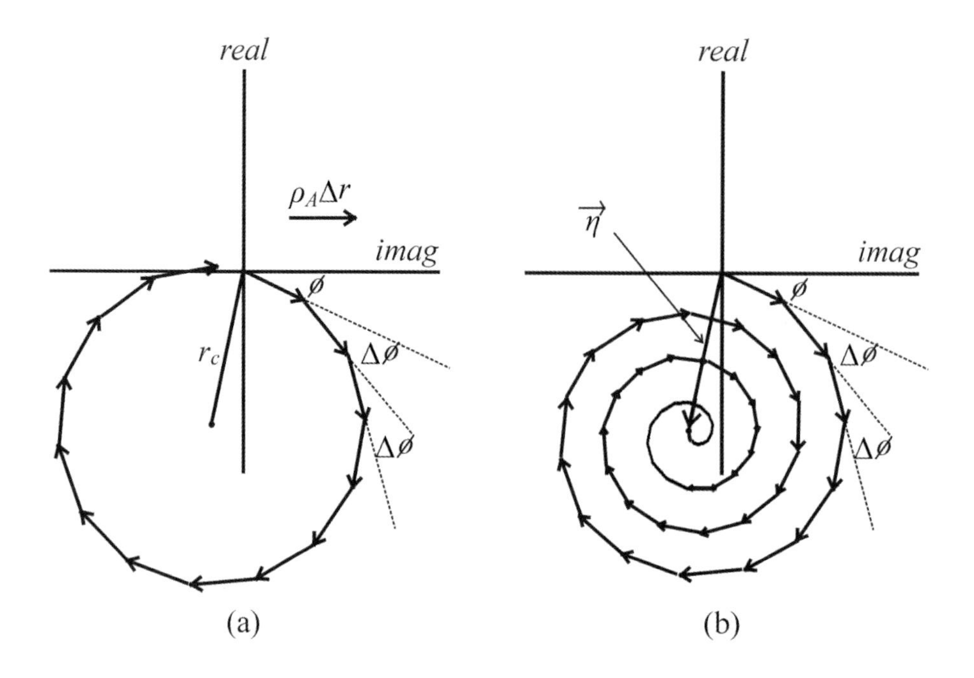

Figure 5.29 Numerical evaluation of $\int_x^{r(max)} \rho_A e^{-i\omega(r/c)}\, dr$ (a) when ρ_A remains constant and (b) when ρ_A diminishes near the edges of the crystal.

$\vec{\eta_0}$ is a vector of length $\rho_A \Delta r$ with phase angle ϕ, $\vec{\eta_1}$ is a vector of length $\rho_A \Delta r$ with phase angle $(\phi + \Delta\phi)$, $\vec{\eta_2}$ is a vector of length $\rho_A \Delta r$ with phase angle $(\phi + \Delta\phi + \Delta\phi)$, and so forth. The results of the integration are shown in Fig. 5.29(a) The constant change in phase angle results in the creation of a polygon as each subsequent vector of length $\rho_A \Delta r$ is added to the previous one rotated by the same angle, $\Delta\phi$. The vector sum will follow a circle with circumference r_c until the last term in the integration is evaluated. The resultant vector will be the vector from the origin to the last vector in the sum, which is arbitrary, depending on the last value of r.

However, while ρ_A will remain constant in the interior of the crystal, as we approach the edges of the crystal the number of oscillators will begin to tail off. Since there are many oscillators on the molecular scale, this tail-off will not be abrupt, but will be gradual, such that ρ_A will slowly diminish to zero. The consequences of the eventual diminution of $\rho_A \Delta r$ are illustrated in Fig. 5.29(b). After many "constant cycles" the length of $\rho_A \Delta r$ gradually begins to get smaller and smaller. The changes in the phase angles, $\Delta\phi$, remain constant, but the shorter vectors result in the formation of a spiral that eventually converges into the center of the circle. The resultant vector, $\vec{\eta}$, is therefore the radius vector of the circle prior to the decrease in $\rho_A \Delta r$.

If the magnitude of Δr is small, then the polygon in Fig. 5.29(a) is essentially a circle, and the length of the resultant vector will be the radius of the circle with circumference $2\pi r_c$. For a single cycle around the circle:

$$\begin{aligned} 2\pi r_c &= \sum \rho_A \Delta r = \frac{c\rho_A}{\omega}\sum \Delta\phi \\ 2\pi r_c &= 2\pi\frac{c\rho_A}{\omega} \Longrightarrow r_c = \frac{c\rho_A}{\omega}, \end{aligned} \tag{5.331}$$

since $\sum \Delta\phi = 2\pi$. Furthermore, if Δr is small, then the initial vector in the superposition (at phase angle ϕ) will be perpendicular to the vector $\overrightarrow{\eta}$, and the phase angle of the resultant vector will be $\phi + 90°$. The integral therefore evaluates to

$$\overrightarrow{\eta} = \frac{c\rho_A}{\omega}e^{i(\phi+\pi/2)} = \frac{c\rho_A}{\omega}\left(\cos\left(\phi+\frac{\pi}{2}\right) + i\sin\left(\phi+\frac{\pi}{2}\right)\right) \tag{5.332}$$

$$\begin{aligned} \overrightarrow{\eta} &= \frac{c\rho_A}{\omega}(-\sin\phi + i\cos\phi) = \frac{c\rho_A}{-i\omega}(\cos\phi + i\sin\phi) \\ &= \frac{c\rho_A}{-i\omega}e^{i\phi} = \frac{c\rho_A}{-i\omega}e^{-i\omega x/c} \\ &= i\frac{c\rho_A}{\omega}e^{-i\omega x/c}. \end{aligned} \tag{5.333}$$

Substituting in Eqn. 5.329 gives us the net electric field vector at the detector:

$$\begin{aligned} \overrightarrow{\mathcal{E}_d} &= \left(-\kappa q_e \omega^2 z_o e^{i\omega t} 2\pi\right)\left(i\frac{c\rho_A}{\omega}e^{-i\omega x/c}\right) \\ &= (-2\pi\kappa q_e c\rho_A)\, i\omega z_o e^{i\omega(t-x/c)}. \end{aligned} \tag{5.334}$$

From Eqn. 5.321,

$$\overrightarrow{\mathcal{E}_d} = (-2\pi\kappa q_e c\rho_A)\, v(t - x/c). \tag{5.335}$$

The electric field at the detector is directly proportional to the velocity of the oscillating electrons.

Having determined the effects on the detected electric field arising from the oscillating electrons in a crystal, we now focus on the relationship between the incident radiation and the detected radiation from these oscillating electrons. In particular we are interested in the phase changes that occur as a result of the retardation effects created by the interaction of the electric field with the bound oscillators.

The electrons are moving about their respective nuclei, subject to the attractive forces of the nuclei. In order to treat the interactions of these electrons with the external electric field of the incident radiation we will make some apparently crude approximations here, which turn out to be amazingly good approximations in the end. We treat the electron's average motion toward and away from the nucleus as if it was subjected to a linear restoring force. While the force is actually an inverse square force over the short distances inside the atom, the changes over small distances vary approximately in proportion to the displacement from the average position of the electron (i.e., they are behaving as harmonic oscillators). Within this approximation, the electrons are "attached" to the nucleus by a spring

with force constant, k, "vibrating" with a natural resonant frequency of $2\pi\nu_o = \omega_o$.*
The force on the electron due to the "spring" is given by Hooke's Law and Newton's second law (simultaneously):

$$F_n(t) = -kz(t) = m_e\, a(t) = m_e \frac{d^2z(t)}{dt^2}. \qquad (5.336)$$

In the absence of an external field the electron is "vibrating" at its resonant frequency. Substituting the expression for the electron oscillator acceleration (Eqn. 5.322),

$$\begin{aligned} -kz(t) &= m_e(-\omega_o^2 z_o e^{i\omega t}) = -m_e\omega_o^2 z(t) \text{ and} \\ \frac{k}{m_e} &= \omega_o^2. \end{aligned} \qquad (5.337)$$

Quantum mechanically, the unperturbed electron is in a stationary state (a conservative state), and does not radiate energy.

We now subject the electron to the oscillating electric field of the incident radiation with frequency $2\pi\nu = \omega$. The electron experiences a second "springlike" force, $F_I(t) = q_e\mathcal{E}_o e^{i\omega t}$, pulling it away, then pushing it toward the nucleus. As the electron oscillates it constantly re-radiates the energy absorbed from the incident radiation. If the incident beam is turned off, the electron will quickly release its excess energy and return to the ground state, again oscillating at ω_o. In order to model this classically, we treat the electron as if was subjected to an additional frictional "drag" force — a damping force — such as that experienced by an automobile that suddenly runs out of gasoline. By analogy, we set this frictional force proportional to the speed of the electron, $F_d(t) = -Ddz/dt$. The net force on the electron is

$$F(t) = -kz(t) + F_d(t) + F_I(t) = m_e \frac{d^2z(t)}{dt^2}, \qquad (5.338)$$

resulting in

$$\frac{d^2z(t)}{dt^2} + \frac{kz(t)}{m_e} + \frac{D}{m_e}\frac{dz}{dt} = \frac{F_I(t)}{m_e}. \qquad (5.339)$$

We define D/m_e as the *damping factor*, γ. Substituting $i\omega z(t)$ (Eqn. 5.321) for $dz(t)/dt$ results in the differential equation of motion of the electron behaving as a *forced oscillator*:

$$\frac{d^2z(t)}{dt^2} + \omega_o^2 z(t) + i\gamma\omega z(t) - \frac{q_e\mathcal{E}_o e^{i\omega t}}{m_e} = 0. \qquad (5.340)$$

The solution to this differential equation (verified by substituting the solution into the equation) is

$$z(t) = \frac{q_e\mathcal{E}_o e^{i\omega t}}{m_e(\omega_o^2 - \omega^2 + i\gamma\omega)}. \qquad (5.341)$$

*The equations derived here will be classical, but they also result from a quantum mechanical treatment of the atom. The "resonant frequencies" of the electron in this classical treatment become those of the transition energies of the atom between its stationary states in the quantum mechanical treatment.

This is the displacement of the oscillating electron as a function of time resulting from the combination of the nuclear attractive force, the damping force, and the force due to the electric field of the incident radiation. The velocity of the oscillating electron is therefore

$$v(t) = \frac{dz(t)}{dt} = i\,\omega \frac{q_e \mathcal{E}_o e^{i\,\omega t}}{m_e(\omega_o^2 - \omega^2 + i\gamma\omega)}. \tag{5.342}$$

Substituting $v(t - x/c)$ and $\kappa = 1/(4\pi\epsilon_o c^2)$ in Eqn. 5.335, the electric field at the detector resulting from the oscillating electrons driven by the electric field of the incident radiation is

$$\begin{aligned} \overrightarrow{\mathcal{E}_d} &= \frac{-2\pi q_e c \rho_A}{4\pi\epsilon_o c^2}\, i\,\omega \frac{q_e \mathcal{E}_o e^{i\,\omega(t-x/c)}}{m_e(\omega_o^2 - \omega^2 + i\gamma\omega)} \\ &= -i\,\omega \frac{q_e^2 \rho_A}{2\epsilon_o c m_e} \frac{1}{(\omega_o^2 - \omega^2 + i\gamma\omega)}\, \mathcal{E}_o\, e^{i\,\omega(t-x/c)}. \end{aligned} \tag{5.343}$$

The resulting electric field is a complex number; it is out of phase with respect to the incident radiation. The damping factor is much smaller than the frequency of the incident radiation. When the frequency of the incident radiation is substantially greater than that of the resonant frequency of the oscillator, i.e., $\omega \gg \omega_o$, $(\omega_o^2 - \omega^2 + i\gamma\omega \approx -\omega^2)$ and

$$\overrightarrow{\mathcal{E}_d} \approx i \frac{q_e^2 \rho_A}{2\epsilon_o c m_e \omega}\, \mathcal{E}_o\, e^{i\,\omega(t-x/c)}. \tag{5.344}$$

In the absence of the crystal the electric field due to the incident radiation at the detector would be $\overrightarrow{\mathcal{E}_d} = \mathcal{E}_o e^{i\omega(t-x/c)}$. Measured at $(t - x/c) = 0$, this wave vector is coincident with the real axis in the complex plane (Fig. 5.23). When $\omega \gg \omega_o$, the scattered radiation from the crystal at $(t - x/c) = 0$ lies entirely along the imaginary axis — it is 90° out of phase with respect to the incident radiation. When $\omega \ll \omega_o$ the electric field from the oscillating electron is −90° out of phase with respect to the incident radiation:

$$\overrightarrow{\mathcal{E}_d} \approx -i \frac{q_e^2 \rho_A \omega}{2\epsilon_o c m_e \omega_o^2}\, \mathcal{E}_o\, e^{i\,\omega(t-x/c)}. \tag{5.345}$$

When the frequency of the incident radiation differs significantly from the resonant frequency of an electron oscillator in the crystal the radiation behaves "normally", shifted by ±90°, as we previously determined when considering the extinction phenomenon.

As ω approaches ω_o we can no longer consider $\gamma\omega$ negligible in comparison to $\omega_o^2 - \omega^2$, and the phase shift begins to deviate from 90°. When the frequency of the incident radiation becomes equal to the oscillator frequency, the electric field at the detector becomes:

$$\overrightarrow{\mathcal{E}_d} \approx -\frac{q_e^2 \rho_A}{2\epsilon_o c m_e \gamma}\, \mathcal{E}_o\, e^{i\,\omega(t-x/c)}. \tag{5.346}$$

At $(t - x/c) = 0$ the entire wave vector is aligned along the negative real axis, indicating a 180° change of phase.

For a range of incident beam frequencies beginning at $\omega \gg \omega_o$ and decreasing to zero, the phase of the scattered wave vector changes only slightly until $\omega_o^2 - \omega^2$ gets close to zero, then dramatically undergoes a phase shift from 90°, through 180° to 270°. This sudden change in phase was considered an anomaly when it was first observed, and the phenomenon is still called *anomalous scattering*, or *anomalous dispersion*, even though there is really nothing "anomalous" about it!

In order to examine dispersion effects on the electric field at the detector as the wavelength varies, we return to the representation of the retarded wave in terms of the refractive index derived in the discussion of primary extinction. From Eqn. 5.259, the electric field resulting from the scattered radiation is

$$-i\omega \frac{(n_r - 1)T}{c} \mathcal{E}_o e^{i\omega(t-x/c)}. \tag{5.347}$$

The expression for the electric field in Eqn. 5.343 is for the radiation produced by the oscillators, i.e., it also describes the scattered radiation. Setting the two expressions equal to one another generates the following relationship:

$$(n_r - 1)T = \frac{q_e^2 \rho_A}{2\epsilon_o m_e} \frac{1}{(\omega_o^2 - \omega^2 + i\gamma\omega)}. \tag{5.348}$$

Since ρ_A is the number of oscillators per unit area, and T is the thickness of the crystal, ρ_A/T is the number of oscillators per unit volume of the crystal, N_V. Dividing by T and solving for n_r results in an expression for the refractive index as a function of the frequency of the incident radiation — a dispersion equation:

$$n_r = 1 + \frac{q_e^2}{2\epsilon_o m_e} \frac{N_V}{(\omega_o^2 - \omega^2 + i\gamma\omega)}. \tag{5.349}$$

The crystal is actually composed of multi-electronic atoms. Classically, each electron in an atom has its own set of resonant frequencies, resulting in different electron oscillators with different resonant frequencies. The electric field and refractive index are the superpositions of the effects from each oscillator:

$$n_r = 1 + \frac{q_e^2}{2\epsilon_o m_e} \sum_k \frac{N_k}{(\omega_{o,k}^2 - \omega^2 + i\gamma\omega)}, \tag{5.350}$$

where N_k is the number of oscillators with resonant frequency $\omega_{o,k}$ per unit volume. A more rigorous (but less intuitive) quantum mechanical treatment produces essentially the same equation, except that each atom is treated as a "multiple oscillator," with each of its resonant frequencies corresponding to the difference in energies between its electronic states — its transition energies. Quantum mechanics tells us that there is a certain probability that each transition will take place, and the resulting electric field and refractive index depends on the sum over all of the waves emanating from each oscillator for each frequency, weighted by the transition probability for each resonant frequency:

$$\overrightarrow{\mathcal{E}_d} = -i\,\omega \frac{q_e^2 \rho_A}{2\epsilon_o c m_e} \sum_a \sum_k \frac{p_k N_a}{(\omega_o^2 - \omega^2 + i\gamma\omega)} \mathcal{E}_o\, e^{i\,\omega(t-x/c)} \quad \text{and} \tag{5.351}$$

$$n_r = 1 + \frac{q_e^2}{2\epsilon_o m_e} \sum_a \sum_k \frac{p_k N_a}{(\omega_{o,k}^2 - \omega^2 + i\gamma\omega)}, \tag{5.352}$$

where the first summation in each expression is over all the atoms in the crystal, the second is over all of the electronic transitions for each atom, and N_a is the number of atoms of atom type a per unit volume.

Transforming the denominator in Eqns. 5.351 and 5.352 separates each expression into its real and imaginary components:

$$\left(\frac{1}{(\omega_o^2 - \omega^2 + i\gamma\omega)}\right)\left(\frac{(\omega_o^2 - \omega^2 - i\gamma\omega)}{(\omega_o^2 - \omega^2 - i\gamma\omega)}\right)$$
$$= \frac{\omega_o^2 - \omega^2}{(\omega_o^2 - \omega^2)^2 + \gamma^2\omega^2} - i\frac{\gamma\omega}{(\omega_o^2 - \omega^2)^2 + \gamma^2\omega^2}. \quad (5.353)$$

The refractive index then takes on the form $n_r = n_r' - i\,n_r''$. Substituting this expression for the refractive index into Eqn. 5.257, the equation for the electric field at the detector *when it is aligned with the incident beam*, results in

$$\overrightarrow{\mathcal{E}_d'(t)} = e^{-\omega n_r'' T/c}\mathcal{E}_o e^{i\omega(t-r/c)} e^{-i\omega(n_r'-1)(T/c)}. \quad (5.354)$$

Squaring both sides of the equation and multiplying each side by the proportionality constant $\epsilon_o\,c$ (recall that $I = \epsilon_o\,c\,\mathcal{E}^2$) illustrates the effect of the damping factor on the intensity observed at the detector:

$$I = I_0 e^{-(2\omega/c\,n_r'')\,x}. \quad (5.355)$$

The thickness of the crystal, T, has been replaced with x, the distance that the beam travels through the crystal. The exponential term decreases as the beam travels through the crystal — attenuating the incident beam intensity. Comparing Eqn. 5.355 with Eqn. 5.187 leads us to conclude that $2\omega/c\,n_r''(\gamma) = \mu_l(\omega)$, *giving us the wavelength dependence of the linear absorption coefficient.* The damping factor represents the energy dissipation mechanism that allows the oscillators to return to their ground states after interacting with the incident radiation — the energy loss that the incident beam undergoes as it passes through the crystal. Thus the imaginary term in Eqns. 5.351 and 5.352 is the term responsible for the absorption of the X-radiation as it passes through the crystal. The absorption maxima illustrated in Fig. 5.12 occur when the frequency of the incident radiation approaches a resonant frequency of the atom ($\omega \approx \omega_o$), and the imaginary term, n_r'', approaches its maximum value.

The attenuation of the beam intensity can be compensated for with an adequate absorption correction, but the effect of dispersion on the relative phase of the diffracted beam cannot. The contribution to this phase shift is different for each atom type. In Sec. 3.2.4 we determined that a spherical atom could be treated as if it's electrons were at the nucleus, scattering with an efficiency determined by the scattering factor, which was attenuated due to destructive interference from scattering from various locations inside the atomic sphere. The relative destructive interference resulted in a diminution of the scattering efficiency at increasing diffraction angles, illustrated in Figs. 3.37 and 4.32. The spherical atom model assumes that the net interference of the waves scattering from each atom depends only on their relative locations in the crystal. This interference is the fundamental phenomenon that allows for the determination of the atom locations. If the wave from a given atom undergoes an additional phase shift as a consequence of dispersion, this change in phase must be modeled into the calculated structure factor in order to match it with the observed structure factor. Because the dispersion effect depends on each atom type, the atomic scattering factor is the logical entity on which to base this model, and we now determine the way in which dispersion affects it.

For simplicity, we consider the simplest atom, a hydrogen atom, with a scattering factor of 1 at $2\theta = 0$. We treat it classically, with a resonant frequency of ω_o, so that the electric field observed at the detector is described by Eqn. 5.343. Recalling that the scattering factor for an atom is the electric field observed at the detector relative to that due to a hypothetical unbound electron scattering from the center of the atom, we note that for a free electron $\omega_o = 0$ and $\gamma = 0$. For the free electron, Eqn. 5.343 becomes

$$\vec{\mathcal{E}}_{d,\,free} = f_e = -i\,\omega \frac{q_e^2 \rho_A}{2\epsilon_o c m_e} \frac{1}{-\omega^2} \mathcal{E}_o\, e^{i\,\omega(t-x/c)}. \tag{5.356}$$

The ratio of $\vec{\mathcal{E}_d}$ and $\vec{\mathcal{E}}_{d,\,free}$ is the atomic scattering factor for the (classical) hydrogen atom:

$$f_{\rm H} = \frac{\vec{\mathcal{E}_d}}{f_e} = \frac{-\omega^2}{\omega_o^2 - \omega^2 + i\gamma\omega} = \frac{\omega^2}{\omega^2 - \omega_o^2 - i\gamma\omega}. \tag{5.357}$$

Transforming the denominator as in Eqn. 5.353 separates the scattering factor into its real and imaginary components:

$$f_{\rm H} = \frac{\omega^2(\omega^2 - \omega_o^2)}{(\omega^2 - \omega_o^2)^2 + \gamma^2\omega^2} + i\frac{\gamma\omega^3}{(\omega^2 - \omega_o^2)^2 + \gamma^2\omega^2} = f'_H + i f''_H. \tag{5.358}$$

Rigorously, all atoms must be treated quantum mechanically, but the results are qualitatively the same; dispersion effects transform the scattering factor for every atom into a complex number, resulting in a change in both amplitude and phase for the structure factor. The structure factor expression in Eqn. 3.101 now becomes

$$\vec{F}_{\mathbf{h}} = \sum_{j=1}^{n} f'_j e^{-2\pi\, i(\mathbf{h}\cdot\,\mathbf{r}_j)} + i\sum_{j=1}^{n} f''_j e^{-2\pi\, i(\mathbf{h}\cdot\,\mathbf{r}_j)}. \tag{5.359}$$

A practical problem arises in determining the scattering factor for each atom, since it now varies with the wavelength as well as the diffraction angle. An analysis of Eqn. 5.358, *which only accounts for the effects of dispersion on the scattering factor*, provides clues to the solution of this dilemma. When $\omega \gg \omega_o$, since γ is small, the real component of the scattering factor approaches unity, while the imaginary component tends toward zero. Since the expression for the scattering factor in Eqn. 5.358 has no 2θ dependence, it does not account for destructive interference due to scattering from different parts of the electron cloud. If we ignore the 2θ dependence, for wavelengths far from resonance we can treat the electron as if it is scattering essentially "dispersion free" from the center of the atom. For $\omega \gg \omega_o$ the real part of Eqn. 5.358 becomes approximately equal to $\omega^4/\omega^4 = 1$ (the number of electrons in the atom), while the imaginary part approaches zero.

The same results are obtained for the calculation of scattering factors for multi-electronic atoms, depicted graphically in Fig. 5.30.* When $\omega \gg \omega_o$ — when dispersion effects are negligible — an atom scatters as if its electrons were concentrated at

*The actual quantum mechanical functions for f' and f'' generally vary substantially from those generated from this simple classical treatment — in which the atom is approximated by Z independent classical dipole oscillators.[91]

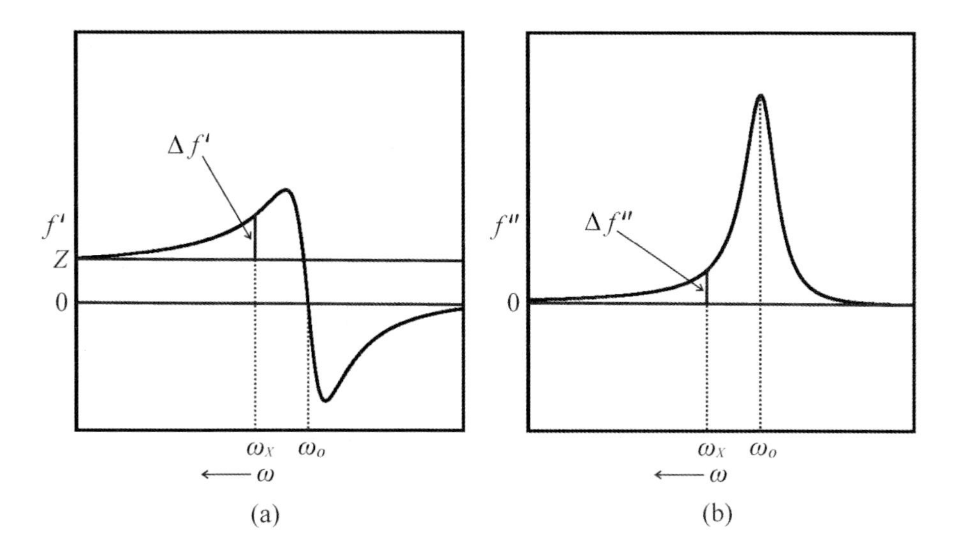

Figure 5.30 (a) Real part of the scattering factor at $2\theta = 0$ for an atom of atomic number Z as a function of frequency. (b) Imaginary part of the scattering factor at $2\theta = 0$ for the atom as a function of frequency.

the nucleus. The real part of the scattering factor under these conditions becomes equal to the atomic number, which is the scattering factor at $2\theta = 0$. As ω approaches ω_o the scattering factor begins to deviate from this value due to dispersion effects. The real part of the scattering factor then becomes $Z + \Delta f'$, as indicated in Fig. 5.30(a). The imaginary part becomes $\Delta f''$, illustrated in Fig. 5.30(b). When diffraction occurs at a specific value of 2θ, the scattering factor in the absence of dispersion decreases to a value less than Z, which we will denote with the symbol f_θ to indicate its θ dependence. The real part of the scattering factor then becomes $f_\theta + \Delta f'$ and the imaginary part remains $\Delta f''$. The overall scattering factor is the sum of the "dispersion free" part and the changes that result from the effects of dispersion:

$$f = f_\theta + \Delta f' + i\Delta f'' \quad \text{and} \tag{5.360}$$

$$\vec{F}_{\mathbf{h}} = \sum_{j=1}^{n} (f_{\theta,j} + \Delta f'_j + i\Delta f''_j)\, e^{-2\pi i(\mathbf{h}\cdot\mathbf{r_j})}. \tag{5.361}$$

Values of $\Delta f'$ and $\Delta f''$ for the elements are listed in the *International Tables for Crystallography* for commonly used X-ray wavelengths.

Dispersion effects cause only small perturbations in the observed intensities unless the frequency of the radiation approaches a transition energy in the atom. Most X-radiation is much higher in energy than the energies of the state-to-state transitions between atomic energy levels in the crystal; the highest absorption probabilities are ordinarily for those transitions resulting in the ejection of the most tightly bound core electrons — the $1s$ electrons in the K levels of atoms with $Z \geq 20$. It is especially important to correct for dispersion in these cases, since the phase shift can be very large in the "anomalous" scattering regime.

An important consequence of dispersion is its influence on the relative intensities of reflections with indices hkl and $\bar{h}\bar{k}\bar{l}$. In Sec. 3.2.5 we determined that the structure factors $\overrightarrow{F}_{hkl}$ and $\overrightarrow{F}_{\bar{h}\bar{k}\bar{l}}$ had the same magnitude, with phases that were negatives of one another. This resulted in the observed intensities, I_{hkl} and $I_{\bar{h}\bar{k}\bar{l}}$, being equal to one another — the Friedel Law. Because of dispersion the real scattering factor in Eqn. 3.103 must be replaced with a complex scattering factor — a vector: $\mathbf{f} = f' + if'' = fe^{i\phi}$, $f' = f_\theta + \Delta f'$ and $f'' = \Delta f''$. Setting $\xi_j = -2\pi(hx_j + ky_j + lz_j)$,

$$\begin{aligned}\overrightarrow{F}_{hkl} &= \sum_{j=1}^{n} \mathbf{f}_j e^{i\xi_j} = \sum_{j=1}^{n} f_j e^{i\phi_j} e^{i\xi_j} = \sum_{j=1}^{n} f_j e^{i(\phi_j+\xi_j)} = F_{hkl} e^{i\varphi_{hkl}} \\ &= \sum_{j=1}^{n} f_j \cos(\phi_j + \xi_j) + i \sum_{j=1}^{n} f_j \sin(\phi_j + \xi_j) \\ &= A_{hkl} + i\, B_{hkl}.\end{aligned} \tag{5.362}$$

The amplitude and phase of $\overrightarrow{F}_{hkl}$ are therefore

$$\begin{aligned}\varphi_{hkl} &= \arctan\left(\frac{B_{hkl}}{A_{hkl}}\right) \\ F_{hkl} &= \left(A_{hkl}^2 + B_{hkl}^2\right)^{\frac{1}{2}}.\end{aligned}$$

The structure factor expression for $\overrightarrow{F}_{\bar{h}\bar{k}\bar{l}}$ is determined by substituting $\bar{h}\bar{k}\bar{l}$ for hkl:

$$\begin{aligned}\overrightarrow{F}_{\bar{h}\bar{k}\bar{l}} &= \sum_{j=1}^{n} f_j e^{i\phi_j} e^{-i\xi_j} = \sum_{j=1}^{n} f_j e^{i(\phi_j-\xi_j)} \\ &= \sum_{j=1}^{n} f_j \cos(\phi_j - \xi_j) + i \sum_{j=1}^{n} f_j \sin(\phi_j - \xi_j) \\ &= A'_{\bar{h}\bar{k}\bar{l}} + i\, B'_{\bar{h}\bar{k}\bar{l}},\end{aligned} \tag{5.363}$$

with amplitude and phase

$$\varphi_{\bar{h}\bar{k}\bar{l}} = \arctan\left(\frac{B'_{\bar{h}\bar{k}\bar{l}}}{A'_{\bar{h}\bar{k}\bar{l}}}\right) \neq \arctan\left(\frac{-B_{hkl}}{A_{hkl}}\right) \tag{5.364}$$

$$F_{\bar{h}\bar{k}\bar{l}} = \left(A'^2_{\bar{h}\bar{k}\bar{l}} + B'^2_{\bar{h}\bar{k}\bar{l}}\right)^{\frac{1}{2}} \neq \left(A_{hkl}^2 + B_{hkl}^2\right)^{\frac{1}{2}}. \tag{5.365}$$

Note that if all of phase shifts resulting from dispersion (the ϕ_j s) are set to zero, we obtain the Friedel Law. The contribution to the phase from the atom locations changes sign when the indices are reversed, while the contribution due to dispersion retains its sign. The magnitudes of $\overrightarrow{F}_{hkl}$ and $\overrightarrow{F}_{\bar{h}\bar{k}\bar{l}}$ are no longer equal, and their phases are no longer negatives of one another. Because of this, in principle, $I_{hkl} \neq$

$I_{\bar{h}\bar{k}\bar{l}}$ — *the Friedel Law no longer holds.* Incorporating the effects of dispersion into the structure factor,

$$\begin{aligned}
\vec{F}_{hkl} &= \sum_{j=1}^{n}(f'_j + if''_j)e^{i\xi_j} \\
&= \sum_{j=1}^{n} f'_j e^{i\xi_j} + i\sum_{j=1}^{n} f''_j e^{i\xi_j} \\
&= \sum_{j=1}^{n} f'_j \cos(\xi_j) + i\sum_{j=1}^{n} f'_j \sin(\xi_j) + i\sum_{j=1}^{n} f''_j \cos(\xi_j) + i^2\sum_{j=1}^{n} f''_j \sin(\xi_j) \\
A_{hkl} &= \sum_{j=1}^{n}((f_{j,\theta} + \Delta f'_j)\cos(\xi_j) - \Delta f''_j \sin(\xi_j)) \qquad (5.366) \\
B_{hkl} &= \sum_{j=1}^{n}((f_{j,\theta} + \Delta f'_j)\sin(\xi_j) + \Delta f''_j \cos(\xi_j)). \qquad (5.367)
\end{aligned}$$

For a centrosymmetric structure, there is an atom at $(-x, -y, -z)$ for every atom at (x, y, z). Thus for every $\xi_j = -2\pi(hx_j + ky_j + lz_j)$ in the structure factor expression, there is a $-\xi_j = -2\pi(-hx_j - ky_j - lz_j) = 2\pi(hx_j + ky_j + lz_j)$:

$$\begin{aligned}
\vec{F}_{hkl} &= \sum_{j=1}^{n/2} f_j e^{i\phi_j} e^{i\xi_j} + \sum_{j=1}^{n/2} f_j e^{i\phi_j} e^{-i\xi_j} \\
&= \sum_{j=1}^{n/2} f_j e^{i\phi_j}(e^{i\xi_j} + e^{-i\xi_j}) \\
&= \sum_{j=1}^{n/2} f_j e^{i\phi_j}(\cos\xi_j + i\sin\xi_j + \cos(-\xi_j) + i\sin(-\xi_j)) \\
&= \sum_{j=1}^{n/2} 2f_j e^{i\phi_j}\cos\xi_j; \qquad (5.368)
\end{aligned}$$

The centrosymmetric structure factor with negative indices,

$$\vec{F}_{\bar{h}\bar{k}\bar{l}} = \sum_{j=1}^{n/2} f_j e^{i\phi_j} e^{-i\xi_j} + \sum_{j=1}^{n/2} f_j e^{i\phi_j} e^{i\xi_j}, \qquad (5.369)$$

is obviously identical — *the Friedel Law is rigorously valid for centrosymmetric structures.* Note that if $\phi = 0$ (if there are no dispersion effects) the centrosymmetric structure factor vector is aligned along the real axis in the complex plane, giving it a phase angle of either 0 or π.

For non-centrosymmetric structures *except in the anomalous dispersion regime*, the effects of dispersion on the relative intensities are usually small, and hkl and $\bar{h}\bar{k}\bar{l}$ are generally considered to be *Friedel equivalent reflections*, even though their intensities may vary slightly. The variations in amplitude and phase and between Friedel equivalent reflections due to anomalous dispersion are often useful in determining the absolute configuration of a chiral structure (Chapter 8), and can also be exploited in structure determination, as we shall observe in Chapter 6.

The Displacement Correction

The spherical atom approximation (Sec. 3.2.4) models the crystal structure with overlapping spheres of electron density. In this approximation each atom in the unit cell is treated as if its electrons were located at the nucleus, scattering with a fraction of their total electron density, depending on the degree of destructive interference in the electron cloud at a given diffraction angle. This fraction is the atomic scattering factor, which, as we determined in the previous section, depends on the diffraction angle, 2θ, and the effects of dispersion. We will denote this scattering factor as $f_{\theta d} = f_\theta + \Delta f' + i\Delta f''$:

$$\vec{F}_{\mathbf{h}} = \sum_{j=1}^{n} f_{j,\theta d}\, e^{-2\pi i(\mathbf{h}\cdot\mathbf{r}_j)} \tag{5.370}$$

where $\mathbf{r}_j$ is the location of the center of atom j in the unit cell in fractional coordinates. When the atoms are displaced from their average positions, either because they are in constant motion (which they always are!) or because "equivalent" atoms in each unit cell do not occupy the same locations in all unit cells, the intensity of the diffracted beam is attenuated. It is therefore necessary to correct intensities for the effects of thermal motion and other atom displacements.

The Effects of Thermal Motion on The Intensity. Implicit in Eqn. 5.370 is the assumption that $\mathbf{r}_j$ is fixed, since the spherical atom scattering factors are based on electron density calculated with respect to a fixed nucleus at the center of the atom. In reality the atoms are in constant motion at any temperature above absolute zero as a consequence of their thermal energies. The electrons about the atoms move much more rapidly than their more massive nuclei — *the electron density distribution of each atom remains effectively spherical with its atomic sphere in constant vibrational motion.* The frequencies of the vibrations are the natural resonant frequencies of the atoms bonded to one another — we will treat the atoms in a crystal as if they were attached to one another with springs – vibrating as *harmonic oscillators.* The amplitudes of these vibrations – the degree of displacement of each atom – depends on the temperature. These motions result in the electron density of each atom occupying a larger volume than it would at absolute zero, and if the motions do not average out to be a sphere (*isotropic*), the average electron density distribution for the atom is now non-spherical (*anisotropic*). Since many vibrations occur during the collection of a single intensity datum, we observe only the average intensities resulting from scattering from each atom as it oscillates about its average position. The larger effective volumes of the atoms result in more destructive interference than predicted by the "fixed atom" scattering factor, resulting in lower intensities. Because the effective atomic volumes increase as the temperature increases, observed intensities decrease with an increase in temperature. The effect of temperature on the observed intensities must therefore be modeled, and since it depends on atom location, it is logically included in the expression for the calculated structure factor.

We begin by denoting the instantaneous location of the atomic center of atom j as $\mathbf{r}'_j = \mathbf{r}_j + \vec{\delta}_j$, where $\vec{\delta}_j$ is the instantaneous displacement vector of the atom from its equilibrium (average) position, $\mathbf{r}_j$. The instantaneous structure factor is then

$$\overrightarrow{F}'_{\mathbf{h}} = \sum_{j=1}^{n} f_{\theta d,\, j}\; e^{-2\pi\, i(\mathbf{h}\cdot\, \mathbf{r}'_j\,)}. \tag{5.371}$$

The observed intensity depends on the incident beam intensity, and is also affected by the other factors (e.g. polarization and absorption) discussed previously in this chapter. In order to compare observed and calculated intensities, we must scale the relative observed intensities: $I_{o,hkl} = S\, I_{rel,hkl}$, where S is the scale factor, yet to be determined (ironically, it is the displacement correction that provides the means for determining S). For the moment, we will assume that the observed intensities are on the same scale as those calculated from the atom locations in Eqn. 5.371. The resultant instantaneous intensity is then

$$\begin{aligned} \overrightarrow{I}'_{\mathbf{h}} &= \left(\sum_{j=1}^{n} f_{j,\,\theta d}\; e^{-2\pi\, i(\mathbf{h}\cdot\, \mathbf{r}'_j\,)}\right)\left(\sum_{m=1}^{n} f^{*}_{m,\,\theta d}\; e^{2\pi\, i(\mathbf{h}\cdot\, \mathbf{r}'_j\,)}\right) \\ &= \sum_{j=1}^{n}\sum_{m=1}^{n} f_{j,\,\theta d}\; f^{*}_{m,\,\theta d}\; e^{-2\pi\, i(\mathbf{h}\cdot\, \mathbf{r}'_j\,)} e^{2\pi\, i(\mathbf{h}\cdot\, \mathbf{r}'_m\,)}. \end{aligned} \tag{5.372}$$

Experimentally, we observe only a time-averaged intensity:

$$\begin{aligned} \langle \overrightarrow{I}'_{\mathbf{h}} \rangle &= \left\langle \sum_{j=1}^{n}\sum_{m=1}^{n} f_{j,\,\theta d}\; f^{*}_{m,\,\theta d}\; e^{-2\pi\, i(\mathbf{h}\cdot\, (\mathbf{r}_j + \overrightarrow{\delta_j}\,))} e^{2\pi\, i(\mathbf{h}\cdot\, (\mathbf{r}_m + \overrightarrow{\delta_m}\,))} \right\rangle \\ &= \left\langle \sum_{j=1}^{n}\sum_{m=1}^{n} f_{j,\,\theta d}\; f^{*}_{m,\,\theta d}\; e^{2\pi\, i(\mathbf{h}\cdot\, (\mathbf{r}_m - \mathbf{r}_j\,))} e^{2\pi\, i(\mathbf{h}\cdot\, (\overrightarrow{\delta_m} - \overrightarrow{\delta_j}\,))} \right\rangle. \end{aligned} \tag{5.373}$$

The only variables in Eqn. 5.373 are the displacements due to vibration:

$$\langle \overrightarrow{I}'_{\mathbf{h}} \rangle = \sum_{j=1}^{n}\sum_{m=1}^{n} f_{j,\,\theta d}\; f^{*}_{m,\,\theta d}\; e^{2\pi\, i(\mathbf{h}\cdot\, (\mathbf{r}_m - \mathbf{r}_j\,))} \left\langle e^{2\pi\, i(\mathbf{h}\cdot\, (\overrightarrow{\delta_m} - \overrightarrow{\delta_j}\,))} \right\rangle. \tag{5.374}$$

The average displacement term in Eqn. 5.374 is the term responsible for the effects of thermal vibration and static atom displacements on the intensity.

As previously discussed in Sec. 5.1.1, the average value of the function $g(x)$ is the sum of the values of the function at each value of x, weighted by the probability of observing that particular value of x. For a discrete function, if we make z measurements of the function, then the number of times that we observe a given value (the ith value), $g(x_i)$, is the fraction of the total number of times that we observe x_i times the total number of observations. The fraction of the times that x_i is observed is the probability of observing it, $Pr(x_i)$. The number of times that x_i is observed is therefore $Pr(x_i) \cdot z$, and the average value of $g(x)$ is obtained by multiplying each value of $g(x_i)$ by the number of times that it is observed and dividing the sum of these products by the total number of observations:

$$\langle g(x) \rangle = \frac{1}{z} \sum_{i} Pr(x_i) z \cdot g(x_i) = \sum_{i} Pr(x_i) g(x_i). \tag{5.375}$$

For a continuous function,

$$\langle g(x)\rangle = \int_{-\infty}^{\infty} Pr(x)g(x)\,dx. \tag{5.376}$$

Again, this is just an average of all the values of $g(x)$, each weighted by the probability that x will be observed.

The average value that we wish to determine is $\langle e^{ix}\rangle$, where $x = 2\pi(\mathbf{h}\cdot(\overrightarrow{\delta_m} - \overrightarrow{\delta_j}))$. It is reasonable to assume that the probability for observing a given displacement for atom j, $\overrightarrow{\delta_j}$, would be centered around its average value and would tail off quickly as the size of the displacement increases (in other words, we would expect the atoms to vibrate about their average positions). If we approximate the bonds between atoms as "springs", then we can treat pairs of atoms as harmonic oscillators. Intuitively, we might expect that the probability for a given displacement from the mean location of an atom would be characterized by a normal (Gaussian) distribution; the probability that a displacement would be observed would be largest at the average value, and would decrease rapidly as the atom moves farther from its equilibrium position. In 1928 Nobel laureate Felix Bloch developed a theorem (*Bloch's Theorem*) that shows that the probability for a given displacement for a harmonic oscillator follows a Gaussian probability distribution.* Since the probabilities of observing $\overrightarrow{\delta_j}$ and $\overrightarrow{\delta_m}$ are both determined (approximately) by Gaussian probability distributions, so will the probability of observing a specific value of x:

$$Pr(x) = \frac{1}{\sigma\sqrt{2\pi}}\, e^{-x^2/(2\sigma^2)} \quad \text{and} \tag{5.377}$$

$$\begin{aligned}\langle e^{ix}\rangle &= \int_{-\infty}^{\infty} Pr(x)e^{ix}\,dx \\ &= \int_{-\infty}^{\infty} Pr(x)\cos x\,dx + i\int_{-\infty}^{\infty} Pr(x)\sin x\,dx.\end{aligned} \tag{5.378}$$

where σ is the standard deviation of the distribution of x. Because a harmonic oscillator has an equal probability of displacement in either direction, $\overrightarrow{\delta_m} - \overrightarrow{\delta_j}$ and $\overrightarrow{\delta_j} - \overrightarrow{\delta_m}$ are equally probable and $Pr(x)$ is an even function — that is, $Pr(x) = P(-x)$. On the other hand, $\sin x$ is an odd function and $\sin(-x) = -\sin x$. The integrand in the second term is therefore an odd function that integrates to zero and

$$\langle e^{ix}\rangle = \int_{-\infty}^{\infty} \cos x\, \frac{1}{\sigma\sqrt{2\pi}}\, e^{-x^2/(2\sigma^2)}\,dx. \tag{5.379}$$

To evaluate this integral we employ two definite integrals:

$$\int_{-\infty}^{\infty} \cos bx\, e^{-a^2x^2}\,dx = 2\left(\frac{\sqrt{\pi}\,e^{-b^2/(4a^2)}}{2a}\right) \tag{5.380}$$

$$\int_{-\infty}^{\infty} x^{2n}e^{-a^2x^2}\,dx = 2\left(\frac{1\cdot 3\cdot 5\cdots(2n-1)}{2^{n+1}a^{2n}}\,\frac{\sqrt{\pi}}{a}\right). \tag{5.381}$$

*An excellent proof of this assertion is given in Appendix E of Prince's book on mathematical techniques in crystallography[92].

Setting $a = 1/(\sigma\sqrt{2})$ and $b = 1$ in Eqn. 5.380 gives

$$\langle e^{ix} \rangle = 2\frac{a}{\sqrt{\pi}} \frac{\sqrt{\pi}\, e^{-1/(4a^2)}}{2a} = e^{-1/(4a^2)}. \tag{5.382}$$

Setting $n = 1$ and $a = 1/(\sigma\sqrt{2})$ in Eqn. 5.381 also allows us to determine $\langle x^2 \rangle$:

$$\langle x^2 \rangle = \int_{-\infty}^{\infty} Pr(x) x^2\, dx = \int_{-\infty}^{\infty} x^2 \frac{1}{\sigma\sqrt{2\pi}} e^{-x^2/(2\sigma^2)}\, dx \tag{5.383}$$

$$= 2\frac{a}{\sqrt{\pi}} \frac{1}{2^2 a^2} \frac{\sqrt{\pi}}{a} = \frac{1}{2a^2} \Longrightarrow e^{-1/(4a^2)} = e^{-\langle x^2 \rangle/2}. \tag{5.384}$$

It follows that the average value of e^{ix} depends on the average value of x^2:

$$\langle e^{ix} \rangle = e^{-\langle x^2 \rangle/2}. \tag{5.385}$$

Substituting $x = 2\pi(\mathbf{h} \cdot (\overrightarrow{\delta_m} - \overrightarrow{\delta_j}))$ and Eqn. 5.385 into Eqn. 5.374,

$$\langle \overrightarrow{I'_{\mathbf{h}}} \rangle = \sum_{j=1}^{n} \sum_{m=1}^{n} f_{j,\theta d}\, f^*_{m,\theta d}\; e^{2\pi i(\mathbf{h}\cdot(\mathbf{r}_m - \mathbf{r}_j))} e^{-2\pi^2 \langle(\mathbf{h}\cdot(\overrightarrow{\delta_m} - \overrightarrow{\delta_j})^2)\rangle}. \tag{5.386}$$

Expanding the average term in the exponential gives:

$$\langle(\mathbf{h} \cdot (\overrightarrow{\delta_m} - \overrightarrow{\delta_j})^2)\rangle = \langle \mathbf{h} \cdot \overrightarrow{\delta_m} \rangle^2 - 2\langle(\mathbf{h} \cdot \overrightarrow{\delta_m})(\mathbf{h} \cdot \overrightarrow{\delta_j})\rangle + \langle(\mathbf{h} \cdot \overrightarrow{\delta_j}\rangle^2.$$

If the atoms are moving independently, then for every $\overrightarrow{\delta_m}$ there is a $\overrightarrow{\delta_j}$ and a $-\overrightarrow{\delta_j}$, the cross term averages to zero,* and

$$\begin{aligned}
\langle \overrightarrow{I'_{\mathbf{h}}} \rangle &= \sum_{j=1}^{n} \sum_{m=1}^{n} f_{j,\theta d}\, f^*_{m,\theta d}\; e^{2\pi i(\mathbf{h}\cdot(\mathbf{r}_m - \mathbf{r}_j))} e^{-2\pi^2 (\langle(\mathbf{h}\cdot\overrightarrow{\delta_m})^2\rangle + \langle(\mathbf{h}\cdot\overrightarrow{\delta_j})^2\rangle)} \\
&= \sum_{m=1}^{n} f_{m,\theta d} e^{2\pi i(\mathbf{h}\cdot\mathbf{r}_m)} e^{-2\pi^2 \langle(\mathbf{h}\cdot\overrightarrow{\delta_m})^2\rangle} \\
&\quad \times \sum_{j=1}^{n} f_{j,\theta d} e^{-2\pi i(\mathbf{h}\cdot\mathbf{r}_j)} e^{-2\pi^2 \langle(\mathbf{h}\cdot\overrightarrow{\delta_j})^2\rangle}
\end{aligned} \tag{5.387}$$

The average instantaneous intensity *is* the observed intensity, and since $I_{hkl} = F^*_{hkl} F_{hkl}$, we immediately identify the structure factor in Eqn. 5.387 as

$$F_{\mathbf{h}} = \sum_{j=1}^{n} f_{j,\theta d} e^{-2\pi i(\mathbf{h}\cdot\mathbf{r}_j)} e^{-2\pi^2 \langle(\mathbf{h}\cdot\overrightarrow{\delta_j})^2\rangle}. \tag{5.388}$$

In the absence of atom displacement, $\delta_j = 0$, and the second exponential term equals one. As the thermal displacements increase, because $(\mathbf{h} \cdot \overrightarrow{\delta_j})^2$ is always positive, the negative exponent in the second exponential results in an attenuation

*This is only approximately true. To some extent vibrational motions are coupled to one another, giving rise to a non-zero term with a large distribution of values, resulting in a diminution of the observed intensity. Some energy is lost from the diffracted beam, which becomes part of the background radiation. The phenomenon is known as *thermal diffuse scattering*.

of the structure factor and therefore the intensity. This attenuation can be viewed as a decrease in the scattering power of the atom, due to a modification of the effective volume and shape of the electron cloud. As a result, $e^{-2\pi^2 \langle (\mathbf{h}\cdot \overrightarrow{\delta_j})^2\rangle}$ is often incorporated into the scattering factor:

$$F_{\mathbf{h}} = \sum_{j=1}^{n} f_{j,\theta d} e^{-2\pi i(\mathbf{h}\cdot\mathbf{r}_j)} e^{-T_j} = \sum_{j=1}^{n} f_{j,\theta dT} e^{-2\pi i(\mathbf{h}\cdot\mathbf{r}_j)}, \tag{5.389}$$

where $f_{j,\theta d}$ is the scattering factor for the spherical atom at rest, $f_{j,\theta dT} = f_{j,\theta d}e^{-T_j}$ includes the effects of thermal motion, and $T_j = 2\pi^2 \langle (\mathbf{h}\cdot \overrightarrow{\delta_j})^2\rangle$. We will refer to T_j as the *atomic displacement term*,* which can be explicitly evaluated in terms of the direct lattice basis:

$$\begin{aligned}
\mathbf{h} &= h\mathbf{a}^* + k\mathbf{b}^* + l\mathbf{c}^* \\
\mathbf{r}_j &= x_j\mathbf{a} + y_j\mathbf{b} + z_j\mathbf{c} \\
\overrightarrow{\delta_j} &= b_{1j}\,\mathbf{a} + b_{2j}\,\mathbf{b} + b_{3j}\,\mathbf{c}. \\
T_j &= 2\pi^2\langle[(h\mathbf{a}^* + k\mathbf{b}^* + l\mathbf{c}^*)\cdot(b_{1j}\,\mathbf{a} + b_{2j}\,\mathbf{b} + b_{3j}\,\mathbf{c})]^2\rangle \\
&= 2\pi^2\langle(hb_{1j} + kb_{2j} + lb_{3j})^2\rangle,
\end{aligned} \tag{5.390}$$

since $\mathbf{a}^*\cdot\mathbf{a} = \mathbf{b}^*\cdot\mathbf{b} = \mathbf{c}^*\cdot\mathbf{c} = 1$ and $\mathbf{a}^*\cdot\mathbf{b} = \mathbf{a}^*\cdot\mathbf{c} = \mathbf{b}^*\cdot\mathbf{c} = 0$. Thus,

$$\begin{aligned}
T_j &= 2\pi^2\langle(h^2b_{1j}b_{1j} + k^2b_{2j}b_{2j} + l^2b_{3j}b_{3j} + 2hkb_{1j}b_{2j} + 2hlb_{1j}b_{3j} + 2klb_{2j}b_{3j})\rangle. \\
&= 2\pi^2(h^2\langle b_{1j}b_{1j}\rangle + k^2\langle b_{2j}b_{2j}\rangle + l^2\langle b_{3j}b_{3j}\rangle \\
&\quad + 2hk\langle b_{1j}b_{2j}\rangle + 2hl\langle b_{1j}b_{3j}\rangle + 2kl\langle b_{2j}b_{3j}\rangle).
\end{aligned}$$

Setting $\beta_{ik} = \langle b_{ij}b_{kj}\rangle$ (for atom j),

$$T_j = 2\pi^2(h^2\beta_{11} + k^2\beta_{22} + l^2\beta_{33} + 2hk\beta_{12} + 2hl\beta_{13} + 2kl\beta_{23}). \tag{5.391}$$

For computational convenience $2\pi^2$ is often incorporated into β_{ik}:

$$T_j = h^2\beta'_{11} + k^2\beta'_{22} + l^2\beta'_{33} + 2hk\beta'_{12} + 2hl\beta'_{13} + 2kl\beta'_{23}, \tag{5.392}$$

where $\beta'_{ik} = 2\pi^2\beta_{ik}$.

The atomic displacement vector, $\overrightarrow{\delta_j}$, has components in the direct lattice basis, and the six parameters required to define the atomic displacement term for each atom are known as its *atomic displacement parameters*, commonly referred to as "ADPs". $\beta_{11} = \langle b_{1j}^2\rangle$, $\beta_{22} = \langle b_{2j}^2\rangle$, and $\beta_{33} = \langle b_{3j}^2\rangle$ are the averages of the squares of the atomic displacements *in fractional coordinates* along each of the direct cell basis vectors. Because these numbers convey little information, it is useful to describe the displacement vectors in a new parallel basis by defining $b_{1j} = u_{1j}\,a^*$, $b_{2j} = u_{2j}\,b^*$, and $b_{3j} = u_{3j}\,c^*$, so that

$$\overrightarrow{\delta_j} = \mathbf{u}_j = (u_{1j}\,a^*)\,\mathbf{a} + (u_{2j}\,b^*)\,\mathbf{b} + (u_{3j}\,c^*)\,\mathbf{c}. \tag{5.393}$$

*It would appear that "thermal displacement term" would be more descriptive. However, other atomic displacements produce the same effects as those due to thermal motion. The most common of these occurs whenever an atom can occupy more than one site in the unit cell. This is a static phenomenon know as *static displacement disorder*, and has the same effect on the intensities that vibrational motion does. In both cases the apparent volume of the atom is increased and the intensities are subsequently decreased whenever an atom can occupy more than one location – either statically or dynamically.

For crystal lattices with axes coincident with the crystal Cartesian basis (orthorhombic, tetragonal, and cubic lattices) the magnitudes of the components provide direct physical information. For these lattices, $\cos\alpha = \cos\beta = \cos\gamma = 0$. Similarly, $\cos\alpha^* = \cos\beta^* = \cos\gamma^* = 0$, and the metric tensors for the direct (Eqn. 4.46) and reciprocal (Eqn. 4.44) lattices become

$$\mathbf{G} = \begin{bmatrix} a^2 & 0 & 0 \\ 0 & b^2 & 0 \\ 0 & 0 & c^2 \end{bmatrix} \qquad \mathbf{G}^* = \begin{bmatrix} a^{*2} & 0 & 0 \\ 0 & b^{*2} & 0 \\ 0 & 0 & c^{*2} \end{bmatrix}. \tag{5.394}$$

Since $\mathbf{GG}^* = \mathbf{I}$, $a = 1/a^*$, $b = 1/b^*$ and $c = 1/c^*$. In addition, the unit cell axes are coincident with the Cartesian axes such that $\mathbf{a} = a\mathbf{i}$, etc., and the displacement vector, *in Cartesian coordinates* is

$$\begin{aligned} \overrightarrow{\delta_j} = \mathbf{u}_j &= (u_{1j}\, a^*)\, a\, \mathbf{i} + (u_{2j}\, b^*)\, b\, \mathbf{j} + (u_{3j}\, c^*)\, c\, \mathbf{k} \\ &= u_{1j}\, (a^*\, a)\, \mathbf{i} + u_{2j}\, (b^*\, b)\, \mathbf{j} + u_{3j}\, (c^*\, c)\, \mathbf{k} \\ &= u_{1j}\, \mathbf{i} + u_{2j}\, \mathbf{j} + u_{3j}\, \mathbf{k}. \end{aligned} \tag{5.395}$$

The components of the vector are now the Cartesian coordinates of the displacement vector, with magnitude $u_j = \sqrt{u_{1j}^2 + u_{2j}^2 + u_{3j}^2}$, *in angstroms.*

If the lattices have inter-axial angles that are not 90°, then the components are the distances in angstroms in the direction of the unit cell axes, and do not give the magnitude of the displacement vector directly (although they often provide a useful approximation for unit cells with angles that do not deviate significantly from 90°). The thermal displacement term can now be expressed in terms of displacements along the direct lattice basis vectors:

$$\begin{aligned} T_j &= 2\pi^2 \langle (\mathbf{h} \cdot \overrightarrow{\delta_j})^2 \rangle = 2\pi^2 \langle (ha^* u_{1j} + kb^* u_{2j} + lc^* u_{3j})^2 \rangle \\ &= 2\pi^2 (h^2 a^{*2} \langle u_{1j} u_{1j} \rangle + k^2 b^{*2} \langle u_{2j} u_{2j} \rangle + l^2 c^{*2} \langle u_{3j} u_{3j} \rangle \\ &\qquad + 2hka^* b^* \langle u_{1j} u_{2j} \rangle + 2hla^* c^* \langle u_{1j} u_{3j} \rangle + 2klb^* c^* \langle u_{2j} u_{3j} \rangle) \\ &= 2\pi^2 (h^2 a^{*2} U_{11} + k^2 b^{*2} U_{22} + l^2 c^{*2} U_{33} \\ &\qquad + 2hka^* b^* U_{12} + 2hla^* c^* U_{13} + 2klb^* c^* U_{23}), \end{aligned} \tag{5.396}$$

where $U_{ik} = \langle u_{ij} u_{kj} \rangle$ for atom j, and $U_{11} = \langle u_{1j}^2 \rangle$, $U_{22} = \langle u_{2j}^2 \rangle$, and $U_{33} = \langle u_{3j}^2 \rangle$ are the averages of the squares of the atomic displacements in angstroms along the $\mathbf{a}$, $\mathbf{b}$, and $\mathbf{c}$ axes, respectively.

In order to correct the intensities for displacement effects it is necessary to determine the average products of the atom displacements, manifest in either β_{ik} or U_{ik}. The choice of β_{ik} or U_{ik} depends on the particular software package employed in structural refinement. Comparing Eqn. 5.396 with Eqn. 5.391 provides the relationship between the two choices:

$$U_{ik} = \frac{\beta_{ik}}{v_i v_k}, \quad i = 1, 2, 3, \tag{5.397}$$

where $v_1 = a^*$, $v_2 = b^*$, and $v_3 = c^*$.

The anisotropic displacement parameters are used to model the effects of vibration and other atomic displacements in the structure factor equation, correcting

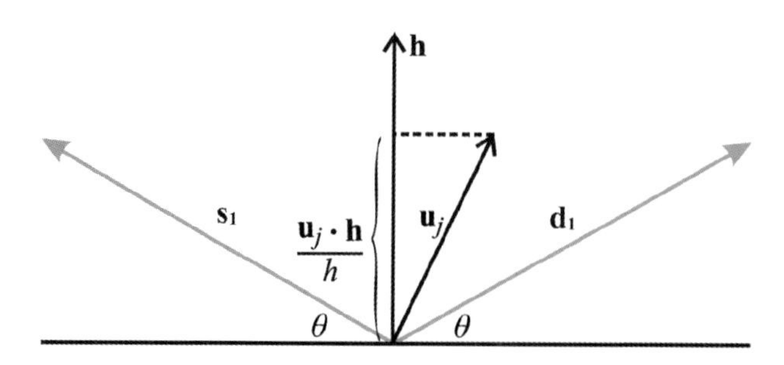

Figure 5.31 Projection of an instantaneous atom displacement vector in the direction of the diffraction vector.

the experimental errors that result from such displacements. However, there is no experimental probe that will provide these values. They are ordinarily calculated as parameters in crystal structure refinement — after the atomic positions in the unit cell have been determined. Unfortunately, the effects of thermal motion on the intensities can significantly impact the structural solution, and we must therefore discover a way to approximate them — *even before we know the locations of the atoms in the structure.* The process begins by rearranging the atomic displacement term and dividing and multiplying by the square of the magnitude of the diffraction vector:

$$T_j = 2\pi^2 \langle (\mathbf{h} \cdot \overrightarrow{\delta_j})^2 \rangle = 2\pi^2 \left\langle \left(\frac{\mathbf{u}_j \cdot \mathbf{h}}{h} \right)^2 \right\rangle h^2 = 2\pi^2 \left\langle u_{\mathbf{h}j}^2 \right\rangle h^2. \tag{5.398}$$

Referring to Fig. 5.31, the term in parentheses can be seen to be the square of the projection of the displacement vector (in angstroms) onto the diffraction vector. If the atom is displaced anisotropically, then the average squared value of $u_{\mathbf{h}j}$ will depend on the direction of **h**, but — if the average atom displacement is isotropic, then its average squared displacement in all directions will be the same, and $\langle u_{\mathbf{h}j}^2 \rangle$ will be independent of the direction of the diffraction vector: $\langle u_{\mathbf{h}j}^2 \rangle = \langle u_j^2 \rangle$. The *isotropic displacement term* then depends only on the average squared displacement of the atom and the magnitude of the diffraction vector, $h = 2\sin\theta/\lambda$:

$$T_{j,\,\mathrm{iso}} = 2\pi^2 \left\langle u_j^2 \right\rangle h^2 = 8\pi^2 \left\langle u_j^2 \right\rangle \frac{\sin^2\theta}{\lambda^2}. \tag{5.399}$$

The quality of the least-squares refinement of a crystal structure depends on having enough experimental measurements of sufficient accuracy to determine each of the parameters in the refinement. *Anisotropic refinement* adds six parameters to the three positional parameters for each atom, while *isotropic refinement, which ignores any anisotropy*, adds only one parameter, $\langle u_j^2 \rangle$. With structural determinations in which the number of reflections (or quality of the data set) is limited in comparison to the number of positional parameters, it is often necessary to refine such structures isotropically.

Prior to the determination of the atom locations, we can assign only an average overall value for $T_{j,\,\mathrm{iso}}$ for each atom. This factor, known as the overall Debye–Waller factor, can be estimated from the number and identity of the atoms in the unit cell by scaling the observed reflections to determine the scale factor, S. We will take this up in Sec. 5.2.

The Displacement Ellipsoid. The displacement ellipsoid, especially its graphical representation, is perhaps the most recognizable aspect of a reported crystal structure in the literature. To many non-crystallographers, the term "crystal structure" is often nearly synonymous with the term "ORTEP", the *Oak Ridge Thermal Ellipsoid Plot Program* created by Carroll K. Johnson[23]. The displacement ellipsoid is logically treated as a component of structural refinement, but the necessary mathematics to describe it were developed in the previous section, and we will consider it here, while the ideas are still fresh. The material in this section can be deferred until anisotropic refinement is discussed without loss of continuity.

As the atom vibrates, its nucleus maps a displacement volume traced out by the vector, $\overrightarrow{\delta_j}$. If we were able to observe the instantaneous location of the atomic nucleus we could sample a large enough number of locations to visualize this displacement volume.* Intuitively, if the displacement vector has an equal probability of being found instantaneously in any direction (isotropic vibration), we would expect the average displacement to be spherical. If not (anisotropic vibration), then the sphere would be "distorted" in some manner. In order to discover the nature of this distortion we will find it convenient to express the displacement term in matrix form. The instantaneous displacement vector for atom j is $\overrightarrow{\delta_j} = [b_{1j}\; b_{2j}\; b_{3j}]$, and we define the symmetric matrix $[\mathcal{B}]_j$ as

$$\mathcal{B}t_j = \left\langle \begin{bmatrix} b_{1j} \\ b_{2j} \\ b_{3j} \end{bmatrix} \begin{bmatrix} b_{1j} & b_{2j} & b_{3j} \end{bmatrix} \right\rangle = \langle \overrightarrow{\delta_j}\,\overrightarrow{\delta_j}^T \rangle \tag{5.400}$$

$$= \begin{bmatrix} \langle b_{1,j}^2 \rangle & \langle b_{1,j}\, b_{2,j} \rangle & \langle b_{1,j}\, b_{3,j} \rangle \\ \langle b_{1,j}\, b_{2,j} \rangle & \langle b_{2,j}^2 \rangle & \langle b_{2,j}\, b_{3,j} \rangle \\ \langle b_{1,3}\, b_{3,j} \rangle & \langle b_{2,j}\, b_{3,j} & \langle b_{3,j}^2 \rangle \end{bmatrix} = \begin{bmatrix} \beta_{11} & \beta_{12} & \beta_{13} \\ \beta_{12} & \beta_{22} & \beta_{23} \\ \beta_{13} & \beta_{23} & \beta_{33} \end{bmatrix}.$$

The displacement term is then

$$T_j = 2\pi^2 \begin{bmatrix} h & k & l \end{bmatrix} \begin{bmatrix} \beta_{11} & \beta_{12} & \beta_{13} \\ \beta_{12} & \beta_{22} & \beta_{23} \\ \beta_{13} & \beta_{23} & \beta_{33} \end{bmatrix} \begin{bmatrix} h \\ k \\ l \end{bmatrix} = 2\pi^2 \mathbf{h}^T [\mathcal{B}]_j\, \mathbf{h}, \tag{5.401}$$

and the structure factor is

$$F_{\mathbf{h}} = \sum_{j=1}^{n} f_{j,\,\theta d} e^{-2\pi\, i(\mathbf{h}\cdot\mathbf{r}_j)} e^{-2\pi^2 \mathbf{h}^T [\mathcal{B}]_j\, \mathbf{h}}. \tag{5.402}$$

*Since the X-rays are scattering from electrons in the atoms, the displacement of an atom is actually that of the center of its electron "cloud", which is assumed to constitute a sphere with the nucleus at its center.

The instantaneous displacement vectors for an atom have a common origin at the equilibrium position of the atom, and are (approximately) normally distributed with respect to that position, i.e., they average to zero. *The terms in the* $[\mathcal{B}]_j$ *matrix are therefore the elements of the covariance matrix for the trivariate probability distribution* (Eqn. 5.138): $f_G(\overrightarrow{\delta_j}) = f_G(b_{1j}, b_{2j}, b_{3j})$, developed earlier in this chapter. To describe the displacement volume of the atom we can think of it in terms of a contour map in three dimensions. A contour in this case is a three-dimensional surface on which the probability of finding the center of the atom per unit volume, *the probability density*, is a constant:†

$$f_G(b_{1j}, b_{2j}, b_{3j}) = f_G(\overrightarrow{\delta_j}) = \frac{1}{(2\pi)^{3/2}\,|\mathcal{B}|_j^{1/2}}\, e^{-\frac{1}{2}\,(\overrightarrow{\delta_j})^T\,[\mathcal{B}]_j^{-1}\,(\overrightarrow{\delta_j})} = C'. \tag{5.403}$$

The pre-exponential term in Eqn. 5.403 is a constant, which we will denote as C''. To simplify the equation we take the logarithm of both sides:

$$\begin{aligned} \ln C'' + \ln\left(e^{-\frac{1}{2}\,(\overrightarrow{\delta_j})^T\,[\mathcal{B}]_j^{-1}\,(\overrightarrow{\delta_j})}\right) &= \ln C' \\ (\overrightarrow{\delta_j})^T\,[\mathcal{B}]_j^{-1}\,(\overrightarrow{\delta_j}) &= C, \end{aligned} \tag{5.404}$$

where $C = 2(\ln C'' - \ln C')$.

The shape of the displacement volume will be described by the subset of the displacement vectors that satisfy Eqn. 5.404 — these vectors will define a surface of constant probability density. The vectors in the distribution are in the non-orthogonal direct lattice basis, and in order to determine the nature of this surface the displacement vectors are transformed from the unit cell basis into Cartesian coordinates:

$$\mathbf{B}\overrightarrow{\delta_j} = \mathbf{B}\begin{bmatrix} b_{1j} \\ b_{2j} \\ b_{3j} \end{bmatrix} = \begin{bmatrix} u_{1j}^c \\ u_{2j}^c \\ u_{3j}^c \end{bmatrix} = \mathbf{u_c}. \tag{5.405}$$

The superscript "c" has been added to differentiate these Cartesian displacements from the displacements along the unit cell axes discussed in the previous section. u_{1j}^c, u_{2j}^c, and u_{3j}^c are the components of $\overrightarrow{\delta_j}$ along the $\mathbf{i}$, $\mathbf{j}$, and $\mathbf{k}$ axes of the crystal Cartesian coordinate system (Sec. 1.5.2).

The covariance matrix in the Cartesian system can now be determined :

$$\begin{aligned} \mathbf{B}\overrightarrow{\delta_j} &= \mathbf{u_c} \\ \overrightarrow{\delta_j}^T\mathbf{B}^T &= \mathbf{u_c}^T \\ \mathbf{B}\overrightarrow{\delta_j}\,\overrightarrow{\delta_j}^T\mathbf{B}^T &= \mathbf{u_c}\mathbf{u_c}^T. \end{aligned} \tag{5.406}$$

The average values for the expressions are determined be adding the matrices for each displacement and dividing by the number of displacements. Since the sums of the matrices are obtained by adding their individual elements, it follows that

$$\langle \mathbf{u_c}\mathbf{u_c}^T \rangle = \mathbf{B}\langle \overrightarrow{\delta_j}\,\overrightarrow{\delta_j}^T \rangle \mathbf{B}^T = \mathbf{B}\,[\mathcal{B}]_j\,\mathbf{B}^T = \begin{bmatrix} U_{11}^c & U_{12}^c & U_{13}^c \\ U_{21}^c & U_{22}^c & U_{23}^c \\ U_{31}^c & U_{32}^c & U_{33}^c \end{bmatrix} = \mathbf{U_c}, \tag{5.407}$$

†This is entirely analogous to the probability density plots for atoms and molecules in elementary chemistry and physics books, except that the illustrations in those books are *electron* probability density plots rather than the "nucleus" probability density plots discussed here.

where $U^c_{ik} = \langle u^c_{ij} u^c_{kj} \rangle$. Eqn. 5.404 can now be expressed in Cartesian coordinates:

$$\begin{aligned}
[\mathcal{B}]_j &= \mathbf{B^{-1}U_c(B^{\mathit{T}})^{-1}} \\
[\mathcal{B}]_j^{-1} &= \mathbf{B}^T\mathbf{U_c^{-1}B} \\
\overrightarrow{\delta_j}^T [\mathcal{B}]_j^{-1} \overrightarrow{\delta_j} &= \overrightarrow{\delta_j}^T \mathbf{B}^T\mathbf{U_c^{-1}B}\, \overrightarrow{\delta_j} \\
&= \mathbf{u_c}^T\mathbf{U_c^{-1}u_c} = C.
\end{aligned} \tag{5.408}$$

The constant probability density surface is revealed by expressing the covariance matrix in terms of its unit eigenvectors and eigenvalues (Sec. 1.4.7):

$$\begin{aligned}
\mathbf{U_c e}_i &= \lambda_i \mathbf{e_i}, \quad i = 1, 2, 3; \\
\mathbf{U_c E} &= \mathbf{E}\,[\lambda].
\end{aligned} \tag{5.409}$$

where $[\lambda]$ is a diagonal matrix with λ_1, λ_2 and λ_3 along the diagonal, and $\mathbf{E}$ is a matrix with the eigenvectors of $\mathbf{U_c}$ as its columns (the modal matrix of $\mathbf{U_c}$). $\mathbf{U_c}$ is a symmetric matrix. In Sec. 1.4.7 it was shown that the modal matrix of a symmetric matrix is orthogonal. It follows that $\mathbf{E}$ is an orthogonal matrix — the eigenvectors of $\mathbf{E}$ are perpendicular to one another and the inverse of $\mathbf{E}$ is its transpose. Thus,

$$\begin{aligned}
\mathbf{U_c E E}^T &= \mathbf{E}\,[\lambda]\mathbf{E}^T, \\
\mathbf{U_c} &= \mathbf{E}\,[\lambda]\mathbf{E}^T \\
\mathbf{U_c}^{-1} &= (\mathbf{E}^T)^T[\lambda]^{-1}\mathbf{E}^T \;=\; \mathbf{E}\,[\lambda]^{-1}\mathbf{E}^T \quad \text{and} \\
\mathbf{u_c}^T\mathbf{U_c^{-1}u_c} &= \mathbf{u_c}^T\mathbf{E}\,[\lambda]^{-1}\mathbf{E}^T\mathbf{u_c} = C.
\end{aligned} \tag{5.410}$$

Expanding Eqn. 5.410 and *noting that the inverse of a diagonal matrix is formed from the inverses of its diagonal elements* results in an equation for the surface in terms of the displacement vectors:

$$\begin{aligned}
&\mathbf{u_c}^T \begin{bmatrix} \mathbf{e_1} & \mathbf{e_2} & \mathbf{e_3} \end{bmatrix} \begin{bmatrix} 1/\lambda_1 & 0 & 0 \\ 0 & 1/\lambda_2 & 0 \\ 0 & 0 & 1/\lambda_3 \end{bmatrix} \begin{bmatrix} \mathbf{e_1} \\ \mathbf{e_2} \\ \mathbf{e_3} \end{bmatrix} \mathbf{u_c} = C, \\
&\begin{bmatrix} \dfrac{(\mathbf{u_c}\cdot\mathbf{e_1})}{\lambda_1} & \dfrac{(\mathbf{u_c}\cdot\mathbf{e_2})}{\lambda_2} & \dfrac{(\mathbf{u_c}\cdot\mathbf{e_3})}{\lambda_3} \end{bmatrix} \begin{bmatrix} \mathbf{u_c}\cdot\mathbf{e_1} \\ \mathbf{u_c}\cdot\mathbf{e_2} \\ \mathbf{u_c}\cdot\mathbf{e_3} \end{bmatrix} = C, \\
&\frac{(\mathbf{u_c}\cdot\mathbf{e_1})^2}{\lambda_1} + \frac{(\mathbf{u_c}\cdot\mathbf{e_2})^2}{\lambda_2} + \frac{(\mathbf{u_c}\cdot\mathbf{e_3})^2}{\lambda_3} = C.
\end{aligned} \tag{5.411}$$

The scalar products in Eqn. 5.411 are the projections of the displacement vector, $\mathbf{u_c}$, onto the unit eigenvectors of $\mathbf{U_c}$, illustrated in Fig. 5.32(a):

$$\frac{p_1^2}{\lambda_1} + \frac{p_2^2}{\lambda_2} + \frac{p_3^2}{\lambda_3} = C. \tag{5.412}$$

Before analyzing this equation further we pause to consider a more general equation known as the *quadratic form.* We do this because terminology related to the quadratic form is used routinely in the discussion of atomic displacements in crystallography. A quadratic form is a second degree polynomial in which the sum

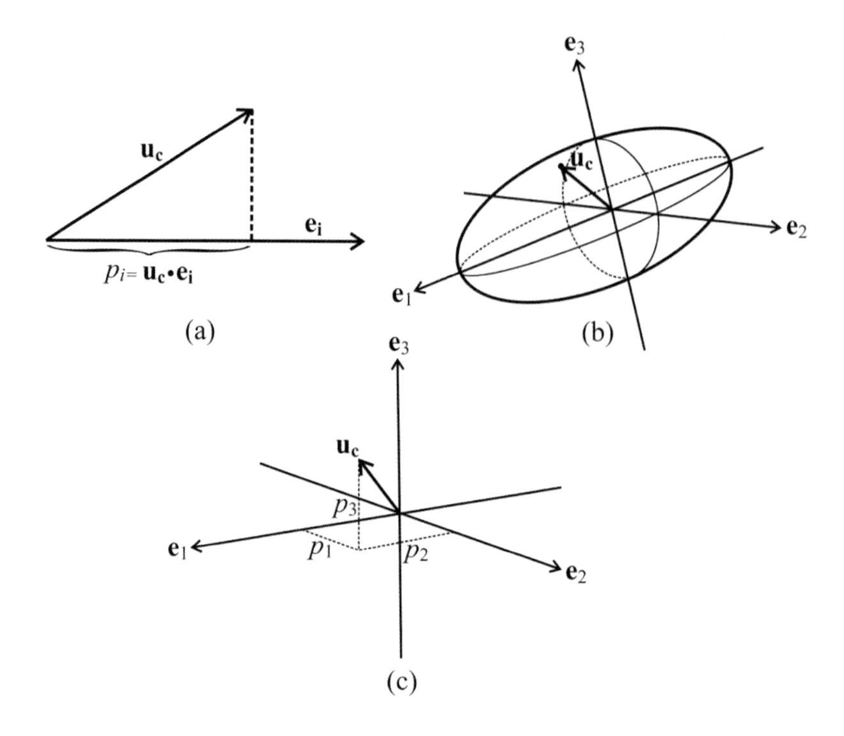

Figure 5.32 (a) Projection of an instantaneous atom displacement vector in the direction of a unit eigenvector of $\mathbf{U_c}$. (b) The displacement vectors for a constant probability density define an ellipsoid. (c) The components of the displacement vector along the eigenvectors of $\mathbf{U_c}$ have maximum values in the directions of the eigenvectors.

of the exponents in each term is constant – a *homogeneous* polynomial. A homogeneous polynomial can be written as an expression incorporating the symmetric matrix of its coefficients:

$$a_{11}x^2 + a_{22}y^2 + a_{33}z^2 + 2a_{12}xy + 2a_{13}xz + 2a_{23}yz$$
$$= \begin{bmatrix} x & y & z \end{bmatrix} \begin{bmatrix} a_{11} & a_{12} & a_{13} \\ a_{12} & a_{22} & a_{23} \\ a_{13} & a_{23} & a_{33} \end{bmatrix} \begin{bmatrix} x \\ y \\ z \end{bmatrix} = \mathbf{x}^T\mathbf{A}\mathbf{x}. \tag{5.413}$$

The matrices defining the quadratic forms are characterized by their eigenvalues and eigenvectors. If all of the eigenvalues of a quadratic form matrix are positive, then the matrix is said to be a *positive definite matrix.* If, on the other hand, the matrix has one or more eigenvalues less than or equal to zero, then the matrix is characterized as *non-positive definite.* When the quadratic forms are set equal to constants the resulting polynomial equations define surfaces known as *quadratic surfaces.* Examples of quadratic surfaces include the paraboloid, the hyperboloid, and the *ellipsoid.* The general equation for an ellipsoid in Cartesian (x, y, z) coordinates is

$$\frac{x^2}{a^2} + \frac{y^2}{b^2} + \frac{z^2}{c^2} = 1, \tag{5.414}$$

where a, b, and c are the lengths of the semi-axes of the ellipsoid.

The matrix $\mathbf{U_c}^{-1}$ is symmetric since $\mathbf{U_c}$ is symmetric (Sec. 1.81). Thus $\mathbf{U_c}^{-1}$ is a quadratic form matrix, and $\mathbf{u_c}^T\mathbf{U_c}^{-1}\mathbf{u_c}$ is a quadratic form. Rearranging Eqn. 5.412 gives

$$\frac{p_1^2}{(\sqrt{C\,\lambda_1})^{\,2}} + \frac{p_2^2}{(\sqrt{C\,\lambda_2})^{\,2}} + \frac{p_3^2}{(\sqrt{C\,\lambda_3})^{\,2}} = 1. \tag{5.415}$$

$\mathbf{u_c}^T\mathbf{U_c}^{-1}\mathbf{u_c} = C$ *is therefore an equation of an ellipsoid surface* with semi-axial lengths of $\sqrt{C\,\lambda_1}$, $\sqrt{C\,\lambda_2}$, and $\sqrt{C\,\lambda_3}$, and principal axes along the eigenvectors of $\mathbf{U_c}$, as illustrated in Fig. 5.32(b). The unit eigenvectors of $\mathbf{U_c}$ are a "natural" basis for the ellipsoid defined by $\mathbf{U_c}$, and since p_1, p_2, and p_3 are the projections of $\mathbf{u_c}$ onto these coordinates, they are the components of the *constant probability density displacement vector* in this basis: $\overrightarrow{\delta_j} = \mathbf{u}'_\mathbf{c} = [p_1\ p_2\ p_3]$, as shown in Fig. 5.32(c).

The significance of the term "non-positive definite" now becomes apparent. The displacement ellipsoid is not determined directly from experiment, but rather is defined during the refinement of the crystal structure. The anisotropic displacement parameters are determined by adjusting them in a least squares fit of the structural model to the diffraction data. If one of the eigenvalues of the orthogonalized U_{ij} tensor becomes negative, then the semi-axial length of the displacement ellipsoid corresponding to the eigenvalue becomes imaginary — a physically impossible situation. The resulting imaginary ellipsoid is often referred to as an "npd."

In structural determinations in which the data are very good, the thermal ellipsoids can tell us something of the nature of the displacements in the structure, such as those resulting from thermal motion.* On the other hand, when the intensity data set is limited either in size or quality, the anisotropic displacement parameters adjust to overcome these limitations (to achieve the best fit to the model). Changing static (equilibrium) atomic positions will affect the calculated intensities considerably – since they alter the locations of the scattering centers — adjusting the displacement parameters will have much less of an effect on the intensities. Thus, the "most adjustable" parameters are the displacement parameters, and the occurrence of non-positive definite ellipsoids in the refinement process is ordinarily a clear indicator of the limitations of the data set. Even in cases where there are no npds, the shapes of the resulting ellipsoids can be revealing — *if a displacement ellipsoid appears as if it is about to become non-positive definite — displaying one or more very short semi-axial lengths, then the data set is likely to be of insufficient quality to warrant the refinement of the anisotropic displacement parameters.*

Earlier in the chapter (Sec. 5.1.1) we noted that the covariance matrix for a set of independent variables could appear to correlate the variables, but a suitable rotation of the coordinate system would "uncorrelated" them. We now see how this is accomplished. Eqn. 5.412 can be written in matrix form as $\mathbf{u}'^{\,T}_\mathbf{c}[\lambda]^{-1}\mathbf{u}'_\mathbf{c} = C$. By rotating the coordinate system so that it is coincident with the eigenvectors of the $\mathbf{U_c}$ matrix, *the new covariance matrix is a diagonal matrix of the eigenvalues of* $\mathbf{U_c}$, and $\mathbf{u_c}^T\mathbf{U_c}^{-1}\mathbf{u_c} \equiv \mathbf{u}'^{\,T}_\mathbf{c}[\lambda]^{-1}\mathbf{u}'_\mathbf{c} = C$, describing the quadratic surface in the crystal Cartesian coordinate system and the "ellipsoid" coordinate system, respectively.

*The harmonic oscillator/Gaussian approximation is just that – an approximation. When more complex displacements are considered, i.e., when the distribution function is not Gaussian, the treatment becomes much more involved (c.f., Coppens[93]).

The eigenvalues are therefore the components of the mean squared displacement of the atom — the variances of p_1, p_2, and p_3 — in the ellipsoid coordinate system:

$$\begin{aligned}
\mathbf{u}_\mathbf{c}'^T[\lambda]^{-1}\mathbf{u}_\mathbf{c}' &= \begin{bmatrix} p_1 & p_2 & p_3 \end{bmatrix} \begin{bmatrix} 1/\lambda_1 & 0 & 0 \\ 0 & 1/\lambda_2 & 0 \\ 0 & 0 & 1/\lambda_3 \end{bmatrix} \begin{bmatrix} p_1 \\ p_1 \\ p_1 \end{bmatrix} \\
&= \begin{bmatrix} p_1 & p_2 & p_3 \end{bmatrix} \begin{bmatrix} 1/\langle p_1^2\rangle & 0 & 0 \\ 0 & 1/\langle p_2^2\rangle & 0 \\ 0 & 0 & 1/\langle p_3^2\rangle \end{bmatrix} \begin{bmatrix} p_1 \\ p_1 \\ p_1 \end{bmatrix} \\
&= \begin{bmatrix} p_1 & p_2 & p_3 \end{bmatrix} \begin{bmatrix} 1/\sigma_1^2 & 0 & 0 \\ 0 & 1/\sigma_2^2 & 0 \\ 0 & 0 & 1/\sigma_3^2 \end{bmatrix} \begin{bmatrix} p_1 \\ p_1 \\ p_1 \end{bmatrix} \\
&= \mathbf{u}_\mathbf{c}'^T\mathbf{S}^{-1}\mathbf{u}_\mathbf{c}' = C.
\end{aligned} \tag{5.416}$$

The constant probability density represented by the ellipsoid surface is therefore

$$\begin{aligned}
f_G(p_1, p_2, p_3) &= \frac{1}{(2\pi)^{3/2}|\mathbf{S}|} e^{-\frac{1}{2}\mathbf{u}_\mathbf{c}'^T\mathbf{S}^{-1}\mathbf{u}_\mathbf{c}'} \\
&= \frac{1}{(2\pi)^{3/2}\sigma_1\,\sigma_2\,\sigma_3} e^{-\frac{1}{2}(p_1^2/\sigma_1^2 + p_2^2/\sigma_2^2 + p_3^2/\sigma_3^2)},
\end{aligned} \tag{5.417}$$

where p_1, p_2, and p_3 are subject to the constraint that $\mathbf{u}_\mathbf{c}' = [p_1\ p_2\ p_3]$ must lie on the surface of an ellipsoid: $p_1^2/\sigma_1^2 + p_2^2/\sigma_2^2 + p_3^2/\sigma_3^2 = C$.

If we define a vector, $\mathbf{t} = [t_1\ t_2\ t_3] = [(p_1/\sigma_1)\ (p_2/\sigma_2)\ (p_3/\sigma_3)]$, then $C = t^2$, the square of the magnitude of $\mathbf{t}$. For a given displacement vector, $[p_1^c\ p_2^c\ p_3^c]$ to fall on the surface of the ellipsoid, $t_c = \sqrt{C}$. For a vector, $[p_1\ p_2\ p_3]$, to fall within the surface defined by the ellipsoid, t will be less than t_c (qualitatively, p_1, p_2 and/or p_3 will be smaller). Thus, to determine the probability that a displacement vector will fall on or within the ellipsoid, we must integrate over the probability density of all the vectors that satisfy $\|[(p_1/\sigma_1)\ (p_2/\sigma_2)\ (p_3/\sigma_3)]\| \le t_c$:

$$Pr(p_1^c, p_2^c, p_3^c) = \frac{1}{(2\pi)^{3/2}\sigma_1\,\sigma_2\,\sigma_3} \int_0^{p_1^c}\int_0^{p_2^c}\int_0^{p_3^c} e^{-\frac{1}{2}(p_1^2/\sigma_1^2 + p_2^2/\sigma_2^2 + p_3^2/\sigma_3^2)}\, dp_1 dp_2 dp_3$$

Since $t_1 = p_1/\sigma_1$, etc., $dp_1 = \sigma_1 dt_1$, $dp_2 = \sigma_2 dt_2$, and $dp_3 = \sigma_3 dt_3$. Thus, the probability that a displacement vector will fall on or within the ellipsoid surface is

$$Pr(\mathbf{t_c}) = \frac{1}{(2\pi)^{3/2}\sigma_1\,\sigma_2\,\sigma_3} \int_0^{t_1^c}\int_0^{t_2^c}\int_0^{t_3^c} e^{-t^2/2}\sigma_1\,\sigma_2\,\sigma_3\, dt_1 dt_2 dt_3.$$

Since the only variable in the integral is the magnitude of $\mathbf{t}$,

$$Pr(t_c) = \frac{1}{(2\pi)^{3/2}} \int_0^{t_1^c}\int_0^{t_2^c}\int_0^{t_3^c} e^{-t^2/2}\, dt_1 dt_2 dt_3.$$

Again, because the integral does not explicitly contain the components of $\mathbf{t}$, it can be simplified significantly by converting to spherical polar coordinates. Referring to Fig. 5.22, integrating over t_1, t_2, and t_3 is equivalent to integrating v from 0 to 2π, η from 0 to π, and t from 0 to t_c, where the differential volume in polar coordinates is $dt_1\, dt_2\, dt_3 = t^2\, dt\, \sin\eta\, d\eta\, dv$:

$$\begin{aligned} Pr(t_c) &= \frac{1}{(2\pi)^{3/2}} \int_0^{t_c} \int_0^{\pi} \int_0^{2\pi} t^2\, e^{-t^2/2}\, dv\, \sin\eta\, d\eta\, dt \\ &= \frac{1}{\sqrt{(2/\pi)}} \int_0^{t_c} t^2 e^{-t^2/2}\, dt. \end{aligned} \tag{5.418}$$

Note that the integral in Eqn. 5.418 was created by dividing the integration variables in the Gaussian distribution by their standard deviations, just as they were when the error function was defined in Sec. 5.1.1. The equation above is the trivariate counterpart of the error function; similarly, it has been evaluated numerically and tabulated in collections of mathematical tables and software packages. The tables contain a list of values of t_c and the corresponding probability that a displacement vector will lie within an ellipsoid defined by that value of t_c.

Consulting the table for a specific probability, for example, 0.5, gives a value of t_c that will generate $Pr(t_c) = 0.5$, which turns out to be 1.5382; 50% of the displacement vectors will have $t \leq t_c$. Since $\sqrt{C} = t_c$, the ellipsoid in which the center of the atom should be found 50% of the time will then have semi-axial lengths of $\sigma_1 t_c$, $\sigma_2 t_c$, and $\sigma_3 t_c$, respectively. The anisotropic displacement parameters describe an ellipsoid with its major axes along the eigenvectors of the orthogonalized displacement matrix (tensor), $\mathbf{U_c}$. The eigenvalues define the mean squared atom displacements in the directions of the principal axes, and the square roots of the eigenvalues determine the standard deviations of the displacements — the semi-axial lengths of the ellipsoid. When displacement ellipsoids with $t_c = 1.5382$ are published as illustrations in the literature they are said to have been plotted "at the 50% probability level."

The Equivalent Isotropic Displacement Parameter. The displacement volume of the atom is described by six anisotropic displacement parameters. It is often difficult to interpret these parameters directly, and if the specific shape of the displacement volume is not important, it is convenient to describe the atom as if its displacement was isotropic. In this case the displacement volume would be a sphere, and we would expect the average squared displacement to be approximated as the average of the squared displacements along the axes of *an orthogonal coordinate system.* If the eigenvalues of $\mathbf{U_c}$ have been determined, then this radius will be $1/3(\lambda_1 + \lambda_2 + \lambda_3) = 1/3(\langle p_1^2\rangle + \langle p_2^2\rangle + \langle p_3^2\rangle) \simeq u_{iso}^2$. Since the diagonal elements of $\mathbf{U_c}$ are mean squared Cartesian displacements of the same vector described in a rotated coordinate system, their average will be the same as for the diagonalized matrix so that $u_{iso}^2 \simeq 1/3(\langle u_1^{c2}\rangle + \langle u_2^{c2}\rangle + \langle u_3^{c2}\rangle) = 1/3(U_{11}^c + U_{22}^c + U_{33}^c)$. The average squared isotropic displacement parameter is the same for all three matrices, and a displacement matrix more or less "equivalent" to the orthogonalized anisotropic displacement matrix would have the average values of the diagonal terms of the displacement matrix (tensor) along its diagonal:

$$\mathbf{U_{eq}} = \begin{bmatrix} U_{eq} & 0 & 0 \\ 0 & U_{eq} & 0 \\ 0 & 0 & U_{eq} \end{bmatrix} = \begin{bmatrix} \frac{U_{11}^c+U_{22}^c+U_{33}^c}{3} & 0 & 0 \\ 0 & \frac{U_{11}^c+U_{22}^c+U_{33}^c}{3} & 0 \\ 0 & 0 & \frac{U_{11}^c+U_{22}^c+U_{33}^c}{3} \end{bmatrix}.$$

The diagonal elements of $\mathbf{U_c}$ determine the squared atom displacements, and the off-diagonal elements determine the directions of the ellipsoid axes. Since there is no preferred direction, $\mathbf{U_{eq}}$ is a diagonal matrix. The isotropic displacement is described by a single isotropic displacement parameter, U_{eq}, with $u_{iso} \simeq \sqrt{U_{eq}}$. The sum of the diagonal elements of a matrix is known as its *trace*, and the *equivalent isotropic displacement parameter*, U_{eq}, is commonly described in the literature as "one third of the trace of the orthogonalized U_{ij} tensor."

Symmetry and the Displacement Ellipsoid. Atoms in symmetry-equivalent positions are in locally identical environments, and these environments are related by the symmetry of the crystal lattice. For example, if an atom is symmetric to another through a mirror plane, as illustrated in Fig. 5.33, then the displacement vectors and displacement ellipsoids must also be related by the mirror plane. Translating a given displacement vector will not change its orientation, but rotations or reflections will — the displacement ellipsoids for two symmetry-equivalent atoms are transformed by the non-translational symmetry operators that link the two atoms.

The general symmetry operation results in the transformation $\mathbf{v_f}' = \mathbf{R}\mathbf{v_f} + \mathbf{t}$, where $\mathbf{v_f}$ is a general vector in the unit cell fractional coordinates, $\mathbf{R}$ is the point symmetry operator, and $\mathbf{t}$ is the translation operator. A displacement vector

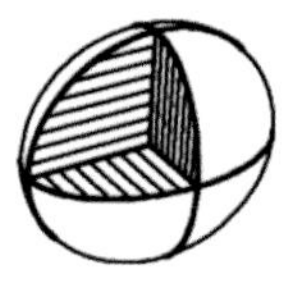

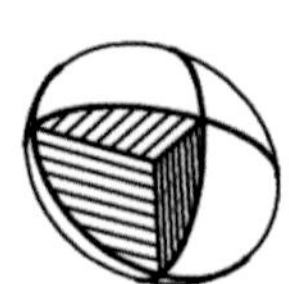

Figure 5.33 ORTEP displacement ellipsoids for two atoms related by a mirror plane.

$\overrightarrow{\delta_j} = [b_{1j}\ b_{2j}\ b_{3j}]$ is therefore transformed by $\mathbf{R}$,* creating a transformed anisotropic displacement tensor:

$$\begin{aligned}
&\mathbf{R}\overrightarrow{\delta_j} = \overrightarrow{\delta'_j} \\
&\overrightarrow{\delta_j}^T \mathbf{R}^T = \overrightarrow{\delta'_j}^T \\
&\mathbf{R}\overrightarrow{\delta_j}\,\overrightarrow{\delta_j}^T \mathbf{R}^T = \overrightarrow{\delta'_j}\,\overrightarrow{\delta'_j}^T \\
&\mathbf{R}\langle\overrightarrow{\delta_j}\,\overrightarrow{\delta_j}^T\rangle \mathbf{R}^T = \langle\overrightarrow{\delta'_j}\,\overrightarrow{\delta'_j}^T\rangle \\
&\mathbf{R}[\mathcal{B}]\mathbf{R}^T = [\mathcal{B}]'.
\end{aligned} \tag{5.419}$$

In the example given here, a mirror plane or glide plane in a monoclinic unit cell parallel to the ac plane would transform the anisotropic displacement matrix (containing either β_{ij}s or U_{ij}s) as follows:

$$\begin{bmatrix} 1 & 0 & 0 \\ 0 & -1 & 0 \\ 0 & 0 & 1 \end{bmatrix} \begin{bmatrix} \beta_{11} & \beta_{12} & \beta_{13} \\ \beta_{12} & \beta_{22} & \beta_{23} \\ \beta_{13} & \beta_{23} & \beta_{33} \end{bmatrix} \begin{bmatrix} 1 & 0 & 0 \\ 0 & -1 & 0 \\ 0 & 0 & 1 \end{bmatrix} = \begin{bmatrix} \beta_{11} & -\beta_{12} & \beta_{13} \\ -\beta_{12} & \beta_{22} & -\beta_{23} \\ \beta_{13} & -\beta_{23} & \beta_{33} \end{bmatrix}. \tag{5.420}$$

Note that the mean squared displacements (along the diagonal) have remained the same; only the directions of the ellipsoid axes have been altered. Note also that an inversion does not change the matrix:

$$\begin{bmatrix} -1 & 0 & 0 \\ 0 & -1 & 0 \\ 0 & 0 & -1 \end{bmatrix} \begin{bmatrix} \beta_{11} & \beta_{12} & \beta_{13} \\ \beta_{12} & \beta_{22} & \beta_{23} \\ \beta_{13} & \beta_{23} & \beta_{33} \end{bmatrix} \begin{bmatrix} -1 & 0 & 0 \\ 0 & -1 & 0 \\ 0 & 0 & -1 \end{bmatrix} = \begin{bmatrix} \beta_{11} & \beta_{12} & \beta_{13} \\ \beta_{12} & \beta_{22} & \beta_{23} \\ \beta_{13} & \beta_{23} & \beta_{33} \end{bmatrix}. \tag{5.421}$$

When the symmetry operator corresponding to a special position in the unit cell is applied, an atom in the special position must have invariant anisotropic displacement parameters, i.e., $[\mathcal{B}] = [\mathcal{B}]'$. This places special constraints on the displacement parameters. An atom on a mirror or glide plane, for example, requires that $\beta_{12} = -\beta_{12}$ and $\beta_{23} = -\beta_{23}$, and that is possible only if β_{12} and $\beta_{23} = 0$.

5.2 Scaling the Intensity Data

X-ray crystal structure determination begins with a set of relative integrated intensities, $\{I_{rel,\mathbf{h}}\}$, ideally corrected for Lorentz, polarization, and absorption effects:

$$I_{rel,\mathbf{h}} = \frac{\langle I_d \rangle_\mathbf{h}}{(LPT)_\mathbf{h}}. \tag{5.422}$$

The scattered waves responsible for these intensities are described by the scattering vector – the structure factor – with amplitude $F_{rel,\mathbf{h}}$:

$$\mathbf{F}_{rel,\mathbf{h}} = F_{rel,\mathbf{h}}\, e^{i\varphi_\mathbf{h}} \tag{5.423}$$

$$I_{rel,\mathbf{h}} = \mathbf{F}^*_{rel,\mathbf{h}} \mathbf{F}_{rel,\mathbf{h}} = F^2_{rel,\mathbf{h}}. \tag{5.424}$$

*Although the ellipsoids do not ordinarily lie on symmetry elements, since their orientations do not depend on translations the operation can be envisioned as a translation of the ellipsoid to the symmetry element, followed by the symmetry operation and a subsequent translation to the symmetry-equivalent position. See Fig. 6.23 and the related discussion for an example in another context.

The structure is solved by determining the locations of atoms and displacement parameters in the unit cell that will predict (calculate) the set of observed intensities:

$$\mathbf{F}_{c,\mathbf{h}} = \sum_{j=1}^{n} f_j(h) e^{-2\pi i(\mathbf{h}\cdot \mathbf{r}_j)} e^{-T_{j,\mathbf{h}}} \tag{5.425}$$

$$I_{c,\mathbf{h}} = \mathbf{F}^*_{c,\mathbf{h}} \mathbf{F}_{c,\mathbf{h}} = F^2_{c,\mathbf{h}}. \tag{5.426}$$

The scattering factor, $f_{j,\theta d}$, has been replaced here with $f_j(h)$ to emphasize that it depends on the magnitude of the diffraction vector, $h = 2\sin\theta/\lambda$.

The magnitudes of the experimental intensities depend on the size of the crystal and the incident beam intensity. The magnitudes of the calculated intensities, however, do not depend on the experimental parameters. The calculated structure factors predict the electron density distribution,

$$\rho(\mathbf{r}) = \frac{1}{V_{cell}} \sum_{\mathbf{h}} \mathbf{F}_{c,\mathbf{h}} e^{-2\pi i(\mathbf{h}\cdot\mathbf{r})}, \tag{5.427}$$

where $\rho(\mathbf{r})$ is a probability density function, giving the probability per unit volume of finding an electron at a location between $\mathbf{r}$ and $\mathbf{r} + d\mathbf{r}$. The function represents the fraction of the electron charge that occupies a unit volume in this region, and has dimensions of electrons per cubic angstrom, $e/^3$. Since $1/V_{cell}$ has dimensions of $1/^3$, the structure factor, and hence the scattering factor, formally has dimensions of "electrons", or more appropriately, "electron scattering power". This is consistent with the definition of the scattering factor, which has a scattering power equal to its atomic number in the absence of destructive interference within the atom, i.e. at $2\theta = 0$ (Sec. 3.2.4). A hypothetical structure factor, calculated at $2\theta = 0$, with $(h, k, l) = (0, 0, 0)$ *gives the numbers of electrons in the unit cell*:

$$\mathbf{F}_{000} = F_{000} = \sum_{j=1}^{n} f_j e^{-2\pi i(0x+0y+0z)} = \sum_{j=1}^{n} f_j e^0 = \sum_{j=1}^{n} f_j. \tag{5.428}$$

In order to compare calculated structure factor magnitudes, F_c, with experimental structure factor magnitudes, we must find a way to put F_{rel} and I_{rel} on the same scale as F_c and I_c. These scaled values will be the experimental observations used in crystal structure solution and refinement, and we will refer to them as F_o and I_o:

$$F_o = S' F_{rel} \text{ and} \tag{5.429}$$

$$I_o = F_o^2 = S'^2 F^2_{rel} = K' I_{rel}. \tag{5.430}$$

As we shall see below, the intensity scale factor, K', can only be approximated without a structural model; it is not determined accurately until it becomes a parameter (i.e., a component of the model) in structural refinement. In refinement the scale factor takes the form of a variable, along with the calculated structure factors (from the atomic positions and displacements). Because of this, it is more appropriate to work with the reciprocals of the scale factors, $K = 1/K'$ and $S = 1/S'$:

$$F_{rel} = SF_o \approx SF_c \quad \text{and} \quad I_{rel} = KI_o \approx KI_c. \tag{5.431}$$

With a knowledge of the atomic positions and displacement parameters, the scale factor, K, can easily be calculated from the average values of the relative and calculated intensities:

$$K = \frac{\langle I_{rel} \rangle_{\mathbf{h}}}{\langle I_c \rangle_{\mathbf{h}}}. \tag{5.432}$$

The **h** subscripts indicate that the averages are over all of the reflections. We have no knowledge of either the atom locations or their displacement parameters, and we must investigate the possibility of determining $\langle I_c \rangle_{\mathbf{h}}$ without them. Beginning with Eqn. 5.425, we insert the isotropic displacement parameter(Eqn. 5.399) into the equation, since we know nothing of anisotropic displacements:

$$\begin{aligned} \mathbf{F}_{c,\mathbf{h}} &= \sum_{j=1}^{n} f_j(h) e^{-2\pi\, i(\mathbf{h}\cdot\mathbf{r}_j)} e^{-2\,\pi^2\,\langle u_j^2 \rangle h^2} & (5.433) \\ &= \sum_{j=1}^{n} f_j(h) e^{-2\pi\, i(\mathbf{h}\cdot\mathbf{r}_j)} e^{-8\,\pi^2\,\langle u_j^2 \rangle \sin^2\theta/\lambda}. & (5.434) \end{aligned}$$

$8\,\pi^2\,\langle u_j^2 \rangle$ is often abbreviated in the literature as B_j. Because we are interested here in the form of the equation involving h, we will abbreviate $2\,\pi^2\,\langle u_j^2 \rangle$ as B'_j:

$$\mathbf{F}_{c,\mathbf{h}} = \sum_{j=1}^{n} f_j(h) e^{-2\pi\, i(\mathbf{h}\cdot\mathbf{r}_j)} e^{-\mathrm{B}'_j h^2}. \tag{5.435}$$

Finally, since at this stage we do not even know the locations of the atoms, we replace the isotropic displacement term for each atom with an average (overall) isotropic displacement term, B':

$$\mathbf{F}_{c,\mathbf{h}} = \sum_{j=1}^{n} f_j(h) e^{-2\pi\, i(\mathbf{h}\cdot\mathbf{r}_j)} e^{-\mathrm{B}' h^2}. \tag{5.436}$$

The sum is over the atoms in the unit cell, and therefore requires some knowledge of the unit cell contents. The number of formula units/molecules per unit cell can usually be determined from the volume of the unit cell (Sec. 1.5.4), and the mass of a formula unit. The mass of the unit cell is an integral multiple of the mass of the asymmetric unit — and that multiple must be consistent with the Bravais lattice of the crystal. For example, a monoclinic unit cell will ordinarily contain either two, four, or eight asymmetric units. If $W = m_{asu}/V_{cell}$, where m_{asu} is the mass of an asymmetric unit, then the calculated density will be either $2W$, $4W$, or $8W$, and only one of these will provide a realistic density.

Using Eqn. 5.436, the average calculated intensity is

$$\begin{aligned} \langle I_{c,\mathbf{h}} \rangle_{\mathbf{h}} &= \langle \mathbf{F}^*_{c,\mathbf{h}}\, \mathbf{F}_{c,\mathbf{h}} \rangle_{\mathbf{h}} \\ &= \left\langle \sum_{j=1}^{n} f_j(h) e^{2\pi\, i(\mathbf{h}\cdot\mathbf{r}_j)} e^{-\mathrm{B}' h^2} \sum_{m=1}^{n} f_m(h) e^{-2\pi\, i(\mathbf{h}\cdot\mathbf{r}_m)} e^{-\mathrm{B}' h^2} \right\rangle_{\mathbf{h}}. & (5.437) \end{aligned}$$

There are two sets of terms in the product, one set for $j = m$ and the other for $j \neq m$:

$$\begin{aligned}\langle I_{c,\mathbf{h}}\rangle_{\mathbf{h}} &= \langle \mathbf{F}^*_{c,\mathbf{h}}\,\mathbf{F}_{c,\mathbf{h}}\rangle_{\mathbf{h}} \\ &= \left\langle \sum_{j=1}^{n} \left(f_j(h) e^{-\mathrm{B}' h^2}\right)^2 e^{2\pi\, i(\mathbf{h}\cdot\,\mathbf{r}_j)} e^{-2\pi\, i(\mathbf{h}\cdot\,\mathbf{r}_j)} \right\rangle_{\mathbf{h}} \\ &\quad + \left\langle \sum_{j=1}^{n}\sum_{m=1}^{n} f_j(h) f_m(h) \left(e^{-\mathrm{B}' h^2}\right)^2 e^{2\pi\, i(\mathbf{h}\cdot\,(\mathbf{r}_j - \mathbf{r}_m))} \right\rangle_{\mathbf{h}} . \end{aligned} \tag{5.438}$$

The exponential products in the first term are all equal to $e^0 = 1$. The first term is therefore independent of the atomic positions. The second term still depends on atomic positions, and since $f_j(h)$ and $f_m(h)$ are functions of the diffraction angle (and therefore h), each term in the sum contains variables that are *not* independent of one another; the terms cannot be factored additionally in order to deal with the expression containing $\mathbf{r}_j - \mathbf{r}_m = \Delta\mathbf{r}_{jm}$.

Rather than attempting to determine the average over the entire set of reflections, consider instead calculating the average intensity for a series of subsets of the reflections for which the change in the magnitude of the diffraction vector is relatively small, as illustrated in Fig. 5.34. This will allow us to replace the atomic scattering factors in each region by their (constant) averages in the region. For a given region, the second term in Eqn. 5.438 then becomes

$$\begin{aligned}&\sum_{j=1}^{n}\sum_{m=1}^{n} \langle f_j(h)\rangle \langle f_m(h)\rangle \left(e^{-\mathrm{B}'\langle h\rangle^2}\right)^2 \left\langle e^{2\pi\, i(\mathbf{h}\cdot\,\Delta\mathbf{r}_{jm})}\right\rangle_{\mathbf{h}} \\ &= \sum_{j=1}^{n}\sum_{m=1}^{n} C\, \left(\langle \cos(2\pi\,\mathbf{h}\cdot\,\Delta\mathbf{r}_{jm}\,)\rangle_{\mathbf{h}} + i\, \langle \sin(2\pi\,\mathbf{h}\cdot\,\Delta\mathbf{r}_{jm}\,)\rangle_{\mathbf{h}}\right) . \end{aligned} \tag{5.439}$$

Each region creates a spherical shell in reciprocal space with a central radius equal to $\langle\, h\rangle$; every diffraction vector within a shell will point in a different direction. Fig. 5.35 illustrates the resulting vectors, $Ce^{2\pi\, i(\mathbf{h}\cdot\,\Delta\mathbf{r})}$, for an arbitrarily chosen value of $\Delta\mathbf{r} = (0.345, -0.191, -0.208)$ in a triclinic unit cell with $a = 10.139$ Å, $b = 10.583$ Å, $c = 9.911$ Å, $\alpha = 75.86°$, $\beta = 77.41°$, and $\gamma = 65.93°$, for values of h between 0.25 and 0.35, with $\langle h\rangle = 0.31$. If the shell is sufficiently large there will be enough vectors so that for every value of $\mathbf{h}$ creating an argument in the cosine term of $2\pi\, i\,(\mathbf{h}\cdot\,\Delta\mathbf{r}_{jm})$, there will be another of approximately equal magnitude but at an angle of (about) π to it. The cosine terms will then tend to average to zero. There will also be a vector at approximately $-2\pi\, i\,(\mathbf{h}\cdot\,\Delta\mathbf{r}_{jm})$ for every $2\pi\, i\,(\mathbf{h}\cdot\,\Delta\mathbf{r}_{jm})$, and the sine terms will also average toward zero. In this example, for a single value of $\Delta\mathbf{r}$, the average value of the cosine term was 0.09, and the average value of the sine term was 0.02. If only positive indices are chosen, and the terms are averaged in this region for $\Delta\mathbf{r} = (\Delta x, \Delta y, \Delta z)$, where Δx, Δy, and Δz vary form zero to one in increments of 0.1, the cosine terms average to 0.007 and the sine terms average to less than 0.001. If Friedel equivalents are included, then there will be a $-\mathbf{h}$ for every $\mathbf{h}$, and the sine terms will cancel one another exactly.

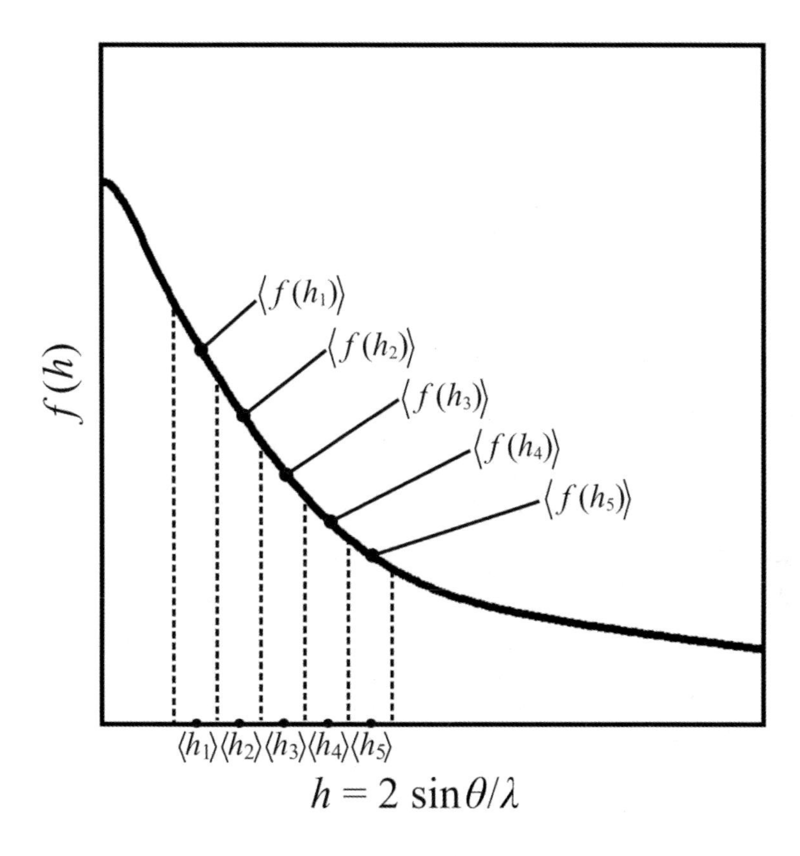

Figure 5.34 Atomic scattering factor curve divided into regions in which Δh is relatively small.

It follows that the second term in Eqn. 5.438 approaches zero for each $\langle h\rangle$, and the average calculated intensity for each shell can be considered to be approximately independent of atom locations:

$$\langle I_{c,\mathbf{h}}\rangle_{\mathbf{h}}(\langle h\rangle) = \sum_{j=1}^{n}\langle f_j(h)\rangle^2 e^{-2\mathrm{B}'\langle h\rangle^2} = e^{-2\mathrm{B}'\langle h\rangle^2}\sum_{j=1}^{n}\langle f_j(h)\rangle^2. \tag{5.440}$$

The scale factor, *for each region* (which should, in principle be the same number) can now be determined independently of the structural model,

$$K = \frac{\langle I_{rel}\rangle_{\mathbf{h}}}{\langle I_c\rangle_{\mathbf{h}}}(\langle h\rangle) = \frac{\langle I_{rel}\rangle_{\mathbf{h}}}{e^{-2\mathrm{B}'\langle h\rangle^2}\sum_{j=1}^{n}\langle f_j(h)\rangle^2.}. \tag{5.441}$$

provided that we know the value for B$'$.

In 1942, A. J. C. Wilson published a paper in *Nature*[94] in which he demonstrated that rearranging Eqn. 5.441, a function of the average values of h in each region,

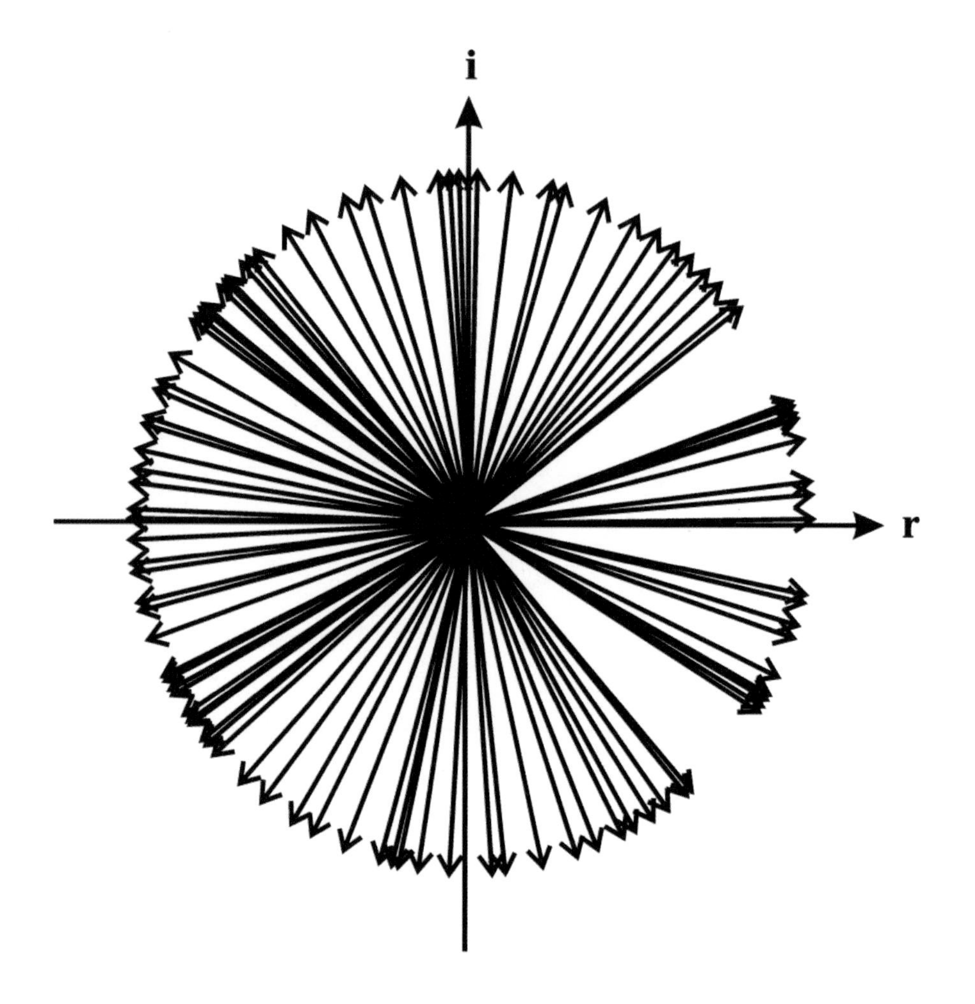

Figure 5.35 108 vectors, $Ce^{2\pi\, i(0.345\, h - 0.191\, k - 0.208\, l)}$, plotted in the complex plane in the range between $h = 0.25$ and $h = 0.35$.

would generate a linear plot to provide a "best-fit" scale factor *and* the overall isotropic temperature factor at the same time:

$$\frac{\langle I_{rel}\rangle_{\mathbf{h}}}{\sum_{j=1}^{n}\langle f_j(h)\rangle^2.} = e^{-2\mathrm{B}'\langle h\rangle^2}K \tag{5.442}$$

$$\ln\left(\frac{\langle I_{rel}\rangle_{\mathbf{h}}}{\sum_{j=1}^{n}\langle f_j(h)\rangle^2.}\right) = -2\,\mathrm{B}'\,\langle h\rangle^2 + \ln K. \tag{5.443}$$

This is the equation of a line with an intercept of ln K and a slope of $-2\,\mathrm{B}'$. A plot of $(\ln\langle I_{rel}\rangle)/(\sum\langle f_j(h)\rangle^2)$ versus $\langle h\rangle^2$ and the resulting least-squares line is known as

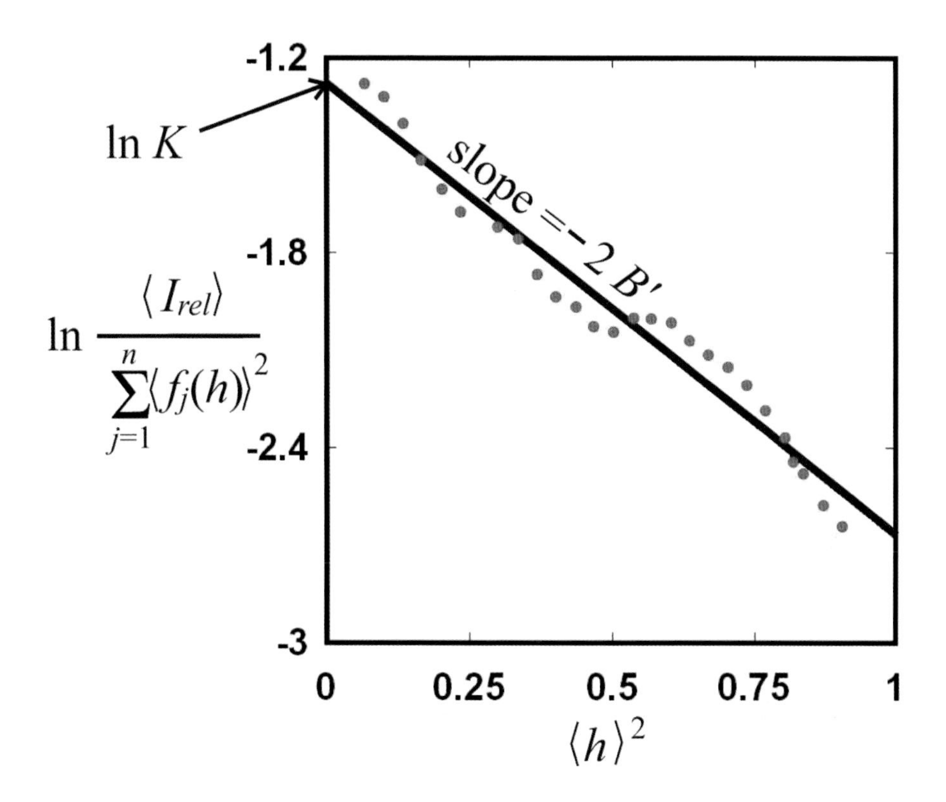

Figure 5.36 Wilson plot of intensity data for $CrC_{40}H_{30}O_4P_2$.

a *Wilson plot.* The $\mathbf{h}$ subscript indicates that I_{rel} is averaged over the indices of all the reflections within a specific shell. An example of a Wilson plot for intensity data collected from a crystal of an organometallic compound, $Cr((C_6H_5)_3P)_2(CO)_4$, is illustrated in Fig. 5.36.

In this example, $\ln K = -1.277$, $K = 0.28$, and $K' = 1/K = 3.58$. The slope of the line is -1.392, which determines B′, and the average atomic displacement: $-2\,\mathrm{B}' = -1.392$, $\mathrm{B}' = 0.696 = 2\pi^2\langle u\rangle^2$, and $\langle u\rangle^2 = 0.035^2 \Rightarrow \langle u\rangle \approx 0.19$. In the literature, the independent variable in Wilson plots is routinely $\langle\sin\theta/\lambda\rangle^2 = \langle h/2\rangle^2$ rather than $\langle h\rangle^2$. The slope in these cases is $-2\,\mathrm{B}$, where $\mathrm{B}= 8\pi^2\langle u\rangle^2$.

5.3 Determining the Space Group

The set of corrected and scaled observed intensities, $\{I_o\}$, and structure factor magnitudes, $\{F_o\}$, provide the starting point for the determination of the crystal structure model. The task at hand (still a formidable one!) is to determine the relative phases, $\varphi_{\mathbf{h}}$, of the scattered waves, and consequently, the electron density distribution in the unit cell:

$$\rho(\mathbf{r}) = \frac{1}{V_c}\sum_{\mathbf{h}} F_{o,\mathbf{h}}\, e^{i\varphi_{\mathbf{h}}}\, e^{-2\pi\, i(\mathbf{r}\cdot\mathbf{h})}. \tag{5.444}$$

The electron density is modeled from the locations of atoms/ions within the unit cell, and the structural solution process is centered around the determination of their locations. Since the procedure is not a trivial one, it is best to simplify it as much as possible — by determining the smallest number of atomic locations necessary to define the structure. It follows that we should take advantage of the space-group symmetry of the unit cell — reducing the problem to finding the locations of the atoms in an asymmetric unit.

Furthermore, the structural solution and refinement methods covered in the next three chapters are intimately linked to the lattice symmetry, and it is therefore important (and very often necessary) to assign the space group of the lattice prior to attempting a structural solution. Beginning with the Laue group, we must now determine the *point and translational symmetry operators* of the lattice in order to establish its space group.

5.3.1 Systematic Absences

In (Sec. 4.6.3), rotation axes and mirror planes were seen to determine the symmetry of the diffraction pattern. With a complete set of intensities, and a need to determine as much symmetry as possible, we now consider the possibility that there are similar effects that result from the translational operators as well.

Screw Axes

Recall that a two fold rotation axis parallel to the b axis requires that for every atom located at (x_f, y_f, z_f), there is a symmetry-equivalent atom at $(-x_f, y_f, -z_f)$.* Half of the atoms will be at locations (x_f, y_f, z_f), and the other half will be at symmetry-equivalent locations,$(-x_f, y_f, -z_f)$. The resulting structure factor for a general reflection will then be

$$\mathbf{F}_{c,hkl} = \sum_{j=1}^{n/2} f_j e^{-2\pi\, i(hx_j+ky_j+lz_j)} + \sum_{m=1}^{n/2} f_m e^{-2\pi\, i(-hx_m+ky_m-lz_m)}. \tag{5.445}$$

The structure factor for the reflection with indices $(\bar{h}, k, \bar{l})$ is identical (after switching the dummy variables, j and m):

$$\begin{aligned}\mathbf{F}_{c,\bar{h}k\bar{l}} &= \sum_{j=1}^{n/2} f_j e^{-2\pi\, i(\bar{h}x_j+ky_j+\bar{l}z_j)} + \sum_{m=1}^{n/2} f_m e^{-2\pi\, i(-\bar{h}x_m+ky_m-\bar{l}z_m)} \\ &= \sum_{j=1}^{n/2} f_j e^{-2\pi\, i(-hx_j+ky_j-lz_j)} + \sum_{m=1}^{n/2} f_m e^{-2\pi\, i(hx_m+ky_m+lz_m)}, \end{aligned} \tag{5.446}$$

and, as we observed in Chapter 4, for a lattice with a two-fold axis parallel to b, $I_{\bar{h}k\bar{l}} = I_{hkl}$.

*To determine the effects of symmetry on the diffraction pattern in Chapter 4, the symmetry of the *observed* intensities was shown to be a consequence of the electron density symmetry in the direct lattice. We take an alternative (but equivalent) approach here in determining the effects of symmetry on the *calculated* diffraction pattern.

Replacing the two-fold rotation axis with a two-fold screw axis places equivalent atoms at (x_f, y_f, z_f) and $(-x_f, (y_f + \frac{1}{2}), -z_f)$. The structure factor for a general reflection now becomes

$$\begin{aligned}\mathbf{F}_{c,hkl} &= \sum_{j=1}^{n/2} f_j e^{-2\pi\, i(hx_j+ky_j+lz_j)} + \sum_{m=1}^{n/2} f_m e^{-2\pi\, i(-hx_m+k(y_m+1/2)-lz_m)} \\ &= \sum_{j=1}^{n/2} f_j e^{-2\pi\, i(hx_j+ky_j+lz_j)} + \sum_{m=1}^{n/2} f_m e^{-2\pi\, i(-hx_m+ky_m-lz_m)}\, e^{-k\pi i}.\end{aligned} \tag{5.447}$$

If the integer, k, is even, $e^{-k\pi i} = \cos(-k\pi) + i\sin(-k\pi) = 1$, Eqn. 5.447 is identical to Eqn. 5.445, and $I_{\bar{h}k\bar{l}} = I_{hkl}$. If k is odd, $e^{-k\pi i} = -1$, and

$$\begin{aligned}\mathbf{F}_{c,hkl} &= \sum_{j=1}^{n/2} f_j e^{-2\pi\, i(hx_j+ky_j+lz_j)} - \sum_{m=1}^{n/2} f_m e^{-2\pi\, i(-hx_m+ky_m-lz_m)} \\ \mathbf{F}_{c,\bar{h}k\bar{l}} &= \sum_{j=1}^{n/2} f_j e^{-2\pi\, i(\bar{h}x_j+ky_j+\bar{l}z_j)} - \sum_{m=1}^{n/2} f_m e^{-2\pi\, i(-\bar{h}x_m+ky_m-\bar{l}z_m)} \\ &= \sum_{j=1}^{n/2} f_j e^{-2\pi\, i(-hx_j+ky_j-lz_j)} - \sum_{m=1}^{n/2} f_m e^{-2\pi\, i(hx_m+ky_m+lz_m)} \\ &= -\mathbf{F}_{c,hkl}.\end{aligned} \tag{5.448}$$

Again, since $I_{hkl} = \mathbf{F}^*_{hkl}\mathbf{F}_{hkl}$, $I_{\bar{h}k\bar{l}} = I_{hkl}$. A diad screw axis has the same effect *on the intensity of a general reflection* that a diad rotation axis has. Both create a two-fold rotation axis in the diffraction pattern. In general, n-fold rotation axes and n-fold screw axes produce n-fold rotation axes in the intensity-weighted reciprocal lattice and *the general symmetry of the reciprocal lattice cannot distinguish between the two.*

Fortunately, all is not lost here! There are certain classes of reflections that *are* affected by the translation of symmetry-equivalent scattering centers. For these reflections, scattering from the two centers in the absence of the translation occurs in phase, while waves scattered from the translated centers are 180° out of phase. Again, consider diad rotation and screw axes parallel to the b axis. We now focus on reflections with $(0k0)$ indices. For the rotation axis, Eqn. 5.445 becomes

$$\mathbf{F}_{c,0k0} = \sum_{j=1}^{n/2} f_j e^{-2\pi\, i\, ky_j} + \sum_{m=1}^{n/2} f_m e^{-2\pi\, i\, ky_m} = 2\sum_{j=1}^{n/2} f_j e^{-2\pi\, i\, ky_j}. \tag{5.449}$$

Because $y_j = y_m$ for every atom, j, and its symmetry-equivalent, m, their contributions to the phase are identical; the atoms scatter in phase. For the diad screw axis, Eqn. 5.447 becomes

$$\begin{aligned}\mathbf{F}_{c,0k0} &= \sum_{j=1}^{n/2} f_j e^{-2\pi\, i\, ky_j} + \sum_{m=1}^{n/2} f_m e^{-2\pi\, ik(y_m+1/2)} \\ &= \sum_{j=1}^{n/2} f_j e^{-2\pi\, i\, ky_j} + \sum_{m=1}^{n/2} f_m e^{-2\pi\, i\, ky_m}\, e^{-k\pi i}.\end{aligned} \tag{5.450}$$

As before, when k is even, Eqn. 5.450 is identical to Eqn. 5.449, and the resultant waves scattered from symmetry-equivalent atoms are in phase with one another. *However*, when k is odd, Eqn. 5.450 becomes

$$\mathbf{F}_{c,0k0} = \sum_{j=1}^{n/2} f_j e^{-2\pi\, i\, ky_j} - \sum_{m=1}^{n/2} f_m e^{-2\pi\, i\, ky_m}. \tag{5.451}$$

For every atom in the first term, there is an identical atom in the second term, with $f_m = f_j$, and $y_m = y_j$. The terms cancel one another — the scattering from atom j is extinguished by the scattering from atom m; i.e., the waves scatter 180° out of phase and undergo complete destructive interference. Thus, for $0k0$ reflections with k odd ($k = 2n+1,\ n = 0, 1, 2, \ldots$), $\mathbf{F}_{c,0k0} = \mathbf{0}$ and $I_{0k0} = 0$. The reflection has an intensity of zero (ideally) and is said to be *systematically absent.* With parallel arguments it is easily shown that $h00$ reflections with h odd and $00l$ reflections with l odd result from two-fold screw axes parallel to a and c, respectively.

An analysis of the list of the $0k0$ (or $h00$ or $00l$) reflections will, in principle, allow us to distinguish between a two-fold rotation axis and a two-fold screw axis. If reflections are systematically observed for $k = 2n$, and systematically absent for $k = 2n+1$, then the presence of a diad screw axis is likely. There is a caveat here — the number of such reflections is often limited, and vibrational motions and other displacements can allow for some constructive interference; there may be some intensity for the formally absent reflections. If the systematically observed reflections are also weak, the assignment may be difficult. In such cases it may be necessary to attempt separate structural solutions incorporating a rotation axis in one case, and a screw axis in the other.

All screw axes exhibit systematic absences in the diffraction pattern, derivable by the method outlined above. The *International Tables For Crystallography* tabulates *reflection conditions*, rather than "absence" conditions. For a diad screw axis parallel to c, the reflection conditions for the $00l$ reflections require that $l = 2n,\ n = 1, 2, \ldots$; observed $00l$ reflections will have l evenly divisible by 2. For triad screw axes parallel to c, the $00l$ reflections are observed for $l = 3n$; l must be evenly divisible by 3. For tetrad and hexad screw axes parallel to c, reflection conditions require $l = 4n$ and $l = 6n$, respectively. Similar reflection conditions exist for screw axes parallel to a (for $h00$), and b (for $0k0$).

Glide Planes

The presence of a mirror plane perpendicular to the b axis and parallel to the ac plane (for example) has symmetry-equivalent atoms at (x_f, y_f, z_f) and $(x_f, -y_f, z_f)$, resulting in the structure factor expression for general indices:

$$\mathbf{F}_{c,hkl} = \sum_{j=1}^{n/2} f_j e^{-2\pi\, i(hx_j+ky_j+lz_j)} + \sum_{m=1}^{n/2} f_m e^{-2\pi\, i(hx_m-ky_m+lz_m)}. \tag{5.452}$$

The structure factor expression for the reflection with indices $(h, \bar{k}, l)$ is identical (again, after switching dummy indices):

$$
\begin{aligned}
\mathbf{F}_{c,h\bar{k}l} &= \sum_{j=1}^{n/2} f_j e^{-2\pi i(hx_j+\bar{k}y_j+lz_j)} + \sum_{m=1}^{n/2} f_m e^{-2\pi i(hx_m-\bar{k}y_m+lz_m)} \\
&= \sum_{j=1}^{n/2} f_j e^{-2\pi i(hx_j-ky_j+lz_j)} + \sum_{m=1}^{n/2} f_m e^{-2\pi i(hx_m+ky_m+lz_m)}, \qquad (5.453)
\end{aligned}
$$

and $I_{h\bar{k}l} = I_{hkl}$. Adding a glide translation parallel to the a axis results in symmetry-equivalent atoms at (x_f, y_f, z_f) and $((x_f + \frac{1}{2}), -y_f, z_f)$, with

$$
\begin{aligned}
\mathbf{F}_{c,hkl} &= \sum_{j=1}^{n/2} f_j e^{-2\pi i(hx_j+ky_j+lz_j)} + \sum_{m=1}^{n/2} f_m e^{-2\pi i(h(x_m+1/2)-ky_m+lz_m)} \\
&\qquad (5.454) \\
&= \sum_{j=1}^{n/2} f_j e^{-2\pi i(hx_j+ky_j+lz_j)} + \sum_{m=1}^{n/2} f_m e^{-2\pi i(hx_m-ky_m+lz_m)} e^{-h\pi i}.
\end{aligned}
$$

For k even Eqn. 5.454 is identical to Eqn. 5.453, and $I_{h\bar{k}l} = I_{hkl}$. For k odd, incorporating the same arguments as those used for the screw axis, $\mathbf{F}_{h\bar{k}l} = -\mathbf{F}_{hkl}$, and again, $I_{h\bar{k}l} = I_{hkl}$. Just as we observed with the screw axis, glide planes and mirror planes have the same effect on the diffraction pattern — they create a mirror plane in the intensity-weighted reciprocal lattice. The general reflections in the reciprocal lattice do not provide the information necessary to distinguish between glide and mirror planes in the direct lattice.

As before, a subset of reflections will assist us in determining the presence of a glide plane in the lattice. Setting k to zero eliminates the differentiating term due to the reflection (just as setting h and l eliminated the terms that changed sign during rotation when we considered screw axes), and we now consider the effect of the $h0l$ reflections on the diffraction pattern. For the mirror plane,

$$
\begin{aligned}
\mathbf{F}_{c,h0l} &= \sum_{j=1}^{n/2} f_j e^{-2\pi i\,(hx_j+lz_j)} + \sum_{m=1}^{n/2} f_m e^{-2\pi i\,(hx_m+lz_m)} \\
&= 2\sum_{j=1}^{n/2} f_j e^{-2\pi i\,(hx_j+lz_j)}, \qquad (5.455)
\end{aligned}
$$

since $(x_j, z_j) = (x_m, z_m)$ and $f_j = f_m$ for each symmetry-equivalent pair. For the a-glide plane,

$$
\mathbf{F}_{c,h0l} = \sum_{j=1}^{n/2} f_j e^{-2\pi i(hx_j+lz_j)} + \sum_{m=1}^{n/2} f_m e^{-2\pi i(hx_m+lz_m)} e^{-h\pi i}. \qquad (5.456)
$$

Just as we observed with the screw axis, when h is even, $e^{-h\pi i} = 1$, and Eqns. 5.456 and 5.455 are the same. The waves scattering from each atom and its symmetry-equivalent are in phase and they will constructively interfere. For h odd, $e^{-h\pi i} =$

-1, and for every atom in the first term, there is an identical atom in the second term, with $f_m = f_j$, and $(x_j, z_j) = (x_m, z_m)$:

$$\mathbf{F}_{c,h0l} = \sum_{j=1}^{n/2} f_j e^{-2\pi\, i(hx_j+lz_j)} - \sum_{m=1}^{n/2} f_m e^{-2\pi\, i(hx_m+lz_m)} = 0. \tag{5.457}$$

Waves scattered from pairs of symmetry-equivalent atoms destructively interfere, and $I_{h0l} = 0$ for h odd. For an a-glide plane perpendicular to the b axis and parallel to the ac plane, the $h0l$ reflections for $h = 2n + 1$ are systematically absent, and the refection conditions require that $h = 2n$. Similarly, for a c-glide plane parallel to the ac plane, the reflection conditions for the $h0l$ reflections require that $l = 2n$. There is usually a much larger number of reflections in the data set affected by the presence of a glide plane, and the assignment is ordinarily much more certain than it is in the case of a screw axis.

An n-glide plane parallel to the ac plane has symmetry-equivalent atoms at (x_f, y_f, z_f) and $((x_f + \frac{1}{2}), -y_f, (z_f + \frac{1}{2}))$. The resulting expression for $\mathbf{F}_{c,h0l}$ is identical to Eqn. 5.456, except that the exponential term is $e^{-h\pi i} e^{-l\pi i} = e^{-(h+l)\pi i}$. Reflection conditions for the n-glide plane are therefore $h + l = 2n$ for the $h0l$ reflections (see the treatment of centered lattices below for a more detailed proof). Reflection conditions for glide planes perpendicular to a (for $0kl$) and c (for $hk0$) are derived in the same manner as those for glide planes perpendicular to the b axis.

Centered Lattices

The presence of centering in a lattice is initially indicated during the cell reduction process (Sec. 4.6.1). Cell reduction is not always unambiguous, however, and the occurrence of systematic absences in the diffraction pattern can serve as a valuable check on the assignment of the Bravais lattice. Unlike translations in lattices described by primitive unit cells, translations resulting from centering alter the intensities of reflections with general indices. Consider, for example, a C-centered unit cell, containing symmetry-equivalent atoms at (x_f, y_f, z_f) and $((x_f + \frac{1}{2}), (y_f + \frac{1}{2}), z_f)$. The structure factor for a *general* reflection is

$$\begin{aligned}\mathbf{F}_{c,hkl} &= \sum_{j=1}^{n/2} f_j e^{-2\pi\, i(hx_j+ky_j+lz_j)} + \sum_{m=1}^{n/2} f_m e^{-2\pi\, i(h(x_m+1/2)+k(y_m+1/2)+lz_m)} \\ &= \sum_{j=1}^{n/2} f_j e^{-2\pi\, i(hx_j+ky_j+lz_j)} + \sum_{m=1}^{n/2} f_m e^{-2\pi\, i(hx_m+ky_m+lz_m)}\, e^{-(h+k)\pi i}.\end{aligned} \tag{5.458}$$

When $h + k$ is odd, $e^{-(h+k)\pi i} = -1$, $\mathbf{F}_{c,hkl} = \mathbf{0}$ and $I_{hkl} = 0$. The reflection conditions for a general reflection if the lattice is C centered therefore require that $h + k = 2n$. Similarly, for an A-centered lattice, only $k + l = 2n$ reflections can be observed, and for a B-centered lattice, only $k + l = 2n$ reflections can be observed. Observed reflections for an I-centered lattice are limited to those for which $h + k + l = 2n$. For a reflection to be observed from an F-centered lattice $h + k$, $k + l$, and $h + l$ must all be even, which constrains h, k, and l to all be either even or odd.

When data are collected from a crystal with a centered lattice, a very large number of reflections are systematically absent. When a serial detector is employed data collection is usually limited to those reflections that satisfy the reflection conditions since virtually half of the reflections will be unobserved.

Reflection Conditions

The International Tables For Crystallography lists reflection conditions for every space group. The entry for the $C\,2/c$ space group (Sec. 2.4.4) illustrates the general format:

Generators selected (1); t(1,0,0); t(0,1,0); t(0,0,1); t($\frac{1}{2}$,$\frac{1}{2}$,0); (2); (3)

Positions

Multiplicity, Wyckoff letter, Site Symmetry			Coordinates			Reflection conditions
			(0,0,0)+	($\frac{1}{2}$,$\frac{1}{2}$,0)+		General:
8	f	1	(1) (x, y, z)	(2) $(\bar{x}, y, \bar{z}+\frac{1}{2})$		
				(3) $(\bar{x}, \bar{y}, \bar{z})$	(4) $(x, \bar{y}, z+\frac{1}{2})$	$hkl : h+k=2n$
						$h0l : h, l=2n$
						$0kl : k=2n$
						$hk0 : h+k=2n$
						$0k0 : k=2n$
						$h00 : h=2n$
						$00l : l=2n$
						Special: as above, plus
4	e	2	$(0, y, \frac{1}{4})$	$(0, \bar{y}, \frac{3}{4})$		no extra conditions
4	d	$\bar{1}$	$(\frac{1}{4}, \frac{1}{4}, \frac{1}{2})$	$(\frac{3}{4}, \frac{1}{4}, 0)$		$hkl : k+l=2n$
4	c	$\bar{1}$	$(\frac{1}{4}, \frac{1}{4}, 0)$	$(\frac{3}{4}, \frac{1}{4}, \frac{1}{2})$		$hkl : k+l=2n$
4	b	$\bar{1}$	$(0, \frac{1}{2}, 0)$	$(0, \frac{1}{2}, \frac{1}{2})$		$hkl : l=2n$
4	a	$\bar{1}$	$(0, 0, 0)$	$(0, 0, \frac{1}{2})$		$hkl : l=2n$

The **Generators selected** line refers to the symmetry operations that are sequentially applied to generate all of the points that are symmetry-equivalent to a point in a general position, (x, y, z), denoted with a (1) label in the table. The translationally-equivalent points are generated with the "t" operations (e.g., $t(1, 0, 0)$), followed by the centering translation, t($\frac{1}{2}$,$\frac{1}{2}$,0), the diad screw operation (2) $(\bar{x}, y, \bar{z}+\frac{1}{2})$, and the inversion, (3)$(\bar{x}, \bar{y}, \bar{z})$. The remaining operation, (4) is redundant, having been generated by the combination of (2) and (3).

The **Positions** heading refers to the locations in the lattice of symmetry-equivalent points, and points on special positions within the lattice, that is, points that lie on symmetry elements. These positions were determined and tabulated by R. W. G. Wyckoff,[95] and are often referred to as *Wyckoff positions.* The two sets of coordinates followed by the "+" sign directly above the position vectors indicate that each position is determined by adding each of the translation vectors in parentheses to a position vector. For a general position, the diad axis applied to (x, y, z) produces position (2), and the inversion operation applied to positions (1) and (2) produces positions (3) and (4) respectively. Adding $(0, 0, 0)$ to these positions leaves them unchanged, and adding $(\frac{1}{2},\frac{1}{2},0)$ to each of these positions generates four more symmetry-equivalent positions due to centering. The symbols to the left of the general position vectors indicate the number of symmetry-equivalent locations for a general position, in this case 8. This number is referred to as the *multiplicity* of the general position. The italicized letter is the symbol assigned by Wyckoff for the type of position, and provides a shorthand reference to a specific position. The symbol that follows is the point symmetry of the position. For the general position this is always 1.

The reflection conditions are listed in the column to the right of the position vectors. In addition to the constraint imposed by centering on a general reflection, $h + k = 2n$, the c-glide plane requires $h = 2n$, and the n-glide plane requires $h + l = 2n$ for the $h0l$ reflections. For $h + l$ to be even, l must be even, since h is even. Thus the reflection condition requires $h, l = 2n$ for a reflection with $h0l$ indices. The $0kl$ reflections with $k = n$ are a subset of the general reflections with $h = 0$, as are the $hk0$ reflections with $h + k = 2n$. The diad screw axis parallel to b restricts the observed $0k0$ reflections to those with $k = 2n$, and the $h00$, $h = 2n$, and $00l$, $l = 2n$ conditions are two more subsets of the reflection conditions for the general hkl reflection.

Following the information related to the general position vectors, additional details regarding locations at special positions are listed in the table. All of the special positions in the $C2/c$ space group exhibit a multiplicity of 4; the symmetry operations bring a point back onto itself at each position. The e position is a point on the diad axis; the remaining special positions lie on inversion centers. The diad axis imposes no more constraints on the observed reflections, but the inversion centers do. For example, the wave scattered from an atom located at $(0, 0, 0)$ with indices $hkl, l = 2n+1$ will destructively interfere with a wave scattered from a symmetry-equivalent atom at $(0, 0, \frac{1}{2})$. Ordinarily, this will not result in a complete loss of intensity, since atoms that do not lie on the symmetry element will scatter waves that continue to interfere "normally". However, heavy atoms (atoms containing many electrons) often dominate the scattering from a crystal and if such an atom is located at a special position, many of the reflections can exhibit very low intensities. In this case, every other reflection could be affected. We will soon discover that the statistical distribution of the intensities is important in space group determination. Later on, in Chapter 7, we will make extensive use of reflection statistics for structural solution employing *direct methods.* The location of a heavy atom in a special position can significantly alter the intensity statistics, and can therefore influence the space group assignment *and* the structural solution.

5.3.2 Intensity Statistics

The most difficult symmetry element to detect is the inversion center. The Friedel Law imposes a center of symmetry on the diffraction pattern, regardless of whether or not it exists in the direct lattice. Because it is a point group operation, there are no special subgroups that might indicate its presence or absence, as there were for the glide planes and screw axes discussed above. Since there are no direct indicators of the inversion center, we take a more global view, focusing on the entire set of intensities, in order to assess whether or not there are any differences in the way they are *distributed* in non-centrosymmetric structures versus centrosymmetric structures. If there are any such differences, they will be reflected in their respective probability distributions.

Inversion Center

Rather than determining the probability distributions of the intensities for a data set, we will determine the probability density functions (Sec. 5.1.1) of the square roots of the intensities — the structure factor amplitudes, F_h. A general structure factor for a *non-centrosymmetric* structure containing n atoms is

$$\begin{aligned} \mathbf{F_h} &= \sum_{j=1}^{n} f_j \cos(-2\pi\mathbf{h}\cdot\mathbf{r}_j) + i\sum_{j=1}^{n} f_j \sin(-2\pi\mathbf{h}\cdot\mathbf{r}_j) \\ &= \sum_{j=1}^{n} X_j + i\sum_{j=1}^{n} Y_j \\ &= A_\mathbf{h} + i\,B_\mathbf{h}, \end{aligned} \tag{5.459}$$

where the atomic scattering factor, $f_j \equiv f_{j,\theta dT}$, includes the effects of atom displacement and dispersion.

There are two basic approaches to determining the probability distribution of F_h. The one most commonly seen in the literature[17] assumes that there is an equal probability of finding an atom at any point in the unit cell, creating random variables X_j and Y_j, such that the structure factor population that determines the distribution occurs from random selections of the values for $\mathbf{r}_j$, with $\mathbf{h}$ fixed. The resulting probability distribution is then assumed to be the same as the distribution for the actual population of structure factors observed for n fixed atoms, when $\mathbf{h}$ varies, rather than $\mathbf{r}_j$. A more intuitive approach, and one that more directly establishes the distribution of interest, is to treat the atom locations as fixed, in which case samples of the random variables are generated by selecting different values of $\mathbf{h}$.

The random variables will follow some kind of distribution; each variable X_j will have mean $\langle X_j\rangle$ and variance $\sigma^2(X_j)$; each variable Y_j will have mean $\langle Y_j\rangle$ and variance $\sigma^2(Y_j)$. Although we do not know the nature of these distributions, $\mathbf{F_h}$, $A_\mathbf{h}$, and $B_\mathbf{h}$ are all sums of these random variables, and therefore, according to the central limit theorem (Sec. 5.1.1), *each of them is normally distributed.* Since $\mathbf{F_h}$ is determined by $A_\mathbf{h}$ and $B_\mathbf{h}$ simultaneously, the probability density function for $\mathbf{F_h}$ is given by the bivariate normal distribution (Sec. 5.1.1), and since $A_\mathbf{h}$ and $B_\mathbf{h}$ are independent, from Eqn. 5.135,

$$\begin{aligned} f_G(\mathbf{F_h}) &= f_G(A_\mathbf{h}, B_\mathbf{h}) = \\ & \frac{1}{\sqrt{2\pi}\,\sigma_A} e^{-(A_\mathbf{h}-\langle A_\mathbf{h}\rangle)^2/2\sigma_A^2} \frac{1}{\sqrt{2\pi}\,\sigma_B} e^{-(B_\mathbf{h}-\langle B_\mathbf{h}\rangle)^2/2\sigma_B^2}. \end{aligned} \tag{5.460}$$

$A_\mathbf{h}$ has mean $\langle A_\mathbf{h} \rangle$, averaged over the all indices (denoted by the subscript, $\mathbf{h}$):

$$\begin{aligned} \langle A_\mathbf{h} \rangle_\mathbf{h} &= \left\langle \sum_{j=1}^{n} f_j \cos(-2\pi \mathbf{h} \cdot \mathbf{r}_j) \right\rangle_\mathbf{h} \\ &= \sum_{j=1}^{n} \langle f_j \cos(-2\pi \mathbf{h} \cdot \mathbf{r}_j) \rangle_\mathbf{h} . \end{aligned} \tag{5.461}$$

Similarly,

$$\langle B_\mathbf{h} \rangle_\mathbf{h} = \sum_{j=1}^{n} \langle f_j \sin(-2\pi \mathbf{h} \cdot \mathbf{r}_j) \rangle_\mathbf{h} . \tag{5.462}$$

Because f_j depends on $\mathbf{h}$, the average product in each term in the sum cannot be expressed as the product of the average values. However, *if we limit f_j to a region of relatively constant values for the magnitude of the diffraction vector*, as we did when scaling the relative intensities (Sec. 5.2), the cosine and sine terms will both average to approximately zero:

$$\langle A_\mathbf{h} \rangle_\mathbf{h} = 0 \tag{5.463}$$

$$\langle B_\mathbf{h} \rangle_\mathbf{h} = 0. \tag{5.464}$$

In this case, the subscript $\mathbf{h}$ indicates that the probability distribution is limited to diffraction vectors within a spherical shell in reciprocal space with a specific mean radius, h. In order to determine σ_A^2 and σ_B^2 within this region, we apply Eqn. 5.80:

$$\sigma_A^2 = \langle A_\mathbf{h}^2 \rangle - \langle A_\mathbf{h} \rangle^2 = \langle A_\mathbf{h}^2 \rangle . \tag{5.465}$$

$$\begin{aligned} &= \left\langle \sum_{j=1}^{n} f_j \cos(-2\pi \mathbf{h} \cdot \mathbf{r}_j) \sum_{m=1}^{n} f_m \cos(-2\pi \mathbf{h} \cdot \mathbf{r}_m) \right\rangle_\mathbf{h} \\ &= \left\langle \sum_{j=1}^{n} f_j^2 \cos^2(-2\pi \mathbf{h} \cdot \mathbf{r}_j) \right\rangle_\mathbf{h} \\ &\quad + \left\langle \sum_{j=1}^{n} \sum_{m=1}^{n} f_j f_m \cos(-2\pi \mathbf{h} \cdot \mathbf{r}_j) \cos(-2\pi \mathbf{h} \cdot \mathbf{r}_m) \right\rangle_\mathbf{h} . \end{aligned} \tag{5.466}$$

Setting $\eta_j = \cos(-2\pi \mathbf{h} \cdot \mathbf{r}_j)$ and $\eta_m = \cos(-2\pi \mathbf{h} \cdot \mathbf{r}_m)$, and incorporating Euler's expression for the cosine,

$$\begin{aligned} \cos\eta_j \cos\eta_m &= \frac{e^{i\eta_j} + e^{-i\eta_j}}{2} \frac{e^{i\eta_m} + e^{-i\eta_m}}{2} \\ &= \frac{1}{4}\left(e^{i(\eta_j+\eta_m)} + e^{i(\eta_j-\eta_m)} + e^{i(\eta_m-\eta_j)} + e^{-i(\eta_j+\eta_m)} \right) . \end{aligned} \tag{5.467}$$

Thus, the second term in Eqn. 5.466 for f_j and f_m approximately constant and $f_j \neq f_m$ becomes

$$\frac{1}{4} \sum_{j=1}^{n} \sum_{m=1}^{n} f_j f_m \left\langle \left(e^{i(\eta_j+\eta_m)} + e^{i(\eta_j-\eta_m)} + e^{i(\eta_m-\eta_j)} + e^{-i(\eta_j+\eta_m)} \right) \right\rangle_\mathbf{h} .$$

The second term now consists of four sets of sums of averages. The first of these is

$$\sum_{j=1}^{n}\sum_{m=1}^{n} f_j\, f_m \left\langle e^{-2\pi\, i\,(\mathbf{h}\cdot(\mathbf{r}_j+\mathbf{r}_m))}\right\rangle_{\mathbf{h}}. \tag{5.468}$$

Using arguments identical to those made in Sec. 5.2, each of these exponential terms will average to zero. It follows that the second term in Eqn. 5.466 averages to zero, and, for f_j approximately constant,

$$\sigma_A^2 = \sum_{j=1}^{n} f_j^2 \left\langle \cos^2(-2\pi\mathbf{h}\cdot\mathbf{r}_j)\right\rangle_{\mathbf{h}}. \tag{5.469}$$

Since we expect $\mathbf{h}$ to take on random orientations, $-2\pi\mathbf{h}\cdot\mathbf{r}_j$ will have values that will essentially vary continuously from zero to 2π, and we can approximate the average of $\cos^2(-2\pi\mathbf{h}\cdot\mathbf{r}_j)$ by considering $-2\pi\mathbf{h}\cdot\mathbf{r}_j$ to be a continuous variable, η_j.

The average value of $\cos^2\eta$ can be evaluated by integrating $\eta\cos^2\eta$ as we have shown previously for other functions, but this integral is not trivial to evaluate, and we take another approach to determining the average. Consider a function, $g(x)$. If we have n discrete values of $g(x)$ then the average value of the function for these values is

$$\langle g(x)\rangle = \frac{g(x_1)+g(x_2)+\cdots+g(x_n)}{n}. \tag{5.470}$$

If the average is between two limits, a and b, then we consider the x_i to be the values at the end of the interval between x_i and $x_{i+1} = x_i + \Delta x$, where $\Delta x = (b-a)/n$ (we have divided the interval into n equal increments). Since $n = (b-a)/\Delta x$,

$$\langle g(x)\rangle = \frac{(g(x_1)+g(x_2)+\cdots+g(x_n))\,\Delta x}{b-a} = \frac{1}{b-a}\sum_{i=1}^{n} g(x_i)\,\Delta x. \tag{5.471}$$

In the limit, as $n\to\infty$ and $\Delta x\to dx$,

$$\langle g(x)\rangle = \frac{1}{b-a}\int_a^b g(x_i)\,dx. \tag{5.472}$$

The integral of $\cos^2\eta$ is evaluated by substituting the identity $\cos^2\eta = \frac{1}{2}(1+\cos 2\eta)$. The interval over which we wish to average is between the limits $a = 0$ and $b = 2\pi$. Thus,

$$\begin{aligned}\int_0^{2\pi}\cos^2\eta\,d\eta &= \frac{1}{2}\int_0^{2\pi} d\eta + \frac{1}{2}\int_0^{2\pi}\cos 2\eta\,d\eta \\ &= \frac{2\pi}{2} - \frac{0}{2} + \frac{\sin 4\pi}{4} - \frac{\sin 0}{4} \\ &= \pi.\end{aligned} \tag{5.473}$$

The average value is then

$$\langle\cos^2\eta\rangle = \frac{1}{2\pi-0}\int_0^{2\pi}\cos^2\eta\,d\eta = \frac{\pi}{2\pi} = \frac{1}{2}. \tag{5.474}$$

Thus the variance of $A_\mathbf{h}$ in the region of approximately constant f_j reduces to

$$\sigma_A^2 = \frac{1}{2}\sum_{j=1}^{n} f_j^2. \tag{5.475}$$

Substituting $\mathbf{B_h}$ for $\mathbf{A_h}$, $\sin(-2\pi\mathbf{h}\cdot\mathbf{r}_j)$ for $\cos(-2\pi\mathbf{h}\cdot\mathbf{r}_j)$, and Euler's relations for the sine function results in the same value for the variance of $B_\mathbf{h}$*:

$$\sigma_B^2 = \frac{1}{2}\sum_{j=1}^{n} f_j^2. \tag{5.476}$$

The probability density function for the distribution of $\mathbf{F_h}$ for a spherical shell with radius $\approx h$ for which f_j is approximately constant is

$$\begin{aligned} f_G(\mathbf{F_h}) &= \frac{1}{\sqrt{2\pi}\,(\frac{1}{2}\sum f_j^2)^{1/2}}\, e^{-A_\mathbf{h}^2/2\frac{1}{2}\sum f_j^2} \\ &\quad\times \frac{1}{\sqrt{2\pi}\,(\frac{1}{2}\sum f_j^2)^{1/2}}\, e^{-B_\mathbf{h}^2/2\frac{1}{2}\sum f_j^2} \\ &= \frac{1}{\pi\sum f_j^2}\, e^{-(A_\mathbf{h}^2+B_\mathbf{h}^2)/\sum f_j^2}. \end{aligned} \tag{5.477}$$

The square of the magnitude of the structure factor is $F_\mathbf{h}^2 = A_\mathbf{h}^2 + B_\mathbf{h}^2$, giving the normal probability density function for the structure factor in terms of its magnitude:

$$f_G(\mathbf{F_h}) = \frac{1}{\pi\sum f_j^2}\, e^{-F_\mathbf{h}^2/\sum f_j^2}. \tag{5.478}$$

Prior to the solution of the crystal structure we have only experimentally determined structure factor amplitudes. While $\mathbf{F_h}$ is normally distributed, $F_\mathbf{h}$ need not be, since it is not represented as a sum of random variables. Given the normal probability density function, $f_G(\mathbf{F_h})$, we wish to determine $f(F_\mathbf{h})$. Eqn. 5.478 gives the probability per unit area of finding a structure factor vector, $\mathbf{F_h}$, at the location $(A_\mathbf{h}, B_\mathbf{h})$ in the complex plane. As illustrated in Fig. 5.37(a), the probability of a structure factor vector falling within the differential area, $dA_\mathbf{h}dB_\mathbf{h}$ is the probability per unit area times the area, $f_G(\mathbf{F_h})dA_\mathbf{h}dB_\mathbf{h}$. Any structure factor vector with the same magnitude will lie somewhere on a circular ring of radius $F_\mathbf{h}$ and thickness $dF_\mathbf{h}$, as depicted in Fig. 5.37(b). Because the $f_G(\mathbf{F_h})$ probability density depends on $F_\mathbf{h}$, it will be uniform for the entire ring. The probability that a vector with magnitude $F_\mathbf{h}$ will be found somewhere on the ring is the probability per unit area times the area of the ring, which is just its circumference times its thickness: $A_{ring} = 2\pi\, F_\mathbf{h}dF_\mathbf{h}$. Thus the probability of observing a structure factor with a magnitude $F_\mathbf{h}$ is

$$\begin{aligned} Pr_1(F_\mathbf{h}) &= \left(\frac{1}{\pi\sum f_j^2}\, e^{-F_\mathbf{h}^2/\sum f_j^2}\right) 2\pi\, F_\mathbf{h}dF_\mathbf{h} \\ &= \frac{2}{\sum f_j^2}\, F_\mathbf{h}\, e^{-F_\mathbf{h}^2/\sum f_j^2}\, dF_\mathbf{h}, \end{aligned} \tag{5.479}$$

*We also determine, by substituting $\sin^2\eta = \frac{1}{2}(1-\cos 2\eta)$, that $\langle\sin^2\eta\rangle = 1/2$.

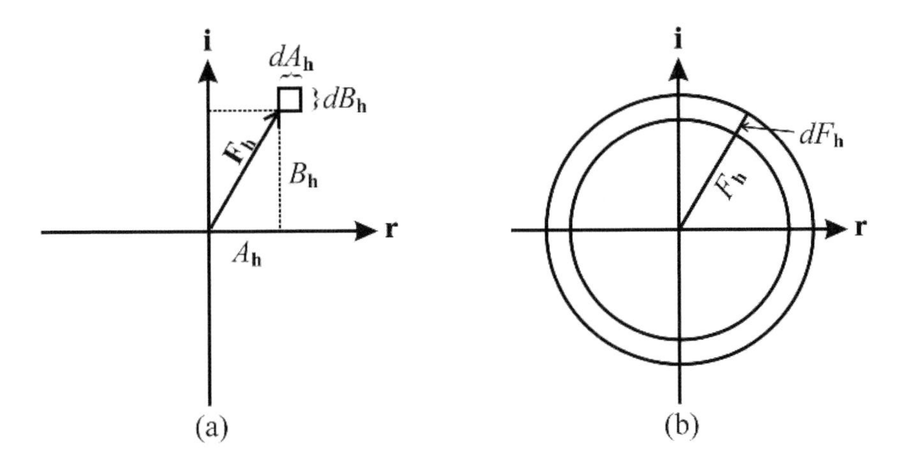

Figure 5.37 (a) $Pr_1(\mathbf{F_h})$ is the probability that there is a structure factor vector in the population $\{\mathbf{F_h}\}$ terminating somewhere within the area $dA_{\mathbf{h}}dB_{\mathbf{h}}$. (b) $Pr_1(F_{\mathbf{h}})$ is the probability that there is a structure factor vector with magnitude $F_{\mathbf{h}}$ that will terminate somewhere within the circular ring with area $2\pi F_{\mathbf{h}}dF_{\mathbf{h}}$.

and the probability density function for the structure factor magnitude from a non-centrosymmetric crystal is

$$f_1(F_{\mathbf{h}}) = \frac{2}{\sum f_j^2} F_{\mathbf{h}}\, e^{-F_{\mathbf{h}}^2/\sum f_j^2}. \tag{5.480}$$

The subscript "1" refers to the lack of an inversion center in the crystal.

There are constraints on the directions of the structure factor vectors from a centrosymmetric structure in the complex plane that are not suffered by those from a non-centrosymmetric structure, and we might therefore suspect that their probability distributions will differ. When an inversion center is present, for every $\mathbf{r}_j$ there is a $-\mathbf{r}_j$ and

$$\begin{aligned}\mathbf{F_h} &= \sum_{j=1}^{n/2} f_j \cos(-2\pi\mathbf{h}\cdot\mathbf{r}_j) + i\sum_{j=1}^{n/2} f_j \sin(-2\pi\mathbf{h}\cdot\mathbf{r}_j)\\ &\quad + \sum_{j=1}^{n/2} f_j \cos(+2\pi\mathbf{h}\cdot\mathbf{r}_j) + i\sum_{j=1}^{n/2} f_j \sin(+2\pi\mathbf{h}\cdot\mathbf{r}_j).\end{aligned} \tag{5.481}$$

Since $\cos(-\eta) = \cos\eta$ and $\sin(-\eta) = -\sin\eta$, the sine terms cancel and

$$\mathbf{F_h} = \sum_{j=1}^{n/2} 2f_j \cos(-2\pi\mathbf{h}\cdot\mathbf{r}_j). \tag{5.482}$$

There is no imaginary component for the centrosymmetric structure factor vector; it must point in either the positive or negative direction along the real axis, with

$\mathbf{F_h} = \pm F_\mathbf{h}$. The probability density function for the centrosymmetric structure factor is therefore described by the univariate normal distribution:

$$f_G(\mathbf{F_h}) = \frac{1}{\sqrt{2\pi}\,\sigma_F} e^{-(F_\mathbf{h}-\langle F_\mathbf{h}\rangle)^2/2\sigma_F^2} \text{ and} \tag{5.483}$$

$$f_G(F_\mathbf{h}) = 2\,f_G(\mathbf{F_h}) = \frac{2}{\sqrt{2\pi}\,\sigma_F} e^{-(F_\mathbf{h}-\langle F_\mathbf{h}\rangle)^2/2\sigma_F^2}. \tag{5.484}$$

The factor of 2 in Eqn. 5.484 occurs because for every $\mathbf{F_h}$ there is an equally probable $-\mathbf{F_h}$, both with magnitude $F_\mathbf{h}$.

The cosine terms average to zero for $F_\mathbf{h}$ just as they did for $A_\mathbf{h}$ in the non-centrosymmetric case, and

$$\langle F_\mathbf{h}\rangle = 0. \tag{5.485}$$

The variance is given by

$$\sigma_F^2 = \langle F_\mathbf{h}^2\rangle - \langle F_\mathbf{h}\rangle^2 = \langle F_\mathbf{h}^2\rangle. \tag{5.486}$$

$$\tag{5.487}$$

As before, the cross-terms in $F_\mathbf{h}^2$ average to zero, and for approximately constant values of f_j and h,

$$\begin{aligned}\sigma_F^2 &= \sum_{j=1}^{n/2} (2f_j)^2 \left\langle \cos^2(-2\pi\mathbf{h}\cdot\mathbf{r}_j)\right\rangle_\mathbf{h} \\ &= 4\frac{1}{2}\sum_{j=1}^{n/2} f_j^2 = 2\sum_{j=1}^{n/2} f_j^2 \\ &= \sum_{j=1}^{n} f_j^2.\end{aligned} \tag{5.488}$$

Thus the probability density function for the structure factor magnitude from a centrosymmetric crystal *is normally distributed*:

$$f_{\bar{1}}(F_\mathbf{h}) = \frac{2}{\sqrt{2\pi}(\sum f_j^2)^{1/2}} e^{-F_\mathbf{h}^2/2\sum f_j^2}. \tag{5.489}$$

Fig. 5.38 illustrates the probability density functions for a selected value of $\sum f_j^2$ for the distributions of structure factor amplitudes from a non-centrosymmetric crystal (in black) and a centrosymmetric crystal (in red). The distribution function for the centrosymmetric structure is the positive half of the symmetric Gaussian curve, centered on its mean, zero; the distribution function is described as a *centric distribution.* The distribution function for a noncentrosymmetric structure is not centered about its mean, and is termed an *acentric distribution.* These terms are *not* interchangeable with the structural properties to which they refer. The terms *centric* and *acentric* always refer to probability distributions.

These distribution functions are quite different, and it now appears that a plot of the fractional (probability) distribution of structure factor amplitudes will tell us whether or not a crystal is centrosymmetric. However, the distribution functions

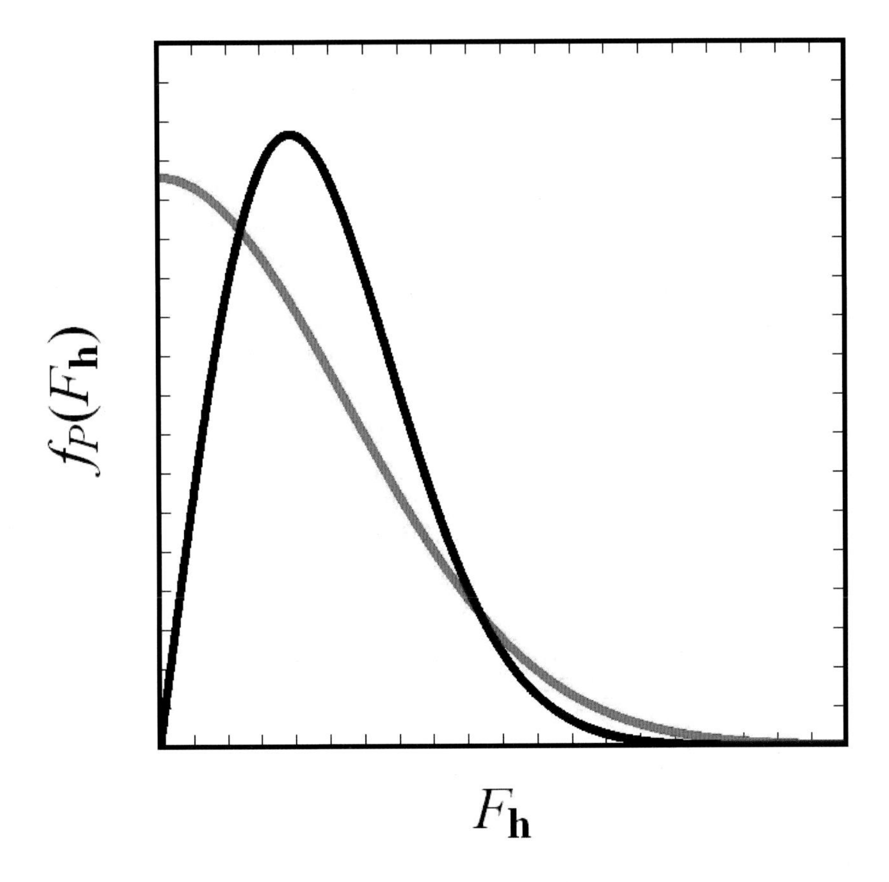

Figure 5.38 Probability density functions for an acentric probability distribution of $F_{\mathbf{h}}$, $f_1(F_{\mathbf{h}})$, (black) and a centric probability distribution of $F_{\mathbf{h}}$, $f_{\bar{1}}(F_{\mathbf{h}})$, (red).

contain a parameter that depends on the diffraction angle, $\sum f_j^2$. In turn, $\sum f_j^2$ depends on the number and types of atoms in the unit cell *and* the thermal (and other) displacements of those atoms. It is therefore difficult to establish statistical criteria to test for an inversion center from the structure factor amplitudes themselves, since they are dependent on different parameters for every crystal. Thus it appears reasonable to attempt to find a way to modify the observed structure factors in a manner that will render them independent of the diffraction angle and atomic displacements, yet still be comparable to calculated structure factors based upon a model of the unit cell containing atom locations. The model will therefore also necessarily be independent of diffraction angle and atom displacements.

The calculated structure factor, based on the spherical atom model, is

$$\mathbf{F}_{c,\mathbf{h}} = \sum_{j=1}^{n} f_j \, e^{-2\pi i (\mathbf{h} \cdot \mathbf{r}_j)}, \tag{5.490}$$

where, again, the atomic scattering factor, $f_j \equiv f_{j,\theta dT}$, includes the effects of atom displacement and dispersion. Because of the finite size of the atoms and their thermal motions, the scattering factors diminish as the diffraction angle increases, due to destructive interference from different regions within each atom. Now, consider what would happen if the electrons for each atom in a crystal were all concentrated to a point at the center of the atom, and that these *point atoms* were at absolute zero, unable to undergo thermal motion. Of course no such structure is possible, but if one did exist there would be no destructive interference from different regions of the atom; each atom would scatter every reflection with the full power of all its electrons at any diffraction angle. For point atom j, the scattering factor, f_j, would simply be the atomic number of the atom, Z_j. The resulting calculated structure factors would no longer depend on the diffraction angle — they would be independent of $\sum f_j^2$:

$$\mathbf{F}^p_{c,\mathbf{h}} = \sum_{j=1}^{n} Z_j \, e^{-2\pi i\,(\mathbf{h}\cdot\mathbf{r}_j)}. \tag{5.491}$$

To create observed structure factors that are independent of the diffraction angle we make the assertion that the effects of destructive interference due to atom size and displacement for a reflection collected within a spherical shell in reciprocal space defined by h will be closely approximated by the effects on the average intensity within that region. In other words, if the average intensity diminishes by a certain fraction at a given diffraction angle, since each intensity depends on the same scattering factors at the same angle, each reflection will be diminished by approximately the same fraction. If, for example, an intensity $I_o(h)$ is diminished by a fraction, g, compared to the intensity that it would have if it could be measured at $h = 0$, $I_o(0)$, then the average value for all of the reflections with the same h, $\langle I_o(h)\rangle$, will also be diminished by the same fraction:

$$\begin{aligned} I_o(h) &= g\, I_o(0) \\ \langle I_o(h)\rangle &= g\,\langle I_o(0)\rangle \\ \frac{I_o(h)}{\langle I_o(h)\rangle} &= \frac{g\, I_o(0)}{g\,\langle I_o(0)\rangle} = \frac{I_o(0)}{\langle I_o(0)\rangle} = I_E. \end{aligned} \tag{5.492}$$

Thus, dividing each intensity by the average value of the intensity for each value of h creates an entity that is independent of the value of h, i.e., does not depend on the diffraction angle. Dividing by the average value in the region also places these entities on a common scale. The result is a set of new "intensities", I_E, that are independent of the diffraction angle and are more or less "normalized." In order to determine their properties we consider the experimental intensities and normalized intensities in terms of their corresponding structure factors. $\mathbf{F}_{o,\mathbf{h}}$ is the scaled, experimentally observed structure factor; $I_{o,\mathbf{h}} = F^2_{o,\mathbf{h}}$. We define $\mathbf{E}_{o,\mathbf{h}}$ as the *normalized structure factor*,* $I_{nor,\mathbf{h}} = E^2_{o,\mathbf{h}}$. $\mathbf{E}_{o,\mathbf{h}}$ is a vector pointing in the same direction as $\mathbf{F}_{o,\mathbf{h}}$ (i.e., they both have the same phase), with magnitude $E^2_{o,\mathbf{h}}$.

*In order for I_E to be rigorously normalized, its Euclidean norm, $\sqrt{\sum I_E^2}$, would be unity. Instead, $\langle I_E\rangle = 1$ and I_E is more appropriately described as *quasi-normalized.* We will, however, use the simpler terminology, with the more formal description considered to be implied. The square root of I_E is not normalized at all, but $E^2 = I_E$ is quasi-normalized, and common usage refers to E as a normalized (or quasi-normalized) structure factor.

From to Eqn. 5.440, the average intensity for a specific region is given by

$$\begin{aligned} \langle I_{o,\mathbf{h}}\rangle &= e^{-2\mathrm{B}'\langle h\rangle^2}\sum_{j=1}^{n}\langle f_{j,\theta\, d}\rangle^2 \\ &= \sum_{j=1}^{n} f_{j,\theta\, d}^2 e^{-2\mathrm{B}'h^2} = \sum_{j=1}^{n} f_{j,\theta\, dT}^2 = \sum_{j=1}^{n} f_j^2 = \sigma_{F_\mathbf{h}}^2. \end{aligned} \tag{5.493}$$

Eqn. 5.493 gives the average intensity for a general reflection, but reflections with intensities affected by symmetry must be given special consideration. For example, consider a unit cell with a mirror plane perpendicular to the b axis. The structure factor for the hkl reflection is

$$\mathbf{F}_{hkl} = \sum_{j=1}^{n/2} f_j e^{-2\pi\, i(hx+ky+lz)} + \sum_{j=1}^{n/2} f_j e^{-2\pi\, i(hx-ky+lz)}. \tag{5.494}$$

For the $h0l$ reflection it is

$$\begin{aligned} \mathbf{F}_{hkl} = 2\sum_{j=1}^{n/2} f_j e^{-2\pi ß(hx+lz)} &= \sum_{j=1}^{n/2}(2f_j)\cos(-2\pi(hx+lz)) \\ &+ i\sum_{j=1}^{n/2}(2f_j)\sin(-2\pi(hx+lz)). \end{aligned}$$

From Eqn. 5.469,

$$\begin{aligned} \sigma_A^2 &= \sum_{j=1}^{n/2}(2f_j)^2\langle\cos^2(-2\pi ß(hx+lz))\rangle = \sum_{j=1}^{n/2} 4f_j^2\left(\frac{1}{2}\right) \\ &= 2\sum_{j=1}^{n/2} f_j^2 = \sum_{j=1}^{n} f_j^2. \end{aligned} \tag{5.495}$$

Similarly,

$$\begin{aligned} \sigma_B^2 &= \sum_{j=1}^{n/2} 4f_j^2\langle\sin^2(-2\pi ß(hx+lz))\rangle = \sum_{j=1}^{n/2} 4f_j^2\left(\frac{1}{2}\right) = \sum_{j=1}^{n} f_j^2 \quad \text{and} \\ \sigma_{F_\mathbf{h}}^2 &= \sigma_A^2+\sigma_B^2 = \sum_{j=1}^{n} f_j^2 + \sum_{j=1}^{n} f_j^2 \Longrightarrow \langle I_{o,\mathbf{h}}\rangle = 2\sum_{j=1}^{n} f_j^2. \end{aligned} \tag{5.496}$$

In general,

$$\langle I_{o,\mathbf{h}}\rangle = \epsilon\sum_{j=1}^{n} f_j^2, \tag{5.497}$$

where $\epsilon = 1$ for the general reflection. For special classes of reflections ϵ varies from 1 to 8, depending on the space group symmetry.[96] Eqn. 5.492 can now be expressed in terms of structure factor magnitudes:

$$E_{o,\mathbf{h}}^2 = \frac{F_{o,\mathbf{h}}^2}{\epsilon\sum_{j=1}^{n} f_j^2}. \tag{5.498}$$

Referring to Fig. 4.32, we note that when the scattering factors for various atoms tail off as the diffraction angle increases, their ratios remain about the same. For heavy atoms, this is a rough approximation — the tail-off does not occur as rapidly as it does for light atoms since the core electrons of the heavy atom are subject to a higher effective nuclear charge, and are more tightly bound to the nucleus (less spread out). For a structure with light atoms the approximation becomes a relatively good one, and in the limit, for an equal-atom structure, it is, of course, exact. For the sake of the argument here, we will assume that the ratios of the scattering factors remain constant and equal to the ratios of the atomic numbers of the atoms (where they begin at $h = 0$). Thus, if $f_{\mathrm{H}}(h)$ is the scattering factor for the hydrogen atom at a given diffraction angle, then $f_{\mathrm{He}}(h) \approx 2f_{\mathrm{H}}(h)$, $f_{\mathrm{Li}}(h) \approx 3f_{\mathrm{H}}(h)$, $f_{\mathrm{C}}(h) \approx 6f_{\mathrm{H}}(h)$, $f_{\mathrm{Cl}}(h) \approx 17f_{\mathrm{H}}(h)$, etc. In general, for atom j, we assume $f_{j,\theta\,d}(h) = Z_j f_{\mathrm{H}}(h)$. Adding in the effects of temperature using the overall isotropic displacement factor gives:

$$f_j(h) = Z_j\, f_{\mathrm{H}}\, e^{-B'h^2} \quad \text{and} \tag{5.499}$$

$$\begin{aligned} E^2_{o,\mathbf{h}} &= \frac{\left(\displaystyle\sum_{j=1}^{n} Z_j\, f_{\mathrm{H}}\, e^{-B'h^2} e^{-2\pi\, i(\mathbf{h}\cdot\,\mathbf{r}_j)}\right)^2}{\epsilon \displaystyle\sum_{j=1}^{n} Z_j^2\, f_{\mathrm{H}}^2\, e^{-2B'h^2}} \\ &= \frac{f_h^2\, e^{-2B'h^2} \left(\displaystyle\sum_{j=1}^{n} Z_j\, e^{-2\pi\, i(\mathbf{h}\cdot\,\mathbf{r}_j)}\right)^2}{f_h^2\, e^{-2B'h^2} \epsilon \displaystyle\sum_{j=1}^{n} Z_j^2} \\ &= \frac{\left(\displaystyle\sum_{j=1}^{n} Z_j\, e^{-2\pi\, i(\mathbf{h}\cdot\,\mathbf{r}_j)}\right)^2}{\epsilon \displaystyle\sum_{j=1}^{n} Z_j^2}. \end{aligned} \tag{5.500}$$

The expression in the numerator is identical to that in Eqn. 5.491, and the square of the magnitude of the normalized structure factor is seen to be the square of the magnitude of the point-atom structure factor divided by a constant:

$$E^2_{o,\mathbf{h}} = \frac{(F^p_{o,\mathbf{h}})^2}{\epsilon \displaystyle\sum_{j=1}^{n} Z_j^2}. \tag{5.501}$$

We are now in a position to generate a set of calculated normalized point-atom structure factors from a model based on atom locations. From Eqn. 5.491,

$$\mathbf{E}_{c,\mathbf{h}} = \frac{\mathbf{F}^p_{c,\mathbf{h}}}{\left(\epsilon \sum_{j=1}^{n} Z_j^2\right)^{1/2}} \quad \text{and} \quad E_{c,\mathbf{h}} = (\mathbf{E}^*_{c,\mathbf{h}} \mathbf{E}_{c,\mathbf{h}})^{1/2}. \tag{5.502}$$

The calculated structure factor magnitudes can be then compared to observed normalized point-atom structure factor magnitudes generated from the diffraction data set:

$$E_{o,\mathbf{h}} = \left(\frac{F_{o,\mathbf{h}}^2}{\epsilon \sum_{j=1}^{n} f_j^2} \right)^{1/2}. \tag{5.503}$$

The main accomplishment here (perhaps lost in the derivation) is that we now have structure factor magnitudes that do not depend on the diffraction angle. *The statistical distribution of $E_{\mathbf{h}}$ can be analyzed for the entire data set.*

For a noncentrosymmetric structure, the point-atom structure factor is

$$\begin{aligned} \mathbf{F}_{\mathbf{h}}^{p} &= \sum_{j=1}^{n} Z_j \cos(-2\pi \mathbf{h} \cdot \mathbf{r}_j) + i \sum_{j=1}^{n} Z_j \sin(-2\pi \mathbf{h} \cdot \mathbf{r}_j) \\ &= A_{\mathbf{h}}^{p} + i\, B_{\mathbf{h}}^{p}, \end{aligned} \tag{5.504}$$

Following identical arguments to those given previously for $\mathbf{F}_{\mathbf{h}}$, $\langle A_{\mathbf{h}}^{p} \rangle = 0$ and $\langle B_{\mathbf{h}}^{p} \rangle = 0$. In addition,

$$\sigma_{A^p}^2 = \langle (A_{\mathbf{h}}^{p})^2 \rangle = \frac{1}{2} \sum_{j=1}^{n} Z_j^2 \quad \text{amd} \quad \sigma_{B^p}^2 = \langle (B_{\mathbf{h}}^{p})^2 \rangle = \frac{1}{2} \sum_{j=1}^{n} Z_j^2. \tag{5.505}$$

The normalized structure factor* is

$$\mathbf{E}_{\mathbf{h}} = \frac{\mathbf{F}_{\mathbf{h}}^{p}}{\left(\sum Z_j^2\right)^{1/2}} = \frac{A_{\mathbf{h}}^{p}}{\left(\sum Z_j^2\right)^{1/2}} + i \frac{B_{\mathbf{h}}^{p}}{\left(\sum Z_j^2\right)^{1/2}} = A_{\mathbf{h}}^{E} + i\, B_{\mathbf{h}}^{E}. \tag{5.506}$$

It follows that $\langle A_{\mathbf{h}}^{E} \rangle = 0$ and $\langle B_{\mathbf{h}}^{E} \rangle = 0$, and the variances of $A_{\mathbf{h}}^{E}$ and $B_{\mathbf{h}}^{E}$ are

$$\sigma_A^2 = \langle (A_{\mathbf{h}}^{E})^2 \rangle = \frac{\langle (A_{\mathbf{h}}^{p})^2 \rangle}{\sum Z_j^2} = \frac{1}{2} \quad \text{and} \tag{5.507}$$

$$\sigma_B^2 = \langle (B_{\mathbf{h}}^{E})^2 \rangle = \frac{\langle (B_{\mathbf{h}}^{p})^2 \rangle}{\sum Z_j^2} = \frac{1}{2}. \tag{5.508}$$

The bivariate normal distribution for $\mathbf{E}_{\mathbf{h}}$ gives

$$f_G(\mathbf{E}_{\mathbf{h}}) = \frac{1}{\sqrt{2\pi}\, \sigma_A} e^{-(A_{\mathbf{h}}^{E} - \langle A_{\mathbf{h}}^{E} \rangle)^2 / 2\sigma_A^2} \frac{1}{\sqrt{2\pi}\, \sigma_B} e^{-(B_{\mathbf{h}}^{E} - \langle B_{\mathbf{h}}^{E} \rangle)^2 / 2\sigma_B^2} \tag{5.509}$$

$$= \frac{1}{\pi} e^{-((A_{\mathbf{h}}^{E})^2 + (B_{\mathbf{h}}^{E})^2)} \tag{5.510}$$

$$= \frac{1}{\pi} e^{-E_{\mathbf{h}}^2}. \tag{5.511}$$

By direct analogy with Eqn. 5.479, the probability of observing a normalized structure factor with a magnitude $E_{\mathbf{h}}$ is

$$\begin{aligned} Pr_1(E_{\mathbf{h}}) &= \left(\frac{1}{\pi} e^{-E_{\mathbf{h}}^2} \right) 2\pi\, E_{\mathbf{h}}\, dE_{\mathbf{h}} \\ &= 2\, E_{\mathbf{h}}\, e^{-E_{\mathbf{h}}^2}\, dE_{\mathbf{h}}, \end{aligned} \tag{5.512}$$

*The symmetry factor, ϵ, will be assumed to correspond to the general reflection ($\epsilon = 1$) for discussions throughout the remainder of the text.

and the probability density function for the normalized structure factor magnitudes from a non-centrosymmetric crystal is

$$f_1(E_{\mathbf{h}}) = 2\,E_{\mathbf{h}}\,e^{-E_{\mathbf{h}}^2}. \tag{5.513}$$

For a centrosymmetric structure the point-atom structure factor is

$$\mathbf{F}_{\mathbf{h}}^p = \sum_{j=1}^{n} Z_j \cos(-2\pi\mathbf{h}\cdot\mathbf{r}_j), \tag{5.514}$$

$\langle F_{\mathbf{h}}^p\rangle = 0$, and by analogy with Eqn. 5.488, $\sigma_{F^p}^2 = \langle(F_{\mathbf{h}}^p)^2\rangle = \sum_{j=1}^{n} Z_j^2$. Thus, $\langle E_{\mathbf{h}}\rangle = 0$ and

$$\sigma_E^2 = \langle E_{\mathbf{h}}^2\rangle = \frac{\langle(F_{\mathbf{h}}^p)^2\rangle}{\sum Z_j^2} = 1. \tag{5.515}$$

The probability density function for the normalized structure factor magnitudes from a centrosymmetric crystal is normally distributed:

$$f_{\bar{1}}(E_{\mathbf{h}}) = \frac{2}{\sqrt{2\pi}\sqrt{1}}\,e^{-E_{\mathbf{h}}^2/2(1)} = \frac{2}{\sqrt{2\pi}}\,e^{-E_{\mathbf{h}}^2/2}. \tag{5.516}$$

It is important to note here that *the probability distributions of the normalized structure factors for both centrosymmetric and noncentrosymmetric structures are completely general.* It is in this sense that the Es are said to be normalized, since their probability distributions are ideally independent of the details of any particular structure. The distributions are plotted in Fig. 5.39.

According to Eqn. 5.515, for centrosymmetric structures, $\langle E_{\mathbf{h}}^2\rangle = 1$. For a non-centrosymmetric structure,

$$\langle E_{\mathbf{h}}^2\rangle = \langle(A_{\mathbf{h}}^E)^2 + (B_{\mathbf{h}}^E)^2\rangle = \langle(A_{\mathbf{h}}^E)^2\rangle + \langle(B_{\mathbf{h}}^E)^2\rangle = \frac{1}{2} + \frac{1}{2} = 1. \tag{5.517}$$

While $\langle E_{\mathbf{h}}^2\rangle$ does not serve to differentiate between centric and acentric distributions, in theory, $\langle E_{\mathbf{h}}\rangle$ does. For an acentric distribution,

$$\langle E_{\mathbf{h}}\rangle_1 = \int_0^\infty E_{\mathbf{h}}\left(2\,E_{\mathbf{h}}\,e^{-E_{\mathbf{h}}^2}\right)dE_{\mathbf{h}} = 2\int_0^\infty E_{\mathbf{h}}^2\,e^{-E_{\mathbf{h}}^2}\,dE_{\mathbf{h}}. \tag{5.518}$$

Letting $u = E_{\mathbf{h}}^2$ such that $dE_{\mathbf{h}} = du/2E_{\mathbf{h}}$ puts the integral in the form of a standard definite integral:

$$\langle E_{\mathbf{h}}\rangle_1 = 2\int_0^\infty u\,e^{-u}\frac{du}{2E_{\mathbf{h}}} = \int_0^\infty \frac{u}{\sqrt{u}}\,e^{-u}\,du,$$

which evaluates to

$$\langle E_{\mathbf{h}}\rangle_1 = \int_0^\infty \sqrt{u}\,e^{-u}\,du = \frac{\sqrt{\pi}}{2} = 0.886. \tag{5.519}$$

For a centric distribution,

$$\langle E_{\mathbf{h}}\rangle_{\bar{1}} = \int_0^\infty E_{\mathbf{h}}\left(\frac{2}{\sqrt{2\pi}}\,e^{-E_{\mathbf{h}}^2/2}\right)dE_{\mathbf{h}} = \frac{2}{\sqrt{2\pi}}\int_0^\infty E_{\mathbf{h}}\,e^{-E_{\mathbf{h}}^2/2}\,dE_{\mathbf{h}}. \tag{5.520}$$

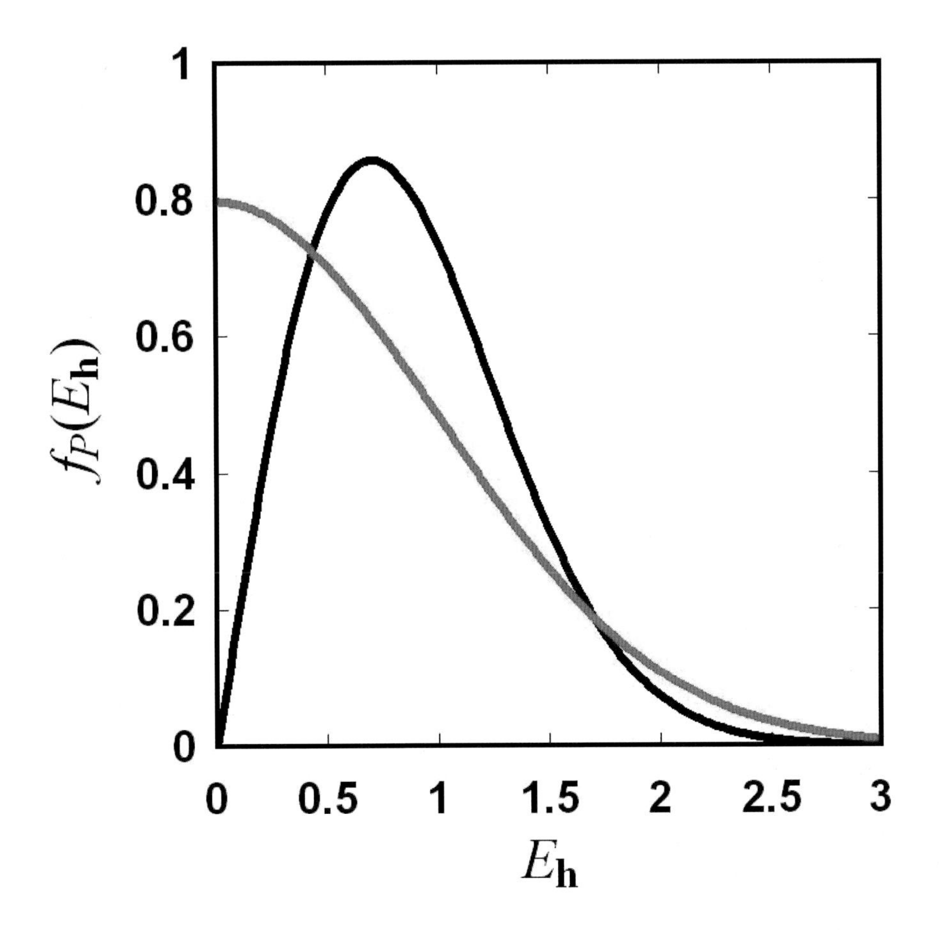

Figure 5.39 Probability density functions for an acentric probability distribution of $E_{\mathbf{h}}$, $f_1(E_{\mathbf{h}})$, (black) and a centric probability distribution of $E_{\mathbf{h}}$, $f_{\bar{1}}(E_{\mathbf{h}})$, (red).

Letting $u = E_{\mathbf{h}}^2/2$ gives $du = E_{\mathbf{h}}\,dE_{\mathbf{h}}$ and

$$\begin{aligned}\langle E_{\mathbf{h}}\rangle_{\bar{1}} &= \frac{2}{\sqrt{2\pi}}\int_0^{\infty} e^{-u}\,du \\ &= -\frac{2}{\sqrt{2\pi}}e^{-u}\Big|_0^{\infty} = \frac{2}{\sqrt{2\pi}} = 0.798. \qquad (5.521)\end{aligned}$$

In principle, then, $\langle E_{\mathbf{h}}\rangle$ should allow us to detect the presence of an inversion center. However, the theoretical averages are relatively close to one another, and it is often best to evaluate the entire distribution of the normalized structure factors. One of the most straightforward ways to do this is to analyze their *cumulative probability distribution*, that is, to determine for each possible value of $E_{\mathbf{h}} = t$, the probability that there will be an $E_{\mathbf{h}}$ less than or equal to t. This will be the fraction of all the $E_{\mathbf{h}}$s that fall within the range $0 \rightarrow t$:

$$Pr(0 \rightarrow t) = \int_0^t f(E_{\mathbf{h}})\,dE_{\mathbf{h}}. \qquad (5.522)$$

For an acentric distribution,

$$Pr_1(0 \rightarrow t) = 2\int_0^t E_{\mathbf{h}}\, e^{-E_{\mathbf{h}}^2}\, dE_{\mathbf{h}}. \tag{5.523}$$

This integral is of the form

$$\int x\, e^{-ax^2}\, dx = -\frac{1}{2a}\, e^{-ax^2}, \tag{5.524}$$

with $x = E_{\mathbf{h}}^2$ and $a = 1$, so that

$$Pr_1(0 \rightarrow t) = 2\left(-\frac{1}{2}e^{-E_{\mathbf{h}}^2}\right)\Bigg|_0^t = 1 - e^{-t^2}. \tag{5.525}$$

For a centric distribution,

$$Pr_{\bar{1}}(0 \rightarrow t) = \frac{2}{\sqrt{2\pi}}\int_0^t e^{-E_{\mathbf{h}}^2/2}\, dE_{\mathbf{h}}. \tag{5.526}$$

This is a normal probability distribution integrated between limits. Setting $u = E_{\mathbf{h}}/\sqrt{2}$ and $du = dE_{\mathbf{h}}/\sqrt{2}$, then integrating from zero to $t/\sqrt{(2)}$ gives,

$$Pr_{\bar{1}}(0 \rightarrow t) = \sqrt{2}\frac{2}{\sqrt{2\pi}}\int_0^{t/\sqrt{2}} e^{-u^2}\, du = \frac{2}{\sqrt{\pi}}\int_0^{t/\sqrt{2}} e^{-u^2}\, du. \tag{5.527}$$

Referring to Eqn. 5.70, this is just $\mathrm{erf}(t/\sqrt{2})$, the error function, which can be calculated by numerical integration or obtained readily from published mathematical tables. Thus

$$Pr_{\bar{1}}(0 \rightarrow t) = \mathrm{erf}\left(\frac{t}{\sqrt{2}}\right). \tag{5.528}$$

Fig. 5.40 illustrates the cumulative probability distributions for theoretical distributions of normalized structure factors for a non-centrosymmetric structure (in blue), a centrosymmetric structure (black), and for the actual centrosymmetric structure of an organometallic complex, $CrC_{40}H_{30}O_4P_2$, which crystallizes in the $P\bar{1}$ space group. * This example is unequivocal, but in many cases the distinction is not as clear. A measure of the goodness of fit is often useful in deciding which distribution best represents the experimental data:

$$g_1 = \left(\frac{1}{m}\sum_m\left(1 - e^{-E_{\mathbf{h}}^2} - \frac{n_E}{N_E}\right)^2\right)^{1/2} \tag{5.529}$$

$$g_{\bar{1}} = \left(\frac{1}{m}\sum_m\left(\mathrm{erf}\left(\frac{E_{\mathbf{h}}}{\sqrt{2}}\right) - \frac{n_E}{N_E}\right)^2\right)^{1/2}, \tag{5.530}$$

where n_E is the number of normalized structure factors with values less than or equal to $E_{\mathbf{h}} = t$, N_E is the total number of normalized structure factors, and m

*see Fig. 7.24 for an example of an experimental cumulative probability distribution for a non-centrosymmetric structure.

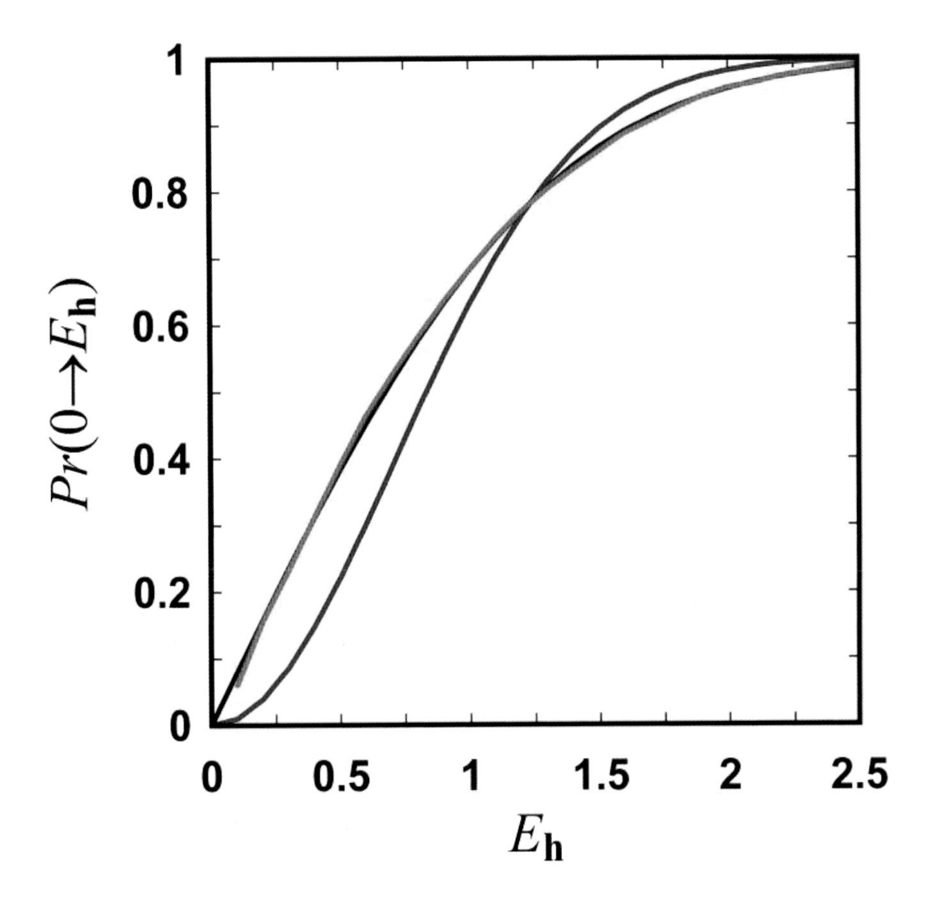

Figure 5.40 Cumulative probability distributions for an ideal acentric probability distribution of $E_{\mathbf{h}} = t$, $1 - e^{-E_{\mathbf{h}}^2}$ (blue), and an ideal centric probability distribution of $E_{\mathbf{h}} = t$, $\mathrm{erf}(E_{\mathbf{h}}/\sqrt{2})$ (black); the fraction of the total number of $E_{\mathbf{h}}$s less than or equal to each specific $E_{\mathbf{h}}$ on the x axis for $CrC_{40}H_{30}O_4P_2$ is shown in red.

is the number of $E_{\mathbf{h}} = t$ values. g_1 and $g_{\bar{1}}$ are the root-mean-square deviations of the fit to each distribution; the smaller number represents the better fit. In the example in Fig. 5.40, $g_1 = 0.4167$ and $g_{\bar{1}} = 0.0085$. Current software packages use a number of variations of the cumulative distribution, all leading to essentially the same analysis.

The distribution of normalized structure factors is also influenced by other symmetry in the structure. For example, a centrosymmetric structure containing a molecule with its own center of symmetry that does not lie on a crystallographic inversion center will often produce a distribution with even greater deviations from the acentric distribution. Such distributions are often referred to as *hypercentric*. It is important to recall that the cumulative distribution functions were derived based on the assumption that the ratios of the scattering factors remain constant and equal to the ratios of the atomic numbers of the atoms. For heavy atoms, this can be a relatively crude approximation, and the deviation from statistical parameters for idealized structures can be substantial. Other factors, including the location of atoms in special positions and non-crystallographic symmetry can also affect the statistical analysis of the intensity data. The analysis of reflection statistics often determines the existence of an inversion center, but in may cases it does not, and it may be necessary to proceed with structural determination by attempting a solution assuming various space groups.

Exercises

1. In a diffraction experiment the integrated intensity of a reflection is measured repeatedly and found to have an average value of 7500 counts per second (cps). Determine the probability that a single measurement of the intensity will provide a value less than or equal to (a) 7450 cps, (b) 7400 cps, (c) 7300 cps, and (d) 7200 cps.

2. Determine the diffraction angles, 2θ, for which the polarization correction will be at a maximum and a minimum and the values of the correction factor at these angles. Ignore secondary polarization effects such as those arising from a monochromator.

3. The orientation matrix and X-ray wavelength for the data collected for the mercaptopyridine crystal discussed in Chapter 1 are given in Exercise 2 of Chapter 4. (a) Using the orientation matrix, determine the Lorentz correction factor for the $(2\,\bar{4}\,11)$ reflection for the case in which the integrated intensity is collected with a serial detector. For comparative purposes, use the form of the factor that includes the wavelength. (b) Determine the Lorentz correction factor for the reflection when its intensity is measured using an area detector at $\omega = 9.41°$, with χ fixed at 54.73° and φ set at 270°.

4. The mercaptopyridine molecule, C_5H_4NSH, is a common ligand in organometallic chemistry. It often occurs as a bridging ligand, attaching itself to more than one metal center in a complex. For example, in the complex $Re_2Os_2\ ((C_6H_5)_3P)_2(CO)_{11}(\mu\text{-}SNC_5H_4)$, the ligand bonds to an osmium atom and a rhenium atom through its sulfur atom, and another rhenium atom via its nitrogen atom (the "μ" indicates that the sulfur atom "bridges" two metal atoms). The unit cell for a crystal of the ligand has a volume of 542.3 Å^3; the space group is $P2_1/n$. The unit cell of a crystal of the complex has a volume of 4920.66 Å^3 in the $P2_1/c$ space group. The table below gives the mass absorption coefficients ($\mu_m = \mu_l/\rho$ (cm^2/g)) for the elements in these compounds for both copper and molybdenum radiation:

Element	Cu $K\alpha$	Mo $K\alpha$	Element	Cu $K\alpha$	Mo $K\alpha$
H	0.435	0.380	S	89.1	9.55
C	4.60	0.625	P	77.28	7.87
N	7.52	0.916	Re	178.1	98.7
O	11.5	1.31	Os	181.8	100.2

(a) Determine the linear absorption coefficients of crystals of the ligand for Cu $K\alpha$ and Mo $K\alpha$ radiation. (b) Determine the linear absorption coefficients of crystals of the complex for Cu $K\alpha$ and Mo $K\alpha$ radiation.

5. Iron pyrite, FeS_2, packs in a simple cubic lattice with four formula units per unit cell and $a = 5.4187$ Å. The crystals are often found in nature in the form of cubes. A cube of FeS_2 with edge length d is mounted between a Mo $K\alpha$ X-ray source and a detector that intercepts the beam. We assume that the beam has a comparatively negligible width, so that its cross-section uniformly passes through the cube as it is rotated. The mass absorption coefficients for Fe and S are 17.74 cm^2/g and 9.63 cm^2/g respectively. Calculate the ratio of the minimum and maximum intensities observed as the crystal is rotated in the beam for (a) $d = 0.1$ mm, (b) $d = 0.5$ mm, and (c) $d = 1.0$ mm.

6. Although the carbon monoselenide molecule, CSe, is known only in the gas phase, its analog, carbon monosulfide, CS, tends to form $(CS)_n$ polymers. Consider a hypothetical crystal of CSe containing a single molecule in a triclinic unit cell. The carbon atom is at the origin, and the selenium atom is at (0.375, 0,375, 0.375). Using the dispersion and scattering factor data below,

atom	f_{110}	$\Delta f'$(Cu $K\alpha$)	$\Delta f''$(Cu $K\alpha$)	$\Delta f'$(Mo $K\alpha$)	$\Delta f''$(Mo $K\alpha$)
C	5.4	0.017	0.009	0.002	0.002
Se	30.6	−0.879	1.139	−0.178	2.223

(a) determine the amplitudes and phases for $\mathbf{F}_{110}$ and $\mathbf{F}_{\bar{1}\bar{1}0}$ for Cu $K\alpha$ radiation, and (b) for Mo $K\alpha$ radiation.

7. A terminal oxygen atom in a typical metal carbonyl complex has the following displacement parameters when refined from data collected at 298 K: $U_{11} = 0.06396$, $U_{22} = 0.08275$, $U_{33} = 0.07527$, $U_{12} = -0.04411$, $U_{13} = -0.00590$, and $U_{23} = 0.01226$. A typical metal carbonyl complex with data collected at 100 K provides the following refined displacement parameters: $U_{11} = 0.04448$, $U_{22} = 0.03012$, $U_{33} = 0.03846$, $U_{12} = 0.01193$, $U_{13} = 0.01169$, and $U_{23} = 0.00365$. (a) Compute the equivalent squared isotropic displacement, u^2_{iso}, for the oxygen atoms observed at both temperatures. (b) For the intensity of a reflection collected at $2\theta = 35°$ with Mo $K\alpha$ radiation ($\lambda = 0.71073$) determine the percentage of the "scattering power" that is lost due to vibration by comparing the effective atomic scattering factor at both temperatures to that of a hypothetical oxygen at rest. The scattering factor parameters for oxygen (Eqn. 3.100) are given in Exercise 8 at the end of Chapter 3.

8. The mercaptopyridine unit cell contains four C_5H_4NSH molecules. The table on the following page summarizes the intensity data for the crystal structure, collected with Mo $K\alpha$ radiation and divided into regions of approximately constant values of the diffraction angle, 2θ. The average intensity is listed for each region, along with the value of the atomic scattering factor, $f(\langle 2\theta \rangle)$, for each type of atom in the unit cell for each region. Create a Wilson plot of the data and determine the scale factor and the average atomic displacement.

Table 5.3 Data for Exercise 8.

$\langle 2\theta \rangle$	$\langle I_{rel} \rangle$	f_H	f_C	f_N	f_S
7.71	1247.8	0.83	5.19	6.26	14.34
10.04	921.59	0.73	4.74	5.83	13.45
12.52	982.03	0.63	4.26	5.33	12.48
14.97	345.16	0.53	3.8	4.83	11.57
17.31	247.5	0.45	3.4	4.37	10.8
19.8	247.41	0.37	3.04	3.93	10.1
22.23	273.53	0.3	2.74	3.55	9.52
24.67	202.97	0.25	2.49	3.21	9.03
27.18	116.46	0.21	2.28	2.92	8.61
29.62	104.61	0.17	2.12	2.68	8.27
32.12	79.41	0.14	1.99	2.47	7.96
34.59	93.92	0.12	1.88	2.29	7.69
37.09	52.02	0.1	1.8	2.15	7.44
39.67	42.16	0.08	1.73	2.02	7.2
42.11	35.15	0.07	1.68	1.92	6.98
44.58	30.18	0.06	1.63	1.84	6.76
47.15	19.9	0.05	1.59	1.77	6.54
49.51	22.17	0.04	1.55	1.71	6.33

9. Derive the reflection conditions for (a) an I-centered crystal, (b) a crystal with a triad screw axis parallel to the c axis.

10. The relatively small difference between the expected values of $\langle E_{\mathbf{h}} \rangle$ for the centric and acentric E distributions has prompted the use of other "E-statistics" to assist in ascertaining the presence or absence of a center of symmetry in the unit cell. In particular, expected values of $\langle |E_{\mathbf{h}}^2 - 1| \rangle$ are commonly employed for this purpose. The absolute value is taken, since $E_{\mathbf{h}}^2 - 1$ can negative. (a) Determine the expected value of $\langle |E_{\mathbf{h}}^2 - 1| \rangle_1$ for the acentric distribution. (Note: Because $E_{\mathbf{h}}^2 - 1$ is negative for $0 \leq E_{\mathbf{h}} < 1$, the probability integral must be evaluated separately from 0 to 1, then added to the integral from 1 to ∞.) (b) Determine the expected value of $\langle |E_{\mathbf{h}}^2 - 1| \rangle_{\bar{1}}$ for the centric distribution by evaluating the probability integral numerically, or estimating it graphically using trapezoidal integration.

Chapter 6

Crystal Structure Solution: Experimental

In Chapter 3 the relationship between structure factors and the structural model was developed in the form of a Fourier pair. Given the locations of the atoms from the model, the scaled magnitudes of the scattered waves along with their relative phases can be calculated for each reflection ($f_j(hkl) \equiv f_{j,\theta\, d\, T}$):

$$\vec{F}_{hkl} = \mathbf{F}_{hkl} = \sum_{j=1}^{n} f_j(hkl) e^{-2\pi\, i(hx_j + ky_j + lz_j)}. \tag{6.1}$$

Given the set of scaled scattered waves and their relative phases, the electron density of the crystal can be calculated and the atoms located at positions of electron density maxima:

$$\rho(x, y, z) = \frac{1}{V_c} \sum_{hkl} \mathbf{F}_{hkl} e^{-2\pi i(hx+ky+lz)}. \tag{6.2}$$

The scaled magnitudes of the structure factors are available from the diffraction experiment. The relative phases are not. In this chapter, and the one that follows, we will consider various approaches to the solution of ***the phase problem***, the fundamental problem in crystal structure determination. These approaches have been divided somewhat artificially into those that make use of the effects of atomic scattering on experimental relationships between observed intensities — the subject of this chapter, and those that are based on statistical relationships between them — the subject of the next chapter.

6.1 The Patterson Function

Although the lack of phase information precludes the direct interpretation of the intensity data from the diffraction experiment, there is, after all, structural information in the relative intensities, since they depend on the locations of the scattering centers. To determine how we might extract this information, consider the hypothetical one-dimensional crystal of BeClF illustrated in Fig. 3.30 and represented as

Understanding Single-Crystal X-Ray Crystallography. Dennis W. Bennett

ISBN: 978-3-527-32677-8 (HC), 978-3-527-32794-2 (SC)

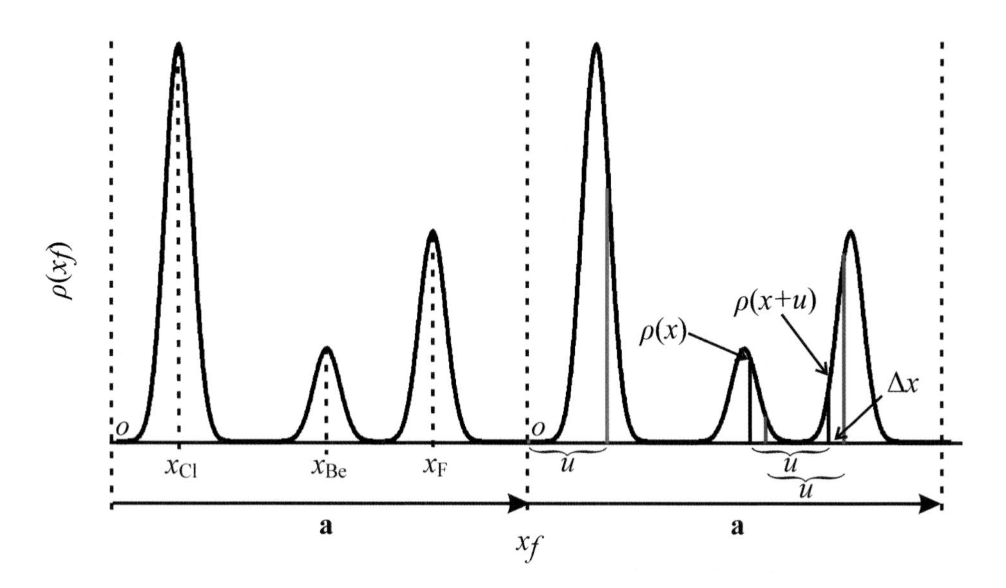

Figure 6.1 Electron density map of a hypothetical one-dimensional crystal of BeClF.

an *electron density map* in Fig. 6.1. Beginning at the origin, we select an increment along the a axis, which we denote u, illustrated by the red vertical lines in Fig. 6.1. We compute the product of the electron densities at the beginning and end of the increment, $\rho(0)\rho(u)$. In the case shown here this will be a very small number, despite the large value of $\rho(u)$, since $\rho(0) \approx 0$. We now move the u displacement a small increment, Δx, along the a axis and again compute the electron density product, $\rho(x)\rho(x+u)$. We repeat this process until a series of such products have been accumulated for the entire unit cell. The black vertical lines and the blue vertical lines represent the process for two of the translations in sequence. Once we have accumulated the products for a given value of the increment, u, we add them all together, each multiplied (weighted) by the translation increment, Δx. There will be more products for a smaller value of Δx, and we weight the terms in the product so that the resulting function of u approximates the same function, known as *the Patterson function*, for different values of Δx:

$$P(u) = \sum_{x=0}^{1} \rho(x)\rho(x+u)\Delta x. \tag{6.3}$$

Considering u as a variable, we can determine values of $P(u)$ for u varying continuously from zero to one. The most important property of this function is that there are large contributions to the sum only when the electron density at some x and the electron density at $x+u$ have significant values *simultaneously*. The consequences of this constraint are seen in Fig. 6.2. In Fig. 6.2(a) the displacement increment, u_a, is the distance between the electron density maxima for two atoms, i.e., it is the distance between the atoms. As the displacement vector moves in increments of Δx through both peaks, there will be large values of the product that contribute to the sum. In Fig. 6.2(b) the displacement increment, u_b, does not correspond to the

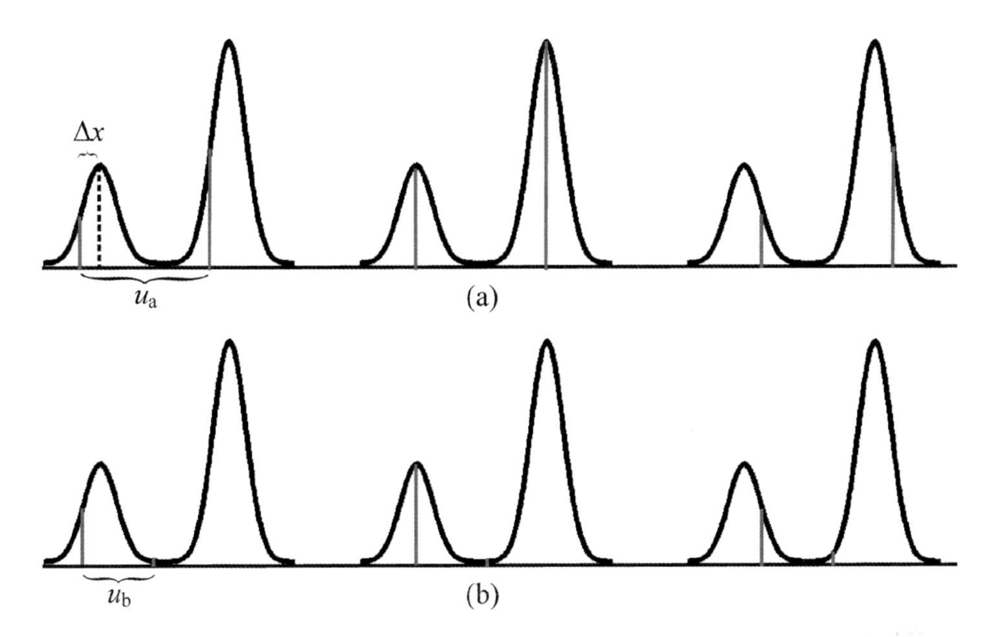

Figure 6.2 Contributions to the Patterson sum for two different values of the displacement variable, u_a and u_b, moved in sequential increments of Δx along the a axis.

interatomic distance, and each translation results in a small value of either $\rho(x)$, $\rho(x+u)$, or both — the *products* will generally be very small numbers. We would therefore expect $P(u_a)$ to be much greater than $P(u_b)$. *Furthermore, we should anticipate that $P(u)$ will have maximum values whenever u is the distance (vector) between two atoms.* The Patterson function becomes a more sensitive indicator of this as Δx becomes smaller, and optimally, when it becomes infinitesimal*:

$$P(u) = \int_0^1 \rho(x)\rho(x+u)\,dx. \tag{6.4}$$

Fig. 6.3 compares the electron density function, $\rho(x)$, for the one-dimensional BeClF structure with the Patterson function, $P(u)$ for the structure. A number of features of the Patterson function are illustrated in this simple example:

1. The values of u vary from 0 to 1 along the a axis, and therefore are described by unit cell fractional coordinates; the unit cell coordinate system for the Patterson function is the same as the coordinate system for the electron density function.

2. When $u = 0$ the value of the function is the integral of the square of the electron density throughout the unit cell. This is effectively the value of the function for a displacement vector between every atom and itself, and is a very large number. The result is a large peak (the largest peak) at the origin for the Patterson function.

*The Patterson function was developed formally by Arthur Lindo Patterson in 1934[97] as the Fourier transform of the intensities, from which the integral arises naturally.

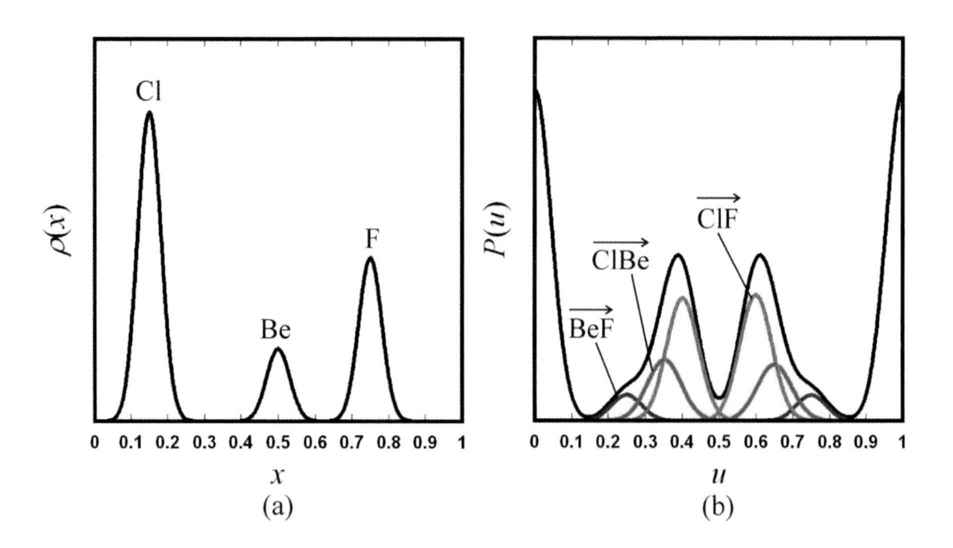

Figure 6.3 (a) Electron density function for the hypothetical BeClF structure. (b) Patterson function for the BeClF structure.

3. The Patterson function is centrosymmetric. This can be understood by referring to Fig. 6.1. The Cl, Be, and F atoms are at $x_{\text{Cl}} = 0.15$, $x_{\text{Be}} = 0.50$, and $x_{\text{F}} = 0.75$. The Cl – F internuclear vector (in fractional coordinates) is $\overrightarrow{\text{ClF}} = 0.75 - 0.15$, and we expect a maximum in the *Patterson map* at $u_f = 0.60$. There is also a chlorine atom at $x_f = 1.15$ in the adjacent unit cell, with $\overrightarrow{\text{FCl}} = 1.15 - 0.75 = 0.40$. This will create an identical maximum at $u_f = 0.40 \equiv 1 - 0.60 \equiv -0.60$ in the Patterson map. In general, $P(u) = P(-u)$.

4. There are many more peaks in the Patterson map in comparison with the electron density map — there is a maximum and its centrosymmetric equivalent for every interatomic vector in the structure. In addition, the Patterson peaks are broader than the electron density peaks as a consequence of the simultaneous contributions of both electron density functions. As can be seen in Fig. 6.3, this results in significant overlap of the numerous peaks in the Patterson map, and a subsequent loss in resolution. The maxima shown in red, blue and green in the figure are those that would have been observed if there were no other atoms in the structure except those responsible for each maximum. Even in this simple structure, the only peaks resolved are those for the Cl-F vector, and even in this case, the presence of the other atoms skews these peaks from their expected locations.

5. The magnitude of a peak in the Patterson map is proportional to the product of the electron densities of the pair of atoms responsible for the peak. *The largest peaks are observed when both atoms are electron-rich heavy atoms.*

The reader may wonder at this stage why we have developed a function that seems to require a knowledge of the electron density function, providing peaks at the locations of interatomic vectors, if we already know the locations of the atoms, (and hence the interatomic vectors). To answer this, we expand the Patterson function to three dimensions and express it in terms of the structure factor representation of the electron density. The displacement variable in the Patterson function now becomes a 3-vector, $\mathbf{u} = [u_f \; v_f \; w_f] \equiv [u \; v \; w]$.*

$$P(\mathbf{u}) = \int_0^1 \int_0^1 \int_0^1 \rho(\mathbf{r})\rho(\mathbf{r}+\mathbf{u}) \, dxdydz. \tag{6.5}$$

$$\begin{aligned}
\rho(\mathbf{r}) &= \frac{1}{V_c} \sum_{\mathbf{h}} \mathbf{F_h} e^{-2\pi i \, (\mathbf{r}\cdot\mathbf{h})} \\
\rho(\mathbf{r}+\mathbf{u}) &= \frac{1}{V_c} \sum_{\mathbf{h}'} \mathbf{F_{h'}} e^{-2\pi i \, ((\mathbf{r}+\mathbf{u})\cdot\mathbf{h}')} \\
P(\mathbf{u}) &= \int_0^1 \int_0^1 \int_0^1 \frac{1}{V_c} \sum_{\mathbf{h}} \mathbf{F_h} e^{-2\pi i \, (\mathbf{r}\cdot\mathbf{h})} \frac{1}{V_c} \sum_{\mathbf{h}'} \mathbf{F_{h'}} e^{-2\pi i \, ((\mathbf{r}+\mathbf{u})\cdot\mathbf{h}')} \, dx \, dy \, dz \\
&= \sum_{\mathbf{h}} \sum_{\mathbf{h}'} \mathbf{F_h F_{h'}} \, e^{-2\pi i (\mathbf{u}\cdot\mathbf{h}')} \frac{1}{V_c^2} \int_0^1 \int_0^1 \int_0^1 e^{-2\pi i (\mathbf{r}\cdot(\mathbf{h}+\mathbf{h}'))} \, dx \, dy \, dz.
\end{aligned}$$

In the general case, $(\mathbf{r} \cdot (\mathbf{h} + \mathbf{h}')) = (h + h')x + (k + k')y + (l + l')z$. If at least one of these terms is not zero (e.g., $h \neq -h'$), then $h + h' = m$, an integer, and

$$\int_0^1 e^{-2\pi i (h+h')x} \, dx = \int_0^1 e^{-2\pi \, imx} \, dx = -\frac{1}{2\pi m} e^{-2\pi \, imx} \Big|_0^1 = 0. \tag{6.6}$$

However, for those terms in the sum in which $h = -h'$, $k = -k'$, *and* $l = -l'$, the exponent is zero and the integral evaluates to the volume of the unit cell (Eqn. 3.80):

$$\frac{1}{V_c^2} \int_0^1 \int_0^1 \int_0^1 e^0 \, dx \, dy \, dz \equiv \frac{1}{V_c^2} \int_0^{V_c} dV = \frac{1}{V_c}. \tag{6.7}$$

The only non-zero terms in the sum occur when $\mathbf{h}' = \bar{\mathbf{h}}$:

$$P(\mathbf{u}) = \frac{1}{V_c} \sum_{\mathbf{h}} \mathbf{F_h F_{\bar{h}}} \, e^{-2\pi i (\mathbf{u}\cdot\bar{\mathbf{h}})}. \tag{6.8}$$

But the Friedel law tells us that $\mathbf{F_h F_{\bar{h}}} = F_{\mathbf{h}} e^{i \varphi_{\mathbf{h}}} F_{\mathbf{h}} e^{-i \varphi_{\mathbf{h}}} = F_{\mathbf{h}}^2$. Furthermore, *the squares of the magnitudes of the structure factors are the observed intensities*:

$$\begin{aligned}
P(\mathbf{u}) &= \frac{1}{V_c} \sum_{\mathbf{h}} F_{\mathbf{h}}^2 \, e^{-2\pi i (\mathbf{u}\cdot\bar{\mathbf{h}})} &\quad (6.9) \\
&= \frac{1}{V_c} \sum_{\mathbf{h}} I_{o,\mathbf{h}} \, e^{-2\pi i (\mathbf{u}\cdot\bar{\mathbf{h}})}. &\quad (6.10)
\end{aligned}$$

The Patterson function is a Fourier series in which the coefficients are intensities rather than structure factors. An experimental Patterson map can be created

*For the entirety of the next two chapters the variables will always be in fractional coordinates, and we will drop the subscript for simplicity.

directly from the data collected in the diffraction experiment. While it does not create a recognizable image of the actual structure, it does create maxima that can, in principle, provide every interatomic vector in the structure!

For every $\mathbf{h}$ in the sum, there is a $-\mathbf{h}$. If we represent the sum in terms of these $(\mathbf{h}, -\mathbf{h})$ pairs,

$$\begin{aligned} P(\mathbf{u}) &= \frac{1}{V_c}\sum_{+\mathbf{h}} I_{o,\mathbf{h}}\, e^{-2\pi\, i(\mathbf{u}\cdot\bar{\mathbf{h}})} + \frac{1}{V_c}\sum_{-\mathbf{h}} I_{o,\mathbf{h}}\, e^{-2\pi\, i(\mathbf{u}\cdot\bar{\mathbf{h}})} \\ &= \frac{1}{V_c}\sum_{+\mathbf{h}} I_{o,\mathbf{h}} \cos(-2\pi(\mathbf{u}\cdot\bar{\mathbf{h}})) + i\frac{1}{V_c}\sum_{+\mathbf{h}} I_{o,\mathbf{h}} \sin(-2\pi(\mathbf{u}\cdot\bar{\mathbf{h}})) \\ &\quad + \frac{1}{V_c}\sum_{-\mathbf{h}} I_{o,\mathbf{h}} \cos(-2\pi(\mathbf{u}\cdot\bar{\mathbf{h}})) - i\frac{1}{V_c}\sum_{+\mathbf{h}} I_{o,\mathbf{h}} \sin(-2\pi(\mathbf{u}\cdot\bar{\mathbf{h}})) \\ &= \frac{1}{V_c}\sum_{\mathbf{h}} I_{o,\mathbf{h}} \cos(2\pi(\mathbf{u}\cdot\mathbf{h})). \end{aligned} \tag{6.11}$$

A major difficulty in interpreting the Patterson function arises from the sheer number of maxima that occur in the Patterson map. If there are n atoms in the unit cell, for every atom there are $n-1$ interatomic vectors to each of the other atoms in the unit cell, resulting in a total of $n(n-1)$ maxima in the Patterson map (n^2 if the origin peak is included as well). Fig. 6.4(a) shows four unit cells of a hypothetical two-dimensional structure of the planar phenol molecule, each unit cell containing two C_6H_5OH asymmetric units related by a center of symmetry. The locations of the maxima in the Patterson map are shown in Fig. 6.4(b); all vectors to the hydrogen atoms (the maxima with the lowest intensities) have been omitted for clarity in the figure. The black circles represent the locations of the termini of the C–C vectors, the red circles are the locations of the O–O vectors, and the blue circles represent the positions of the C–O vectors. Note that the six-fold symmetry of the benzene ring is reflected in the six-interatomic C–C vectors centered at the origin. Note also that the vectors representing the C–O bonds are nearly superimposed onto C–C vectors; resulting in unresolved peaks of higher intensity in the Patterson map.

Using the two-dimensional analog of Eqn. 6.1, structure factors were generated and the corresponding electron density and Patterson functions were calculated for the two-dimensional C_6H_5OH structure. The functions are represented graphically in Fig. 6.5. Note that the hydrogen atom peaks in the electron density plot are about the same size as the "ripples" in the plot. These ripples occur because the Fourier series must be calculated with a finite number of terms, and those that are left out fail to remove the residual waves that would otherwise add to zero. The locations of the carbon and oxygen atoms are readily ascertained from the plot. The Patterson map is clearly more complex that the electron density map — even for this simple two-dimensional structure. The origin peak is larger than all the others, but it is still more-or-less resolved from the peaks near the origin. This is important, since the shorter inter-atomic vectors are those that represent the bonds between atoms, and are most useful in deducing the structure from the Patterson map. As already mentioned, the origin peak is essentially the vector between every atom and itself. It therefore has a magnitude that scales as the sum of the squares of the electron densities of all of the atoms in the unit cell. For actual three-dimensional crystal structures the relative size of the origin peak can virtually obliterate peaks

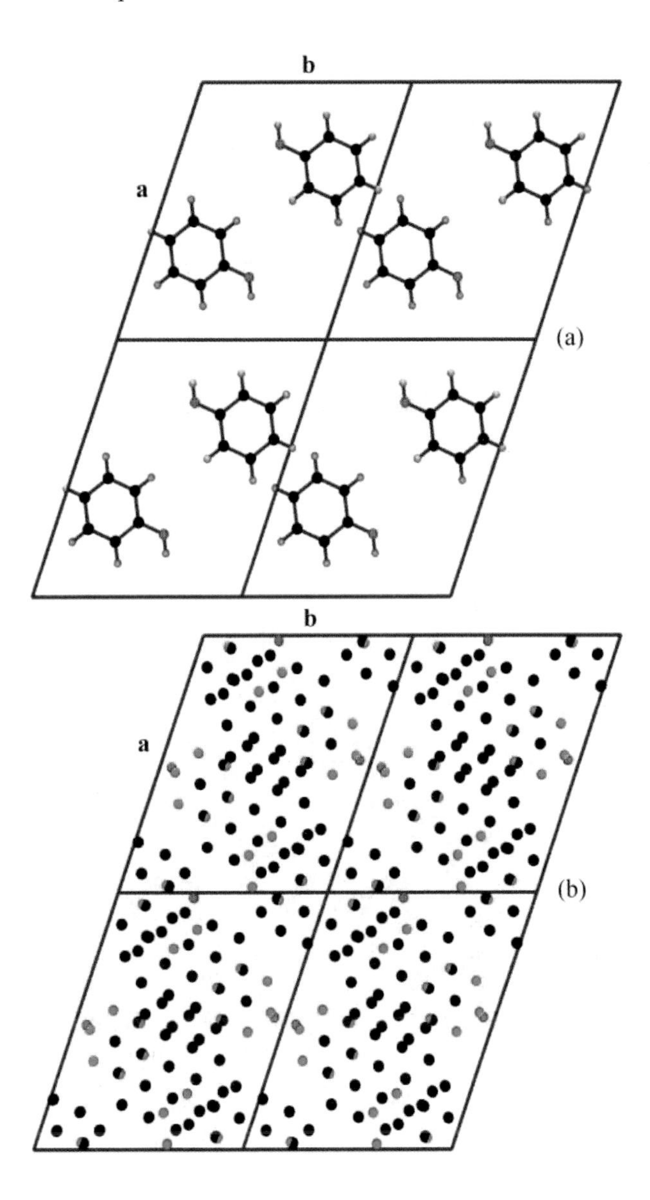

Figure 6.4 (a) Atom locations for the hypothetical centrosymmetric two-dimensional C_6H_5OH structure. (b) Patterson maxima locations for the two-dimensional C_6H_5OH structure (X-H maxima and origin peak not shown).

due to the interatomic vectors near the origin. This effect is exacerbated when the actual experimental Patterson map is calculated, since random error in the data will tend to broaden all of the peaks.

The origin peak can be effectively removed from the experimental Patterson map by subtraction of that portion of each intensity that does not depend on the

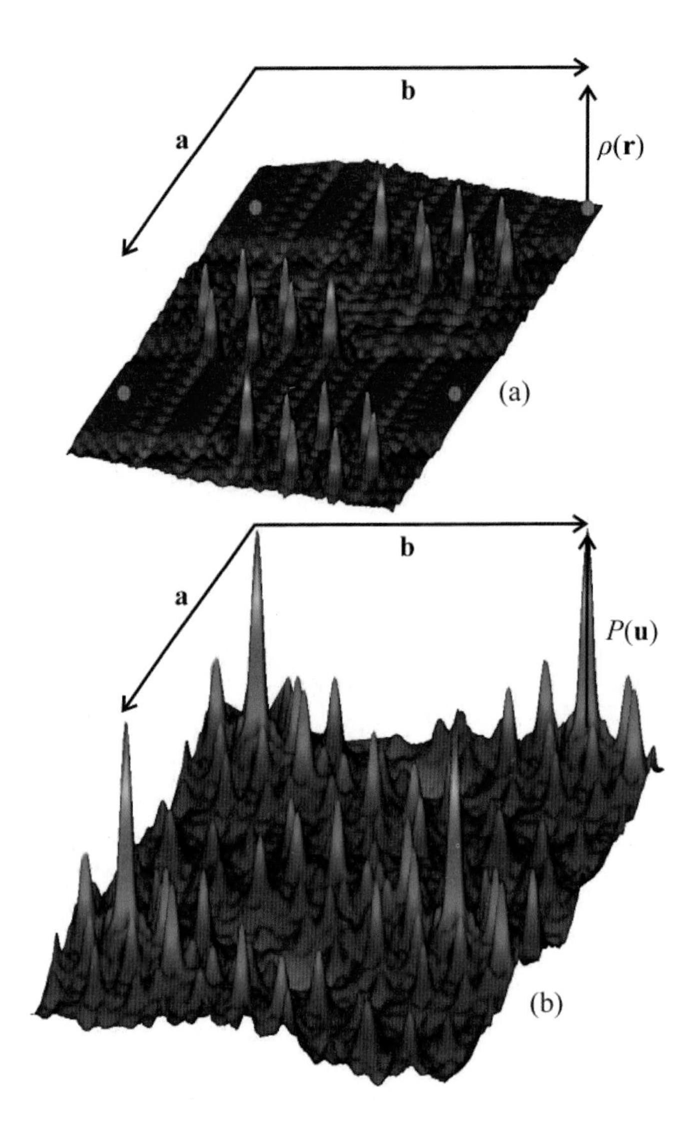

Figure 6.5 (a) Electron density function for the hypothetical two-dimensional C_6H_5OH structure. The red circles signify the origin locations. (b) Patterson function for the hypothetical two-dimensional C_6H_5OH structure.

relative locations of the atoms (i.e., the contributions to the intensities of the atoms "with themselves"). The intensities can be factored into two components:

$$
\begin{aligned}
I_{\mathbf{h}} &= F_{\mathbf{h}}^2 = \mathbf{F}_{\mathbf{h}}^* \mathbf{F}_{\mathbf{h}} \\
&= \left(\sum_{j=1}^{n} f_j(hkl) e^{+2\pi\, i(hx_j+ky_j+lz_j)} \right) \left(\sum_{m=1}^{n} f_m(hkl) e^{-2\pi\, i(hx_m+ky_m+lz_m)} \right) \\
&= \sum_{j=1}^{n} f_j^2(hkl) + \sum_{j \neq m}^{n} \sum^{n} f_j(hkl) f_m(hkl) e^{-2\pi\, i(\mathbf{h}\cdot(\mathbf{r}_m - \mathbf{r}_j))}. \qquad (6.12)
\end{aligned}
$$

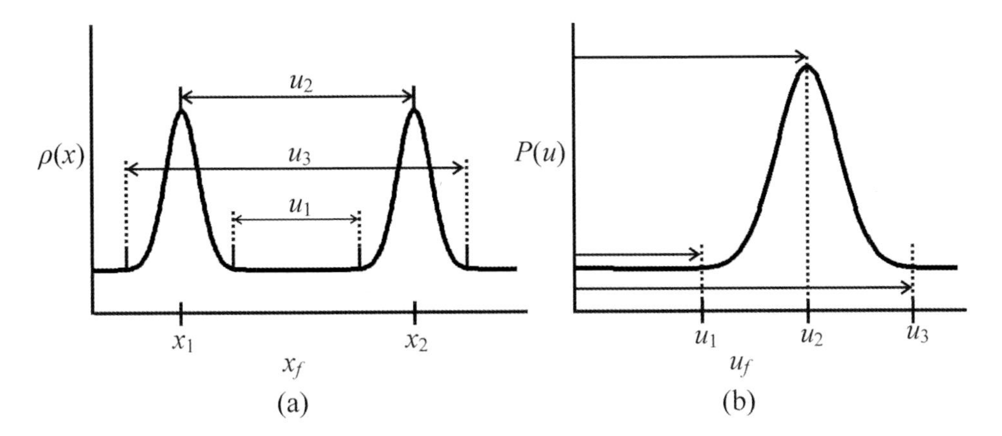

Figure 6.6 (a) One-dimensional electron density peaks for two atoms separated by u_2. (b) One-dimensional Patterson peak resulting from the two atoms. The relative heights are arbitrary.

The second sum accounts for all of the terms for which $j \neq m$ — the contributions to the intensity that depend on the locations of the atoms with respect to one another. When summed over all of the intensities, these components create the Patterson maxima at various locations in the unit cell. The first sum, for $j = m$, depends only on the scattering power of each atom independently, and is the component that gives rise to the origin peak. The creation of modified "intensities", in which this contribution is subtracted out, should therefore produce a Patterson map with the origin peak removed:

$$I'_{o,\mathbf{h}} = I_{o,\mathbf{h}} - \sum_{j=1}^{n} f_j^2(\mathbf{h}) \tag{6.13}$$

$$P'(\mathbf{u}) = \frac{1}{V_c} \sum_{\mathbf{h}} I'_{o,\mathbf{h}} \cos(2\pi(\mathbf{u} \cdot \mathbf{h})). \tag{6.14}$$

The overlap of peaks in the Patterson map presents a more serious problem. The origin of this overlap is illustrated in Fig. 6.6. Fig. 6.6(a) represents two atoms with equal electron density distributions separated by a distance, u_2. The Patterson peak resulting from these two atoms is shown in Fig. 6.6(b). Contributions to the Patterson peak occur for any distance greater than u_1 and less than u_3. The width of the Patterson peak is therefore $u_3 - u_1$; *the sum of the widths of the electron density peaks.* For the three-dimensional Patterson function of a structure with a unit cell containing n atoms, superposition of the $n(n-1)$ peaks, each approximately twice as broad as the electron density peaks, generally creates a Patterson map that is nearly impossible to resolve into individual interatomic vectors.

The overlap problem is a result of the electron density distribution of the atoms, and would not exist if the electron density of each atom was concentrated at its nucleus *and* if the atoms were not vibrating. In the example above, this would result in $u_1 \approx u_2 \approx u_3$, creating significantly sharper peaks in the Patterson map. In the previous chapter we determined that dividing the observed intensities by the sum of the squares of the scattering factors resulted in modified point-atom intensities — normalized intensities that would (approximately) be observed if the atoms were stationary and had electron densities concentrated at the nucleus:

$$I_{E,\mathbf{h}} = E_{o,\mathbf{h}}^2 = \frac{F_{o,\mathbf{h}}^2}{\sum_{j=1}^{n} f_j^2(\mathbf{h})} = \frac{I_{o,\mathbf{h}}}{\sum_{j=1}^{n} f_j^2(\mathbf{h})}. \tag{6.15}$$

The derivations in Chapter 5 hold equally for the modified intensities generated to remove the origin. Substituting $I'_{o,\mathbf{h}}$ in Eqn. 6.15,

$$I'_{E,\mathbf{h}} = \frac{I_{o,\mathbf{h}} - \sum_{j=1}^{n} f_j^2(\mathbf{h})}{\sum_{j=1}^{n} f_j^2(\mathbf{h})} = \frac{I_{o,\mathbf{h}}}{\sum_{j=1}^{n} f_j^2(\mathbf{h})} - 1 = E_{o,\mathbf{h}}^2 - 1. \tag{6.16}$$

Using values of $E_{o,\mathbf{h}}^2 - 1$ as modified intensities in the Patterson Fourier series significantly enhances the resolution, creating *a sharpened Patterson map with the origin peak removed*:

$$P'_E(\mathbf{u}) = \frac{1}{V_c} \sum_{\mathbf{h}} I'_{E,\mathbf{h}} \cos(2\pi(\mathbf{u} \cdot \mathbf{h})). \tag{6.17}$$

6.1.1 Patterson Solution: Structures Without Heavy Atoms.

Patterson Vector Superposition

The Patterson map contains maxima corresponding to every interatomic vector in the unit cell, with all of the vectors translated to the origin. If the structure is known, all of the vectors to a given atom can be generated by translating the unit cell contents so that the atom lies at the origin of the unit cell. It follows that the maxima in the entire Patterson map can be generated by sequentially translating the unit cell contents so that each of the n atoms in the unit cell lies at the origin. The hypothetical two-dimensional unit cell in Fig. 6.4(a) serves to illustrate this. Ignoring the hydrogen atoms, the maxima in the Patterson map resulting from the vectors to an oxygen atom are determined by translating the structure so that the atom is at the origin, as illustrated in Fig. 6.7(a). The O–C maxima (vector heads) are indicated with blue circles, and the O–O maxima are indicated with red circles. Figs. 6.7(b) and (c) show the maxima resulting from the vectors to two different carbon atoms, C and C'. The C–C maxima are indicated with black circles. *For each translation an image of the entire structure is created in the Patterson unit cell.* In addition, for every maximum generated in this manner there is a centrosymmetric equivalent; these have been omitted from the figure for illustrative purposes.

Suppose that two copies of the Patterson map are created, and superimposed upon one another such that the origin of one Patterson map lies atop a peak in the second map. For example, consider the case in which the origin of the Patterson map in Fig. 6.7(a) is translated to the head of the C–O vector in Fig. 6.7(b), illustrated by the dashed arrow on the left. This will place two images of the

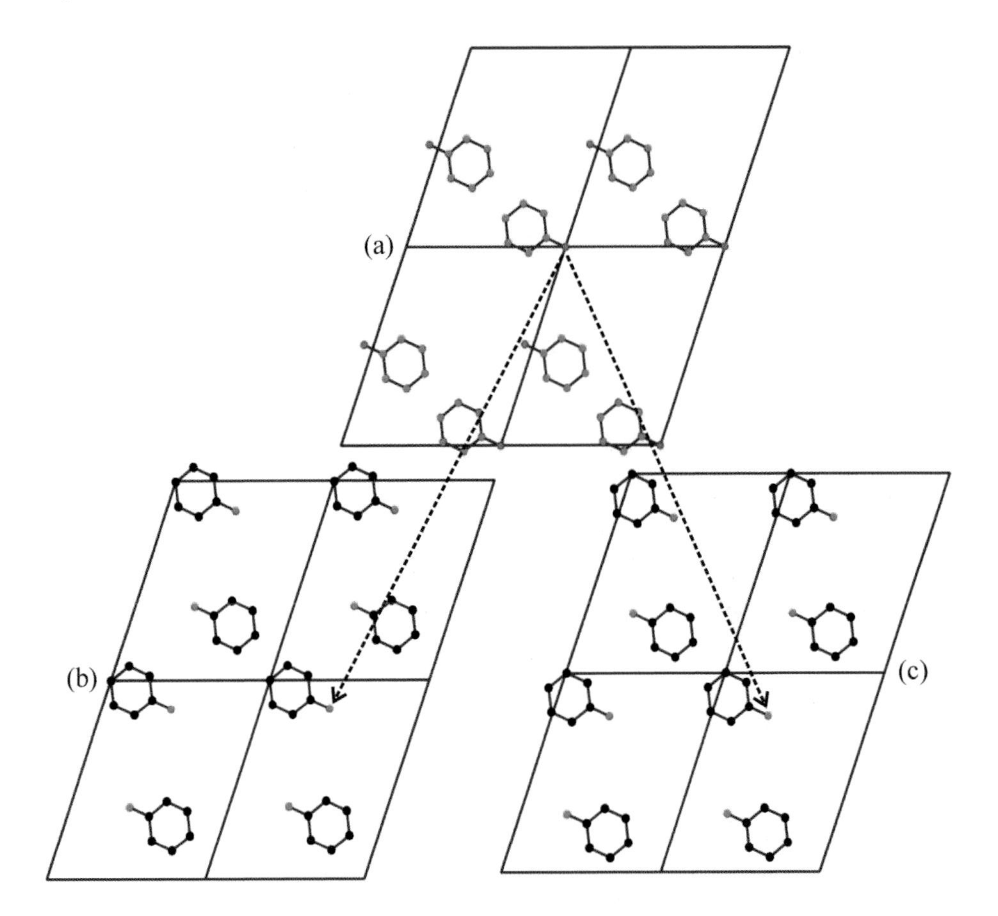

Figure 6.7 (a) Patterson maxima for O–X vectors. (b) Patterson maxima for C–X vectors. (c) Patterson maxima for C'–X vectors. The heads of C–C vectors, O–O vectors, and C–O/O–C vectors are denoted with black circles, red circles, and blue circles, respectively. The peaks are connected to indicate the recurrence of structural images in the Patterson map.

structure atop one another, superpositioning the image created by the vectors to the oxygen atom with the image created by those to the carbon atom. A different superposition will result if the map origin is translated instead to the head of the C'–O vector in Fig. 6.7(c). In general, if the origin of one copy of a Patterson map is translated to a peak in a second copy of the map, two images of the structure will be superimposed, and the magnitudes of the peaks at the overlapping positions will be proportional to the sum of the intensities of the individual peaks in each copy. *In principle, if there are no other fortuitous overlaps, identifying these peaks should be sufficient to determine the relative positions of the atoms in a centrosymmetric unit cell.*

If the unit cell is non-centrosymmetric the overlapping peaks will correspond to an image of the structure *and* a second image related to the first by an inversion center (due to the centrosymmetric nature of the Patterson function). This is illustrated in Fig. 6.8 for a hypothetical two-dimension structure of the planar

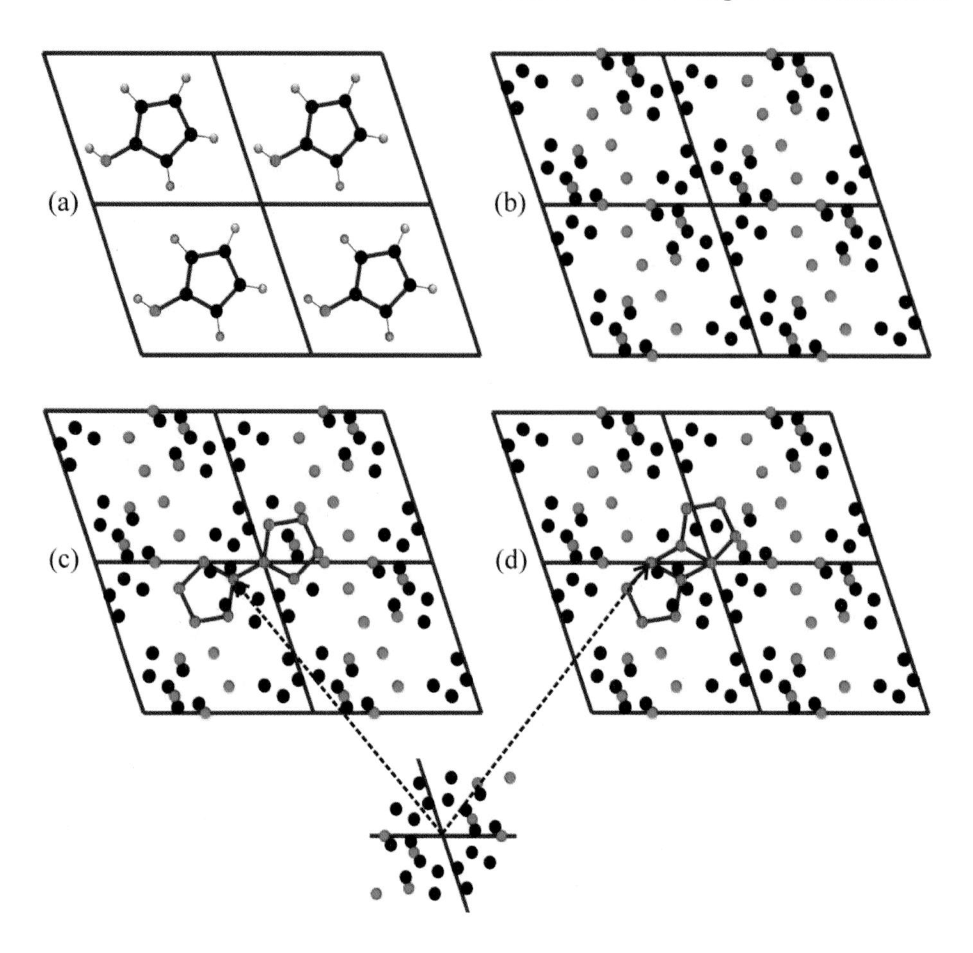

Figure 6.8 (a) Atom locations for the hypothetical non-centrosymmetric two-dimensional C_5H_4OH structure. (b) Patterson maxima locations for the two-dimensional C_5H_4OH structure (X-H maxima and origin peak not shown). (c) Overlapping peaks (shown in red) resulting from a superposition of two copies of the Patterson map. (d) Overlapping peaks for a second superposition. Non-overlapping peaks from the second copies in (c) and (d) are not shown.

furanol molecule, C_5H_4OH. The overlapping maxima resulting from translation of the origin of a copy of the Patterson map (insert) to a C–O maximum in the original map is shown in Fig. 6.8(c). The red circles indicate the locations of the overlapping maxima; the remaining peaks from the second map have been omitted for clarity. The resulting image is seen to consist of the actual structure connected to its inversion. In this optimal case, no other overlapping peaks are observed. A second superposition is shown in Fig. 6.8(d), again resulting in the asymmetric unit connected to its inversion — but in a different manner (the selected peak serves as a connection point in each case). Combining these two superpositions results in coincident peaks solely for the asymmetric unit. *In theory, a judiciously selected second superposition is generally sufficient to provide the relative positions of the atoms in a non-centrosymmetic unit cell.*

For these simple two-dimensional examples it is possible to determine vector superpositions visually — but in three dimensions this task must be left to the computer. Overlapping peaks are found by searching through superimposed Patterson maps for maxima that are approximately doubled in magnitude. This search is complicated if there are parallel vectors of equal length in the unit cell, since they will also appear as doubly-weighted peaks in the Patterson map. This can occur, for example, if a molecule in the unit cell has a benzene ring. The hexagonal structure of the ring contains three pairs of parallel carbon-carbon bonds, creating three doubly weighted peaks in the Patterson map when the vectors are translated to the origin. For non-centrosymmetric unit cells the occurrence of doubly weighted peaks is observed only in special (or fortuitous) circumstances, but for centrosymmetric unit cells a large proportion of the vectors are parallel to one another. This occurs since ever X–Y vector in the cell is relate to a Y–X vector by the inversion center. Each (X–Y, Y–X) pair produces a (Y–X, X–Y) pair in the Patterson map, resulting in doubly weighted peaks for nearly every vector in the direct unit cell. Only those direct unit cell vectors that pass through the origin will produce singly-weighted Patterson maxima, since they are redundant (parallel to themselves only). A centrosymmetric unit cell contains $2n$ atoms with half of them related to the other half by an inversion center. There are $2n-1$ vectors between each atom and all the others, resulting in $2n(2n-1)$ vectors in the direct unit cell. Of these, there is a vector for each atom and its inversion equivalent, passing through the origin and resulting in $2n$ singly-weighted peaks. The remaining $(2n(2n-1)-2n))/2) = 2n(n-1)$ maxima are therefore doubly-weighted in the Patterson map of a centrosymmetric structure. The doubly-weighted peaks result from the overlapping of two images of the structure, and selecting a doubly-weighted peak for the superposition will result in the creation of two images of the structure. As with the double image in the non-centrosymmetric unit cell, this can be resolved by performing a second superposition.

If the superposition exercise is successful, the relative locations of the atoms (or at least a significant portion of them), including their orientation with respect to the unit cell axes, will be known. However, since the interatomic vectors have all been translated to the Patterson cell origin, they still must be placed in their proper locations in the unit cell. To accomplish this we can perform a translational search through the unit cell, seeking a match between the experimentally observed structure factor magnitudes, and those calculated from Eqn. 6.1. The atom locations, (x_j, y_j, z_j), determined from vector superposition are referenced to an arbitrary origin $(0, 0, 0)$. Translating the unit cell contents is effected by shifting this origin to some point $p = (x_p, y_p, z_p)$ in the unit cell. The predicted structure factors for the unit cell contents resulting from the translation are therefore

$$
\begin{aligned}
\mathbf{F}_c &= \sum_{j=1}^{n} f_j(hkl) e^{-2\pi i(h(x_p+x_j)+k(y_p+y_j)+l(z_p+z_j))} \\
&= e^{-2\pi i(hx_p+ky_p+lz_p)} \sum_{j=1}^{n} f_j(hkl) e^{-2\pi i(hx_j+ky_j+lz_j)}
\end{aligned}
$$

$$= (\cos(-2\pi\, i(hx_p + ky_p + lz_p)) + i\sin(-2\pi\, i(hx_p + ky_p + lz_p)))$$
$$\times \left(\sum_{j=1}^{n} f_j(hkl)\cos(-2\pi\, i(hx_j + ky_j + lz_j)) \right.$$
$$\left. + i \sum_{j=1}^{n} f_j(hkl)\sin(-2\pi\, i(hx_j + ky_j + lz_j)) \right). \quad (6.18)$$

Thus,

$$\begin{aligned} \mathbf{F}_c &= (a_p + i\, b_p)(A + i\, B) \\ &= a_p A + i\, a_p B + i\, b_p A + i^2\, b_p B \\ &= (a_p A - b_p B) + i\,(a_p B + b_p A) \end{aligned} \quad (6.19)$$

The magnitude of the predicted structure factor is determined from $\mathbf{F}_c^*\mathbf{F}_c$:

$$\begin{aligned} F_c^2 &= I_c = (a_p A - b_p B)^2 + (a_p B + b_p A)^2 \\ &= (a_p^2 + b_p^2)(A^2 + B^2). \end{aligned} \quad (6.20)$$

The origin point, (x_p, y_p, z_p), is translated systematically in small increments, Δx_p, Δy_p, and Δz_p, each time computing F_c from the atom locations resulting from the translation. Because the relative positions of the atoms do not change, their relative contribution to the phase and magnitude of the structure factor is constant, with this contribution shifted by the change in position of the origin. It is therefore necessary to compute the sums defining $A(hkl)$ and $B(hkl)$ only once. Each trial structure factor is then determined from $a_p(hkl)$ and $b_p(hkl)$ alone, requiring the evaluation of only a single sine and cosine.

The goal of the search is to obtain the location of the origin point that provides the closest agreement between observed and calculated structure factor magnitudes. There are a number of ways to assess this; a logical one is to treat the differences between calculated and observed structure factors as residual errors, using the minimization of the sum of the squares of the residuals (Sec. 4.5) as the criterion for the best fit. The residuals are $F_c(hkl) - F_o(hkl)$, and the location of the origin is determined from the minimum value of D obtained in the search:

$$D = \sum_{hkl} w_{hkl}(F_c(hkl) - F_o(hkl))^2. \quad (6.21)$$

The w_{hkl} term is a factor that weights each squared residual according to the reliability of the observed structure factor. At this stage of the structural determination process the weights are ordinarily the reciprocals of the variances of each observed structure factor magnitude, $w_{hkl} = 1/\sigma^2(F_o(hkl))$.* This is qualitatively reasonable since small variances should result in more reliable, and therefore more heavily weighted structure factor magnitudes.

*The subject of weighting will be treated more formally in Chapter 8, since structural refinement is also based on least-squares criteria. $w_{hkl} = 1/\sigma^2(F_o(hkl))$ is the proper statistical weight for a normal distribution of structure factors with no other errors in the experiment. Since many structure factor distributions are acentric, and since other errors can affect the reliability, various empirical weighting schemes are employed to take these deviations into account. Prior to the final stages of structural refinement the reciprocal of the variance is a suitable approximation for w_{hkl}.

It would appear that we have arrived at a point where structural solution has been reduced to a few Patterson superpositions and a translation search. Unfortunately the use of Patterson superposition alone rarely results in a structural solution. If there is no initial model for the structure, it is nearly impossible to randomly select vectors in the Patterson map that will overlap the same image. Furthermore, *even with sharpening and origin removal*, the three-dimensional Patterson map ordinarily contains a very large number of overlapping peaks, and the lack of resolution in the map usually results in a number of multiple images that are very difficult to sort out, even with repeated superpositions. *However, the chances for successful extraction of a structural image from the Patterson map are markedly improved if the location of one or more atoms in the unit cell is known.*

Light Atom Location: Patterson Symmetry

For two atoms in the unit cell located at (x_1, y_1, z_1) and (x_2, y_2, z_2), the corresponding maximum in the Patterson unit cell occurs at

$$\begin{aligned} u_{12} &= x_2 - x_1 \\ v_{12} &= y_2 - y_1 \\ w_{12} &= z_2 - z_1. \end{aligned} \tag{6.22}$$

If the two atoms are related by symmetry, then Patterson maxima are located as a result of that symmetry. For example, if the atoms are related by a two-fold rotation axis parallel to the b axis, as depicted in Fig. 6.9(a), then

$$\begin{aligned} x_2 &= -x_1 \\ y_2 &= y_1 \\ z_2 &= -z_1 \\ u_{12} &= 2x_2 = -2x_1 \\ v_{12} &= y_1 - y_1 = 0 \\ w_{12} &= 2z_2 = -2z_1. \end{aligned} \tag{6.23}$$

Since each atom at (x, y, z) in the unit cell is related to another by the two-fold rotation, there will be a large concentration of Patterson maxima at $(2x, 0, 2z)$ and $(-2x, 0, -2z)$ in the Patterson unit cell — on the ac plane (Fig. 6.9(b)). On the other hand, if the atoms are related by a two-fold screw axis parallel to the b axis as in Fig. 6.9(c), then

$$\begin{aligned} x_2 &= -x_1 \\ y_2 &= \frac{1}{2} + y_1 \\ z_2 &= -z_1 \\ u_{12} &= 2x_2 = -2x_1 \\ v_{12} &= \frac{1}{2} + y_1 - y_1 = \frac{1}{2} \\ w_{12} &= 2z_2 = -2z_1. \end{aligned} \tag{6.24}$$

The diad screw axis in the direct unit cell creates a diad rotation axis in the Patterson unit cell. Because all Patterson vectors are translated to the origin, *the*

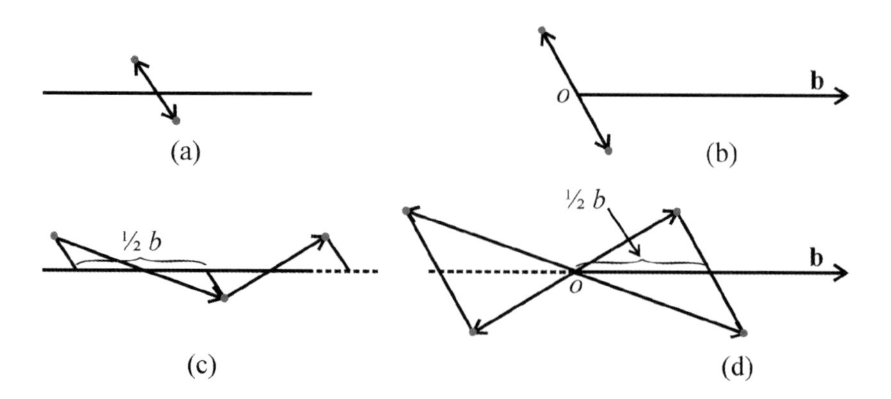

Figure 6.9 (a) Atoms related by a two-fold rotation axis parallel to the b axis. (b) Patterson maxima resulting from atoms related by a two-fold rotation axis parallel to the b axis. (c) Atoms related by a two-fold screw axis parallel to the b axis. (d) Patterson maxima resulting from atoms related by a two-fold screw axis parallel to the b axis.

Patterson lattice no longer reflects the translational symmetry of the direct lattice. In this case there will be a large concentration of Patterson maxima at $(2x, 1/2, 2z)$ and $(-2x, 1/2, -2z)$, on a plane parallel to the ac plane, but passing through the center of the Patterson unit cell at $v = 1/2$ (Fig. 6.9(d)).

If the atoms in the direct unit cell are related by either a mirror plane or a glide plane, then the Patterson maxima resulting from this symmetry will accumulate along a line. For example, in Fig. 6.10(a) the atoms are related by a mirror plane perpendicular to the b axis and parallel to the ac plane. The Patterson vectors are therefore given by

$$\begin{aligned}
x_2 &= x_1 \\
y_2 &= -y_1 \\
z_2 &= z_1 \\
u_{12} &= x_1 - x_1 = 0 \\
v_{12} &= 2y_2 = -2y_1 \\
w_{12} &= z_1 - z_1 = 0.
\end{aligned} \qquad (6.25)$$

The Patterson maxima due to symmetry-equivalent atoms have coordinates $(0, \pm 2y, 0)$, lying along the b axis of the Patterson unit cell. If the mirror plane is

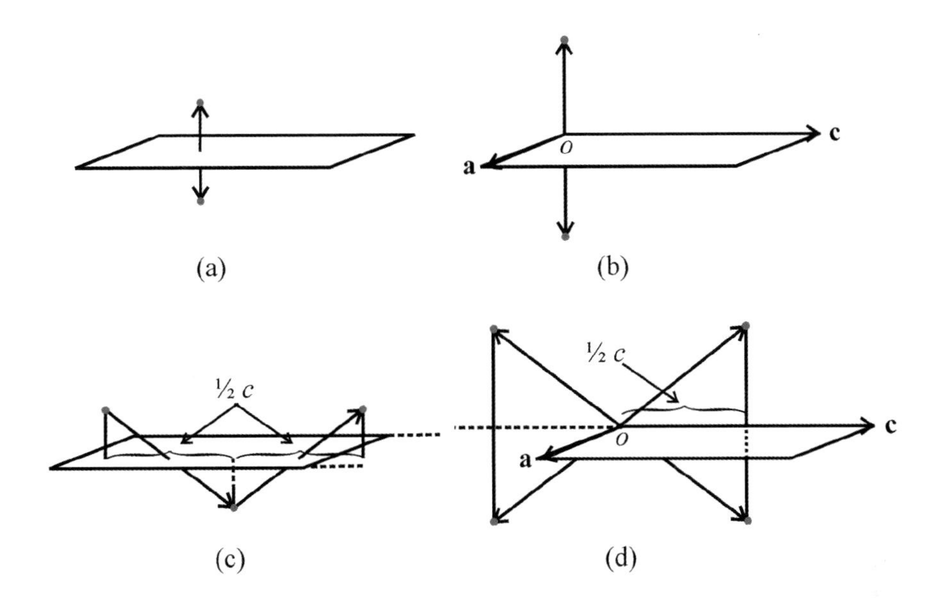

Figure 6.10 (a) Atoms related by a mirror plane perpendicular to the b axis. (b) Patterson maxima resulting from atoms related by a mirror plane perpendicular to the b axis. (c) Atoms related by a c-glide plane perpendicular to the b axis. (d) Patterson maxima resulting from atoms related by a c-glide plane perpendicular to the b axis.

replaced by a c-glide plane perpendicular to the b axis, as depicted in Fig. 6.10(c), the Patterson vectors become

$$
\begin{aligned}
x_2 &= x_1 \\
y_2 &= -y_1 \\
z_2 &= \frac{1}{2} + z_1 \\
u_{12} &= x_1 - x_1 = 0 \\
v_{12} &= 2y_2 = -2y_1 \\
w_{12} &= \frac{1}{2} + z_1 - z_1 = \frac{1}{2}.
\end{aligned}
\tag{6.26}
$$

The Patterson maxima now accumulate at $(0, \pm 2y, 1/2)$, along a line parallel to the b axis which now bisects the c axis at $w = 1/2$ (Fig. 6.10(d)). Again, the direct lattice translational symmetry is lost in the Patterson lattice. The transformation of screw axes into rotation axes and all symmetry planes into lines reduces the number of possible space groups in the Patterson lattice to 24.

The observation that specific symmetry elements in the direct lattice result in specific lines and planes in the Patterson lattice was made in the 1930's by David Harker, one of the founding fathers of crystallography, while still a graduate student.[98] The lines and planes that exhibit an accumulation of Patterson maxima, known as *Harker lines* and *Harker sections*, are capable of differentiating between rotational and screw axes and/or mirror planes and glide planes, *and are therefore often very useful in space group determination.*

There is logical relationship — an "implication" — between the location of atoms in the direct unit cell and the locations of maxima in Harker lines and planes in the Patterson unit cell. In 1946 Martin Buerger[99] defined this relationship in the form of an *implication function*, $I(x, y, z)$, where x, y, z are direct cell fractional coordinates. The concept was expanded by William Lipscomb (later to be awarded the Nobel Prize in Chemistry), into a form that allowed for the search for atomic positions in the Patterson map via a function which Lipscomb described as the *Symmetry Minimum Function.*[100] The implication function is based on the notion that an atom and its symmetry-equivalent will produce a maximum in the Patterson map related to the symmetry element, as we have just observed. For an atom at $\mathbf{r} = (x, y, z)$, there will be a maximum in the Patterson map at $P(\mathbf{u}(\mathbf{r})) = P(\mathbf{r} - \mathbf{S}_i\mathbf{r})$, where $\mathbf{S}_i$ is the ith symmetry operator of the space group. The search for atoms is effected by varying $\mathbf{r}$ sequentially throughout the direct unit cell and evaluating the implication function in the Patterson map for each value of $\mathbf{r}$:

$$I_i(\mathbf{r}) = P(\mathbf{r} - \mathbf{S}_i\mathbf{r}). \tag{6.27}$$

If $\mathbf{S}_i$ is the operator for an $\bar{n}$ operation, then a maximum in the implication function species all of the coordinates of an atom. For example, consider an atom in space group $P2_1/c$, with an inversion center, a two-fold screw axis, and a glide plane. For $\mathbf{S}_i = \bar{1}$ the implication function is

$$I_{\bar{1}}(x, y, z) = P(x - (-x), y - (-y), z - (-z)) = P(2x, 2y, 2z), \tag{6.28}$$

and a maximum in the function provides the coordinates of the atom at (x, y, z). However, the implication functions for the screw axis and glide plane are

$$I_{2_1}(x, y, z) = P\left(2x, \frac{1}{2}, \left(\frac{1}{2} - 2z\right)\right) \quad \text{and} \tag{6.29}$$

$$I_c(x, y, z) = P\left(0, \left(\frac{1}{2} - 2y\right), \frac{1}{2}\right), \tag{6.30}$$

neither of which provides all of the coordinates of the atom. In addition, the evaluation of any one of these functions independently of the others suffers from a number of additional ambiguities resulting from "satellite" peaks from other symmetry operators, overlapping peaks in the Patterson map, and the common occurrence of "general" peaks lying close to those from symmetry related atoms. Because of these (and other) ambiguities, the implication function alone is of limited use in determining atom locations.

However, if *all* of the implication functions can be compared simultaneously, the ambiguities are often minimized, and reasonable trial atom positions can be obtained. The most common approach is embodied in the Symmetry Minimum Function, which is simply the smallest value for any one of the implication functions, evaluated at each search point in the direct unit cell:

$$\begin{aligned} SMF(x, y, z) &= M_{i=1}^{p} I_i(x, y, z)/m_i \\ &= M_{i=1}^{p} P((x, y, z) - \mathbf{S}_i(x, y, z))/m_i. \end{aligned} \tag{6.31}$$

M is a "minimization operator" that evaluates each of the p implication functions at point (x, y, z) and selects the minimum value. Each implication function is divided by the multiplicity of the operation in order to weight each function equally. The screw axis and glide planes produce doubly-weighted peaks due to the redundant coordinates at $v = \pm 1/2$ and $w = \pm 1/2$, respectively — imposed by the center of symmetry in the Patterson lattice. Thus $m_{2_1} = 2$, $m_c = 2$ and $m_i = 1$ since the inversion is singly-weighted. $SMF(x, y, z)$ will be small unless all three implication functions have large values for the same value of (x, y, z). This is, of course, the criterion for the location of an actual atom, since it is subjected to the three symmetry conditions simultaneously. It follows that maxima in the Symmetry Minimum Function tend to correspond to reasonable trial atom locations.

Once one or more trial atom locations are determined from the Symmetry Minimum Function, it is generally necessary to include the entire Patterson map in the search for more atoms in order to find an image of the structure in the map. This is usually accomplished by expanding the "simultaneity" principle of the Symmetry Minimum Function to all of the Patterson vectors (rather than limiting the search to those related by symmetry). If a trial atom is located at $\mathbf{r}_t = (x_y, y_t, z_t)$, with symmetry-equivalent atoms at $\mathbf{S}_i\mathbf{r}_t$, then an *Image Seeking Minimum Function** can be defined as

$$ISMF(x, y, z) = M_{i=1}^{p}\, P((x, y, z) - \mathbf{S}_i(x_t, y_t, z_t)). \tag{6.32}$$

This function tests for the presence of an atom at point (x, y, z) be searching for Patterson maxima for vectors between the point and a trial atom *and* its symmetry-equivalents. As before, the function will have a large value only if all of the maxima have large values simultaneously.

The advantage of seeking images in this manner is that the variables are coordinates in the direct unit cell, rather than the Patterson cell. Thus a successfully extracted image of the structure will already be located properly in the unit cell. In most cases, Patterson methods provide only a partial structure, and it is usually necessary to complete the structure using Fourier methods, which we will discuss later on in the chapter.

6.1.2 Patterson Solution: Structures With Heavy Atoms.

The magnitude of a Patterson maximum is proportional to the product of the electron densities of the two atoms responsible for the Patterson vector. Structures that consist entirely of light atoms produce Patterson maps with numerous maxima of comparable magnitude. Because of this, the superposition and search strategies outlined in the previous section are often difficult to apply in practice. The severity of this problem is underscored by noting that the 1985 Nobel Prize in Chemistry was awarded to Herbert A. Hauptman amd Jerome Karle in 1985 "for their outstanding achievements in the development of direct methods for the determination of crystal structures." Before direct methods, which do not require the use of the Patterson function, the structure of many light atom structures simply went unsolved.

On the other hand, a large number of crystal structures containing one or more heavy atoms in the unit cell *were* solved, at least in part, by Patterson methods, prior to the development of direct methods. Direct methods are based on statistical relationships between reflection intensities; they assume that the ratio of the

*Prior to the advent of direct methods for structural determination (the subject of Chapter 7), Patterson methods provided the only systematic means for solving "light atom" crystal structures. Many approaches to the extraction of structural images evolved during this period, including a number of variations of the Image Seeking Minimum Function described here.

scattering factors at various diffraction angles is approximately independent of the diffraction angle. For atoms of nearly equal electron density this is a reasonable approximation, but for structures containing heavy atoms, the approximation often breaks down. In these cases direct methods sometimes fail, and it is in such cases that Patterson methods can still provide a viable structural solution. While direct methods work best when all atoms have about the same electron density distributions, Patterson methods work best when they do not.

Heavy Atom Location: Patterson Symmetry

The major difficulty in Patterson searches for light atom locations arises from the existence of a large number of peaks of approximately equal intensity. The use of symmetry is often helpful, but rarely unambiguous, and light atom structures using Patterson methods generally involve a significant trial-and-error effort before a reasonable structural image emerges from the map. However, if the structure contains an electron-rich "heavy atom" the selection of a trial atom to begin structural determination becomes much less ambiguous.

Fig. 6.11(a) is a representation of a hypothetical centrosymmetric two-dimensional structure of C_6H_5I, in which the OH groups in Fig. 6.4 has been replaced with heavy iodine atoms. The resulting Patterson maxima are shown in Fig. 6.11(b). The termini of the C–C vectors are represented with black circles, the blue circles are the locations of the C–I vectors, and the red circles represent the positions of the I–I vectors. Unlike the light atom case, there is no need to perform a symmetry search to find the I–I maxima. The magnitudes of the C–C peaks are proportional to $Z_{\mathrm{C}}^2 = 6^2 = 36$, the magnitudes of the C–I peaks are proportional to $Z_{\mathrm{C}}Z_{\mathrm{I}} = 6 \times 53 = 318$, and the magnitudes of the I–I peaks are proportional to $Z_{\mathrm{I}}^2 = 53^2 = 2809$. The I–I vectors illustrated in Fig. 6.11(a) therefore create easily recognizable maxima at the locations shown in red in Fig. 6.11(b). The selection of either of these vectors as (u, v) will place the iodine atom at $(x, y) = (u/2, v/2)$ as a consequence of the center of symmetry. The selection of one of the vectors to define the position of the iodine atom will shift the origin of the unit cell to the midpoint of the a axis in the figure, the other to the b axis; any inversion center can serve as the unit cell origin.

The predominant peaks in the Patterson map of a structure containing heavy atoms result from the interatomic vectors between the heavy atoms. These vectors can be located by direct inspection of the Patterson map, and the heavy atoms can generally be placed in the unit cell by application of the space group symmetry operators. Furthermore, as we shall discover later in the chapter, *the location of a heavy atom that accounts for a substantial fraction of the X-ray scattering may be sufficient to provide for a complete solution of the crystal structure.* The solution of heavy atom structures is historically the most common application of Patterson methods, and we will consider illustrative examples of the three most common space groups for small molecule crystal structures.* An example of a rarely occurring space group is also considered in order to demonstrate a problem that arises for space groups that do not have uniquely defined origins.

1. Space Group $P\bar{1}$: A heavy atom in this triclinic centrosymmetric space group will be located at the each of the symmetry-equivalent positions $\mathbf{r}_1 = (x, y, z)$ and

*Approximately 60% of all small molecule space groups are accounted for by the $P\bar{1}$, $P2_1/c$, and $P2_12_12_1$ space groups.

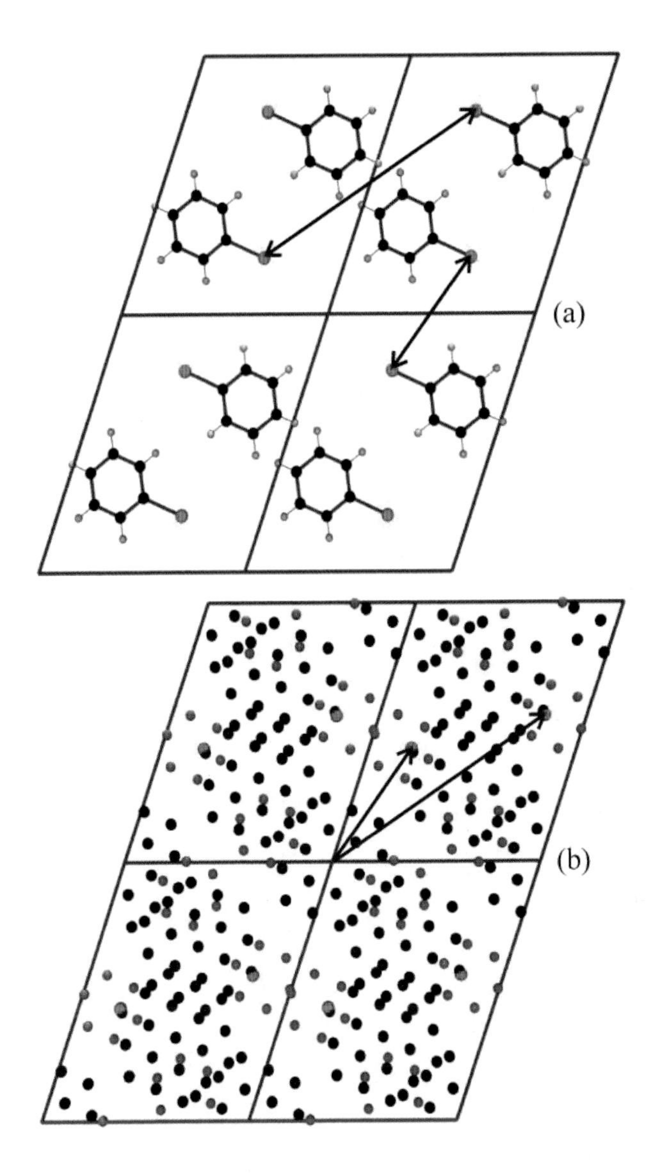

Figure 6.11 (a) Atom locations for the hypothetical centrosymmetric two-dimensional C_6H_5I structure. (b) Patterson maxima locations for the two-dimensional C_6H_5I structure (X-H maxima and origin peak not shown).

$\mathbf{r}_2 = (\bar{x}, \bar{y}, \bar{z})$, resulting in two symmetry-equivalent vectors in the Patterson map at

$$\mathbf{u}_{12} = \mathbf{r}_2 - \mathbf{r}_1 = (\bar{x} - x, \bar{y} - y, \bar{z} - z) = (-2x, -2y, -2z) \quad \text{and} \tag{6.33}$$

$$\mathbf{u}_{21} = \mathbf{r}_1 - \mathbf{r}_2 = (x - \bar{x}, y - \bar{y}, z - \bar{z} = (2x, 2y, 2z). \tag{6.34}$$

The organometallic complex, $CrC_{40}H_{30}O_4P_2$, crystallizes in the $P\bar{1}$ space group. A contour plot of a planar section of the 3-dimensional Patterson map for the

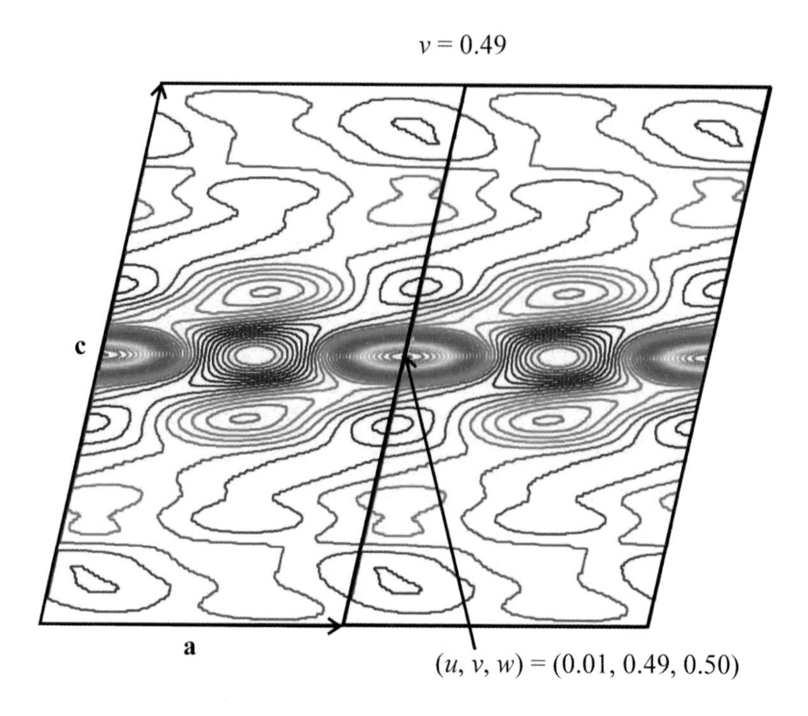

Figure 6.12 Contour plot of the $(u, 0.49, w)$ section of the Patterson map of the organometallic complex, $CrC_{40}H_{30}O_4P_2$, crystallized in the $P\bar{1}$ space group.

complex, parallel to the ac plane at $v = 0.49$ along b, is shown in Fig. 6.12. The colors indicate relative magnitudes, with warmer colors indicating higher values of the Patterson function. A maximum resulting from the Cr–Cr vector is found in this section at $(u, v, w) = (2x, 2y, 2z) = (0.01,\ 0.49,\ 0.50)$. Beginning with a chromium atom at $(x, y, z) = (0.005,\ 0.245,\ 0.250)$ as an approximation to the asymmetric unit, three iterative cycles of Fourier difference synthesis (Sec. 6.3) coupled with least-squares refinement (Chapter 8) resulted in a determination of the locations of the remainder of the atoms in the asymmetric unit. The chromium atom was determined to reside at $(x, y, z) = (0.016,\ 0.245,\ 0.248)$ in the refined structure, illustrated in Fig. 6.13.

2. Space Group $P2_1/c$: In this centrosymmetric monoclinic space group a heavy atom will be located at each of the symmetry-equivalent positions $\mathbf{r}_1 = (x, y, z)$, $\mathbf{r}_2 = (\bar{x}, \bar{y}, \bar{z})$, $\mathbf{r}_3 = (\bar{x}, \frac{1}{2} + y, \frac{1}{2} - z)$, and $\mathbf{r}_4 = x, \frac{1}{2} - y, \frac{1}{2} + z)$. The glide planes and screw axes in $P2_1/c$ do not pass through the center of symmetry and as a result the intense peaks found in the Harker planes due to the screw axes are offset by 1/2 from $w = 2z$ (Eqn. 6.24), and those found along the Harker lines will be offset by 1/2 from $v = 2y$ (Eqn. 6.26). In the Harker planes created from the screw axes,

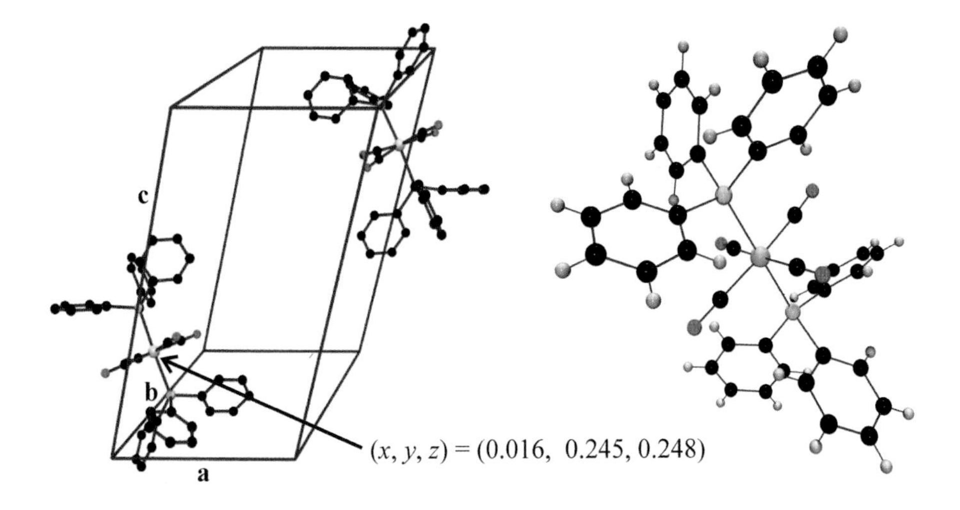

Figure 6.13 $P\bar{1}$ unit cell contents and molecular structure of $Cr(P(C_6H_5)_3)_2(CO)_4$.

the related Patterson vectors, $\mathbf{u}_{ij}$ and $\mathbf{u}_{ji}$, result in maxima in the Harker plane related by a center of symmetry lying in the plane:

$$\mathbf{u}_{13} = \mathbf{r}_3 - \mathbf{r}_1 = (-2x,\ \tfrac{1}{2},\ \tfrac{1}{2} - 2z), \tag{6.35}$$
$$\mathbf{u}_{31} = \mathbf{r}_1 - \mathbf{r}_3 = (2x,\ -\tfrac{1}{2},\ -\tfrac{1}{2} + 2z), \tag{6.36}$$
$$\mathbf{u}_{24} = \mathbf{r}_4 - \mathbf{r}_2 = (2x,\ \tfrac{1}{2},\ \tfrac{1}{2} + 2z), \text{ and} \tag{6.37}$$
$$\mathbf{u}_{42} = \mathbf{r}_2 - \mathbf{r}_4 = (-2x,\ -\tfrac{1}{2},\ -\tfrac{1}{2} - 2z). \tag{6.38}$$

Since $v = -\frac{1}{2} \equiv \frac{1}{2}$ (*recall that adding or subtracting 1 simply places the coordinate in an adjacent unit cell at an equivalent position*), $\mathbf{u}_{13}$ produces a peak at $(-2x,\ \frac{1}{2},\ \frac{1}{2} - 2z) = (-u, \frac{1}{2}, -w)$ and $\mathbf{u}_{31}$ generates a peak at $(2x,\ \frac{1}{2},\ -\frac{1}{2} + 2z) = (u, \frac{1}{2}, w)$. Furthermore, since $w = -\frac{1}{2} + 2z \equiv \frac{1}{2} + 2z$, and $w = -\frac{1}{2} - 2z \equiv \frac{1}{2} - 2z$, $\mathbf{u}_{24} = \mathbf{u}_{31}$ and $\mathbf{u}_{42} = \mathbf{u}_{13}$. The two screw-related pairs of atoms produce the same maxima in the Patterson map, thus doubling the intensity of the peaks. The glide planes create maxima in Harker lines at

$$\mathbf{u}_{14} = \mathbf{r}_4 - \mathbf{r}_1 = (0,\ \tfrac{1}{2} - 2y,\ \tfrac{1}{2}), \tag{6.39}$$
$$\mathbf{u}_{41} = \mathbf{r}_1 - \mathbf{r}_4 = (0,\ -\tfrac{1}{2} + 2y,\ -\tfrac{1}{2}), \tag{6.40}$$
$$\mathbf{u}_{23} = \mathbf{r}_3 - \mathbf{r}_2 = (0,\ \tfrac{1}{2} + 2y,\ \tfrac{1}{2}), \text{ and} \tag{6.41}$$
$$\mathbf{u}_{32} = \mathbf{r}_2 - \mathbf{r}_3 = (0,\ -\tfrac{1}{2} - 2y,\ -\tfrac{1}{2}), \tag{6.42}$$

with $\mathbf{u}_{14} = \mathbf{u}_{32}$ and $\mathbf{u}_{41} = \mathbf{u}_{23}$, again doubling the peak intensities at the two locations on the Harker line. The inversion center in the $P2_1/c$ unit cell and the $2/m$ symmetry of the Patterson function produce single-intensity heavy atom peaks at

$$\mathbf{u}_{12} = \mathbf{r}_2 - \mathbf{r}_1 = (-2x,\ -2y,\ -2z), \tag{6.43}$$
$$\mathbf{u}_{21} = \mathbf{r}_1 - \mathbf{r}_2 = (2x,\ 2y,\ 2z), \tag{6.44}$$
$$\mathbf{u}_{34} = \mathbf{r}_4 - \mathbf{r}_3 = (2x,\ -2y,\ 2z), \text{ and} \tag{6.45}$$
$$\mathbf{u}_{43} = \mathbf{r}_3 - \mathbf{r}_4 = (-2x,\ 2y,\ -2z). \tag{6.46}$$

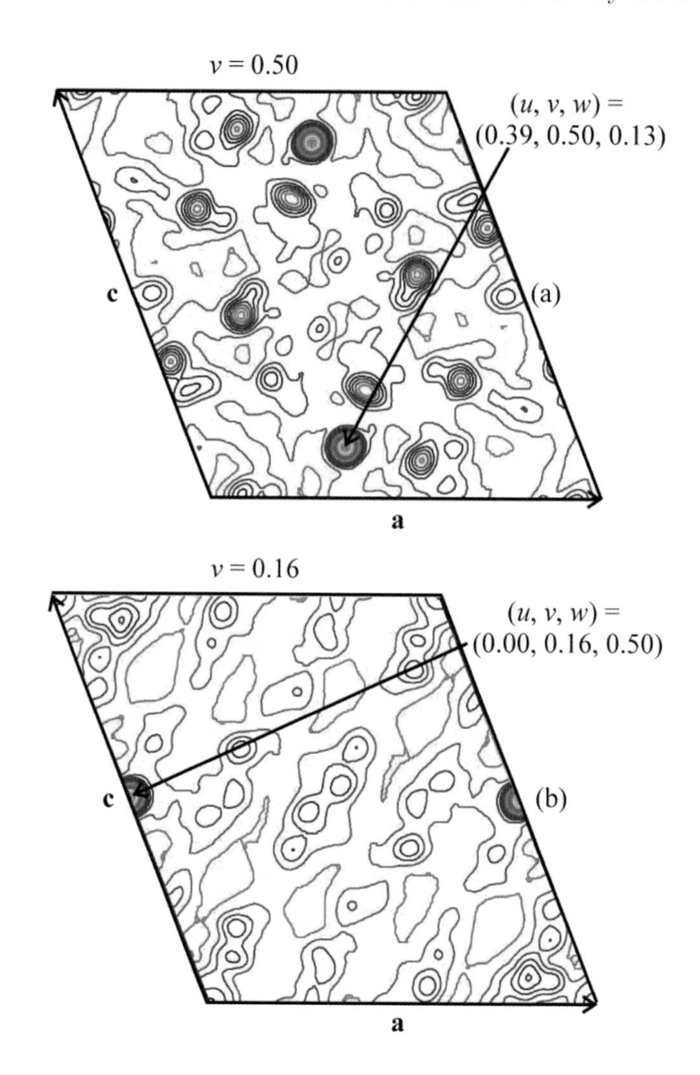

Figure 6.14 (a) Contour plot of the $(u, 0.50, w)$ section of the Patterson map of the organometallic complex, $FeC_{12}H_{12}O_5$, crystallized in the $P2_1/c$ space group. (b) Contour plot of the $(u, 0.16, w)$ section of the Patterson map of the complex.

$FeC_{12}H_{12}O_5$ is an organometallic complex that crystallizes in the $P2_1/c$ space group. The most intense peaks are expected to be those arising from the screw axes and glide planes. A contour plot of a planar section of the 3-dimensional Patterson map for the complex, parallel to the ac plane at $v = 0.50$ along b, is shown in Fig. 6.14(a). This section is the Harker plane at $v = \frac{1}{2}$, containing a maximum at $(u, \frac{1}{2}, w) = (0.39, 0.50, 0.13)$. The x and z coordinates of the Fe atom are determined from $2x = 0.39 \Rightarrow x = 0.195$ and $2z - \frac{1}{2} = 0.13 \Rightarrow z = 0.315$. A second maximum is found in the $v = 0.16$ contour plot in Fig. 6.14(b) on the Harker line at $(0, v, \frac{1}{2}) = (0.00, 0.16, 0.50)$. This provides the y coordinate of the Fe atom; $\frac{1}{2} - 2y = 0.16 \Rightarrow y = 0.170$. Beginning with an iron atom at $(0.195, 0.170, .315)$ as the initial asymmetric unit, as in the previous example, three cycles of difference

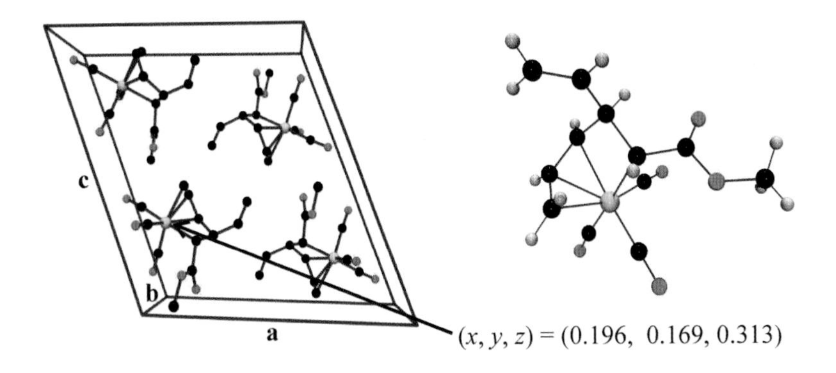

Figure 6.15 $P2_1/c$ unit cell contents and molecular structure of $Fe(\eta^4\text{-}(C_9H_{12}O_2))(CO)_3$.

syntheses and refinement resulted in a complete solution of the structure with the iron atom at $(x, y, z) = (0.196,\ 0.169,\ 0.313)$, illustrated in Fig. 6.15.

3. Space Group $P2_12_12_1$: This orthorhombic space group is the most common space group lacking a center of symmetry. Diad screw axes parallel to the three unit cell axes constitute its symmetry elements, and result in three Harker planes. The screw axes do not intersect, nor do they pass through the origin. Consequently, as in the $P2_1/c$ space group, the maxima in the Patterson map are related by a center of symmetry in each Harker plane, displaced by 1/2 from $u = 2x$, $v = 2y$ and $w = 2z$ for screw axes parallel to a, b, and c, respectively. Heavy atoms are located at $\mathbf{r}_1 = (x, y, z)$, $\mathbf{r}_2 = (\frac{1}{2} - x,\ \bar{y},\ \frac{1}{2} + z)$, $\mathbf{r}_3 = (\frac{1}{2} + x,\ \frac{1}{2} - y,\ -z)$, and $\mathbf{r}_4 = (\bar{x},\ \frac{1}{2} + y,\ \frac{1}{2} - z)$. Because the screw axes do not pass through the origin, each Harker plane contains four maxima related by a center of symmetry in the plane resulting from the vectors between the heavy atoms. In the $(u, v, \frac{1}{2})$ plane,

$$\mathbf{u}_{12} = \mathbf{r}_2 - \mathbf{r}_1 = (\tfrac{1}{2} - 2x,\ -2y,\ \tfrac{1}{2}), \tag{6.47}$$

$$\mathbf{u}_{21} = \mathbf{r}_1 - \mathbf{r}_2 = (-\tfrac{1}{2} + 2x,\ 2y,\ -\tfrac{1}{2}) \equiv (\tfrac{1}{2} + 2x,\ 2y,\ \tfrac{1}{2}), \tag{6.48}$$

$$\mathbf{u}_{34} = \mathbf{r}_4 - \mathbf{r}_3 = (-\tfrac{1}{2} - 2x,\ 2y,\ \tfrac{1}{2}) \equiv (\tfrac{1}{2} - 2x,\ 2y,\ \tfrac{1}{2}), \text{ and} \tag{6.49}$$

$$\mathbf{u}_{43} = \mathbf{r}_3 - \mathbf{r}_4 = (\tfrac{1}{2} + 2x,\ -2y,\ -\tfrac{1}{2}) \equiv (\tfrac{1}{2} + 2x,\ -2y,\ \tfrac{1}{2}). \tag{6.50}$$

In the $(\frac{1}{2}, v, w)$ plane,

$$\mathbf{u}_{13} = \mathbf{r}_3 - \mathbf{r}_1 = (\tfrac{1}{2},\ \tfrac{1}{2} - 2y,\ -2z), \tag{6.51}$$

$$\mathbf{u}_{31} = \mathbf{r}_1 - \mathbf{r}_3 = (\tfrac{1}{2},\ \tfrac{1}{2} + 2y,\ 2z), \tag{6.52}$$

$$\mathbf{u}_{24} = \mathbf{r}_4 - \mathbf{r}_2 = (\tfrac{1}{2},\ \tfrac{1}{2} + 2y,\ -2z), \text{ and} \tag{6.53}$$

$$\mathbf{u}_{42} = \mathbf{r}_2 - \mathbf{r}_4 = (\tfrac{1}{2},\ \tfrac{1}{2} - 2y,\ 2z), \tag{6.54}$$

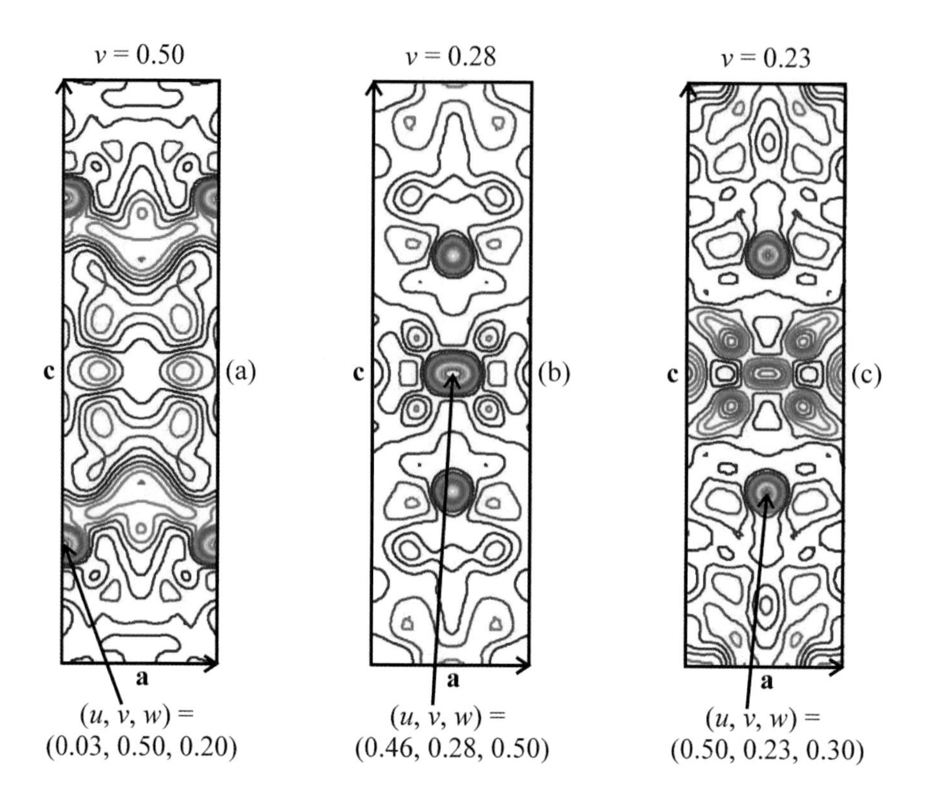

Figure 6.16 (a) Contour plot of the $(u, 0.50, w)$ section of the Patterson map of the organometallic complex, $CrC_{15}H_{16}O_4$, crystallized in the $P2_12_12_1$ space group. (b) Contour plot of the $(u, 0.28, w)$ section of the Patterson map of the complex. (c) Contour plot of the $(u, 0.23, w)$ section of the Patterson map of the complex.

and in the $(u, \frac{1}{2}, w)$ plane,

$$\mathbf{u}_{14} = \mathbf{r}_4 - \mathbf{r}_1 = (-2x,\ \tfrac{1}{2},\ \tfrac{1}{2} - 2z), \tag{6.55}$$
$$\mathbf{u}_{41} = \mathbf{r}_1 - \mathbf{r}_4 = (2x,\ \tfrac{1}{2},\ \tfrac{1}{2} + 2z), \tag{6.56}$$
$$\mathbf{u}_{23} = \mathbf{r}_3 - \mathbf{r}_2 = (2x,\ \tfrac{1}{2},\ \tfrac{1}{2} - 2z), \quad \text{and} \tag{6.57}$$
$$\mathbf{u}_{32} = \mathbf{r}_2 - \mathbf{r}_3 = (-2x,\ \tfrac{1}{2},\ \tfrac{1}{2} + 2z). \tag{6.58}$$

The organometallic complex, $CrC_{15}H_{16}O_4$, crystallizes in the $P2_12_12_1$ space group. Three planar sections of the Patterson map parallel to the ac plane are shown in Fig. 6.16. The $(u, \frac{1}{2}, v)$ section in Fig. 6.16(a) shows the four maxima resulting from the vectors defined in Eqns. 6.55–6.58. Setting $\mathbf{u}_{14} = (0.03, 0.50, 0.20)$ provides the x and z coordinates for the chromium atom:

$$\begin{aligned}
-2x &= 0.03 \\
x &= -0.02 = 1 - 0.02 = 0.98 \quad \text{and} \\
\tfrac{1}{2} - 2z &= 0.20 \\
z &= 0.15.
\end{aligned}$$

The y coordinate is determined from the peaks in the $(u, v, \frac{1}{2})$ section (Fig. 6.16(b)), or the $(\frac{1}{2}, v, w)$ (Fig. 6.16(c)). Selecting the peak at $(0.46, 0.28, 0.50)$ and setting it equal to $\mathbf{u}_{12}$ results in $\frac{1}{2} - 2x = 0.46$ and $x = 0.02$, which is the x coordinate of a symmetry-related atom. However, a consistent y coordinate is found by assigning the peak to $\mathbf{u}_{12}$:

$$\begin{aligned} 2y &= 0.28 \\ y &= 0.14 \quad \text{and} \\ \tfrac{1}{2} + 2x &= 0.46 \\ x &= -0.02 = 0.98. \end{aligned}$$

If the peak at $(0.50, 0.23, 0.30)$ in Fig. 6.16(c) is selected, it must to be assigned to $\mathbf{u}_{42}$ to be consistent:

$$\begin{aligned} 2z &= 0.30 \\ z &= 0.15 \quad \text{and} \\ \tfrac{1}{2} + 2y &= 0.23 \\ y &= 0.14. \end{aligned}$$

Using a chromium atom at $(0.98, 0.14, 0.15)$ as an approximation to the asymmetric unit resulted in a complete solution of the structure, illustrated in Fig. 6.17(a); the chromium atom position refined to $(0.966, 0.135, 0.150)$.

The structure depicted in Fig. 6.17(b) illustrates an important caveat whenever a noncentrosymmetric space group is involved. If we had selected $\mathbf{u}_{41} = (0.03, 0.50, 0.20)$ as the (x, z)-defining peak in the Harker plane, the x and z coordinates would be different than those generated from the $\mathbf{u}_{14}$ vector:

$$\begin{aligned} 2x &= 0.03 \\ x &= 0.02 \quad \text{and} \\ \tfrac{1}{2} + 2z &= 0.20 \\ z &= 0.85. \end{aligned}$$

As previously determined, the $\mathbf{u}_{12}$ vector is consistent with this assignment, resulting in $-2y = 0.28$ and $y = 0.86$. Placing a chromium atom at $(0.02, 0.86, 0.85) \equiv (-0.98, -0.14, -0.15)$ results in a complete structural solution with the chromium atom at $(0.034, 0.865, 0.850)$, *but it is not the same structure!* The coordinates of the atoms in the unit cell in Fig. 6.17(b) are the inverses of those in Fig. 6.17(a), and the resulting molecular structures are seen to be mirror images of one another. In other words, *depending on the initial choice of Patterson vectors, we determine one of the enantiomorphic structures of the complex.* While this may seem a trivial distinction, different enantiomers exhibit different physical and chemical properties, and it would appear that X-ray diffraction is incapable of distinguishing between the two structures. Fortunately, with good data, it is possible to exploit the anomalous dispersion phenomenon during the refinement process in order to determine which of the enantiomorphs is the correct one — in this case the one depicted in Fig. 6.17(a). We will defer this topic until Chapter 8.

4. Space Group Pn: While many space groups exhibit origin locations constrained by symmetry (as in the previous three examples), a number of noncentrosymmetric space groups do not. Such space groups are rather rare for small-molecule crystal structures, but they are much more common in macromolecular

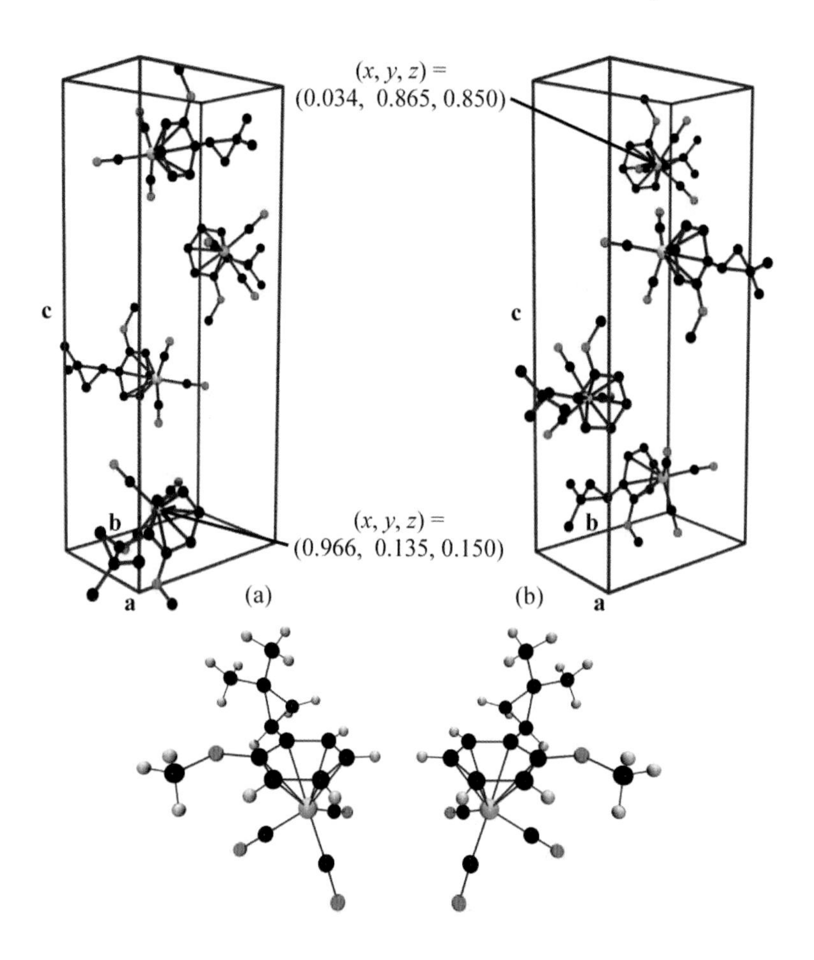

Figure 6.17 (a) $P2_12_12_1$ unit cell contents of an enantiomorph and molecular structure of the corresponding enantiomer of $Cr(\eta^6\text{-}(C_6H_4C_5H_9OCH_3))(CO)_3$. (b) Unit cell contents of the other enantiomorph and molecular structure of the other enantiomer.

structures. For these space groups the selection of one or more of the origin coordinates is arbitrary. For heavy atom structure solutions this can result in the imposition of a center of symmetry in the structure solution, and the resulting atom locations will be those of the structure *and* its mirror image. The Pn space group (or its equivalent in another setting, Pc) is typical in this regard.* The glide direction is along the ac diagonal with the reflection in the b direction, and the origin can be located at any position along the diagonal. The position of a heavy atom will require only that it reflects to a position equidistant from a plane parallel to the ac plane along b. The plane passes through $y = 0$ and the y coordinate for the heavy atom is constrained to be reflected across that plane. There are no such

*The most common space group of this type is the monoclinic space group, $P2_1$, observed very rarely for organometallic and inorganic structures, rarely for organic structures, but quite often in protein crystallography. The origin in the $P2_1$ lattice can be placed at any position along the unique axis.

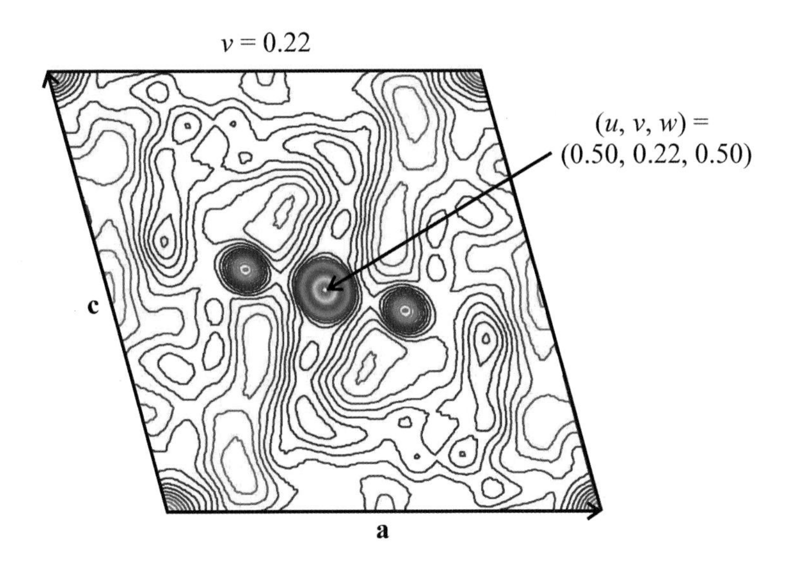

Figure 6.18 Contour plot of the $(u, 0.22, w)$ section of the Patterson map of the organometallic complex, $FeC_{26}H_{24}F_6O_2P_2$, crystallized in the *Pn* space group.

restrictions on its x and z positions (which will define the position of the origin, once selected); they can be assigned arbitrarily.

For the *Pn* space group, a heavy atom at $\mathbf{r}_1 = (x, y, z)$ generates another at $\mathbf{r}_2 = (\frac{1}{2} + x,\ \bar{y},\ \frac{1}{2} + z)$. The Patterson map contains a single Harker plane at $(\frac{1}{2},\ v,\ \frac{1}{2})$ with heavy atom vectors

$$\mathbf{u}_{12} = \mathbf{r}_2 - \mathbf{r}_1 = (\tfrac{1}{2},\ -2y,\ \tfrac{1}{2}), \text{ and} \tag{6.59}$$

$$\mathbf{u}_{21} = \mathbf{r}_1 - \mathbf{r}_2 = (\tfrac{1}{2},\ 2y,\ \tfrac{1}{2}). \tag{6.60}$$

$FeC_{26}H_{24}F_6O_2P_2$ is an organometallic complex that crystallizes in the *Pn* space group. As indicated in Fig. 6.18, a maximum is found in the Patterson map at $(\frac{1}{2}, v, \frac{1}{2}) = (0.50, 0.22, 0.50)$, giving $2y = 0.22$ and $y = 0.11$. x and z can be assigned arbitrarily; it is instructive to set them at $x = 0.25$ and $z = 0.25$. This puts an iron atom at (0.25, 0.11, 0.25), and the other at (0.75, 0.89, 0.75). *In the absence of other atoms, the structure appears centrosymmetric.* If Fourier methods are used to complete the structure, the first step is the generation of a Fourier map based solely on the positions of the iron atoms. Because these atoms are related by a quasi center of symmetry, they will generate structure factors with phases of either 0 or π. These phases are subsequently assigned to a subset of the observed experimental intensities to create an electron density map in order to search for more atoms in the structure (Sec. 6.3). It follows that the calculated electron density map will be centrosymmetric, even though there is no center of symmetry in the structure. The electron density map will therefore contain both the asymmetric unit and its mirror image.

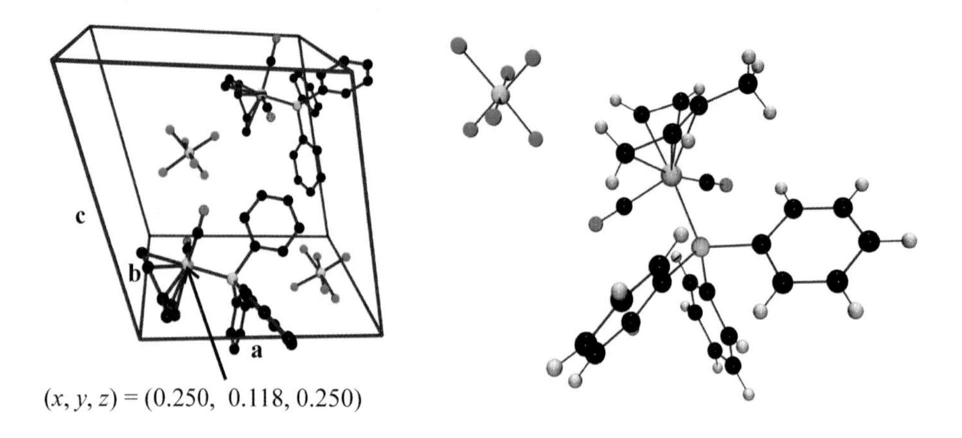

Figure 6.19 *Pn* unit cell contents and molecular structure of $Fe(\eta^5\text{-}(C_5H_6CH_3))(P(C_6H_5)_3)(CO)_2^+PF_6^-$.

In this example, the difference Fourier map created by "phasing" the reflections based on an iron atom at $(0.25, 0.11, 0.25)$ contains two large peaks (of equal intensity), one at $(x, y, z) = (0.420, 0.110, 0.203)$ and the other at $(x', y', z') = (0.080, 0.110, 0.297)$. These intense peaks arise from phosphorus atoms, since they represent substantially more electron density that any of the remaining atoms in the structure. A phosphorus atom at $(0.420, 0.110, 0.203)$ generates a symmetry-equivalent one at $(0.920, -0.110, 0.703)$. If the unit cell was centrosymmetric, there would be another phosphorus atom at $(-0.920, 0.110, -0.703) = (0.080, 0.110, 0.297)$, at the location of the second "phosphorus" peak. These peaks are collinear with the iron atom and equidistant from it, as if the complex itself was centrosymmetric. If we know that the complex does not contain phosphorus atoms in this configuration (structural solution often involves a significant amount of "chemical intuition"), then we can select one of the phosphorus atoms, ignore the other, and proceed with the structural solution.

Inclusion of a phosphorus atom at $(0.420, 0.110, 0.203)$ serves to locate the phosphorus atom in the PF_6^- anion in the next solution cycle, and two subsequent iterations complete the structure, illustrated in Fig. 6.19. If the peak in the Fourier map at $(0.080, 0.110, 0.297)$ is selected for the phosphorus atom position the enantiomeric structure is obtained. As in the previous example, the determination of the correct structure must await the final stages of the refinement process.

The structures in these four examples each contain a transition metal atom with enough electron density to allow them to be completed using Fourier techniques. In structures that contain "lighter" heavy atoms, the electron density of the heavy atom is often insufficient to determine the remainder of the structure in the iterative Fourier process. In these cases it is possible to use its position at a starting point for Patterson vector superposition — or as an initial atom in the expansion of the structure using direct methods, the subject of Chapter 7.

6.1.3 Patterson Solution: Search Methods.

In some cases the structure of a molecule in the unit cell is known prior to crystallographic investigation. More commonly, the structure of a fragment of the molecule is known. In both cases, the known entity can be used as a model to be located in its proper position in the unit cell. As before, the atomic positions of the model serve as a starting point for completion of the entire structure.

Patterson maps exhibit maxima corresponding to vectors between atoms in the unit cell, all translated to the origin. In molecular crystals these atom-atom vectors can be divided into two subsets: *intra*molecular vectors — those between atoms in the same molecule, and *inter*molecular vectors — those between atoms in different molecules.

Fig. 6.20(a) represents the location of a fragment in the hypothetical two-dimensional furanol structure depicted in Fig. 6.8. The maxima in the Patter-

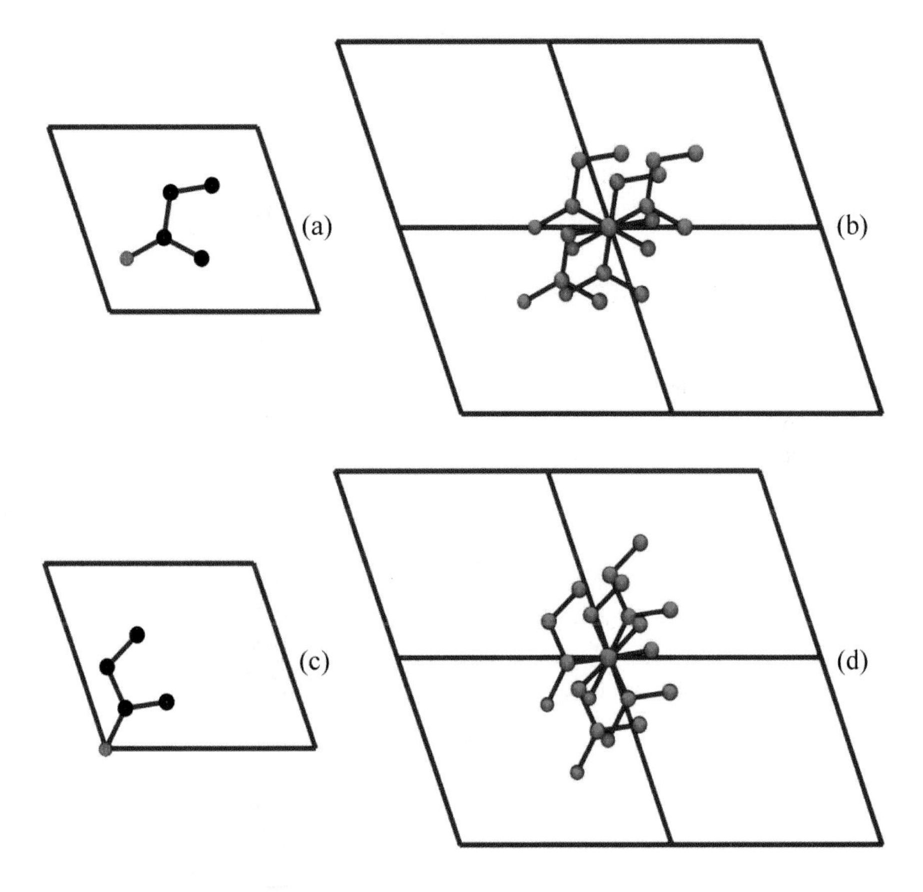

Figure 6.20 (a) Fragment of the hypothetical two-dimensional C_5H_4OH structure in its "natural" position in the unit cell. (b) Patterson maxima resulting from the intramolecular vectors in the properly oriented fragment. (c) Fragment in an arbitrary position and orientation in the unit cell. (d) Patterson maxima resulting from the intramolecular vectors in the arbitrarily oriented fragment.

son map due to the intramolecular vectors of the fragment can be generated by sequentially translating it so that each of its atoms is at the origin, as shown in Fig. 6.20(b). The "bonds" are included in the figure to illustrate each image of the fragment in the Patterson map.

Suppose that we know the structure of the fragment, but do not know its position or orientation in the unit cell. Since all of the atom-atom vectors are translated to the origin in the Patterson map, the translational position of the fragment in the unit cell does not affect the locations of the intramolecular Patterson maxima. Furthermore, if the fragment model is placed in an arbitrary orientation in the unit cell, its intramolecular vectors will be rotated by some angle φ with respect to the "natural" unit cell. Fig. 6.20(c) depicts the known fragment placed at an arbitrary position in the unit cell, rotated by $\varphi = 35°$. The resulting Patterson maxima are shown in Fig. 6.20(d); the Patterson vectors are rotated by the same angle. The correct orientation of the fragment in the unit cell can be determined by placing the fragment in an arbitrary location in the unit cell and rotating it in small steps, each time calculating putative intramolecular contributions to the Patterson map. For each rotation an intramolecular Patterson map is calculated from the rotated model fragment and compared to the experimental Patterson map, looking for the best agreement between the two maps. The search region is limited to the region containing the intramolecular vectors to avoid fortuitous overlap with intermolecular vectors in the map.

Once the orientation of the fragment is established, it can be located in its proper position in the unit cell by incorporating a translation search. The intermolecular Patterson vectors — those between molecules — are determined by the relative positions of symmetry-equivalent fragments in the unit cell. The translational search proceeds stepwise along the unit cell axes in small increments, this time including the region in the Patterson map containing the intermolecular vectors. The search criteria are the same — the correct translation is determined by the best agreement between the maxima created from the fragment model and those in the experimental Patterson map.

For a three-dimensional structure, the rotational search begins with the placement of the model in the unit cell at a convenient location and orientation. The rotation of the model must now be conducted in three dimensions, a task readily accomplished using the Euler angles – just as we did when orienting a single crystal with the diffractometer. All possible orientations of the model can be accessed by converting each model vector, $\mathbf{r}_M$, to Cartesian coordinates and sequentially rotating it about the z axis by φ_1, the x axis by φ_2 and the z axis (again) by φ_3, as illustrated in Fig. 6.21. In matrix form,

$$\mathbf{R} = \begin{bmatrix} \cos\varphi_3 & \sin\varphi_3 & 0 \\ -\sin\varphi_3 & \cos\varphi_3 & 0 \\ 0 & 0 & 1 \end{bmatrix} \begin{bmatrix} 1 & 0 & 0 \\ 0 & \cos\varphi_2 & \sin\varphi_2 \\ 0 & -\sin\varphi_2 & \cos\varphi_2 \end{bmatrix} \begin{bmatrix} \cos\varphi_1 & \sin\varphi_1 & 0 \\ -\sin\varphi_1 & \cos\varphi_1 & 0 \\ 0 & 0 & 1 \end{bmatrix} \tag{6.61}$$

$$\mathbf{r}'_M = \mathbf{B}^{-1}\,\mathbf{R}\,\mathbf{B}\,\mathbf{r}_M, \tag{6.62}$$

where $\mathbf{B}$ is the transformation matrix from fractional to Cartesian coordinates; $\mathbf{B}^{-1}$ converts the rotated vector back into fractional coordinates.

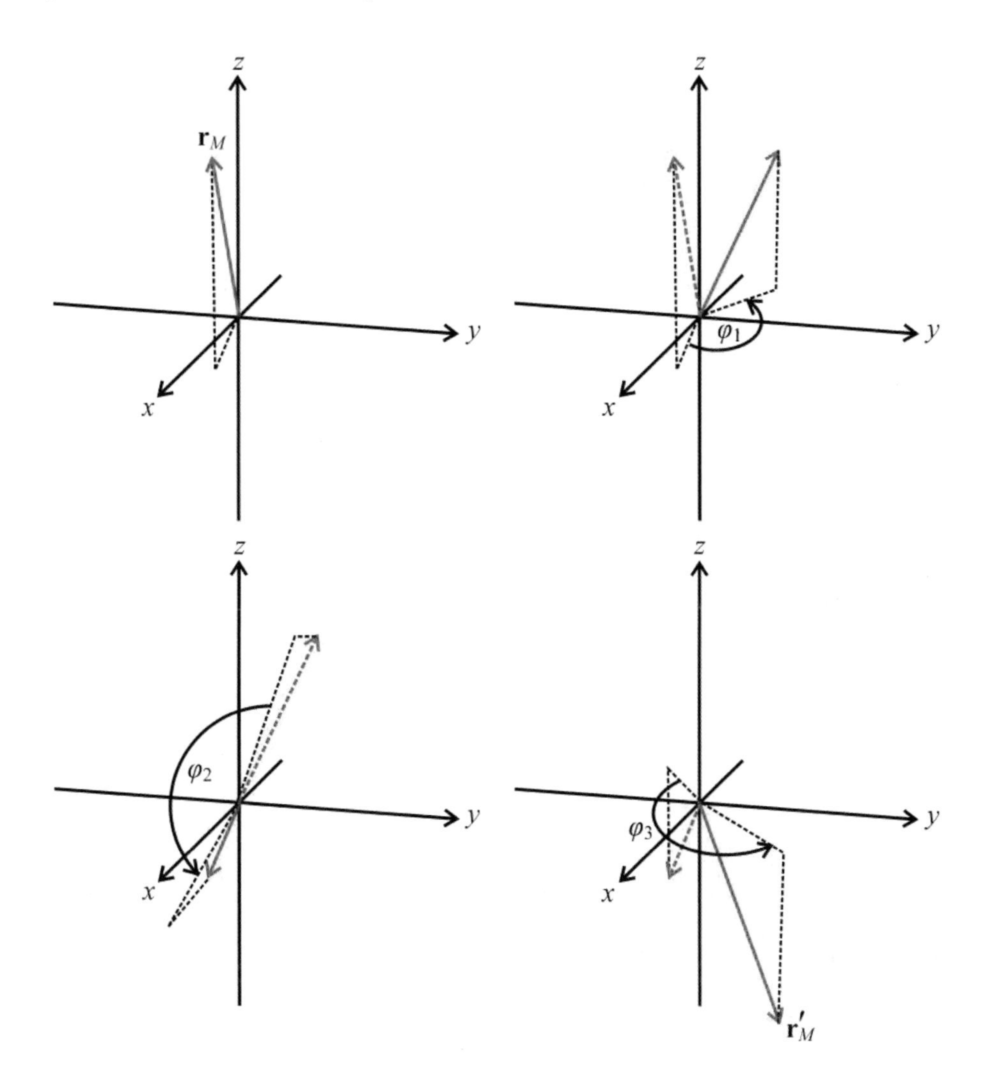

Figure 6.21 Eulerian rotation of a vector by a φ_1 rotation about the z axis, a φ_2 rotation about the x axis, and a second φ_3 rotation about the z axis.

For each orientation of the model, "structure factors" are calculated from the m model vectors in the unit cell:

$$\mathbf{F}_M(\mathbf{h}) = \sum_{j=1}^{m} f_j \, e^{-2\pi i(\mathbf{h}\cdot\mathbf{r}'_{M,j})}. \tag{6.63}$$

While the overall phase of each of these quasi structure factors depends on the location (translation) of the model in the unit cell, their magnitudes do not. If the model is translated by a vector $\mathbf{t}$, then

$$\mathbf{F}'_M(\mathbf{h}) = \sum_{j=1}^{m} f_j \, e^{-2\pi i(\mathbf{h}\cdot(\mathbf{r}'_{M,j}+\mathbf{t}))}. \tag{6.64}$$

Setting $\eta_j = -2\pi i(\mathbf{h}\cdot\mathbf{r}'_{M,j})$ and $\alpha = -2\pi i(\mathbf{h}\cdot\mathbf{t})$,

$$F_M^2 = \left(\sum_j f_j \cos\eta_j\right)^2 + \left(\sum_j f_j \sin\eta_j\right)^2, \quad \text{and} \tag{6.65}$$

$$F_M'^2 = \left(\sum_j f_j \cos(\eta_j+\alpha)\right)^2 + \left(\sum_j f_j \sin(\eta_j+\alpha)\right)^2. \tag{6.66}$$

Expanding Eqn. 6.66 and employing the trigonometric identities, $\cos^2\alpha + \sin^2\alpha = 1$, $\cos(\eta_j+\alpha) = (\cos\eta_j \cos\alpha - \sin\eta_j \sin\alpha)$ and $\sin(\eta_j+\alpha) = (\cos\eta_j \sin\alpha + \sin\eta_j \cos\alpha)$, it is readily shown that

$$F_M'^2 = \left(\sum_j f_j \cos\eta_j\right)^2 + \left(\sum_j f_j \sin\eta_j\right)^2. \tag{6.67}$$

Translating the model adds a constant phase shift to the overall phase of $\mathbf{F}_M$, but does not affect its magnitude. It follows that a model Patterson map, $P_M(\mathbf{u}, \varphi_1, \varphi_2, \varphi_3)$, calculated from these quasi structure factors (e.g., from Eqn. 6.11, 6.14, or 6.17) will depend only on the orientation of the model in the unit cell — determined by the Euler angles. For a given orientation, the agreement between the model Patterson map, $P_M(\mathbf{u})$, and the experimental Patterson map, $P_E(\mathbf{u})$, can be evaluated by computing the product of the value of both functions at each point in the region encompassed by the intramolecular vectors. $P_E(\mathbf{u})P_M(\mathbf{u})$ will have a numerically significant value at $\mathbf{u}$ only if both functions have significant values there. If we integrate (sum) the products over the region, we can define a rotational *overlap function* that will tend to maximize as the maps become coincident with one another:

$$\mathcal{O}_r(\varphi_1, \varphi_2, \varphi_3) = \int_{\mathbf{u}(min)}^{\mathbf{u}(max)} P_E(\mathbf{u})\, P_M(\mathbf{u}, \varphi_1, \varphi_2, \varphi_3)\, d\mathbf{u}. \tag{6.68}$$

Once an optimal value for $\mathcal{O}_r(\varphi_1, \varphi_2, \varphi_3)$ is obtained and the model is oriented in the unit cell, we have only to determine the translation of the model with respect to the unit cell origin. One way to accomplish this is to employ an incremental translational search, translating the model and using Eqn. 6.20 to determine the best fit between F_c and F_o. This requires the generation of calculated structure factors for each increment in the search. A more straightforward approach is to employ the Patterson function. Although, as we have just illustrated, the translation of a model does not alter the contributions to the Patterson Function for the vectors between atoms within the model, translation *does* alter the vectors between the atoms within the model and those related by symmetry in a symmetry-equivalent model within the unit cell. The location of the molecule is defined by the symmetry operations in the unit cell and the vector relationships between symmetry-equivalent "copies" of the model will serve to define its position with respect to the origin. Beginning at a convenient location in the unit cell, symmetry-equivalent atoms are generated by applying the space group symmetry operators. For symmetry operation $\mathbf{S}_j$ and atom i in the model:

$$\mathbf{r}'_i = \mathbf{S}_j\mathbf{r}_i = \mathbf{R}_j\mathbf{r}_i + \mathbf{t}_j. \tag{6.69}$$

The search region is now limited to that containing the intermolecular vectors (known as cross-Patterson vectors), and the search is conducted by translating the model, each time generating symmetry-equivalent atoms and calculating a Patterson map, $P_M(\mathbf{u}, \mathbf{s})$, where $\mathbf{s} = (x_s, y_s, z_s)$ is the vector from the origin of the unit cell to a local origin for the model (the local origin moves with the model as it is translated). This time a translational overlap function is employed to determine the location of the model in the unit cell:

$$\mathcal{O}_t(x_s, y_s, z_s) = \int_{\mathbf{u}'(min)}^{\mathbf{u}'(max)} P_E(\mathbf{u})\, P_M(\mathbf{u}, x_s, y_s, z_s)\, d\mathbf{u}. \tag{6.70}$$

Alternatively, the translation can be conducted by moving the origin and using the structure factors directly, as described in Sec. 6.1.1.

For small molecule structural determinations, Patterson searches have been replaced with direct methods to a large extent.* In contrast, Patterson searches are employed routinely in the solution of macromolecular crystal structures since there are currently no general direct methods for that purpose. The strategy for macromolecular structure solution differs somewhat from that for small molecules. The three-dimensional structures of very large molecules are dictated to a great extent by the way in which their long polymeric chains tend to fold back upon one another, due to local interactions within the polymer. This, in turn, is strongly influenced by the nature of the sequence of monomeric units in the polymer. Polymers with similar monomeric sequences tend to form molecules with similar structures. A large number of macromolecular crystal structures, especially those of proteins (polymers consisting of amino acids), have been elucidated by using the known structure of a molecule as the model for a Patterson search in the crystal lattice of a molecule with an unknown structure, but with a similar (homologous) monomeric sequence. These search techniques are therefore referred to by macromolecular crystallographers as *molecular replacement* methods.[102] In the best cases, the model molecule is almost identical to the target molecule, each differing from the other in only a few regions.*

The Rotation Function

For every rotation in Eqn. 6.68, a new set of model structure factors must be computed. The large number of reflections required for a typical macromolecular structure determination makes the direct application of Eqn. 6.68 very inefficient. As an alternative to the process of rotating the model structure within the experimental unit cell coordinate system, we can consider the unit cells of the model and experimental Patterson maps to share a common origin, effecting the rotational search by rotating the Patterson map of the model unit cell with respect to the experimental Patterson map. If the model unit cell is rotated with respect to the experimental unit cell with the rotation matrix $\mathbf{R}$ (e.g., through the Euler angles),

*Patterson searches are still useful in troublesome cases, especially when combined with direct methods.[101]

*In the worst cases, molecular replacement can result in an incorrect structural determination. A thorough understanding of the fundamentals of crystallography is necessary to in order to avoid such mistakes. While molecular replacement methods are powerful and important tools for the determination of macromolecular crystal structures, they must be used with caution.[103]

then each point defined by $\mathbf{u}$ in the experimental Patterson map will correspond to a point at $\mathbf{u}'$ in the model Patterson map:

$$\mathbf{u}' = \mathbf{B}^{-1}\,\mathbf{R}\,\mathbf{B}\,\mathbf{u} = \mathbf{C}\mathbf{u}. \tag{6.71}$$

From Eqn. 6.9, the Patterson function for the experimental data is

$$P_E(\mathbf{u}) = \frac{1}{V_c}\sum_{\mathbf{h}} F_o^2(\mathbf{h})\, e^{-2\pi\, i(\mathbf{h}\cdot\mathbf{u})}. \tag{6.72}$$

For the model structure, it is

$$\begin{aligned} P_M(\mathbf{u}') &= \frac{1}{V_c}\sum_{\mathbf{h}'} F_M^2(\mathbf{h}')\, e^{-2\pi\, i(\mathbf{h}'\cdot\mathbf{u}')} \\ &= \frac{1}{V_c}\sum_{\mathbf{h}'} F_M^2(\mathbf{h}')\, e^{-2\pi\, i(\mathbf{h}'\cdot\mathbf{C}\mathbf{u})}. \end{aligned} \tag{6.73}$$

If the constant factor $1/V_c^2$ is incorporated into the rotational overlap function, it becomes

$$\mathcal{O}_r(\varphi_1, \varphi_2, \varphi_3) = \int_{\mathbf{u}(min)}^{\mathbf{u}(max)} \sum_{\mathbf{h}} F_o^2(\mathbf{h})\, e^{-2\pi\, i(\mathbf{h}\cdot\mathbf{u})} \sum_{\mathbf{h}'} F_M^2(\mathbf{h}')\, e^{-2\pi\, i(\mathbf{h}'\cdot\mathbf{C}\mathbf{u})}\, d\mathbf{u}. \tag{6.74}$$

The inner product in the second sum in the equation can be transformed so that *the rotational variable is the reciprocal lattice vector*:

$$\mathbf{h}'\cdot\mathbf{C}\mathbf{u} \equiv \mathbf{h}'^T\mathbf{C}\mathbf{u} = (\mathbf{u}^T\mathbf{C}^T\mathbf{h}')^T = \mathbf{u}\cdot\mathbf{C}^T\mathbf{h}', \quad \text{and} \tag{6.75}$$

$$\mathcal{O}_r(\varphi_1, \varphi_2, \varphi_3) = \sum_{\mathbf{h}}\sum_{\mathbf{h}'} F_o^2(\mathbf{h})\, F_M^2(\mathbf{h}') \int_{\mathbf{u}(min)}^{\mathbf{u}(max)} e^{-2\pi\, i(\mathbf{u}\cdot(\mathbf{h}+\mathbf{C}^T\mathbf{h}'))} d\mathbf{u} \tag{6.76}$$

$$= \sum_{\mathbf{h}}\sum_{\mathbf{h}'} F_o^2(\mathbf{h})\, F_M^2(\mathbf{h}') \int_{\mathbf{u}(min)}^{\mathbf{u}(max)} e^{-2\pi\, i(\mathbf{u}\cdot\mathbf{g})} d\mathbf{u}, \tag{6.77}$$

where $\mathbf{g} = \mathbf{h} + \mathbf{C}^T\mathbf{h}'$. The integral form in Eqn. 6.77 has been widely used to describe the interference between scattered waves, and is often referred to as an *interference function.* In this application, it registers the interference between "Patterson waves" in the Fourier series representations of the experimental and model Patterson functions. The representation of the overlap function that incorporates the interference function is known as the *Rossmann-Blow* rotation function, named for its developers[104]. The major advantage of representing the rotational overlap function in this manner is that the variability in the function is encompassed entirely in the integral. It is therefore necessary to compute structure factors for the model structure only once to generate $F_M^2(\mathbf{h}')$; the rotational search is accomplished by rotating the initial reciprocal lattice vectors, $\mathbf{h}'$. Furthermore, if the integration volume is selected appropriately, the integral can be evaluated analytically, leading to a functional form for the overlap function.

In particular, if the integral variable $\mathbf{u}$ is converted to Cartesian coordinates, and then expressed in spherical polar coordinates such that $\mathbf{u} = (u, \eta, \upsilon)$ and $d\mathbf{u} =$

$u^2 \sin\eta\, d\eta\, dv\, du$ — and if $\mathbf{g}$ is aligned along the z (polar) axis (see Fig. 5.22, lower right) then the angle between $\mathbf{u}$ and $\mathbf{g}$ is η and $\mathbf{u}\cdot\mathbf{g} = ug\cos\eta$. Selecting a spherical search region of radius r_s and integrating over the sphere gives

$$\begin{aligned}
\int_{\mathbf{u}} e^{-2\pi\, i(\mathbf{u}\cdot\mathbf{g})}\, d\mathbf{u} &= \int_0^{r_s}\int_0^{\pi}\int_0^{2\pi} e^{-2\pi\, i\,(ug\cos\eta)} u^2 \sin\eta\, d\eta\, dv\, du \\
&= 2\pi \int_0^{r_s} u^2 \left(\int_0^{\pi} e^{-2\pi\, i\,(ug\cos\eta)} \sin\eta\, d\eta\right) du \\
&= \frac{2}{g}\int_0^{r_s} u\sin(2\pi g u)\, du. \qquad (6.78)
\end{aligned}$$

The integral in Eqn. 6.78 is of the form

$$\int x\sin(ax)\, dx = \frac{\sin(ax)}{a^2} - \frac{x\cos(ax)}{a}, \qquad (6.79)$$

where $u \equiv x$ and $2\pi g \equiv a$. Thus

$$\frac{2}{g}\int_0^{r_s} u\sin(2\pi g u)\, du = \frac{2}{g}\left(\frac{\sin(2\pi g r_s)}{(2\pi g)^2} - \frac{r_s\cos(2\pi g r_s)}{2\pi g}\right). \qquad (6.80)$$

The volume of the integration sphere is $V_s = 4/3\pi r_s^3$. Multiplying the first term in Eqn. 6.80 by V_s/V_s and the second term by $2\pi g V_s/2\pi g V_s$ gives

$$\begin{aligned}
\frac{2}{g}\int_0^{r} u\sin(2\pi g u)\, du &= V_s \frac{3(\sin(2\pi g r_s) - 2\pi g r_s\cos(2\pi g r_s))}{(2\pi g r_s)^3} \qquad (6.81) \\
&= V_s G_{\mathbf{hh}'} \text{ and} \qquad (6.82) \\
\mathcal{O}_r(\varphi_1,\varphi_2,\varphi_3) &= V_s \sum_{\mathbf{h}}\sum_{\mathbf{h}'} F_o^2(\mathbf{h})\, F_M^2(\mathbf{h}') G_{\mathbf{hh}'}(\varphi_1,\varphi_2,\varphi_3). \qquad (6.83)
\end{aligned}$$

Fig. 6.22 shows $G_{\mathbf{hh}'}(\varphi_1,\varphi_2,\varphi_3)$ plotted versus $g(\varphi_1,\varphi_2,\varphi_3)r_s$. The function has a maximum value of 1, and contributes very little to the overlap function for values of $gr_s > 1$. Thus only those values of $g(\varphi_1,\varphi_2,\varphi_3)$ that produce $gr_s \leq 1$ need to be calculated, further improving the efficiency of the rotational search. In addition to these improvements the computational speed and accuracy of the spherical expansion can be improved dramatically by expanding the Fourier series using spherical harmonics (Eqn. 5.244) [105].

The Translation Function

The "small molecule" translation overlap function, Eqn. 6.70, also requires the computation of a set of structure factors for each translation of the model, and as it was with Eqn. 6.68, the large number of atoms in a macromolecular structure makes this process cumbersome and time-consuming. As with the rotation function, the process can be expedited considerably if we can create an overlap function that requires only a single determination of the structure factors for the model.

The goal of the translation search is to map the maxima resulting from the intermolecular vectors between the translated model and its symmetry-equivalents onto the experimental Patterson map. The model Patterson map for these cross vectors results from the integral of the product of electron density functions, $\rho(\mathbf{r})$

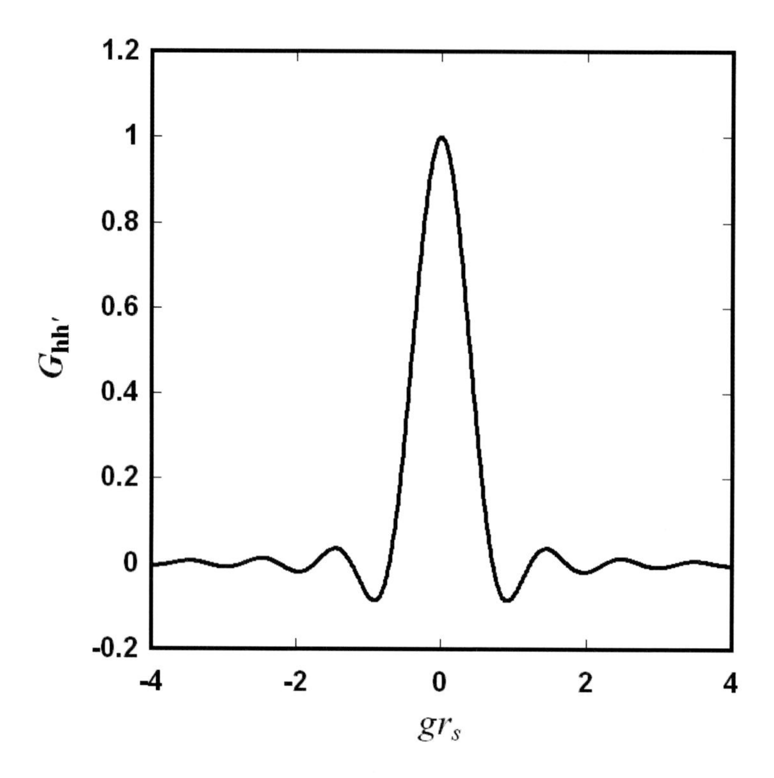

Figure 6.22 The interference function, $G_{\mathbf{hh}'}(\varphi_1, \varphi_2, \varphi_3)$, plotted as a function of $g(\varphi_1, \varphi_2, \varphi_3)r_s$.

at points within the model structure, and $\rho(\mathbf{r}')$ at points within a second model structure related by a symmetry transformation to the first, illustrated in Fig. 6.23:

$$P_X(\mathbf{u}) = \int_{V_c} \rho(\mathbf{r})\,\rho(\mathbf{r}')\,d\mathbf{r} = \int_{V_c} \rho(\mathbf{r})\,\rho(\mathbf{r}+\mathbf{u})\,d\mathbf{r}. \tag{6.84}$$

In order to determine structure factors for the model, we place the origin of the unit cell at a local origin for the model, o_M:

$$\mathbf{F}_M(\mathbf{h}) = \sum_{j=1}^{n} f_j e^{-2\pi\, i(\mathbf{h}\cdot\,\mathbf{r}_{M,j})}, \tag{6.85}$$

where $\mathbf{r}_{M,j}$ is the position of atom j in the model with respect to the local origin. The electron density at some point $\mathbf{r}_M$ in the model can now be expanded as a Fourier series using these calculated structure factors:

$$\rho_M(\mathbf{r}_M) = \frac{1}{V_c}\sum_{\mathbf{h}} F_M(\mathbf{h})e^{-2\pi\, i(\mathbf{h}\cdot\,\mathbf{r}_M)}. \tag{6.86}$$

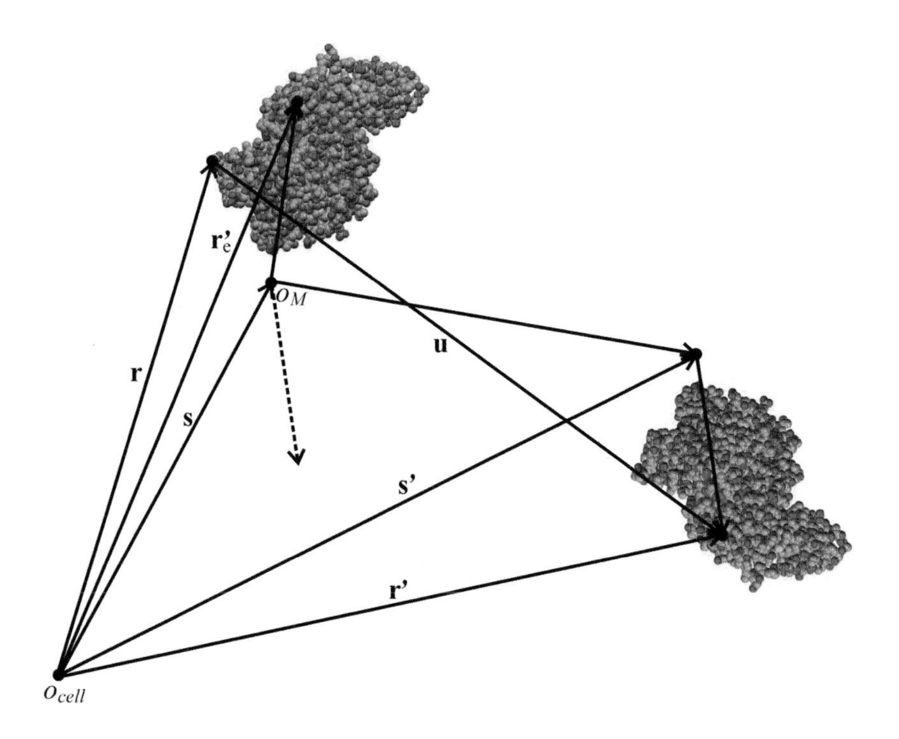

Figure 6.23 Cross-Patterson vector, **u**, between a point within a model structure and a point within a symmetry-equivalent model structure in the unit cell.

If the model is now translated by moving the local origin to a point in the unit cell at the end of the vector **s**, and **r** is at the translated point $\mathbf{r}_M$, then $\mathbf{r}_M = \mathbf{r} - \mathbf{s}$ and

$$\rho(\mathbf{r}) = \rho_M(\mathbf{r} - \mathbf{s}) = \frac{1}{V_c} \sum_{\mathbf{h}} F_M(\mathbf{h}) e^{-2\pi\, i(\mathbf{h}\cdot\, (\mathbf{r}-\mathbf{s}))}. \tag{6.87}$$

A symmetry-equivalent model can be positioned in the unit cell by applying the symmetry operator, $\mathbf{S} = \mathbf{R} + \mathbf{t}$ to every point in the model, *or* the transformation can be applied solely to the local origin:

$$\mathbf{s}' = \mathbf{R}\mathbf{s} + \mathbf{t}. \tag{6.88}$$

The model is then transformed locally with respect to o_M using the point symmetry operator, **R**, followed by translation of the origin from **s** to $\mathbf{s}'$. This equivalency can be seen in Fig. 6.23, where the two molecules are seen to be related by either a horizontal two-fold screw axis or a glide plane, midway between the molecules. If the symmetry operator is moved vertically to the local origin and the first molecule is rotated (or reflected), a translation that centers the origin at $\mathbf{s}'$ will place the molecule in its symmetry-equivalent position.

In order to determine the electron density at the point $\mathbf{r}'$ within the symmetry-equivalent model, we perform the inverse operation to determine the electron density at the equivalent point, $\mathbf{r}'_{\mathbf{e}}$, within the original model. Subtracting $\mathbf{s}'$ from $\mathbf{r}'$ creates a local vector, $\mathbf{r}'_{\mathbf{M}}$, translated to the local origin, as indicated by the dashed arrow in the figure. The vector is then operated upon by $\mathbf{R}^{-1}$, transforming it to a symmetry-equivalent vector within the original model:

$$\mathbf{r}'_{M\mathbf{e}} = \mathbf{R}^{-1}(\mathbf{r}' - \mathbf{s}') = \mathbf{R}^{-1}(\mathbf{r} + \mathbf{u} - \mathbf{R}\mathbf{s} - \mathbf{t})). \tag{6.89}$$

The electron density at $\mathbf{r}'$ is therefore

$$\rho(\mathbf{r}+\mathbf{u}) = \rho_M(\mathbf{R}^{-1}(\mathbf{r}' - \mathbf{s}')) = \frac{1}{V_c}\sum_{\mathbf{h}'} F_M(\mathbf{h}')e^{-2\pi\, i(\mathbf{h}'\cdot\, (\mathbf{R}^{-1}(\mathbf{r}+\mathbf{u}-\mathbf{R}\mathbf{s}-\mathbf{t})))}. \tag{6.90}$$

Substituting the Fourier series expressions for the electron density gives us a cross-Patterson vector function based on the model to be translated that explicitly contains $\mathbf{s}$, the location of the local origin of the model in the unit cell:

$$\begin{aligned} P_X(\mathbf{u}, \mathbf{s}) &= \frac{1}{V_c^2}\int_{V_c}\sum_{\mathbf{h}} F_M(\mathbf{h})e^{-2\pi\, i(\mathbf{h}\cdot\, (\mathbf{r}-\mathbf{s}))}\sum_{\mathbf{h}'} F_M(\mathbf{h}')e^{-2\pi\, i(\mathbf{h}'\cdot\, (\mathbf{R}^{-1}(\mathbf{r}+\mathbf{u}-\mathbf{R}\mathbf{s}-\mathbf{t})))}d\mathbf{r} \\ &= \frac{1}{V_c^2}\sum_{\mathbf{h}}\sum_{\mathbf{h}'} F_M(\mathbf{h})F_M(\mathbf{h}')\, e^{2\pi\, i(\mathbf{h}\cdot\,\mathbf{s}\, +\, \mathbf{h}'\cdot\, (\mathbf{R}^{-1}(\mathbf{R}\mathbf{s}\, +\, \mathbf{t})))} \\ &\quad \times\ e^{-2\pi\, i(\mathbf{h}'\cdot\, \mathbf{R}^{-1}\mathbf{u})}\int_{V_c} e^{-2\pi\, i(\mathbf{h}\,\cdot\,\mathbf{r}\ +\ \mathbf{h}'\,\cdot\,\mathbf{R}^{-1}\mathbf{r})}\, d\mathbf{r}. \end{aligned} \tag{6.91}$$

The argument in the exponential in the integrand contains the inner product of reciprocal lattice vectors with integer components, $\mathbf{h} = [hkl]$ and $\mathbf{h}' = [h'k'l']$, with $\mathbf{h}\cdot\mathbf{r}\ +\ \mathbf{h}'\cdot\mathbf{R}^{-1}\mathbf{r} = (\mathbf{h}^T + \mathbf{h}'^T\mathbf{R}^{-1})\mathbf{r}$. But $\mathbf{h}'^T\mathbf{R}^{-1} = [(\mathbf{R}^{-1})^T\mathbf{h}'] = (\mathbf{R}\mathbf{h}')^T$, where $\mathbf{R}$ is a point symmetry operator in the lattice, generating symmetry-equivalent points in both direct space — often in conjunction with a translation operator — and in reciprocal space — where there is no translational symmetry. It follows that $\mathbf{R}\mathbf{h}' = \mathbf{h}''$, another reciprocal lattice vector with integer components. The argument therefore becomes $(\mathbf{h}^T + \mathbf{h}''^T)\mathbf{r} = (\mathbf{h} + \mathbf{h}'')\cdot\mathbf{r}$, which, as we have already observed, causes the integral to vanish (Eqn. 6.6) except when $\mathbf{h} = -\mathbf{h}'' = -\mathbf{R}\mathbf{h}'$, in which case the integral evaluates to the volume of the unit cell, V_c (Eqn. 6.7). Since $\mathbf{R}$ is an orthogonal transformation matrix, $\mathbf{h}' = -\mathbf{R}^{-1}\mathbf{h} = -\mathbf{R}^T\mathbf{h}$, and the arguments in the exponentials in Eqn. 6.91 reduce to

$$\begin{aligned} &\mathbf{h}\cdot\,\mathbf{s} + \mathbf{h}'\cdot\,(\mathbf{R}^{-1}(\mathbf{R}\mathbf{s} + \mathbf{t}))) \ = \ \mathbf{h}^T\mathbf{s} - (\mathbf{R}^T\mathbf{h})^T\mathbf{s} - (\mathbf{R}^T\mathbf{h})^T(\mathbf{R}^T\mathbf{t}) \\ &= \ \mathbf{h}^T\mathbf{s} - \mathbf{h}^T\mathbf{R}\mathbf{s} - \mathbf{h}^T\mathbf{R}\mathbf{R}^T\mathbf{t} \ = \ \mathbf{h}^T(\mathbf{s} - \mathbf{R}\mathbf{s} - \mathbf{t}) \\ &= \ \mathbf{h}\cdot(\mathbf{s} - \mathbf{R}\mathbf{s} - \mathbf{t}) \ \text{ and} \end{aligned} \tag{6.92}$$

$$\begin{aligned} &\mathbf{h}'\cdot\,(\mathbf{R}^{-1}\mathbf{u}) \ = \ -\mathbf{h}^T\mathbf{R}\mathbf{R}^T\mathbf{u} \ = \ -\mathbf{h}^T\mathbf{u} \\ &= \ -\mathbf{h}\cdot\mathbf{u}. \end{aligned} \tag{6.93}$$

The model cross-Patterson vector function becomes

$$P_X(\mathbf{u}, \mathbf{s}) = \frac{1}{V_c}\sum_{\mathbf{h}} F_M(\mathbf{h})F_M(-\mathbf{R}^T\mathbf{h})\, e^{2\pi\, i(\mathbf{h}\cdot\, (\mathbf{s}\, -\, \mathbf{R}\mathbf{s}\, -\, \mathbf{t}))}\, e^{2\pi\, i(\mathbf{h}\cdot\,\mathbf{u})}. \tag{6.94}$$

The translational overlap function between the cross vector function and the experimental Patterson function is

$$\mathcal{O}_t(x_s, y_s, z_s) = \mathcal{O}_t(\mathbf{s}) = \int_{V_c} P_E(\mathbf{u})\, P_X(\mathbf{u}, \mathbf{s})\, d\mathbf{u}$$

$$= \int_{V_c} \sum_{\mathbf{h}'} F_o^2(\mathbf{h}') e^{-2\pi\, i(\mathbf{h}'\cdot\mathbf{u})} \frac{1}{V_c} \sum_{\mathbf{h}} F_M(\mathbf{h}) F_M(-\mathbf{R}^T\mathbf{h})\, e^{2\pi\, i(\mathbf{h}\cdot\,(\mathbf{s}\,-\,\mathbf{Rs}\,-\,\mathbf{t}))}\, e^{2\pi\, i(\mathbf{h}\cdot\,\mathbf{u})} d\mathbf{u}$$

$$= \frac{1}{V_c} \sum_{\mathbf{h}'} \sum_{\mathbf{h}} F_o^2(\mathbf{h}') F_M(\mathbf{h}) F_M(-\mathbf{R}^T\mathbf{h})\, e^{2\pi\, i(\mathbf{h}\cdot\,(\mathbf{s}\,-\,\mathbf{Rs}\,-\,\mathbf{t}))} \int_{V_c} e^{-2\pi\, i((\mathbf{h}'-\mathbf{h})\cdot\,\mathbf{u})}\, d\mathbf{u}.$$

Again, the integral vanishes unless $\mathbf{h}' = \mathbf{h}$, and evaluates to the cell volume whenever $\mathbf{h}' = \mathbf{h}$, resulting in an analytical form for the translational overlap function:

$$\mathcal{O}_t(x_s, y_s, z_s) = \mathcal{O}_t(\mathbf{s}) = \sum_{\mathbf{h}} F_o^2(\mathbf{h}) F_M(\mathbf{h}) F_M(-\mathbf{R}^T\mathbf{h})\, e^{2\pi\, i(\mathbf{h}\cdot\,(\mathbf{s}\,-\,\mathbf{Rs}\,-\,\mathbf{t}))}. \quad (6.95)$$

As the model is translated throughout the unit cell the overlap function is evaluated for values of $\mathbf{s} - \mathbf{Rs} - \mathbf{t} = \mathbf{s} - \mathbf{s}'$, the vector between the local origin of the model structure and its symmetry-equivalent for a selected symmetry operator in the space group. A maximum value for $\mathcal{O}_t(\mathbf{s})$ is indicative of an optimal translation. The translation function, originally authored by Crowther and Blow[106], can be modified to remove the intramolecular vectors from the experimental Patterson map, since they are unnecessary, and only serve to increase random error in the computation of the Fourier sum that determines $\mathcal{O}_t(\mathbf{s})$. The function can be further modified to include all of the symmetry operators simultaneously.

6.2 Other Experimental Methods

6.2.1 The Isomorphous Replacement Method

Although the molecular replacement method described in the previous section is the most commonly used strategy for the determination of macromolecular structures, it requires model structures; at some point these structures must be determined *ab initio* — from the beginning.

While heavy atoms can usually be incorporated into macromolecules, they do not account for a large enough percentage of the scattering to generate a structural solution from ordinary Patterson or Fourier methods. However, it is often possible to obtain crystals of the macromolecule that contain different atoms in one or more positions, but are otherwise effectively structurally identical (*isomorphous*). The structures are viewed as having one atom replaced with another (or an empty site in one structure filled with an atom in another). The method is therefore known as *isomorphous replacement*.

While it is possible to infuse replacement atoms directly into macromolecular crystals, isomorphic crystals containing small molecules are more difficult to obtain. Since the formation of crystals depends on subtle intermolecular/interionic interactions, very similar molecules can pack in very different unit cells. The method therefore has seen only limited use in small-molecule crystallography, since it depends on somewhat fortuitous circumstances. Nevertheless, in the early days of

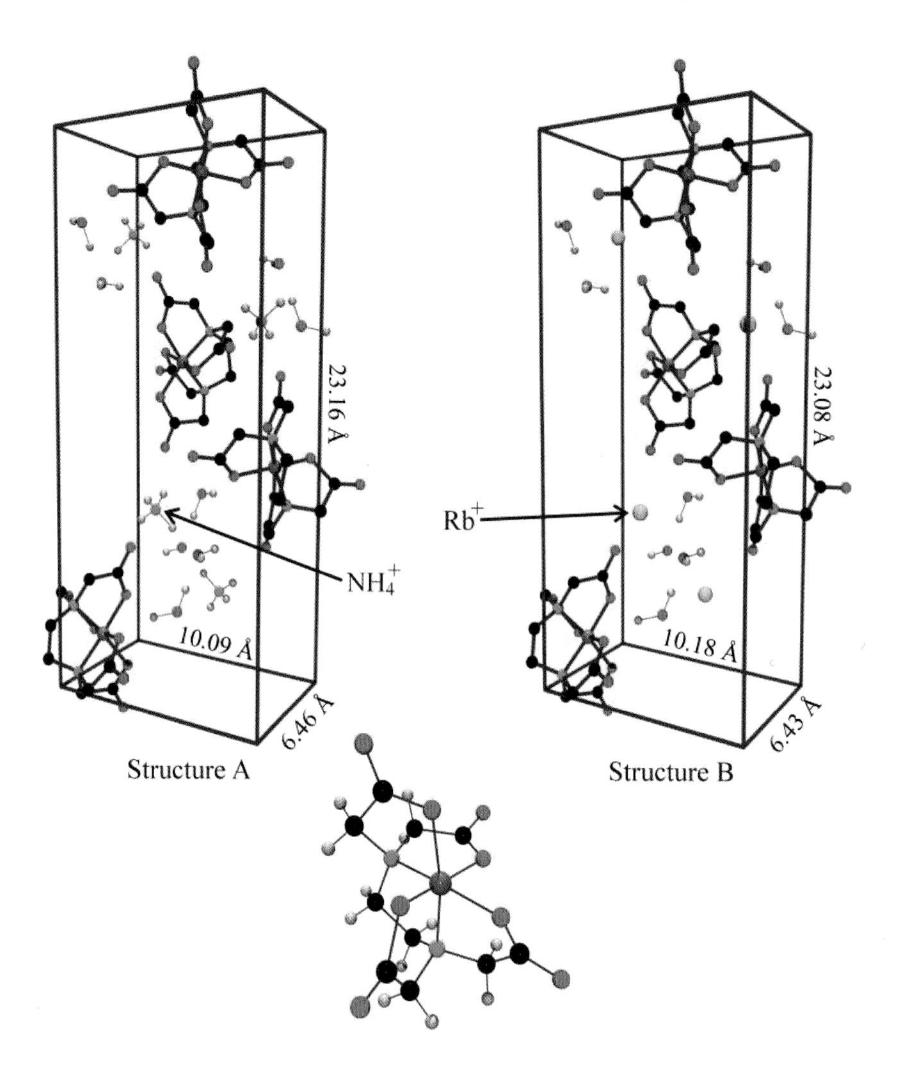

Figure 6.24 Structure A: Ammonium Ethylenediaminetetraacetatocobaltate(III). Structure B: Rubidium Ethylenediaminetetraacetatocobaltate(III).

molecular structure determination, a number of crystal structures were solved by isomorphous replacement. An example of such a structure, reported in 1959[107], is illustrated in Fig. 6.24. The unit cell for Structure A in the figure contains the $NH_4Co(III)(EDTA){\cdot}2H_2O$ asymmetric unit, while the unit cell for Structure B contains $RbCo(III)(EDTA){\cdot}2H_2O$. Although the ammonium ion and the rubidium ion may seem to be quite different, the effective ionic radius of NH_4^+ is very close to the ionic radius of Rb^+. This similarity results in nearly identical unit cells, both in space group $P2_12_12_1$. While the two salts are structurally isomorphous, the nitrogen atom* of the ammonium ion contributes significantly less to X-ray scattering

*We ignore the hydrogen atoms since their electron density is diffuse and they contribute only a small amount to the scattering.

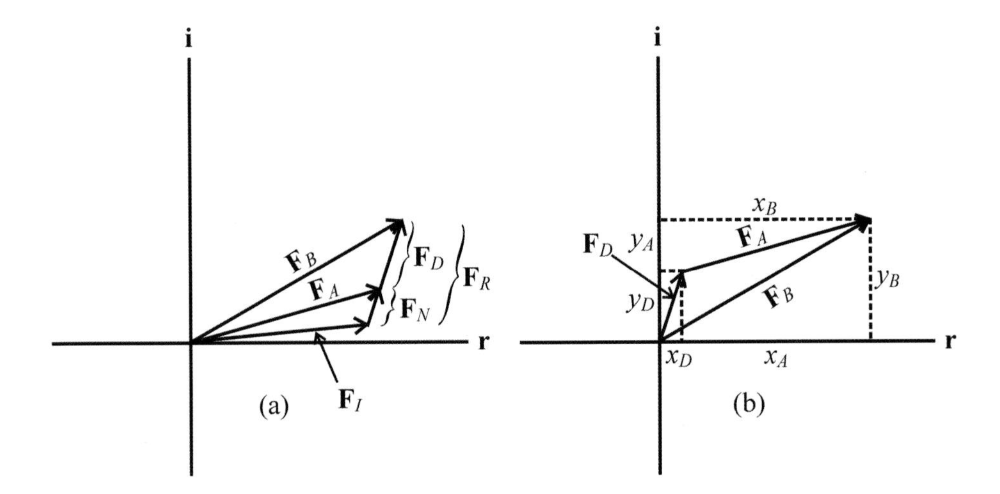

Figure 6.25 (a) Structure factor vectors for reflection **h** scattering from isomorphous crystals A and B. (b) Relationship between the structure factors for the reflection.

than the rubidium ion does, and *it is this difference that allows us to determine the phases of the reflections from both structures.*

The structural solution begins with a Patterson map of Structure B, from which four symmetry-equivalent heavy rubidium ions are located. Four ammonium ions are then assumed to be in the same locations in the isomorphic unit cell of Structure A. For a given **h** the structure factors for the reflection from each crystal are given by

$$\mathbf{F}_A(\mathbf{h}) = \sum_{j=1}^{n} f_j e^{-2\pi i(\mathbf{h}\cdot\mathbf{r}_j)} + f_{\mathrm{N}} \sum_{k=1}^{4} e^{-2\pi i(\mathbf{h}\cdot\mathbf{r}_{\mathrm{N},k})} = \mathbf{F}_I(\mathbf{h}) + \mathbf{F}_N(\mathbf{h}) \tag{6.96}$$

$$\mathbf{F}_B(\mathbf{h}) = \sum_{j=1}^{n} f_j e^{-2\pi i(\mathbf{h}\cdot\mathbf{r}_j)} + f_{\mathrm{Rb}} \sum_{k=1}^{4} e^{-2\pi i(\mathbf{h}\cdot\mathbf{r}_{\mathrm{Rb},k})} = \mathbf{F}_I(\mathbf{h}) + \mathbf{F}_R(\mathbf{h}), \tag{6.97}$$

where the first sum in each expression corresponds to the n atoms in each unit cell that are in effectively the same locations, and $\mathbf{r}_{\mathrm{N}}$ and $\mathbf{r}_{\mathrm{Rb}}$ are the locations of the ammonium ions and rubidium ions, respectively. These vectors are shown in Fig. 6.25(a). Because $\mathbf{r}_{\mathrm{N},k} = \mathbf{r}_{\mathrm{Rb},k}$ the exponential terms in Eqns. 6.96 and 6.97 are identical, and $\mathbf{F}_N(\mathbf{h})$ and $\mathbf{F}_R(\mathbf{h})$ have the same phase angle, differing only in magnitude. Because we know the locations of the rubidium and ammonium ions, $\mathbf{F}_N(\mathbf{h})$ and $\mathbf{F}_R(\mathbf{h})$ can be determined, but we know only the magnitudes $F_A(\mathbf{h})$ and $F_B(\mathbf{h})$; we seek to determine the phase angles, $\varphi_A(\mathbf{h})$ and $\varphi_B(\mathbf{h})$.

Solving for the constant contribution to both structure factors, $\mathbf{F}_I(\mathbf{h})$, gives

$$\mathbf{F}_A(\mathbf{h}) - \mathbf{F}_N(\mathbf{h}) = \mathbf{F}_B(\mathbf{h}) - \mathbf{F}_R(\mathbf{h}) \tag{6.98}$$

$$\begin{aligned}\mathbf{F}_B(\mathbf{h}) &= \mathbf{F}_A(\mathbf{h}) + (\mathbf{F}_R(\mathbf{h}) - \mathbf{F}_N(\mathbf{h})) \\ &= \mathbf{F}_A(\mathbf{h}) + \mathbf{F}_D(\mathbf{h}),\end{aligned} \tag{6.99}$$

where $\mathbf{F}_D(\mathbf{h})$ is known from the location of the ammonium and rubidium ions. Fig. 6.25(b) illustrates the vectors in Eqn. 6.99 in the complex plane in which

$$\mathbf{F}_A(\mathbf{h}) = x_A + i\,y_A \tag{6.100}$$
$$\mathbf{F}_B(\mathbf{h}) = x_B + i\,y_B \tag{6.101}$$
$$\mathbf{F}_D(\mathbf{h}) = x_D + i\,y_D \tag{6.102}$$

The known magnitudes of the structure factors are

$$F_A^2(\mathbf{h}) = x_A^2 + y_A^2 \tag{6.103}$$
$$\begin{aligned} F_B^2(\mathbf{h}) &= x_B^2 + y_B^2 \\ &= (x_A + x_D)^2 + (y_A + y_D)^2 \\ &= x_A^2 + y_A^2 + 2\,x_D\,x_A + 2\,y_D\,y_A + x_D^2 + y_D^2 \\ &= x_A^2 + y_A^2 + D\,x_A + E\,y_A + G, \end{aligned} \tag{6.104}$$

where $D = 2x_D$, $E = 2y_D$, and $G = x_D^2 + y_D^2$, all known quantities. Subtracting Eqn. 6.103 from Eqn. 6.104, and setting $H = F_B^2(\mathbf{h}) - F_A^2(\mathbf{h})$ (a known quantity),

$$H = F_B^2(\mathbf{h}) - F_A^2(\mathbf{h}) = D\,x_A + E\,y_A + G, \tag{6.105}$$

providing x_A in terms of y_A:

$$x_A = \frac{1}{D}(H - E\,y_A - G) = \frac{1}{D}(I - E\,y_A), \tag{6.106}$$

where $I = H - G$. Substituting the expression for x_A into Eqn. 6.103 results in a quadratic equation:

$$\left(\frac{E^2}{D^2} + 1\right) y_A^2 - \left(\frac{2IE}{D^2}\right) y_A + \left(\frac{I^2}{D^2} - F_A^2\right) = 0. \tag{6.107}$$

Solving the equation for y_A and substituting into Eqn. 6.106 to obtain x_A *should* provide the phase angles for the reflections from both crystals:

$$\varphi_A(\mathbf{h}) = \arctan\left(\frac{y_A}{x_A}\right) \tag{6.108}$$
$$\varphi_B(\mathbf{h}) = \arctan\left(\frac{y_B}{x_B}\right) = \arctan\left(\frac{y_A + y_D}{x_A + x_D}\right). \tag{6.109}$$

However, the quadratic equation yields two roots, and therefore two possible values of $\varphi_A(\mathbf{h})$ and $\varphi_B(\mathbf{h})$ for every reflection. This ambiguity can be seen graphically in Fig. 6.26(a) by noting that Eqn. 6.103 is the equation of a circle of radius F_A centered at the origin, and Eqn. 6.104 is the equation of a circle of radius F_B centered at $-\mathbf{F}_D = (-x_D, -y_D)$ (with x_A and y_A as the variables). The points of intersection of the two circles are the solutions to the simultaneous equations; each solution produces two vectors with magnitudes F_A and F_B, but at different phase angles.

There have been several approaches to the resolution of the ambiguities that occur when a *single isomorphous replacement* is made. One approach is to use *both* possible structure factor vectors for each reflection in a Fourier series, with the

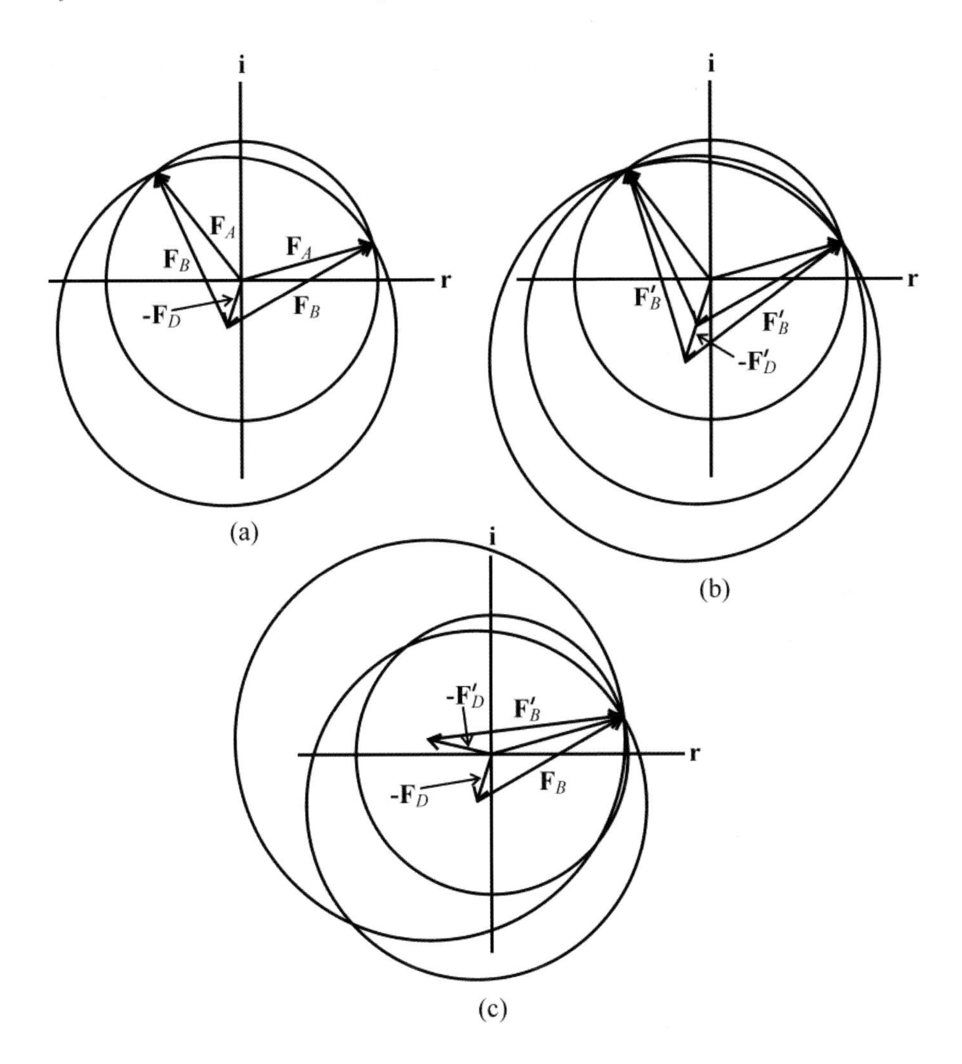

Figure 6.26 (a) Possible structure factor vectors for reflection **h** scattering from isomorphous crystals A and B. (b) Possible structure factor vectors for reflection **h** scattering from isomorphous crystals A, B, and B', with the "replacement atoms" located in the same position. (c) Possible structure factor vectors for reflection **h** scattering from isomorphous crystals A, B, and B', with the "replacement atoms" located in the different positions.

assumption that the vectors with correct phases will interfere to produce maxima at atomic locations, while those with incorrect phases will interfere randomly, and tend to appear as background noise in the Fourier map. Unfortunately, this background noise is usually substantial, and the resulting Fourier map (Known as a SIR map), is often too noisy to produce an unequivocal structure.

A more systematic line of attack is to create one or more additional isomorphic crystals containing different heavy atoms, an approach known as *multiple isomorphous replacement.* Fig. 6.26(b) illustrates an important constraint on the additional isomorphic structure (Structure B′) — the "replacement atoms" must reside in different locations from those in Structure B. Suppose that this is not the case — that the structure contains cesium atoms in place of the rubidium atoms. Because the heavier cesium atoms are in the same locations, they will scatter with a different magnitude than the rubidium atoms, but with the same phase. The difference vector, $\mathbf{F}'_D$, will be collinear with $\mathbf{F}_D$, with $x'_D = k\,x_D$ and $y'_D = k\,y_D$, resulting in $D' = kD$, $E' = kE$ and $I' = H' - G' = D'\,x_A + E'\,y_A = kI$. Substituting D', E' and I' into Eqn. 6.107 results in an identical equation with identical solutions — the three circles intersect at the same two points and the ambiguity remains. Suppose, instead, that an isomorphic crystal is obtained which contains a rhodium atom in place of the cobalt atom in the EDTA complex. In this case $\mathbf{F}'_D$ will *not* be collinear with $\mathbf{F}_D$. Eqn. 6.107 will now generate a second set of solutions, but only one set will be consistent with the solutions generated from Structures A and B. This is illustrated graphically in Fig. 6.26(c); the circles share only a single point in common and the assignment of phases to all three reflections is now unambiguous.

The solution of small-molecule structures using multiple isomorphous replacement requires the formation of three or more isomorphous crystals with heavy "replacement atoms" in at least two different locations in the unit cell. The probability that such crystals will form with the same cell parameters, space group symmetry, and atomic locations is extremely small. The method is therefore virtually never employed for small molecule structure determination. On the other hand, while the probability that multiple isomorphous macromolecular crystals will form is equally unlikely, *the crystals of macromolecules can often be modified to incorporate heavy atoms at specific sites in the molecules within the crystals.* Biological macromolecules generally co-crystallize with a large number of water molecules; they are often described as being partly "in solution" within the crystal. If crystals of the native molecule (without heavy metals) are soaked in an aqueous solution containing a heavy metal salt, the metal ions can diffuse into the crystal, eventually finding themselves at preferential coordination sites within the macromolecule. Alkali and alkaline earth metal ions tend to coordinate in sites where ionic bonding dominates, while transition metal ions tend to prefer coordinate-covalent bonding. It is therefore possible to select metal ions that will occupy different sites in the crystal.

The solution of macromolecular structures using multiple isomorphous replacement still requires locating the heavy atoms, a task that cannot generally be performed with an ordinary Patterson map because of the tremendous number of overlapping Patterson vectors. Fortunately, the availability of data from isomorphous crystals allows us to determine the heavy atom locations by creating a *difference Patterson map,* which eliminates the Patterson peaks common to both structures. The difference-Patterson map is computed with the intensities from two isomorphous structures:

$$P_\delta(\mathbf{u}) = \frac{1}{V_c}\sum_{\mathbf{h}} I_B(\mathbf{h})e^{-2\pi\, i(\mathbf{u}\cdot\mathbf{h})} - \frac{1}{V_c}\sum_{\mathbf{h}} I_A(\mathbf{h})e^{-2\pi\, i(\mathbf{u}\cdot\mathbf{h})} \tag{6.110}$$

$$= \frac{1}{V_c}\sum_{\mathbf{h}} (F_B^2(\mathbf{h}) - F_A^2(\mathbf{h}))\, e^{-2\pi\, i(\mathbf{u}\cdot\mathbf{h})}. \tag{6.111}$$

In principle, the vectors between identical atoms in identical positions — those not involving the replacement atoms — should subtract from one another, leaving behind a Patterson function with a significant reduction in the number of vectors, especially for a macromolecular crystal. The equations for the small molecule example are perfectly general, and we will retain the notation, using $\mathbf{F}_N(\mathbf{h})$ and $\mathbf{F}_R(\mathbf{h})$ to indicate the replacement atom contributions to the corresponding structure factors. The coefficients in the difference-Patterson Fourier series are devoid of contributions to vectors between atoms other than those involving the replacement atoms:

$$\begin{aligned} F_B^2(\mathbf{h}) - F_A^2(\mathbf{h}) &= \mathbf{F}_B(\mathbf{h})\mathbf{F}_B^*(\mathbf{h}) - \mathbf{F}_A(\mathbf{h})\mathbf{F}_A^*(\mathbf{h}) && (6.112)\\ &= (\mathbf{F}_I(\mathbf{h}) + \mathbf{F}_R(\mathbf{h}))\,(\mathbf{F}_I^*(\mathbf{h}) + \mathbf{F}_R^*(\mathbf{h}))\\ &\quad -(\mathbf{F}_I(\mathbf{h}) + \mathbf{F}_N(\mathbf{h}))\,(\mathbf{F}_I^*(\mathbf{h}) + \mathbf{F}_N^*(\mathbf{h})) && (6.113)\\ &= (F_R^2(\mathbf{h}) - F_N^2(\mathbf{h})) + (\mathbf{F}_R(\mathbf{h}) - \mathbf{F}_N(\mathbf{h}))\mathbf{F}_I^*(\mathbf{h})\\ &\quad +(\mathbf{F}_R^*(\mathbf{h}) - \mathbf{F}_N^*(\mathbf{h}))\mathbf{F}_I(\mathbf{h}). && (6.114) \end{aligned}$$

The $\mathbf{F}_I(\mathbf{h})\mathbf{F}_I^*(\mathbf{h})$ terms cancel, leaving only contributions for vectors between replacement atoms in the two structures (the first term in Eqn. 6.114) and between replacement atoms and the atoms in the "identical" portions of the structures (the second and third terms). The difference-Patterson map will contain peaks at the locations of the replacement atom vectors (the heavy atom vectors) with magnitudes largely dependent on the difference between $F_R^2(\mathbf{h})$ and $F_N^2(\mathbf{h})$.

The difference-Patterson map still contains a large number of peaks, and if $F_R^2(\mathbf{h})$ and $F_N^2(\mathbf{h})$ are not too different, the heavy atom difference vectors may be difficult to discern from those due to other vectors between heavy atoms and lighter atoms. Ideally, the difference-Patterson function would generate only those vectors due to the replacement atoms in the structure. If the structures of the isomorphic crystals were known, it would be possible to calculate such a function. A "difference" electron density function would have

$$\begin{aligned} \Delta\rho &= \rho_B(\mathbf{r}) - \rho_A(\mathbf{r}) = \frac{1}{V_c}\sum_{\mathbf{h}}(\mathbf{F}_B(\mathbf{h}) - \mathbf{F}_A(\mathbf{h}))\,e^{-2\pi i(\mathbf{r}\cdot\mathbf{h})}\\ &= \frac{1}{V_c}\sum_{\mathbf{h}}\mathbf{F}_D(\mathbf{h})\,e^{-2\pi i(\mathbf{r}\cdot\mathbf{h})}, && (6.115) \end{aligned}$$

where $\Delta\rho$ would have substantial values only at the positions of the replacement atoms, since the remainder of the electron density is the same in both structures. Referring to Eqn. 6.9, this difference structure would generate a Patterson function with coefficients that are the squared magnitudes of the differences between the structure factors for the two isomorphic crystals,

$$\begin{aligned} P_\Delta(\mathbf{u}) &= \frac{1}{V_c}\sum_{\mathbf{h}}|\mathbf{F}_B(\mathbf{h}) - \mathbf{F}_A(\mathbf{h})|^2 e^{-2\pi i(\mathbf{u}\cdot\mathbf{h})}\\ &= \frac{1}{V_c}\sum_{\mathbf{h}}|\mathbf{F}_D(\mathbf{h})|^2 e^{-2\pi i(\mathbf{u}\cdot\mathbf{h})} = \frac{1}{V_c}\sum_{\mathbf{h}}F_D^2 e^{-2\pi i(\mathbf{u}\cdot\mathbf{h})}, && (6.116) \end{aligned}$$

yielding vectors exclusively between the replacement atom positions.

Since the phases of $\mathbf{F}_A(\mathbf{h})$ and $\mathbf{F}_B(\mathbf{h})$ are unknown, we do not have the values of F_D necessary to generate the "ideal" difference-Patterson function. Fig. 6.27 shows the relationships between the structure factor vectors for reflection $\mathbf{h}$ necessary to

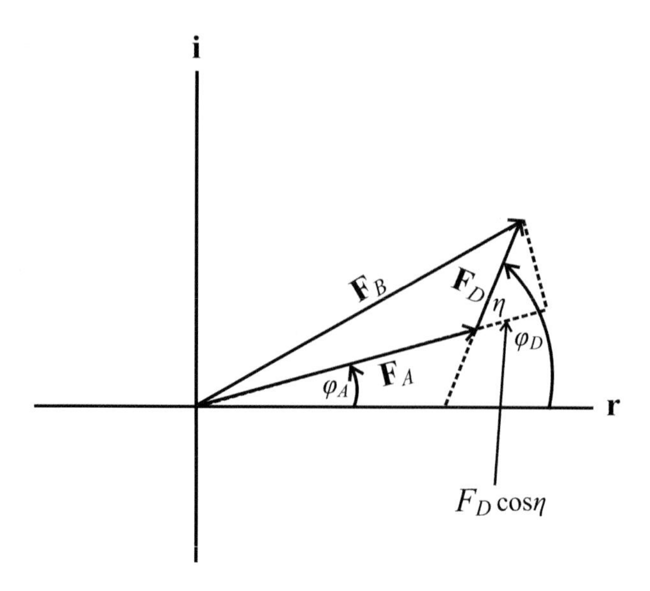

Figure 6.27 Phase relations between structure factor vectors for reflection **h** scattering from isomorphous crystals A and B.

create such a function. For macromolecules, the major contribution to the structure factors will come from the many atoms in both structures that are identical, and the vectors are therefore reasonably close to parallel. Because of this a perpendicular dropped from the head of $\mathbf{F}_B(\mathbf{h})$ to a line collinear with $\mathbf{F}_A(\mathbf{h})$ will intersect the line at a length nearly equal to F_B, such that

$$F_B \simeq F_A + F_D \cos\eta \quad \text{and} \quad (F_B - F_A)^2 \simeq F_D^2 \cos^2 \eta, \tag{6.117}$$

where $\eta = \varphi_D - \varphi_A$, the difference between the phase angles of $\mathbf{F}_D(\mathbf{h})$ and $\mathbf{F}_A(\mathbf{h})$. Employing the trigonometric identity for $\cos^2 \eta$ gives

$$(F_B - F_A)^2 \simeq F_D^2 \left(\frac{1 + \cos 2\eta}{2}\right) = \frac{1}{2}F_D^2 + \frac{1}{2}F_D^2 \cos 2\eta. \tag{6.118}$$

A Patterson function can now be generated using the differences between structure factor amplitudes (*moduli*) rather than the differences between the structure factors themselves:

$$\begin{aligned} P'_\Delta(\mathbf{u}) &= \frac{1}{V_c}\sum_{\mathbf{h}}(F_B(\mathbf{h}) - F_A(\mathbf{h}))^2 e^{-2\pi\, i(\mathbf{u}\cdot\mathbf{h})} && (6.119)\\ &\simeq \frac{1}{V_c}\sum_{\mathbf{h}}\left(\frac{1}{2}F_D^2 + \frac{1}{2}F_D^2\cos(2\varphi_D - 2\varphi_A)\right) e^{-2\pi\, i(\mathbf{u}\cdot\mathbf{h})}. && (6.120)\end{aligned}$$

This type of difference-Patterson function is known as a *modulus difference squared function*, or more commonly, as an *isomorphous difference-Patterson function.* There is no special relationship between φ_A and φ_D, and the second term containing $\cos(2\varphi_D - 2\varphi_A)$ takes on more or less random values, adding to the noise in the map. Thus,

$$P'_{\Delta}(\mathbf{u}) = \frac{1}{2}\frac{1}{V_c}\sum_{\mathbf{h}} F_D^2\, e^{-2\pi\, i(\mathbf{u}\cdot\mathbf{h})} + \text{ noise}, \tag{6.121}$$

providing a noisier, less intense approximation to the ideal difference-Patterson function and emphasizing the vectors between heavy atoms in the isomorphous crystals. Among macromolecular crystallographers, the isomorphous difference-Patterson function is the one most commonly used for the determination of heavy atom positions.

6.2.2 The Anomalous Dispersion Method

The isomorphous replacement method exploits the differences in the contributions to the structure factors of two or more different atoms occupying the same site in two or more isomorphous crystals. The success of the method is dependent on the degree to which the unit cells and contents of the isomorphic crystals are otherwise identical. Unfortunately, the infusion of heavy atoms from the solvent can distort the crystal by changing the volume of the unit cell, or changing the orientation and/or the geometry of the molecules as the heavy atoms coordinate to sites within them. If these distortions are significant then the structures are no longer isomorphous, and the method breaks down.

These problems could be avoided if there was some way to change the X-ray scattering at specific atomic sites in the same crystal (effectively turning the atoms into "replacement atoms"), leaving scattering from the remainder of the atoms essentially unchanged. In general, the relative scattering efficiencies of the atoms measured at different wavelengths do not vary significantly, but there are special conditions in which they do. We addressed this "anomalous" scattering phenomenon in Chapter 5 as an experimental error caused by the wavelength-dependent dispersion of the radiation scattered from different atoms. The error was corrected by including these effects in the scattering factor for each type of atom. For atom j,

$$f_j = f_{\theta,\,j} + \Delta f'_j + i\Delta f''_j \quad \text{and} \tag{6.122}$$

$$\mathbf{F}(\mathbf{h}) = \sum_{j=1}^{n} (f_{\theta,\,j} + \Delta f'_j + i\Delta f''_j)\; e^{-2\pi\, i(\mathbf{h}\cdot\,\mathbf{r_j}\,)}, \tag{6.123}$$

where the real part of the scattering factor for the atom is $f_{\theta,\,j} + \Delta f'_j$ and the imaginary part is $\Delta f''_j$. Referring to Fig. 5.30, these effects become prominent only when the X-ray wavelength approaches the binding energies of the $1s$ electrons in atoms with $Z \geq 20$, although dispersion effects from lighter atoms (e.g., sulfur) are often substantial enough to be of use in structural determination.

If intensity data are collected from a crystal at two different wavelengths, one chosen so that a specific atom type in the unit cell (m of them) is undergoing significant anomalous dispersion at one wavelength ($\lambda = Y$), and comparatively negligible anomalous dispersion at another ($\lambda = X$), with the remainder of the atoms (n of them) undergoing comparatively negligible anomalous dispersion at both wavelengths, the structure factors will vary between the two data sets as a result of the anomalous scattering:

$$\begin{aligned}
\mathbf{F}_X(\mathbf{h}) &\simeq \sum_{j=1}^{n} f_{\theta,j}\, e^{-2\pi i(\mathbf{h}\cdot\mathbf{r_j})} + \sum_{k=1}^{m} f_{\theta,k}\, e^{-2\pi i(\mathbf{h}\cdot\mathbf{r_k})} \\
&= \mathbf{F}_I(\mathbf{h}) + \mathbf{F}_H(\mathbf{h}) \qquad (6.124) \\
\mathbf{F}_Y(\mathbf{h}) &\simeq \sum_{j=1}^{n} f_{\theta,j}\, e^{-2\pi i(\mathbf{h}\cdot\mathbf{r_j})} + \sum_{k=1}^{m} f_{\theta,k}\, e^{-2\pi i(\mathbf{h}\cdot\mathbf{r_k})} \\
&\quad + \sum_{k=1}^{m} \Delta f_k'\, e^{-2\pi i(\mathbf{h}\cdot\mathbf{r_k})} + i\sum_{k=1}^{m} \Delta f_k''\, e^{-2\pi i(\mathbf{h}\cdot\mathbf{r_k})} \\
&= \mathbf{F}_I(\mathbf{h}) + \mathbf{F}_H(\mathbf{h}) + \mathbf{F}_Y'(\mathbf{h}) + \mathbf{F}_Y''(\mathbf{h}) \qquad (6.125) \\
&= \mathbf{F}_I(\mathbf{h}) + \mathbf{F}_H(\mathbf{h}) + \mathbf{F}_D(\mathbf{h}), \qquad (6.126)
\end{aligned}$$

where $\mathbf{F}_H(\mathbf{h})$ is the "non-dispersion" contribution to the structure factor of the atoms exhibiting the difference in anomalous dispersion, which we will refer to as the "anomalous scatterers," and $\mathbf{F}_I$ is the contribution from the remaining atoms. Combining Eqns. 6.124 and 6.126 results in an equation entirely analagous to Eqn. 6.99:

$$\mathbf{F}_Y(\mathbf{h}) = \mathbf{F}_X(\mathbf{h}) + \mathbf{F}_D(\mathbf{h}). \qquad (6.127)$$

If the locations of the anomalous scatterers (the heavy atoms) are known, then we know $\mathbf{F}_D(\mathbf{h})$, F_X and F_Y. Setting $H = F_Y^2(\mathbf{h}) - F_X^2(\mathbf{h})$ results in Eqn. 6.107, allowing for the determination of phases of the reflections from both data sets. The situation is identical to that for single isomorphous replacement; we are faced with the dilemma of two possible phases for each reflection.

In order to resolve the ambiguity in the single isomorphous replacement case, it was necessary to locate a second heavy atom at a different location in the unit cell so that $\mathbf{F}_D(\mathbf{h})$ and $\mathbf{F}_D'(\mathbf{h})$ were not parallel. This same effect can be duplicated by judiciously sorting the reflection data collected at the "significant dispersion" wavelength into two sets, one consisting of a set of reflections with indices $\{\mathbf{h}\}$, the other consisting of the set of reflections with the negatives of these indices, $\{\bar{\mathbf{h}}\}$. This allows us to take advantage of the effects of anomalous dispersion on the structure factors $\mathbf{F}_Y(\mathbf{h})$ and $\mathbf{F}_Y(\bar{\mathbf{h}})$ which would otherwise exhibit identical magnitudes. In the previous chapter (Eqns. 5.364 and 5.365) we determined that both the phases and amplitudes of $\mathbf{F}_Y(\mathbf{h})$ and $\mathbf{F}_Y(\bar{\mathbf{h}})$ are no longer identical as a result of dispersion effects. These effects are ordinarily very small, except in the regime where the wavelength of the X-radiation is close to the energy of a core electron for a specific type of atom, the case that we are considering here.

The phase relationships for $\mathbf{F}_Y(\mathbf{h})$, illustrated in Fig. 6.28, can be deduced from Eqn. 6.125:

$$\begin{aligned}
\mathbf{F}_Y(\mathbf{h}) &= \mathbf{F}_I(\mathbf{h}) + \mathbf{F}_H(\mathbf{h}) + \mathbf{F}_Y'(\mathbf{h}) + \mathbf{F}_Y''(\mathbf{h}) \\
&= \mathbf{F}_I(\mathbf{h}) + \mathbf{F}_H(\mathbf{h}) + \mathbf{F}_Y'(\mathbf{h}) + i\,\mathbf{F}_{Real}''(\mathbf{h}) \qquad (6.128)
\end{aligned}$$

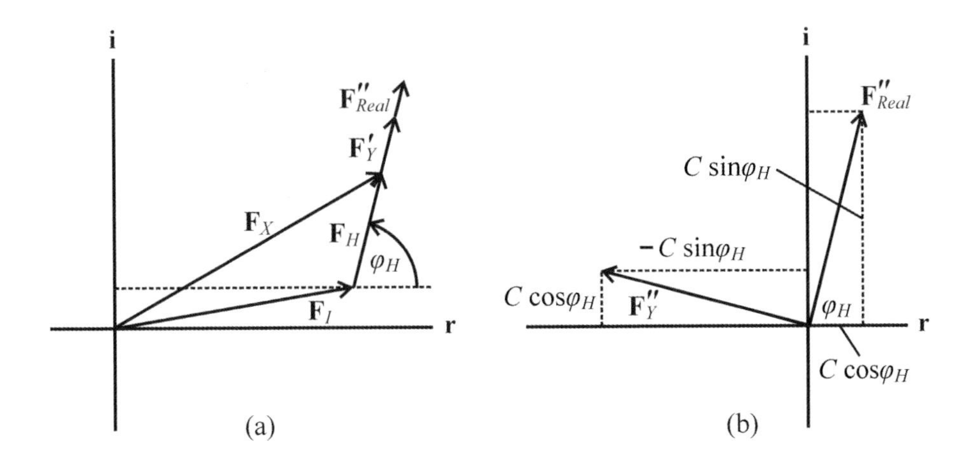

Figure 6.28 (a) Contributions to the structure factor of a reflection undergoing significant anomalous scattering. $\mathbf{F}''_{Real}$ is the imaginary contribution rotated clockwise by 90°. (b) Imaginary contribution to the structure factor, $\mathbf{F}''_Y$, rotated counterclockwise by $\varphi_H + 90°$.

$$\mathbf{F}_H(\mathbf{h}) = f_\theta \sum_{k=1}^{m} e^{-2\pi\, i(\mathbf{h}\cdot\,\mathbf{r_k})} = Ae^{i\,\varphi_H} \tag{6.129}$$

$$\mathbf{F}'_Y(\mathbf{h}) = \Delta f' \sum_{k=1}^{m} e^{-2\pi\, i(\mathbf{h}\cdot\,\mathbf{r_k})} = Be^{i\,\varphi_H} \tag{6.130}$$

$$\mathbf{F}''_{Real}(\mathbf{h}) = \Delta f'' \sum_{k=1}^{m} e^{-2\pi\, i(\mathbf{h}\cdot\,\mathbf{r_k})} = Ce^{i\,\varphi_H}, \tag{6.131}$$

where $A = f_\theta$, $B = \Delta f'$, and $C = \Delta f''$ for the anomalous scatterers.

To determine the phase of the imaginary contribution from anomalous scattering to $\mathbf{F}_Y(\mathbf{h})$, we consider the vector $\mathbf{F}''_{Real}(\mathbf{h})$, with $\mathbf{F}''_Y(\mathbf{h}) = i\,\mathbf{F}''_{Real}(\mathbf{h})$. As depicted in Fig. 6.28(a), $\mathbf{F}_H(\mathbf{h})$, $\mathbf{F}'_Y(\mathbf{h})$ and $\mathbf{F}''_{Real}(\mathbf{h})$ are parallel to one another, rotated through angle φ_H in the complex plane. The effect of multiplying $\mathbf{F}''_{Real}(\mathbf{h})$ by i is determined from Euler's relations,

$$\begin{aligned}\mathbf{F}''_Y(\mathbf{h}) &= i\,Ce^{i\,\varphi_H} = i\,C\cos\varphi_H + i^2\,C\sin\varphi_H \\ &= -C\sin\varphi_H + i\,C\cos\varphi_H,\end{aligned} \tag{6.132}$$

as illustrated in Fig. 6.28b. *The imaginary contribution due to dispersion is rotated counterclockwise through an angle of* 90° *with respect to the phase angle of the real contribution due to dispersion.* As shown in Fig. 6.29(a) $\mathbf{F}'_Y(\mathbf{h})$ has a phase angle of φ_H, and $\mathbf{F}''_Y(\mathbf{h})$ is 90° out of phase with $\mathbf{F}_H(\mathbf{h})$, rotated counterclockwise in the complex plane to phase angle $\varphi_H + 90°$. $\mathbf{F}'_Y(\bar{\mathbf{h}})$ has a phase angle of $-\varphi_H$, and $\mathbf{F}''_Y(\bar{\mathbf{h}})$ is 90° out of phase with $\mathbf{F}_H(\bar{\mathbf{h}})$, *also rotated counterclockwise in the complex plane* to phase angle $-\varphi_H + 90°$. As illustrated in Fig. 6.29(b), the contributions

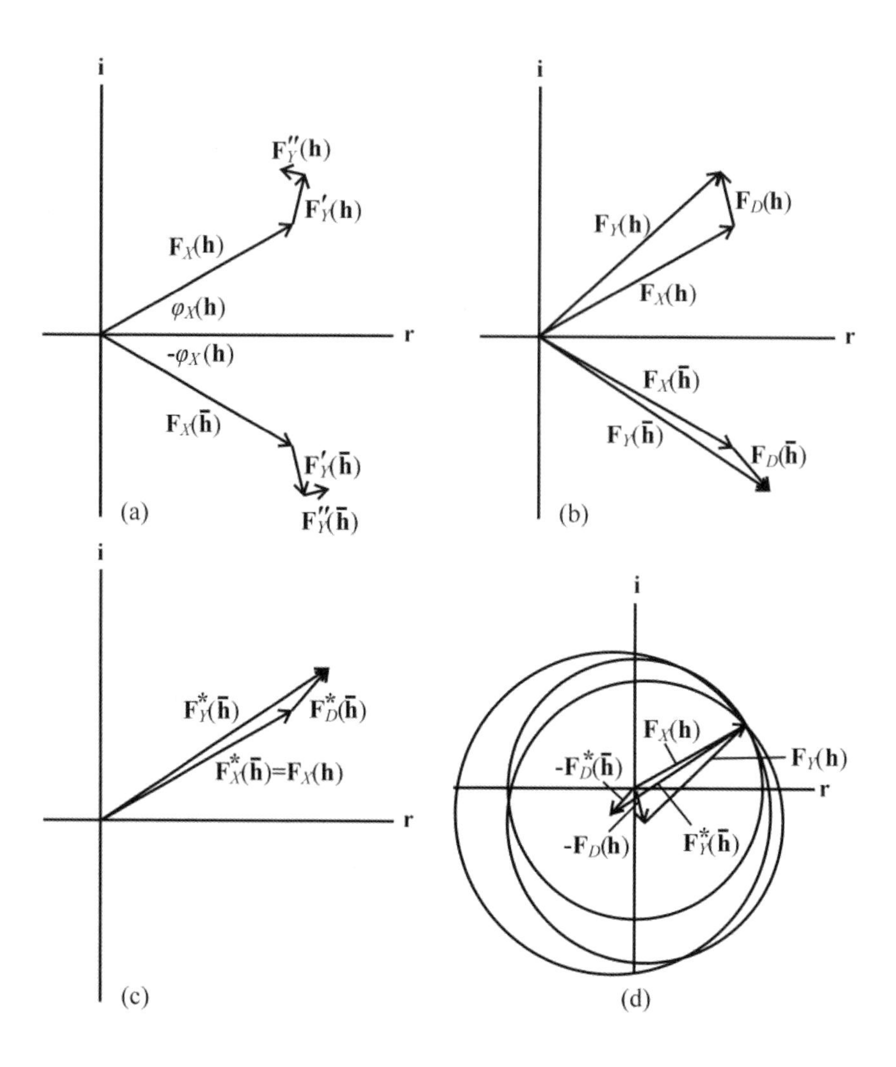

Figure 6.29 (a) Contributions to the structure factors $\mathbf{F}_Y(\mathbf{h})$ and $\mathbf{F}_Y(\bar{\mathbf{h}})$ for reflections from a crystal undergoing significant anomalous dispersion. (b) Resulting structure factors $\mathbf{F}_Y(\mathbf{h})$ and $\mathbf{F}_Y(\bar{\mathbf{h}})$. (c) Complex conjugate vectors for $\mathbf{F}_Y(\bar{\mathbf{h}})$. (d) Unique solution for $\mathbf{F}_X(\mathbf{h})$, $\mathbf{F}_Y(\mathbf{h})$, and $\mathbf{F}_Y^*(\bar{\mathbf{h}})$.

due to dispersion exhibit different magnitudes, with phase angles that are ***not*** the negatives of one another, creating structure factors with different magnitudes and phase angles that are ***not*** the negatives of one another.

Thus the data from the two sets of reflections can serve the same purpose as data sets from two heavy atoms at different locations in the unit cell. As with the multiple isomorphous replacement method, each reflection set must be referenced to a third, in this case consisting of structure factors for the wavelength for which dispersion effects are negligible, $\mathbf{F}_X(\mathbf{h})$. However, the reflection set consisting of the negative indices is referenced to $\mathbf{F}_X(\bar{\mathbf{h}})$. The complex conjugates of $\mathbf{F}_X(\bar{\mathbf{h}})$,

$\mathbf{F}'_Y(\bar{\mathbf{h}})$, and $\mathbf{F}''_Y(\bar{\mathbf{h}})$ yield vectors of the same magnitude – while negating the phase angles – as illustrated in Fig. 6.29(c). Since $\mathbf{F}^*_X(\bar{\mathbf{h}}) = \mathbf{F}_X(\mathbf{h})$, $\mathbf{F}^*_D(\bar{\mathbf{h}})$ and $F^*_Y(\bar{\mathbf{h}})$ provide a third set of values for Eqn. 6.107. Since $\mathbf{F}^*_D(\bar{\mathbf{h}})$ has a different magnitude and direction than $\mathbf{F}_D(\mathbf{h})$, the situation is the same as for multiple isomorphous replacement, providing two pairs of solutions, but only a single set consistent between the pairs. This is illustrated graphically in Fig. 6.29(d) showing circles with radii $F_X(\mathbf{h})$, $F_Y(\mathbf{h})$, and $F^*_Y(\bar{\mathbf{h}})$, centered on the origin, $-\mathbf{F}_D(\mathbf{h})$, and $-\mathbf{F}^*_D(\bar{\mathbf{h}})$, respectively, and intersecting in a single common point.

The heavy atoms that facilitate a multiple isomorphous replacement solution are also those that exhibit dispersion effects of a magnitude that makes them useful for anomalous dispersion solutions. While the approaches are rarely used for small molecule solutions, they are currently the only methods available for the *ab initio* solutions of macromolecular structures. Both methods allow for the assignment of phases based on the locations of heavy atoms, and are often used in combination.

6.3 Completion of the Structural Solution: Fourier Methods

In rare cases it is possible to determine the entire structure from one of the experimental methods described in this chapter, or from the statistical methods described in the next chapter. However, the solution of most structures is an iterative process in which a known portion of the structure is utilized to find the remainder of the atoms in the structure.

The amplitude and phase of each reflection is determined from a superposition of the waves scattered from each atom in the unit cell. Given a partial structure, we can calculate structure factor amplitudes and phases as if the structure consisted solely of the atoms in the partial structure. For a noncentrosymmetric n atom structure, with m known atomic positions, the *calculated structure factors* are given by

$$\mathbf{F}_c(\mathbf{h}) = \sum_{j=1}^{m \leq n} f_j(\mathbf{h}) e^{-2\pi\, i(\mathbf{h}\cdot\mathbf{r}_j)} = F_c(\mathbf{h})\, e^{i\,\varphi_c}. \tag{6.133}$$

If enough of the electron density in the unit cell is accounted for by the m atoms so that the phases are approximately correct, then these phases can be applied to the observed structure factor amplitudes to create *observed structure factors* from Eqns. 3.103, 3.105, and 3.104:

$$\mathbf{F}_o(\mathbf{h}) = F_o(\mathbf{h})\, e^{i\,\varphi_c}. \tag{6.134}$$

Since there is no imaginary term for centrosymmetric structure factors (Eqn. 3.113), the phase angle for $\mathbf{F}_o$ is either 0, with $\mathbf{F}_o$ pointing in the positive direction along the real axis — or π, with $\mathbf{F}_o$ pointing in the negative direction. The calculated structure factors are therefore given by

$$\mathbf{F}_c(\mathbf{h}) = 2 \sum_{j=1}^{m' \leq n'} f_j(\mathbf{h}) \cos(-2\pi\, i(\mathbf{h}\cdot\mathbf{r}_j)) = F_c(\mathbf{h})\, S_c, \tag{6.135}$$

$m' = m/2$, $n' = n/2$, and S_c is the sign of the calculated structure factor. For $\varphi_c = 0$, $S_c = +1$; for $\varphi_c = \pi$, $S_c = -1$. The observed structure factors then become

$$\mathbf{F}_o(\mathbf{h}) = F_o(\mathbf{h})\, S_c. \tag{6.136}$$

6.3.1 Electron Density Synthesis

If the phases calculated from Eqn. 6.133 are reasonable approximations to the actual phases of the reflections with amplitudes F_o, then the resulting structure factors should be improvements over the calculated structure factors (since they have more accurate amplitudes) and an electron density map calculated from them should create a better model of the structure, providing the locations of more of its atoms:

$$\begin{aligned}\rho(x,y,z) &= \frac{1}{V_c}\sum_{\mathbf{h}} \mathbf{F}_o(\mathbf{h}) e^{-2\pi i(hx+ky+lz)} \\ &= \frac{1}{V_c}\sum_{hkl} F_{o,hkl}\, e^{i\,\varphi_c} e^{-2\pi i(hx+ky+lz)}.\end{aligned} \tag{6.137}$$

Using the experimental structure factor magnitudes to compute Eqn. 6.137 presents us with a bit of a dilemma. The average electron density should equal the number of electrons in the unit cell divided by its volume. However, expanding the equation *appears to indicate that the average electron density in the unit cell is zero*:

$$\begin{aligned}\langle\rho(x,y,z)\rangle &= \frac{1}{V_c}\sum_{hkl} F_{o,hkl}\, e^{i\,\varphi_c}\langle\cos(-2\pi(kx+ky+lz))\rangle \\ &\quad + i\frac{1}{V_c}\sum_{hkl} F_{o,hkl}\, e^{i\,\varphi_c}\langle\sin(-2\pi(kx+ky+lz))\rangle.\end{aligned} \tag{6.138}$$

The sines and cosines in each term in the Fourier series are periodic functions oscillating between positive and negative values; they average to zero when evaluated over every x, y, z in the unit cell.

The reason for this odd result is that we are unable to provide the one experimental structure factor that occurs in a term in the Fourier series that does *not* depend on the locations of the atoms, $\mathbf{F}_{000}$. The reciprocal lattice vector $\mathbf{h} = (0\,0\,0)$ is the null vector with zero magnitude, $\mathrm{h} = 0$. The corresponding diffraction angle is given by $\lambda\,\mathrm{h} = 2\sin\theta = 0$, with $\theta = 0$, in the direction of the incident beam. At this angle the radiation is scattered without destructive interference in the direction of the beam (Fig. 3.34) with the full scattering power of every electron in the unit cell. While we are unable to measure the intensity of this reflection, we can calculate it:

$$\mathbf{F}_{000} = \sum_{j=1}^{n} f_j(0,0,0)e^{-2\pi\, i(0x+0y+0z)} = \sum_{j=1}^{n} f_j(0,0,0) = \sum_{j=1}^{n} Z_j, \tag{6.139}$$

where Z_j is the atomic number for atom j, the magnitude of its scattering factor at $\theta = 0$. Thus $\mathbf{F}_{000} = F_{000} =$ the number of electrons in the unit cell. Inclusion

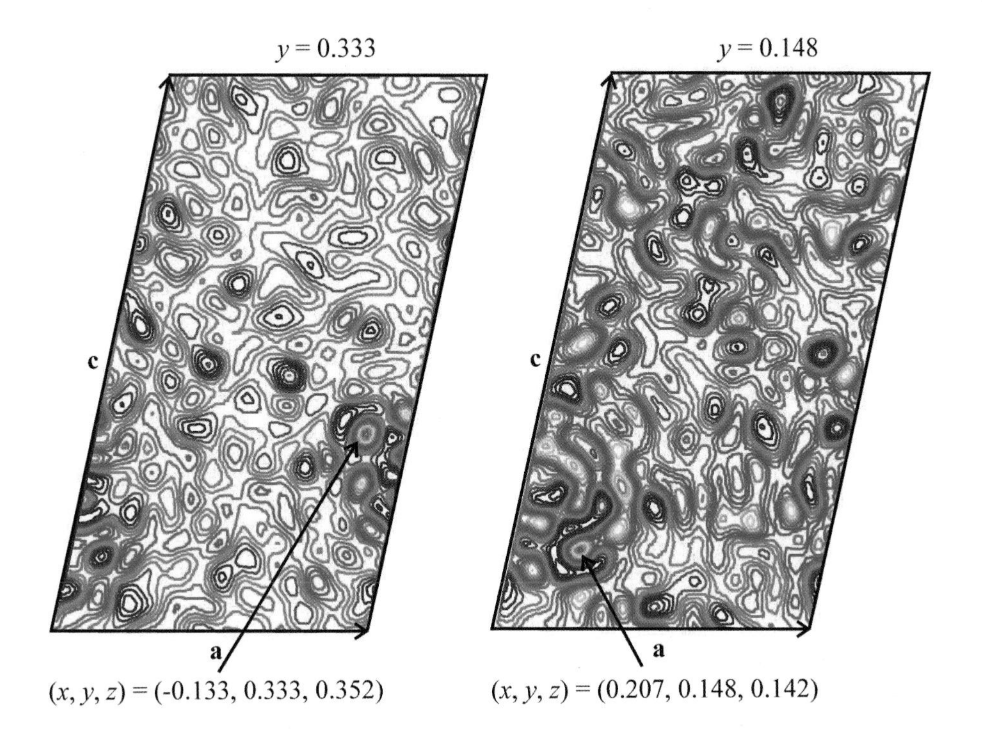

Figure 6.30 Contour plots of two sections of an electron density map for $Cr(P(C_6H_5)_3)_2(CO)_4$ with phases based solely on the locations of the chromium atoms in the unit cell. Warmer colors represent larger magnitudes for contour values.

of $\mathbf{F}_{000}$ in the Fourier series now averages the electron density to the number of electrons in the unit cell divided by its volume:

$$\rho(x, y, z) = \frac{1}{V_c}\left(F_{000}e^{-2\pi i(0x+0y+0z)} + \sum_{hkl} F_{o,hkl}\, e^{i\,\varphi_c} e^{-2\pi i(hx+ky+lz)}\right). \quad (6.140)$$

$$\rho(x, y, z) = \frac{1}{V_c}\left(F_{000} + \sum_{hkl} F_{o,hkl}\, e^{i\,\varphi_c} e^{-2\pi i(hx+ky+lz)}\right) \quad (6.141)$$

$$\langle\rho(x, y, z)\rangle = \frac{F_{000}}{V_c}. \quad (6.142)$$

For structures with a sufficiently heavy atom (e.g., a transition metal), it is often possible to use the heavy atom as the only atom in the Fourier series to compute initial phases. In a previous example, the chromium atom in the $Cr(P(C_6H_5)_3)_2(CO)_4$ structure was determined from a Patterson map to be located at (0.005, 0.245, 0.250) in a centrosymmetric triclinic unit cell (Fig. 6.13), and refined (Chapter 8) to fractional coordinates (0.016, 0.245, 0.248). There are two molecules in the $P\bar{1}$ unit cell, with a = 10.139, b = 10.583, c = 17.911, α = 75.86°, β = 77.41° γ = 65.93°, V_c = 1686.33^3, and F_{000} = 712 electrons. Since the structure is centrosymmetric, structure factors ($\mathbf{F}_c = F_c S_c$) are calculated from Eqn. 6.135, and the electron density is computed from Eqn. 3.114:

$$\rho(x,y,z) = \frac{1}{V_c}\left(F_{000} + \sum_{hkl} F_{o,hkl} S_c \cos(2\pi(hk+ky+lz))\right). \tag{6.143}$$

For every observed reflection, a structure factor is calculated and its sign determined from

$$\mathbf{F}_c(\mathbf{h}) = 2f_{Cr}(\mathbf{h})\cos(-2\pi\, i(0.016h + 0.245k + 0.248l)) = F_c(\mathbf{h})\, S_c.$$

These signs are incorporated into Eqn. 6.143 and the resulting electron density map is searched for maxima indicative of atom positions. Fig. 6.30 shows two contour plots of sections of the map parallel to the ac plane at $y = 0.333$ and $y = 0.148$. Although the two chromium atoms account for less than 7% of the electron density in the unit cell, two relatively large maxima are located in these two sections at (-0.133, 0.333, 0.352) and (0.207, 0.148, 0.142). The map is seen to be relatively "flat"and noisy — and the second peak is not as resolved as the first. Nevertheless, analysis of the locations of these peaks indicates that they both lie on a line passing through the chromium atom and are approximately equidistant from it – each at a distance of ca. 2.3 Å. The covalent radii of chromium and phosphorus are 1.24 Å and 1.10 Å, respectively; a Cr-P bond is expected to have a length approximately equal to the sum of the covalent radii, 2.34 Å. Adding the positions of the two phosphorus atoms and refining the positions of all three atoms places Cr, P(1), and P(2) in the unit cell at (0.0217, 0.2423, 0.2421), (-0.1549, 0.3422, 0.3589) and (0.1841, 0.1492, 0.1517), respectively. New structure factors and their signs are calculated from these positions,

$$\begin{aligned}
\mathbf{F}_c(\mathbf{h}) &= 2f_{Cr}(\mathbf{h})\cos(-2\pi\, i(0.0217h + 0.2423k + 0.2421l)) \\
&+ 2f_{P}(\mathbf{h})\cos(-2\pi\, i(-0.1549h + 0.3422k + 0.3589l)) \\
&+ 2f_{P}(\mathbf{h})\cos(-2\pi\, i(0.1841h + 0.1492k + 0.1517l)) = F_c(\mathbf{h})\, S_c,
\end{aligned}$$

and the signs are again incorporated into Eqn. 6.143. Successive cycles of atom location, refinement, sign calculation and Fourier electron density (F_o) map calculation results in a determination of the positions of all of the atoms in the structure. One of the major reasons for the lack of resolution in the Fourier map when phases are determined from a partial structure arises from our inability to discern which of the structure factors are more or less correctly phased, and which are not. The major contributors to the Fourier sum are the strong reflections, and a relatively crude, but simple approach to the problem is simply to omit the weaker reflections. One way to accomplish this is to omit a certain fraction of the high angle reflections, since they are generally the weakest. A somewhat more systematic approach is to assign phases only to those reflections for which F_o and F_c are close to one another. (i.e., F_c/F_o lies within some specified range). While this seems a reasonable approach, the actual selection process turns out to be more difficult than it might first appear. If only those reflections for which $F_o \approx F_c$ are selected, the electron density map will be unlikely to contain anything new:

$$\begin{aligned}\rho(\mathbf{r}) &= \frac{1}{V_c}\left(F_{000} + \sum_{\mathbf{h}} F_{o,\mathbf{h}}\, e^{i\,\varphi_c} e^{-2\pi i(\mathbf{h}\cdot\mathbf{r})}\right)\\ &\approx \frac{1}{V_c}\left(F_{000} + \sum_{\mathbf{h}} F_{c,\mathbf{h}}\, e^{i\,\varphi_c} e^{-2\pi i(\mathbf{h}\cdot\mathbf{r})}\right),\end{aligned}$$

creating maxima in the positions from which the structure factors were calculated in the first place (i.e., $\rho(\mathbf{r})$ is simply the Fourier transform of $\mathbf{F}_c(\mathbf{h})$).

The appearance of new peaks in the Fourier map therefore depends on the fortuitous assignment of correct phases to those reflections for which F_o and F_c are reasonably different. Selection of the appropriate range is somewhat arbitrary, and often difficult — it is generally advisable to exclude only those reflections for which the difference between F_o and F_c is substantial.

Weighting

Ideally, the Fourier sum for the calculated electron density would include all of the structure factors, each multiplied by a "weighting" factor that would correct for the difference between the "observed phase", $\varphi_o(\mathbf{h})$ (the phase that would be observed if it was possible to do so), and the calculated phase, $\varphi_c(\mathbf{h})$:

$$\rho(\mathbf{r}) = \frac{1}{V_c}\left(F_{000} + \sum_{\mathbf{h}} w(\mathbf{h}) F_{o,\mathbf{h}}\, e^{i\,\varphi_c(\mathbf{h})} e^{-2\pi i(\mathbf{h}\cdot\mathbf{r})}\right). \quad (6.144)$$

The observed structure factor, is the resultant of waves scattered from each of the atoms in the unit cell, each wave with a magnitude determined by the number of electrons in the atom and a phase determined by the atom location. Fig. 6.31(a) is a representation of this, with the contributions to the structure factor from the known atoms shown in black, and the contributions from the unknown atoms shown in red. The observed structure factor, $\mathbf{F}_o = F_o\, e^{i\varphi_o}$, is the vector sum of the structure factor calculated from the known positions, $\mathbf{F}_c = F_c\, e^{i\varphi_c}$, and the difference vector, $\mathbf{F}_u$, as illustrated in Fig. 6.31(b):

$$\mathbf{F}_u = \mathbf{F}_o - \mathbf{F}_c = F_o\, e^{i\varphi_o} - F_c\, e^{i\varphi_c}. \quad (6.145)$$

The calculated structure factors are in error by the angle $\phi = \varphi_o - \varphi_c$. With $\varphi_o = \phi + \varphi_c$,

$$\begin{aligned}\mathbf{F}_u &= F_o e^{i\phi} e^{i\varphi_c} - F_c e^{i\varphi_c}\\ &= (F_o e^{i\phi} - F_c) e^{i\varphi_c}.\end{aligned} \quad (6.146)$$

If we knew ϕ, then $e^{i\phi} = w(\mathbf{h})$ would serve as the weighting factor alluded to above:

$$\begin{aligned}\mathbf{F}_u &= (w(\mathbf{h})F_o - F_c)e^{i\varphi_c} \quad \text{and}\\ \mathbf{F}_o &= \mathbf{F}_c + (w(\mathbf{h})F_o - F_c)e^{i\varphi_c} = w(\mathbf{h})F_o e^{i\varphi_c}.\end{aligned} \quad (6.147)$$

Since we are unable to measure φ_o, we are clearly unable to determine ϕ. Suppose, however, that we had some way of estimating ϕ. For example if we knew

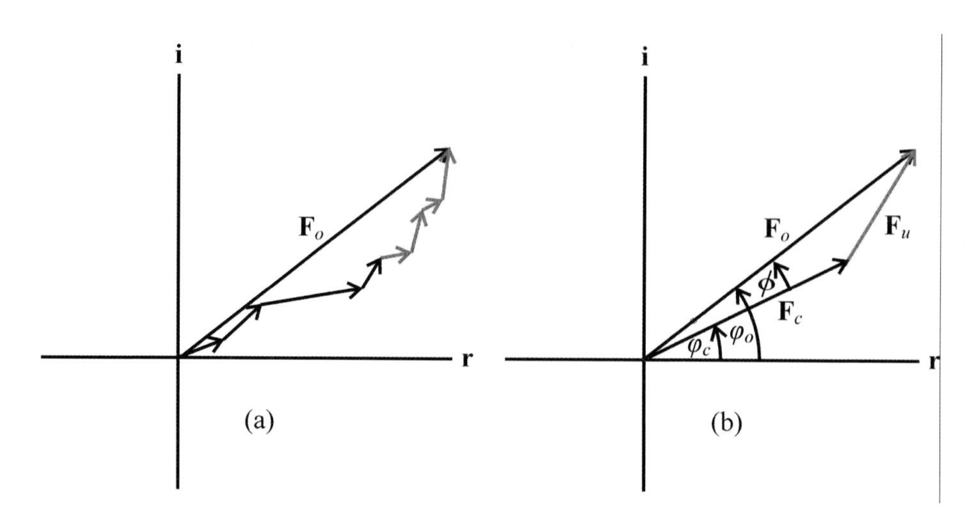

Figure 6.31 (a) The observed structure factor ($\mathbf{F}_o$) represented as the superposition of waves from each atom in the unit cell. Black vectors indicate contributions from known atomic positions; red vectors are contributions from unknown atomic positions. (b) Observed structure factor represented as the sum of the net contribution from known atomic positions ($\mathbf{F}_c$) and unknown atomic positions ($\mathbf{F}_u$).

ϕ for a number of reflections with observed magnitudes near F_o (between F_o and $F_o + dF_o$), and corresponding calculated values near F_c (between F_c and $F_c + dF_c$) then we could use the most probable value of ϕ from these reflections as an approximation to ϕ. The most probable value of ϕ is its *expected value*, $\langle\phi\rangle$, which is an average of all the ϕ's, each one weighted by the probability of observing it. Then

$$w(\mathbf{h}) = e^{i\langle\phi\rangle} = \langle e^{i\phi}\rangle = \langle\cos\phi\rangle + i\langle\sin\phi\rangle. \tag{6.148}$$

Assuming that ϕ can be equally positive or negative, $\langle\sin\phi\rangle$=0 and

$$w(\mathbf{h}) = \langle\cos\phi\rangle. \tag{6.149}$$

While we do not have the necessary values of ϕ to determine its expected value, we note from Eqn. 6.147 that for specific F_c, φ_c and F_o (all known values), $\mathbf{F}_u$ depends only on ϕ, and the probability of finding a value of ϕ between ϕ and $\phi + d\phi$ is the same as the probability of finding the corresponding $\mathbf{F}_u$ between $\mathbf{F}_u$ and $\mathbf{F}_u + d\mathbf{F}_u$. $\mathbf{F}_u$ and ϕ must therefore have the same probability distribution functions. For a non-centrosymmetric crystal the probability for $\mathbf{F}_u$ is given by Eqn. 5.478:

$$Pr_1(\mathbf{F}_u) = f_G(\mathbf{F_u})\,d\mathbf{F}_u = \frac{1}{\pi\sum f_j^2}\, e^{-F_u^2/\sum f_j^2}\, d\mathbf{F}_u, \tag{6.150}$$

where $\sum f_j^2$ is the sum over all of the atoms with *unknown* locations ($\mathbf{F}_u$ *would* be the structure factor if the unit cell contained only those atoms). The squared magnitude for F_u is determined from the triangle in Fig. 6.31(b) using the law of cosines:

$$F_u^2 = F_o^2 + F_c^2 - 2F_oF_c\cos\phi \quad \text{and} \tag{6.151}$$

$$\begin{aligned} Pr_1(\mathbf{F}_u) &= \frac{1}{\pi\sum f_j^2} e^{-(F_o^2+F_c^2-2F_oF_c\cos\phi)/\sum f_j^2}\, d\mathbf{F}_u \\ &= \frac{1}{\pi\sum f_j^2} e^{-(F_o^2+F_c^2)/\sum f_j^2} e^{(2F_oF_c\cos\phi)/\sum f_j^2}\, d\mathbf{F}_u. \end{aligned} \tag{6.152}$$

The only variable is ϕ. Accumulating the terms before the second exponential into a constant, K, and setting $X(\mathbf{h}) = (2F_oF_c)/\sum f_j^2 = (2F_o(\mathbf{h})F_c(\mathbf{h}))/\sum_{j=1}^{k} f_j(\mathbf{h})^2$ for k unknown atoms, gives

$$\begin{aligned} Pr_1(\mathbf{F}_u) &= K\, e^{X\cos\phi}\, d\mathbf{F}_u, \quad \text{and} \\ Pr_1(\phi) &= K\, e^{X\cos\phi}\, d\phi. \end{aligned} \tag{6.153}$$

The expected value of a function is its weighted average value, determined by summing each value of the function weighted by the probability that the value will be observed, and dividing this sum by the sum of the probabilities. For a specific value of F_o and F_c (and thus a specific value of X) the expected value for $\cos\phi$ is therefore

$$\langle\cos\phi\rangle_X = \frac{\int_o^{2\pi} K\, e^{X\cos\phi}\cos\phi d\phi}{\int_o^{2\pi} K\, e^{X\cos\phi}\, d\phi} = \frac{\int_o^{2\pi} e^{X\cos\phi}\cos\phi d\phi}{\int_o^{2\pi} e^{X\cos\phi}\, d\phi}. \tag{6.154}$$

The desired weighting factor can now be obtained from the ratio of two definite integrals. While these integrals could be evaluated numerically, it would be necessary to do so for each reflection (each value of X). Fortunately the integrals take the form of modified Bessel functions, and can be computed as converging power series. From Appendix F, the *modified Bessel function of the first kind of order n* has both series and integral forms:*

$$\begin{aligned} I_n(x) &= \sum_{m=0}^{\infty} \frac{1}{m!(m+n)!}\left(\frac{x}{2}\right)^{2m+n} = \frac{1}{\pi}\int_0^{\pi} e^{x\cos\phi}\cos n\phi\, d\phi \\ &= \frac{1}{2\pi}\int_0^{2\pi} e^{x\cos\phi}\cos n\phi\, d\phi. \end{aligned} \tag{6.155}$$

It follows that

$$\langle\cos\phi\rangle_X = w(\mathbf{h}) = w(X) = \frac{\int_o^{2\pi} e^{X\cos\phi}\cos(1)\phi d\phi}{\int_o^{2\pi} e^{X\cos\phi}\cos(0)\phi d\phi} = \frac{I_1(X)}{I_0(X)}. \tag{6.156}$$

It is now possible to create a table of values of the ratio $I_1(X)/I_0(X)$ as a function of $X = (2F_oF_c)/\sum f_j^2$, providing a relatively efficient means for the determination of improved weights for a Fourier electron density calculation. This scheme was devised by Sim[108] for non-centrosymmetric structures, and is commonly known as *Sim weighting.*

*The use of "$I_n(x)$" for the modified Bessel function is nearly universal. The same terminology is also used to indicate an intensity, but the context in which the term is employed should preclude any confusion.

The reflections measured from a centrosymmetric crystal exhibit a different probability distribution than those from a noncentrosymmetric crystal, and are limited to only two possible possible phases. In order to derive $w(\mathbf{h})$ for centrosymmetric structures, we again seek a way to estimate the contribution of the unknown atomic positions, $\mathbf{F}_u$, to the observed structure factor,

$$\mathbf{F}_o = \mathbf{F}_u + \mathbf{F}_c \approx \langle \mathbf{F}_u \rangle + \mathbf{F}_c, \tag{6.157}$$

with $\langle \mathbf{F}_u \rangle$ the expected value of $\mathbf{F}_u$. Recalling that there is no imaginary term for centrosymmetric structure factors (Eqn. 3.113), the phase angles for $\mathbf{F}_o$, $\mathbf{F}_u$, and $\mathbf{F}_c$ are either 0, pointing in the positive direction along the real axis, or π, pointing in the negative direction. Thus $\mathbf{F}_o = \pm F_o$, $\mathbf{F}_c = \pm F_c$, and

$$\mathbf{F}_u = \mathbf{F}_o - \mathbf{F}_c = S_o F_o - S_c F_c, \tag{6.158}$$

where S_o is the sign of $\mathbf{F}_o$ and S_c is the sign of $\mathbf{F}_c$. Unlike the noncentrosymmetric case (with infinitely many possible phase angles), there are only two possibilities to be considered in determining $\mathbf{F}_u$ — either the signs of $\mathbf{F}_o$ and $\mathbf{F}_c$ are equal, or they are not. If $S_o = S_c$, then

$$\mathbf{F}_u(eq) = S_c F_o - S_c F_c = S_c(F_o - F_c), \tag{6.159}$$

and if $S_o = -S_c$, then

$$\mathbf{F}_u(neq) = -S_c F_o - S_c F_c = -S_c(F_o + F_c). \tag{6.160}$$

If we have a set of reflections with the same F_o, F_c, and S_c, then the average (expected) value of $\mathbf{F}_u$ will be the average value of the $\mathbf{F}_u$ for reflections for which the signs are the same plus the average value of $\mathbf{F}_u$ for the reflections with opposite signs, weighted by the probability of observing each, P_{eq} and P_{neq}:

$$\langle \mathbf{F}_u \rangle = P_{eq} \langle \mathbf{F}_u(eq) \rangle + P_{neq} \langle \mathbf{F}_u(neq) \rangle, \tag{6.161}$$

with $P_{eq} + P_{neq}=1$. For a centrosymmetric crystal the probability for $\mathbf{F}_u$ is given by Eqn. 5.489:

$$Pr_{\bar{1}}(\mathbf{F}_u) = \frac{2}{\sqrt{2\pi}(\sum f_j^2)^{1/2}} e^{-F_u^2/2\sum f_j^2} \, d\mathbf{F}_u = K \, e^{-F_u^2/2\sum f_j^2} \, d\mathbf{F}_u. \tag{6.162}$$

$\mathbf{F}_u(eq)$ and $\mathbf{F}_u(neq)$ are distributed in the same manner, and

$$\begin{aligned} P_{eq} &= K \, e^{-F_u(eq)^2/2\sum f_j^2} \, d\mathbf{F}_u = K \, e^{-S_c^2(F_o - F_c)^2/2\sum f_j^2} \, d\mathbf{F}_u \\ &= K \, e^{-(F_o^2 - 2F_oF_c + F_c^2)/2\sum f_j^2} \, d\mathbf{F}_u \\ &= K \, e^{-(F_o^2 + F_c^2)/2\sum f_j^2} e^{2F_oF_c/2\sum f_j^2} \, d\mathbf{F}_u. \end{aligned} \tag{6.163}$$

Similarly,

$$\begin{aligned} P_{neq} &= K \, e^{-F_u(neq)^2/2\sum f_j^2} \, d\mathbf{F}_u = K \, e^{-(-S_c)^2(F_o + F_c)^2/2\sum f_j^2} \, d\mathbf{F}_u \\ &= K \, e^{-(F_o^2 + F_c^2)/2\sum f_j^2} e^{-2F_oF_c/2\sum f_j^2} \, d\mathbf{F}_u. \end{aligned} \tag{6.164}$$

Taking the ratio of these probabilities,

$$\frac{P_{eq}}{P_{neq}} = \frac{e^{F_oF_c/\sum f_j^2}}{e^{-F_oF_c/\sum f_j^2}} = \frac{e^{X/2}}{e^{-X/2}}, \tag{6.165}$$

where $X = (2F_oF_c)/\sum f_j^2$, as defined for the noncentrosymmetric case. In order to determine the actual probabilities, constrained so that their sum is equal to 1, we make use of the hyperbolic tangent function:

$$\tanh(y) = \frac{e^y - e^{-y}}{e^y + e^{-y}}. \tag{6.166}$$

Setting $y = X/2$,

$$\begin{aligned}
e^y &= \frac{1}{2}(e^y + e^{-y}) + \frac{1}{2}(e^y - e^{-y}) \\
e^{-y} &= \frac{1}{2}(e^{-y} + e^y) + \frac{1}{2}(e^{-y} - e^y) = \frac{1}{2}(e^y + e^{-y}) - \frac{1}{2}(e^y - e^{-y}) \\
\frac{P_{eq}}{P_{neq}} &= \frac{e^y}{e^{-y}} = \frac{\frac{e^y}{e^{-y}+e^y}}{\frac{e^{-y}}{e^{-y}+e^y}} = \frac{\frac{1}{2}\frac{e^y+e^{-y}}{e^y+e^{-y}} + \frac{1}{2}\frac{e^y-e^{-y}}{e^y+e^{-y}}}{\frac{1}{2}\frac{e^y+e^{-y}}{e^y+e^{-y}} - \frac{1}{2}\frac{e^y-e^{-y}}{e^y+e^{-y}}} \\
&= \frac{\frac{1}{2}+\frac{1}{2}\tanh(y)}{\frac{1}{2}-\frac{1}{2}\tanh(y)}.
\end{aligned} \tag{6.167}$$

This is the *ratio* of the probabilities: $P_{eq} = C(\frac{1}{2}+\frac{1}{2}\tanh(y))$ and $P_{neq} = C(\frac{1}{2} - \frac{1}{2}\tanh(y))$. But $C(\frac{1}{2}+\frac{1}{2}\tanh(y)) + C(\frac{1}{2}-\frac{1}{2}\tanh(y)) = C = 1$, and

$$P_{eq} = \frac{1}{2} + \frac{1}{2}\tanh(y) \tag{6.168}$$

$$P_{neq} = \frac{1}{2} - \frac{1}{2}\tanh(y). \tag{6.169}$$

For specific F_o, F_c, and S_c, $\langle \mathbf{F}_u(eq)\rangle = S_c(F_o - F_c)$ and $\langle \mathbf{F}_u(neq\rangle) = -S_c(F_o+F_c)$. The expected value of $\mathbf{F}_u$ is given by

$$\begin{aligned}
\langle \mathbf{F}_u \rangle &= P_{eq}(S_c(F_o - F_c)) + P_{neq}(-S_c(F_o + F_c)) \\
&= S_cF_o\tanh(y) - S_cF_c \\
&= S_cF_o\tanh(y) - \mathbf{F}_c.
\end{aligned} \tag{6.170}$$

From Eqn. 6.157 we obtain the centrosymmetric weighting factor, originally derived by Woolfson[109]:

$$\mathbf{F_o} \approx S_cF_o\tanh(y) - \mathbf{F}_c + \mathbf{F}_c = \tanh(X/2)F_oS_c, \quad \text{and} \tag{6.171}$$

$$w(\mathbf{h}) = w(X) = \tanh(X/2), \tag{6.172}$$

allowing for an improved electron density map:

$$\rho(\mathbf{r}) = \frac{1}{V_c}\left(F_{000} + \sum_{\mathbf{h}} w(\mathbf{h})F_{o,\mathbf{h}}S_c(\mathbf{h})\cos(2\pi(\mathbf{h}\cdot\mathbf{r}))\right). \tag{6.173}$$

6.3.2 Difference Electron Density Synthesis

For a partially known structure, the observed structure factor, $\mathbf{F}_o$, is the vector sum of the contribution from atoms with known locations, $\mathbf{F}_c$ — which we can calculate — and the contribution from the atoms with unknown locations, $\mathbf{F}_u$ —

which we must determine to complete the structure. *If* the phases of the observed structure factors were known, then an electron density map calculated from the known portion of the structure, subtracted from the electron density map calculated from the observed structure factors would create *a difference electron density* map with maxima at the positions of the remaining atoms in the structure:

$$\begin{aligned}
\rho_u(x,y,z) &= \rho_o(x,y,z) - \rho_c(x,y,z) \\
&= \frac{1}{V_c}\left(F_{000} + \sum_{hkl} F_{o,hkl}\, e^{i\varphi_o} e^{-2\pi i(hx+ky+lz)}\right) \\
&\quad -\frac{1}{V_c}\left(F_{000} + \sum_{hkl} F_{c,hkl}\, e^{i\varphi_c} e^{-2\pi i(hx+ky+lz)}\right) \\
&= \frac{1}{V_c}\sum_{hkl}(F_{o,hkl}\, e^{i\varphi_o} - F_{c,hkl}\, e^{i\varphi_c})e^{-2\pi i(hx+ky+lz)} \qquad (6.174) \\
&= \frac{1}{V_c}\sum_{hkl}(\mathbf{F}_o - \mathbf{F}_c)e^{-2\pi i(\mathbf{h}\cdot\mathbf{r})} \\
&= \frac{1}{V_c}\sum_{hkl}\mathbf{F}_u(h,k,l)e^{-2\pi i(\mathbf{h}\cdot\mathbf{r})}. \qquad (6.175)
\end{aligned}$$

As before, the absence of phases for the observed structure factors makes it necessary to approximate the difference electron density by assuming that $\varphi_o \approx \varphi_c$ for each reflection:

$$\begin{aligned}
\rho_u(x,y,z) &\approx \frac{1}{V_c}\sum_{hkl}(F_{o,hkl}\, e^{i\varphi_c} - F_{c,hkl}\, e^{i\varphi_c})e^{-2\pi i(hx+ky+lz)} \qquad (6.176) \\
&= \frac{1}{V_c}\sum_{hkl}(F_o - F_c)\, e^{i\varphi_c} e^{-2\pi i(\mathbf{h}\cdot\mathbf{r})} = \frac{1}{V_c}\sum_{hkl}(\Delta F)\, e^{i\varphi_c} e^{-2\pi i(\mathbf{h}\cdot\mathbf{r})} \\
&= \frac{1}{V_c}\sum_{hkl}\mathbf{F}'_u(h,k,l)e^{-2\pi i(\mathbf{h}\cdot\mathbf{r})}. \qquad (6.177)
\end{aligned}$$

While this may appear to be no more than a restatement of the electron density synthesis formalism, constructing the Fourier series in this manner offers a number of distinct advantages, the most obvious being a "cleaner" map resulting from to the absence of the large peaks at the known atom locations that remain in a conventional electron density (F_o) map. Furthermore, the difference Fourier (ΔF) map is positive in regions where the electron density has not been accounted for, and is negative in regions where there is more electron density assigned than there should be. Thus, if there is small positional error in an atom location, the difference map will exhibit a negative peak at the selected location of an atom, and a positive peak in the region next to it, indicating the direction in which the atom should be moved. This is illustrated in Fig. 6.32(a) for a hypothetical one-dimensional structure. Fig. 6.32(b) indicates a situation in which an atom can occupy more than one location in the unit cell. This phenomenon, known as *static displacement disorder*, results in a diminution of the scattering intensity from each position, *depending on the relative occupancy of the atom in each site.* In the example shown here an atom at the calculated position occupies only 60% of those sites in the crystal, and scatters on the average as if there was only 0.6 of an atom at that location in the

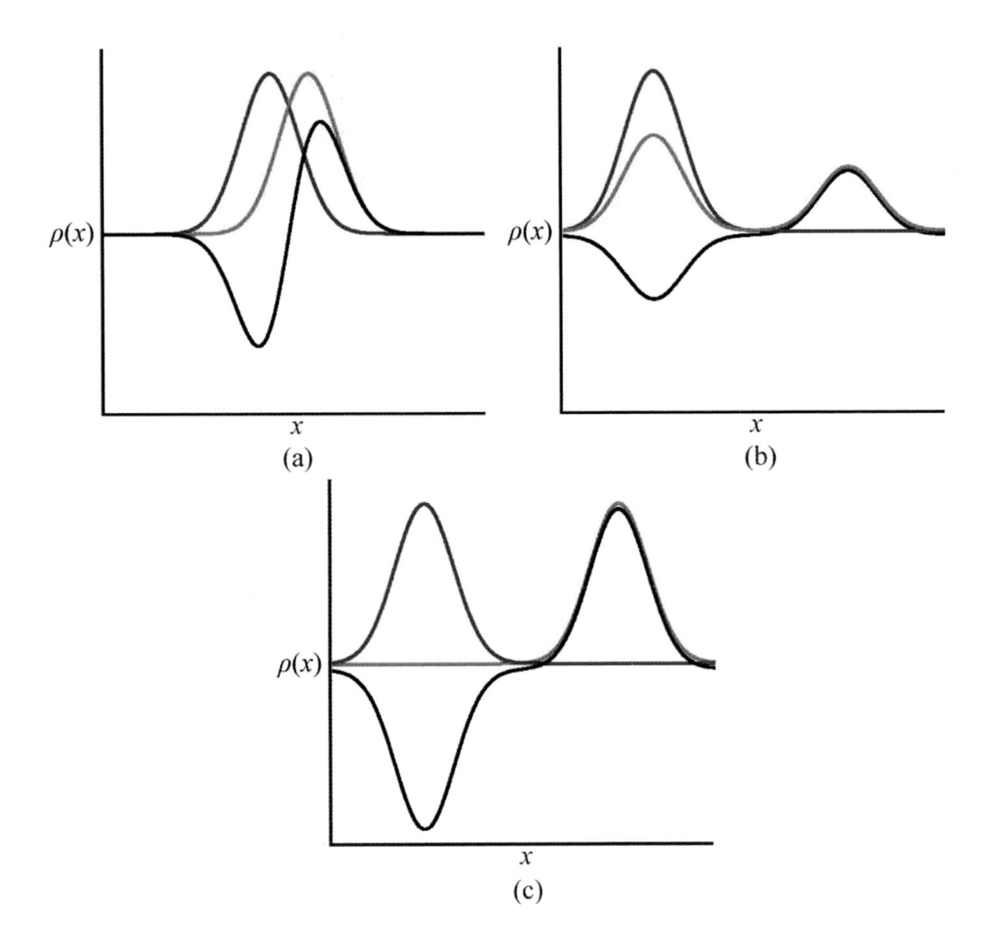

Figure 6.32 Difference electron density plots for a hypothetical one-dimensional structure with: (a) a positional error in the placement of an atom; (b) static displacement disorder; (c) an atom placed incorrectly. The actual electron density is shown in red, the calculated electron density (from atom placement) in blue, and the difference electron density in black.

unit cell. Thus there is a negative peak (hole) at the assigned location, corresponding to an excess of 0.4 of an atom. A positive peak corresponding to a 40% site occupancy is observed at the location where the atom is found when it is not in the calculated position. Fig. 6.32(c) represents an extension of this, in which an atom has simply been placed incorrectly, creating a large negative peak at the incorrect location, and a large positive peak at the correct location.

In addition to its diagnostic advantages, the difference electron density synthesis can actually provide a more accurate Fourier map. There are two major reasons for this. The first involves a problem that arises with all Fourier series calculations — although they are formally infinite series, at some point they must be terminated; truncation of the series necessarily leaves out higher frequency sinusoidal components. Referring to Fig. 3.30, as the higher frequency terms necessary to model

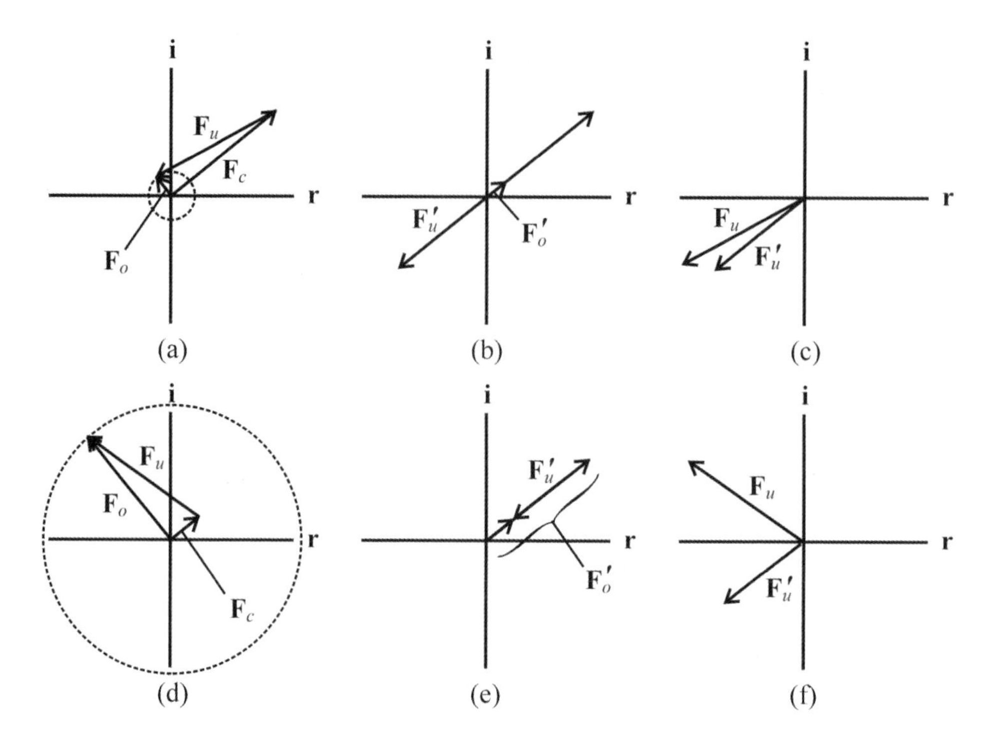

Figure 6.33 (a)–(c) Comparison of $\mathbf{F}_u = F_o e^{i\varphi_o} + F_c e^{i\varphi_c}$ with $\mathbf{F}'_u = F_o e^{i\varphi_c} + F_c e^{i\varphi_c}$ for $F_o \ll F_c$. (d)–(f) Comparison of $\mathbf{F}_u = F_o e^{i\varphi_o} + F_c e^{i\varphi_c}$ with $\mathbf{F}'_u = F_o e^{i\varphi_c} + F_c e^{i\varphi_c}$ for $F_o \gg F_c$.

the election density are included in the series they create residual ripples in regions next to the actual peaks, regions where there is often little electron density. The inclusion of components of even higher frequency will tend to smooth these ripples by providing destructive interference in the appropriate regions. If these terms are not included in the Fourier series, the ripples remain, contributing to the noise in the electron density Fourier map (Fig. 6.30). Fourier series based on $F_o e^{i\varphi_c}$ and $F_c e^{i\varphi_c}$ will be truncated at the same frequency limit, and will tend to contain the same residual frequencies. Subtracting the two Fourier series will consequentially tend to remove the ripples, providing increased resolution in the difference Fourier map.

The second reason for more accuracy in the difference Fourier synthesis occurs because, unlike the electron density synthesis, the assignment of approximately correct phases to reflections with large differences in F_o and F_c is no longer simply fortuitous, provided that $F_o \ll F_c$. This can be seen in Fig. 6.33. Fig. 6.33(a) represents the case in which the observed structure factor has a very different phase from the phase of the calculated structure factor, with $F_o \ll F_c$. The difference vector with the correct phase is $\mathbf{F}_u$. Fig. 6.33(b) illustrates the consequences of assigning the calculated phase to F_o, creating a putative approximation to the observed structure factor, $\mathbf{F}'_o$ and the resulting difference vector $\mathbf{F}'_u$. Although $\mathbf{F}'_o$ is a very poor approximation to $\mathbf{F}_o$, $\mathbf{F}'_u$ is a good approximation to $\mathbf{F}_u$, both in magnitude and in phase, as indicated in Fig. 6.33(c). Indeed, any orientation of $\mathbf{F}_o$ will produce essentially the same result, since the $\mathbf{F}_o$ vector is constrained to a small circle of radius F_o, and $\mathbf{F}_u$ is limited by being forced to terminate at some point on the circle. Furthermore, the approximation improves as F_o becomes smaller and smaller, and is actually best if a reflection with a relatively large calculated structure factor has an observed structure factor that is close to zero. Physically, this represents contributions from atoms in undetermined locations that scatter in a destructively interfering manner with respect to those in known positions, and therefore have a substantial influence in determining $\mathbf{F}_u$. Unlike electron density synthesis, difference Fourier synthesis should be performed with the weakest of reflections for which $F_o \ll F_c$, especially those that are labeled as "unobserved" by virtue of having intensities with approximately the same values as their standard deviations (e.g., $I_\mathbf{h} < 2\sigma(I_\mathbf{h})$). In the electron density synthesis the weaker reflections are only minor contributors to the Fourier sum, and are often omitted to remove noise from the electron density map. In the difference electron density synthesis, those reflections for which F_o is much less than F_c add large terms to the sum and are arguably the most important contributors to the Fourier series.

A large difference between F_o and F_c can also occur when an observed structure factor magnitude is much larger than the calculated magnitude, $F_o \gg F_c$. While the assignment of correct phases for these reflections would be equally useful, Fig. 6.33(a) illustrates that there are no constraints on possible phases for $\mathbf{F}_u$, since it can terminate at any point on the large circle of radius F_o. When $F_o \gg F_c$, Fig. 6.33(c) demonstrates that $\mathbf{F}'_u$ is *not* necessarily a good approximation to $\mathbf{F}_u$. Because ΔF for these reflections will introduce large terms in the Fourier series in which the phase of $\mathbf{F}'_u$ may be significantly in error, it is probably advisable to omit them from the difference electron density calculation, or at least limit them to cases where the difference between F_o and F_c is not too large.

A third possibility occurs when $F_o \approx F_c$. As indicated in the previous section, this provides little additional information to the electron density Fourier map, as it tends to reproduce the known portion of the model. These reflections can, however, produce significant errors in the map, especially for large values of F_o and F_c for which there is a substantial difference between φ_o and φ_c, since the terms will be weighted heavily in the Fourier series. In the difference Fourier calculation, $F_o - F_c \approx 0$, and subsequent error is reduced significantly since the contributions from these reflections in the Fourier sum are minimized.

From these considerations, it would appear that a difference electron density calculation should incorporate those reflections for which $F_o \leq F_c$, and omit those with $F_o > kF_c$, where the value of k establishes a threshold for the selection of reflections to omit. Two contour plots of sections from a difference electron density map for $Cr(P(C_6H_5)_3)_2(CO)_4$, based solely on the assignment of the chromium atom positions, are shown in Fig. 6.34. In comparing these plots with those from the electron density map for the same structure in Fig. 6.30, it is immediately clear that the resolution in the difference map has been enhanced and the noise decreased; the locations of the two phosphorus atoms in the molecule are easily determined from the plots.

Prior to the availability of modern crystallography software and fast computers, crystal structures were solved by creating contour plots similar to those shown here, often drawn by hand on plastic sheets and stacked into a three-dimensional model in order to locate the atoms in the Fourier map. Today the contouring is ordinarily

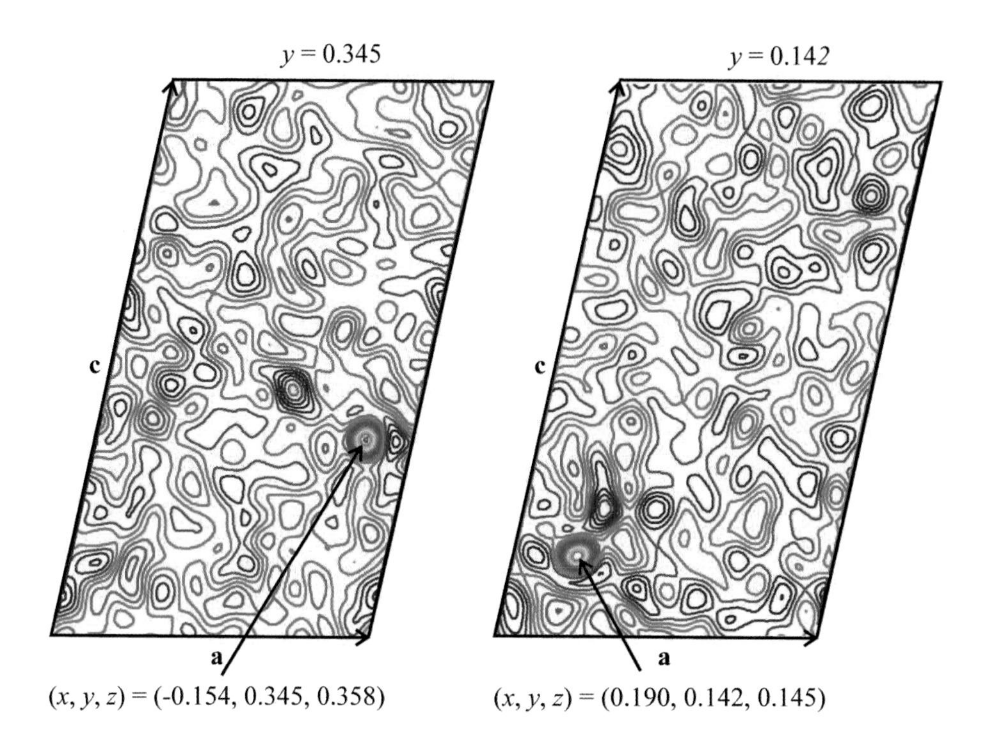

Figure 6.34 Contour plots of two sections of a difference electron density map for $Cr(P(C_6H_5)_3)_2(CO)_4$ with phases based solely on the locations of the chromium atoms in the unit cell. Warmer colors represent larger magnitudes for contour values.

performed by the computer — calculating the Fourier series sum at points on a three-dimensional grid and either searching this grid for maxima (for small molecule structures) or creating three-dimensional contour maps (for macromolecular structures). Small molecule structures are usually completed by an iterative process in which atoms are located from maxima in the difference map, their positions and thermal parameters refined (either isotropically or anisotropically), and a new difference map generated in order to find more atoms. The process is illustrated graphically in Fig. 6.35 for the $Cr(P(C_6H_5)_3)_2(CO)_4$ structure. In the first cycle, a difference map based solely on the position of the chromium atom is created. After least squares refinement of the chromium atom position and isotropic temperature factor, the peak search routine locates the peaks in the difference map with the largest magnitudes. The positions of these peaks relative to the chromium atom are shown in Fig. 6.35(a), indicating a structure with the expected pseudo-octahedral geometry about the chromium atom. The two largest maxima, for peaks **1** and **2** are found at $(-0.1542, 0.3470, 0.3538)$ and $(0.1901, 0.1423, 0.1454)$ with magnitudes of 12.37 e/Å^{3}* and 11.28 e/Å^3, respectively. These are the maxima observed in the contour sections in Fig. 6.34, and are almost certainly due to the two phosphorus atoms bonded to the chromium atom. Peaks **3**, **4**, **5**, and **7** have maxima in the range of 3.88 e/Å^3 - 3.21 e/Å^3, and appear to represent the oxygen atoms of the

*The magnitudes are in electron density units — the number of electrons per cubic angstrom.

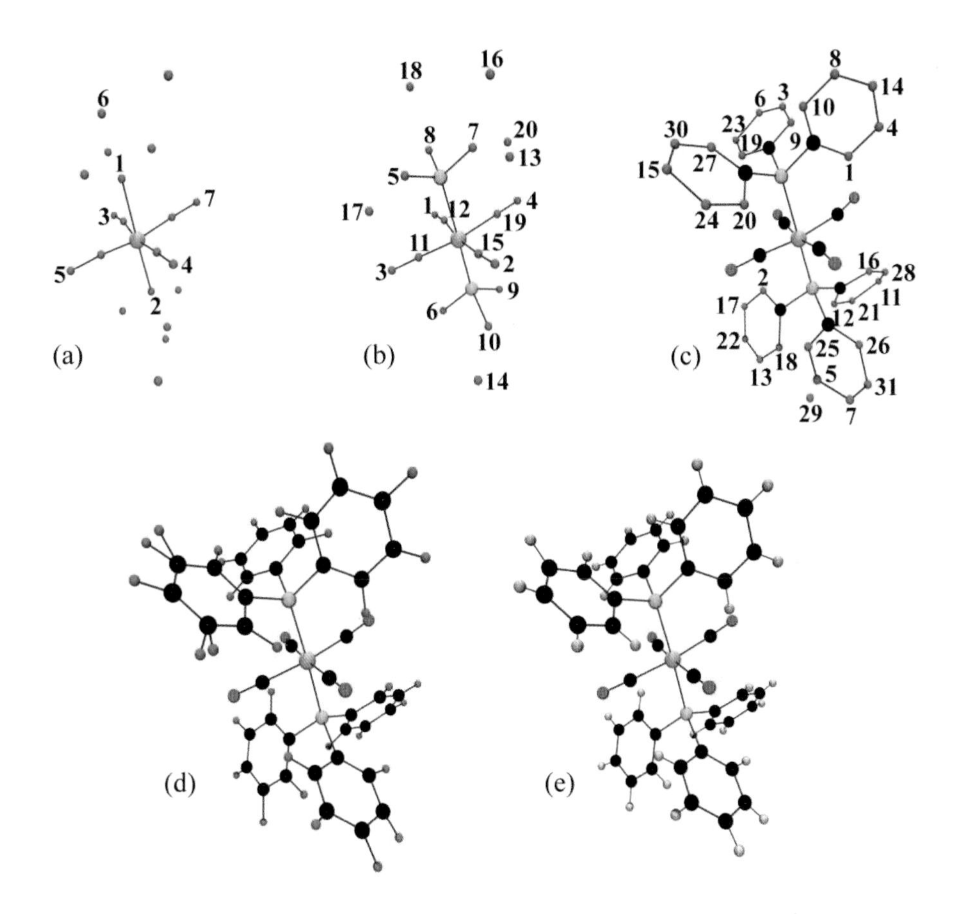

Figure 6.35 Difference Fourier solution of the $Cr(P(C_6H_5)_3)_2(CO)_4$ structure: (a) Cycle 1. (b) Cycle 2. (c) Cycle 3. (d) Cycle 4. (e) Completed structure. The blue spheres represent locations of maxima in the difference electron density Fourier map. The integers represent the order of the magnitudes of the maxima (**1** is the highest).

carbonyl (CO) groups in the molecule. In the early states of structural solution it is often advisable to be conservative about atom placement, at least until the heavier atoms are located. Thus only the two phosphorus atoms are added to the structure, and after isotropic refinement of the three atomic positions, the resulting difference map produces the peaks in Fig. 6.35(b). The CO peaks have returned (if they don't something is usually awry), and the carbon atoms bonded to the phosphorus atoms creating the expected pseudo-tetrahedral bonding environment for phosphorus show up clearly in the difference map. These assignments are verified with the calculation of bond lengths and angles for the various atoms (Sec. 1.5.3). The assignment of these atoms to the structure in a third cycle generates the remaining carbon atom locations in the difference map, completing the hexagonal phenyl rings in the structure, as illustrated in Fig. 6.35(c). With the inclusion of all the non-hydrogen atoms, refined anisotropically, the difference map produces the locations of the weakly scattering (one-electron) hydrogen atoms. As can be seen from Fig. 6.35(d), all of the hydrogen atoms are located in the structure, although two of them appear as double peaks. *The hydrogen atom intensities in the difference map are generally only just slightly above the noise in the map*, and a ripple in the wrong place will tend to "split" a contour. The assignment of one of the two peaks in each case and final positional refinement of the hydrogen atoms with fixed isotropic temperature factors, along with anisotropic refinement of the non-hydrogen atoms, results in the completed structure of the molecule — modeled in Fig. 6.35(e). It should be noted here that the location and refinement of hydrogen atom positions depends on the relative contributions of the hydrogen atoms to the overall scattering and the quality of the data set, among a number of things. In many cases it is not possible to locate the hydrogen atoms in a structure, especially if the scattering is dominated by one or more heavy atoms. Hydrogen atoms are often placed in their calculated positions and refined as if they were "riding" on the atom to which they are bonded.

Charge Density Distributions

The difference electron density synthesis has its ultimate application in determining the actual electron density distribution in the crystal structure. Indeed, this has been a goal of crystallographers since the beginning of X-ray crystallography. Darwin's 1922 paper entitled *The reflexion of x-rays from imperfect crystals*[83] begins with "The recent work of James, Bragg, and Bosanquet,[110,111] on the reflexion of X-rays from rock salt crystals is of extreme importance in that it promises more directly than any other method to supply information about the actual positions which the electrons occupy in the atom." In the early days of X-ray diffraction intensities were only crudely estimated — much too crudely to reflect the "positions" of the electrons. Most of the electrons in a structure reside in the atomic cores, which have spherically symmetric charge distributions, and it is only the electron density of the valence electrons that participates in bonding and is therefore polarized toward the regions between the atomic nuclei. This is a small fraction of the average electron density in the crystal, placing extreme demands on the quality of the intensity data if such *non-spherical* charge density distributions are to be determined experimentally.

If the intensity data are very accurate, and the systematic errors (polarization, absorption, extinction, etc.) are handled appropriately, then the final difference Fourier map represents the difference between the "spherical atom" electron density model and the experimental electron density. The resulting *deformation density map* should indicate positive regions where the valence elections tend to concentrate (into the bonding regions) and corresponding negative difference electron densities where they do not.[112] The determination of charge density distributions remains an active area of investigation in crystallography. The deformation density approach has been largely replaced by a more direct method pioneered by Philip Coppens[113] and coworkers which models the atomic electron density as an expansion of spherical harmonic functions centered on each atom (coined *pseudo-atoms* by their originator, Robert F. Stewart[114]), followed by the formulation of atomic scattering factors for the resulting *non-spherical* charge distributions. Coppens, along with coworker, Niels K. Hansen,[115] modeled the pseudo-atom electron density as

$$\begin{aligned} \rho_{atom}(\mathbf{r}) &= P_{core}\rho(r)_{core} + P_{valence}\kappa^3\rho(\kappa\mathbf{r})_{valence} \\ &+ \sum_{l=0}^{lmax} \kappa'^3 R_l(\kappa' r) \sum_{m=0}^{l} P_l^m y_l^m(\eta, \upsilon). \end{aligned} \tag{6.178}$$

P_{core} and $P_{valence}$ are the core and valence electron populations of the atom. $\rho(r)_{core}$ and $\rho(r)_{valence}$ are the electron densities of the core and valence electrons determined from Hartree-Fock-Slater calculations[43–45], each normalized to a single electron. The κ and κ' terms vary the radial extension of the electron density to account for differences in electronegativity and hence electron-electron repulsion. The $y_l^m(\eta, \upsilon)$ terms are real spherical harmonics (see Fig. 5.22 and Eqn. 5.248) representing the "shapes" of the atomic orbital basis set for each atom. The P_l^m and R_l terms are the populations and radial functions of each orbital on the atom; the radial functions are the same as the functions used in electronic structure calculations. P_{core} is usually fixed at the core electron population, while $P_{valence}$, P_l^m, κ, and κ' are *charge density variables* incorporated into the least-squares refinement of the structure (See Chapter 8). This is accomplished by generating scattering factors based on the atomic electron density distributions given by Eqn. 6.178 for each cycle of the iterative refinement process (Eqn. 3.99 expanded to three dimensions, with $\rho_{atom}(\mathbf{r}) \equiv \psi^2_{atom}(\mathbf{r})$).

Since linear combinations of the spherical harmonic functions can generate non-spherical distributions with various contributions for each atom, those familiar with molecular orbital calculations will note the close similarity to the linear combination of atomic orbitals (*LCAO*) formalism[46] used extensively in the theoretical determination of charge density distributions.

6.3.3 The Completion of Macromolecular Structures

The completion of macromolecular structures is a more involved process. Fourier contour maps are generally created in three dimensions and viewed with the aid of specially designed graphics software. The large number of atoms and the relative lack of resolution generally requires the visual manipulation of a model in order to map it onto the contour map. This process is coupled with least-squares refinement, which may include additional geometric and energetic constraints applied to the structure. We will defer a discussion of this process until Chapter 8. The two approaches to structure completion described in this chapter, electron density synthesis and difference electron density synthesis, occur in various forms, often in combination. The weights applied to F_o in Sim/Woolfson weighting can also be applied in difference electron density synthesis, and various combinations of these approaches are often used, especially in macromolecular crystallography[116].

Exercises

1. Representing each peak in a Fourier map as a single point, determine the "peak density" (the average number of peaks per unit volume) for an electron density map (D_e) and a Patterson map (D_P) for each of the following (ignore the hydrogen atoms): (a) C_5H_4NSH, $V_{cell} = 542.3$ Å^3, $Z = 4$, (b) Re_2Os_2 $((C_6H_5)_3P)_2(CO)_{11}(\mu\text{-}SNC_5H_4)$, $V_{cell} = 4920.7$ Å^3, $Z = 4$, and (c) Rat Short Chain Acyl-CoA Dehydrogenase (RSCAD)[117], asymmetric unit consisting of 6330 non-hydrogen atoms, $V_{cell} = 1,383,302$ Å^3, $Z = 6$.

2. A two dimensional unit cell for the *very* hypothetical ion pair, H^+ $C_5H_5^-$ (ignoring the hydrogen atoms and ion) is shown below.

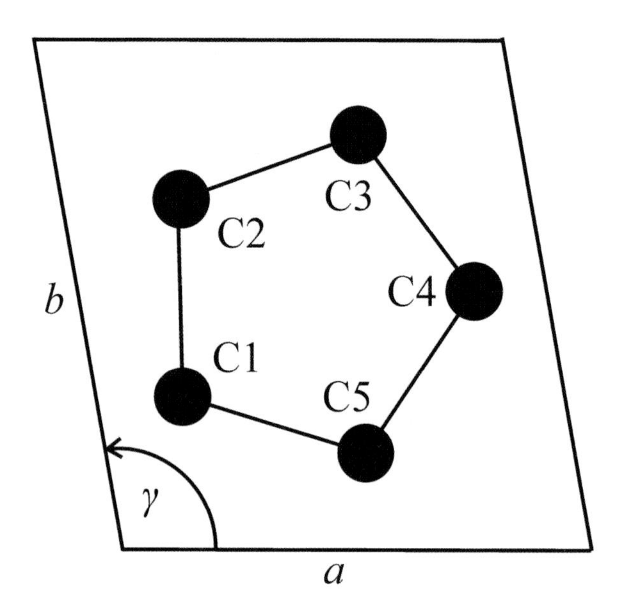

The unit cell parameters are $a = 3.5$ Å, $b = 3.8$ Å, and $\gamma = 100°$. The fractional coordinates of the carbon atoms: C1(0.1951, 0.3110), C2(0.2586, 0.6847), C3(0.6595, 0.8065), C4(0.8438, 0.5079) and C5(0.5568, 0.2017). Calculate the Cartesian components of the Patterson vectors and the four origins that encompass the unit cell. Plot the predicted locations of the Patterson maxima.

3. Predict the lines and/or planes in the Patterson unit cell upon which Patterson maxima tend to accumulate (in Harker lines and sections) for the following space groups: (a) $P2/c$, (b) $P2_1/c$, (c) $P2_12_12$, (d) $P2_12_12_1$, (e) $P3_1$, and (f) $P4_1$.

4. Extend the Patterson map created in Exercise 2 so that there are four unit cells arranged about a common origin (e.g., Fig. 6.8). Overlay the plot with a sheet of paper that is transparent enough to provide a view of the points underneath and carefully mark an "x" at the center of each point on the plot. Use this template to demonstrate by Patterson superposition that there are numerous images of the H^+ $C_5H_5^-$ structure in the Patterson map.

5. A sharpened Patterson map for the mercaptopyridine structure, space group $P2_1/n$, revealed the following peak locations, heights, and vector lengths:

peak	u_f	v_f	w_f	height	length
1	0.0000	0.0000	0.0000	999	0.00
2	0.5000	0.3730	0.5000	339	7.58
3	0.9764	0.5000	0.3130	282	5.51
4	0.0666	0.5000	0.1778	184	4.03
5	0.5189	0.1243	0.1880	144	4.44
6	0.7005	0.5000	0.2653	143	5.53
7	0.3194	0.0000	0.4201	141	5.94
8	0.8408	0.0000	0.0399	132	1.22
9	0.5136	0.5000	0.0481	123	4.49
10	0.5000	0.0000	0.5000	119	7.20

Determine the location of the sulfur atom in the unit cell and verify that its location is consistent with that listed in Sec. 1.5.3, found by direct methods.

6. The unit cell parameters for the hypothetical two-dimensional C_5H_4OH structure illustrated in Fig. 6.8(a) are $a = 7$ Å, $b = 6$ Å, and $\gamma = 110°$. The following table lists the $6 \times 5 = 30$ Patterson vectors, centered about the origin, due to the six non-hydrogen atoms:

vector	u_f	v_f	vector	u_f	v_f
$\overrightarrow{\mathrm{O1C1}}$	0.2130	0.1136	$\overrightarrow{\mathrm{C1O1}}$	−0.2130	−0.1136
$\overrightarrow{\mathrm{O1C2}}$	0.3164	0.3589	$\overrightarrow{\mathrm{C2O1}}$	−0.3164	−0.3589
$\overrightarrow{\mathrm{O1C3}}$	0.5253	0.3976	$\overrightarrow{\mathrm{C3O1}}$	−0.5253	−0.3976
$\overrightarrow{\mathrm{O1C4}}$	0.5511	0.1763	$\overrightarrow{\mathrm{C4O1}}$	−0.5511	−0.1763
$\overrightarrow{\mathrm{O1C5}}$	0.3581	0.0008	$\overrightarrow{\mathrm{C5O1}}$	−0.3581	−0.0008
$\overrightarrow{\mathrm{C1C2}}$	0.1034	0.2453	$\overrightarrow{\mathrm{C2C1}}$	−0.1034	−0.2453
$\overrightarrow{\mathrm{C1C3}}$	0.3123	0.2840	$\overrightarrow{\mathrm{C3C1}}$	−0.3123	−0.2840
$\overrightarrow{\mathrm{C1C4}}$	0.3381	0.0627	$\overrightarrow{\mathrm{C4C1}}$	−0.3381	−0.0627
$\overrightarrow{\mathrm{C1C5}}$	0.1451	−0.1128	$\overrightarrow{\mathrm{C5C1}}$	−0.1451	0.1128
$\overrightarrow{\mathrm{C2C3}}$	0.2089	0.0387	$\overrightarrow{\mathrm{C3C2}}$	−0.2089	−0.0387
$\overrightarrow{\mathrm{C2C4}}$	0.2347	−0.1826	$\overrightarrow{\mathrm{C4C2}}$	−0.2347	0.1826
$\overrightarrow{\mathrm{C2C5}}$	0.0417	−0.3581	$\overrightarrow{\mathrm{C5C2}}$	−0.0417	0.3581
$\overrightarrow{\mathrm{C3C4}}$	0.0258	−0.2213	$\overrightarrow{\mathrm{C4C3}}$	−0.0258	0.2213
$\overrightarrow{\mathrm{C3C5}}$	−0.1672	−0.3968	$\overrightarrow{\mathrm{C5C3}}$	0.1672	0.3968
$\overrightarrow{\mathrm{C4C5}}$	−0.1930	−0.1755	$\overrightarrow{\mathrm{C5C4}}$	0.1930	0.1755

A model Patterson map can be created by orienting a model of the molecule in a convenient location in the unit cell. The orientation of the "actual" molecule in the structure can then be determined by superimposing the origin of the two maps and rotating the model map until the maxima in both maps are superimposed — emulating a rotation search. Using a C–C bond length of 1.400 Å, and a C–O bond length of 1.411 Å, create a model Patterson map by placing the center of the five carbon pentagon at the origin and the oxygen atom coincident with the a (x) axis. Plot the Patterson maxima for both maps on semi-transparent paper and determine the angle at which the molecule is rotated in the unit cell with respect to the orientation of the model.

7. Intensities for the (1 2 2) reflection are collected from two isomorphous crystals containing cobalt amine transition metal complex ions, one with ammonium ions, Crystal A, the other with rubidium ions, Crystal B, located in the same site (site 1). The intensity from a third isomorphous crystal, crystal B', containing an ammonium ion in site 1, is formed from an iridium amine complex, in which an iridium atom has replaced the cobalt atom inside the complex (at site 2 in the crystal). After scaling, the structure factor amplitudes of the reflection from each crystal are: $F_A = 43.23$, $F_B = 76.16$, $F_{B'} = 103.81$. From a Patterson map generated from data collected from Crystal B, the rubidium ion is located at $\mathbf{r}_R$; a Patterson map generated from data collected from Crystal B' provides the location of the iridium atom at $\mathbf{r}_I$. Based on these locations and their three symmetry-equivalents, the structure factor contributions for each of these heavy atoms can be calculated:

$$\mathbf{F}_{R,122} = f_{Rb} \sum_{i=1}^{4} e^{-2\pi i(\mathbf{h}_{122} \cdot \mathbf{r}_{R,i})}, \quad F_R = 46.25, \ \varphi_R = 188^\circ$$

$$\mathbf{F}_{I,122} = f_{Ir} \sum_{i=1}^{4} e^{-2\pi i(\mathbf{h}_{122} \cdot \mathbf{r}_{I,i})}, \quad F_R = 97.88, \ \varphi_R = 185^\circ$$

Determine the phase of the (1 2 2) reflection for Crystal A.

8. Given the following scattering factor parameters for Eqn. 3.100,

	a_1	b_1	a_2	b_2	a_3	b_3	a_4	b_4	c
H	0.4899	20.6593	0.2620	7.7404	0.1968	49.5519	0.0499	2.2016	0.0013
C	2.3100	20.8439	1.0200	10.2075	1.5886	0.5687	0.8650	51.6512	0.2156
N	12.2126	0.0057	3.1322	9.8933	2.0125	28.9975	1.1663	0.5826	−11.5290
O	3.0485	13.2771	2.2868	5.7011	1.5463	0.3239	0.8670	32.9089	0.2508
S	6.9053	1.4679	5.2034	22.2151	1.4379	0.2536	1.5863	56.1720	0.8669

determine the Fourier weighting factors, $w(\mathbf{h})$, following location of the sulfur atom from a Patterson map for: (a) The (1 2 1) reflection for 2-dimethylsulfuranylidene–indan 1,3 dione, $C_{11}H_{10}O_2S$, which packs in the $P2_12_12_1$ space group. The observed structure factor magnitude for the (1 2 1) reflection, collected at $2\theta = 25.21°$ with CuKα radiation ($\lambda = 1.5418$), is $F_o = 76.19$. The sulfur atom alone produces a calculated structure factor magnitude of $F_c = 19.92$ for the reflection. (The modified Bessel functions needed are discussed in Appendix F); (b) the (1 1 2) reflection for the 2-mercaptopyridine structure, collected at $2\theta = 11.76°$ with MoKα radiation ($\lambda = 0.71073$). The observed structure factor is $F_o = 120.30$, and the calculated structure factor, based on the sulfur atom alone is $F_c = 40.73$.

Chapter 7

Crystal Structure Solution: Statistical

7.1 Direct Methods

In the previous chapter, methods for the assignment of structure factor phases required a partial model of the structure, initially determined from an experimental Fourier map. Such methods were loosely categorized as "experimental methods". While these methods were very often successful for the determination of structures containing heavy atoms, structures containing only lighter atoms – characterized (approximately) as "equal atom" structures – were often virtually impossible to solve by such methods. In the late 1940's John S. Kasper and David Harker were interested in the structural solution of molecules containing only boron hydrides, for which the bonding models of the time did not predict realistic stereochemical models. Composed of only light boron and hydrogen atoms, crystals of these molecules often evaded structural determination. In order to circumvent the need for initial atomic positions, Kasper and Harker sought methods that would assign the structure factor phases directly. These *direct methods* for phase assignment were based on relationships among the phases of reflections with large amplitudes. While the resulting inequality relationships[118] have been supplanted by more general and robust "statistical methods," the underlying principles upon which they are based remain the same – the phases of reflections are not independent of one another – the relationships among them are constrained to provide physically meaningful electron densities. That is, the structure factor waves, which we will refer to as *density waves**, must superposition in the Fourier series so that

(a) The electron density is positive.
(b) The electron density is localized (atomic).

This is illustrated in Fig. 7.1 for a hypothetical centrosymmetric one-dimensional

*The structure factors consist of sinusoidal electromagnetic waves that superposition to create an image of the electron density distribution in the crystal. Because of this they are often referred to as density waves; *density image waves* would be a more rigorous characterization.

Understanding Single-Crystal X-Ray Crystallography. Dennis W. Bennett

ISBN: 978-3-527-32677-8 (HC), 978-3-527-32794-2 (SC)

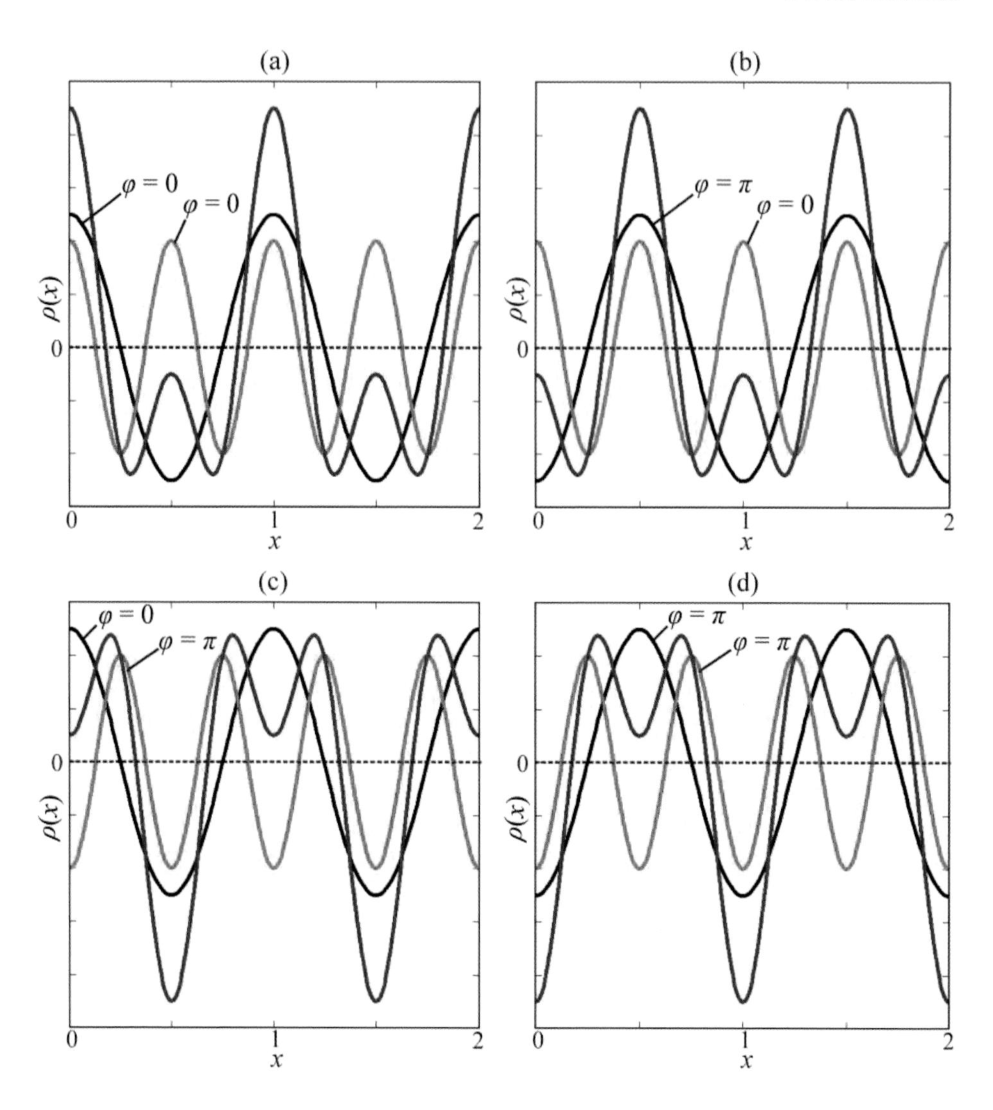

Figure 7.1 Contribution of structure factors $\mathbf{F}_1$ (plotted in black) and $\mathbf{F}_2$ (plotted in red), both with large amplitudes, to the electron density of a hypothetical centrosymmetric one-dimensional crystal: (a) $\varphi_1 = 0$, $\varphi_2 = 0$, (b) $\varphi_1 = \pi$, $\varphi_2 = 0$, (c) $\varphi_1 = 0$, $\varphi_2 = \pi$, and (d) $\varphi_1 = \pi$, $\varphi_2 = \pi$. The net contribution, $\mathbf{F}_1 + \mathbf{F}_2$, is plotted in blue.

crystal for which F_h and F_{2h}, in this case, F_1 and F_2, both have large magnitudes and therefore both contribute significantly in the determination of $\rho(x)$. Because F_1 and F_2 are both large, they will *probably* superposition to make the electron density positive. Since the structure is centrosymmetric, the phases of the reflections are restricted to either 0 or π, creating the four possible scenarios depicted in Figs. 7.1(a)–(d). The frequency of $\mathbf{F}_2$ is twice that of $\mathbf{F}_1$, and as depicted in Fig. 7.1(a) and Fig. 7.1(b), if the phase of $\mathbf{F}_2$ is 0, then $\mathbf{F}_2$ will tend to add to $\mathbf{F}_1$ so that the electron density tends to be positive, *regardless of the phase of* $\mathbf{F}_1$.

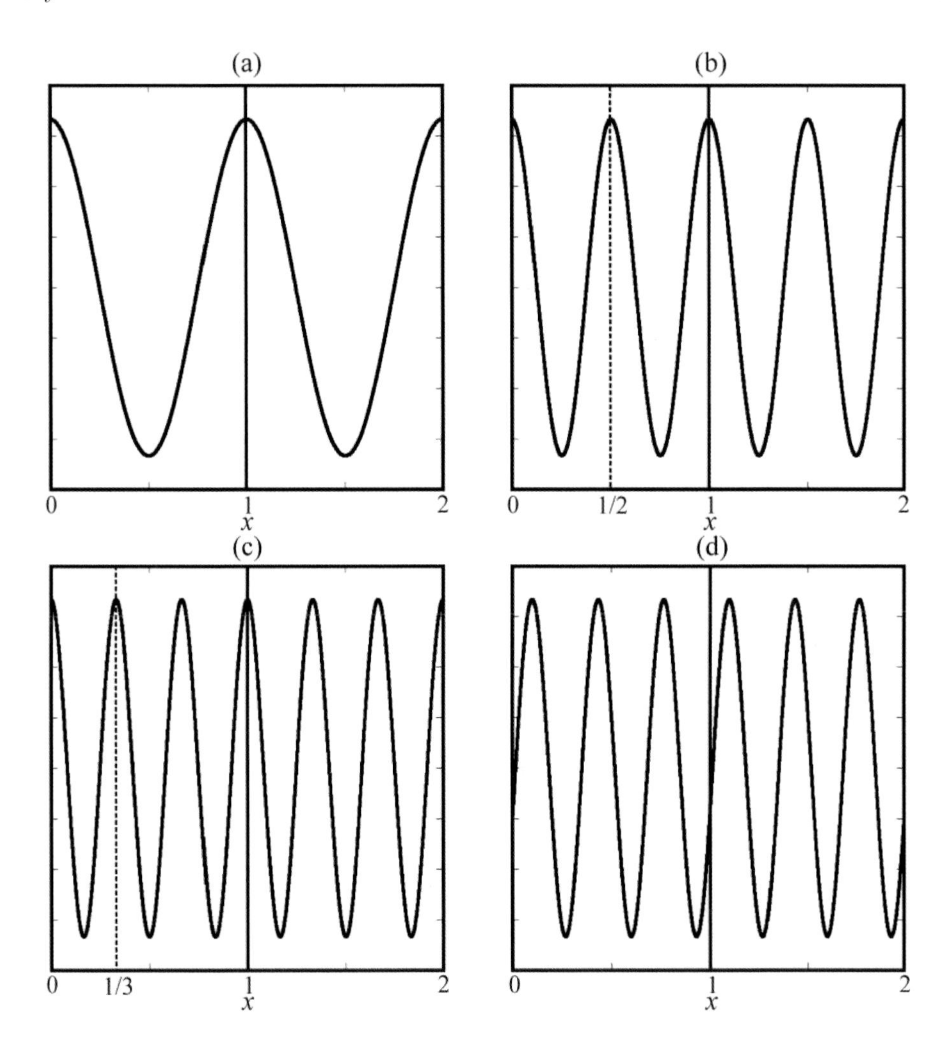

Figure 7.2 Structure factors (density waves) for a hypothetical one-dimensional crystal: (a) $\mathbf{F}_1$, $\varphi = 0$. (b) $\mathbf{F}_2$, $\varphi = 0$. (c) $\mathbf{F}_3$, $\varphi = 0$. and (d) $\mathbf{F}_3$, $\varphi \neq 0$.

Figs. 7.1(c) and (d) indicate the opposite effect when $\varphi_2 = \pi$: the electron density tends to be negative – which is physically unreasonable. It follows that φ_{2h} is *more likely* to be 0 than π if F_h and F_{2h} are both large, because the electron density must be positive. While it is not possible to assign individual phases based on intensities, the tendency for reflections that have large magnitudes to have structure factors (density waves) with phases leading to positive electron density prompts us to seek relationships for the phases of intense reflections that will assist us in the assignment of those phases.

7.1.1 Probability Methods: Structure Invariants

The structure factors must clearly have identical patterns in each unit cell, with frequencies that depend on the reflection indices, as illustrated in one dimension in

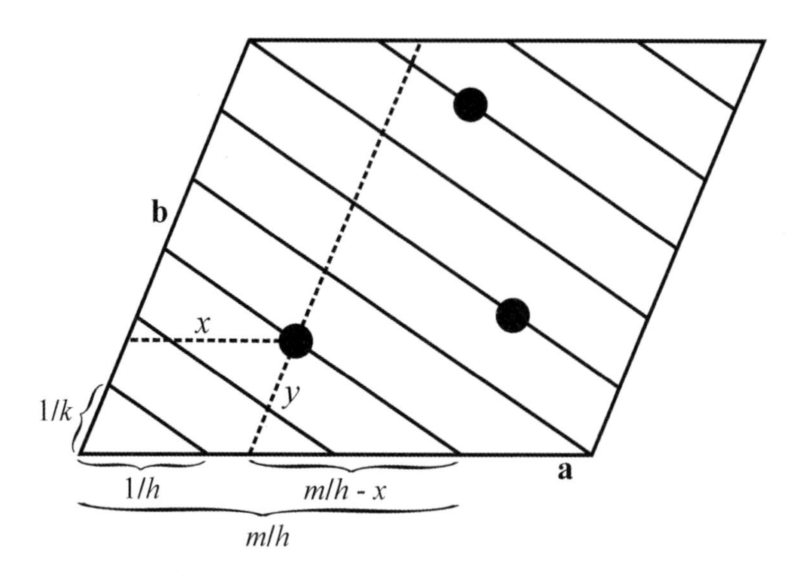

Figure 7.3 Two-dimensional unit cell with three identical atoms coincident with hk lattice lines.

Fig. 7.2 for (a) $h = 1$, $\varphi = 0$, (b) $h = 2$, $\varphi = 0$, (c)$h = 3$, $\varphi = 0$, and (d) $h = 3$, $\varphi \neq 0$. In this simple example the maxima are separated by the distances between the intersection points, $d_h = 1$, $d_h = 1/2$, and $d_h = 1/3$ for $\mathbf{F}_1$, $\mathbf{F}_2$, and $\mathbf{F}_3$, respectively. In two dimensions the density waves for $\mathbf{F}_{hk}$ have maxima that are separated by the distance between the lines, d_{hk}; in three dimensions the density waves for $\mathbf{F}_{hkl}$ have maxima that are separated by the distance between the planes, d_{hkl}.

In many elementary texts Bragg's Law is easily derived by treating the structure factors as if they were simply reflections arising from parallel planes separated by distances determined by their indices (see Appendix A). The actual scattering arises from electrons in atoms, and intuitively it would appear that a reflection would be strong if there were atoms on (or close to) the planes responsible for the reflection, and weak if there were not. To verify this, we consider the two-dimensional unit cell in Fig. 7.3, containing three identical atoms, each atom lying on an hk line in the lattice. The lattice lines will divide the a axis into h equal segments and the b axis into k equal segments. An atom lying on a line at (x, y) creates a triangle with edges $(m/h) - x$ and y, where m is an integer. The triangle is similar to the triangle with edges $1/h$ and $1/k$ so that

$$\frac{\frac{m}{h} - x}{y} = \frac{1/h}{1/k} = \frac{k}{h} \tag{7.1}$$

$$hx + ky = m. \tag{7.2}$$

The structure factor for this equal-atom structure is

$$\mathbf{F}_{hk} = f \sum_{j=1}^{3} e^{-2\pi\, i(hx_j + ky_j)}, \tag{7.3}$$

where f is the scattering factor for each atom. For any one of the atoms,

$$\begin{aligned} e^{-2\pi\, i(hx_j+ky_j)} &= e^{-2\pi\, i\, m_j} = \cos(-2\pi\, m_j) + i \sin(-2\pi\, m_j) = 1; \\ \Longrightarrow \mathbf{F}_{hk} &= F_{hk} = 3f. \end{aligned}$$

All of the atoms are scattering in phase,* and will contribute maximally to $\mathbf{F}_{hk}$. If one of the atoms is moved away from a line, then $hx_j + ky_j$ for that atom will no longer be an integer and $-2\pi\, i(hx_j + ky_j) = i\,\eta \neq -2\pi\, i\, m_j$. In this case,

$$\begin{aligned} \mathbf{F}_{hk} &= f \sum_{j=1}^{2} 1 + f\, e^{i\,\eta} = 2f + f\, \cos\eta + i\, f\, \sin\eta, \\ F_{hk}^2 &= (f(2+\cos\eta))^2 + (f\, \sin\eta))^2 \\ &= f^2(4 + 4\cos\eta + \cos^2\eta + \sin^2\eta) = f^2(5 + 4\cos\eta) \quad \text{and} \\ F_{hk} &= f\sqrt{5 + 4\cos\eta}. \end{aligned} \tag{7.4}$$

The maximum value for $\cos\eta$ is 1, and that is when $\eta = -2\pi\, m_j$ — when the atom is on a line — and $F_{hk} = f\sqrt{9} = 3f$. For any other location of the atom, $\cos\eta < 1$ and $F_{hk} < 3f$. Fig. 7.4 illustrates the effect on F_{hk} resulting from moving an atom in Fig. 7.3 along the dashed line parallel to the b axis (x fixed) from $y = 0$ to $y = 1$. The maximum value of $3f$ occurs each time an atom crosses one of the hk lines. The structure factor amplitude has a minimum value of $1f$ when the atom is midway between the two lines. In this case the atom is scattering 180° out of phase with the two atoms still residing on the planes, the destructive interference canceling out the contribution of one of the atoms on the planes. As the atom moves, its contribution to the structure factor increases as it approaches an hk line, and becomes maximum when it resides on the line.

It follows that structure factors for reflections for which atoms lie on or near the lattice planes of the reflection will tend to have large amplitudes. The converse is also true — if a structure factor is large then it is highly likely that many more atoms will be close to the reflection planes, with fewer in other positions (since scattering from them would tend to diminish the intensity). Thus, *reflections with high intensities tend to arise from lattice planes with large concentrations of electron density on or near them, and much less electron density in regions in-between.*

The electron density map — the structure — depends on the relative phases of the structure factors with respect to one another, and are independent of the scattering reference — the origin. On the other hand, the *absolute* phases depend on our choice of an origin. This can be seen in Fig. 7.5. The parallel lines represent lattice planes for an intense reflection; each plane will probably contain a large amount of electron density and consequentially a density wave that will be likely to have its maximum on or near the plane. If we place the origin at point A on one

*The scattering from a point located on any *hkl* plane in a 3-dimensional lattice will be in phase with the scattering from a point on any other plane with the same indices. This will be seen to be very useful when we consider structure semi-invariants.

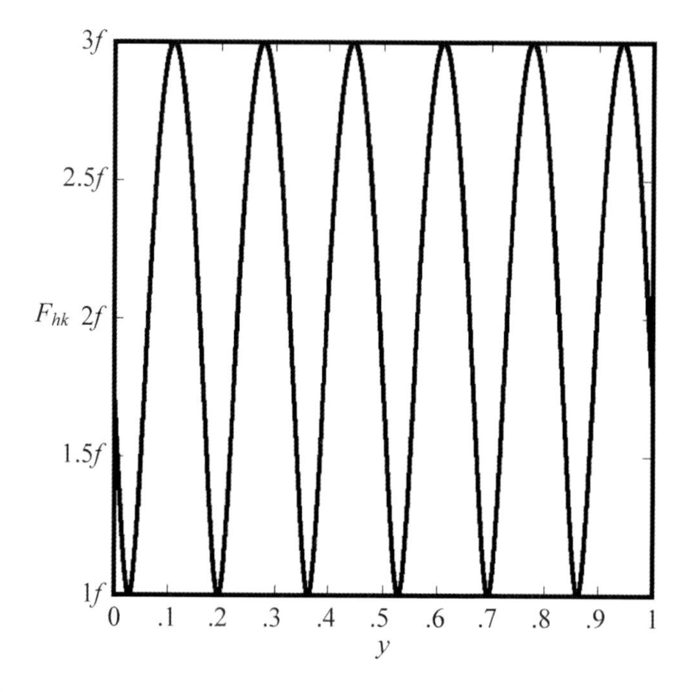

Figure 7.4 Variation in the structure factor amplitude, F_{hk}, as an atom is moved along y in a two-dimensional unit cell.

of the planes, the phase will be ≈ 0 at that point. On the other hand if we place it at the arbitrary point o, the phase of the structure factor will be shifted by some phase angle, φ. Extending the Ao line segment to A' places the origin again at a maximum, now with a phase of $\approx 2\pi \equiv 0$.

For the single reflection in Fig. 7.5 we could initially choose to set the phase to zero, thus restricting the origin to one of the planes, but this would tell us nothing about the phases of other reflections. Since the structure does not depend on the choice of an origin, neither should the relationships between the phases that depend on the structure (and vice versa). We are therefore looking for phase relationships that do not depend on the choice of an origin — *structure invariant* relationships.*

Consider the projection in Fig. 7.6(a), representing the intersection of two sets of lattice planes for intense reflections **h** and **k**. We expect that the density waves resulting from reflections from these two sets of planes will have maxima on or near the planes. If the maximum of each density wave resided on the plane, and the electron density was spread uniformly throughout each plane, then, with certainty, we could conclude that there would be an even larger concentration of electron density at their lines of intersection (perpendicular to the page), indicated by the dashed circles. In Fig. 7.6(b) the blue lines indicate the lattice planes defined by $-\mathbf{h}-\mathbf{k}$. If the reflection from this set of planes is also intense, if the maximum of its density wave lies on the plane, and if the electron density is uniformly distributed

*The more descriptive term, *origin invariant*, has been suggested by H. Schenk,[119] the author of the geometric approach used here to describe the relationships between phases.

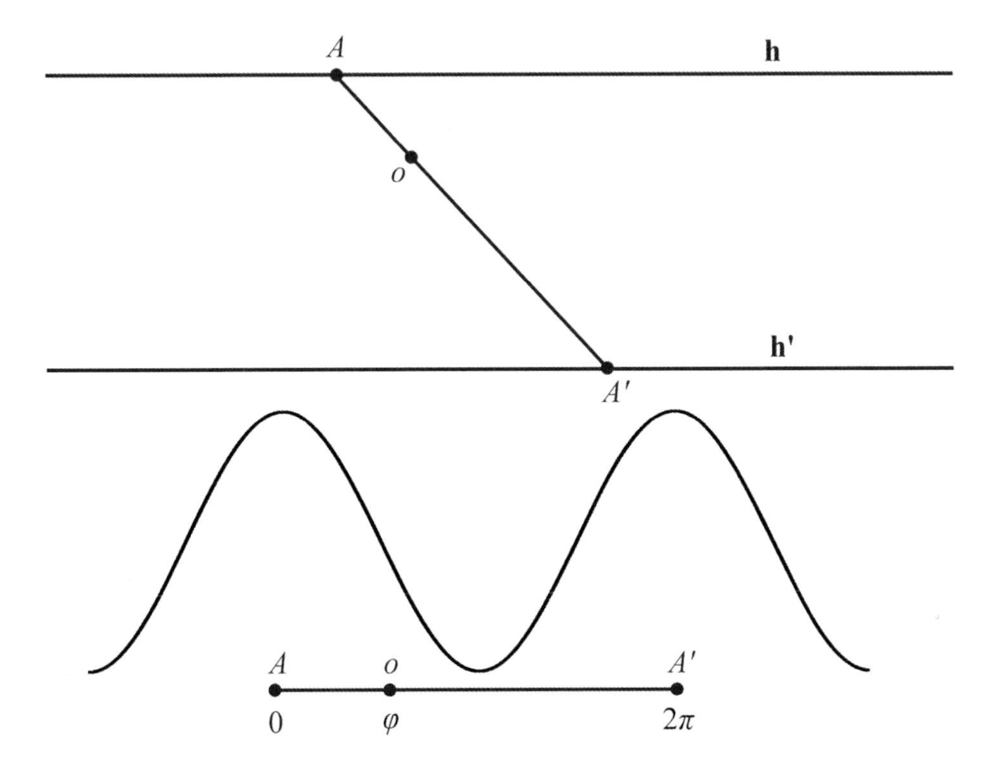

Figure 7.5 Top: Parallel lattice planes for an intense reflection with the origin chosen at an arbitrary point o. Bottom: Structure factor (density wave) for the reflection with maxima on the lattice planes.

on the plane, then the maxima of the density waves for the three reflections will coincide at their points of intersection. *Because the electron density on each set of planes must be positive*, there will be an even larger concentration of electron density at the points where the planes intersect. In reality, the maxima may be on or near the planes, and the electron density is localized and may not coincide exactly with an intersection point. *Large concentrations of electron density are therefore likely, but not certain, to reside at the points of intersection of the* **h**, **k**, *and* $-\mathbf{h}-\mathbf{k}$ *planes.*

Fig. 7.6(c) represents the placement of an origin at an arbitrary point in the lattice. Since the points of intersection are also likely to be the points of coincidence for the density wave maxima of the three intense reflections, geometric arguments allow us to determine a probable relationship for the phases of the three reflections. From Fig. 7.5, we note that $\overline{Ao}/\overline{AA'} \approx \varphi_{\mathbf{h}}/2\pi$, where the "approximately equal to" symbol "≈" is used to indicate "probably equal to". Similarly, $\overline{Bo}/\overline{BB'} \approx \varphi_{\mathbf{k}}/2\pi$ and $\overline{Co}/\overline{CC'} \approx \varphi_{-\mathbf{h}-\mathbf{k}}/2\pi$. The dotted line segments in the figure are parallel to

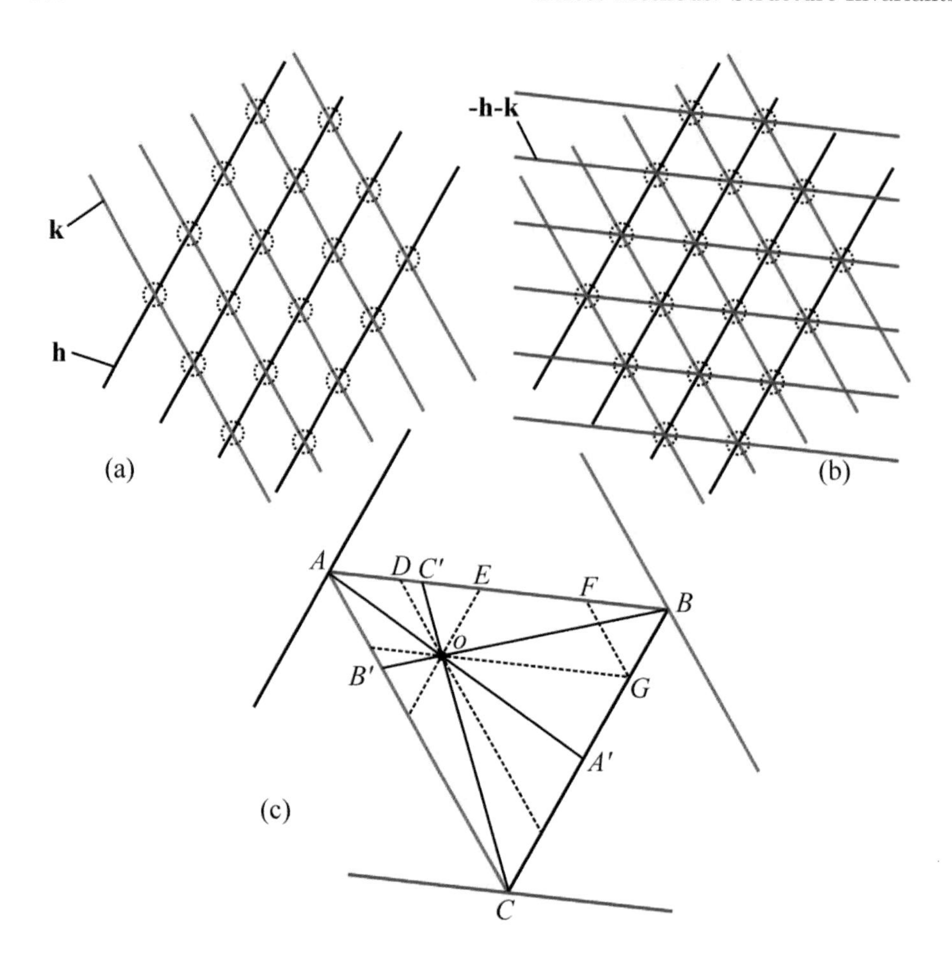

Figure 7.6 (a) Density wave maxima concentrating at the lines of intersection of the lattice planes corresponding to the intense reflections **h** and **k**. (b) Density wave maxima concentrating at the points of intersection of the lattice planes corresponding to the intense reflections **h**, **k**, and −**h**−**k**. (c) Arbitrary origin located between three intersection points.

the lattice planes and create a set of similar triangles:

$$AA'B \sim AoE \tag{7.5}$$

$$BB'A \sim BoD \tag{7.6}$$

$$CC'B \sim CoG \tag{7.7}$$

$$CAB \sim GFB \tag{7.8}$$

$$oDE \sim GFB \tag{7.9}$$

From 7.5, $\overline{Ao}/\overline{AA'} = \overline{AE}/\overline{AB}$, and from 7.6, $\overline{Bo}/\overline{BB}' = \overline{BD}/\overline{AB}$. From 7.7, $\overline{Co}/\overline{CC}' = \overline{CG}/\overline{GB}$ and from 7.8, $\overline{CG}/\overline{GB} = \overline{AF}/\overline{AB}$; thus, $\overline{Co}/\overline{CC}' = \overline{AF}/\overline{AB}$.

The sum of the probable phase ratios can now be expressed with a common denominator:

$$\frac{\overline{Ao}}{\overline{AA'}} + \frac{\overline{Bo}}{\overline{BB'}} + \frac{\overline{Co}}{\overline{CC'}} = \frac{\overline{AE} + \overline{BD} + \overline{AF}}{\overline{AB}} \approx \frac{\varphi_{\mathbf{h}}}{2\pi} + \frac{\varphi_{\mathbf{k}}}{2\pi} + \frac{\varphi_{-\mathbf{h}-\mathbf{k}}}{2\pi}. \tag{7.10}$$

$oDFG$ is a parallelogram, so that $\overline{oD} = \overline{GF}$; thus from 7.9, $\overline{DE} = \overline{FB}$, and the numerator is equal to $2\overline{AB}$:

$$\begin{aligned}\overline{AE} + \overline{BD} + \overline{AF} &= (\overline{AD} + \overline{DE}) + (\overline{DE} + \overline{EF} + \overline{FB}) + (\overline{AD} + \overline{DE} + \overline{EF}) \\ &= \overline{AD} + \overline{DE} + \overline{DE} + \overline{EF} + \overline{FB} + \overline{AD} + \overline{FB} + \overline{EF}\end{aligned}$$

$$\begin{aligned} &= 2(\overline{AD} + \overline{DE} + \overline{EF} + \overline{FB}) \\ &= 2\overline{AB}. \end{aligned} \tag{7.11}$$

Because the origin location is arbitrary, the structure invariant relationship for the phases of these three intense reflections is

$$\begin{aligned}&\frac{\varphi_{\mathbf{h}}}{2\pi} + \frac{\varphi_{\mathbf{k}}}{2\pi} + \frac{\varphi_{-\mathbf{h}-\mathbf{k}}}{2\pi} \approx \frac{2\overline{AB}}{\overline{AB}} = 2 \Longrightarrow \\ &\varphi_{\mathbf{h}} + \varphi_{\mathbf{k}} + \varphi_{-\mathbf{h}-\mathbf{k}} \approx 4\pi \equiv 0.\end{aligned} \tag{7.12}$$

The relationship for the three phases in 7.12 is known as a *triplet invariant.* The ability to correctly assess the probability of the validity of these relationships lies at the heart of direct methods structural solutions. Other invariants will come into play as well (e.g., *quartet* invariants), along with *semi-invariants* that *do* depend on the selection of the origin, but remain unchanged when the origin is shifted to one that is symmetry-equivalent. Before we apply probability methods to the triplet relationship, we first establish it on more rigorous ground.

Sayre's Equation: Triplet Invariants

In the previous section the triplet relationship arose from an intuitive argument related to the requirement that electron density is a positive quantity. In 1952 David Sayre[120] published an approach to phase determination that leads to the same relationship for intense reflections, but is based on the knowledge that electron density is localized around atoms — the second criterion stated at the beginning of the chapter.

Sayre's treatment approximates the structure by considering that it contains equal atoms — a good approximation for "lighter atom" compounds.* The derivation is based on the observation that the function $\rho(\mathbf{r})^2$ is very similar to the $\rho(\mathbf{r})$ function. Indeed, the squared electron density function generates peaks much like those from the electron density function, localized at the same centers, and actually sharper, tending to tail off more rapidly because the squares of numbers less than 1 are smaller than the numbers themselves. This is illustrated in Fig. 7.7 for a hypothetical one-dimensional unit cell. It is therefore reasonable to conclude

*The approximation is often not as good when the structure contains a heavy atom that dominates the scattering — and therefore skews the relationship.

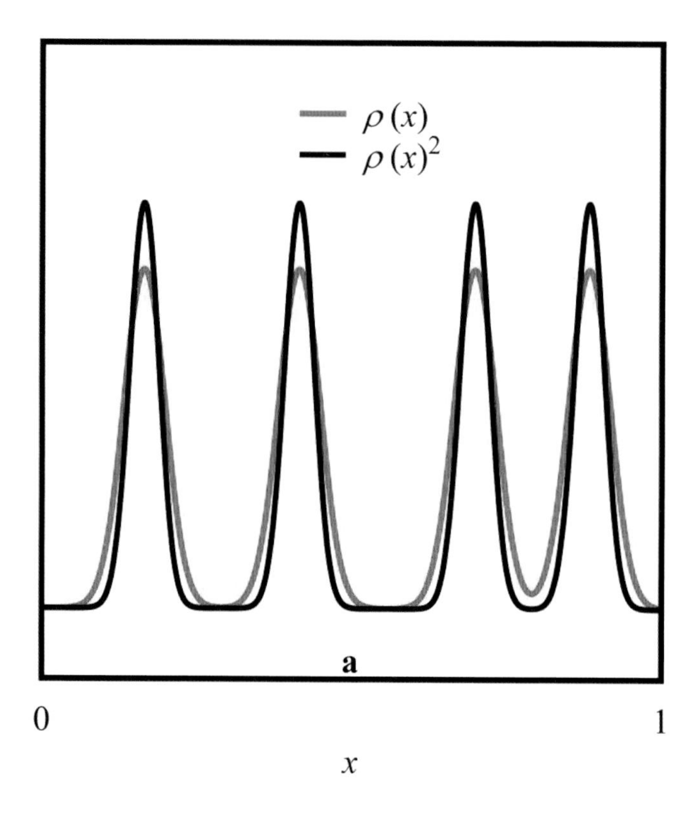

Figure 7.7 Electron density and squared electron density for a hypothetical one-dimensional unit cell containing four identical atoms.

that the $\rho(\mathbf{r})^2$ function can be represented as a Fourier series similar to the $\rho(\mathbf{r})$ representation:

$$\rho(\mathbf{r})^2 = \frac{1}{V_c}\sum_{\mathbf{h}} \mathbf{G}_{\mathbf{h}} e^{-2\pi\, i(\mathbf{h}\cdot\mathbf{r})}, \tag{7.13}$$

where the density-squared structure factor is

$$\mathbf{G}_{\mathbf{h}} = \sum_{j=1}^{n} g_j e^{-2\pi\, i(\mathbf{h}\cdot\mathbf{r})} \tag{7.14}$$

with g_j the "scattering factor" for the squared electron density. The $\rho(\mathbf{r})^2$ function can also be generated from the electron density Fourier series:

$$\rho(\mathbf{r}) = \frac{1}{V_c}\sum_{\mathbf{l}} \mathbf{F}_{\mathbf{l}} e^{-2\pi\, i(\mathbf{l}\cdot\mathbf{r})}$$

$$\begin{aligned}\rho(\mathbf{r})^2 &= \left(\frac{1}{V_c}\sum_{\mathbf{l}} \mathbf{F}_{\mathbf{l}} e^{-2\pi\, i(\mathbf{l}\cdot\,\mathbf{r})}\right)\left(\frac{1}{V_c}\sum_{\mathbf{l}'} \mathbf{F}_{\mathbf{l}'} e^{-2\pi\, i(\mathbf{l}'\cdot\,\mathbf{r})}\right) \\ &= \frac{1}{V_c^2}\sum_{\mathbf{l}}\sum_{\mathbf{l}'} \mathbf{F}_{\mathbf{l}}\mathbf{F}_{\mathbf{l}'} e^{-2\pi\, i((\mathbf{l}+\mathbf{l}')\cdot\,\mathbf{r})}. \end{aligned} \tag{7.15}$$

The summation variables can be redefined as $\mathbf{k} = \mathbf{l}'$ and $\mathbf{h} = \mathbf{l} + \mathbf{l}' = \mathbf{l} + \mathbf{k}$; both $\mathbf{h}$ and $\mathbf{k}$ remain independent since $\mathbf{l}$ and $\mathbf{l}'$ are independent. Thus,

$$\rho(\mathbf{r})^2 = \frac{1}{V_c^2}\sum_{\mathbf{h}-\mathbf{k}}\sum_{\mathbf{k}} \mathbf{F}_{\mathbf{h}-\mathbf{k}}\mathbf{F}_{\mathbf{k}} e^{-2\pi\, i((\mathbf{h}-\mathbf{k}+\mathbf{k})\cdot\,\mathbf{r})}$$

There is a value of $\mathbf{h}-\mathbf{k}$ for every value of $\mathbf{h}$ in the summations, and the summation variable can be replaced with $\mathbf{h}$. Thus,

$$\rho(\mathbf{r})^2 = \frac{1}{V_c^2}\sum_{\mathbf{h}}\sum_{\mathbf{k}} \mathbf{F}_{\mathbf{h}-\mathbf{k}}\mathbf{F}_{\mathbf{k}} e^{-2\pi\, i(\mathbf{h}\cdot\,\mathbf{r})}, \tag{7.16}$$

and from Eqn. 7.13, the squared density structure factor can be identified as

$$\mathbf{G}_{\mathbf{h}} = \frac{1}{V_c}\sum_{\mathbf{k}} \mathbf{F}_{\mathbf{h}-\mathbf{k}}\mathbf{F}_{\mathbf{k}}. \tag{7.17}$$

For an equal-atom structure, the electron density structure factor is

$$\mathbf{F}_{\mathbf{h}} = f\sum_{j=1}^{n} e^{-2\pi\, i(\mathbf{h}\cdot\mathbf{r})}, \tag{7.18}$$

and the squared electron density structure factor is

$$\mathbf{G}_{\mathbf{h}} = g\sum_{j=1}^{n} e^{-2\pi\, i(\mathbf{h}\cdot\mathbf{r})}, \tag{7.19}$$

resulting in a simple relationship between $\mathbf{G}_{\mathbf{h}}$ and $\mathbf{F}_{\mathbf{h}}$:

$$\frac{\mathbf{F}_{\mathbf{h}}}{f} = \frac{\mathbf{G}_{\mathbf{h}}}{g}. \tag{7.20}$$

This relationship gives an expression for every structure factor as the sum of products of the structure factors:

$$\mathbf{F}_{\mathbf{h}} = \frac{f}{g}\mathbf{G}_{\mathbf{h}} = \frac{f}{g}\frac{1}{V_c}\sum_{\mathbf{k}} \mathbf{F}_{\mathbf{h}-\mathbf{k}}\mathbf{F}_{\mathbf{k}}. \tag{7.21}$$

Eqn. 7.21, known as *Sayre's Equation*, is exact for equal-atom structures, and a very good approximation for "lighter-atom" structures (almost equal-atom). In terms of the amplitudes and phases of the structure factors, Sayre's equation becomes

$$\begin{aligned} F_{\mathbf{h}} e^{i\varphi_{\mathbf{h}}} &= \frac{f}{g}\frac{1}{V_c}\sum_{\mathbf{k}} F_{\mathbf{h}-\mathbf{k}}\, e^{i\varphi_{\mathbf{h}-\mathbf{k}}}\, F_{\mathbf{k}} e^{i\varphi_{\mathbf{k}}} \\ &= \frac{f}{g}\frac{1}{V_c}\sum_{\mathbf{k}} F_{\mathbf{k}} F_{\mathbf{h}-\mathbf{k}}\, e^{i(\varphi_{\mathbf{k}}+\varphi_{\mathbf{h}-\mathbf{k}})}. \end{aligned} \tag{7.22}$$

Now suppose that our data contains three intense reflections with indices $\mathbf{h}$, $\mathbf{k}'$, and $\mathbf{h}-\mathbf{k}'$, for a specific reflection identified by $\mathbf{k}'$. The $F_{\mathbf{k}}F_{\mathbf{h}-\mathbf{k}}$ factors are the weighting factors in the sum, and if the term for $\mathbf{k}' = \mathbf{k}$ dominates the sum then we can probably ignore the remainder of the terms:

$$F_{\mathbf{h}}e^{i\varphi_{\mathbf{h}}} \approx \frac{f}{g}\frac{1}{V_c}F_{\mathbf{k}}F_{\mathbf{h}-\mathbf{k}}\, e^{i(\varphi_{\mathbf{k}}+\varphi_{\mathbf{h}-\mathbf{k}})}, \tag{7.23}$$

with

$$F_{\mathbf{h}} \approx \frac{f}{g}\frac{1}{V_c}F_{\mathbf{k}}F_{\mathbf{h}-\mathbf{k}}, \tag{7.24}$$

and

$$e^{i\varphi_{\mathbf{h}}} \approx e^{i(\varphi_{\mathbf{k}}+\varphi_{\mathbf{h}-\mathbf{k}})}$$

$$\varphi_{\mathbf{h}} \approx \varphi_{\mathbf{k}} + \varphi_{\mathbf{h}-\mathbf{k}} \tag{7.25}$$

$$\varphi_{\mathbf{h}} - \varphi_{\mathbf{k}} - \varphi_{\mathbf{h}-\mathbf{k}} \approx 0. \tag{7.26}$$

From Eqn. 7.25 for strong reflections $-\mathbf{h}$, $\mathbf{k}$, and $-\mathbf{h}-\mathbf{k}$,

$$-\varphi_{-\mathbf{h}} + \varphi_{\mathbf{k}} + \varphi_{-\mathbf{h}-\mathbf{k}} \approx 0. \tag{7.27}$$

The Freidel law requires that the $\mathbf{h}$ reflection is also intense, and that $-\varphi_{-\mathbf{h}} = \varphi_{\mathbf{h}}$ (ignoring the usually small effects of anomalous dispersion), providing the triplet invariant relation for the phases of three intense reflections derived in the previous section from more intuitive arguments:

$$\varphi_{\mathbf{h}} + \varphi_{\mathbf{k}} + \varphi_{-\mathbf{h}-\mathbf{k}} \approx 0. \tag{7.28}$$

For a centrosymmetric structure, Eqn. 7.21 becomes

$$F_{\mathbf{h}}S_{\mathbf{h}} = \frac{f}{g}\frac{1}{V_c}\sum_{\mathbf{k}} F_{\mathbf{h}-\mathbf{k}}S_{\mathbf{h}-\mathbf{k}}\, F_{\mathbf{k}}S_{\mathbf{k}}, \tag{7.29}$$

leading to

$$F_{\mathbf{h}}S_{\mathbf{h}} \approx \frac{f}{g}\frac{1}{V_c}F_{\mathbf{h}-\mathbf{k}}S_{\mathbf{h}-\mathbf{k}}\, F_{\mathbf{k}}S_{\mathbf{k}} \tag{7.30}$$

for three intense reflections with the appropriate indices. This gives

$$\frac{S_{\mathbf{h}}}{S_{\mathbf{h}-\mathbf{k}}S_{\mathbf{k}}} \approx \frac{fF_{\mathbf{h}-\mathbf{k}}\, F_{\mathbf{k}}}{gV_cF_{\mathbf{h}}}. \tag{7.31}$$

Since $S_{\mathbf{k}} = \pm 1$, $S_{\mathbf{k}} = 1/S_{\mathbf{k}}$, etc., the term on the left becomes $S_{\mathbf{h}}S_{\mathbf{h}-\mathbf{k}}S_{\mathbf{k}} = \pm 1$. All of the variables in the term on the right are positive numbers, and the triplet relationship for centrosymmetric structures becomes

$$S_{\mathbf{h}}S_{\mathbf{k}}S_{\mathbf{h}-\mathbf{k}} \approx +1. \tag{7.32}$$

This relationship, relatively simple compared to that for non-centrosymmetric structures, was published by William Cochran in a paper in 1952 in *Acta Crystallographica*[121] immediately following the Sayre paper. The signs of the structure factors are limited to ± 1, and it is not surprising that the first attempts to develop direct methods focused on the determination of structure factor signs for centrosymmetric structures. These relationships are seen in various forms, all of which are equivalent. For example, for intense reflections with indices $-\mathbf{h}$, $\mathbf{k}$, and $-\mathbf{h}-\mathbf{k}$,

$$S_{-\mathbf{h}} S_{\mathbf{k}} S_{-\mathbf{h}-\mathbf{k}} \approx +1. \tag{7.33}$$

For centrosymmetric reflections, $S_{-\mathbf{h}} = S_{\mathbf{h}}$, since the reflections have the same phases ($-0 \equiv 0, -\pi \equiv \pi$), and

$$S_{\mathbf{h}} S_{\mathbf{k}} S_{-\mathbf{h}-\mathbf{k}} \approx +1, \tag{7.34}$$

consistent with Eqn. 7.28.

Although we are not there yet, the triplet relations point the way to obtaining phases directly. If we begin with a starting set of phases, and have at our disposal a set of triplet relations that have a high probability of being valid, then we should be able to use the relationships to determine new phases. In other words, if we know $\varphi_{\mathbf{k}}$ and $\varphi_{\mathbf{h}-\mathbf{k}}$, and are reasonably certain that $\varphi_{\mathbf{h}} - \varphi_{\mathbf{k}} - \varphi_{\mathbf{h}-\mathbf{k}} \approx 0$, then we can calculate $\varphi_{\mathbf{h}}$. This new phase can then be incorporated into another triplet relationship to determine yet another phase, and so forth. The caveat here is that an incorrect phase will lead to more incorrect phases – and will ordinarily *not* lead to a solution. *We will assess the relative likelihood that a triplet relationship is valid by determining its probability distribution.*

Triplet Probabilities

In order to compare structure factor magnitudes, it is useful to have the structure factors in a form that is independent of the diffraction angle. In addition, we wish to establish criteria that do not depend on the actual magnitudes of the structure factors, since they scale with the number of electrons in the unit cell. The quasi-normalized structure factors used to describe intensity statistics in Chapter 5 were chosen for the same reasons, and we will make use of them here. Dividing each structure factor in Eqn. 7.21 by $(\sum f_j^2)^{1/2} = \sqrt{n} f$ for an n equal-atom structure, Sayre's equation becomes

$$\mathbf{E}_{\mathbf{h}} = \frac{f}{g} \frac{\sqrt{n} f}{V_c} \sum_{\mathbf{k}} \mathbf{E}_{\mathbf{k}} \mathbf{E}_{\mathbf{h}-\mathbf{k}} \tag{7.35}$$

For three reflections with indices $\mathbf{h}$, $\mathbf{k}$, and $\mathbf{h}-\mathbf{k}$ and large values of E,

$$E_{\mathbf{h}} \approx \frac{f}{g} \frac{\sqrt{n} f}{V_c} E_{\mathbf{k}} E_{\mathbf{h}-\mathbf{k}}, \tag{7.36}$$

and again,

$$\varphi_{\mathbf{h}} - \varphi_{\mathbf{k}} - \varphi_{\mathbf{h}-\mathbf{k}} \approx 0. \tag{7.37}$$

We now turn to the fundamental question for determining phases by direct methods: *Given the three reflections in a triplet relationship, if we know the magnitudes (E values) of the three reflections, and two of the phases, what is the probability that the third phase is correctly determined by* Eqn. 7.37? The probability is known as a *conditional probability* and is formally represented as $Pr(A|B)$, which is read "given B, what is the probability of A?" In our case, the conditional probability that we are aiming for is $Pr(\varphi_{\mathbf{h}}|E_{\mathbf{h}}, \mathbf{E_k}, \mathbf{E_{h-k}})$.

We begin by analyzing the probability distribution of $\mathbf{E_h}$. From Eqn. 5.502,

$$\mathbf{E_h} = \frac{\mathbf{F}_{\mathbf{h}}^p}{\left(\sum_{j=1}^{n} Z_j^2\right)^{1/2}} = \frac{\sum_{j=1}^{n} Z_j\, e^{-2\pi\, 1(\mathbf{h}\cdot\mathbf{r}_j)}}{\left(\sum_{j=1}^{n} Z_j^2\right)^{1/2}}. \tag{7.38}$$

The analysis is simplified by assuming an equal-atom structure with n atoms in the unit cell, for which $Z_j = Z$, and

$$\frac{Z}{\left(\sum_{j=1}^{n} Z^2\right)^{1/2}} = \frac{Z}{Z\sqrt{n}} = \frac{1}{\sqrt{n}} \tag{7.39}$$

$$\mathbf{E_h} = \frac{1}{\sqrt{n}} \sum_{j=1}^{n} e^{-2\pi\, i(\mathbf{h}\cdot\,\mathbf{r}_j)} \tag{7.40}$$

$$= \frac{1}{\sqrt{n}} \sum_{j=1}^{n} \cos(-2\pi\mathbf{h}\cdot\mathbf{r}_j) + i \sum_{j=1}^{n} \sin(-2\pi\mathbf{h}\cdot\mathbf{r}_j) \tag{7.41}$$

$$= A_{\mathbf{h}} + i\, B_{\mathbf{h}}.$$

$A_{\mathbf{h}}$ and $B_{\mathbf{h}}$ can also be expressed in terms of the phase and magnitude of $\mathbf{E_h}$:

$$\mathbf{E_h} = E_{\mathbf{h}} e^{i\varphi_{\mathbf{h}}} = E_{\mathbf{h}} \cos\varphi_{\mathbf{h}} + i\, E_{\mathbf{h}} \sin\varphi_{\mathbf{h}} = A_{\mathbf{h}} + i\, B_{\mathbf{h}}. \tag{7.42}$$

As we did in Chapter 5, we will consider the atoms in fixed locations, with the indices as random variables. This is entirely equivalent to treating the atom locations as equally probable, with fixed indices. In either case $\mathbf{E_h}$ is the sum of random variables, and according to the central limit theorem (Section 5.1.1), $\mathbf{E_h}$ is normally distributed with probability density function,

$$f_G(\mathbf{E_h}) = \frac{1}{\sqrt{2\pi}\,\sigma_{\mathbf{h}}} e^{-(\mathbf{E_h}-\langle\mathbf{E_h}\rangle)^2/2\sigma_{\mathbf{h}}^2}. \tag{7.43}$$

Since $\mathbf{E_h}$ is determined by $A_{\mathbf{h}}$ and $B_{\mathbf{h}}$ simultaneously, the probability density function for $\mathbf{E_h}$ is given by the bivariate normal distribution (Sec. 5.1.1), and since $A_{\mathbf{h}}$ and $B_{\mathbf{h}}$ are independent, from Eqn. 5.135,

$$f_G(\mathbf{E_h}) = f_G(A_{\mathbf{h}}, B_{\mathbf{h}}) = \frac{1}{\sqrt{2\pi}\,\sigma_A} e^{-(A_{\mathbf{h}}-\langle A_{\mathbf{h}}\rangle)^2/2\sigma_A^2} \frac{1}{\sqrt{2\pi}\,\sigma_B} e^{-(B_{\mathbf{h}}-\langle B_{\mathbf{h}}\rangle)^2/2\sigma_B^2}. \tag{7.44}$$

In order to express $\mathbf{E_h}$ in terms of $\mathbf{E_k}$ and $\mathbf{E_{h-k}}$, we add and subtract $\mathbf{k}$ in the exponential term:

$$\mathbf{E_h} = \frac{1}{\sqrt{n}} \sum_{j=1}^{n} e^{-2\pi\, i((\mathbf{h}+\mathbf{k}-\mathbf{k})\cdot\,\mathbf{r}_j)} = \frac{1}{\sqrt{n}} \sum_{j=1}^{n} e^{-2\pi\, i(\mathbf{k}\cdot\,\mathbf{r}_j)}\, e^{-2\pi\, i((\mathbf{h}-\mathbf{k})\cdot\,\mathbf{r}_j)}. \tag{7.45}$$

Because **k** is independent of **h** the average of the product is the product of the averages, and

$$\langle \mathbf{E_h} \rangle = \frac{1}{\sqrt{n}} \sum_{j=1}^{n} \langle e^{-2\pi i(\mathbf{k}\cdot \mathbf{r}_j)} \rangle \langle e^{-2\pi i((\mathbf{h}-\mathbf{k})\cdot \mathbf{r}_j)} \rangle. \tag{7.46}$$

If we now establish the constraint that $E_{\mathbf{h}}$, $\mathbf{E_k}$ and $\mathbf{E_{h-k}}$ are all known (fixed), the exponentials in the expressions for each sum can be replaced with their average values:

$$\mathbf{E_k} = \frac{1}{\sqrt{n}} \sum_{j=1}^{n} e^{-2\pi i(\mathbf{k}\cdot \mathbf{r}_j)} = \frac{1}{\sqrt{n}} \sum_{j=1}^{n} \langle e^{-2\pi i(\mathbf{k}\cdot \mathbf{r}_j)} \rangle = \frac{n}{\sqrt{n}} \langle e^{-2\pi i(\mathbf{k}\cdot \mathbf{r}_j)} \rangle;$$

$$\mathbf{E_{h-k}} = \frac{n}{\sqrt{n}} \langle e^{-2\pi i((\mathbf{h}-\mathbf{k})\cdot \mathbf{r}_j)} \rangle. \tag{7.47}$$

Substituting the average values in Eqn. 7.46 results in

$$\begin{aligned}\langle \mathbf{E_h} \rangle &= \frac{1}{\sqrt{n}} \sum_{j=1}^{n} \frac{\sqrt{n}}{n} \mathbf{E_k} \frac{\sqrt{n}}{n} \mathbf{E_{h-k}} = \frac{n}{\sqrt{n}} \frac{\sqrt{n}}{n} \frac{\sqrt{n}}{n} \mathbf{E_k E_{h-k}} \\ &= \frac{1}{\sqrt{n}} \mathbf{E_k E_{h-k}} = \frac{1}{\sqrt{n}} E_{\mathbf{k}} E_{\mathbf{h-k}} e^{i(\varphi_{\mathbf{k}} + \varphi_{\mathbf{h-k}})} \qquad (7.48) \\ &= \frac{1}{\sqrt{n}} E_{\mathbf{k}} E_{\mathbf{h-k}} \cos(\varphi_{\mathbf{k}} + \varphi_{\mathbf{h-k}}) \\ &\qquad + i \frac{1}{\sqrt{n}} E_{\mathbf{k}} E_{\mathbf{h-k}} \sin(\varphi_{\mathbf{k}} + \varphi_{\mathbf{h-k}}) \qquad (7.49) \\ &= \langle A_{\mathbf{h}} \rangle + i \langle B_{\mathbf{h}} \rangle.\end{aligned}$$

The variances of $A_{\mathbf{h}}$ and $B_{\mathbf{h}}$ can now be determined from Eqn. 5.78:

$$\sigma_A^2 = \langle A_{\mathbf{h}}^2 \rangle - \langle A_{\mathbf{h}} \rangle^2 \tag{7.50}$$

$$= \frac{1}{n} \left\langle \left(\sum_{j=1}^{n} \cos(-2\pi \mathbf{h}\cdot \mathbf{r}_j) \right)^2 \right\rangle - \frac{1}{n} \left\langle \left(\sum_{j=1}^{n} \cos(-2\pi \mathbf{h}\cdot \mathbf{r}_j) \right) \right\rangle^2 \tag{7.51}$$

$$= \frac{1}{n} \sum_{j=1}^{n} \left(\langle \cos^2(-2\pi \mathbf{h}\cdot \mathbf{r}_j) \rangle - \langle \cos(-2\pi \mathbf{h}\cdot \mathbf{r}_j) \rangle^2 \right). \tag{7.52}$$

The averages are taken over the atom locations in the unit cell with **h** fixed. If they were over all possible locations (positive and negative) and all indices, cosine-squared terms would average to 0.5 (Eqn. 7.174) and cosine terms would average to zero. Instead, cosine terms average to small numbers (e.g., 0.02). Squaring the $\langle A_{\mathbf{h}}^2 \rangle$ term in Eqn. 7.51,

$$\left\langle \left(\sum_{j=1}^{n} \cos(-2\pi \mathbf{h}\cdot \mathbf{r}_j) \right)^2 \right\rangle$$

$$= \left\langle \left(\sum_{j1=1}^{n} \cos(-2\pi \mathbf{h}\cdot \mathbf{r}_{j1}) \right) \right\rangle \left\langle \left(\sum_{j2=1}^{n} \cos(-2\pi \mathbf{h}\cdot \mathbf{r}_{j2}) \right) \right\rangle, \tag{7.53}$$

results in two types of terms in the sum:

$$\langle \cos^2(-2\pi \mathbf{h} \cdot \mathbf{r}_j) \rangle \quad \text{for} \quad j1 = j2 \quad \text{and} \tag{7.54}$$

$$\langle \cos(-2\pi \mathbf{h} \cdot \mathbf{r}_{j1}) \rangle \langle \cos(-2\pi \mathbf{h} \cdot \mathbf{r}_{j2}) \rangle \quad \text{for} \quad j1 \neq j2. \tag{7.55}$$

The average cosine terms are very small, making the cross terms negligible. The squared terms are evaluated using the trigonometric identity $\cos^2 \eta = (1+\cos 2\eta/2)$:

$$\langle \cos^2(-2\pi \mathbf{h} \cdot \mathbf{r}_j) \rangle = \frac{1 + \langle \cos(-4\pi \mathbf{h} \cdot \mathbf{r}_j) \rangle}{2} \simeq \frac{1}{2}. \tag{7.56}$$

The average value of the cosine is given by

$$\begin{aligned} \langle \cos(-2\pi \mathbf{h} \cdot \mathbf{r}_j) \rangle &= \frac{1}{n} \sum_{j=1}^{n} \cos(-2\pi \mathbf{h} \cdot \mathbf{r}_j) \quad \text{and} \\ \langle \cos(-2\pi \mathbf{h} \cdot \mathbf{r}_j) \rangle^2 &= \frac{1}{n^2} \left(\sum_{j=1}^{n} \cos(-2\pi \mathbf{h} \cdot \mathbf{r}_j) \right)^2 . \end{aligned} \tag{7.57}$$

The result is an already small number made even smaller by squaring it (e.g., 10^{-4}). The second term in the variance is therefore negligible compared to the first, and

$$\sigma_A^2 \simeq \frac{1}{n} \sum_{j=1}^{n} \frac{1}{2} = \frac{1}{2}. \tag{7.58}$$

Similarly, using $\sin^2 \eta = (1 - \sin 2\eta/2)$,

$$\sigma_B^2 \simeq \frac{1}{2}. \tag{7.59}$$

Substituting the variances in Eqn. 7.44 gives the conditional probability density function

$$f_G(s|E_{\mathbf{h}}, \mathbf{E_k}, \mathbf{E_{h-k}}) = \frac{1}{\pi} e^{-(A_{\mathbf{h}} - \langle A_{\mathbf{h}} \rangle)^2 - (B_{\mathbf{h}} - \langle B_{\mathbf{h}} \rangle)^2}, \tag{7.60}$$

where s, which defines $A_{\mathbf{h}}$ and $B_{\mathbf{h}}$, is a location on the edge of a circle of radius $E_{\mathbf{h}}$ in the complex plane, as illustrated in Fig. 7.8. The probability that $\mathbf{E_h}$ terminates on a circle with radius $E_{\mathbf{h}}$ between s and $s + ds$, or equivalently, that $\mathbf{E_h}$ has a phase between $\varphi_{\mathbf{h}}$ and $\varphi_{\mathbf{h}} + d\varphi_{\mathbf{h}}$ is

$$Pr(s|E_{\mathbf{h}}, \mathbf{E_k}, \mathbf{E_{h-k}}) = \frac{1}{\pi} e^{-(A_{\mathbf{h}} - \langle A_{\mathbf{h}} \rangle)^2 - (B_{\mathbf{h}} - \langle B_{\mathbf{h}} \rangle)^2} ds. \tag{7.61}$$

Since $ds = E_{\mathbf{h}} \, d\varphi_{\mathbf{h}}$,

$$Pr(\varphi_{\mathbf{h}}|E_{\mathbf{h}}, \mathbf{E_k}, \mathbf{E_{h-k}}) = \frac{1}{\pi} e^{-(A_{\mathbf{h}} - \langle A_{\mathbf{h}} \rangle)^2 - (B_{\mathbf{h}} - \langle B_{\mathbf{h}} \rangle)^2} E_{\mathbf{h}} \, d\varphi_{\mathbf{h}}, \tag{7.62}$$

where $A_{\mathbf{h}} = E_{\mathbf{h}} \cos \varphi_{\mathbf{h}}$ and $B_{\mathbf{h}} = E_{\mathbf{h}} \sin \varphi_{\mathbf{h}}$. With substitutions and the trigonometric identities $\cos^2 \alpha + \sin^2 \alpha = 1$ and $\cos \alpha \cos \beta + \sin \alpha \sin \beta = \cos(\alpha - \beta)$,

$$\begin{aligned} &Pr(\varphi_{\mathbf{h}}|E_{\mathbf{h}}, \mathbf{E_k}, \mathbf{E_{h-k}}) = \\ &\frac{E_{\mathbf{h}}}{\pi} e^{-(E_{\mathbf{h}} \cos \varphi_{\mathbf{h}} - \frac{1}{\sqrt{n}} E_{\mathbf{k}} E_{\mathbf{h-k}} \cos(\varphi_{\mathbf{k}} + \varphi_{\mathbf{h-k}}))^2 - (E_{\mathbf{h}} \sin \varphi_{\mathbf{h}} - \frac{1}{\sqrt{n}} E_{\mathbf{k}} E_{\mathbf{h-k}} \sin(\varphi_{\mathbf{k}} + \varphi_{\mathbf{h-k}}))^2} d\varphi_{\mathbf{h}} \\ &= \frac{E_{\mathbf{h}}}{\pi} e^{-E_{\mathbf{h}}^2 + \frac{1}{n}(E_{\mathbf{k}} E_{\mathbf{h-k}})^2} e^{\frac{2}{\sqrt{n}} E_{\mathbf{h}} E_{\mathbf{k}} E_{\mathbf{h-k}} \cos(\varphi_{\mathbf{h}} - \varphi_{\mathbf{k}} - \varphi_{\mathbf{h-k}})} d\varphi_{\mathbf{h}} \\ &= K e^{\frac{2}{\sqrt{n}} E_{\mathbf{h}} E_{\mathbf{k}} E_{\mathbf{h-k}} \cos(\varphi_{\mathbf{h}} - \varphi_{\mathbf{k}} - \varphi_{\mathbf{h-k}})} d\varphi_{\mathbf{h}}. \end{aligned} \tag{7.63}$$

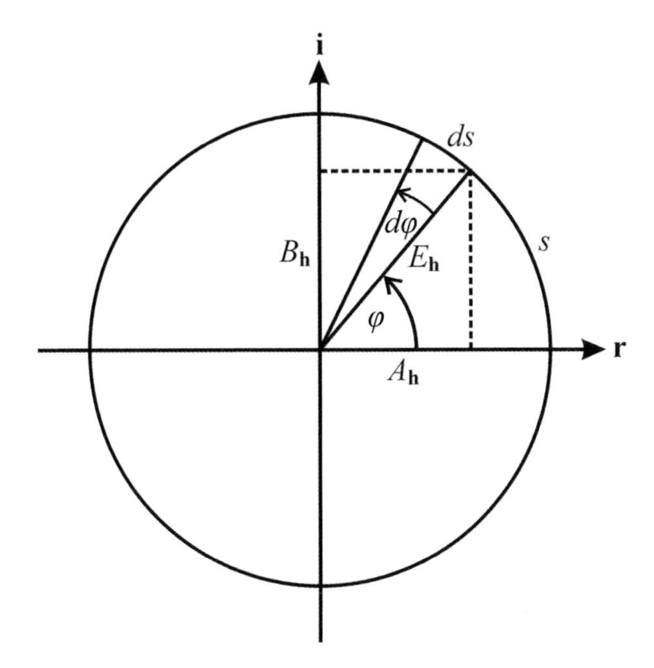

Figure 7.8 The normalized structure factor vector $\mathbf{E_h}$ with known magnitude $E_{\mathbf{h}}$ will terminate somewhere on a circle of radius $E_{\mathbf{h}}$ in the complex plane.

The terms in K are constants since they are fixed by the conditional constraints; to assure that the probability that $\mathbf{E_h}$ terminates *somewhere* on the circle is unity, we must multiply the (un-normalized) probability by another constant, C, such that $KC = N$, a normalization constant. The products of the structure factor magnitudes are also known; for convenience we define $X_{\mathbf{hk}} = \frac{2}{\sqrt{n}} E_{\mathbf{h}} E_{\mathbf{k}} E_{\mathbf{h-k}}$, giving

$$Pr(\varphi_{\mathbf{h}}|E_{\mathbf{h}}, \mathbf{E_k}, \mathbf{E_{h-k}}) = N e^{X_{\mathbf{hk}} \cos(\varphi_{\mathbf{h}} - \varphi_{\mathbf{k}} - \varphi_{\mathbf{h-k}})} d\varphi_{\mathbf{h}}. \tag{7.64}$$

$$\int_0^{2\pi} N e^{X_{\mathbf{hk}} \cos(\varphi_{\mathbf{h}} - \varphi_{\mathbf{k}} - \varphi_{\mathbf{h-k}})} d\varphi_{\mathbf{h}} = 1. \tag{7.65}$$

$$N = \left(\int_0^{2\pi} e^{X_{\mathbf{hk}} \cos(\varphi_{\mathbf{h}} - \varphi_{\mathbf{k}} - \varphi_{\mathbf{h-k}})} d\varphi_{\mathbf{h}} \right)^{-1}. \tag{7.66}$$

Because $\varphi_{\mathbf{k}}$ and $\varphi_{\mathbf{h-k}}$ are fixed, $d(\varphi_{\mathbf{h}} - \varphi_{\mathbf{k}} - \varphi_{\mathbf{h-k}}) = d\varphi_{\mathbf{h}}$, and from Appendix F, the integral is of the form

$$I_0(x) = \frac{1}{2\pi} \int_0^{2\pi} e^{x \cos\phi} \cos 0 \; d\phi = \frac{1}{2\pi} \int_0^{2\pi} e^{x \cos\phi} \; d\phi, \tag{7.67}$$

the integral representation of the modified Bessel function of the first kind of order 0. The integral is conveniently represented in this form since it can be evaluated as a series expansion. Thus $N = 1/(2\pi I_0(X_{\mathbf{hk}}))$ and

$$Pr(\varphi_{\mathbf{h}}|E_{\mathbf{h}}, \mathbf{E_k}, \mathbf{E_{h-k}}) = \frac{e^{X_{\mathbf{hk}} \cos(\varphi_{\mathbf{h}} - \varphi_{\mathbf{k}} - \varphi_{\mathbf{h-k}})} d\varphi_{\mathbf{h}}}{2\pi I_0(X_{\mathbf{hk}})}, \tag{7.68}$$

giving the probability that $\varphi_{\mathbf{h}}$ lies between $\varphi_{\mathbf{h}}$ and $\varphi_{\mathbf{h}} + d\varphi_{\mathbf{h}}$, subject to the constraints of the triplet relationship.[122]

The factor, $X_{\mathbf{hk}}$, determines the nature of the $\varphi_{\mathbf{h}}$ probability distribution. We derived it based on the assumption that the structure consists of equal atoms. If the structure does not, then $1/\sqrt{n}$ will not factor out of the expression for E_h in Eqn. 7.40, and we might expect that some other approximate factor would replace $1/\sqrt{n}$ in $X_{\mathbf{hk}}$ when the structure does not contain equal atoms. The derivation of the factor requires some rather tedious mathematics[123]; we will forego that for a less rigorous approach – determining an expression that reduces to $1/\sqrt{n}$ in the equal-atom case. Given that $E_{\mathbf{k}} = E_{-\mathbf{k}}$ and $E_{\mathbf{h-k}} = E_{-(\mathbf{h-k})}$, the structure factor form of the triplet invariant is

$$\mathbf{E_h E_{-k} E_{-(h-k)}} = E_{\mathbf{h}} e^{i\varphi_{\mathbf{h}}} E_{\mathbf{k}} e^{-i\varphi_{\mathbf{h}}} E_{\mathbf{h-k}} e^{-i\varphi_{\mathbf{h-k}}} \approx E_{\mathbf{h}} E_{\mathbf{k}} E_{\mathbf{h-k}}. \tag{7.69}$$

Defining σ_m* as

$$\sigma_m = \sum_{j=1}^{n} Z_j^m \Longrightarrow \sigma_2 = \sum_{j=1}^{n} Z_j^2, \tag{7.70}$$

the product of the structure factors is

$$\mathbf{E_h E_{-k} E_{-(h-k)}} =$$
$$\sum_{j1=1}^{n} \frac{Z_{j1}}{(\sigma_2)^{1/2}} e^{-2\pi\,(\mathbf{h}\cdot\mathbf{r}_{j1})} \sum_{j2=1}^{n} \frac{Z_{j2}}{(\sigma_2)^{1/2}} e^{+2\pi\,(\mathbf{k}\cdot\mathbf{r}_{j2})} \sum_{j3=1}^{n} \frac{Z_{j3}}{((\sigma_2)^{1/2}} e^{+2\pi\,((\mathbf{h-k})\cdot\mathbf{r}_{j3})}.$$

$$\mathbf{E_h E_{-k} E_{-(h-k)}} =$$
$$\sum_{j1=1}^{n}\sum_{j2=1}^{n}\sum_{j3=1}^{n} \frac{Z_{j1}Z_{j2}Z_{j3}}{(\sigma_2)^{3/2}} e^{-2\pi\,(\mathbf{h}\cdot(\mathbf{r}_{j1}-\mathbf{r}_{j3})+(\mathbf{k}\cdot(\mathbf{r}_{j1}-\mathbf{r}_{j2}))}. \tag{7.71}$$

We now consider some specific terms in the sum. For $j1 = j2 = j3 = j$, the terms are

$$\sum_{j=1}^{n} \frac{Z_j^3}{(\sigma_2)^{3/2}} e^0 = \frac{\sigma_3}{(\sigma_2)^{3/2}}. \tag{7.72}$$

For $j1 = j3 \neq j2$, we have

$$\sum_{j1=1}^{n}\sum_{j2=1}^{n} \frac{Z_{j1}^2 Z_{j2}}{(\sigma_2)^{3/2}} e^{-2\pi\,(\mathbf{h}\cdot(\mathbf{r}_{j1}-\mathbf{r}_{j2}))} \tag{7.73}$$

*The symbol "σ" is not to be confused with the standard deviation. It is used here in lieu of a less ambiguous symbol because of its common occurrence in the direct methods literature.

The pre-exponential factor in the sum can be approximated by assuming that the atomic number of atom $j1$ does not differ too greatly from the atomic number of the other atoms such that

$$\sigma_3 = \sum_{j=1}^{n} Z_j^3 \simeq \sum_{j=1}^{n} Z_{j1} Z_j^2 = Z_{j1}\sigma_2 \quad \text{and} \tag{7.74}$$

$$\begin{aligned}\frac{Z_{j1}^2 Z_{j2}}{(\sigma_2)^{3/2}} &= \frac{\sigma_3}{(\sigma_2)^{3/2}} \frac{Z_{j1}^2 Z_{j2}}{\sigma_3} \simeq \frac{\sigma_3}{(\sigma_2)^{3/2}} \frac{Z_{j1} Z_{j1} Z_{j2}}{Z_{j1}\sigma_2} \\ &= \frac{\sigma_3}{(\sigma_2)^{3/2}} \frac{Z_{j1}}{(\sigma_2)^{1/2}} \frac{Z_{j2}}{(\sigma_2)^{1/2}}. \end{aligned} \tag{7.75}$$

Expression 7.73 becomes

$$\frac{\sigma_3}{(\sigma_2)^{3/2}} \sum_{j1=1}^{n} \sum_{j2=1}^{n} \frac{Z_{j1}}{(\sigma_2)^{1/2}} \frac{Z_{j2}}{(\sigma_2)^{1/2}} e^{-2\pi\,(\mathbf{h}\cdot(\mathbf{r}_{j1}-\mathbf{r}_{j2}))} \tag{7.76}$$

Noting that

$$\begin{aligned} E_{\mathbf{h}}^2 &= \mathbf{E_h E_h^*} = \sum_{j1=1}^{n} \frac{Z_{j1}}{(\sigma_2)^{1/2}} e^{-2\pi\,(\mathbf{h}\cdot(\mathbf{r}_{j1}))} \sum_{j2=1}^{n} \frac{Z_{j2}}{(\sigma_2)^{1/2}} e^{2\pi\,(\mathbf{h}\cdot(\mathbf{r}_{j2}))} \quad (7.77) \\ &= \sum_{j=1}^{n} \frac{Z_j^2}{(\sigma_2)} + \sum_{j1=1\neq j2}^{n} \sum_{j2=1\neq j1}^{n} \frac{Z_{j1}}{(\sigma_2)^{1/2}} \frac{Z_{j2}}{(\sigma_2)^{1/2}} e^{-2\pi\,(\mathbf{h}\cdot(\mathbf{r}_{j1}-\mathbf{r}_{j2}))} \\ &= 1 + \sum_{j1=1}^{n} \sum_{j2=1}^{n} \frac{Z_{j1}}{(\sigma_2)^{1/2}} \frac{Z_{j2}}{(\sigma_2)^{1/2}} e^{-2\pi\,(\mathbf{h}\cdot(\mathbf{r}_{j1}-\mathbf{r}_{j2}))}. \\ E_{\mathbf{h}}^2 - 1 &= \sum_{j1=1}^{n} \sum_{j2=1}^{n} \frac{Z_{j1}}{(\sigma_2)^{1/2}} \frac{Z_{j2}}{(\sigma_2)^{1/2}} e^{-2\pi\,(\mathbf{h}\cdot(\mathbf{r}_{j1}-\mathbf{r}_{j2}))}. \quad (7.78) \end{aligned}$$

Thus, expression 7.73 reduces to

$$\frac{\sigma_3}{(\sigma_2)^{3/2}}(E_{\mathbf{h}}^2 - 1). \tag{7.79}$$

For an equal atom structure, Eqn. 7.78 becomes

$$E_{\mathbf{h}}^2 - 1 = \sum_{j1=1}^{n} \sum_{j2=1}^{n} \frac{1}{\sqrt{n}} \frac{1}{\sqrt{n}} e^{-2\pi\,(\mathbf{h}\cdot(\mathbf{r}_{j1}-\mathbf{r}_{j2}))} = \frac{1}{n} \sum_{j1=1}^{n} \sum_{j2=1}^{n} e^{-2\pi\,(\mathbf{h}\cdot(\mathbf{r}_{j1}-\mathbf{r}_{j2}))},$$

and expression 7.73 becomes

$$\begin{aligned} \sum_{j1=1}^{n} \sum_{j2=1}^{n} \left(\frac{1}{\sqrt{n}}\right)^2 \frac{1}{\sqrt{n}} e^{-2\pi\,(\mathbf{h}\cdot(\mathbf{r}_{j1}-\mathbf{r}_{j2}))} &= \frac{1}{n} \frac{1}{\sqrt{n}} \sum_{j1=1}^{n} \sum_{j2=1}^{n} e^{-2\pi\,(\mathbf{h}\cdot(\mathbf{r}_{j1}-\mathbf{r}_{j2}))}, \\ &= \frac{1}{\sqrt{n}}(E_{\mathbf{h}}^2 - 1). \end{aligned} \tag{7.80}$$

For an equal atom structure, expression 7.72 is

$$\sum_{j=1}^{n} \left(\frac{1}{\sqrt{n}}\right)^3 e^0 = \frac{n}{\sqrt{n}^3} = \frac{1}{\sqrt{n}}. \tag{7.81}$$

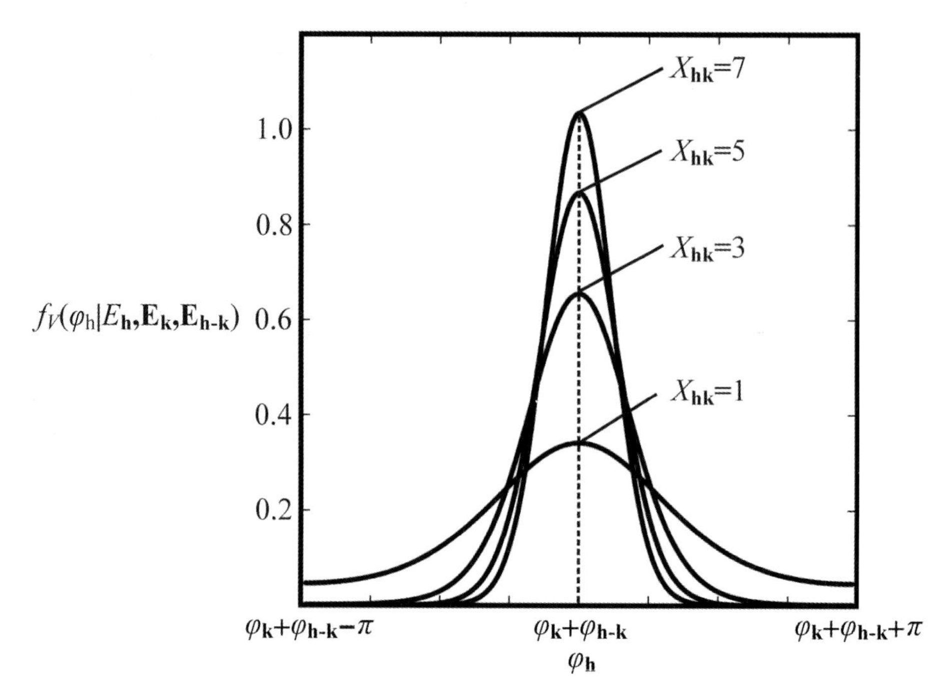

Figure 7.9 Probability density function $f_V(\varphi_\mathbf{h}|E_\mathbf{h}, \mathbf{E_k}, \mathbf{E_{h-k}})$ for different values of $X_\mathbf{hk}$.

It is evident that the term that we are seeking to replace $1/\sqrt{n}$ for structures without identical atoms is $\sigma_3/(\sigma_2)^{3/2}$. Indeed, for equal atoms,

$$\frac{\sigma_3}{(\sigma_2)^{3/2}} = \frac{\sum Z^3}{(\sum Z^2)^{3/2}} = \frac{nZ^3}{(nZ^2)^{3/2}} = \frac{nZ^3}{n^{3/2}Z^3} = \frac{1}{\sqrt{n}}. \tag{7.82}$$

For an unequal atom structure,

$$X_\mathbf{hk} = \frac{2\sigma_3}{(\sigma_2)^{3/2}} E_\mathbf{h} E_\mathbf{k} E_\mathbf{h-k}. \tag{7.83}$$

Since $\varphi_\mathbf{h} \approx \varphi_\mathbf{k} + \varphi_\mathbf{h-k}$, with $\varphi_\mathbf{k}$ and $\varphi_\mathbf{k} + \varphi_\mathbf{h-k}$ fixed, the probability density function, $f_V(\varphi_\mathbf{h}|E_\mathbf{h}, \mathbf{E_k}, \mathbf{E_{h-k}})$,* will be centered on its expected value, $\varphi_\mathbf{k} + \varphi_\mathbf{h-k}$. Fig. 7.9 is a plot of the probability density function from $\varphi_\mathbf{h} = \varphi_\mathbf{k} + \varphi_\mathbf{h-k} - \pi$ to $\varphi_\mathbf{h} = \varphi_\mathbf{k} + \varphi_\mathbf{h-k} + \pi$ for various values of $X_\mathbf{hk}$. Recalling that the triplet relationship was derived from Sayre's equation by assuming that the reflections had structure factors large enough to dominate the summation over all of the structure factors (Eqn. 7.24), the triplet relationship should become increasingly valid as $E_\mathbf{h}$, $E_\mathbf{k}$, and $E_\mathbf{h-k}$ collectively increase in magnitude. The graph demonstrates clearly that as $X_\mathbf{hk}$ increases, the variance of the distribution decreases, consequently increasing the probability that $\varphi_\mathbf{h}$ will be close to its most probable value, as expected.

*The V subscript indicates that the probability distribution is a normal distribution for circular variables, known as a von Mises[124] distribution.

For centrosymmetric structures the phases are constrained to 0 and π, and the conditional probability becomes $Pr(S_{\mathbf{h}}^{+}|E_{\mathbf{h}}, \mathbf{E_k}, \mathbf{E_{h-k}})$, where the "+" sign indicates the probability that the sign of $\mathbf{E_h}$ is positive, given the magnitudes of the three structure factors and the signs of $\mathbf{E_k}$ and $\mathbf{E_{h-k}}$. The normalized structure factor for an equal atom centrosymmetric structure is

$$\mathbf{E_h} = \frac{1}{\sqrt{n}}\sum_{j=1}^{n}\cos(-2\pi\mathbf{h}\cdot\mathbf{r}_j) = \frac{1}{\sqrt{n}}\sum_{j=1}^{n/2}2\cos(-2\pi\mathbf{h}\cdot\mathbf{r}_j). \tag{7.84}$$

$\mathbf{E_h}$ is again the sum of random variables, and from the central limit theorem, is normally distributed about its most probable (expected) value, $\langle\mathbf{E_h}\rangle$, given by Eqn. 7.43.

Following the same procedure as in the non-centrosymmetic case, adding and subtracting $\mathbf{k}$ in the cosine term results in

$$\langle\mathbf{E_h}\rangle = \frac{1}{\sqrt{n}}\mathbf{E_k}\mathbf{E_{h-k}}. \tag{7.85}$$

The variance is determined with the same procedure that we used to determine σ_A^2. Because the average cosine terms are close to zero, $\langle\mathbf{E_h}\rangle^2 \simeq 0$, $\langle\cos^2(-2\pi\mathbf{h}\cdot\mathbf{r}_j)\rangle \simeq 1/2$ and

$$\sigma_{E_{\mathbf{h}}}^2 = \langle{\mathbf{E_h}}^2\rangle - \langle\mathbf{E_h}\rangle^2 \tag{7.86}$$

$$\simeq \frac{1}{n}\left\langle\left(\sum_{j=1}^{n/2}2\cos(-2\pi\mathbf{h}\cdot\mathbf{r}_j)\right)^2\right\rangle = \frac{1}{n}\sum_{j=1}^{n/2}2^2\frac{1}{2} = 1. \tag{7.87}$$

The conditional probability density function for the centrosymmetic distribution of $\mathbf{E_h}$ is therefore

$$f_G(\mathbf{E_h}|\mathbf{E_k}, \mathbf{E_{h-k}}) = \frac{1}{\sqrt{2\pi}}e^{-(\mathbf{E_h}-\frac{1}{\sqrt{n}}\mathbf{E_k}\mathbf{E_{h-k}})^2/2} \Longrightarrow$$

$$f_G(S_{\mathbf{h}}E_{\mathbf{h}}|S_{\mathbf{k}}E_{\mathbf{k}}, S_{\mathbf{h-k}}E_{\mathbf{h-k}}) = \frac{1}{\sqrt{2\pi}}e^{-(S_{\mathbf{h}}E_{\mathbf{h}}-\frac{1}{\sqrt{n}}S_{\mathbf{k}}E_{\mathbf{k}}S_{\mathbf{h-k}}E_{\mathbf{h-k}})^2/2}, \tag{7.88}$$

where $S_{\mathbf{h}}$ is the sign of $\mathbf{E_h}$, etc. The probability density function for the case when $S_{\mathbf{h}}$ is positive is given by

$$f_G(S_{\mathbf{h}}^{+}|E_{\mathbf{h}}, S_{\mathbf{k}}E_{\mathbf{k}}, S_{\mathbf{h-k}}E_{\mathbf{h-k}}) = \frac{1}{\sqrt{2\pi}}e^{-(+E_{\mathbf{h}}-\frac{1}{\sqrt{n}}S_{\mathbf{k}}E_{\mathbf{k}}S_{\mathbf{h-k}}E_{\mathbf{h-k}})^2/2}, \tag{7.89}$$

and that $S_{\mathbf{h}}$ is negative,

$$f_G(S_{\mathbf{h}}^{-}|E_{\mathbf{h}}, S_{\mathbf{k}}E_{\mathbf{k}}, S_{\mathbf{h-k}}E_{\mathbf{h-k}}) = \frac{1}{\sqrt{2\pi}}e^{-(-E_{\mathbf{h}}-\frac{1}{\sqrt{n}}S_{\mathbf{k}}E_{\mathbf{k}}S_{\mathbf{h-k}}E_{\mathbf{h-k}})^2/2}. \tag{7.90}$$

The ratio of the probabilities is the ratio of the probability density functions, and

$$\frac{Pr(S_{\mathbf{h}}^{+})}{Pr(S_{\mathbf{h}}^{-})} = \frac{e^{-(+E_{\mathbf{h}}-\frac{1}{\sqrt{n}}S_{\mathbf{k}}E_{\mathbf{k}}S_{\mathbf{h-k}}E_{\mathbf{h-k}})^2/2}}{e^{-(-E_{\mathbf{h}}-\frac{1}{\sqrt{n}}S_{\mathbf{k}}E_{\mathbf{k}}S_{\mathbf{h-k}}E_{\mathbf{h-k}})^2/2}}, \tag{7.91}$$

$$= e^{\frac{2}{\sqrt{n}}E_{\mathbf{h}}E_{\mathbf{k}}E_{\mathbf{h-k}}S_{\mathbf{k}}S_{\mathbf{h-k}}} \tag{7.92}$$

$$= e^{X_{\mathbf{hk}}S_{\mathbf{h}}S_{\mathbf{h-k}}}. \tag{7.93}$$

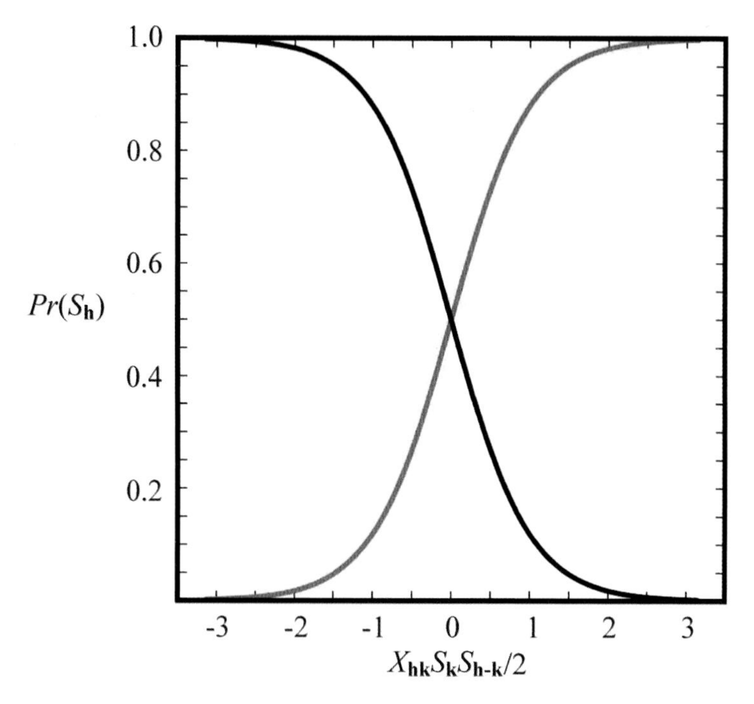

Figure 7.10 Sign probability for a centrosymmetric structure as a function of $X_{\mathbf{hk}}S_{\mathbf{h}}S_{\mathbf{h-k}}/2$. $Pr(S_{\mathbf{h}}^{+})$ is shown in red; $Pr(S_{\mathbf{h}}^{-})$ is shown in black.

Setting $y = X_{\mathbf{hk}}S_{\mathbf{h}}S_{\mathbf{h-k}}/2$, from Eqns. 6.166–6.169,

$$\frac{Pr(S_{\mathbf{h}}^{+})}{Pr(S_{\mathbf{h}}^{-})} = e^{2y} = \frac{e^{y}}{e^{-y}} = \frac{\frac{1}{2} + \frac{1}{2}\tanh(y)}{\frac{1}{2} - \frac{1}{2}\tanh(y)}, \tag{7.94}$$

$$Pr(S_{\mathbf{h}}^{+}) = \frac{1}{2} + \frac{1}{2}\tanh(X_{\mathbf{hk}}S_{\mathbf{h}}S_{\mathbf{h-k}}/2) \quad \text{and} \tag{7.95}$$

$$Pr(S_{\mathbf{h}}^{-}) = \frac{1}{2} - \frac{1}{2}\tanh(X_{\mathbf{hk}}S_{\mathbf{h}}S_{\mathbf{h-k}}/2). \tag{7.96}$$

Fig. 7.10 illustrates, in red, the probability that the sign of $\mathbf{E_h}$ will be positive — and in black, the probability that the sign will be negative. When $S_{\mathbf{h}}$ and $S_{\mathbf{h-k}}$ both have the same sign, $Pr(S_{\mathbf{h}}^{+})$ is 1/2 when $E_{\mathbf{h}}E_{\mathbf{k}}E_{\mathbf{h-k}}$ is zero, increasing rapidly to 1 as $E_{\mathbf{h}}E_{\mathbf{k}}E_{\mathbf{h-k}}$ increases. Again, large values of the product of the magnitudes of all three reflections indicate a high probability that the sign of $\mathbf{E_h}$ is positive. When $S_{\mathbf{h}}$ and $S_{\mathbf{h-k}}$ have opposite signs, the sign of $\mathbf{E_h}$ approaches -1 as $E_{\mathbf{h}}E_{\mathbf{k}}E_{\mathbf{h-k}}$ increases.

The formula for $Pr(S_{\mathbf{h}}^{+})$, originally derived by William Cochran and Michael M. Woolfson,[123] allows for the use of triplet relationships to predict new structure factor phases (signs) from known amplitudes and phases. The formula is rarely used in practice; *the tangent formula, discussed in the next section, can be used for the determination of phases for both centrosymmetric and non-centrosymmetric structures.*

The Tangent Formula

A single triplet relationship may not be reliable enough to assign an unknown phase, but if there are several relationships for the same reflection, it is possible to combine them so that the predicted phase becomes much more certain. Assuming that the probability distribution for each triplet is independent of the others, the conditional probability of observing $Pr(\varphi_{\mathbf{h}}) = Pr(\varphi_{\mathbf{h}}|E_{\mathbf{h}}, \mathbf{E}_{\mathbf{k}1}, \mathbf{E}_{\mathbf{h}-\mathbf{k}1}, \mathbf{E}_{\mathbf{k}2}, \mathbf{E}_{\mathbf{h}-\mathbf{k}2}\dots)$ is the product of the individual probabilities, multiplied by a normalization constant (since each individual probability is independently normalized). For m independent triplet relationships involving $\varphi_{\mathbf{h}}$, for a unit phase angle the conditional probability of observing $\varphi_{\mathbf{h}}$ (the probability density) is given by

$$f_V(\varphi_{\mathbf{h}}) = f_1(\varphi_{\mathbf{h}})f_2(\varphi_{\mathbf{h}})\dots f_m(\varphi_{\mathbf{h}}) = \prod_{j=1}^{m} f_j(\varphi_{\mathbf{h}}) \tag{7.97}$$

$$= \prod_{j=1}^{m} e^{X_{\mathbf{hk}_j}\cos(\varphi_{\mathbf{h}}-\varphi_{\mathbf{k_j}}-\varphi_{\mathbf{h}-\mathbf{k_j}})}$$

$$= \prod_{j=1}^{m} e^{X_{\mathbf{hk}_j}\cos(\varphi_{\mathbf{h}}-\alpha_j)} \tag{7.98}$$

$$= e^{\sum_{j=1}^{m} X_{\mathbf{hk}_j}\cos(\varphi_{\mathbf{h}}-\alpha_j)}; \tag{7.99}$$

$\alpha_j = \varphi_{\mathbf{k_j}} + \varphi_{\mathbf{h}-\mathbf{k_j}}$, the expected value for $\varphi_{\mathbf{h}}$ for the jth triplet relationship. The probability of observing $\varphi_{\mathbf{h}}$ between $\varphi_{\mathbf{h}}$ and $\varphi_{\mathbf{h}} + d\varphi_{\mathbf{h}}$ is then

$$Pr(\varphi_{\mathbf{h}}) = N\, e^{\sum_{j=1}^{m} X_{\mathbf{hk}_j}\cos(\varphi_{\mathbf{h}}-\alpha_j)}d\varphi_{\mathbf{h}}, \tag{7.100}$$

where N is the normalization constant. The exponent in Eqn. 7.100 is a sum of cosine functions, which is itself a cosine function:

$$\sum_{j=1}^{m} X_{\mathbf{hk}_j}\cos(\varphi_{\mathbf{h}} - \alpha_j) = Y_{\mathbf{h}}\cos\eta. \tag{7.101}$$

Defining the angle $\beta_{\mathbf{h}}$ as the difference between $\varphi_{\mathbf{h}}$ and η, such that $\eta = \varphi_{\mathbf{h}} - \beta_{\mathbf{h}}$, and employing $\cos(a - b) = \cos a \cos b + \sin a \sin b$,

$$\sum_{j=1}^{m} X_{\mathbf{hk}_j}\cos(\varphi_{\mathbf{h}} - \alpha_j) = Y_{\mathbf{h}}\cos(\varphi_{\mathbf{h}} - \beta_{\mathbf{h}}) \tag{7.102}$$

$$\cos(\varphi_{\mathbf{h}})\sum_{j=1}^{m} X_{\mathbf{hk}_j}\cos(\alpha_j) + \sin(\varphi_{\mathbf{h}})\sum_{j=1}^{m} X_{\mathbf{hk}_j}\sin(\alpha_j) =$$

$$Y_{\mathbf{h}}\cos\varphi_{\mathbf{h}}\cos\beta_{\mathbf{h}} + Y_{\mathbf{h}}\sin\varphi_{\mathbf{h}}\sin\beta_{\mathbf{h}}. \tag{7.103}$$

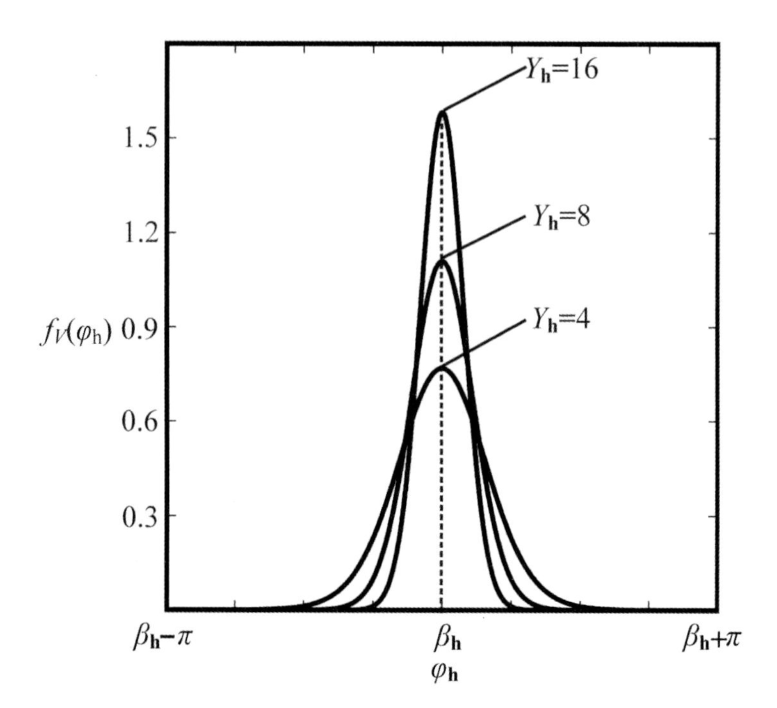

Figure 7.11 Probability density function $f_V(\varphi_{\mathbf{h}}) = Ne^{Y_{\mathbf{h}}\cos(\varphi_{\mathbf{h}}-\beta_{\mathbf{h}})}$ for different values of $Y_{\mathbf{h}}$, illustrating the decrease in variance as $Y_{\mathbf{h}}$ increases.

It follows that

$$Y_{\mathbf{h}}\cos\beta_{\mathbf{h}} = \sum_{j=1}^{m} X_{\mathbf{hk}_j}\cos(\alpha_j) \tag{7.104}$$

$$Y_{\mathbf{h}}\sin\beta_{\mathbf{h}} = \sum_{j=1}^{m} X_{\mathbf{hk}_j}\sin(\alpha_j) \tag{7.105}$$

$$Y_{\mathbf{h}}^2\cos^2\beta_{\mathbf{h}} + Y_{\mathbf{h}}^2\sin^2\beta_{\mathbf{h}} = \left(\sum_{j=1}^{m} X_{\mathbf{hk}_j}\cos(\alpha_j)\right)^2 + \left(\sum_{j=1}^{m} X_{\mathbf{hk}_j}\sin(\alpha_j)\right)^2$$

$$Y_{\mathbf{h}}^2(\cos^2\beta_{\mathbf{h}} + \sin^2\beta_{\mathbf{h}}) = \left(\sum_{j=1}^{m} X_{\mathbf{hk}_j}\cos(\alpha_j)\right)^2 + \left(\sum_{j=1}^{m} X_{\mathbf{hk}_j}\sin(\alpha_j)\right)^2$$

$$Y_{\mathbf{h}}^2 = \left(\sum_{j=1}^{m} X_{\mathbf{hk}_j}\cos(\alpha_j)\right)^2 + \left(\sum_{j=1}^{m} X_{\mathbf{hk}_j}\sin(\alpha_j)\right)^2$$

$$Y_{\mathbf{h}} = \left(\left(\sum_{j=1}^{m} X_{\mathbf{hk}_j}\cos(\alpha_j)\right)^2 + \left(\sum_{j=1}^{m} X_{\mathbf{hk}_j}\sin(\alpha_j)\right)^2\right)^{1/2}. \tag{7.106}$$

Furthermore, $Y_{\mathbf{h}} \sin\beta_{\mathbf{h}} / Y_{\mathbf{h}} \cos\beta_{\mathbf{h}} = \tan\beta_{\mathbf{h}}$, and

$$\tan\beta_{\mathbf{h}} = \frac{\sum_{j=1}^{m} X_{\mathbf{hk}_j} \sin(\alpha_j)}{\sum_{j=1}^{m} X_{\mathbf{hk}_j} \cos(\alpha_j)}$$

$$= \frac{\sum_{j=1}^{m} 2\sigma_3/\sigma_2^{3/2} E_{\mathbf{h}} E_{\mathbf{k}_j} E_{\mathbf{h}-\mathbf{k}_j} \sin(\varphi_{\mathbf{k}_j} + \varphi_{\mathbf{h}-\mathbf{k}_j})}{\sum_{j=1}^{m} 2\sigma_3/\sigma_2^{3/2} E_{\mathbf{h}} E_{\mathbf{k}_j} E_{\mathbf{h}-\mathbf{k}_j} \cos(\varphi_{\mathbf{k}_j} + \varphi_{\mathbf{h}-\mathbf{k}_j})}$$

$$\tan\beta_{\mathbf{h}} = \frac{\sum_{j=1}^{m} E_{\mathbf{k}_j} E_{\mathbf{h}-\mathbf{k}_j} \sin(\varphi_{\mathbf{k}_j} + \varphi_{\mathbf{h}-\mathbf{k}_j})}{\sum_{j=1}^{m} E_{\mathbf{k}_j} E_{\mathbf{h}-\mathbf{k}_j} \cos(\varphi_{\mathbf{k}_j} + \varphi_{\mathbf{h}-\mathbf{k}_j})}. \tag{7.107}$$

We can now determine $Y_{\mathbf{h}}$ and $\beta_{\mathbf{h}}$ from the known magnitudes and phases; the probability that the phase of $\mathbf{E}_{\mathbf{h}}$ lies between $\varphi_{\mathbf{h}}$ and $\varphi_{\mathbf{h}} + d\varphi_{\mathbf{h}}$ is

$$Pr(\varphi_{\mathbf{h}}) = N\, e^{\sum_{j=1}^{m} X_{\mathbf{hk}_j} \cos(\varphi_{\mathbf{h}} - \alpha_j)} d\varphi_{\mathbf{h}} = N e^{Y_{\mathbf{h}} \cos(\varphi_{\mathbf{h}} - \beta_{\mathbf{h}})} d\varphi_{\mathbf{h}}. \tag{7.108}$$

The normalization constant is $N = 1/(2\pi I_0(Y_{\mathbf{h}}))$ where $I_0(Y_{\mathbf{h}})$ is a modified Bessel function (cf. Eqns. 7.65 – 7.68) and

$$Pr(\varphi_{\mathbf{h}}) = \frac{e^{Y_{\mathbf{h}} \cos(\varphi_{\mathbf{h}} - \beta_{\mathbf{h}})} d\varphi_{\mathbf{h}}}{2\pi I_0(Y_{\mathbf{h}})}. \tag{7.109}$$

The probability has the same form as the probability for a single triplet (Eqn. 7.68). The probability density function, $f_V(\varphi_{\mathbf{h}})$ is illustrated for three values of $Y_{\mathbf{h}}$ in Fig. 7.11. The distribution is centered at $\beta_{\mathbf{h}}$, ***rendering*** $\beta_{\mathbf{h}}$ ***the most probable value of*** $\varphi_{\mathbf{h}}$. Eqn. 7.107, which provides this value, is the *tangent formula* of Herbert A. Hauptmann and Jerome Karle[125], who were awarded the Nobel Prize in 1985 for their contributions to direct methods. As observed in the single triplet case for $X_{\mathbf{hk}}$, as $Y_{\mathbf{h}}$ increases, the variance decreases, thus increasing the reliability of $\varphi_{\mathbf{h}} \approx \beta_{\mathbf{h}}$.

The role of increasing the number of triplet relationships in determining the phase of a reflection can be seen graphically by referring to Fig. 7.12. Eqns. 7.104 and 7.105 are indicative of a vector of magnitude $Y_{\mathbf{h}}$ at angle $\beta_{\mathbf{h}}$ in the complex plane, with components along the real and imaginary axes given by the sum of vectors of magnitude $X_{\mathbf{hk}_j}$ and angle $\alpha_j = \varphi_{\mathbf{k}_j} + \varphi_{\mathbf{h}-\mathbf{k}_j}$. α_j is the value of $\varphi_{\mathbf{h}}$ predicted by each triplet. $\beta_{\mathbf{h}}$ is the value of $\varphi_{\mathbf{h}}$ predicted by the tangent formula from all of the triplets. Even with a great deal of variance in the phases predicted for each triplet, the tangent formula tends toward a reliable value of the phase. As the number of triplets increase, the influence of any one triplet on $\beta_{\mathbf{h}}$ decreases. The variance of the phase estimate, $\varphi_{\mathbf{h}} \approx \beta_{\mathbf{h}}$, is a measure of its reliability:

$$\sigma^2_{\varphi_{\mathbf{h}}} = \langle \varphi^2_{\mathbf{h}} \rangle - \langle \varphi_{\mathbf{h}} \rangle^2 = \langle \varphi^2_{\mathbf{h}} \rangle - \beta^2_{\mathbf{h}}. \tag{7.110}$$

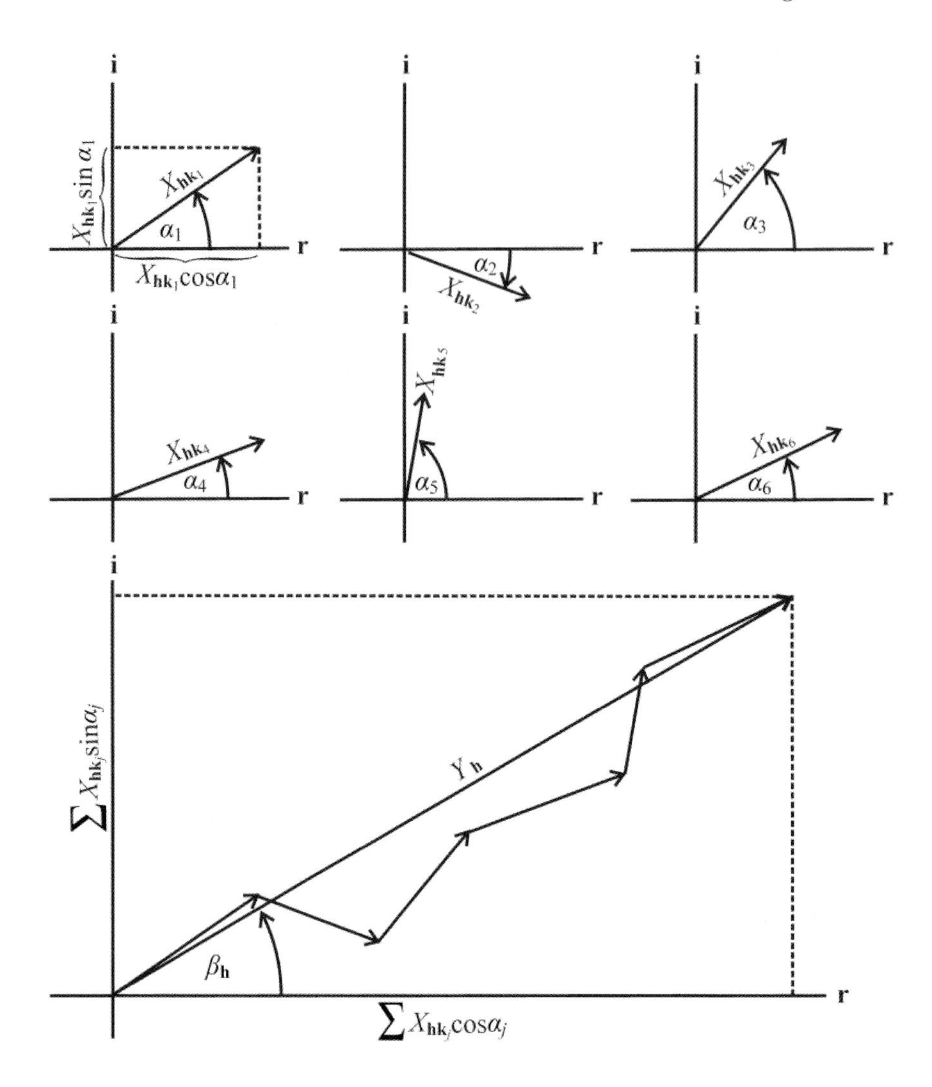

Figure 7.12 The combination of six triplet relationships for the reflection with normalized structure factor $\mathbf{E_h}$ as a sum of vectors with magnitudes $X_{\mathbf{hk}_j}$ and phases angles $\alpha_j = \varphi_{\mathbf{k}_j} + \varphi_{\mathbf{h}-\mathbf{k}_j}$, the expected value of $\varphi_\mathbf{h}$ for each triplet.

The average value of $\varphi_\mathbf{h}^2$ is the sum of each value, weighted by the probability of observing it (Eqn. 5.39):

$$\begin{aligned}\langle\varphi_\mathbf{h}^2\rangle &= \int_{\beta_\mathbf{h}-\pi}^{\beta_\mathbf{h}+\pi} \varphi_\mathbf{h}^2 f_V(\varphi_\mathbf{h})d\varphi_\mathbf{h} \\ &= (2\pi I_0(Y_\mathbf{h}))^{-1}\int_{\beta_\mathbf{h}-\pi}^{\beta_\mathbf{h}+\pi} \varphi_\mathbf{h}^2 e^{Y_\mathbf{h}\cos(\varphi_\mathbf{h}-\beta_\mathbf{h})}d\varphi_\mathbf{h}. \qquad (7.111)\end{aligned}$$

To evaluate this integral, we define $q = \varphi_{\mathbf{h}} - \beta_{\mathbf{h}}$:

$$\langle \varphi_{\mathbf{h}}^2 \rangle = (2\pi I_0(Y_{\mathbf{h}}))^{-1} \int_{-\pi}^{\pi} (q + \beta_{\mathbf{h}})^2 e^{Y_{\mathbf{h}} \cos q} dq$$
$$= 2\pi I_0(Y_{\mathbf{h}}))^{-1} \left(\int_{-\pi}^{\pi} q^2 e^{Y_{\mathbf{h}} \cos q} dq + 2\beta_{\mathbf{h}} \int_{-\pi}^{\pi} q e^{Y_{\mathbf{h}} \cos q} dq + \beta_{\mathbf{h}}^2 \int_{-\pi}^{\pi} e^{Y_{\mathbf{h}} \cos q} dq \right).$$

The third integral is in the integral form of a modified Bessel function (cf. Eqn. 7.67) and the integrand in the second integral is an odd function, thus integrating to zero:

$$\langle \varphi_{\mathbf{h}}^2 \rangle = (2\pi I_0(Y_{\mathbf{h}}))^{-1} \int_{-\pi}^{\pi} q^2 e^{Y_{\mathbf{h}} \cos q} dq + (2\pi I_0(Y_{\mathbf{h}}))^{-1} \beta_{\mathbf{h}}^2 (2\pi I_0(Y_{\mathbf{h}}))$$
$$= (2\pi I_0(Y_{\mathbf{h}}))^{-1} \int_{-\pi}^{\pi} q^2 e^{Y_{\mathbf{h}} \cos q} dq + \beta_{\mathbf{h}}^2. \tag{7.112}$$

The variance therefore becomes

$$\sigma_{\varphi_{\mathbf{h}}}^2 = \langle \varphi_{\mathbf{h}}^2 \rangle - \beta_{\mathbf{h}}^2 = (2\pi I_0(Y_{\mathbf{h}}))^{-1} \int_{-\pi}^{\pi} q^2 e^{Y_{\mathbf{h}} \cos q} dq. \tag{7.113}$$

The exponential in the integrand can be expanded as a series of modified Bessel functions (Eqn. F.23):

$$\sigma_{\varphi_{\mathbf{h}}}^2 = (2\pi I_0(Y_{\mathbf{h}}))^{-1} \int_{-\pi}^{\pi} q^2 \left(I_0(Y_{\mathbf{h}}) + 2 \sum_{n=1}^{\infty} I_n(Y_{\mathbf{h}}) \cos nq \right) dq. \tag{7.114}$$
$$= \frac{1}{2\pi} \int_{-\pi}^{\pi} q^2 dq + (\pi I_0(Y_{\mathbf{h}}))^{-1} \sum_{n=1}^{\infty} I_n(Y_{\mathbf{h}}) \int_{-\pi}^{\pi} q^2 \cos nq \, dq. \tag{7.115}$$

From the identity,

$$\int x^2 (\cos ax) \, dx = \frac{2x \cos ax}{a^2} + \frac{a^2 x^2 - 2}{a^3} \sin ax, \tag{7.116}$$

noting that $\sin n\pi = 0$,

$$\int_{-\pi}^{\pi} q^2 \cos nq \, dq = \left. \frac{2q \cos nq}{n^2} \right|_{-\pi}^{\pi} = \frac{4\pi \cos n\pi}{n^2} = (-1)^n \frac{4\pi}{n^2}. \tag{7.117}$$

The $(-1)^n$ factor occurs because $\cos n\pi = \pm 1$ is positive for n even, and negative for n odd. The variance of the phase estimate[126] is therefore

$$\sigma_{\varphi_{\mathbf{h}}}^2 = \frac{\pi^2}{3} + 4 \sum_{n=1}^{\infty} \frac{(-1)^n}{n^2} \frac{I_n(Y_{\mathbf{h}})}{I_0(Y_{\mathbf{h}})}; \tag{7.118}$$

the variance depends solely on $Y_{\mathbf{h}}$. Fig. 7.13 is a plot of the standard deviation of $\varphi_{\mathbf{h}}$, $\sigma_{\varphi_{\mathbf{h}}} = \sqrt{\sigma_{\varphi_{\mathbf{h}}}^2}$, from $\beta_{\mathbf{h}}$ as a function of $Y_{\mathbf{h}}$. For values of $Y_{\mathbf{h}}$ above about 4 the standard deviation tails off slowly in a nearly linear fashion (indicating a more reliable estimate). For values below 3–4 the standard deviation increases significantly.

The tangent formula was derived based upon a continuous probability distribution for non-centrosymmetric structures, yet it can be applied to centrosymmetric

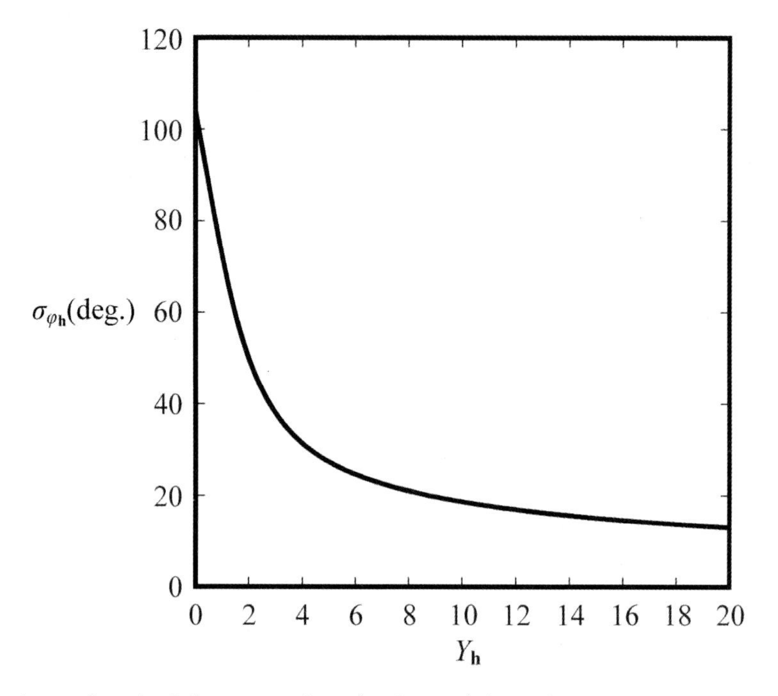

Figure 7.13 Standard deviation of $\varphi_{\mathbf{h}}$ (in degrees) from $\beta_{\mathbf{h}}$, the expected value of $\varphi_{\mathbf{h}}$, as a function of $Y_{\mathbf{h}}$.

structures as well. This may at first appear impossible, since, for phases of 0 and π, $\tan\beta_{\mathbf{h}}$ will always be equal to its expected value of zero! An alternative derivation of the tangent formula[127] may serve to clarify the way in which it is used to determine both centrosymmetric and non-centrosymmetric phases. *In principle, if we begin by assuming nothing about restrictions on the phase, a general solution would eventually arrive at those phases, even if they are restricted.*

Given a reflection with an unknown phase, $\mathbf{E_h}$, and a set of m triplet-related reflections $\{\mathbf{E}_{\mathbf{k}_j}, \mathbf{E}_{\mathbf{h}-\mathbf{k}_j}\}$, with $\Phi_{\mathbf{k}_j} = \varphi_{\mathbf{h}} - \varphi_{\mathbf{k}_j} - \varphi_{\mathbf{h}-\mathbf{k_j}} \approx 0$, the "structure factor form" of the triplet relationship is a product of the normalized structure factors:

$$\begin{aligned}\mathbf{E_h}\mathbf{E}^*_{\mathbf{k}_j}\mathbf{E}^*_{\mathbf{h}-\mathbf{k}_j} &= E_{\mathbf{h}}e^{i\varphi_{\mathbf{h}}}E_{-\mathbf{k}_j}e^{-i\varphi_{\mathbf{k}_j}}E_{-(\mathbf{h}-\mathbf{k}_j)}e^{-i\varphi_{\mathbf{h}-\mathbf{k}}}\\ &= E_{\mathbf{h}}e^{i\varphi_{\mathbf{h}}}E_{\mathbf{k}_j}e^{-i\varphi_{\mathbf{k}_j}}E_{\mathbf{h}-\mathbf{k}_j}e^{-i\varphi_{\mathbf{h}-\mathbf{k}_j}} = E_{\mathbf{h}}E_{\mathbf{k}_j}E_{\mathbf{h}-\mathbf{k}_j}e^{i(\varphi_{\mathbf{h}}-\varphi_{\mathbf{k}_j}-\varphi_{\mathbf{h}-\mathbf{k}_j})}\\ &\approx E_{\mathbf{h}}E_{\mathbf{k}_j}E_{\mathbf{h}-\mathbf{k}_j}e^{i(0)} \approx E_{\mathbf{h}}E_{\mathbf{k}_j}E_{\mathbf{h}-k_j}.\end{aligned} \quad (7.119)$$

We now consider the sum of these relationships as a function of the continuous variable, $\varphi_{\mathbf{h}}$:

$$Z(\varphi_{\mathbf{h}}) = \sum_{\mathbf{h}}\sum_{j=1}^{m}\mathbf{E_h}\mathbf{E}^*_{\mathbf{k}_j}\mathbf{E}^*_{\mathbf{h}-\mathbf{k}_j}, \quad (7.120)$$

where all of the reflections and their Friedel equivalents are included in the sum. If we allow $\varphi_{\mathbf{h}}$ to vary from 0 to 2π, we expect $Z(\varphi_{\mathbf{h}})$ to reach its maximum value when $\varphi_{\mathbf{h}}$ approaches its expected value, $\beta_{\mathbf{h}}$, and the triplet relationships approach zero. The function can be divided into two sums, one with $-\varphi_{\mathbf{h}}$, and all pairs ($\varphi_{\mathbf{k}}$, $\varphi_{\mathbf{h-k}}$), the other with $\varphi_{\mathbf{h}}$, and all pairs ($-\varphi_{\mathbf{k}}$, $-\varphi_{\mathbf{h-k}}$):

$$
\begin{aligned}
Z(\varphi_{\mathbf{h}}) &= \frac{1}{2}\left(\sum_{j=1}^{m} \mathbf{E}_{\mathbf{h}}^{*}\mathbf{E}_{\mathbf{k}_j}\mathbf{E}_{\mathbf{h}-\mathbf{k}_j}\right) + \frac{1}{2}\left(\sum_{j=1}^{m} \mathbf{E}_{\mathbf{h}}\mathbf{E}_{\mathbf{k}_j}^{*}\mathbf{E}_{\mathbf{h}-\mathbf{k}_j}^{*}\right) && (7.121) \\
&= \frac{1}{2}\left(E_{\mathbf{h}}\sum_{j=1}^{m} E_{\mathbf{k}_j}E_{\mathbf{h}-\mathbf{k}_j} e^{-i\varphi_{\mathbf{h}}} e^{i\varphi_{\mathbf{k}_j}} e^{i\varphi_{\mathbf{h}-\mathbf{k}_j}}\right) \\
&\quad + \frac{1}{2}\left(E_{\mathbf{h}}\sum_{j=1}^{m} E_{\mathbf{k}_j}E_{\mathbf{h}-\mathbf{k}_j} e^{i\varphi_{\mathbf{h}}} e^{-i\varphi_{\mathbf{k}_j}} e^{-i\varphi_{\mathbf{h}-\mathbf{k}_j}}\right) \\
&= E_{\mathbf{h}}\sum_{j=1}^{m} E_{\mathbf{k}_j}E_{\mathbf{h}-\mathbf{k}_j} \frac{e^{i(\varphi_{\mathbf{h}}-\varphi_{\mathbf{k}_j}-\varphi_{\mathbf{h}-\mathbf{k}_j})} + e^{-i(\varphi_{\mathbf{h}}-\varphi_{\mathbf{k}_j}-\varphi_{\mathbf{h}-\mathbf{k}_j})}}{2} && (7.122) \\
&= E_{\mathbf{h}}\sum_{j=1}^{m} E_{\mathbf{k}_j}E_{\mathbf{h}-\mathbf{k}_j} \cos(\varphi_{\mathbf{h}} - \varphi_{\mathbf{k}_j} - \varphi_{\mathbf{h}-\mathbf{k}_j}). && (7.123)
\end{aligned}
$$

Now, suppose that we begin with $\varphi_{\mathbf{h}} = 0$ and observe the behavior of $Z(\varphi_{\mathbf{h}})$ as $\varphi_{\mathbf{h}}$ increases until it reaches 2π. As $\varphi_{\mathbf{h}}$ approaches its expected value, $\beta_{\mathbf{h}}$, each $\Phi_{\mathbf{k}_j}$ approaches zero and each $\cos(\Phi_{\mathbf{k}_j})$ approaches its maximum value of one; we therefore expect that $Z(\varphi_{\mathbf{h}})$ will have a maximum value when $\varphi_{\mathbf{h}} = \beta_{\mathbf{h}}$. The function will be at a maximum when its derivative is zero:

$$
\begin{aligned}
\frac{\partial Z(\varphi_{\mathbf{h}})}{\partial \varphi_{\mathbf{h}}} &= -E_{\mathbf{h}}\sum_{j=1}^{m} E_{\mathbf{k}_j}E_{\mathbf{h}-\mathbf{k}_j} \sin(\varphi_{\mathbf{h}} - \varphi_{\mathbf{k}_j} - \varphi_{\mathbf{h}-\mathbf{k}_j}) && (7.124) \\
\left(\frac{\partial Z(\varphi_{\mathbf{h}})}{\partial \varphi_{\mathbf{h}}}\right)_{\beta_{\mathbf{h}}} &= -E_{\mathbf{h}}\sum_{j=1}^{m} E_{\mathbf{k}_j}E_{\mathbf{h}-\mathbf{k}_j} \sin(\beta_{\mathbf{h}} - (\varphi_{\mathbf{k}_j} + \varphi_{\mathbf{h}-\mathbf{k}_j})) && (7.125) \\
&= -E_{\mathbf{h}}\sin\beta_{\mathbf{h}}\sum_{j=1}^{m} E_{\mathbf{k}_j}E_{\mathbf{h}-\mathbf{k}_j} \cos(\varphi_{\mathbf{k}_j} + \varphi_{\mathbf{h}-\mathbf{k}_j})) \\
&\quad + E_{\mathbf{h}}\cos\beta_{\mathbf{h}}\sum_{j=1}^{m} E_{\mathbf{k}_j}E_{\mathbf{h}-\mathbf{k}_j} \sin(\varphi_{\mathbf{k}_j} + \varphi_{\mathbf{h}-\mathbf{k}_j})) = 0. && (7.126)
\end{aligned}
$$

Thus,

$$
\frac{\sin\beta_{\mathbf{h}}}{\cos\beta_{\mathbf{h}}} = \tan\beta_{\mathbf{h}} = \frac{\sum_{j=1}^{m} E_{\mathbf{k}_j}E_{\mathbf{h}-\mathbf{k}_j} \sin(\varphi_{\mathbf{k}_j} + \varphi_{\mathbf{h}-\mathbf{k}_j})}{\sum_{j=1}^{m} E_{\mathbf{k}_j}E_{\mathbf{h}-\mathbf{k}_j} \cos(\varphi_{\mathbf{k}_j} + \varphi_{\mathbf{h}-\mathbf{k}_j})}.
$$

The tangent formula provides the values of the phases that maximize $Z(\varphi_{\mathbf{h}})$ for each reflection. An approach to the determination of these phases might be to begin with a subset of all the reflections with large E values (so that the triplet relations are likely to be valid) and guess the initial phases of these reflections. We assume, for the sake of argument, that these values are not too far from the actual values. For each reflection, sequentially, we would accumulate all of the triplet relationships, then vary $\varphi_{\mathbf{h}}$ from zero to 2π, seeking the maximum value for $Z(\varphi_{\mathbf{h}})$. Each time a new (and improved) phase is determined it is used in the subsequent tangent formulas for the remaining reflections. The new phases would then be used for a second cycle of the process, and so forth. The process would be repeated iteratively, creating improved values of the phases during each iteration until the changes in all of the phases were negligible — providing a self-consistent solution. While this approach would be impractical (or at least tremendously time-consuming), from the derivation above we can see that the iterative application of the tangent formula will accomplish the same result, providing phases that maximize $Z(\varphi_{\mathbf{h}})$.

7.1.2 Probability Methods: Initial Phases

With the tangent formula we have a powerful tool for determining unknown phases from known phases. The process of structural determination begins with a few phases, using the tangent formula to generate new phases, then using those phases to generate more phases, and so on until all of the reflections have been assigned phases. The next step is therefore to determine reliable phases for a few reflections in order to initiate the process.

Earlier in the chapter it was noted that the structure is solved by determining the relative phases of the reflections. The structure factors – the density waves – create an image of the structure – and it useful to consider how this image is formed. If it was possible to detect all of the reflections simultaneously with their amplitudes and phases, we could (with an appropriate "eye") combine them (in a Fourier series), reconstructing an image of the crystal structure immediately. The waves are all traveling waves, and their actual phases would depend on the instant in time that we were observing them, but their relative phases would always be the same. A snapshot at any instant would reveal all of the waves with their relative phases. To define their actual phases we would pick a reference point and assign phases to enough reflections (usually three) to establish the phases of the remaining reflections and provide the starting phases that we are looking for. The reference point that we have chosen is the origin of the unit cell. The actual phases therefore depend on the location of the origin. If the origin is moved, the phases of all the reflections change accordingly.

There is often some confusion in attempts to explain the effects of origin location on phases; points of reference often become confused. In order to avoid this confusion we will take our snapshot of the density waves referenced to an initial origin at a fixed location in the lattice. The phases of the reflections are therefore defined with respect to the original origin location. If we move to another "reference" point in the unit cell and determine the phases of all of the density waves relative to that point, the phases will be different — undergoing shifts relative to the phases referenced to the origin. *It follows that the assignment of these shifted phases to all of the reflections will serve to determine the location of the reference point.* This is equally true of the initial origin. If the unshifted phases are assigned to the reflections they will serve to define the original origin location. A reference point defined by the reflection phases will be referred to as a *scattering reference.*

The location of the origin in the unit cell is determined by the requirement that it reflects the point symmetry of the space group. Any point in the unit cell that meets this criterion is a *permissible origin*; we will limit the locations of potential scattering reference points to the permissible origins in the unit cell. Because the symmetry operations in the unit cell are defined with respect to the permissible origins, the electron density, and hence the structure factors — must have the same symmetry with respect to those origins. If the scattering references are located at these points of symmetry in the unit cell, there are often constraints imposed by that symmetry on the phases of classes of reflections referenced to them. *Conversely, assigning appropriate phases to reflections within these classes can serve to define the phase reference point, and at the same time establish initial phases for structural solution.* Our goal will therefore be to assign enough reflection phases to uniquely define a scattering "origin" to which all of the remaining phases will be referenced. Once the phases of all of these reflections have been assigned, translating the original origin to the location of the selected scattering reference will maintain both the symmetry and the relationships between the phases — it is in this sense that we will have "defined the origin." In this context, *origin definition* and *determination of a scattering reference* are synonymous, and we refer to them interchangeably.

Initial Phases Without Symmetry Constraints

Consider the structure factors from a unit cell with an origin originally at o, depicted in Fig. 7.14:

$$\mathbf{F_h} = \sum_{j=1}^{n} f_j e^{-2\pi\, i(\mathbf{h}\cdot\mathbf{r}_j)} = F e^{i\,\varphi_\mathbf{h}}.$$

If the space group is $P1$ (or if we ignore any symmetry*), there are no constraints on the location of the origin — the scattering reference can therefore be placed at any one of infinitely many permissible origin locations, o'. Since this location is also an origin, we can determine new reflection phases referenced to o' by translating the origin along the vector $\mathbf{v}_o$. In general, the changes in phases that occur when a scattering reference is defined can be determined in this manner, and *we will refer to the specification of a scattering reference point as an "origin shift."* The atoms originally at $\mathbf{r}_j$ are located at $\mathbf{r}'_j = \mathbf{r}_j - \mathbf{v}_o$ in the translated unit cell and the structure factors become

$$\mathbf{F}'_\mathbf{h} = \sum_{j=1}^{n} f_j e^{-2\pi\, i(\mathbf{h}\cdot\mathbf{r}'_j)} = \sum_{j=1}^{n} f_j e^{-2\pi\, i(\mathbf{h}\cdot(\mathbf{r}_j - \mathbf{v}_o))} \tag{7.127}$$

$$= e^{2\pi\, i\mathbf{h}\cdot\mathbf{v}_o} \sum_{j=1}^{n} f_j e^{-2\pi\, i(\mathbf{h}\cdot\mathbf{r}_j)} = F e^{i\,\varphi_\mathbf{h}} e^{2\pi\, i\mathbf{h}\cdot\mathbf{v}_o} = F e^{i\,(\varphi_\mathbf{h} + 2\pi\mathbf{h}\cdot\mathbf{v}_o)}. \tag{7.128}$$

The origin shift creates a phase shift of $2\pi\mathbf{h}\cdot\mathbf{v}_o$ for all of the structure factors, leaving their amplitudes (and resultant intensities) unaltered. In Sec. 7.1.1 we showed that an atom positioned at any point on an hk line in a two dimensional unit cell would scatter in phase with an atom at another point on an hk line – even the

*It is nearly always advisable to solve the structure in its assigned space group. However, it is sometimes useful in the early stages of structural elucidation to ignore the symmetry, and solve the structure in the $P1$ space group.

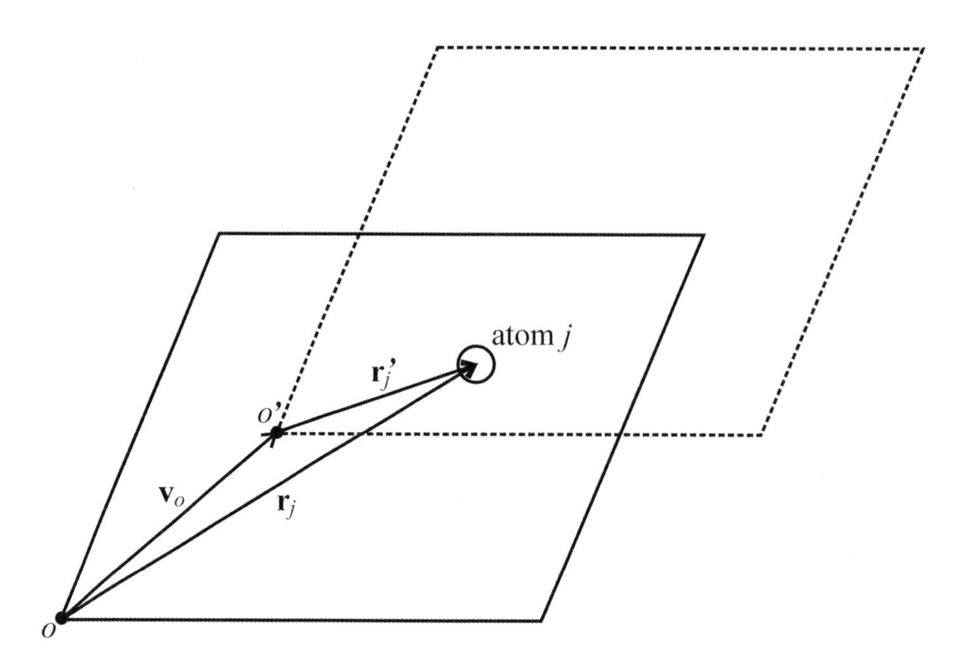

Figure 7.14 The effect of an origin shift on the coordinates of the atoms in the unit cell.

same line. This came about because, for a point (x, y) on an hk line, $hx + ky = m$, where m is an integer. We now consider a similar situation, depicted in Fig. 7.15. The scattering reference is placed at an arbitrary location, $o' = (x_o, y_o)$, along a line parallel to an hk line in a two-dimensional unit cell. The structure factor, $\mathbf{F}_{hk}$ experiences a phase shift relative to this location (along with the structure factors for all of the reflections):

$$\mathbf{F}'_{hk} = F_{hk}e^{i\,(\varphi_{\mathbf{h}}+2\pi\mathbf{h}\cdot\mathbf{v}_o)} = F_{hk}e^{i\,(\varphi_{\mathbf{h}}+2\pi(hx_o+ky_o))}. \tag{7.129}$$

The parallel line intersects the original a axis at the point $m/h - q$, creating similar triangles, one with edges of length $m/h - x_o - q$ and y_o, the other with lengths $1/h$ and $1/k$. Thus,

$$\frac{\frac{m}{h} - x_o - q}{y_o} = \frac{1/h}{1/k} = \frac{k}{h} \tag{7.130}$$

$$hx_o + ky_o = m - hq = C, \quad \text{and}$$

$$\varphi'_{\mathbf{h}} = \varphi_{\mathbf{h}} + C, \tag{7.131}$$

where C is a constant for any location on the parallel line. *This specific reflection will have the same phase if the scattering reference is located at any point along a line parallel to the hk line.* C determines the distance of the parallel line from the hk line; the selection of any value of C will satisfy the same "equal phase" conditions. Turning this around, *if the hk reflection is assigned an arbitrary phase,* $\varphi_{\mathbf{h}} + C$*, the scattering reference will be constrained to lie somewhere along a line parallel to the hk lines in the lattice.*

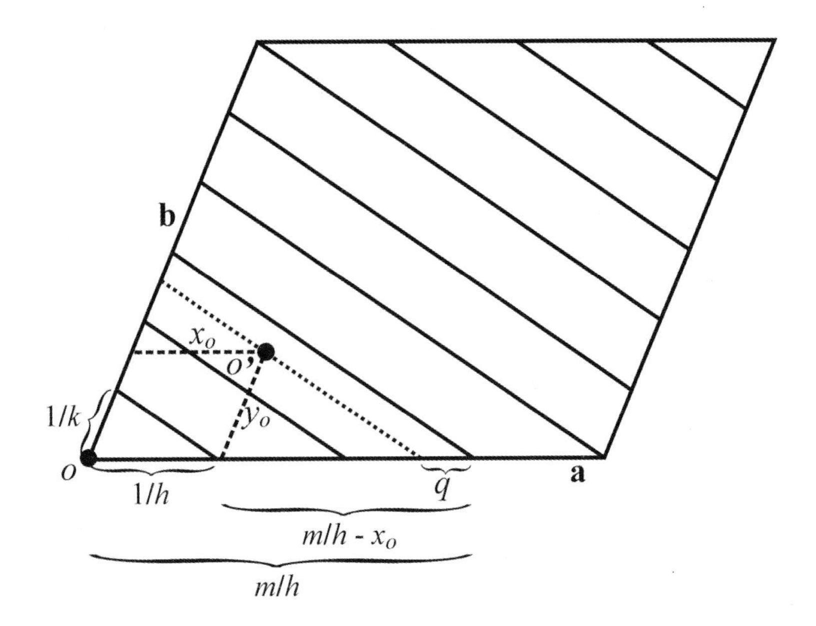

Figure 7.15 Origin shift to a line parallel to an hk line in a two-dimensional unit cell.

Assigning an arbitrary phase to a second reflection with indices (h' k') will constrain the scattering reference to a line parallel to the $h'k'$ lines in the lattice. Thus assigning phases to two reflections restricts the scattering reference to the points of intersection of the sets of lines parallel to the hk and $h'k'$ lines, illustrated in Fig. 7.16. Each intersection marks a point at which the phases of the hk and $h'k'$ reflections will be the same as the phases of the two reflections at any other intersection point. Each point is a potential scattering reference. However, the phases of the remaining reflections will depend on which of these scattering references that we select – *the assignment of two arbitrary phases to two arbitrary reflections does not define the location of the scattering reference unequivocally.*

To guarantee that the scattering reference is assigned unambiguously, the reflections that we select must generate parallelograms that do not create more than one intersection point within the unit cell. This will be the case only if the area of the parallelogram equals the area of the unit cell. The corresponding reciprocal parallelogram must therefore also have an area equal to the area of the reciprocal unit cell. This places significant limits on those reflections that can be assigned arbitrary phases in order to define the origin.

In an extension of the treatment above to three dimensions, it is readily shown that an origin shift to any point on a plane parallel to an hkl plane will result in a constant phase shift:

$$hx_o + ky_o + lz_o = C. \tag{7.132}$$

This specific reflection will have the same phase if the scattering reference is located at any point on a plane parallel to the hkl plane. Conversely, if an arbitrary

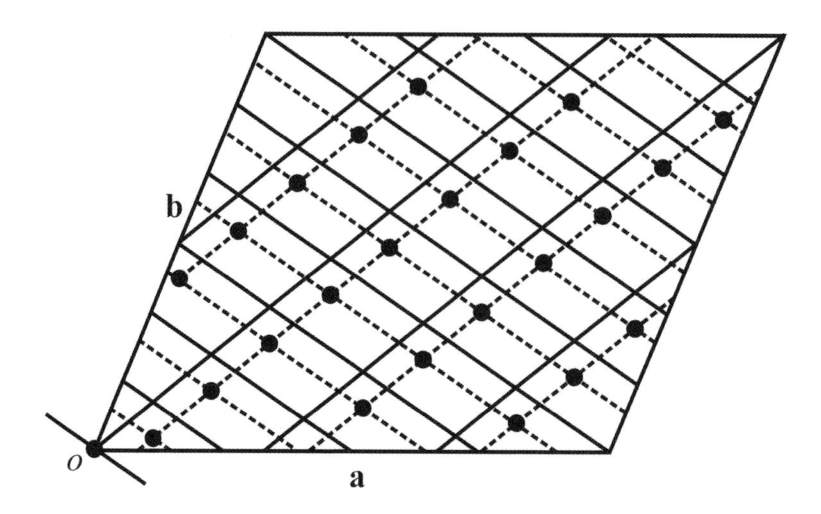

Figure 7.16 Possible scattering reference locations defined by the selection of phases for the $(h\ k) = (4\ 6)$ and $(h'\ k') = (2\ \bar{3})$ reflections.

phase is assigned to the reflection, the scattering reference will be constrained to lie somewhere on a plane parallel to an $h_1k_1l_1$ plane. Assigning an arbitrary phase to a second reflection will constrain the scattering reference to a second set of planes parallel to an $h_2k_2l_2$ plane; the scattering reference must therefore lie on one of the parallel lines created by the intersection of the two planes. The assignment of a phase to a third reflection will constrain the scattering reference to a third set of planes parallel to the $h_3k_3l_3$ planes. The scattering reference must consequently be located on one of the points resulting from the intersection of these planes with the parallel lines. As it was for two dimensions, the three reflections for which we are allowed to assign arbitrary phases in order to define the scattering reference location must assign that location unambiguously. This will occur only if the scattering reference is defined by reflections that produce parallelepipeds with only a single intersection point within the unit cell. To guarantee this, the parallelepipeds resulting from the intersections of the planes must have the same volume as the unit cell, and the reciprocal parallelepipeds created by the selected reflections must have the same volume as the reciprocal unit cell. The direct lattice parallelepipeds are determined by the vectors connecting the parallel planes, and the reciprocal lattice parallelepipeds are therefore defined by the reciprocal lattice vectors for the reflections (the simplest example being the direct unit cell and the corresponding [1 0 0], [0 1 0], and [0 0 1] reciprocal lattice vectors defining the reciprocal unit cell). The vectors defining the reciprocal parallelepiped are

$$\begin{aligned}
\mathbf{h}_1 &= h_1\,\mathbf{a}^* + k_1\,\mathbf{b}^* + l_1\,\mathbf{c}^* \\
\mathbf{h}_2 &= h_2\,\mathbf{a}^* + k_2\,\mathbf{b}^* + l_2\,\mathbf{c}^* \\
\mathbf{h}_3 &= h_3\,\mathbf{a}^* + k_3\,\mathbf{b}^* + l_3\,\mathbf{c}^*.
\end{aligned} \tag{7.133}$$

The reciprocal cell volume is given by $V^* = \mathbf{a}^* \cdot (\mathbf{b}^* \times \mathbf{c}^*)$ (Eqn. 1.135). The reciprocal parallelepiped volume is $V_p = \mathbf{h}_1 \cdot (\mathbf{h}_2 \times \mathbf{h}_3)$:

$$\begin{aligned}
V_p &= (h_1\,\mathbf{a}^* + k_1\,\mathbf{b}^* + l_1\,\mathbf{c}^*) \\
&\quad \cdot (h_2\,\mathbf{a}^* + k_2\,\mathbf{b}^* + l_2\,\mathbf{c}^*) \times (h_3\,\mathbf{a}^* + k_3\,\mathbf{b}^* + l_3\,\mathbf{c}^*) \\
&= (h_1\,\mathbf{a}^* + k_1\,\mathbf{b}^* + l_1\,\mathbf{c}^*) \\
&\quad \cdot (h_2\,h_3\,\mathbf{a}^*\times\mathbf{a}^* + h_2\,k_3\,\mathbf{a}^*\times\mathbf{b}^* + h_2\,l_3\,\mathbf{a}^*\times\mathbf{c}^* \\
&\quad + k_2\,h_3\,\mathbf{b}^*\times\mathbf{a}^* + k_2\,k_3\,\mathbf{b}^*\times\mathbf{b}^* + k_2\,l_3\,\mathbf{b}^*\times\mathbf{c}^* \\
&\quad + l_2\,h_3\,\mathbf{c}^*\times\mathbf{a}^* + l_2\,k_3\,\mathbf{c}^*\times\mathbf{b}^* + l_2\,l_3\,\mathbf{c}^*\times\mathbf{c}^*).
\end{aligned}$$

The first term in the expansion of the scalar product is (See Eqn. 1.136)

$$h_1h_2h_3\,(\mathbf{a}^* \cdot (\mathbf{a}^*\times\mathbf{a}^*)) = h_1h_2h_3 \begin{vmatrix} a_x^* & a_y^* & a_z^* \\ a_x^* & a_y^* & a_z^* \\ a_x^* & a_y^* & a_z^* \end{vmatrix} = 0. \tag{7.134}$$

This term is equal to zero because the determinant contains two identical rows (in this case, three identical rows). For any term containing two $\mathbf{a}^*$, $\mathbf{b}^*$ or $\mathbf{c}^*$ vectors, the resulting determinant will contain two identical rows, and will therefore equal zero. The only non-zero terms will contain a scalar triple product containing all three reciprocal axial vectors — equal to the reciprocal cell volume, e.g.,

$$h_1k_2l_3\,(\mathbf{a}^* \cdot (\mathbf{b}^*\times\mathbf{c}^*)) = h_1k_2l_3 \begin{vmatrix} a_x^* & a_y^* & a_z^* \\ b_x^* & b_y^* & b_z^* \\ c_x^* & c_y^* & c_z^* \end{vmatrix} = h_1k_2l_3\,V^*. \tag{7.135}$$

Noting that $(\mathbf{a}^*\times\mathbf{b}^*) = -(\mathbf{b}^*\times\mathbf{a}^*)$, etc.,

$$\begin{aligned}
V_p &= h_1k_2l_3\,V^* + h_2k_3l_1\,V^* + h_3k_1l_2\,V^* \\
&\quad - h_3k_2l_1\,V^* - h_1k_3l_2\,V^* - h_2k_1l_3\,V^*.
\end{aligned} \tag{7.136}$$

Factoring out V^*, the sum is seen to consist of a signed permutation of the indices, conveniently expressed in a determinant:

$$V_p = \begin{vmatrix} h_1 & k_1 & l_1 \\ h_2 & k_2 & l_2 \\ h_3 & k_3 & l_3 \end{vmatrix} V^* = |\mathbf{H}|V^*. \tag{7.137}$$

Only the reflections with indices that have $V_p = V^*$ are allowable reflections for defining the scattering reference, requiring that $|\mathbf{H}| = 1$. Since switching a row in the determinant will change its sign, but will not alter the volume of the parallelepiped, three reflections with $|\mathbf{H}| = \pm 1$ can be assigned arbitrary phases to define the location of the scattering reference and be utilized to begin phase determination. The (1 0 0), (0 1 0), and (0 0 1) reflections clearly satisfy this condition. The (2 2 1), (1 2 2) and (1 1 1) reflections are another example. While it may appear obvious that the axial reflections would always be chosen, recall that the purpose of assigning phases to the origin-defining reflections is to employ them as starting phases for triplet evaluation. As we shall soon see, reflections with large estimated $Y_{\mathbf{h}}$ values are ordinarily chosen to define the origin, and these may or may not include the axial reflections.

Initial Phases With Symmetry Constraints

Space group symmetry limits the choice of origins to specific locations in the unit cell, and the choice of origin-defining reflections must constrain the phases so that they are referenced to one of these locations. The assignment of arbitrary phases to arbitrarily selected general reflections will not establish one of these locations as an unambiguous origin position. However, for every space group there exist specific classes of reflections that are allowed only certain phases in order for the density waves (and therefore the electron density) to be identical before and after application of the symmetry operator. Assignment of appropriate phases to these reflections *will* require the phases to be in sync with the symmetry operations, serving to limit the origin to locations that are consistent with the symmetry. To identify these special classes we must first determine the way in which symmetry operations in direct space affect the structure factors in reciprocal space. From Chapter 2, the general symmetry operator is $\mathbf{S} = \mathbf{R} + \mathbf{t}$, such that for location $\mathbf{r}_j$ in the unit cell, $\mathbf{r}'_j = \mathbf{R}\mathbf{r}_j + \mathbf{t}$. The application of this symmetry operator leaves the electron density in the unit cell unchanged, and must therefore leave the density waves — the structure factors — unchanged in the sense that they have the same amplitudes and, if they undergo a phase shift, that shift must be an integer multiple of 2π ($\varphi_{\mathbf{h}}^{\mathbf{S}} = \varphi_{\mathbf{h}} + 2m\pi,\ m = 0, 1, 2, \ldots$):

$$\begin{aligned}
\mathbf{F}_{\mathbf{h}} &= \sum_{j=1}^{n} f_j e^{-2\pi\, i(\mathbf{h}\cdot\mathbf{r}_j)} = F_{\mathbf{h}} e^{i\varphi_{\mathbf{h}}} \\
\mathbf{F}_{\mathbf{h}}^{\mathbf{S}} &= \sum_{j=1}^{n} f_j e^{-2\pi\, i(\mathbf{h}\cdot\mathbf{R}\mathbf{r}_j + \mathbf{t})} = \sum_{j=1}^{n} f_j e^{-2\pi\, i(\mathbf{h}\cdot\mathbf{R}\mathbf{r}_j + \mathbf{h}\cdot\mathbf{t})} \\
&= e^{-2\pi\, i(\mathbf{h}\cdot\mathbf{t})} \sum_{j=1}^{n} f_j e^{-2\pi\, i(\mathbf{h}\cdot\mathbf{R}\mathbf{r}_j)} = F_{\mathbf{h}} e^{-2\pi\, i(\mathbf{h}\cdot\mathbf{t})} e^{i(\varphi_{\mathbf{h}}^{\mathbf{R}})} \\
&= F_{\mathbf{h}} e^{i(\varphi_{\mathbf{h}}^{\mathbf{R}} - 2\pi\, i\, \mathbf{h}\cdot\mathbf{t})} = F_{\mathbf{h}} e^{i\varphi_{\mathbf{h}}^{\mathbf{S}}}.
\end{aligned} \tag{7.138}$$

Thus,

$$\varphi_{\mathbf{h}} + 2m\pi = \varphi_{\mathbf{h}}^{\mathbf{R}} - 2\pi\, \mathbf{h}\cdot\mathbf{t}, \tag{7.139}$$

where $\varphi_{\mathbf{h}}^{\mathbf{R}}$ is the phase of the reflection after the point symmetry operator is applied (before translation to a symmetry-equivalent location).

Now, suppose for a certain class of reflections, the sign of the indices are reversed for $\mathbf{h}\cdot\mathbf{R}\mathbf{r}_j$, such that $\mathbf{h}^T\mathbf{R} = -\mathbf{h}$, and $\mathbf{h}\cdot\mathbf{R}\mathbf{r}_j = -\mathbf{h}\cdot\mathbf{r}_j$. This will occur, for example, with centrosymmetric space groups for all of the reflections. This will change the sigh of the contribution for every atom location in the Fourier series, resulting in $\varphi_{\mathbf{h}}^{\mathbf{R}} = -\varphi_{\mathbf{h}}$ and

$$\begin{aligned}
\varphi_{\mathbf{h}} &= -\varphi_{\mathbf{h}} - 2\pi\, \mathbf{h}\cdot\mathbf{t} - 2m\pi \\
2\varphi_{\mathbf{h}} &= -2\pi\, \mathbf{h}\cdot\mathbf{t} = 2\pi\, \mathbf{h}\cdot\mathbf{t} - 2m\pi \\
\varphi_{\mathbf{h}} &= \pi\, \mathbf{h}\cdot\mathbf{t} - m\pi,
\end{aligned} \tag{7.140}$$

thus establishing the phase for the specific reflections in the class.

Origin-defining phases in centrosymmetric space groups. As noted above, in space groups other than $P1$, the origin is selected to reflect point symmetry of the lattice. The location of the origin is no longer arbitrary, and phases must be assigned with values that will define the origin at one of its allowed positions. In centrosymmetric unit cells the permissible origins are the inversion centers. The point symmetry operator is the inversion operator, which changes the sign of *all* of the reflections in the set: $\mathbf{h}^T\mathbf{R}(\bar{1}) = -\mathbf{h}^T$. For the inversion operation, $\mathbf{t}^T = [0\ 0\ 0]$, and, from Eqn. 7.140,

$$\varphi_{\mathbf{h}} \quad = \quad -m\pi \equiv m\pi. \tag{7.141}$$

The phases of all of the reflections are restricted to 0 when m is even, and to π when m is odd. Assigning either one of these phases to three reflections will serve to constrain the scattering reference to one or more centers of symmetry. Unlike in the $P1$ case, other points of intersection in the unit cell are inconsequential, since the remaining reflections would have phases other than 0 or π if one of the "forbidden" points could be chosen as an origin. Thus, while the assignment reflections would have the same phase at any of the intersection points, the remaining reflections must always have only 0 and π phases, and possible origins will be only those permitted for the space group. The origin is defined by the phases of three appropriate reflections *and* restrictions on the phases of the class of reflections constrained by symmetry — in this case all of the reflections.

As before, the assignment of phases must be made so that the origin definition is unique. This will require the selection of specific sets of reflections. We will again begin by considering the simpler case of a two-dimensional unit cell, illustrated in Fig. 7.17. There are four unique inversion centers in the unit cell (the local environments around each one can be different), indicated by the numbers 1–4 in the figure. To determine the relative phases for different reflections, we examine the phase shifts that will occur when origin 1 is translated along $\mathbf{v}_o$ to each of the other permissible origins. If the origin is moved to location 2, $(x_o, y_o) = (1/2, 1/2)$ and the phase shift is $\Delta\varphi_{hk} = 2\pi((h/2 + k/2) = \pi(h + k)$. If $h + k = 2m$ is even, then $\Delta\varphi_{hk} = 2m\pi \equiv 0$; the reflection has the same phase at both locations. If $h + k = 2m + 1$ is odd, then $\Delta\varphi_{hk} = 2m\pi + \pi \equiv \pi$, and the phase of the reflection is shifted by 180°. For a translation to origin 3, $(x_o, y_o) = (1/2, 0)$ and $\Delta\varphi_{hk} = 2\pi((h/2) = \pi h$. If $h = 2m$ is even, the phase remains the same, and if $h = 2m + 1$ is odd then the phase shifts by 180°. A translation to origin 4 has $(x_o, y_o) = (0, 1/2)$, resulting in a zero phase shift for k even and a 180° phase shift for k odd.

We now divide the $\{h, k\}$ reflection set into four groups, known as *parity groups*, depending on whether each of the two indices is odd or even. Using the symbol "e" to denote even, and "o" to denote odd, the groups can be labeled "ee", "oo", "eo", and "oe", examples of which are shown in the figure. For reflections with both indices even, $h+k$, h, and k are all even. The phases of these reflections are invariant with respect to a shift in permissable origins. These phases are *not* structure invariants since they depend on the origin location, and are instead known as *structure semi-invariants. Structure semi-invariants are phases or linear combinations of phases that do not change if the origin is shifted to any of the allowable origins (origins with the same point symmetry) in the lattice.* Fig, 7.17(a) illustrates the lattice lines in the unit cell of an ee reflection. The structure factor wave has a phase of 0 at origin 1, the original origin, with maxima coincident with the lattice lines (See Fig. 7.2).

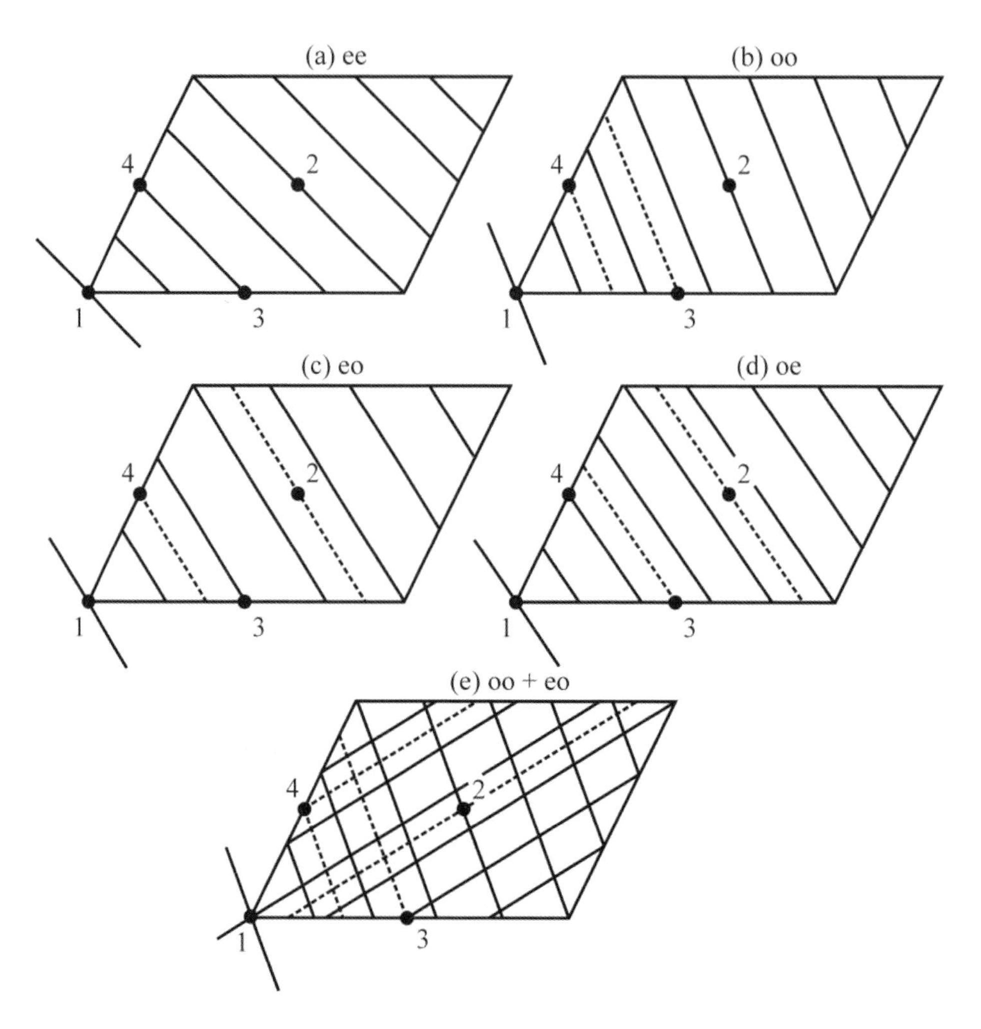

Figure 7.17 Lattice lines in a two-dimensional unit cell with an inversion center. (a) Lattice lines for $(h\ k) = (4\ 4)$. (b) Lattice lines for $(h\ k) = (5\ 3)$. (c) Lattice lines for $(h\ k) = (4\ 3)$. (d) Lattice lines for $(h\ k) = (5\ 4)$. (e) Lattice lines for $(h\ k) = (5\ 3)$ and $(h\ k) = (4\ \bar{3})$.

All four permissible origins lie on the lattice lines; the phase of the reflection will be the same if the scattering reference is chosen at any one of these locations. It follows that a phase assigned to one of these reflections will place no new constraints on the scattering reference location — *structure semi-invariants place no restrictions on the possible choices for the location of the scattering reference, and therefore are not useful for origin definition.*

Fig. 7.17(b) is an example of the lattice lines of an oo reflection. Note that only origins 1 and 2 intersect the lattice lines. For these reflections $h + k$ is even, and h and k are odd. There will be no change in phase if the origin is moved to location 2, but the phase will shift to π (indicated by the dashed lines) if the

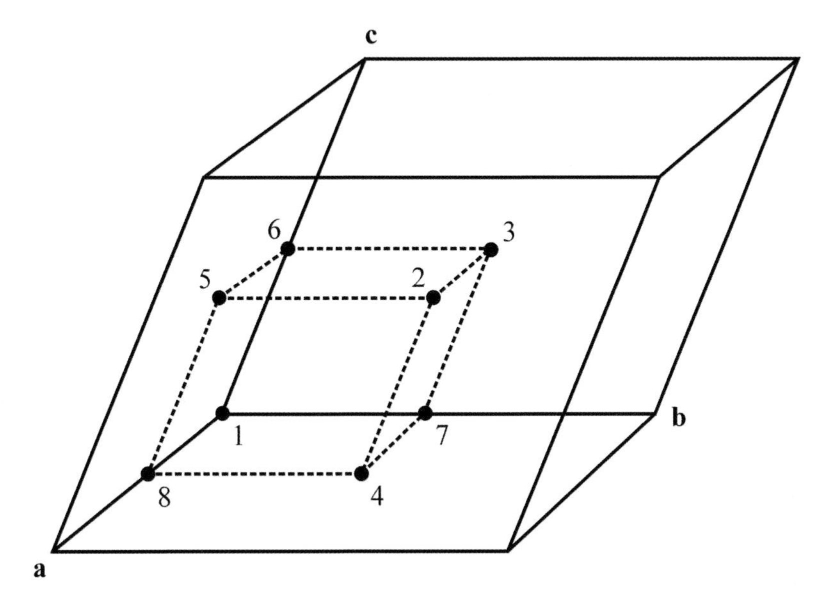

Figure 7.18 Permissible origins for a primitive centrosymmetric unit cell.

origin is moved to either location 3 or location 4. The assignment of a phase of 0 to an oo reflection restricts the scattering reference to location 1 or location 2, while assignment of a phase of π restricts the scattering reference to location 3 or location 4. The lattice lines for an eo reflection are depicted in Fig. 7.17(c); only origins 1 and 3 lie on a lattice line. This is consistent with $h+k$ odd, h even, and k odd. The assignment of a phase of 0 to an eo reflection thus restricts the scattering reference to location 1 or location 3 — and locations 2 or 4 if π is assigned. Finally, as indicated in Fig. 7.17(d), an oe reflection has origins 1 and 4 on the lattice lines, consistent with $h + k$ odd, h odd, and k even. The assignment of a phase of 0 to an oe reflection thus restricts the scattering reference to location 1 or location 4 — and locations 2 or 3 if π is assigned.

The origin can be uniquely defined by assigning phases of 0 (or π) to appropriately selected reflections chosen from different parity groups. For example, a phase assignment of 0 to an oo reflection will restrict the origin to location 1 or 2. The subsequent assignment of a phase of 0 to an eo reflection eliminates location 2 as a scattering reference and defines the origin at location 1, illustrated in Fig. 7.17(e).

Fig. 7.18 illustrates the eight permissible origin locations in a primitive centrosymmetric unit cell in three-dimensions. The reflections are now separated into eight parity groups, depending on their indices: eee, eeo, eoe, eoo, oee, oeo, ooe, and ooo. The phase changes for each parity group and origin combination can be determined in the same manner as they were in the two-dimensional case. For example, moving the origin to $(1/2, 1/2, 1/2)$ results in $\Delta\varphi_{hk} = 2\pi((h/2+k/2+l/2) = \pi(h + k + l)$. If $h + k + l$ is even (for eee, eoo, oeo, and ooe), there is no change in phase; if the sum is odd (for eeo, eoe, oee, and ooo), there is a phase shift of π. Table 7.1 is the result of an analysis of the phase changes induced by origin shifts for the reflections in each of the parity groups.

Table 7.1 Phase shifts resulting from changes in the scattering reference location in a centosymmetric lattice for all parity groups.

	Parity Group Phase Shift							
Origin	eee	eeo	eoe	eoo	oee	oeo	ooe	ooo
(1) 0, 0, 0	0	0	0	0	0	0	0	0
(2) $\frac{1}{2}$, $\frac{1}{2}$, $\frac{1}{2}$	0	π	π	0	π	0	0	π
(3) 0, $\frac{1}{2}$, $\frac{1}{2}$	0	π	π	0	0	π	π	0
(4) $\frac{1}{2}$, $\frac{1}{2}$, 0	0	0	π	π	π	π	0	0
(5) $\frac{1}{2}$, 0, $\frac{1}{2}$	0	π	0	π	π	0	π	0
(6) 0, 0, $\frac{1}{2}$	0	π	0	π	0	π	0	π
(7) 0, $\frac{1}{2}$, 0	0	0	π	π	0	0	π	π
(8) $\frac{1}{2}$, 0, 0	0	0	0	0	π	π	π	π

The judicious assignment of phases to three reflections from different parity groups will serve to define a unique scattering reference at the intersection of the lattice planes corresponding to the reflections (if the phase is 0) or planes midway between the lattice planes (if the phase is π). The eee reflections are structure semi-invariants, and cannot be used for origin definition.* Assigning phases of 0 to eeo, oee, and ooe reflections will, sequentially, restrict the scattering reference to locations 1, 4, 7, and 8, then 1 and 7, and finally, location 1. Assigning 0, π, and 0 to the same reflections will restrict the scattering reference to locations 1, 4, 7, and 8, then 4 and 8, and finally location 4. Other combinations of 0 and π will place the scattering reference at different locations. As an example, the $(2, 2, 3)$, $(1, 2, 2)$, and $(1, 1, 0)$ reflections are suitable origin-defining reflections for a centrosymmetric lattice.

Note that not all combinations of parity groups result in unambiguous origin definition. If phases of 0 are assigned to eeo, oee, and oeo reflections, the first two reflections will restrict the scattering reference to locations 1 and 7, but the third reflection cannot distinguish between these locations. If phases of 0, π, and 0 are assigned, the first two reflections restrict the scattering reference to locations 4 and 8, but the third reflection does not restrict the scattering reference to either of these locations. In order to develop a systematic method for selecting those reflections

*Note that an eee reflection will have lattice planes that pass through the origin and the center of the unit cell. They will also bisect each of the unit cell axes and pass through the centers of the plane faces of the cell. The lattice planes will therefore contain all of the permissible origins; the eee reflection will have the same phase with the scattering reference located at any of these origins. *In order for the phase of a reflection to be semi-invariant, its lattice planes must contain all of the permissible origins in the unit cell.*

that define a unique origin we consider the general phase shift for a given shift in the origin. Expanding Eqn. 7.129,

$$\begin{aligned}\mathbf{F}'_{hkl} &= F_{hk}e^{i(\varphi_{\mathbf{h}}+2\pi\mathbf{h}\cdot\mathbf{v}_o)} = F_{hk}e^{i(\varphi_{\mathbf{h}}+2\pi(hx_o+ky_o+lz_o))}.\\ \Delta\varphi &= \varphi'_{\mathbf{h}} - \varphi_{\mathbf{h}} = 2\pi(hx_o+ky_o+lz_o)\\ &= 2\pi\begin{bmatrix} h & k & l\end{bmatrix}\begin{bmatrix} x_o\\ y_o\\ z_o\end{bmatrix}.\end{aligned} \tag{7.142}$$

For three reflections,

$$2\pi\begin{bmatrix} h_1 & k_1 & l_1\\ h_2 & k_2 & l_2\\ h_3 & k_3 & l_3\end{bmatrix}\begin{bmatrix} x_o\\ y_o\\ z_o\end{bmatrix} = \begin{bmatrix}\Delta\varphi_1\\ \Delta\varphi_2\\ \Delta\varphi_3\end{bmatrix}. \tag{7.143}$$

$$\begin{bmatrix} x_o\\ y_o\\ z_o\end{bmatrix} = \frac{1}{2\pi}\begin{bmatrix} h_1 & k_1 & l_1\\ h_2 & k_2 & l_2\\ h_3 & k_3 & l_3\end{bmatrix}^{-1}\begin{bmatrix}\Delta\varphi_1\\ \Delta\varphi_2\\ \Delta\varphi_3\end{bmatrix}. \tag{7.144}$$

The solution for a unique origin shift defined by $[x_o\ y_o\ z_o]$ is possible only if the matrix of indices, **H**, has an inverse — and that is possible only if the determinant of the matrix is not zero.

The analysis can be further simplified by noting that each index can be written as $h = 2m + p$, where m is an integer; $p = 1$ if h is odd and $p = 0$ if h is even. The same holds for k and l, and

$$\begin{aligned}\Delta\varphi &= 2\pi((2m+p_h)x_o + (2m'+p_k)y_o + (2m''+p_l)z_o\\ &= (4\pi m + 2\pi p_h)x_o + (4\pi m' + 2\pi p_k)y_o + (4\pi m'' + 2\pi p_l)z_o\\ &= 2\pi(p_h x_o + p_k y_o + p_l z_o).\end{aligned} \tag{7.145}$$

Thus,

$$\begin{bmatrix} x_o\\ y_o\\ z_o\end{bmatrix} = \frac{1}{2\pi}\begin{bmatrix} p_{h_1} & p_{k_1} & p_{l_1}\\ p_{h_2} & p_{k_2} & p_{l_2}\\ p_{h_3} & p_{k_3} & p_{l_3}\end{bmatrix}^{-1}\begin{bmatrix}\Delta\varphi_1\\ \Delta\varphi_2\\ \Delta\varphi_3\end{bmatrix}. \tag{7.146}$$

$[x_o\ y_o\ z_o]$ has a unique solution only if

$$\begin{vmatrix} p_{h_1} & p_{k_1} & p_{l_1}\\ p_{h_2} & p_{k_2} & p_{l_2}\\ p_{h_3} & p_{k_3} & p_{l_3}\end{vmatrix} \neq 0. \tag{7.147}$$

Because the 3×3 matrix contains only ones and zeros (known as a binary matrix), its determinant can have values of either -1, 0, or 1. Each row in the matrix represents the parity of the indices of the reflection, and we will refer to it as a *parity vector*. We therefore need only know the parity of the set of potential origin defining reflections in order to assess whether or not they will define a unique origin. For example, the eeo, oee, and ooe set is represented by [0 0 1], [1 0 0], and [1 1 0];

The eeo, oee, and oeo set is represented by [0 0 1], [1 0 0], and [1 0 1]. If we create a matrix from the first set of vectors and evaluate its determinant, we find that the determinant is non-zero,

$$\begin{vmatrix} 0 & 0 & 1 \\ 1 & 0 & 0 \\ 1 & 1 & 0 \end{vmatrix} = 1, \qquad (7.148)$$

indicating that the vectors are linearly independent. For the second set of vectors we can readily see that this is not the case, since [0 0 1] + [1 0 0] = [1 0 1] and:

$$\begin{vmatrix} 0 & 0 & 1 \\ 1 & 0 & 0 \\ 1 & 0 & 1 \end{vmatrix} = 0. \qquad (7.149)$$

Thus, *the assignment of phases for three reflections will serve to define the origin if the vectors representing the parity groups of the reflections are linearly independent*, easily checked by calculating a determinant of the matrix of the parity vectors. Table 7.1 is general for all primitive centrosymmetric unit cells of orthorhombic symmetry or lower. For centered space groups, some of the permissible origins are related by translational symmetry and there are fewer independent origins to choose between. The parity rules outlined above are modified to take this into account.

Origin- and enantiomorph-defining phases in non-centrosymmetric space groups. As in the centrosymmetric case, the permissible origins in non-centrosymmetric space groups are determined by symmetry; the assigned phases must define a scattering reference located at one of these origins. While there are no phase restrictions on a general reflection in a non-centrosymmetric space group, there are classes of reflections with phases that depend on the symmetry, and assignment of phases to these reflections will serve to locate the scattering reference in positions consistent with the symmetry. A simple two-dimensional example is illustrated in Fig. 7.19. The unit cell has two-fold axial symmetry illustrated by the location of two-fold screw axes parallel to the b axis (the substitution of two-fold rotational axes will produce an identical result). The symmetry operator for the two-fold screw operation in two-dimensions is

$$\mathbf{S}(2_1) = \mathbf{R}(2) + \mathbf{t} = \begin{bmatrix} -1 & 0 \\ 0 & 1 \end{bmatrix} + \begin{bmatrix} 0 \\ \frac{1}{2} \end{bmatrix}. \qquad (7.150)$$

Permissible origins can be placed at any point on one of the 2_1 axes; the x coordinates of potential scattering references are limited to 0 and 1/2, with no limits on their y coordinates. While $\mathbf{h}^T\mathbf{R} \neq -\mathbf{h}^T$ for the general hk reflection, $\mathbf{h}^T\mathbf{R} = -\mathbf{h}^T$ for the $h0$ class of reflections. From Eqn. 7.140, $\varphi_{\mathbf{h}} = \pi\,\mathbf{h}\cdot\mathbf{t} - m\pi = -m\pi = m\pi$ for this class of reflections. Fig. 7.19(a) illustrates the case when h is odd. Placing a scattering reference on the screw axis passing through the origin will set the phase of the reflection to 0; placing it on the screw axis passing through the center of the unit cell will set the phase to π. Assigning a phase of 0 to the reflection will therefore restrict the scattering reference to the screw axis passing through the original origin, while assigning a phase of π will place the reference on the screw axis at

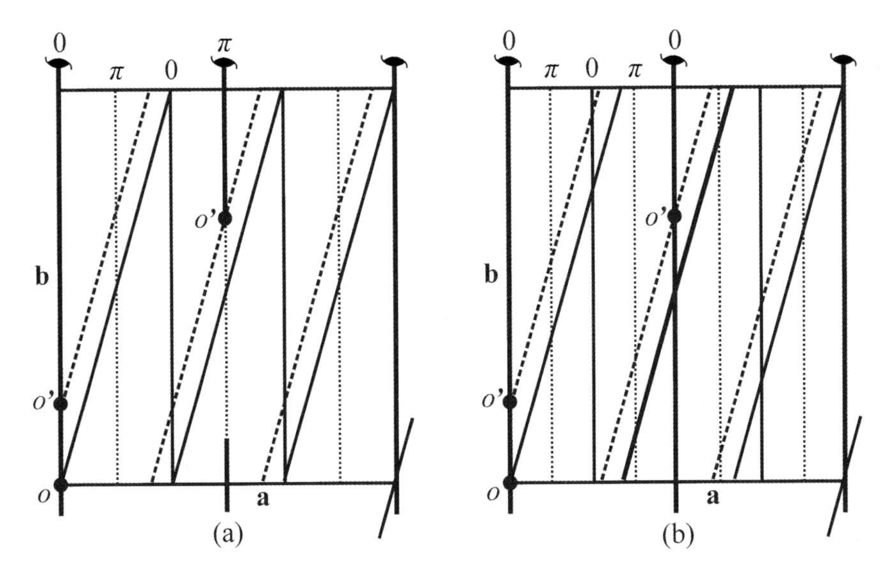

Figure 7.19 Lattice lines, phases, and potential scattering references in a two-dimensional unit cell with two-fold rotational symmetry for (a) $(h\ k) = (3\ 0)$; $(h'\ k') = (3\ \bar{1})$ and (b) $(h\ k) = (4\ 0)$; $(h'\ k') = (3\ \bar{1})$.

$x = 1/2$. A second reflection can be assigned an arbitrary phase, unequivocally establishing the location of the scattering reference at the intersection of the two sets of parallel lines corresponding to each reflection. The only restriction on this reflection is that it must cross the b axis only once — otherwise there would be more than one potential scattering reference in the unit cell. Restrictions such as this will always occur when there is a continuum of potential origin locations in the unit cell, as we have already observed in the $P1$ case. Thus the arbitrary phase must be assigned to a reflection with $(h'\ 1)$ or $(h'\ \bar{1})$ indices.

Fig. 7.19(b) illustrates the case when h is even. Placing the scattering reference on either of the screw axes will set the phase of the reflection to 0. The phase of this reflection does not change with an origin shift, and the reflection is therefore a structure semi-invariant. It cannot be used for origin definition.

In three dimensions, a two-fold axis coincident with the b axis results in space groups $P2$ or $P2_1$, depending on whether the axis is a rotational axis or a screw axis. A $P2_1$ unit cell is illustrated in Fig. 7.20. The two-fold axes are parallel to the b axis, one coincident with it, two others bisecting the a and c axes, respectively, and a fourth passing through the midpoint of the ac plane. The permissible origin locations are located at $(0, y, 0)$, $(1/2, y, 0)$, $(0, y, 1/2)$ and $(1/2, y, 1/2)$ with y arbitrary. The $(h\ 0\ l)$ class of reflections have $\mathbf{R}(2)[h\ 0\ l] = [-h\ 0\ -l]$, resulting in $\varphi_{\mathbf{h}} = \pi\,[h\ 0\ l][0\ 1/2\ 0] - m\pi = m\pi$. A shift to origin 1 has $\mathbf{v}_o = [0\ y\ 0]$, with $\Delta\varphi_{\mathbf{h}} = 2\pi[h\ 0\ l][0\ y\ 0] = 0$, as expected. An origin shift to 2 results in $\Delta\varphi_{\mathbf{h}} = 2\pi[h\ 0\ l][1/2\ y\ 1/2] = (h + l)\pi$. There is no shift in phase when $h + l$ is even, and a phase shift of π when $h + l$ is odd. An origin shift to 3 results in $\Delta\varphi_{\mathbf{h}} = 2\pi[h\ 0\ l][0\ y\ 1/2] = l\pi$; zero when l is even and π when l is odd. Finally, a shift to origin 4 results in a phase shift of zero for h even and π for l odd. Since $k = 0$ for this class of reflections there are four possible parity groups, eee, oee, eeo,

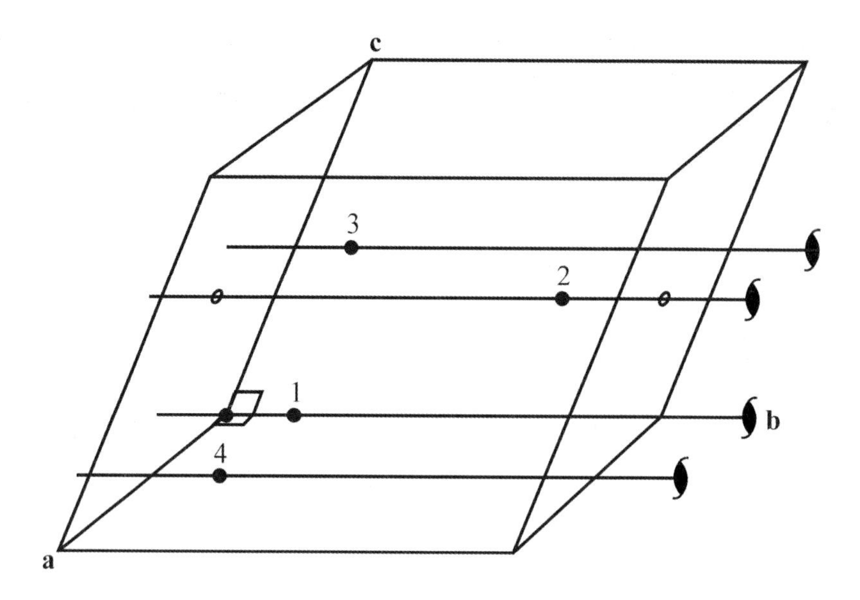

Figure 7.20 Permissible origins for a primitive non-centrosymmetric unit cell in the $P2_1$ space group. The origin must be located on one of the independent screw axes in the unit cell. The location on the axis in the direction parallel to the b axis (y) is arbitrary.

and oeo. The effects on the phase of each parity group for shifts to each origin are summarized in Table 7.2. Reflections with *eee* parity are structure semi-invariants.

Table 7.2 Phase shifts resulting from changes in the scattering reference location in a $P2$ or $P2_1$ unit cell for $h0l$ parity groups.

	Parity Group Phase Shift			
Origin	eee	oee	eeo	oeo
(1) $0,\ y,\ 0$	0	0	0	0
(2) $\frac{1}{2}, y, \frac{1}{2}$	0	π	π	0
(3) $0,\ y,\ \frac{1}{2}$	0	0	π	π
(4) $\frac{1}{2}, y, 0$	0	π	0	π

The parity vectors for oee, eeo, and oeo are [1 0 0], [0 0 1] and [1 0 1], respectively. Any two of these vectors are linearly independent, and can be used to select the axis upon which the origin resides, but a third vector from the remaining parity group will be a linear combination of the other two vectors, and therefore will be unable to establish the location of the scattering reference along that axis. Thus, for example, assigning a phase of 0 to the $(3\ 0\ \bar{2})$ reflection and π to the $(\bar{4}\ 0\ 1)$ reflection will locate the scattering reference on the screw axis bisecting the c axis in the unit cell. In order to fix the position of the scattering reference at a specific location along the axis, a third reflection with a linearly independent parity vector must be assigned a phase. The phase in this case will be arbitrary, serving to establish the position of the scattering reference along the axis. If the phase of this reflection is set to zero it will place the reference on the c axis. The reflection planes for the third reflection must cross the b axis only once to avoid multiple scattering references in the same unit cell, and must therefore be a member of the set of reflections with $k = 1$, e.g., $(5\ 1\ \bar{3})$.

Before turning to the exercise of phase determination, we must first resolve another dilemma presented by non-centrosymmetric structures, one that we have already encountered in the previous chapter. Structures without a center of symmetry can exist in one of two enantiomorphic forms. The asymmetric units of the two unit cells are mirror images of one another (enantiomers), and the coordinates of the atoms in one unit cell are the negatives of the atom coordinates in the other. This is illustrated in Fig. 7.21 for the D and L forms of glucose. Except for small differences due to anomalous dispersion, $F_{\mathbf{h}} = F_{\bar{\mathbf{h}}}$, and the intensity data cannot discriminate between the two structures. For the D enantiomer,

$$\mathbf{F_h}^D = \sum_{j=1}^{n} f_j e^{-2\pi i\,(\mathbf{h}\cdot\mathbf{r})} = F_h e^{i\varphi_{\mathbf{h}}}, \tag{7.151}$$

and for the L enantiomer,

$$\mathbf{F_h}^L = \sum_{j=1}^{n} f_j e^{-2\pi i\,(-\mathbf{h}\cdot\mathbf{r})} = F_h e^{-i\varphi_{\mathbf{h}}}. \tag{7.152}$$

If the assignment of phases is initiated without further restraints, then subsequent triplet relationships will be consistent with either enantiomorph. The result will be a mixture of structure factors phased for both enantiomorphs and a structure containing both enantiomers, often overlapping, and very difficult to interpret in a Fourier map. The origin-defining reflections can also suffer this fate.

In order to get around this we must assign a phase to a fourth reflection — a phase that we do not know — in order to *fix the enantiomorph.* If the selection is made appropriately then the triplet relations will remain consistent with the phase of this reflection and will not generate the enantiomorphic structure. Once a scattering reference has been selected it is possible to transfer it to the location of any one of the permissible origins, maintaining the symmetry and the relative phases of all the reflections. The actual phases for the origin-defining reflections will change, along with the phases of the remaining reflections, unless the phases are structure invariants or semi-invariants. These new phases must be consistent with the structure – one enantiomorph can't be allowed to switch to another simply because the scattering reference is changed. It follows that the enantiomorph will be fixed by the assignment of a fourth phase only if a shift in the origin will not serve to alter the enantiomorph selected by the assignment. A convenient way to guarantee that the origin-fixing reflections and the enantiomorph-fixing reflection do not simultaneously undergo phase changes to those for the enantiomorph when the scattering reference is changed, is to assign the phase of a structure invariant or semi-invariant, thus guaranteeing that it will not undergo any change when the origin shift occurs. For this space group the simplest invariant is the triplet. For simplicity, let us assume that phases of 0 have been assigned to the origin-defining

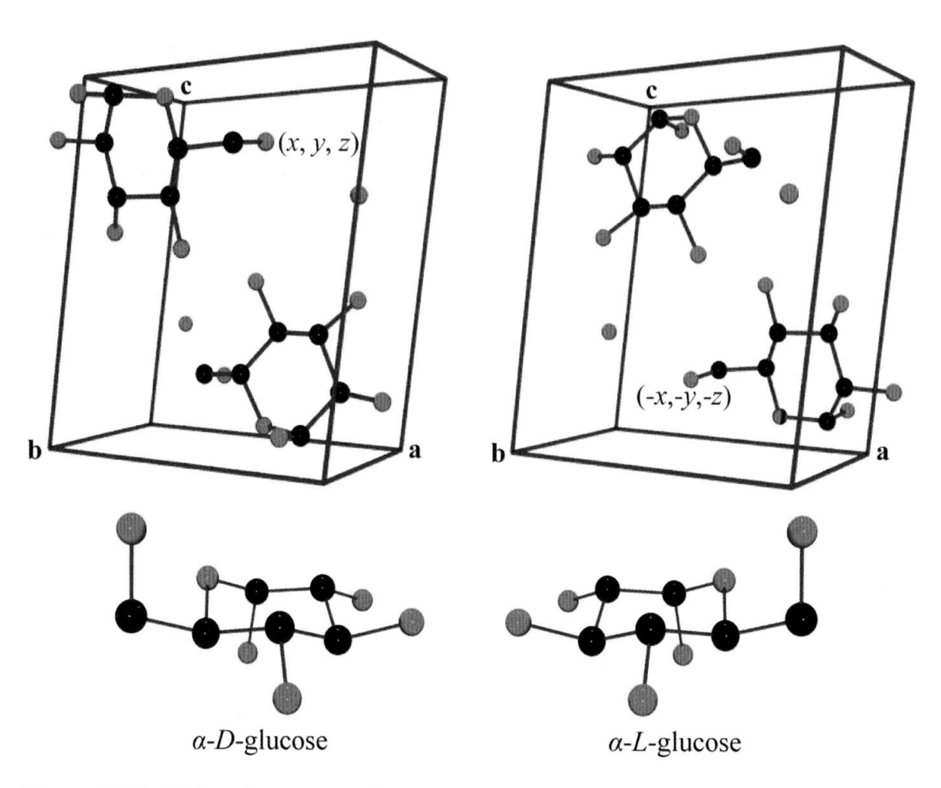

Figure 7.21 Unit cell contents of enantiomeric forms of α-glucose monohydrate in the $P2_1$ space group (hydrogen atoms not shown).

reflections $(3\ 0\ \bar{2})$, $(\bar{4}\ 0\ 1)$, and $(5\ 1\ \bar{3})$. Further, suppose that the $(1\ 1\ \bar{2})$ reflection is intense, so that the triplet relationship

$$\Phi = \varphi_{11\bar{2}} - \varphi_{51\bar{3}} - \varphi_{\bar{4}01} \approx 0 \tag{7.153}$$

is highly probable. Since all phases change signs when the enantiomorph selection is changed, all structure invariants and semi-invariants change sign. *Fixing the enantiomorph is tantamount to establishing the signs of the structure invariants and semi-invariants.* Assigning a value of zero to $\varphi_{11\bar{2}}$ would be consistent with its expected value, but would not serve to fix the enantiomorph, since $-\Phi = \Phi$; Φ must be restricted so that it remains positive (or negative). Because $\varphi_{11\bar{2}}$ is only approximately zero, it might appear that assigning a phase of a few degrees would be useful. However, a small phase angle will probably not be sufficient to ensure that the triplet relationships that arise from this reflection will not "cross over" to the other enantiomorph. As noted earlier, a single incorrect phase does not have a great effect on the tangent formula, and to some degree, we are free to assign a fourth phase arbitrarily. Values in increments of $\pi/4$ are often used; $\pi/4$ might be selected in this instance since it is closest to the expected value of the reflection. The value is allowed to change during the phase-refinement process, but is constrained (along with the value of Φ) to remain in the interval $0 < \varphi_{11\bar{2}} < \pi$. As the phase converges toward its expected value (probably close to zero), there will (hopefully) be enough additional phases established to maintain the chirality of the structure.

The assignment of a fourth phase with one sign will generate new phases consistent with the structure of one enantiomorph, while assignment of a phase with the same magnitude but with the opposite sign will generate new phases consistent with the other enantiomorph. Thus, the assignment of an arbitrary phase to a fourth reflection *fixes the enantiomorph, but it does not tell us if our choice is the correct one.* The determination of the correct enantiomorph must await the final refinement of the crystal structure. This is the same dilemma that we encountered in the solution of non-centrosymmetric structures using the Patterson function in the previous chapter. A further problem is encountered if the fourth reflection does not turn out to be involved in many triplet relationships. If this is the case, it is again possible that the phase of another reflection can be assigned a phase with the incorrect sign, creating phases for the other enantiomorph in the triplet relationships in which it is involved, again propagating the phases of the enantiomorphic structure.

Symmetry constraints do not always result in phases of 0 or π for reflection classes. For small molecules, the most common non-centrosymmetric space group is $P2_12_12_1$, which we will use as an example (the treatment will be general for any member of the 222 point group). Again, we begin with a two-dimensional analog of the space group, with two-fold screw axes parallel to the a and b axes as illustrated in Fig. 7.22. These screw axes cross the perpendicular unit cell axes at $x = 1/4$; $x = 3/4$ and $y = 1/4$; $y = 3/4$, respectively. A general point at (x, y) is transformed to $(1/2 - x, y)$ with a rotation about an axis parallel to $\mathbf{b}$, and to $(x, 1/2 - y)$ with a rotation about an axis parallel to $\mathbf{a}$. In order for the origin to be symmetrically disposed with respect to both axes it must lie at the midpoint of the rectangle formed by the points of interjection of all four screw axes in the unit cell. There are therefore four permissible origins in the unit cell, identified by the numbers 1–4 in the figure. Note that these are at the same locations as they are in the two-dimensional centrosymmetric unit cell in Fig. 7.17. The symmetry operator for the two-fold screw operation parallel to $\mathbf{b}$ is

$$\mathbf{S}(2_1) = \mathbf{R}(2) + \mathbf{t} = \begin{bmatrix} -1 & 0 \\ 0 & 1 \end{bmatrix} + \begin{bmatrix} \frac{1}{2} \\ \frac{1}{2} \end{bmatrix}. \tag{7.154}$$

For the $h0$ class of reflections $\mathbf{h}^T\mathbf{R}(2) = -\mathbf{h}^T$, and the allowed phases for this class of reflections are

$$\varphi_{\mathbf{h}} = \pi\,\mathbf{h}\cdot\mathbf{t} - m\pi = \pi\frac{h}{2} - m\pi = \pi\left(\frac{h - 2m}{2}\right). \tag{7.155}$$

If h is even $(h = 2n)$, then

$$\varphi_{\mathbf{h}} = \pi\left(\frac{2n - 2m}{2}\right) = (n - m)\pi. \tag{7.156}$$

If $n - m$ is even $(n - m = 2m')$, then $\varphi_{\mathbf{h}} \equiv 2m'\pi \equiv 0$; if $n - m$ is odd, then $\varphi_{\mathbf{h}} \equiv (2m' + 1)\pi \equiv \pi$. Thus, for h even, $\varphi_{\mathbf{h}}$ is allowed by symmetry to have a value of 0 or π. If h is odd $(h = 2n + 1)$, then

$$\varphi_{\mathbf{h}} = \pi\left(\frac{2n + 1 - 2m}{2}\right) = m'\frac{\pi}{2}, \tag{7.157}$$

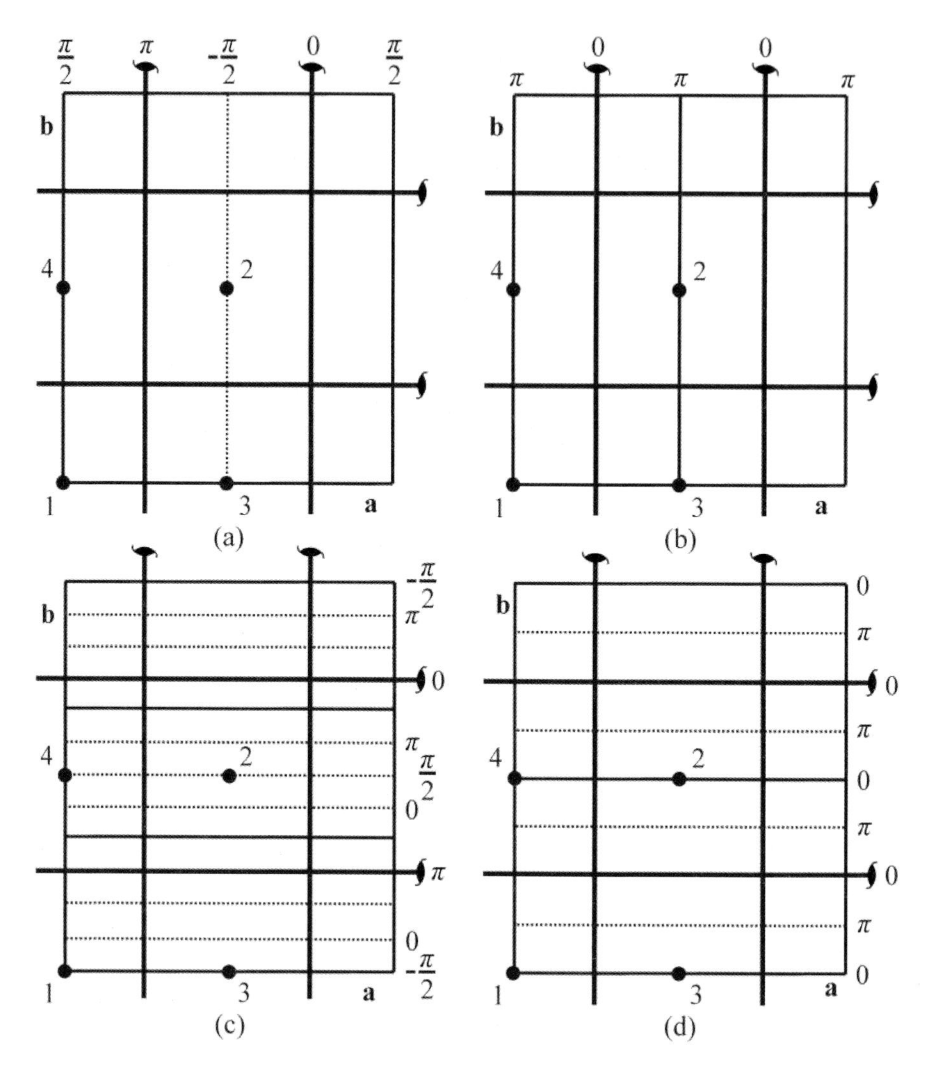

Figure 7.22 Lattice lines, phases, and potential scattering references in a two-dimensional unit cell with two-fold rotational symmetry in directions parallel to the a and b axes for (a) $(h\ k) = (1\ 0)$, (b) $(h\ k) = (2\ 0)$, (c) $(h\ k) = (0\ 3)$, and $(h\ k) = (0\ 4)$. 0 phase is the location of the density wave maximum; π is the location of the density wave minimum.

where m' is an odd integer. Thus, for h odd, $\varphi_{\mathbf{h}}$ is allowed by symmetry to have a value of $\pi/2$ or $3\pi/2 \equiv -\pi/2$. The symmetry operator for the two-fold screw operation parallel to $\mathbf{b}$ is

$$\mathbf{S}(2_1) = \mathbf{R}(2) + \mathbf{t} = \begin{bmatrix} 1 & 0 \\ 0 & -1 \end{bmatrix} + \begin{bmatrix} \frac{1}{2} \\ \frac{1}{2} \end{bmatrix}, \tag{7.158}$$

leading to the same results for k for the $0k$ class of reflections. Fig. 7.22(a) depicts the lattice lines for the (1 0) reflection, with a phase of $\pi/2$. A shift to origin 2 or 3

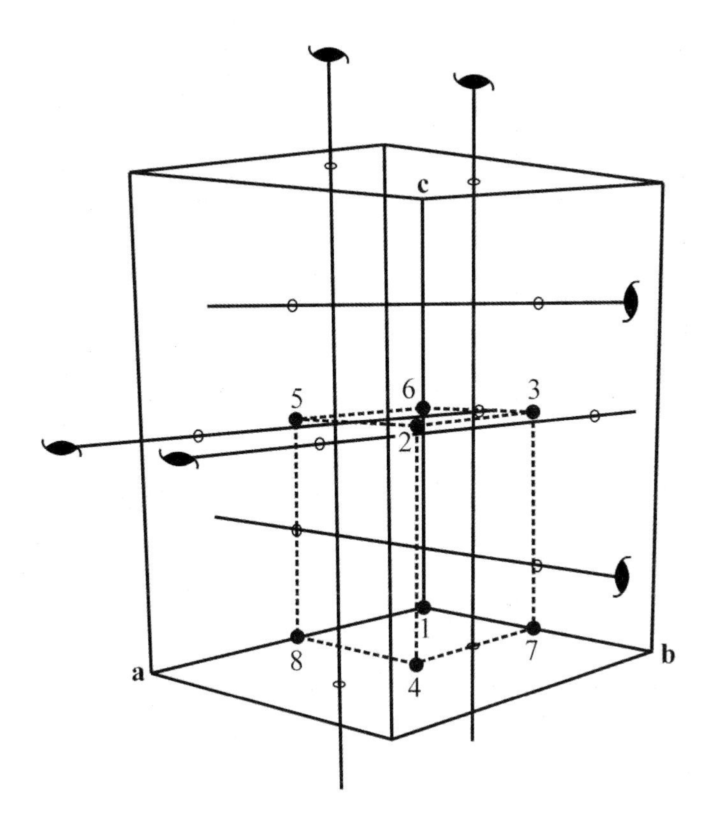

Figure 7.23 Permissible origins for a primitive non-centrosymmetric unit cell in the $P2_12_12_1$ space group.

will result in a phase change of π for this reflection, to $-\pi/2$. The (2 0) reflection, illustrated in Fig. 7.22(b), is a structure semi-invariant; all four permissible origins lie on lattice lines, and there is no change in the phase of π resulting from an origin shift to any of the origins. The (0 3) reflection, shown in Fig. 7.22(c), experiences a phase shift of π, from $-\pi/2$ to $\pi/2$ when the scattering reference is located on origins 2 or 4. Fig. 7.22(d) indicates that the (0 4) reflection is also a structure semi-invariant. Assigning phases of $\pm\pi/2$ to the reflections with oe and eo parity will therefore serve to define the origin.

The permissible origins of the $P2_12_12_1$ unit cell are shown in Fig. 7.23. They are at the intersections of lines midway between the screw axes, at the same fractional coordinates as those of the centrosymmetric unit cell (Fig. 7.18). There are now three sets of screw axes, parallel to each of the unit cell axes. The three dimensional symmetry operator for the screw operation parallel to $\mathbf{c}$ is

$$\mathbf{S}(2_1) = \mathbf{R}(2) + \mathbf{t} = \begin{bmatrix} -1 & 0 & 0 \\ 0 & -1 & 0 \\ 0 & 0 & 1 \end{bmatrix} + \begin{bmatrix} \frac{1}{2} \\ 0 \\ \frac{1}{2} \end{bmatrix}. \tag{7.159}$$

For the $hk0$ class of reflections, $\mathbf{h}^T\mathbf{R}(2) = -\mathbf{h}^T$ and the results are identical to those for the two dimensional case (Eqns. 7.155–7.157). For h even, $\varphi_{\mathbf{h}} = 0$ or π; for h odd, $\varphi_{\mathbf{h}} = \pm\pi/2$. For the screw operation parallel to $\mathbf{a}$, the phases of the $0kl$ reflections are restricted to $\varphi_{\mathbf{h}} = 0$ or π for k even and $\varphi_{\mathbf{h}} = \pm\pi/2$ for k odd. For the screw operation parallel to $\mathbf{b}$, the phases of the $h0l$ reflections are restricted to $\varphi_{\mathbf{h}} = 0$ or π for l even and $\varphi_{\mathbf{h}} = \pm\pi/2$ for l odd. In order to determine which combinations of these reflections to use to define the origin, we need only to determine which origin shifts render phase shifts, and which do not. The phase shifts are due solely to the locations of the permissible origins in the unit cell and the parity of the indices for each reflection — the result is identical to the centrosymmetric case, since the permissible origins are at the same locations. Thus Table 7.1 serves the purpose of selecting origin-defining reflections for the $P2_12_12_1$ space group, as well as other members of the 222 point group.

The enantiomorph is readily fixed in this space group as it was in the $P2_1$, by assigning a phase to a general reflection that will constrain the sign of a triplet invariant. The parity vectors of the $\mathbf{h}_1 = (5\ 2\ 0)$, $\mathbf{h}_2 = (3\ 1\ 0)$, and $\mathbf{h}_3 = (1\ 0\ 5)$ reflections are linearly independent,

$$\begin{vmatrix} 1 & 1 & 1 \\ 0 & 1 & 0 \\ 0 & 0 & 1 \end{vmatrix} = 1, \tag{7.160}$$

and serve to define the origin by assigning $\varphi_{\mathbf{h}_1} = \pi/2$, $\varphi_{\mathbf{h}_2} = -\pi/2$, and $\varphi_{\mathbf{h}_3} = \pi/2$. The sign of the structure invariant

$$\Phi = \varphi_{625} - \varphi_{520} - \varphi_{105} \approx 0 \tag{7.161}$$

can be set positive to fix the enantiomorph by

$$\begin{aligned} &0 < \left(\varphi_{625} - \frac{\pi}{2} - \frac{\pi}{2}\right) < \pi; \\ &\pi < \varphi_{625} < 2\pi. \end{aligned} \tag{7.162}$$

The expected value of φ_{625} is π; a reasonable initial assignment would be $\varphi_{625} = 5\pi/4$.

For space groups such as $P2_12_12_1$ — space groups that have reflection classes with phases constrained by symmetry — it is also possible to fix the enantiomorph unequivocally by assigning the phase of an appropriate reflection in one of those classes. Since the phase of this reflection will not change, it will maintain the enantiomorph choice as new triplet relationships come in to play, provided that it is involved in enough of them. These reflections are not generally structure semi-invariants, and we must be certain that they do not switch enantiomorphs when an origin shift occurs. According to Table 7.1, a shift to origin 8 results in phase shifts of π for each of the origin-defining reflections listed above. If the enantiomorph-fixing reflection also undergoes a phase shift of π when the origin is shifted, then the phases of all four reflections will have changed sign — the origin shift will have created phases consistent with the other enantiomorph. It is therefore necessary to select a reflection with a phase that will not change sign when the scattering reference is moved to origin 8. If $\mathbf{h}_4 = (0\ 3\ 7)$, with $\varphi_{\mathbf{h}_4} = -\pi/2$, then a shift to origin 8 will result in phases of $\varphi_{\mathbf{h}_1} = -\pi/2$, $\varphi_{\mathbf{h}_2} = \pi/2$, $\varphi_{\mathbf{h}_3} = -\pi/2$, and

$\varphi_{\mathbf{h}_4} = -\pi/2$. The reflection is suitable for fixing the enantiomorph. On the other hand, if $\mathbf{h}_4 = (5\ 0\ 7)$, with $\varphi_{\mathbf{h}_4} = -\pi/2$, then a phase shift to origin 8 will result in $\varphi_{\mathbf{h}_1} = -\pi/2$, $\varphi_{\mathbf{h}_2} = \pi/2$, $\varphi_{\mathbf{h}_3} = -\pi/2$, and $\varphi_{\mathbf{h}_4} = \pi/2$. The reflection cannot be used to fix the enantiomorph.

As the space group symmetry relationships become more complex, so do the criteria for defining the origin and fixing the enantiomorph. Fortunately, these criteria have been derived systematically by Hauptman and Karle.[128–131] Their results are summarized in the *International Tables for Crystallography.* More detailed discussions can be found in books by Giacovazzo[132] and Ladd and Palmer[133].

7.1.3 Probability Methods: Solving the Structure

Once the phases of the origin- and enantiomorph-defining reflections are assigned, the strategy for structural solution now appears to be straightforward. A triplet relationship consisting of two assigned phases and one unknown phase is used to determine the unknown phase, which, in turn, is used to generate additional phases from other triplet relationships. These new phases provide even more new phases; the process continues sequentially until all of the phases have been assigned. Structural solution would therefore appear to be a relatively trivial task.

Unfortunately, as we have discussed at length, any given triplet relationship will provide only an approximate phase, with a variance depending on the value of $X_{\mathbf{hk}} = 2\sigma_3/(\sigma_2)^{3/2}E_{\mathbf{h}}E_{\mathbf{k}}E_{\mathbf{h-k}}$. We can improve our chances of making correct assignments by beginning with triplets with the largest values of $X_{\mathbf{hk}}$, but it takes only one incorrect phase assignment in the sequence to generate incorrect phases from all of the triplet relationships involving the incorrect phase. Unless the incorrect assignment(s) occur relatively late in the process, the likelihood that a structural solution will result becomes prohibitively small.

Phase Determination: Symbolic Addition

In 1966 Jerome and Isabella Karle published a breakthrough method for phase determination[126] that delayed the assignment of numerical phases until enough triplet relationships had been established so that any one incorrect assignment would be unlikely to lead to a large number of incorrect phases. In this method, rather than computing numerical values for initial phases, phases for a few reflections with large values of $E_{\mathbf{h}}$ are treated as variables. The symbols assigned to these variables are carried through the phase assignment process with new phases becoming linear combinations of the symbols – *symbolic addition.* At some stage in the process the relationships between these linear combinations are used to determine values for the symbols, providing the numerical phases necessary to define the structure.

The crystal structure of the mineral bikitaite, $Li[AlSi_2O_6]\cdot H_2O$, was solved with the symbolic addition procedure in the early mid-1970's by Kocman, Gait, and Rucklidge[134]. It was common practice at that time for structural papers to include a table of the observed and calculated structure factors for a crystal structure, along with the refined phases. The reader will find it a useful and informative exercise to access these data and attempt to mimic the structural solution,* an outline of which will be given here.

*Although the "manual" solution of a crystal structural is rarely (if ever) necessary with the availability of today's powerful direct methods software, the essential relationships between phases

Table 7.3 180 reflections with the largest values of E_{hkl} for $Li[AlSi_2O_6]{\cdot}H_2O$.

h	k	l	E	h	k	l	E	h	k	l	E	h	k	l	E	h	k	l	E
$\bar{6}$	0	2	2.40	4	4	0	1.80	8	2	0	1.59	$\bar{6}$	3	7	1.48	$\bar{1}$	2	5	1.39
$\bar{5}$	2	10	2.40	$\bar{8}$	1	5	1.79	$\bar{7}$	4	6	1.59	5	5	3	1.47	2	3	5	1.38
$\bar{5}$	0	5	2.39	$\bar{5}$	6	4	1.76	$\bar{6}$	0	4	1.58	$\bar{8}$	1	7	1.46	$\bar{2}$	1	6	1.38
7	1	0	2.38	$\bar{6}$	6	3	1.76	$\bar{1}$	0	5	1.58	$\overline{10}$	1	8	1.46	$\bar{5}$	4	8	1.38
$\bar{2}$	0	4	2.38	$\bar{8}$	5	4	1.75	$\bar{9}$	4	3	1.58	7	5	1	1.46	0	4	1	1.37
0	6	1	2.34	4	2	2	1.73	1	1	3	1.58	$\bar{1}$	6	4	1.46	$\bar{2}$	2	2	1.37
$\bar{9}$	2	3	2.31	$\bar{2}$	1	8	1.73	11	0	0	1.58	$\bar{9}$	3	4	1.45	10	2	0	1.37
$\overline{11}$	0	7	2.28	$\bar{3}$	2	9	1.72	6	2	3	1.57	$\bar{4}$	4	1	1.45	6	1	2	1.37
$\bar{7}$	0	7	2.19	1	2	8	1.72	$\bar{3}$	2	6	1.57	2	2	4	1.45	6	3	2	1.35
4	0	2	2.13	$\bar{9}$	2	8	1.72	1	3	0	1.56	$\bar{5}$	2	3	1.45	$\bar{2}$	6	3	1.35
$\bar{8}$	1	10	2.12	4	2	7	1.71	$\bar{3}$	1	7	1.56	3	1	5	1.45	$\bar{9}$	2	4	1.35
$\bar{9}$	0	3	2.10	1	1	0	1.70	6	0	3	1.56	$\overline{11}$	0	6	1.45	$\bar{2}$	6	2	1.35
$\bar{1}$	0	10	2.09	$\bar{3}$	1	1	1.70	5	1	2	1.55	8	0	2	1.45	$\bar{1}$	0	7	1.34
0	0	2	2.05	$\bar{7}$	4	7	1.70	5	0	2	1.54	$\bar{2}$	2	4	1.44	5	0	0	1.34
$\bar{5}$	0	3	2.04	$\bar{8}$	3	5	1.69	$\overline{10}$	1	7	1.54	5	5	2	1.44	2	1	0	1.33
$\bar{5}$	2	5	2.02	$\overline{11}$	2	5	1.68	$\bar{2}$	1	9	1.54	$\bar{9}$	4	4	1.44	$\bar{2}$	5	7	1.33
$\bar{2}$	0	2	2.00	3	5	4	1.67	$\bar{8}$	1	8	1.54	$\bar{7}$	1	5	1.43	5	1	3	1.33
$\bar{7}$	2	7	1.98	$\bar{3}$	5	6	1.67	$\bar{8}$	0	5	1.53	$\overline{10}$	2	1	1.43	$\bar{3}$	4	6	1.33
$\bar{1}$	4	4	1.97	$\bar{4}$	1	10	1.66	$\bar{4}$	1	5	1.53	1	0	8	1.43	8	3	1	1.33
4	0	7	1.96	$\bar{6}$	1	10	1.65	2	1	1	1.53	$\bar{7}$	4	8	1.43	$\bar{6}$	3	2	1.32
$\bar{5}$	0	10	1.96	$\bar{1}$	5	4	1.65	$\bar{6}$	2	2	1.53	1	1	7	1.43	$\bar{3}$	1	5	1.32
$\bar{3}$	3	2	1.94	$\bar{6}$	5	2	1.65	$\bar{6}$	6	1	1.53	7	3	0	1.42	$\bar{4}$	5	6	1.32
4	4	1	1.92	7	1	5	1.63	$\bar{9}$	3	2	1.53	$\bar{1}$	4	6	1.42	$\bar{5}$	1	3	1.32
$\bar{9}$	0	8	1.92	4	2	0	1.63	8	1	3	1.52	3	1	7	1.42	$\bar{5}$	4	2	1.31
5	0	6	1.92	4	0	0	1.62	2	4	4	1.52	$\bar{2}$	4	3	1.42	$\bar{9}$	2	1	1.31
$\bar{3}$	3	1	1.91	2	1	5	1.62	$\bar{3}$	5	1	1.52	0	0	7	1.41	1	6	2	1.31
1	0	3	1.90	6	0	0	1.62	5	5	1	1.51	0	0	5	1.41	$\bar{4}$	1	3	1.31
3	3	0	1.90	2	1	3	1.62	$\bar{5}$	3	4	1.51	6	1	1	1.41	7	1	2	1.31
$\bar{8}$	1	3	1.89	$\bar{8}$	3	4	1.62	$\bar{1}$	4	5	1.51	$\bar{7}$	5	3	1.41	5	4	3	1.30
0	2	0	1.87	$\bar{7}$	1	2	1.61	6	0	5	1.51	$\bar{5}$	2	9	1.41	6	5	0	1.30
$\bar{8}$	3	6	1.87	$\bar{3}$	0	9	1.61	$\bar{4}$	3	5	1.50	$\bar{9}$	2	5	1.41	0	4	6	1.30
$\overline{11}$	0	5	1.82	$\bar{5}$	2	8	1.60	$\bar{1}$	1	5	1.50	$\bar{5}$	4	4	1.41	$\bar{9}$	3	3	1.30
$\overline{10}$	2	2	1.81	2	3	0	1.59	0	2	5	1.49	$\bar{5}$	4	6	1.40	$\bar{3}$	2	1	1.30
8	2	2	1.81	$\bar{6}$	4	3	1.59	$\bar{6}$	1	8	1.48	8	2	1	1.40	1	5	3	1.30
$\bar{2}$	1	10	1.80	$\bar{2}$	1	1	1.59	$\bar{7}$	5	1	1.48	2	3	1	1.39	2	3	7	1.30
4	6	1	1.80	4	4	2	1.59	$\bar{8}$	3	7	1.48	2	5	0	1.39	2	5	2	1.30

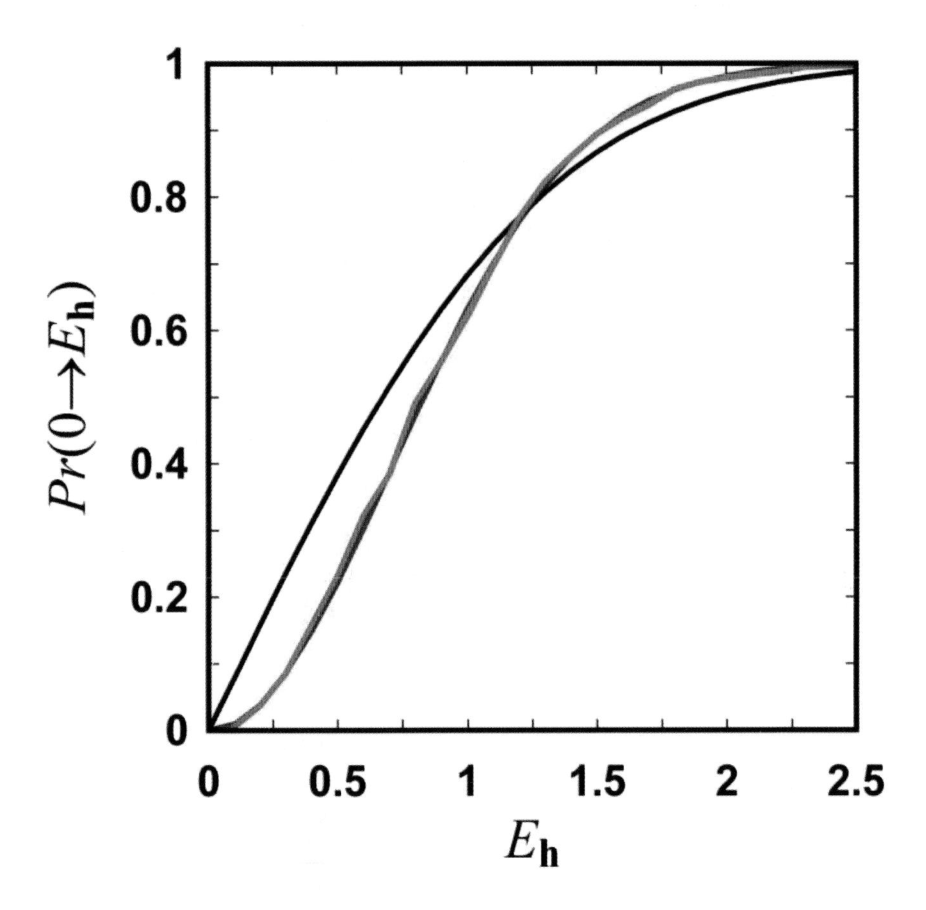

Figure 7.24 Cumulative probability distributions for an ideal acentric probability distribution of $E_{\mathbf{h}} = t$, $1 - e^{-E^2_{\mathbf{h}}}$ (blue), and an ideal centric probability distribution of $E_{\mathbf{h}} = t$, $\mathrm{erf}(E_{\mathbf{h}}/\sqrt{2})$ (black); the fraction of the total number of $E_{\mathbf{h}}$s less than or equal to each specific $E_{\mathbf{h}}$ on the x axis for $Li[AlSi_2O_6]{\cdot}H_2O$ is shown in red.

The mineral from Bikita, Rhodesia, crystallizes in a monoclinic unit cell with cell parameters $a = 8.613(4)$, $b = 4.962(2)$, $c = 7.600(4)$, and $\beta = 111.45(1)°$. 947 reflections exhibited systematic absences for the $0k0$ reflections for odd k, consistent with the non-centrosymmetric space group $P2_1$ and the centrosymmetric space group $P2_1/m$. The initial step in all direct methods strategies is the generation of the quasi-normalized structure factor magnitudes from the scaled values of the observed structure factor amplitudes. The cumulative probability distribution of these magnitudes for the bikitaite data set are shown in Fig. 7.24, strongly indicating the $P2_1$ space group (cf. Fig. 5.40).

Triplet relationships with the largest values of $E_{\mathbf{h}}E_{\mathbf{k}}E_{\mathbf{h-k}}$ have the highest probability of producing correct phases. Table 7.3 lists the 180 largest E_{hkl} values – the values used for the structural determination. Structural solution begins with

that determine the structure, as well as the sources of potential errors, are readily observed with the determination of symbolic phases.

a list of all possible triplet relationships for these reflections, accumulated into a list commonly referred to as a Σ_2 list.*

Although there are only 180 E_{hkl} values in the reflection list, there are three additional symmetry-related reflections for each value of E_{hkl} for which we can determine the phase, based on Eqn. 7.139. Since the lattice is monoclinic, $E_{hkl} = E_{h\bar{k}l} = E_{\bar{h}\bar{k}\bar{l}} = E_{\bar{h}k\bar{l}}$. Expanding the reflection list to include these reflections provides over 2000 triplet relationships.

The phases of the symmetry-related reflections depend on the parity of the k index. The operator for the screw operation parallel to the b axis in the $P2_1$ space group is

$$\mathbf{S}(2_1) = \mathbf{R}(2) + \mathbf{t} = \begin{bmatrix} -1 & 0 & 0 \\ 0 & 1 & 0 \\ 0 & 0 & -1 \end{bmatrix} + \begin{bmatrix} 0 \\ \frac{1}{2} \\ 0 \end{bmatrix}, \tag{7.163}$$

giving

$$\varphi_{hkl} + 2m\pi = \varphi_{\bar{h}k\bar{l}} - 2\pi\left(\frac{k}{2}\right) \quad \text{and} \tag{7.164}$$

$$\varphi_{\bar{h}k\bar{l}} = \varphi_{hkl} + (2m+k)\pi. \tag{7.165}$$

Since $\varphi_{\bar{h}\bar{k}\bar{l}} = -\varphi_{hkl}$,

$$\varphi_{h\bar{k}l} = -\varphi_{hkl} - (2m+k)\pi. \tag{7.166}$$

For k even ($k = 2n$), $(2m+k)\pi \equiv 0$ and

$$\varphi_{hkl} = \varphi_{\bar{h}k\bar{l}} = -\varphi_{\bar{h}\bar{k}\bar{l}} = -\varphi_{h\bar{k}l}. \tag{7.167}$$

For k odd ($k = 2n+1$), $(2m+k)\pi \equiv \pi$ and

$$\varphi_{hkl} = \pi + \varphi_{\bar{h}k\bar{l}} = -\varphi_{\bar{h}\bar{k}\bar{l}} = \pi - \varphi_{h\bar{k}l}. \tag{7.168}$$

The $h0l$ reflections are a special class of reflections in the $P2_1$ space group:

$$\begin{aligned} \varphi_{h0l} + 2m\pi &= \varphi_{\bar{h}0\bar{l}} - 2\pi\left(\frac{0}{2}\right) \\ &= -\varphi_{h0l} \\ 2\varphi_{h0l} &= -2\pi\, m = 2\pi\, m \quad \text{and} \\ \varphi_{h0l} &= m\,\pi. \end{aligned} \tag{7.169}$$

For m even, $\varphi_{h0l} = 0$, and for m odd, $\varphi_{h0l} = \pi$; the phases of the $h0l$ reflections are constrained to values of 0 or π, and are often termed *centric reflections.*

*The first monograph on direct methods by Hauptmann and Karle[128] focused specifically on centrosymmetric structures. The authors showed that the sign of $E_{\mathbf{h}}$ was the sign of a series of sums: $S_{\mathbf{h}} = \Sigma_1 + \Sigma_2 + \Sigma_3 + \Sigma_4$. The Σ_2 contribution resulted from the triplet relationship for centrosymmetric structures (Eqn. 7.32): $S_{\mathbf{h}} \approx \sum_i S_{\mathbf{k}_i} S_{\mathbf{h}-\mathbf{k}_i}$, and triplet relations in general became known as "Σ_2 relationships."

Assignment of symbolic phases. In order to optimize the chances for success, the triplet list is scanned for reflections that occur in a large number of relationships for which $E_{\mathbf{h}}E_{\mathbf{k}}E_{\mathbf{h-k}}$ is also large. Three of these reflections are selected as origin-defining reflections and a small number of reflections, in this case three, are assigned symbolic phases:[†]

h	k	l	E_{hkl}	φ_{hkl}
$\bar{5}$	0	5	2.39	0
4	0	7	1.97	0
7	1	5	1.63	0
$\bar{6}$	0	2	2.40	s_1
$\bar{3}$	3	1	1.91	s_2
$\bar{5}$	2	10	2.40	s_3

This *starting set* of reflections is now used to generate new symbolic phases by selecting triplet relationships with large values of $X_{\mathbf{hk}} = E_{\mathbf{h}}E_{\mathbf{k}}E_{\mathbf{h-k}}$ containing pairs of these reflections. Initial assignments are generally restricted to the most probable triplets, those with $X_{\mathbf{hk}}$ greater than some threshold value, in this case $X_{\mathbf{hk}} >= 4.2$. A few examples will serve to illustrate the process:

(1)

h	k	l	π	s_1	s_2	s_3
$\bar{5}$	0	5	0	0	0	0
$\bar{6}$	0	2	0	1	0	0
1	0	3		φ_{103}		
1	0	3	0	−1	0	0

(2)

h	k	l	π	s_1	s_2	s_3
$\bar{5}$	0	5	0	0	0	0
3	$\bar{3}$	$\bar{1}$	0	0	−1	0
$\bar{8}$	3	6		$\varphi_{\bar{8}36}$		
$\bar{8}$	3	6	0	0	1	0

(3)

h	k	l	π	s_1	s_2	s_3
$\bar{5}$	0	5	0	0	0	0
$\bar{5}$	2	10	0	0	0	1
0	$\bar{2}$	$\bar{5}$		$-\varphi_{025}$		
0	2	5	0	0	0	1

(4)

h	k	l	π	s_1	s_2	s_3
$\bar{6}$	0	2	0	1	0	0
$\bar{3}$	3	1	0	0	1	0
$\bar{3}$	$\bar{3}$	1	1	0	−1	0
$s_1 = \pi$						

(5)

h	k	l	π	s_1	s_2	s_3
$\bar{5}$	0	5	0	0	0	0
6	0	$\bar{2}$	0	−1	0	0
$\overline{11}$	0	7		$\varphi_{\overline{11}07}$		
$\overline{11}$	0	7	0	1	0	0

(6)

h	k	l	π	s_1	s_2	s_3
$\bar{6}$	0	2	0	1	0	0
$\bar{5}$	2	10	0	0	0	1
$\bar{1}$	$\bar{2}$	$\bar{8}$		$-\varphi_{128}$		
1	2	8	0	−1	0	1

(7)

h	k	l	π	s_1	s_2	s_3
$\bar{5}$	2	10	0	0	0	1
$\bar{3}$	3	1	0	0	1	0
$\bar{2}$	$\bar{1}$	9		$\pi - \varphi_{\bar{2}19}$		
$\bar{2}$	1	9	1	0	1	−1

(8)

h	k	l	π	s_1	s_2	s_3
$\overline{11}$	0	7	0	1	0	0
$\bar{3}$	3	1	0	0	1	0
$\bar{8}$	$\bar{3}$	6	1	0	−1	0
$s_1 = \pi$						

(9)

h	k	l	π	s_1	s_2	s_3
4	0	7	0	0	0	0
$\bar{5}$	2	10	0	0	0	1
9	$\bar{2}$	$\bar{3}$		$-\varphi_{\bar{9}23}$		
$\bar{9}$	2	3	0	0	0	1

(10)

h	k	l	π	s_1	s_2	s_3
$\bar{5}$	0	5	0	0	0	0
$\bar{9}$	2	3	0	0	0	1
4	$\bar{2}$	2		$-\varphi_{422}$		
4	2	2	0	0	0	1

(11)

h	k	l	π	s_1	s_2	s_3
$\bar{5}$	2	10	0	0	0	1
4	2	2	0	0	0	1
$\bar{9}$	0	8		$\varphi_{\bar{9}08}$		
$\bar{9}$	0	8	0	0	0	0

(12)

h	k	l	π	s_1	s_2	s_3
$\bar{8}$	$\bar{3}$	6	1	0	−1	0
3	$\bar{3}$	$\bar{1}$	0	0	−1	0
$\overline{11}$	0	7		$\varphi_{\overline{11}07}$		
$\overline{11}$	0	7	1	0	0	0

In general, a symbolic phase is a sum of symbols of the form $\varphi_{hkl} = c_0 s_0 + c_1 s_1 + c_2 s_2 + c_3 s_3 + \cdots c_n s_n$, where s_0 is a constant, and $c_0 \cdots c_n$, are the integer multipliers for each of n symbols. In our case, $c_0 \cdots c_3$ are the multipliers for the constant, π,

[†]The values of E_{hkl} given here differ somewhat from those in the literature. Differences in scaling and scattering factor functions are largely responsible for the deviations.

and s_1, s_2, and s_3, listed in the columns for each triplet. The triplets are listed in the order **h**, **k**, and **h** − **k**. After all symbolic phase assignments for a given $X_{\mathbf{hk}}$ threshold have been made, the threshold is lowered and additional phases are generated, based on the more probable initial phase assignments, until each reflection in the "large E" subset has been assigned a symbolic phase.

Assignment of numerical phases. Numerical values for some or all of the symbols can often be determined from recurring relationships that appear during the symbolic addition process. For triplet (1), the assigned phases, $\varphi_{\bar{5}05} = 0$ and $\varphi_{\bar{6}02} = s_1$ result in the triplet relationship, $s_1 + \varphi_{103} \approx 0$ and $\varphi_{103} \approx 0\pi - 1s_1 + 0s_2 + 0s_3 = -s_1$. For triplet (2), $\varphi_{33\bar{1}} \approx -\varphi_{\bar{3}11} \approx -s_2$, and the triplet relationship $\varphi_{\bar{8}36} - s_2 \approx 0$ gives $\varphi_{\bar{8}36} \approx s_2$. Similarly, triplet (3) results in the assignment of s_3 to φ_{025}. Triplet (4) is a relationship with three assigned symbolic phases; since $\varphi_{3\bar{3}\bar{1}} = \pi - \varphi_{\bar{3}31}$, the triplet relationship is $\pi - s_2 + s_2 \approx s_1$, providing a strong indication that $s_1 \approx \pi$. This is consistent with the phase of the assigned $\bar{6}02$ reflection, which must have a value of either 0 or π. Triplets (5), (6), and (7) produce symbolic phases $\varphi_{\bar{1}107} \approx s_1$, $\varphi_{128} \approx s_3 - s_1$, and $\varphi_{\bar{2}19} \approx s_2 - s_3$, respectively. Triplet (8) contains three assigned phases with the same relationship as triplet (4), again indicating that $s_1 \approx \pi$. Triplets (9), (10) indicate that $\varphi_{\bar{9}23} \approx s_3$ and $\varphi_{422} \approx s_3$. φ_{422} provides the triplet relationship for (11): $\varphi_{\bar{9}08} + s_3 \approx s_3$, indicating that $\varphi_{\bar{9}08} \approx 0$. The triplet relationship for (12) is $\varphi_{\bar{1}107} - s_2 \approx \pi - s_2$, indicating that $\varphi_{\bar{1}107} \approx \pi$. For triplet (5) we assigned s_1 to $\varphi_{\bar{1}107}$, providing yet another indication that $s_1 \approx \pi$.

For most structures solved by the symbolic addition method, some or all of the relationships lead to more ambiguous conclusions. In our case, the assignment of numerical phases to s_2 and s_3 is not as straightforward as it is for s_1. These phases are not constrained to specific values, and relationships involving them indicate several possible values for each symbol. In such cases it is usually necessary to attempt to determine the correct structure by assigning a sequence of values for each symbol, ordinarily in phase increments of 45°, applying the values to the symbolic phases and testing each set of phases as a potential solution. The numerical values for each trial phase set are generally only rough approximations to the actual phases; those with large $X_{\mathbf{hk}}$ having a higher probability of being correct.

It is here that the tangent formula comes into play, with each phase refined from all of the triplet relationships in which it is involved. The process is iterative, since each new value of $\varphi_{\mathbf{h}}(new) = \beta_{\mathbf{h}}$ in Eqn. 7.107 alters all of the triplet relationships that contain it. After the phase of the first reflection has been determined from the tangent formula, the new phase is incorporated into any of the triplets further down in the list that include it. The same holds for the second reflection, and so forth. For a given reflection in the reflection list, the tangent formula improves the phases of reflections further down in the list, but has no effect on those above it. Thus it is necessary to apply the tangent formula repeatedly, with each cycle producing an improved set of phases. This *tangent refinement* generally proceeds until a cycle produces no changes – until the process has become self-consistent.

Determining the structure: The E map. One or more of the refined phase sets is a potential structural solution. In order to determine if a given phase set

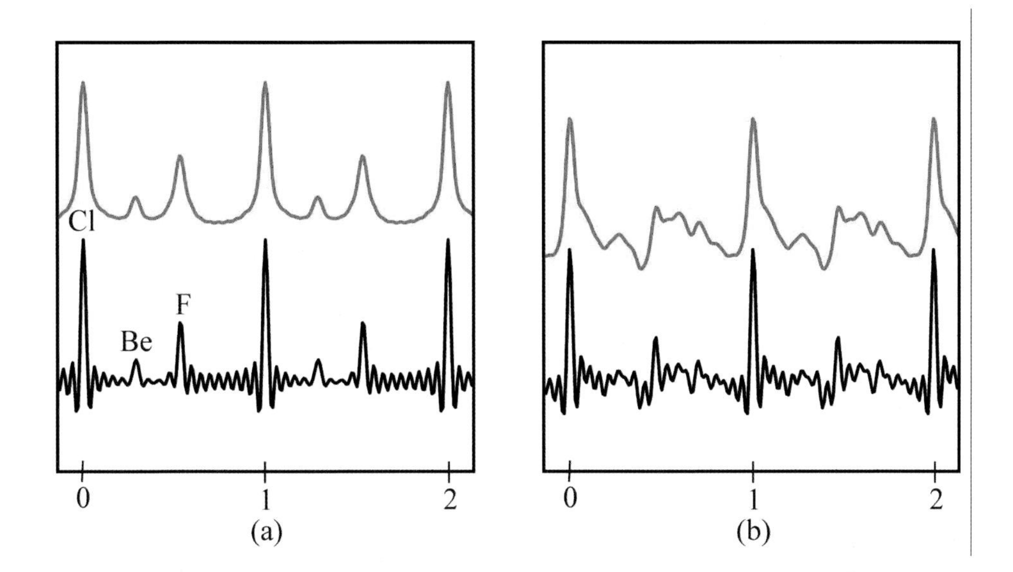

Figure 7.25 (a) F map for the hypothetical one-dimensional BeClF structure with correct phases (red) and the corresponding E map (black). (b) F map (red) and E map (black) for the BeClF structure with phase errors.

produces a reasonable structure, a Fourier map based on the observed structure factors (an F map) can be generated and searched for atomic positions:

$$\rho(\mathbf{r}) = \frac{1}{V_c} \sum_{\mathbf{h}} F_{o,\mathbf{h}}\, e^{i\varphi_{\mathbf{h}}}\, e^{\mathbf{h}\cdot\mathbf{r}}. \tag{7.170}$$

An F map for the hypothetical BeClF structure (Fig. 3.30) is shown in red in Fig. 7.25(a).* The peaks in F map are relatively broad, due in large part to the electron distributions and thermal motion of the atoms. If the phases are correct, the F map reveals the location of all of the atoms in the structure. However, the phases are only approximate; if the symbols are assigned in increments of $\pi/4$, then even the phase set for the correct structure can have phase errors in the neighborhood of this increment. An F map for the structure with errors in the phases is shown in red in Fig. 7.25(b).† The broad peaks tend to coalesce into one another, making it difficult to determine the location of the peak for either the fluorine atom or the beryllium atom.

If the atoms had their electrons located at their nuclei, and if the atoms were not in constant motion, we might expect that the peaks in the map would be sharper, perhaps making the effects of phase errors less pronounced. In Chapter 5 we developed the normalized structure factors that have been used throughout this chapter as approximations to structure factors that would result from such

*The origin has been moved to the center of the chlorine atom. Structure factors were generated for $h = -20$ to $h = 20$ with the scattering factor for atom $j = f_j(h)\, e^{-T_j}$; $T_j = 2\pi^2\,(0.1)\, h^2$.

†The errors were introduced by adding $r \cdot \pi/4$ to each phase, where r is a random number between -1 and 1.

hypothetical "point-atoms" at rest (Eqn. 5.503). A Fourier map based on these structure factors is known as an E map:

$$\epsilon(\mathbf{r}) = \sum_{\mathbf{h}} \left(\frac{F^2_{o,\mathbf{h}}}{\sum_j f_j^2(\mathbf{h})} \right)^{1/2} e^{i\varphi_{\mathbf{h}}} e^{\mathbf{h}\cdot\mathbf{r}} = \sum_{\mathbf{h}} E_{o,\mathbf{h}} e^{i\varphi_{\mathbf{h}}} e^{\mathbf{h}\cdot\mathbf{r}}. \tag{7.171}$$

One-dimensional E maps for the BeClF structure with correct phases and phase errors are shown in black in Figs. 7.25(a) and Fig. 7.25(b), respectively. The beryllium atom (with only three electrons) cannot be discerned in either map in Fig. 7.25(b), but the fluorine maximum is still resolved in the E map. The corresponding peak in the F map would be difficult to distinguish from its neighbors.

For the bikitaite structure, a trial phase set consisting of refined phases initially assigned with $s_1 = \pi$, $s_2 = \pi$ and $s_3 = \pi/2^*$ results in a three-dimensional E map that reveals the entire structure. After refinement, s_2 refines to 145.4° and s_3 refines to 79.6°. Both of these values deviate from their assigned values to an extent where their linear combinations produce ambiguous symbolic phases when they are first generated from the triplet relationships. For example, as the errors propagate, a number of $h0l$ reflections. which are required to have phases of 0 or π, are assigned phases that would equal $\pi/2$, based on the initial assignments of s_2 and s_3. Such ambiguities are common with the "manual" approach to symbolic addition. Success in applying the procedure depends critically on the selection of origin-defining reflections and the reflections that are assigned the initial symbolic phases. While these reflections are ordinarily those with large E values that appear in a substantial number of triplets, their specific selection is subjective. Although a significant number of structures were solved by the method, many were not, and investigators sought more robust approaches to direct structure determination. Two separate approaches arose from these attempts, both designed to turn the major decisions in the structural solution over to the computer. The first approach was the *multi-solution* method, treated later in the chapter. This method was remarkably successful, but its utility was often limited by the small memory and slow speed of the computers of the day (circa 1970), and the symbolic addition method was revisited as a computationally efficient approach to structural determination.

Determination of the starting set: Convergence. One of the principals in the search for an automated symbolic addition procedure, Henk Schenk, in collaboration with a number of co-workers, combined innovations made in the multisolution approach with the use of additional structure invariants and an algorithm for the assignment of numerical phases into the computer program *SIMPEL*.[135] A major contribution to the program was a modified version of a method for determining an optimal starting set, developed by Germain, Main and Woolfson[136] for the original multisolution program, *MULTAN*.[137] As we have already noted, origin-defining and enantiomorph-fixing phases are rarely sufficient to solve the structure. The structural solution depends on the sequential generation of new phases from those already assigned, and with only three or four reflections to begin with, there is a high probability that a phase will be incorrectly defined in such a manner as to

*Although this value produces a reasonable solution of the structure, the triplet relationships themselves indicated that a value of π for s_3 would be more appropriate. Automatic solution of the structure by the program *SIMPEL* uses π as the value for the symbol assigned to the $\bar{5}\,2\,10$ reflection.

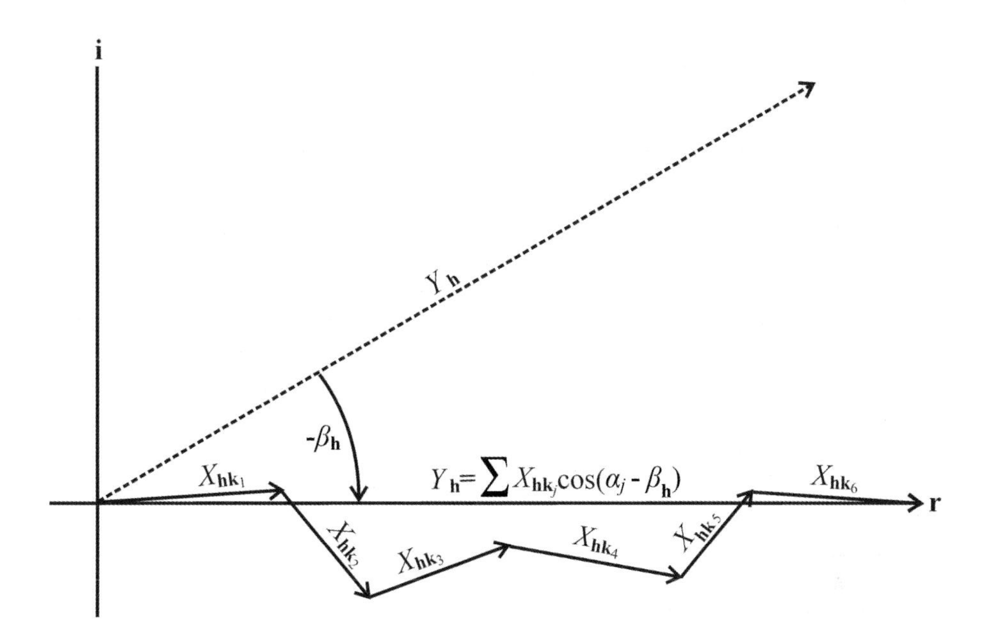

Figure 7.26 The resultant vector for the combination of the six triplet relationships in Fig. 7.12, rotated by angle $-\beta_{\mathbf{h}}$ to make it coincident with the real axis in the complex plane.

cause the solution pathway to divert to an incorrect solution. It is therefore necessary to add more phases to the starting set — from reflections which have the highest probability of keeping the phase-determining process on track. In the "manual" symbolic addition procedure the selection of these reflections was more-or-less arbitrary; failure to generate a good starting set often resulted in failure to solve the structure.

Reflections with relatively large E values that are involved in a large number of triplet relationships are likely to provide the most reliable phases, since both of these criteria tend to minimize the variance of the phases predicted by the tangent formula. Ideally, the starting set of phases will consist of those with the smallest variances, which, as we have previously noted (Fig. 7.13), decrease as the magnitude of the resultant of the vector sum of the triplet vectors for each reflection, $Y_{\mathbf{h}}$, increases. The optimal starting set will therefore consist of a subset of reflections with the largest values of $Y_{\mathbf{h}}$. The determination of $Y_{\mathbf{h}}$ is simplified by rotating the $\overrightarrow{Y_{\mathbf{h}}}$ vector (see Fig. 7.12) through the angle $-\beta_{\mathbf{h}}$ onto the real axis in the complex plane, as depicted in Fig. 7.26. $Y_{\mathbf{h}}$ is then expressed as the sum of the projections of each of the rotated triplet vectors onto the real axis:

$$\begin{aligned} Y_{\mathbf{h}} &= \sum_{j=1}^{m} X_{\mathbf{hk}_j} \cos(\alpha_j - \beta_{\mathbf{h}}) = \sum_{j=1}^{m} X_{\mathbf{hk}_j} \cos(\beta_{\mathbf{h}} - \alpha_j) \\ &= \sum_{j=1}^{m} X_{\mathbf{hk}_j} \cos(\beta_{\mathbf{h}} - \varphi_{\mathbf{k}_j} - \varphi_{\mathbf{h}-\mathbf{k}_j}) = \sum_{j=1}^{m} X_{\mathbf{hk}_j} \cos(\Phi_j), \end{aligned} \tag{7.172}$$

where $\Phi_j = \beta_{\mathbf{h}} - \varphi_{\mathbf{k}_j} - \varphi_{\mathbf{h}-\mathbf{k}_j}$. At the beginning of the phase-determining process we do not know any of the phases necessary to determine the $\cos(\Phi_j)$ values in order to obtain $Y_{\mathbf{h}}$ for each reflection. However, we can estimate the value of each $Y_{\mathbf{h}}$ if we are able to determine the expected value of each $\cos(\Phi_j)$ in the sum. In order to accomplish this we first note that for the triplet relationship, $\Phi_{\mathbf{hk}} = \varphi_{\mathbf{h}} - \varphi_{\mathbf{k_j}} - \varphi_{\mathbf{h}-\mathbf{k_j}}$, the probability distribution of the triplet is the same as the probability distribution of the unknown phase (because $\varphi_{\mathbf{k_j}}$ and $\varphi_{\mathbf{h}-\mathbf{k_j}}$ are fixed): $Pr(\Phi_{\mathbf{hk}}|E_{\mathbf{h}}, \mathbf{E_k}, \mathbf{E_{h-k}}) = Pr(\varphi_{\mathbf{h}}|E_{\mathbf{h}}, \mathbf{E_k}, \mathbf{E_{h-k}})$. It follows that $(\beta_{\mathbf{h}} - \varphi_{\mathbf{k_j}} - \varphi_{\mathbf{h}-\mathbf{k_j}}) \simeq (\varphi_{\mathbf{h}} - \varphi_{\mathbf{k_j}} - \varphi_{\mathbf{h}-\mathbf{k_j}})$ and $\Phi_j \simeq \Phi_{\mathbf{hk}}$, since $\beta_{\mathbf{h}} = \langle\varphi_{\mathbf{h}}\rangle$, resulting in the probability distribution of Φ_j from Eqn. 7.68:

$$Pr(\Phi_j) = \frac{e^{X_{\mathbf{hk}_j}\cos(\Phi_j)}d\Phi_j}{2\pi I_0(X_{\mathbf{hk}_j})}. \tag{7.173}$$

The expected value of $\cos(\Phi_j)$ is its probability-weighted average over all possible values,

$$\begin{aligned}\langle\cos(\Phi_j)\rangle &= \int_0^{2\pi}\cos(\Phi_j)Pr(\Phi_j)d\Phi_j \\ &= \frac{1}{2\pi I_0(X_{\mathbf{hk}_j})}\int_0^{2\pi} e^{X_{\mathbf{hk}_j}\cos(\Phi_j)}\cos(\Phi_j)d\Phi_j.\end{aligned} \tag{7.174}$$

The integral is in the form of a modified Bessel function of order 1, giving

$$\langle\cos(\Phi_j)\rangle = \frac{2\pi I_1(X_{\mathbf{hk}_j})}{2\pi I_0(X_{\mathbf{hk}_j})} = \frac{I_1(X_{\mathbf{hk}_j})}{I_0(X_{\mathbf{hk}_j})}. \tag{7.175}$$

Substituting these expected values into Eqn. 7.172 provides an expected value for $Y_{\mathbf{h}}$:

$$\langle Y_{\mathbf{h}}\rangle = \sum_{j=1}^{m} X_{\mathbf{hk}_j}\left(\frac{I_1(X_{\mathbf{hk}_j})}{I_0(X_{\mathbf{hk}_j})}\right). \tag{7.176}$$

Recalling that

$$X_{\mathbf{hk}_j} = \frac{2\sigma_3}{(\sigma_2)^{3/2}}E_{\mathbf{h}}E_{\mathbf{k}_j}E_{\mathbf{h}-\mathbf{k}_j}, \tag{7.177}$$

we can estimate $Y_{\mathbf{h}}$ for each reflection from the magnitudes of the normalized structure factors for the triplets associated with **h** and the values of the Bessel functions, which can be determined quickly by expanding the power series form of the functions (Appendix F) – or by looking the values up in published tables.

The procedure to generate a starting set begins with the creation of a list of (ca. 150–300) reflections with the largest values of E. All triplet relationships for each reflection in the list are determined, followed by calculation of $\langle Y_{\mathbf{h}}\rangle$ for each one, creating a ranked list of reflections from which the starting set is to be chosen by elimination of all reflections but those with the most reliably predicted triplet relationships – those with the largest values of $\langle Y_{\mathbf{h}}\rangle$. The process of selecting the starting subset is still not as straightforward as it first might appear. Each value of $\langle Y_{\mathbf{h}}\rangle$ depends on all of the triplet relationships that define it, and elimination of a reflection from the set impacts on all of the values of $\langle Y_{\mathbf{h}}\rangle$ in which it forms triplets. Thus, each time a reflection is removed from the list, the remaining values of $\langle Y_{\mathbf{h}}\rangle$ must be reevaluated to reflect the absence of the removed reflection. A procedure developed by Germain[136] provides a systematic way of accomplishing this so that the starting set consists of those reflections with the strongest links to one another.

The procedure, coined *convergence* by Germain (since it tends to "converge" toward the most strongly interrelated reflections), involves a series of cycles of an elimination process in which a single reflection is added to the starting set at the end of each cycle. The process is continued until the starting set contains a selected number of reflections. Each cycle begins with a calculation of $\langle Y_{\mathbf{h}}\rangle$ for each reflection and the removal of the reflection with the smallest value of $\langle Y_{\mathbf{h}}\rangle$ for all reflections not in the starting set. This reflection is temporarily eliminated as a candidate for a reflection in the starting set, provided that there are enough reflections with larger values of $\langle Y_{\mathbf{h}}\rangle$ that can define the origin. If not — if the reflection is required in order to define the origin — it is retained, since the origin *must* be defined.* After removal of this reflection, the remaining triplet relationships are all updated and new values of $\langle Y_{\mathbf{h}}\rangle$ are computed; all the triplets containing the phase of the removed reflection are now absent. The process is repeated sequentially, each time eliminating the reflection with the smallest value of $\langle Y_{\mathbf{h}}\rangle$, with the aim of retaining only the reflection with the largest number of mutual triplet relationships. The "surviving" reflection in each cycle is the one predicted to have the largest number of reliable triplet relationships and therefore the strongest links to the reflections already assigned to the starting set — the reflections that are optimal for the phase-determining process.

Additional structure invariants. Although the triplet invariant is the cornerstone of direct methods, other structure invariants also provide structural information. We first consider a special case of the triplet relationship, $\varphi_{\mathbf{h}}-\varphi_{\mathbf{k}}-\varphi_{\mathbf{h}-\mathbf{k}}\approx 0$, when $\mathbf{h}=-\mathbf{k}$:

$$\begin{aligned}
\varphi_{\mathbf{h}}-\varphi_{-\mathbf{h}} &\approx \varphi_{2\mathbf{h}} \\
\varphi_{\mathbf{h}}+\varphi_{\mathbf{h}} &\approx \varphi_{2\mathbf{h}} && (7.178)\\
\varphi_{\mathbf{h}} &\approx \frac{\varphi_{2\mathbf{h}}}{2}. && (7.179)
\end{aligned}$$

The relationship in Eqn. 7.179 is often referred to as a Σ_1 relationship.† The probability that the relationship holds is dependent on the triplet probability. If $E_{\mathbf{h}}=E_{\mathbf{k}}$ and $E_{2\mathbf{h}}=E_{\mathbf{h}-\mathbf{k}}$ are both large, then the phase of one reflection can serve to determine the phase of the other. We have already encountered an example of this for centrosymmetric structures. Since $\varphi_{\mathbf{h}}$ is 0 or π, $\varphi_{2\mathbf{h}}$ is likely to equal zero since $2\cdot 0\equiv 2\pi\equiv 0$; for large $E_{\mathbf{h}}$ and $E_{2\mathbf{h}}$, the sign of $E_{2\mathbf{h}}$ is likely to be positive, regardless of the sign of $E_{\mathbf{h}}$ (see Fig. 7.1 and the accompanying discussion).

The success of the triplet relationship in providing structural information leads us to ask if there are other structure invariant relationships that might be useful.

*Note that this way of defining the origin provides origin-defining reflections with the strongest possible triplet relationships.

†In the Hauptmann and Karle monograph on centrosymmetric direct methods[128] the centrosymmetric analog of this formula was the first term in the series of sums: $S_{\mathbf{h}}=\Sigma_1+\Sigma_2+\Sigma_3+\Sigma_4$. The explicit form of the Σ_1 relationship depends on the space group symmetry.

Recalling that a structure invariant relationship is defined as a relationship that remains unchanged when the origin is shifted, we consider the structure factor itself, with the origin shifted from o to o' along the vector $\mathbf{v}_o$:

$$\begin{aligned}\mathbf{F}_{\mathbf{h}} &= \sum_{j=1}^{n} f_j e^{-2\pi i(\mathbf{h}\cdot\mathbf{r}_j)} = F_h e^{i\varphi_{\mathbf{h}}} \\ \mathbf{F}'_{\mathbf{h}} &= \sum_{j=1}^{n} f_j e^{-2\pi i(\mathbf{h}\cdot(\mathbf{r}_j-\mathbf{v}_o))} \\ &= \mathbf{F}_{\mathbf{h}} e^{2\pi i(\mathbf{h}\cdot\mathbf{v}_o)} = F_h e^{i\varphi_{\mathbf{h}}} e^{2\pi i(\mathbf{h}\cdot\mathbf{v}_o)} = F_h e^{i\varphi'_{\mathbf{h}}}. \end{aligned} \tag{7.180}$$

The phase of the structure factor changes with an origin shift, while its magnitude does not; the magnitude of the structure factor is a structure invariant. Note that for $\mathbf{h} = \mathbf{0}$, $\mathbf{F}_{\mathbf{h}} = \mathbf{F}'_{\mathbf{h}}$, since $\varphi = \varphi' = 0$. The $\mathbf{F}_{000}$ reflection, equal to the number of electrons in the unit cell, *is* a structure invariant. It follows that any product of structure factors for which the net phase is zero will also be invariant. For

$$\begin{aligned}\mathbf{F}_{\mathbf{h}1}\mathbf{F}_{\mathbf{h}2}\ldots\mathbf{F}_{\mathbf{h}m} &= F_{\mathbf{h}_1} e^{i\varphi_1} F_{\mathbf{h}_2} e^{i\varphi_2}\ldots F_{\mathbf{h}_m} e^{i\varphi_m} \\ &= F_{\mathbf{h}_1} F_{\mathbf{h}_2}\ldots F_{\mathbf{h}_m} e^{i(\varphi_1+\varphi_2+\cdots+\varphi_m)}, \end{aligned} \tag{7.181}$$

an origin shift results in

$$\begin{aligned}\mathbf{F}'_{\mathbf{h}1}\mathbf{F}'_{\mathbf{h}2}\ldots\mathbf{F}'_{\mathbf{h}m} &= F_{\mathbf{h}_1} e^{i\varphi_1} e^{2\pi i(\mathbf{h}_1\cdot\mathbf{v}_o)} F_{\mathbf{h}_2} e^{i\varphi_2} e^{2\pi i(\mathbf{h}_2\cdot\mathbf{v}_o)}\ldots F_{\mathbf{h}_m} e^{i\varphi_m} e^{2\pi i(\mathbf{h}_m\cdot\mathbf{v}_o)} \\ &= F_{\mathbf{h}_1} F_{\mathbf{h}_2} F_{\mathbf{h}_3}\ldots F_{\mathbf{h}_m} e^{i(\varphi_1+\varphi_2+\cdots+\varphi_m)} e^{2\pi i((\mathbf{h}_1+\mathbf{h}_2+\cdots+\mathbf{h}_m)\cdot\mathbf{v}_o)} \\ &= \mathbf{F}_{\mathbf{h}1}\mathbf{F}_{\mathbf{h}2}\ldots\mathbf{F}_{\mathbf{h}m} e^{2\pi i((\mathbf{h}_1+\mathbf{h}_2+\cdots+\mathbf{h}_m)\cdot\mathbf{v}_o)}. \end{aligned} \tag{7.182}$$

The structure factor product will remain unchanged with an origin shift only if $\mathbf{h}_1 + \mathbf{h}_2 + \cdots + \mathbf{h}_m = \mathbf{0}$.

For $m = 2$, $\mathbf{h}_1 + \mathbf{h}_2 = 0$ holds for $\mathbf{h} + (-\mathbf{h}) = 0$; giving the invariant "doublet", $(\varphi_{\mathbf{h}} + \varphi_{-\mathbf{h}})$, and the product

$$\mathbf{F}_{\mathbf{h}1}\mathbf{F}_{\mathbf{h}2} = \mathbf{F}_{\mathbf{h}}\mathbf{F}_{\bar{\mathbf{h}}} = F_{\mathbf{h}}^2 e^{(\varphi_{\mathbf{h}}-\varphi_{\mathbf{h}})} = F_{\mathbf{h}}^2, \tag{7.183}$$

the scaled intensity of the reflection (also a structure invariant).

For $m = 3$, $\mathbf{h}_1 + \mathbf{h}_2 + \mathbf{h}_3 = 0$ is satisfied by $\mathbf{h} - \mathbf{h} + \mathbf{k} - \mathbf{k} \equiv \mathbf{h} - \mathbf{k} - (\mathbf{h} - \mathbf{k}) = 0$, resulting in the triplet invariant, $(\varphi_{\mathbf{h}} - \varphi_{\mathbf{k}} - \varphi_{\mathbf{h-k}})$. For large values of $F_{\mathbf{h}}F_{\mathbf{k}}F_{\mathbf{h-k}}$ we have already discovered that this results in the triplet probability relationship, $\varphi_{\mathbf{h}} - \varphi_{\mathbf{k}} - \varphi_{\mathbf{h-k}} \approx 0$ and

$$\mathbf{F}_{\mathbf{h}1}\mathbf{F}_{\mathbf{h}2}\mathbf{F}_{\mathbf{h}3} = \mathbf{F}_{\mathbf{h}}\mathbf{F}_{\mathbf{k}}\mathbf{F}_{\mathbf{h-k}} = F_{\mathbf{h}}F_{\mathbf{k}}F_{\mathbf{h-k}} e^{(\varphi_{\mathbf{h}}-\varphi_{\mathbf{k}}-\varphi_{\mathbf{h-k}})} \approx F_{\mathbf{h}}F_{\mathbf{k}}F_{\mathbf{h-k}}. \tag{7.184}$$

The probability that $\Phi_{\mathbf{hk}} = \varphi_{\mathbf{h}} - \varphi_{\mathbf{k}} - \varphi_{\mathbf{h-k}} = 0$ is dependent on the magnitudes, $E_{\mathbf{h}}$, $E_{\mathbf{k}}$, and $E_{\mathbf{h-k}}$. The probability distribution is the same as that for Eqn. 7.68, but is centered around zero, the expected value of $\Phi_{\mathbf{hkl}}$:

$$f(\Phi_{\mathbf{hkl}}|E_{\mathbf{h}}, E_{\mathbf{k}}, E_{\mathbf{h-k}}) = \frac{e^{X_{\mathbf{hk}}\cos\Phi_{\mathbf{hk}}}}{2\pi I_0(X_{\mathbf{hk}})}, \tag{7.185}$$

where $X_{\mathbf{hk}} = (2/\sqrt{n})\, E_{\mathbf{h}}E_{\mathbf{k}}E_{\mathbf{h-k}}$ for an equal atom structure. The triplet distribution is often referred to as a *three magnitude distribution.*

For $m = 4$, $\mathbf{h}+\mathbf{k}+\mathbf{l}+(-\mathbf{h}-\mathbf{k}-\mathbf{l}) = 0$ satisfies $\mathbf{h}_1 + \mathbf{h}_2 + \mathbf{h}_3 + \mathbf{h}_4 = 0$ and the sum of phases $(\varphi_{\mathbf{h}} + \varphi_{\mathbf{k}} + \varphi_{\mathbf{l}} + \varphi_{-\mathbf{h}-\mathbf{k}-\mathbf{l}})$ becomes a *quartet* invariant. In order to determine if quartets can be useful for determining phases we ask a question similar to that asked for the triplet relationship: *Given the four reflections in a quartet relationship, if we know the magnitudes (E values) of the four reflections, and three of the phases, what is the probability that the fourth phase is correctly determined by*

$$\Phi_{\mathbf{hkl}} = \varphi_{\mathbf{h}} + \varphi_{\mathbf{k}} + \varphi_{\mathbf{l}} + \varphi_{-\mathbf{h}-\mathbf{k}-\mathbf{l}} \approx 0? \tag{7.186}$$

The answer to this question is encompassed in the conditional probability,

$$Pr(\varphi_{\mathbf{h}}|E_{\mathbf{h}}, \mathbf{E}_{\mathbf{k}}, \mathbf{E}_{\mathbf{l}}, \mathbf{E}_{-\mathbf{h}-\mathbf{k}-\mathbf{l}}). \tag{7.187}$$

The derivation parallels that for the triplet probability. We will again assume an "equal atom" structure. In order to express $\mathbf{E}_{\mathbf{h}}$ in terms of $\mathbf{E}_{\mathbf{k}}$, $\mathbf{E}_{\mathbf{l}}$ and $\mathbf{E}_{-\mathbf{h}-\mathbf{k}-\mathbf{l}}$, we add and subtract $\mathbf{k}$ and $\mathbf{l}$ in the exponential term in Eqn. 7.40 for $\mathbf{E}_{\bar{\mathbf{h}}}$:

$$\begin{aligned}\mathbf{E}_{\bar{\mathbf{h}}} &= \frac{1}{\sqrt{n}}\sum_{j=1}^{n} e^{-2\pi\, i((-\mathbf{h}+\mathbf{k}+\mathbf{l}-\mathbf{k}-\mathbf{l})\cdot\,\mathbf{r}_j)} \\ &= \frac{1}{\sqrt{n}}\sum_{j=1}^{n} e^{-2\pi\, i(\mathbf{k}\cdot\,\mathbf{r}_j)}\, e^{-2\pi\, i(\mathbf{l}\cdot\,\mathbf{r}_j)}\, e^{-2\pi\, i((-\mathbf{h}-\mathbf{k}-\mathbf{l})\cdot\,\mathbf{r}_j)}.\end{aligned} \tag{7.188}$$

The average value of $\mathbf{E}_{\bar{\mathbf{h}}}$ is then (cf. Eqn. 7.49)

$$\begin{aligned}\langle\mathbf{E}_{\bar{\mathbf{h}}}\rangle &= \frac{n}{\sqrt{n}}\frac{\sqrt{n}}{n}\frac{\sqrt{n}}{n}\frac{\sqrt{n}}{n}\;\mathbf{E}_{\mathbf{k}}\mathbf{E}_{\mathbf{l}}\mathbf{E}_{-\mathbf{h}-\mathbf{k}-\mathbf{l}} \\ &= \frac{1}{n}\;E_{\mathbf{k}}E_{\mathbf{l}}E_{-\mathbf{h}-\mathbf{k}-\mathbf{l}}\cos(\varphi_{\mathbf{k}} + \varphi_{\mathbf{l}} + \varphi_{-\mathbf{h}-\mathbf{k}-\mathbf{l}}) \\ &\qquad + i\,\frac{1}{n}\;E_{\mathbf{k}}E_{\mathbf{l}}E_{-\mathbf{h}-\mathbf{k}-\mathbf{l}}\sin(\varphi_{\mathbf{k}} + \varphi_{\mathbf{l}} + \varphi_{-\mathbf{h}-\mathbf{k}-\mathbf{l}}) \\ &= \langle A_{\bar{\mathbf{h}}}\rangle + i\,\langle B_{\bar{\mathbf{h}}}\rangle.\end{aligned} \tag{7.189}$$

This results in the conditional probability for $\varphi_{\bar{\mathbf{h}}}$:

$$Pr(\varphi_{\bar{\mathbf{h}}}|E_{\mathbf{h}}, \mathbf{E}_{\mathbf{k}}, \mathbf{E}_{\mathbf{l}}, \mathbf{E}_{-\mathbf{h}-\mathbf{k}-\mathbf{l}}) = \frac{1}{\pi}\, e^{-(A_{\bar{\mathbf{h}}}-\langle A_{\bar{\mathbf{h}}}\rangle)^2-(B_{\bar{\mathbf{h}}}-\langle B_{\bar{\mathbf{h}}}\rangle)^2}\, E_{\mathbf{h}}\, d\varphi_{\bar{\mathbf{h}}}.$$

The variances of $A_{\bar{\mathbf{h}}}$ and $B_{\bar{\mathbf{h}}}$ are determined from Eqns. 7.51–7.59, and since $A_{\bar{\mathbf{h}}} = E_{\mathbf{h}}\cos\varphi_{\bar{\mathbf{h}}} \equiv E_{\mathbf{h}}\cos\varphi_{\mathbf{h}}$ and $B_{\bar{\mathbf{h}}} = E_{\mathbf{h}}\sin\varphi_{\bar{\mathbf{h}}}$ (cf. Eqn. 7.63),

$$Pr(\varphi_{\bar{\mathbf{h}}}|E_{\mathbf{h}}, \mathbf{E}_{\mathbf{k}}, \mathbf{E}_{\mathbf{l}}, \mathbf{E}_{-\mathbf{h}-\mathbf{k}-\mathbf{l}}) = K' e^{\frac{2}{n}E_{\mathbf{h}}E_{\mathbf{k}}E_{\mathbf{l}}E_{-\mathbf{h}-\mathbf{k}-\mathbf{l}}\cos(\varphi_{\mathbf{h}}+\varphi_{\mathbf{k}}+\varphi_{\mathbf{l}}+\varphi_{-\mathbf{h}-\mathbf{k}-\mathbf{l}})}\, d\varphi_{\mathbf{h}},$$

where

$$K' = \frac{E_{\mathbf{h}}}{\pi} e^{-E_{\mathbf{h}}^2+\frac{1}{n^2}(E_{\mathbf{k}}E_{\mathbf{l}}E_{-\mathbf{h}-\mathbf{k}-\mathbf{l}})^2}. \tag{7.190}$$

Because $\varphi_{\mathbf{h}}$ has the same probability distribution as $\varphi_{\bar{\mathbf{h}}}$, setting

$$W_{\mathbf{hkl}} = (2/n)\, E_{\mathbf{h}}E_{\mathbf{k}}E_{\mathbf{l}}E_{-\mathbf{h}-\mathbf{k}-\mathbf{l}} \tag{7.191}$$

and multiplying $K' \cdot C' = N'$ to normalize the probability distribution (cf. Eqns. 7.65–7.68) results in the quartet probability function,

$$Pr(\varphi_{\mathbf{h}}|E_{\mathbf{h}}, \mathbf{E_k}, \mathbf{E_l}, \mathbf{E_{-h-k-l}}) = \frac{e^{W_{\mathbf{hkl}} \cos(\varphi_{\mathbf{h}}+\varphi_{\mathbf{k}}+\varphi_{\mathbf{l}}+\varphi_{-\mathbf{h}-\mathbf{k}-\mathbf{l}})} d\varphi_{\mathbf{h}}}{2\pi I_0(W_{\mathbf{hkl}})}, \quad (7.192)$$

giving us the probability that $\varphi_{\mathbf{h}}$ lies between $\varphi_{\mathbf{h}}$ and $\varphi_{\mathbf{h}} + d\varphi_{\mathbf{h}}$, subject to the constraints of the quartet relationship. The triplet and quartet probabilities are very similar in form. For non-equal atom structures

$$W_{\mathbf{hkl}} = \frac{2\sum Z^4}{(\sum Z^2)^2} E_{\mathbf{h}} E_{\mathbf{k}} E_{\mathbf{l}} E_{-\mathbf{h}-\mathbf{k}-\mathbf{l}} = \frac{2\sigma_4}{\sigma_2^2} E_{\mathbf{h}} E_{\mathbf{k}} E_{\mathbf{l}} E_{-\mathbf{h}-\mathbf{k}-\mathbf{l}}. \quad (7.193)$$

The expected value of $\varphi_{\mathbf{h}}$ is $-\varphi_{\mathbf{k}} - \varphi_{\mathbf{l}} - \varphi_{-\mathbf{h}-\mathbf{k}-\mathbf{l}}$; the expected value of $\Phi_{\mathbf{hkl}}$ is zero. The probability distribution of $\Phi_{\mathbf{hkl}}$ depends on the magnitudes $E_{\mathbf{h}}$, $E_{\mathbf{k}}$, $E_{\mathbf{l}}$, and $E_{-\mathbf{h}-\mathbf{k}-\mathbf{l}}$:

$$f(\Phi_{\mathbf{hkl}}|E_{\mathbf{h}}, E_{\mathbf{k}}, E_{\mathbf{k}}, E_{-\mathbf{h}-\mathbf{k}-\mathbf{l}}) = \frac{e^{W_{\mathbf{hk}} \cos \Phi_{\mathbf{hk}}}}{2\pi I_0(W_{\mathbf{hk}})}, \quad (7.194)$$

and is referred to as a *four magnitude distribution.* The quartet relationship should, in principle, be capable of providing the same type of phase information as the triplet relationship. The primary distinction between the two probability distributions is the exponential term. In the triplet relationship (for an equal atom structure), $X_{\mathbf{hk}} = (2/\sqrt{n}) E_{\mathbf{h}} E_{\mathbf{k}} E_{\mathbf{h}-\mathbf{k}}$, while in the quartet relationship, $W_{\mathbf{hkl}} = (2/n) E_{\mathbf{h}} E_{\mathbf{k}} E_{\mathbf{l}} E_{-\mathbf{h}-\mathbf{k}-\mathbf{l}}$. As with $X_{\mathbf{hk}}$, the larger the value of $W_{\mathbf{hkl}}$, the more likely that $\Phi_{\mathbf{hkl}} = 0$ and that $\varphi_{\mathbf{h}}$ is predicted by the quartet relationship. However, the $1/n$ versus $1/\sqrt{n}$ factor diminishes the quartet probabilities in comparison with triplet probabilities; quartet relationships are much less reliable phase predictors.

In 1973 Schenk "rescued" the quartet as a structural-determination tool when he noted that a phase predicted by a quartet relationship is much more likely to be correct if $E_{\mathbf{h}+\mathbf{k}}$, $E_{\mathbf{h}+\mathbf{l}}$, and $E_{\mathbf{k}+\mathbf{l}}$ are also large. Fig. 7.27 (a) illustrates a scenario in which the $\mathbf{h}$, $\mathbf{k}$, $\mathbf{l}$, and $-\mathbf{h} - \mathbf{k} - \mathbf{l}$ planes all intersect at a common point of high electron density. As a result the density waves will tend to be in phase, and an extension of the geometric argument for the triplet phase relationship (Fig. 7.6) would yield the quartet phase relationship. Fig. 7.27(b) illustrates the case in which $E_{\mathbf{h}+\mathbf{k}}$ is also large, providing further evidence that electron density is concentrated at the points of intersection. The addition of the magnitudes $E_{\mathbf{h}+\mathbf{k}}$, $E_{\mathbf{h}+\mathbf{l}}$, and $E_{\mathbf{k}+\mathbf{l}}$ to the evaluation of the quartet relationship resulted in the term *seven magnitude quartet.*

Unlike the triplet relationship, the quartet relationship relies on the intersection of four sets of planes, and they need not all intersect at the same point. Fig. 7.28(a) illustrates a case in which all four planes intersect regions of high electron density with the $-\mathbf{h} - \mathbf{k} - \mathbf{l}$ planes bisecting the lines of intersection of the three other planes. In this case, the density waves from the $\mathbf{h}$, $\mathbf{k}$ and $\mathbf{l}$ planes will tend to be in phase, while the wave from the $-\mathbf{h} - \mathbf{k} - \mathbf{l}$ planes will tend to be 180° out of phase, resulting in

$$\Phi_{\mathbf{hkl}} = \varphi_{\mathbf{h}} + \varphi_{\mathbf{k}} + \varphi_{\mathbf{l}} + \varphi_{-\mathbf{h}-\mathbf{k}-\mathbf{l}} \approx \pi. \quad (7.195)$$

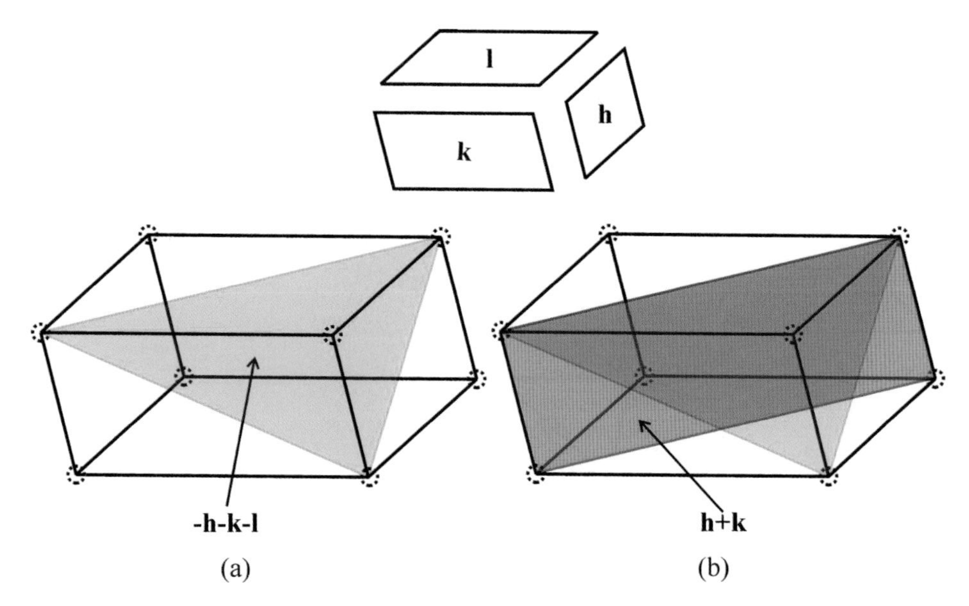

Figure 7.27 (a) Large values of $E_{\mathbf{h}}$, $E_{\mathbf{k}}$, $E_{\mathbf{l}}$, and $E_{-\mathbf{h}-\mathbf{k}-\mathbf{l}}$ indicate that electron density is concentrated at points of intersection of the **h**, **k**, **l**, and $-\mathbf{h}-\mathbf{k}-\mathbf{l}$ planes. (b) An additional large value of $E_{\mathbf{h}+\mathbf{k}}$ increases the probability that electron density resides at the points of intersection.

As illustrated in Fig. 7.28(b), the situation is different for the $\mathbf{h}+\mathbf{k}$ reflections than it was in the previous example. In this idealized case, the electron density resides alternatively on the $\mathbf{h}+\mathbf{k}$ planes and between them. Thus the scattered waves waves from each region will be out of phase with respect to one another, resulting in destructive interference. The seven magnitude quartets for which $\varphi_{\mathbf{h}}+\varphi_{\mathbf{k}}+\varphi_{\mathbf{l}}+\varphi_{-\mathbf{h}-\mathbf{k}-\mathbf{l}} \approx \pi$ exhibit *weak* intensities for the $\mathbf{h}+\mathbf{k}$, $\mathbf{h}+\mathbf{l}$, and $\mathbf{k}+\mathbf{l}$ reflections.

Quartets for which $\Phi_{\mathbf{hkl}} \approx 0$ are termed *positive quartets* because the cosine of $\Phi_{\mathbf{hkl}}$ is positive. Quartets for which $\Phi_{\mathbf{hkl}} \approx \pi$ have $\cos\Phi_{\mathbf{hkl}} < 0$, and are known as *negative quartets*. The intuitive arguments outlined above, originally put forth by Schenk,[138–140] were placed on a firm theoretical footing with independent derivations of the seven magnitude conditional probability distribution by Hauptman[141,142] and Giacovazzo[143].* For an equal atom structure,

$$f(\Phi_{\mathbf{hkl}}|E_{\mathbf{h}}, E_{\mathbf{k}}, E_{\mathbf{l}}, E_{-\mathbf{h}-\mathbf{k}-\mathbf{l}}, E_{\mathbf{h}+\mathbf{k}}, E_{\mathbf{h}+\mathbf{l}}, E_{\mathbf{k}+\mathbf{l}}) =$$

$$Ne^{(-4/n)E_{\mathbf{h}}E_{\mathbf{k}}E_{\mathbf{l}}E_{-\mathbf{h}-\mathbf{k}-\mathbf{l}}\cos\Phi_{\mathbf{hkl}}} \times I_0(E_{\mathbf{h}+\mathbf{k}}Z_{hk})I_0(E_{\mathbf{h}+\mathbf{l}}Z_{hl})I_0(E_{\mathbf{k}+\mathbf{l}}Z_{kl}), \quad (7.196)$$

*While the principles of the derivations are essentially the same as those used for the derivation of the triplet and quartet probability distributions, they require the evaluation of a 14-fold integral — a rather tedious undertaking. The interested reader will find an outline of the derivations in the cited references.

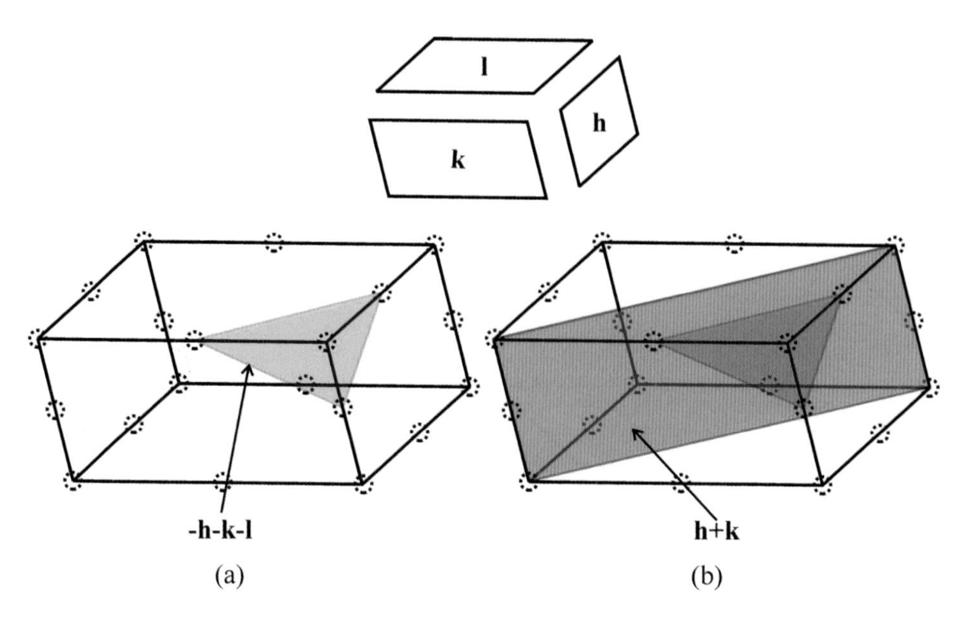

Figure 7.28 (a) Large values of $E_{\mathbf{h}}$, $E_{\mathbf{k}}$, $E_{\mathbf{l}}$, and $E_{-\mathbf{h}-\mathbf{k}-\mathbf{l}}$ are observed when three of the planes in the quartet intersect at points of high electron density and one of the planes in the quartet intersects regions of high electron density that is not at the intersection points of the other three planes. (b) The $\mathbf{h}+\mathbf{k}$ reflection scatters from electron density on the planes *and* electron density between them, weakening the intensity of the reflection.

where N is a normalization constant, $I_0(EZ)$ is a zero-order modified Bessel function with argument EZ, and

$$Z_{hk} = \frac{2}{\sqrt{n}}(E_{\mathbf{h}}^2 E_{\mathbf{k}}^2 + E_{\mathbf{l}}^2 E_{-\mathbf{h}-\mathbf{k}-\mathbf{l}}^2 + 2E_{\mathbf{h}}E_{\mathbf{k}}E_{\mathbf{l}}E_{-\mathbf{h}-\mathbf{k}-\mathbf{l}}\cos\Phi_{\mathbf{hkl}})^{\frac{1}{2}}, \quad (7.197)$$

$$Z_{hl} = \frac{2}{\sqrt{n}}(E_{\mathbf{h}}^2 E_{\mathbf{l}}^2 + E_{\mathbf{k}}^2 E_{-\mathbf{h}-\mathbf{k}-\mathbf{l}}^2 + 2E_{\mathbf{h}}E_{\mathbf{k}}E_{\mathbf{l}}E_{-\mathbf{h}-\mathbf{k}-\mathbf{l}}\cos\Phi_{\mathbf{hkl}})^{\frac{1}{2}}, \quad (7.198)$$

$$Z_{kl} = \frac{2}{\sqrt{n}}(E_{\mathbf{k}}^2 E_{\mathbf{l}}^2 + E_{\mathbf{h}}^2 E_{-\mathbf{h}-\mathbf{k}-\mathbf{l}}^2 + 2E_{\mathbf{h}}E_{\mathbf{k}}E_{\mathbf{l}}E_{-\mathbf{h}-\mathbf{k}-\mathbf{l}}\cos\Phi_{\mathbf{hkl}})^{\frac{1}{2}}. \quad (7.199)$$

The normalization constant is determined by setting

$$\int_{-\pi}^{\pi} f(\Phi_{\mathbf{hkl}})d\Phi_{\mathbf{hkl}} = 1 \quad (7.200)$$

and solving for N (usually numerically, since it varies for each seven magnitude quartet).

In the special case in which $E_{\mathbf{h}+\mathbf{k}}$, $E_{\mathbf{h}+\mathbf{l}}$, and $E_{\mathbf{k}+\mathbf{l}}$ are all zero, the Bessel functions all evaluate to $I_0(EZ) = 1$ (Appendix F). The exponential term in Eqn. 7.196 will be a large negative number when $\Phi_{\mathbf{hkl}} = 0$, minimizing the probability, and a large positive number when $\Phi_{\mathbf{hkl}} = \pi$; the probability is maximized when $\cos\Phi_{\mathbf{hkl}} = -1$ (a negative quartet). For the case in which the "cross terms" are all large, the exponential term is minimized at $\Phi_{\mathbf{hkl}} = 0$, but the Bessel functions more than compensate with $\cos\Phi_{\mathbf{hkl}} = 1$, and the probability is a maximum at $\Phi_{\mathbf{hkl}} = 0$ (a positive quartet).

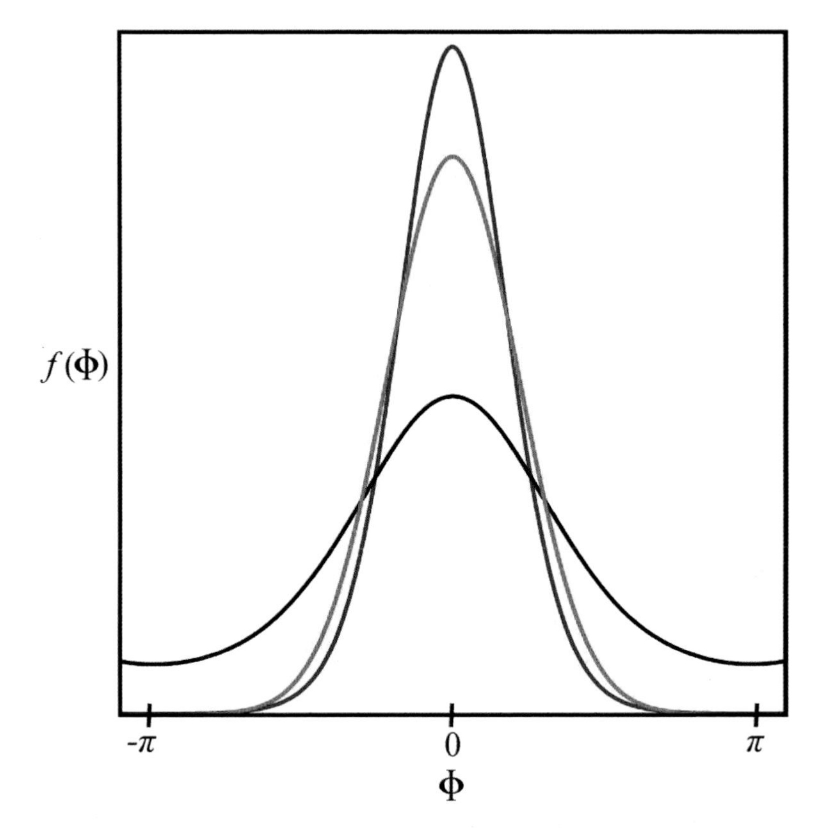

Figure 7.29 Selected probability distributions for the $Li[AlSi_2O_6]{\cdot}H_2O$ structure. Blue — three magnitude (triplet): $f(\Phi_{\mathbf{hk}}|E_{061}, E_{461}, E_{\bar{4}00})$. Black — four magnitude (quartet): $f(\Phi_{\mathbf{hkl}}|E_{061}, E_{\bar{6}02}, E_{\bar{4}00}, E_{\bar{2}63})$. Red — seven magnitude (positive quartet): $f(\Phi_{\mathbf{hkl}}|E_{061}, E_{\bar{6}02}, E_{\bar{4}00}, E_{\bar{2}63}, E_{\bar{6}63}, E_{461}, E_{\bar{2}02})$.

Fig. 7.29 illustrates selected three, four, and seven magnitude probability distributions for the bikitaite data set. The triplet probability distribution, shown in blue, results from three reflections with $E_{\mathbf{h}} = E_{061} = 2.67$, $E_{\mathbf{k}} = E_{461} = 2.64$, and $E_{\mathbf{h-k}} = E_{\bar{4}00} = 1.08$. The quartet probability distribution, in black, arises from $E_{\mathbf{h}} = E_{061} = 2.67$, $E_{\mathbf{k}} = E_{\bar{6}02} = 2.04$, $E_{\mathbf{l}} = E_{\bar{4}00} = 1.08$, and $E_{-\mathbf{h-k-l}} = E_{\bar{2}63} = 1.72$. The variance for the quartet distribution is significantly larger than for the triplet distribution, and is therefore unlikely to be a reliable predictor for $\varphi_{\mathbf{h}}$. Adding $E_{\mathbf{h+k}} = E_{\bar{6}63} = 2.85$, $E_{\mathbf{h+l}} = E_{461} = 2.64$, and $E_{\mathbf{k+l}} = E_{\bar{2}02} = 1.22$ to the quartet reflections results in the seven magnitude probability distribution shown in red in Fig. 7.29. The variance has decreased substantially, now approaching the variance (and hence the reliability) of the triplet, thus enhancing the prediction that $\Phi_{hkl} \approx 0$ and that $\varphi_{\mathbf{h}} \approx -\varphi_{\mathbf{k}} - \varphi_{\mathbf{l}} - \varphi_{-\mathbf{h-k-l}}$.

Fig. 7.30 illustrates a four magnitude probability distribution for the bikitaite data set (in black) with $E_{\mathbf{h}} = E_{600} = 1.52$, $E_{\mathbf{k}} = E_{\bar{1}101} = 1.20$, $E_{\mathbf{l}} = E_{002} = 1.24$, and $E_{-\mathbf{h-k-l}} = E_{50\bar{3}} = 1.21$, and the corresponding seven magnitude probability distribution with small magnitudes for the cross-terms: $E_{\mathbf{h+k}} = E_{\bar{5}01} = 0.19$,

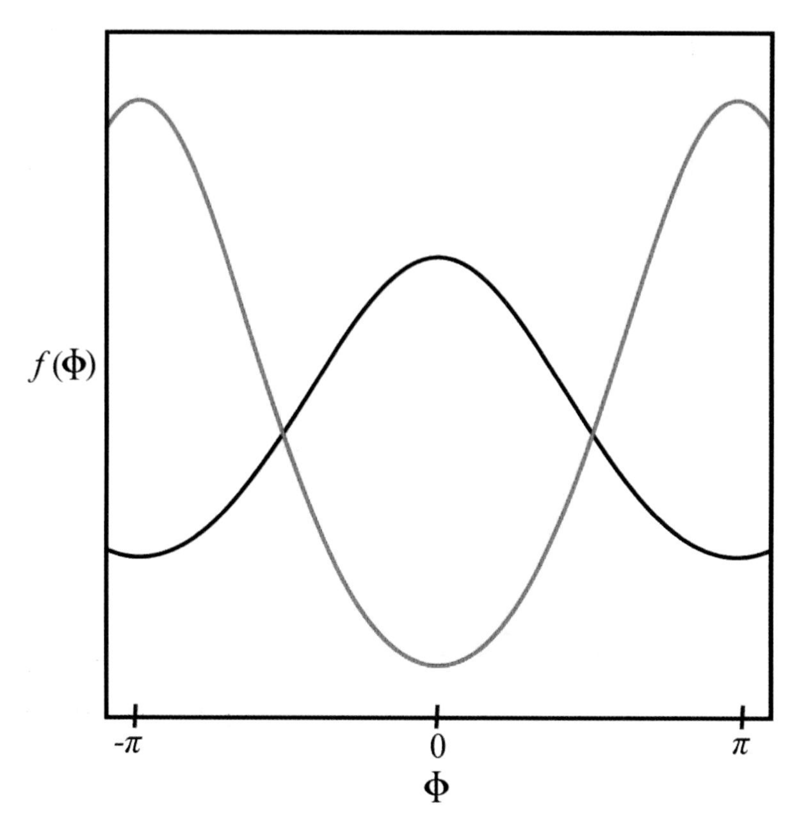

Figure 7.30 Selected probability distributions for the $Li[AlSi_2O_6]{\cdot}H_2O$ structure. Black — four magnitude (quartet): $f(\Phi_{\mathbf{hkl}}|E_{600}, E_{\bar{1}\bar{1}01}, E_{002}, E_{50\bar{3}})$. Red — seven magnitude (negative quartet): $f(\Phi_{\mathbf{hkl}}|E_{600}, E_{\bar{1}\bar{1}01}, E_{002}, E_{50\bar{3}}, E_{\bar{5}01}, E_{602}, E_{\bar{1}\bar{1}03})$.

$E_{\mathbf{h+l}} = E_{602} = 0.17$, and $E_{\mathbf{k+l}} = E_{\bar{1}\bar{1}03} = 0.16$. The seven magnitude negative quartet has an expected value very close to π. In general, the variance of a negative quartet is seen to be larger than a comparative positive quartet (e.g., with the same four E values), since the values of the modified Bessel functions are smaller for the negative quartet. In addition, the number of negative quartets for a given data set is usually very much smaller than the number of triplets or positive quartets.

The convergence procedure to determine an optimal starting set can be enhanced by combining large-cross-term quartet relationships with triplet relationships. A parallel tangent formula exists for quartets, which is readily derived by considering a reflection with an unknown phase, $\mathbf{E_h}$, and a set of n quartet-related reflections $\{\mathbf{E}_{\mathbf{k}_j}, \mathbf{E}_{\mathbf{l}_j}, \mathbf{E}_{-\mathbf{h}-\mathbf{k}_j-\mathbf{l}_j}\}$, with $\Phi_{\mathbf{k}_j} = \varphi_{\mathbf{h}} + \varphi_{\mathbf{k}_j} + \varphi_{\mathbf{l}_j} + \varphi_{-\mathbf{h}-\mathbf{k}_j-\mathbf{l}_j} \approx 0$. Substituting

$\mathbf{E_h E_{k_j} E_{l_j} E_{-h-k_j-l_j}}$ for $\mathbf{E_h E^*_{k_j} E^*_{h-k_j}}$ in Eqn. 7.119, and following the procedure in Eqns. 7.120 - 7.126 gives

$$\frac{\sin\beta_{\mathbf{h}}}{\cos\beta_{\mathbf{h}}} = \tan\beta_{\mathbf{h}} = \frac{\sum_{j=1}^{n} E_{\mathbf{k}_j} E_{\mathbf{l}_j} E_{-\mathbf{h}-\mathbf{k}_j-\mathbf{l}_j} \sin(\varphi_{\mathbf{k}_j} + \varphi_{\mathbf{l}_j} + \varphi_{-\mathbf{h}-\mathbf{k}_j-\mathbf{l}_j})}{\sum_{j=1}^{n} E_{\mathbf{k}_j} E_{\mathbf{l}_j} E_{-\mathbf{h}-\mathbf{k}_j-\mathbf{l}_j} \cos(\varphi_{\mathbf{k}_j} + \varphi_{\mathbf{l}_j} + \varphi_{-\mathbf{h}-\mathbf{k}_j-\mathbf{l}_j})}. \quad (7.201)$$

In order to make practical use of the formula, the expressions in the sums must be weighted according to their reliability, dictated by the magnitudes of the cross terms in the seven magnitude quartet:

$$\tan\beta_{\mathbf{h}} = \frac{\sum_{j=1}^{n} w\, E_{\mathbf{k}_j} E_{\mathbf{l}_j} E_{-\mathbf{h}-\mathbf{k}_j-\mathbf{l}_j} \sin(\varphi_{\mathbf{k}_j} + \varphi_{\mathbf{l}_j} + \varphi_{-\mathbf{h}-\mathbf{k}_j-\mathbf{l}_j})}{\sum_{j=1}^{n} w\, E_{\mathbf{k}_j} E_{\mathbf{l}_j} E_{-\mathbf{h}-\mathbf{k}_j-\mathbf{l}_j} \cos(\varphi_{\mathbf{k}_j} + \varphi_{\mathbf{l}_j} + \varphi_{-\mathbf{h}-\mathbf{k}_j-\mathbf{l}_j})}, \quad (7.202)$$

where w is a weighting factor, e.g.,[19]

$$w = E^2_{\mathbf{h+k}} + E^2_{\mathbf{h+l}} + E^2_{\mathbf{k+l}} - 2. \quad (7.203)$$

This suggests that formulas for the triplet and seven magnitude quartets might be combined (cf. Eqn. 7.172) by substituting $W_{\mathbf{hkl}}$ in the quartet probability function with $W'_{\mathbf{hkl}} = wW_{\mathbf{hkl}}$ in order to incorporate the contribution of the seven magnitude quartets into the convergence procedure:

$$Y_{\mathbf{h}} = \sum_{i=1}^{m} X_{\mathbf{hk}_i} \cos(\Phi^t_i) + \sum_{j=1}^{n} W'_{\mathbf{hk}_j\mathbf{l}_j} \cos(\Phi^q_j), \quad (7.204)$$

where $\Phi^t_i = \beta^t_{\mathbf{h}} - \varphi_{\mathbf{k}_i} - \varphi_{\mathbf{h}-\mathbf{k}_i}$ for the m triplet relationships, and $\Phi^q_j = \beta^q_{\mathbf{h}} + \varphi_{\mathbf{k}_j} + \varphi_{\mathbf{l}_j} + \varphi_{-\mathbf{h}-\mathbf{k}_j-\mathbf{l}_j}$ for the n quartet relationships. This gives

$$Pr(\Phi^q_j) \simeq \frac{e^{W'_{\mathbf{hk}_j\mathbf{l}_j} \cos(\Phi_j)} d\Phi_j}{2\pi I_0(W'_{\mathbf{hk}_j\mathbf{l}_j})}. \quad (7.205)$$

From Eqn. 7.174,

$$\langle\cos(\Phi^q_j)\rangle = \frac{2\pi I_1(W'_{\mathbf{hk}_j\mathbf{l}_j})}{2\pi I_0(W'_{\mathbf{hk}_j\mathbf{l}_j})} = \frac{I_1(W'_{\mathbf{hk}_j\mathbf{l}_j})}{I_0(W'_{\mathbf{hk}_j\mathbf{l}_j})} \quad \text{and} \quad (7.206)$$

$$\langle Y_{\mathbf{h}}\rangle = \sum_{i=1}^{m} X_{\mathbf{hk}_i} \left(\frac{I_1(X_{\mathbf{hk}_i})}{I_0(X_{\mathbf{hk}_i})}\right) + \sum_{j=1}^{n} W'_{\mathbf{hk}_j\mathbf{l}_j} \left(\frac{I_1(W'_{\mathbf{hk}_j\mathbf{l}_j})}{I_0(W'_{\mathbf{hk}_j\mathbf{l}_j})}\right). \quad (7.207)$$

Automated symbolic addition. The automated symbolic addition procedure employs seven magnitude quartet invariants in addition to triplet relationships in determining the optimal starting set. As with the "manual" process, it begins with a subset of the reflections with E_{hkl} above a certain value. At the completion of the convergence procedure, the origin-defining reflections are assigned numerical phases and several of the remaining reflections in the starting set are assigned symbolic phases. For the bikitaite data set, the E threshold is 1.24, resulting in 199 reflections to be used for structural solution. The starting set generated by *SIMPEL* is

h	k	l	E_{hkl}	φ_{hkl}
7	1	0	1.84	0
1	0	$\bar{10}$	2.65	0
11	0	$\bar{5}$	1.98	0
6	6	$\bar{3}$	2.85	s_1
6	0	$\bar{2}$	2.04	s_2
11	0	$\bar{7}$	2.76	s_3
5	2	$\bar{10}$	2.90	s_4
9	2	$\bar{3}$	1.94	s_5

The $6\,6\,\bar{3}$ reflection is selected to be the enantiomorph-defining reflection; once a numerical phase is determined for s_1 the phase will be constrained to retain its sign during tangent refinement. Note that only two of these reflections are the same as those selected in the original structural solution.

After the starting set has been determined, a limited symbolic addition process using both triplets and quartets is undertaken in order to expand the starting set. Only those reflections generating values of $\langle Y_h \rangle$ above a relatively high threshold are accepted, subject to the additional constraint that all of the assigned symbolic phases must be linearly independent (so that the new combinations of phases behave effectively as new symbols). For the bikitaite data, the starting set is expanded to include the following reflections:

h	k	l	E_{hkl}	φ_{hkl}
0	6	1	2.67	$s_1 + s_2$
5	6	$\bar{4}$	2.13	$s_1 + s_3$
5	0	$\bar{5}$	1.85	$s_2 + s_3$
4	0	7	2.68	$-s_4 + s_5$
7	5	1	2.39	$\pi + s_1 + s_2$
8	1	$\bar{10}$	2.88	0

In order to determine whether or not the expanded starting set is capable of generating symbolic phases for all of the reflections in the "large E" subset, the convergence procedure is reversed — becoming a *divergence* process. Beginning with the starting set, $\langle Y_{\mathbf{h}} \rangle$ is calculated for all of the reflections and the reflection with the largest $\langle Y_{\mathbf{h}} \rangle$ is added back into the reflection set. Since each addition affects all of the values of $\langle Y_{\mathbf{h}} \rangle$, the process is conducted in cycles, with a new cycle beginning after a certain minimum value of $\langle Y_h \rangle$ has been reached, thus guaranteeing that the most reliable reflections are accessed first. This minimum value is reduced for each cycle until all of the reflections have been added back into reflection set *or* a limiting value of the threshold is reached. In the latter case, if more than a few (e.g., 10) of the reflections cannot be accessed from the reflections added to the reflection set, then additional symbols are added and the process is re-initiated until the phases of all of the reflections can be generated from assigned numerical or symbolic phases.

Once the starting set is seen to be capable of providing phases for all of the reflections, the symbolic addition procedure is expanded in order to assign symbolic phases to the entire "large E" reflection set. *These phases are determined from triplet relations only*; they are assigned systematically so that more reliable triplets generate phases for less reliable triplets. In the beginning, only phases for which $\langle Y_{\mathbf{h}} \rangle$ is above a relatively high threshold will be accepted. At the end of the first cycle, only a few reflections have been added to the list of phased reflections. The threshold is then lowered, and a second cycle of symbolic addition is initiated. This cyclic process continues with a lower threshold for each cycle, until symbolic phases have been assigned to all of the reflections. *In the event that a reflection is assigned two or more inconsistent symbolic phases, the reflection is omitted from the list.*

The assignment of numerical phases: Figures of Merit. In the "manual" symbolic addition procedure, the assignment of numerical phases was accomplished by searching for unambiguous relationships between assigned phases that would yield a numerical value for a given symbol. For those symbols that could not be assigned a numerical phase, it was necessary to select sets of trial values for the symbols, refine the resulting numerical phases with the tangent formula, create an E map for each trial phase set, and search the map for a reasonable structure. Even with small numbers of symbols this process was a cumbersome one; with larger numbers of symbols it became a formidable one. In order to determine the trial numerical values of the symbols most likely to produce a correct structure, a number of formulas to generate values indicative of the most probable phase sets — *figures of merit* — have been devised (many had their origin in the multi-solution approach).* We will consider here the more common ones used in automated symbolic addition, and will take up others when multiple solution methods are discussed.

The *absolute figure of merit*[137] is a measure of the internal consistency of the assigned phases. A randomly distributed set of phases would be expected to contain no structural information. Even if this is the case, there will be a finite value for $Y_{\mathbf{h}}$. If the phases are randomly distributed, so are the values of α in Eqn. 7.106. The cosine and sine terms can then be replaced with their average values. Expanding the "cosine" portion of equation produces terms containing $< \cos(\alpha_i) >< \cos(\alpha_j) >= 0$, and terms containing $< \cos^2(\alpha_i) >= 1/2$. The results are identical for the "sine" portion of the equation. Thus

$$Y_{\mathbf{h}}(rand) = \left(\left(\sum_{j=1}^{m} \frac{1}{2} X^2_{\mathbf{hk}_j} \right) + \left(\sum_{j=1}^{m} \frac{1}{2} X^2_{\mathbf{hk}_j} \right) \right)^{1/2} = \left(\sum_{j=1}^{m} X^2_{\mathbf{hk}_j} \right)^{1/2} . \quad (7.208)$$

*Current direct methods software packages employ a number of practical variations of the figures of merit discussed in this section. Their presentation here has been designed to clarify the principles upon which they are based.

The absolute figure of merit formula compares the deviation of the computed value of $Y_{\mathbf{h}}$ from $Y_{\mathbf{h}}(rand)$ with the deviation of the expected value of $Y_{\mathbf{h}}$, $\langle Y_{\mathbf{h}}\rangle$ (computed during the convergence process) from $Y_{\mathbf{h}}(rand)$:

$$Z_{abs} = \frac{\sum_{\mathbf{h}}(Y_{\mathbf{h}} - Y_{\mathbf{h}}(rand))}{\sum_{\mathbf{h}}(\langle Y_{\mathbf{h}}\rangle - Y_{\mathbf{h}}(rand))}. \tag{7.209}$$

Z_{abs} is essentially a measure of the internal consistency among the triplet relationships. It will be close to zero if the determined phases are randomly distributed, and close to unity if the phases render all of the values of $Y_{\mathbf{h}}$ approximately equal to their expected values. Thus values of Z_{abs} that are close to unity tend to represent correct phase sets. A closely related absolute figure of merit is also employed in direct methods:

$$M_{abs} = \frac{\sum_{\mathbf{h}} Y_{\mathbf{h}}}{\sum_{\mathbf{h}}\langle Y_{\mathbf{h}}\rangle}. \tag{7.210}$$

M_{abs} is also a measure of the internal consistency of the triplet relationships, and is expected to be close to unity when the calculated values of $Y_{\mathbf{h}}$ are close to the expected values, again indicating a correct phase set.

A measure of the degree of deviation of $Y_{\mathbf{h}}$ from its statistically expected value is given by a similar figure of merit:

$$R_Y = \frac{\sum_{\mathbf{h}}(Y_{\mathbf{h}} - \langle Y_{\mathbf{h}}\rangle)}{\sum_{\mathbf{h}}\langle Y_{\mathbf{h}}\rangle}. \tag{7.211}$$

R_Y [126] should approach zero for a correct set of phases.

All of the figures of merit discussed above are based to some degree on the self-consistency of the phases assigned from the large $E_{\mathbf{h}}$ values; all tend to indicate the same thing since they depend on the criteria used to determine the phases themselves. It would be therefore appear useful to establish a figure of merit that depended on relationships separate from those involving the determined phases in the "large E" subset. To accomplish this we return to Sayres Equation (Eqn. 7.35), — valid for all reflections — *including those with small values of* $E_{\mathbf{h}}$:

$$\mathbf{E_h} = C\sum_{\mathbf{k}} \mathbf{E_k}\mathbf{E_{h\text{-}k}} \quad \text{and}$$

$$E_{\mathbf{h}} = C\left|\sum_{\mathbf{k}} E_{\mathbf{k}}E_{\mathbf{h\text{-}k}}e^{i(\varphi_{\mathbf{k}}+\varphi_{\mathbf{h-k}})}\right|. \tag{7.212}$$

In order to assign phases, we have focussed on triplets for which $E_{\mathbf{h}}$, $E_{\mathbf{k}}$, and $E_{\mathbf{h-k}}$ are all large. Suppose, instead, that we collect those triplets with assigned phases $\varphi_{\mathbf{k}}$ and $\varphi_{\mathbf{h-k}}$ for which $E_{\mathbf{k}}$, and $E_{\mathbf{h-k}}$ are both large, but for which $E_{\mathbf{h}} \simeq 0$. If $\varphi_{\mathbf{k}}$ and $\varphi_{\mathbf{h-k}}$ have been correctly assigned for all of the triplets associated with $E_{\mathbf{h}}$, then the terms must cancel in the sum in Eqn. 7.212 in order to correctly predict $E_{\mathbf{h}}$. On the other hand, if they are not assigned correctly it is unlikely that they will fortuitously cancel one another. Thus we consider a figure of merit based on these special triplets:[144]

$$\psi_0 = \frac{\sum_{\mathbf{h}}\left(\left|\sum_{\mathbf{k}} E_{\mathbf{k}}E_{\mathbf{h\text{-}k}}e^{i(\varphi_{\mathbf{k}}+\varphi_{\mathbf{h-k}})}\right|\right)}{\sum_{\mathbf{h}}\left(\left|\sum_{\mathbf{k}} E_{\mathbf{k}}E_{\mathbf{h\text{-}k}}\right|^2\right)^{1/2}}, \tag{7.213}$$

where $\mathbf{h}$ corresponds only to those reflections with values of $E_{\mathbf{h}}$ close to zero; ψ_0 should therefore be a small value for a correct set of phases.

Another important figure of merit that is independent of the triplet relationships used to assign phases is based on quartet invariants, since they are not used for final phase assignment. Positive quartets are *not* independent of the triplets, however. Given large values of $E_{\mathbf{h}}$ and $E_{\mathbf{k}}$, a large value of $E_{\mathbf{h}+\mathbf{k}}$ results in the triplet relationship $\varphi_{\mathbf{h}+\mathbf{k}} - \varphi_{\mathbf{h}} - \varphi_{\mathbf{k}} \approx 0$. In turn, large values of $E_{\mathbf{l}}$ and $E_{-\mathbf{h}-\mathbf{k}-\mathbf{l}}$ result in the triplet relationship $\varphi_{\mathbf{h}+\mathbf{k}} - \varphi_{\bar{\mathbf{l}}} - \varphi_{\mathbf{h}+\mathbf{k}+\mathbf{l}} \approx 0$. Another pair of triplet relationships will result from large values of $E_{\mathbf{h}+\mathbf{l}}$, and another from large values of $E_{\mathbf{k}+\mathbf{l}}$. Thus the existence of a seven magnitude quartet with large values of $E_{\mathbf{h}} \equiv E_{\bar{\mathbf{h}}}, E_{\mathbf{k}} \equiv E_{\bar{\mathbf{k}}}, E_{\mathbf{l}} \equiv E_{\bar{\mathbf{l}}}, E_{-\mathbf{h}-\mathbf{k}-\mathbf{l}} \equiv E_{\mathbf{h}+\mathbf{k}+\mathbf{l}}, E_{\mathbf{h}+\mathbf{k}}, E_{\mathbf{h}+\mathbf{l}}$, and $E_{\mathbf{k}+\mathbf{l}}$, will predict three pairs of related triplets. Conversely, the existence of the three triplet pairs with strong E values will predict the positive quartet relationship. While positive quartet relationships are not independent of triplet relationships, negative quartets are characterized by small values of $E_{\mathbf{h}+\mathbf{k}}$, $E_{\mathbf{h}+\mathbf{l}}$, and $E_{\mathbf{k}+\mathbf{l}}$, which are generally *not* contained in triplets used to assigned phases. Thus negative quartet relationships can be employed as an independent metric to determine whether or not the phases assigned from triplets will predict the expected values of $\sim \pi$ for $\varphi_{\mathbf{h}}+\varphi_{\mathbf{k}}+\varphi_{\mathbf{l}}+\varphi_{-\mathbf{h}-\mathbf{k}-\mathbf{l}}$. If $\varphi_{\mathbf{h}}$ is the phase of reflection $\mathbf{h}$ determined from triplet relationships, and $\phi_{\mathbf{h}} = \pi - \varphi_{\mathbf{k}} - \varphi_{\mathbf{k}} - \varphi_{-\mathbf{h}-\mathbf{k}-\mathbf{l}}$ is the phase predicted from a negative quartet relationship (with phases determined from triplets), then a straightforward figure of merit, summed over all negative quartet relationships is

$$N_{fom} = \frac{\sum_{\mathbf{h}} Y_{\mathbf{h}} |\phi_{\mathbf{h}} - \varphi_{\mathbf{h}}|}{\sum_{\mathbf{h}} Y_{\mathbf{h}}}, \tag{7.214}$$

where $Y_{\mathbf{h}}$ weights each term in proportion to the reliability of the triplet-determined phase. N_{fom} is expected to approach zero for a correctly assigned phase set. Negative quartets have been shown to be a powerful discriminator between phase sets for which other figures of merit often fail to provide a reasonable structure.

While each of these figures of merit has its own intrinsic value, it is convenient to combine them in some manner to provide a single measure of the likelihood that a phase set is reasonable — a *combined figure of merit.* This is generally a weighted sum of the individual figures of merit:

$$C_{fom} = w_1 \frac{1}{Z_{abs}} + w_2 R_Y + w_3 \psi_0 + w_4 N_{fom}. \tag{7.215}$$

C_{fom} should be a small value for a correct phase set. The weights are often unit weights by default, but can be adjusted by the investigator to emphasize a specific figure of merit. In practice the combined figures of merit from a selected number of trial sets with the smallest figures of merit are ordinarily divided by the maximum value of C_{fom} so that they are scaled to be less than or equal to unity.

The *SIMPEL* program generates potential phase sets by creating trial sets with each of the initial symbols assigned phases in increments of 45 degrees. Restricted reflections are assigned phases consistent with the restrictions (e.g., 0 or π). With modern computers it would be possible at this stage to subject each of these possible solutions to tangent refinement (see below) and select the sets with the best figures of merit. However, recalling that *SIMPEL* has it roots in a search for optimal computational efficiency, the program evaluates possible phase sets prior to tangent refinement. Furthermore, in order to limit the number of phase sets to a manageable number, it is possible to estimate the internal consistency of each set by comparing $\sum_{\mathbf{k}} X_{\mathbf{hk}}$ to $Y_{\mathbf{h}}$ for each reflection. Referring to Fig. 7.12, if the individual phases predicted from each triplet are identical, then $Y_{\mathbf{h}}$ will equal $\sum_{\mathbf{k}} X_{\mathbf{hk}}$. We would therefore expect

$$Q = \sum_{\mathbf{h}} \left(\sum_{\mathbf{k}} X_{\mathbf{hk}} - Y_{\mathbf{h}} \right) \tag{7.216}$$

to be a minimum when the phases calculated from each triplet for a given reflection are close to one another. Schenk devised a closely related (and somewhat less rigorous) figure of merit, Q_{fom}, that is calculated much more rapidly than Q.[145] It is used in the *SIMPEL* program, both to limit the number of trial phase sets to those with the smallest values of Q_{fom}, and to use it as a figure of merit, incorporated as a term in C_{fom}, which is then scaled so that its maximum value is 1.0.

For the 199 reflections from the bikitaite data set with assigned symbolic phases, this process resulted in the following sets of trial phases, ranked by decreasing internal consistency (increasing Q_{fom}):

Set	Q_{fom}	s_1	s_2	s_3	s_4	s_5	$1/Z_{abs}$	R_Y	ψ_0	N_{fom}	C_{fom}
1	0.50	π	π	0	π	0	2.18	0.41	0.37	0.00	0.59
2	0.51	π	π	0	0	π	5.36	0.60	0.28	0.00	1.15
3	0.55	$\frac{3\pi}{4}$	π	0	π	0	6.74	0.63	0.30	0.00	1.40
4	0.55	$\frac{3\pi}{4}$	π	0	π	0	6.74	0.63	0.30	0.00	1.40
5	0.56	$\frac{3\pi}{4}$	π	0	0	π	31.03	0.71	0.23	0.00	5.57
6	0.56	$-\frac{3\pi}{4}$	π	0	0	π	31.03	0.71	0.23	0.00	5.57
7	0.67	$\frac{\pi}{2}$	π	0	π	0	2.13	0.40	0.42	0.00	0.62
8	0.67	$-\frac{\pi}{2}$	π	0	π	0	2.13	0.40	0.42	0.00	0.62
9	0.67	$\frac{\pi}{2}$	π	0	0	π	5.18	0.60	0.29	0.00	1.15
10	0.67	$-\frac{\pi}{2}$	π	0	0	π	5.18	0.60	0.29	0.00	1.15
11	0.78	$\frac{\pi}{4}$	π	0	0	π	24.42	0.71	0.25	0.00	4.48
12	0.78	$-\frac{\pi}{4}$	0	π	0	π	24.42	0.71	0.25	0.00	4.48
13	0.78	$\frac{\pi}{4}$	π	0	π	0	5.32	0.60	0.27	0.00	1.19
14	0.78	$-\frac{\pi}{4}$	π	0	π	0	5.32	0.60	0.27	0.00	1.19
15	0.83	0	π	0	0	π	5.87	0.61	0.26	0.00	1.29
16	0.83	0	π	0	π	0	2.37	0.43	0.38	0.00	0.68

Although phase sets 1 and 16 appear to be reasonable candidates, they were rejected by the program because the average value of $\varphi_{\mathbf{h}}$ for each phase set was over 45° from the predicted average value of the phase for the structure, computed by averaging phases determined from Eqn. 7.175. With the exception of phase sets 7 and 8, the remaining sets were rejected because of the large value of $1/Z_{abs}$. There were (apparently) insufficient negative quartets for the program to determine N_{fom}.

Once the phase sets deemed most likely to provide a solution are selected, the phases are refined using the tangent formula. In this *tangent refinement*, the tangent formula is applied sequentially to the multiple triplet relationships for each reflection, generating an improved set of phases. As the process proceeds through the reflection list, those reflections later on in the list will be evaluated based on improved phases from preceding reflections. Thus another cycle through the list is required to subject those reflections earlier on in the list to improved phases, and so forth. As the phase estimate for a specific reflection improves, Y_h for the reflection increases. The tangent refinement cycles continue until there is no longer any change in $\sum_{\mathbf{h}} Y_h^2$.

Phase set 7 was chosen as the best candidate (phase set 8 will have all the phases negated – corresponding to the enantiomorph) and subjected to tangent refinement, followed by computation of an E map and a search for peaks. The following peaks and their maximum values were located in the map:

Peak	x_f	y_f	z_f	max
1	0.3856	0.9632	0.4336	30.556
2	0.1049	0.9608	0.5884	26.036
3	0.1085	0.4816	0.5881	22.439
4	0.1050	0.0258	0.0000	21.512
5	0.4640	0.9112	0.5530	19.209
6	0.3851	0.4685	0.4362	19.115
7	0.3920	0.5454	0.9970	17.697
8	0.0654	0.1962	0.9453	14.277
9	0.1899	0.5942	0.1200	14.188
10	0.3953	0.0297	0.9812	14.037

A three dimensional plot of these peaks, illustrated in Fig. 7.31(a), reveals virtually no structural information. Molecular crystal structures often contain groups that are easily identifiable in such a plot, but the structures of extended solids such as minerals are generally more difficult to discern. $Li[AlSi_2O_6]\cdot H_2O$ is a member of a large class of minerals known as aluminosilicates. These materials can be envisioned as derivatives of silica, SiO_2, which contains silicon atoms, formally existing as Si^{4+} ions, surrounded by four tetrahedrally arranged oxygen atoms (formally O^{2-}). The minerals are formed by the replacement of a Si^{4+} ion with an Al^{3+} ion in a tetrahedral site; another positive ion must therefore occupy a position in the lattice, in this case the Li^+ ion, in order to maintain charge balance. The silicon atoms in SiO_2 are separated by 3.1 Å; Peaks 1 and 2 are separated by 3.1 Å, as are peaks 2 and 4. Peak 3 is 3.8 Å from peak 1, 2.4 Å from peak 2, and 4.1 Å from peak 4, and is unlikely to be a Si or Al atom. The peaks with the largest values in the map are likely to be the heavier Al and Si atoms, which should occupy tetrahedral sites and be approximately 3.1 Å apart. It therefore seems seasonable to assign two Si atoms and one Al atom to the positions 1, 2, and 4 (Fig. 7.31(b))* and attempt to locate the remaining atoms by difference synthesis. This exercise underscores an important practical axiom for crystal structure determination – the more that one knows about the contents of the crystals under investigation, the greater the chances for a solution.

*At this juncture the choice of the specific type of atom to assign to each site is somewhat arbitrary since Si and Al differ by only a single electron, and will scatter X-rays similarly.

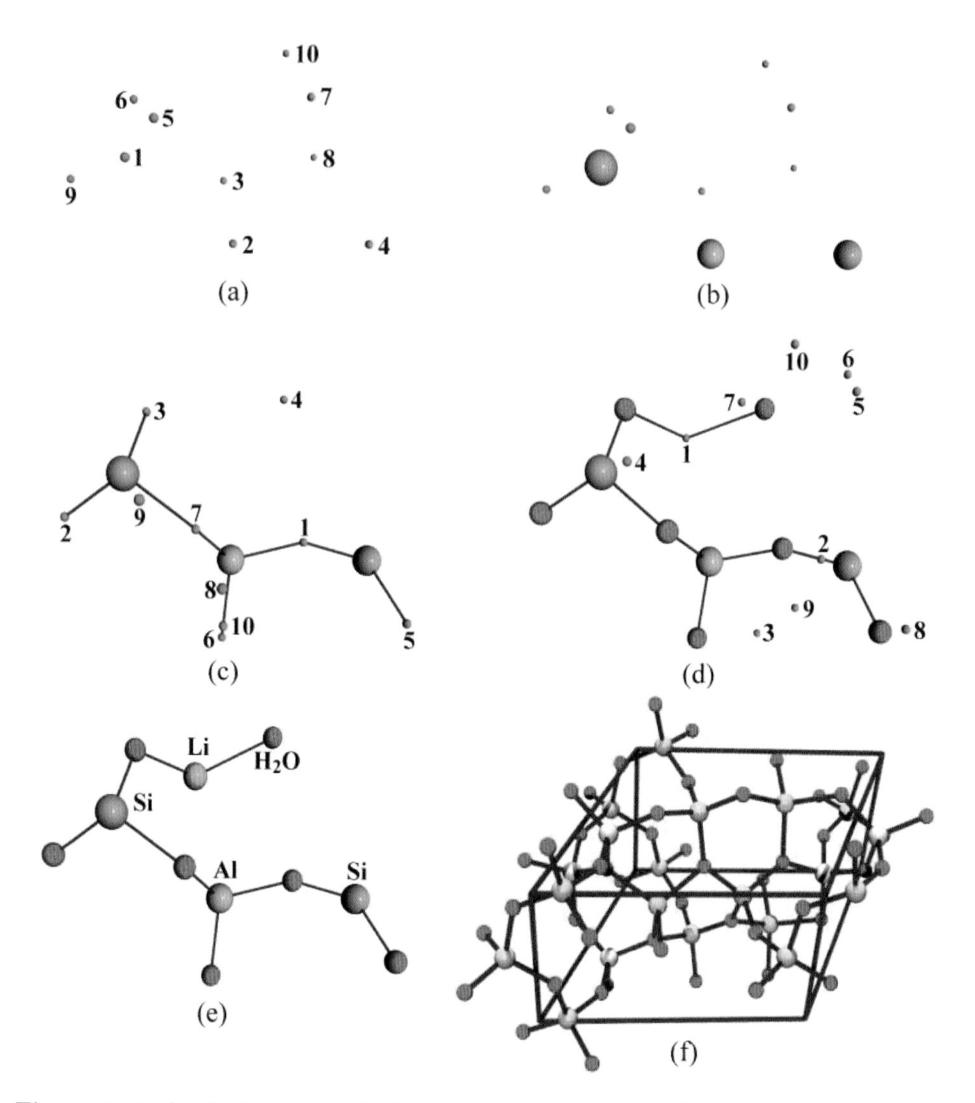

Figure 7.31 Symbolic addition/difference Fourier solution of the Li[$AlSi_2O_6$]·H_2O structure: (a) E map. (b) Assignment of heavy atoms. (c) First ΔF map. (d) Second ΔF map. (e) Completed structure. (f) Packing diagram. The small blue spheres represent locations of maxima in the Fourier maps.

Four cycles of least-squares refinement, followed by the generation of a ΔF map revealed the peaks shown in Fig. 7.31(c). The peak numbers correspond to their relative magnitudes. Peaks 1–7 had magnitudes of 7.51–10.75 $e/\text{Å}^3$; peaks 8–10 had magnitudes of less than 4.54 $e/\text{Å}^3$. The seven peaks of largest magnitude were assigned as oxygen atoms, resulting in Fig. 7.31(d). A second cycle produced a single large peak (peak 1) of magnitude 4.01 $e/\text{Å}^3$; the remaining 9 peaks had magnitudes of less than 0.99 $e/\text{Å}^3$. Peak 1 was assigned as a lithium atom; completing the structure as shown in Fig. 7.31(e). The structure of the mineral is revealed in the packing diagram in Fig. 7.31(f), clearly showing the arrangement of the tetrahedral sites. The authors of the original structural report argued that two of the three different tetrahedral sites within the lattice are disordered, occupied equally by Al and Si atoms, while the third is occupied solely by Si atoms. We will revisit this in the discussion of refinement in the next chapter.

Although we have spent a considerable amount of time discussing symbolic addition, very few crystal structures are solved by this method today. Faster computers and more powerful multisolution algorithms solve most small molecule structures. Even so, symbolic addition can be a useful alternative when these methods fail. Furthermore, it is an elegantly logical approach to structural solution that allows the student/investigator to observe many of the underpinnings of direct methods that are much more difficult to discern in today's "black box" software packages.

Phase Determination: Multiple Solutions

Although the symbolic addition procedure provided the first systematic approach to direct structure determination, it suffered from a number of drawbacks. The manual assignment of phases required significant expertise, and was tedious, time-consuming, and easily subject to error during the symbol-assignment process. A more serious defect arose from the use of linear combinations of symbolic phases to determine new phases, which are fundamentally cyclic in nature. This resulted in the assignment of ambiguous phases, forcing the investigator to either decide which of the assignments was valid (based on the relative magnitudes of $X_{\mathbf{hk}}$, $\langle Y_{\mathbf{h}} \rangle$, or some other criterion), or ignore the reflections altogether (as in *SIMPEL*). For example, during the "manual" assignment of symbols in the bikitaite structure, the following triplet relationships occurred:

(1)							(2)							(3)							(4)						
h	k	l	π	s_1	s_2	s_3	h	k	l	π	s_1	s_2	s_3	h	k	l	π	s_1	s_2	s_3	h	k	l	π	s_1	s_2	s_3
$\bar{8}$	3	6	0	0	1	0	$\bar{9}$	2	3	0	0	0	1	0	2	5	0	0	0	1	$\bar{3}$	2	1	0	−1	0	1
$\bar{5}$	2	5	1	0	0	1	$\bar{3}$	2	6	1	1	0	1	5	2	$\bar{5}$	1	0	0	2	2	0	$\bar{4}$	0	−1	0	−1
$\bar{3}$	1	1	1	−1	0	2	$\bar{6}$	0	$\bar{3}$	1	−1	0	−1	$\bar{5}$	0	10	1	−1	0	−1	$\bar{5}$	2	5	1	0	0	2
$s_3 = \frac{1}{2}s_1$							$s_3 = 2\pi \equiv 0$							$s_1 = 2\pi \equiv 0$							$\pi = 0$!!						

At first glance triplets (1), (2), and (3) appear to be consistent with one another. However, as noted earlier, a number of strong relationships (with large values of $X_{\mathbf{hk}}$) indicated that s_1 is almost certainly equal to π; triplet (3) is ambiguous, and probably incorrect. Thus triplet (1) indicates that $s_3 \approx \pi/2$, while triplet (2) indicates that $s_3 \approx 0$. Triplet (4) leads us to the untenable conclusion that $\pi \approx 0$!

The "symbol inconsistency" problem arises because the symbols from different predicted phases must combine in integer increments of one another. This lead Michael Woolfson and Gabriel Germain[146] to propose that the phases assigned to a reflection from different triplet relationships would combine more naturally if the phases were initially assigned numerical values by direct application of the tangent formula. The symbolic addition method had already demonstrated that trial phase sets with phases assigned to the reflections in the starting set in increments of $\pi/4$ would lead to reasonable solutions, and Germain and Woolfson proposed that phases for the starting set be assigned all possible permutations of the four phases

$\pm\pi/4$ and $\pm 3\pi/4$ (*quadrant permutation*). The tangent formula could then be applied systematically to each of the trial starting sets in order to generate phases for the remaining reflections in the "large E" subset. Each of the resulting trial phase sets would then be tested as a plausible solution – a *multisolution* approach to the problem. Based upon initial success with the approach, Germain and Woolfson collaborated with Peter Main to create the multiple-solution tangent-phasing program known as *MULTAN*.[147]

As with the symbolic addition procedure, the multisolution process begins with a "large E" subset, with phases assigned from triplet relationships. To limit the possibility that an incorrectly assigned phase will lead to failure, the starting set is determined by the *convergence method* previously described for the symbolic addition procedure (the method was originally devised for use in *MULTAN*). Recall that convergence creates a starting set by sequentially rejecting reflections with larger predicted variances (smaller values of $\langle Y_{\mathbf{h}} \rangle$), indicating fewer connections via triplets to other reflections and/or "weaker" triplet relationships (smaller $X_{\mathbf{hk}}$ values). Following the determination of the starting set, numerical phases are assigned to each member of the set, $\pm\pi/4$ and $\pm 3\pi/4$ for general reflections (*quadrant permutation*), and appropriate values assigned to "special" reflections (e.g., 0 or π; $\pm\pi/2$, etc.). This creates a large number of trial phase sets (depending on the number of reflections in the starting set), each of which must be evaluated independently. If there are four general reflections in the starting set in addition to those defining the origin and enantiomorph (assigned initially), each of which can take on any one of four values, there will be 4^4 trial starting sets, resulting in 256 trial phase sets to evaluate.

In the symbolic addition procedure, the convergence process was reversed – becoming *divergence* – to test the ability of the starting set to generate phases for the remaining reflections. *In the multiple solution approach divergence is utilized to systematically assign numerical phases.* During convergence, each cycle adds another reflection to the starting set, and as the selected number of reflections in the starting set is approached, the final reflections that are eliminated are those that will be most related to the reflections in the starting set. These reflections provide an essential link to the more weakly related reflections; the last reflection to be temporarily eliminated is thus added back to the starting set to provide an appropriate bridge to the reflections that have been rejected from inclusion in the starting set. The longer that a reflection remains in the list, the stronger its links to the remaining reflections. The order of the reflections removed during convergence is therefore stored so that it can be utilized in the divergence process.

Divergence begins by effectively reversing the convergence process. Starting with the last removed reflection – the reflection most strongly linked to the starting set – the starting set phases are applied to the tangent formula in order to determine its phase. This new phase, along with the starting set phases, is then used by the tangent formula to compute the phase of the "next best" reflection and so forth. Thus each phase is determined only from those reflections in the starting set and those reflections with stronger links. As each phase is assigned, $Y_{\mathbf{h}}$, is computed. In the early stages of phase development phase assignments are limited to a selected subset of the reflections, e.g., the last 60–100 to be eliminated during convergence. Only those phases with $Y_{\mathbf{h}}$ above a given threshold (i.e., those that are most reliably predicted) are accepted. In this manner the reflections with the weakest links are assigned phases at the end of the process, when they have the lowest probability of sending the phase-determining process down an unproductive path. After the phases of the subset have been assigned, the remaining reflections are added back into the list. Starting from the beginning of the list, the phases are redetermined using the tangent formula, now incorporating all of the assigned phases. This tangent refinement is then repeated cyclically to self-consistency, generating a complete set of phases for the trial set. *MULTAN* employed triplet and Σ_1 relations exclusively in convergence and divergence, but later modifications also incorporated quartet relationships.

The weighted tangent formula. The tangent formula expressed in Eqn. 7.107 gives equal weight to the phase contribution from each triplet. However, the phases from each contributor in the sums are not equally reliable, and it would clearly be appropriate to weight each $E_{\mathbf{k}}$ and $E_{\mathbf{h}-\mathbf{k}}$ by a parameter related to its respective variance. Thus a weighted tangent formula takes the form

$$\tan\beta_{\mathbf{h}} = \frac{\sum_{j=1}^{m} w_{\mathbf{k}_j} E_{\mathbf{k}_j} w_{\mathbf{h}-\mathbf{k}_j} E_{\mathbf{h}-\mathbf{k}_j} \sin(\varphi_{\mathbf{k}_j} + \varphi_{\mathbf{h}-\mathbf{k}_j})}{\sum_{j=1}^{m} w_{\mathbf{k}_j} E_{\mathbf{k}_j} w_{\mathbf{h}-\mathbf{k}_j} E_{\mathbf{h}-\mathbf{k}_j} \cos(\varphi_{\mathbf{k}_j} + \varphi_{\mathbf{h}-\mathbf{k}_j})}, \tag{7.217}$$

where $w_{\mathbf{k}}$ is a weighting factor that becomes smaller as the expected variance of $\varphi_{\mathbf{k}}$ increases. Since $Y_{\mathbf{k}}$ becomes smaller as the variance increases, Germain[147] suggested that a suitable weighting factor would assume an approximately linear relationship between the reciprocal of the variance and $Y_{\mathbf{k}}$: $w_{\mathbf{k}} = c\,Y_k$. Germain proposed a value of $c = 0.2$, with each weight limited to a maximum value, $w_{max} = 1$.

For space groups without translational symmetry, known as *symmorphic* space groups, this simple linear weighting scheme often works too well! For such space groups (e.g., $P\bar{1}$), when all of the phases approach zero, the tangent formula becomes self-consistent (every phase of zero produces another phase of zero), and each $Y_{\mathbf{h}}$ approaches a maximum (see Fig. 7.26):

$$Y_{\mathbf{h}} = \sum_{j=1}^{m} X_{\mathbf{hk}_j} \cos(\beta_{\mathbf{h}} - \varphi_{\mathbf{k}_j} - \varphi_{\mathbf{h}-\mathbf{k}_j}) = \sum_{j=1}^{m} X_{\mathbf{hk}_j}. \tag{7.218}$$

During the assignment of phases, as more and more phases are set equal to zero, more weights take on their maximum values, tending to drive the phase assignment process toward an unlikely solution – with all of the phases equal to zero! In order to avoid this problem Hull and Irwin[148] devised a weighting scheme in which $w_{\mathbf{k}}$ would steadily increase with increasing $Y_{\mathbf{h}}$ until $w_{\mathbf{k}}$ reached a value of one, after which $w_{\mathbf{k}}$ would decrease, thus avoiding the problem presented by too many large values of $Y_{\mathbf{k}}$. The scheme was based on the argument that the most reliable phases were those with values of $Y_{\mathbf{k}}$ closest to their expected values: $Y_{\mathbf{k}} \simeq \langle Y_{\mathbf{k}}\rangle \Rightarrow Y_{\mathbf{k}}/\langle Y_{\mathbf{k}}\rangle \simeq 1$. The weighting function incorporated the ratio of $Y_{\mathbf{k}}$ to its expected value as the independent variable, with the weight increasing until $Y_{\mathbf{k}}/\langle Y_{\mathbf{k}}\rangle$ reaches one. The function in this region is approximately Gaussian and decreases more gradually outside of the "Gaussian" region (as the integral becomes larger):

$$w_{\mathbf{k}} = C\,e^{-(Y_{\mathbf{k}}/\langle Y_{\mathbf{k}}\rangle)^2} \int_0^{Y_{\mathbf{k}}/\langle Y_{\mathbf{k}}\rangle} e^{-x^2}\,dx. \tag{7.219}$$

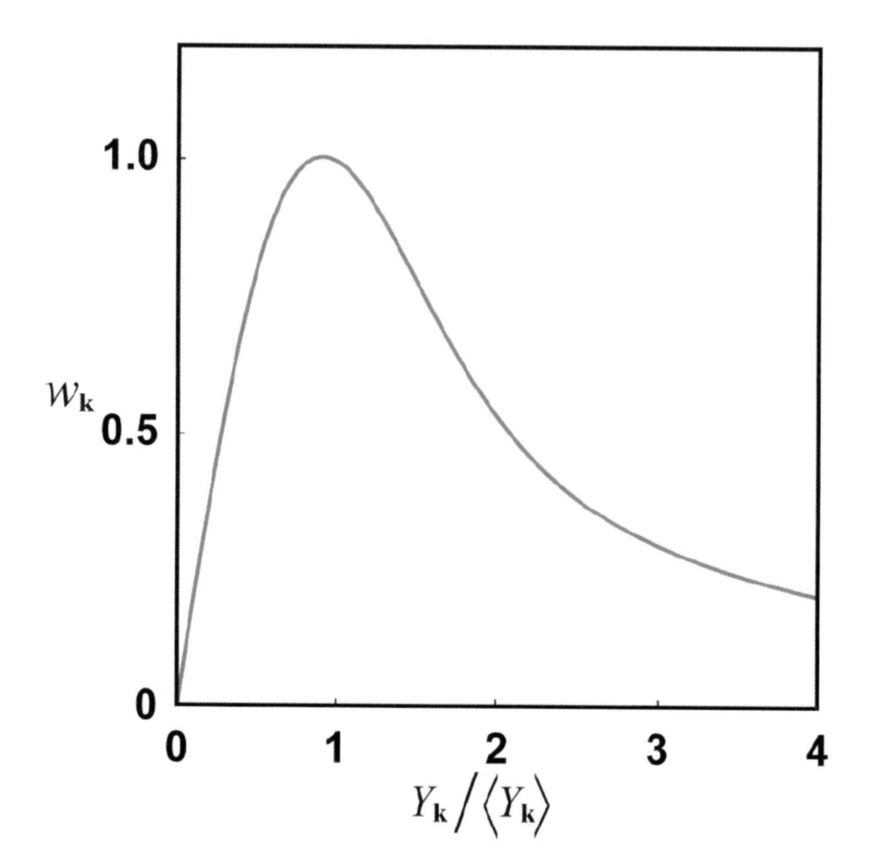

Figure 7.32 Hull-Irwin weights as a function of $Y_k/\langle Y_k\rangle$.

The function has a maximum value at $Y_{\mathbf{k}}/\langle Y_{\mathbf{k}}\rangle = 1$; $C \simeq 0.58$ is a scaling constant that renders the maximum value of $w_{\mathbf{k}}$ equal to unity. The Hull-Irwin weighting function is plotted in Fig. 7.32. There are a number of variations of this weighting scheme incorporated in current direct methods software.

Selecting the most likely solution(s): Figures of Merit. In automated symbolic addition the most likely structural solutions are selected by comparing figures of merit for the trial phase sets. The multisolution approach generates many more trial phase sets, and in most cases it would be virtually impossible to evaluate each phase set by producing an E map and searching for a structure. Thus figures of merit take on an even more important role in multiple solution procedures. Most direct methods algorithms employ an internal consistency "absolute" figure of merit such as Z_{abs} (Eqn. 7.209) or M_{abs} (Eqn. 7.210). In addition, *MULTAN* used the ψ_0 figure of merit (Eqn. 7.213), R_Y (Eqn. 7.211), and a combined figure of merit

$$C_{fom} = w_1 \frac{Z_{abs} - Z_{abs}^{min}}{Z_{abs}^{max} - Z_{abs}^{min}} + w_2 \frac{\psi_0^{max} - \psi_0}{\psi_0^{max} - \psi_0^{min}} + w_3 \frac{R_Y^{max} - R_Y}{R_Y^{max} - R_Y^{min}}. \tag{7.220}$$

The minimum and maximum values in each ratio create larger values for the optimally smaller values of ψ_0 and R_Y; higher values of C_{fom} represent probable solutions. Weights of 0.6, 1.2, and 1.2, respectively, scale C_{fom} so that the highest value is 3.0 and the lowest is 0.0.

Current multisolution programs also incorporate some form of a negative quartet figure of merit as a metric for the comparison of trial phase sets, since the negative quartets (like ψ_0) are independent of the triplet relationships used for phase assignment. A simple criterion for estimating the integrity of a trial phase set involves a sum over all quartets with small cross-terms[149]:

$$N_{qest} = \frac{\sum_{\mathbf{h}}\sum_{\mathbf{k}}\sum_{\mathbf{l}} \frac{2\sigma_4}{\sigma_2^2} E_{\mathbf{h}} E_{\mathbf{k}} E_{\mathbf{l}} E_{-\mathbf{h}-\mathbf{k}-\mathbf{l}} \cos(\varphi_{\mathbf{h}} + \varphi_{\mathbf{k}} + \varphi_{\mathbf{l}} + \varphi_{-\mathbf{h}-\mathbf{k}-\mathbf{l}})}{\sum_{\mathbf{h}}\sum_{\mathbf{k}}\sum_{\mathbf{l}} \frac{2\sigma_4}{\sigma_2^2} E_{\mathbf{h}} E_{\mathbf{k}} E_{\mathbf{l}} E_{-\mathbf{h}-\mathbf{k}-\mathbf{l}}}$$

$$= \frac{\sum_{\mathbf{h}}\sum_{\mathbf{k}}\sum_{\mathbf{l}} W_{\mathbf{hkl}} \cos(\varphi_{\mathbf{h}} + \varphi_{\mathbf{k}} + \varphi_{\mathbf{l}} + \varphi_{-\mathbf{h}-\mathbf{k}-\mathbf{l}})}{\sum_{\mathbf{h}}\sum_{\mathbf{k}}\sum_{\mathbf{l}} W_{\mathbf{hkl}}}. \quad (7.221)$$

For a correct structure, $\Phi_{\mathbf{hkl}} = \varphi_{\mathbf{h}} + \varphi_{\mathbf{k}} + \varphi_{\mathbf{l}} + \varphi_{-\mathbf{h}-\mathbf{k}-\mathbf{l}} \approx \pi$ for the negative quartets, and N_{qest} will therefore tend to be negative, with a minimum value of -1. The maximum value of N_{qest} is +1, indicating $\Phi_{\mathbf{hkl}} \approx 0$, inconsistent with the negative quartets of a viable phase set.

E maps are generated from the phased reflections of the trial sets with the most favorable figures of merit and searched for maxima. The set with the highest combined figure of merit often provides the correct solution, but many times the actual solution is determined from a phase set with less favorable figures of merit.

Expanding the starting set: Magic Integers. The expansion of the starting set beyond the initially assigned phases was necessitated by the tendency for a single "bad" phase relationship to lead to an incorrect structure unless there are a sufficient number of early "good" phase relationships to keep the phase assignment process on a productive path. It follows that increasing the number of reflections in the starting set would tend to obviate this problem, as we have already observed in the automatic symbolic addition procedure. Unfortunately, quadrant permutation results in 4^n trial phase sets for n general reflections in the starting set. For example, if we were to include the additional reflections included in the starting set for the symbolic addition solution of the bikitaite structure, the number of trial phase sets would be $4^8 = 65536$! Each of these sets would require phase assignments and tangent refinement. This would be a tremendous task for the fast computers of today; in the early 1970's, it was simply beyond the reach of most computer systems.

This led Woolfson and White to search for a way to assign starting set phases that would systematically lead to the most probable trial phase sets[150].* The procedure had it roots in symbolic addition, but instead of assigning symbols to phases in the starting set, a number of phases are assigned simultaneously by combining

*Woolfson and White initially proposed this approach as a phase-determining procedure.

a single symbol with a number of integers, which the authors coined *magic integers*. We will find it convenient here to express phases in terms of cycles rather than radians or degrees (e.g., $\pi/4 \equiv 1/8$). For three arbitrarily chosen phases, we consider the following three relationships, which arise from the cyclic nature of the phases — the difference between a phase and an "equivalent" phase will always be an integer number of cycles:

$$\varphi_1 - m_1 s = i_1 \quad (7.222)$$

$$\varphi_2 - m_2 s = i_2 \quad (7.223)$$

$$\varphi_3 - m_3 s = i_3, \quad (7.224)$$

where s is a non-integer symbol between 0 and 1 (cycle) and m_1, m_2, m_3, i_1, i_2, and i_3 are integers. Thus, for example, φ_1 and $m_1 s$ are "equivalent" phases, differing by i_1 cycles. While there will always be some value of s that will satisfy any one of the equations above, the authors demonstrated that, for n such expressions, a single symbol could be found that would *approximately* satisfy all of the expressions simultaneously. This occurs because the phases are cyclic, so that a single number can be found that will put each phase in a set of phases in the vicinity of its equivalent value if the number is multiplied by an appropriate (magic) integer. While this is not entirely intuitive (and the details of the theory are beyond our scope[151]), two examples from the original paper will serve as illustrations:

Selecting integers $m_1 = 3$, $m_2 = 4$, and $m_3 = 5$, we consider $\varphi_1 = 0.3$, $\varphi_2 = 0.2$ and $\varphi_3 = 0.7$ and search for a value of s that will predict the phases with the least average error. For $s = 0.766$,

$$\begin{aligned}
&\varphi_1 - m_1 s = 0.3 - 3 \times 0.766 = 0.3 - 2.298 \simeq 0.3 - (\sim 2.3) = -2 \\
&\varphi_1 - 2.298 \simeq -2 \Longrightarrow \varphi_1 \simeq 0.298 \\
&\varphi_2 - m_2 s = 0.2 - 4 \times 0.766 = 0.2 - 3.064 \simeq 0.2 - (\sim 3.2) = -3 \\
&\varphi_2 \simeq 0.064 \\
&\varphi_3 - m_3 s = 0.7 - 5 \times 0.766 = 0.7 - 3.830 \simeq 0.7 - (\sim 3.7) = -3 \\
&\varphi_3 \simeq 0.830.
\end{aligned}$$

The predicted phases are in error by 0.002, 0.136, and 0.130 cycles for φ_1, φ_2, and φ_3, respectively. Using the same integers, we select the phases $\varphi_1 = 0.8$, $\varphi_2 = 0.6$ and $\varphi_3 = 0.0$, and find the value of the symbol that minimizes the average error: $s = 0.622$, and

$$\begin{aligned}
&\varphi_1 - m_1 s = 0.8 - 3 \times 0.622 = 0.1 - 1.866 \simeq 0.8 - (\sim 1.8) = -1 \\
&\varphi_1 - 1.866 \simeq -1 \Longrightarrow \varphi_1 \simeq 0.866 \\
&\varphi_2 - m_2 s = 0.6 - 4 \times 0.622 = 0.6 - 2.488 \simeq 0.6 - (\sim 2.6) = -2 \\
&\varphi_2 \simeq 0.488 \\
&\varphi_3 - m_3 s = 0.0 - 5 \times 0.622 = 0.0 - 3.110 \simeq 0.0 - (\sim 3.0) = -3 \\
&\varphi_3 \simeq 0.110,
\end{aligned}$$

with errors in the predicted phases of 0.066, 0.111, and 0.111 cycles, respectively. Having selected the "magic" integers 3,4, and 5, we can approximate three arbitrarily chosen phases with a single symbol, s.

Suppose that we expand the starting set to include three sets of three reflections with phases related by a large number of triplet relationships – reflections eliminated during the final steps of the convergence process. Using the same three magic integers,

$$\begin{aligned} \varphi_1 &= i_1 + 3s_1 & \varphi_4 &= i_4 + 3s_2 & \varphi_7 &= i_7 + 3s_3 \\ \varphi_2 &= i_2 + 4s_1 & \varphi_5 &= i_5 + 4s_2 & \varphi_8 &= i_8 + 4s_3 \\ \varphi_3 &= i_3 + 5s_1 & \varphi_6 &= i_6 + 5s_2 & \varphi_9 &= i_9 + 5s_3 \,. \end{aligned}$$

Let us further suppose that one of the triplet relationships is

$$\varphi_1 + \varphi_4 + \varphi_6 \approx 0 \equiv 2\pi,$$

which, in cycles is

$$\varphi_1 + \varphi_4 + \varphi_6 \approx k',$$

where k' represents an integral number of cycles. The relationship is approximated by

$$\begin{aligned} &(i_1 + 3s_1) + (i_4 + 3s_2) + (i_6 + 5s_2) \approx k' \\ &3s_1 + 3s_2 + 5s_2 = k' - i_1 - i_4 - i_6 \approx k \\ &3s_1 + 8s_2 \approx k, \end{aligned}$$

another integral number of cycles. If s_1, s_2, and s_3 are expressed in radians,

$$\begin{aligned} &3s_1 + 3s_2 + 5s_2 \approx 2k\pi \equiv 0 \\ &3s_1 + 8s_2 \approx 2k\pi \equiv 0. \end{aligned}$$

Recall from symbolic addition that a general triplet includes a constant term, s_0 (often 0, π, or $\pi/2$) to reflect space group symmetry. Thus a general relationship for triplets containing any three of these 9 phases in terms of the three symbols s_1, s_2, and s_3 is given by

$$Is_1 + I's_2 + I''s_3 + s_0 \approx 2k\pi. \tag{7.225}$$

This implies that

$$\cos(2k\pi) \approx \cos(Is_1 + I's_2 + I''s_3 + s_0) \approx 1, \tag{7.226}$$

with $\cos(Is_1 + I's_2 + I''s_3 + s_0)$ approaching unity, but ≤ 1. The likelihood that this relationship holds is measured by the likelihood that the triplet relationship holds, indicated by the magnitude $X_{\mathbf{hk}}$. It is also measured by just how close $\cos(Is_1 + I's_2 + I''s_3 + s_0)$ is to unity. It follows that the sum over all of the triplets involving these 9 reflections,

$$S(s_0, s_1, s_2, s_3) = \sum_i E_{\mathbf{h},i} E_{\mathbf{k},i} E_{\mathbf{h}-\mathbf{k},i} \cos(Is_1 + I's_2 + I''s_3 + s_0), \tag{7.227}$$

should be a maximum for the best values of s_1, s_2 and s_3. By numerically evaluating $S(s_0, s_1, s_2, s_3)$ $(0 < s_1, s_2, s_3 < 2\pi)$ in small increments of s_1, s_2, and s_3, we can search for maxima. The values of s_1, s_2, and s_3 for these maxima allow us to establish the initial phases of starting sets for phase assignment with the tangent formula and subsequent tangent refinement (in which the starting set phases themselves can be refined). Each maximum represents a possible solution, resulting in significantly fewer trial solutions to analyze, compared to those generated by quadrant permutation with an expanded starting set.

The employment of magic integers in expanding the starting set was often unsuccessful in early versions of the approach, due to errors in the phase approximations. Subsequent improvements in the method, however, resulted in a program known as *MAGIC* that was able to solve structures that could not be solved with *MULTAN*.

Expanding the starting set: Random Phases. In 1977 Woolfson reported an improved method for phase refinement that did not involve the tangent formula[152]. It arose naturally from the "integer cycle" concept that produced the magic integers. For a general triplet relationship,

$$\pm\varphi_p \pm \varphi_q \pm \varphi_r + s_0 \approx 2k\pi, \tag{7.228}$$

where s_0 is a constant to allow for the effects of translational symmetry on equivalent reflections, and k is an integer. For φ_p, φ_q, φ_r, and s_0 expressed in cycles, the triplet relationships become

$$\pm\varphi_p \pm \varphi_q \pm \varphi_r \approx k - s_0. \tag{7.229}$$

Each of these expressions is an independent linear equation. In general, we have n reflections (phases) and m triplet relationships, with $m > n$. If all of the triplet relationships were correct (which, of course, they are not), and we knew the values of k for each triplet, we could immediately solve for the phases since we would replace the "≈" with "=" and solve the m simultaneous linear equations in n unknowns. Although n and m are often on the order 50–100 and 1000–2000, as a simple example, suppose that we have four phases ($n = 4$) involved in six triplet relationships ($m = 6$):

$$\begin{aligned}
1\varphi_1 + 1\varphi_2 + 1\varphi_3 + 0\varphi_4 &= k_1 - s_{0,1} = c_1 \\
0\varphi_1 + 1\varphi_2 + 1\varphi_3 + 1\varphi_4 &= k_2 - s_{0,2} = c_2 \\
1\varphi_1 + 0\varphi_2 - 1\varphi_3 + 1\varphi_4 &= k_3 - s_{0,3} = c_3 \\
-1\varphi_1 + 0\varphi_2 + 1\varphi_3 - 1\varphi_4 &= k_4 - s_{0,4} = c_4 \\
1\varphi_1 - 1\varphi_2 + 0\varphi_3 + 1\varphi_4 &= k_5 - s_{0,5} = c_5 \\
-1\varphi_1 + 0\varphi_2 - 1\varphi_3 - 1\varphi_4 &= k_6 - s_{0,6} = c_6
\end{aligned}$$

In matrix form this becomes

$$\begin{bmatrix} 1 & 1 & 1 & 0 \\ 0 & 1 & 1 & 1 \\ 1 & 0 & -1 & 1 \\ -1 & 0 & 1 & -1 \\ 1 & -1 & 0 & 1 \\ -1 & 0 & -1 & -1 \end{bmatrix} \begin{bmatrix} \varphi_1 \\ \varphi_2 \\ \varphi_3 \\ \varphi_4 \end{bmatrix} = \begin{bmatrix} c_1 \\ c_2 \\ c_3 \\ c_4 \\ c_5 \\ c_6 \end{bmatrix},$$

Which can be generalized for m triplets involving n phases as

$$\mathbf{A}_1\vec{\varphi} = \mathbf{c}, \tag{7.230}$$

Where $\mathbf{A}_1$ is an $m \times n$ matrix and $\vec{\varphi}$ and $\mathbf{c}$ are n- and m-dimensional vectors, respectively.

If the integers creating the $\mathbf{c}$ vector are known (they generally aren't) the phase vector, $\vec{\varphi}$, containing a *least-squares estimate* of the phases, can now be obtained as a matrix solution (we will take this up formally in the next chapter):

$$\vec{\varphi} = (\mathbf{A}_1^T\mathbf{A}_1)^{-1}\mathbf{A}_1^T\mathbf{c}. \tag{7.231}$$

The solution is only approximate, since the triplet relationships are only approximations – approximations that vary considerably in their reliability. To improve on the phase estimates we might consider weighting each triplet expression by an indicator of its reliability. We have already employed the product of the E values as a measure of the variance of the triplet, suggesting that, for the ith triplet, $X_i = (\sigma_3/\sigma_2^{3/2})E_pE_qE_r$, would be a suitable weighting factor:

$$X_i(\pm\varphi_p \pm \varphi_q \pm \varphi_r) \approx X_ic_i. \tag{7.232}$$

For our simple example this would result in

$$\begin{bmatrix} X_1 & X_1 & X_1 & 0 \\ 0 & X_2 & X_2 & X_2 \\ X_3 & 0 & -X_3 & X_3 \\ -X_4 & 0 & X_4 & -X_4 \\ X_5 & -X_5 & 0 & X_5 \\ -X_6 & 0 & -X_6 & -X_6 \end{bmatrix} \begin{bmatrix} \varphi_1 \\ \varphi_2 \\ \varphi_3 \\ \varphi_4 \end{bmatrix} = \begin{bmatrix} X_1c_1 \\ X_2c_2 \\ X_3c_3 \\ X_4c_4 \\ X_5c_5 \\ X_6c_6 \end{bmatrix} = \begin{bmatrix} C_1 \\ C_2 \\ C_3 \\ C_4 \\ C_5 \\ C_6 \end{bmatrix},$$

$$\mathbf{A}\vec{\varphi} = \mathbf{C}, \tag{7.233}$$

with the solution providing a *weighted least-squares* estimate of the phases.

Since the $\mathbf{C}$ vector is not known, a direct solution for the phases is ordinarily not possible using this approach. However, *if a set of estimated phases already exists* (e.g., from the divergence procedure), then $\mathbf{C}$ can be estimated and approximations to its component *integers*, $k_i = (C_i - X_is_{0,i})/X_i$, can be determined. Setting the k_i equal to the integers closest to them (phase errors will render some of them incorrect) allows for a determination of an improved $\mathbf{C}$ vector, subsequently allowing for a least-squares solution, providing a better approximation to the phases. This should, in turn, generate more k_i estimates closer to their correct integer values, and so forth. This phase refinement process continues until there are no changes in the integers.

The integer estimates determined in each iteration of the refinement process suggest another way to weight the triplet relationships in the process. Intuitively, an estimated value of k_i close to an integer indicates a high probability that the phases in the triplet are correct, while a significant deviation from an integer value implies a less reliable triplet. This indicates that a weight corresponding to the difference between the estimated k value and the closest integer would be appropriate. Woolfson proposed that for $-0.5 < \alpha < 0.5$, where α is the difference between k_{est} and the nearest integer, a weight $w(\alpha)$ could be applied such that $w(\alpha)$ approaches unity as α approaches zero, and zero as $|\alpha|$ approaches 0.5. Combining both weighting factors results in an overall weighting factor for each triplet: $w_i = X_i w(\alpha_i)$.

With some modifications to the weighting method, the least-squares approach to phase refinement turned out to provide convergence to correct structures, even when the initial phases were only moderately accurate. This prompted Ricardo Baggio, in collaboration with Woolfson, Germain and Declercq, to investigate the limitations of the method. To do so, correct phases were modified by adding random numbers to them, so that the phase errors were normally distributed. The magnitude of the random numbers was varied so that each test used initial phases with a greater average error, measured by the standard deviation of the modified phases with respect to the correct phases. This investigation produced a remarkable result – *even with standard deviations as high as 90° there was a very high probability that the phases would converge to the correct values.* Although not every set of randomly modified phases resulted in a correct solution, enough of them did that Baggio, et. al., considered the possibility that the 35–100 starting reflections chosen from the last reflections rejected in the convergence process could be *assigned random phases* to generate a number of trial phase sets. Each of these randomly assigned phase sets would then be refined by the least-squares method and tested with figures of merit to determine whether or not it constituted a probable solution. The method proved surprisingly successful, and resulted in the program *YZARC*.[153] One of the major advantages of the random-phase approach is that it generates a large starting set in which all of the initial phases are assigned at the same time, diminishing the chances that a particular phase assignment will result in failure to solve the structure.

The success of the *YZARC* method led Yao Jia-Xing to investigate the use of random starting phases applied to the weighted tangent refinement of *MULTAN* in lieu of the least-squares phase refinement process described above.[154] In Jia-Xing's *RANTAN* procedure, rather than assigning initial phases from triplet relationships, random phases are assigned to a substantial number of reflections, creating a large trial starting set with as many as 300–400 phases. The size of the starting set and the simultaneous assignment of phases makes it unnecessary to preselect the most strongly linked reflections – obviating the need for the convergence procedure. The resulting phases are then assigned the following initial weights:

origin-fixing phase	$w = 1.00$
enantiomorph-fixing phase with special value	$w = 0.99$
enantiomorph-fixing phase with general value	$w = 0.85$
random phase	$w = 0.25$.

The "special value" phases are those constrained by space group symmetry, such as the centric $h0l$ reflections in the $P2_1$ space group. Each trial phase set is subjected to weighted tangent refinement using Eqn. 7.217. The refinement process is essentially that of *MULTAN*; after the first cycle weights are assigned on the basis of the variances of the phases, with the exception that the random phases retain their weights of 0.25 until they achieve a weight greater than that value. After a weight becomes greater than 0.25 it is allowed to vary along its normal refinement path. When each trial phase set has refined to self-consistency, standard figures of merit are used to evaluate whether or not is likely to provide a correct solution. The

RANTAN procedure proved to be so successful that it was eventually incorporated into *MULTAN* as the default procedure.

A random walk alternative to tangent refinement: Phase Annealing. Tangent refinement begins with a set of phases that are iteratively adjusted to self-consistency, whereupon figures of merit, especially the combined figure of merit, C_{fom}, are determined in order to assess whether or not the phase set is a viable one. There is, however, no *a priori* reason to conclude that the resulting figures of merit are optimal – tangent refinement does not necessarily produce phases with figures of merit that represent the highest likelihood that those phases will provide a correct structure. To attain this end, an alternative approach to tangent refinement (and perhaps even structural solution) could involve making small adjustments to a set of phases in order to optimize a figure of merit directly. One way to accomplish this is to make small random changes in each phase and assess whether or not a specific change results in a better figure of merit. If the change produces an improved figure of merit, the change is accepted – otherwise it is ignored. After each phase is altered and tested, a second cycle of changes is made, and so forth until there are no further changes in the figure of merit. A simple example of this method of optimization – a *Monte Carlo random walk* – is given in Appendix G. In the simple case discussed there, a least-squares line is obtained from a set of data by allowing the slope and intercept of the line to vary incrementally, using the sum of the squares of the residual errors as a figure of merit. Unfortunately, the extrapolation to phase refinement is not straightforward. There are many phases, and it is almost certain that the random walk will find a figure of merit at a local minimum – requiring a walk "uphill" to get back on the path to the global minimum (the actual minimum value).* Since a given change is accepted only if it lowers the figure of merit we will find ourselves "trapped" at the local minimum and we must find a way to systematically increase the figure of merit until we can safely proceed toward the global minimum.

To resolve this dilemma, we draw on a thermodynamic analogy. Consider a sample of molten aluminum at 2800 K that we wish to cool and crystalize. The aluminum atoms have a large amount of kinetic energy, allowing them to move about relatively freely, so that the atomic arrangement at any instant can be in any one of a very large number of configurations (one of which is the configuration of the crystal). The aluminum atoms in each of the configurations have the same kinetic energy on the average (the temperature is a measure of their average kinetic energy), so that the total kinetic energy of any one configuration is the same as for any other. On the other hand, the forces of interaction are different for different arrangements of the atoms; the configurations have different potential energies, with the crystal configuration being the lowest. Let us now suppose that we cool the sample to room temperature. As the kinetic energy is removed from the atoms, the higher potential energy configurations will become less accessible (the atoms will not have sufficient kinetic energy to rearrange into these configurations), and the sample will tend to find itself in a lower potential energy configuration. If the sample is cooled slowly, the number of higher energy configurations available at a given temperature will be sufficient to allow the sample to gradually find a pathway to the global

*We make the assumption here that we are seeking a minimum figure of merit. If we are attempting to arrive at a maximum figure of merit we can simply search for its reciprocal.

energy minimum – that of the crystal. However, if the kinetic energy of the atoms is removed rapidly, the atoms will find themselves trapped in a configuration with insufficient kinetic energy to get back out – a local minimum. Because the sample is in a higher potential energy state it may find a mechanical pathway to a lower energy – it may crack with only a small external perturbation.

To remove this "thermal stress"we must reverse the effects of rapid cooling. We do this by raising the temperature of the sample to some temperature, T, just below its melting point, where the atoms have substantial kinetic energy and are able to diffuse into different locations. At this temperature, the Boltzmann distribution function gives us the probability that the sample will be in a configuration with potential energy E_1,

$$Pr(E_1) = Z(T)^{-1} e^{-E_1/kT}, \tag{7.234}$$

where T is the absolute temperature, k is Boltzmann's constant, and $Z(T)$ is a sum over all the probabilities (the partition function). Since we must attain a higher potential energy to get out of the local minimum, we now consider the probability that atoms in the sample will be in a new configuration with higher potential energy, E_2, such that

$$Pr(E_2) = Z(T)^{-1} e^{-E_2/kT}. \tag{7.235}$$

The relative probability that we will find the system at potential energy E_2 rather than potential energy E_1 is proportional to the ratio of these two probabilities:

$$Pr(\Delta E) = \frac{Pr(E_2)}{Pr(E_1)} = e^{-(E_2-E_1)/kT} = e^{-(\Delta E)/kT}. \tag{7.236}$$

For a given $|\Delta E|$, the relative probability will always be larger (usually much larger) when $E_2 < E_1$. As the sample is cooled slowly, it will generally continue to find itself in configurations of lower potential energy as the atoms move about, but a small fraction of those movements will put the sample in a higher potential energy configuration, thus providing a means to get the sample out of a local energy minimum. As T is lowered $Pr(\Delta E)$ decreases, diminishing the probability that *any* change will occur and allowing the sample to gradually walk toward the global energy minimum in a process known as *annealing*.

We now consider how we might simulate this annealing process. Assuming that we can compute the potential energy of a given configuration of atoms as a function of the locations of the atoms, we can use the potential energy as a figure of merit that we wish to minimize. Beginning with a random configuration at potential energy E_1, we change the position of an atom with a small random increment and calculate the potential energy, E_2. If the potential energy is lower ($\Delta E < 0$) then we accept the new position. If $\Delta E > 0$, then we assess the relative probability that the system might end up in a higher energy configuration by computing $Pr(\Delta E)$. We are now faced with the dilemma of determining whether or not the system will actually find itself in the new configuration. In order not to bias the decision we generate a random number between 0 and 1. If $Pr(\Delta E)$ is greater than the random number the new position is accepted — otherwise it is ignored; the larger the value of $Pr(\Delta E)$, the greater the probability that the new higher energy conformation will be accepted. After each atom has been allowed a chance to change its position, the temperature is lowered and another cycle is initiated. The "cooling" process continues until there is no change in the potential energy of the sample. This allows the system to randomly walk toward a potential energy minimum, along a pathway that includes states of higher potential energy, as it does in the actual annealing process. This *simulated annealing* approach to optimization was first described by Nicholas Metropolis, Arianna Rosenbluth, Marshall Rosenbluth, Augusta Teller, and Edward Teller, and is often referred to as the *Metropolis Algorithm.*[155] Although it was initially designed to determine the low energy configurations of thermodynamic systems, simulated annealing has been applied to a variety of optimization situations. In such cases the "temperature" is a parameter that is selected empirically, and the changes in energies are paralleled with the changes in the parameters necessary to optimize the figure of merit applicable to a given situation.

To apply the simulated annealing method to phase optimization rigorously, we would treat phases as analogs of atom positions. To optimize the reliability of the phases we then might seek values of the phases that would minimize the average variance of the phases, which would occur when $\mathcal{E} = \sum_{\mathbf{h}} Y_{\mathbf{h}}$ is at a maximum. $\mathcal{E}$ is the analog of the potential energy of a configuration — except that we wish to maximize it. A phase is altered by making a small random change, $\Delta\varphi_{\mathbf{h}}$, resulting in a corresponding"potential energy" change, $\Delta\mathcal{E}$. If $\Delta\mathcal{E} > 0$ we would accept the change, otherwise we would assess the probability,

$$Pr(\Delta\mathcal{E}) = e^{\Delta\mathcal{E}/k\mathcal{T}}, \tag{7.237}$$

that a smaller $\mathcal{E}$ (larger variance) would be tolerated, and accept the change if $Pr(\Delta\mathcal{E})$ was greater than a random number between 0 and 1. Since $\mathcal{E}$ is not a thermodynamic quantity, we would need to find an initial value of $\mathcal{T}$ that would scale the $Pr(\Delta\mathcal{E})$ values so that they remained less than 1 (adjusting the magnitude empirically to attain changes appropriate to the scale of the "energies"). Other figures of merit might also be utilized. A logical choice would be C_{fom}, which we would attempt to minimize or maximize, depending on the way that the figure of merit was defined.

The artificer of the concept of applying simulated annealing to phase optimization is George M. Sheldrick, who, in the latest version of his popular software package *SHELX**, uses a modified simulated annealing approach to refine phases of small molecule structures. He coined the term *phase annealing*[156] to characterize the process. Sheldrick's phase annealing was first applied to centrosymmetric structures, and then extended to treat non-centrosymmetric structures. For a centrosymmetric structure the phase is restricted to values of 0 or π, so that $\Delta\varphi_{\mathbf{h}} = \pi$ for any change in phase. Initial phases are determined by the tangent formula, *modified to included negative quartets.* We consider the case for $\varphi_{\mathbf{h}} = 0$; the other case, for $\varphi_{\mathbf{h}} = \pi$, will parallel the treatment exactly. For a given set of phases, C_{fom}

*A number of excellent software packages currently exist for structural solution and refinement. These packages all contain special features that distinguish them from others and practicing crystallographers generally have several of them in their arsenal. *SHELX*, however, deserves special mention here. In 1976 Sheldrick provided the first truly portable and "user friendly" crystallography software package, *SHELX*-76, making structural solution possible for a large number of investigators (including this author). Over the years he has continued to incorporate new solution and refinement methods, adding many of his own innovations (such as phase annealing). Following its initial release *SHELX* soon became the most-used software package for small molecule crystallography; it has remained so for over 30 years.[1]

is calculated, then each phase is changed sequentially, followed by a new calculation of C_{fom}. If C_{fom} is lower, then the change is accepted. If it is higher, then we must assess the probability that such an event will occur. For a single triplet, Eqn. 7.93 gives the relative probability that $\varphi'_{\mathbf{h}} = \pi$:

$$\frac{Pr(\varphi'_{\mathbf{h}})}{Pr(\varphi_{\mathbf{h}})} = \frac{1 - Pr(\varphi_{\mathbf{h}})}{Pr(\varphi_{\mathbf{h}})} = Pr(\Delta\phi_{\mathbf{h}}) = e^{-X_{\mathbf{hk}} S_{\mathbf{h}} S_{\mathbf{h-k}}}. \tag{7.238}$$

For n triplets contributing to $\varphi_{\mathbf{h}}$, the net probability is the product of the individual probabilities:

$$\begin{aligned} Pr(\Delta\phi_{\mathbf{h}}) &= e^{-X_{\mathbf{hk_1}} S_{\mathbf{h}} S_{\mathbf{h-k_1}}} e^{-X_{\mathbf{hk_2}} S_{\mathbf{h}} S_{\mathbf{h-k_2}}} \ldots e^{-X_{\mathbf{hk_n}} S_{\mathbf{h}} S_{\mathbf{h-k_n}}} \\ &= e^{-\sum_{i=1}^{n} X_{\mathbf{hk_i}} S_{\mathbf{h}} S_{\mathbf{h-k_i}}}. \end{aligned} \tag{7.239}$$

Referring to Fig. 7.12, α_i will be either 0 or π, and $\mathbf{Y_h}$ will lie along the real axis. It follows that

$$\begin{aligned} \sum_{i=1}^{n} X_{\mathbf{hk_i}} S_{\mathbf{h}} S_{\mathbf{h-k_i}} &= Y_{\mathbf{h}}, \quad \text{and} \\ Pr(\Delta\varphi_{\mathbf{h}}) &= e^{-Y_{\mathbf{h}}}. \end{aligned} \tag{7.240}$$

To make this probability "thermodynamic", we introduce a "temperature", $\mathcal{T}$:

$$Pr(\Delta\varphi_{\mathbf{h}}) = e^{-Y_{\mathbf{h}}/k\mathcal{T}}. \tag{7.241}$$

Ordinarily, a random number would be generated and the "higher energy" change would be made if $Pr(\Delta\phi_{\mathbf{h}})$ was greater than the random number. Sheldrick incorporated a more stringent probability ratio for the decision criterion for a centrosymmetric structure:

$$Pr(\Delta\varphi_{\mathbf{h}}) = \frac{1}{2} e^{-Y_{\mathbf{h}}/k\mathcal{T}}. \tag{7.242}$$

For non-centrosymmetric structures, Sheldrick postulated that Eqn. 7.240 would also apply to unrestricted phases. This makes qualitative sense — Y_h is computed for the configuration before the change and a larger value of Y_h corresponds to a smaller variance in the calculated phase, indicating that it is less likely that it should be changed. The negative argument in the exponential reflects this — as Y_h increases, $Pr(\Delta\varphi_{\mathbf{h}})$ decreases. Rather than creating random changes in the phases directly, Sheldrick empirically tested a number of formulae for determining $\Delta\varphi_{\mathbf{h}}$, settling on

$$\Delta\varphi_{\mathbf{h}} = \text{Arccos}\left(\frac{4Y_{\mathbf{h}}/k\mathcal{T} + \ln(R)}{4Y_{\mathbf{h}}/k\mathcal{T} - \ln(R)}\right). \tag{7.243}$$

R is a random number between 0 and 1 and the sign of $\Delta\varphi_{\mathbf{h}}$ is assigned randomly (e.g., by generating a random number between 0 and 1 and setting the sign positive if it is less than 0.5, and negative if it is greater). Using this formula $\Delta\varphi_{\mathbf{h}}$ approaches 0 as $4Y_{\mathbf{h}}/k\mathcal{T}$ becomes large, as expected (as $\mathcal{T}$ decreases and Y_h increases).

To scale the initial "temperature" appropriately, we recall that there is a finite probability that the phases will be entirely random. For each reflection, Eqn. 7.208

provides a value of $Y_{\mathbf{h}}$ for a random distribution of phases. We would expect that the probability of a change toward a random phase distribution would be relatively small. It was found (again empirically) that

$$Pr(\Delta\varphi_{\mathbf{h}}) = e^{-\langle Y_{\mathbf{h}}(rand)\rangle/k\mathcal{T}} = 0.3 \tag{7.244}$$

provides a suitable starting value for $\mathcal{T}$, which is then lowered by reducing $\mathcal{T}$ by 5% for each cycle and slowly "cooling" the phase configuration to a figure-of-merit minimum.

Sheldrick's phase annealing algorithm has been shown to be quite capable of determining an *ab initio* solution, arriving at a global minimum for C_{fom} from a random set of initial phases. However, this approach has proved to be somewhat inefficient, and it is quite possible that its direct application will be unable to distinguish between two minima that are very close to one another, but representative of very different configurations in "phase space." Thus the phase annealing algorithm in *SHELX* retains the multisolution approach, generating random phase sets and refining them using the tangent formula (as in *RANTAN*), then "annealing" the resulting phase sets to optimize the figure of merit. This additional step has made it possible to solve a number of quite large small molecule structures that had previously evaded solution.

The evaluation of the E map generated from a given solution (usually the one with the smallest C_{fom}) is also enhanced in *SHELX* by analyzing the $1.3m$ highest peaks in the E map, where m is the number of atoms in the asymmetric unit. In this *peaklist optimization*, peaks are temporarily removed sequentially, beginning with the peak of lowest magnitude. After each removal, atoms are assigned to the remaining peaks and new phases are calculated from Eqn. 3.108 for the "large E" reflections (e.g., $E > 1.2$). The average residual difference between the observed and calculated larger E values,

$$R_E = \frac{\sum_{\mathbf{h}} |E_{\mathbf{h},o} - E_{\mathbf{h},c}|}{E_{\mathbf{h},c}}, \tag{7.245}$$

is then evaluated and the peak is eliminated if R_E increases. After all of the peaks have been tested, the remaining peaks are assigned as atoms and a new set of phases is computed. An E map is calculated and the $1.3m$ highest peaks are again evaluated. This procedure is conducted cyclically until no changes are observed in R_E.

A solution of the bikitaite structure illustrates the power of the combination of random phase tangent refinement, phase annealing, and peaklist optimization. Recall that the automated symbolic addition procedure produced an E map in which only the three heavier atoms could be discerned. It was necessary to complete the structure with two cycles of difference-Fourier synthesis. A "multisolution" attempt to solve the structure with an earlier version of *MULTAN* (without random phase starting sets) proved unsuccessful. Fig. 7.33(a) illustrates the maxima in the E map produced from *SHELX* using default values for the input parameters. Fig. 7.33(b) reveals that the entire structure has been determined. The small blue spheres represent the locations of the maxima in the E map. The numbers are indicative of the relative amplitudes of the maxima: the three peaks with the largest magnitudes correspond to the three heavier atoms; the seven next largest

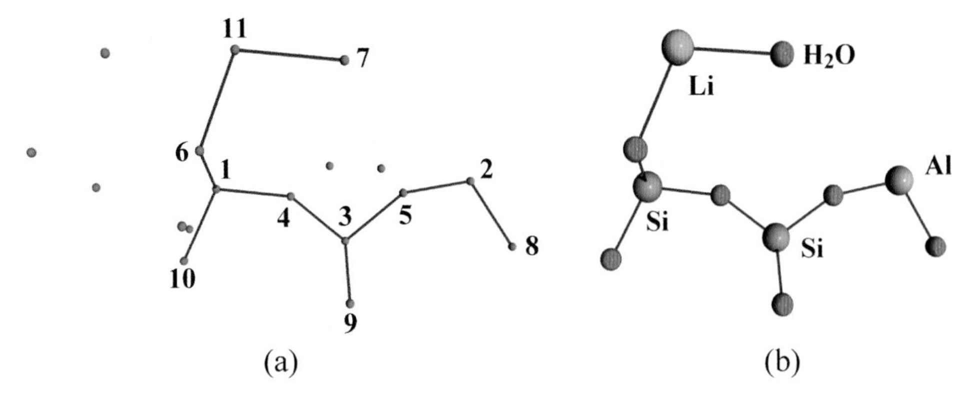

Figure 7.33 Multiple solution/phase annealing elucidation of the $Li[AlSi_2O_6]{\cdot}H_2O$ structure: (a) E map. (b) Completed structure. The small blue spheres represent locations of maxima in the E map, numbered in accordance with their relative values.

peaks corresponded to the oxygen atoms in the asymmetric unit, and the eleventh peak locates the lightest non-hydrogen atom — the lithium atom.*

7.2 Other Direct Methods

Small molecule crystal structures containing up to about 100 atoms in the asymmetric unit are easily solved today, primarily using random phase starting sets incorporated into software packages such as *SHELX* or *SIR*[157]. In many cases solution and refinement using these programs require only minimal user intervention and small-molecule structure determination has become routine, provided that the data are collected at a resolution limit (Eqn. 4.121) that will allow for the determination of the location of individual atoms (ca., 1.1–1.2 Å). The success of these methods arises chiefly from the redundancy of the structural information inherent in the diffraction data — there are ordinarily many more experimental measurements (reflections), than there are structural parameters to be determined in the small molecule diffraction experiment. Determining a single parameter requires *at least* one experimental measurement; locating 100 atoms requires establishing 300 positional parameters (x, y, z). A typical data set for a small molecule structure contains ~1000–50000 reflections (depending on the nature of the data collection). The data are said to be *over-determined.* Direct solution methods usually begin with a subset of these reflections — those with high E values — but there are ordinarily more than enough for structure elucidation.

Macromolecular crystals are very often smaller than small molecule crystals, and most of them contain solvent molecules (water) — the large portion of which exhibit little or no long-range order in the crystal. These and other physical constraints result in resolution limits of 2 Å or more, such that individual atomic positions are no longer discernable. In addition, these molecules can contain from a thousand to

*Note that this is a different asymmetric unit than the one determined in the symbolic addition solution. The origin-defining reflections in each solution produced different origin locations. Both asymmetric units produce the same structure. The data were of insufficient quality to allow for the determination of the hydrogen atom locations.

several thousand atoms; the "high-E" data are generally *under-determined*, often severely so. The combination of having fewer measurements than parameters, along with the experimental constraints imposed by smaller crystals that diffract poorly, ordinarily renders established direct methods algorithms ineffective in determining the structures of macromolecular crystals. While there are currently no universal approaches to direct macromolecular structure determination, progress has been made on several fronts; a brief description of two approaches will be given here.

7.2.1 Dual-Space Iteration

In 1993, Hauptman[158] and coworkers proposed the basis of an algorithm in which it might be possible to adjust atom locations in order to optimize the agreements between calculated triplets and quartets and their expected values. The atom locations are constrained to reflect chemical bonding — atoms can get no closer than the sum of their covalent radii – the approximate length of the chemical bond between them. If the atoms are in their correct positions we would expect the average residual differences (squared) between the triplets and quartets and their expected values to be at a minimum; this would also be true of their cosines. It follows that atoms in correct positions would tend to minimize the following function (note the similarity to $\sum \epsilon^2$, the sum of the squares of the residuals, in the least-squares optimization procedure described in Appendix G):

$$R_\Phi = \frac{\sum\limits_{\mathbf{hk}} X_{\mathbf{hk}} \left(\cos \Phi_{\mathbf{hk}} - \langle \cos \Phi_{\mathbf{hk}} \rangle\right)^2 + \sum\limits_{\mathbf{hkl}} |W'_{\mathbf{hkl}}| \left(\cos \Phi_{\mathbf{hkl}} - \langle \cos \Phi_{\mathbf{hkl}} \rangle\right)^2}{\sum\limits_{\mathbf{hk}} X_{\mathbf{hk}} + \sum\limits_{\mathbf{hkl}} |W'_{\mathbf{hkl}}|}, \tag{7.246}$$

where n is the number of atoms in the unit cell,

$$\Phi_{\mathbf{hk}} = \varphi_{\mathbf{h}} + \varphi_{\mathbf{k}} + \varphi_{-\mathbf{h}-\mathbf{k}}, \tag{7.247}$$

$$\Phi_{\mathbf{hkl}} = \varphi_{\mathbf{h}} + \varphi_{\mathbf{k}} + \varphi_{\mathbf{l}} + \varphi_{-\mathbf{h}-\mathbf{k}-\mathbf{l}}, \tag{7.248}$$

$$X_{\mathbf{hk}} = \frac{2}{\sqrt{n}} E_{\mathbf{h}} E_{\mathbf{k}} E_{-\mathbf{h}-\mathbf{k}}, \quad \text{and} \tag{7.249}$$

$$W'_{\mathbf{hkl}} = \frac{2}{n} E_{\mathbf{h}} E_{\mathbf{k}} E_{\mathbf{l}} E_{-\mathbf{h}-\mathbf{k}-\mathbf{l}} (E^2_{\mathbf{h}+\mathbf{k}} + E^2_{\mathbf{h}+\mathbf{l}} + E^2_{\mathbf{k}+\mathbf{l}} - 2). \tag{7.250}$$

This is seen to be a weighted sum of the squares of the residuals between the cosines of the triplets and quartets and their expected values. The weighting factors are assigned to reflect the reliabilities of the structure invariants; $W'_{\mathbf{hkl}}$ reflects the seven magnitude quartet variance by including the cross-terms (cf. Eqn. 7.204). Its absolute value is taken since it can become negative when the cross terms are small (i.e., for negative quartets). The expected values of the triplets and quartets are determined from Eqns. 7.175 and 7.206:

$$\langle \cos(\Phi_{\mathbf{hk}}) \rangle = \frac{I_1(X_{\mathbf{hk}})}{I_0(X_{\mathbf{hk}})},$$

$$\langle \cos(\Phi_{\mathbf{hkl}}) \rangle = \frac{I_1(W'_{\mathbf{hkl}})}{I_0(W'_{\mathbf{hkl}})},$$

where I_0 and I_1 are modified Bessel functions.

The resulting algorithm[159] that incorporates this minimization function begins in direct space with a random distribution of the atoms in the unit cell, with locations subject to bonding constraints. Phases are calculated from these locations with a Fourier series (Eqn. 3.108). The basis of the algorithm is to adjust each phase (in reciprocal space) by adding and subtracting a fixed increment (e.g., $\pi/2$), then evaluating R_{Φ}, accepting any change that reduces it. After each phase has been provided the opportunity to change, the increment is divided by 2 and another cycle is initiated (note the similarities to Appendix G). Following a fixed number of cycles (e.g., 5), the algorithm returns to direct space, computing a Fourier map and selecting the m highest peaks (and symmetry-equivalents), where $m <= n$, subject to bonding constraints. These new atom locations are used to determine a new set of phases in reciprocal space, which are then subjected to another round of R_{Φ} minimization. When R_{Φ} has reached a minimum, an E map is created to determine if a reasonable structure has been determined. The algorithm forms the basis of the program *SnB*.[160]* Sheldrick has created a version of this approach in his *SHELXD* program that uses the tangent formula to generate and improve the phases and peaklist optimization to select atom locations.[161]

The dual-space approach has resulted in the structural determination of small protein molecules containing in the neighborhood of 1000 atoms. While this success has allowed for the determination of much larger structures by direct methods, the method still appears to require near-atomic resolution.

7.2.2 Maximum Entropy

The phase problem is typical of many in the physical sciences in which we attempt to determine a function on the basis of things that we know with certainty (the structure factor magnitudes), and things that we do not (the phases, and therefore the electron density distribution). In seeking a function that will allow us to determine $\rho(\mathbf{r})$, we expect it to meet two criteria: (a) It must account for the things that we know ($F_{\mathbf{h}}$) — and represent the things that we don't know ($\rho(\mathbf{r})$) *without bias*. To simplify this abstract treatment here let us divide the unit cell into small identical volumes centered on points on a uniform grid, with each volume centered on a discrete vector to each grid point. The probability of finding an electron in the jth volume element is measured by $\rho(\mathbf{r}_j)$,

$$\rho(\mathbf{r}_j) = \frac{1}{V_c} \sum_{\mathbf{h}} F_{\mathbf{h}} e^{i\varphi_{-\mathbf{h}}} e^{-2\pi i(\mathbf{h}\cdot\mathbf{r}_j)}. \tag{7.251}$$

For the discrete values $(\mathbf{r}_1, \mathbf{r}_2, \ldots, \mathbf{r}_n)$, the corresponding (relative) probabilities are $(\rho(\mathbf{r}_1), \rho(\mathbf{r}_2), \ldots, \rho(\mathbf{r}_n))$. For such a situation, Shannon[162] showed that the relative uncertainty in the overall probability distribution is given by

$$S(\rho(\mathbf{r}_1), \rho(\mathbf{r}_2), \ldots, \rho(\mathbf{r}_n)) = k \sum_{j=1}^{n} \rho(\mathbf{r}_j) \ln \rho(\mathbf{r}_j). \tag{7.252}$$

This measure of uncertainty is the analog of the entropy derived in statistical thermodynamics[163] (when k is Boltzmann's constant it *is* the entropy; for our purposes

*"SnB" stands for "Shake and Bake". The shaking occurs in reciprocal space and the baking is conducted in direct space.

k can be set to unity). The least biased calculated electron density distribution is one that maximizes this entropy, while still accounting for the knowledge that we have a portion of the information necessary to determine $\rho(\mathbf{r})$. In other words, in our simplified discrete treatment we seek to determine the set $\{\rho(\mathbf{r})\}$, such that

$$\frac{\partial S(\rho(\mathbf{r}_1), \rho(\mathbf{r}_2), \ldots, \rho(\mathbf{r}_n))}{\partial \rho(\mathbf{r}_j)} = 0, \quad j = 1, 2, \ldots, n, \tag{7.253}$$

subject to the constraints that the $F_{\mathbf{h}_j}$ are known. This is accomplished mathematically using the method of undetermined multipliers (see Appendix H), by maximizing

$$L(\rho(\mathbf{r}_1), \rho(\mathbf{r}_2), \ldots \lambda_{\mathbf{h}_1}, \lambda_{\mathbf{h}_2} \ldots) = S(\rho(\mathbf{r}_1), \rho(\mathbf{r}_2), \ldots) + \lambda_{\mathbf{h}_1} g_1(F_{\mathbf{h}_1}) + \lambda_{\mathbf{h}_2} g_2(F_{\mathbf{h}_2}) \ldots \tag{7.254}$$

for each $\rho(\mathbf{r}_j)$. The $\lambda_{\mathbf{h}}$ are known as *Lagrange multipliers*, and the $g(F)$ functions are constraining functions determined by the known magnitudes of the structure factors. The problem is now "reduced" to substituting Eqn. 7.251 into Eqn. 7.252, creating L (the "Lagrangian"), and determining the Lagrange multipliers and $\rho(\mathbf{r}_j)$ that maximize it.*

Maximum entropy found its first practical physical applications in the field of radio astronomy, where it was shown to be capable of creating an image from an incomplete data set. The seminal paper by Gull and Daniell[165] was entitled *Image reconstruction from incomplete and noisy data.* This problem parallels the general problem in macromolecular crystallography, in which we seek to create an image of the atomic arrangement in the crystal with less than the necessary data, often at poor resolution. Because of this, it is conceivable that maximum entropy algorithms could prove to be the most likely to provide *ab initio* macromolecular structural solutions. Unfortunately, the actual determination of the Lagrange multipliers and the subsequent generation of the electron density function present very difficult challenges, both theoretically and computationally. It is fair to state at the time of the writing of this book that the jury is still out.

7.3 Completion of the Structural Solution: Probability Methods

Experimental or statistical methods often provide only a partial structural solution, with the resulting Fourier map exhibiting either a recognizable fragment or a number of peaks of substantially greater intensities than the remainder. In both cases, the remaining atoms can very often be located with a series of difference Fourier syntheses, as we have already observed. There are, however, instances in which the located atoms do not represent a sufficient portion of the electron density to produce calculated phases that will provide directly resolvable peaks in a Fourier

*The maximization is accomplished variationally – by considering the changes in S and $g(F)$, δS and $\delta g(F)$, when $\rho(r)$ is changed by $\delta\rho(r)$. The brief discussion here is neither complete nor rigorous — it was designed to provide the reader with the general principles underlying the maximum entropy concept. A thorough treatment of the maximum entropy principle applied to crystallography can be found in the comprehensive paper by Bricogne[164].

map. In these cases the use of the phases generated from assigned atom positions as starting sets for phase expansion can serve as a viable alternative to Fourier synthesis.

7.3.1 Phases from Fourier Structure Factors

Phase expansion using structural information was first proposed by Jerome Karle in 1968[166]. Karle's procedure was designed to add atoms in cycles, each time utilizing the tangent formula to generate new phases and a new E map after determining the location of additional atoms. The procedure became known as *Karle recycling.* The recycling process begins by generating calculated phases with a Fourier series from an initial set of m atom locations, which we will refer to as a *fragment*:

$$\mathbf{F}_{c,\,\mathbf{h}} = \sum_{j=1}^{m} f_j(\mathbf{h}) e^{-2\pi\, i(\mathbf{h}\cdot\mathbf{r}_j)}. \tag{7.255}$$

The contribution of the fragment to the normalized structure factor is then

$$\mathbf{E}_{c,\,\mathbf{h}} = \left(\frac{F^2_{c,\mathbf{h}}}{\sum_j f_j^2(\mathbf{h})} \right)^{1/2} e^{i\varphi_{c,\,\mathbf{h}}}. \tag{7.256}$$

As with ab initio solutions, only those reflections with the largest observed $E_{o,\,\mathbf{h}}$ values are utilized as candidates for inclusion in the starting set (e.g., $E_{o,\,\mathbf{h}} > 1.4$). The resulting phases are assumed to be (approximately) correct if the structure factor amplitude, $F_{c,\,\mathbf{h}}$, calculated in the Fourier series, is of sufficient magnitude to represent the relative scattering power of the fragment atoms that are responsible for $F_{c,\,\mathbf{h}}$. In other words, if $F_{c,\,\mathbf{h}} \geq p\, F_{o,\,\mathbf{h}}$, then $\varphi_{c,\,\mathbf{h}}$ becomes a phase in the starting set; p is the ratio of the number of electrons in the fragment to the total number of electrons in the asymmetric unit, i.e., the relative scattering power of the fragment. The starting set obtained in this manner is utilized in the tangent formula and following tangent refinement, a new E map is generated and searched for additional peaks. The process is continued iteratively until the structure is complete.

The Karle recycling procedure requires user intervention after each cycle is completed in order to determine the atom locations for the next cycle. In 1978 *MULTAN* was modified to make use of known structural information more automatically, by using initial phases generated from the fragment to provide an initial starting set, followed by a weighted tangent refinement. In order to incorporate the known structural information into the procedure, the weights assigned to each reflection in Eqn. 7.217 should correct for the error in the phases arising from the lack of complete structural information. For example, if $\mathbf{E}_{c,\,\mathbf{k}} = E_{\mathbf{c},\,\mathbf{k}} e^{i\varphi_{c,\,\mathbf{k}}}$, then the weight, $w_{\mathbf{k}}$, should provide an approximation to the correct ("observed") structure factor $\mathbf{E}_{o,\,\mathbf{k}}$:

$$\mathbf{E}_{o,\,\mathbf{k}} = E_{o,\,\mathbf{k}}\, e^{i\varphi_{o,\mathbf{k}}} \simeq E_{o,\,\mathbf{k}} w_{\mathbf{k}}\, e^{i\varphi_{c,\,\mathbf{k}}}. \tag{7.257}$$

It follows that in the tangent formula, which contains the calculated phases, $E_{o,\,\mathbf{k}}$ should be replaced with $w_{\mathbf{k}}\, E_{o,\,\mathbf{k}}$.

The observed and calculate normalized structure factors are shown in Fig. 7.34, in which the contribution of the unknown portion of the structure to the phase

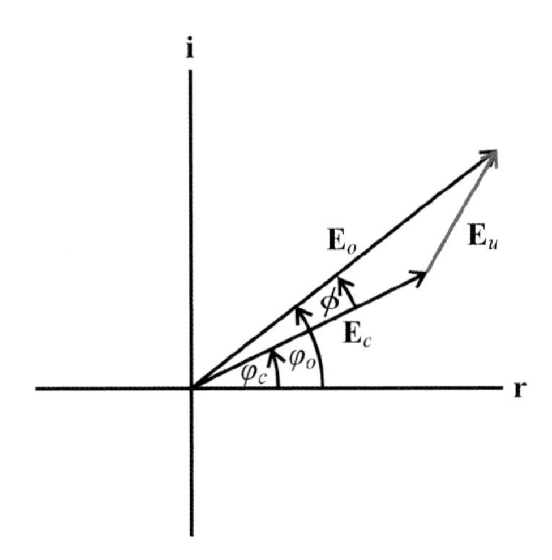

Figure 7.34 "Observed" normalized structure factor represented as the sum of the net contribution from known atomic positions ($\mathbf{E}_c$) and unknown atomic positions ($\mathbf{E}_u$).

is given by $\phi = \varphi_o - \varphi_c$ and the corresponding contribution to the normalized structure factor is given by

$$\mathbf{E}_{u,\,\mathbf{k}} = \mathbf{E}_{o,\,\mathbf{k}} - \mathbf{E}_{c,\,\mathbf{k}} = E_{o,\,\mathbf{k}}\, e^{i\varphi_{o,\mathbf{k}}} - E_{c,\,\mathbf{k}}\, e^{i\varphi_{c,\mathbf{k}}}. \tag{7.258}$$

The calculated phase generated by the fragment atoms differs from the actual phase by the angle $\phi = \varphi_o - \varphi_c$. With $\varphi_o = \phi + \varphi_c$,

$$\begin{aligned}\mathbf{E}_{u,\,\mathbf{k}} &= E_{o,\,\mathbf{k}}\, e^{i\phi}\, e^{i\varphi_{c,\mathbf{k}}} - E_{c,\,\mathbf{k}} e^{i\varphi_c} \\ &= (E_{o,\,\mathbf{k}} e^{i\phi} - E_{c,\,\mathbf{k}}) e^{i\varphi_c}.\end{aligned} \tag{7.259}$$

This will be seen to parallel Eqn. 6.149, in which the weighting factor was formulated in order to determine the correction factor needed to approximate observed structure factors from those calculated from a partial structure. The treatment following Eqn. 6.149 can be duplicated here, simply by substituting normalized structure factors for "regular" structure factors, resulting in the probability distribution function for ϕ and the subsequent weighting factor for each normalized structure factor in the tangent formula,

$$\begin{aligned}Pr_1(\phi) &= K\, e^{X^E_{\mathbf{k}} \cos\phi} d\phi \quad \text{and} \\ w_{\mathbf{k}} &= \frac{I_1(X^E_{\mathbf{k}})}{I_0(X^E_{\mathbf{k}})},\end{aligned} \tag{7.260}$$

where $X^E_{\mathbf{k}} = E_{o,\mathbf{k}} E_{c,\,\mathbf{k}} / \sum_{j=1}^{k} f_j^2(\mathbf{k})$ and I_1 and I_0 are modified Bessel functions (Appendix F). In this statistically weighted expansion scheme reflections for the starting set are generated as described above, but only the most reliable of the phases, those with $X^E_{\mathbf{k}} > 4.0$, are chosen to begin tangent refinement. These reflections are given initial weights defined by Eqn. 7.260, which are then refined using the weighted tangent formula (only the starting set is refined). The starting

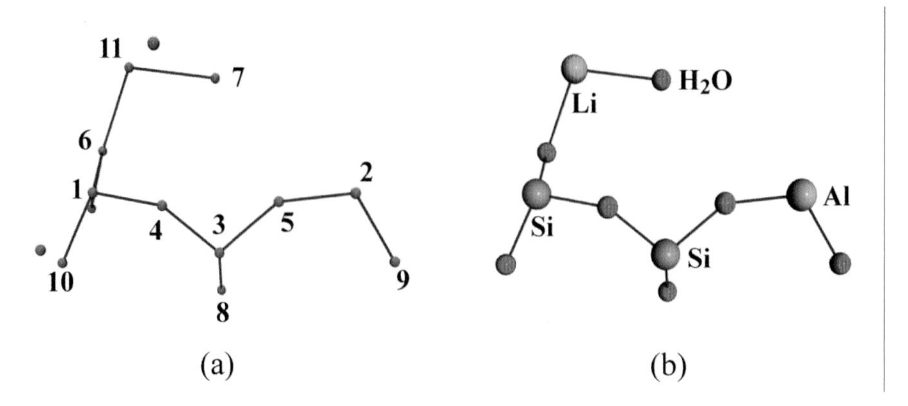

Figure 7.35 Symbolic addition/phase expansion solution of the $Li[AlSi_2O_6]{\cdot}H_2O$ structure with initial phases based on the positions of the silicon and aluminum atoms: (a) E map. (b) Completed structure. The small blue spheres represent locations of maxima in the E map, numbered in accordance with their relative values.

set phases are then fixed and the reflection list is expanded to include those phases with $X^E_{\mathbf{k}} > 2.4$. These phases are refined and accepted if $Y^2_{\mathbf{h}}$ is greater than 2.0. Following refinement, an E map is generated and searched for peaks. As an example, recall the symbolic addition solution of bikitaite, which only revealed the positions of the silicon and aluminum atoms in the structure. Using these three atoms as a fragment, one cycle of phase expansion provides an E map with the entire structure, as illustrated in Fig. 7.35.

While the entire structure is often determined in a single cycle of the statistically weighted expansion procedure, in some instances it is necessary to repeat the process more than once in order to locate all of the atoms in the structure. In order to enhance the probability that the entire structure is revealed without outside intervention, Jia-Xing[167] drew on the success of the "random phase" approach to develop a multisolution method for partial structure expansion. In this approach the initial phases are assigned and refined as above, but in order to treat problems in which only a small fraction of the scattering power results from the fragment electrons, "reduced" weights are assigned to the starting set reflections: $w^r_{\mathbf{k}} = 2w_{\mathbf{k}} - 1$. Instead of proceeding directly into tangent expansion, several sets of random phases are generated and assigned to the reflections that are not in the starting set. Phases for all of the reflections in each trial set are then refined, as in the *RANTAN* algorithm, using figure of merit criteria to assess the trial solutions. The starting set phases are only allowed to change if the calculated weights based on $Y_{\mathbf{k}}$ (e.g., Eqn. 7.219) become greater than the initially assigned weights. *SHELX* provides a powerful variation of this approach, using a starting set generated by phases determined from a fragment, followed by phase annealing and peaklist optimization. Since the phase annealing algorithm is itself a random walk to the final structure, the generation of multiple trial solutions from random phases becomes unnecessary.

7.3.2 Phases from Difference Fourier Structure Factors

At the end of Chapter 6 it was observed that if a portion of the structure is known, a Fourier series using difference structure factors as Fourier coefficients ($\Delta F = F_o - F_c$,), produces a Fourier map with better resolution than the traditional map created using only the magnitudes of the observed structure factors. Referring to Eqn. 6.175, if $\mathbf{F}_{c,\mathbf{h}}$ is the structure factor calculated on the basis of a known fragment, and $\mathbf{F}_{o,\mathbf{h}}$ is the "observed" structure factor, generated by all of the atoms in the structure, then $\mathbf{F}_{u,\mathbf{h}} = \mathbf{F}_{o,\mathbf{h}} - \mathbf{F}_{c,\mathbf{h}}$ is the contribution of the atoms with unknown locations to $\mathbf{F}_{o,\mathbf{h}}$, and a Fourier series using the difference coefficients, $\mathbf{F}_{u,\mathbf{h}}$, should produce a difference map containing peaks corresponding to the unknown portion of the structure. Since we do not know the phase for $\mathbf{F}_{o,\mathbf{h}}$, $\varphi_{o,\mathbf{h}}$, we assign the calculated phase from the Fourier series generated from the fragment in the structure, $\varphi_{c,\mathbf{h}}$ to $\mathbf{F}_{o,\mathbf{h}}$. This results in a modified difference Fourier coefficient, $\mathbf{F}'_{u,\mathbf{h}}$:

$$\mathbf{F}_{u,\mathbf{h}} = F_{o,\mathbf{h}} e^{i\varphi_{o,\mathbf{h}}} - F_{c,\mathbf{h}} e^{i\varphi_{c,\mathbf{h}}}; \tag{7.261}$$

$$\mathbf{F}'_{u,\mathbf{h}} = F_{o,\mathbf{h}} e^{i\varphi_{c,\mathbf{h}}} - F_{c,\mathbf{h}} e^{i\varphi_{c,\mathbf{h}}} = (F_{o,\mathbf{h}} - F_{c,\mathbf{h}}) e^{i\varphi_{c,\mathbf{h}}}. \tag{7.262}$$

In difference Fourier synthesis, a Fourier map is generated using $\mathbf{F}'_{u,\mathbf{h}}$ in place of $\mathbf{F}_{u,\mathbf{h}}$, using only those reflections for which we have some confidence that $\mathbf{F}'_{u,\mathbf{h}} \approx \mathbf{F}_{u,\mathbf{h}}$. The process of locating the remainder of the structure usually results in several cycles of the procedure, each time requiring the generation of a ΔF map and the search for atom locations.

A direct methods alternative to difference Fourier synthesis was created in the Crystallography Laboratory at the University of Nijmegen[168]. In this method quasi-normalized *difference* structure factors are employed in the tangent formula for phase expansion. As with normal phase expansion this requires an assessment of the relative probabilities of the starting phases, that is, whether or not $\mathbf{F}'_{u,\mathbf{h}} \approx \mathbf{F}_{u,\mathbf{h}}$, which will be true if $\varphi_{c,\mathbf{h}} \approx \varphi_{o,\mathbf{h}}$. In order to assess the likelihood of this relationship, we compare the assumption that $\mathbf{F}_{c,\mathbf{h}}$ and $\mathbf{F}_{o,\mathbf{h}}$ are in phase with the assumption that they are completely out of phase:

$$\mathbf{F}''_{u,\mathbf{h}} = -F_{o,\mathbf{h}} e^{i\varphi_{c,\mathbf{h}}} - F_{c,\mathbf{h}} e^{i\varphi_{c,\mathbf{h}}} = (-F_{o,\mathbf{h}} - F_{c,\mathbf{h}}) e^{i\varphi_{c,\mathbf{h}}}. \tag{7.263}$$

In order to assess the relative probabilities of $\mathbf{F}'_{u,\mathbf{h}}$ and $\mathbf{F}''_{u,\mathbf{h}}$ and to incorporate $\mathbf{F}'_{u,\mathbf{h}}$ in the tangent formula both entities must first be converted to quasi-normalized structure factors. A modified two-dimensional Wilson plot is used to calculate separate overall isotropic displacement factors for the known and unknown portions of the structure.[169] From Eqn. 5.442,

$$\langle I_{rel} \rangle_{\mathbf{h}} = K \left(\sum_{j=1}^{m} \langle f_j(h) \rangle^2 e^{-2\mathrm{B}_u \langle h \rangle^2} + \sum_{l=1}^{n} \langle f_l(h) \rangle^2 e^{-2\mathrm{B}_c \langle h \rangle^2} \right), \tag{7.264}$$

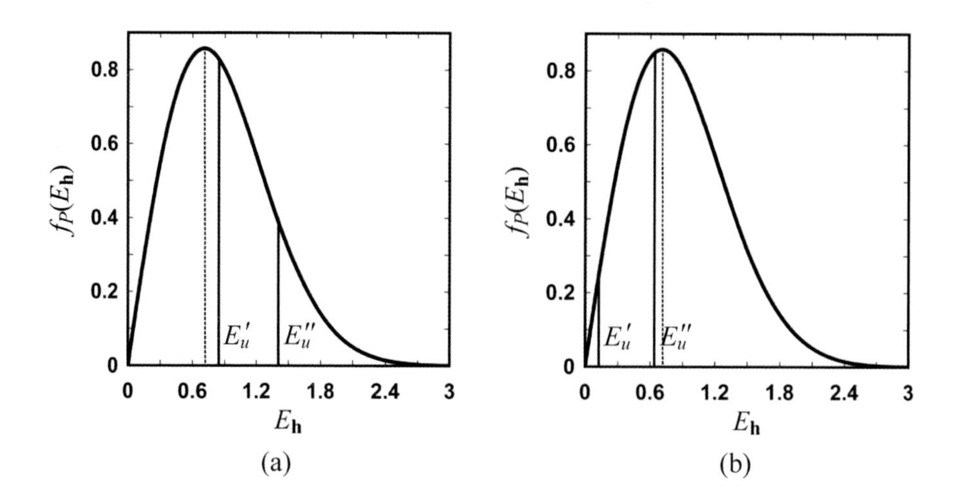

Figure 7.36 Comparative probabilities for E'_u and E''_u (a) when both have magnitudes greater than $E_{max} = \sqrt{2}/2$ (dashed line), and (b) when both have magnitudes less than E_{max}.

where there are n atoms in the fragment and m unknown atom positions. The temperature corrected scattering factors are then used to "normalize" $\mathbf{F}'_{u,\mathbf{h}}$ and $\mathbf{F}''_{u,\mathbf{h}}$:

$$\mathbf{E}'_{u,\,\mathbf{h}} = \left(\frac{(F_{o,\,\mathbf{h}} - F_{c,\,\mathbf{h}})^2}{\sum_j f_j^2(\mathbf{h})e^{-2\mathrm{B}_u h^2}}\right)^{1/2} e^{i\varphi_{c,\,\mathbf{h}}}; \tag{7.265}$$

$$\mathbf{E}''_{u,\,\mathbf{h}} = \left(\frac{(-F_{o,\,\mathbf{h}} - F_{c,\,\mathbf{h}})^2}{\sum_j f_j^2(\mathbf{h})e^{-2\mathrm{B}_u h^2}}\right)^{1/2} e^{i\varphi_{c,\,\mathbf{h}}}. \tag{7.266}$$

From these equations it is clear that $E'_{u,\,\mathbf{h}} < E''_{u,\,\mathbf{h}}$. The probability distribution for both of these entities is given by Eqn. 5.480.* The function has a maximum value, determined by setting $\partial(f_1(E_{\mathbf{h}}))/\partial E_{\mathbf{h}} = 0$, resulting in $E_{\mathbf{h}}(max) = \sqrt{2}/2$.

We are seeking reflections that have a higher probability of having $\mathbf{F}_{c,\,\mathbf{h}}$ and $\mathbf{F}_{o,\,\mathbf{h}}$ in phase. From Fig. 7.36(a), those reflections that have both $E'_{u,\,\mathbf{h}}$ and $E''_{u,\,\mathbf{h}}$ greater than $E_{\mathbf{h}}(max)$ will all meet that criterion, while those with $E'_{u,\,\mathbf{h}}$ and $E''_{u,\,\mathbf{h}}$ less than $E_{\mathbf{h}}(max)$ will not, as depicted in Fig. 7.36(b). Reflections with a value on either side of the line will have similar probabilities. We therefore limit reflections to those with both values greater than $E_{\mathbf{h}}(max)$. The larger the difference between the E values for these two reflections, the more likely that $\varphi_{c,\,\mathbf{h}} \approx \varphi_{o,\,\mathbf{h}}$ — and that $\varphi_{c,\,\mathbf{h}}$ is a suitable phase for $E'_{u,\,\mathbf{h}}$. This is made quantitative by determining the relative probability,

$$Pr_u = \frac{f_1(E'_{u,\,\mathbf{h}})\,dE}{f_1(E'_{u,\,\mathbf{h}})\,dE + f_1(E''_{u,\,\mathbf{h}})\,dE} = \frac{f_1(E'_{u,\,\mathbf{h}})}{f_1(E'_{u,\,\mathbf{h}}) + f_1(E''_{u,\,\mathbf{h}})}. \tag{7.267}$$

*For centrosymmetric structures the probability distribution is given by Eqn. 5.489.

The $E'_{u,\mathbf{h}}$ magnitudes of reflections in the selected set with large E values (e.g., $E > 1.2$) are assigned initial phases $\varphi_{c,\mathbf{h}}$ and weights based on Pr_u (e.g., $w_{\mathbf{h}} = (2Pr_u - 1)^2$). These initial phases are then incorporated into the weighted tangent formula (Eqn. 7.217) and the reflection list is expanded to include reflections from the selected set with smaller values of $E'_{u,\mathbf{h}}$ (e.g., $E > 0.9$). The remaining phases are determined from the tangent formula, followed by tangent refinement that includes the phases of the starting set. At the end of the refinement, the normalized difference structure factors have taken on their own phases, $\mathbf{E}'_{u,\mathbf{h}} = E'_{u,\mathbf{h}} e^{i\varphi_{u,\mathbf{h}}}$. These phases are assigned to the "regular" difference structure factors,

$$\mathbf{F}_{u,\mathbf{h}} \simeq F''_{u,\mathbf{h}} e^{i\varphi_{u,\mathbf{h}}}, \tag{7.268}$$

and equation 7.261 is used to determine the phases of the observed structure factors by adjusting $F''_{u,\mathbf{h}} = F_{u,\mathbf{h}}$, so that $|\mathbf{F}_{o,\mathbf{h}}| = |\mathbf{F}_{u,\mathbf{h}} + \mathbf{F}_{c,\mathbf{h}}|$ and

$$F_{o,\mathbf{h}} e^{i\varphi_{o,\mathbf{h}}} = F_{u,\mathbf{h}} e^{i\varphi_{o,\mathbf{h}}} + F_{c,\mathbf{h}} e^{i\varphi_{c,\mathbf{h}}}. \tag{7.269}$$

This direct method for difference structures has been incorporated into the program *DIRDIF*. It has been shown to be especially successful in solving structures for which the known fragment represents only a small fraction of the scattering power.

Exercises

1. The triplet phase relationship (TPR) for strong reflections, $\varphi_{\mathbf{h}} \approx \varphi_{\mathbf{k}} + \varphi_{\mathbf{h}-\mathbf{k}}$, is seen in the literature in various forms. (a) Show that $\varphi_{-\mathbf{h}} + \varphi_{\mathbf{k}} + \varphi_{\mathbf{h}-\mathbf{k}} \approx 0$ is equivalent. (b) Show that $\varphi_{\mathbf{h}} + \varphi_{\mathbf{k}} - \varphi_{\mathbf{h}+\mathbf{k}} \approx 0$ is equivalent.

2. Triplet relationships are analyzed with conditional probabilities since the phase of a given reflection is *not* independent of the phases of the other reflections in the relationship. To gain an idea of the difference between a simple probability and a conditional probability, solve the following two problems: (a) You are told that your new neighbor has two children. What is the probability that the older child's sibling is a brother? (b) You discover that the older child's name is John. What is the probability that the older child's sibling is a brother?

3. Show that the tangent formula provides the expected value for a single triplet relationship.

4. The E values for the strong reflections for the bikitaite structure, $Li[AlSi_2O_6]{\cdot}H_2O$, are given in Table 7.3. The (7 1 0) reflection is found in a number of triplet relationships. During the structural solution process the (4 0 7) and ($\bar{3}$ 1 7) are assigned phases of 0° and 30°, respectively. (a) Determine the phase of the (7 1 0) reflection predicted by the triplet relationship among these reflections. (b) Based on this single relationship, determine the probability that φ_{710} lies within ±10° of the predicted phase.

5. As the solution of the bikitaite structure described in Exercise 4 continues, the phases of a number of reflections involved in triplet relationships with the (7 1 0) reflection emerge (listed in (**h**, **h** − **k**) pairs, one for each triplet):

h	k	l	E	φ	h	k	l	E	φ	h	k	l	E	φ	h	k	l	E	φ
4	0	7	1.96	0	4	4	1	1.92	24	$\bar{3}$	3	2	1.94	120	3	3	0	1.90	288
$\bar{3}$	1	7	1.56	30	$\bar{3}$	3	1	1.91	144	4	2	2	1.73	74	4	2	0	1.63	69
$\bar{5}$	2	5	2.02	226	$\bar{8}$	1	10	2.12	24	$\bar{9}$	0	8	1.92	0	$\bar{3}$	3	2	1.94	120
2	1	5	1.62	188	$\bar{1}$	0	10	2.09	0	$\bar{2}$	1	8	1.73	202	$\overline{10}$	2	2	1.81	254

(a) Determine the expected value of φ_{710} based upon the assigned phases of these reflections. (b) Determine the probability that φ_{710} lies within $\pm 10°$ of the predicted phase (Note: The modified Bessel function that determines the normalization constant tends to require a large number of terms in the series expansion when the argument becomes large. Since the probability integral itself must be evaluated numerically, in such cases it is much easier to evaluate the normalization integral numerically as well.)

6. Referring to Table 7.3 in order of descending values of E: (a) show that the first three reflections in the table do *not* constitute a potential set of origin-defining reflections; (b) proceeding down the table, determine the first set of three reflections that would be suitable for origin-definition.

7. Assigning phases of 0 to the ($\bar{5}$ 0 5), ($\overline{11}$ 0 7), and (7 1 0) reflections in Table 5.3 (defining the origin), and symbolic phases, s_1, s_2, and s_3 to the ($\bar{2}$ 0 2), ($\bar{5}$ 2 5), and ($\bar{3}$ 3 2) reflections, respectively, assign the symbolic phases of the (a) ($\bar{6}$ 0 2), (b) ($\bar{7}$ 0 7), (c) ($\bar{8}$ 3 5), and (d) ($\bar{6}$ 2 2) reflections.

8. Using the origin-defining and symbolic phases assigned in the previous exercise, show that triplet relationships involving the ($\bar{5}$ 0 5), ($\bar{5}$ 2 5), (7 1 0), and (0 2 0) reflections predict that s_2 is approximately π.

Chapter 8

Crystal Structure Refinement

The structural solution methods described in the previous two chapters result in the determination of the locations of some or all of the atoms in the unit cell. These locations are determined from maxima in a Fourier map, and are only approximate. A *model* of the structure (or a portion of the structure) is created by placing spherical atoms at these locations, allowing for the determination of calculated structure factors:

$$F_{c,\,\mathbf{h}}e^{i\varphi_{\mathbf{h}}} = \sum_{j=1}^{n} f_{j,\,\theta d}e^{-2\pi\, i(hx_j+ky_j+lz_j)}e^{-T_j}. \tag{8.1}$$

The structure factors derived from this model predict reflection intensities, $I_{c,\,\mathbf{h}}$ from the squares of the magnitudes of the structure factors calculated from Eqn. 8.1, $\mathbf{F}^*_{c,\,\mathbf{h}}\mathbf{F}_{c,\,\mathbf{h}} = F^2_{c,\,\mathbf{h}}$. The experimental entities upon which the structural solution is based are the observed reflection intensities, I_o, scaled to "electron scattering" units. The parameters of the model that determine the calculated structure factor magnitudes are the atom locations, (x_j, y_j, z_j), and the temperature factors, T_j. In order to complete the structure with as much accuracy as possible we must refine these parameters to generate calculated intensities that are as close as possible to those observed experimentally. This is accomplished by employing the method of least squares, which we encountered earlier in the refinement of unit cell parameters in Chapter 4 and in the description of phase annealing in Chapter 7 (also see Appendix G).

8.1 Linear Least Squares

Before we take up the *non-linear least squares* methods used for crystal structure refinement, we will find it useful to discuss *linear least squares*, which involves the solution of simultaneous linear equations. In order to link this treatment to the crystallography problem, we consider a similar situation that involves refining a model to obtain the best fit to a set of experimental observations.

The top of Fig. 8.1 represents a plot of the absorption spectrum of a solution containing differing amounts of seven components. The spectrum has been collected on an instrument in which the absorption of light is measured in small

Understanding Single-Crystal X-Ray Crystallography. Dennis W. Bennett

ISBN: 978-3-527-32677-8 (HC), 978-3-527-32794-2 (SC)

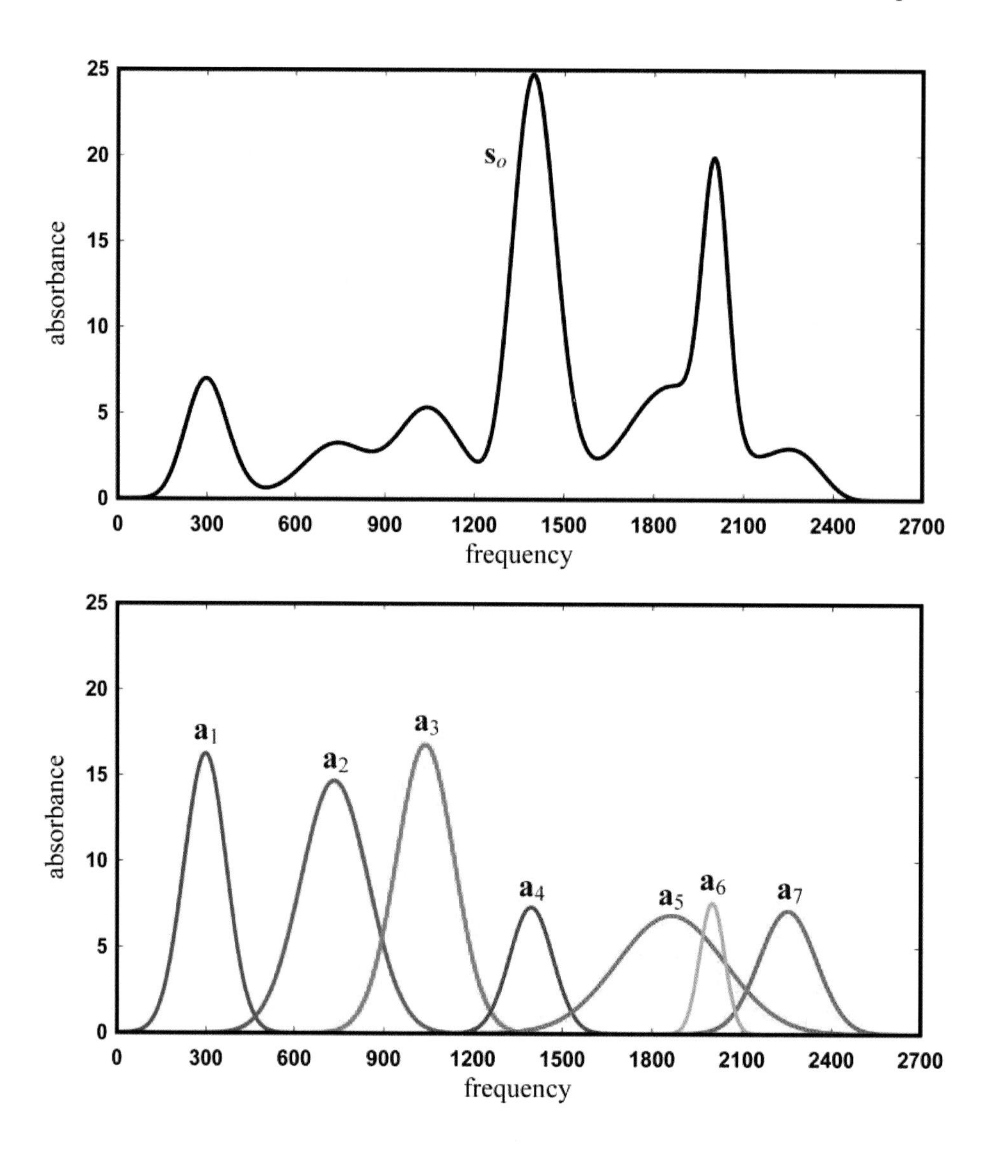

Figure 8.1 Top: Hypothetical observed absorption spectrum, $\mathbf{s}_o$, of a solution containing a mixture of components (units are arbitrary). Bottom: Individual spectra, $\mathbf{a}_j$, of each component in the mixture, all at unit concentration (e.g., 1 mol/liter).

equal increments of the frequency of the absorbed radiation; we will refer to the absorbance measured at each increment as a "data point". Each data point constitutes a separate experimental observation. If there are m data points collected, then the *observed* spectrum is represented by the vector, $\mathbf{s}_o = [s_1^o \; s_2^o \; \dots \; s_m^o]$, where s_i is the absorption measured for data point i. The plot at the bottom of Fig. 8.1 represents the spectra of the individual components, all at the same concentration (1 mol/liter). The spectra are different for each component because each absorbs radiation at a different frequency and with its own efficiency (some materials absorb light more effectively than others). Our goal is to determine the contribution of each of the components to the observed spectrum — and hence the concentrations of each component in the sample. The concentration of each component, x_j, is related linearly to its absorbance; the spectrum is the sum of the contributions of each of the components. Each component spectrum is collected with the same number of data points at the same frequencies as the observed spectrum; the jth spectrum is represented by $\mathbf{a}_j = [a_{1,j}\ a_{2,j}\ \ldots\ a_{m,j}]$. If the concentrations of each component in the sample are known, then the observed spectrum can be modeled with a *calculated* spectrum, created as a *linear combination* of the component spectra, $\mathbf{s}_c = x_1\mathbf{a}_1 + x_2\mathbf{a}_2 + \ldots + x_n\mathbf{a}_n$. For the ith data point,

$$s_i^c = x_1 a_{i,1} + x_2 a_{i,2} + \ldots + x_n a_{i,n}, \tag{8.2}$$

where $n = 7$ in our example. The calculated spectrum, $\mathbf{s}_c$, is therefore

$$\begin{bmatrix} a_{11} & a_{12} & \cdots & a_{1n} \\ a_{21} & a_{22} & \cdots & a_{2n} \\ a_{31} & a_{32} & \cdots & a_{3n} \\ \vdots & \vdots & & \vdots \\ \vdots & \vdots & & \vdots \\ a_{m11} & a_{m2} & \cdots & a_{mn} \end{bmatrix} \begin{bmatrix} x_1 \\ x_2 \\ \vdots \\ x_n \end{bmatrix} = \begin{bmatrix} s_1^c \\ s_2^c \\ s_3^c \\ \vdots \\ \vdots \\ s_m^c \end{bmatrix}, \tag{8.3}$$

which is the matrix equation

$$\mathbf{Ax} = \mathbf{s}_c. \tag{8.4}$$

The $\mathbf{A}$ matrix, the matrix of the column vectors of the components, is often referred to as the *design matrix*.

If the model is perfect it will generate the experimental spectrum exactly such that $\mathbf{s}_c = \mathbf{s}_o$, and we can calculate the concentrations directly from the observed spectrum by creating an invertible (square) matrix, $\mathbf{A}^T\mathbf{A}$:

$$\begin{aligned} \mathbf{Ax} &= \mathbf{s}_o \\ \mathbf{A}^T\mathbf{Ax} &= \mathbf{A}^T\mathbf{s}_o \\ (\mathbf{A}^T\mathbf{A})^{-1}\mathbf{A}^T\mathbf{Ax} &= (\mathbf{A}^T\mathbf{A})^{-1}\mathbf{A}^T\mathbf{s}_o \\ \mathbf{x} &= (\mathbf{A}^T\mathbf{A})^{-1}\mathbf{A}^T\mathbf{s}_o. \end{aligned} \tag{8.5}$$

The model, however, is rarely perfect (there may, for example, be interactions among the components in the mixture), and the calculated model spectrum will not be identical to the observed spectrum. We can still get approximate concentrations by solving Eqn. 8.5, which we might intuitively assume would provide the best possible values of x_i. In order to clarify the nature of these "best" values we take a closer look at the relationship between the observed and calculated spectrum. For every data point there is a difference between the observed absorbance and the calculated absorbance. If we subtract the calculated absorbance from the observed absorbance for the ith data point, the difference, the quantity remaining after subtraction, is the *residual*, $r_i = s_i^o - s_i^c$, resulting in the residual m vector, $\mathbf{r}$:

$$\mathbf{r} = \mathbf{s}_o - \mathbf{s}_c = \mathbf{s}_o - \mathbf{Ax}. \tag{8.6}$$

The "best" values of $\mathbf{x}$ be will be those that create a model spectrum as close as possible to the observed spectrum — those that make the average magnitude of the difference between the observed and calculated absorbances as small as possible. These values will therefore make the length of the residual vector as small as possible. The length of $\mathbf{r}$ is determined by the square root of its inner product, $\mathbf{r}^T\mathbf{r}$.* Since minimizing the square root also minimizes the inner product, we consider the values of x_i that minimize:

$$\mathbf{r}^T\mathbf{r} = (\mathbf{s}_o - \mathbf{A}\mathbf{x})^T(\mathbf{s}_o - \mathbf{A}\mathbf{x}) = r_1^2 + r_2^2 + \cdots + r_m^2, \tag{8.7}$$

the *sum of the squares of the residuals.* The process of obtaining these values is known as *least-squares minimization.* $\mathbf{r}^T\mathbf{r}$ will be at a minimum when its derivative with respect to each value of x_i is zero:

$$\frac{\partial(\mathbf{r}^T\mathbf{r})}{\partial x_i} = 2r_1\frac{\partial r_1}{\partial x_i} + 2r_2\frac{\partial r_2}{\partial x_i} + \ldots + 2r_m\frac{\partial r_m}{\partial x_i} = 0, \qquad i = 1, 2, \ldots, n. \tag{8.8}$$

The jth component of the residual vector is

$$r_j = s_j^o - s_j^c = s_j^o - a_{j1}x_1 - a_{j2}x_2 - \ldots - a_{ji}x_i \ldots - a_{jn}x_n, \tag{8.9}$$

and its derivative with respect to x_i is therefore

$$\frac{\partial r_j}{\partial x_i} = -a_{ji}. \tag{8.10}$$

It follows that the derivative of $\mathbf{r}^T\mathbf{r}$ is

$$\begin{aligned}\frac{\partial(\mathbf{r}^T\mathbf{r})}{\partial x_i} &= -2r_1a_{1i} - 2r_2a_{2i} - \ldots - 2r_ma_{ni} \\ &= -2(\mathbf{a}_i^T\mathbf{r}) = 0, \qquad i = 1, 2, \ldots, n.\end{aligned} \tag{8.11}$$

Taking the derivative for each x_i produces n simultaneous linear equations:

$$\begin{aligned}\frac{\partial(\mathbf{r}^T\mathbf{r})}{\partial x_1} &= -2(\mathbf{a}_1^T\mathbf{r}) = 0 \\ \frac{\partial(\mathbf{r}^T\mathbf{r})}{\partial x_2} &= -2(\mathbf{a}_2^T\mathbf{r}) = 0 \\ &\vdots \qquad\quad \vdots \qquad \vdots \\ \frac{\partial(\mathbf{r}^T\mathbf{r})}{\partial x_n} &= -2(\mathbf{a}_n^T\mathbf{r}) = 0,\end{aligned}$$

resulting in the matrix equation

$$\mathbf{A}^T\mathbf{r} = \mathbf{A}^T(\mathbf{s}_o - \mathbf{A}\mathbf{x}) = \mathbf{0}. \tag{8.12}$$

Rearranging this equation gives

$$\mathbf{A}^T\mathbf{A}\mathbf{x} = \mathbf{A}^T\mathbf{s}_o, \tag{8.13}$$

*The "length" of an abstract vector in m space in known formally as its *Euclidean norm.* For a residual vector with $m = 3$, this would be the length of the vector with its components treated as Cartesian coordinates, $|\mathbf{r}| = \sqrt{r_1^2 + r_2^2 + r_3^2}$.

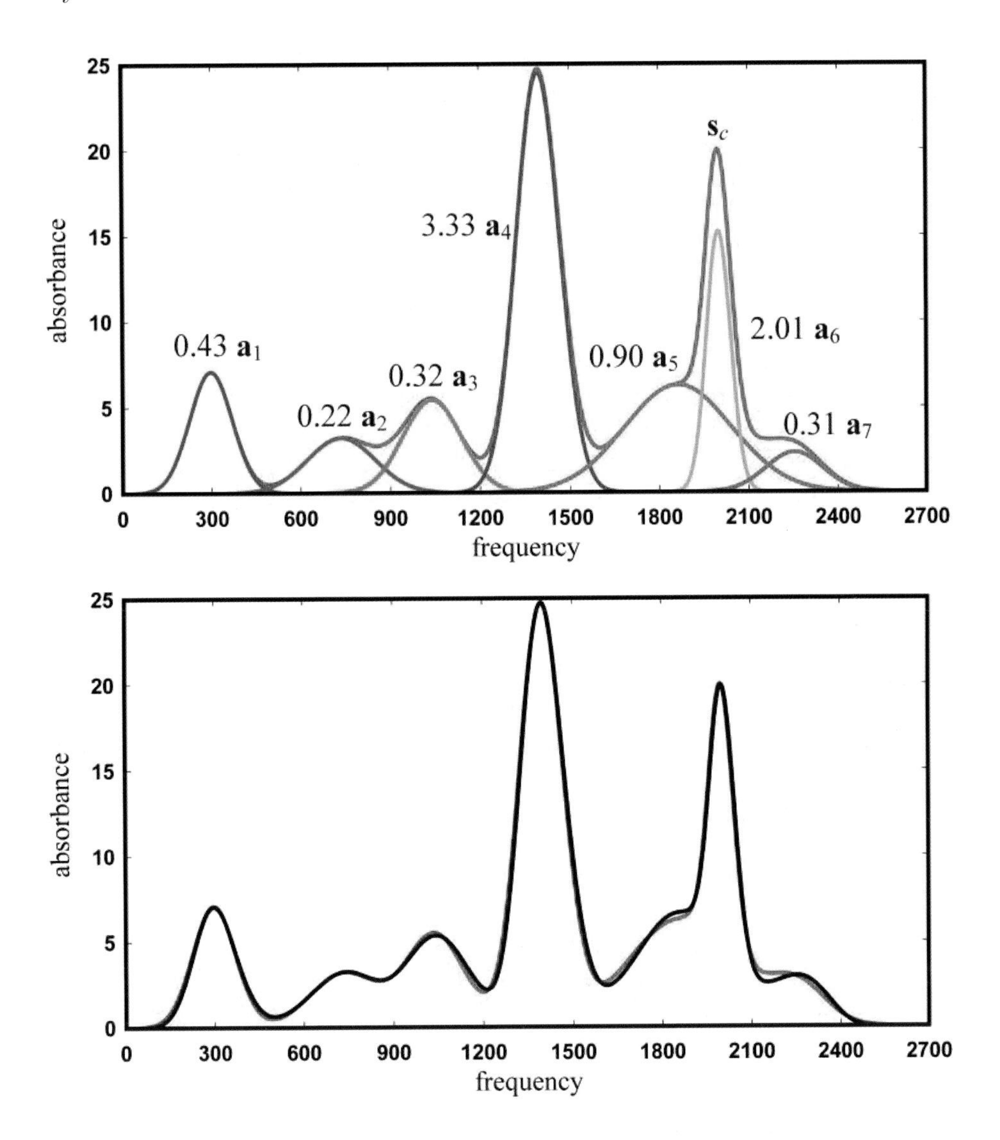

Figure 8.2 Top: Calculated absorption spectrum, $\mathbf{s}_c$, from a least-squares determination of the concentrations of the individual components. Bottom: Calculated spectrum (red) compared to the observed spectrum (black).

and

$$\mathbf{x} = (\mathbf{A}^T\mathbf{A})^{-1}\mathbf{A}^T\mathbf{s}_o.$$

The linear equations that are represented by Eqn. 8.13 are known as the *normal equations.* The solution of the normal equations is identical to the solution given by Eqn. 8.5— it is the solution that minimizes the sum of the squares of the residuals – the *least-squares solution.* Fig. 8.2 shows the results of the least-squares solution for the concentrations in our example, giving $\mathbf{x} = [0.43\ 0.22\ 0.32\ 3.33\ 0.90\ 2.01\ 0.31]$.

8.1.1 Weighted Least Squares

The simple spectroscopy example was based on having equal concentrations for the component spectra. In addition, we presumed that the data collected from different frequencies were equally reliable. If the concentrations differ, then the resulting values of x_i can be corrected by multiplying each x_i by the concentration of the ith component. For example, if the concentration of component 2 was actually 1.3 mol/liter when the component spectrum was measured then we would correct x_2 by multiplying it by 1.3. The second assumption is likely to be reasonable in our particular experiment if all of the spectra are collected on the same instrument — and if the instrument detector does not vary over the frequency range. However, it is often the case that spectrometer detectors are "noisier" in some regions of the spectrum than they are in other regions. In such an instance it would be prudent to weight the residuals in the least-squares determination so that the influence of the residuals on the final solution correspond to the reliability of the measurements that determine them. In other words, the residual vector to be minimized becomes $\mathbf{r}_w = [w_1' r_1 \; w_2' r_2 \; \ldots w_m' r_m]$, where w_i' is a measure of the reliability of r_i, which in our case, is a measure of the reliability of the absorbance collected for data point i. In order to generate a weight for each data point we might consider recording the spectrum k times, and evaluating the degree to which the absorbance for each data point varies with respect to its average value. Provided that the data vary because of random error (i.e., the errors are normally distributed – and are said to be *unbiased*), this number is reflected in its standard deviation. σ_i is obtained for each data point by measuring the values of the deviations from the mean value, and determining the average square root of the squared deviations (the root-mean-square deviation). From Eqn. 5.2,

$$\sigma_i = \left(\frac{1}{k-1} \sum_{j=1}^{k} (s_{i,j}^o - \langle s_{i,j}^o \rangle)^2 \right)^{\frac{1}{2}} . \qquad (8.14)$$

The larger the value of σ_i the less reliable the measurement, and a reasonable value for w_i' is $1/\sigma_i$. Thus the *weighted least squares* solution minimizes

$$(\mathbf{r}^T\mathbf{r})_w = \begin{bmatrix} w_1' r_1 & w_2' r_2 & \cdots & w_m' r_m \end{bmatrix} \begin{bmatrix} w_1' r_1 \\ w_2' r_2 \\ \vdots \\ w_m' r_m \end{bmatrix} \qquad (8.15)$$

$$\begin{aligned} &= w_1'^2 r_1^2 + w_2'^2 r_2^2 + \cdots + w_m'^2 r_m^2 \\ &= \frac{1}{\sigma_1^2} r_1^2 + \frac{1}{\sigma_2^2} r_2^2 + \cdots + \frac{1}{\sigma_m^2} r_m^2 \\ &= w_1 r_1^2 + w_2 r_2^2 + \cdots + w_m r_m^2, \end{aligned} \qquad (8.16)$$

where the unbiased least-squares weight for data point i is $1/\sigma_i^2$, *the reciprocal of its variance.*

Differentiating with respect to x_i now becomes

$$\begin{aligned}\frac{\partial(\mathbf{r}^T\mathbf{r})_w}{\partial x_i} &= -2w_1r_1a_{1i} - 2w_2r_2a_{2i} - \ldots - 2w_mr_ma_{ni}\\ &= -2(\mathbf{a}_i^T\mathbf{W}\mathbf{r}) = 0, \qquad i = 1, 2, \ldots, n,\end{aligned} \tag{8.17}$$

where $\mathbf{W}$ is the diagonal matrix

$$\mathbf{W} = \begin{bmatrix} w_1 & 0 & \cdots & 0 \\ 0 & w_2 & \cdots & 0 \\ \vdots & \vdots & \ddots & \vdots \\ 0 & 0 & \cdots & w_m \end{bmatrix}. \tag{8.18}$$

The weighted least-squares solution parallels the previous un-weighted solution:

$$\mathbf{A}^T\mathbf{W}\mathbf{r} = \mathbf{A}^T\mathbf{W}(\mathbf{s}_o - \mathbf{A}\mathbf{x}) = \mathbf{0} \tag{8.19}$$

$$\mathbf{A}^T\mathbf{W}\mathbf{A}\mathbf{x} = \mathbf{A}^T\mathbf{W}\mathbf{s}_o \tag{8.20}$$

$$\mathbf{x} = (\mathbf{A}^T\mathbf{W}\mathbf{A})^{-1}\mathbf{A}^T\mathbf{W}\mathbf{s}_o. \tag{8.21}$$

8.1.2 Estimation of Parameter Errors

In the crystallography experiment the parameters that we determine from the least-squares method are the coordinates and displacement factors for each atom. The positions will give us bond lengths and angles, and it will be important to assess the estimated errors in these quantities. These errors are reflected in the errors inherent in the experimental data — the intensities — and we must therefore determine how to estimate the parameter errors in terms of the errors in the data.

The spectroscopy example again serves as a simpler case. The observed spectrum, $\mathbf{s}_o$, is approximated with a model spectrum $\mathbf{s}_c$, by minimizing the sum of the squares of the residuals, $\mathbf{r}$, which represent the differences between the observed spectrum and the model. The actual errors in the observations, however, are those between the observed spectrum and an ideal spectrum without errors, which we refer to as $\mathbf{s}_e$ (e for *exact*), such that $s_i^e = s_i^o + \epsilon_i$, where ϵ_i is the error in the ith observation. If we were able to measure this "perfect" spectrum, the least-squares method would produce the parameters x_r^e instead of the x_r actually determined. Intuitively, we might expect that the residuals, r_i are related to the errors, ϵ_i — and if we can determine this relationship we should be able to utilize the measured residuals to estimate the variances of the parameters.

Expanding Eqn. 8.20 provides the normal equations:

$$\begin{bmatrix} \sum_{i=1}^{m} w_i a_{i1} a_{i1} & \sum_{i=1}^{m} w_i a_{i1} a_{i2} & \cdots & \sum_{i=1}^{m} w_i a_{i1} a_{in} \\ \sum_{i=1}^{m} w_i a_{i2} a_{i1} & \sum_{i=1}^{m} w_i a_{i2} a_{i2} & \cdots & \sum_{i=1}^{m} w_i a_{i2} a_{in} \\ \vdots & \vdots & \vdots & \vdots \\ \sum_{i=1}^{m} w_i a_{in} a_{i1} & \sum_{i=1}^{m} w_i a_{in} a_{i2} & \cdots & \sum_{i=1}^{m} w_i a_{in} a_{in} \end{bmatrix} \begin{bmatrix} x_1 \\ x_2 \\ \vdots \\ x_n \end{bmatrix} =$$

$$\begin{bmatrix} w_1 a_{11} & w_2 a_{21} & \cdots & \cdots & w_m a_{m1} \\ w_1 a_{12} & w_2 a_{22} & \cdots & \cdots & w_m a_{m2} \\ \vdots & \vdots & \vdots & \vdots & \vdots \\ w_1 a_{1n} & w_2 a_{2n} & \cdots & \cdots & w_m a_{mn} \end{bmatrix} \begin{bmatrix} s_1^o \\ s_2^o \\ \vdots \\ \vdots \\ s_m^o \end{bmatrix}. \tag{8.22}$$

The resulting jth normal equation is

$$x_1 \sum_{i=1}^{m} w_i a_{ij} a_{i1} + x_2 \sum_{i=1}^{m} w_i a_{ij} a_{i2} + \ldots + x_n \sum_{i=1}^{m} w_i a_{ij} a_{in} = \sum_{i=1}^{m} w_i a_{ij} s_i^o. \tag{8.23}$$

Letting

$$c_{rk} = c_{kr} = \sum_{i=1}^{m} wia_{ir} a_{ik} \quad \text{and} \tag{8.24}$$

$$v_r = \sum_{i=1}^{m} w_i a_{ir} s_i^o \tag{8.25}$$

simplifies the n normal equations:

$$\begin{aligned} c_{11} x_1 + c_{12} x_2 + \ldots + c_{1n} x_n &= v_1 \\ c_{21} x_1 + c_{22} x_2 + \ldots + c_{2n} x_n &= v_2 \\ \vdots \qquad \vdots \qquad\qquad \vdots \;\; &\vdots \;\; \vdots \\ c_{n1} x_1 + c_{n2} x_2 + \ldots + c_{nn} x_n &= v_n. \end{aligned}$$

The coefficient matrix provides the least-squares solution:

$$\begin{bmatrix} c_{11} & c_{12} & \cdots & c_{1n} \\ c_{21} & c_{21} & \cdots & c_{2n} \\ \vdots & \vdots & \vdots & \vdots \\ c_{n1} & c_{n2} & \cdots & c_{nn} \end{bmatrix} \begin{bmatrix} x_1 \\ x_2 \\ \vdots \\ x_n \end{bmatrix} = \begin{bmatrix} v_1 \\ v_2 \\ \vdots \\ v_n \end{bmatrix}. \tag{8.26}$$

Thus,

$$\mathbf{Cx} = \mathbf{A}^T \mathbf{W} \mathbf{A} \mathbf{x} = \mathbf{v} \tag{8.27}$$

$$\mathbf{x} = \mathbf{C}^{-1} \mathbf{v}. \tag{8.28}$$

For reasons that will soon become evident, the inverse of the normal equation coefficient matrix,

$$\mathbf{C}^{-1} = (\mathbf{A}^T\mathbf{W}\mathbf{A})^{-1} = \begin{bmatrix} c^i_{11} & c^i_{12} & \cdots & c^i_{1n} \\ c^i_{21} & c^i_{21} & \cdots & c^i_{2n} \\ \vdots & \vdots & \vdots & \vdots \\ c^i_{n1} & c^i_{n2} & \cdots & c^i_{nn} \end{bmatrix}, \tag{8.29}$$

is known as the *variance-covariance matrix*, or more commonly as the *covariance* matrix.

The rth parameter (of n parameters) is therefore determined by

$$x_r = \sum_{k=1}^{n} c^i_{rk} v_k = \sum_{k=1}^{n} c^i_{rk} \left(\sum_{i=1}^{m} w_i a_{ik} s^o_i \right) = \sum_{i=1}^{m} \left(\sum_{k=1}^{n} c^i_{rk} a_{ik} \right) w_i s^o_i. \tag{8.30}$$

Let $Q_{ri} = \sum_{k=1}^{n} c^i_{rk} a_{ik}$, so that

$$x_r = \sum_{i=1}^{m} w_i Q_{ri} s^o_i. \tag{8.31}$$

For the error-free spectrum the value of the rth parameter would be,

$$x^e_r = \sum_{i=1}^{m} w_i Q_{ri} s^e_i = \sum_{i=1}^{m} w_i Q_{ri} (s^o_i + \epsilon_i), \tag{8.32}$$

and

$$x^e_r - x_r = \sum_{i=1}^{m} w_i Q_{ri} \epsilon_i. \tag{8.33}$$

If the errors are normally distributed, then x^e_r will be the average value of x_r (say, for 100 measured spectra), and the variance of x_r will be

$$\begin{aligned} \sigma^2_r &= \langle (x_r - \langle x_r \rangle) \rangle^2 = \langle (x_r - x^e_r)^2 \rangle \\ &= \langle (w_1 Q^2_{r1})(w_1 \epsilon^2_1) + (w_2 Q^2_{r2})(w_2 \epsilon^2_2) + \ldots + (w_m Q^2_{rm})(w_m \epsilon^2_m) \\ &\quad -2 w_1 w_2 Q_{r1} Q_{r2} \epsilon_1 \epsilon_2 - 2 w_1 w_3 Q_{r1} Q_{r3} \epsilon_1 \epsilon_3 - \cdots \rangle. \end{aligned} \tag{8.34}$$

Since $w_i = 1/\sigma^2_i$ represents the scatter of the observed values of s^o_i about their mean, and ϵ_i represents the mean deviations between the observed values and the "error-free" values, we would expect their behavior to parallel one another — as $1/\sigma^2_i$ gets smaller, ϵ^2_i gets larger. Thus the values of $w_i \epsilon^2_i$ are approximately constant, and we can replace them with their average value, $\langle w_i \epsilon^2_i \rangle$. Furthermore, the cross-terms containing $\epsilon_i \epsilon_j$ should be equally distributed (positive and negative) and average to zero. Thus,

$$\sigma^2_r = \langle w_i \epsilon^2_i \rangle \sum_{i=1}^{m} w_i Q^2_{ri}. \tag{8.35}$$

The sum becomes

$$\sum_{i=1}^{m} w_i Q_{ri}^2 = \sum_{i=1}^{m} w_i Q_{ri} \left(\sum_{k=1}^{n} c_{rk}^i a_{ik} \right) = \sum_{k=1}^{n} \left(\sum_{i=1}^{m} w_i Q_{ri} a_{ik} \right) c_{rk}^i. \tag{8.36}$$

Substituting the expression for Q_{ri} into the sum in parentheses in the expression on the right gives

$$\sum_{i=1}^{m} w_i Q_{ri} a_{ik} = \sum_{i=1}^{m} w_i \left(\sum_{j=1}^{n} c_{rj}^i a_{ij} \right) a_{ik} = \sum_{j=1}^{n} c_{rj}^i \sum_{i=1}^{m} w_i a_{ij} a_{ik}. \tag{8.37}$$

From Eqn. 8.22, the second sum is the c_{jk} matrix element of the $\mathbf{C}$ matrix and

$$\sum_{i=1}^{m} w_i Q_{ri} a_{ik} = \sum_{j=1}^{n} c_{rj}^i c_{jk}. \tag{8.38}$$

This is the inner product of the rth row of the $\mathbf{C}^{-1}$ matrix with the kth column of the $\mathbf{C}$ matrix. Since $\mathbf{C}^{-1}\mathbf{C} = \mathbf{I}$, the sum is unity for $k = r$ and zero for $k \neq r$. It follows from Eqn. 8.36 that

$$\sum_{i=1}^{m} w_i Q_{ri}^2 = c_{rr}^i, \tag{8.39}$$

the rth element on the diagonal of $\mathbf{C}^{-1}$, and

$$\sigma_r^2 = c_{rr}^i \langle w_i \epsilon_i^2 \rangle. \tag{8.40}$$

Because we do not have access to the error-free spectrum we are not able to assess $\langle w_i \epsilon_i^2 \rangle$ and we can't use this relationship directly. We must therefore find a way to relate the errors to the residual differences between the model spectrum and the observed spectrum, which we can calculate:

$$r_i = s_i^o - s_i^c = s_i^o - \sum_{k=1}^{n} x_k a_{ik}, \qquad i = 0, 1, 2, \ldots, m. \tag{8.41}$$

The errors are given by $\epsilon_i = s_i^e - s_i^o$, and the difference between the error-free spectrum and the calculated spectrum is

$$s_i^e - s_i^c = \epsilon_i + r_i = \sum_{k=1}^{n} (x_k^e - x_k) a_{ik} \tag{8.42}$$

$$= \sum_{k=1}^{n} a_{ik} \sum_{j=1}^{m} w_j Q_{kj} \epsilon_j. \tag{8.43}$$

Multiplying both sides of Eqn. 8.42 by $w_i r_i$,

$$w_i \epsilon_i r_i + w_i r_i^2 = \sum_{k=1}^{n} (x_k^e - x_k) w_i a_{ik} r_i$$

and summing from $i = 1$ to $i = m$ results in

$$\sum_{i=1}^{m} w_i \epsilon_i r_i + \sum_{i=1}^{m} w_i r_i^2 = \sum_{k=1}^{n} (x_k^e - x_k) \sum_{i=1}^{m} a_{ik} w_i r_i. \tag{8.44}$$

From Eqn. 8.17, the sum on the right is the minimization criterion for weighted least squares,

$$\sum_{i=1}^{m} a_{ik} w_i r_i = \mathbf{a}_k^T \mathbf{W} \mathbf{r} = 0, \tag{8.45}$$

allowing for expression of the errors in terms of the residuals:

$$\sum_{i=1}^{m} w_i \epsilon_i r_i = -\sum_{i=1}^{m} w_i r_i^2. \tag{8.46}$$

Multiplying both sides of Eqn. 8.43 by $w_i \epsilon_i$,

$$w_i \epsilon_i^2 + w_i \epsilon_i r_i = \sum_{k=1}^{n} \sum_{j=1}^{m} w_i a_{ik} Q_{kj} w_j \epsilon_i \epsilon_j,$$

using Eqn. 8.46, and again summing over i and gives

$$\sum_{i=1}^{m} w_i \epsilon_i^2 - \sum_{i=1}^{m} w_i r_i^2 = \sum_{k=1}^{n} \sum_{i=1}^{m} \sum_{j=1}^{m} w_i a_{ik} Q_{kj} w_j \epsilon_i \epsilon_j. \tag{8.47}$$

If the equation is averaged over all the observations, the terms containing $\langle \epsilon_i \epsilon_j \rangle$ will average to zero, *except* when $i = j$. Substituting $\langle w_i \epsilon_i^2 \rangle$ for $w_i \epsilon_i^2$ results in

$$m\langle w_i \epsilon_i^2 \rangle - \sum_{i=1}^{m} w_i r_i^2 = \sum_{k=1}^{n} \left(\sum_{i=1}^{m} w_i Q_{kj} a_{ik} \right) \langle w_i \epsilon_i^2 \rangle. \tag{8.48}$$

From Eqn. 8.38, the expression in parentheses is the inner product of the kth row of the $\mathbf{C}^{-1}$ matrix with the kth column of the $\mathbf{C}$ matrix, which is unity. Thus,

$$m\langle w_i \epsilon_i^2 \rangle - \sum_{i=1}^{m} w_i r_i^2 = n\langle w_i \epsilon_i^2 \rangle. \tag{8.49}$$

and

$$\langle w_i \epsilon_i^2 \rangle = \frac{1}{m-n} \sum_{i=1}^{m} w_i r_i^2. \tag{8.50}$$

From Eqn. 8.40, the estimated variance of the rth parameter (x_r), is therefore

$$\sigma_r^2 = c_{rr}^i \left(\frac{1}{m-n} \sum_{i=1}^{m} w_i r_i^2 \right), \tag{8.51}$$

where m is the number of observations, n is the number of parameters, and r_i is the residual between the ith observed data point and its calculated value. As expected, the variance increases as the number of parameters increases, compared to the number of observations. It takes at least one experimental measurement to define a parameter, so that the number of "extra" measurements is $m - n$, which is often called the number of *degrees of freedom*. Experiments with positive degrees of freedom are considered to be *overdetermined*, those with negative degrees of freedom are deemed to be *underdetermined*.

If the covariance between parameter r and parameter s (Eqn. 5.74) is substituted into Eqn. 8.34, a similar result is obtained for σ_{rs}:

$$\sigma_{rs} = \langle (x_r - x_r^e)(x_s - x_s^e) \rangle = c_{rs}^i \left(\frac{1}{m-n} \sum_{i=1}^{m} w_i r_i^2 \right), \tag{8.52}$$

where c_{rs}^i is the *rsth* off-diagonal element of the $\mathbf{C}^{-1} = (\mathbf{A}^T \mathbf{W} \mathbf{A})^{-1}$ matrix. This matrix (along with the residuals) provides the variances and covariances of the parameters, and is therefore known as the *variance-covariance* matrix. The covariances are expected to be small (close to zero), *except* when the variables are correlated.

Estimation of Errors in the Lattice Parameters

In Chapter 4 the method of least squares was employed to determine the orientation matrix and unit cell parameters from the observed diffractometer angles for a set of m indexed reflections. The treatment in Chapter 4 emphasized the creation of the orientation matrix, and it is useful to reconsider it here more formally in order to estimate the errors in the lattice parameters determined from the orientation matrix. Eqns. 4.52–4.54 can be rewritten in terms of the Cartesian components of the reciprocal axes:

$$a_x^* h_j + a_y^* k_j + a_z^* l_j = x_j^o \tag{8.53}$$
$$b_x^* h_j + b_y^* k_j + b_z^* l_j = y_j^o \tag{8.54}$$
$$c_x^* h_j + c_y^* k_j + c_z^* l_j = z_j^o, \quad j = 1, 2, \ldots, m. \tag{8.55}$$

There are m simultaneous equations for each axis. x_j^o, y_j^o, and z_j^o are the Cartesian components of the reciprocal lattice vector for the jth reflection in the diffractometer reference frame, determined experimentally from the diffractometer angles (Eqn. 4.15):

$$\begin{bmatrix} x_j^o \\ y_j^o \\ z_j^o \end{bmatrix} = \frac{2 \sin \theta}{\lambda} \begin{bmatrix} \cos \varphi \sin(\omega - \theta) - \sin \varphi \cos \chi \cos(\omega - \theta) \\ \sin \varphi \sin(\omega - \theta) + \cos \varphi \cos \chi \cos(\omega - \theta) \\ \sin \chi \cos(\omega - \theta) \end{bmatrix}. \tag{8.56}$$

These same components can be calculated from the least-squares orientation matrix:

$$\mathbf{A}^* \begin{bmatrix} h_j \\ k_j \\ l_j \end{bmatrix} = \begin{bmatrix} x_j^c \\ y_j^c \\ z_j^c \end{bmatrix}. \tag{8.57}$$

Since the axes are independent they can be treated separately. We will consider the b^* axis here; the treatments for the a^* and c^* axes will be identical. The m equations are therefore:

$$\begin{aligned}
h_1 b_x^* + k_1 b_y^* + l_1 b_z^* &= y_1^o \\
h_2 b_x^* + k_2 b_y^* + l_2 b_z^* &= y_2^o \\
\vdots \qquad \vdots \qquad \vdots \\
h_m b_x^* + k_m b_y^* + l_m b_z^* &= y_m^o.
\end{aligned} \tag{8.58}$$

The residuals are $r_j = y_j^o - y_j^c$. Denoting the $m \times 3$ design matrix as $\mathbf{H}$ and the m vector on the right as $\mathbf{y}$,

$$\begin{aligned}
\mathbf{Hb} &= \mathbf{y}, \\
\mathbf{H}^T\mathbf{Hb} &= \mathbf{H}^T\mathbf{y}, \quad \text{and} \\
\mathbf{b}^* &= (\mathbf{H}^T\mathbf{H})^{-1}\mathbf{H}^T\mathbf{y}.
\end{aligned} \tag{8.59}$$

Comparing this result with Eqn. 4.62, $\mathbf{H}^T\mathbf{H} \equiv \mathbf{T}$, $\mathbf{b}^* \equiv [a_{21}\ a_{22}\ a_{23}]$, and $\mathbf{H}^T\mathbf{y} \equiv [w_{21}\ w_{22}\ w_{23}]$:

$$\begin{bmatrix} t_{11} & t_{12} & t_{13} \\ t_{21} & t_{22} & t_{23} \\ t_{31} & t_{32} & t_{33} \end{bmatrix} \begin{bmatrix} b_x^* \\ b_y^* \\ b_z^* \end{bmatrix} = \begin{bmatrix} w_{21} \\ w_{22} \\ w_{23} \end{bmatrix}. \tag{8.60}$$

The corresponding variances of the Cartesian components of the $\mathbf{b}^*$ vector are given by Eqn. 8.51:

$$\sigma_r^2 = t_{rr}^i \left(\frac{1}{m-3} \sum_{j=1}^{m} (y_j^o - y_j^c)^2 \right), \qquad r = 1, 2, 3, \tag{8.61}$$

where the t_{rr}^i elements are the first, second and third elements along the diagonal of the inverse of the normal equation coefficient matrix, $\mathbf{T}^{-1}$ — for b_x^*, b_y^*, and b_z^*, respectively. The covariances of the $\mathbf{b}^*$ components are

$$\sigma_{rs} = t_{rs}^i \left(\frac{1}{m-3} \sum_{j=1}^{m} (y_j^o - y_j^c)^2 \right), \qquad \text{etc.} \tag{8.62}$$

In order to estimate the variances of the direct lattice cell parameters, we must now see how errors in the reciprocal lattice *propagate* into the direct lattice. The relationships between direct and reciprocal cell parameters were developed in Eqns. 3.19–3.24. In general, any direct lattice parameter D, will be a function of the nine reciprocal axis vector components. The goal here is to estimate the variance of $D(a_x^*, a_y^*, a_z^*, b_x^*, b_y^*, b_z^*, c_x^*, c_y^*, c_z^*)$, $\sigma^2(D)$, from the estimated variances of the reciprocal lattice components.

While an estimate of the variance of each of these parameters is determined from Eqn. 8.61, a more rigorous way to obtain the variances would be to collect the reflection set n times, determine the parameters for each set of reflections, and compute their variances directly, e.g.,

$$\sigma^2(b_x^*) = \frac{1}{n-1} \sum_{i=1}^{n} (b_{x,i}^* - \langle b_x^* \rangle)^2 = \frac{1}{n-1} \sum_{i=1}^{n} (\Delta b_{x,i}^*)^2. \tag{8.63}$$

Similarly, the covariances are determined from, e.g.,

$$\sigma(b_x^*, b_y^*) = \frac{1}{n}\sum_{i=1}^{n}(b_{x,i} - \langle b_x^* \rangle)(b_{y,i}^* - \langle b_y^* \rangle) = \frac{1}{n}\sum_{i=1}^{n}(\Delta b_{x,i}^*)(\Delta b_{y,i}^*). \tag{8.64}$$

The variance in the direct lattice parameter, D, would also be

$$\sigma^2(D) = \frac{1}{n}\sum_{i=1}^{n}(D_i - \langle D \rangle)^2 = \sum_{i=1}^{n}(\Delta_i D)^2. \tag{8.65}$$

If b_x^* (for example) changes by a small amount, ∂b_x^*, then D will change by ∂D, and the change in D per unit change in b_x^* is $\partial D/\partial b_x^*$. Thus if b_x^* varies from its mean value by Δb_x^*, D will vary from its mean value by $(\partial D/\partial b_x^*)\Delta b_x^*$. It follows that the total change in D for the ith set of reflections is the sum of the changes due to each of the parameters*:

$$\Delta D_i = \frac{\partial D}{\partial a_x^*}\Delta a_{x,i}^* + \frac{\partial D}{\partial a_y^*}\Delta a_{y,i}^* + \frac{\partial D}{\partial a_z^*}\Delta a_{z,i}^* + \frac{\partial D}{\partial b_x^*}\Delta b_{x,i}^* + \ldots. \tag{8.66}$$

The variance of D is therefore approximately

$$\sigma^2(D) \simeq \frac{1}{n}\sum_{i=1}^{n}\left(\frac{\partial D}{\partial a_x^*}\Delta a_{x,i}^* + \frac{\partial D}{\partial a_y^*}\Delta a_{y,i}^* + \frac{\partial D}{\partial a_z^*}\Delta a_{z,i}^* + \frac{\partial D}{\partial b_x^*}\Delta b_{x,i}^* + \ldots\right)^2. \tag{8.67}$$

When the expression on the right is squared, two types of terms result:

$$\frac{1}{n}\sum_{i=1}^{n}\left(\frac{\partial D}{\partial a_x^*}\Delta a_{x,i}^*\right)^2 \quad \text{and} \quad \frac{1}{n}\sum_{i=1}^{n}\left(\frac{\partial D}{\partial a_x^*}\Delta a_{x,i}^*\right)\left(\frac{\partial D}{\partial b_y^*}\Delta b_{y,i}^*\right). \tag{8.68}$$

These correspond to the variances and covariances of the reciprocal Cartesian components, and

$$\begin{aligned}\sigma^2(D) &= \left(\frac{\partial D}{\partial a_x^*}\right)^2\sigma^2(a_x^*) + \left(\frac{\partial D}{\partial a_y^*}\right)^2\sigma^2(a_y^*) + \ldots \\ &+ \left(\frac{\partial D}{\partial a_x^*}\right)\left(\frac{\partial D}{\partial a_y^*}\right)\sigma(a_x^*, a_y^*) + \ldots\end{aligned} \tag{8.69}$$

The axial components are usually considered to be independent of one another (*uncorrelated*) such that $\Delta a_{x,i}^*$ and $\Delta a_{y,i}^*$, etc., each have equal probability of being negative or positive; the covariances all will be approximately equal to zero. Thus, for the unit cell parameters,

$$\sigma^2(D) \simeq \left(\frac{\partial D}{\partial a_x^*}\right)^2\sigma^2(a_x^*) + \left(\frac{\partial D}{\partial a_y^*}\right)^2\sigma^2(a_y^*) + \ldots \tag{8.70}$$

We are now able to determine the variance of any direct cell parameter "simply" by differentiating it with respect to each of the components of the reciprocal axes. For example, the magnitude of the b axis is given by

$$b = \frac{a^* c^* \sin\beta^*}{V^*}.$$

*For a more rigorous treatment see the discussion of the total differential in Appendix H.

The variance in b is then

$$\begin{aligned}\sigma^2(b) &= \left(\frac{\partial b}{\partial a^*}\right)^2 \sigma^2(a^*) + \left(\frac{\partial b}{\partial c^*}\right)^2 \sigma^2(c^*) \\ &\quad + \left(\frac{\partial b}{\partial \beta^*}\right)^2 \sigma^2(\beta^*) + \left(\frac{\partial b}{\partial V^*}\right)^2 \sigma^2(V^*). \end{aligned} \tag{8.71}$$

The variance in a^* is given by

$$\begin{aligned} a^* &= (a_x^{*\,2} + a_y^{*\,2} + a_z^{*\,2})^{1/2}; \\ \sigma^2(a^*) &= \left(\frac{\partial a^*}{\partial a_x^*}\right)^2 \sigma^2(a_x^*) + \left(\frac{\partial a^*}{\partial a_y^*}\right)^2 \sigma^2(a_y^*) + \left(\frac{\partial a^*}{\partial a_z^*}\right)^2 \sigma^2(a_z^*) \\ &= \frac{1}{a^{*\,2}}\,(a_x^{*\,2}\,\sigma^2(a_x^*) + a_y^{*\,2}\,\sigma^2(a_y^*) + a_z^{*\,2}\,\sigma^2(a_z^*)). \end{aligned} \tag{8.72}$$

Similarly,

$$\sigma^2(c^*) = \frac{1}{c^{*\,2}}\,(c_x^{*\,2}\,\sigma^2(c_x^*) + c_y^{*\,2}\,\sigma^2(c_y^*) + c_z^{*\,2}\,\sigma^2(c_z^*)). \tag{8.73}$$

To obtain the variance in β^*, we must determine its cosine from the scalar product of $a*$ and $c*$:

$$\begin{aligned} \cos\beta^* &= \frac{a_x^* c_x^* + a_y^* c_y^* + a_z^* c_z^*}{a * c*}; \\ \sigma^2(\cos\beta^*) &= \left(\frac{c_x^*}{a^* c^*}\right)^2 \sigma^2(a_x^*) + \left(\frac{a_x^*}{a^* c^*}\right)^2 \sigma^2(c_x^*) + \left(\frac{c_y^*}{a^* c^*}\right)^2 \sigma^2(a_y^*) \\ &\quad + \left(\frac{a_y^*}{a^* c^*}\right)^2 \sigma^2(c_y^*) + \left(\frac{c_z^*}{a^* c^*}\right)^2 \sigma^2(a_z^*) + \left(\frac{a_z^*}{a^* c^*}\right)^2 \sigma^2(c_z^*) \\ &\quad + \left(\frac{a_x^* c_x^* + a_y^* c_y^* + a_z^* c_z^*}{a^{*\,2} c^*}\right)^2 \sigma^2(a^*) \\ &\quad + \left(\frac{a_x^* c_x^* + a_y^* c_y^* + a_z^* c_z^*}{a^* c^{*\,2}}\right)^2 \sigma^2(c^*). \end{aligned} \tag{8.74}$$

The variance in the angle itself is determined by considering the relationship between a change in the cosine of the angle and the angle itself:

$$\begin{aligned} \frac{d\cos\beta^*}{d\beta^*} &= -\sin\beta^* \\ d\beta^* &= -\frac{d\cos\beta^*}{\sin\beta^*} \\ \sigma(\beta^*) &= -\frac{\sigma(\cos\beta^*)}{\sin\beta^*} \\ \sigma^2(\beta^*) &= \frac{\sigma^2(\cos\beta^*)}{\sin^2\beta^*}. \end{aligned} \tag{8.75}$$

The variances in b^*, α^* and γ^* are determined in the same manner. Finally, the variance in V^* is determined from

$$V^* = a^* b^* c^* (1 - \cos^2\alpha^* - \cos^2\beta^* - \cos^2\gamma^* + 2\cos\alpha^*\cos\beta^*\cos\gamma^*)^{1/2}. \tag{8.76}$$

Thus,

$$\begin{aligned}\sigma^2(V^*) &= \left(\frac{a^*}{V^*}\right)^2 \sigma^2(a^{*\,2}) + \left(\frac{b^*}{V^*}\right)^2 \sigma^2(b^{*\,2}) + \left(\frac{c^*}{V^*}\right)^2 \sigma^2(c^{*\,2}) \\ &+ \frac{(a^*b^*c^*)^4}{V^{*\,2}}(\sin\alpha(\cos\alpha^* - \cos\beta^*\cos\gamma^*))^2\sigma^2(\alpha^*) \\ &+ \frac{(a^*b^*c^*)^4}{V^{*\,2}}(\sin\beta(\cos\beta^* - \cos\alpha^*\cos\gamma^*))^2\sigma^2(\beta^*) \\ &+ \frac{(a^*b^*c^*)^4}{V^{*\,2}}(\sin\gamma(\cos\gamma^* - \cos\alpha^*\cos\beta^*))^2\sigma^2(\gamma^*). \end{aligned} \tag{8.77}$$

Substituting these variances into Eqn. 8.71,

$$\begin{aligned}\sigma^2(b) &= \left(\frac{c^*\sin\beta^*}{V^*}\right)^2 \sigma^2(a^*) + \left(\frac{a^*\sin\beta^*}{V^*}\right)^2 \sigma^2(c^*) \\ &+ \left(\frac{a^*c^*\cos\beta^*}{V^*}\right)^2 \sigma^2(\beta^*) + \left(\frac{a^*b^*c^*}{V^{*\,2}}\right)^2 \sigma^2(V^*). \end{aligned} \tag{8.78}$$

As is evident, the variance for each of the direct cell parameters is a complex function of the reciprocal axis components and their variances. Because we are only estimating the variances, we might consider numerical differentiation as an alternative. For example, to determine $\partial b/\partial a_x^*$, we make a small change in a_x^*: $a_x^{*'} = a_x^* + \delta(a_x^*)$. Then

$$\begin{aligned} a^{*'} &= ((a_x^* + \delta(a_x^*))^2 + a_y^{*2} + a_z^{*2})^{1/2}, \\ \beta^{*'} &= \arccos\left(\frac{\mathbf{a}^{*'}\cdot\mathbf{c}^*}{a^{*'}c^*}\right), \\ V^{*'} &= \mathbf{a}^{*'}\cdot(\mathbf{b}^*\times\mathbf{c}^*), \\ b' &= \frac{a^{*'}c^*\sin\beta^{*'}}{V^{*'}}, \\ \delta(b) &= b' - b \quad \text{and} \\ \frac{\partial b}{\partial a_x^*} &\simeq \frac{\delta(b)}{\delta(a_x^*)}, \quad \text{etc.} \end{aligned} \tag{8.79}$$

The variance in b is then

$$\sigma^2(b) \simeq \left(\frac{\delta b}{\delta a_x^*}\right)^2 \sigma^2(a_x^*) + \left(\frac{\delta b}{\delta a_y^*}\right)^2 \sigma^2(a_y^*) + \ldots, \text{etc.} \tag{8.80}$$

8.1.3 Constrained Least Squares

In the least-squares refinement of the crystal structure it is often necessary to constrain a parameter to a specific value, or to restrict some of the variable parameters so that they are related to one another. For example, the location of the origin in a $P2_1$ unit cell is arbitrary along the b axis, and it is necessary to fix the z location of one atom (thus establishing the location of the origin) to remove the ambiguity. Since the refinement involves the adjustment of parameters, it is necessary to assign the initial parameters unambiguously. Another common occurrence in structural refinement arises from situations in which an atom or atoms can occupy more than one site in the lattice. Since the X-rays "see" only an average structure, the atom will appear to partially occupy each of the sites, with some fraction of the atom in a particular site. The atomic scattering factor accounts for the atom's electrons, and the refinement of the atoms in each position is effected by multiplying the scattering factor by the fractional occupancy of the atom in each location. This fraction is known as the *site occupancy factor.** Since we do not know the fractional occupancies of the atom in each site they are treated as variable parameters in refinement. It is often useful, and sometimes necessary, to constrain these fractional parameters, so that they add to unity – to account for the single atom distributed among the sites.

A situation similar to the $P2_1$ ambiguity would arise in our spectroscopy example if two of the component spectra had the same line shape and sat atop one another (or were too close to be resolved). There would then be infinitely many combinations of the component spectra that would produce the same contribution to the model spectrum, and it would be impossible to determine the concentrations of either component from a least-squares treatment. Mathematically the situation is even more dire. The two overlapping spectra would produce parallel vectors in the design matrix since each absorbance in one vector would just be a constant times the corresponding absorbance in the other vector. This would result in parallel columns and rows in the normal equation $\mathbf{C}$ matrix, $(\mathbf{A}^T\mathbf{W}\mathbf{A})$, giving it a zero determinant *with no inverse* (a singular matrix). A way around this might be to determine the concentration of one of the components independently (from another experiment) and constrain it in the least-squares minimization so that only the other component is a variable parameter. The "occupancy factor" problem is mimicked in the spectroscopy example if some of the concentrations are dependent on others. Suppose that we have the following situation: Compound 1 is known to have been added to the solution so that its concentration is 0.5 M ($x_1 = 0.5$). Furthermore, Compound 6, initially added to the solution at a concentration of 2.5 M, is known to decompose into Compound 2, which does not exist initially in the solution ($x_2 + x_6 = 2.5$). We know nothing about the rest of the concentrations. We now wish to determine concentrations, $\mathbf{x}$, that will minimize $f(\mathbf{x}) = \mathbf{r}^T\mathbf{r}$, *subject to the constraints on the variables*:

$$q_1(\mathbf{x}) = 1x_1 + 0x_2 + 0x_3 + 0x_4 + 0x_5 + 0x_6 + 0x_7 = 0.5; \tag{8.81}$$

$$q_2(\mathbf{x}) = 0x_1 + 1x_2 + 0x_3 + 0x_4 + 0x_5 + 1x_6 + 0x_7 = 2.5. \tag{8.82}$$

These constraints mandate that from all of the values of $\mathbf{x}$ that are possible for the unconstrained solution, $\mathbf{x}_u$, only those values that satisfy these constraining equations, $\mathbf{x}_c$, will be allowed for the constrained solution. Referring to Appendix H,† a function is optimized, subject to constraints, if the *gradient vector* of the *Lagrangian* is equal to $\mathbf{0}$. The gradient vector of a function is the vector with components equal to the partial derivatives of the function with respect to each of the

*For an atom in a general position the site occupancy factor is ordinarily equal to unity. For atoms in special positions, symmetry operators will duplicate an atom at the same site. In these special cases site occupancy factors are fractions determined by the multiplicity of the symmetry operation. For example, an atom on a three-fold rotational axis would be assigned a site occupancy factor of 1/3.

†The vector $\mathbf{x}_c$ corresponds to $\mathbf{r}$ in the appendix, not to be confused with the residual vector here.

variables, and is indicated by the symbol, ∇ (the gradient of $f(\mathbf{x}_c)$ is $\nabla f(\mathbf{x}_c)$). The Lagrangian is a function combined with the constraining equations, each multiplied by an undetermined multiplier, λ':

$$L(\mathbf{x}_c, \overrightarrow{\lambda}) = f(\mathbf{x}_c) + \lambda_1'(z_1 - q_1(\mathbf{x}_c)) + \lambda_2'(z_2 - q_2(\mathbf{x}_c)) + \ldots. \tag{8.83}$$

The criterion that $f(\mathbf{x}_c)$ is minimized, subject to the limitations on the variables imposed by the constraining equations, is that $\nabla L(\mathbf{x}_c, \lambda) = 0$ (we will assume unit weights for simplicity). In our example, the constraints are all linear equations, and are represented conveniently with a matrix equation:

$$\begin{bmatrix} 1 & 0 & 0 & 0 & 0 & 0 & 0 \\ 0 & 1 & 0 & 0 & 0 & 1 & 0 \end{bmatrix} \begin{bmatrix} x_1 \\ x_2 \\ x_3 \\ x_4 \\ x_5 \\ x_6 \\ x_7 \end{bmatrix} = \begin{bmatrix} 0.5 \\ 2.5 \end{bmatrix}; \tag{8.84}$$

$$\mathbf{Q}\mathbf{x}_c = \mathbf{z}. \tag{8.85}$$

For convenience we set $\lambda = \lambda'/2$. $f(\mathbf{x}_c) = \mathbf{r}^T\mathbf{r} = (\mathbf{s}_o - \mathbf{A}\mathbf{x}_c)^T(\mathbf{s}_o - \mathbf{A}\mathbf{x}_c)$, and the Lagrangian is

$$\begin{aligned} L(\mathbf{x}_c, \overrightarrow{\lambda}) &= \mathbf{r}^T\mathbf{r} + \lambda_1'(0.5 - q_1(\mathbf{x}_c)) + \lambda_2'(2.5 - q_2(\mathbf{x}_c)) \\ &= \mathbf{r}^T\mathbf{r} + 2\overrightarrow{\lambda}(\mathbf{z} - \mathbf{Q}\mathbf{x}_c), \end{aligned} \tag{8.86}$$

where $\overrightarrow{\lambda}$ is the column vector of undetermined multipliers, $[\lambda_1\ \lambda_2]$. The values of $\mathbf{x}_c$ that minimize the Lagrangian are those for which

$$\nabla(\mathbf{r}^T\mathbf{r}) + 2\overrightarrow{\lambda}\nabla(\mathbf{z} - \mathbf{Q}\mathbf{x}_c) = 0. \tag{8.87}$$

The first gradient is the vector with components

$$\frac{\partial(\mathbf{r}^T\mathbf{r})}{x_i} = \mathbf{r}^T\frac{\partial\mathbf{r}}{x_i} + \mathbf{r}\frac{\partial\mathbf{r}^T}{x_i} = 2\mathbf{r}^T\frac{\partial\mathbf{r}}{x_i}. \tag{8.88}$$

Thus,

$$\begin{aligned} \nabla(\mathbf{r}^T\mathbf{r}) &= 2\mathbf{r}^T\nabla\mathbf{r} = 2(\mathbf{s}_o - \mathbf{A}\mathbf{x}_c)^T\nabla(\mathbf{s}_o - \mathbf{A}\mathbf{x}_c) \\ &= -2(\mathbf{s}_o - \mathbf{A}\mathbf{x}_c)^T\mathbf{A}\nabla\mathbf{x}_c \end{aligned} \tag{8.89}$$

since $\nabla\mathbf{s}_o = 0$ (consisting of the derivatives of a constant vector). The second gradient is

$$2\overrightarrow{\lambda}\nabla(\mathbf{z} - \mathbf{Q}\mathbf{x}_c) = 2\overrightarrow{\lambda}\nabla\mathbf{z} - 2\overrightarrow{\lambda}\mathbf{Q}\nabla\mathbf{x}_c = -2\overrightarrow{\lambda}\mathbf{Q}\nabla\mathbf{x}_c. \tag{8.90}$$

Substituting Eqns. 8.89 and 8.90 into Eqn. 8.87 gives

$$\begin{aligned} (\mathbf{s}_o - \mathbf{A}\mathbf{x}_c)^T\mathbf{A}\nabla\mathbf{x}_c + \overrightarrow{\lambda}\mathbf{Q}\nabla\mathbf{x}_c &= 0 \\ \mathbf{s}_o^T\mathbf{A} - \mathbf{x}_c^T\mathbf{A}^T\mathbf{A} + \overrightarrow{\lambda}\mathbf{Q} &= 0 \\ \overrightarrow{\lambda}\mathbf{Q} &= \mathbf{x}_c^T\mathbf{A}^T\mathbf{A} - \mathbf{s}_o^T\mathbf{A}. \end{aligned} \tag{8.91}$$

The $\mathbf{A}^T\mathbf{A}$ matrix is the normal equation coefficient matrix, $\mathbf{C}$, a symmetric matrix. From Eqn. 8.5, the *unconstrained* least-squares solution is

$$\begin{aligned}
\mathbf{x}_u &= (\mathbf{A}^T\mathbf{A})^{-1}\mathbf{A}^T\mathbf{s}_o = \mathbf{C}^{-1}\mathbf{A}^T\mathbf{s}_o \\
\mathbf{x}_u^T &= \mathbf{s}_o^T\mathbf{A}(\mathbf{C}^{-1})^T = \mathbf{s}_o^T\mathbf{A}\mathbf{C}^{-1} \\
\mathbf{s}_o^T\mathbf{A} &= \mathbf{x}_u^T\mathbf{C}.
\end{aligned} \tag{8.92}$$

Thus,

$$\vec{\lambda}\mathbf{Q} = \mathbf{x}_c^T\mathbf{C} - \mathbf{x}_u^T\mathbf{C} = (\mathbf{x}_c^T - \mathbf{x}_u^T)\mathbf{C}. \tag{8.93}$$

Post-multiplying both sides by $\mathbf{C}^{-1}\mathbf{Q}^T$,

$$\vec{\lambda}\mathbf{Q}\mathbf{C}^{-1}\mathbf{Q}^T = (\mathbf{x}_c^T - \mathbf{x}_u^T)\mathbf{C}\mathbf{C}^{-1}\mathbf{Q}^T = \mathbf{x}_c^T\mathbf{Q}^T - \mathbf{x}_u^T\mathbf{Q}^T, \tag{8.94}$$

and inserting the constraint relationship (Eqn. 8.85), $\mathbf{x}_c^T\mathbf{Q}^T = \mathbf{z}^T$,

$$\vec{\lambda}\mathbf{Q}\mathbf{C}^{-1}\mathbf{Q}^T = \mathbf{z}^T - \mathbf{x}_u^T\mathbf{Q}^T = (\mathbf{z} - \mathbf{Q}\mathbf{x}_u)^T, \tag{8.95}$$

and

$$\vec{\lambda} = (\mathbf{z} - \mathbf{Q}\mathbf{x}_u)^T(\mathbf{Q}\mathbf{C}^{-1}\mathbf{Q}^T)^{-1}. \tag{8.96}$$

This produces the Lagrange multiplier vector, which provides the constrained least-squares solution when substituted into Eqn. 8.93:

$$\begin{aligned}
(\mathbf{z} - \mathbf{Q}\mathbf{x}_u)^T(\mathbf{Q}\mathbf{C}^{-1}\mathbf{Q}^T)^{-1}\mathbf{Q} &= (\mathbf{x}_c^T - \mathbf{x}_u^T)\mathbf{C} \\
\mathbf{x}_c^T - \mathbf{x}_u^T &= (\mathbf{z} - \mathbf{Q}\mathbf{x}_u)^T(\mathbf{Q}\mathbf{C}^{-1}\mathbf{Q}^T)^{-1}\mathbf{Q}\mathbf{C}^{-1} \\
\mathbf{x}_c &= \mathbf{x}_u + (\mathbf{C}^{-1})^T\mathbf{Q}^T((\mathbf{Q}\mathbf{C}^{-1}\mathbf{Q}^T)^{-1})^T(\mathbf{z} - \mathbf{Q}\mathbf{x}_u) \\
\mathbf{x}_c &= \mathbf{x}_u + \mathbf{C}^{-1}\mathbf{Q}^T(\mathbf{Q}\mathbf{C}^{-1}\mathbf{Q})^{-1}(\mathbf{z} - \mathbf{Q}\mathbf{x}_u).
\end{aligned} \tag{8.97}$$

The matrix expression on the right then becomes a "correction vector", providing the values necessary to force the variables to satisfy the constraining equations. For example, in our specific case, if $x_{1,u} = 0.52$, then the expression on the right will create a vector with its ith component equal to -0.02, so that $x_{1,c} = 0.52 - .02 = 0.5$.

Note that although this derivation began with a specific example, the proof is general — and correct for any number of variables and constraints. Statistically, the addition of a constraint is effectively the same as the addition of a parameter, and the variance of the parameters for the constrained least-squares minimization determined by adding these "parameters" to Eqn. 8.51 is

$$\sigma_r^2 = c_{rr}^i\left(\frac{1}{m-(n+k)}\sum_{i=1}^{m} w_i, r_i^2\right), \tag{8.98}$$

where m is the number of observations, n is the number of parameters, and k is the number of constraints. The variance generally increases since the constrained variables are no longer free to take on values that would minimize the residuals to the extent that they would with an unconstrained fit.

8.1.4 Restrained Least Squares

The use of undetermined multipliers restricts the "variable space" to those values that rigorously satisfy the constraining equations. In structural refinement, there are instances in which it is desirable to relax these conditions, so that the variables are allowed to take on values that can deviate from the constraints, but are still restrained so that they remain in the neighborhood of the limits imposed by the equations.

Returning to the spectroscopy example, suppose that we knew that there was some experimental error in the measuring of the volumes of compound 1 and compound 6, so that the $k = 1, 2$ constraining equations become $q_k(\mathbf{x}) = z_k - \delta_k$, where δ_k is the difference between the value of the constraining equation from the actual concentrations and, z_k, the value of equation if the measurement was not in error. This creates a "constraint residual", $\delta_k = z_k - q_k(\mathbf{x})$. As an alternative to applying rigid constraints with undetermined multipliers, we might consider incorporating these new "residuals" directly into the least squares process, generating concentrations that tend to minimize these residuals and keep the concentrations *in the vicinity* of the values necessary to rigorously satisfy the constraining equations. Thus the constraints are "softened" somewhat, and we refer to them as *restraints*.

The new weighted residual vector becomes

$$\mathbf{r}_r = \begin{bmatrix} w'_1 r_1 \\ w'_2 r_2 \\ \vdots \\ w'_m r_m \\ w'^c_1 \delta_1 \\ w'^c_2 \delta_2 \\ \vdots \\ w'^c_o \delta_o \end{bmatrix} \tag{8.99}$$

for m data points and o restraints, and the least-squares solution minimizes

$$\mathbf{r}_r^T \mathbf{r}_r = w_1 r_1^2 + w_2 r_2^2 + \ldots + w_m r_m^2 + w_1^c \delta_1^2 + w_2^c \delta_2^2 + \ldots + w_o^c \delta_o^2. \tag{8.100}$$

Thus

$$\begin{aligned}
\frac{\partial(\mathbf{r}_r^T \mathbf{r}_r)}{x_i} &= -2\mathbf{a}_i^T \mathbf{W}\mathbf{r} + \sum_{k=1}^{o} w_k^c \frac{\partial \delta_k^2}{\partial x_i} \\
&= -2\mathbf{a}_i^T \mathbf{W}\mathbf{r} - 2\sum_{k=1}^{o} w_k^c (z_k - q(\mathbf{x}_r)_k) \frac{\partial q(\mathbf{x}_r)_k}{\partial x_i} \\
&= -2\mathbf{a}_i^T \mathbf{W}(\mathbf{s}_o - \mathbf{A}\mathbf{x}_r) - 2\sum_{k=1}^{o} w_k^c (z_k - q(\mathbf{x}_r)_k) \frac{\partial q(\mathbf{x}_r)_k}{\partial x_i} \\
&= \mathbf{a}_i^T \mathbf{W}(\mathbf{s}_o - \mathbf{A}\mathbf{x}_r) + \sum_{k=1}^{o} w_k^c (z_k - q(\mathbf{x}_r)_k) \frac{\partial q(\mathbf{x}_r)_k}{\partial x_i} = 0.
\end{aligned} \tag{8.101}$$

The "restraint" expression on the right depends on the nature of the constraining equations. If it evaluates to $p_i(\mathbf{x}_r)$ for the ith parameter, then $\mathbf{p}(\mathbf{x}_r) = [p_1(\mathbf{x}_r)\ p_2(\mathbf{x}_r)\ \cdots\ p_n(\mathbf{x}_r)]$, and

$$\mathbf{A}^T\mathbf{W}(\mathbf{s}_o - \mathbf{A}\mathbf{x}_r) + \mathbf{p}(\mathbf{x}_r) = \mathbf{0} \tag{8.102}$$

$$\mathbf{A}^T\mathbf{A}\mathbf{x}_r = \mathbf{A}^T\mathbf{W}\mathbf{s}_o + \mathbf{p}(\mathbf{x}_r)$$

$$\mathbf{x}_r = (\mathbf{A}^T\mathbf{A})^{-1}\mathbf{A}^T\mathbf{W}\mathbf{s}_o + (\mathbf{A}^T\mathbf{A})^{-1}\mathbf{p}(\mathbf{x}_r). \tag{8.103}$$

Written in this form, it is clear that a general solution for $\mathbf{x}_r$ does not exist, since it occurs on both sides of the equation. However, if we assume that the initial values for the parameters are reasonably close to their optimal values, Eqn. 8.103 can be solved iteratively. Approximate values for the "restraint vector," $\mathbf{p}(\mathbf{x}_r)$, are computed from the original parameters, and a new parameter vector, $\mathbf{x}_r$, is calculated. These new parameters then create an improved restraint vector, and so forth until the parameters no longer change.

The degree to which a specific term in the least-squares refinement affects the values of the parameters depends on its weighting factor. The weights of the experimental residuals are assigned as functions of their variances, and it is reasonable to weight the restraining residuals in the same manner, by selecting reasonable values for their variances. Thus we might let $w_k^c = K/\sigma_k^2$, where σ_k is an estimate of the standard deviation that we expect for the restraint condition (the root mean square value of δ_k), and K is a constant that puts the observational residuals and the restraint residuals on the same scale. Note that if the standard deviation approaches zero, the weight approaches infinity. *Thus very small values of the selected standard deviation will tend to give the restraining equation sufficient weight in the solution to have it serve effectively as a constraint.* On the other hand, increasing values of the standard deviation allows the restraint to have more and more latitude, until, in the limit the restraint equals zero – returning the parameters to their unrestrained values.

8.2 Non-linear Least Squares: Structure Refinement

The experimental observations in the X-ray diffraction experiment are the relative intensities, I_{rel}, multiplied by a scale factor (see Sec. 5.2) in order to place the intensities calculated from a model, I_c, and the observed intensities, I_o, on the same scale: $I_o = KI_{rel}$. The scale factor, K, is only a rough estimate, and needs to be updated in order to provide the closest fit between I_o and I_c. The scale factor can be modified by a multiplier, K', such that $K'KI_{rel} = K'I_o \simeq I_c$, where K' is adjusted to provide the best fit between the calculated and pre-scaled observed intensities. In the least squares process, experimental observations are compared to those predicted by the model, and the experimental observations clearly should remain constant (in practice, if the experimental intensities are allowed to vary along with the calculated intensities the least-squares method will optimize by minimizing the observed intensities). It follows that the scale factor, K', should be varied as a parameter in the model: $I_o \simeq (1/K')I_c = p_1 I_c$. The least-squares residual for the observed and calculated intensity for a given reflection is therefore $r_{\mathbf{h}} = (I_{o,\mathbf{h}} - p_1 I_{c,\mathbf{h}})$, resulting in the weighted residual vector, $\mathbf{r} = [w'_{\mathbf{h}_1} r_{\mathbf{h}_1}\ w'_{\mathbf{h}_2} r_{\mathbf{h}_2}\ \ldots\ w'_{\mathbf{h}_m} r_{\mathbf{h}_m}]$, and the optimal least-squares fit of the calculated intensities to the observed intensities is obtained by minimizing*

$$\begin{aligned} R(\mathbf{p}) &= \mathbf{r}^T\mathbf{r} = \sum_{\mathbf{h}} w_{\mathbf{h}}(I_{o,\mathbf{h}} - p_1\, I_{c,\mathbf{h}}(p_2, p_3, \ldots, p_n))^2 \\ &= \sum_{\mathbf{h}} w_{\mathbf{h}}(I_{o,\mathbf{h}} - I_{c,\mathbf{h}}(\mathbf{p}))^2, \end{aligned} \tag{8.105}$$

where $\mathbf{p}$ is the vector of parameters, — the scale factor, p_1, the atom positions, $\mathbf{r}_j$, the displacement parameters, T_j, and *the site occupancy factors*, s_j.

The site occupancy factors allow for the partial occupancy of an atomic site. This can occur, for example, if a molecule in the crystal has access to two or more locations in which it is able to position a specific atom — often with one location more likely to be occupied than the others. The result is known as displacement disorder, and the average structure effectively contains only a "partial atom" at each location. This is reflected in the intensities as scattering from an atom in a given site with a fractional scattering power equal to the fraction of the sites occupied by the atom. The scattering of an atom is predicted by its scattering factor, and it follows that an atom (atom j) that occupies a fraction, s_j, of the locations x_j, y_j, z_j will behave as if it is an atom located at x_j, y_j, z_j with scattering factor $s_j\, f_{j,\theta d}$. In the absence of disorder, the site occupancy factors are ordinarily equal to unity. However, if the structure is disordered, the occupancy factors are often unknown, and must be determined as variable parameters in the least-squares fit.

The intensity calculated from the parameters $\mathbf{p}$ for reflection $\mathbf{h}$ is

$$I_{c,\mathbf{h}} = F^2_{c,\mathbf{h}} = \left(\sum_j s_j\, f_{j,\theta d} e^{-2\pi\, i(\mathbf{h}\cdot\mathbf{r}_j)} e^{-T_j}\right)^* \left(\sum_j s_j\, f_{j,\theta d} e^{-2\pi\, i(\mathbf{h}\cdot\mathbf{r}_j)} e^{-T_j}\right). \tag{8.106}$$

The minimization of Eqn. 8.105 is more difficult than the minimization of Eqn. 8.7 for the linear least-squares problem, since the calculated intensities from Eqn. 8.106 are *not* linear combinations of the atom positions and thermal parameters (cf., Eqn. 8.2). Fortunately, the initial parameters from the structural solution are reasonably close to the optimized parameters, and the function can be expanded as a Taylor series evaluated at the initial values. Referring to Appendix I, the local

*It must be noted here that for many years refinements were traditionally based on the magnitude of the structure factors rather than the intensities, by minimizing

$$R'(\mathbf{p}) = \sum_{\mathbf{h}} w'_{\mathbf{h}}(F_{o,\mathbf{h}} - \sqrt{p_1} F_{c,\mathbf{h}})^2, \tag{8.104}$$

which should provide the same refined parameters. However, it is experimentally possible to obtain negative intensities — in cases where the background exceeds the measurement. Since it is impossible to determine the magnitudes of negative intensities (the square roots of negative numbers) these intensities must be set to zero when refinement is based on F rather than I. In addition to negative intensities, in the early days of crystal structure analysis, Fourier series calculations were very demanding exercises, and it was traditional to use only those reflections with intensities greater than some multiple (e.g. 2 or 3) of their standard deviations — setting the "unobserved" intensities to zero. However, there is no experimental or statistical justification for setting these observations to a fixed number (zero), and doing so introduces a systematic bias into the data.[170] We will therefore consider only so-called F^2 refinement, incorporating all of the observed data, noting only that the treatment here is readily applied to refinement based on F.

behavior of a function at specific point can be approximated as an infinite series of the function and its derivatives evaluated at that point. Thus if $\mathbf{p}_o$ is the initial parameter vector, then, in the vicinity of the n initial parameters,

$$\begin{aligned} I_{c,\mathbf{h}}(\mathbf{p}) \simeq\; & I_{c,\mathbf{h}}(\mathbf{p})_{\mathbf{p}_o} + \frac{1}{1!}\sum_{i=1}^{n}\left(\frac{\partial I_{c,\mathbf{h}}(\mathbf{p})}{\partial p_i}\right)_{\mathbf{p}_o}(p_i - p_{i,o}) \\ & + \frac{1}{2!}\sum_{i=1}^{n}\left(\frac{\partial^2 I_{c,\mathbf{h}}(\mathbf{p})}{\partial p_i^2}\right)_{\mathbf{p}_o}(p_i - p_{i,o})^2 + \cdots . \end{aligned} \tag{8.107}$$

The $\mathbf{p}_o$ subscript indicates that the expression is a number – generated by evaluating the expression with the initial values. If the variables of interest are all reasonably close to these initial values, then the function can be approximated as the first two terms in the Taylor series — the sum of value of the function evaluated with the original parameters, $p_{o,i}$, and the first derivatives of the function evaluated with the original parameters, each multiplied by $\Delta p_i = p_i - p_{i,o}$:

$$I_{c,\mathbf{h}}(\mathbf{p}) \simeq I_{c,\mathbf{h}}(\mathbf{p})_{\mathbf{p}_o} + \sum_{i=1}^{n}\left(\frac{\partial I_{c,\mathbf{h}}(\mathbf{p})}{\partial p_i}\right)_{\mathbf{p}_o}\Delta p_i . \tag{8.108}$$

This is an equation that approximates the calculated intensities as linear equations with respect to the parameters to be refined. Setting $\Delta I_{\mathbf{h}} = I_{o,\mathbf{h}} - (I_{c,\mathbf{h}}(\mathbf{p}))_{\mathbf{p}_o}$ (a number with a zero derivative), Eqn. 8.105 becomes

$$R(\mathbf{p}) = \sum_{\mathbf{h}} w_{\mathbf{h}}\left(\Delta I_{\mathbf{h}} - \sum_{i=1}^{n}\left(\frac{\partial(I_{c,\mathbf{h}}(\mathbf{p}))}{\partial p_i}\right)_{\mathbf{p}_o}\Delta p_i\right)^2 , \tag{8.109}$$

the expression that we seek to minimize with respect to the parameters by differentiation with respect to each parameter and setting each $\partial R(\mathbf{p})/\partial p_j$ to zero. The derivative of the expression inside the parentheses before it is squared is

$$-\sum_{i=1}^{n}\left(\frac{\partial(I_{c,\mathbf{h}}(\mathbf{p}))}{\partial p_i}\right)_{\mathbf{p}_o}\left(\frac{\partial(p_i - p_{i,o})}{\partial p_j}\right) = -\left(\frac{\partial(I_{c,\mathbf{h}}(\mathbf{p}))}{\partial p_j}\right)_{\mathbf{p}_o} , \tag{8.110}$$

since the only non-zero term in the sum is for $j = i$, in which case $\partial\Delta p_j/\partial p_j = 1$. This results in n simultaneous equations ($j = 1, 2, \ldots, n$):

$$\frac{\partial R(\mathbf{p})}{\partial p_j} = -2\sum_{\mathbf{h}} w_{\mathbf{h}}\left(\Delta I_{\mathbf{h}} - \sum_{i=1}^{n}\left(\frac{\partial(I_{c,\mathbf{h}}(\mathbf{p}))}{\partial p_i}\right)_{\mathbf{p}_o}\Delta p_i\right)\left(\frac{\partial(I_{c,\mathbf{h}}(\mathbf{p}))}{\partial p_j}\right)_{\mathbf{p}_o} = 0.$$

Rearrangement results in the normal equations,

$$\begin{aligned} & \Delta p_1 \sum_{\mathbf{h}} w_{\mathbf{h}}\left(\frac{\partial(I_{c,\mathbf{h}}(\mathbf{p}))}{\partial p_1}\frac{\partial(I_{c,\mathbf{h}}(\mathbf{p}))}{\partial p_j}\right)_{\mathbf{p}_o} \\ & + \Delta p_2 \sum_{\mathbf{h}} w_{\mathbf{h}}\left(\frac{\partial(I_{c,\mathbf{h}}(\mathbf{p}))}{\partial p_2}\frac{\partial(I_{c,\mathbf{h}}(\mathbf{p}))}{\partial p_j}\right)_{\mathbf{p}_o} + \cdots \\ & + \Delta p_n \sum_{\mathbf{h}} w_{\mathbf{h}}\left(\frac{\partial(I_{c,\mathbf{h}}(\mathbf{p}))}{\partial p_n}\frac{\partial(I_{c,\mathbf{h}}(\mathbf{p}))}{\partial p_j}\right)_{\mathbf{p}_o} \\ & \qquad = \sum_{\mathbf{h}} w_{\mathbf{h}}\Delta I_{\mathbf{h}}\left(\frac{\partial(I_{c,\mathbf{h}}(\mathbf{p}))}{\partial p_j}\right)_{\mathbf{p}_o} . \end{aligned} \tag{8.111}$$

Setting

$$t_{ji} = \sum_{\mathbf{h}} w_{\mathbf{h}} \left(\frac{\partial(I_{c,\mathbf{h}}(\mathbf{p}))}{\partial p_i} \frac{\partial(I_{c,\mathbf{h}}(\mathbf{p}))}{\partial p_j} \right)_{\mathbf{p}_o}, \tag{8.112}$$

and

$$d_j = \sum_{\mathbf{h}} w_{\mathbf{h}} \Delta I_{\mathbf{h}} \left(\frac{\partial(p_1 I_{c,\mathbf{h}}(\mathbf{p}))}{\partial p_j} \right)_{\mathbf{p}_o}, \tag{8.113}$$

gives n simultaneous linear equations in n unknowns – *the differences between the optimized parameters and their initial values*:

$$\begin{aligned}
t_{11}\Delta p_1 + t_{12}\Delta p_2 + t_{13}\Delta p_3 + \cdots + t_{1n}\Delta p_n &= d_1 \\
t_{21}\Delta p_1 + t_{22}\Delta p_2 + t_{23}\Delta p_3 + \cdots + t_{2n}\Delta p_n &= d_2 \\
\vdots \qquad\quad \vdots \qquad\quad \vdots \qquad\qquad\quad \vdots \quad &\quad \vdots \\
t_{n1}\Delta p_1 + t_{n2}\Delta p_2 + t_{n3}\Delta p_3 + \cdots + t_{nn}\Delta p_n &= d_n.
\end{aligned} \tag{8.114}$$

The matrix equation for the solution is

$$\mathbf{T}\boldsymbol{\Delta}\mathbf{p} = \mathbf{d} \quad \text{and} \quad \boldsymbol{\Delta}\mathbf{p} = \mathbf{T}^{-1}\mathbf{d}. \tag{8.115}$$

The variance-covariance matrix is $\mathbf{T}^{-1}$ (cf., Eqns. 8.26–8.29); the least-squares parameter values are determined from $\mathbf{p} = \mathbf{p}_o + \boldsymbol{\Delta}\mathbf{p}$. The matrix elements of $\mathbf{T}$ and the components of $\mathbf{d}$ are determined from the partial derivatives of the calculated intensities evaluated with the initial parameters. The scale factor, p_1, is a multiplicative factor, and the derivative is straightforward:

$$\left(\frac{\partial(p_1 I_{c,\mathbf{h}}(p2, p3, \ldots, pn))}{\partial p_1} \right)_{\mathbf{p}_o} = I_{c,\mathbf{h}}(p_2, p_3, \ldots, p_n)_{\mathbf{p}_o}. \tag{8.116}$$

Although this might not appear intuitive, note that if p_1 is the *only* parameter that varies, then Eqn. 8.108 becomes

$$\begin{aligned}
I_{c,\mathbf{h}}(\mathbf{p}) &\simeq p_{1,o} I_{c,\mathbf{h}}(p_2, p_3, \ldots, p_n)_{\mathbf{p}_o} + I_{c,\mathbf{h}}(p_2, p_3, \ldots, p_n)_{\mathbf{p}_o}(p_1 - p_{1,o}) \\
&= p_1 I_{c,\mathbf{h}}(p_2, p_3, \ldots, p_n)_{\mathbf{p}_o},
\end{aligned} \tag{8.117}$$

the original calculated intensity multiplied by a new least-squares scale factor. To obtain the remaining derivatives we must consider the explicit nature of the calculated intensity. From Eqn. 8.1,

$$\begin{aligned}
F_{c,\mathbf{h}} e^{i\varphi_{\mathbf{h}}} &= F_{c,\mathbf{h}} \cos\varphi_{\mathbf{h}} + i\, F_{c,\mathbf{h}} \sin\varphi_{\mathbf{h}} \\
&= \sum_{j=1}^{m} s_j\, f_{j,\theta d} \cos(-2\pi\, i(hx_j + ky_j + lz_j))\; e^{-T_{j,\mathbf{h}}} \\
&\qquad + i \sum_{j=1}^{m} s_j\, f_{j,\theta d} \sin(-2\pi\, i(hx_j + ky_j + lz_j))\; e^{-T_{j,\mathbf{h}}} \\
&= A_{\mathbf{h}} + i\, B_{\mathbf{h}},
\end{aligned} \tag{8.118}$$

and

$$I_{c,\mathbf{h}} = A_{\mathbf{h}}^2 + B_{\mathbf{h}}^2. \tag{8.119}$$

$T_{j,\mathbf{h}}$ from Eqn. 5.396 is the anisotropic displacement factor for atom j:

$$T_{j,\mathbf{h}} = 2\pi^2(h^2a^{*2}U_{11}^j + k^2b^{*2}U_{22}^j + l^2c^{*2}U_{33}^j + 2hka^*b^*U_{12}^j + 2hla^*c^*U_{13}^j + 2klb^*c^*U_{23}^j). \quad (8.120)$$

The derivative with respect to a general parameter, p_i, is

$$\left(\frac{\partial(I_{c,\mathbf{h}}(\mathbf{p}))}{\partial p_i}\right)_{\mathbf{p}_o} = \left(2A_{\mathbf{h}}(\mathbf{p})\frac{\partial A_{\mathbf{h}}(\mathbf{p})}{\partial p_i} + 2B_{\mathbf{h}}(\mathbf{p})\frac{\partial B_{\mathbf{h}}(\mathbf{p})}{\partial p_i}\right)_{\mathbf{p}_o} \quad (8.121)$$

$$= \left(2\cos\varphi_{\mathbf{h}}\frac{\partial A_{\mathbf{h}}(\mathbf{p})}{\partial p_i} + 2\sin\varphi_{\mathbf{h}}\frac{\partial B_{\mathbf{h}}(\mathbf{p})}{\partial p_i}\right)_{\mathbf{p}_o}. \quad (8.122)$$

The derivatives will vary in form, depending on whether the parameter is a fractional coordinate or an element of the displacement matrix. Suppose, for example, that $p_j \equiv y_q$ is the y coordinate for the qth atom. Then

$$\left(\frac{\partial A_{\mathbf{h}}(\mathbf{p})}{\partial p_i}\right)_{\mathbf{p}_o} \equiv \left(\frac{\partial A_{\mathbf{h}}(\mathbf{p})}{\partial y_q}\right)_{\mathbf{p}_o}$$

$$= p_{1,o}\sum_{j=1}^{m} f_{j,\theta d}\left(s_j\, e^{-T_{j,\mathbf{h}}}\frac{\partial\left(\cos(-2\pi\, i(hx_j + ky_j + lz_j))\right)}{\partial y_q}\right)_{\mathbf{p}_o}.$$

The only non-zero term in the sum has $y_j = y_q$, and

$$\left(\frac{\partial A_{\mathbf{h}}(\mathbf{p})}{\partial y_q}\right)_{\mathbf{p}_o} = 2\pi\, k\, p_{1,o}\, f_{q,\theta d}\left(s_q\, e^{-T_{q,\mathbf{h}}}\sin(-2\pi\, i(hx_q + ky_q + lz_q))\right)_{\mathbf{p}_o}. \quad (8.123)$$

Similarly,

$$\left(\frac{\partial B_{\mathbf{h}}(\mathbf{p})}{\partial y_q}\right)_{\mathbf{p}_o} = -2\pi\, k\, p_{1,o}\, f_{q,\theta d}\left(s_q\, e^{-T_{q,\mathbf{h}}}\cos(-2\pi\, i(hx_q + ky_q + lz_q))\right)_{\mathbf{p}_o}. \quad (8.124)$$

If, on the other hand, $p_i \equiv U_{12}^q$, one of the displacement parameters for the qth atom, then

$$\left(\frac{\partial A_{\mathbf{h}}(\mathbf{p})}{\partial U_{12}^q}\right)_{\mathbf{p}_o} = p_{1,o}\sum_{j=1}^{m} f_{j,\theta d}\left(s_j\,\cos(-2\pi\, i(\mathbf{h}\cdot\mathbf{r}_j))\,\frac{\partial(e^{-T_{j,\mathbf{h}}})}{\partial U_{12}^q}\right)_{\mathbf{p}_o}.$$

Again the only non-zero term in the sum is for $U_{12}^j = U_{12}^q$, and

$$\left(\frac{\partial A_{\mathbf{h}}(\mathbf{p})}{\partial U_{12}^q}\right)_{\mathbf{p}_o} = -2\pi^2\, h\, k\, a^*\, b^*\, p_{1,o}\, f_{q,\theta d}\left(s_q\, e^{-T_{q,\mathbf{h}}}\cos(-2\pi\, i(\mathbf{h}\cdot\mathbf{r}_q))\right)_{\mathbf{p}_o}. \quad (8.125)$$

Similarly,

$$\left(\frac{\partial B_{\mathbf{h}}(\mathbf{p})}{\partial U_{12}^q}\right)_{\mathbf{p}_o} = -2\pi^2\, h\, k\, a^*\, b^*\, p_{1,o}\, f_{q,\theta d}\left(s_q\, e^{-T_{q,\mathbf{h}}}\sin(-2\pi\, i(\mathbf{h}\cdot\mathbf{r}_q))\right)_{\mathbf{p}_o}. \quad (8.126)$$

If the temperature factor for atom q is isotropic, then, from Eqn. 5.399,

$$T_{q,\,\mathrm{iso}} = 8\pi^2\left\langle u_q^2\right\rangle\frac{\sin^2\theta_{\mathbf{h}}}{\lambda^2} = 8\pi^2\, U_{iso}^q\frac{\sin^2\theta_{\mathbf{h}}}{\lambda^2}, \quad (8.127)$$

resulting in

$$\left(\frac{\partial A_{\mathbf{h}}(\mathbf{p})}{\partial U^q_{iso}}\right)_{\mathbf{p}_o} = -8\pi^2 \frac{\sin^2\theta_{\mathbf{h}}}{\lambda^2} p_{1,o} f_{q,\theta d} \left(s_q e^{-T_{q,\,iso}} \cos(-2\pi\, i(\mathbf{h}\cdot\mathbf{r}_q))\right)_{\mathbf{p}_o}, \quad (8.128)$$

and

$$\left(\frac{\partial B_{\mathbf{h}}(\mathbf{p})}{\partial U^q_{iso}}\right)_{\mathbf{p}_o} = -8\pi^2 \frac{\sin^2\theta_{\mathbf{h}}}{\lambda^2} p_{1,o} f_{q,\theta d} \left(s_q e^{-T_{q,\,iso}} \sin(-2\pi\, i(\mathbf{h}\cdot\mathbf{r}_q))\right)_{\mathbf{p}_o}. \quad (8.129)$$

The site occupancy factors are multiplicative terms. For the derivative with respect to $p_i = s_q$, the only non-zero term in the sums is for $s_q = s_j$, giving

$$\left(\frac{\partial A_{\mathbf{h}}(\mathbf{p})}{\partial s_q}\right)_{\mathbf{p}_o} = p_{1,o} f_{q,\theta d} \left(e^{-T_{q,\,\mathbf{h}}} \cos(-2\pi\, i(hx_q + ky_q + lz_q))\right)_{\mathbf{p}_o}, \quad (8.130)$$

and

$$\left(\frac{\partial B_{\mathbf{h}}(\mathbf{p})}{\partial s_q}\right)_{\mathbf{p}_o} = p_{1,o} f_{q,\theta d} \left(e^{-T_{q,\,\mathbf{h}}} \sin(-2\pi\, i(hx_q + ky_q + lz_q))\right)_{\mathbf{p}_o}. \quad (8.131)$$

These derivatives provide specific formulae for the matrix elements for **T** and the vector components for **d**, allowing for a least squares solution for the parameters, based upon their initial values.

It is important to note here that the entire treatment above was based on the assumption that the initial parameter values were sufficiently close to the optimized parameter values that calculated intensities could be modeled as linear equations in the region of the initial values of the parameters. If the initial structure solution is largely incorrect, then refinement will almost certainly fail to provide a corrected structure. Because the solution is based on a linear approximation to the calculated intensity function, the least squares values of the parameters should represent an improvement over the initial values, but will not ordinarily be optimal after a single solution of the system of linear equations. Thus, it is necessary to repeat the process iteratively, using the new parameters generated in each iteration as initial values for the next cycle. The refinement process is generally repeated for the number of cycles sufficient to render negligible the relative parameter shifts, measured as fractions of the parameter values (i.e., $\Delta p_i/p_i$).

8.2.1 Weights in Refinement

In the discussion of weighted refinement in linear least squares it was noted that the best "unbiased" weight for each observation was the reciprocal of the variance of the observation. The estimated standard deviation of the observed intensity, $\sigma(I_{o,\mathbf{h}})$, is determined from counting statistics (Poisson statistics – see Chapter 5), so that the weight for each reflection is given by

$$w_{\mathbf{h}} = \frac{1}{\sigma^2(I_{o,\mathbf{h}})}. \quad (8.132)$$

The estimated standard deviations based on counting statistics do not account for other random errors or systematic errors that are either unknown, or may not have

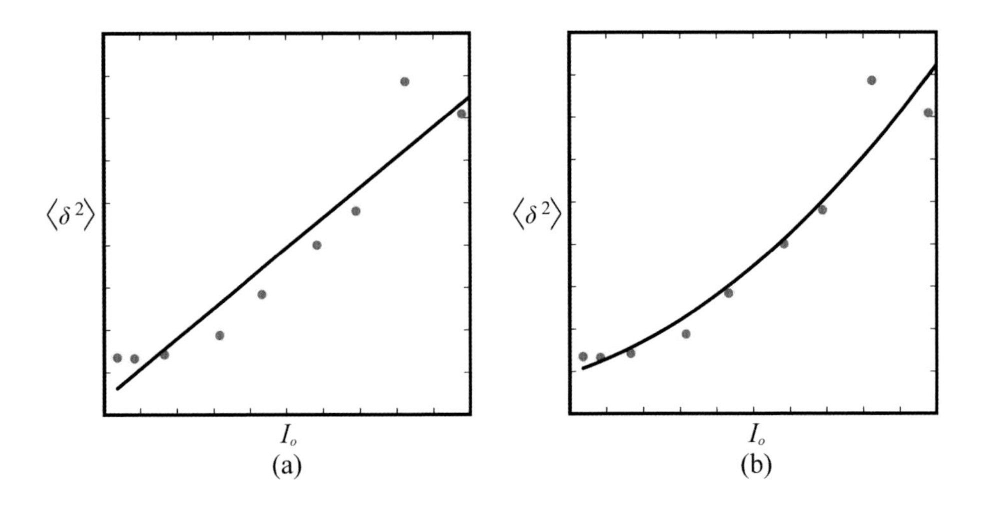

Figure 8.3 (a) Linear least-squares fit of $\langle\delta^2\rangle$ as a function of I_0. (b) Second degree polynomial least-squares fit of $\langle\delta^2\rangle$ as a function of I_0.

been modeled appropriately (e.g., absorption or extinction errors). Since the systematic errors are dependent on I_o, an alternative weighting function that includes them would incorporate the intensity:

$$w_{\mathbf{h}} = f(I_o), \tag{8.133}$$

where the function, $f(I_o)$, is evaluated by comparing the observed and calculated intensities before each cycle of refinement. If there is systematic error that depends on I_o in the data then the deviation of the observed intensities from those calculated from the model, $\delta_{\mathbf{h}} = I_{o,\mathbf{h}} - I_{c,\mathbf{h}}$, will vary with $I_{o,\mathbf{h}}$, and will reflect all of the error in the data. We therefore might consider replacing the standard deviation with $\delta_{\mathbf{h}}$,

$$w_{\mathbf{h}} = \frac{1}{\delta^2(I_{o,\mathbf{h}})}, \tag{8.134}$$

but since $w_{\mathbf{h}}\delta^2_{\mathbf{h}}$ is the function being optimized in the least-squares fit the result would be an attempt to minimize $\sum \delta^2_{\mathbf{h}}/\delta^2_{\mathbf{h}}$. A way around this is to approximate $\delta^2(I_{o,\mathbf{h}})$ with a simple function.

The approach is illustrated in Fig. 8.3 for a typical data set. The intensity data are divided into "regions" and the average value of $\langle\delta^2\rangle$ is determined and plotted versus I_o in the center of each region. In Fig. 8.3, the points are fit to a line, $\langle\delta^2\rangle = a + bI_o$, resulting in the weighting function

$$w_{\mathbf{h}} = \frac{1}{a + bI_{o,\mathbf{h}}}. \tag{8.135}$$

While such a weighting function would appear to be reasonable, Wilson[170] showed that the dependence on the observed intensity introduced a bias into the data that becomes especially important for values of the intensity close to zero. For illustrative purposes, suppose that $a = 1$, $b = 0.2$, and the actual value of I_o is zero, so that

$w_{\mathbf{h}} = 1$. If a random fluctuation renders $I_o = 2$, then $w_{\mathbf{h}} = 0.711$. If an equally probable fluctuation occurs in the negative direction, $I_o = -2$, then $w_{\mathbf{h}} = 1.666$ — equally probable fluctuations result in unequal changes in the weights. The subsequent introduction of bias into the parameter values can be seen by considering the least-squares minimization of $R(\mathbf{p})$ with respect to some parameter, p_i. To simplify the terminology in the analysis we let $p = p_i - p_{i,o} = \Delta p_i$, $R(\mathbf{p}) = R$, $I_o = I_{o,\mathbf{h}}$ and $I_c = I_{c,\mathbf{h}}(\mathbf{p})$, which includes the scale factor, p_1. Thus, we considerer minimizing

$$R = \sum_{\mathbf{h}} w_{\mathbf{h}}(I_o - I_c)^2 \tag{8.136}$$

with respect to p. R can be expanded as a Taylor series in the vicinity of $p = 0$:

$$R = R_o + R'_o p + \frac{1}{2} R''_o p^2 + \cdots, \tag{8.137}$$

where R' and R'' are the first and second derivatives of R with respect to p, and the subscript indicates that these are the values at $p = 0$. Since p is very small (close to zero) the higher order terms are negligible. Ignoring these terms, R is minimized with respect to p by setting its derivative to zero:

$$\frac{\partial R}{\partial p} = R'_o + R''_o p = 0. \tag{8.138}$$

The value of p that minimizes R is therefore

$$p = -\frac{R'_o}{R''_o}. \tag{8.139}$$

If the weighting function includes the observed intensity, then a random fluctuation in I_o, ϵ, will produce a change in $w_{\mathbf{h}}$, and subsequent changes in R'_o, R''_o, and p. If the weighting function does not introduce bias into the parameter values, then we would expect that, on the average, $\langle p \rangle$, will not depend on $\langle \epsilon^2 \rangle$ (ϵ will average to zero, but its magnitude will not). Beginning with the evaluation of R'_o and R''_o,

$$R'_o = \left(\frac{\partial S}{\partial p}\right)_o = \sum_{\mathbf{h}} \left(\frac{\partial w_{\mathbf{h}}}{\partial p}(I_o - I_c)^2 - 2w_{\mathbf{h}}(I_o - I_c)\frac{\partial I_c}{\partial p} \right)_o. \tag{8.140}$$

From the chain rule for differentiation,

$$R'_o = \left(\frac{\partial S}{\partial p}\right)_o = \sum_{\mathbf{h}} \left(\frac{\partial w_{\mathbf{h}}}{\partial I_c}\frac{\partial I_c}{\partial p}(I_o - I_c)^2 - 2w_{\mathbf{h}}(I_o - I_c)\frac{\partial I_c}{\partial p} \right)_o, \tag{8.141}$$

and

$$\begin{aligned} R''_o &= \left(\frac{\partial^2 S}{\partial p^2}\right)_o = \sum_{\mathbf{h}} \left(\frac{\partial^2 w_{\mathbf{h}}}{\partial p^2}(I_o - I_c)^2 - 4\frac{\partial w_{\mathbf{h}}}{\partial p}(I_o - I_c)\frac{\partial I_c}{\partial p} \right. \\ &\qquad \left. + 2w_{\mathbf{h}}\frac{\partial^2 I_c}{\partial p^2} - 2w_{\mathbf{h}}(I_o - I_c)\frac{\partial^2 I_c}{\partial p^2} \right)_o. \end{aligned} \tag{8.142}$$

As refinement progresses $I_o - I_c$ approaches zero, and only one term in the sum remains:

$$R_o'' = 2\sum_{\mathbf{h}} w_{\mathbf{h}} \left(\frac{\partial^2 I_c}{\partial p^2}\right)_o . \tag{8.143}$$

At the average value of I_o, $I_o - I_c = \delta$. Now, suppose that a statistical fluctuation, $-\epsilon$, occurs in the observed intensity initially at its average value, such that $I_o - I_c = \delta - \epsilon$. Then R_o' becomes

$$R_o' = \sum_{\mathbf{h}} \left(\frac{\partial w_{\mathbf{h}}}{\partial I_c}(\delta^2 - 2\delta\epsilon + \epsilon^2) - 2w_{\mathbf{h}}(\delta - \epsilon)\right)_o \left(\frac{\partial I_c}{\partial p}\right)_o . \tag{8.144}$$

We are considering changes in $w_{\mathbf{h}}$ as a function of ϵ and δ, and can therefore expand $w_{\mathbf{h}}(\delta, \epsilon)$ as a Taylor series in the vicinity of the initial values of the variables, δ and $-\epsilon$ (ignoring terms higher than second order):

$$w_{\mathbf{h}} = w_{\mathbf{h},o} - \left(\frac{\partial w_{\mathbf{h}}}{\partial \epsilon}\right)_o \epsilon + \frac{1}{2}\left(\frac{\partial^2 w_{\mathbf{h}}}{\partial \epsilon^2}\right)_o \epsilon^2 + \left(\frac{\partial w_{\mathbf{h}}}{\partial \delta}\right)_o \delta + \frac{1}{2}\left(\frac{\partial^2 w_{\mathbf{h}}}{\partial \delta^2}\right)_o \delta^2 . \tag{8.145}$$

Before substituting this expression for $w_{\mathbf{h}}$ into Eqn. 8.144, we note the following: All of the second order terms will contain third order products such as ϵ^3 or $\epsilon^2\delta$, and since ϵ and δ are small numbers, these products will be negligible. Furthermore, if we consider an average value of R_o', $\langle R_o' \rangle$, (from repeated measurements of I_o), then $\langle \epsilon \rangle$ and $\langle \epsilon\delta \rangle$ will vanish since they will have equally probable positive and negative values. Thus the first term in the expression for the expected value of R_o' is

$$\left(\frac{\partial w_{\mathbf{h}}}{\partial I_c}\langle \delta^2 \rangle + \frac{\partial w_{\mathbf{h}}}{\partial I_c}\langle \epsilon^2 \rangle - 2w_{\mathbf{h}}\langle \delta \rangle - 2\frac{\partial w_{\mathbf{h}}}{\partial \epsilon}\langle \epsilon^2 \rangle - 2\frac{\partial w_{\mathbf{h}}}{\partial \delta}\langle \delta^2 \rangle\right)_o . \tag{8.146}$$

Since δ changes when I_c changes, and ϵ changes when I_o changes, $\partial w_{\mathbf{h}}/\partial \delta = \partial w_{\mathbf{h}}/\partial I_c$, and $\partial w_{\mathbf{h}}/\partial \epsilon = \partial w_{\mathbf{h}}/\partial I_o$. The average value for the change in parameter p_i, p, is now obtained by substituting these derivatives into Eqn. 8.146 and applying Eqn. 8.139:

$$\begin{aligned} \langle p \rangle &= -\sum_{\mathbf{h}} \left(\left(-\frac{\partial w_{\mathbf{h}}}{\partial I_c}\langle \delta^2 \rangle + \frac{\partial w_{\mathbf{h}}}{\partial I_c}\langle \epsilon^2 \rangle - 2w_{\mathbf{h}}\langle \delta \rangle - 2\frac{\partial w_{\mathbf{h}}}{\partial I_o}\langle \epsilon^2 \rangle\right)\left(\frac{\partial I_c}{\partial p}\right)\right)_o \\ &\quad \times \left(2\sum_{\mathbf{h}} w_{\mathbf{h}} \left(\frac{\partial^2 I_c}{\partial p^2}\right)_o\right)^{-1} . \end{aligned} \tag{8.147}$$

Random fluctuations, ϵ, affect weighting functions involving observed intensities in a manner that introduces bias into the parameter values of magnitude

$$\left(\frac{\partial w_{\mathbf{h}}}{\partial I_c}\right)_o \langle \epsilon^2 \rangle - 2\left(\frac{\partial w_{\mathbf{h}}}{\partial I_o}\right)_o \langle \epsilon^2 \rangle = \left(\frac{\partial w_{\mathbf{h}}}{\partial I_c} - 2\frac{\partial w_{\mathbf{h}}}{\partial I_o}\right)_o \langle \epsilon^2 \rangle. \tag{8.148}$$

This bias will be eliminated if the coefficient of $\langle \epsilon^2 \rangle$ is zero, i.e., if

$$\left(\frac{\partial w_{\mathbf{h}}}{\partial I_c} = 2\frac{\partial w_{\mathbf{h}}}{\partial I_o}\right)_o . \tag{8.149}$$

It follows that the weighting function must contain contributions from both the observed and calculated intensities, such that the derivative of the weight for each satisfies Eqn. 8.149. Substituting $P_{\mathbf{h}} = (1/3)I_{o,\mathbf{h}} + (2/3)I_{c,\mathbf{h}}$ for $I_{o,\mathbf{h}}$* in Eqn. 8.135 accomplishes this:

$$w_{\mathbf{h}} = (a + bP_{\mathbf{h}})^{-1} = \left(a + b\frac{1}{3}I_{o,\mathbf{h}} + b\frac{2}{3}I_{c,\mathbf{h}}\right)_o^{-1} \tag{8.150}$$

$$\frac{\partial w_{\mathbf{h}}}{\partial I_{o,\mathbf{h}}} = -\frac{1}{3}b\left(a + b\frac{1}{3}I_{o,\mathbf{h}} + b\frac{2}{3}I_{c,\mathbf{h}}\right)_o^{-2} \tag{8.151}$$

$$\frac{\partial w_{\mathbf{h}}}{\partial I_{c,\mathbf{h}}} = -\frac{2}{3}b\left(a + b\frac{1}{3}I_{o,\mathbf{h}} + b\frac{2}{3}I_{c,\mathbf{h}}\right)_o^{-2}. \tag{8.152}$$

To employ this function a linear least-squares fit of $\langle\delta^2\rangle$ versus $P_{\mathbf{h}}$ would produce the values of a and b.

The weighting function does not have to depend on a linear equation. Cruickshank[171] suggested that a second degree polynomial would provide a more accurate representation of errors in the weighting function (Fig. 8.3(b)):

$$w_{\mathbf{h}} = \frac{1}{a + bI_{o,\mathbf{h}} + cI_{o,\mathbf{h}^2}}. \tag{8.153}$$

If $P_{\mathbf{h}}$ is substituted for $I_{o,\mathbf{h}}$ in a general polynomial in the denominator,

$$w_{\mathbf{h}} = (a + \sum_j b_j P^j)_o^{-1} \tag{8.154}$$

$$\frac{\partial w_{\mathbf{h}}}{\partial I_{o,\mathbf{h}}} = \frac{1}{3}\left(a + \sum_j b_j P^j\right)^{-2}\left(\sum_k k\, b_k P^{k-1}\right)_o \tag{8.155}$$

$$\frac{\partial w_{\mathbf{h}}}{\partial I_{c,\mathbf{h}}} = \frac{2}{3}\left(a + \sum_j b_j P^j\right)^{-2}\left(\sum_k k\, b_k P^{k-1}\right)_o, \tag{8.156}$$

satisfying Eqn. 8.149. Thus the *Cruikshank weights* can be corrected for bias as well:

$$w_{\mathbf{h}} = \left(\frac{1}{a + bP_{\mathbf{h}} + cP_{\mathbf{h}}^2}\right)_o. \tag{8.157}$$

If there is no systematic error, $\langle\delta^2\rangle$ would not vary with P, and would represent the average random error in the intensity data. In the weighting function, a would equal $\langle\delta^2\rangle$, and b and c would equal zero. If the random error is then considered to arise completely from counting statistics, then $a = \langle\sigma_{\mathbf{h}}^2\rangle$. Of course, in this case, a more reasonable weighting scheme would allow $a \equiv a_{\mathbf{h}}$ to represent the variance of each intensity by employing it as a traditional weight, $w_{\mathbf{h}} = 1/\sigma_{\mathbf{h}}^2$. In order to

*Most of the derivations in this chapter closely parallel those for which F_o and F_c are substituted for I_o and I_c. There is, however, no straightforward parallel for determining unbiased weights when refinement is based on F. This is yet another reason for preferring refinement based on $F^2 \equiv I$.

make use of the variances of each reflection *and* treat the data for systematic error, Sheldrick[1] employs a hybrid weighting scheme:

$$w_{\mathbf{h}} = \left(\frac{1}{\sigma_{\mathbf{h}}^2 + bP_{\mathbf{h}} + cP_{\mathbf{h}}^2}\right)_o. \tag{8.158}$$

8.2.2 Estimation of Parameter Errors in Refinement

The normal equations for refinement, Eqns. 8.114 and 8.115, are based on changes in the parameter values, Δp_i, and the elements in the variance-covariance matrix, $\mathbf{T}^{-1}$, correspond to those changes. The variances in the changes, however, are identical to the variances in the parameters:

$$\sigma^2(\Delta p_i) = \left(\frac{\partial p_i}{\partial p_i}\right)\sigma^2(p_i) - \left(\frac{\partial p_{i,o}}{\partial p_i}\right)\sigma^2(p_i) = \sigma^2(p_i). \tag{8.159}$$

From Eqn. 8.51, the estimated variance of the rth parameter (p_r), is therefore

$$\sigma^2(p_r) = \sigma_{rr} = t_{rr}^i\left(\frac{1}{m-n}\sum_{\mathbf{h}} w_{\mathbf{h}}(I_{o,\mathbf{h}} - I_{c,\mathbf{h}}(\mathbf{p}))^2\right), \tag{8.160}$$

where m is the number of reflections, n is the number of parameters, and t_{rr}^i is the rth diagonal element of the variance-covariance matrix. The covariance for parameter r and parameter s is given by Eqn. 8.52:

$$\sigma_{rs} = t_{rs}^i\left(\frac{1}{m-n}\sum_{\mathbf{h}} w_{\mathbf{h}}(I_{o,\mathbf{h}} - I_{c,\mathbf{h}}(\mathbf{p}))^2\right), \tag{8.161}$$

where t_{rs}^i is the *rsth* off-diagonal element of the $\mathbf{T}^{-1}$ matrix.

The variances and covariances in the refined parameters are reflected in the quantities derived from them, such as interatomic distances and angles. In order to estimate the errors in such quantities it is necessary to propagate the parameter errors through the equations used for the derivation. In Sec. 8.1.2 the treatment of errors in the lattice parameters produced a general expression (Eqn. 8.69) that is equally applicable to the structural parameters. Because the coordinate system upon which the parameters are defined is also determined experimentally, the variances of the unit cell parameters must also be included in the analysis. Let us consider a general function defining an entity derived from the refined parameters:

$$D(q_1, q_2, \ldots, q_k, u_1, u_2, u_3, u_4, u_5, u_6), \tag{8.162}$$

with q_i, the ith of k atomic parameters in the derivation, and $u_1 \ldots u_6$ the six unit cell parameters.

Unlike the unit cell parameters, which are ordinarily (but not rigorously) consider to be uncorrelated, the refined parameters may be correlated significantly — the changes in one parameter may parallel the changes in another if the parameters are linked to one another — either physically, or mathematically. If this happens the cross terms corresponding to these parameters (cf., 8.68) will not cancel one another, and the covariances for these parameters will *not* be negligible. A general expression for the variance in D is therefore

$$\sigma^2(D) = \sum_{j=1}^{k}\sum_{i=i}^{k}\left(\frac{\partial D}{\partial q_i}\right)\left(\frac{\partial D}{\partial q_j}\right)\sigma_{ij} + \sum_{j=1}^{6}\sum_{i=i}^{6}\left(\frac{\partial D}{\partial u_i}\right)\left(\frac{\partial D}{\partial u_j}\right)\sigma_{ij}, \qquad (8.163)$$

which includes the covariances of the lattice parameters. If they are ignored (set to zero) then

$$\sigma^2(D) = \sum_{j=1}^{k}\sum_{i=i}^{k}\left(\frac{\partial D}{\partial q_i}\right)\left(\frac{\partial D}{\partial q_j}\right)\sigma_{ij} + \sum_{j=1}^{6}\left(\frac{\partial D}{\partial u_j}\right)^2\sigma_j^2. \qquad (8.164)$$

Errors in Interatomic Distances

In order to determine the length of a vector in fractional coordinates, $\mathbf{v}_f = [x_f\ y_f\ z_f]$, it must first be converted to Cartesian coordinates. Using Eqn. 1.130,

$$\mathbf{B}\mathbf{v}_f = \begin{bmatrix} a_x & b_x & c_x \\ a_y & b_y & c_y \\ a_z & b_z & c_z \end{bmatrix}\begin{bmatrix} x_f \\ y_f \\ z_f \end{bmatrix} = \begin{bmatrix} x_c \\ y_c \\ z_c \end{bmatrix} = \mathbf{v}_c; \qquad (8.165)$$

$$v^2 = \mathbf{v}_c^T\mathbf{v}_c = \begin{bmatrix} x_c & y_c & z_c \end{bmatrix}\begin{bmatrix} x_c \\ y_c \\ z_c \end{bmatrix} = x_c^2 + y_c^2 + z_c^2. \qquad (8.166)$$

Thus,

$$v^2 = \mathbf{v}_f^T\mathbf{B}^T\mathbf{B}\mathbf{v}_f = \mathbf{v}_f^T\begin{bmatrix} \mathbf{a}\cdot\mathbf{a} & \mathbf{a}\cdot\mathbf{b} & \mathbf{a}\cdot\mathbf{c} \\ \mathbf{b}\cdot\mathbf{a} & \mathbf{b}\cdot\mathbf{b} & \mathbf{b}\cdot\mathbf{c} \\ \mathbf{c}\cdot\mathbf{a} & \mathbf{c}\cdot\mathbf{b} & \mathbf{c}\cdot\mathbf{c} \end{bmatrix}\mathbf{v}_f = \mathbf{v}_f^T\mathbf{G}\mathbf{v}_f \qquad (8.167)$$

$$= \begin{bmatrix} x_f & y_f & z_f \end{bmatrix}\begin{bmatrix} a^2 & ab\cos\gamma & ac\cos\beta \\ ab\cos\gamma & b^2 & bc\cos\alpha \\ ac\cos\beta & bc\cos\alpha & c^2 \end{bmatrix}\begin{bmatrix} x_f \\ y_f \\ z_f \end{bmatrix} \qquad (8.168)$$

$$= x_f^2a^2 + y_f^2b^2 + z_f^2c^2 + 2x_fy_fab\cos\gamma + 2x_fz_fac\cos\beta + 2y_fz_fbc\cos\alpha. \qquad (8.169)$$

Fig. 8.4 depicts three atoms in the unit cell, at locations $\mathbf{v}_{f,a}$, $\mathbf{v}_{f,b}$, and $\mathbf{v}_{f,c}$ in fractional coordinates. The distance between atom a and atom c is the length of the vector $\mathbf{v}_{f,ac} = [(x_{f,c} - x_{f,a})\ (y_{f,c} - y_{f,a})\ (z_{f,c} - z_{f,a})]$:

$$\begin{aligned} v_{ac}^2 = {} & (x_{f,c} - x_{f,a})^2a^2 + (y_{f,c} - y_{f,a})^2b^2 + (z_{f,c} - z_{f,a})^2c^2 \\ & + 2(x_{f,c} - x_{f,a})(y_{f,c} - y_{f,a})\,ab\cos\gamma \\ & + 2(x_{f,c} - x_{f,a}(z_{f,c} - z_{f,a})\,ac\cos\beta \\ & + 2(y_{f,c} - y_{f,a})(z_{f,c} - z_{f,a})\,bc\cos\alpha. \end{aligned} \qquad (8.170)$$

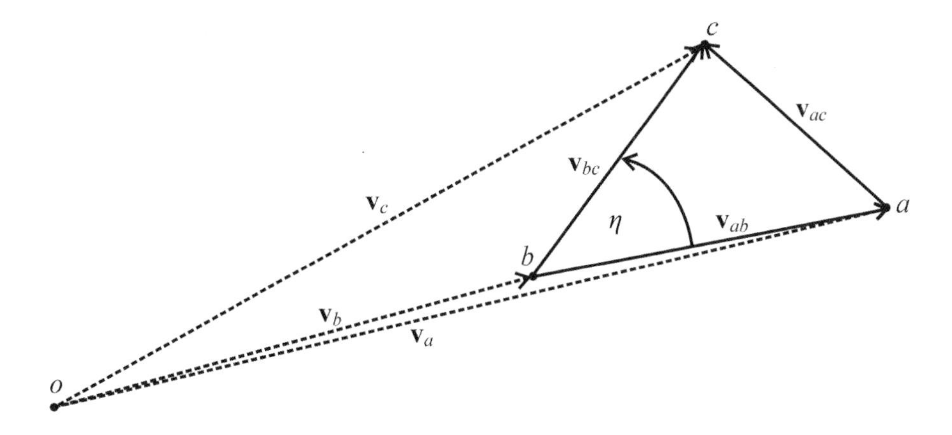

Figure 8.4 Vectors to three atom locations in the unit cell: $\mathbf{v}_{f,a} = [x_{f,a}\ y_{f,a}\ z_{f,a}]$, $\mathbf{v}_{f,b} = [x_{f,b}\ y_{f,b}\ z_{f,b}]$, $\mathbf{v}_{f,c} = [x_{f,c}\ y_{f,c}\ z_{f,c}]$.

The estimated error in the squared magnitude of the vector (assuming that the covariances of the unit cell parameters are equal to zero) is given by

$$\begin{aligned}\sigma^2(v_{ac}^2) &= \left(\frac{\partial v_{ac}^2}{\partial x_{f,a}}\right)_o^2 \sigma^2(x_{f,a}) + \left(\frac{\partial v_{ac}^2}{\partial x_{f,c}}\right)_o^2 \sigma^2(x_{f,c}) + \left(\frac{\partial v_{ac}^2}{\partial y_{f,a}}\right)_o^2 \sigma^2(y_{f,a}) + \ldots \\ &+ \left(\frac{\partial v_{ac}^2}{\partial a}\right)_o^2 \sigma^2(a) + \left(\frac{\partial v_{ac}^2}{\partial b}\right)_o^2 \sigma^2(b) + \left(\frac{\partial v_{ac}^2}{\partial \gamma}\right)_o^2 \sigma^2(\gamma) + \ldots \\ &+ 2\left(\frac{\partial v_{ac}^2}{\partial x_{f,a}}\right)_o \left(\frac{\partial v_{ac}^2}{\partial x_{f,c}}\right)_o \sigma(x_{f,a}, x_{f,c}) + \ldots \quad . \end{aligned} \tag{8.171}$$

Since $dv_{ac}^2/dv_{ac} = 2v_{ac}$, $\sigma(v_{ac}) = \sigma(v_{ac}^2)/(2v_{ac})$ — and $\sigma^2(v_{ac}) = \sigma^2(v_{ac}^2)/(4v_{ac}^2)$. Differentiating Eqn. 8.170 with respect to each of the twelve parameters, and inserting the derivatives into the 27 terms in Eqn. 8.171 provides an explicit expression for the estimated error in v_{ac}. Equations for errors in more complex quantities, such as interatomic angles, torsional angles, or dihedral angles, will be even more complicated.

A more general treatment is possible if the derivatives are estimated numerically. This is readily accomplished by calculating $D(p_1, \ldots p_i, \ldots)$ — then making a small change in parameter p_i, e.g., $\delta p_i = 0.01\,\sigma(p_i)$, and calculating $\delta D = D(p_1, \ldots (p_i + \delta p_i), \ldots) - D(p_1, \ldots p_i, \ldots)$. The estimated error is then determined by substituting the approximate derivatives for all parameters involved in D,

$$\frac{\partial D}{\partial p_i} \simeq \frac{\delta D}{\delta p_i}, \tag{8.172}$$

directly into Eqn. 8.164 and setting all terms exclusively containing parameters not involved in D to zero.

Errors in Interatomic Angles

Referring to Fig. 8.4, the interatomic angle, η, is determined from the dot product of the Cartesian components of $\mathbf{v}_{f,ab} = [(x_{f,a} - x_{f,b})\ (y_{f,a} - y_{f,b})\ (z_{f,a} - z_{f,b})] = [x_{f,ab}\ y_{f,ab}\ z_{f,ab}]$ and $\mathbf{v}_{f,bc} = [(x_{f,c} - x_{f,b})\ (y_{f,c} - y_{f,b})\ (z_{f,c} - z_{f,b})] = [x_{f,bc}\ y_{f,bc}\ z_{f,bc}]$:

$$\begin{aligned}
\cos\eta &= \frac{\mathbf{v}_{ab}^T\mathbf{v}_{bc}}{v_{ab}v_{bc}} = \frac{\mathbf{v}_{f,ab}^T\mathbf{B}^T\mathbf{B}\mathbf{v}_{f,bc}}{v_{ab}v_{bc}} = \frac{\mathbf{v}_{f,ab}^T\mathbf{G}\mathbf{v}_{f,bc}}{v_{ab}v_{bc}} = \\
&\frac{1}{v_{ab}v_{bc}}\begin{bmatrix} x_{f,ab} & y_{f,ab} & z_{f,ab}\end{bmatrix}\begin{bmatrix} a^2 & ab\cos\gamma & ac\cos\beta \\ ab\cos\gamma & b^2 & bc\cos\alpha \\ ac\cos\beta & bc\cos\alpha & c^2\end{bmatrix}\begin{bmatrix} x_{f,bc} \\ y_{f,bc} \\ z_{f,bc}\end{bmatrix} \\
&\frac{1}{v_{ab}v_{bc}}(a^2 x_{f,ab}x_{f,bc} + b^2 y_{f,ab}y_{f,bc} + c^2 z_{f,ab}z_{f,bc} \\
&\quad + ab(x_{f,ab}y_{f,bc} + y_{f,ab}x_{f,bc})\cos\gamma + ac(x_{f,ab}z_{f,bc} + z_{f,ab}x_{f,bc})\cos\beta \\
&\quad + bc(y_{f,ab}z_{f,bc} + z_{f,ab}y_{f,bc})\cos\alpha).
\end{aligned} \tag{8.173}$$

The variables above are replaced by their refined-parameter equivalents (e.g., $x_{f,ab} = x_{f,a} - x_{f,b}$ — including the expressions for v_{ab} and v_{bc}), and $\sigma^2(\cos\eta)$ is then evaluated, either explicitly, by taking the appropriate partial derivatives, or numerically, as described above:

$$\begin{aligned}
\sigma^2(\cos\eta) &= \left(\frac{\partial\cos\eta}{\partial x_{f,a}}\right)_o^2\sigma^2(x_{f,a}) + \left(\frac{\partial\cos\eta}{\partial x_{f,b}}\right)_o^2\sigma^2(x_{f,b}) + \left(\frac{\partial\cos\eta}{\partial y_{f,a}}\right)_o^2\sigma^2(y_{f,a}) + \ldots \\
&+ \left(\frac{\partial\cos\eta}{\partial a}\right)_o^2\sigma^2(a) + \left(\frac{\partial\cos\eta}{\partial b}\right)_o^2\sigma^2(b) + \left(\frac{\partial\cos\eta}{\partial\gamma}\right)_o^2\sigma^2(\gamma) + \ldots \\
&+ 2\left(\frac{\partial\cos\eta}{\partial x_{f,a}}\right)_o\left(\frac{\partial\cos\eta}{\partial x_{f,b}}\right)_o\sigma(x_{f,a},x_{f,b}) + \ldots \quad .
\end{aligned} \tag{8.174}$$

The estimated error in the interatomic angle is then determined from (cf., Eqn. 8.75):

$$\begin{aligned}
d\eta &= -\frac{d\cos\eta}{\sin\eta} \\
\sigma(\eta) &= -\frac{\sigma(\cos\eta)}{\sin\eta}.
\end{aligned} \tag{8.175}$$

Another angle that is important in molecular structural analysis is the torsional angle, which corresponds to the rotational angle about a specific chemical bond. Fig. 8.5 illustrates this for the $S_2O_5^{2-}$ anion. If the *ab* and *cd* bonds are initially aligned in a projection looking down the *bc* bond, then ν is the torsional angle. To determine ν, normal vectors to the *abc* and *bcd* planes are constructed and the angle between these two vectors is then the dihedral angle between the two planes, which is also the torsional angle. The locations of the four atoms that determine the angle are $\mathbf{v}_{f,a} = [x_{f,a}\ y_{f,a}\ z_{f,a}]$, $\mathbf{v}_{f,b} = [x_{f,b}\ y_{f,b}\ x_{f,b}]$, $\mathbf{v}_{f,c} = [x_{f,c}\ y_{f,c}\ z_{f,c}]$, and $\mathbf{v}_{f,d} = [x_{f,d}\ y_{f,d}\ z_{f,d}]$. The normal to the *abc* plane is generated from the vector product of a vector along the *ab* bond, and a vector along the *bc* bond; the normal to the *bcd* plane is generated from the vector product of a vector along the *cb* bond

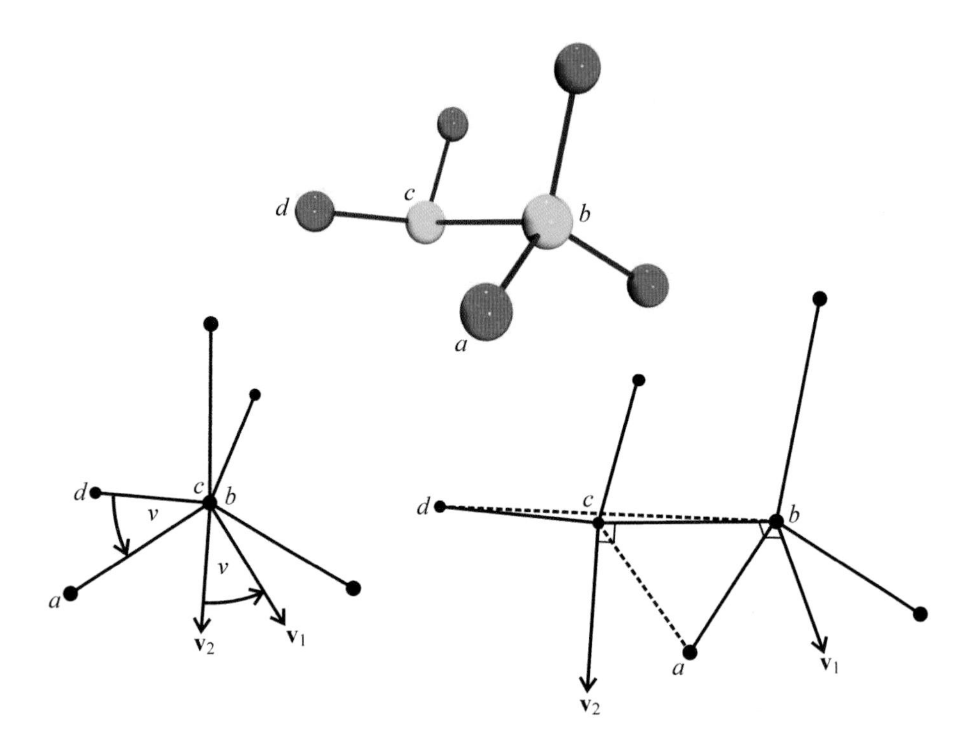

Figure 8.5 The torsional angle for rotation about the sulfur-sulfur bond in $S_2O_5^{2-}$.

and a vector along the *cd* bond. The following sequence of operations produces the cosine of the torsional angle:

$$\mathbf{Bv}_{f,\,a} = \mathbf{v}_{c,\,a} \quad \mathbf{Bv}_{f,\,b} = \mathbf{v}_{c,\,b} \quad \mathbf{Bv}_{f,\,c} = \mathbf{v}_{c,\,c} \quad \mathbf{Bv}_{f,\,d} = \mathbf{v}_{c,\,d}$$

$$\mathbf{v}_{ba} = \mathbf{v}_{c,\,a} - \mathbf{v}_{c,\,b} \qquad \mathbf{v}_{bc} = \mathbf{v}_{c,\,c} - \mathbf{v}_{c,\,b} \qquad \mathbf{v}_{dc} = \mathbf{v}_{c,\,d} - \mathbf{v}_{c,\,c}$$

$$\mathbf{v}_1 = \mathbf{v}_{ba} \times \mathbf{v}_{bc} = \begin{bmatrix} y_{ba}z_{bc} - y_{bc}z_{ba} \\ z_{ba}x_{bc} - z_{bc}x_{ba} \\ x_{ba}y_{bc} - x_{bc}y_{ba} \end{bmatrix}$$

$$\mathbf{v}_2 = -\mathbf{v}_{bc} \times \mathbf{v}_{dc} = - \begin{bmatrix} y_{bc}z_{dc} - y_{dc}z_{bc} \\ z_{bc}x_{dc} - z_{dc}x_{bc} \\ x_{bc}y_{dc} - x_{dc}y_{bc} \end{bmatrix}$$

$$v_1 = ((y_{ba}z_{bc} - y_{bc}z_{ba})^2 + (z_{ba}x_{bc} - z_{bc}x_{ba})^2 + (x_{ba}y_{bc} - x_{bc}y_{ba})^2)^{1/2}$$
$$v_2 = ((y_{bc}z_{dc} - y_{dc}z_{bc})^2 + (z_{bc}x_{dc} - z_{dc}x_{bc})^2 + (x_{bc}y_{dc} - x_{dc}y_{bc})^2)^{1/2}$$
$$\cos\nu = \frac{\mathbf{v}_1^T\mathbf{v}_2}{v_1\,v_2}. \tag{8.176}$$

The vector products are complex functions of all the coordinates of the four atoms *and* the unit cell parameters. Although it is possible to produce a general formula for $\cos\nu$ from the scenario above, the derivatives render the function even more complex. Numerical derivatives are the logical choice in this instance. Changing cell parameters will be reflected in the **B** matrix; changing positional parameters will be reflected in their transformations throughout the sequence of operations. Each small change, δp results in a change in the computed value of $\cos\nu$, $\delta\cos\nu$, and

$$\begin{aligned}\sigma^2(\cos\nu) \simeq & \left(\frac{\delta\cos\nu}{\delta x_{f,a}}\right)_o^2 \sigma^2(x_{f,a}) + \left(\frac{\delta\cos\nu}{\delta x_{f,b}}\right)_o^2 \sigma^2(x_{f,b}) + \left(\frac{\delta\cos\nu}{\delta y_{f,a}}\right)_o^2 \sigma^2(y_{f,a}) + \dots \\ & + \left(\frac{\delta\cos\nu}{\delta a}\right)_o^2 \sigma^2(a) + \left(\frac{\delta\cos\nu}{\delta b}\right)_o^2 \sigma^2(b) + \left(\frac{\delta\cos\nu}{\delta\gamma}\right)_o^2 \sigma^2(\gamma) + \dots \\ & + 2\left(\frac{\delta\cos\nu}{\delta x_{f,a}}\right)_o \left(\frac{\delta\cos\nu}{\delta x_{f,b}}\right)_o \sigma(x_{f,a}, x_{f,b}) + \dots \quad , \end{aligned} \tag{8.177}$$

with $\sigma(\nu) = (\sigma^2(\cos\nu))^{1/2}/\sin\nu$. Errors in other calculated quantities can be evaluated in the same manner.

8.2.3 Refinement Figures of Merit

In least-squares optimization, the "best fit" criterion is the minimization of the weighted sum of the squares of the residual differences between the scaled observed intensities and those calculated from the structural model. A logical measure of the validity of the model is the degree to which it decreases the Euclidean norm (length) of the residual vector, $|\mathbf{r_h}| = (R(\mathbf{p}))^{1/2}$, reflected in the weighted root-mean-square deviation,

$$R_{rms} = \frac{1}{n}\left(\sum_{\mathbf{h}} w_{\mathbf{h}}(I_{o,\mathbf{h}} - I_{c,\mathbf{h}})^2\right)^{1/2} , \tag{8.178}$$

where n is the number of reflections and $I_{c,\mathbf{h}}$ includes the scale factor, p_1. This number is difficult to use as a refinement figure of merit, since it scales with the average magnitude of the intensities, which depends on the contents of the unit cell. It can be rendered "content independent" by dividing the weighted mean-square deviation by the weighted mean-square intensity:

$$R_I = \left(\frac{\frac{1}{n}\sum_{\mathbf{h}} w_{\mathbf{h}}(I_{o,\mathbf{h}} - I_{c,\mathbf{h}})^2}{\frac{1}{n}\sum_{\mathbf{h}} w_{\mathbf{h}} I_o^2}\right)^{1/2} = \left(\frac{\sum_{\mathbf{h}} w_{\mathbf{h}}(I_{o,\mathbf{h}} - I_{c,\mathbf{h}})^2}{\sum_{\mathbf{h}} w_{\mathbf{h}} I_o^2}\right)^{1/2} . \tag{8.179}$$

R_I is known as the *reliability index.** R_I values in the neighborhood of 0.10 ± 0.08 are commonly observed for small molecule crystal structures. A related scale-independent measure of the reliability of the least squares fit is the correlation coefficient,

$$\rho_I = \frac{\langle (I_o - \langle I_o\rangle)(I_c - \langle I_c\rangle)\rangle}{(\langle (I_o - \langle I_o\rangle)^2\rangle)^{1/2}(\langle (I_c - \langle I_c\rangle)^2\rangle)^{1/2}}, \tag{8.180}$$

which approaches unity for a reliable fit. Another value often cited as a figure of merit is the *goodness-of-fit*, S, defined as the squared magnitude of the residual vector, $R(\mathbf{p})$, divided by its expected value, $\langle R(\mathbf{p})\rangle$,

$$S = \frac{R(\mathbf{p})}{\langle R(\mathbf{p})\rangle} = \frac{\sum_{\mathbf{h}} w_{\mathbf{h}}(I_o - I_c)^2}{\sum_{\mathbf{h}} \langle w_{\mathbf{h}}\rangle \langle (I_o - I_c)^2\rangle}. \tag{8.181}$$

*The reliability index based on the intensities is often referred to as the "crystallographic R-factor based on F^2," and is commonly denoted as $wR2$.

If there are no systematic errors, then the unbiased weights are the variances of the intensities, and

$$\langle w_h \rangle = \frac{1}{\langle \sigma_{\mathbf{h}}^2 \rangle} = \frac{1}{\langle (I_o - I_c)^2 \rangle}. \tag{8.182}$$

Thus,

$$\langle R(\mathbf{p}) \rangle = \sum_{\mathbf{h}} \frac{\langle (I_o - I_c)^2 \rangle}{\langle (I_o - I_c)^2 \rangle} = \sum_{\mathbf{h}} 1 = m, \tag{8.183}$$

where m is the number of reflections. Wilson[172] has shown that the actual expected value is somewhat smaller than m, and should rigorously be replaced by the number of degrees of freedom, $m - n$, where n is the number of parameters, giving

$$S = \frac{\sum_{\mathbf{h}} w_{\mathbf{h}} (I_o - I_c)^2}{m - n}. \tag{8.184}$$

If the weighting function truly compensates for systematic errors and reflects random errors correctly, then the numerator above will approach its expected value, and S should approach unity. The *goodness-of-fit* value is often cited as an indicator of the validity of the weighting function.

For many years refinements were routinely based on the magnitudes of the structure factors rather than the scaled intensities. For refinement based on F, the reliability index is

$$R_F = \left(\frac{\sum_{\mathbf{h}} w'_{\mathbf{h}} (F_{o,\mathbf{h}} - F_{c,\mathbf{h}})^2}{\sum_{\mathbf{h}} w'_{\mathbf{h}} F_o^2} \right)^{1/2}. \tag{8.185}$$

The relatively recent trend toward refinements based on intensities has created some problems in making literature comparisons, since a large body of the literature contains R factors based on structure factor refinement. These values are ordinarily smaller than R_I for the same structure by a factor of two or three (e.g, 0.05 ± 0.04). This has become especially important in the peer review process, since the reliability index is often the major criterion used for the evaluation of the structure. Because of this, it is also useful to calculate a "conventional" R factor in addition to R_I. In order to report R_F in addition to R_I, it would be necessary to repeat the refinement based on F in order to redetermine the weighting function, since, as we have already noted, there is no straightforward relationship between the weights based on each type of refinement. In general, a "non-weighted" reliability index will suffice for comparative purposes:

$$R_1 = \left(\frac{\sum_{\mathbf{h}} (F_{o,\mathbf{h}} - F_{c,\mathbf{h}})^2}{\sum_{\mathbf{h}} F_o^2} \right)^{1/2}. \tag{8.186}$$

In addition, R_F was usually reported for "observed" reflections only, those for which the observed intensity was greater than its estimated standard deviation by some factor (e.g., $I_o \geq 2\sigma(I_o)$). Values for R_I and R_1 are routinely reported in the literature for all of the reflections *and* "observed" reflections.

Although a large percentage of small molecule structures are refined today based on intensities, some small molecule refinements are still based on F. On the other hand, macromolecular structures are routinely refined based on structure factor magnitudes, arguably because the small statistical biases that are introduced are far outweighed by other approximations. Because of this there is little impetus to modify currently used macromolecular crystallography refinement programs. R_F is the common reliability index employed in macromolecular refinement.

8.2.4 Constrained Refinement

There may be good reasons in structural refinement to constrain selected variables in a specific manner. A common example occurs for an atom in a special position. In such cases the site occupancy factors are constrained to specific values (the reciprocal of the site multiplicity), and the displacement parameters are constrained by symmetry. If the atom is on a mirror plane or a glide plane, for example, referring to Sec. 4.4, this would require the simple constraining equations: $s = 0.5$, $U_{12} = 0$ and $U_{23} = 0$. These parameters can be constrained by fixing their values and eliminating them from the normal equations (using them only for generating calculated intensities). However, it may be necessary to constrain relationships between parameters while allowing them to vary. This is illustrated in another common example, in which a single atom is able to occupy two separate sites in the structure, site a and site b. This can be modeled with two sets of freely refined positional parameters, one for each site, with the fractional occupancy of each site characterized by its site occupancy factor. However, the site occupancies can take on only a subset of all possible values, since they must sum to unity (a single atom). The atomic coordinates, $(x_{f,a}, y_{f,a}, z_{f,a})$ and $(x_{f,b}, yx_{f,b}, z_{f,b})$, and the site occupancy factors, s_a and s_b, are therefore refined (along with the remaining parameters), subject to the linear constraint, $q'(\mathbf{p}) = s_a + s_b = 1$. Applying the constraining equations for both of these examples can seen to be analogous to the discussion of linear least-squares constraints in Sec. 8.1.3, indicating that the method of undetermined multipliers can be utilized to minimize the magnitude of the residual vector, subject to the constraining equations, $q'_i(\mathbf{p}) = z$, $i = 1, 2, \ldots, k$. The Lagrangian is (See Appendix H and Eqn. 8.86):

$$L(\mathbf{p}_c, \overrightarrow{\lambda}) = \sum_{\mathbf{h}} w_{\mathbf{h}}(I_{o,\mathbf{h}} - I_{c,\mathbf{h}}(\mathbf{p}_c))^2 + \sum_{i=1}^{k} \lambda_i \, q(\mathbf{p}_c)_i, \tag{8.187}$$

where $q(\mathbf{p}_c)_i = z - q'_i(\mathbf{p}_c) = 0$ and $\mathbf{p}_c$ is the subset of possible parameter values that satisfy the constraining equations.

Unlike the discussion in Sec. 8.1.3, the constraining equations in structural refinement might not be linear equations. For example, suppose that we wish to keep a certain bond length between atom a and atom b, d_{ab}, fixed to a specific value throughout the refinement process. In this case $\mathbf{d}_{ab} = [(x_{f,b} - x_{f,a}) \; (y_{f,b} - y_{f,a}) \; (z_{f,b} - z_{f,a})]$, and the coordinates are refined, subject to the constraining equation,

$$\begin{aligned} q'(\mathbf{p}_c) &= \begin{bmatrix} (x_{f,b} - x_{f,a}) & (y_{f,b} - y_{f,a}) & (z_{f,b} - z_{f,a}) \end{bmatrix} \mathbf{B}^T\mathbf{B} \begin{bmatrix} (x_{f,b} - x_{f,a}) \\ (y_{f,b} - y_{f,a}) \\ (z_{f,b} - z_{f,a}) \end{bmatrix} \\ &= \mathbf{d}_{ab}^T \mathbf{G} \mathbf{d}_{ab} = (x_{c,b} - x_{c,a})^2 + (y_{c,b} - y_{c,a})^2 + (z_{c,b} - z_{c,a})^2 \\ &= d_{ab}^2. \end{aligned}$$

The constraining equations, like the squares of the residuals themselves, can be approximated as linear equations by expanding them as Taylor series about the initial parameter values:

$$q(\mathbf{p}_c)_i \simeq q(\mathbf{p}_c)_o + \sum_{i=1}^{n} \left(\frac{\partial q(\mathbf{p}_c)}{\partial p_i} \right)_o \Delta p_i, \tag{8.188}$$

and the Lagrangian becomes

$$\begin{aligned} L(\mathbf{p}_c, \overrightarrow{\lambda}) &= \sum_{\mathbf{h}} w_{\mathbf{h}} \left(\Delta I_{\mathbf{h}} - \sum_{i=1}^{n} \left(\frac{\partial (I_{c,\mathbf{h}}(\mathbf{p}_c))}{\partial p_i} \right)_o \Delta p_i \right)^2 \\ &\quad + \sum_{l=1}^{k} \lambda_j \left(q_l(\mathbf{p}_c)_o + \sum_{i=1}^{n} \left(\frac{\partial q_l(\mathbf{p}_c)}{\partial p_i} \right)_o \Delta p_i \right). \end{aligned} \tag{8.189}$$

The normal equations are created by differentiating the Lagrangian with respect to the parameters and undetermined multipliers, and setting the derivatives to zero:

$$\begin{aligned} \frac{\partial L(\mathbf{p}_c, \overrightarrow{\lambda})}{\partial p_j} &= -2 \sum_{\mathbf{h}} w_{\mathbf{h}} \left(\Delta I_{\mathbf{h}} - \sum_{i=1}^{n} \left(\frac{\partial (I_{c,\mathbf{h}}(\mathbf{p}))}{\partial p_i} \right)_o \Delta p_i \right) \left(\frac{\partial (I_{c,\mathbf{h}}(\mathbf{p}))}{\partial p_j} \right)_o \\ &\quad + \sum_{l=1}^{k} \lambda_l \left(\frac{\partial q_l(\mathbf{p}_c)}{\partial p_j} \right)_o = 0 \end{aligned} \tag{8.190}$$

$$\frac{\partial L(\mathbf{p}_c, \overrightarrow{\lambda})}{\partial \lambda_l} = q_l(\mathbf{p}_c)_o + \sum_{i=1}^{n} \left(\frac{\partial q_l(\mathbf{p}_c)}{\partial p_i} \right)_o \Delta p_i = 0. \tag{8.191}$$

Rearrangement creates n equations containing the parameters and multipliers as variables:

$$\begin{aligned} &\Delta p_1 \sum_{\mathbf{h}} w_{\mathbf{h}} \left(\frac{\partial (I_{c,\mathbf{h}}(\mathbf{p}))}{\partial p_1} \frac{\partial (I_{c,\mathbf{h}}(\mathbf{p}))}{\partial p_j} \right)_o + \Delta p_2 \sum_{\mathbf{h}} w_{\mathbf{h}} \left(\frac{\partial (I_{c,\mathbf{h}}(\mathbf{p}))}{\partial p_2} \frac{\partial (I_{c,\mathbf{h}}(\mathbf{p}))}{\partial p_j} \right)_o + \cdots \\ &-\frac{1}{2}\lambda_1 \left(\frac{\partial q_1(\mathbf{p}_c)}{\partial p_j} \right)_o - \cdots - \frac{1}{2}\lambda_k \left(\frac{\partial q_k(\mathbf{p}_c)}{\partial p_j} \right)_o = \sum_{\mathbf{h}} w_{\mathbf{h}} \Delta I_{\mathbf{h}} \left(\frac{\partial (I_{c,\mathbf{h}}(\mathbf{p}))}{\partial p_j} \right)_o, \end{aligned} \tag{8.192}$$

and k equations with non-zero coefficients for the parameters alone:

$$-\frac{1}{2}\Delta p_1 \left(\frac{\partial q_l(\mathbf{p}_c)}{\partial p_1} \right)_o - \frac{1}{2}\Delta p_2 \left(\frac{\partial q_l(\mathbf{p}_c)}{\partial p_2} \right)_o + \cdots + 0\,\lambda_1 + 0\,\lambda 2 + \cdots = \frac{1}{2} q_l(\mathbf{p}_c)_o. \tag{8.193}$$

Using the defined entities from Eqns. 8.112 and 8.113, and setting

$$t_{j,n+l} = t_{n+l,j} = -\frac{1}{2} \left(\frac{\partial q_l(\mathbf{p}_c)}{\partial p_j} \right)_0 \tag{8.194}$$

and

$$d_{n+l} = \frac{1}{2} q_l(\mathbf{p}_c)_o, \tag{8.195}$$

gives $n + k$ simultaneous linear equations in $n + k$ unknowns:

$$\begin{aligned}
t_{11}\Delta p_1 + t_{12}\Delta p_2 + \cdots + t_{1,n+l}\lambda_1 + \cdots + t_{1,n+k}\lambda_k &= d_1 \\
t_{21}\Delta p_1 + t_{22}\Delta p_2 + \cdots + t_{2,n+l}\lambda_1 + \cdots + t_{2,n+k}\lambda_k &= d_2 \\
\vdots \qquad \vdots \qquad \vdots \qquad \vdots \qquad \vdots & \\
t_{n+1,1}\Delta p_1 + t_{n+1,2}\Delta p_2 + \cdots + 0\lambda_1 + \cdots + 0\lambda_k &= d_{n+1} \\
\vdots \qquad \vdots \qquad \vdots \qquad \vdots \qquad \vdots & \\
t_{n+k,1}\Delta p_1 + t_{n+k,2}\Delta p_2 + \cdots + 0\lambda_1 + \cdots + 0\lambda_k &= d_{n+k}.
\end{aligned} \tag{8.196}$$

The matrix equation for the solution is

$$\begin{bmatrix} t_{11} & t_{12} & \cdots & t_{1,n+1} & \cdots & t_{1,n+k} \\ t_{21} & t_{22} & \cdots & t_{2,n+1} & \cdots & t_{2,n+k} \\ \vdots & \vdots & \vdots & \vdots & \vdots & \vdots \\ t_{n+1,1} & t_{n+1,2} & \cdots & 0 & \cdots & 0 \\ \vdots & \vdots & \vdots & \vdots & \vdots & \vdots \\ t_{n+k,1} & t_{n+k,2} & \cdots & 0 & \cdots & 0 \end{bmatrix} \begin{bmatrix} \Delta p_1 \\ \Delta p_2 \\ \vdots \\ \lambda_1 \\ \vdots \\ \lambda_k \end{bmatrix} = \begin{bmatrix} d_1 \\ d_2 \\ \vdots \\ d_{n+1} \\ \vdots \\ d_{n+k} \end{bmatrix}. \tag{8.197}$$

The estimated variance of the rth parameter (p_r), is therefore

$$\sigma^2(p_r) = \sigma_{rr} = t^i_{rr}\left(\frac{1}{m-(n+k)}\sum_{\mathbf{h}} w_{\mathbf{h}}(I_{o,\,\mathbf{h}} - I_{c,\,\mathbf{h}}(\mathbf{p}))^2\right), \tag{8.198}$$

where m is the number of reflections, n is the number of parameters, and t^i_{rr} is the rth diagonal element of the variance-covariance matrix, the inverse of the matrix in Eqn. 8.197. The covariance for parameter r and parameter s is given by :

$$\sigma_{rs} = t^i_{rs}\left(\frac{1}{m-(n+k)}\sum_{\mathbf{h}} w_{\mathbf{h}}(I_{o,\,\mathbf{h}} - I_{c,\,\mathbf{h}}(\mathbf{p}))^2\right). \tag{8.199}$$

In constrained linear least squares, a solution is obtained from the normal equations in a single step by conveniently incorporating undetermined multipliers. As we have just observed, this results in an increase in the number of normal equations. The addition of the multipliers to the already large number of parameters in crystallographic refinements has lead crystallographers to avoid their use by using the relationships between the parameters to limit the number of normal equations, taking advantage of the iterative nature of non-linear least squares. This is relatively straightforward for simple constraining equations. Returning to the site occupancy example, since $s_b = s_a - 1$, initial values of s_a and s_b (e.g., 0.5) can be used to determine $I_{c,\,\mathbf{h}}$, but only s_a is refined. s_b is computed from $1 - s_a$ after each cycle of the refinement before $I_{c,\,\mathbf{h}}$ is recalculated.

The application of more complex constraining equations using this approach is not as straightforward. For example, consider the distance constraint,

$$d^2_{ab} = (x_{c,\,b} - x_{c,\,a})^2 + (y_{c,\,b} - y_{c,\,a})^2 + (z_{c,\,b} - z_{c,\,a})^2 = x^2_{ab} + y^2_{ab} + z^2_{ab}.$$

Rather than refining all six positional parameters, in the unconstrained case, we can equivalently refine $x_{f,\,a}$, $y_{f,\,a}$, $z_{f,\,a}$, $u = x_{ab}^2$, $v = y_{ab}^2$, and $w = z_{ab}^2$ after evaluating the derivatives of the last three expressions with respect to the calculated intensities for incorporation into the normal equations, e.g.,

$$\frac{\partial I_{c,\,\mathbf{h}}}{\partial u} = \frac{\partial I_{c,\,\mathbf{h}}}{\partial x_{f,\,a}}\frac{\partial x_{f,\,a}}{\partial u} + \frac{\partial I_{c,\,\mathbf{h}}}{\partial x_{f,\,b}}\frac{\partial x_{f,\,b}}{\partial u}, \tag{8.200}$$

with $\partial x_{f,\,a}/\partial u$ and $\partial x_{f,\,b}/\partial u$ evaluated either analytically or numerically.

After solution, new values of $x_{c,\,b}$, $y_{c,\,b}$ and $z_{c,\,b}$ can be determined,

$$\begin{aligned} x_{c,\,b} &= x_{c,\,a} \pm \sqrt{u} \\ y_{c,\,b} &= x_{c,\,a} \pm \sqrt{v} \\ z_{c,\,b} &= z_{c,\,a} \pm \sqrt{w}, \end{aligned} \tag{8.201}$$

and converted to fractional coordinates. The $\pm$ indicates the original signs of x_{ab}, etc. Although this gains us nothing in the absence of the constraint, application of the constraint relationships reduces the number of independent parameters by one (the number of independent parameters will always be reduced by the number of constraints):

$$\begin{aligned} u &= d_{ab}^2 - (y_{c,\,b} - y_{c,\,a})^2 - (z_{c,\,b} - z_{c,\,a})^2 \\ v &= d_{ab}^2 - (x_{c,\,b} - x_{c,\,a})^2 - (z_{c,\,b} - z_{c,\,a})^2 \end{aligned} \tag{8.202}$$

The third parameter, w, is uniquely determined from u and v:

$$w = d_{ab}^2 - u - v. \tag{8.203}$$

The five parameters, $x_{f,\,a}$, $y_{f,\,a}$, $z_{f,\,a}$, $u = x_{ab}^2$ and $v = y_{ab}^2$ are included in the normal equation matrix and after the solution the sixth parameter, $w = z_{ab}^2$, is calculated from Eqn. 8.203. Following conversion of u, v, and w to $x_{f,\,b}$, $y_{f,\,b}$ and $z_{f,\,b}$, new intensities are calculated and another cycle of refinement is initiated. The refinement will then converge to the optimal least-squares minimum while maintaining the distance d_{ab} between the two atoms. A major advantage of this approach is that it actually *reduces* the number of normal equations.

As an example of typical constraints incorporated into the refinement process we return to the solution and refinement of the bikitaite structure discussed in detail in the previous chapter. In the original paper, the authors noted that the aluminum atom exhibited a decidedly smaller displacement ellipsoid in comparison with the silicon atom ellipsoids, when refined in any one of the three sites that it could conceivably occupy, as illustrated in Figs. 8.6(a)–(c). Note that the displacement ellipsoids are at the 99% probability level for clarity. The silicon and aluminum atoms differ by only a single electron, and should exhibit similar displacement ellipsoids. The authors noted that the displacement ellipsoid for the aluminum atom in site 1 appeared smaller than the silicon ellipsoid in site 2, while the silicon atom in site 3 displayed a larger ellipsoid than the one in site 2. A smaller-than-expected ellipsoid is indicative of the existence of unaccounted-for electron density at a specific site, and the authors concluded that the aluminum and silicon atoms in site 1 and site 3 were disordered. They subsequently assigned a "hybrid atom" to each site, using the average value of the scattering factors for aluminum and

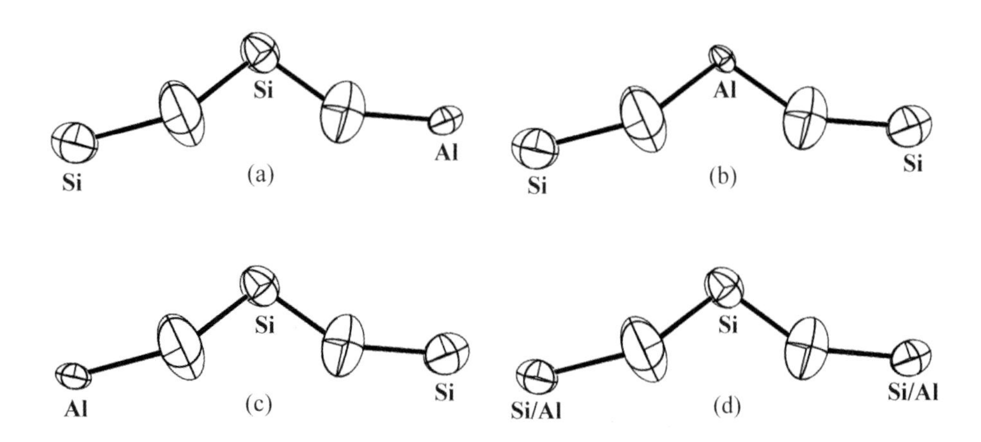

Figure 8.6 Refined displacement ellipsoids for the heavier atom portion of the bikitaite structure with various site occupancies: (a) Si-O-Si-O-Al, (b) Si-O-Al-O-Si, (c) Al-O-Si-O-Si, and (d) Al/Si-O-Si-Al/Si. Ellipsoids are shown at the 99% probability level.

silicon atoms — effectively placing half a silicon atom and half an aluminum atom in each site. At the time of the refinement, constraints were not a routine part of the algorithms in the software of the day. With modern software, the assignment made by the authors can be tested by refining the structure with constrained site occupancy factors, by placing an aluminum and silicon atom at each location such that $s(\mathrm{Al}_1) = s(\mathrm{Si}_3)$, and $s(\mathrm{Al}_3) = 1 - s(\mathrm{Al}_1) = s(\mathrm{Si}_1) = 1 - s(\mathrm{Si}_3)$. In addition, the atomic coordinates and anisotropic temperature factors of the atoms at site 1 are constrained to be equal, as are those for site 3. The resulting site occupancy factors for Al refine to 0.491(3) for site 1 and 0.509(3) for site 3, and 0.509(3) and 0.491(3) for silicon in sites 1 and 3, respectively. The numbers in parentheses indicate standard deviations of 0.003 for each factor. The displacement ellipsoids are plotted in Fig. 8.6 (d), indicating a more physically reasonable structure, and one essentially identical to that reported by the authors.

Rigid Group Refinement

Under ordinary circumstances, the refined parameters in small-molecule structures are significantly over-determined, with many more observations (reflections) than parameters, resulting in an accurate structural model. There are, however, instances when the number of observations is insufficient to provide accurate parameter values. In such cases it is often possible to employ known structural information to reduce the number of refined parameters.* Many molecules have atoms in known conformations, e.g., carbon atoms in a phenyl ring (a hexagon) or in a cyclopentadienide ligand (a pentagon), oxygen atoms in a sulfate ion (a tetrahedron), etc. In these cases it is possible to constrain a group of atoms to adopt a known conformation and refine the atoms collectively as a *rigid group*.

*Macromolecular structures are routinely under-determined, and it is generally *necessary* to incorporate known structural information into the refinement.

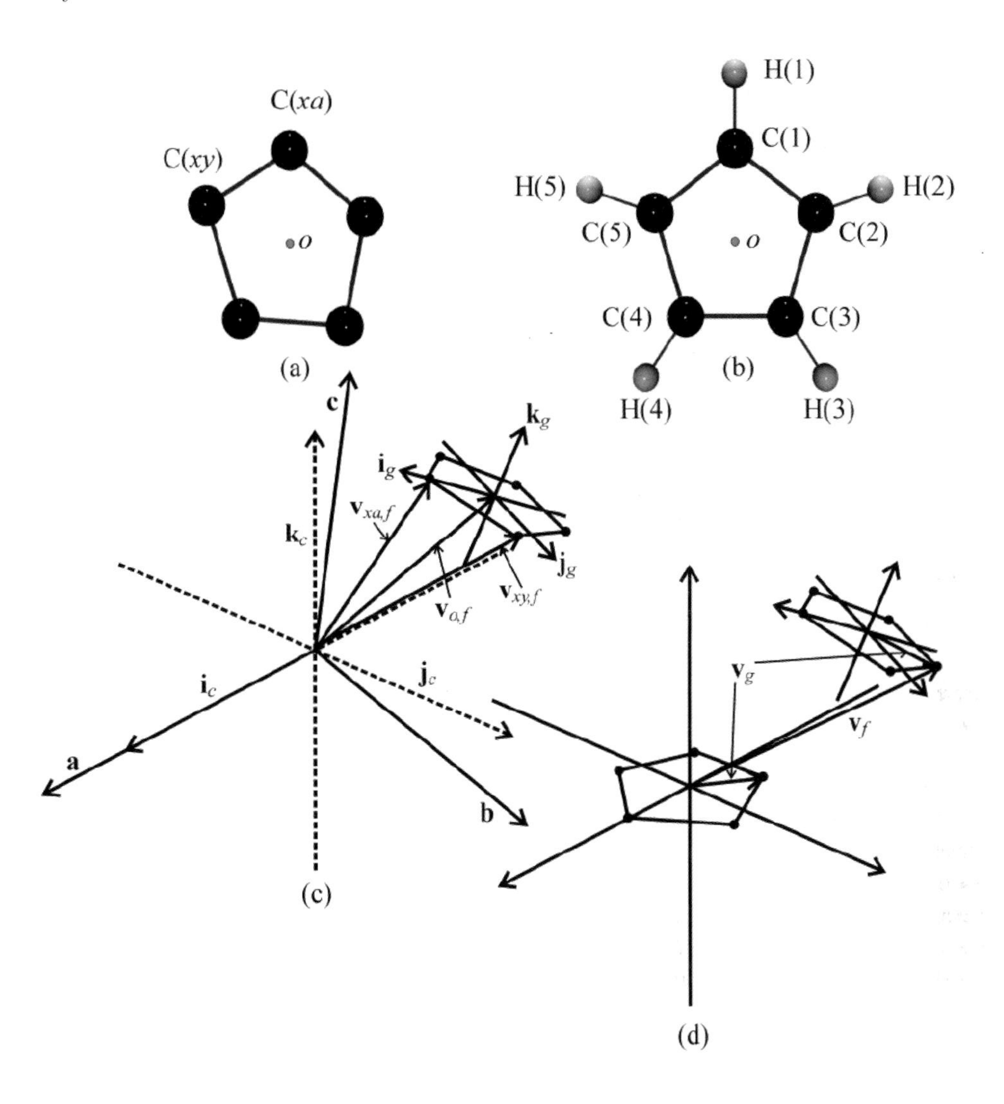

Figure 8.7 (a) Cyclopentadienide carbon atoms from a structural solution. (b) Idealized cyclopentadienide rigid group. (c) Rigid group with local coordinate system origin at $\mathbf{v}_{o,f}$ in the unit cell. The standard Cartesian coordinate system is illustrated with dashed axes. (d) Local coordinate system of the rigid group rotated and translated onto the standard coordinate axes.

Defining the group to be refined begins with the construction of a local Cartesian coordinate system based on the positions of atoms that define the group. A convenient way to accomplish this is to select three non-collinear locations in the unit cell in order to define a plane that determines the position and orientation of the group. Referring to Fig. 8.7(a) as an example, the five carbon atoms in a group known to be a cyclopentadienide ion are shown as they are initially found in a difference electron density Fourier map. Fig. 8.7(b) is the idealized structure of the ion, which will replace the crude structure during refinement.* The initial origin of the group is assigned as the average of the fractional coordinates of the five carbon atoms in the group. The local coordinate system is defined by selecting a vector from the origin through C(xa) as the x axis and a second vector from the origin through C(xy) to delineate the xy plane. The fractional coordinates of these positions are $\mathbf{v}_{o,f}$, $\mathbf{v}_{xa,f}$, and $\mathbf{v}_{xy,f}$, respectively, illustrated in Fig. 8.7(c). The coordinate system is created in the following manner:

1. $\mathbf{v}_{o,f}$, $\mathbf{v}_{xa,f}$ and $\mathbf{v}_{xy,f}$ are converted to Cartesian coordinates, $\mathbf{v}'_o$, $\mathbf{v}'_{xa}$, and $\mathbf{v}'_{xy}$, in the standard system: $\mathbf{B}\mathbf{v}_f = \mathbf{v}'_c$. Any other coordinates to be retained as part of the group are transformed as well (e.g., amino acid coordinates in proteins).

2. The coordinates defined by these vectors are translated to the origin of the standard coordinate system, setting $\mathbf{v}_o = \mathbf{v}'_o - \mathbf{v}'_o = \mathbf{0}$, $\mathbf{v}_{xa} = \mathbf{v}'_{xa} - \mathbf{v}'_o$, $\mathbf{v}_{xy} = \mathbf{v}'_{xy} - \mathbf{v}'_o$; in general, $\mathbf{v}_g = \mathbf{v}'_c - \mathbf{v}'_o$.

3. A unit vector in the local x direction is determined along $\mathbf{v}_{xa}$: $\mathbf{i}_g = \mathbf{v}_{xa}/v_{xa}$.

4. A unit vector in the local z direction, perpendicular to the local xy plane, is now created from the vector product of $\mathbf{i}_g$ and$\mathbf{v}_{xy}$: $\mathbf{v}_z = \mathbf{i}_g \times \mathbf{v}_{xy}$; $\mathbf{k}_g = \mathbf{v}_z/v_z$.

5. A unit vector in the local y direction is created from the vector product of $\mathbf{k}_g$, and $\mathbf{i}_g$: $\mathbf{j}_g = \mathbf{k}_g \times \mathbf{i}_g$.

In order to rotate and translate the group it must be referenced to the standard Cartesian system. This is conveniently accomplished by determining the Eulerian angles (Fig. 6.21) that will rotate the group coordinate system (already translated to the origin) so that it is coincident with the standard coordinate system. All manipulations can then be performed in the standard system, after which the resulting locations can be returned to the original fractional coordinate system by reversing the sequence of operations. These two systems are brought into coincidence in three steps:

1. The $\mathbf{k}_g$ axis has coordinates (k_x, k_y, k_z) in the standard system. The projection of this axis onto the $\mathbf{i}_c\mathbf{j}_c$ plane has coordinates $(k_x, k_y, 0)$, and the angle necessary to rotate this axis about the $\mathbf{k}_c$ axis onto the $\mathbf{j}_c\mathbf{k}_c$ plane is

$$\varphi_1 = -\arctan\left(\frac{k_x}{k_y}\right), \tag{8.204}$$

and

$$\mathbf{R}_1\mathbf{v}_g = \begin{bmatrix} \cos\varphi_1 & \sin\varphi_1 & 0 \\ -\sin\varphi_1 & \cos\varphi_1 & 0 \\ 0 & 0 & 1 \end{bmatrix} \mathbf{v}_g = \mathbf{v}'_g. \tag{8.205}$$

*The initial location and orientation of the group often requires the locations of only a portion of its atoms. This illustrates another use of rigid group refinement. During structural solution, missing atoms are located by assigning electron density distributions or phases from the positions of atoms already determined. In the intermediate stages of refinement it is useful to refine these positions, and inclusion of atom locations from known structural information can improve the resulting Fourier map or phase assignment.

2. The rotated $\mathbf{k}_g$ axis now has coordinates $(0, k'_y, k'_z)$ in the standard system, and the angle necessary to rotate this axis about the $\mathbf{i}_c$ axis so that it is collinear with the $\mathbf{k}_c$ axis is

$$\varphi_2 = \arctan\left(\frac{k'_y}{k'_z}\right), \tag{8.206}$$

and

$$\mathbf{R_2 v}'_g = \mathbf{R_2 R_1 v}_g = \begin{bmatrix} 1 & 0 & 0 \\ 0 & \cos\varphi_2 & \sin\varphi_2 \\ 0 & -\sin\varphi_2 & \cos\varphi_2 \end{bmatrix} \mathbf{v}'_g = \mathbf{v}''_g. \tag{8.207}$$

3. The twice rotated $\mathbf{i}_g$ axis now lies in the $\mathbf{i}_c\mathbf{j}_c$ plane with coordinates $(i''_x, i''_y, 0)$ in the standard system, and the angle necessary to rotate this axis about the $\mathbf{k}_c$ axis so that it is collinear with the $\mathbf{i}_c$ axis is

$$\varphi_3 = -\arctan\left(\frac{i''_y}{i''_x}\right), \tag{8.208}$$

and

$$\mathbf{R_3 v}''_g = \mathbf{R_3 R_2 R_1 v}_g = \mathbf{R v}_g = \begin{bmatrix} \cos\varphi_3 & \sin\varphi_3 & 0 \\ -\sin\varphi_3 & \cos\varphi_3 & 0 \\ 0 & 0 & 1 \end{bmatrix} \mathbf{v}''_g = \mathbf{v}'''_g. \tag{8.209}$$

Fig. 8.7(d) illustrates the results of the translation and rotation of the local coordinate system of the cyclopentadienide group so that it coincides with the standard Cartesian coordinate system. Rigid group coordinates are now created in this transformed local coordinate system for all of the atoms in Fig. 8.7(b): $\{\mathbf{v}_r\} \equiv \mathbf{v}_{\mathrm{C}(1)}, \mathbf{v}_{\mathrm{H}(1)}, \ldots, \mathbf{v}_{\mathrm{C}(5)}, \mathbf{v}_{\mathrm{H}(5)}$. These vectors are stored since they will be used for each cycle of the refinement.

The rigid group coordinates are now transformed back to coordinates in the unit cell by reversing the rotation and translation of the local coordinate system (since $\mathbf{R}$ is an orthogonal matrix $\mathbf{R}^{-1} = \mathbf{R}^T$):

$$\mathbf{v}^r_c = \mathbf{R}^T\mathbf{v}_r + \mathbf{v}'_o. \tag{8.210}$$

Fractional coordinates for the rigid group atoms are now generated in preparation for refinement:

$$\mathbf{v}^r_f = \mathbf{B}^{-1}\mathbf{v}^c_r. \tag{8.211}$$

The group can be rotated in the unit cell by changing the angles in $\mathbf{R}$ and applying Eqns. 8.210 and 8.211. It can be translated by changing the group origin location, $\mathbf{v}_{o,f}$, generating $\mathbf{v}'_o = \mathbf{B}\mathbf{v}_{o,f}$, and applying the equations. The group atom locations are incorporated into Eqn. 8.106 to calculate $I_{c,\mathbf{h}}$ and $S(\mathbf{p})$, but the three Eulerian angles and the fractional coordinates of the group origin are the parameters refined in the least-squares process.

To incorporate these parameters into the normal equations it is necessary to determine the appropriate derivatives. A change in the origin location, $\Delta\mathbf{v}_{o,f} =$

$[\Delta x_o\ \Delta y_o\ \Delta z_o]$ results in an identical shift for all of the atoms in the group, i.e. $\Delta x_i = \Delta x_o$, etc. The calculated intensity changes are due to the contribution from each shifted atom, and

$$\frac{\partial I_{c,\mathbf{h}}}{\partial x_o} = \sum_{i=1}^{N} \frac{\partial I_{c,\mathbf{h}}}{\partial x_i}\frac{\partial x_i}{\partial x_o} = \sum_{i=1}^{N} \frac{\partial I_{c,\mathbf{h}}}{\partial x_i}, \quad \text{etc,} \tag{8.212}$$

where the derivatives for each of N atoms in the group are determined from Eqns. 8.123 and 8.124. A change in an Eulerian angle will result in rotations of all of the vectors in the group, altering all $3N$ fractional coordinates differently:

$$\frac{\partial I_{c,\mathbf{h}}}{\partial \varphi_j} = \sum_{i=1}^{N} \frac{\partial I_{c,\mathbf{h}}}{\partial x_i}\frac{\partial x_i}{\partial \varphi_j} + \sum_{i=1}^{N} \frac{\partial I_{c,\mathbf{h}}}{\partial y_i}\frac{\partial y_i}{\partial \varphi_j} + \sum_{i=1}^{N} \frac{\partial I_{c,\mathbf{h}}}{\partial z_i}\frac{\partial z_i}{\partial \varphi_j}, \quad j = 1, 2, 3. \tag{8.213}$$

The derivatives of the angles with respect to each coordinate can be determined analytically by expanding and differentiating Eqns. 8.210 and 8.211, but numerical derivatives are a clear choice here, for which a small change, $\delta\varphi_j$, is made and $\delta I_{c,\mathbf{h}}$ is evaluated, such that $\partial I_{c,\mathbf{h}}/\partial\varphi_j \simeq \delta I_{c,\mathbf{h}}/\delta\varphi_j$.

The least-squares solution from the normal equations creates new values of the origin coordinates and the rotational angles. After transformation of the new origin location to Cartesian coordinates, the values are substituted into Eqns. 8.210 and 8.211 for the next refinement cycle.

Hydrogen Atom Refinement

The refinement of hydrogen atom locations presents a special problem. Having only a single electron, hydrogen atoms contribute only a fraction of the scattering power of most of the atoms in the unit cell. Because of this, their locations are not as well resolved as the locations of the heavier atoms in a Fourier map based on intensity data. This is particularly true for structures that contain heavier metals — in such cases the hydrogen atoms provide only a very small portion of the scattering power, and are often impossible to locate during structural solution.

A second problem with modeling hydrogen atom locations involves the spherical atom approximation – which assumes that the charge density about each atom is symmetrically disposed. For most atoms this is an adequate approximation, since the internuclear electron density in the chemical bonds does not significantly alter the spherical distribution. However, for bonds to the hydrogen atom this is not the case. The effective nuclear charge of most atoms bonded to hydrogen tends to overcome the influence of the single proton in the hydrogen nucleus, polarizing the electron density on the hydrogen atom into the internuclear region. An approximate hydrogen charge sphere in the model will be localized away from the hydrogen nucleus in the direction of the bond — the apparent internuclear distance is less than the actual internuclear distance.

Even in instances in which hydrogen atoms are located in the structural solution, their ideal positions are very often more accurate than those determined from the Fourier map. It is therefore reasonable to incorporate the positions of idealized hydrogen atoms whenever it is prudent to do so. In a situation similar to that of the rigid group refinement described above, it is possible to consider a case in which the position of one atom is refined, and a related atom is assumed to move in concert

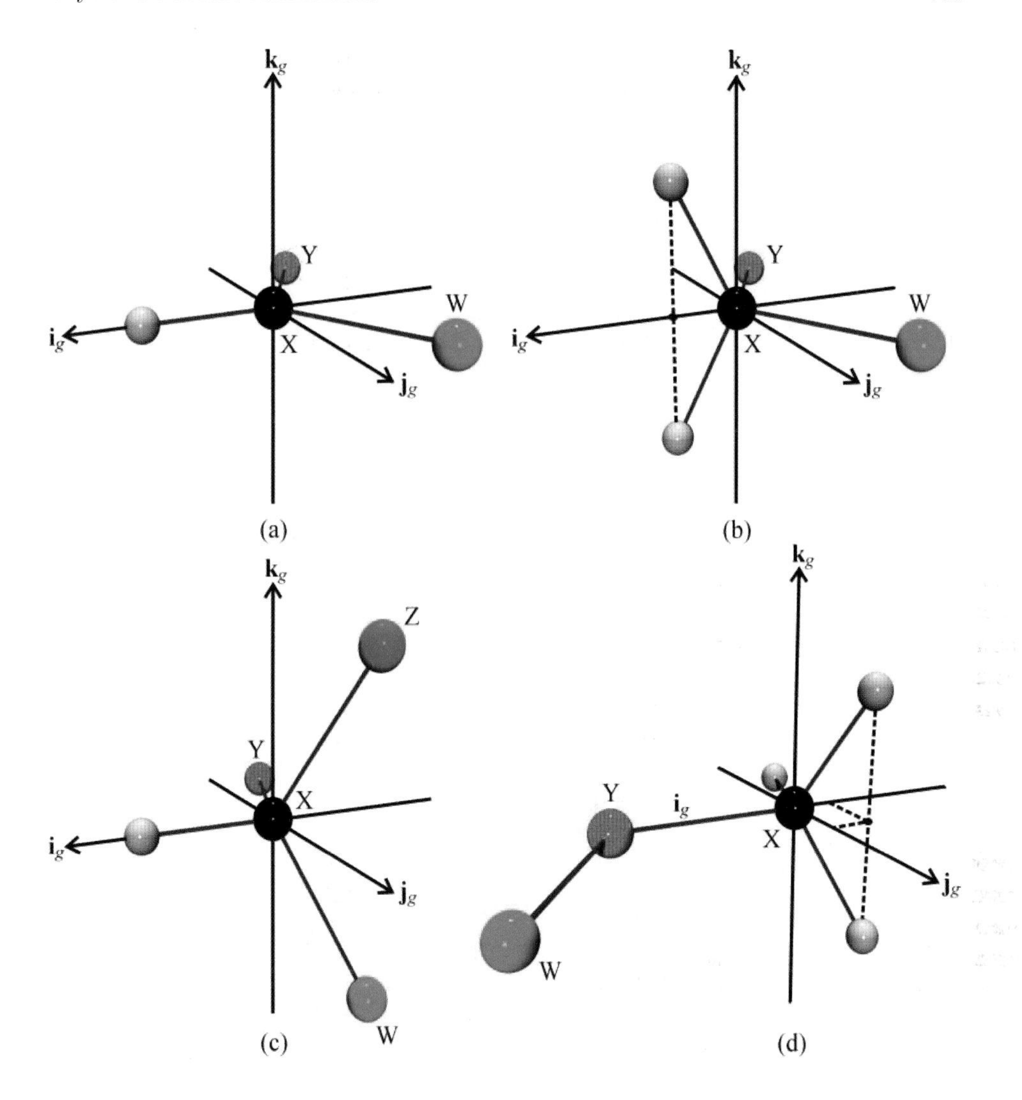

Figure 8.8 Hydrogen atoms in typical arrangements: (a) Planar WXHY-group. (b) WXH_2Y group. (c) WXHYZ group. (d) Terminal $WYXH_n$ group ($n = 1$, 2, or 3).

with the refined atom. In this case the interatomic vector between the two atoms translates with the refined atom – the second atom appears to be "riding" on the first atom. This type of constraint is physically most reasonable when the "riding" atom has substantially less mass than the refined atom and is tightly bound to it; the *riding model* is used routinely for the refinement of hydrogen atom positions.

Typical hydrogen atom arrangements are shown in Fig. 8.8. Fig. 8.8(a) depicts a planar arrangement often observed when the atom to which the hydrogen atom is bonded is considered to be "sp^2" hybridized. Idealized coordinates for this atom are determined by considering it to lie a distance d_H (the apparent X–H distance described above) along a vector bisecting the WXY angle. The atom is placed in this location in a manner similar to that for defining the atoms in a rigid group. The fractional coordinates of the W, X, and Y atoms are converted to Cartesian coordinates: $\mathbf{B}\mathbf{v}_{X,f} = \mathbf{v}'_{X,c}$, etc.. The X atom is chosen as the origin, and the three atoms are translated to the standard coordinate system origin : $\mathbf{v}_{W,c} = \mathbf{v}'_{W,c} - \mathbf{v}'_{X,c}$, etc. The average of the unit vectors,

$$\mathbf{v}_x = \frac{1}{2}\left(\frac{\mathbf{v}_{W,c}}{v_{W,c}} + \frac{\mathbf{v}_{Y,c}}{v_{Y,c}}\right), \tag{8.214}$$

creates a vector in the WXY plane bisecting the WXY angle. The x axis of the local coordinate system is now defined as a unit vector in the negative $\mathbf{v}_x$ direction: $\mathbf{i}_g = -\mathbf{v}_x/v_x$. The $\mathbf{v}_{W,c}$ vector defines the xy plane in the local system, and $\mathbf{k}_g$ and $\mathbf{j}_g$ are created as they were for the rigid group. The hydrogen atom is now assigned coordinates $(d_H, 0, 0)$ in this coordinate system. To transform these coordinates back into fractional coordinates we note that transformation matrix to convert the coordinates of the local coordinate axes in the standard coordinate system ($\mathbf{i}_g = [i_x\ i_y\ i_z]$, etc.) into the local system is generated from

$$\begin{bmatrix} c_{11} & c_{12} & c_{13} \\ c_{21} & c_{22} & c_{23} \\ c_{31} & c_{32} & c_{33} \end{bmatrix} \begin{bmatrix} i_x & j_x & k_x \\ i_y & j_y & k_y \\ i_z & j_z & k_z \end{bmatrix} = \begin{bmatrix} 1 & 0 & 0 \\ 0 & 1 & 0 \\ 0 & 0 & 1 \end{bmatrix}, \tag{8.215}$$

and in general, $\mathbf{C}\mathbf{v}_c = \mathbf{v}_g$, where

$$\begin{bmatrix} c_{11} & c_{12} & c_{13} \\ c_{21} & c_{22} & c_{23} \\ c_{31} & c_{32} & c_{33} \end{bmatrix} = \begin{bmatrix} i_x & j_x & k_x \\ i_y & j_y & k_y \\ i_z & j_z & k_z \end{bmatrix}^{-1}. \tag{8.216}$$

The hydrogen atom location is therefore converted into Cartesian coordinates in the standard system by: $\mathbf{v}_{H,c} = \mathbf{C}^{-1}[d_H\ 0\ 0] = [\mathbf{i}_g\ \mathbf{j}_g\ \mathbf{i}_g][d_H\ 0\ 0]$. It is then translated to its location in the standard coordinate system: $\mathbf{v}'_{H,c} = \mathbf{v}_{H,c} + \mathbf{v}'_{X,c}$ and converted to fractional coordinates: $\mathbf{v}_{H,f} = \mathbf{B}^{-1}\mathbf{v}'_{H,c}$.

Fig. 8.8(b) illustrates an approximately tetrahedral arrangement observed when two hydrogen atoms are bonded to an atom that is considered to be "sp^3" hybridized. The local coordinate system and transformations are identical to those described above, except that two hydrogen atoms are generated with local coordinates $(d_H \cos\eta, d_H \sin\eta, 0)$ and $(d_H \cos\eta, -d_H \sin\eta, 0)$, where η is either 54.735° (tetrahedral), or an angle generated such that the average of the HXH and WXW angles is 109.471°. The second choice is designed to compensate for occasions when the WXW angle is substantially less than the tetrahedral angle. When atom X is "sp^3" hybridized and bonded to three heavy atoms, as illustrated in Fig. 8.8(c), the single hydrogen atom is positioned in a local coordinate system in which the $\mathbf{i}_g$ axis is created from the average of the unit vectors to the three bonded atoms, and the xy plane is created from any one of the bonded atom locations.

In each of the three scenarios just described the hydrogen atom can be refined with the riding model. The refinement is identical to that for a rigid group for which there is no rotation. All atoms "move" in the refinement with an identical displacement vector. The heavy atom, X, and the hydrogen atoms attached do it are incorporated into Eqn. 8.106 to calculate $I_{c,\mathbf{h}}$ and $S(\mathbf{p})$. The displacements of a single positional vector are refined in least squares, $\Delta\mathbf{v}_{XH}$, such that $\mathbf{v}_X = \mathbf{v}_{X,o} + \Delta\mathbf{v}_{XH}$, and $\mathbf{v}_H = \mathbf{v}_{H,o} + \Delta\mathbf{v}_{XH}$ for all m hydrogen atoms bonded to X. The derivatives of the three parameters of the displacement vector are determined as they are from the contributors to the intensity of the rigid group atoms (noting that $\Delta x_X = \Delta x_H = \Delta x_{XH}$, etc.):

$$\frac{\partial I_{c,\mathbf{h}}}{\partial x_{XH}} = \frac{\partial I_{c,\mathbf{h}}}{\partial x_X}\frac{\partial x_X}{\partial x_{XH}} + \sum_{i=1}^{m}\frac{\partial I_{c,\mathbf{h}}}{\partial x_i}\frac{\partial x_i}{\partial x_{XH}} = \frac{\partial I_{c,\mathbf{h}}}{\partial x_X} + \sum_{i=1}^{m}\frac{\partial I_{c,\mathbf{h}}}{\partial x_i}, \text{ etc.} \tag{8.217}$$

Given that the positional parameters of the hydrogen atoms in a structure are difficult to determine accurately, it should not be surprising that their displacement parameters are even less accurately determined, and they are rarely refined. In the riding model a reasonable approximation is to assign an isotropic displacement parameter that is about 20% larger than the equivalent isotropic displacement factor of the atom to which the hydrogen atom is bonded (e.g., $U_{iso}(\mathrm{H}){=}1.2U_{eq}(\mathrm{X})$).

The riding model maintains the X–H distances and orientation of the XH_m group, but the group can translate in the unit cell, altering the other angles to hydrogen (e.g., YXH). A useful alternative to the riding model is found in the refinement software package, *CRYSTALS*.* Rather than refining the hydrogen atom locations, they are simply fixed after their locations are computed and the intensities and residuals have been calculated. The idealized positions are then recalculated from the heavy atom locations after a number of refinement cycles and the process is repeated until the positions no longer change. *SHELX* combines the riding model with this approach, computing idealized positions after each cycle of "riding model" refinement.

Hydrogen atoms bonded to a "terminal" atom present special difficulties in assigning idealize positions. As illustrated in Fig. 8.8(d) a group containing hydrogen atoms bonded to a terminal X atom is more-or-less free to rotate about the X–Y bond. In many cases this can lead to rotationally disordered hydrogen atoms. In these cases the group can often be modeled with a number of conformations, each with the appropriate number of hydrogen atoms with varying site occupancy factors. If X is a carbon atom, then the group is a methyl group with hydrogen atoms in a tetrahedral arrangement. The rotation of the group is hindered to some degree by repulsive interactions with the atoms bonded to the Y atom, and the methyl group will often tend to adopt a conformation that has one of the hydrogen atoms in the same plane as the atom with the shortest W–Y bond (the closest atom to the hydrogen atoms), but rotated 180° away from it. The other two hydrogen atoms lie equidistant above and below the plane, as illustrated. To assign these positions a local coordinate system is generated by assigning the X atom as the origin, a vector from the origin to the Y atom as the x axis, and the xy plane is determined from a vector from the origin to the W atom. In this local system, the vector to the hydrogen atom in the XYW plane, $\mathbf{v}_H$, has coordinates $(d_H\cos\eta, -d_H\sin\eta, 0)$, $\eta = 109.471°$. Substituting $\varphi = 120°$ and $\varphi = 240°$ into Eqn. 1.54 and applying the resulting rotation matrices to $\mathbf{v}_H$ will generate the coordinates of the other two hydrogen atoms by rotation about the x axis in the local system. Following this initial assignment of hydrogen positions, it may be advisable to refine the methyl

*There are many variations on the refinement themes discussed in this chapter. The software package, *CRYSTALS*,[173] authored by David J. Watkin and associates at Oxford University was designed specifically for structural refinement, and contains many of these variations.

group as a rigid group. In this case rotation can be restricted to the torsional angle about the X–Y bond. If the terminal group is an amino group ($-NH_2$) or a hydroxyl group ($-OH$), then the determination of a sterically-favored orientation of the hydrogen atoms(s) is less unequivocal, and refinement of the X–H_m moiety as a rigid group may be more appropriate. One alternative to this is employed in *SHELX*, in which a Fourier map is computed and the orientation of the hydrogen atoms are located by determining the torsional angle about the bond that maximizes the sum of the electron densities computed at the locations for each of the hydrogen atoms.

8.2.5 Restrained Refinement

In lieu of constraining specific parameters in refinement, it may be more reasonable to allow them to vary within limits, such that their refinement is *restrained*, rather than constrained. As an example, let us return to the internuclear distance constraint requiring that

$$\begin{aligned} d_{ab}^2 &= \mathbf{d}_{ab}^T\mathbf{G}\mathbf{d}_{ab} = q'(\mathbf{p}_c) \quad \text{and} \\ q(\mathbf{p}_c) &= d_{ab}^2 - q'(\mathbf{p}_c) = 0 \end{aligned} \tag{8.218}$$

for $\mathbf{d}_{ab} = [(x_{f,b} - x_{f,a})\ (y_{f,b} - y_{f,a})\ (z_{f,b} - zx_{f,a})]$. Instead of requiring this rigorously, we wish to impose the "softer" constraint,

$$q(\mathbf{p}_r) = d_{ab}^2 - q'(\mathbf{p}_r) = \delta_{ab}, \tag{8.219}$$

allowing the internuclear distance, d_{ab}, to vary within the limits imposed by δ_{ab} — which we intend to keep small by adjusting the parameters appropriately. In discussing restraints in linear least squares, we viewed such expressions as observational equations in which parameter values were determined that would include these "constraint residuals" in the weighted residual vector (Eqn. 8.99), so that the parameters would be those that minimized the sum of the squares of all of the residuals. For structural refinement, if $\delta_k = q_k - q_k'(\mathbf{p}_r)$ is the soft-constraint residual for the kth restraining equation of o restraints, the least squares fit minimizes

$$R(\mathbf{p}_r) = \sum_{\mathbf{h}} r_{\mathbf{h}}^T r_{\mathbf{h}} = \sum_{\mathbf{h}} w_{\mathbf{h}}(I_{o,\mathbf{h}} - I_{c,\mathbf{h}}(\mathbf{p}_r))^2 + \sum_{k=1}^{o} w_k(q_k - q_k'(\mathbf{p}_r))^2, \tag{8.220}$$

where q_k is the target value (e.g., d_{ab}^2) for $q_k'(\mathbf{p}_r)$.

The second sum in Eqn. 8.220 may not be linear, and in order to generate the normal equations, both $I_{c,\mathbf{h}}(\mathbf{p}_r))$ and $q_k'(\mathbf{p}_r)$ must be "linearized" by expansion as Taylor series about the initial parameter values. The first expansion is given by Eqn. 8.108 and the second by

$$q_k'(\mathbf{p}_r) \simeq q_k'(\mathbf{p}_r)_{\mathbf{p}_o} + \sum_{i=1}^{n} \left(\frac{\partial q_k'(\mathbf{p}_r)}{\partial p_i}\right)_{\mathbf{p}_o} \Delta p_i. \tag{8.221}$$

Setting $\Delta I_{\mathbf{h}} = I_{o,\mathbf{h}} - (I_{c,\mathbf{h}}(\mathbf{p}_r))_{\mathbf{p}_o}$ and $\Delta q_k = q_k - q_k'(\mathbf{p}_r)_{\mathbf{p}_o}$, Eqn. 8.220 becomes

$$\begin{aligned} R(\mathbf{p}_r) &= \sum_{\mathbf{h}} w_{\mathbf{h}} \left(\Delta I_{\mathbf{h}} - \sum_{i=1}^{n} \left(\frac{\partial(I_{c,\mathbf{h}}(\mathbf{p}_r))}{\partial p_i}\right)_{\mathbf{p}_o} \Delta p_i\right)^2 \\ &\quad + \sum_{k=1}^{o} w_k \left(\Delta q_k - \sum_{i=1}^{n} \left(\frac{\partial(q_k'(\mathbf{p}_r))}{\partial p_i}\right)_{\mathbf{p}_o} \Delta p_i\right)^2, \end{aligned} \tag{8.222}$$

which is minimized by differentiating with respect to the n parameters and setting each of the derivatives to zero. This results in n simultaneous linear equations:

$$\frac{\partial R(\mathbf{p})}{\partial p_j} = \sum_{\mathbf{h}} w_{\mathbf{h}} \left(\Delta I_{\mathbf{h}} - \sum_{i=1}^{n} \left(\frac{\partial (I_{c,\mathbf{h}}(\mathbf{p}))}{\partial p_i} \right)_{\mathbf{p}_o} \Delta p_i \right) \left(\frac{\partial (I_{c,\mathbf{h}}(\mathbf{p}))}{\partial p_j} \right)_{\mathbf{p}_o}$$
$$+ \sum_{k=1}^{o} w_k \left(\Delta q_k - \sum_{i=1}^{n} \left(\frac{\partial (q'_k(\mathbf{p}_r))}{\partial p_i} \right)_{\mathbf{p}_o} \Delta p_i \right) \left(\frac{\partial (q'_k(\mathbf{p}_r))}{\partial p_j} \right)_{\mathbf{p}_o} = 0. \quad (8.223)$$

Rearrangement results in n normal equations — the jth normal equation is:

$$\sum_{i=1}^{n} \left(\sum_{\mathbf{h}} w_{\mathbf{h}} \left(\frac{\partial (I_{c,\mathbf{h}}(\mathbf{p}))}{\partial p_i} \frac{\partial (I_{c,\mathbf{h}}(\mathbf{p}))}{\partial p_j} \right)_{\mathbf{p}_o} + \sum_{k=1}^{o} w_k \left(\frac{\partial (q'_k(\mathbf{p}_r))}{\partial p_i} \frac{\partial (q'_k(\mathbf{p}_r))}{\partial p_j} \right)_{\mathbf{p}_o} \right) \Delta p_i$$
$$= \sum_{\mathbf{h}} w_{\mathbf{h}} \Delta I_{\mathbf{h}} \left(\frac{\partial (I_{c,\mathbf{h}}(\mathbf{p}))}{\partial p_j} \right)_{\mathbf{p}_o} + \sum_{k=1}^{o} w_k \Delta q_k \left(\frac{\partial (q'_k(\mathbf{p}_r))}{\partial p_j} \right)_{\mathbf{p}_o}. \quad (8.224)$$

As with the rigorous application of the equations as "constraining equations" using undetermined multipliers, the derivatives of the restraining equations with respect to the parameters must be evaluated. This is ordinarily a straightforward process since the equations are usually relatively simple functions of specific parameters. When they are not, numerical differentiation is an option. Note that the number of normal equations remains the same when the equations are treated as "restraining equations," which was not the case when they were treated as "constraining equations" using Lagrange multipliers.

The normal equations take the form of Eqns. 8.114 and 8.115, providing the vector of parameter shifts, $\mathbf{\Delta p}$, and the covariance matrix, $\mathbf{T}^{-1}$. It is important to note here that the covariances between variables that are restrained with respect to one another will be large, and *must* be included in the analysis of parameter errors (Eqn. 8.164).

The choice of weights for the restraint conditions is not as straightforward as it is for intensity data, and since they are included in the same equations they must be put on approximately the same scale. The *goodness-of-fit* index (Eqn. 8.184) provided a test for the validity of the weights applied to the reflection residuals; for reasonable weights,

$$S = \frac{\sum_{\mathbf{h}} w_{\mathbf{h}} (I_o - I_c)^2}{m - n} \simeq 1. \quad (8.225)$$

We therefore might expect that a similar index for restraint residuals would be

$$S' = \frac{\sum_k w_k (q_k - q'_k(\mathbf{p}_r))^2}{o} \simeq S \simeq 1, \quad (8.226)$$

for o restraining equations. For a selected single restraint, since we are interested only in the scale, we approximate $(q_k - q'_k(\mathbf{p}_r))^2$ with its estimated expected value, σ_k^2, and assume that

$$\frac{\sum_{\mathbf{h}} w_{\mathbf{h}} (I_o - I_c)^2}{m - n} \simeq \frac{w_k \sigma_k^2}{1}. \quad (8.227)$$

The scaled weight for then kth restraint is then dependent on $1/\sigma_k^2$:

$$w_k \simeq \frac{K_r}{\sigma_k^2}, \quad K_r = \frac{\sum_{\mathbf{h}} w_{\mathbf{h}}(I_o - I_c)^2}{m - n}. \tag{8.228}$$

Eqn. 8.228 provides a suitable scale factor, but the assigned values of the estimated standard deviations of the restraint conditions are the values that determine the actual weights. The smaller the estimate of σ_k^2, the larger the weight, and the more that the restraint condition influences the restrained parameter values toward attainment of the target value. If σ_k^2 becomes very small (approaching zero) the weight becomes very large (approaching infinity) — in the limit the restraining equation becomes a constraining equation. In practice, optimal values for the restraint weights must be determined empirically.

8.2.6 The Refinement of Twinned Structures

Crystallographers are often confronted with multiple crystals that appear macroscopically as if they were a single crystal. For example, crystals can tend to grow as very thin sheets or needles, and these individual crystals will tend to stack together into something that looks very much like a single crystal, but is actually a polycrystalline aggregate. In other instances, very small crystals can pack together along similar faces, but with varying alignments, again creating a polycrystalline aggregate. There is generally no relationship between the orientation of one of the crystals in the aggregate and the others. If one of these aggregates is rotated in the X-ray beam a rotation image generally exhibits multiple diffraction spots for each reflection, often with low intensities, since each crystal in the aggregate generally diffracts weakly. Many times there are so many such spots that they coalesce into a diffraction ring, similar to that from a finely ground powder rotated in the beam (a *powder pattern*). Polycrystalline aggregates are generally easy to identify, and are ordinarily rejected as candidates for crystallographic analysis.

There is, however, a type of multiple crystal, known as a *twin*, that not only may look like a single crystal, but that diffracts like one as well. The formation of a twin, known as *twinning*, results from conditions in the dynamic process of crystal growth that causes a crystal to alter its growth pattern, creating two (or more) lattices that are geometrically identical, but with unit cells *systematically* rotated, translated, or reflected with respect to one another. Twinning tends to occur when the components of the twin, the *twin domains*, are thermodynamically equivalent and there is a small energy barrier required for the lattice to begin growing in a different manner. For example, if a monoclinic crystal has $\beta = 90° + \delta \simeq 90°$, it may be relatively easy for a mirror-image of the structure to begin growing with $\beta = 90° - \delta$. If this happens, the physical interactions of the molecules at the interface of these two lattices will be different (often weaker), and a *twin boundary* results.

In general, the unit cells of each twin domain, and therefore the reciprocal lattices, are related by a rotational and/or translational and/or reflection operation known as the *twin law*, e.g., for $I_{o,\mathbf{h_1}}$ and $I_{o,\mathbf{h_2}}$ from twin components 1 and 2, respectively,

$$\begin{bmatrix} z_{11} & z_{12} & z_{13} \\ z_{21} & z_{22} & z_{23} \\ z_{31} & z_{32} & z_{33} \end{bmatrix} \begin{bmatrix} h_1 \\ k_1 \\ l_1 \end{bmatrix} = \begin{bmatrix} h_2 \\ k_2 \\ l_2 \end{bmatrix}. \tag{8.229}$$

The **Z** matrix is known as the *twin law matrix.* In some instances the twin operation is a symmetry operation for the lattice (the crystal system), but not for the contents of the unit cells. This type of twin, which usually arises from a higher symmetry lattice (orthorhombic or higher), is known as a *merohedral* twin. One important type of merohedral twinning, known as *racemic twinning*, appears with lattices of all symmetries, and occurs when the twin operator belongs to the Laue group of the crystal system but not its point group. This can occur when crystallization occurs from a racemic mixture of the enantiomers of a chiral compound, with each twin component consisting of one of the enantiomorphs. The lattice constants of both enantiomorphs are identical, and if the interaction energy between the two enantiomorphic unit cells is not too different from the interaction energies of each enantiomorph with itself, then the lattice can contain an approximate 50/50 mixture of each unit cell, and the crystal can appear centrosymmetric. The lattice points of two twin domains from a lower symmetry lattice (triclinic or monoclinic) can also superimpose if they are not related by an operation of the crystal system point group, but collectively appear to exist (approximately) in a higher symmetry lattice. These twins are known as *pseudo-merohedral twins*— the monoclinic twin with $\beta \approx 90°$ is an example — the combined lattice appears orthorhombic.

For both merohedral and pseudo-merohedral twins, the direct lattice points from each component of the twin superimpose upon one another (at least approximately), and the reciprocal lattices are therefore coincident. The indices of the overlapping reciprocal lattice points, however, are not the same (they are related by the twin law), and the intensity for each reciprocal lattice point as it passes through the sphere of reflection is the sum of the intensities from different reflections, one arising from each component of the twin. For a twinned crystal with two components,

$$I_{o,\mathbf{h_1},\mathbf{h_2}} = t_1 I_{o,\mathbf{h_1}} + t_2 I_{o,\mathbf{h_2}}, \tag{8.230}$$

where t_1 and t_2 are the relative contributions from each of the components of the twin, which, in turn, depend on the relative volumes of each component.

The solution and refinement of a twinned structure for merohedral and pseudo-merohedral twins depends on the determination of the twin law and the ability to deconvolute overlapping intensities. In some cases the twin law must be known in order to solve the structure. In these cases it is often necessary to estimate t_1 and t_2 in Eqn. 8.230, allowing for the calculation of the intensities for each of the twin components. The twin law is then determined by trial and error – by testing all possible twin matrices with the appropriate symmetry. For each trial twin law, the intensities from each of the components is used to determine orientation matrices, unit cells, indexed reflections, and the correct space group, from which a solution for one of the twins can be obtained. In other cases, the structure will solve, at least partially, in the unit cell determined from the collective reflections from both twin components. In either situation the twin law and correct space group must be determined before the structure can be refined. Once these are known, the intensity contributions from each crystal can be approximated and combined to create a calculated intensity for least-squares refinement:

$$I_{c,\mathbf{h_1},\mathbf{h_2}}(\mathbf{p}) = p_1(t_1 I_{c,\mathbf{h_1}} + t_2 I_{c,\mathbf{h_2}}), \tag{8.231}$$

where p_1 is the overall scale factor. The refinement process includes $t1$ and $t2$ as parameters, subject to the constraint that their sum must be unity.

A third type of twin has the unit cells related by an operation that is not a symmetry operation for the unit cell or the lattice. In this case, the overlap of reciprocal lattice points is fortuitous, and depends on the relative proximity of the reciprocal lattice points of each twin (with reciprocal lattice points with lower indices closer together and therefore more likely to overlap). This type of twin is known as a *non-merohedral* twin. Once the twin law is determined (which itself may be problematic), many of the reflections from the components will not overlap, and an orientation matrix for each component will allow for an attempted solution and refinement of the structure. The fundamental problem that arises here is that some, perhaps many, of the reflections overlap, and handling such reflections is often difficult. One approach is to eliminate those reflections that exhibit very large disagreements between the calculated and observed intensities, provided, of course, that the structure can be solved in the first place.

8.2.7 The Refinement of Chiral Structures

Because the intensity-weighted reciprocal lattice is very nearly centrosymmetric, structural solutions do not ordinarily distinguish between the two enantiomorphs of a non-centrosymmetric structure. In such cases an arbitrary selection of one of the enantiomorphs usually proves sufficient to solve the structure.

In Chapter 5 we determined that dispersion phenomena cause phase shifts in the reflections that are not due to atom locations, and that these phase shifts result in changes in the observed intensities. These effects are ordinarily small, except for cases in which the X-ray frequency is close to that of an electronic transition in one of the atoms in the structure. The phase shift for each atom type depends only on the diffraction angle, and can be incorporated into the atomic scattering factor by expressing it as a vector: $\mathbf{f} = f' + i\,f'' = fe^{i\phi}$. The result is a breakdown of the Friedel Law, rendering I_{hkl} and $I_{\bar{h}\bar{k}\bar{l}}$, unequal. In Sec. 5.1.2 the intensity weighted reciprocal lattice was shown to be centrosymmetric only for centrosymmetric structures. For non-centrosymmetric structures, the calculated structure factors for one of the enantiomorphs (labeled with a "+") are given by

$$\mathbf{F}^+_{hkl} = \sum_{j=1}^{n} f_j e^{i(\phi_j - 2\pi(hx_j+ky_j+lz_j))}; \tag{8.232}$$

$$\mathbf{F}^+_{\bar{h}\bar{k}\bar{l}} = \sum_{j=1}^{n} f_j e^{i(\phi_j + 2\pi(hx_j+ky_j+lz_j))}. \tag{8.233}$$

Thus a model of the structure with one enantiomorph will have different values for $I^+_{c,\mathbf{h}}$ and $I^+_{c,\bar{\mathbf{h}}}$. The structure factors for the other enantiomorph (labeled with a "−")

are calculated by inverting the structure — changing the signs of all of the atom coordinates:

$$\mathbf{F}^{-}_{hkl} = \sum_{j=1}^{n} f_j e^{i(\phi_j + 2\pi(hx_j + ky_j + lz_j))}; \tag{8.234}$$

$$\mathbf{F}^{-}_{\bar{h}\bar{k}\bar{l}} = \sum_{j=1}^{n} f_j e^{i(\phi_j - 2\pi(hx_j + ky_j + lz_j))}. \tag{8.235}$$

This results in a different set of calculated intensities: $I^{-}_{c,\mathbf{h}} = I^{+}_{c,\bar{\mathbf{h}}}$ and $I^{-}_{c,\bar{\mathbf{h}}} = I^{+}_{c,\mathbf{h}}$ Only one of these two sets should map with the observed intensities, $I_{o,\mathbf{h}}$ and $I_{o,\bar{\mathbf{h}}}$, and if the dispersion effects are large enough it should be possible to refine the structure with both enantiomorphs and select the structure with the smaller value of the reliability index, R_I. An equivalent and convenient way to generate both sets of intensities is to utilize the same indices and atom positions, and use $\mathbf{f}^{+} = f' + i\,f'' = fe^{i\phi}$ and $\mathbf{f}^{-} = f' - i\,f'' = fe^{-i\phi}$ to calculate $I^{+}_{c,\mathbf{h}}$ and $I^{-}_{c,\mathbf{h}}$ respectively.

When the dispersion effects are small, refinement of the incorrect enantiomorph will still tend to minimize the residual. The only way that the refinement can compensate for the phase errors introduced by using ϕ with the wrong signs is for it to alter the positions of the atoms in the model. It is unlikely that the shifts will be able to compensate completely for the error, but the value of R_I for the incorrect structure will tend to approach the value for the correct structure — *at the expense of introducing systematic errors into the atomic positions.* It is therefore always important, whenever it is possible, to refine the structure with the correct configuration.

Because R^{+}_{I} and R^{-}_{I} are not very different under ordinary circumstances, Hamilton[174] suggested that the ratio of the two residuals could be evaluated statistically using linear hypothesis testing to determine whether or not the correct enantiomorph (often called the *absolute structure*) could be discerned from ratio. The validity of the test relies on having normally distributed intensity data with no systematic error. This is rarely (if ever) a valid assumption. Based on this, and what appear to be flaws in the use of the linear hypothesis model, Rogers[175] suggested an alternative way to determine the absolute structure. In Rogers' approach, only a single refinement is undertaken, but the imaginary contribution to the scattering factor is allowed to vary. Thus the scattering factor vector becomes $\mathbf{f} = f' + i\eta\,f''$, and η is treated as a variable parameter in the least-squares refinement. If the atom locations for the refined structure are correct η would be expected to refine to a number close to $+1$. If they should actually be inverted, η should refine to a number close to -1 in order to effectively invert the structure.

While the η parameter provided the first attempt to utilize a refinement variable to determine the absolute configuration, Flack[176] pointed out that the functional behavior of $I_{c,\mathbf{h}}$ with respect to η could result in an assignment of the incorrect enantiomorph. This led him to propose an alternative parameter for refinement, now in common use, and known as x — *the Flack parameter.* In this approach, the initial structure is assumed to be a racemic twin (see the previous section), and is refined as such. Eqn. 8.231 becomes

$$I_{c,\mathbf{h}}(\mathbf{p}, x) = p_1((1-x)\,I^{+}_{c,\mathbf{h}} + x\,I^{-}_{c,\mathbf{h}}). \tag{8.236}$$

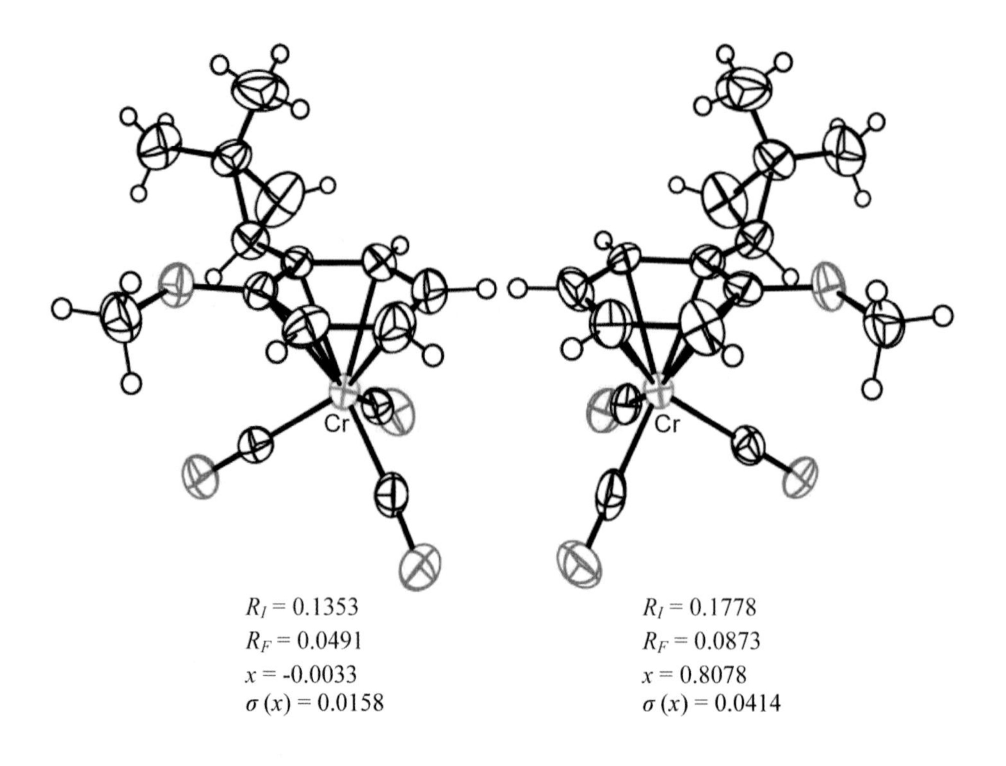

Figure 8.9 Enantiomers of $CrC_{15}H_{16}O_4$ and the results of refinement of model enantiomorphs containing each enantiomer.

If the putative "twin domain" with the original structure is the correct structure, x should refine favoring intensities calculated from the original atom positions — to a number in the vicinity of zero. If the other putative domain is the correct structure, then the refinement will tend to use the calculated intensities of the inverted structure, and x will approach unity.

The structural solution of the organometallic complex, $CrC_{15}H_{16}O_4$ (Fig. 6.17), was described In Chapter 6. The complex crystallizes in the noncentrosymmetric space group, $P2_12_12_1$. It was noted in the discussion that it was necessary to select an enantiomorph in order to solve the structure, but that the determination of the correct enantiomorph would have to await refinement. Fig. 8.9 illustrates the two possible enantiomers and the refinement results for each. The enantiomorph containing the enantiomer on the left refines to a reliability index of $R_I = 0.1353$, while the other enantiomer has a higher R factor of $R_I = 0.1778$. The original solution provided the structure on the right, with a Flack parameter of $x = 0.8078(0.0414)$, reasonably close to 1, and indicative of the likelihood that the other enantiomorph was likely to be the correct one. Inversion of the structure and refinement generated the enantiomer on the left with a Flack parameter of $x = -0.0033(0.0158)$. If the refinement of one of the enantiomorphs does not produce a value of x within a specified number of estimated standard deviations (e.g., $3\sigma(x)$) of zero, then the data may be insufficient to discern between the two enantiomorphs. On the other hand, the crystal may actually be a racemic twin!

8.3 Macromolecular Refinement

Small molecule structure solution often involves a cyclic process in which the Fourier map from a partial structure is improved by refinement of known atomic positions in order to locate more atoms — until the complete structure is revealed. The interaction between solution and refinement is generally straightforward, and often demands only minimal intervention by the investigator seeking the solution. In contrast, the interplay between the initial location of atomic positions and the completion of the model structure for macromolecules is an essential part of the process, such that the solution of a macromolecular crystal structure is inexorably linked to its refinement. The process of macromolecular structural solution ordinarily requires constant and intimate involvement by the investigator. Proteins constitute the vast majority of macromolecular structure determinations, both past and present, and we will concentrate on these biologically important amino acid polymers in the discussion that follows.

8.3.1 Heavy Atom Solutions

With the exception of a only a very few "small" macromolecules with data collected at or near atomic resolution, macromolecular crystals do not lend themselves to structural solution with current direct methods. A macromolecular structure that is solved "from the beginning" almost universally begins with a determination of the location of heavy atoms that have been infused into the crystals, usually from a modulus difference squared synthesis (Eqn. 6.121). The isomorphous replacement or anomalous dispersion methods described in detail in Chapter 6 are then used to generate initial phases for the reflections. The most common approach involves at least three data sets, one from the protein without heavy atoms (the *native protein*), and two others, each containing a different heavy atom selected to coordinate to a different type of binding site within the protein (in order to resolve the phase ambiguity problem).

The initial location of the heavy atom positions is relatively crude, and just as with small molecule structures, it is important to begin the phasing process with those positions as accurately determined as possible. Fig. 8.10 illustrates the relationship between the structure factors of a native protein and its heavy atom derivative. A reflection wave scattered from the heavy atom derivative is determined from a superposition of the waves scattered from the heavy atoms in the protein ($\mathbf{F}_H$) and the native structure (without the heavy atoms):

$$\mathbf{F}_{HP} = \mathbf{F}_H + \mathbf{F}_P. \tag{8.237}$$

For small molecule structures the heavy atom positions can be refined by minimizing*

$$R(\mathbf{p}) = \sum_{\mathbf{h}} w_{\mathbf{h}}(F_{o,\mathbf{h}} - F_{c,\mathbf{h}}(H))^2, \tag{8.238}$$

where $F_{c,\mathbf{h}}(H)$ are the magnitudes of the structure factors calculated solely from the heavy atom locations, and it is assumed that the heavy atoms dominate the phases

*Recall that macromolecular structure refinements are usually based on F rather than $I \equiv F^2$.

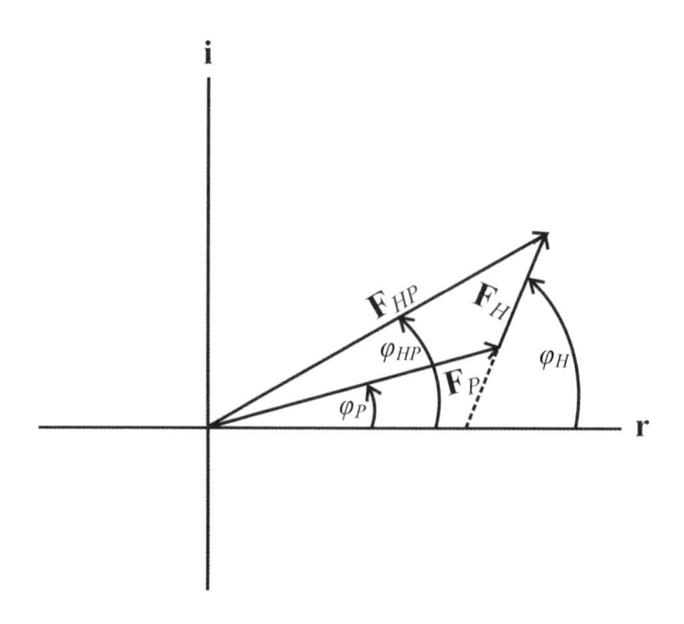

Figure 8.10 Relationship between the structure factors of a native protein ($\mathbf{F}_P$) and a heavy atom derivative of the protein ($\mathbf{F}_{HP}$).

and magnitudes of the observed structure factors. Heavy atoms in macromolecular crystals constitute a substantially smaller fraction of the electron density in the crystal, and they do not dominate the scattering to the extent that contributions from all of the remaining atoms in the native protein can be ignored – Eqn. 8.238 is not suitable for the refinement of heavy atom positions in macromolecules. However, if the phases of the native protein are not altered too much by the heavy atoms, then observed structure factor magnitudes for the heavy atom contribution can be crudely approximated by assuming that $\mathbf{F}_H$, $\mathbf{F}_P$, and $\mathbf{F}_{HP}$ are parallel, and

$$F_{o,\mathbf{h}}(H) \simeq |F_{o,\mathbf{h}}(HP) - F_{o,\mathbf{h}}(P)|. \tag{8.239}$$

This allows for refinement of the heavy atom positions by minimizing[177]

$$R(\mathbf{p}) = \sum_{\mathbf{h}} w_{\mathbf{h}}(F_{o,\mathbf{h}}(H) - F_{c,\mathbf{h}}(H))^2. \tag{8.240}$$

Instead of attempting to refine the heavy atom positions based only on their contributions to the phase of the heavy atom derivative, the phases of the derivative itself can be used by allowing the native protein phases to vary, $\varphi_P(\mathbf{h})$, refining them as parameters in addition to the heavy atom positions, and using the observed structure factor amplitudes to calculate $\mathbf{F}_{c,\mathbf{h}}(P)$. Thus

$$\mathbf{F}_{c,\mathbf{h}}(HP) = F_{o,\mathbf{h}}(P)\, e^{i\varphi_P} + \mathbf{F}_{c,\mathbf{h}}(H) \tag{8.241}$$

and

$$F_{c,\mathbf{h}}(HP) = |F_{o,\mathbf{h}}(P)\, e^{i\,\varphi_P(\mathbf{h})} + \mathbf{F}_{c,\mathbf{h}}(H)|. \tag{8.242}$$

This *phase refinement* minimizes

$$R(\mathbf{p}) = \sum_{\mathbf{h}} w(F_{o,\mathbf{h}}(HP) - F_{c,\mathbf{h}}(HP))^2. \tag{8.243}$$

The refinement is initiated with values of the phases for all of the reflections calculated from the initial heavy atom positions. The weight, w, is selected to reflect the difference in errors in the intensities at various diffraction angles,

$$w = \frac{1}{\langle (F_{c,\mathbf{h}}(HP) - F_{c,\mathbf{h}}(HP))^2 \rangle}, \tag{8.244}$$

calculated in ranges of $h = 2\sin\theta/\lambda$; the weight for a given reflection is that for the range in which h for that reflection is found. This method of heavy atom refinement makes no assumptions (crude or otherwise) about relationships between the phases of $\mathbf{F}_H$, $\mathbf{F}_P$, and $\mathbf{F}_{HP}$.

Assuming that multiple isomorphous replacement is the solution method chosen (it is the most common), we focus on the heavy atom positions of the first heavy atom derivative. The heavy atom contribution to the phases of the derivative are given by

$$\mathbf{F}_H(\mathbf{h}) = \sum_{j=1}^{k} f_j e^{-2\pi\, i(\mathbf{h}\cdot\mathbf{r}_j)} = x_H + i\, y_H, \tag{8.245}$$

where $\mathbf{r}_j$ is the vector to the jth of k heavy atoms in the unit cell, and

$$x_H = \sum_{j=1}^{k} f_j \cos(-2\pi\mathbf{h}\cdot\mathbf{r}_j) \quad \text{and} \tag{8.246}$$

$$y_H = \sum_{j=1}^{k} f_j \sin(-2\pi\mathbf{h}\cdot\mathbf{r}_j). \tag{8.247}$$

The experimental intensities provide the magnitudes of the structure factors for the native protein, $F_P(\mathbf{h})$ and the heavy atom derivative, $F_{HP}(\mathbf{h})$, for which we wish to determine the phases from

$$\mathbf{F}_P(\mathbf{h}) = x_P + i\, y_P \quad \text{and} \tag{8.248}$$

$$\mathbf{F}_{HP}(\mathbf{h}) = \mathbf{F}_P(\mathbf{h}) + \mathbf{F}_H(\mathbf{h}) = x_{HP} + i\, y_{HP}. \tag{8.249}$$

The quantities, $D = 2x_H$, $E = 2y_H$, $G = x_H^2 + y_H^2$, $H = F_{HP}^2(\mathbf{h}) - F_P^2(\mathbf{h})$, and $I = H - G$ are determined from the known values, and From Eqn. 6.106, x_P can now be expressed in terms of y_P,

$$x_P = \frac{1}{D}(I - E\, y_P), \tag{8.250}$$

and from Eqn. 6.107,

$$\left(\frac{E^2}{D^2} + 1\right) y_P^2 - \left(\frac{2IE}{D^2}\right) y_P + \left(\frac{I^2}{D^2} - F_P^2\right) = 0. \tag{8.251}$$

Solving the quadratic equation gives two roots for y_P, two pairs of x_P, y_P values, and two possible phases for each reflection from the native protein:

$$\varphi_P(\mathbf{h}) = \arctan\left(\frac{y_P}{x_P}\right). \tag{8.252}$$

The phase of the heavy atom derivative is then

$$\varphi_{HP}(\mathbf{h}) = \arctan\left(\frac{y_P + y_H}{x_P + x_H}\right). \tag{8.253}$$

An alternative geometric derivation is often found in protein crystallography texts. Referring to Fig. 8.10, the angle between $\mathbf{F}_P$ and $\mathbf{F}_H$ is $\pi - (\varphi_H - \varphi_P)$. From the law of cosines,

$$\begin{aligned} F_{HP}^2 &= F_P^2 + F_H^2 - 2F_H F_P \cos(\pi - (\varphi_H - \varphi_P)) \\ &= F_P^2 + F_H^2 + 2F_H F_P \cos(\varphi_H - \varphi_P) \\ \varphi_P &= \varphi_H - \arccos\left(\frac{F_{HP}^2 - F_P^2 - F_H^2}{2F_H F_P}\right). \end{aligned} \tag{8.254}$$

Although it is not quite so obvious, this solution also leads to a phase ambiguity, since there are always two angles with the same cosine, again creating two phases for each reflection. This phase ambiguity is removed by repeating the phase determination with a second heavy atom derivative containing heavy atoms in different locations.* The solution results in a second set of pairs of phases, but for every reflection, each pair will have only one phase in common with the other pair (Fig. 6.26). This initial phase set will be used to initiate the final stages of the solution and refinement process, which we will take up after discussing the other common method for structural solution known as molecular replacement.

8.3.2 Molecular Replacement Solutions

In the first few pages of this book it was noted that molecules without symmetry of their own could still form symmetric crystalline arrays. There is perhaps no clearer case for this than the existing number of reported protein crystal structures, now numbering well over 10,000. Most of these structures have been determined in the last two decades, and the number of new structures is certain to increase even more rapidly with new improvements in data collection, crystal growth, and modeling techniques. One of the major reasons for the rapid growth of macromolecular crystallography is the use of Patterson search methods described in Chapter 6. Patterson searches are successful for small molecules when the geometry of the molecule, or a fragment of it is known. Since Patterson search methods require a model structure with which to perform the search, it would at first appear hopeless to attempt such an undertaking for macromolecular structures. However, many biological macromolecules occur in nature in similar forms, and the known structure

*In practice, the infusion of heavy metals into the protein crystal is an imperfect process. There are usually a number of coordination sites in the protein for a given metal ion, and none of these sites are completely occupied. *Multiple isomorphous replacement* (MIR), and/or its anomalous dispersion counterpart, *multiple anomalous dispersion* (MAD), ordinarily require more than two heavy atom derivatives.

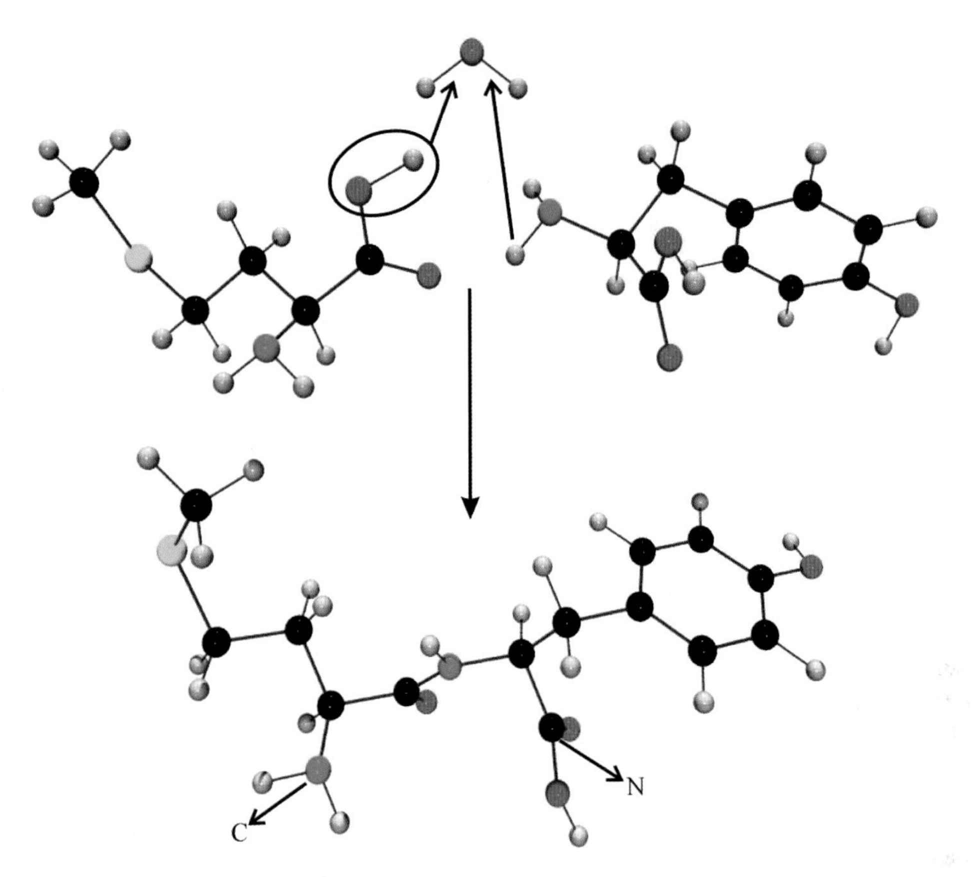

Figure 8.11 Dimer created from the amino acids methionine and tyrosine by dehydration synthesis. The arrows to C and N indicate the next C–N bonds created in polypeptide formation.

of one molecule can often be used as a model structure for another — *molecular replacement.**

The vast majority of all of the macromolecular crystal structures that have been reported are those of the most diverse of biological compounds – proteins. Proteins are large polymers made up of sequences of the 20 naturally occurring amino acids. In these poly-amino acids, known as *polypeptides*, the amino acid momomers are linked by the loss of a hydroxyl moiety from the carboxyl group (–COOH) of one amino acid, and a hydrogen atom from the amino group (–NH_2) of another, creating a C–N bond with the subsequent elimination of water (dehydration). The formation of a *dipeptide* dimer from the amino acids methionone and tyrosine is illustrated in Fig. 8.11. Further *dehydration synthesis* extends the polypeptide chain in both directions. Each amino acid has a characteristic side group, and it is the variation in

*The original concept of the molecular replacement method was introduced by Michael Rossmann and David Blow in the early 1960's. Rossmanns' book entitled *The Molecular Replacement Method* (1972),[178], and his detailed update in a review article with the same title published in *Acta Crystallographica*(1990),[179] contain excellent expositions of the details of the method. A description of its main elements can be found in Chapter 6.

these side groups that makes each protein unique. As the polypeptides form, they tend to "fold" into three dimensional conformations that, if not thermodynamically the most stable, are kinetically the most robust — every protein with the same sequence of amino acids adopts approximately the same structure as another with the same sequence. This tendency to form conformationally equivalent molecules is markedly important to protein crystallographers, since there would be no protein crystals without it. Nucleic acids also tend to form crystals, but other biological polymers, such as polysaccharides, tend to form amorphous materials. It is indeed fortunate that proteins crystallize, since an understanding of their functions in biological systems often depends critically upon a knowledge of their three-dimensional structures.

Many proteins serve as biological catalysts – enzymes. The catalyzed reactions generally involve the interaction with a substrate molecule in order to facilitate a reaction involving that molecule, followed by release of the product. The substrate-enzyme interaction generally takes place in a specialized and structurally sensitive region of the protein known as the active site. Enzymes that catalyze similar reactions, or the same reaction in various species (or both), generally have similar amino acid sequences (*sequence homology*) and subsequently adopt similar three-dimensional conformations. In these cases, the known molecular structure of one molecule (perhaps determined with heavy atom techniques) can often be used as a template for an unknown structure using the rotation and translation search techniques previously discussed in Chapter 6.

Somewhat surprisingly, many proteins with only modest similarities in their amino acid sequences, often catalyzing different classes of reactions, will tend to form similar structures. It is often possible to use a molecule with a limited sequence homology to model another, and a number of structures have been successfully determined in this manner. However, an important caveat must be mentioned in the employment of the molecular replacement method for proteins with limited similarities. Protein crystal structures are nearly always underdetermined,* and are rarely refined from diffraction data at atomic resolution. Their refinement often results in large R factors, with no reliable figure of merit to determine whether or not the structural solution is correct. Molecular replacement in these cases must be used with great care and significant trepidation — especially if the structure appears to contradict other reported experimental data.

8.3.3 Completion of the Model

Both heavy atom and molecular replacement methods establish an initial set of phases than result in a crude electron density map for the native protein. This map will ordinarily be quite difficult to interpret, and a simpler map is usually generated at the outset using the more reliable low angle (and low resolution) reflections. A subset of reflections in a resolution range of e.g., 5–6 Å generally creates an "envelope structure" of the protein in the unit cell. The initial low-resolution map will serve

*"Underdetermined" is used somewhat loosely here. The minimum number of measurements necessary to define the atomic positions is $3n$, where n is the number of atoms. The number of reflections collected with an area detector often exceeds this number, even with macromolecular crystals. However, in order to *accurately* define molecular parameters, the ratio of observations to variables must ordinarily be greater than 5, with optimal values closer to 10. This is rarely the case for macromolecules.

to identify the location of the molecule in the unit cell, but will not be sufficiently resolved to reveal details of the structure.

Unlike small molecule structures, protein crystals are ordinarily full of water. Some of the water molecules are in specific locations in the unit cell, usually a part of the protein structure, while many others are more-or-less disordered. If the initial outline of the protein is evident in the low-resolution map, then the phases can be improved by creating a weighted electron density function in which regions outside of the protein envelope are given smoothed average electron densities that are weighted according to the solvent/protein ratio in the crystal (either known from density measurements or estimated). New phases are then calculated from this *solvent flattened* electron density map.

Once the protein envelope is identified in the unit cell, the higher resolution data are added and a new Fourier map is calculated. The data are now at a resolution that should allow for portions of the polypeptide chain to be located inside the protein envelope. The amino acid sequence of the protein is generally known, and it can prove invaluable at this stage, since the general structure of an amino acid attached to one already located in the map will be known, and can be placed in its proper conformation, based on the positions of its identifiable side groups (e.g., a phenyl ring). The process is performed with sophisticated computer graphics software that creates a three-dimension contour plot of the electron density. The unit cell is divided into a three-dimensional grid and the points that have the same electron density in each grid volume(specified by the user) are connected by lines, creating a "wire-mesh" contour plot, as illustrated in the top portion of Fig. 8.12.

As amino acids are added to the structure they are represented as stick-figures, making it easy to superimpose the centers of the heavy atoms in each amino acid onto the Fourier map, illustrated in the bottom portion of Fig. 8.12. The software allows the user to identify groups of atoms to create a rigid group that can be manually rotated and translated. The process is the same as that for rigid group refinement, in which the group is transformed to a local coordinate system, manipulated, then transformed back to the fractional coordinate system. Amino acids are generally added sequentially in approximate locations, then rotated and translated visibly within the electron density contours.

As more amino acids are modeled into the structure, new electron density maps are calculated in order to continue to improve the phases and determine more of the structure. At some stage in this process, however, the phases will be dominated by that portion of the model already assigned, and the model will cease to improve the electron density map. At this juncture the logical option is to turn to difference electron density synthesis, described in detail in Sec. 6.3.2. In positive regions the $\Delta F = F_o - F_c$ map provides the location of the electron density not accounted for by the model, and in negative regions it indicates excessive electron density — i.e., something is where it doesn't belong. For example, if the difference electron density at the location of an amino acid side chain is all negative, it may indicate that another amino acid belongs at that position in the chain. There may also be a large negative region at the location of the side chain, and a positive region adjacent to it, indicating that the conformation of group should be changed.

While the $F_o - F_c$ map is most sensitive to differences between the model and the observations, it is often difficult to interpret, since the model electron density is eliminated from the map. The map also tends to be noisy, since it contains all

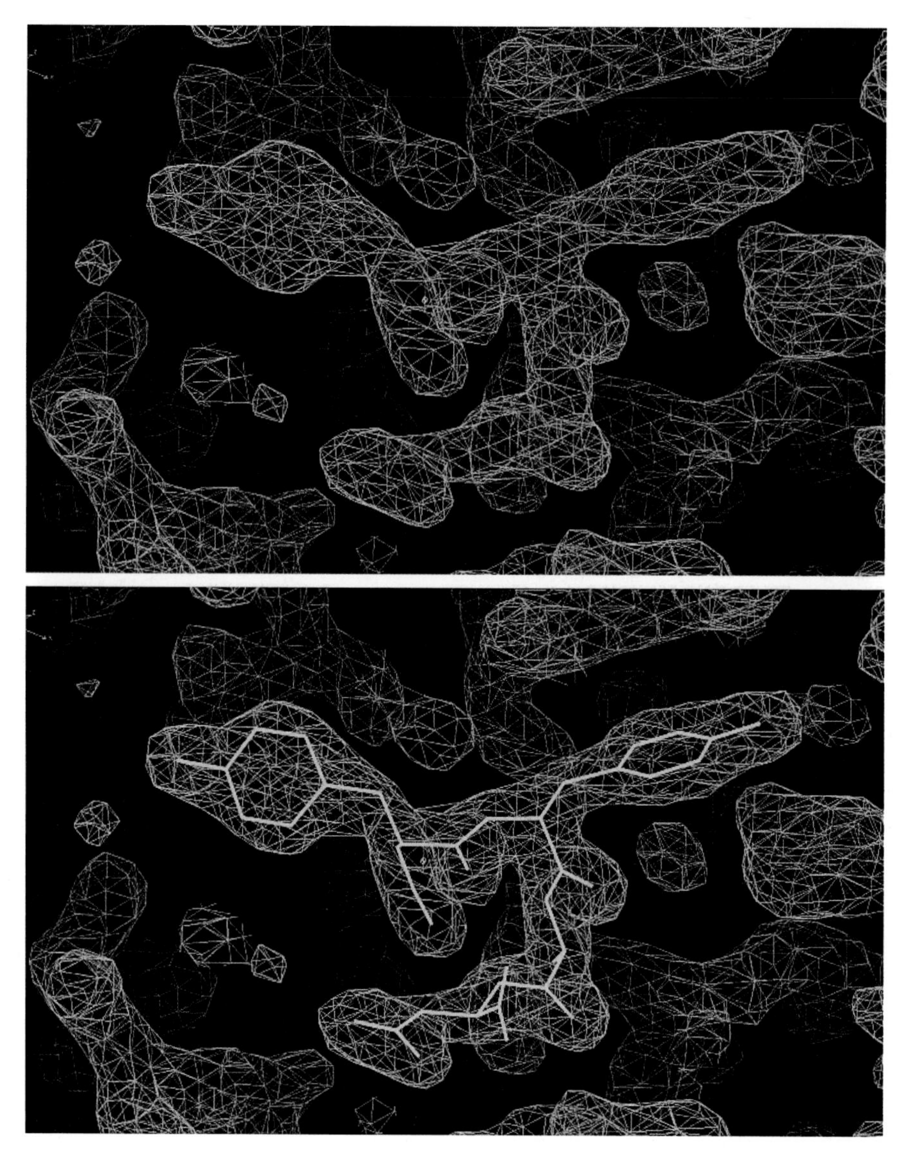

Figure 8.12 Top: $2F_o - F_c$ electron density contour map. Bottom: Refined amino acid positions and conformations superimposed onto the map.

of the errors in the model. As an alternative, macromolecular crystallographers employ a number of different "difference map" schemes in the intermediate stages of model construction. The most common is the $2F_o - F_c$ synthesis, which produces difference electron density, but retains the original model electron density in the map. As the structure improves and F_c approaches F_o for all of the reflections, the map becomes an approximation to a conventional F_o map.

8.3.4 Refinement of the Model

Least Squares.

Once the macromolecular model is complete and has been matched manually with the electron density map, the model must be refined, just as it is in small molecule crystallography. The method of least squares is to be employed, but as we have already noted, this can't be accomplished for all of the atomic parameters (coordinates, site occupancy factors, and displacement parameters) since there are far too many atomic parameters for the number of observed reflections. Fortunately, a large number of crystal structures of small polypeptides, refined at atomic resolution, reveal that the bond lengths in a given amino acid vary little from one polypeptide to the next. Furthermore, the C–N bond in the amide portion of each dipeptide moiety (known as a *peptide* bond) has an occupied π molecular orbital that creates a rotational barrier about the C–N bond of over 50 kJ/mol. The nitrogen and carbon atoms in the amide portion are effectively sp^2 hybridized, and the amide portion of the dipeptide remains planar, as illustrated in the top portion of Fig. 8.13.

A primary goal in macromolecular refinement is the reduction of the number of refined variables. As we discussed earlier on in the chapter, this can be accomplished by refining a portion of a molecule as an idealized rigid group. A modified version of rigid group refinement is employed in macromolecular refinement. As an example, consider the dipeptide illustrated in the bottom portion of Fig. 8.13. The hydrogen atoms shown in the figure are not incorporated into the refinement, due to their low scattering power. The carboxyl and amino groups at the bottom of the figure would be linked to amino acids via peptide bonds inside the polypeptide — unless the dipeptide terminated the chain. Hypothetically, if the dimer was *actually* a rigid group, it could be treated in a manner identical to the cyclopentadienide group previously discussed. That is, three atoms would be selected to create a local coordinate system on the dimer. Logical choices would be the carbon and nitrogen atoms connected by the C–N bond, and the carbonyl oxygen atom in the amide plane – since these are constant for all of the dimers in the protein. The remaining atoms in the dimer would then be referenced to this coordinate system, which would then be transported to the origin of the crystal Cartesian system and rotated into coincidence with it. The atom locations in this orientation ($\{\mathbf{v}_r\}$) would then be saved for use as the rigid group coordinates, along with the Eulerian angles necessary to rotate the local coordinate axes onto the crystal Cartesian axes. After transforming the dimer back to its original position, structure factors would be calculated from the atom positions, displacement parameters*, and possible site occupancy factors. The positional parameters to be refined would now be reduced from over 50 for the $3n$ atomic coordinates to six — the coordinates of the origin atom and the rotational angles of the group. After each cycle of refinement, the new values for the group origin coordinates and rotational angles would be substituted into Eqns. 8.210 and 8.211, new structure factors calculated, and the next refinement cycle would be initiated.

The dimer, of course, is not a rigid group. The internuclear distances in the dimer remain effectively constant, but there is essentially free rotation about each

*An overall displacement factor in the early stages of refinement, and isotropic displacement parameters for all of the atoms in later stages.

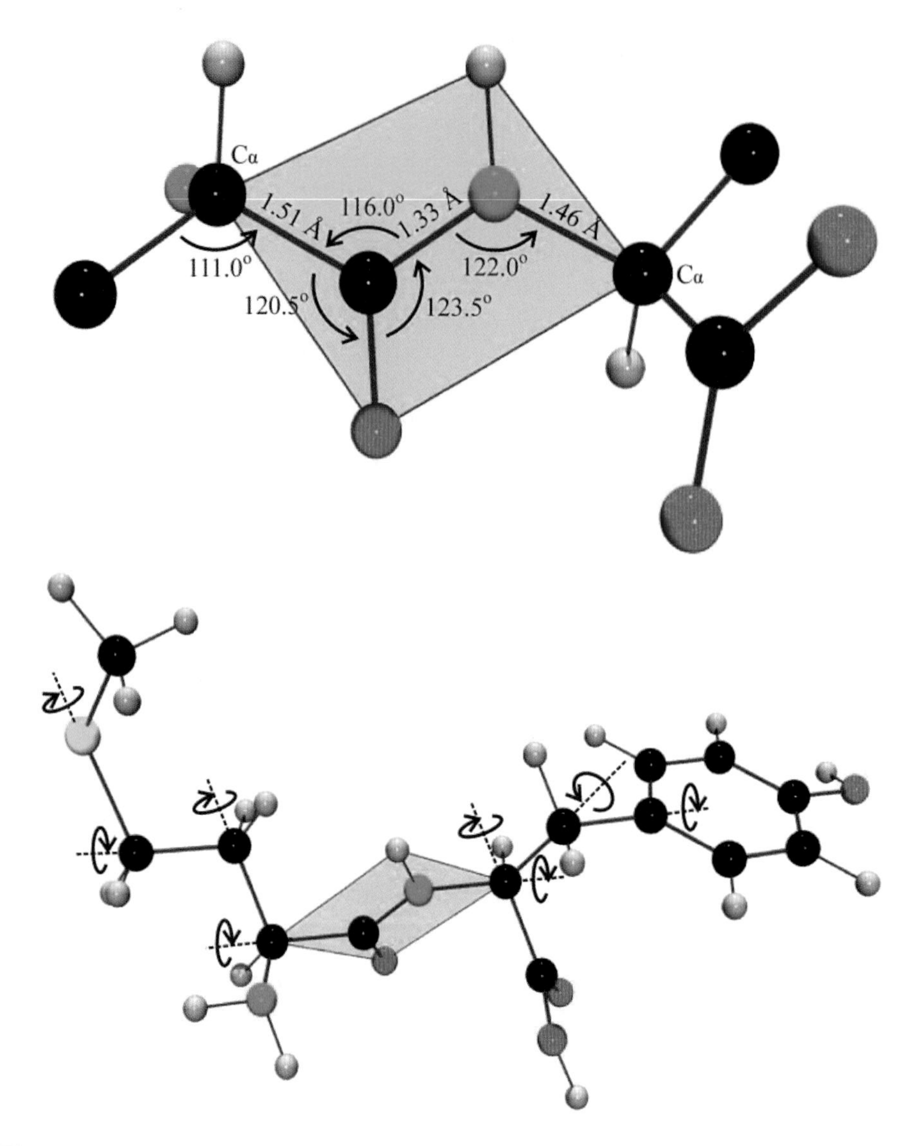

Figure 8.13 Top: Planar amide region in a dipeptide. Bottom: Variable rotational angles in a dipeptide. Hydrogen atoms are not included in the refinement.

of the bonds indicated with a dashed line in Fig. 8.13. The refinement of the dimer as a rigid group involves rotating and translating the entire vector set, $(\{\mathbf{v}_r\})$. It is also possible to rotate or translate portions of it. Consider the carbon-carbon bond, C_1–C_2, in the figure. Rotation about this bond will change the coordinates of the terminal carbon atom, C_3 and the sulfur atom. We will denote the vectors to each of these fours atoms (in the rigid group coordinate system) as $\mathbf{v}_{C_1}$, $\mathbf{v}_{C_2}$, $\mathbf{v}_S$ and $\mathbf{v}_{C_3}$. In order to accomplish the rotation, $\mathbf{v}_{C_1}$ is subtracted all of the vectors, translating the vectors to the group origin, e.g., $\mathbf{v}'_{C_2} = \mathbf{v}_{C_2} - \mathbf{v}_{C_1}$. $\mathbf{v}'_{C_2}$ is now the rotation axis, which we denote as $\mathbf{r}$, with magnitude r, and coordinates (r_x, r_y, r_z). The rotation matrix for a vector about this axis (see Eqn. 2.46) is

$$\mathbf{R}_{\phi_2} =$$

$$\begin{bmatrix} \frac{r_x^2+(r_y^2+r_z^2)\cos\phi_2}{r^2} & \frac{r_x r_y(1-\cos\phi_2)-r_z r\sin\phi_2}{r^2} & \frac{r_x r_z(1-\cos\phi_2)+r_y r\sin\phi_2}{r^2} \\ \frac{r_x r_y(1-\cos\phi_2)+r_z r\sin\phi_2}{r^2} & \frac{r_y^2+(r_x^2+r_z^2)\cos\phi_2}{r^2} & \frac{r_y r_z(1-\cos\phi_2)-r_x r\sin\phi_2}{r^2} \\ \frac{r_x r_z(1-\cos\phi_2)-r_y r\sin\phi_2}{r^2} & \frac{r_y r_z(1-\cos\phi_2)+r_x r\sin\phi_2}{r^2} & \frac{r_z^2+(r_x^2+r_y^2)\cos\phi_2}{r^2} \end{bmatrix}.$$

The rotation is accomplished with $\mathbf{v}''_{\mathrm{S}} = \mathbf{R}_{\phi_2}\mathbf{v}'_{\mathrm{S}}$ and $\mathbf{v}''_{\mathrm{C}_3} = \mathbf{R}_{\phi_2}\mathbf{v}'_{\mathrm{C}_3}$.

Rotation matrices can be generated for all eight rotational angles in the dimer, and the angles, initially set to zero, can be refined as parameters. The refinement is based on F, and the normal equations become, for $j = 1 \ldots, n$,

$$\sum_{i=1}^{n}\sum_{\mathbf{h}} w_{\mathbf{h}} \left(\frac{\partial(F_{c,\mathbf{h}}(\mathbf{p}))}{\partial p_i}\frac{\partial(F_{c,\mathbf{h}}(\mathbf{p}))}{\partial p_j}\right)_{\mathbf{p}_o} \Delta p_i = \sum_{\mathbf{h}} w_{\mathbf{h}}\Delta F_{\mathbf{h}} \left(\frac{\partial(F_{c,\mathbf{h}}(\mathbf{p}))}{\partial p_j}\right)_{\mathbf{p}_o}.$$

Evaluation of the derivatives for the group origin coordinates and Eulerian angles for rotation of the group parallel the derivations based on $I_{c,\mathbf{h}}$ described earlier in the chapter. As with the Eulerian angles, the derivatives for the bond rotation angles are evaluated numerically, by making a small change, $\delta\phi_i$, and after translating the vectors to the local origin, rotating the vectors to S and C_3, then adding the components of the C_1 vector back to each rotated vector, recalculating $F_{c,\mathbf{h}}$, and determining $\delta F_{c,\mathbf{h}}$ — from which $\partial(F_{c,\mathbf{h}}(\mathbf{p}))/\partial\phi_i \simeq \delta(F_{c,\mathbf{h}}(\mathbf{p}))/\delta\phi_i$. For the twenty atoms refined in this portion of the protein, assuming site occupancy factors of 1.0, there would be 80 parameters to refine without constraints, 60 positional coordinates and 20 displacement parameters. With constraints the number has been reduced to 34 — six parameters defining the origin and rotation angles of the dimer, and 8 more for the rotational angles within the dimer.

It is also possible to refine soft constraints as we did with small molecules. Soft constraints (restraints) can be incorporated to allow selected distances, angles, volumes, or other parameters to vary be an amount determined by the restraint weight. Recall that restraints are incorporated as observations in the normal equations, and do not increase the size of the normal equation matrix.

Molecular Dynamics.

For most macromolecular structures, the actual locations of the atoms are known only approximately; to determine their actual locations we must know more than the X-ray data can tell us — we need additional information.

An atom in the crystal interacts with the atoms surrounding it via attractive "bonding" forces and repulsive "contact forces". If we began with a crude structure and could determine the interaction forces exactly, we could, in principle, predict the structure. A common method for predicting structures in molecular systems based on interactive forces is known as *molecular dynamics*. An atom in motion with some initial velocity will experience forces from surrounding atoms dependent upon their relative positions (usually with respect to an equilibrium position). These forces will result in adding an acceleration vector to each atom in the direction of the force, generating a new configuration of atoms each with a new velocity and new forces, and so forth, with the structure evolving in time. If the configurations are sampled in very small time increments, (e.g., picoseconds), then the structure will appear as if it is in a movie, bouncing around its equilibrium configuration. The rate at which this happens depends on the selection of initial velocities of the atoms, which in turn depends on their kinetic energies, and therefore the temperature.

Suppose that we have a dynamic macromolecule with atomic coordinates $\{\mathbf{r}(t)\} = \{\mathbf{r}_1(t), \mathbf{r}_2(t), \ldots\}$, at some time t. Since it is dynamic each atom will have a velocity, $\{\mathbf{v}(t)\} = \{\mathbf{v}_1(t), \mathbf{v}_2(t), \ldots\}$, the average of which will depend on the temperature, and a corresponding momentum $\{\mathbf{p}(t)\} = \{m_1\mathbf{v}_1(t), m_2\mathbf{v}_2(t), \ldots\} = \{\mathbf{p}_1(t), \mathbf{p}_2(t), \ldots\}$. At time t, each atom experiences a net force due to the interactions with surrounding atoms, and (from $\mathbf{F} = m\mathbf{a}$), is accelerated in the direction of the force vector. After a time interval, Δt, short enough to presume a constant force (e.g., a picosecond), the atoms will move to a new configuration with new momenta (by applying the accelerations due to the forces at the end of the interval):

$$\begin{aligned}
\mathbf{r}(t+\Delta t) &= \mathbf{r}(t) + \mathbf{v}(t)\Delta t \\
&= \mathbf{r}(t) + \frac{\mathbf{p}(t)}{m}\Delta t && (8.255) \\
\mathbf{p}(t+\Delta t) &= \mathbf{p}(t) + m\Delta\mathbf{v} \\
&= \mathbf{p}(t) + \frac{m\Delta\mathbf{v}}{\Delta t}\Delta t \\
&= \mathbf{p}(t) + m\mathbf{a}\Delta t = \mathbf{p}(t) + \mathbf{F}\Delta t && (8.256) \\
\mathbf{v}(t+\Delta t) &= \mathbf{p}(t+\Delta t)/m. && (8.257)
\end{aligned}$$

The new momenta result in new velocity vectors, with the atoms in new locations. Because of their displacements, the atoms are subjected to new forces; incorporating the new positions, momenta, velocities, and forces into Eqns. 8.255–8.257 results in a new configuration with new momenta. The forces for each configuration are "restoring forces", tending to decrease the energy of the system to its equilibrium (lowest energy) conformation. If we were able to determine the exact forces for any configuration we would simply set the atoms in motion by assigning initial velocities to each atom, followed by the application of continual cycles of Eqns. 8.255–8.257 until it is seen to be oscillating around a minimum energy conformation – the analog of a molecule vibrating at equilibrium. This is ordinarily accomplished by generating a random probability, and selecting the velocity from a Boltzmann probability distribution at a specified temperature. (The lower the temperature, the slower the atoms move). A sequence of short time intervals would show various conformations of the molecule, all in the proximity of the equilibrium conformation. The step-by-step pathway of the molecular conformation is known as a conformational *trajectory*.

It would therefore appear that we need only a crude structure, after which we would use molecular dynamics to refine the structure to its equilibrium configuration, making no further use of the intensities. This would always be the case if we had the ability to accurately calculate all of the interaction energies – but this is an impossible task — rooted in rigorous quantum mechanical calculations of the

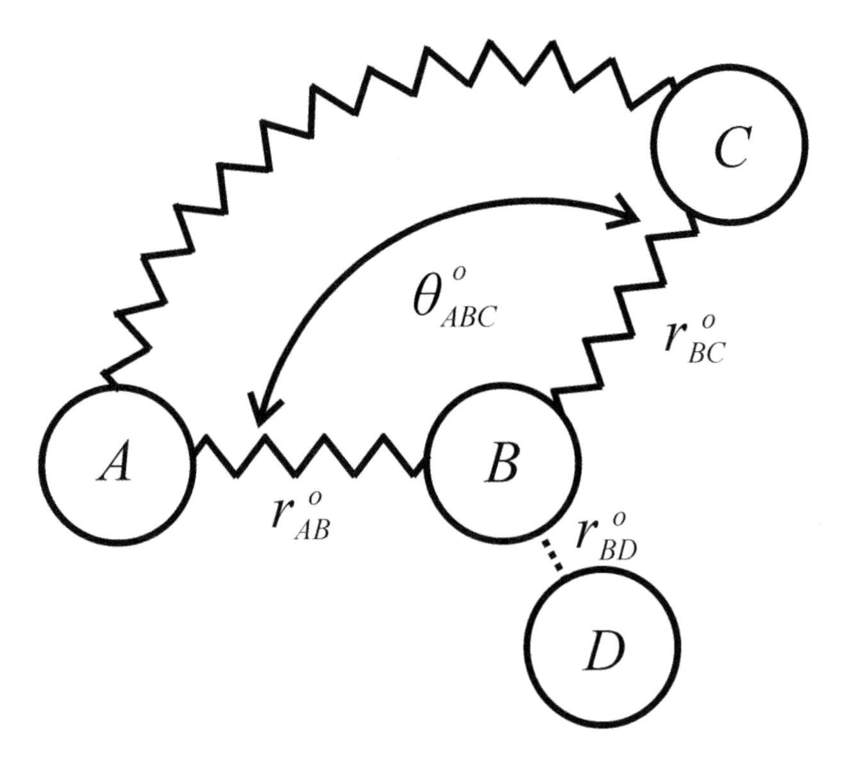

Figure 8.14 Bonding interactions approximated as springs between atoms A, B, and C. Non-bonded interactions occur between atoms B and D.

electronic structure of molecules that are far too large to model computationally. A more reasonable option is to approximate the major interaction energies and combine them in some way with the intensity data, so that the trajectory can move the molecule toward a conformation that simultaneously has a low potential energy *and* a small structure factor residual.[180]

We consider two types of interaction energies – those between bonded atoms, and those between non-bonded atoms. Although the energy changes in a non-linear fashion when an internuclear distance between two bonded atoms changes, we approximate it with a linear response, treating the bond as a spring. In Fig. 8.14 bonding interactions are shown between atoms A, B, and C. Atoms B and D interact through non-bonding forces. The figure shows the atoms in their initial (reference) positions. If, during the motion of the atoms, the internuclear distance between atom A and atom B increases, the force on the atoms (relative to the initial positions) will be that of a stretched spring with force constant k_{AB}:

$$F_{AB} = -k_{AB}(r_{AB} - r^o_{AB}). \tag{8.258}$$

The contribution to the potential energy due to this displacement will be

$$E_{AB} = \frac{1}{2}k_{AB}(r_{AB} - r^o_{AB})^2. \tag{8.259}$$

If the bond angle, θ^o_{ABC}, changes, its effect can also be approximated with a curved spring with force constant k_{ABC}:

$$F_{ABC} = -k_{ABC}(\theta_{ABC} - \theta^o_{ABC}) \tag{8.260}$$

$$E_{ABC} = \frac{1}{2}k_{ABC}(\theta_{ABC} - \theta^o_{ABC})^2. \tag{8.261}$$

Similar forces and potential energy contributions can be derived for torsional angles and other motions of bonded atoms.

Atoms that are not bonded do not interact until they get very close to one another. As they approach, dispersion forces cause the atoms to attract one another (the force is proportional to r^{-7}) until the electron clouds of the two atoms penetrate and overcome the attractive force with electron-electron repulsion. The black curve in Fig. 8.15 illustrates this behavior for two argon atoms, measured experimentally. The potential energy minimum is negative, implying that all neutral atoms attract one another in the equilibrium structure. The combination of the attractive and repulsive non-bonding forces was modeled by John Lennard-Jones in 1931 [181]:

$$F_{BD} = -24\epsilon\left(2\left(\frac{\sigma^{12}}{r^{13}_{BD}}\right) - \left(\frac{\sigma^6}{r^7_{BD}}\right)\right), \tag{8.262}$$

where ϵ is the depth of the potential energy well and σ is the effective distance at which the potential is zero. The contribution to the potential energy of the conformation is

$$E_{BD} = 4\epsilon\left(\left(\frac{\sigma}{r_{BD}}\right)^{12} - \left(\frac{\sigma}{r_{BD}}\right)^6\right). \tag{8.263}$$

Electrostatic interactions can also be included by assigning charges to the atoms:

$$F_{AB} = \frac{1}{4\pi\epsilon_o}\frac{q_A q_B}{r^2_{AB}}; \tag{8.264}$$

$$E_{AB} = \frac{1}{2\pi\epsilon_o}\frac{q_A q_B}{r_{AB}}. \tag{8.265}$$

The forces resulting from the approximated interactions are used to define a net force on each atom and a subsequent acceleration. At the end of each time increment the forces are recalculated in preparation for the next set of motions. The trajectory is followed until the sum of the potential energies is minimized. The aim of molecular dynamics is to find a conformation that minimizes the molecular dynamic potential energy of the form:

$$\begin{aligned} E_{MD} &= \sum_{AB}\frac{1}{2}k_{AB}(r_{AB} - r^o_{AB})^2 + \sum_{ABC}\frac{1}{2}k_{ABC}(\theta_{ABC} - \theta^o_{ABC})^2 \\ &= \sum_{BD}4\epsilon\left(\left(\frac{\sigma}{r_{BD}}\right)^{12} - \left(\frac{\sigma}{r_{BD}}\right)^6\right) + \sum_{AB}\frac{1}{2\pi\epsilon_o}\frac{q_A q_B}{r_{AB}} + \ldots \end{aligned} \tag{8.266}$$

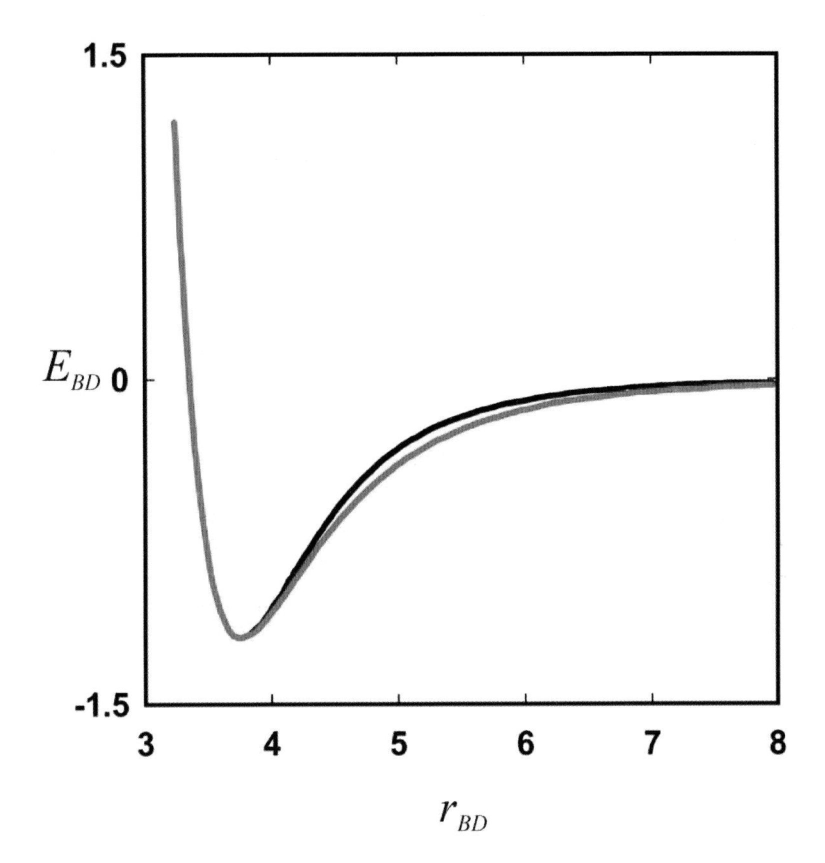

Figure 8.15 Non-bonded interaction between two argon atoms. The black curve (which is nearly coincident with much of the red curve) is the experimentally measured interaction potential energy. The red curve is the interaction energy modeled with the Lennard-Jones potential. The internuclear distance is in angstroms, and the potential energy is in kJ/mol.

The refinement of the X-ray intensity data leads to a conformation that minimizes the sum of the squares of the residuals much as if they represented an "effective potential energy:"

$$E_{Xray} = \sum_{\mathbf{h}} w_{\mathbf{h}}(F_{o,\mathbf{h}} - \sqrt{p_1}F_{c,\mathbf{h}})^2. \tag{8.267}$$

In order to incorporate molecular dynamics information into the structural refinement, we minimize

$$E_{total} = E_{Xray} + KE_{MD}, \tag{8.268}$$

where K is a factor that scales the two "energies" and weights the relative amount of each in determining the final conformation. The molecular dynamics trajectory is now followed sequentially until a minimum "energy" conformation is attained.

Because of the multidimensional nature of the problem it is very likely that a false minimum will be obtained, for which the atoms will have insufficient velocities to cross over into a new conformation. One way around this is to begin with a Boltzmann distribution of velocities at a high temperature (with many large velocities in the distribution), then, after minimization, "cool" the molecule by selecting a Boltzmann distribution at a lower temperature. This can be seen to be very similar to the simulated annealing approach to optimization discussed in Chapter 7. Modern macromolecular refinement programs such as *X-PLOR*[182] make use of sophisticated simulated annealing molecular dynamics algorithms.

Evaluating the Model.

Small molecule crystal structures generally have a large measurement-to-parameter ratio, and the model is almost always accurate. While there are exceptions, a low reliability index (e.g., 0.03–0.10) generally indicates a correct structure. On the other hand, macromolecular data are plagued by a large number of parameters compared to the number of reflections, and reliability indices are often between 0.20 and 0.30; the R factor alone is insufficient to assess whether or not the model is correct. If the model is correct, then refinement of the model with a subset of the reflections removed from the refinement should correctly predict the structure factor magnitudes of the reflections in the subset. If this is correct then a reliability index including only those reflections not used in the refinement, R_{free}, should remain in the vicinity of the reliability index for the entire structure. While R_{free} will be higher than R_F, a determination of several values of R_{free} can be useful in evaluating the model. Most importantly, the molecule must make chemical sense. It must be structurally reasonable, with bond lengths within 0.01 Å–0.02 Å and bond angles within 2°–3° of their expected values. If most of the model is reasonable geometrically, then a portion that is not may indicate a need for "manual" correction and re-refinement.

Exercises

1. Consider a spring suspended from a horizontal support beam with a weight with mass m attached to the end of the spring. The weight is pulled downward and released, setting it into oscillatory (harmonic) motion, oscillating at frequency ν. If the mass of the spring is negligible, the frequency is determined by the force constant of the spring, k, and the mass of the weight:

$$\nu = \frac{1}{2\pi}\sqrt{\frac{k}{m}}.$$

The period — the time that it takes for a single oscillation to occur — is the reciprocal of the frequency:

$$T = 2\pi\sqrt{\frac{m}{k}}.$$

Since the spring has mass, in order to be rigorous, the formula must also contain the "effective mass" of the spring, m_s:

$$T = 2\pi\sqrt{\frac{m + m_s}{k}}.$$

In a series of experiments, various weights are attached to the spring and the oscillation frequency is measured in each case:

m (g)	10	20	30	40	50	60	70	80	90	100
ν (s^{-1})	1.778	1.463	1.263	1.137	1.036	0.950	0.899	0.838	0.795	0.759

(a) Derive a relationship that is linear in m (the equation of a line with m as the independent variable). (b) Use the relationship and the method of least squares to determine the effective mass and force constant of the spring (assume that the masses are exact and that the values of the dependent variable are equally weighted).

2. The unweighted linear least-squares solution of Exercise 1 can be refined by including appropriate weights for each of the residuals. If the error in the measurement of each frequency is entirely random, then these weights are the reciprocals of the variances of the dependent variables, $y_i = T_i^2$. The weighted sum of the squares of the residuals, $\sum_1^n w_i(sm_i + y_0 - y_i)^2$, is minimized, where $w_i = 1/\sigma^2(y_i)$. (a) Derive a matrix solution for the slope and intercept of a weighted least-squares line for the data in Exercise 1. (b) Assuming that the variance of the frequency (and therefore the period) is the same for each data point (independent of ν and T), show that the weights in part (a) are *not* independent of T. (c) Determine the weighted least-squares values for the force constant and effective mass of the spring in Exercise 1. (d) Given the following repeated measurements of the frequency of the oscillating 50 g weight (in s^{-1}): 1.038, 1.033, 1.029, 1.029, 1.034, 1.027, 1.028, 1.037, 1.028, 1.034, 1.032, 1.030, 1.037, 1.031, 1.033, 1.035, 1.031, 1.031, 1.035, and 1.035, determine the estimated standard deviations of the force constant and effective mass of the spring in Exercise 1.

3. In a chemical kinetics experiment a compound, compound X, is known to form slowly along a zero-order pathway, and decompose more rapidly along an independent first-order pathway. The concentration of X can be modeled as a function of time as

$$[\mathrm{X}](t) = p_1 t + p_2 e^{-p_3 t}.$$

In the experiment, [X] is measured in intervals of 0.1 h, and we seek the values of the parameters, p_1, p_2, and p_3 that will provide a "best fit" (a least-squares fit) to the data:

t (h)	0.0	0.1	0.2	0.3	0.4	0.5	0.6	0.7	0.8	0.9	1.0
[X](mM)	190.6	109.1	65.7	40.7	35.7	7.9	18.8	12.1	1.7	9.9	5.3

(a) Determine initial guesses for p_1, p_2, and p_3 by selecting data points with appropriate values of t. (b) Using a Taylor series approximation to the function, determine optimal least-squares values for p_1, p_2, and p_3. (c) Plot the least-squares model function and the data in the table.

4. The pyridine ring in Fig. 1.32 can be approximated by a hexagon with an edge length of 1.36 Å. (a) Given the fractional coordinates of S(1), C(1), and N(1), calculate the fractional coordinates of the remaining atoms based on this approximation, retaining the original locations of C(1) and S(1).

5. Given the following multiple isomorphous replacement data for the (0 6 0) reflection from a protein crystal,[7] estimate φ_P for the native protein:
   native protein: $F_P = 858$,

   Pt derivative: $F_{HP} = 756$, $F_H = 141$, $\varphi_H = 78°$,
   U derivative: $F_{HP} = 856$, $F_H = 154$, $\varphi_H = 63°$,
   I derivative: $F_{HP} = 940$, $F_H = 100$, $\varphi_H = 146°$.

Appendix A

A Geometric Derivation of Bragg's Law

When a crystal is subjected to X-rays, the radiation waves behave as if they are reflecting from parallel planes in the crystal (the hkl planes). For a given set of lattice planes with indices hkl, the incident beam approaches the planes at an angle θ and is reflected at the same angle. The wave reflecting from a lower plane must travel an additional distance compared to the wave above it. This additional

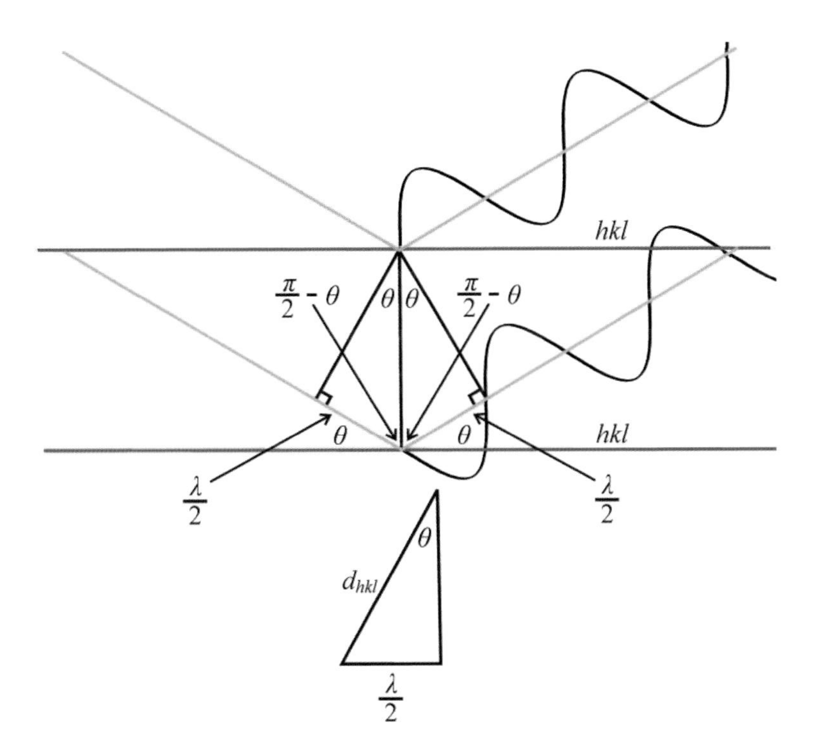

Figure A.1 Constructively interfering waves reflecting from adjacent parallel planes in Bragg diffraction.

Understanding Single-Crystal X-Ray Crystallography. Dennis W. Bennett

ISBN: 978-3-527-32677-8 (HC), 978-3-527-32794-2 (SC)

distance will tend to render the waves out of phase with respect to one another, and in general, they will destructively interfere – no diffraction will be observed. As depicted in Fig. A.1, this distance will vary with the incident angle, θ. When the waves are reflected they all undergo a 90° phase shift. Because the phases of the waves scattered from each plane are the same, the arguments made here do not depend on the relative phases of the incident and scattered beams; we can ignore this shift in phase (a phase shift of 180° has been depicted in the figure for illustrative purposes). If θ is adjusted so that the additional distance traveled is exactly one wavelength, then the reflected beams will constructively interfere, and a diffracted beam will be observed. When this occurs the relationship between the angles and the distance between the planes is illustrated in the triangle at the bottom of the figure, resulting in

$$\sin\theta_{hkl} = \frac{\lambda/2}{d_{hkl}}, \tag{A.1}$$

and

$$\lambda = 2d_{hkl}\sin\theta_{hkl}. \tag{A.2}$$

Appendix B

The Fourier Transform: Electron Density & The Structure Factor

Fourier demonstrated rigorously that any function, $f(\theta)$, continuous in the region $-\pi \leq 0 \leq \pi$ can be expanded as an infinite superposition of sinusoidal functions with varying phases and amplitudes (phases shift the functions "back and forth", while amplitudes change the "size" of the contribution of each function):

$$f(\theta) = \sum_{n=-\infty}^{n=\infty} A(n)e^{in\theta}$$

where n is an integer. As n increases the wavelength of the component waves gets shorter and shorter.

Given the coefficients, the function can be easily reproduced as an infinite sum. Note that this tells us nothing of the exact function – it just models its behavior. Suppose, on the other hand, that we know the function and wish to express it as an infinite series. How do we obtain the coefficients to model the function?

Consider the coefficient for integer m. Multiplying both sides of the above equation by $e^{-im\theta}$ (note the minus sign) and integrating over the interval $-\pi$ to π:

$$\int_{-\pi}^{\pi} f(\theta)e^{-im\theta}d\theta = \int_{-\pi}^{\pi} e^{-im\theta}\left(\sum_{n=-\infty}^{n=\infty} A(n)e^{in\theta}\right)d\theta.$$

There are two types of terms in the sum on the right. When $m = n$ the term becomes

$$\int_{-\pi}^{\pi} A(n)e^{i(n-m)\theta}d\theta = \int_{-\pi}^{\pi} A(n)e^{0}d\theta = A(n)2\pi.$$

When $n \neq m \Longrightarrow n - m = k$, the terms in the sum are of the form

$$\int_{-\pi}^{\pi} A(n)e^{ik\theta}d\theta.$$

Understanding Single-Crystal X-Ray Crystallography. Dennis W. Bennett

ISBN: 978-3-527-32677-8 (HC), 978-3-527-32794-2 (SC)

From Euler's relations, $e^{ik\theta} = \cos(k\theta) + i\sin(k\theta)$, and each of these terms integrates to zero. Thus (n and m are any integer),

$$A(n) = \frac{1}{2\pi}\int_{-\pi}^{\pi} f(\theta)e^{-in\theta}\, d\theta. \tag{B.1}$$

Now we have a pair of functions. Given one member of the pair we can generate the other. However, the variable ranges for these functions are different, and one of the "functions" is actually just an infinite series representation of that function. Thus we wish to generalize this to get a pair of functions in the same variable space which can be represented in terms of one another. We wish to turn $f(\theta)$ into an integral, then determine the behavior of both functions from $-\infty$ to ∞. To do this we must play some games with variables. These are no more than "tricks" to get us where we want to go.

We begin by replacing θ with the variable $\pi x/L$. Since $\theta = \pi x/L$ we have done nothing more than redefine x as the independent variable. The L gives us a way to modify the variable θ, resulting in

$$f(x) = \sum_{n=-\infty}^{n=\infty} A(n)e^{in\pi x/L} \qquad \text{and}$$

$$A(n) = \frac{1}{2L}\int_{-L}^{L} f(x)e^{-in\pi x/L}\, dx. \tag{B.2}$$

We now wish to see what happens as we increase L, eventually letting it go to infinity. To do this we will compute a series of integrals with increasing values of L. Since L is in the denominator, we simplify the process by defining a variable in the numerator which will change as L changes, $k_n = (n\pi)/L$. For every n there will be a different value of k_n, but each value will vary by a constant amount, depending on L: $\Delta k = k_{n+1} - k_n = \pi/L \Longrightarrow k_n = n\Delta k$. Since $A(n)$ is a function of n, and k_n is just a constant times n, we can now define $A(n)$ in terms of a function of k_n rather than n. We write it in a specific form, incorporating a constant $(1/\sqrt{2\pi})$ to allow us to determine what will happen as L goes to infinity:

$$A(n) = \left(\frac{1}{L}\right)\left(\frac{\pi}{2}\right)^{\frac{1}{2}} g(k_n) = \left(\frac{1}{L}\right)\left(\frac{\pi}{2}\right)^{\frac{1}{2}} g(n\Delta k) \tag{B.3}$$

This may appear to be an odd construction, but we have simply redefined a variable related to $A(n)$'s "natural variable", and assumed that some function of that new variable will produce $A(n)$. This new definition gives

$$f(x) = \sum_{n=-\infty}^{n=\infty} \left(\frac{1}{L}\right)\left(\frac{\pi}{2}\right)^{\frac{1}{2}} g(n\Delta k)e^{in\Delta kx}.$$

We now substitute $\pi/\Delta k$ for L, since as $L \longrightarrow \infty$, Δk goes to zero. The reason we are doing this is that wish to turn Δk into dk to make an integral. Substitution gives us

$$f(x) = \left(\frac{1}{2\pi}\right)^{\frac{1}{2}} \sum_{n=-\infty}^{n=\infty} g(n\Delta k)e^{in\Delta kx}\Delta k.$$

Combining Equations B.2 and B.3 gives

$$g(n\Delta k) = \left(\frac{1}{2\pi}\right)^{\frac{1}{2}} \int_{-L}^{L} f(x)e^{-in\Delta kx}\, dx. \tag{B.4}$$

As $L \longrightarrow \infty$, $\pi/L = \Delta k \longrightarrow 0$. As Δk goes to zero, $k_{n+1} \longrightarrow k_n \longrightarrow k_{n-1} \longrightarrow k$. Since $k_n = n\Delta k$, $n\Delta k$ collapses to k for all n. Thus, as $L \longrightarrow \infty$,

$$g(k) = \left(\frac{1}{2\pi}\right)^{\frac{1}{2}} \int_{-\infty}^{\infty} f(x)e^{-ikx}\, dx, \tag{B.5}$$

and as Δk becomes infinitesimally small,

$$f(x) = \left(\frac{1}{2\pi}\right)^{\frac{1}{2}} \sum_{n=-\infty}^{n=\infty} g(n\Delta k)e^{in\Delta kx}\Delta k \longrightarrow$$

$$f(x) = \left(\frac{1}{2\pi}\right)^{\frac{1}{2}} \int_{-\infty}^{\infty} g(k)e^{ikx}dk. \tag{B.6}$$

We now have two functions, Eqn. B.5 and Eqn. B.6, which are collectively called a *Fourier pair*. They are essentially the same function expressed with two different dynamic variables. Given the function in terms of one variable, we can determine its behavior with respect to the other variable by doing a Fourier transform. That is, given the function $g(k)$ (telling us how the function behaves when we vary k), in order to get f(x) (telling us how the function behaves when we vary x), we choose an x, and integrate over $g(k)e^{ikx}$, and determine $f(x)$ for that x, and so on.

From Chapter 3, we derived an expression for the structure factor (Eqn. 3.79):

$$\vec{F}_{hkl} = \int_0^{V_c} \rho(x_f, y_f, z_f)e^{-2\pi i(hx_f+ky_f+lz_f)}\, dV. \tag{B.7}$$

The Fourier transform gives the electron density function (the integration is over the reciprocal lattice vectors from 0 to V_c^*):

$$\rho(x_f, y_f, z_f) = \int_0^{V_c^*} \vec{F}_{hkl}e^{2\pi i(hx_f+ky_f+lz_f)}\, dV^*. \tag{B.8}$$

The variables in the integral are not continuous – they have integer values at the reciprocal lattice points (h, k, l). The integral therefore becomes a sum:

$$\rho(x_f, y_f, z_f) = \sum_{h=-\infty}^{\infty}\sum_{k=-\infty}^{\infty}\sum_{l=-\infty}^{\infty} \vec{F}_{hkl}e^{-2\pi i(hx_f+ky_f+lz_f)}\Delta V^*. \tag{B.9}$$

To determine ΔV^* we consider the simpler one-dimensional case:

$$\rho(x) = \sum_{h'=-\infty}^{\infty} \vec{F}_{h'}e^{2\pi i(h'x_f)}\Delta x^*$$

$$\vec{F}_h = \int_0^a \rho(x)e^{-2\pi i(hx/a)}\, dx, \tag{B.10}$$

where $x_f = x/a$. Substituting,

$$\begin{aligned}
\vec{F}_h &= \int_0^a \left(\sum_{h'=-\infty}^{\infty} \vec{F}_{h'} e^{2\pi i (h' x_f)} \Delta x^* \right) e^{-2\pi i (hx/a)}\, dx \\
&= \sum_{h'=-\infty}^{\infty} \int_0^a \left(\vec{F}_{h'} e^{2\pi i (h' x_f)} \Delta x^* \right) e^{-2\pi i (hx/a)}\, dx \\
&= \sum_{h'=-\infty}^{\infty} \int_0^a \left(\vec{F}_{h'} e^{2\pi i (h'-h) x/a} \Delta x^* \right) dx \qquad \text{(B.11)}
\end{aligned}$$

There are two types of terms in the sum, one for $h = h'$ and the other for $h \neq h'$. For $h = h'$,

$$\int_0^a \left(\vec{F}_{h'} e^{2\pi i (h'-h) x/a} \Delta x^* \right) dx = \int_0^a \vec{F}_{h'} \Delta x^* e^0 dx = \vec{F}_h \Delta x^* a. \qquad \text{(B.12)}$$

For $h \neq h'$

$$\begin{aligned}
\int_0^a \left(\vec{F}_{h'} e^{2\pi i (h'-h) x/a} \Delta x^* \right) dx &= \frac{a}{2\pi i (h'-h)} e^{2\pi i (h'-h) x/a} \Big|_0^a \\
&= \frac{a}{2\pi i (h'-h)} \left(e^{2\pi i (h'-h)(1)} - e^{2\pi i (h'-h)(0)} \right) \\
&= \frac{a}{2\pi i (h'-h)} (1-1) = 0. \qquad \text{(B.13)}
\end{aligned}$$

The only non-zero term in the sum has $h = h'$, and

$$\begin{aligned}
\vec{F}_h &= \vec{F}_h \Delta x^* a \Rightarrow \\
\Delta x^* &= \frac{1}{a}. \qquad \text{(B.14)}
\end{aligned}$$

Extrapolating to three dimensions gives $\Delta V^* = 1/V_c$, and we obtain Eqn. 3.83,

$$\rho(x_f, y_f, z_f) = \frac{1}{V_c} \sum_{h=-\infty}^{\infty} \sum_{k=-\infty}^{\infty} \sum_{l=-\infty}^{\infty} \vec{F}_{hkl} e^{-2\pi i (h x_f + k y_f + l z_f)}. \qquad \text{(B.15)}$$

Appendix C

Determination of the Phase Parameter in the Amplitude Reflectivity Ratio

As the X-ray beam crosses adjacent planes in the crystal the diffracted wave undergoes a phase shift characterized by the parameter, η:

$$\mathbf{S}_{j+1} = \mathbf{S}_j e^{i\pi} e^{-\eta} \tag{C.1}$$

$$\mathbf{S}_j = \mathbf{S}_{j+1} e^{-i\pi} e^{\eta} = \mathbf{S}_{j+1} e^{i\pi} e^{\eta}$$

$$\mathbf{S}_{j-1} = \mathbf{S}_j e^{i\pi} e^{\eta}. \tag{C.2}$$

Eqns. 5.261 and 5.262 contain both transmitted and reflected amplitudes, and to determine η we must first combine these relationships into an equation that contains reflected wave vectors exclusively. Rearranging Eqn. 5.261 and noting that it is equally valid for $\mathbf{T}_j$ and $\mathbf{T}_{j-1}$ gives

$$i\sigma\mathbf{T}_j = \mathbf{S}_{j+1} e^{i\varphi} e^{-i\gamma} - \mathbf{S}_j \tag{C.3}$$

$$i\sigma\mathbf{T}_{j-1} = \mathbf{S}_j e^{i\varphi} e^{-i\gamma} - \mathbf{S}_{j-1}. \tag{C.4}$$

Multiplying Eqn. 5.262 by $i\sigma$ and substituting the reflection wave vector expressions above for $i\sigma\mathbf{T}_j$ and $i\sigma\mathbf{T}_{j-1}$ results in

$$(\mathbf{S}_{j+1} e^{i\varphi} e^{-i\gamma} - \mathbf{S}_j) e^{-i\varphi} = (\mathbf{S}_j e^{i\varphi} e^{-i\gamma} - \mathbf{S}_{j-1}) e^{-i\gamma} + \sigma^2 \mathbf{S}_j e^{i\varphi}. \tag{C.5}$$

Expanding, collecting terms, and multiplying each side by $e^{-i\varphi}$ gives

$$(\mathbf{S}_{j+1} + \mathbf{S}_{j-1}) e^{-i\varphi} e^{-i\gamma} = (\sigma^2 + (e^{-i\gamma})^2 + e^{-i2\varphi}) \mathbf{S}_j. \tag{C.6}$$

Substituting from Eqns. C.1 and C.2, and noting that $\varphi = \pi + \delta$ and $e^{-i2\pi} = 1$,

$$(\mathbf{S}_j e^{i\pi} e^{-\eta} + \mathbf{S}_j e^{i\pi} e^{\eta}) e^{-i\varphi} e^{-i\gamma} = (\sigma^2 + (e^{-i\gamma})^2 + e^{-i2\varphi}) \mathbf{S}_j \quad \text{and}$$

$$(e^{-i\delta} e^{-i\gamma})(e^{-\eta} + e^{\eta}) = \sigma^2 + (e^{-i\gamma})^2 + e^{-i2\delta}. \tag{C.7}$$

Since γ is small, $e^{-i\gamma} \approx (1 - i\gamma)$ and

$$(1 - i\gamma) e^{-i\delta} (e^{-\eta} + e^{\eta}) = \sigma^2 + (1 - i\gamma)^2 + e^{-i2\delta}. \tag{C.8}$$

Understanding Single-Crystal X-Ray Crystallography. Dennis W. Bennett

ISBN: 978-3-527-32677-8 (HC), 978-3-527-32794-2 (SC)

Expanding the exponentials as power series (Eqn. 5.258) to second order,

$$
\begin{aligned}
e^{-i\delta} &= 1 - i\delta - \frac{1}{2}\delta^2 && \text{(C.9)}\\
e^{-\eta} + e^{\eta} &= 1 - \eta + \frac{1}{2}\eta^2 + 1 + \eta + \frac{1}{2}\eta^2 = 2 + \eta^2 && \text{(C.10)}\\
e^{-i2\delta} &= 1 - i2\delta - 2\delta^2, && \text{(C.11)}
\end{aligned}
$$

resulting in

$$(1 - i\delta - \frac{1}{2}\delta^2 - i\gamma - \gamma\delta - \frac{1}{2}i\gamma\delta^2)(2 + \eta^2) = \sigma^2 + 1 - i2\gamma - \gamma^2 + 1 - i2\gamma - 2\delta^2. \quad \text{(C.12)}$$

Setting the real parts of the expressions on each side equal to one another gives

$$
\begin{aligned}
(1 - \frac{1}{2}\delta^2 - \gamma\delta)(2 + \eta^2) &= 2 + \sigma^2 - \gamma^2 - 2\delta^2\\
2 + \eta^2 - \delta^2 - \frac{1}{2}\delta^2\eta^2 - 2\gamma\delta - \gamma\delta\eta^2 &= 2 + \sigma^2 - \gamma^2 - 2\delta^2.
\end{aligned}
$$

The terms $\frac{1}{2}\delta^2\eta^2$ and $\gamma\delta\eta^2$ are multiple products of numbers much less than unity, and are negligible, compared to the other terms. Neglecting these terms results in an expression for η^2:

$$
\begin{aligned}
\eta^2 &= \sigma^2 - \delta^2 + 2\delta\gamma - \gamma^2\\
&= \sigma^2 - (\delta - \gamma)^2. && \text{(C.13)}
\end{aligned}
$$

Appendix D

Reflection From a Single Plane

In the description of dynamic diffraction in Chapter 5 an expression for the integrated intensity was derived by taking into account the phase relationships resulting from diffraction from successive planes as the transmitted beam passes through the crystal. Central to the discussion was the factor, σ, relating the transmitted beam amplitude (the intensity of the incident beam) to the amplitude of the diffracted beam from a single plane in the crystal lattice:

$$\mathbf{S} = -i\sigma\mathbf{T} \tag{D.1}$$

$$Se^{i\varphi} = \sigma Te^{i\varphi}e^{-i\pi/2}$$

$$S = \sigma T$$

$$\sigma = \frac{S}{T} = \frac{\mathcal{E}_d}{\mathcal{E}_I}, \tag{D.2}$$

where $\mathbf{S}$ and $\mathbf{T}$ differ in phase by 90° and $\mathcal{E}_I$ and $\mathcal{E}_d$ are the amplitudes of the incident and diffracted beams, respectively.

Scattering from a single layer of atoms.

For simplicity we consider a hypothetical crystal consisting of layers of unit cells, each cell with volume, V_c and containing a single atom. We will first consider diffraction from the sets of planes containing the atoms, separated by the distance d. Here we are interested in the diffraction from a single plane in the set. Fig. D.1 represents such a plane, with each atom treated as a scattering center at some point p. Classically, the incident radiation emanating from the source at s drives a secondary oscillator at p, producing radiation 90° out of phase with respect to the incident radiation that travels to the detector at point d. The net amplitude of the scattered radiation at the detector is determined by a superposition of the wavelets from each scatterer in the plane.

The relative phase of each wavelet depends on the distance that the beam has traveled, spd, which differs in length from the direct path, sod, by $\varepsilon = \varepsilon_1 + \varepsilon_2$. The triangle osq is an isosceles triangle, dividing $\angle qos$ into two equal angles of magnitude $\upsilon/2$. The perpendicular bisector creates two right triangles and $\angle soq =$

Understanding Single-Crystal X-Ray Crystallography. Dennis W. Bennett

ISBN: 978-3-527-32677-8 (HC), 978-3-527-32794-2 (SC)

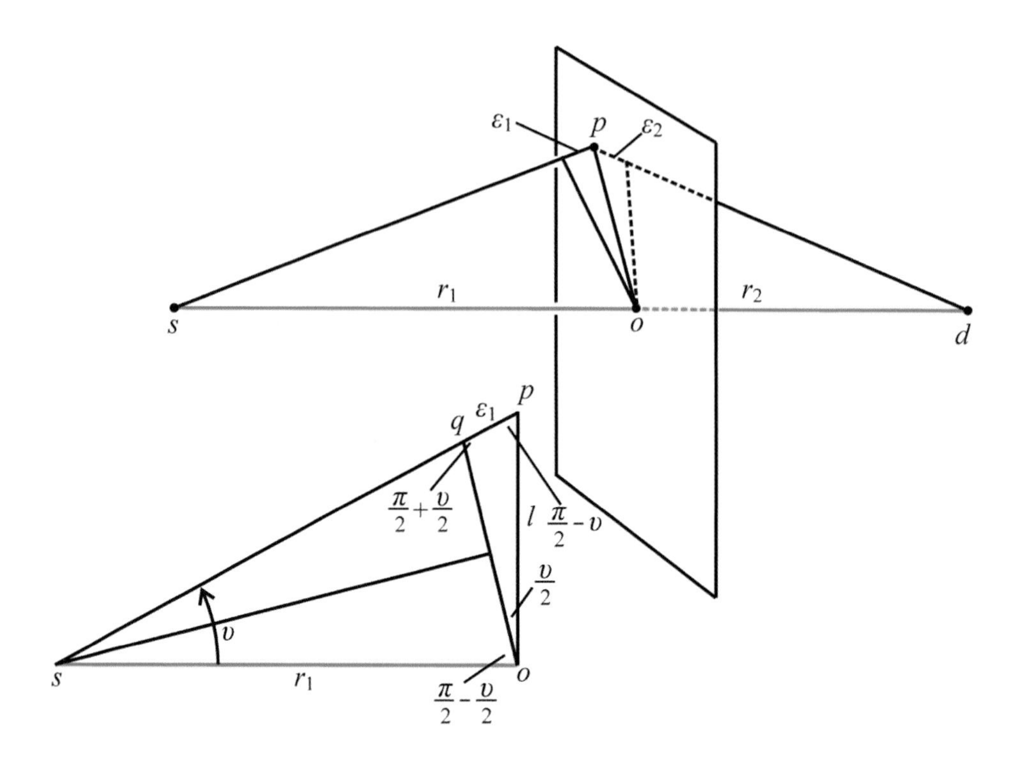

Figure D.1 Geometric relationship between a radiation source at point s, a plane of secondary oscillators at points p, and a detector at point d.

$\angle sqo = (\pi/2 - \upsilon/2)$. The result is the triangle qop with angles $\upsilon/2$, $(\pi/2 + \upsilon/2)$, and $(\pi/2 - \upsilon)$ and edge lengths l and ε_1. From the law of sines,

$$\frac{\varepsilon_1}{\sin(\upsilon/2)} = \frac{l}{\sin(\pi/2 + \upsilon/2)} = \frac{l}{\cos(\upsilon/2)};$$

$$\varepsilon_1 = l\,\frac{\sin(\upsilon/2)}{\cos(\upsilon/2)} = l\,\tan(\upsilon/2). \tag{D.3}$$

The angle υ is much smaller than its depiction in the figure, since the source and detector are generally located at greater relative distances from a very small crystal. For very small angles, the tangent of the angle is very nearly equal to the angle, leading to the approximation

$$\tan(\upsilon/2) \approx \frac{1}{2}\upsilon \approx \frac{1}{2}\tan\upsilon = \frac{1}{2}\frac{l}{r_1}. \tag{D.4}$$

From Eqn. D.3,

$$\varepsilon_1 \approx \frac{l^2}{2r_1}; \qquad \text{similarly,} \qquad \varepsilon_2 \approx \frac{l^2}{2r_2}. \tag{D.5}$$

The wavelet emitted from the atom at point p travels an additional distance and therefore arrives at the detector at a different phase than it would have if it em-

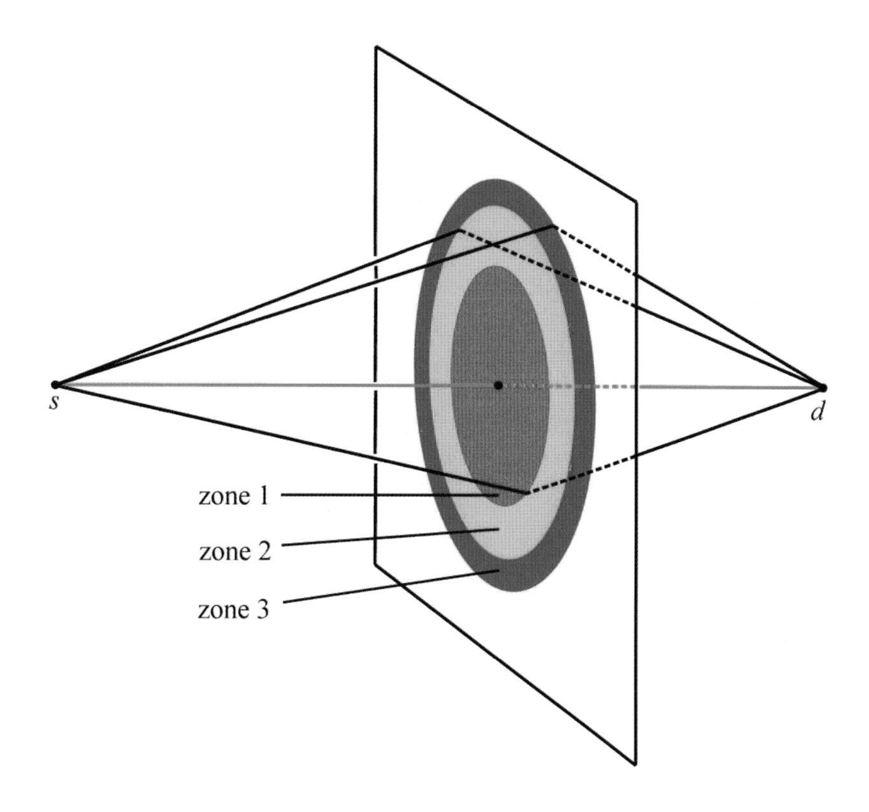

Figure D.2 Plane of secondary oscillators separated into Fresnel zones.

anated from point o, behaving as if it had been "retarded" at the plane for a time ε/c. The extra distance, ε, is often referred to as the *retardation*:

$$\begin{aligned}\varepsilon &= \varepsilon_1 + \varepsilon_2 \approx \frac{l^2}{2r_1} + \frac{l^2}{2r_2} \\ &= \frac{l^2}{2}\left(\frac{1}{r_1} + \frac{1}{r_2}\right) = \frac{l^2}{2}\left(\frac{r_1 + r_2}{r_1 r_2}\right).\end{aligned} \tag{D.6}$$

The plane of atoms is normal to the incident beam and it follows that a wavelet from any atom that lies at a point on a circle with radius l will arrive at the detector in phase with a wavelet from any other atom on the circle.

In order to determine the resultant amplitude at the detector, we divide the atomic plane into concentric circles, depicted in Fig. D.2. The first circle has a radius that gives $\varepsilon = \lambda/2$, where λ is the radiation wavelength. A wavelet from any point on the circle will therefore have a relative phase of π at d. The area inside the circle defines a zone known as a *Fresnel zone*, in this case, the first Fresnel zone, with retardations from 0 to $\lambda/2$. The second Fresnel zone has retardations

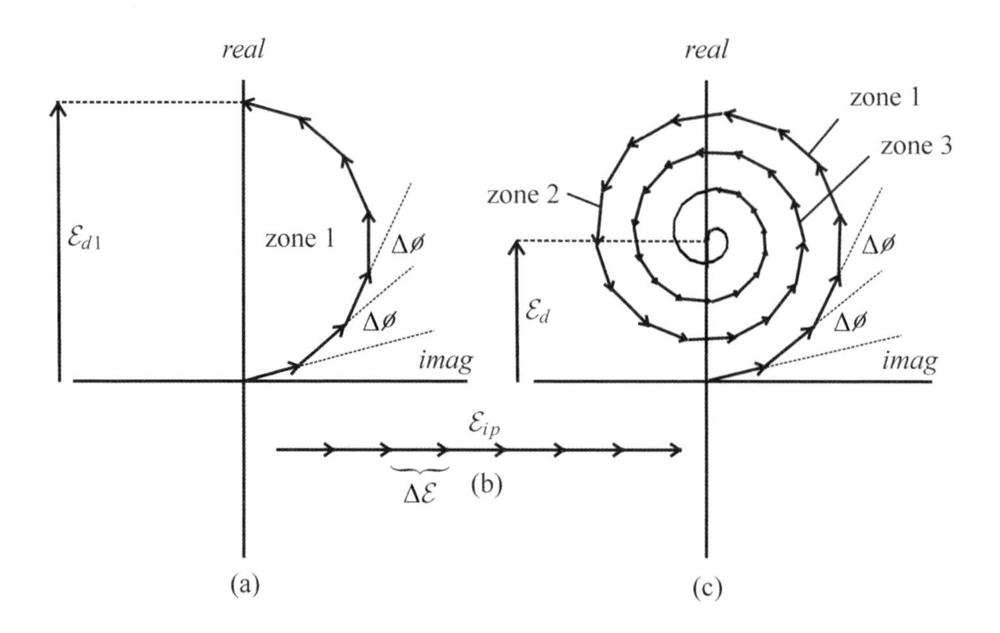

Figure D.3 (a) Resultant amplitude of the wave observed at the detector from scatterers in Fresnel zone 1. (b) Resultant amplitude from all of the Fresnel zones.

from $\lambda/2$ to $2\lambda/2$, and so forth. The outer edge of the nth Fresnel zone is therefore defined by

$$\varepsilon_n = n\frac{\lambda}{2}, \tag{D.7}$$

and the area of the nth zone is the area of the ring,

$$\begin{aligned} A_n &= \pi_n^2 - \pi_{n-1}^2 \approx 2\pi\varepsilon_n\left(\frac{r_1r_2}{r_1+r_2}\right) - 2\pi\varepsilon_{n-1}\left(\frac{r_1r_2}{r_1+r_2}\right) \\ &= 2\pi(n-(n-1)\frac{\lambda}{2}\left(\frac{r_1r_2}{r_1+r_2}\right) \\ A_n &\approx \pi\lambda\left(\frac{1}{r_1}+\frac{1}{r_2}\right)^{-1}. \end{aligned} \tag{D.8}$$

Thus, based on the approximation that $\tan v \approx v$, etc., the areas of the Fresnel zones are approximately equal.

We now divide the first Fresnel zone into a series of concentric rings, each making up a subzone. We make the rings thin enough that we can consider the retardation from each subzone to remain effectively constant. The radii of the rings are selected so that the difference between the retardations from one subzone to the next gives the same variation in phase, $\Delta\phi$, at the detector. The relationship between the change in retardation and the change in phase between each zone is $\Delta\varepsilon/\Delta\phi = \lambda/2\pi$, and

$$\Delta\phi = \frac{\Delta\varepsilon}{\lambda}2\pi. \tag{D.9}$$

Because $\Delta\phi$ is the same from one subzone to the next, $\Delta\varepsilon$ will also be the same, and the areas of the subzones and consequently the number of scatterers will be approximately the same. It follows that amplitude of the net wave from each subzone will be approximately equal to the amplitude from any another subzone. The resultant wave from the first Fresnel zone is then the superposition of waves from each zone, each with the about same amplitude, $\Delta\mathcal{E}$, with the resultant vector, $\overrightarrow{\mathcal{E}_{d1}}$, sequentially shifting in phase by $\Delta\phi$ from $\Delta\phi$ for the first subzone to π for all of the subzones, as depicted in Fig. D.3(a). The phase of π of the resultant vector follows since $\sum\Delta\varepsilon = \lambda/2$ (to the edge of Fresnel zone 1), and

$$\phi = \sum\Delta\phi = \frac{\sum\Delta\varepsilon}{\lambda}2\pi = \frac{\lambda/2}{\lambda}2\pi = \pi. \tag{D.10}$$

The subzones can be made continually smaller, with the superposition of the vectors eventually turning into a semicircle. If there had been no change in the phase of the vectors, their net amplitude would be $\mathcal{E}_{ip} = \sum\Delta\mathcal{E}$ as in Fig. D.3(b). The circle will have a circumference of $2\mathcal{E}_{ip} = \pi\mathcal{E}_{d1}$, and

$$\mathcal{E}_{d1} = \frac{2}{\pi}\mathcal{E}_{ip} = \frac{2}{\pi}\sum\Delta\mathcal{E}. \tag{D.11}$$

The semicircle that results from the superposition of wavelets from zone 1 is only an approximation since the radius from the center to the edge of the zone is approximated by $l = \varepsilon_1/\tan(\upsilon/2) \approx \varepsilon_1/\upsilon$. In reality, for all but the smallest angles, this approximation breaks down, with $\tan\upsilon$ becoming larger than υ with increasing rapidity as υ increases. If all of the Fresnel zones are subdivided, for a given change in phase, Δl decreases as υ increases, slowly at first, then at an increasing rate as the subzones get farther from the center. The areas of the subzones are therefore not actually the same, but decrease at the same rate as l^2. As a consequence, the amplitudes of the waves from each subzone decrease at this rate, and the superposition of the amplitude vectors from each subzone creates a spiral that converges into the center of a circle, as illustrated in Fig. D.3(c). The effect is very small for zone 1, and the circle is essentially the same as that in Fig. D.3(a). It follows that the resultant amplitude from all of subzones on the plane is determined by the radius of the circle, and equal to half of the resultant amplitude from the first Fresnel zone:

$$\mathcal{E}_{d} = \frac{1}{2}\mathcal{E}_{d1}. \tag{D.12}$$

Reflection from a single layer of unit cells.

In the diffraction experiment the plane of atomic oscillators is no longer normal to the incident beam, but is tilted so that is lies at angle θ with respect to the beam, as illustrated in Fig. D.4(a). In addition, the path of the scattered radiation is diverted to d', at an angle of 2θ with respect to the incident radiation. Nevertheless, the path to the detector at d' is simply the reflection of the path to d across the scattering plane, and the distance and phase relationships are therefore the same.

Consider a point p on the plane before it is tilted to the diffraction angle. If p is projected onto the plane after it is tilted – to p' – the oscillator at p' will produce

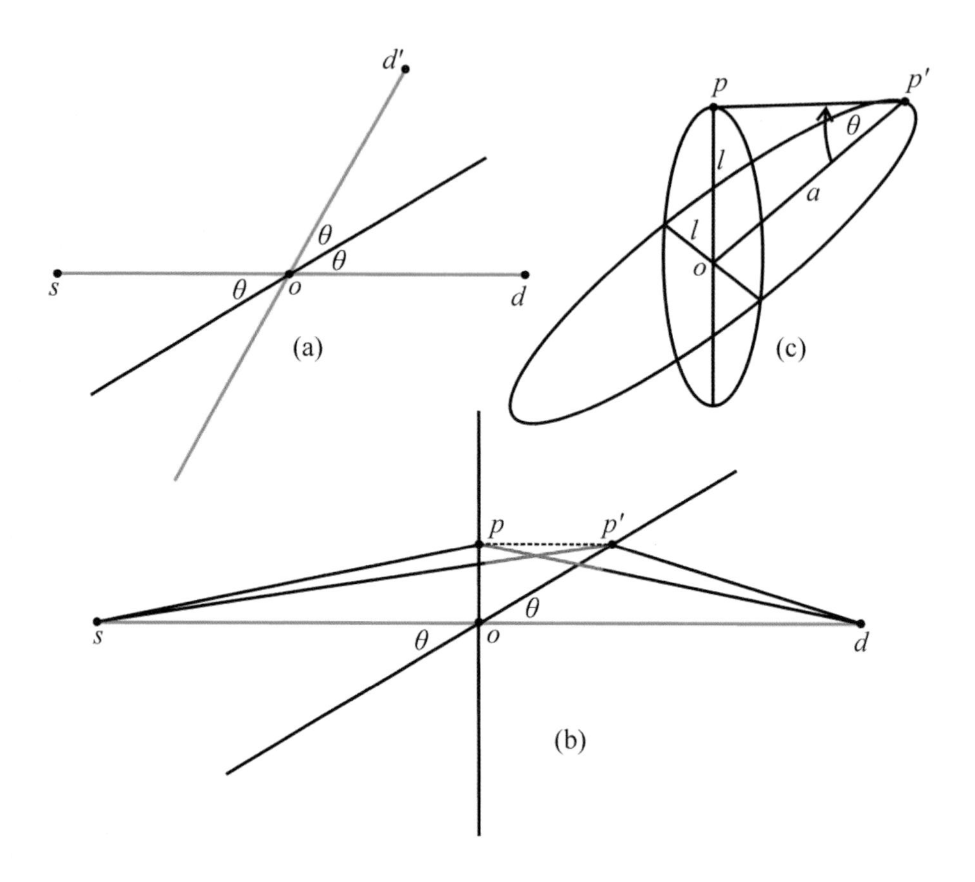

Figure D.4 (a) Symmetry between a diffracted beam with the detector at d' and a scattered beam with the detector at d. (b) Comparison of path lengths from a secondary scatterer located at point p on a plane normal to the beam and a scatterer located at the projection of p, p', onto a tilted plane. The red segments indicate the increase and decrease of the distance that the wave travels from source to detector. (c) Circular Fresnel zone on a normal plane of scatterers and its elliptical projection onto a tilted plane of scatterers.

a wavelet with the same phase at the oscillator at p. Referring to Fig. D.4(b), this happens because the change in path length on the source side of the plane is almost exactly compensated for by a change in path length on the detector side, $spd = sp'd$. The projection of the Fresnel circle onto the tilted plane becomes the ellipse shown in Fig. D.4(c), and the area of the elliptical Fresnel zone is π multiplied by the lengths of the major and minor axes of the ellipse, $A_\theta = \pi\, l\, a$, where $l = a \sin\theta$:

$$A_\theta = \frac{\pi\, l^2}{\sin\theta}. \tag{D.13}$$

The area of the Fresnel circle is $\pi\, l^2$, and from Eqn. D.8,

$$A_\theta = \frac{\pi\lambda}{\sin\theta}\left(\frac{1}{r_1}+\frac{1}{r_2}\right)^{-1}. \tag{D.14}$$

We now consider a plane of unit cells with a single atom in each unit cell. Because there are no other atoms in the unit cell we can place the origin at the center of the atom and treat it as if was a point scatterer consisting of f electrons, where f is the atomic scattering factor at the diffraction angle. According to Eqn. 5.167 the intensity observed at the detector for a single electron oscillator in the plane is

$$I_e = \left(\frac{1}{4\pi\epsilon_0 c^2}\right)^2\left(\frac{I_0 q_e^2}{r_2\, m_e}\right)^2 P = \mathcal{I}_I\left(\frac{\kappa_o}{r_2}\right)^2 P, \tag{D.15}$$

and the amplitude of the wave observed at the detector for the electron oscillator is

$$\mathcal{E}_e = \frac{1}{4\pi\epsilon_0 c^2}\frac{\mathcal{E}_I q_e^2}{r_2\, m_e}P^{1/2} = \mathcal{E}_I\frac{\kappa_o}{r_2}P^{1/2}, \tag{D.16}$$

where $\kappa_o = q_e^2/4\pi\epsilon_o m_e c^2$, r_2 is the distance from the plane to the detector, and P is the polarization factor, $(1+\cos^2 2\theta)/2$. The observed amplitude of the wave from each atom is therefore

$$\mathcal{E}_a = f\,\mathcal{E}_I\frac{\kappa_o}{r_2}P^{1/2}. \tag{D.17}$$

In order to make use of the area of the Fresnel zone amplitude we must know the amplitude of the wave from a unit area on the plane, and must therefore divide the amplitude per atom by the area accounted for by the atom, which is the area of the unit cell, A_c:

$$\mathcal{E}_u = f\,\mathcal{E}_I\frac{\kappa_o}{A_c\, r_2}P^{1/2}. \tag{D.18}$$

From Eqn. D.11, the amplitude from the first Fresnel zone is therefore

$$\mathcal{E}_{d1} = \frac{2}{\pi}\sum \mathcal{E}_u = \frac{2}{\pi}\mathcal{E}_u A_\theta, \tag{D.19}$$

where A_θ is the elliptical area of the first Fresnel zone. Thus,

$$\mathcal{E}_{d1} = \frac{2\lambda}{\sin\theta}\mathcal{E}_u\left(\frac{1}{r_1}+\frac{1}{r_2}\right)^{-1}. \tag{D.20}$$

The amplitude at the detector is half of the amplitude from the first Fresnel zone:

$$\mathcal{E}_d = \frac{\lambda}{\sin\theta}\mathcal{E}_u\left(\frac{1}{r_1}+\frac{1}{r_2}\right)^{-1} = \frac{\mathcal{E}_I\lambda\kappa_o P^{1/2}}{A_c\sin\theta\, r_2}f\left(\frac{1}{r_1}+\frac{1}{r_2}\right)^{-1}. \tag{D.21}$$

The volume of the unit cell is $V_c = A_c\, d$, where d is the distance between planes. In the diffraction experiment the distance from the source, r_1, is so large that $(1/r_1 \approx 0)$. Multiplying by d/d, the amplitude observed at the detector is therefore

$$\mathcal{E}_d = \frac{\mathcal{E}_I\lambda\kappa_o\, dP^{1/2}\, r_2}{A_c\, d\sin\theta\, r_2}f = \frac{\mathcal{E}_I\lambda\kappa_o\, dP^{1/2}}{V_c\sin\theta}f. \tag{D.22}$$

Now, suppose that the unit cell contains a second atom. If the atom is located so that it scatters in phase with the first atom, then the amplitude observed at the detector will be the sum of the amplitudes from each set of atoms:

$$\mathcal{E}_d = \mathcal{E}_1 + \mathcal{E}_2 = \frac{\mathcal{E}_I \lambda \kappa_o \, d P^{1/2}}{V_c \sin\theta} f_1 + \frac{\mathcal{E}_I \lambda \kappa_o \, d P^{1/2}}{V_c \sin\theta} f_2. \tag{D.23}$$

In general, the atoms are located at positions $\mathbf{r}_1$ and $\mathbf{r}_2$, where they scatter out of phase; the waves scattered from each atom interfere with one another. We must therefore consider the relative phases of the waves scattered from each set of atoms:

$$\begin{aligned}
\vec{\mathcal{E}_1} &= \frac{\mathcal{E}_I \lambda \kappa_o \, d P^{1/2}}{V_c \sin\theta} f_1 \, e^{-2\pi \, i(\mathbf{h}\cdot\mathbf{r}_1)} \\
\vec{\mathcal{E}_2} &= \frac{\mathcal{E}_I \lambda \kappa_o \, d P^{1/2}}{V_c \sin\theta} f_2 \, e^{-2\pi \, i(\mathbf{h}\cdot\mathbf{r}_1)} \\
\vec{\mathcal{E}_d} &= \frac{\mathcal{E}_I \lambda \kappa_o \, d P^{1/2}}{V_c \sin\theta} (f_1 \, e^{-2\pi \, i(\mathbf{h}\cdot\mathbf{r}_1)} + f_2 \, e^{-2\pi \, i(\mathbf{h}\cdot\mathbf{r}_2)}).
\end{aligned} \tag{D.24}$$

For n atoms in the unit cell,

$$\vec{\mathcal{E}_d} = \frac{\mathcal{E}_I \lambda \kappa_o \, d P^{1/2}}{V_c \sin\theta} \sum_{j=1}^{n} f_j \, e^{-2\pi \, i(\mathbf{h}\cdot\mathbf{r}_j)}, \tag{D.25}$$

and we immediately recognize the sum as the structure factor for the reflection:

$$\vec{\mathcal{E}_d} = \mathcal{E}_d e^{i\varphi} = \frac{\mathcal{E}_I \lambda \kappa_o \, d P^{1/2}}{V_c \sin\theta} \mathbf{F}_{\mathbf{h}} = \frac{\mathcal{E}_I \lambda \kappa_o \, d P^{1/2}}{V_c \sin\theta} F_{\mathbf{h}} e^{i\varphi}. \tag{D.26}$$

The amplitude of the wave observed at the detector is therefore

$$\mathcal{E}_d = \frac{\mathcal{E}_I \lambda \kappa_o \, d P^{1/2}}{V_c \sin\theta} F_{\mathbf{h}}. \tag{D.27}$$

For a given reflection, the reflecting efficiency* of a single plane of atoms, σ, can now be determined:

$$\sigma = \frac{\mathcal{E}_d}{\mathcal{E}_I} = \frac{\lambda \kappa_o \, d P^{1/2}}{V_c \sin\theta} F_{\mathbf{h}} = \frac{\lambda \, d}{\sin\theta} \left(\frac{q_e^2}{4\pi\epsilon_o m_e c^2 V_c} \right) P^{1/2} F_{\mathbf{h}}. \tag{D.29}$$

---

*It should be noted here that if the crystal is not in the diffraction condition, the intensity will be observed at d rather than at d'. It will be greater than the intensity observed at d' because there will be no destructive interference resulting from scattering from different locations within each atom (i.e., $f_j = Z_j$). Thus the electric field at d will be

$$\mathcal{E}_d = \frac{\mathcal{E}_I \lambda \kappa_o \, d P^{1/2}}{V_c \sin\theta} F_{000}. \tag{D.28}$$

In this case σ will represent the coherent scattering efficiency of the plane of atoms.

Appendix E

A Discussion of Kinematical Models for Extinction

In 1922 C.G Darwin published a paper in *Philosophical Magazine*[83] entitled "The Reflexion of X-rays from Imperfect Crystals" in which he applied his dynamical theory of X-ray diffraction to describe the loss of intensity in the diffracted beam — "primary extinction" — and a kinematical treatment to describe the loss of intensity in the incident beam — "secondary extinction." Darwin's dynamical model, which explicitly considers phase relationships as the transmitted beam propagates through the crystal in the diffraction condition, was verified independently by Ewald[183]. It correctly describes the extinction phenomenon, but is based on a perfect and infinite crystal, and does not lend itself to the analysis of small single crystals of arbitrary shape. It has yet to be applied in routine crystal structure analysis. Dynamical diffraction is described in Chapter 5.

The kinematical model ignores phase relationships completely, and instead is based upon energy conservation. It is formulated from the apparently obvious proposition that energy lost in the transmitted beam must equal energy gained in the diffracted beam, and vice versa. The result is a pair of coupled simultaneous differential equations:

$$\frac{\partial I_0}{x_1} = -\sigma I_0 + \sigma I_d \tag{E.1}$$

$$\frac{\partial I_d}{x_2} = -\sigma I_d + \sigma I_0, \tag{E.2}$$

where x_1 and x_2 are the penetration distances for incident and diffracted beams, and σ is the fraction of radiation scattered per unit volume from either beam (Eqn. D.29). Solutions to these equations result in extinction correction factors which are applicable to various geometries and crystal shapes.[87,184] Indeed, the equations appear intuitively obvious. The incident beam loses a portion of its intensity, $-\sigma I_0$, which is gained by the diffracted beam. The same fraction of the diffracted beam is re-diffracted, and the diffracted beam loses a portion of its intensity, $-\sigma I_d$. The beam from this second diffraction propagates in the same direction as the incident beam and the energy lost from the diffracted beam is

Understanding Single-Crystal X-Ray Crystallography. Dennis W. Bennett

ISBN: 978-3-527-32677-8 (HC), 978-3-527-32794-2 (SC)

apparently regained by the incident beam. Based upon these assumptions, the overall change in intensity (energy) is

$$\frac{\partial I_0}{x_1} + \frac{\partial I_d}{x_2} = -\sigma I_0 + \sigma I_d - \sigma I_d + \sigma I_0 = 0. \tag{E.3}$$

Although Darwin used the model to describe secondary extinction, others extended it to incorporate primary extinction as well. In addition to its adaptability, energy is apparently conserved in the model, and although investigators do not point this out explicitly in the literature, in his introductory book on X-ray crystallography Woolfson notes that the dynamical theory apparently does not conserve energy[17]:

> This theory, which is the one usually given in elementary texts, suffers from the drawback that energy seems not to be conserved. The primary beam is reduced in intensity and so is the diffracted beam and one might well ask where all the energy has gone.

The kinematical theory for extinction is more readily adaptable to real crystals, and has been cited as a limiting case of the dynamical theory[26], but a careful analysis demonstrates that this clearly cannot be the case – *the dynamical and kinematical models for extinction are mutually contradictory.*

Energy conservation would appear to render the kinematical model correct, implying some fundamental flaw in the dynamical model, which predicts that destructive interference will reduce the intensity of the incident beam. Recalling Eqn. 5.264, as the incident beam passes from layer j to layer $j+1$ in the crystal, the intensity of the beam decreases:

$$\begin{aligned} \mathbf{T}_{j+1}e^{-i\varphi} &\simeq \mathbf{T}_j(1-\sigma^2) \\ (\mathbf{T}_{j+1}e^{-i\varphi})(\mathbf{T}_{j+1}e^{-i\varphi})^* &\simeq \mathbf{T}_j\mathbf{T}_j^*(1-\sigma^2)^2 \\ I_{j+1} &\simeq I_j(1-\sigma^2)^2. \end{aligned} \tag{E.4}$$

This occurs because the diffracted beam is 90° out of phase with the transmitted beam, and the radiation from the second diffraction undergoes another 90° phase shift, sending it in the direction of the transmitted beam, *but* 180° *out of phase with it.* The resulting destructive interference reduces the intensity of the beam. A third diffraction of this radiation sends a fraction of the radiation back in the direction of the diffracted beam that is 180° out of phase with it, *slightly* reducing its intensity as well (the "fraction of a fraction of a fraction" is a very small number). The diffracted beam is reduced in intensity primarily because of the attenuation of the transmitted beam — both beams are diminished in intensity.

In order to analyze the premise that energy is not conserved in the dynamical model, it is useful to consider absorption and extinction collectively from a photon scattering perspective. Consider a crystal in an X-ray beam out of the diffraction condition. As the X-ray photons pass through the crystal most of them do not encounter an electron, and pass through unencumbered. Those photons that do encounter an electron suffer one of three fates. Some are absorbed completely, with the energy of the photon exciting the electron into the vacuum – out of the atom in which it was initially bound. The electron is a photoelectron, with a kinetic energy that represents the difference between the energy of the photon and the binding energy of the electron. Other electrons are excited into stationary states in the atom. In this case a new photon is created with an energy that is the difference between the transition energy in the atom and the energy of the original photon. These photons are scattered in all directions becoming part of the background radiation. Because they have a different frequency than the incident photons they are said to have been *incoherently scattered.* This type of scattering is known as *Compton scattering.* These two phenomena are often collectively referred to as "true absorption."

The photons can also suffer a third fate. Instead of being absorbed into a stationary state, they can be absorbed into a "virtual" time-dependent state. After a short time delay (retardation) a new photon is created with the same energy as the absorbed photon. These photons are also scattered in all directions*. Because they have the same frequency as incident photons they are said to have been *coherently scattered.* This type of scattering is known as *Thompson scattering.* The coherently scattered photons also end up as part of the background radiation. Experimental absorption coefficients are determined by measuring the attenuation of the beam due to all of the effects listed above. *This includes the loss of intensity due to coherent scattering.* We now consider the results of rotating the crystal into the diffraction condition. The kinematical model assumes that in this condition a fraction of the photons in the beam are suddenly extracted from the beam and coherently scattered in phase toward a diffraction maximum. The removal of these photons from the beam is the putative basis for secondary extinction. This is the fundamental flaw in the kinematical model of extinction. *Photons do not scatter coherently because they diffract. They diffract because they scatter coherently.* Diffraction does not create coherently scattered photons — it redirects the photons created from coherent scattering toward diffraction maxima. These photons are created by independent events that occur whether or not the crystal happens to be in the diffraction condition. The loss in intensity "due to diffraction" is actually the loss in intensity due to coherent scattering, which is proportional to the intensity of the incident radiation. It is accounted for completely in the experimental absorption coefficient.

In his 1929 paper Darwin justified the existence of secondary extinction with an equation relating the intensity of the transmitted beam after traversing m planes to the initial intensity of the incident beam and the intensity of the diffracted beam:

$$\begin{aligned} T_m^2 + S_0^2 &= T_0^2; \\ I_m &= I_0 - I_d. \end{aligned} \tag{E.5}$$

This equation implies that the transmitted beam intensity is diminished by the intensity of the diffracted beam. From Eqn. 5.267

$$\begin{aligned} \mathbf{S}_1 &= \mathbf{S}_0 e^{i\pi} e^{-\eta} \\ \mathbf{S}_2 &= \mathbf{S}_1 e^{i\pi} e^{-\eta} = \mathbf{S}_0 e^{i\pi} e^{-\eta} e^{i\pi} e^{-\eta} = \mathbf{S}_0 e^{i2\pi} e^{-2\eta} \\ \mathbf{S}_3 &= \mathbf{S}_0 e^{i3\pi} e^{-3\eta} \\ &\vdots \\ \mathbf{S}_m &= \mathbf{S}_0 e^{im\pi} e^{-m\eta} \end{aligned}$$

*This statement is not quite true. The scattering probability is a function of the cosine of the angle with respect to the incident beam, resulting in a nodal plane orthogonal to the beam. The result is the polarization effect discussed in Chapter 5.

From Eqn. 5.272,

$$\mathbf{R}_A = \frac{\mathbf{S}_m}{\mathbf{T}_m} = \frac{\mathbf{S}_0}{\mathbf{T}_0}, \tag{E.6}$$

and

$$\begin{aligned}
\mathbf{T}_m &= \frac{\mathbf{S}_m \mathbf{T}_0}{\mathbf{S}_0} = \frac{\mathbf{S}_0 e^{im\pi} e^{-m\eta} \mathbf{T}_0}{\mathbf{S}_0}, \\
\mathbf{T}_m \mathbf{T}_m^* &= e^{2im\pi} (e^{-m\eta})(e^{-m\eta})^* \mathbf{T}_0 \mathbf{T}_0^*, \\
T_m^2 &= T_0^2 e^{(-m(\eta_r + i\eta_i))} e^{(-m(\eta_r - i\eta_i))}, \\
I_m &= I_0 e^{-2m\eta_r},
\end{aligned} \tag{E.7}$$

where $\eta = \eta_r + i\eta_i$; η_r is the real portion of the complex quantity, η. It follows that the diminution in the transmitted beam intensity is not related to the intensity of the diffracted beam, but arises solely from interference phenomena. No additional photons are removed from the transmitted beam when primary diffraction occurs, but they are removed from the transmitted beam when secondary diffraction occurs — as a consequence of destructive interference.

The contradiction between the dynamical and kinematical models now becomes clear. The kinematical model for extinction adds the photons from secondary diffraction back into the transmitted beam, while the dynamical model removes them. In the dynamical model, these photons are found neither in the incident nor the diffracted beam. Just as with diffraction, interference has directed a portion of the photons away from either beam. *Interference does not destroy photons — it alters their trajectory.* The redirection of the plane wave for these photons must alter their paths by only a few degrees to render them incapable of undergoing diffraction. None of the energy is lost – it is simply unaccounted for in either the incident or diffracted beams.

The kinematical model for extinction therefore fails on two counts. First, it is based on the assumption that diffraction reduces the intensity of the incident beam and second, it does not account for the phase changes that are an inherent part of the scattering process. The dynamical model suffers from neither of these problems, and is clearly the proper way to treat extinction. In addition, all extinction is "primary" extinction, resulting from destructive interference. "Secondary extinction" is due to nothing more than the coherent scattering of photons. It is taken care of in the absorption correction.

Appendix F

Probability Integrals: The Modified Bessel Function

The determination of expected (average) values in probability theory generally involves the summation (integration) of a function at each value of the independent variable, weighted by the probability that the value will be observed:

$$\langle f(x) \rangle = \int_{x_1}^{x_2} Pr(x) f(x)\, dx. \tag{F.1}$$

The nature of the integrand often makes evaluation of the integral difficult; it is frequently necessary to determine its value numerically.

In crystallography we are concerned with expected values related to the phases assigned to reflections, which often take the form

$$\int_0^{2\pi} e^{x \cos\phi} \cos(n\phi)\, d\phi. \tag{F.2}$$

Fortunately, such integrals can be evaluated with a series expansion, known as *a modified Bessel function of the first kind.* Bessel functions are attributed to their discoverer, the astronomer Freidrich Wilhelm Bessel, who published a paper in 1824 entitled "Investigation of the Part of Planetary Perturbation which Arises from the Motion of the Sun." The examination of such perturbations often begins with a differential equation describing the perturbation, in this case, *Bessel's equation*:

$$x^2 \frac{d^2(J_n(x))}{dx^2} + x\frac{d(J_n(x))}{dx} + (x^2 - n^2)J_n(x) = 0, \tag{F.3}$$

where the solution, $J_n(x)$, is a function that describes the perturbed motion for a given value of the integer, n.*

The second order differential equation can be solved by the Frobenius method, in which a power series solution that satisfies the equation has the form:

$$J_n(x) = x^k \sum_{j=0}^{\infty} a_j x^j = \sum_{j=0}^{\infty} a_j x^{j+k}. \tag{F.4}$$

*Bessel functions can take on non-integer orders as well. We will consider only those functions of integer order here.

Understanding Single-Crystal X-Ray Crystallography. Dennis W. Bennett

ISBN: 978-3-527-32677-8 (HC), 978-3-527-32794-2 (SC)

To determine the solution we insert Eqn. F.4 into Eqn. F.3 and determine the values of the coefficients, a_j, that satisfy the equation:

$$x^2 \sum_{j=0}^{\infty}(k+j)(k+j-1)a_j x^{k+j-2} + x \sum_{j=0}^{\infty}(k+j)a_j x^{k+j-1}$$
$$+ x^2 \sum_{j=0}^{\infty} a_j x^{k+j} - n^2 \sum_{j=0}^{\infty} a_j x^{k+j} = 0$$
$$\sum_{j=0}^{\infty}(k+j)(k+j-1)a_j x^{k+j} + \sum_{j=0}^{\infty}(k+j)a_j x^{k+j}$$
$$+ \sum_{j=2}^{\infty} a_{j-2} x^{k+j} - n^2 \sum_{j=0}^{\infty} a_j x^{k+j} = 0 \qquad \text{(F.5)}$$

After expansion, the coefficient of each power of x must be zero, since the entire sum equals zero. For $j = 0$ and $x^{j+k} = x^k$, $(k(k-1) + k - n^2)a_0 x^k = 0$, and

$$a_0(k^2 - n^2) = 0 \Longrightarrow k = \pm n, \qquad \text{(F.6)}$$

since $a_0 \neq 0$. For $j = 1$ and $x^{j+k} = x^{k+1}$, we obtain

$$a_1((k+1)^2 - n^2) = 0. \qquad \text{(F.7)}$$

Since $k = \pm n$, $(k+1)^2 - n^2$ cannot be zero, and it follows that a_1=0. For $j \geq 2$,

$$([(k+j)(k+j-1) + (k+j) - n^2]a_j + a_{j-2})x^{j+k} = 0, \qquad \text{(F.8)}$$

and

$$a_j = \frac{-a_{j-2}}{(k+j)^2 - n^2}. \qquad \text{(F.9)}$$

For $j = 3$, $a_{j-2} = a_1 = 0$ and $a_3 = 0$. Similarly, when $j = 5$, $a_{j-2} = a_3$, and $a_5 = 0$. Following the same logic, $a_j = 0$ for all odd values of j, i.e., $j = 2m + 1$, $m = 1, 2, \ldots$; the only non-zero values have $j = 2m$. Selecting $n = k$ gives

$$\begin{aligned}
a_{2m} &= \frac{-a_{2m-2}}{(n+2m)^2 - n^2} = \frac{-a_{2m-2}}{4m(m+n)} \qquad \text{(F.10)} \\
a_2 &= \frac{-a_0}{4(1)(1+n)} = \frac{-a_0}{2^2(1)(1+n)} \\
a_4 &= \frac{-a_2}{4(2)(2+n)} = \frac{+a_0}{2^2(1)\cdot 2^2(2)(2+n)(1+n)} \\
a_6 &= \frac{-a_4}{4(3)(3+n)} = \frac{-a_0}{2^2(1)\cdot 2^2(2)\cdot 2^2(3)(3+n)(2+n)(1+n)} \\
a_8 &= \frac{-a_6}{4(4)(4+n)} = \frac{+a_0}{2^2(1)\cdot 2^2(2)\cdot 2^2(3)\cdot 2^2(4)(4+n)(3+n)(2+n)(1+n)},
\end{aligned}$$

etc. Each coefficient is generated by the preceding one, a relationship known as a *recursion relationship*:

$$a_{2m} = \frac{(-1)^m a_0}{(2^2)^m \, m!(m+n)!} = \frac{(-1)^m a_0}{2^{2m} \, m!(m+n)!}. \qquad \text{(F.11)}$$

a_0 is an arbitrary constant (any value of a_0 will provide a solution to the equation). For convenience we set a_0 to 2^{-n}, giving

$$a_{2m} = \frac{(-1)^m}{m!(m+n)!}\frac{1}{2^{2m+n}}. \tag{F.12}$$

Setting $k = n$ in Eqn. F.4, the solutions to Bessel's equation therefore become

$$J_n(x) = \sum_{m=0}^{\infty} a_{2m} x^{2m+n} = \sum_{m=0}^{\infty} \frac{(-1)^m}{m!(m+n)!}\left(\frac{x}{2}\right)^{2m+n}. \tag{F.13}$$

For a given value of n, the resulting function, $J_n(x)$ is known as a *Bessel function of the first kind of order n*. While the functions were first derived to solve the Bessel differential equation, they have important properties that allow them to be used for a large variety of applications. In cylindrical coordinate systems they turn out to be the analogs of sinusoidal functions, and since they exhibit oscillatory behavior (every term in the sum changes sign) Bessel functions can be used to expand arbitrary functions, just as sinusoidal functions are employed in a Fourier series.

The functions can also be altered to form solutions for a more extensive range of differential equations. For our purposes, the most important alteration produces the *modified Bessel function of the first kind*:

$$I_n(x) = i^{-n} J_n(ix). \tag{F.14}$$

The effect of the inclusion of the imaginary argument in this fashion is to remove the oscillatory behavior:

$$\begin{aligned} I_n(x) &= i^{-n} \sum_{m=0}^{\infty} \frac{(-1)^m}{m!(m+n)!}\left(\frac{ix}{2}\right)^{2m+n} = \sum_{m=0}^{\infty} \frac{(-1)^m}{m!(m+n)!}\left(\frac{x}{2}\right)^{2m+n} (i^2)^m \\ &= \sum_{m=0}^{\infty} \frac{(-1)^m(-1)^m}{m!(m+n)!}\left(\frac{x}{2}\right)^{2m+n} = \sum_{m=0}^{\infty} \frac{1}{m!(m+n)!}\left(\frac{x}{2}\right)^{2m+n}. \end{aligned} \tag{F.15}$$

Note that although $I_n(x)$ is created by incorporating an imaginary argument into the real function $J_n(x)$, the i^{-n} term keeps $I_n(x)$ real.

The modified Bessel function for $-n$ is given by

$$I_{-n}(x) = \sum_{m=0}^{\infty} \frac{1}{m!(m-n)!}\left(\frac{x}{2}\right)^{2m-n}. \tag{F.16}$$

Although it is an infinite series, the terms for $(m - n) < 0$ are zero. To prove this, we consider the following:

$$\begin{aligned}
n! &= \frac{(n+1)!}{n+1} \\
2! &= \frac{3!}{3} = 2 \\
1! &= \frac{2!}{2} = 1 \\
0! &= \frac{1!}{1} = 1 \\
-1! &= \frac{0!}{0} = \infty \\
-2! &= \frac{-1!}{-1} = \infty, \text{etc.}
\end{aligned}$$

In order to handle these terms we set $m = m' + n$, giving

$$\begin{aligned}
I_{-n}(x) &= \sum_{(m'+n)=0}^{\infty} \frac{1}{(m'+n)!m'!} \left(\frac{x}{2}\right)^{2m'+n} \\
&= \sum_{m'=-n}^{-1} \frac{1}{m'!(m'+n)!} \left(\frac{x}{2}\right)^{2m'+n} + \sum_{m'=0}^{\infty} \frac{1}{m'!(m'+n)!} \left(\frac{x}{2}\right)^{2m'+n}
\end{aligned}$$

The terms in the first sum are all zero since each contains $1/m'! = 1/\infty = 0$. The second sum begins with $m' = 0$ and we can now drop the prime:

$$I_{-n}(x) = \sum_{m=0}^{\infty} \frac{1}{m!(m+n)!} \left(\frac{x}{2}\right)^{2m+n} = I_n(x). \tag{F.17}$$

When summed from $-\infty$ to $+\infty$, for every $I_n(x)$ there is an equal value for $I_{-n}(x)$.*

In addition to their many physical applications, it is fortuitous that integrals which arise in crystallographic probability distributions can be evaluated as the *integral forms of modified Bessel functions.* In order to establish these forms, we first consider the function

$$e^{(x/2)(t+1/t)} = e^{xt/2} e^{x/2t}. \tag{F.18}$$

The function e^q can be expanded as a Maclaurin series:

$$e^q = q^0 + \frac{q^1}{1} + \frac{q^2}{2} + \frac{q^3}{6} + \cdots = \sum_{n=0}^{\infty} \frac{q^n}{n!}.$$

Thus,

$$\begin{aligned}
e^{xt/2} e^{x/2t} &= \left(\sum_{k=0}^{\infty} \left(\frac{x}{2}\right)^k \left(\frac{t^k}{k!}\right)\right) \left(\sum_{m=0}^{\infty} \left(\frac{x}{2}\right)^m \left(\frac{t^{-m}}{m!}\right)\right). \\
&= \sum_{k=0}^{\infty} \sum_{m=0}^{\infty} \frac{1}{m!k!} \left(\frac{x}{2}\right)^{k+m} t^{k-m}.
\end{aligned} \tag{F.19}$$

*Using the same treatment it is easy to show that the magnitudes of $J_n(x)$ and $J_{-n}(x)$ are the same, but they have opposite signs for odd n: $J_{-n}(x) = (-1)^n J_n(x)$.

We now focus on the exponent of t, $n = k - m$. Since k and m each cover the range from zero to infinity, there will be a value of n for every integer from $-\infty$ to $+\infty$. We now wish to collect all of the terms in the sum in Eqn. F.19 for a specific value of n. For every value of m there will be a corresponding value of k such that $k - m = n$ and $k = n + m$, creating a term in the sum

$$\frac{1}{m!(m+n)!}\left(\frac{x}{2}\right)^{2m+n} t^n. \tag{F.20}$$

Since there is a term for every m, the net contribution to the sum for each n is

$$\sum_{m=0}^{\infty}\frac{1}{m!(m+n)!}\left(\frac{x}{2}\right)^{2m+n} t^n = I_n(x)t^n. \tag{F.21}$$

Eqn. F.19 is then the sum over all of the values of n:

$$e^{(x/2)(t+1/t)} = \sum_{n=-\infty}^{\infty} I_n(x)t^n, \tag{F.22}$$

an expansion of the function in terms of the modified Bessel functions. In order to derive the integral representation of $I_n(x)$ we consider the case for $t = e^{i\phi}$:

$$\begin{aligned}
(x/2)(t+1/t) &= x\left(\frac{e^{i\phi}+e^{-i\phi}}{2}\right) = x\cos\phi \\
e^{(x/2)(t+1/t)} &= e^{x\cos\phi} \\
e^{x\cos\phi} &= I_0(x)e^0 + I_1(x)e^{i\phi} + I_{-1}(x)e^{-i\phi} + I_2(x)e^{i2\phi} + I_{-2}(x)e^{-i2\phi} \\
&\quad + I_3(x)e^{i3\phi} + I_{-i3}(x)e^{-i3\phi} + I_4(x)e^{i4\phi} + I_{-i4}(x)e^{-i4\phi} + \ldots \\
&= I_0(x) + I_1(x)(e^{i\phi}+e^{-i\phi}) + I_2(x)(e^{i2\phi}+e^{-i2\phi}) \\
&\quad + I_3(x)(e^{i3\phi}+e^{-i3\phi}) + I_4(x)(e^{i4\phi}+e^{-i4\phi}) + \ldots \\
&= I_0(x) + 2(I_1(x)\cos\phi) + 2(I_2(x)\cos 2\phi) \\
&\quad + 2(I_3(x)\cos 3\phi) + 2(I_4(x)\cos 4\phi) + \ldots \\
&= I_0(x) + 2\sum_{n=1}^{\infty} I_n(x)\cos n\phi.
\end{aligned} \tag{F.23}$$

For a particular value of n, say m, we now create the following integral:

$$\int_0^{\pi} e^{x\cos\phi}\cos m\phi\, d\phi = \int_0^{\pi}\left(I_0(x) + 2\sum_{n=1}^{\infty} I_n(x)\cos n\phi\right)\cos m\phi\, d\phi. \tag{F.24}$$

Any term in the sum (except for the term involving $I_0(x)$) will have the following form:

$$2I_n(x)\int_0^{\pi}\cos n\phi\,\cos m\phi\, d\phi. \tag{F.25}$$

For $m \neq n$, from the trigonometric identity for $\cos\alpha\,\cos\beta$ (Eqn. 3.58),

$$\begin{aligned}\int_0^\pi \cos n\phi\ \cos m\phi\ d\phi &= \frac{1}{2}\int_0^\pi \cos((m+n)\phi)\ d\phi + \frac{1}{2}\int_0^\pi \cos((m-n)\phi)\ d\phi \\ &= \frac{1}{2}\int_0^\pi \cos k\phi\ d\phi + \frac{1}{2}\int_0^\pi \cos l\phi\ d\phi \\ &= \frac{1}{2}\left(\frac{\sin k\phi}{k} + \frac{\sin l\phi}{l}\right)\Bigg|_0^\pi = 0, \quad m \neq n.\end{aligned}$$

If $m = n \neq 0$, then from the integral for $\cos^2\alpha$ (Eqn. 3.57),

$$\int_0^\pi \cos^2 n\phi\ d\phi = \left(\frac{n\phi}{2n} + \frac{\sin 2n\phi}{4n}\right)\Bigg|_0^\pi = \frac{\pi}{2}. \tag{F.26}$$

Thus the only non-zero term in the sum has $n = m$, and

$$\int_0^\pi e^{x\cos\phi}\cos n\phi\ d\phi = 2I_n(x)\frac{\pi}{2} = \pi I_n(x),$$

resulting in *the integral representation of the modified Bessel function of the first kind of order* n:

$$I_n(x) = \frac{1}{\pi}\int_0^\pi e^{x\cos\phi}\cos n\phi\ d\phi. \tag{F.27}$$

Note that in the case where $m = 0$,

$$\int_0^\pi e^{x\cos\phi} d\phi = \int_0^\pi \left(I_0(x) + 2\sum_{n=1}^{\infty} I_n(x)\cos n\phi\right) d\phi = \pi I_0(x), \tag{F.28}$$

since each term with a cosine integrates to zero. Thus

$$I_0(x) = \frac{1}{\pi}\int_0^\pi e^{x\cos\phi}\ d\phi = \frac{1}{\pi}\int_0^\pi e^{x\cos\phi}\cos 0\ d\phi, \tag{F.29}$$

and Eqn. F.27 is valid for all values of n.

Appendix G

Monte Carlo Optimization – A Simple Example

Phase annealing, described in Chapter 7, is a method of determining optimized phases by allowing a starting set of phases to undergo changes that result in a random walk toward an optimal figure of merit. The fundamental concept is nicely illustrated by considering a simpler optimization process – that of determining the "best line" to a set of data. This line is a "least-squares" or "linear regression" line, one that can easily be determined analytically. The example shown here is therefore an unnecessary one, but the method can readily be extended to fit data to any function.

We consider a set of n experimental absorbance data, $\{A_{E,1}, A_{E,2}, \ldots, A_{E,n}\}$, collected with a spectrometer from a set of samples of compound X at concentrations $\{c_1, c_2, \ldots, c_n\}$. Furthermore, we note that the windows of our instrument have been contaminated with an inadvertent spill of compound X, so that a blank sample records an absorbance of A_0. If there is no error in our measurements we anticipate the data to fit exactly on a straight line, $A_{L,i} = sc_i + A_0$, which we can plot, connect the points and use as a calibration curve to determine the concentration for any value of the absorbance. However, our data are experimental, and therefore subject to errors. In this case we have a systematic error represented by A_0, which adds to every value of the absorbance, and random errors (normally distributed errors) $\{\epsilon_1, \epsilon_2, \ldots, \epsilon_n\}$, due to instrument fluctuations and lack of accuracy in preparing the samples. The "best fit" calibration line is one that minimizes the average (root mean square) error, which is at its minimum (optimal) value when $\sum_i^n \epsilon_i^2$ is at a minimum, where $\epsilon_i = A_{L,i} - A_{E,i}$, the residual difference between the predicted absorbance from the line, $A_{L,i}$ and the experimentally measured absorbance $A_{E,i}$.

Understanding Single-Crystal X-Ray Crystallography. Dennis W. Bennett

ISBN: 978-3-527-32677-8 (HC), 978-3-527-32794-2 (SC)

The least-squares line is determined by finding a value of the slope, s and the intercept, A_0, that minimizes $\sum_i^n \epsilon_i^2 = \sum_i^n (sc_i + A_0 - A_{E,i})^2$. This will be the values of s and A_0 for which

$$\frac{\partial \sum_i^n (sc_i + A_0 - A_{E,i})^2}{\partial s} = 0 \quad \text{and} \quad \frac{\partial \sum_i^n (sc_i + A_0 - A_{E,i})^2}{\partial A_0} = 0 \Longrightarrow$$

$$\left(\sum_{i=1}^{n} c_i^2\right) s + \left(\sum_{i=1}^{n} c_i\right) A_0 = \sum_{i=1}^{n} c_i A_{E,i} \tag{G.1}$$

$$\left(\sum_{i=1}^{n} c_i\right) s + nA_0 = \sum_{i=1}^{n} A_{E,i} \tag{G.2}$$

The result is two linear equations in two unknowns; the optimal values of s and A_0 are readily determined:

$$\begin{bmatrix} \sum_{i=1}^n c_i^2 & \sum_{i=1}^n c_i \\ \sum_{i=1}^n c_i & n \end{bmatrix} \begin{bmatrix} s \\ A_0 \end{bmatrix} = \begin{bmatrix} \sum_{i=1}^n c_i A_{E,i} \\ \sum_{i=1}^n A_{E,i} \end{bmatrix}$$

$$\begin{bmatrix} s \\ A_0 \end{bmatrix} = \begin{bmatrix} \sum_{i=1}^n c_i^2 & \sum_{i=1}^n c_i \\ \sum_{i=1}^n c_i & n \end{bmatrix}^{-1} \begin{bmatrix} \sum_{i=1}^n c_i A_{E,i} \\ \sum_{i=1}^n A_{E,i} \end{bmatrix}. \tag{G.3}$$

A simple alternative to the analytical solution would be to continually guess values of s and A_0, seeking values that would minimize $\sum_i^n \epsilon_i^2$. A Monte Carlo random walk provides a way to make systematic guesses in a manner that continues to lead us toward an optimal solution. We begin by making an initial guess of the values of s and A_0 and using these values to calculate $\sum_i^n \epsilon_i^2$, which we will utilize as a figure of merit. Our goal will be to find values of s and A_0 that minimize this figure of merit, as we did above in the analytical solution. To achieve this goal we modify the initial values and determine whether or not the figure of merit has improved. This is accomplished by generating a random number, r, between -1 and 1, and multiplying a preset increment, δ, by r. This creates a random number somewhere between $-\delta$ and δ. δr is added to s and the figure of merit is calculated with this modified value. If the figure of merit is smaller, then the new value of the slope is retained; otherwise the change is ignored. A_0 is then modified and tested in the same manner. This process is continued iteratively until A_0 and s no longer undergo changes; *until they have randomly walked to their optimal values.*

A practical strategy is to begin with a large value for the increment δ, which will quickly put s and A_0 in the neighborhood of the optimal solution. Once there, the large incremental changes will be unable to improve the values of the parameters. If a large number of attempts (say, 100) fail to produce a change, δ is halved and the procedure is repeated. Each time that a value of δ fails to produce a change it is halved until the value of the increment reaches a pre-selected tolerance value, a small number that represents the degree of precision required in the values of the slope and intercept. As an example, consider the following data:

Concentration	Absorbance
0.0	0.1074
0.1	0.2576
0.2	0.2142
0.4	0.5011
0.5	0.5105
0.6	0.5560
0.7	0.6047
0.8	0.7597
0.9	0.8982
1.0	1.0289

A plot of these data, along with the least-squares line determined analytically, with $s = 0.8417$ and $A_0 = 0.1105$, is illustrated in Fig. G.1. A random walk to the same least-squares line begins with an initial guess of a line along the c axis with $s = 0$ and $A_0 = 0$, resulting in an initial value for the figure of merit

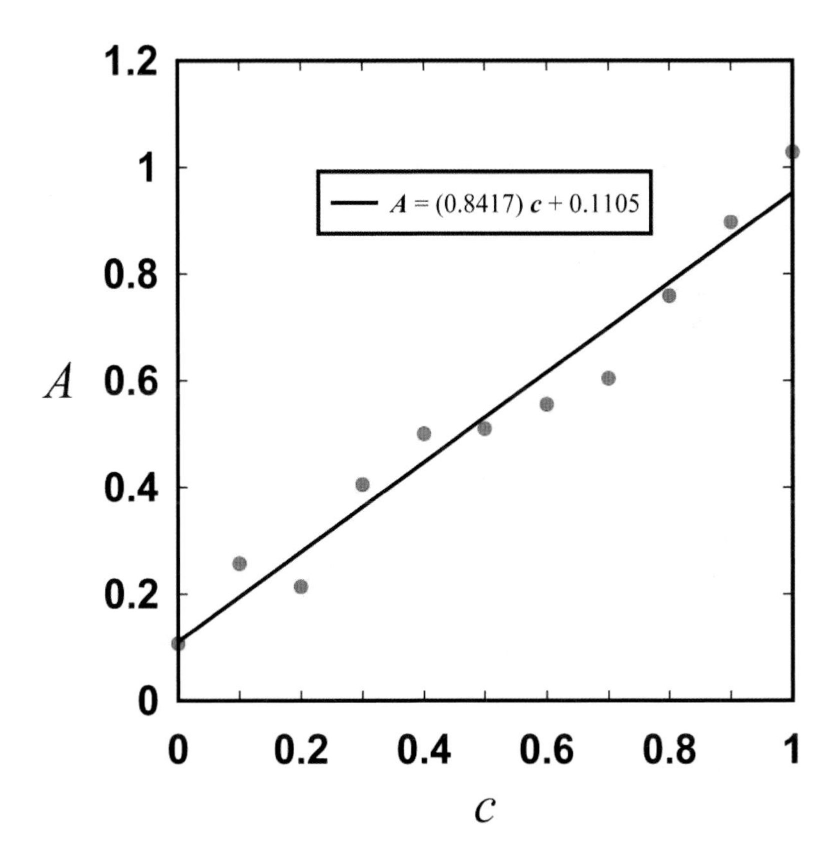

Figure G.1 Plot of experimental absorbances (A) versus concentration (c) and analytically determined least-squares line.

of $\sum \epsilon^2 = 3.9176$. The following table lists sequential values of the maximum increment, slope, intercept, and figure of merit for the changes that resulted in a lower figure of merit, illustrating that the algorithm rapidly converges to the same slope and intercept as the ones determined analytically:

δ	s	A_0	$\sum \epsilon^2$	δ	s	A_0	$\sum \epsilon^2$
0.5000	0.0000	0.0000	3.9175869	0.1250	0.8548	0.1025	0.0334954
0.5000	0.0991	0.0000	3.1925928	0.1250	0.8548	0.1030	0.0334811
0.5000	0.0991	0.2445	1.2589406	0.1250	0.8529	0.1030	0.0334592
0.5000	0.0991	0.2983	1.0099601	0.1250	0.8529	0.1062	0.0334417
0.5000	0.0991	0.5914	0.7721838	0.1250	0.8484	0.1062	0.0333427
0.5000	0.1224	0.5914	0.7643392	0.1250	0.8484	0.1076	0.0333353
0.5000	0.1224	0.4568	0.6043431	0.1250	0.8445	0.1076	0.0333157
0.5000	0.2959	0.4568	0.4203054	0.1250	0.8445	0.1087	0.0332934
0.5000	0.2959	0.4096	0.3685208	0.0625	0.8445	0.1091	0.0332923
0.5000	0.5233	0.4096	0.3603494	0.0625	0.8441	0.1091	0.0332904
0.5000	0.5233	0.1745	0.2443707	0.0313	0.8436	0.1091	0.0332895
0.5000	0.6110	0.1745	0.1207874	0.0313	0.8436	0.1096	0.0332878
0.5000	0.7621	0.1745	0.0467025	0.0313	0.8425	0.1096	0.0332867
0.5000	0.7543	0.1745	0.0462202	0.0313	0.8428	0.1096	0.0332861
0.5000	0.7543	0.1469	0.0422671	0.0156	0.8428	0.1101	0.0332855
0.5000	0.8197	0.1469	0.0409360	0.0156	0.8424	0.1101	0.0332842
0.5000	0.8197	0.0990	0.0393353	0.0156	0.8422	0.1101	0.0332840
0.5000	0.8910	0.0990	0.0378961	0.0156	0.8422	0.1101	0.0332840
0.5000	0.8749	0.0990	0.0347946	0.0156	0.8422	0.1101	0.0332840
0.5000	0.8743	0.0990	0.0347177	0.0156	0.8422	0.1103	0.0332839
0.5000	0.8737	0.0990	0.0346460	0.0156	0.8418	0.1103	0.0332837
0.5000	0.8601	0.0990	0.0337104	0.0156	0.8418	0.1103	0.0332837
0.5000	0.8587	0.0990	0.0336956	0.0156	0.8418	0.1103	0.0332836
0.1250	0.8587	0.0993	0.0336818	0.0156	0.8418	0.1104	0.0332836
0.1250	0.8587	0.1001	0.0336395	0.0156	0.8417	0.1104	0.0332835
0.1250	0.8548	0.1001	0.0336296	0.0078	0.8417	0.1105	0.0332835

Appendix H

Constrained Optimization

Crystal structure refinement often includes constraining certain parameters in the least-squares method. Because the parameters are determined from matrix algebra, they are not treated as individual entities, and restricting specific parameters becomes problematic. The most straightforward method to accomplish this makes use of Lagrange's method of undetermined multipliers. Before considering this method it is necessary to establish some preliminary concepts from multivariate calculus, since we are dealing with more than one variable in the refinement process.

The Derivative Mean Value. Consider the single-valued function $f(x)$ in the region $x_a < x < x_b$, illustrated in Fig. H.1(a). The *mean value theorem* states that there is at least one point on the curve characterizing the function, $f(x_o)$, at

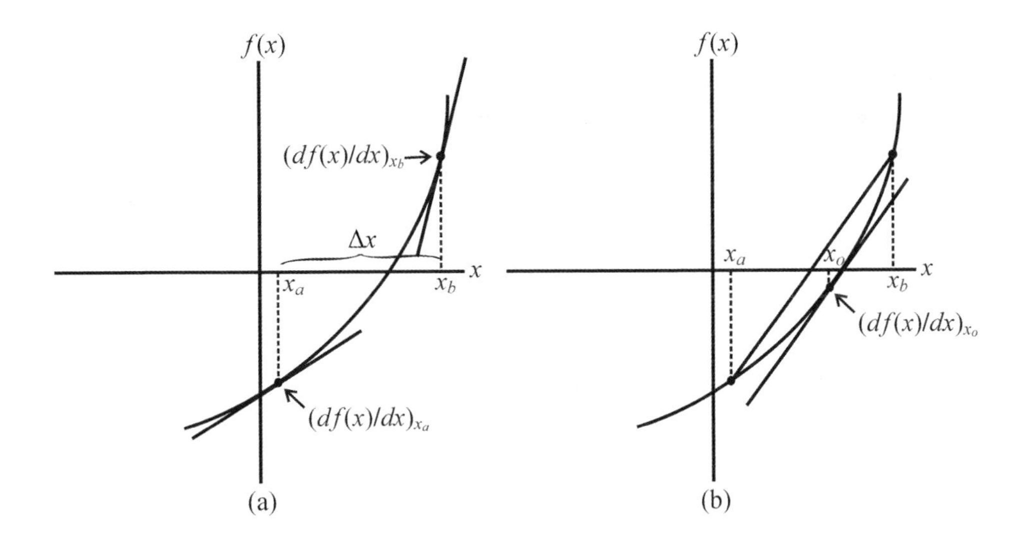

Figure H.1 (a) Function $f(x)$ with derivatives at x_a and x_b. (b) Secant between $(x_a, f(x_a))$ and $(x_b, f(x_b))$ and derivative at x_o parallel to the secant.

Understanding Single-Crystal X-Ray Crystallography. Dennis W. Bennett

ISBN: 978-3-527-32677-8 (HC), 978-3-527-32794-2 (SC)

which the derivative of the function is equal to the "average" derivative within the interval.

The derivative changes over the interval, and at some point on the curve, x_o, the tangent to the curve will be parallel to the secant connecting $(x_a, f(x_a))$ with $(x_b, f(x_b))$, as illustrated in Fig. H.1(b). For this specific "mean" value of the derivative,

$$\frac{f(x_b) - f(x_a)}{x_b - x_a} = \frac{f(x_a + \Delta x) - f(x_a)}{\Delta x} = \left(\frac{df(x)}{dx}\right)_{x=x_o}, \quad \text{and} \tag{H.1}$$

$$f(x_a + \Delta x) - f(x_a) = \Delta x \left(\frac{df(x)}{dx}\right)_{x=x_o}. \tag{H.2}$$

Since x_o is between x_a and x_b, it will be some fraction, δ, of Δx greater than x_a: $x_o = x_a + \delta \Delta x$, $0 \leq \delta \leq 1$. Thus,

$$f(x_a + \Delta x) - f(x_a) = \Delta x \left(\frac{df(x)}{dx}\right)_{x=x_a+\delta\Delta x}. \tag{H.3}$$

The Total Differential. Now consider a function of two variables, $u = f(x, y)$. If we begin at some point, (x_a, y_a) and move to another point (x_b, y_b), we will have $x_b = x_a + \Delta x$ and $y_b = y_a + \Delta y$, with

$$\Delta u = f(x_a + \Delta x, y_a + \Delta y) - f(x_a, y_a). \tag{H.4}$$

Adding and subtracting $f(x_a, y_a + \Delta y)$ gives

$$\Delta u = (f(x_a+\Delta x, y_a+\Delta y) - f(x_a, y_a)+\Delta y)) + (f(x_a, y_a+\Delta y - f(x_a, y_a))). \tag{H.5}$$

Δu can now be treated as the sum of two expressions. In the first expression only x changes, and we can treat it as a single valued function. Similarly, only y changes in the second expression, and we can also treat it as a single-valued function. Provided that the derivatives of each expression are constrained to the variable that changes (partial derivatives), the mean value theorem holds for each expression ($0 \leq \delta_x, \delta_y \leq 1$):

$$f(x_a + \Delta x, y_a + \Delta y) - f(x_a, y_a) = \Delta x \left(\frac{\partial f(x, y)}{\partial x}\right)_{x=x_a+\delta_x\Delta x} \tag{H.6}$$

$$f(x_a, y_a + \Delta y - f(x_a, y_a)) = \Delta x \left(\frac{\partial f(x, y)}{\partial y}\right)_{y=y_a+\delta_y\Delta y}, \tag{H.7}$$

and

$$\Delta u = \left(\frac{\partial f(x, y)}{\partial x}\right)_{x=x_a+\delta_x\Delta x} \Delta x + \left(\frac{\partial f(x, y)}{\partial y}\right)_{y=y_a+\delta_y\Delta y} \Delta y. \tag{H.8}$$

In the limit, as Δ_x and Δ_y approach zero, the partial derivatives approach their values at $x = x_a$ and $y = y_b$. As the interval between x_a and x_b shrinks toward x_a and becomes very small, the partial derivatives become

$$\left(\frac{\partial f(x, y)}{\partial x}\right)_{x=x_a} + \epsilon_x \quad \text{and} \quad \left(\frac{\partial f(x, y)}{\partial y}\right)_{y=y_a} + \epsilon_y, \tag{H.9}$$

where ϵ_x and ϵ_y are infinitesimal quantities that go to zero as the interval shrinks to the point where x_b and x_a, and y_b and y_a, coincide. Since x_a and y_a are arbitrary, we drop the subscripts, giving

$$\begin{aligned} \Delta u &= \left(\frac{\partial f(x,y)}{\partial x} + \epsilon_x\right)\Delta x + \left(\frac{\partial f(x,y)}{\partial y} + \epsilon_y\right)\Delta y \\ &= \frac{\partial f(x,y)}{\partial x}\Delta x + \frac{\partial f(x,y)}{\partial y}\Delta y + \epsilon_x\Delta x + \epsilon_y\Delta y. \end{aligned} \tag{H.10}$$

In the limit, as Δx and Δy approach zero, $\epsilon_x\Delta x$ and $\epsilon_y\Delta y$ are the products of two very small numbers, and approach zero much more rapidly than Δx, Δy, and Δu. Thus when Δx, Δy, and Δu become infinitesimal quantities, dx, dy and du, the products have already vanished:

$$du = \frac{\partial f(x,y)}{\partial x}\,dx + \frac{\partial f(x,y)}{\partial y}\,dy. \tag{H.11}$$

The differential, du is known as *the total differential.* For small Δx and Δy it is a good approximation to Δu. The extrapolation to three (or more) dimensions is obvious. For $u = f(x,y,z)$,

$$du = \frac{\partial f(x,y,z)}{\partial x}\,dx + \frac{\partial f(x,y,z)}{\partial y}\,dy + \frac{\partial f(x,y,z)}{\partial z}\,dz; \tag{H.12}$$

$$\Delta u \simeq \frac{\partial f(x,y,z)}{\partial x}\Delta x + \frac{\partial f(x,y,z)}{\partial y}\Delta y + \frac{\partial f(x,y,z)}{\partial z}\Delta z. \tag{H.13}$$

The Gradient. For the function $u = f(x,y,z)$, every point $P = (x,y,z)$ defines a vector $\mathbf{v} = [x\ y\ z]$ from the origin to P, for which there is a value of the function, $f(x,y,z)$. We now generate a displacement from point P to point $Q = (x+\Delta x, y+\Delta y, z+\Delta z)$, and consider how $f(x,y,x)$ changes when this displacement occurs:

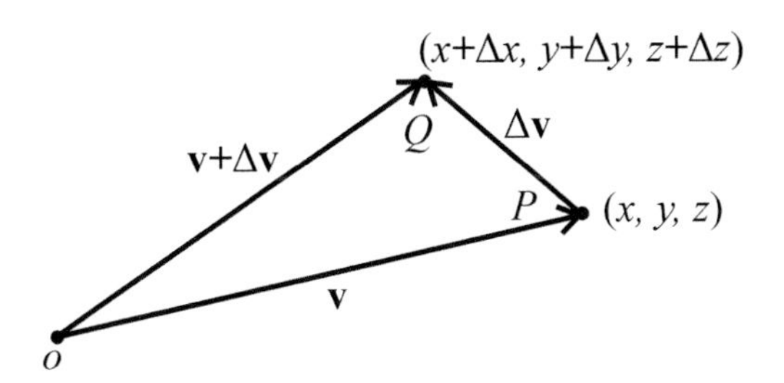

If $\Delta v = |\Delta\mathbf{v}|$, the magnitude of the change in the coordinates, then

$$\frac{\Delta u}{\Delta v} = \frac{\partial u}{\partial x}\frac{\Delta x}{\Delta v} + \frac{\partial u}{\partial y}\frac{\Delta y}{\Delta v} + \frac{\partial u}{\partial z}\frac{\Delta z}{\Delta v}, \tag{H.14}$$

and $\Delta u/\Delta v$ is the change in the function per unit change in the coordinates — the rate of change of the function with respect to the coordinates. As Δx, Δy, and Δz approach zero, $\Delta u/\Delta v$ becomes the limiting value, du/dv. It is the derivative of the function in the direction of the displacement, and in the general case is known as the *directional derivative* of the function:

$$\frac{du}{dv} = \frac{\partial u}{\partial x}\frac{dx}{dv} + \frac{\partial u}{\partial y}\frac{dy}{dv} + \frac{\partial u}{\partial z}\frac{dz}{dv}. \tag{H.15}$$

The directional derivative can be seen to be the dot product of two vectors in orthonormal coordinates:

$$\frac{du}{dv} = \left(\frac{\partial u}{\partial x}\mathbf{i} + \frac{\partial u}{\partial y}\mathbf{j} + \frac{\partial u}{\partial x}\mathbf{k}\right) \cdot \left(\frac{dx}{dv}\mathbf{i} + \frac{dy}{dv}\mathbf{j} + \frac{dz}{dv}\mathbf{k}\right). \tag{H.16}$$

The second vector is the displacement vector $d\mathbf{v} = [dx\ dy\ dz]$ divided by its magnitude, dv, i.e., a unit vector in the direction of the displacement, $\mathbf{v}_u = d\mathbf{v}/dv$. The first vector is known as the *gradient vector* of the function, or more commonly, as the *gradient* of the function, symbolized by ∇. The gradient is the rate of change of the function with respect to the coordinates:

$$\nabla u = \nabla f(x, y, z) = \frac{\partial u}{\partial x}\mathbf{i} + \frac{\partial u}{\partial y}\mathbf{j} + \frac{\partial u}{\partial x}\mathbf{k}. \tag{H.17}$$

The directional derivative for a displacement is therefore the dot product of the gradient of the function and a unit vector in the direction of the displacement:

$$\frac{du}{dv} = \nabla f(x, y, z) \cdot \frac{d\mathbf{v}}{dv} = \nabla u \cdot \mathbf{v}_u. \tag{H.18}$$

Recalling that the dot product of two vectors is the projection of one vector along the other,

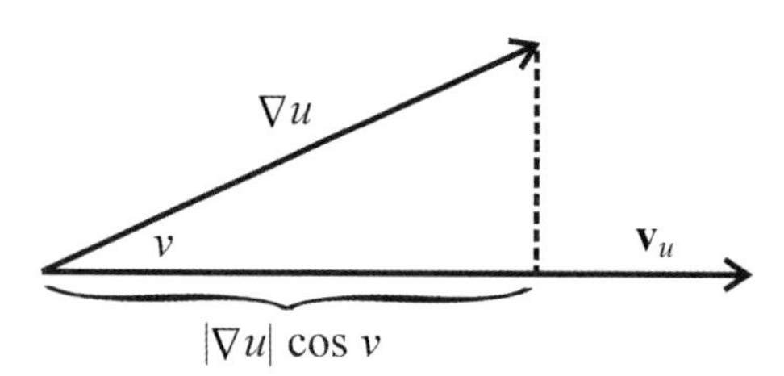

du/dv is the projection of the gradient vector in the direction of the displacement.

Since $\mathbf{v}_u$ is a unit vector $du/dv = v_u|\nabla u|\cos\nu = |\nabla u|\cos\nu$. We now consider different displacement directions to determine along which direction the function will experience the greatest rate of change, du/dv. This rate of change of the function will be at a maximum when $\cos\nu = 1$, that is, when $\nu = 0$. This will be when the displacement vector is coincident with the gradient vector. *The gradient vector at point (x_o, y_o, z_o) extends in the direction of the greatest rate of change of the function at the point.* Furthermore, the magnitude of the gradient vector at (x_o, y_o, z_o) is the rate of change of the function at that point:

$$\left(\frac{du}{dv}\right)_{(x_o,y_o,z_o)} = |\nabla u|\cos\nu = |\nabla u|. \tag{H.19}$$

Since the gradient vector determines the rate of change of the function, we should be able to exploit it to determine at which point the function is at a maximum or a minimum. In order to visualize this we will consider the function $f(x, y)$ so that the function can be plotted with respect to the independent variables. An example is illustrated in Fig. H.2. We focus on a particular point on the surface of the curve, $(x_o, y_o, f(x_o, y_o))$. A contour slice is taken through the point, creating a *level curve* at $f(x, y) = c$, a constant. A tangent to the point in the plane of the level curve is generated, defining a unit vector, $\mathbf{v}_t$, in the direction of the tangent. The level curve is shown in projection in the upper right portion of the figure.

Because $f(x, y)$ is a constant at any point on the level curve, $df(x, y)/dv = 0$ for any point on the curve. The directional derivative in the direction of the tangent at $(x_o, y_o, f(x_o, y_o))$ is

$$\left(\frac{df(x,y)}{dv}\right)_{x_o,y_o} = (\nabla f(x,y))_{x_o,y_o} \cdot \mathbf{v}_t = 0. \tag{H.20}$$

Because the dot product is zero, in an orthonormal coordinate system $\nabla f(x, y)$ *is perpendicular to the tangent to the level surface at the point at which it is evaluated* (Fig. H.3(a)). In three dimensions, the level curve contour for $f(x, y) = c$ becomes

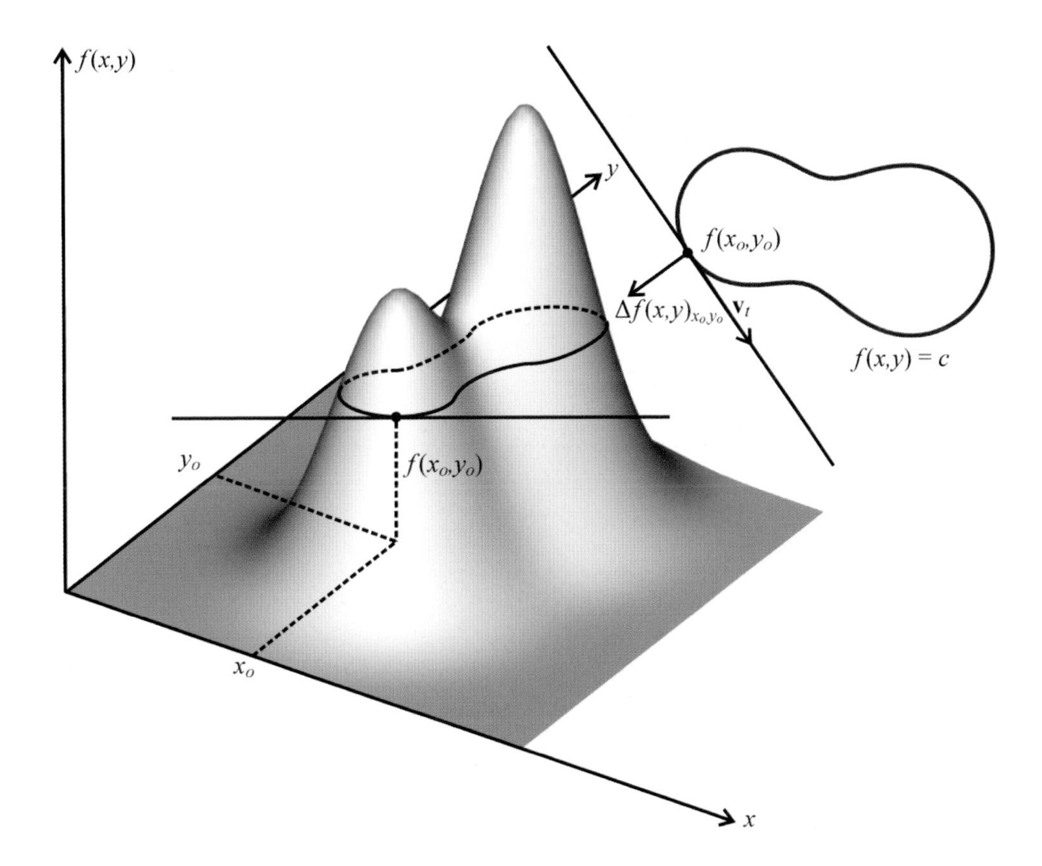

Figure H.2 Plot of $f(x, y)$ and level surface contour through point $(x_o, y_o, f(x_o, y_o))$.

a contour surface for $f(x, y, z) = c$ (Fig. H.3(b)). The gradient is perpendicular to the surface and points in the maximum rate of change of the function at the point at which it is evaluated.

Fig. H.3(c) illustrates a function of two variables with a local extreme (a maximum or a minimum) at (x_o, y_0). For constant $y = y_o$, a curve is created across the surface, $f(x, y_o)$, with a maximum at x_o. Similarly, if x is held at x_o, the curve $f(x_o, y)$ has a maximum at y_o. Thus $(\partial f(x, y)/\partial x)_{x_o} = 0$ in the first instance, and $(\partial f(x, y)/\partial y)_{y_o} = 0$ in the second instance. Since

$$\nabla f(x, y)_{x_o, y_o} = \left(\frac{\partial f(x, y)}{\partial x}\right)_{x_o} \mathbf{i} + \left(\frac{\partial f(x, y)}{\partial y}\right)_{y_o} \mathbf{j}, \tag{H.21}$$

when a local maximum or minimum exists at (x_o, y_0), $\nabla f(x, y) = \mathbf{0}$ at that point. This is a point where the change is going from positive to negative; the vector in

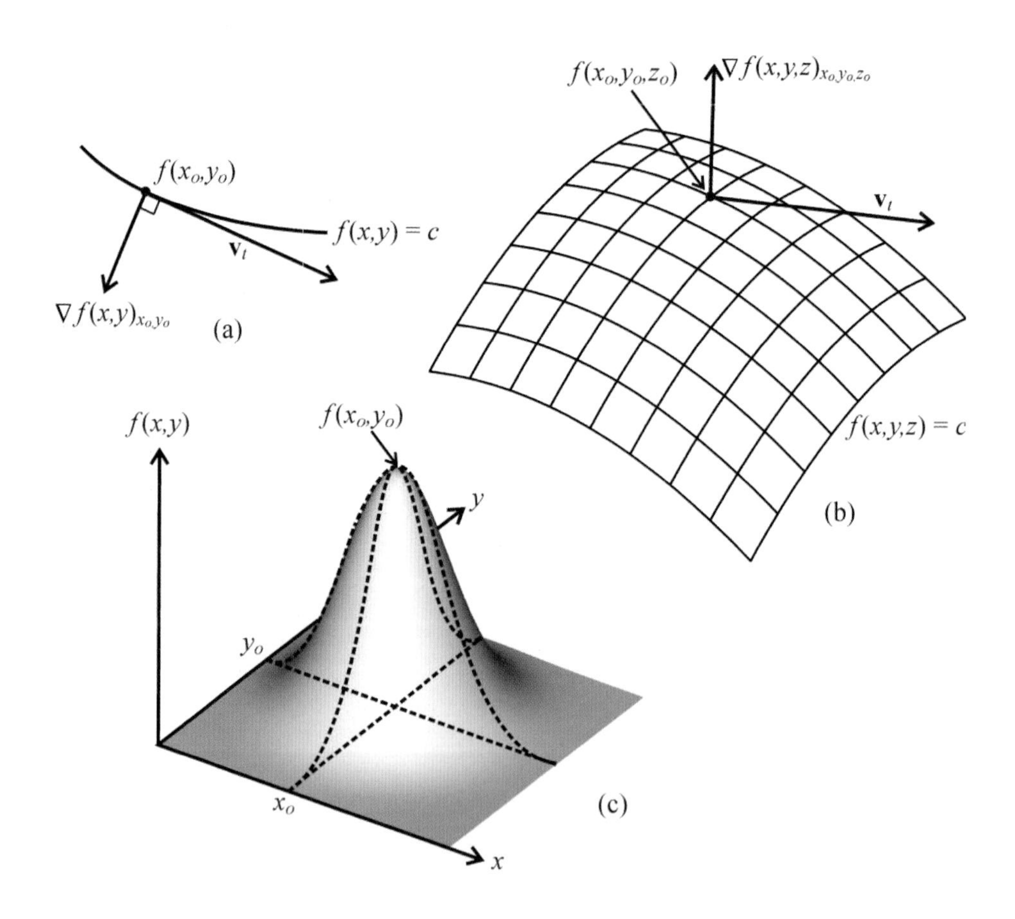

Figure H.3 (a) Relationship between the gradient vector and a tangent unit vector at a point on a level curve. (b) Relationship between the gradient vector and a tangent unit vector at a point on a contour surface. (c) $f(x, y)$ with a maximum at $f(x_o, y_o)$.

the direction of maximum change at this point is therefore $\mathbf{0}$, the null vector. *The function is said to be optimized at a point where the gradient is zero.*

The Method of Lagrange. For a given function, the values of the independent variables (the parameters) that determine an optimal value of the function are those for which the function will have a zero gradient. We now wish to determine the criteria for the optimization of $f(x, y)$ when the variables x and y are subjected to another relationship that constrains them. In general, the variables will be limited by a separate function that is fixed to some constant value (so that only certain values of x and y are allowed). We will find it convenient to denote x, y pairs as $\mathbf{v}$, such that the function $f(x, y) = f(\mathbf{v})$, is subject to the constraints on the variables created by the function $g(x, y) = g(\mathbf{v}) = c$. Although all variable (x, y) vectors, $\{\mathbf{v}\}$ are available to $f(x, y)$, only a subset, $\{\mathbf{r}\}$, satisfy the constraining equation, and we are looking for a vector, $\mathbf{r}_o \equiv \mathbf{v}_o$, in this subset that produces an optimal value (the largest or smallest) of $f(\mathbf{r}) = f(x, y) = f(\mathbf{v})$. In other words, the "sub-function", $f(\mathbf{r})$, with only those values of the function for the x, y pairs that satisfy the constraining equation, will have a maximum at $\mathbf{r}_o = (x_o, y_o)$.

The situation is depicted in Fig. H.4. In this example, the "constraining" function, $g(\mathbf{v})$, lies beneath the general function, $f(\mathbf{v})$, for illustrative purposes, but the only requirement for this function is that the ranges of the independent variables of $f(x, y)$ and $g(x, y) = c$ overlap (they must have some values of x and y in common). The constraining relationship, $g(\mathbf{v}) = c$, creates the level curve on $g(\mathbf{v})$ at height c along the functional axis. The subset of x, y pairs that satisfy the constraining relationship are those that lie on the projection of the level curve on the xy plane, defining $\mathbf{r}$ vectors that satisfy the constraints, and the corresponding values of $f(\mathbf{r})$. There will be some (x_o, y_o), $\mathbf{r}_o$, that will correspond to a maximum value of $f(\mathbf{r})$. This value will be $f(\mathbf{r}_o) = f(\mathbf{v}_o)$ as illustrated in the figure. The projection of the level curve on the $f(\mathbf{v})$ surface, is the "sub function", $f(\mathbf{r})$, and we can ignore the other values of the function, since the constraining equation removes these values from consideration. The function of interested, illustrated with the solid line, are the only values of $f(\mathbf{v})$ allowed by the constraint relationship, and the maximum value on the projected curve is the maximum value of $f(\mathbf{v})$, subject to the constraints imposed by $g(\mathbf{v}) = c$. In our discussion of the gradient we were concerned with the behavior of a function when a displacement of the independent variable vectors occur. In this case we are concerned with the behavior of the "sub function", $f(\mathbf{r})$, as $\mathbf{r}$ undergoes a displacement. The only difference here is that the points P and Q now must lie on the projection of the constraining equation level curve, shown in the lower left of Fig. H.4. As $\Delta\mathbf{r}$ becomes infinitesimal, the unit vector $\mathbf{u}_t$ in direction of the displacement vector, $d\mathbf{r}/dr$, becomes tangent to the projected curve, and Eqn. H.18 becomes

$$\frac{df(\mathbf{r})}{dr} = \nabla f(\mathbf{r}) \cdot \frac{d\mathbf{r}}{dr} = \nabla f(\mathbf{r}) \cdot \mathbf{u}_t. \tag{H.22}$$

Since $f(\mathbf{r})$ varies with $\mathbf{r}$, this derivative is, in general non-zero. However, at $\mathbf{r}_o$, $f(\mathbf{r})$ is passing through a maximum, and $df(\mathbf{r})/dr = 0$ at this point. As can be seen in Fig. H.4, this is the point on the apex of the projection of the level curve on $f(\mathbf{v})$. Thus, at $\mathbf{r}_o$,

$$\left(\frac{df(\mathbf{r})}{dr}\right)_{r_o} = (\nabla f(\mathbf{r}))_{r_o} \cdot (\mathbf{u}_t)_{r_o} = 0. \tag{H.23}$$

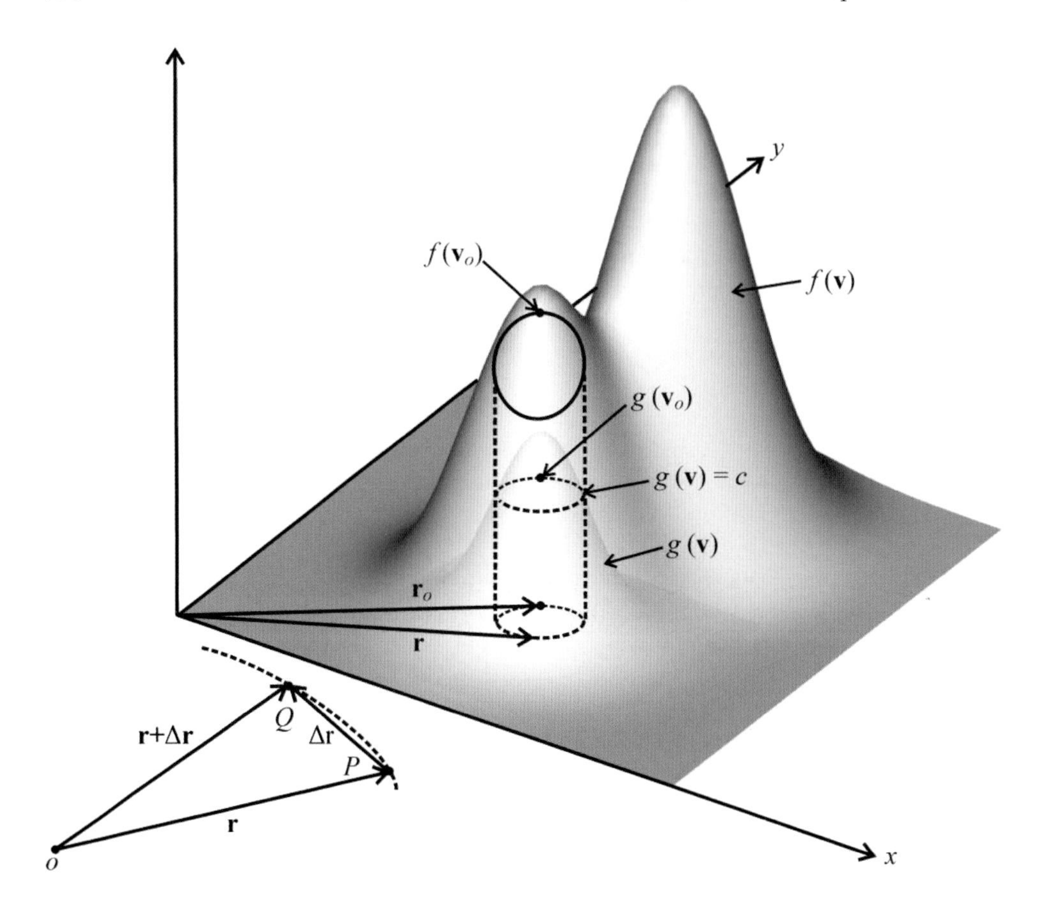

Figure H.4 A general function, $f(x, y)$, the "constraining function", $g(x, y)$, and the level curve corresponding to the constraining relationship, $g(x, y) = c$.

The dot product is equal to zero at this point, and the gradient vector, $\nabla f(\mathbf{r}))_{r_o}$ is therefore perpendicular to the unit vector, $\mathbf{u}_t$, tangent to the projected level curve at the point $(x_o, y_o, 0)$. It follows that $f(\mathbf{r}))_{r_o}$ is also perpendicular to the tangent vector at the point (x_o, y_o, c) on the level curve, as shown on the left in Fig. H.5.

Since $\nabla g(\mathbf{r}))_{r_o}$ is perpendicular to the level curve at every point on the curve (Fig. H.3), it is perpendicular to the curve at (x_o, y_o, c), the point producing the constrained maximum, and the gradient vectors $\nabla f(\mathbf{r}))_{r_o}$ and $\nabla g(\mathbf{r}))_{r_o}$ are parallel to one another, differing only in their magnitude:

$$\begin{aligned}(\nabla f(\mathbf{r}))_{r_o} &= (\lambda \nabla g(\mathbf{r}))_{r_o} \quad \text{and} \\ (\nabla f(\mathbf{r}))_{r_o} &- (\lambda \nabla g(\mathbf{r}))_{r_o} = 0. \end{aligned} \tag{H.24}$$

The scaling factor, λ, is known as a *Lagrange multiplier*, named for the author of this relationship, the famed mathematician, Joseph-Louis Lagrange. Because it is not determined directly by the relationship it is often referred to as an *undetermined multiplier*.

In order to make use of the relationship we define a new function, called the *Lagrangian*:

$$L(\mathbf{r}, \lambda) = f(\mathbf{r}) - \lambda g(\mathbf{r}). \tag{H.25}$$

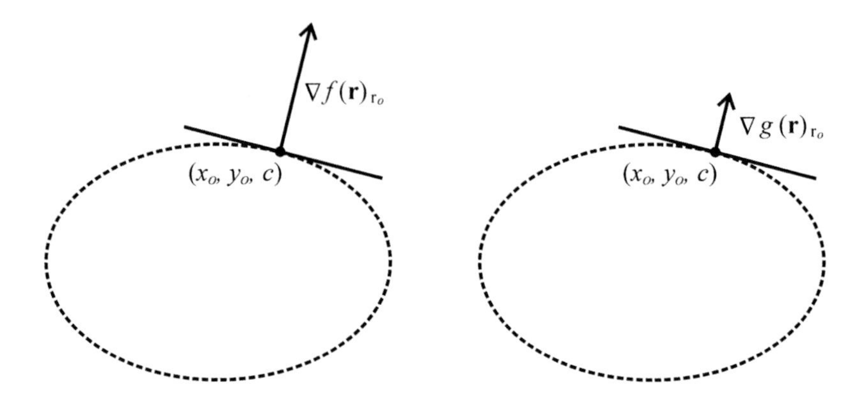

Figure H.5 Relationship of gradient vectors evaluated at the point where a general function, $f(x, y)$, is optimized at (x_o, y_o), subject to constraints on the variables, $g(x, y) = c$. The level curve for $g(x, y) = c$ is indicated with dashes.

When the constraining equation is established, $g(\mathbf{r})$ is a constant, and $f(\mathbf{r})$ will be at an extreme point (maximum or minimum) when $\mathbf{r} = \mathbf{r}_o$. Thus the Lagrangian will be at an extreme point at $\mathbf{r}_o$, and its gradient will be equal to zero:

$$(\nabla L(\mathbf{r}, \lambda))_{r_o} = (\nabla(f(\mathbf{r}) - \lambda g(\mathbf{r})))_{r_o} = (\nabla(f(\mathbf{r})))_{r_o} - (\lambda\nabla(g(\mathbf{r})))_{r_o} = 0. \quad \text{(H.26)}$$

Thus the constrained optimization problem is reduced to finding the values of the variables and Lagrange multiplier(s) that optimize the Lagrangian. Referring to Fig. H.4, the arguments above are also valid for the projection of $g(\mathbf{r}) = c$ onto the xy plane, and indeed, onto any plane parallel to the xy plane. Thus the gradient of $f(r)$ will be parallel to the gradient of a level curve with values $g(\mathbf{r}) - z$, where z is the height of the projection above the xy plane. This allows for the introduction of a constant into the Lagrangian:

$$\begin{aligned} L(\mathbf{r}, \lambda) &= f(\mathbf{r}) - \lambda(g(\mathbf{r}) - z) \\ &= f(\mathbf{r}) + \lambda(z - g(\mathbf{r})). \end{aligned} \quad \text{(H.27)}$$

While the discussion here has been limited to a function of two variables, it is perfectly general; the Lagrangian can be expanded to include any number of variables and/or constraining equations:

$$L(\mathbf{r}, \lambda) = f(\mathbf{r}) + \lambda_1(z_1 - g_1(\mathbf{r})) + \lambda_2(z_2 - g_2(\mathbf{r})) + \ldots. \quad \text{(H.28)}$$

Appendix I

Taylor Series

The Taylor Series, named after the English mathematician Brook Taylor, represents the local behavior of a function, $f(x)$ in the vicinity of a specific point, x_o, as an infinite series of the function and its derivatives evaluated at that point:

$$
\begin{aligned}
f(x) &= f(x)_{x_o} + \frac{1}{1!}\left(\frac{df(x)}{dx}\right)_{x_o}(x - x_o) + \frac{1}{2!}\left(\frac{d^2f(x)}{dx^2}\right)_{x_o}(x - x_o)^2 + \ldots \\
&= \sum_{n=0}^{\infty}\frac{1}{n!}\left(\frac{d^nf(x)}{dx^n}\right)_{x_o}(x - x_o)^n. \qquad \text{(I.1)}
\end{aligned}
$$

Fig. I.1 illustrates the Taylor series expansion of the function $f(x) = e^x$ about the point $x_o = 0.5$. As more terms are added the expansion represents the function at values farther away from x_o. If we wish to model the function in regions relatively close to x_o, then only a few terms (e.g., n=3) are necessary to create a good approximation to $f(x)$ in the vicinity of x_o.

The expansion is especially useful if the values of the function are those with x very close to x_o. In this case, only the first two terms are necessary to approximate the function ($n = 1$ in the figure):

$$
f(x) \simeq mx + (f(x)_{x_o} - m(x_o)) = mx + b, \qquad \text{(I.2)}
$$

a line with a slope, m, equal to the derivative of the function evaluated at x_o, and passing through $(x_o, f(x_o))$. This allows us to approximate a function in the vicinity of a point *as linear function.*

Understanding Single-Crystal X-Ray Crystallography. Dennis W. Bennett

ISBN: 978-3-527-32677-8 (HC), 978-3-527-32794-2 (SC)

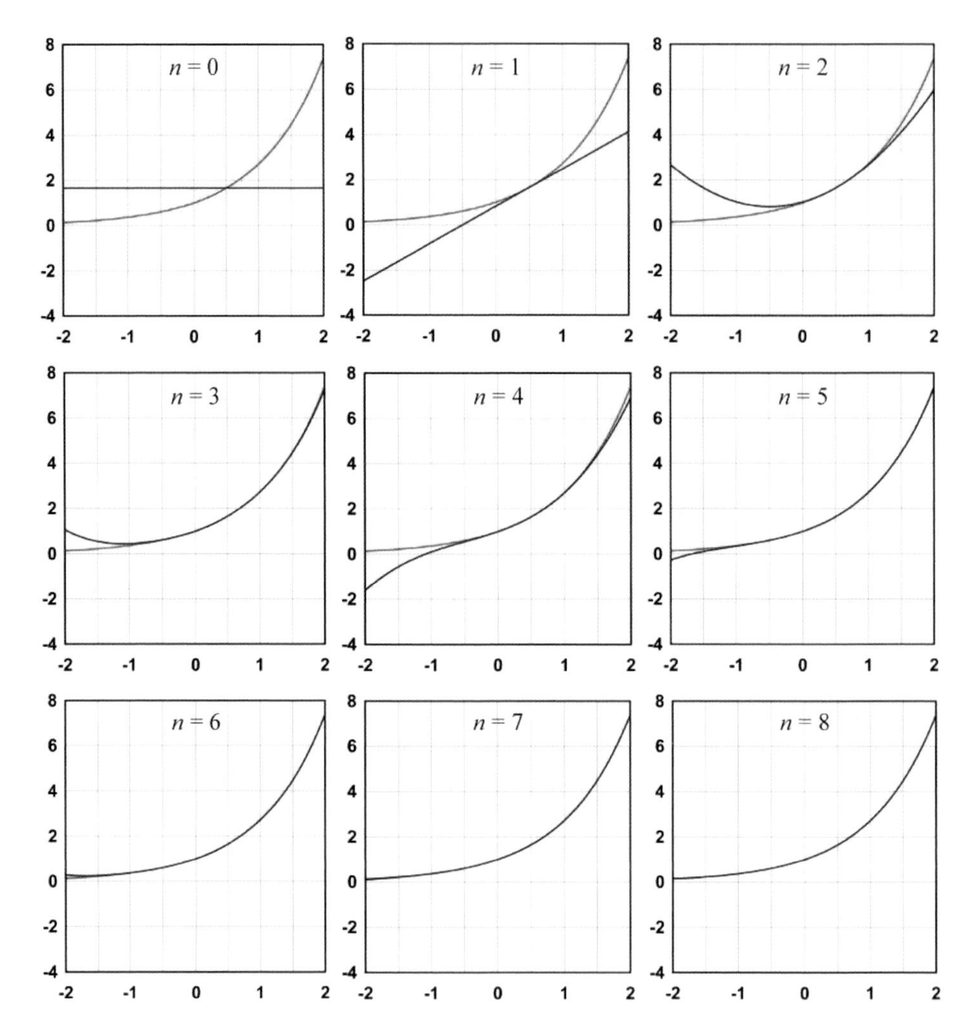

Figure I.1 Taylor series expansion of $f(x) = e^x$ about the point $x_o = 0.5$ for $n = 0$ through $n = 8$.

Bibliography

[1] G. M. Sheldrick. A short history of *SHELX*. *Acta Cryst.*, A64:112, 2008.

[2] D. W. Bennett. *MOLXTL*: Molecular graphics for small molecule crystallography. *J. Appl. Cryst.*, 37:1038, 2004.

[3] W. Massa. *Crystal Structure Determination.* Springer-Verlag, New York, 2nd edition, 2004.

[4] J. Glusker and K. Trueblood. *Crystal Structure Analysis.* Oxford University Press, Oxford, New York, 2nd edition, 1992.

[5] W. Clegg, editor. *Crystal structure analysis : principles and practice.* Number 6 in International Union of Crystallography Texts on Crystallography. Oxford Science Publications, Chester, 2001.

[6] H. Lipson and W. Cochran. *Determination of Crystal Structures.* G. Bell & Sons, Ltd., London, 1966.

[7] M. Ladd and R. Palmer. *Structure Determination by X-ray Crystallography.* Plenum Press, New York, 4th edition, 2003.

[8] G. H. Stout and L. H. Jensen. *X-ray Structure Determination.* John Wiley & Sons, New York, 2nd edition, 1989.

[9] J. Dunitz. *X-ray Analysis and the Structure of Organic Molecules.* Cornell University Press, Ithaca, 1979.

[10] T. L. Blundell and L. Johnson. *Protein Crystallography.* Academic Press, New York, 1976.

[11] C. Giacovazzo, editor. *Fundamentals of Crystallography.* Number 7 in International Union of Crystallography Texts on Crystallography. Oxford University Press, Oxford, New York, 2nd edition, 2002.

[12] D. E. Sands. *Introduction to Crystallography.* Dover Publications, New York, 1994.

[13] G. Rhodes. *Cystallography Made Crystal Clear.* Academic Press, New York, 2006.

[14] W. Clegg. *Crystal Structure Determination.* Oxford University Press, New York, 1998.

Understanding Single-Crystal X-Ray Crystallography. Dennis W. Bennett

ISBN: 978-3-527-32677-8 (HC), 978-3-527-32794-2 (SC)

[15] B.D. Cullity and S.R. Stock. *Elements of X-Ray Diffraction.* Prentice Hall, Upper Saddle River, NJ, 3rd edition, 2001.

[16] P. Luger. *Modern X-ray Analysis on Single Crystals.* W. de Gruyter, New York, 1980.

[17] M. M. Woolfson. *An Introduction to X-ray Crystallography.* Cambridge University Press, Cambridge [Eng], 2nd edition, 1997.

[18] M. M. Woolfson and F. Hai-fu. *Physical and Non-Physical Methods of Solving Crystal Structures.* Cambridge University Press, Cambridge [Eng], 1995.

[19] W. H. Miller. *A Treatise on Crystallography.* J. & J. J. Deighton, Cambridge [Eng], 1829.

[20] H. G. Campbell. *Linear Algebra with Applications.* Appleton-Century-Crofts, New York, 1971.

[21] R. W. D. Nickalls. A new approach to solving the cubic: Cardan's solution revealed. *The Mathematical Gazette*, 77:354, 1993.

[22] F. H. Allen, S. Bellard, M. D. Brice, B. A. Cartwright, A. Doubleday, H. Higgs, T. Hummelink, B. G. Hummelink-Peters, O. Kennard, W. D. S. Motherwell, J. R. Rodgers, and D. G. Watson. The Cambridge Crystallographic Data Centre: Computer-based search, retrieval, analysis and display of information. *Acta Cryst.*, B35:2331, 1979.

[23] M.N. Burnett and C.K. Johnson. *ORTEP-III.* Technical Report ORNL-6895, Oak Ridge National Laboratory, Oak Ridge, Tennessee, 1996.

[24] F. A. Cotton. *Chemical Applications of Group Theory.* Wiley, New York, 3rd edition, 1990.

[25] E. B. Wilson, J. C. Decius, and P. C. Cross. *Molecular Vibrations: The Theory Of Infrared And Raman Vibrational Spectra.* McGraw-Hill, New York, 1955.

[26] W. H. Zachariasen. *Theory of X-ray Diffraction in Crystals.* John Wiley & Sons, New York, 1945.

[27] T. Hahn, editor. *International Tables For Crystallography*, volume A. Kluwer Academic Publishers, Dordrecht, The Netherlands, 1995.

[28] Y. Le Page and G. Donnay. Refinement of the crystal structure of low-quartz. *Acta Cryst.*, B32:2456, 1976.

[29] E. B. Fleischer. X-ray structure determination of cubane. *J. Am. Chem. Soc.*, 86(18):3889, 1964.

[30] M. Vlasse, R. Naslain, J. S. Kasper, and K. Ploog. Crystal structure of tetragonal boron related to α-AlB_{12}. *J. Solid State Chem.*, 28(3):289, 1979.

[31] W. H. Bragg. The crystal structure of ice. *Proc. Phys. Soc. London*, 34:98, 1921.

[32] A. Anderson, T. S. Sun, and M. C. A. Donkersloot. Raman spectra and lattice dynamics of the α phases of nitrogen and carbon monoxide crystals. *Can. J. Physics*, 48:2265, 1970.

[33] D. T. Cromer, D. Schiferl, R. Lesar, and Robert L. Mills. Room-temperature structure of carbon monoxide at 2.7 and 3.6 gpa. *Acta Cryst.*, C39:1146, 1983.

[34] G. N. Ramachandran. Crystal structure of diamond. *Nature*, 156:83, 1945.

[35] M. E. Straumanis and L. S. Yu. Lattice parameters, densities, expansion coefficients and perfection of structure of Cu and of Cu-In α phase. *Acta Cryst.*, A25:676, 1969.

[36] W. L. Bragg. The structure of some crystals as indicated by their diffraction of X-rays. *Proc. Roy. Soc.*, A89:248, 1913.

[37] A. D. Mighell, V. L. Himes, and J. R. Rodgers. Space-group frequencies for organic compounds. *Acta Cryst.*, A39:737, 1983.

[38] H. Eyring, J. Walter, and G. Kimball. *Quantum Chemistry.* John Wiley & Sons, New York, 1944.

[39] M. von Laue. Röntgenstrahlinterferenzen. *Physikalische Zeitschrift*, 14(22/23):1075, 1913.

[40] W.H. Bragg and W. L. Bragg. *X-rays and Crystal Structure.* G. Bell and Sons, London, 191.

[41] P. P. Ewald. Zur theorie der interferenzen der Röntgenstrahlen in kristallen. *Phys. Z.*, 14:465, 1913.

[42] J. Fourier. *The Analytical Theory of Heat.* The University Press, Cambridge [Eng], 1878.

[43] D. R. Hartree. The wave mechanics of an atom with a non-coulomb central field. Part I-theory and methods. *Proc. Royal Soc.*, 24:89, 1928.

[44] V. A. Fock. Näherungsmethode zur lösung des quantenmechanischen mehrkörperproblems. *Zeit. fur Physik*, 61:126, 1930.

[45] J. C. Slater. Note on Hartree's method. *Phys. Rev.*, 35:210, 1930.

[46] A. Szabo and N. S. Ostlund. *Modern Quantum Chemistry: Introduction to Advanced Electronic Structure Theory.* Dover Publications, New York, 1996.

[47] U.W. Arndt and A.J. Wonacott, editors. *The Rotation Method in Crystallography.* North-Holland, Amsterdam, 1977.

[48] M.J. Buerger. *X-ray Crystallography.* John Wiley & Sons, 1942.

[49] M.J. Buerger. *The Precession Method in X-ray Crystallography.* John Wiley & Sons, New York, 1964.

[50] K. Weissenberg. Ein neues röntgengoniometer. *Z. Physik*, 23:229, 1924.

[51] W. F. de Jong and J. Bouman. Das photographieren von reziproken kristallnetzen mittels röntgenstrahlen. *Z. Krist.*, A98:456, 1938.

[52] T. C. Furnas, Jr. and D. Harker. Apparatus for measuring complete single-crystal X-ray diffraction data by means of a geiger counter diffractometer. *Rev. Sci. Instrum.*, 26:449, 1955.

[53] W. Kabsch. Automatic processing of rotation diffraction data from crystals of initially unknown symmetry and cell constants. *J. Appl. Cryst.*, 26(795), 1993.

[54] P. Niggli. *Geometrische Kristallographie des Diskontinuums.* Gebrüder Borntraeger, Leipzig, 1910.

[55] A. Santoro and A.D. Mighell. Determination of reduced cells. *Acta Cryst.*, A26:124, 1970.

[56] I. Krivy and B. Gruber. A unified algorithm for determining the reduced (Niggli) cell. *Acta Cryst.*, A32:297, 1976.

[57] Y. Le Page. The derivation of the axes of the conventional unit cell from the dimensions of the Buerger-reduced cell. *J. Appl. Cryst.*, 15:255, 1982.

[58] G. Friedel. Sur les symétries cristallines que peut révéler la diffraction des rayons Röntgen. *Compt. rend.*, 157:1533, 1913.

[59] J. Als-Neilson and D. McMorrow. *Elements of Modern X-ray Physics.* John Wiley and Sons, New York, 2001.

[60] R. Diamond. Profile analysis in single crystal ditfractometry. *Acta Cryst.*, A25:43, 1969.

[61] W. Kabsch. Evaluation of single-crystal x-ray diffraction data from a position-sensitive detector. *J. Appl. Cryst.*, 21:916, 1988.

[62] R.E. Stenkamp and L.H. Jensen. Resolution revisited: Limit of detail in electron density maps. *Acta Cryst.*, A40:251, 1984.

[63] H.D. Young. *Statistical Treatment of Experimental Data.* McGraw-Hill, New York, 1962.

[64] D. Wells. *The Penguin Dictionary of Curious and Interesting Numbers.* Penguin Books, Middlesex [Eng], 1986.

[65] S. Winitzki. Uniform approximations for transcendental functions. *Proc. ICCSA*, LNCS 2667/2003:962, 2003.

[66] R.P. Feynman, R.B. Leighton, and M. Sands. *Feynman Lectures on Physics*, volume 1. Addison-Wesley, Reading, Mass., 1989.

[67] P. Bouguer. *Essai d'Optique sur la Gradation de la Lumière.* Gauthier-Villars et Cie, Paris, 1729.

[68] J.H. Lambert. *Lambert's Photometrie.* Verlag von Wilhelm Engelmann, Leipzig, 1892. Published originally by Lambert in 1760 as *Photometria sive de mensura et gradibus luminus, colorum et umbrae*; published in German by E. Anding in 1892.

[69] A. Beer. Bestimmung der absorption des rothen lichts in farbigen flüssigkeiten. *Annal. Phys. Chem.*, 86:78, 1852.

[70] W. R. Busing and H. A. Levy. High-speed computation of the absorption correction for single crystal diffraction measurements. *Acta Cryst.*, 10:180, 1957.

[71] W. R. Busing and H. A. Levy. Angle calculations for 3- and 4- circle X-ray and neutron diffraetometers. *Acta Cryst.*, 22:457, 1967.

[72] E. T Whittaker. *The Calculus of Observations;: An Introduction to Numerical Analysis.* Dover Publications, New York, 1967.

[73] J. De Meulenaer and H. Tompa. The absorption correction in crystal structure analysis. *Acta Cryst.*, 19:1014, 1965.

[74] T.C. Furnas. Single crystal orienter instruction manual, 1957. General Electric Co., Milwaukee, Wisconsin.

[75] A. C. T. North, D. C. Phillips, and F. S. Mathews. A semi-empirical method of absorption correction. *Acta Cryst.*, A24:351, 1968.

[76] H. D. Flack. An experimental absorption-extinction correction technique. *Acta Cryst.*, A33:890, 1977.

[77] R. H. Blessing. An empirical correction for absorption anisotropy. *Acta Cryst.*, A51:33, 1995.

[78] A. Paturle and P. Coppens. Normalization factors for spherical harmonic density functions. *Acta Cryst.*, A44, 1988.

[79] N. Walker and D. Stuart. An empirical method for correcting diffractometer data for absorption effects. *Acta Cryst. (1983)*, A39:158, 1983.

[80] S. Parkin, B. Moezzi, and H. Hope. *XABS2*: an empirical absorption correction program. *J. Appl. Cryst.*, 28:53, 1995.

[81] C. G. Darwin. The theory of X-ray reflexion. *Phil. Mag.*, 27:315, 1914.

[82] C. G. Darwin. The theory of X-ray reflexion part II. *Phil. Mag.*, 775:315, 1914.

[83] C. G. Darwin. The reflexion of X-rays from imperfect crystals. *Phil. Mag.*, 43:800, 1922.

[84] P. P. Ewald. Theorie der dispersion. *Ann. Phys. Lpz.*, 49:1, 1916.

[85] P. P. Ewald. Theorie der reflexion und brechung. *Ann. Phys. Lpz.*, 49:117, 1916.

[86] P. P. Ewald. Die kristalloptik der röntgenstrahlen. *Ann. Phys. Lpz.*, 54:519, 1917.

[87] W. H. Zachariasen. A general theory of x-ray diffraction in crystals. *Acta Cryst.*, 23(558), 1967.

[88] P. J. Becker and P. Coppens. Extinction within the limit of validity of the darwin transfer equations. I. General formalisms for primary and secondary extinction and their application to spherical crystals. *Acta Cryst.*, A30:129, 1974.

[89] P. J. Becker and P. Coppens. Extinction within the limit of validity of the darwin transfer equations. II. Refinement of extinction in spherical crystals of SrF_2 and LiF. *Acta Cryst.*, A30:148, 1974.

[90] P. J. Becker and P. Coppens. Extinction within the limit of validity of the darwin transfer equations. III. Non-spherical crystals and anisotropy of extinction. *Acta Cryst.*, A31:417, 1975.

[91] D. T. Cromer and D. Liberman. Relativistic calculations of anomalous scattering factore for X-rays. *J. Chem. Phys.*, 53(5):1891, 1970.

[92] E. Prince. *Mathematical Techniques in Crystallography and Materials Science.* Springer-Verlag, New York, 1982.

[93] P. Coppens. Evidence for systematic errors in X-ray temperature parameters resulting from bonding effects. *Acta Cryst.*, B24:1272, 1968.

[94] A. J. C. Wilson. Determination of absolute from relative X-ray intensity data. *Nature*, 150:152, 1942.

[95] R. W. G. Wyckoff. *The Analytical Expression of the Results of the Theory of Space Groups.* Carnegie Institution of Washington, Washington, D.C., 2nd edition, 1930.

[96] U. Shmueli, editor. *International Tables For Crystallography*, volume B. Kluwer Academic Publishers, Dordrecht, The Netherlands, 1993.

[97] A. L. Patterson. A fourier series method for the determination of the components of interatomic distances. *Phys. Rev.*, 46:372, 1934.

[98] D. Harker. The application of the three-dimensional Patterson method and the crystal structures of proustite, Ag_3AsS_3, and pyrargyrite, Ag_3SbS_3. *J. Chem. Phys.*, 4:381, 1936.

[99] M. J. Buerger. The interpretation of Harker syntheses. *J. App. Phys*, 17:579, 1946.

[100] P. G. Simpson, R. D. Dobrott, and W. N. Lipscomb. The symmetry minimum function: High order image seeking functions in X-ray crystallography. *Acta Cryst.*, 18:169, 1965.

[101] E. Egert and G. M.Sheldrick. Search for a fragment of known geometry by integrated Patterson and direct methods. *Acta Cryst.*, A41:262, 1985.

[102] M. G. Rossmann. The molecular replacement method. *Acta Cryst.*, A46:73, 1990.

[103] B. W. Matthews. Five retracted structure reports: Inverted or incorrect? *Protein Sci.*, 16:1013, 2007.

[104] M. G. Rossmann and D. M. Blow. The detection of sub-units within the crystallographic asymmetric unit. *Acta Cryst.*, 15:24, 1062.

[105] R. A. Crowther. *In The Molecular Replacement Method. A Collection Of Papers On The Use Of Non-Crystallographic Symmetry.* Gordon and Breach, Science Publishers, Inc., New York, 1972. M. G. Rossmann, editor.

[106] R. A. Crowther and D.M. Blow. A method of positioning a known molecule in an unknown crystal structure. *Acta Cryst.*, 23:544, 1967.

[107] H. A. Weakliem and J. L. Hoard. The structures of ammonium and rubidium ethylenediaminetetraacetatocobaltate(III). *J. Am. Chem. Soc.*, 81(3):549, 1959.

[108] G. A. Sim. A note on the heavy-atom method. *Acta Cryst.*, page 511, 1960.

[109] M. M. Woolfson. An improvement of the 'heavy-atom' method of solving crystal structures. *Acta Cryst.*, 9:804, 1956.

[110] W. H. Bragg, R. W. James, and C. H. Bosanquet. The intensity of reflexion of X-rays by rock-salt (xii). *Phil. Mag.*, 41:309, 1924.

[111] W. H. Bragg, R. W. James, and C. H. Bosanquet. The intensity of reflexion of X-rays by rock-salt (xiii). *Phil. Mag.*, 42:1, 1924.

[112] P. Coppens and M. B. Hall, editors. *Electron Distributions and the Chemical Bond.* Plenum Press, New York, 1982.

[113] P. Coppens. Charge-density analysis at the turn of the century. *Acta Cryst.*, A54:779, 1998.

[114] R. F. Stewart. Electron population analysis with rigid pseudoatoms. *Acta Cryst.*, A32:565, 1976.

[115] N. K. Hansen and P. Coppens. Testing aspherical atom refinements on small-molecule data sets. *Acta Cryst.*, A34:909, 1978.

[116] P. Main. A theoretical comparison of β, γ' and $2F_o - F_c$ syntheses. *Acta Cryst.*, A35:779, 1979.

[117] K.P. Battaile, J. Molin-Case, R. Paschke, M. Wang, D.W. Bennett, J. Vockley J.J.P., and Kim. Crystal structure of rat short chain acyl-coa dehydrogenase complexed with acetyletyl-coa. *J. Biol. Chem.*, 227:12200, 2002.

[118] D. Harker and J. S. Kasper. Phases of Fourier coefficients directly from crystal diffraction data. *Acta Cryst.*, page 70, 1948.

[119] H. Schenck. An introduction to direct methods. Published by the University College Cardiff Press for the International Union of Crystallography with the financial assistance of UNESCO Contract No. SC/RP 250.271.

[120] D. Sayre. The squaring method: a new method for phase determination. *Acta Cryst.*, 5:90, 1952.

[121] W. Cochran. A relation between the signs of structure factors. *Acta Cryst.*, 5:65, 1952.

[122] W. Cochran. Relations between the phases of structure factors. *Acta Cryst.*, 8:473, 1955.

[123] W. Cochran and M. M. Woolfson. The theory of sign relations between structure factors. *Acta Cryst.*, 8:1, 1955.

[124] M. Evans, N. Hastings, and B. Peacock. *Statistical Distributions.* Wiley-Interscience, New York, 2000.

[125] H. Hauptman and J. Karle. A theory of phase determination for the four types of non-centrosymmetric space groups $1P222$, $2P22$, $3P_12$, $3P_22$. *Acta Cryst.*, 9:635, 1956.

[126] J. Karle and I. L. Karle. The symbolic addition procedure for phase determination for centrosymmetric and noncentrosymmetric crystals. *Acta Cryst.*, 21:849, 1966.

[127] T. Debaerdemaeker, C. Tate, and M. M. Woolfson. On the application of phase relationships to complex structures. XXIV. The Sayre tangent formula. *Acta Cryst.*, A41:286, 1085.

[128] H. Hauptman and J. Karle. *Solution of the Phase Problem. I. The Centrosymmetric Crystal.* Polycrystal Book Service, New York, 1953. A.C.A. Monograph No. 3.

[129] H. Hauptman and J. Karle. Structure invariants and seminvariants for non-centrosymmetric space groups. *Acta Cryst.*, 9:45, 1956.

[130] H. Hauptman and J. Karle. Seminvariants for centrosymmetric space groups with conventional centered cells. *Acta Cryst.*, 12:93, 1959.

[131] J. Karle and H. Hauptman. Seminvariants for non-centrosymmetric space groups with conventional centered cells. *Acta Cryst.*, 14:217, 1961.

[132] C. Giacovazzo. *Direct Phasing in Crystallography.* Number 8 in IUCr Monographs on Crystallography. IUCr/Oxford Science Publications, Oxford, 1999.

[133] M. C. F. Ladd and R. A. Palmer. *Theory and Practice of Direct Methods in Crystallography.* Plenum Press, New York, 1980.

[134] V. Kocman, R. I. Gait, and J. Rucklidge. The crystal structure of bikitaite, $Li[AlSi_2O_6]\cdot H_2O$. *Amer. Miner.*, 59:71, 1974.

[135] H. Schenck and C. T. Kiers. *Crystallographic Computing 3*. Oxford University Press, Oxford, 1985. G. M. Sheldrick, C. Kruger, and R. Goddard, editors.

[136] G. Germain, P. Main, and M. M. Woolfson. On the application of phase relationships to complex structures II. Getting a good start. *Acta Cryst.*, B26:274, 1970.

[137] P. Main. *Crystallographic Computing 3*. Oxford University Press, Oxford, 1985. G. M. Sheldrick, C. Kruger, and R. Goddard, editors.

[138] H. Schenk and J. G. H. de Jong. A method for direct structure determinations in $P1$ and related groups. *Acta Cryst.*, A29:31, 1973.

[139] H. Schenk. Direct structure determination in $P1$ and other non-centrosymmetric symmorphic space groups. *Acta Cryst.*, A29:480, 1973.

[140] H. Schenk. On the use of negative quartets. *Acta Cryst.*, A30:477, 1974.

[141] H. Hauptman. A joint probability distribution of seven structure factors. *Acta Cryst.*, A31:671, 1975.

[142] H. Hauptman. A new method in the probabilistic theory of the structure invariants. *Acta Cryst.*, A31:680, 1975.

[143] C. Giacovazzo. A probabilistic theory in $P\bar{1}$ of the invariant $E_{\mathbf{h}}E_{\mathbf{k}}E_{\mathbf{l}}E_{\mathbf{h+k+l}}$. *Acta Cryst.*, A31:252, 1975.

[144] W. Cochran and A. S. Douglas. The use of a high-speed digital computer for the direct determination of crystal structures. II. *Proc. Royal Soc. of London*, A243(1233):281, 1957.

[145] H. Schenk. Automation of the non-centrosymmetric symbolic addition. I. Fast determination of the unknown symbols. *Acta Cryst.*, B27:2037, 1971.

[146] G. Germain and M. M. Woolfson. On the application of phase relationships to complex structures. *Acta Cryst.*, B24:91, 1968.

[147] G. Germain, P. Main, and M. M. Woolfson. The application of phase relationships to complex structures III. The optimum use of phase relationships. *Acta Cryst.*, A27:368, 1971.

[148] Hull S. E and M. J. Irwin. On the application of phase relationships to complex structures XIV. The additional use of statistical information in tangent-formula refinement. *Acta Cryst.*, A34:863, 1978.

[149] G. T. De Titta, J. W. Edmonds, D. A. Langs, and H. Hauptman. Use of negative quartet cosine invariants as a phasing figure of merit: *NQEST*. *Acta Cryst.*, A31:472, 1975.

[150] P. S. White and M. M. Woolfson. The application of phase relationships to complex structures. VII. Magic integers. *Acta Cryst.*, A31:53, 1975.

[151] P. Main. On the application of phase relationships to complex structures. XI. A theory of magic integers. *Acta Cryst.*, A33:750, 1977.

[152] M. M. Woolfson. On the application of phase relationships to complex structures. X. MAGLIN - A successor to *MULTAN*. *Acta Cryst.*, A33:219, 1977.

[153] R. Baggio, M. M. Woolfson, J. P. Declercq, and G. Germain. On the application of phase relationships to complex structures. XVI. A random approach to structure determination. *Acta Cryst.*, A34:883, 1978.

[154] Y. Jia-Xing. On the application of phase relationships to complex structures. XVIII. *RANTAN*- random *MULTAN*. *Acta Cryst.*, A37:642, 1981.

[155] N. Metropolis, A. W. Rosenbluth, M. N. Rosenbluth, A. H. Teller, and E. Teller. Equation of state calculations by fast computing machines. *J. Chem. Phys.*, 21:1087, 1953.

[156] G. M. Sheldrick. Phase annealing in *SHELX-90*: Direct methods for larger structures. *Acta Cryst.*, A46:467, 1990.

[157] A. Altomare, M. C. Burla, M. Camalli, G. L. Cascarano, C. Giacovazzo, A. Guagliardi, A. G. G. Moliterni, G. Polidorib, and R. Spagnac. *SIR97*: A new tool for crystal structure determination and refinement. *J. Appl. Cryst.*, 32:115, 1999.

[158] C. M. Weeks, G. T. Detitta, R. Miller, and H. A. Hauptman. Applications of the minimal principle to peptide structures. *Acta Cryst.*, D49:179, 1993.

[159] R. Miller, G. T. DeTitta, R. Jones, D. A. Langs, C. M. Weeks, and H. A. Hauptman. On the application of the minimal principle to solve unknown structures. *Science*, 259(5100):1430, 1993.

[160] R. Miller, S. M. Gallo, H. G. Khalak, and C. M. Weeks. *SnB*: Crystal structure determination via shake-and-bake. *J. Appl. Cryst.*, 27:613, 1994.

[161] I. Usón and G. M. Sheldrick. Advances in direct methods for protein crystallography. *Current Opinion in Structural Biology*, 9:643, 1999.

[162] C. E. Shannon and Weaver W. *The Mathematical Theory of Communication*. University of Illinois Press, Urbana, 1949.

[163] E. T. Jaynes. Information theory and statistical mechanics. *Phys. Rev.*, 57(4), 1957.

[164] G. Bricogne. Maximum entropy and the foundations of direct methods. *Acta Cryst.*, A40:410, 1984.

[165] S.F. Gull and G.J. Daniell. Image reconstruction from incomplete and noisy data. *Nature*, 272:686, 1978.

[166] J. Karle. Partial structural information combined with the tangent formula for noncentrosymmetric crystals. *Acta Cryst.*, B24:182, 1968.

[167] Y. Jia-Xing. On the application of phase relationships to complex structures. XX. *RANTAN* for large structures and fragment development. *Acta Cryst.*, A39:35, 1983.

[168] Th. E. M. Van Den Hark, P. Prick, and P. T. Beurskens. The application of direct methods to non-centrosymmetric structures containing heavy atoms. *Acta Cryst.*, A32:816, 1976.

[169] P. T. Beurskens, Th. E. M. Van Den Hark, and G. Beurskens. Application of direct methods on difference Fourier coefficients for the solution of partially known structures. *Acta Cryst.*, A32:821, 1976.

[170] A.J.C. Wilson. Statistical bias in least-squares refinement. *Acta Cryst.*, A32:994, 1976.

[171] D. W. J. Cruickshank. *In Computing Methods in Crystallography.* Pergamon Press, Oxford, 1965. J. S. Rollett, editor.

[172] A. J. C. Wilson. Testing the hypothesis 'no remaining systematic error' in parameter determination. *Acta Cryst.*, A36:937, 1980.

[173] P. W. Betteridge, J. R. Carruthers, R. I. Cooper, K. Prout, and D. J. Watkin. *CRYSTALS* version 12: software for guided crystal structure analysis. *J. Appl. Cryst.*, 36:1487, 2003.

[174] W. C. Hamilton. Significance tests on the crystallographic R factor. *Acta Cryst.*, 18:502, 1965.

[175] D. Rogers. On the application of Hamilton's ratio test to the assignment of absolute configuration and an alternative test. *Acta Cryst.*, A37:734, 1981.

[176] H. D. Flack. On enantiomorph-polarity estimation. *Acta Cryst.*, A39:876, 1983.

[177] M. G. Rossmann. The accurate determination of the position and shape of heavy-atom replacement groups in proteins. *Acta Cryst.*, 13:221, 1960.

[178] M. G. Rossmann, editor. *The Molecular Replacement Method. A Collection Of Papers On The Use Of Non-Crystallographic Symmetry.* Gordon and Breach, Science Publishers, Inc., New York, 1972.

[179] M. G. Rossmann. The molecular replacement method. *Acta Cryst.*, A46:73, 1990.

[180] A. T. Brünger, J. Kuriyan, and M. Karplus. Crystallographic R factor refinement by molecular dynamics. *Science*, 235(4787):458, 1987.

[181] J. E. Lennard-Jones. Cohesion. *Proceedings of the Physical Society*, 43(240):461, 1931.

[182] A. T. Brünger. *X-PLOR, version 3.1: A system for X-ray crystallography and NMR.* Yale University Press, New Haven, 1992.

[183] P. P. Ewald. Zur begründung der kristalloptik. *Ann. Phys. (Liepzig)*, 54:519, 1917.

[184] P. J. Becker and P. Coppens. Extinction within the limit of validity of the Darwin transfer equations. I. General formalisms for primary and secondary extinction and their application to spherical crystals. *Acta Cryst.*, A30:129, 1974.

Index

Understanding Single-Crystal X-Ray Crystallography. Dennis W. Bennett

ISBN: 978-3-527-32677-8 (HC), 978-3-527-32794-2 (SC)